Russ:

Thank you for your time & effort in reviewing CDMA content.

- Nishith.

October 2014.

Cellular Communications

IEEE Press
445 Hoes Lane
Piscataway, NJ 08854

Technical Reviewers

Yash Vashanda, *Hughes Network Systems*
Matthew Valenti, *West Virginia University*
Yiping Wang, *Award Solutions, Inc.*

Cellular Communications

A Comprehensive and Practical Guide

Nishith D. Tripathi
Award Solutions, Inc.

Jeffrey H. Reed
Virginia Tech University

IEEE SERIES ON DIGITAL & MOBILE COMMUNICATION

IEEE

WILEY

Published by John Wiley & Sons, Inc., Hoboken, New Jersey. All rights reserved.
Published simultaneously in Canada.

For general information on our other products and services or for technical support, please contact our Customer Care Department within the United States at (800) 762-2974, outside the United States at (317) 572-3993 or fax (317) 572-4002.

Wiley also publishes its books in a variety of electronic formats. Some content that appears in print may not be available in electronic format. For information about Wiley products, visit our web site at www.wiley.com.

Library of Congress Cataloging-in-Publication Data is available.

Printed in the United States of America.

10 9 8 7 6 5 4 3 2 1

To our families:
Khyati, Siya, & Shivam
and
Pam, Julia, & Ryan

CONTENTS

Preface xxiii

Acknowledgments xxix

PART I FUNDAMENTAL CONCEPTS OF CELLULAR COMMUNICATIONS

1 Introduction to Cellular Communications 3

1.1 Motivation for Cellular Communications 4
1.2 The History of Wireless Communications and the Birth of a Cellular System 4
1.2.1 Once Upon a Time... A Really Brief History of Communications 5
1.2.2 Frequency Spectrum Allocation 5
1.2.3 Pre-Cellular Mobile Telephone Systems 8
1.2.4 The Advent of Cellular Systems 9
1.3 Concepts of a Cellular System 11
1.3.1 Network Architecture 11
1.3.2 Air Interface 12

1.3.3 Frequency Reuse 13
1.3.4 Sectorization and Its Impact on Quality and Capacity 20
1.4 Concepts of Duplexing and Multiple Access 22
1.4.1 Duplexing Methods 22
1.4.2 Multiple Access Methods 24
1.5 Types of Interference 26
1.5.1 Adjacent Channel Interference 26
1.5.2 Co-Channel Interference 27
1.6 Evolution of Cellular Standards 28
1.6.1 Evolutionary Path of Cellular Standards 28
1.6.2 Spectrum Allocation in the United States 32
1.6.3 Spectrum Allocation Outside the U.S. 34
1.7 Ecosystem of Cellular Systems 34
1.7.1 Service Providers 35
1.7.2 Mobile Device Manufacturers 36
1.7.3 Radio Network Infrastructure Vendors 37
1.7.4 Baseband Chipset Manufacturers 37
1.7.5 Original Equipment Manufacturers 39
1.7.6 Core Network Vendors 39
1.7.7 Design, Optimization, and Testing Tool Manufacturers 40
1.7.8 Appplication Providers 40
1.8 Phases of a Cellular System 42
1.8.1 Phase 1: Specifications of the Standard 42
1.8.2 Phase 2: Prototype Design and Lab Tests 43
1.8.3 Phase 3: Field Trials 43
1.8.4 Phase 4: Commercial Deployments 44
1.8.5 Phase 5: Optimization 44
1.9 Performance of Cellular Systems 45
1.9.1 Sector Capacity and Call Blocking Probability 45
1.9.2 Accessibility, Access Failure Rate, and Paging Failure Rate 47
1.9.3 Retainability and Call-Drop Rate 47
1.9.4 Error Rate and Quality of Service 47
1.9.5 Cell-Edge Reliability and Cell-Area Reliability 48
1.9.6 Peak User Data Rate 49
1.9.7 Average Cell or Sector Throughput and User-Perceived Throughput 49
1.9.8 Latency or Delay 50

1.9.9 Delay Jitter 50
1.10 End-User Services 50
1.11 Points to Remember 52

2 Elements of a Digital Communication System 55

2.1 Overview of a Digital Cellular Communication System 56
2.2 Transceiver Operations in a Digital Cellular System 57
2.3 Information Bits: Construction at the Transmitter and Retrieval at the Receiver 59
2.3.1 Linear Predictive Speech Coding: A Simplistic View 61
2.3.2 Enhanced Variable-Rate Codec 62
2.3.3 Adaptive Multirate Codec 66
2.4 Forward Error Correction Coding and Decoding 70
2.4.1 Convolutional Coding at the Transmitter 72
2.4.2 Example Convolutional Decoding at the Receiver 74
2.4.3 Turbo Coding at the Transmitter 83
2.4.4 Turbo Decoding at the Receiver 85
2.4.5 What Should I Choose: Convolutional or Turbo? 86
2.5 Interleaving and De-Interleaving 87
2.6 Digital Modulation and Demodulation 89
2.6.1 Modulation Schemes 90
2.6.2 Demodulation Approaches 96
2.6.3 Choosing a Modulation Scheme 99
2.7 RF Processor Functions at the Transmitter and the Receiver 100
2.8 Points to Remember 109

3 Radio Frequency Propagation 111

3.1 Radio Frequency Waves 111
3.2 Free-Space Propagation 113
3.3 Cellular Propagation Mechanisms 116
3.3.1 Reflection 116
3.3.2 Diffraction 117
3.3.3 Scattering 119
3.3.4 Absorption and Penetration 120
3.4 Prediction of Received Signal Strength 121
3.4.1 Distance-Based Path Loss 124
3.4.2 Large-Scale Fading 128
3.4.3 Small-Scale Fading 131

3.5 Points to Remember 153

4 IP and Associated Technologies for a Cellular System 155

4.1 Why Internet Protocol? (Or, Perhaps, Why *Not* Internet Protocol?) 156
4.2 Protocol Stack for the Internet 159
4.2.1 Overview of the Transport Layer (Layer 4) Protocols 160
4.2.2 Overview of the Internetwork Layer (Layer 3) Protocols 173
4.2.3 Overview of the Link Layer (Layer 2) Protocols 184
4.3 Routing and Forwarding of IP Packets 193
4.4 Transport of Information within the Network Infrastructure 200
4.5 Voice-Over-IP–Related Protocols 204
4.5.1 Session Initiation Protocol 206
4.5.2 Session Description Protocol 209
4.5.3 Real-Time Transport Protocol and Real-Time Transport Control Protocol 211
4.6 A Potpourri of IP-Centric Protocols 213
4.6.1 Dynamic Host Configuration Protocol 213
4.6.2 Mobile IP 214
4.6.3 Domain Name System 220
4.6.4 RADIUS and Diameter 222
4.6.5 IP Quality-of-Service Implementation: Integrated Services and Differentiated Services 224
4.6.6 IP Tunneling, GTP, and GRE 228
4.6.7 IPsec 231
4.6.8 Robust Header Compression 231
4.7 Essence of IPv6 232
4.8 Points to Remember 235

PART II 2G, 2.5G, AND 3G CELLULAR TECHNOLOGIES

5 Overview of GSM 239

5.1 Introduction 240
5.1.1 The Origin of GSM and GSM Milestones in Two Minutes 240
5.1.2 Highlights of a GSM System 241
5.2 GSM System Architecture 242
5.2.1 Base Station System 244
5.2.2 Core Network 247

5.3 Air-Interface Protocol Stack between MS and BSS 249
5.3.1 High-Level View of MS–BSS Communications 249
5.3.2 Air-Interface Protocol Stack: A Brief Overview 251
5.4 Radio Interface: Frame Structure and Channels 259
5.4.1 Frequency and Time Structures 259
5.4.2 Air-Interface Channels 263
5.4.3 Bursts: An Inside Look 267
5.5 Network Acquisition 269
5.5.1 Beyond Radio Network Acquisition: MS–Core-Network Communications 272
5.6 Voice Call Setup 279
5.6.1 MS-Originated Call Setup 279
5.6.2 MS-Terminated Call Setup and Paging Procedure 281
5.7 Mobility Management in Idle Mode 282
5.8 Mobility Management in Active or Connected Mode 283
5.8.1 What are the Channel Conditions? 284
5.8.2 Making a Handover Decision: Who is in Charge Around Here? 287
5.8.3 Follow the Leader: Handover Completion 293
5.9 Power Control 294
5.9.1 A Detailed Look at Power Control 295
5.9.2 Further Reduction in the Transmit Power Consumption: DTX 297
5.10 Deployment Considerations 298
5.11 Evolution of GSM 302
5.12 Points to Remember 302

6 GPRS and EDGE 305

6.1 Introduction 306
6.1.1 The Motivation for GPRS 306
6.1.2 The Motivation for EDGE 308
6.1.3 GPRS in Two Minutes 308
6.1.4 EDGE in Two Minutes 309
6.2 System Architecture 310
6.2.1 Packet-Domain Core Network Architecture: An Overview 310
6.2.2 A Detailed View of the Components of the Packet-Domain Core Network 312

6.2.3 Interfaces of the Packet-Domain Core Network 321
6.2.4 End-to-End View of the Data Transfer 323
6.3 MS–BSS Air-Interface Communications 324
6.3.1 Physical Layer of GPRS and EDGE 325
6.3.2 RLC/MAC Layer and Its Functions 334
6.4 Radio Interface: Frame Structure and Channels 342
6.4.1 Review of Timing Structures 342
6.4.2 New Logical Channels in GPRS and EDGE 344
6.4.3 Association of Logical and Physical Channels 346
6.5 Network Acquisition 348
6.5.1 Cell Search and Synchronization: A Brief Review 348
6.5.2 The MS's GPRS Attach Procedure: Making the First Contact with the Core Network 349
6.5.3 Getting Ready for Packet Data Transfer: PDP Context Activation 351
6.6 Packet Transfer in GPRS and EDGE 353
6.6.1 MS-Originated Packet Transfer 353
6.6.2 MS-Terminated Packet Transfer 355
6.6.3 Management of Circuit-Switched and GPRS Paging 356
6.7 Mobility Management 358
6.7.1 Routing Area Update: STANDBY State Mobility Management 358
6.7.2 Mobility Management in READY State 359
6.8 Deployment Considerations and Network Evolution 361
6.9 Points to Remember 362

7 Fundamentals of CDMA, WCDMA, and IS-95 363

7.1 A Brief Introduction to CDMA, WCDMA, and IS-95 364
7.2 CDMA or WCDMA: An Efficient Resource-Sharing Technique 366
7.3 Characteristics of CDMA 369
7.3.1 Wideband Transmission 369
7.3.2 Frequency Reuse and Impact of Sectorization 370
7.3.3 Interference and its Control in CDMA 371
7.3.4 Handoff 375
7.3.5 Power Control 381
7.3.6 Capacity of a CDMA System 388
7.4 IS-95 System Architecture 390
7.5 IS-95 MS–Radio-Network Communications: An Overview 391

7.5.1 Duplexing: Simultaneous Downlink and Uplink Transmissions 392
7.5.2 IS-95 Protocol Stack Overview 392
7.5.3 Downlink and Uplink Radio Channels of IS-95 394
7.5.4 Physical Layer (Layer 1) Processing 399
7.5.5 Multiplex Sublayer (Between Layer 1 and Layer 2): An Overview 403
7.5.6 Layer 2 Versus Layer 3 404
7.6 IS-95 Call Setup 405
7.7 IS-95 Mobility Management 407
7.7.1 Power-On Network Acquisition—One-Way Traffic! 407
7.7.2 MS to Network: "Here I Am"—Registration and Other Idle Mode Activities 408
7.7.3 Handoff in Connected or Traffic Mode—Now We Are Getting Really Busy! 411
7.8 IS-95 Power Control 414
7.8.1 Uplink Power Control 415
7.8.2 Downlink Power Control 416
7.9 IS-95: Deployment, Evolution, and Enhancements 418
7.10 Points to Remember 420

8 CDMA2000 1x 423

8.1 A Brief Overview of CDMA2000 424
8.2 CDMA2000 1x Network Architecture 426
8.2.1 Overview of the Network Architecture 426
8.2.2 A Detailed View of the Packet-Switched Core Network 429
8.3 MS–Radio-Access-Network Communications 431
8.3.1 Overview of Air-Interface Protocol Stack 431
8.3.2 Radio Channels and Radio Configurations of 1x 433
8.3.3 Overview of Physical Layer (Layer 1) Processing 441
8.3.4 Overview of Link Layer (Layer 2) Processing 445
8.3.5 A Brief Overview of Upper-Layer Processing 454
8.4 1x Call Setup 454
8.4.1 Voice Call Setup in 1x: An Overview 455
8.4.2 Voice Call Setup in 1x: A Closer Look 456
8.4.3 How About Data Call Setup in 1x? 459
8.4.4 What About MS-Terminated Call Setup? 460

8.5 Over-the-Air Transmission of Traffic for Forward and Reverse Links 462
8.5.1 Forward Link Packet Data Transmission 462
8.5.2 Reverse Link Packet Data Transmission 467
8.6 Mobility Management 468
8.6.1 Registration and Authentication 469
8.6.2 Using the Access Channel 472
8.6.3 Handoff in Connected Mode: 1x Enhancements 477
8.7 1x Power Control 481
8.7.1 Reverse Link Power Control for the Reverse Fundamental Channel 481
8.7.2 Forward Link Power Control for the Forward Fundamental Channel 482
8.8 1x Deployment and Evolution 484
8.9 Points to Remember 485

9 Universal Mobile Telecommunication System Release 99 487

9.1 Universal Mobile Telecommunication System Release 99: Executive Summary 488
9.2 UMTS Network Architecture 489
9.2.1 Radio Network Subsystem 491
9.2.2 Protocol Stacks: From User Equipment to the Edge of the Network 493
9.3 Radio Interface Protocol Stack between UE and UTRAN 495
9.3.1 Radio Interface Protocol Stack: A Brief Overview 495
9.3.2 Physical Layer 499
9.3.3 Medium Access Control Sublayer 503
9.3.4 Radio Link Control Sublayer 504
9.3.5 Packet Data Convergence Protocol Sublayer 510
9.3.6 Broadcast-Multicast Control Sublayer 511
9.3.7 Radio Resource Control Layer 512
9.4 Radio Interface Between the UE and the UTRAN 515
9.4.1 Overview of Frame Structure and Radio Channels 515
9.4.2 Orthogonal Variable Spreading Factor Codes 519
9.4.3 Primary Synchronization Channel 523
9.4.4 Secondary Synchronization Channel 523
9.4.5 Primary Common Pilot Channel 524
9.4.6 Primary Common Control Physical Channel 525

9.4.7 Page Indicator Channel 526
9.4.8 Secondary—Common Control Physical Channel 527
9.4.9 Physical Random Access Channel 528
9.4.10 Acquisition Indicator Channel 533
9.4.11 Uplink Dedicated Physical Data Channel and Dedicated Physical Control Channel 534
9.4.12 Downlink Dedicated Physical Channel and Dedicated Physical Control Channel 535
9.4.13 Combining of the Downlink Channels in a Cell 536
9.4.14 Transmission of Uplink Channels by User Equipment 539
9.5 Cell Search and Synchronization 539
9.5.1 Overview of Power-Up Synchronization and Cell Search 539
9.5.2 Power-Up Synchronization and Cell Search: A Closer Look 540
9.5.3 Beyond Synchronization: User Equipment and Core Network Communications 542
9.6 Voice Call Setup 544
9.6.1 Overview of Voice Call Setup and Teardown 544
9.6.2 Detailed Voice Call Setup 545
9.7 Data Call Setup 553
9.7.1 Overview of the Data Session Setup 553
9.7.2 Detailed Data Session Setup 554
9.8 Mobility Management 557
9.8.1 Idle Mode Mobility: An Overview 558
9.8.2 Connected Mode Mobility: An Overview of Basic Concepts 559
9.8.3 Stages of Intra-Frequency Handover: A Closer Look 562
9.8.4 Intra-UMTS Inter-Frequency and Inter-RAT Handover 568
9.8.5 Compressed Mode 569
9.9 Power Control 573
9.9.1 Uplink Closed-Loop Power Control 574
9.9.2 Downlink Closed-Loop Power Control 579
9.10 Quality-of-Service in the Universal Mobile Telecommunication System 580
9.11 Evolution of the Universal Mobile Telecommunication System 581
9.12 Points to Remember 583

10 1xEvolution–Data-Optimized Revision 0 585

10.1 1xEV-DO: Executive Summary 586
10.2 Overview of Call Setup and Data Transmission 587
10.3 Network Architecture 590
10.4 1xEV-DO Protocol Stack: An Overview 591
10.5 Introduction to Radio Channels and Timing Structure 593
10.6 A Closer Look at Forward Link Radio Channels 596
10.6.1 Pilot Channel 597
10.6.2 Medium Access Control Channel 597
10.6.3 Control Channel 600
10.6.4 Forward Traffic Channel 604
10.6.5 Combining the Forward Link Channels 610
10.7 A Closer Look at Reverse Link Channels 612
10.7.1 Access Channel 612
10.7.2 Pilot Channel and Data Channel 615
10.7.3 Data Rate Control Channel 617
10.7.4 Acknowledgment Channel 619
10.7.5 Combining of the Reverse Link Channels 621
10.8 Call Setup: A Closer Look 623
10.8.1 Session Setup 624
10.8.2 Point-to-Point Protocol Setup 628
10.8.3 Mobile Internet Protocol Setup 628
10.9 Forward Link Data Transmission: A Detailed View 630
10.10 Reverse Link Data Transmission: A Detailed View 633
10.10.1 Stage 1: Configuration of the Access Terminal with Reverse Link Parameters 633
10.10.2 Stage 2: Fast Indications of Reverse Link Interference 635
10.10.3 Stage 3: Execution of the Reverse Link Data Determination Algorithm 635
10.10.4 Stage 4: The Access Terminal's Data Transmission 638
10.11 Mobility Management in 1xEV-DO 640
10.12 A Glimpse of 1xEV-DO Optimization 642
10.12.1 RF Optimization 643
10.12.2 Parameter and Configuration Optimization 644
10.13 1xRTT and 1xEV-DO Interworking 645
10.14 1xEV-DO Link Budget 646
10.15 Points to Remember 649

11 High-Speed Downlink Packet Access 651

11.1 Overview of Third Generation Partnership Project Release 5 652
11.2 A Bird's-Eye View of High-Speed Downlink Packet Access Data Transmission 653
11.3 High-Speed Downlink Packet Access Channels and Their Use: A Deep Dive 655
11.3.1 High-Speed–Physical Downlink Shared Channel 655
11.3.2 High-Speed–Shared Control Channel 659
11.3.3 High-Speed–Dedicated Physical Control Channel 667
11.4 Data Transmission Process 671
11.4.1 Step 1: Reporting of Channel Conditions 671
11.4.2 Step 2: Scheduling of Users and User Packets 672
11.4.3 Step 3: Packet Transmission from the Node B and Packet Reception at UEs 677
11.4.4 Step 4: UE Response to a Received Packet 680
11.5 High-Speed Downlink Packet Access Call Setup 681
11.6 Resource Management of the High-Speed Downlink Packet Access Channels 683
11.7 Mobility Management in High-Speed Downlink Packet Access 685
11.8 Network and UE Architecture 689
11.8.1 Radio Interface Protocol Stack Changes from Release 99 to Release 5 689
11.8.2 High-Speed Downlink Packet Access UE Categories 691
11.8.3 Interaction Between the Node B and the RNC for High-Speed Downlink Packet Access 691
11.8.4 Construction of a Packet at the Node B 695
11.9 Points to Remember 697

12 1x Evolution–Data-Optimized Revision A 699

12.1 Introduction 700
12.2 Overview of Rev. A Enhancements 702
12.3 Characteristics of Physical Layer Subtypes 705
12.3.1 Overview of Physical Layer Subtypes 705
12.3.2 Reverse Link Frame Structure of Subtype 2 Physical Layer 706
12.4 Rev. A Forward Link Channels 707
12.4.1 Overview of FL Channels 707
12.4.2 ARQ Channel: A New Forward Link Channel in Rev. A 715

12.5 Overview of Rev. A Reverse Link Channels 718
12.5.1 Access Channel and RRI Channel Enhancements in Rev. A 721
12.5.2 Auxiliary Pilot Channel and DSC Channel: New Rev. A Channels 723
12.6 Rev. A Enhancement for Call Setup and Session Setup 724
12.6.1 Session Configuration and Multiple Personalities 725
12.6.2 Generic Attribute Update Protocol 728
12.6.3 Multimode Capability Discovery Protocol 728
12.7 Rev. A Forward Link Data Transmission: A Detailed View 728
12.7.1 Overview of Enhancements in Forward Link Transmission 728
12.7.2 DRC Enhancements 731
12.7.3 Multi-User Packets 734
12.7.4 MACIndex Enhancements 737
12.7.5 The AT's Detection of the FL Packet 739
12.7.6 Forward Link H-ARQ 739
12.8 Overview of Rev. A Reverse Link Data Rate Control 740
12.9 Reverse Link Interference Control via T2P 742
12.9.1 Motivation for T2P-Based Rate Control 743
12.9.2 What Is T2P, Anyway? 744
12.10 Subtype 2 RTC MAC Protocol Data Rate Control 745
12.10.1 Terminology for Subtype 2 and 3 RTC MAC Protocol 745
12.10.2 Overview of RL Data Transmission 750
12.10.3 A Comprehensive View of RL Data Transmission 751
12.11 Subtype 3 RTC MAC Protocol Data Rate Control 756
12.11.1 Overview of RL Data Transmission 756
12.11.2 Detailed View of RL Data Transmission 758
12.12 Rev. A QoS Control 761
12.12.1 Air-Interface QoS 762
12.12.2 Network QoS Control on R-P Interface 763
12.13 Handoff within 1xEV-DO 764
12.14 Idle State Enhancements and Signaling Enhancements 766
12.15 1xEV-DO Rev. A Interworking: Rev. 0, Rev. A, and CDMA2000 1x 767
12.16 Points to Remember 767

13 High-Speed Uplink Packet Access 769

13.1 Introduction to HSUPA: An Executive Summary 770
13.2 Basics of Data Transmission: A Brief Overview 771
13.2.1 Simplified HSUPA Call Setup 772
13.2.2 Some Terminology 773
13.2.3 Major Stages of Uplink Data Transmission: An Overview 774
13.3 HSUPA Channels 777
13.3.1 The E-DCH and E-DPDCHs (Uplink) 777
13.3.2 The E-DPCCH (Uplink) 784
13.3.3 The E-AGCH (Downlink) 785
13.3.4 The E-RGCH (Downlink) 787
13.3.5 The E-HICH (Downlink) 788
13.3.6 A Really Close Look at Timing Relationships Among HSUPA Channels 789
13.4 Prior to Data Transmission 790
13.5 Steps of Uplink Data Transmission 793
13.5.1 Scheduling Requests from the UE to the Node Bs 794
13.5.2 The Node B's Determination of Grants 794
13.5.3 Determination of the Serving Grant 798
13.5.4 E-DCH Transmission 800
13.5.5 Node B's Packet Decoding and HARQ Transmission 811
13.6 Physical Layer Retransmissions and HARQ 811
13.7 Support for User Mobility 814
13.8 System Architecture 815
13.8.1 Enhancements to the Existing Architecture 815
13.8.2 Interaction Between the Node B and the RNC for the Management of HSUPA Channels and Resources 817
13.8.3 HSUPA UE Categories 818
13.9 Points to Remember 818

PART III IP MULTIMEDIA SUBSYSTEM AND 4G CELLULAR TECHNOLOGIES

14 IP Multimedia Subsystem 823

14.1 Introduction to IMS: What and Why? 823
14.2 IMS, IP Convergence, and Fixed Mobile Convergence 827
14.3 History of IMS 828
14.4 IMS Architecture 831
14.4.1 High-Level View of the IMS Architecture 831

14.4.2 IMS Architecture: A Closer Look 835
14.5 IMS Communication Protocols for Signaling and Media Transport 841
14.5.1 Using SIP in IMS 842
14.5.2 H.248 in IMS 845
14.6 Identifying the IMS Subscriber and the IMS Elements 848
14.7 IMS Session Setup Scenarios 852
14.7.1 IMS Device to PSTN Call Setup: A Closer Look 852
14.7.2 PSTN-Originated and IMS Device-Terminated VoIP Call Setup: An Overview 856
14.7.3 Multimedia Session Setup for IMS Endpoints: A Brief Overview 858
14.8 IMS Services and Applications 860
14.9 Implementation of Quality-of-Service using IMS 862
14.10 Points to Remember 864

15 Fourth-Generation Technologies 867

15.1 Why 4G Technologies? 867
15.2 Essential Elements of 4G Technologies 869
15.3 Fundamentals of OFDM and OFDMA 871
15.3.1 OFDM and OFDMA: Why and Why Now? 871
15.3.2 OFDM: A High-Level View 872
15.3.3 Construction of Orthogonal Subcarriers 874
15.3.4 OFDM Challenges and Solutions 878
15.3.5 Simplified OFDMA Transceiver 884
15.4 Multiple Antenna Techniques 888
15.4.1 Diversity Techniques: Some Simple and Some Sophisticated! 888
15.4.2 Spatial Multiplexing 892
15.4.3 Beamforming 893
15.4.4 Space-Division Multiple Access 895
15.5 WiMAX Overview 897
15.5.1 Salient Features of WiMAX 897
15.5.2 WiMAX Network Architecture 900
15.5.3 WiMAX Air Interface 902
15.5.4 WiMAX Operations 903
15.6 Overview of LTE 908
15.6.1 Salient Features of LTE 909

15.6.2 LTE Network Architecture 911

15.6.3 LTE Air Interface 914

15.6.4 LTE Operations 915

15.6.5 A Brief Overview of LTE-Advanced 919

15.7 4G Technology Challenges 921

15.8 Points to Remember 922

Appendix A: A Brief Overview of Signaling System 7 923

Appendix B: Erlang-B Table 929

Appendix C: A High-Level Comparison of Third-Generation Technologies 931

Appendix D: HSPA+ Overview 943

D.1 Summary of HSPA+ Features 943

D.2 Data Transmission in HSPA+: A Closer Look 947

D.3 HSPA+: Beyond Release 7 952

References 955

Glossary 977

Index 991

Preface

These are exciting times for cellular communications. Cellular networks have evolved from first-generation analog systems to fourth-generation digital systems in just about three decades. We welcome you to a journey that will take us through cellular fundamentals and a variety of commercial cellular networks. The first-generation cellular systems and certain second-generation systems have been replaced by newer-generation systems. Remaining second-generation systems are expected to be replaced by newer-generation systems in the 2010s. We have included second-generation systems in the book because several concepts from these systems are reused (sometimes with enhancements) in the third-generation and fouth-generation systems.

The book consists of three parts. The first part provides a foundation for cellular communications by explaining fundamentals of cellular communications, radio propagation, digital communications, and IP-related[1] technologies. The second part focuses on specific cellular technologies, many of which are expected to exist for years to come. The third part introduces a services network called IP Multimedia Subsystem (IMS) and fourth-generation (4G) technologies, such as Long-Term Evolution (LTE) and Worldwide Interoperability for Microwave Access (WiMAX). A brief synopsis of each chapter follows.

[1] IP represents Internet Protocol.

Chapter 1 introduces fundamental concepts of a cellular communication system. Motivation for a cellular system is provided. A glimpse of the history of cellular communications is given. A cellular system relies on reuse of the radio resources to serve many users simultaneously in a given area. The concepts of frequency reuse, sectorization, interference, duplexing, and multiple access techniques play an important role in such resource reuse. These concepts are illustrated along with the system architecture. The evolution of cellular systems from first-generation to fourth-generation is described. An overview of the cellular communications industry is given. The lifecycle of a typical cellular system, which originates from the definition of a cellular standard and culminates in a well-optimized commercial deployment, is illustrated. Metrics that quantify how well the cellular system is performing are summarized. The types of services a subscriber can expect from a cellular system are mentioned.

Chapter 2 provides an overview of the fundamental principles for digital cellular systems. An introduction to the digital communication system is followed by a summary of the basic transmitter and receiver processing blocks. The transmitter includes processing blocks, such as speech coding, channel coding, modulation, spreading, and scrambling. It also includes an RF processor containing modules such as a digital-to-analog converter, a frequency upconverter, a power amplifier, and an antenna. The major blocks at the receiver are speech decoding, channel decoding, demodulation, despreading, and descrambling. The receiver also includes an RF processor containing modules such as an analog-to-digital converter, a frequency downconverter, a power amplifier, and an antenna.

Chapter 3 discusses various aspects of the radio propagation at the cellular radio frequencies. The cellular radio signal encounters a dynamic propagation environment as it travels from the transmitter to the receiver. The signals in cellular systems are one type of radio frequency (RF) waves. Hence, RF waves are introduced first. The free-space propagation is described as a starting point. The cellular propagation is more complex than the free-space propagation. Major mechanisms involved in cellular propagation, such as reflection, diffraction, scattering, and absorption, are described. While predicting the signal strength at a given distance away from the transmitter, three major factors are considered: distance-based path loss, large-scale fading, and small-scale fading. These factors are extensively discussed, along with analytical and mathematical formulations that model these factors. Various statistics that quantify the nature of the propagation environment are specified. Finally, examples are provided of the areas in which the knowledge of the cellular propagation is applied.

Chapter 4 provides an overview of IP, which is at the heart of the Internet, and related technologies from the cellular system perspective. A short introduction to IP is followed by a brief history of the Internet. Each layer of the IP stack is discussed. The IP packets contain information, such as web pages and email, and routing protocols influence the path these packets travel. Major routing protocols are summarized. The host that is the final destination of an IP packet usually resides on some physical network, such as an Ethernet network; interworking between IP and Ethernet is explained. Within a cellular network, the transport of information is implemented using mechanisms such as Asynchronous Transfer Mode (ATM) and Multi-Protocol

Label Switching (MPLS); such transport is also described. As the industry transitions from circuit-switched voice to Voice over IP (VoIP), understanding the roles of relevant protocols is quite important. Therefore, the basic operation of such protocols is also covered. While the primary focus is on IP version 4 (IPv4), a brief introduction to IP version 6 (IPv6) is also given.

Chapter 5 presents an overview of Global System for Mobile Communications (GSM), a second-generation (2G) digital cellular system. A brief summary of salient features of GSM is followed by a discussion of an end-to-end system architecture. The GSM core network forms the circuit-switched core network of a third-generation (3G) Universal Mobile Telecommunication System (UMTS). The GSM air interface is discussed next. Various radio channels used for over-the-air communications are described along with the structure of the radio frame. The mobile device's power-on initialization and network acquisition are summarized. The steps of a voice call setup are illustrated. Operations such as handover and power control are discussed. Finally, the evolutionary path of GSM is outlined.

Chapter 6 discusses the main concepts of General Packet Radio Service (GPRS) and Enhanced Data Rates for GSM Evolution (EDGE), which are an evolution of 2G GSM and often considered to be second-and-a-half–generation (2.5G) systems. First, a brief introduction to these 2.5G systems is given. The network architecture is illustrated at length, including discussions of new network elements that did not exist in GSM. An end-to-end packet flow is described. The air-interface protocol stack, consisting of the physical layer, the Radio Link Control layer, and the Medium Access Control layer, is discussed. GPRS and EDGE reuse the basic timing structure of the GSM radio interface; this timing structure is briefly reviewed, followed by a discussion of various new radio channels GPRS and EDGE introduce. The mobile station's network acquisition after power-on is illustrated along with two GPRS-specific procedures called GPRS attach and Packet Data Protocol (PDP) context activation. GPRS and EDGE are focused on efficient packet data services; both mobile-station–originated and mobile-station–terminated packet transfer are summarized. Idle Mode mobility and Active Mode mobility are considered. Deployment of 2.5G systems and the network evolution path beyond 2.5G are summarized.

Chapter 7 consists of two parts. The first part discusses Code Division Multiple Access (CDMA) and wideband CDMA (WCDMA) as multiple access techniques. The second part describes Interim Standard 95 (IS-95), which was the first CDMA-based cellular standard. Various characteristics of CDMA and WCDMA, such as wideband transmission, universal frequency reuse, interference management, soft handoff, and power control, are discussed. An analytical approach to estimate CDMA capacity is narrated. As part of the IS-95 discussion, network architecture is illustrated first. The air-interface protocol stack and various radio channels that enable over-the-air communication between the mobile station and the radio network are explained. The method of generating orthogonal codes, which are critical in a CDMA-based system, is specified. The process of setting up a call in IS-95 is described. The mobile device's network acquisition after power-up as well as mobility management are addressed next. Important deployment considerations are also mentioned. Features of the second version of IS-95, IS-95B, are summarized.

Chapter 8 provides an overview of CDMA2000 1x radio transmission technology (1xRTT), which is a CDMA-based 3G cellular system. CDMA2000 1xRTT evolved from IS-95. A brief history of CDMA2000 1xRTT, often called CDMA2000 1x, is given. The system architecture of CDMA20001x is illustrated. The communication between the mobile station and the network is described by discussing the air-interface protocol stack and the radio channels defined for the forward and reverse links. The call setup procedure is illustrated. The way in which data transfer occurs in the forward link (FL) and the reverse link (RL) is described. Operations such as handoff and power control are discussed. Deployment and evolution of CDMA2000 1x are briefly discussed.

Chapter 9 is on UMTS, which is a 3G cellular system that evolved from 2G GSM and 2.5G GPRS and EDGE. This chapter provides an overview of UMTS Release 99 with a focus on the frequency division duplexing (FDD) mode. First, a brief overview of UMTS is given, followed by the description of an end-to-end system architecture that includes a radio access network and a core network. The air-interface protocol stack implemented in the mobile device and the radio network is discussed. Various radio channels used for over-the-air communications are described. Cell search, a process whereby the user equipment (UE) finds the existence of a UMTS network in its neighborhood, is summarized. Voice call setup and data session setup are illustrated. UMTS' implementation of handover, power control, and quality of service (QoS) are also discussed. Finally, UMTS' evolutionary path is depicted.

Chapter 10 focuses on 1xEvolution–Data-Only or –Data-Optimized (1xEV-DO) Revision 0 (Rev. 0). 1xEV-DO Rev. 0 is a de facto evolution of CDMA2000 1xRTT. Its primary focus is on the FL air interface. The network architecture is introduced, followed by an explanation of the basic call setup. The air-interface protocol stack and FL and RL channels are described. The FL and RL data transmission are discussed in detail. Optimization of 1xEV-DO is considered, and its link budget is analyzed. Finally, interworking between 1xEV-DO and 1x is summarized.

Chapter 11 is all about High-Speed Downlink Packet Access (HSDPA), which is a Third Generation Partnership Project (3GPP) Release 5 feature. The main focus of HSDPA is to provide much higher downlink throughput to users than UMTS Release 99. Unique features and characteristics of HSDPA are summarized first. Data transmission in HSDPA is briefly discussed next. Characteristics of new HSDPA radio channels and their roles in HSDPA data transmission are described. A closer look at the data transmission in the downlink follows. Call-setup–related changes required to support HSDPA are summarized. HSDPA relies upon efficient management of resources, such as downlink power and OVSF codes; an explanation of such resource management is followed by a discussion on the impact of mobility on the HSDPA operation. The architecture of the network is illustrated with a focus on the radio interface protocol stack and interactions between the Node B and the radio network controller (RNC). Characteristics of different categories of HSDPA UEs are also specified.

Chapter 12 provides an overview of 1xEV-DO Revision A (Rev. A), which is an evolution of 1xEV-DO Rev. 0. Physical layer FL and RL radio channels are discussed. Session configuration and associated protocols are presented. Overall FL data transmission is described along with enhancements, such as multiuser packets. Rev. A more directly controls RL interference to achieve higher RL throughput. RL data transmission is illustrated by incorporating details of traffic-to-pilot–power-ratio–based data rate and interference control. QoS control mechanisms that enable real-time services are summarized. Details of handoff enhancements are given. Enhancements for an access terminal (AT) in the Idle State are described. Interworking among 1xEV-DO Rev. A, 1xRTT, and 1xEV-DO Rev. 0 is explained.

Chapter 13 discusses High-Speed Uplink Packet Access (HSUPA), which is a 3GPP Release 6 feature. The main goal of HSUPA is to significantly increase the peak and average uplink data rates. HSUPA is primarily an air-interface enhancement and is an evolution of HSDPA. Unique characteristics and benefits of HSUPA are summarized first. A brief overview of the downlink data transmission, uplink data transmission, and call setup is given. New HSUPA radio channels are introduced, and the uplink data transmission is discussed in detail. HSUPA defines new sublayers in the air-interface protocol stack. These architectural enhancements are briefly touched upon. Six categories of UEs for HSUPA are specified, and their characteristics are outlined.

Chapter 14 provides an overview of IMS, a service delivery platform that could potentially yield numerous benefits to consumers and service operators. Motivation for IMS is given, followed by the history of IMS. The basic components of an IMS network are described. The IP-based protocols IMS exploits are briefly discussed. Various call scenarios are illustrated to explain how IMS acts as a call control architecture. Examples of services IMS offers are given.

Chapter 15 discusses 4G cellular technologies. 4G technologies enable the networks to cost-effectively meet increasing data demands. Essential elements of 4G systems include Orthogonal Frequency Division Multiplexing, advanced antenna techniques, a flat all-IP network architecture, a converged packet-switched core network, and a scalable service delivery platform. This chapter first discusses the essential elements of 4G systems and then dives into specifics of two 4G system contenders, WiMAX and LTE. Finally, the challenges 4G technologies face are summarized.

In addition to the fifteen chapters, the book has four appendices. Appendix A provides a brief overview of Signaling System 7 (SS7), which enables cellular networks to interface with the Public-Switched Telephone Network. Appendix B specifies the Erlang-B table. The Erlang-B formulation plays an important role in capacity planning for voice-centric systems. Appendix C compares various 3G technologies to accelerate and reinforce your learning of one 3G technology if you are familiar with another 3G technology. Finally, Appendix D discusses evolution of HSDPA and HSUPA called HSPA+ (or HSPA-evolved).

In our industry, we cannot escape acronyms, and their count just keeps increasing as new technologies are developed. We have tried to balance readability with using the language of the industry and have included a glossary of acronyms to ease your pain! An acronym can have multiple valid meanings depending on the context, but we've tried to be quite clear about which meaning is intended.

We have also attempted to be consistent in terms of terminology, spelling, and capitalization within a given technology. Nevertheless, we caution that conventions vary among different technologies (and even different standards documents within a technology). We have set a uniform approach within a technology to the extent possible.

The chapter sections are written at two levels: a low-complexity, overview level for those interested in the basics of the technology and a medium-complexity, detailed level for those interested in the specifics. (The high-complexity level would correspond to the cellular standard itself.) ***The titles of detailed sections appear in gray type to separate such sections from overview sections.*** The overview sections are suitable for all audiences, including university students, managers, and professionals seeking main concepts. The detailed sections are for practicing engineers who want to develop expertise in a given area and for students who want to carry out research in a given area. This book is intended to equip you with just the right balance of *what*, *how*, and *why* so you can easily dive into the standard itself if necessary.

Furthermore, this book is designed to be interactive. Questions are sprinkled throughout the book to give you an opportunity to recall important concepts and to reinforce your learning, and answers generally appear in a footnote. Interesting stories are included to make reading the book not only rewarding but also enjoyable. The last thing we want is for the book to be used as a sleep-aid!

When you purchase this book, you are just beginning a relationship with us. Though this book provides a strong foundation of cellular communications and various commercial standards, we will use power of the ubiquitous Internet to make your pursuit of cellular communications knowledge more beneficial and more entertaining. Our website, http://www.thewirelessuniversity.com, includes errata as well as numerous fun activities, such as Jeopardy games. We have discussion forums in which we can share and grow the cellular communications knowledge among us. On the website, you can also submit feedback to help us improve the quality of the book and the teaching and learning process.

We sincerely hope that this book prepares you well for your future or current professional career. We look forward to interacting with you. Enjoy the book. Happy learning!

DR. NISHITH TRIPATHI & DR. JEFFREY REED

Parker, TX, and Blacksburg, VA
September, 2014

Acknowledgments

The comprehensive nature of this book obviously means that we have a long list of people to thank! We appreciate the financial support of Wireless@Virginia Tech's Industrial Affiliates Program that enabled us to hire a professional technical editor, compositor, and project manager. Lori Hughes and her team, including Gail Cocker at ArtTYPE Studio, who created the art, and Kathryn Hargreaves at Dangerous Curve, who helped convert some of the original MS Word files to LaTeX, assisted greatly with the process of preparing this manuscript, and the clarity and beauty of the book is largely due to them.

Our employment at Nortel Networks, Huawei Technologies, Virginia Tech, and Award Solutions, gave us a hands-on experience with commercial incarnations of technologies, which has made the book more practical.

Time is a resource that can never be replenished. We truly appreciate our families' sacrifice of time that gave us an opportunity to work on the book.

We are grateful to our reviewers for making their invaluable time available to review the book chapters. We thank the following reviewers whose time and efforts have enhanced the technical content of the book: Yasir Ahmed (National ICT R&D Fund, Pakistan), Ihsan Akbar (Wireless Innovation Forum), Parag Agashe (Qualcomm), Kenneth Baker (University of Colorado at Boulder), Bob Best (Award Solutions), David Boettger (Consultant), Robert Boyle (Consultant), Jim Bunch (Metaswitch Networks), Gary Cote (Award Solutions), Dave Curtis (Award Solutions), Luiz

DaSilva (Virginia Tech), Juan Deaton (Idaho National Lab), Satyajit Doctor (Award Solutions), Neerav Dalal (Award Solutions), Russ Edwards (Award Solutions), Alan Gatherer (Huawei), Don Hanley (Award Solutions), Bhagwan Khanka (AT&T), Matt Kohl (TowerCo), Bill Krenik (Texas Instruments), Hongyan Lei (AT&T), Gaviphat Lekutai (AT&T), Allen MacKenzie (Virginia Tech), Amit Mate (Qualcomm), Milap Majmundar (AT&T), Nitin Mangalvedhe (Nokia Siemens Networks), Mike McKinley (Award Solutions), Rias Muhammed (AT&T), William Newhall (Ball Aerospace), Jim Noel (Cassidian Communications), Malcolm Oliphant (Award Solutions), Narayan Parmeshwar (Solubiq), Dave Paranchych (Intellectual Ventures), Mark Pecen (Approach Infinity), Kyle Pittman (TowerCo), Paritosh Rai (AT&T), Hooman Razani (Award Solutions), Ramki Rajagopalan (Award Solutions), Ramesh Rajaram (Ericsson), Chris Reece (Award Solutions), Dennis Roberson (Illinois Institute of Technology), Mansoor Shafi (Telecom New Zealand), Ray Sulewski (Award Solutions), John Treichler (Applied Signal Technology), Ashok Trivedi (Award Solutions), Yash Vasavada (Hughes Networks), Anil Vullikanti (Virginia Bioinformatics Institute), and Min Zhang (Alcatel-Lucent).

Last and most importantly, we thank our publisher, IEEE/Wiley, for their incredible patience while we worked to complete the book.

N. D. T. & J. H. R.

PART I

FUNDAMENTAL CONCEPTS OF CELLULAR COMMUNICATIONS

CHAPTER 1

INTRODUCTION TO CELLULAR COMMUNICATIONS

Welcome to the exciting world of cellular communications, one of the most innovative technologies of modern times! The main objective of this chapter is to introduce fundamental concepts of a cellular communication system. Section 1.1 provides the motivation for a cellular system. Section 1.2 looks at the history of wireless communications and discusses how the cellular system was born. A cellular system relies on reuse of precious radio resources to serve many users simultaneously in a given area. The concepts of frequency reuse, sectorization, interference, duplexing, and multiple access techniques play an important role in such resource reuse and are illustrated in Sections 1.3–1.5 along with the cellular network architecture.

A cellular system has evolved from first-generation (1G) systems to fourth-generation (4G) systems, and Section 1.6 describes that evolutionary path. The cellular communications industry is made up of different types of companies, and Section 1.7 provides a glimpse of such an ecosystem. Section 1.8 narrates the life-cycle of a typical cellular system, which originates from the definition of a cellular standard and culminates in a well-optimized commercial deployment. Several metrics that quantify how well the cellular system is performing are summarized in Section 1.9. Section 1.10 lists the types of services a subscriber or consumer can expect from a cellular system. Finally, Section 1.11 concludes the chapter.

Cellular Communications, First Edition. Nishith D. Tripathi and Jeffrey H. Reed

1.1 MOTIVATION FOR CELLULAR COMMUNICATIONS

Just look around you, and you will discover how cellular communications has influenced our day-to-day lives. While waiting in the airport lounge before boarding a flight, we can respond to email and browse the Internet to read the news. We can place an order for food while on the move, and the food will be ready to be picked by the time we reach the restaurant. Our friends and family members can receive high-quality digital pictures from us when we send them our vacation pictures. Emerging cellular technologies will enable numerous users to simultaneously experience different types of services such as high-definition video streaming and interactive video gaming.

The cellular phone was ranked number two (the Internet was number one) of the top 25 innovations that have changed the world [1]. At the end of 2011, mobile subscriptions numbered over 6 billion, representing 87% of the world's population [2]. Cellular phones have been a blessing to many; more than 240,000 emergency calls (i.e., E911) per day are made in the U.S. using cell phones [3]. Imagine what these callers would have to go through in the absence of cellular phones. Some hospitals have started using private and digital cell phone technology to communicate with emergency medical service providers, medical helicopters, and other trauma centers, which expedites treatment for critical patients. The overall response time has been halved, significantly increasing the chances of survival [4]. News and information is now at our fingertips, thanks to the Internet and to the cellular devices that give wireless subscribers anytime and anywhere wireless access to the Internet. Cellular devices can increase productivity and efficiency. Imagine a delayed business traveler waiting for the plane to take off. A cellular device provides this user wireless access to the enterprise emails, and issues requiring immediate attention can be resolved. Imagine that you are in South Korea or Japan and you are thirsty, but you have no change or cash in your wallet to buy a drink. You really don't want to use a credit card for such a minor purchase. Well, you can quench your thirst if you have your mobile device handy—buy your favorite drink using your cell phone [5]! During a long commute in a train in Europe, you can watch news or a video on your hand-held wireless device. For developing countries (e.g., India and China) in which no extensive landline communications infrastructure exists, cellular communications has rapidly enabled communications services.

1.2 THE HISTORY OF WIRELESS COMMUNICATIONS AND THE BIRTH OF A CELLULAR SYSTEM

Cellular communications has changed our lives forever. Cellular communications is one example of a wireless communication system. The history of wireless communications is briefly touched upon in Section 1.2.1. Cellular communications typically operates in licensed frequency spectrums, the examples of which are given in Section 1.2.2. Section 1.2.3 describes some pre-cellular wireless systems. Section 1.2.4 discusses the arrival of cellular systems.

1.2.1 Once Upon a Time... A Really Brief History of Communications

Wireless communications obviously allows users to communicate without wires, which translates into offering users convenience. Numerous types of wireless communication systems are available, such as cellular systems, satellite systems, broadband wireless access systems, and broadcast radio systems.

Table 1.1 outlines some of the major milestones of wireless communications prior to the emergence of cellular systems and summarizes the early work done in the areas related to cellular communications. Table 1.1 captures only some of the major events associated with communications. The reader is encouraged to seek more information for a more in-depth history of communications [6–8].

1.2.2 Frequency Spectrum Allocation

Different countries may have somewhat different spectrum allocations for cellular systems. This is true espcially when the countries have some spectrum occupied by entities such as TV broadcasters and military. Spectrum allocation in the U.S. is briefly discussed next.

Wireless communications relies upon the frequency spectrum for communications. The frequency spectrum is a valuable and scarce resource. The allocation of the frequency spectrum is controlled by governments around the world. For example, the FCC (Federal Communications Commission) is in charge of spectrum management in the U.S. Figure 1.1 shows a simplified view of some of the frequency bands allocated to various wireless services in the U.S. Other works offer a comprehensive list of the frequency bands and associated services [10, 11] and provide the exact frequency spectrum for a given service, e.g., cellular or Personal Communication Services (PCS), for the U.S. [12, 13].

Examples of services include the following.

- **Broadcasting.** AM and FM radio broadcasts and TV broadcasts use the designated bands.
- **Radio Location.** This spectrum is primarily used by the U.S. government for military use and agencies such as NASA to assist in space flights.
- **Radio Astronomy.** These bands are used for continuous observations such as the electromagnetic radiation from the solar objects (e.g., the Sun and Jupiter).
- **Meteorological Use.** Various meteorological data collection and radiosonde operations occur in these bands for weather forecasting.
- **Industrial, Scientific, and Medical Bands.** These bands are designated for industrial, scientific, and medical applications. They are also used for wireless access in homes and enterprises.
- **Aeronautical.** This band is used by air traffic controllers for air-ground communication to control commercial, private, and recreational aviation.

Table 1.1: Selected Milestones of Wireless or Cellular Communications [6]
SOURCE: Quoted by permission from the IEEE History Center, Rutgers University, New Brunswick, NJ, U.S.A., `www.ieeeghn.org/wiki/index.php/IEEE_Communications_Society_History`.

Year	Event
1837	William Cooke and Charles Wheatstone obtain a patent on the telegraph. Samuel Morse publicly demonstrates the telegraph.
1876	Alexander Graham Bell and Elisha Gray independently invent the telephone. Bell was awarded priority because he filed for a patent a few hours earlier than Gray. (What a difference a few hours can make! Just ask those around you, "Who invented the telephone?" and see who is named!)
1887	Heinrich Hertz showed that electromagnetic waves exist.
1895	Guglielmo Marconi proves the feasibility of radio communications by sending and receiving the first radio signals.
1935	Edwin Armstrong invents frequency modulation.
1940s	Pre-cellular mobile phone service appears.
1947	The cellular concept is born at Bell Labs. Handoffs between transmitters occur and frequencies are reused.
1949	Claude Shannon's paper "Communication in the Presence of Noise" [9] is published.
1958	Jack Kilby invents an integrated circuit.
1963	Bolt, Beranek and Newman company (later to become BBN Technologies) develops the first modem.
1968	Y. Okumura publishes radio frequency (RF) propagation studies.
1972	Bell Labs patents its mobile communications system (3663762).
1973	Motorola introduces a small mobile phone, DynaTAC.
1974	Vinton Cerf and Robert Kahn introduce Transmission Control Protocol and use the term *Internet* for the first time.
1977	Experimental cellular systems are launched in Chicago, Washington, D.C., and Baltimore
1979	A microwave telephone system is installed in Saudi Arabia. The first commercial cellular system is installed by NTT in Tokyo.
1983	Advanced Mobile Phone System (AMPS), the world's first standardized and fully automated commercial cellular service, is deployed in Chicago.
1997	The term *smartphone* is used for the first time.

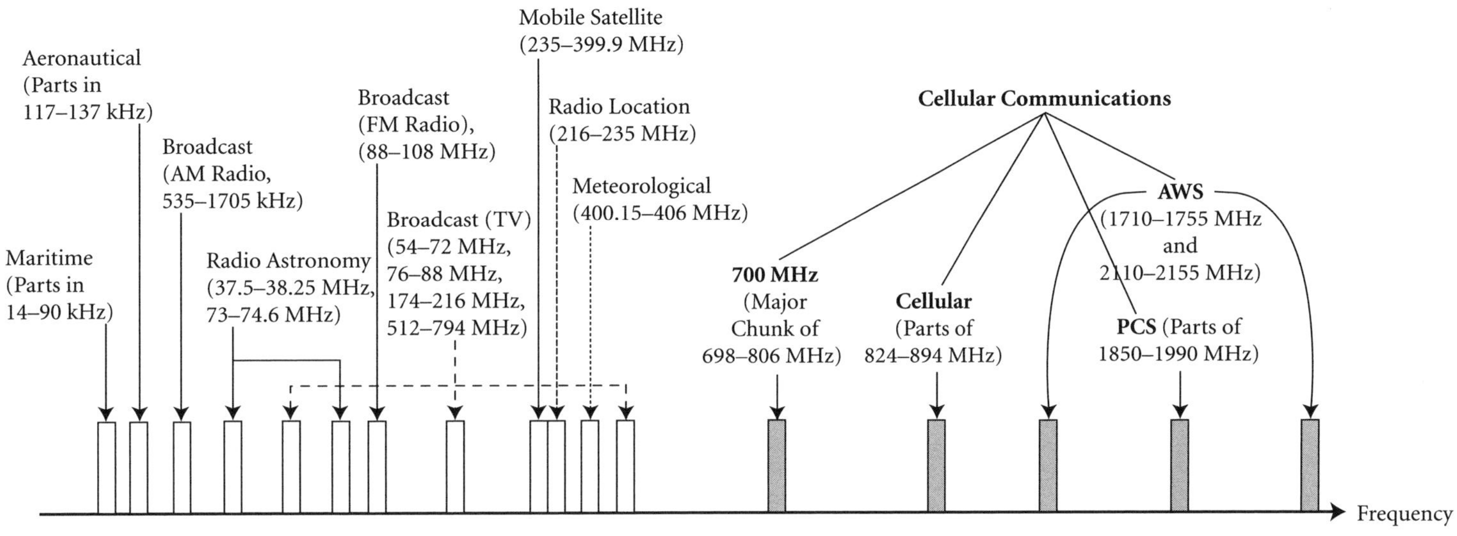

Figure 1.1: Example Usage of Spectrum in the U.S.

- **Maritime Mobile.** These bands are used for ship-to-ship and ship-to-coast communications.
- **Land Mobile Radio.** Various non-military federal agencies use these bands.
- **Mobile Satellite.** These bands are used by the military for training operations and for command and control connectivity among ground, air, and surface and/or sub-surface mobile platforms.
- **Cellular Communications Bands.** Example frequency bands used by cellular communications are cellular (around 850 MHz), PCS (around 1900 MHz), 700 MHz, and AWS (Advanced Wireless Services, at around 1700 and 2100 MHz). We will take a closer look at frequency bands in Sections 1.6.1 and 1.6.2.

1.2.3 Pre-Cellular Mobile Telephone Systems

The mobile telephone systems in existence before cellular systems are characterized by autonomous and non-interacting geographic zones. A base station (BS) with a large power amplifier and a tall tower is used in a given zone. Figure 1.2 illustrates such a mobile system [14].

Each zone is served by a set of frequency channels. In Figure 1.2, the users in Zone 1 are served on channels f_1, f_2, and f_3, and the users in Zone 2 are served on channels f_4, f_5, f_6, and f_7. Mobility across the zones is not supported, and the call drops as the user moves from one zone to the next. As shown in Figure 1.2, User 1 initially communicates with the BS in Zone 1 on f_3. However, as User 1 moves toward Zone 2, the signal quality degrades and the call drops. User 1 now has to re-initiate the call in Zone 2. Examples of pre-cellular mobile systems are Mobile Telephone System, Improved Mobile Telephone Service MJ, and Improved Mobile Telephone

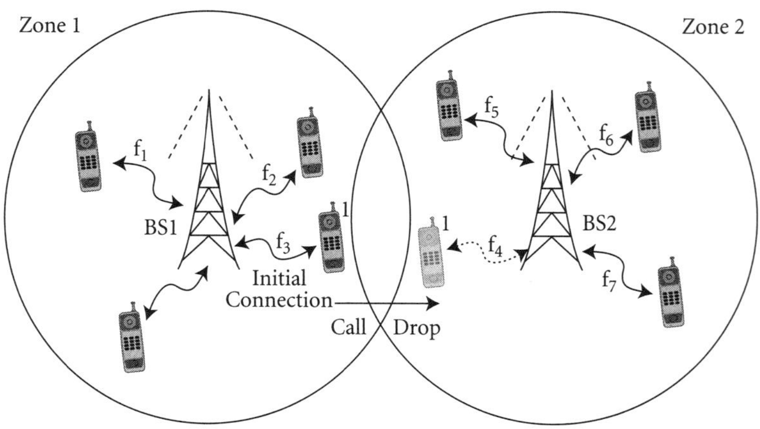

Figure 1.2: Pre-Cellular Mobile Telecommunication Systems

Table 1.2: Pre-Cellular Mobile Systems

System	Frequency Band	Number of Channels
MTS	40 MHz	11
IMTS MJ	150 MHz	11
IMTS MK	450 MHz	12

Service MK[2] systems [14]. Some characteristics of such systems are summarized in Table 1.2.

In 1976, New York City had six IMTS MJ channels and six IMTS MK channels [14]. Major drawbacks of these pre-cellular wireless systems are limited capacity (i.e., the maximum number of users that can be supported) and lack of support for mobility across regions. Since the number of channels is small, very few users can be supported. Furthermore, the lack of coordination between the adjacent regions prevents call continuation as the user moves from one region to the next.

1.2.4 The Advent of Cellular Systems

In 1958, Bell Labs proposed a 75 MHz system (FCC Docket 11997). In December 1971, Bell Labs showed how a cellular system could be designed [15, 16]. The FCC allocated 40 MHz of spectrum for cellular systems in 1974. The frequency spectrum was chosen to be at 800 MHz due to spectrum congestion at lower frequencies. Higher frequencies, such as those above 10 GHz, were avoided due to larger propagation path loss and subsequent reduction in cell sizes. Figure 1.3 illustrates spectrum utilization of the lower end of the spectrum.

The FCC initially allocated 40 MHz of spectrum for one cellular system to be licensed per market. However, to enable competition, the FCC later allowed two carriers (i.e., service providers) per market. The frequencies were divided into 20 MHz

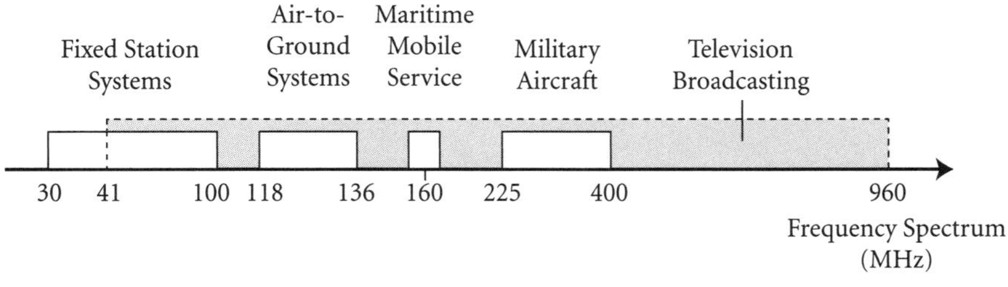

Figure 1.3: Early Frequency Spectrum Utilization

[2]MJ and MK have no further definition.

Table 1.3: Classification of U.S. Markets

Group	Classification
Top 1 to 30 Markets	Very Large
Top 31 to 60 Markets	Large-Sized
Top 61 to 90 Markets	Medium-Sized
Top 91 to 120 Markets	Below Medium-Sized
Top 121 to 150 Markets	Small-Sized

groups called Block A and Block B, which were sometimes referred to as Band A and Band B. One block was reserved for wireline carriers, and the other block was kept for non-wireline carriers. A total of 150 major market areas are defined by the FCC and classified by their populations into five groups as summarized in Table 1.3. Each group has thirty cities.

Table 1.4 summarizes the 1G cellular systems deployed in various countries in the world. These analog systems were gradually replaced by second-generation (2G)

Table 1.4: Initial Analog Cellular Systems [6]
SOURCE: Quoted by permission from the IEEE History Center, Rutgers University, New Brunswick, NJ, U.S.A., `www.ieeeghn.org/wiki/index.php/IEEE_Communications_Society_History`.

Country	Analog Cellular System	Year of Introduction
U.S.	AMPS, first service in Chicago	1983
Japan	800 MHz system by NTT, Tokyo; first commercial cellular system in the world	1979
United Kingdom	TACS by Cellnet	1985
Canada	Aurora at 400 MHz	1987
Denmark, Norway, Sweden, Finland, Spain	NMT at 450 MHz	1982
Netherlands, Luxemburg, Belgium	AFT at 450 MHz	1985
France	Radicom 2000 at 160 MHz, 200 MHz, and 450 MHz	1984
Austria	NMT at 450 MHz	1981
Germany	C-450	1985
Switzerland	NMT 900 MHz by Swiss PT&T	1986
Saudi Arabia, Scandinavia	NMT 450 MHz	1981

digital cellular systems. In the U.S., the cellular service operators were not required to offer AMPS beyond February 18, 2008.

A Bit of Cellular History... Have you ever heard about the first cellular network in the world? An AMPS-type cellular service was first offered in Saudi Arabia and Scandinavia in 1981 [6]. The first commercial cellular service was offered by NTT in Tokyo in 1979 [8, 17].

1.3 CONCEPTS OF A CELLULAR SYSTEM

The fundamental concepts of cellular communications are discussed below. The architecture of a cellular system is depicted in Section 1.3.1. Air interface enables the cellular system to offer wireless communications and is briefly summarized in Section 1.3.2. Frequency reuse enables us to design a commercially viable network. Section 1.3.3 describes frequency reuse. The impact of *sectorization* on quality and capacity is discussed in Section 1.3.4.

1.3.1 Network Architecture

A cellular communication system provides wireless communication services to users in geographical regions called *cells*. Various entities that participate in the communications process in a cellular system are shown in Figure 1.4. The cellular system establishes a wireless communication link between the user device and a fixed infrastructure. This fixed infrastructure connects to the existing landline voice and data networks.

The architecture shown in Figure 1.4 represents a typical third-generation (3G) cellular system. Specific network architectures for various cellular standards are discussed in relevant chapters of the book. The radio access network (RAN) consists of BSs and BS controllers (BSCs). The BSs are referred to as Node Bs in some technologies, and the BSCs are called radio network controllers (RNCs) in some technologies. The BS provides radio interface to the mobile user. The BSC controls multiple BSs and allows mobility between adjacent cells. The core network consists of a circuit-switched core network and a packet-switched core network. The circuit-switched core network consists of components such as the MSC (mobile switching center) and interfaces with the landline phone network such as the PSTN (public-switched telephone network). The MSC controls the voice calls and performs functions such as billing. The packet-switched core network includes gateways.[3] The packet-switched core network interfaces with the Internet and enterprise networks (e.g., networks of

[3]The gateways depend upon specific technologies. Examples of gateways include packet data serving node, Foreign Agent, Home Agent, serving GPRS support node in which GPRS represents General Packet Radio Service, and gateway GPRS support node.

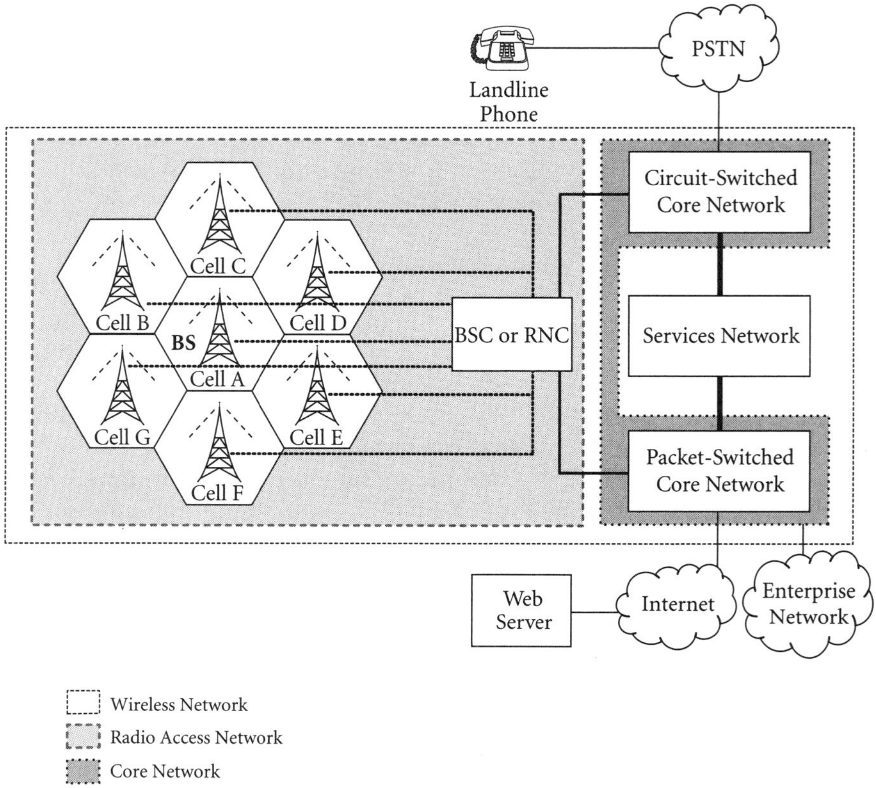

Figure 1.4: Elements of a Cellular System

companies, such as cnn.com). A cellular service provider's system also includes a services network that interfaces with the core networks to provide special services such as prepaid calling, ring-tone downloads, and voice and/or text messaging.

In Figure 1.4, the only link that would always be *wireless* is the communication link between the BS and the mobile device. All other links, such as those between the BSC and the core network, are *wired*. The link between the BS and the BSC or RNC could be wired or wireless; wired links are more common in the U.S., and wireless links are prevalent in Europe. The 4G system eliminates the BSC or RNC from the RAN and replaces two distinct core networks with a single, unified packet-switched core network.

1.3.2 Air Interface

The air interface between the BS and the mobile station (MS) is a critical design component in a cellular system. The communication links between the BS and the MS utilize the frequency bands reserved for cellular communications services. The electromagnetic waves exhibit different propagation properties at different frequencies [18]. For example, in the 3–30 kHz range, the primary mode of propagation is aided by the ionosphere. In the 30–3000 kHz range, propagation occurs via ground waves (i.e., the waves that follow the curvature of the Earth) in addition to the ionosphere. Typical cellular communication frequencies range from 450 MHz to 3 GHz, with widespread deployments around 900 MHz and 1.9 GHz. At these radio frequencies (RFs), the communications occur by line-of-sight and non-line-of-sight propagation. A line-of-sight communication means that the BS antenna and the MS antenna see each other directly with no obstacles between. Non-line-of-sight communication occurs when the path between the BS antenna and the MS antenna is blocked due to an obstacle such as a building.

The radio waves in a cellular system experience a dynamic propagation environment. As the signal propagates through the radio interface from a transmitter to a receiver, its strength changes. The signal strength measured at the receiver is primarily a combination of three major effects or factors: (i) distance-based path loss, (ii) environment-based or topology-based variations, and (iii) mobility-dependent variations. Chapter 3 treats the radio propagation in a comprehensive manner. As the distance between the transmitter and the receiver increases, the signal becomes weaker. Such signal loss is captured in the distance-based path loss, which is a function of parameters such as distance between the transmitter and the receiver; carrier frequency of the signal; and heights and gains of the mobile terminal antenna and the BS antenna. The topology-based signal strength variations reflect the scenario in which the signal strength is different at two separate locations that are equi-distant from the transmitter. These topology-based variations in signal strength are modeled by log-normal shadow fading. The user mobility (more accurately, the relative movement between the user and the immediate environment) introduces short-term fading, modeled by Rayleigh and Ricean fading.

1.3.3 Frequency Reuse

As the communication signal propagates in space, it experiences loss in strength. At a distance far from the BS, the signal becomes very weak. This seemingly simple property allows the same frequency to be reused many times throughout the system. Consider a geographical region shown in Figure 1.5.

The area where service is to be provided, the *service region*, is divided into cells. The size of the cell is specified by the cell radius, which can range from less than 100 m to more than 100 km. While no standard definitions exist for different types of cells, Table 1.5 shows an example classification of cells according to the cell radius [19–21].

A group of cells forms a *cluster*. The available frequency spectrum is reused in every cluster. For example, assume that a service provider has 5.6 MHz of spectrum and that the service operator has chosen a cellular standard in which one radio channel requires 200 kHz of bandwidth. The total number of radio channels that can fit in the 5.6 MHz spectrum is 5.6 MHz $\div$ 200 kHz $=$ 28. Figure 1.6 shows the radio channels for the example under consideration.

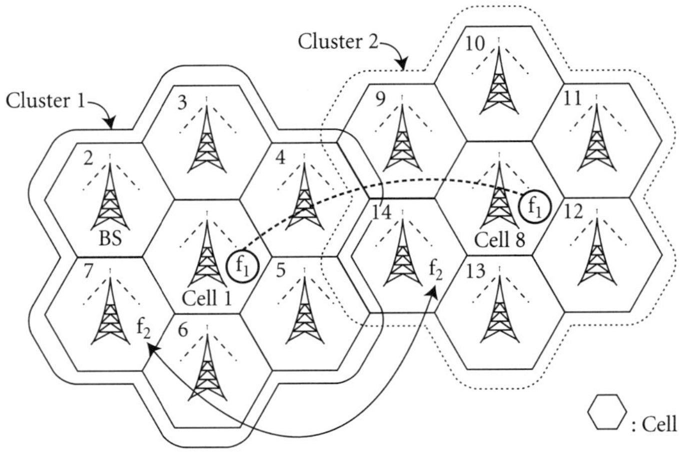

Figure 1.5: Concept of Frequency Reuse

Let's distribute these twenty-eight radio channels in seven cells. Each cell has four radio channels. Consider Cluster 1 in Figure 1.7. Of the twenty-eight channels, the first radio channel, f_1, is allocated to Cell 1 and the second channel, f_2, is allocated to Cell 7. All the twenty-eight channels are used in Cluster 1. The pattern of cells in Cluster 1 is repeated throughout a geographical region. For example, Cluster 2 in Figure 1.7 has the same cell layout as Cluster 1. Hence, channel f_1 is reused in Cell $1'$. Note that the same channel is reused in different clusters. Similarly, channel f_2 is used in Cell 7 of Cluster 1 and Cell $7'$ of Cluster 2. Such frequency reuse enables an operator to use the same spectrum many times in a region, thereby saving significant capital. Without frequency reuse, a service operator would require an excessive amount of spectrum for which there would be no business case. The number of cells in a cluster is called *cluster size*. The inverse of the cluster size is commonly known as the *frequency reuse factor* (FRF). For example, if the cluster

Table 1.5: Cell Classification

Cell Class	Cell Radius	Typical Deployment Scenario
Boomer	>100 km (up to 240 km)	Large, open space (e.g., plains in Australia)
Macro	Typically 1 to 20 km, sometimes up to 100 km	Rural and suburban areas
Micro	100 m to 1 km	Urban areas
Pico	Less than 100 m	Inside buildings
Femto	Few meters	Homes

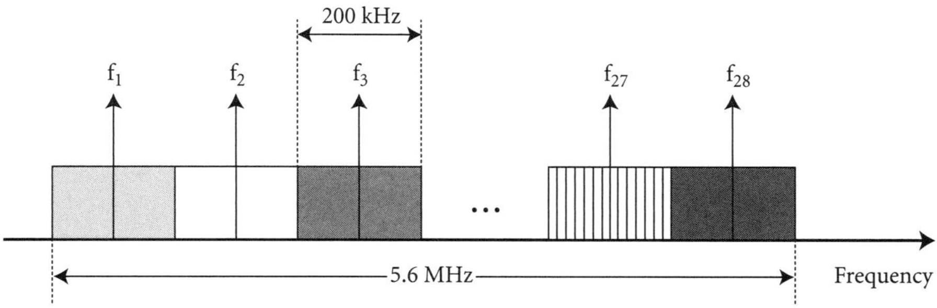

Figure 1.6: Example of Frequency Reuse

size is seven, the cluster has seven cells and the FRF is 1/7. The specific value of the FRF used in a given cellular system depends on the technology. The number of cells in the cluster cannot be arbitrary. A hexagon typically represents a cell, and the hexagon geometry imposes constraints on the FRF values.

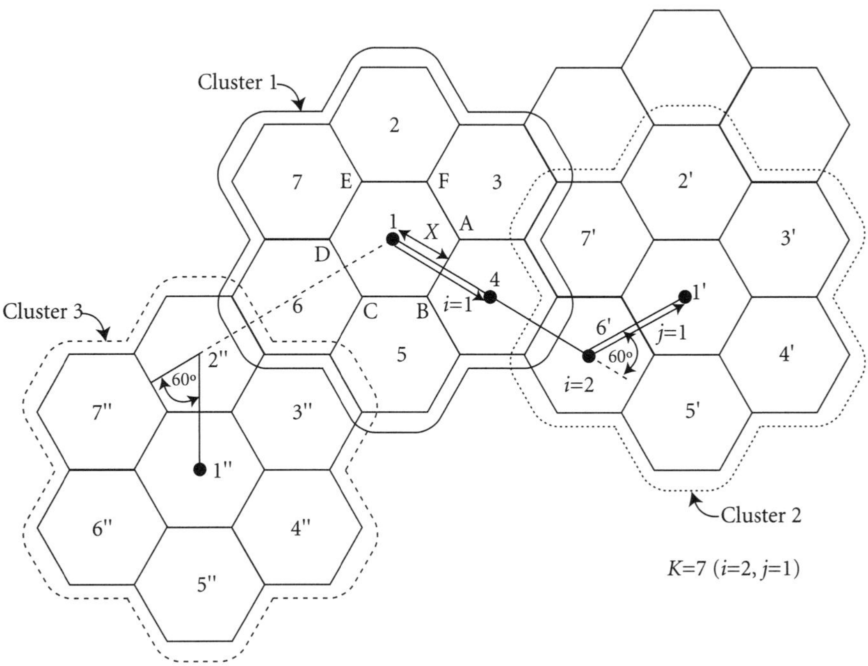

Figure 1.7: Creation of the Frequency Reuse Pattern Using the Cluster Size K

Ever Wondered... Why is a cell depicted as a hexagon? Bell Labs considered three shapes for geometric representation of the cell: triangle, square, and hexagon. The hexagon shape was eventually chosen because it allows the maximum number of mobiles to be connected to the BS in the center at the farthest distance from the BS [8]. Furthermore, the hexagon shape closely mimics the circular signal energy pattern radiating from the BS using an omnidirectional antenna [22]. The choice of the circular shape would have led to either significant overlap (and hence more cells to cover a given area) or areas with no coverage. The hexagonal shape helps achieve an efficient layout for system deployment and analysis.

The number of cells in the cluster, K, is given by the following formula [16, 23].

$$K = i^2 + j^2 + i \cdot j \tag{1.1}$$

where i and j are non-negative integers (i.e., 0 or greater). The values of i and j are used to locate the cell that reuses the same frequency or radio channel. Two cells that use the same radio channel are called *co-channel cells*. Figure 1.7 illustrates how i and j can be used to locate a co-channel cell.

For $i = 2$ and $j = 1$, $K = 7$. Hence, Cluster 1 in Figure 1.7 has seven cells. Now, consider Cell 1 in Cluster 1. Since the cell is a hexagon, it has six sides denoted by AB, BC, CD, DE, EF, and FA. To find a location of the co-channel cell of Cell 1, pick one of the sides, say, side AB. From the center of Cell 1, draw a line passing through the center of AB. Let the distance between the center of the cell and the center of side AB be X. The value of $i = 1$ corresponds to the cell-to-cell distance (i.e., between the centers of the two cells) of $i * 2X = 1 * 2X = 2X$. The value of $i = 2$ corresponds to the cell-to-cell distance of $i * 2X = 2 * 2X = 4X$. Since $i = 2$ for $K = 7$, we reach cell $6'$. Now draw a line at an angle of $60°$ counterclockwise with respect to the line connecting the centers of Cell 1 and Cell $6'$. Along this line, traverse the distance of $j = 1$ corresponding to $j * 2X = 1 * 2X = 2X$. We will then reach Cell $1'$, and we just found the cell that reuses the radio frequency channel used in Cell 1! The same process can be repeated for all the sides of the hexagonal Cell 1. We will then find a total of six co-channel cells since the hexagonal cell has six sides. These six co-channel cells constitute the first tier of co-channel interference for Cell 1.[4] Let D be the center-to-center distance between two co-channel cells. A given cell such as Cell 1 receives co-channel interference from six co-channel cells that are at a distance D from Cell 1. A total of twelve co-channel cells at a distance of $2D$ exist, which represent the second tier of co-channel interference. When analysis is performed, one or two tiers of co-channel interference are typically assumed. Figure 1.7 also shows one more example of finding the cell (which is Cell $1''$) that reuses the frequency used by Cell 1. Table 1.6 lists the allowed values of K for a hexagonal cell geometry. Initial commercial deployments of 1G AMPS and 2G GSM

[4]Interference between the cells due to the use of the same radio channels is called *co-channel interference*. See Section 1.5.2 for details.

Table 1.6: Examples of Allowed Cluster Sizes

i	j	K
0	1	1
1	1	3
2	0	4
2	1	7
2	2	12
3	0	9
3	1	13
3	2	19
3	3	27

(Global System for Mobile Communication) used $K = 7$. GSM deployments later used smaller values of K (e.g., four and lower) as optimization of deployed GSM networks continued. Systems based on code-division multiple access (CDMA) use $K = 1$.

The FRF is one of the key design parameters of cellular technologies. It dictates how often the given spectrum can be reused. The higher the FRF is, the higher the system capacity would be. In our example, when the FRF of $1/7$ is used, as shown in Figure 1.7, one cell receives four channels. When one user is assigned one radio channel, four users can be supported in a given cell. Now, let's assume that we can somehow make the FRF 1. Then, the number of cells in the cluster is one, and the service provider in our example can use its entire 5.6 MHz spectrum in every single cell! Such large spectrum availability in a cell will enable the service operator to support many more than four users in a given cell. Why not always use the FRF of 1? That's where *interference* comes into the picture.

Consider user MS1 in Figure 1.8. The BS in cell C1 is sending the signal to MS1 on the radio channel f1. At the same time, BSs in cells C8 and C15 are sending signals to the MSs (MS2 and MS3) in their own cells on the same radio channel. Recall that frequency reuse is fundamental to cellular communications; we *must* reuse the channels! When MS1 is trying to detect its own signal coming from BS1, the signals from the other BSs propagate through the radio environment and cause interference to MS1. The ratio of the received signal power, C, and the total received interference power, I, is called the C/I ratio, which is the *carrier-to-interference ratio*. The C/I ratio controls the quality of service (QoS), which can be quantified by the error rate. If the C/I ratio is very low, corresponding to strong interference relative to the signal, the error rate is very high and the user-perceived communication quality is poor. If the C/I ratio is high, the signal is stronger than the interference, leading to a low error rate and good user-perceived QoS. Hence, maintaining a good C/I ratio is key to maintaining good QoS. One way to ensure a high C/I ratio is to widely separate the base stations so that the interference signals are weak. When the FRF is one, two adjacent or neighboring BSs are using the same radio channel and interference is

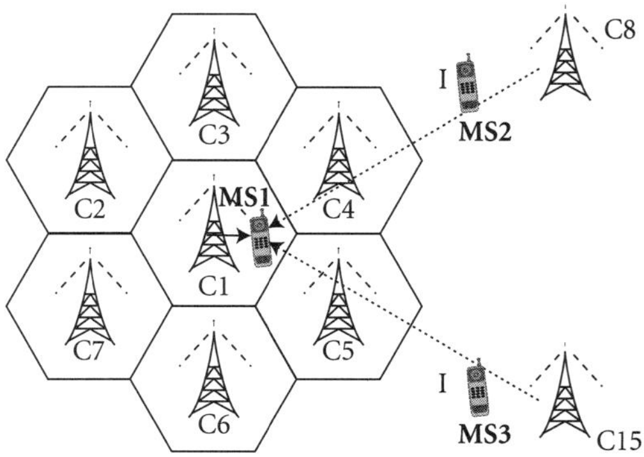

Figure 1.8: Impact of the Frequency Reuse Factor

very high. When the FRF is less than one, say $1/7$, the two BSs that use the same radio channel are about three cells apart, so interference is low.

Some technologies, such as a 1G AMPS system, cannot tolerate a low C/I ratio and require a large separation between two BSs using the same radio channel to achieve a higher C/I ratio. A lower FRF value, such as $1/7$, leads to such higher separation, enabling such systems to meet QoS requirements. Some technologies, such as those based on CDMA, can tolerate a high level of interference and work well even when the FRF is 1. If a high FRF is feasible for a technology, such a technology achieves higher capacity because the same radio channel is used more frequently in a given geographical region.

Let's formulate the relationship between the C/I ratio and the cluster size K. The C/I ratio directly influences the QoS perceived by users. For example, subjective tests for an analog cellular system indicate that a C/I ratio of 18 dB leads to "good" or "excellent" voice quality [16]. Now, the question is how far apart the co-channel cells should be to ensure a good C/I ratio. The C/I ratio at the MS can be calculated as

$$\left(\frac{C}{I}\right) = \frac{\text{Received Signal Power}}{\text{Received Interference Power}} = \frac{S}{\sum_{k=1}^{k=N_{co-ch}} R_k} \tag{1.2}$$

where S is the signal power measured at the MS, R_k is the co-channel interference measured at the MS from the co-channel BS k, and N_{co-ch} is the total number of co-channel BSs. The received signal power is proportional to the transmit signal power and can be simplified as

$$P_r = KP_t d^{-n} \tag{1.3}$$

where P_t is the transmit signal power, d is the distance between the transmitter and the receiver, n is the propagation path loss exponent, and K is the proportionality constant. The value of n usually ranges from 2 to 4, with 4 being the typical value [23].

Eq. (1.2) now becomes

$$\left(\frac{C}{I}\right) = \frac{KP_t d^{-n}}{\sum_{k=1}^{k=N_{co-ch}} KP_{tk} d_k^{-n}} \tag{1.4}$$

where P_t is the transmit power of the desired BS at the MS receiving the desired signal, d is the distance between the desired BS and the MS, P_{tk} is the transmit power of the co-channel BS k, and d_k is the distance between the MS and the co-channel BS k. The first two tiers of co-channel BSs cause most interference, with the first tier of co-channel interference being most significant. Assume that all the BSs are transmitting the same power P_t. The worst-case C/I ratio occurs near the cell edge where C is minimum and I is maximum. Considering the first tier of six co-channel BSs, the distance d_k between such co-channel BSs and the MS are approximated as $(D-R)$, $(D-R)$, $(D-(R/2))$, D, $(D+(R/2))$, and $(D+R)$ [22]. See other works [24] for more accurate d_k values.

Eq. (1.4) can now be further simplified as

$$\begin{aligned}\left(\frac{C}{I}\right) &= (R^{-n})/(2(D-R)^{-n} + (D-(R/2))^{-n} + D^{-4} \\ &\quad + (D+(R/2))^{-n} + (D+R)^{-n}).\end{aligned} \tag{1.5}$$

To derive a much simpler form of Eq. (1.5), let's assume $(D-R) \approx (D+R) \approx (D-(R/2)) \approx (D+(R/2)) \approx D$. Then,

$$\left(\frac{C}{I}\right) = \frac{R^{-n}}{6D^{-n}} = \frac{(D/R)^n}{6} = \frac{Q^n}{6} \tag{1.6}$$

where $Q = (D/R)$ is referred to as the *co-channel interference reduction factor*. Note that as the co-channel distance D increases, co-channel interference decreases and the C/I ratio increases. For the hexagonal geometry,

$$Q = \frac{D}{R} = \sqrt{3K} \tag{1.7}$$

where K is the cluster size.

From Eq. (1.6) and Eq. (1.7), the relationship between the C/I ratio and cluster size K can now be quantified as

$$\left(\frac{C}{I}\right) = \frac{(\sqrt{3K})^n}{6}. \tag{1.8}$$

Note that this relationship assumes six co-channel base stations with D as the co-channel distance, the same cell radius R for all the cells, the same transmit power for all BSs, omnidirectional BSs, and an MS located at the cell edge. If one or more of these assumptions do not hold true, Eq. (1.4) can be used to estimate the C/I ratio.

Initial cellular systems, such as AMPS, utilized $K = 7$. When $K = 7$, use of Eq. (1.8) yields a C/I ratio of 73.5 or $10 * \log 10(73.5) = 18.66$ dB for $n = 4$. Since a C/I ratio of 18 dB offers good voice quality, $K = 7$ appears adequate based on the simplified Eq. (1.8). A more accurate C/I ratio is slightly less than 18 dB for $K = 7$.

1.3.4 Sectorization and Its Impact on Quality and Capacity

When the BS transmits a signal for an MS from an omnidirectional antenna as in Figure 1.8, the signal energy or power is dissipated in all directions and very little energy reaches the MS. Another way to transmit a signal from the BS is to use directional antennas instead of omnidirectional antennas. When 120° sectorization is used, one cell that usually covers 360° is divided into three 120° regions called *sectors*.[5] Figure 1.9 shows an example of 120° sectorization where each hexagon now represents a sector. For example, Cell 1 controlled by the base station BS1 is divided into three sectors, α_1, β_1, and γ_1. Similarly, Cell 2 consists of three sectors, α_2, β_2, and γ_2. When 60° sectorization is used, one cell comprises six sectors, with each sector covering a 60° region. One cell that contains one or more sectors is usually controlled by one physical BS. Multiple omnidirectional antennas can be electrically combined together to create a directional antenna (e.g., 120° or 60°). Since sectorization concentrates signal energy in a specific direction, the immediate benefit is higher signal energy at a given MS location with a sectored system than with an omnidirectional system.

The impact of sectorization varies according to the multiple access technique. We will discuss various multiple access techniques, such as frequency-division multiple access (FDMA), time-division multiple access (TDMA), and CDMA, in Section 1.4. Sectorization in TDMA or FDMA systems results in lower interference and a higher C/I ratio, which results in better signal quality and increased reliability. If one omnidirectional cell has three radio channels (f_1, f_2, and f_3), 120° sectorization assigns one radio channel to each of the three sectors of this cell. Sectorization

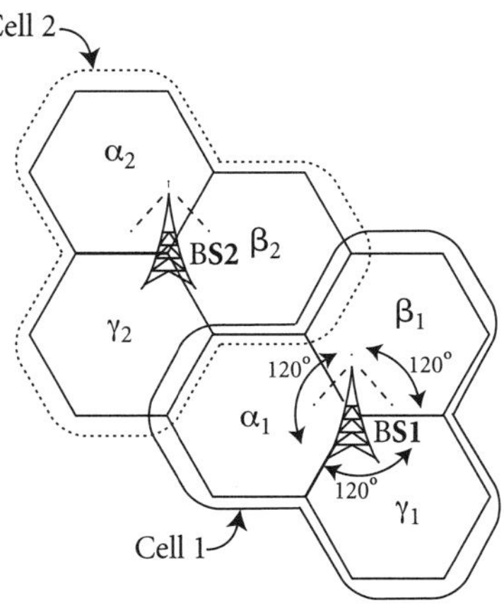

Figure 1.9: Example of 120° Sectorization

[5]In another approach, a hexagon still represents the cell, and three parts of the hexagon represent three sectors.

in TDMA and/or FDMA systems increases capacity only if the FRF increases; otherwise, the cell continues to have the same number of channels after sectorization. If an increase in the C/I ratio results in a higher FRF, the capacity of the TDMA and/or FDMA system increases because more channels become available per cell. When a cluster size is seven cells or the frequency reuse factor is 1/7, sectorization makes it easier to meet the target QoS and reliability. To understand how sectorization changes the interference in TDMA or FDMA systems, compare Figures 1.10 and 1.11. In Figure 1.10, six first-tier BSs cause co-channel interference to MS1 in Cell 1. These BSs include BS2 through BS7. However, MS1 experiences co-channel interference from only two first-tier BSs, BS2 and BS7. In other words, 120° sectorization reduces the number of co-channel interferers from six (for omnidirectional cells) to two. With 60° sectorization, the number of co-channel interferers is just one.

Sectorization immediately increases the capacity of CDMA systems. When 120° sectorization is used, the same radio channel is used in all the three sectors of the same cell in CDMA-based systems. Compared to an omnidirectional system, 120° sectorization theoretically increases the capacity by a factor of three. In practice, however, tri-sectored cells provide a capacity increase of a factor of about 2.5 to 2.6.[6]

In summary, capacity may or may not increase for non-CDMA systems when sectorization is used. For non-CDMA systems, if the improvement in the C/I ratio

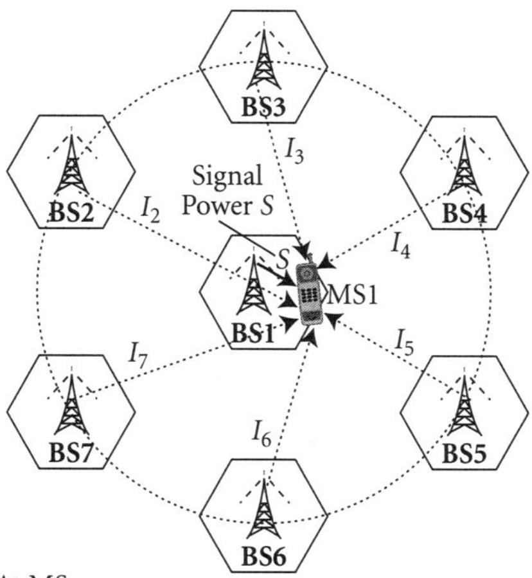

At MS_1
S: Received Signal Power
I_2 through I_7: Received Interference Power

$$\frac{C}{I} = \frac{S}{I_2+I_3+I_4+I_5+I_6+I_7}$$

Figure 1.10: C/I Ratio for an Omnidirectional System

[6]The capacity gain factor of less than three for 120 ° sectorization is attributable to higher interference and more handoffs compared to omnidirectional cells.

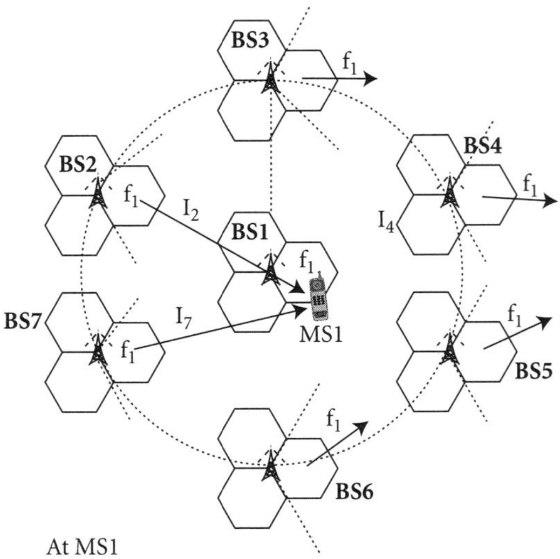

At MS1

S: Received Signal Power from BS1

I_2 and I_7: Received Interference Power from BS2 & BS7

$$\frac{C}{I} = \frac{S}{I_2+I_7}$$

Figure 1.11: C/I Ratio for 120° Sectorization

due to sectorization is large enough to allow an increase in the FRF, the capacity increases at a given reliability; otherwise, the capacity remains the same but reliability improves due to a higher C/I ratio. Sectorization in CDMA-based systems increases the capacity at a given QoS.

1.4 CONCEPTS OF DUPLEXING AND MULTIPLE ACCESS

The concepts of duplexing and multiple access are central to the bidirectional communications between the mobile user and the network and to the sharing of network resources among multiple users. These concepts are introduced below.

1.4.1 Duplexing Methods

The link from the MS to BS is called *reverse link* or *uplink* (UL), and the link from BS to MS is referred to as *forward link* or *downlink* (DL). The simultaneous communications on DL and UL are typically achieved by one of the multiplexing methods: frequency-division duplexing (FDD) or time-division duplexing (TDD). Figure 1.12[7] illustrates the FDD and TDD methods.

[7]Rx in the figure represents "receive," and Tx represents "transmit."

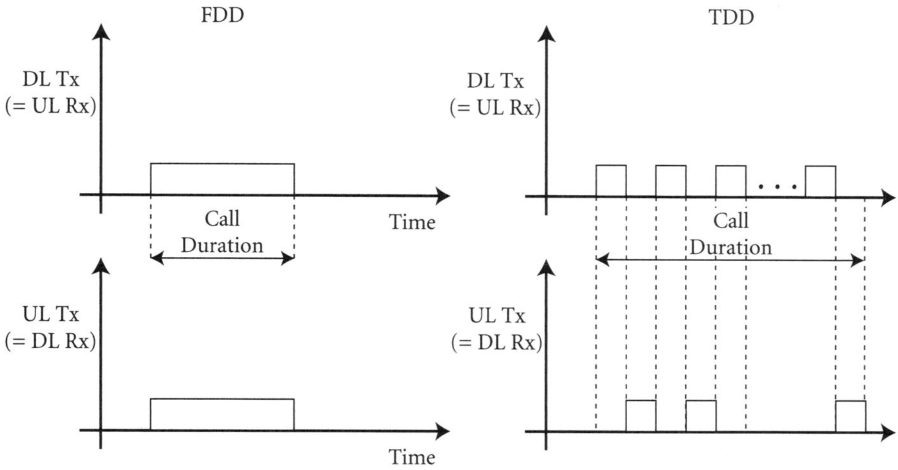

Figure 1.12: Frequency-Division Duplexing and Time-Division Duplexing

In the FDD method, the DL and UL transmissions occur on different frequencies at the same time. In the TDD method, at any given time, *either* (but not both) DL or UL transmission occurs. The advantage of FDD is its simple implementation, and the disadvantage is its requirement for two separate frequency bands. The TDD method obviates the need for two frequency bands but may lead to poor signal quality because of stringent timing requirements. Universal Mobile Telecommunication System (UMTS), a 3G cellular system, supports TDD as one of its options, and this TDD option is deployed by T-Mobile in the Czech Republic and by China Mobile in China. WiMAX (Worldwide Interoperability for Microwave Access), a 4G cellular system, also uses TDD. Most commercial cellular networks utilize FDD due to its ease of deployment and optimization. For example, global deployments of 2G GSM and 3G UMTS typically use FDD.

FDD and TDD are the basic duplexing approaches, and half-FDD is another duplexing approach. Half-FDD is similar to FDD in that one frequency is used for transmission and another frequency is used for reception. However, unlike FDD, half-FDD can either transmit or receive at a given instant. For example, a mobile device that operates using half-FDD transmits on one frequency at one time instant and receives on another frequency at another time instant. Such an half-FDD device cannot transmit and receive at the same time. The attractiveness of half-FDD is the lower cost for the device since such a device needs fewer components. However, the drawback of such a device is lower data rates compared to FDD devices.

1.4.2 Multiple Access Methods

Simultaneous communications between an MS and a BS are achieved through a duplexing method as described earlier. Simultaneous communications between several MSs and one BS in a given direction, DL or UL, are made possible by multiple access techniques. The basic multiple access techniques used in commercial cellular networks are FDMA, TDMA, CDMA, and orthogonal frequency-division multiple access (OFDMA). A combination of these multiple access techniques is also possible. For example, the GSM standard uses a TDMA and FDMA combination but is commonly referred to as a TDMA-based system.

The concept of FDMA is depicted in Figure 1.13. All the users receive signals from the BS simultaneously. However, these signal transmissions occur on different frequencies, such as f_1, f_2, and f_3 in Figure 1.13. The typical bandwidth of a communication channel (DL or UL) for the FDMA system is 30 kHz or 200 kHz.

Figure 1.14 introduces the TDMA system. The same frequency channel f_1 is shared among multiple users. The BS transmits a signal to User 1 in time-slot 1 of the radio frame 1 and time-slot 1 of the radio frame 2. The signal to User 2 is transmitted in time-slot 2 of the radio frames. Similarly, User 3 receives the signal from the BS during time-slot 3 of the radio frames. Note that only one user receives information

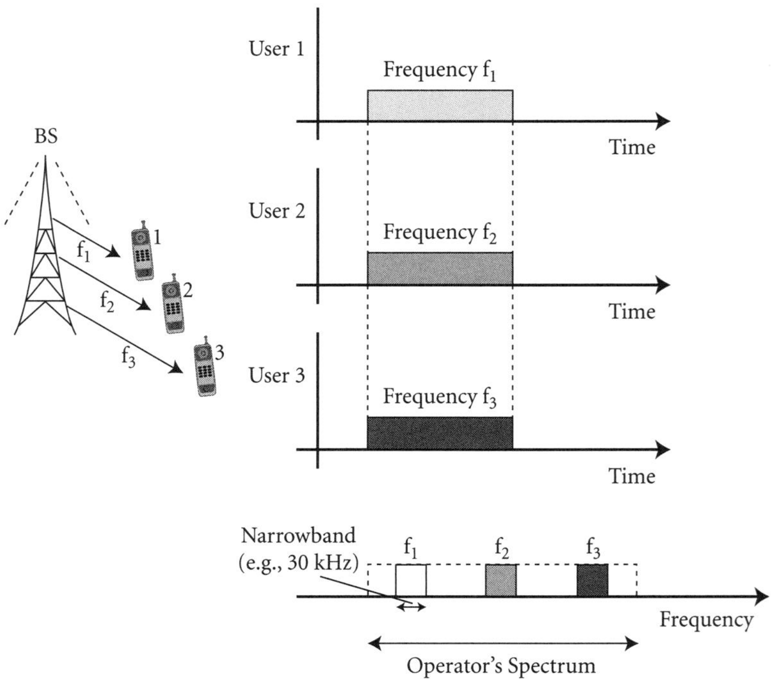

Figure 1.13: Frequency-Division Multiple Access

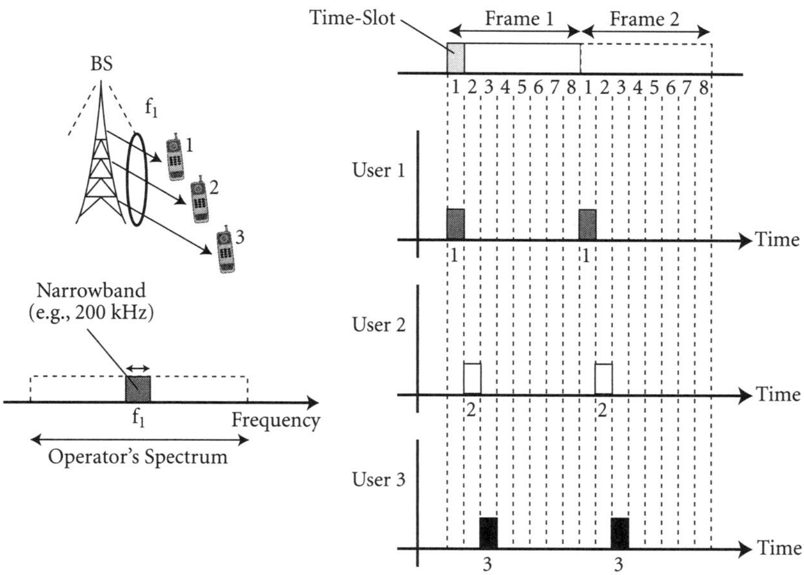

Figure 1.14: Time-Division Multiple Access

at any given time on a given frequency channel. The transmissions from the BS to the MS are synchronized. Figure 1.14 shows the DL operation. The UL operates similar to the DL. The transmission from MSs to the BS is synchronized such that the signal from only one MS is received at a given time on a given frequency channel.

The CDMA method is illustrated in Figure 1.15. The users are assigned different codes, and the bandwidth of each communication channel is wide (e.g., 1.25 MHz or 5 MHz). All the users share the same channel such as f_c in Figure 1.15.

4G cellular systems use OFDMA, which shares some similarities with FDMA. For example, the spectrum is divided into multiple narrowband frequency channels in OFDMA and FDMA. OFDMA refers to these frequency channels as subcarriers. The main differences between FDMA and OFDMA are that the subcarriers are orthogonal to one another and that the users are allocated sets of subcarriers.

OFDMA is similar to the basic FDMA in that different users are allocated different frequency channels. The frequency channels in OFDMA are constructed in a special fashion such that they are orthogonal to one another, and hence the letter "O" at the front of OFDMA. Orthogonality means that no interference exists among the frequency channels. Since interference among frequency channels is of no concern, numerous frequency channels can be squeezed into a limited spectrum bandwidth. A mobile device is usually allocated multiple frequency channels at a given instant for higher data rates.

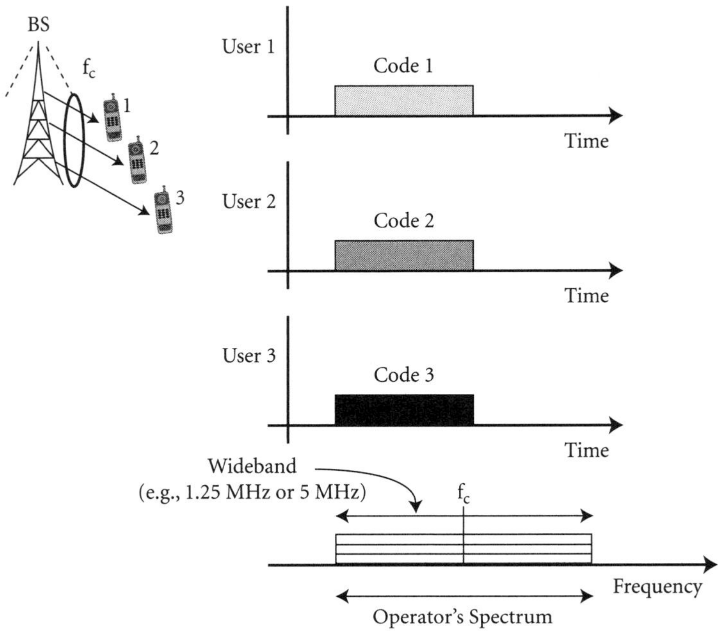

Figure 1.15: Spectrum Sharing in a CDMA System

1.5 TYPES OF INTERFERENCE

The signal captured at a receiver is corrupted by thermal noise (usually expressed in decibel-milliwatts per Hertz) and interference, which in a cellular system is either adjacent channel interference or co-channel interference.

1.5.1 Adjacent Channel Interference

Adjacent channel interference occurs when the signal energy from one radio channel blends into another radio channel. Assume that the focus is on User 1. User 1 receives a signal U_1 that contains information for User 1. The receiver of User 1 receives some energy from signals U_2 and U_3 that are intended for other users. For User 1, U_2 and U_3 represent interference, and this interference causes degradation in the signal quality perceived by User 1. The upper half of Figure 1.16 shows adjacent channel interference for an FDMA system in which significant adjacent channel interference exists between adjacent time-slots. The lower half of Figure 1.16 shows adjacent channel interference for a TDMA system in which significant adjacent channel interference exists among neighboring frequency channels. To reduce adjacent channel interference, the user signals must be separated in the time domain and/or frequency domain. The higher the separation between the channels, the lower the degree of interference and the lower the spectrum utilization efficiency. The spectrum utilization efficiency

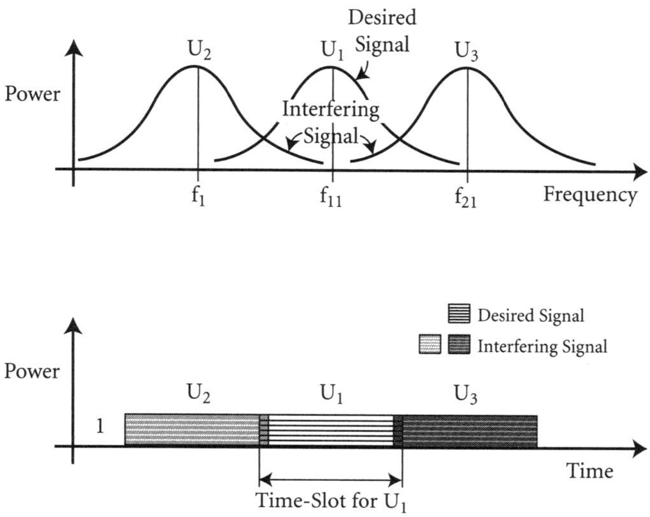

Figure 1.16: Example of Adjacent Channel Interference

can be quantified by the number of users that can be supported in a given spectrum bandwidth. Typically, the channel separation is determined to maximize spectrum utilization efficiency at the minimum tolerable SIR required to meet certain QoS. See Section 1.7 for a discussion on various performance metrics.

1.5.2 Co-Channel Interference

Co-channel interference (CCI) occurs when the user signal and the interference signal coincide in the time domain and frequency domain. Figure 1.17 shows a CCI scenario. The CCI occurs among users in different clusters, rather than in the same cluster, because the same communication channels are reused in different clusters. CCI is reduced by increasing the distance between the cells that use the same frequencies, which increases the number of cells per cluster. In a TDMA or FDMA system, frequency reuse factors of four and seven typically provide an adequate SIR. Let's turn our attention to CDMA.

A Question for You. . . What do you think the frequency reuse factor is for CDMA? See the footnote for the answer.[8]

[8]The frequency reuse factor is one in CDMA.

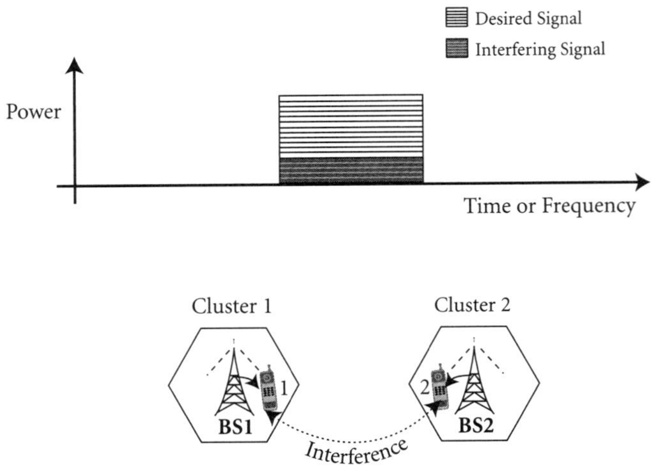

Figure 1.17: Co-Channel Interference Scenario

In the CDMA system, the same wideband channel is used in all the sectors or cells of the system. A CDMA system can tolerate relatively larger CCI compared to non-CDMA systems.

1.6 EVOLUTION OF CELLULAR STANDARDS

Cellular standards have evolved from 1G to 4G in just about three decades. Section 1.6.1 depicts the evolutionary path of cellular standards. Sections 1.6.2 and 1.6.3 specify the spectrum allocation for cellular systems in the U.S. and other parts of the world.

1.6.1 Evolutionary Path of Cellular Standards

Figure 1.18[9] depicts the evolutionary path of cellular systems. Initial cellular systems are referred to as 1G systems. The 1G systems are analog and support voice calls only. Examples of 1G cellular standards are AMPS in the U.S. and European Total Access Communication System (ETACS) in Europe. The main attraction of the 1G cellular systems was the ability to make and receive a call while moving. Such wireless access made the 1G systems quite successful. In Europe, numerous 1G systems emerged, including ETACS and NMT 500. The main drawbacks of the FDMA-based

[9]In Figure 1.18, IP represents Internet Protocol. You'll rarely see the words written out simply because this acronym is so widely used. We'll discuss many of these cellular systems a bit later, but for now we'll just explain that LTE represents Long-Term Evolution, HSPA is High-Speed Packet Access, and HSPA+ means High-Speed Packet Access Evolved.

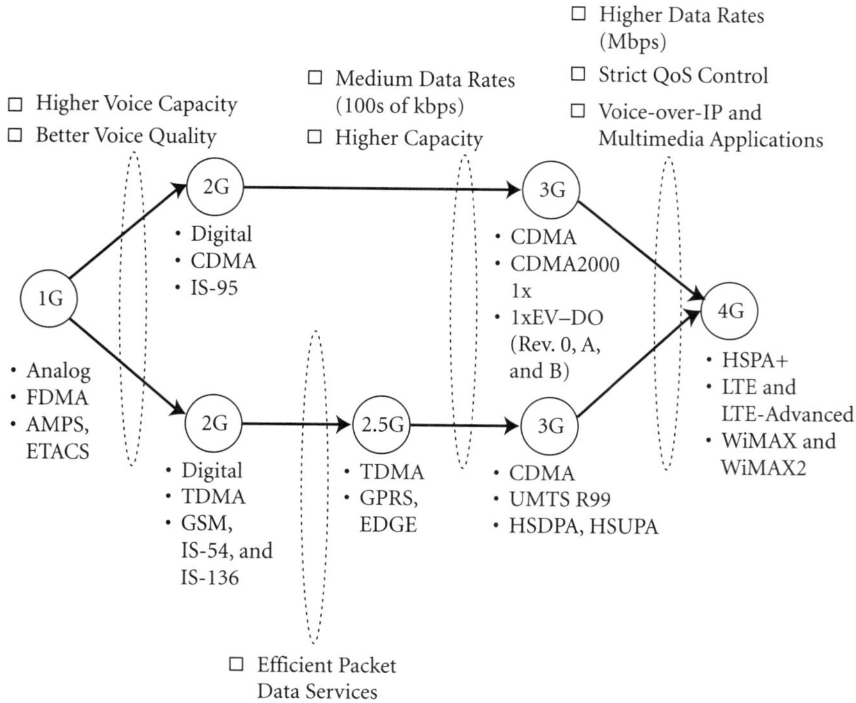

Figure 1.18: Evolution of Cellular Systems

1G systems were limited capacity and poor voice quality during user movement across cells.

Such limitations of 1G systems motivated the development of 2G standards. TDMA-based IS-54 systems were widely deployed in the U.S. and later enhanced to IS-136. TDMA-based GSM systems enabled seamless mobility throughout Europe as numerous analog systems gave way to a common GSM system. GSM is still the most dominant cellular standard in the world, serving more than 80% of wireless subscribers as of early 2011. Another digital 2G standard designed was CDMA-based IS-95. (The term CDMA is often used to denote IS-95 and its successors.) IS-95 systems use a spread-spectrum technique widely used in military communications.

A heated debate followed in the wireless industry around the superiority of GSM versus IS-95. After widespread deployments of GSM and IS-95 occurred, the air-interface efficiency of IS-95 was found to be better than that of GSM, and GSM was found to be better from the perspective of delivering a variety of services, such as short message service (SMS) and seamless roaming. When the time arrived to choose a multiple access technique for a 3G system, CDMA was chosen, which is a testament to its higher spectral efficiency.

The 2G digital systems increased voice capacity, improved voice quality, and started addressing Internet access. However, the data rates for 2G digital systems are quite low (e.g., tens of kilobits per second), and the use of the circuit-switched paradigm introduces many inefficiencies. In the circuit-switched paradigm, expensive resources are reserved in the system for a long time but are actually used only a fraction of the time. The 2G systems evolved to 3G systems to support data services (e.g., Internet access) efficiently and to increase voice capacity. The supported data rates in 3G systems are in excess of 100 kbps. IS-95-based 2G systems typically evolved to 3G CDMA2000 1xRTT (often referred to as 1x) systems, and GSM-based 2G systems evolved to 3G UMTS Release 99 (UMTS R99) systems. Both 1x and UMTS R99 use CDMA as the air-interface multiple access technique. The UMTS systems were relatively slow to take off compared to 1x systems primarily due to lack of device availability and higher cost. Hence, GSM operators needed a transient technology to fill the void on the path to the 3G UMTS system. Second-and-a-half–generation (2.5G) systems, such as GPRS (General Packet Radio Service) and EDGE (Enhanced Data Rates for GSM Evolution), enabled GSM operators to support medium data rates efficiently and compete with 1x operators.

While widespread GSM deployments initially occurred in Europe, early widespread EDGE deployments occurred in the U.S. as the TDMA operators geared up for the competition with CDMA operators and as they transitioned from IS-54 and IS-36 to GSM, GPRS, or EDGE. The focus of 1x and UMTS R99 evolution is on higher data rates and packet data services. 1x systems first evolved to 1x Evolution–Data Optimzed (1xEV-DO) Revision 0 (Rev. 0) systems. Another possible evolution path for the 1x system was to 1x Evolution–Data and Voice (1xEV-DV) systems. However, when two major U.S. operators, Verizon Wireless and Sprint, decided to evolve their 1x networks to 1xEV-DO, work on the design of 1xEV-DV systems stopped. (See Section 1.7.4 for more of this story.)

The initial focus of 1x and UMTS evolution was on non-real-time asymmetric services, such as web-browsing and FTP (File Transfer Protocol) download to the mobile device. Since higher data rates are desirable for the downlink and lower rates are adequate for the uplink for such services, technologies immediately succeeding the 1x and UMTS R99 systems emphasized the downlink performance. Table 1.7 specifies peak data rates for 3G technologies. 1xEV-DO Rev. 0 provides a 2.4 Mbps peak user data rate in the downlink and a 153.6 kbps peak user data rate in the uplink. Similarly, the UMTS R99 evolution, HSDPA (High-Speed Downlink Packet Access, which is a Release 5 feature), provides a 14 Mbps peak user data rate in the downlink and a 2 Mbps peak user data rate in the uplink. These are data rates supported and specified in the standard but may not be supported in commercial deployments. In general, data rates are higher for UMTS-based systems compared to 1x-based systems because the UMTS systems utilize larger spectrum bandwidths. For example, a Third Generation Partnership Project (3GPP) Release 5 FDD system occupies a total of 10 MHz of spectrum, and a 1xEV-DO Rev. 0 system consumes a total of 2.5 MHz spectrum. A need arose to support higher data rates in the uplink for applications, such as sending an email with a large file attached. Furthermore, QoS control became important for efficient support of delay sensitive applications, such as Voice-over-IP

Table 1.7: Peak Data Rates in 3G and 4G Networks

Technology	Peak Downlink Data Rate	Peak Uplink Data Rate	Channel Bandwidth (MHz)
1x	153.6 kbps	153.6 kbps	1.25
UMTS R99	384 kbps (2 Mbps)	128 kbps (2 Mbps)	5
1xEV-DO Rev. 0	2.4 Mbps	153.6 kbps	1.25
HSDPA	14 Mbps	384 kbps (2 Mbps)	5
1xEV-DO Rev. A	3.1 Mbps	1.8 Mbps	1.25
HSUPA	14 Mbps	5.5 Mbps	5
HSPA+ (Release 7)	21 or 28 Mbps	11 Mbps	5
1xEV-DO Rev. B	14.7 Mbps	5.4 Mbps	5
WiMAX (802.16e-2005)	250 Mbps	84 Mbps	20
LTE	300 Mbps	75 Mbps	20
WiMAX2 (802.16m)	1 Gbps (target)	500 Mbps (target)	40
LTE-Advanced	3 Gbps	1.5 Gbps	40

(VoIP), and advanced services, such as broadcast and/or multicast services. To address such issues, 1xEV-DO Rev. 0 evolved to 1xEV-DO Rev. A and HSDPA evolved to HSUPA (High-Speed Uplink Packet Access), which is a 3GPP Release 6 feature. The peak user data rates are 3.07 Mbps in the DL and 1.8 Mbps in the UL in 1xEV-DO Rev. A. The peak user data rates in HSUPA are 14 Mbps in the DL (same as in R5) and 5.5 Mbps in the UL. To achieve data rates comparable to UMTS-based systems, the 1xEV-DO Rev. A evolved to Rev. B by resorting to multiple 1.25-MHz–wide carrier frequencies. A 5-MHz–wide 1xEV-DO Rev. B system supports 14.7 Mbps in the DL and 5.4 Mbps in the UL. HSDPA and HSUPA together are referred to as HSPA. 3GPP Release 7 and beyond enhance the basic HSPA, resulting in HSPA+, which can achieve in the downlink the peak data rates of 21 or 28 Mbps in Release 7, 42 Mbps in Release 8, and 84 Mbps in Release 9. In the uplink, HSPA+ supports 11 Mbps in Releases 7 and 8 and 22 Mbps in Release 9.

Several 4G systems utilize features such as OFDMA, multiple antenna techniques, and IP-based radio and core networks. Examples of such 4G systems include Long-Term Evolution (LTE) and IEEE 802.16e-2005–based WiMAX. A standard called Ultra Mobile Broadband was defined as an evolution of 1xEV-DO Rev. B. However, lack of operator interest led to the demise of Ultra Mobile Broadband. LTE is a 3GPP Release 8 feature and is embraced by UMTS operators, such as AT&T, and 1xEV-DO operators, such as Verizon. LTE can support 300 Mbps in the DL and 75 Mbps in the UL. A WiMAX system with similar features can support about 250 Mbps in the DL and 84 Mbps in the UL. Enhancements to basic LTE and WiMAX, such as LTE-Advanced and WiMAX2, support even higher data rates. 4G systems offer extremely high data rates and tight QoS control. Voice services in LTE and WiMAX are supported via VoIP. LTE and WiMAX exploit the multiple access technique of OFDMA.

1.6.2 Spectrum Allocation in the United States

Figures 1.19, 1.20, and 1.21 specify examples of the bands available for cellular communications in the U.S.[10] The following basic sets of spectrum ranges are available: cellular spectrum in 850 MHz range, PCS spectrum in 1900 MHz range, and AWS spectrum in 1700 or 2100 MHz range and 700 MHz spectrum. Let's take a closer look at the cellular spectrum, which was allocated first for analog cellular systems.

The FCC divided a total of 50 MHz of spectrum at 800 MHz into two 25 MHz blocks, Block A and Block B, as shown in Figure 1.19. Block A was for non-wireline operators or carriers, and Block B was for wireline operators. The FCC initially specified 305 metropolitan statistical areas defined by counties based on the 1980 U.S. census. The Gulf of Mexico Service Area was later added as Market 306. The FCC then created 428 rural service areas using the counties excluded in 306 metropolitan statistical areas. The FCC created a total of 306 + 428 = 734 markets. Each market has a duopoly; it has two competing cellular service providers. AMPS-based cellular systems utilized this 800 MHz cellular spectrum. These AMPS systems were gradually replaced by digital cellular systems. The FCC announced that cellular operators need not provide AMPS service beyond February 2008 [25, 26]. Most cellular networks discontinued the AMPS service.

As cellular communications became popular, an additional spectrum, called the PCS spectrum, was allocated. Digital cellular systems were first deployed at PCS frequencies. Figure 1.20 depicts how the PCS spectrum is allocated by the FCC. The PCS band consists of six frequency blocks, A through F. These blocks are licensed in specific geographical areas called major trading areas and basic trading areas. The fifty-one major trading areas and 493 basic trading areas are defined based on maps [27]. Blocks A and B are licensed in each of the fifty-one major trading areas, and blocks C through F are licensed in each of the 493 basic trading areas.

Though digital cellular systems initially used the 1900 MHz band, they have also been deployed at the 850 MHz band. More spectrum was made available to cellular systems in 2006 as the AWS spectrum to meet the demands of bandwidth-intensive multimedia applications such as full-motion videos and messaging (in addition to regular voice services) [28]. Figure 1.21 specifies the AWS spectrum that can be used for cellular communications.

The FCC allocated licenses for the six blocks as follows:

- Block A: 734 cellular market area licenses;
- Blocks B and C: 176 economic area licenses each;

[10]In the U.S., initial LTE deployments use the 700 MHz spectrum and initial WiMAX deployments use the 2.5 MHz spectrum.

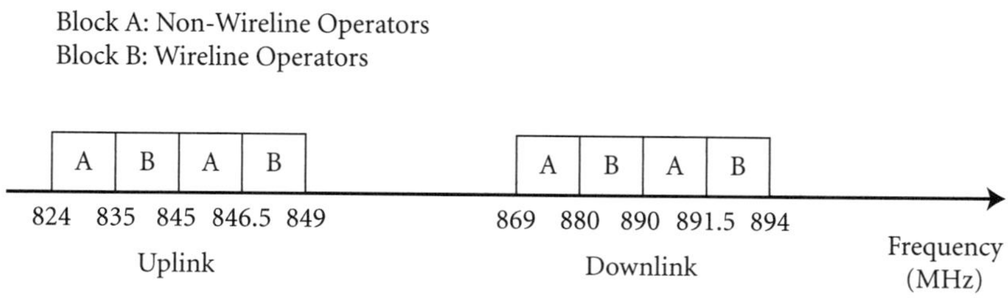

Figure 1.19: Cellular Band for Cellular Communications

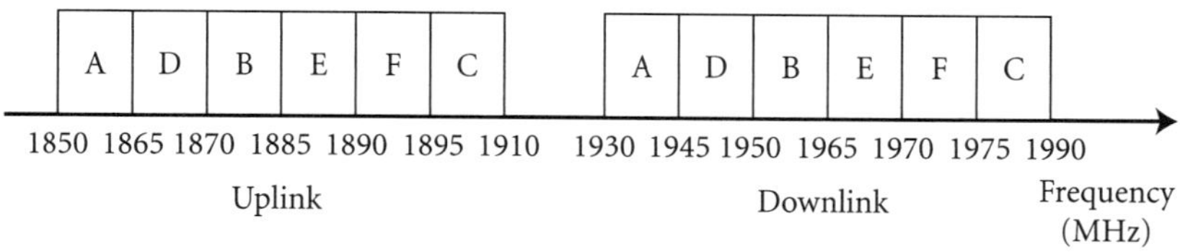

Figure 1.20: PCS Band for Cellular Communications

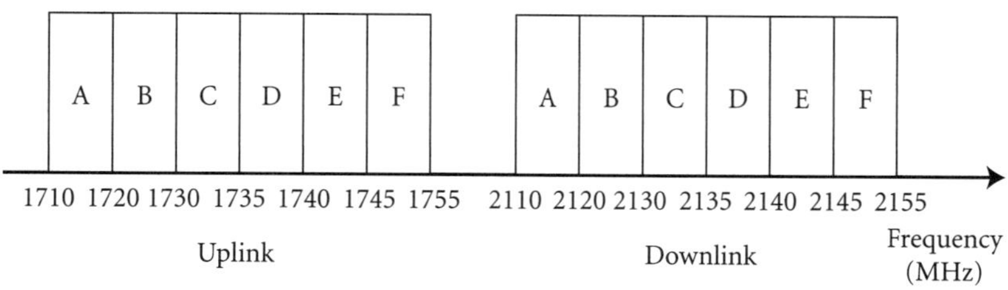

Figure 1.21: AWS Band for Cellular Communications

- Blocks D, E, and F: 12 regional economic area grouping licenses each.

Ever Wondered... What is a multi-band multimode radio? You might have come across the terms *dual-band dual-mode*, *dual-band tri-mode*, and *multi-band multi-mode* phone. *Dual-band* means that the phone supports two frequency bands, e.g., the 850 MHz band and 1900 MHz band. *Tri-mode* means that the phone supports three cellular standards. If a phone supports GSM, EDGE, and UMTS, the phone is said to be tri-mode because it supports three modes or standards.

1.6.3 Spectrum Allocation Outside the U.S.

Different countries have different frequency bands allocated to cellular services, and examples are given below.

- **European Union.** The following 2G spectrum is used in the EU [29]: (i) Extended GSM: 880–890 MHz for the uplink and 925–935 MHz for the downlink, (ii) GSM-900: 890–915 MHz for the uplink and 935–960 MHz for the downlink, and (iii) GSM-1800: 1710–1785 MHz for the uplink and 1805–1880 MHz for the downlink. The following 3G spectrum is allocated to 3G UMTS systems in the EU [30]: (i) 1920–1980 MHz for the uplink and 2110–2170 GHz for the downlink in case of FDD, (ii) 1900–1920 MHz and 2010–2025 MHz in case of TDD.

- **India.** 2G cellular systems are deployed at 800 MHz, 900 MHz, and 1800 MHz bands [31]. Target bands for 3G deployments are 450 MHz, 800 MHz, and 2100 MHz. A total of at least (2×32.5 MHz) of spectrum is expected to be available for 3G cellular systems with five blocks of (2×5 MHz) at 2100 MHz, one block of (2×5 MHz) at 450 MHz, and two blocks of (2×1.25 MHz) at 800 MHz [32]. 20 MHz spectrum at 2.3 GHz was made available for broadband wireless access in early 2010.

- **China.** The frequency bands used by 2G networks in China are as follows: (i) GSM-Railroad: 915–919 MHz for the uplink and 960–964 MHz for the downlink, (ii) GSM-900: 885–915 MHz for the uplink and 930–960 MHz for the downlink, (iii) GSM-1800: 1710–1755 MHz for the uplink and 1805–1850 MHz for the downlink (2×45 MHz), (iv) IS-95 CDMA: 825–835 MHz for the uplink and 870–880 MHz for the downlink.

1.7 ECOSYSTEM OF CELLULAR SYSTEMS

This section provides a glimpse of the types of companies involved in cellular communications. The cellular communications industry involves a variety of players as

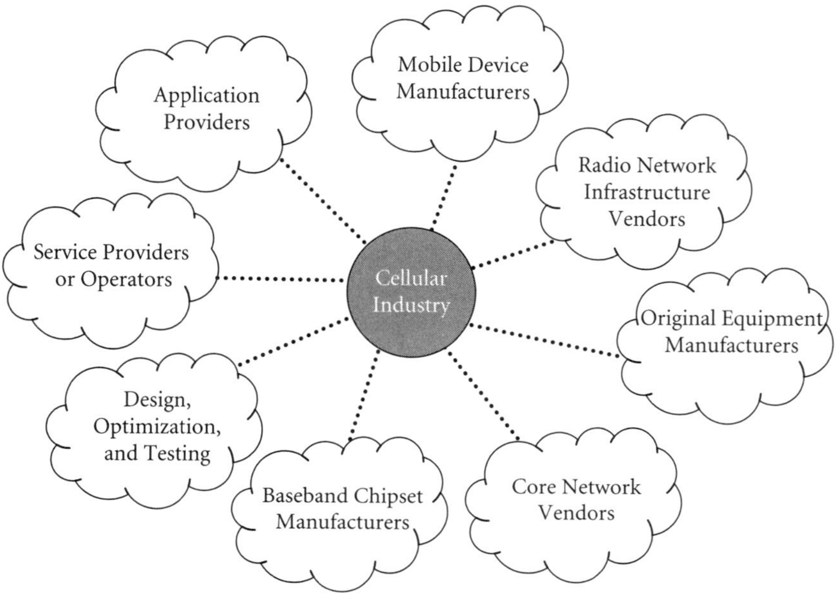

Figure 1.22: Cellular Communications Industry

seen from Figure 1.22. *The lists of companies mentioned in this section are by no means exhaustive or complete.* The intent here is merely to provide some examples. If the company for which your heart is beating is mentioned here, congratulations! Otherwise, please accept the authors' sincere apologies.

1.7.1 Service Providers

Service providers offer cellular services to consumers. They provide a variety of service or subscription plans to residential and/or business subscribers depending upon their business models. Table 1.8 lists some of the major cellular operators around the world and the technologies they utilize.

The service operators have seen both breakups and mergers and acquisitions in a short span of a couple of decades. For example, the monopoly of AT&T resulted in its breakup as mandated by AT&T's settlement with the U.S. Department of Justice, leading to the creation of many "baby Bells" or Regional Bell Operating Companies, such as Ameritech Corporation, Bell Atlantic Corporation, BellSouth Corporation, NYNEX Corporation, Pacific Telesis Group, Southwestern Bell Corporation, and U.S. West [33]. However, in 2001, SBC Communications (Southwestern Bell Corporation, now AT&T) and BellSouth jointly created Cingular Wireless. Furthermore, in December 2006, AT&T became the sole owner of Cingular Wireless after acquiring BellSouth [34].

Table 1.8: Cellular Service Providers

Service Provider	Major Service Regions	Technologies
Vodafone	Europe	GSM, GPRS, UMTS, HSDPA, HSUPA, HSPA+, LTE
Telefonica O2	Europe and Latin America	GSM, GPRS, EDGE, IS-95, 1x, 1xEV-DO, HSPA+, LTE
Orange	Europe	GSM, GPRS, EDGE, UMTS, HSDPA, HSUPA, HSPA+, LTE
Verizon Wireless	U.S.	IS-95, 1x, 1xEV-DO, LTE
AT&T	U.S.	GSM, GPRS, EDGE, UMTS, HSDPA, HSUPA, HSPA+, LTE
T-Mobile	U.S.	GSM, GPRS, EDGE, UMTS, HSDPA, HSUPA, HSPA+
Sprint-Nextel and Clearwire	U.S.	IS-95, 1x, 1xEV-DO, WiMAX
China Mobile	China	GSM, GPRS, EDGE
China Unicom	China	GSM, GPRS, IS-95, 1x, UMTS, HSPA
Bharti Airtel	India	GSM
Bharat Sanchar Nigam Ltd.	India	GSM, UMTS, HSDPA, HSUPA, WiMAX
Reliance	India	GSM, 1x, 1xEV-DO, WiMAX
Hutchison Essar	India	GSM
Tata Teleservices	India	1x, 1xEV-DO, WiMAX

Some companies lease their network from a regular service operator rather than build their networks, and these companies are called *mobile virtual network operators*. For example, Disney is leasing network capacity from Sprint-Nextel to offer services to consumers [35], such as Disney Mobile, which gives parents control over their children's phone (e.g., how and when the phone can be used).

1.7.2 Mobile Device Manufacturers

Cell phone manufacturers produce devices with different capabilities and features. High-end devices typically focus on business customers, tend to be feature-rich and expensive, and exploit the latest technologies. Table 1.9 mentions major mobile device vendors around the globe.

Early mobile devices were just phones because the initial cellular standards focused on the voice service. The advent of the Internet led to multimedia devices (e.g., smartphones) that are no longer just phones but virtually a mobile laptop!

Table 1.9: Mobile Device Manufacturers

Vendors	Headquarters	Technologies
Apple	U.S.	GSM, GPRS, UMTS, EDGE, HSDPA, HSUPA, HSPA+, LTE
Nokia	Finland	GSM, GPRS, UMTS, EDGE, HSDPA, HSUPA, HSPA+, LTE
Motorola Mobility†	U.S.	GSM, GPRS, UMTS, EDGE, HSDPA, HSUPA, 1x, 1xEV-DO, WiMAX, LTE
Samsung	S. Korea	1x, 1xEV-DO, UMTS, HSDPA, HSUPA, WiMAX, LTE
Sony Mobile Communications	U.K.	GSM, GPRS, EDGE, UMTS, HSDPA, HSUPA, HSPA+, LTE
LG	Korea	GSM, GPRS, UMTS, EDGE, HSDPA, HSUPA, HSPA+, 1x, 1xEV-DO, LTE
HTC	Taiwan	GSM, GPRS, EDGE, UMTS, HSDPA, HSUPA, HSPA+, 1x, 1xEV-DO, LTE

† Google acquired Motorola Mobility in 2012 and agreed to sell it to Lenovo in January 2014.

1.7.3 Radio Network Infrastructure Vendors

Radio infrastructure vendors build the RAN components, such as the BSs or Node Bs and the BSCs or RNCs. Mergers of service providers have led to consolidation of the infrastructure vendors as the number of customers (i.e., service operators) has decreased. For example, in the 1990s, major radio network vendors included Lucent Technologies, Motorola, Alcatel, Siemens, Nokia, and Samsung. In the late 2000s, Alcatel acquired Lucent Technologies and bought Nortel's UMTS infrastructure business. Siemens and Nokia's infrastructure units merged. Nokia Siemens Networks acquired most of Motorola's wireless infrastructure in 2010 and later became a wholly owned subsidiary of Nokia. Also, in the late 2000s, Ericsson acquired Nortel's CDMA business. Table 1.10 lists major radio network vendors.

1.7.4 Baseband Chipset Manufacturers

The radio interface protocol stack includes the physical layer as the first layer, frequently called L1. The physical layer processing can be thought of as having two types of processing: baseband processing and RF processing. The baseband processing includes functions such as addition of redundancy to minimize the impact of errors due to a hostile radio environment, and the RF processing includes functions such as signal amplification and frequency upconversion (i.e., translating the baseband signal

Table 1.10: Radio Network Vendors

Radio Network Vendor	Headquarters	Technologies
Ericsson	Sweden	GSM, GPRS, EDGE, UMTS, HSDPA, HSUPA, HSPA+, 1x, 1xEV-DO, LTE
Alcatel-Lucent	France	1x, 1xEV-DO, GSM, GPRS, EDGE, UMTS, HSDPA, HSUPA, HSPA+, LTE, WiMAX
Nokia	Finland	GSM, GPRS, EDGE, UMTS, HSDPA, HSUPA, HSPA+, LTE
Samsung	S. Korea	1x, 1xEV-DO, WiMAX, LTE
ZTE	China	1x, 1xEV-DO, UMTS, HSDPA, HSUPA, HSPA+, WiMAX, LTE
Huawei	China	GSM, GPRS, EDGE, UMTS, 1x, 1xEV-DO, WiMAX, LTE

to the intended carrier frequency, such as 1.9 GHz). The baseband processing is often carried out in ASICs (Application Specific Integrated Circuits) for faster processing. Such an ASIC is called a baseband chipset and is one of the most critical components of the MS and the BS. Very few companies possess the know-how to derive optimal performance from the chipset. The baseband processing is sometimes carried out by DSPs (digital signal processors) and FPGAs (field programmable gate arrays). Table 1.11 lists the chipset manufacturers for various technologies.

Table 1.11: Examples of Chipset Vendors

Vendor	Headquarters	Technologies
Qualcomm	U.S.	GSM, EDGE, UMTS, HSDPA, HSUPA, HSPA+, 1x, 1xEV-DO, LTE
Texas Instruments	U.S.	GSM, EDGE, UMTS, HSDPA, HSUPA, HSPA+, WiMAX, LTE
Broadcom	U.S.	GSM, GPRS, EDGE, UMTS, HSDPA, HSUPA, WiMAX, LTE
Infineon	Germany	GSM, GPRS, EDGE, UMTS, HSDPA, HSUPA, HSPA+, WiMAX, LTE
Ericsson	Switzerland	GSM, EDGE, UMTS, HSDPA, HSUPA, HSPA+, LTE
Sequans	France	WiMAX, LTE

A Bit of Cellular History... Unavailability or delayed availability of the chipsets could even result in the demise of a technology. For example, 1x had two possible evolution paths: 1xEV-DO and 1xEV-DV. IS-95 and its successor technologies have traditionally relied upon Qualcomm for chipset design. Qualcomm made available the 1xEV-DO chipsets while commercial-grade 1xEV-DV chipsets never appeared. Two major CDMA operators in the U.S., Verizon Wireless and Sprint, decided to follow the evolution path to 1xEV-DO because the 1xEV-DO chipset was available and the 1xEV-DV chipset wasn't. Hence, commercial 1x systems evolved to 1xEV-DO.

1.7.5 Original Equipment Manufacturers

Service providers buy the radio network and core network components from the infrastructure vendors. Some vendors may use components from another manufacturer rather than doing the in-house design and development of these components. In such a case, the infrastructure vendor performs integration of the components built by another manufacturer. The companies that design and build the components are called original equipment manufacturers (OEMs). Some factors that motivate a big vendor to partner with OEMs are possibly shorter time-to-market and lack of business case for in-house development. The OEMs run the risk of losing the business if their customer, a big company, decides to pursue in-house development. The benefit to the OEMs, other than the business from the big vendor, is that they need not design the entire radio network that needs to work with legacy networks.

1.7.6 Core Network Vendors

Many radio network equipment manufacturing companies also provide core network components. Let's refresh your memory.

Do You Recall... What are the two types of core networks in a 3G network? Hint: Their names start with C and P. The footnote gives the answer.[11]

Table 1.12 gives examples of the companies that provide core network infrastructure components. In general, the interface between the radio network and the core network is *open*, meaning that the radio network from one vendor can operate with the core network from another vendor. Such interoperability is ensured by specifying the radio-core interface in the cellular standards.

[11]In a 3G network, two core networks exist: a circuit-switched core network and a packet-switched core network.

Table 1.12: Core Network Manufacturers

Core Network Vendor	Headquarters	Components
Ericsson	Sweden	Softswitch, IP Multimedia Subsystem, gateways and routers
Alcatel-Lucent	France	Softswitch, IP Multimedia Subsystem, gateways and routers
Nokia	Finland	IP Multimedia Subsystem, gateways and routers
Samsung	S. Korea	Softswitch, IP Multimedia Subsystem, gateways and routers
ZTE	China	Softswitch, IP Multimedia Subsystem, gateways and routers
Huawei	China	Softswitch, IP Multimedia Subsystem, gateways and routers
Cisco or Starrent	U.S.	IP Multimedia Subsystem, gateways and routers

1.7.7 Design, Optimization, and Testing Tool Manufacturers

Different types of tools are available to facilitate the design of the cellular network elements and the design, deployment, and optimization of cellular networks. Various tools are required to test the operation of the MS, the BS, and other components of a cellular system. These tools can ensure that the component, such as the MS, conforms to the requirements of the standard and can help troubleshoot problems in the lab before the components are used in a widespread commercial deployment. Some tools are useful during the design phase of the components, such as the MS. Some tools facilitate the design and planning of a commercial cellular network. The radio network planning and design phase specifies where the BSs should be located and what settings should be used for various configuration parameters. Table 1.13 lists products for network planning and design. Some tools can be used to collect various logs and measure relevant performance metrics in a pre-commercial network through drive tests. Table 1.13 also gives examples of such drive-test tools. These tools can also be used for optimization of commercial deployments.

1.7.8 Appplication Providers

Application providers develop applications. Apple, Microsoft, BlackBerry Limited, and Nokia are examples of application providers. Thousands of applications are available for just Apple's iPhone alone! Example applications include restaurant reviews by Urbanspoon and iTunes by Apple.

Table 1.13: Design and Test Tool Manufacturers

Tool Vendor	Headquarters	Tool Example	Tool Type	Example Technologies
Qualcomm	U.S.	CAIT	Data collection and troubleshooting	1x, 1xEV-DO, UMTS, HSDPA, HSUPA, HSPA+, LTE
Agilent	U.S.	Protocol Testers		GSM, GPRS, EDGE, UMTS, HSDPA, HSUPA, HSPA+, WiMAX, LTE
Andrew	U.S.	Invex 3G and Invex.NxG	Data collection and optimization	1x, 1xEV-DO, UMTS, GSM, GPRS, HSDPA, LTE
Ascom	Switzerland	TEMS	Data collection and troubleshooting	GSM, UMTS, HSDPA, HSUPA, HSPA+, LTE
InfoVista	France	Mentum Planet and CellPlanner	Network planning and optimization	GSM, CDMA, 1x, UMTS, HSDPA, HSUPA, HSPA+, LTE
Anritsu	Japan	MS and BS Testers	Testing and troubleshooting	1x, 1xEV-DO, UMTS, HSDPA, HSUPA, WiMAX, LTE
Rohde and Schwartz	Germany	Universal radio communication testers	Testing	GSM, GPRS, EDGE, UMTS, HSPDA, 1x, 1xEV-DO
Actix	U.K.	Analyzer	Data collection and optimization	1x, 1xEV-DO, GPRS, EDGE, UMTS, HSDPA, HSUPA, HSPA+, LTE
Tektronix	U.K.	Protocol Analyzers		GSM, GPRS, EDGE, UMTS, HSDPA, HSUPA, HSPA+, WiMAX, LTE
Anite	U.K.	Nemo Analyzer		GSM, GPRS, EDGE, UMTS, HSDPA, HSUPA, HSPA+, WiMAX, LTE

1.8 PHASES OF A CELLULAR SYSTEM

A cellular system is not static; it undergoes changes througout its lifecycle. The lifecycle of the cellular system from conception to evolution involves numerous phases. A cellular system undergoes various stages as shown in Figure 1.23.

1.8.1 Phase 1: Specifications of the Standard

Long before a cellular system can be deployed, the system must first be defined. A set of specifications outlining the requirements is created as part of a standard. The creation of the standard consists of several stages. First, the basic expectations or service requirements are specified, and such requirements are often considered Stage 1 documents of a standard. For example, a Stage 1 requirement may specify that the intended goals are to support voice services and medium-speed packet data services to support applications such as email. A high-level network architecture is then defined, and various components of the network are identified. Stage 2 documents specify the network architecture and high-level signaling flows. The details of the signaling aspects for technical realization or implementation are described in Stage 3 documents. Various layers of the protocol stack, such as the physical layer and medium access control layer, are gradually defined in one or more documents.

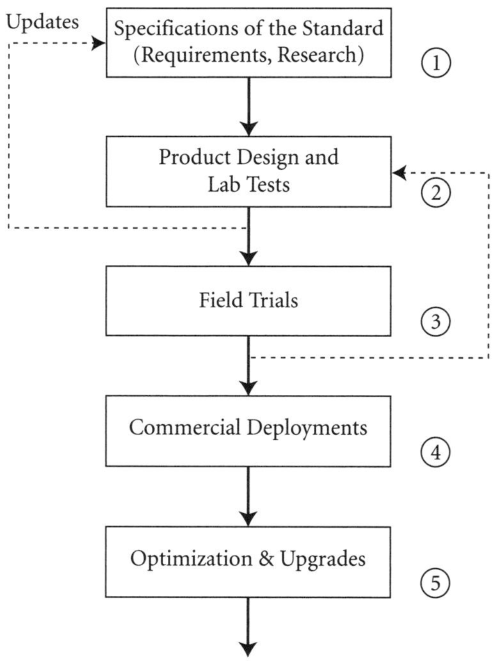

Figure 1.23: Phases of a Cellular Network

Numerous options for a given task, e.g., Quadrature Phase Shift Keying (QPSK) and 16-Quadrature Amplitude Modulation (16-QAM) options for a modulation scheme, are evaluated and selected through voting. Typically, the expected performance of available or feasible options is considered while choosing the specific techniques. In practice, the standardization process becomes political, and different camps advocating different techniques emerge. At the end of the day, democracy (or sometimes the heavy influence of a company, such as Ericsson or Qualcomm!) prevails, and the standard is created. Mandatory features of the standard become sources of revenue for companies that have essential or major patents or IPRs (intellectual property rights). For example, \$1.2 billion out of a total of \$2 billion of revenue for Qualcomm came from IPRs for the first financial quarter in 2007 [36, 37]. Large corporations often give one another rights to use their IPRs, and, in such cases, no exchange of money takes place.

1.8.2 Phase 2: Prototype Design and Lab Tests

As the standard is being created, prototype devices and prototype infrastructure elements are designed and developed. These prototypes serve as proof-of-concepts and are exhibited in telecom tradeshows and conferences. The Cellular Telecommunication and Internet Association (CTIA) is an example of a tradeshow often used by wireless companies to showcase their products and services. Lab tests of prototypes help troubleshoot implementation problems and can sometimes point to changes in the standard. Interoperability of the wireless device and the infrastructure can be tested in the lab tests. Successful lab trials provide a good public relations opportunity to the manufacturing companies. The device and infrastructure manufacturer can draw the attention of service providers through such tests.

1.8.3 Phase 3: Field Trials

Before a wireless network based on a standard is made commercially available to consumers or subscribers, field trials are carried out to evaluate and quantify the performance of the network. Expected capacity, data rates, and coverage can be verified or estimated during this phase.

While theoretical performance for a technology can be quite close to the real-world performance, sometimes we are, well, off-the-mark significantly. For example, the capacity of CDMA-based IS-95 systems was expected to be UL-limited. However, real-world deployments led to exactly the opposite outcome: The capacity was found to be DL-limited. In other words, the system ran out of DL resources before running out of UL resources. Many research papers have focused on improving the UL to increase the capacity. No matter how hard we try to improve the UL, doing so would not increase overall capacity because the downlink would still be the bottleneck. Thus, real-world field trials are very useful, especially when brand-new technologies are deployed for the first time.

1.8.4 Phase 4: Commercial Deployments

Once the standard-based cellular network and mobile device have been tested in lab tests and field trials, we are ready for commercial deployment. A network deployment that is unconstrained by the reuse of existing cell sites is usually referred to as a *green-field deployment*. Such a deployment occurs for a new service operator. In practice, a new technology deployment is usually required to maximize the reuse of existing cell sites. Such a deployment is called an *overlay deployment* and presents numerous challenges. In some cases, even the same antennas must be reused to save costs. The link budget[12] comparison between an existing technology and a new technology may indicate the potential need for extra cell sites to fill the coverage holes.

The network deployment could be *coverage-driven* or *capacity-driven*. The type of deployment primarily depends upon the expected subscriber density. The link budget provides the maximum cell radius and hence the largest possible cell size. When subscriber density is low, the largest possible cell size is used to cover a given geographical area with as few BSs as possible. Such a deployment is referred to as *coverage-driven deployment* in which the focus is on coverage and the cell size is dictated by the link budget. Typical rural and suburban areas are likely to have coverage-driven deployment. In capacity-driven deployment, a high subscriber density necessitates smaller cell sizes because cell capacity is limited and more cells are required to support a large number of subscribers in a given area. Techniques such as down-tilting of antennas and reduction in antenna heights can be used to reduce cell coverage.[13] Since such deployment focuses on capacity rather than coverage, it is called *capacity-driven deployment*. As the subscriber growth continues, capacity must increase. The techniques for meeting the subscriber growth vary depending upon the technologies, such as TDMA and CDMA. Examples of capacity-enhancing techniques are sectorization, the addition of new cell sites, and the addition of radio channels.

1.8.5 Phase 5: Optimization

The initial network deployment is hardly optimal. Optimization of a deployed network is an on-going and strenuous task. The network optimization is carried out to achieve the best possible performance within the given constraints, including non-technical constraints such as cost. An optimal network performance keeps subscribers happy and the wireless business growing. Specific network optimization techniques vary for different technologies; however, general techniques include changes in antenna parameters (e.g., down-tilt, height, azimuth, and beamwidth), changes in the BS power amplifier, and optimization of technology-specific parameters (e.g., the parameters that influence how the mobile device first accesses the wireless network to establish

[12]The link budget is the maximum allowable path loss (e.g., 125 dB) between the transmitter and receiver at a target QoS level at the cell edge. The link budget, when used with a suitable propagation model, corresponds to a certain cell radius (e.g., 3 km).

[13]See Chapter 2 for details of various antenna parameters.

a call). If properly done, network optimization can postpone the addition of a new cell site and/or a carrier frequency, reducing the capital expenditure.

1.9 PERFORMANCE OF CELLULAR SYSTEMS

Good performance of a cellular network leads to satisfaction for the customer and customer loyalty and higher revenue for the service provider. The performance of a network is quantified by numerous performance metrics and measures, which are often referred to as *key performance indicators* (KPIs). Some KPIs quantify cellular service perceived by subscribers, and some indicate how efficiently the network resources are being utilized. Worse than expected KPIs may point to a need for network optimization and/or additional capacity. Several KPIs for a cellular network serving voice and data users are discussed further here.

1.9.1 Sector Capacity and Call Blocking Probability

Remember that sector capacity is the maximum number of users that can be simultaneously supported in the sector. If the sector capacity is twenty, only twenty users can be active (i.e., involved in radio communications with the sector). If the twenty-first user attempts to make a call, that call is blocked. Since the twenty-first user will always be blocked, the blocking probability is 1 (i.e., 100%). The *call blocking probability* is the ratio of the number of calls blocked and the number of successful call attempts. A successful call attempt occurs when the network detects that a user is attempting to access the radio network. The radio network executes the call admission control algorithm, which blocks the call if resources are inadequate to support the call. A cellular network cannot operate at 100% blocking. The cellular network typically operates with a 1–3% blocking probability. The Erlang-B formula describes the relationship between three parameters: the sector capacity, the blocking probability, and the average number of supported users.[14] The average number of users is referred to as the *Erlang capacity* of the sector. It is also called the *offered traffic intensity*. The maximum number of users is equivalent to the number of trunked channels.[15] Eq. (1.9) specifies the Erlang-B formula.

$$P_b = \frac{(E^N)/(N!)}{\sum_{i=0}^{N}((E^i)/(i!))} \tag{1.9}$$

In Eq. 1.9, P_b is the probability of call blocking, N is the number of trunked channels, and E is the Erlang capacity. For example, assume that the maximum number of users

[14]The Erlang-B formulation uses the offered traffic as a parameter. When the call blocking probability is low (e.g., 2% or 0.02), virtually no difference exists between the offered traffic and the carried traffic. Carried traffic is $(1 - P_b) * \mathit{Offered\ Traffic}$ where P_b is the call blocking probability. Due to almost identical values of offered traffic and carried traffic, we are using the average number of supported users as a parameter (which quantifies the carried traffic and approximately represents the offered traffic intensity).

[15]Trunking allows multiple users to share the same resource.

supported by the air interface in a sector is twenty and the call blocking probability is 2% (i.e., 0.02). Then, the Erlang capacity of the sector based on Eq. (1.9) is 13.15 Erlangs.

Several websites (such as `http://www.erlang.com/calculator/erlb/` and `http://www.ansapoint.com/calculator/erlb/`) provide Erlang-B calculators, which allow users to enter values for two parameters and find the third parameter. Some websites also provide Erlang-B tables (e.g., `http://www.umtsworld.com/resources/erlang.htm`). See Appendix B for an Erlang-B table for limited ranges of parameters.

The Erlang capacity is thus the average number of users that can be served simultaneously at a given blocking probability. The call blocking probability is a way of specifying grade of service (GoS). For an optimized network deployment, consistently high call blocking probability indicates the need for additional cell capacity. The number of distinct radio links being used simultaneously in a sector is the *Erlang sector capacity* and can be calculated by taking the time average of the number of simultaneously active radio links. Since all the subscribers do not talk all the time, the actual number of subscribers that can be served is much higher than the Erlang capacity. An approach typically used to estimate the number of supportable subscribers relies on the statistics of busy hour call attempts (BHCA). Assume that a subscriber makes μ calls per minute and talks for H minutes during a call. μ is the call arrival rate per subscriber, and H is the call holding time in minutes. Then, the Erlang traffic generated by one subscriber is μH Erlang. Further assume that the sector capacity is X Erlangs. Hence, the number of subscribers or users that can be supported in a sector is $U = (X/(\mu H))$. U is very useful to service providers because they can provision an adequate number of sectors to cover a given geographical area based on the market predictions of cellular service penetration in such an area.

The Erlang formulation is quite useful for traditional voice calls. However, it is not valid for data calls or sessions (e.g., web-browsing sessions) because the network's resources are no longer tied up completely during the data calls.

Example. When a user makes three calls per hour and the average call duration is two minutes, μ is $3/60$ calls per minute = 0.05 calls per minute, and $H = 2$ minutes. The user traffic intensity $= \mu H$ Erlang $= 0.05 * 2 = 0.1$ Erlang. In the case of $N = 20$ and $P_b = 0.02$, the Erlang capacity is $E = 13.15$ Erlang from the Erlang-B formula or table. The number of subscribers supported in the sector becomes $U = 13.15$ Erlang/0.1 Erlang $= 131.5$. If the total population is 100,000 in a suburban area and cellular service penetration is 70%, a total of $70000/U = 70000/131.5 = 532$ sectors would be required to cover the suburban area. In summary, when cellular subscribers are uniformly distributed among 532 sectors and each subscriber generates 0.1 Erlang of traffic during the busy hour, the call blocking probability would be 2%. Two out of one hundred call attempts would be blocked due to inadequate resources.

1.9.2 Accessibility, Access Failure Rate, and Paging Failure Rate

When the user device has no dedicated radio channel assigned to it by the RAN, it utilizes a common channel called the *access channel* in the uplink. When the RAN cannot detect the MS's access attempt, it sends no acknowledgment, and such a scenario is termed an access failure. The *access failure rate* is the ratio of the number of failed access attempts and the total number of access attempts. An access failure rate of around 2% is frequently the goal of a commercial cellular network. The access failure rate is a measure of accessibility. Access failure rate quantifies the accessibility when the MS is trying to access the RAN. The *paging success or failure rate* measures the accessibility when the RAN is trying to access the MS. When the MS has no dedicated radio channel in the downlink, the RAN utilizes the common downlink channel called the *paging channel* to reach the MS. The paging failure rate is the ratio of the failed page attempts and the total number of page attempts. The target paging failure rate is about 2% in commercial networks.

1.9.3 Retainability and Call-Drop Rate

Once the call has been successfully established and dedicated resources have been allocated to the call, retaining the call becomes important. The call-drop rate is the ratio of the number of dropped calls to the total number of successfully admitted calls. The call-drop rate statistics are a measure of retainability. While call blocking probability, access and paging failure rate, and call-drop rate are all important, call-drop rate is usually considered more critical from the user's perspective. Imagine two scenarios. In one scenario, you are dialing a phone number in an attempt to call someone and the call cannot be connected. In this case, you would re-try and would likely succeed after a few attempts. In another scenario, you are talking to someone and the call drops in the middle of the conversation. Though the degree of annoyance depends upon your personality, you would probably be more annoyed in the second scenario than in the first. Call drops are so important that some cellular operators use their low call-drop rate in advertisements to promote their cellular service.

1.9.4 Error Rate and Quality of Service

The radio environment is usually hostile, and information does get corrupted over the air interface. Information is typically transmitted in units of radio frames or blocks, with each radio frame or block containing a certain number of bits. The cyclic redundancy check in a frame or block enables error detection at the receiver. When one or more bit errors occur in a frame or block, frame or block error is said to occur. The frame error rate or the block error rate is the ratio of the number of bad frames or blocks (i.e., frames containing bit errors) to the total number of frames or blocks. Frame error rate or block error rate specifies the QoS for an active radio link. The target physical layer frame error rate or block error rate for voice service typically ranges from 1–3%.

1.9.5 Cell-Edge Reliability and Cell-Area Reliability

The target frame error rate cannot be achieved everywhere all the time in a cellular system because the radio propagation environment is quite dynamic. Reliability measures specify how often the target QoS is met. *Cell-edge reliability* indicates what percentage of calls made at the cell edge would experience the target QoS. Similarly, *cell-area reliability* indicates what percentage of the cell area experiences the target QoS. The cell-edge reliability is one of the parameters that influences the link budget.

Do You Recall... Do you remember the significance of the link budget? See the footnote for the answer.[16]

The relationship between the cell-edge reliability and cell-area reliability is summarized as shown here [38]. Let the average signal strength x be the normal random variable with mean $\bar{x}$ (measured in decibel-milliwatts) and standard deviation σ dB.[17] Let x_0 be the signal strength threshold above which the signal quality is considered good. The probability density function (pdf) of x is given by

$$p(x) = \frac{e^{((-(x-\bar{x})^2)/(2\sigma^2))}}{\sigma\sqrt{2\pi}}. \tag{1.10}$$

The probability that x exceeds x_0 is

$$P_{x_0} = \int_{x_0}^{x} p(x)dx = \frac{1}{2} + \frac{1}{2}\operatorname{erf}\left(\frac{x_0 - \bar{x}}{\sigma\sqrt{2}}\right) \tag{1.11}$$

where erf(.) is the error function and is related to the Q-function as shown in Eq. (1.12).

$$\mathrm{Q}(z) = \frac{1}{\sqrt{2\pi}}\int_{z}^{\infty} e^{((-x^2)/2)}dx = \frac{1}{2} - \frac{1}{2}\operatorname{erf}\left(\frac{z}{\sqrt{2}}\right) \tag{1.12}$$

Microsoft Excel readily gives the error function erf(.).

Eq. (1.11) provides the cell-edge reliability. For example, if x_0 of -90 dBm is considered a good value of the signal strength, $\bar{x}$ at the cell edge (i.e., at the distance of cell radius R) is -85 dBm, and σ is 10 dB, the cell-edge reliability is

$$P_{x_0} = \frac{1}{2} + \frac{1}{2}\operatorname{erf}\left(\frac{-90 + 85}{10\sqrt{2}}\right) = 0.69. \tag{1.13}$$

[16]The link budget quantifies the maximum tolerable path loss between the transmitter and the receiver for a given QoS level. The link budget essentially dictates the cell radius.

[17]A phenomenon called *fading* causes variations in the signal strength. Chapter 3 discusses different types of fading, including shadow fading, which is usually modeled as a log-normal distribution.

Note that $\text{erf}(-y) = \text{erf}(y)$ in Eq. (1.13) because erf expects a non-negative argument. The cell-area reliability, F_u, indicates the probability that the signal strength exceeds x_0 within the circle of cell-radius R. F_u can be calculated as

$$F_u = \int P_{x_0} dA. \tag{1.14}$$

Assume that the average signal strength $\bar{x}$ is given by

$$\bar{x} = \alpha - 10n \log_{10}(r/R) \tag{1.15}$$

where α depends on parameters such as the transmit power, antenna height, and antenna gains, r is the distance between the transmitter and the receiver, n is the propagation path loss exponent, and R is the cell radius. Let

$$a = \frac{(x_0 - \alpha)}{\sigma\sqrt{2}} \qquad \text{and} \qquad b = \frac{10n \log_{10}(e)}{\sigma\sqrt{2}}. \tag{1.16}$$

Then, the cell-area reliability is

$$F_u = \frac{R^2}{2}\left\{1 + \text{erf}(a) + e^{((2ab+1)/(b^2))}\left[1 - \text{erf}\left(\frac{ab+1}{b}\right)\right]\right\}. \tag{1.17}$$

In Eq. (1.17), erf(.) is the error function [39]. Reudink's work includes a graph representing the relationship among cell-edge reliability, cell-area reliability, and σ/n [38]. As an example, for $\sigma/n = 2$ and a cell-edge reliability of 0.75, the cell-area reliability is approximately 0.90 [40]. For a cell-edge reliability of 0.90, the cell-area reliability is approximately 0.96. The target cell-edge reliability of 75% and 90% are widely used in commercial cellular networks.

1.9.6 Peak User Data Rate

For a system providing data services, such as email and web-browsing, peak data rates are a measure of how fast the data packets can be transferred. The peak user data rates for the downlink and the uplink are relatively easy to specify because the cellular standards and the real-world implementations of the MS and the RAN readily provide their values. These data rates are the maximum possible data rates that a given user can achieve in a given direction during some time period. These peak rates may or may not be the data rates actually perceived by the user depending upon factors such as the prevailing channel conditions and the resource availability for the MS under consideration. For example, the UMTS Release 99 standard specifies a 2 Mbps peak user data rate; however, the commercial realization of Release 99 supported only 384 kbps.

1.9.7 Average Cell or Sector Throughput and User-Perceived Throughput

The average cell or sector throughput is the total number of bits per second in a cell or sector and is measured separately for the downlink and the uplink. The peak user data rate is hardly the throughput perceived by the user. The user-perceived throughput is a relatively difficult parameter to quantify and is the total number of bits per second downloaded or uploaded by a given MS. Specific tests such as FTP download or upload help determine the user-perceived throughput.

1.9.8 Latency or Delay

Latency or *delay* is the average delay experienced by a packet. One way to measure the end-to-end delay is the ping operation for a website, which provides round-trip-time (RTT) for the packets. Since the ping operation includes the total delay between the MS and the web server, server delays and IP network congestion influence the delay statistics. The packet delay between the MS and the radio network and between the MS and the edge of the core network (e.g., a gateway that interfaces with the Internet) can be monitored. Simple tests can be carried out within the network infrastructure to find the delay introduced by the network. Tools from Tektronix and Agilent are widely used in commercial deployments to quantify delay (and other performance metrics). As delay-sensitive applications, such as VoIP and interactive gaming, become popular, latency monitoring and control would be quite important.

1.9.9 Delay Jitter

Delay jitter quantifies the variations in packet delays. One way to measure delay jitter is to measure the time difference between successive packet departures from a network node. Another way to measure delay jitter is to measure the difference in arrival instants for successive IP packets at the MS.

While it is easier to define the performance metrics, determining their values is quite a challenging task. All relevant conditions (e.g., a fully loaded network, uniformly distributed users, and specific applications, such as FTP download) must be specified while quantifying the performance metrics.

1.10 END-USER SERVICES

The ultimate goal of a cellular communication system is to reliably and satisfactorily provide a variety of services to users. Early cellular systems offered only voice services, but recent systems offer much more. Recall that the cellular network consists of the radio network and the core network. The core network interfaces with the PSTN so that the user can communicate with a landline phone; with the Internet so that the user can access the Internet; and with the operator's services network so that special services can be provided to the user. For example, the services network could provide toll-free call service (e.g., an 800 number) and prepaid service (e.g., using a prepaid card). A typical cellular network can offer the following types of services with a suitable mobile device and a service plan.

- **Basic Voice Calls and Enhanced Voice-Related Services.** As of mid-2009, about 73% of revenue comes from voice [41]. In other words, voice is a dominant application. Because cellular networks are designed to have large footprints and adequate capacity, the user can usually make and receive voice calls anytime and anywhere. In addition to the basic voice calls, additional services such as voice mail, call hold, call forwarding, and conference calls are also available. Push-to-talk is another example of a voice-centric service that allows a group of users to talk among themselves. When one person speaks, the remaining group members can simultaneously listen.

- **Tones.** A *ringtone* is the music played by the mobile device instead of normal rings when someone calls the mobile device. It is possible to personalize the ringtone for different callers so that the user of the mobile device knows the identity of the caller without even touching the mobile device. The user can then decide whether to answer the call. Another type of tone is a *ringback* tone. When a person calls the mobile device, such a caller hears music instead of normal rings. This music is referred to as *ringback tones*. It is often possible to customize the ringback tones for different callers.
- **Short Message Service.** Sending a simple text message from a mobile device to an email address or another mobile device is an example of SMS. A text message can also be sent from a website to the mobile device. These messages are short and are typically limited to 160 characters. One operational benefit of the SMS to the service provider is that the message can be sent to the mobile device when the network is less busy. Another benefit is that a low-rate radio channel is adequate and very few resources are consumed.
- **Multimedia Message Service.** This service is an enhancement to the SMS and allows a large amount of data to be transferred. As the name implies, multimedia message service contains data that could be text, picture, or video.
- **Instant Messaging.** While SMS and the multimedia message service are non-real-time, instant messaging allows users to exchange messages in real-time. Instant messaging can also be used as an indicator of the status of the user (e.g., busy, offline, or available).
- **Web-Browsing.** Wireless access to the Internet enables the user to browse the web on the device. 3G and 4G systems allow faster web-browsing due to their efficient support of high data rates.
- **Email.** It is possible to configure the mobile device such that corporate, enterprise, and personal emails are downloaded periodically to the mobile device.
- **Music and Video Downloads.** Music files are downloaded to the mobile device and stored on the device and/or an external memory card. These music files can then be played. Video clips available at the operator's website can also be watched on the mobile device.
- **Gaming and Interactive Gaming.** A variety of games, including 3D games, can be downloaded to the mobile device from the operator's website. Multi-

player or interactive gaming is becoming popular and allows users to play together instead of playing solo on the mobile device [42].

- **Location-Based Services.** The user location can be accurately predicted via some techniques (e.g., with a GPS-enabled phone). Location services enable parents to track their kids [35] and allow emergency service providers to locate and help people in an emergency (e.g., via E911 calls). Turn-by-turn (audible) driving directions and point-of-interest information (e.g., restaurants) are made available to the user [43].

- **Electronic Cash.** Cell phones are used in some countries to make payments [5]. Hence, the only thing that you need is your cell phone—no need to carry cash and no need to run to your bank's ATM. (We're talking about an Automatic Teller Machine here, not Asynchronous Transfer Mode!)

- **Voice-Over-IP.** While cellular networks use a circuit-switched core network for a normal voice call, 4G networks have no circuit-switched core network. These emerging cellular networks would use a packet-switched core network even for a voice call. Voice would be carried in IP packets. These IP packets would then flow to and from the mobile device.

- **Mobile TV and Broadcast and/or Multicast Services.** Some operators are now offering real-time mobile TV services [44]. These services are highly bandwidth-intensive; therefore, special broadcast technologies, such as DVB-H, are used in conjunction with traditional cellular technologies [45, 46].

- **Lawful Interception.** The cellular service providers are required to cooperate with law enforcement agencies to facilitate electronic surveillance. In the U.S., a specific law called the Communications Assistance for Law Enforcement Act was created in 1994 for lawful interception.

- **Applications.** Innovative applications have been emerging at an unimaginable rate. Widespread availability of smartphones, such as iPhones and Droids, has led to thousands of applications being accessible to cellular subscribers. Example applications include restaurant reviews, diet watch, and physical activity monitoring.

1.11 POINTS TO REMEMBER

- The desire to communicate anytime and anywhere without wires led to the arrival of wireless communications. A cellular system is a terrestrial wireless communication system that efficiently reuses scarce and expensive radio resources to serve multiple users simultaneously.

- A given geographical area is divided into cells, and a group of cells constitutes a cluster. The available frequency spectrum is distributed among the cells of a cluster. Since each cluster is using the same radio spectrum, it becomes economically viable to offer cellular services.

- Two types of interference exist in a cellular system: co-channel interference and adjacent channel interference. CCI occurs due to the reuse of the same frequency channel in two different clusters. Adjacent channel interference occurs due to the use of adjacent frequency channels. Sectorization decreases CCI compared to an omnidirectional system. Such interference reduction helps enhance the QoS and/or capacity.

- Duplexing allows simultaneous transmission and reception of signals between the BS and the MS, while multiple access techniques allow multiple MSs to access the BS simultaneously. Examples of multiple access techniques include FDMA, TDMA, CDMA, and OFDMA. Initial 1G analog systems, such as AMPS and ETACS, utilize FDMA. 2G digital systems, such as GSM, use TDMA. 2G systems, such as IS-95, and 3G systems, such as UMTS and 1x, use CDMA. 4G systems, such as LTE and WiMAX, utilize OFDMA.

- A 3G cellular network consists of an RAN and core networks. The 3G RAN includes BSs or Node Bs and RNCs. Two types of core networks are the circuit-switched core network interfacing with the PSTN and the packet-switched core network interfacing with the Internet.

- The ecosystem of the cellular industry includes radio and core infrastructure equipment manufacturers, mobile device manufacturers, baseband chipset manufacturers, service providers, test or design toolset developers, and application providers.

- A cellular system originates from the standard and passes through various phases, such as prototype design of the BS and the MS, lab trials, field trials, commercial deployment, and optimization.

- The performance of a deployed cellular network must be quantified using various metrics. These metrics can point to a need for optimization and/or capacity increase to ensure subscriber satisfaction.

- A cellular system offers a range of services, from basic voice calls to high-speed Internet access. Cellular communications is making wireless communications a necessity rather than a luxury and making wireless communications an integral part of our lives.

- While the voice revenue is still the bulk of the overall revenue, an explosion in data growth is leading to a significant increase in the data revenue. Emerging 4G cellular standards aim to cost-effectively support the rising data traffic demands.

CHAPTER 2

ELEMENTS OF A DIGITAL COMMUNICATION SYSTEM

First-generation cellular systems are analog, but other generations, such as second-generation (2G), third-generation (3G), and fourth-generation (4G), are digital. This chapter provides an overview of the fundamental principles for digital cellular systems. An introduction to the digital communication system identifies major building blocks and is followed by a summary of the basic transmitter and receiver processing blocks. The transmitter includes processing blocks such as speech coding, channel coding, modulation, spreading, and scrambling. It also includes a radio frequency (RF) processor containing modules such as a digital-to-analog converter, a frequency upconverter, a power amplifier, and an antenna. The major processing blocks at the receiver are speech decoding, channel decoding, demodulation, despreading, and descrambling, and the receiver also includes an RF processor containing modules such as an analog-to-digital converter, a frequency downconverter, a power amplifier, and an antenna. The details of a given processing block at the transmitter and the associated processing block at the receiver are discussed together to accelerate understanding of related concepts (e.g., channel coding at the transmitter and channel decoding at the receiver are discussed together).

Cellular Communications, First Edition. Nishith D. Tripathi and Jeffrey H. Reed

2.1 OVERVIEW OF A DIGITAL CELLULAR COMMUNICATION SYSTEM

Consider the scenario shown in Figure 2.1, in which a caller's speech is sent over the air from the mobile station (MS) to the base station (BS). The speech vocoder in the MS converts the caller's speech signal, which is analog, into speech bits. The caller's speech has been transformed into a digital signal now, a set of 1s and 0s. Each cellular standard [e.g., CDMA2000 1x (where CDMA stands for Code-Division Multiple Access) and Universal Mobile Telecommunication System (UMTS)] specifies an air-interface protocol stack. Such a protocol stack helps transmit and receive information such as speech bits and data bits (e.g., email bits or web page bits) over the air. The protocol stack consists of several layers, and the bottom-most layer is the physical layer. This chapter discusses components of a generic physical layer. The physical layer can be thought of as having two building blocks: the baseband processor and the RF processor. Our speech bits pass through various layers of the protocol stack and eventually arrive at the physical layer. Remember that the radio environment is not the late Mother Teresa; it is not kind! The list of atrocities of the radio environment includes weakening of the radio signal as a result of propagation, shadow fading due to buildings, and short-term fading due to the movement of the MS relative to its surroundings.

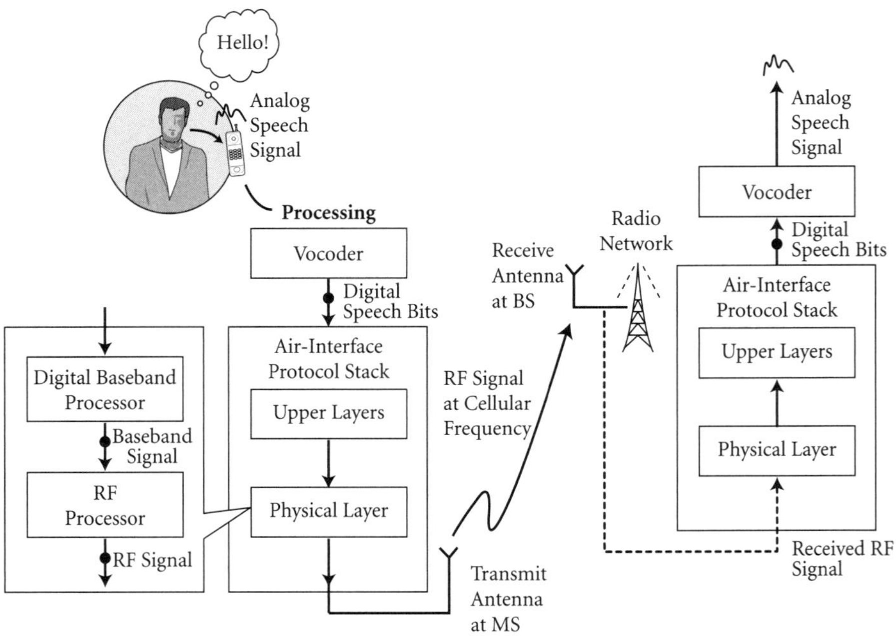

Figure 2.1: High-Level View of Signal Transmission in a Digital Cellular System

The baseband processor converts the speech bits into a signal that can withstand the adversities of the radio environment. Examples of these components at the transmitter include forward error correction (FEC) or *channel coding*, interleaving, digital modulation, spreading (if applicable), and scrambling (if applicable). Because computationally intensive tasks must be completed very quickly in only micro-seconds, the baseband processor is typically implemented in an ASIC (Application-Specific Integrated Circuit) or a digital signal processor. If the signal is observed coming out of the baseband processor using a spectrum analyzer, the signal is at the baseband, occupying the spectrum from 0 to 200 kHz in Global System Mobile (GSM), 0 to 1.2288 MHz in CDMA2000 1x, and 0 to 3.84 MHz in UMTS. Remember that such signals cannot be transmitted over the air because the government authorities [e.g., the Federal Communications Commission (FCC) in the U.S.] have allocated a different spectrum to cellular communications. Hence, *something* must transport our signal from the baseband to the higher frequency cellular spectrum, and that's the job of the RF processor at the physical layer. The RF processor takes the baseband signal as its input and performs frequency upconversion to a suitable frequency, such as 900 MHz, via analog modulation. Now, our speech signal is occupying some bandwidth (e.g., 200 kHz, 1.2288 MHz, or 3.84 MHz for GSM, CDMA2000 1x, and UMTS, respectively) around the cellular carrier frequency. The RF processor at the transmitter amplifies the signal and transmits the resulting signal from an antenna.

The receiver at the BS also implements the physical layer of the protocol stack and attempts to retrieve the speech bits by reversing the processing completed at the transmitter. The RF processor at the receiver accepts the RF signal at a cellular frequency and downconverts the signal to the baseband signal. The baseband processor then carries out tasks such as descrambling (if applicable), despreading (if applicable), demodulation, de-interleaving, and decoding. The bits from the physical layer then pass through upper layers and ideally should resemble the speech bits generated by the speech vocoder at the transmitter.

2.2 TRANSCEIVER OPERATIONS IN A DIGITAL CELLULAR SYSTEM

The physical layer of a transceiver in a digital communication system consists of the major processing blocks shown in Figure 2.2.[18] The transceiver consists of a transmitter and a receiver. Let's consider the transmitter first. A packet or frame containing information bits arrives at the physical layer. Examples of information bits are speech bits and application bits, such as email bits. Speech bits are generated by a speech vocoder, e.g., the Enhanced Variable-Rate Codec (EVRC) and the Adaptive Multirate (AMR) codec. FEC coding or channel coding adds redundancy to the information bits, so the receiver can recover the packet even if the radio channel causes some errors. Convolutional coding and Turbo coding are two major channel coding

[18]Figure 2.2 represents a CDMA-based transceiver. In a non-CDMA-based transceiver, no spreading and scrambling processing blocks exist at the transmitter. Of course, associated despreading and descrambling are absent at the receiver.

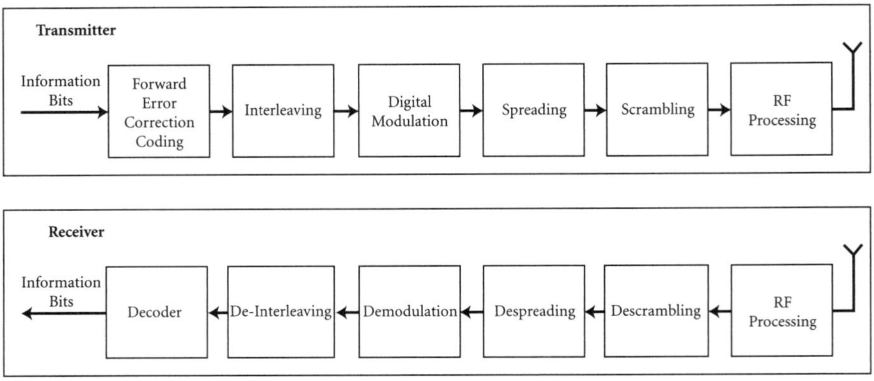

Figure 2.2: Physical Layer Processing in a Digital Transceiver

techniques. The FEC coding techniques generate code symbols. Interleaving helps disperse the consecutive bit errors caused by the radio environment, so the decoder at the receiver experiences non-consecutive bit errors and (potentially) corrects such errors. Interleaved code symbols pass through digital modulation schemes, such as QPSK (Quadrature Phase Shift Keying), and get transformed into modulation symbols. In CDMA-based systems, the modulation symbols pass through spreading and scrambling. Spreading increases the signal bandwidth, and scrambling identifies the signal source and provides some level of resistance to interference. The RF processing amplifies the baseband signal and transfers it to the RF spectrum. The receiver needs to reverse-engineer all the transmitter processing so it can recover the original information bits.

The receiver first carries out RF processing on the received signal to bring the signal to the baseband. When the signal is at the baseband, the receiver descrambles and then despreads the received signal. Next, the receiver performs demodulation and seeks the help of radio channel estimation techniques to facilitate demodulation. After demodulation, de-interleaving is completed. Finally, decoding helps retrieve the original set of transmitted information bits. That's a basic introduction to the overall transmitter and receiver processing blocks.

The next step is to consider these blocks in further detail. Section 2.3 discusses the vocoders utilized by various cellular standards; both speech coding and speech decoding are covered. Section 2.4 describes channel coding carried out at the transmitter and decoding carried out at the receiver. Section 2.5 explains the process of interleaving and de-interleaving. Section 2.6 illustrates various digital modulation methods and briefly touches upon the process of demodulation. (Spreading and despreading and scrambling and descrambling are also major blocks of the transceiver, but these topics are covered in Chapter 7.) Section 2.7 summarizes the RF processing performed at the transmitter and at the receiver.

2.3 INFORMATION BITS: CONSTRUCTION AT THE TRANSMITTER AND RETRIEVAL AT THE RECEIVER

Applications such as email generate bits that eventually reach the physical layer, but speech is an analog signal that needs to be converted into bits. Many different speech coding techniques are available, but the simplest way to convert an analog speech signal would be to sample the signal and represent each sample by a certain number of bits as shown in Figure 2.3.

Do You Recall... Let's test your short-term memory! What is the name of the component that converts speech into a digital signal? See the footnote[19] for the answer.

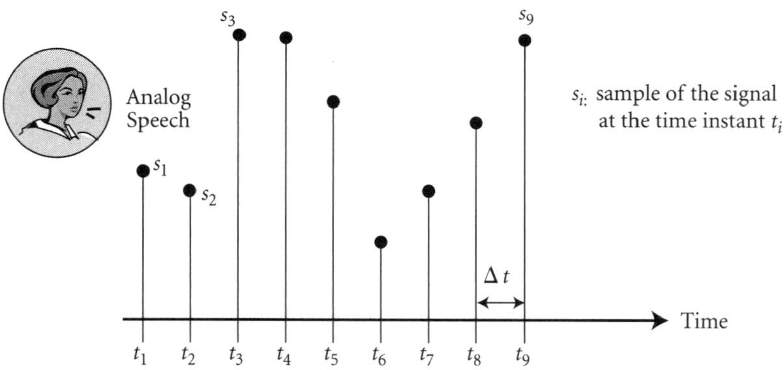

Step 1: Sampling of the Signal

Sampling Frequency, $f_s = \dfrac{1}{\Delta t} \geq 2f_{max}$

f_{max}: Highest Frequency Present in the Signal

Step 2: Quantization of the Samples s_i

Example: s_1 : 0 0 1 1 1 1 0 1

s_2 : 0 0 0 0 1 0 0 1

Figure 2.3: Converting an Analog Speech Signal into a Digital Signal

[19] It is called a speech vocoder (voice coder + decoder) or codec (coder or encoder + decoder).

In Figure 2.3, the speech signal is sampled at regular time intervals at time instants t_1, t_2, and so on. The time interval between successive sampling instants is Δt.

A Question for You... How fast should we sample the signal so that it can be faithfully reconstructed at the receiver? See the footnote[20] for the answer.

Let f_{max} be the highest frequency (bandwidth) present in the analog speech signal. The sampling frequency, f_s, should be $2f_{max}$. While human speech typically contains frequencies ranging from 20 Hz to 20 kHz [47], f_{max} is usually chosen to be 4 kHz. Then, f_s becomes 8 kHz. Δt is $1/8\,\text{kHz} = 0.125\,\text{ms}$. The greater the number of bits that represent a speech sample, the better the quality of the signal becomes because the amplitude of the analog signal is closer to the quantized digital signal. The downside is that a larger buffer is required to store the bits and a higher data rate is required to carry the speech signal. In Figure 2.3, the speech samples s_1 and s_2 are represented by a set of eight bits, `00111101` and `00001001`, respectively. If the speech signal is sampled at 8 kHz and eight bits are used to represent each sample, the required data rate of the channel that can carry speech is $8\,\text{kHz} * 8 = 64$ kbps. Imagine that you are a cellular service operator. You have spent millions of dollars to purchase the frequency spectrum. You will not have a successful business case if one voice call needs 64 kbps of radio channel; radio resources are very expensive! Remember that you must be able to carry many voice calls simultaneously using the acquired spectrum. You need to reduce the data rate to convey the speech signal. If each speech sample is represented by fewer bits (e.g., four or five), the voice quality suffers.

Now, you really need help! You want to maintain good voice quality, but you want to reduce the data rate. Fortunately, efficient speech vocoders can represent the speech signal using low data rates (e.g., 8 kbps and 12.2 kbps) with wireline-type voice quality.[21] Commercial cellular networks utilize vocoders or speech codecs that are based on linear predictive coding (LPC). The speech codecs used in typical commercial deployments of 2G and 3G technologies include the AMR codec for GSM and UMTS standards and the EVRC for IS-95 and 1xRTT standards. Section 2.3.1 introduces the concept of the LPC. Examples of widely used LPC-based speech vocoders are the 8k EVRC and the AMR vocoder. Section 2.3.2 discusses EVRC, and Section 2.3.3 describes the AMR vocoder.

[20]The signal should be sampled at the Nyquist rate, which is twice the bandwidth present in the signal. Such Nyquist sampling prevents aliasing, i.e., a copy of the original signal interfering with the original signal.

[21]By the way, have you ever wondered how to measure the voice quality? Mean opinion score (MOS) is widely used to quantify the quality of reconstructed speech.

2.3.1 Linear Predictive Speech Coding: A Simplistic View

LPC coding attempts to mimic the speech production in humans. Let's look at how humans produce speech because this provides insight on effective coding approaches. The air from the lungs passes through the vocal cords and the vocal tract to create a sound. The vocal tract consists of the throat, the tongue, the nose, and the mouth. The sounds generated by the vocal tract may be voiced (e.g., the sound of the letter b in "book") or unvoiced (e.g., the sound of the letter p in "pool"). The voiced sounds contain distinct resonant frequencies called formant frequencies and typically have high energy levels. The vocal cords vibrate with some periodicity to create air pulses called glottal pulses, leading to voiced sounds. The frequency of the vibrations is the pitch of the sound. Women and children tend to have higher pitch, and men have lower pitch. The unvoiced sounds contain little energy and consist of higher frequencies. The vocal tract stays open and does not vibrate during unvoiced sounds, and the air passes through the cords of the vocal tract in the form of a turbulent flow. As a person speaks, the shape of the vocal tract changes, resulting in different resonant frequencies for different words. The amount of air determines the loudness of the sound.

Figure 2.4 provides a high-level view of the LPC-based speech vocoder. The speech signal is sampled and quantized at the transmitter. The speech encoding algorithm attempts to determine the speech parameters in the form of the input signal for an IIR (Infinite Impulse Response) filter and the filter coefficients such that the application of the input or excitation signal yields the filter output that resembles the sampled input speech signal. *The filter excitation signal is analogous to the air pressure, and the IIR filter coefficients are analogous to the vocal cords and the vocal tract.* Now the transmitter sends out the speech parameters to the receiver instead of the samples of the speech signal. The speech parameters essentially represent the excitation signal and the IIR filter coefficients. Recall that the regular sampling and

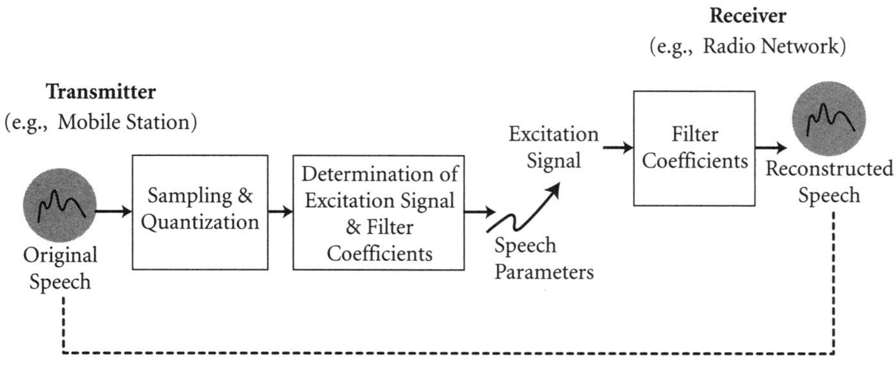

Figure 2.4: Linear Predictive Speech Coding

quantization of the input speech signal would need 64 kbps radio channel, but we want to reduce the over-the-air data rate. The number of bits representing the speech parameters is quite low, requiring low data rates such as (approximately) 8 or 12 kbps. The receiver uses the excitation signal and the filter coefficients received from the transmitter to reconstruct the speech signal. If the speech encoding technique used by the transmitter works well, the receiver can faithfully reconstruct the speech signal. See Bradbury's work for more information on speech production and LPC [48].

Ever Wondered... Why is it called *linear predictive* coding? Since the output of an IIR filter at a given sampling instant depends on a linear combination of the previous samples, the process of speech reconstruction appears as if we were trying to predict the next sample of the speech signal based on previous speech samples. Hence, this speech encoding technique is called *linear predictive* coding.

2.3.2 Enhanced Variable-Rate Codec

2G CDMA-based IS-95 systems supported 13 kbps rate and 8 kbps rate speech codes. The 13 kbps rate gave better voice quality but lower capacity, and the 8 kbps rate yielded higher capacity but lower voice quality. For example, a 13 kbps data rate corresponds to the sector capacity of thirteen users, and an 8 kbps rate supports twenty voice users in a sector. The basic 8 kbps rate codec was then *enhanced* to give better voice quality while preserving the benefit of higher voice capacity. Hence, the codec was termed *enhanced variable-rate codec* or EVRC. CDMA2000 1xRTT systems typically use 8 kbps EVRC as part of their Service Option 3.[22] The 8 kbps EVRC associated with Service Option 2 is briefly summarized in Section 2.3.2.1. Examples of EVRC variations include one corresponding to Service Option 68 and another corresponding to Service Option 70. See the standard for further details regarding EVRC implementations [50].

A speech codec consists of an *encoder* as part of the transmitter and a *decoder* as part of the receiver. Section 2.3.2.1 deals with the speech encoder, and Section 2.3.2.2 focuses on the speech decoder.

2.3.2.1 Speech Encoder for EVRC at the Transmitter

The EVRC works in 20 ms intervals. The speech encoder represents the 20 ms block of speech by a set of parameters, which is then transmitted over the air using the help of the physical layer. These parameters are described in the following paragraphs. The receiver can reconstruct the original speech signal using these parameters. Figure 2.5 depicts the block diagram of the speech encoder that is part of the transmitter.

[22] A service option specifies the characteristics of a service, such as voice service, with a certain data rate. See the standard for a comprehensive list of service options [49].

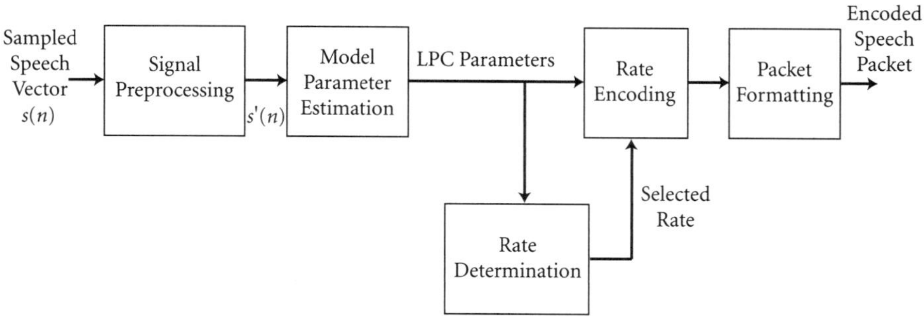

Figure 2.5: Speech Encoder at the Transmitter

Input Speech Signal. A 20 ms block of the Input Speech Signal is sampled at 8 kHz and quantized with the dynamic range of thirteen bits to generate a uniform pulse coded modulation signal. The resulting set of pulse coded modulation bits forms the speech vector $s(n)$ in Figure 2.5.

Signal Preprocessing Module. The Signal Preprocessing Module carries out high-pass filtering and adaptive noise suppression filtering to yield the vector $s'(n)$. The main job of this module is to minimize the impact of excessive low frequency and other background noises that can degrade the voice quality. The cutoff frequency of the high-pass filter is 120 Hz.

Model Parameter Estimation Module. The main objective of the Model Parameter Estimation Module is to determine several LPC parameters that represent the 20 ms speech block. Major inputs for this module are preprocessed speech samples for the current frame, some speech samples from the previous frame, some predicted samples for the next frame, and line spectral pairs (LSPs) and LPC prediction gains from the previous frame. This module carries out LPC analysis and first calculates the LPC coefficients and an optimal pitch delay τ. Furthermore, it converts the LPC coefficients into LSPs, which represent the LPC coefficients in a pseudo-frequency domain and are used because they have better quantization and interpolation properties than the basic LPC coefficients. This module yields the following major outputs: LPCs, LSPs, short-term prediction residual signal $\epsilon(n)$, long-term pitch delay estimate τ, long-term prediction gain β, spectral transition indicator flag, and the bandwidth expanded correlation coefficients R_w.

Rate Determination Module. The Rate Determination Module analyzes the LPC parameters determined by the model parameter estimation module and uses a voice activity detection (VAD) algorithm to decide the data rate to be used for the current 20 ms speech block. The module uses a rate determination algorithm to select one of three encoding rates: Rate 1, Rate 1/2, and Rate 1/8. Active speech is encoded at

Rate 1 or Rate 1/2, and background noise is encoded at Rate 1/8. Major inputs for this module are R_W and τ. The rate determination algorithm compares the current frame energy in each of two frequency bands, 0.3–2.0 kHz and 2.0–4.0 kHz, with the background noise estimates in these respective bands. Different signal-to-noise ratio thresholds are defined for different rates. The highest rate calculated from the two frequency bands is chosen as the encoding rate for the current frame. In general, the higher the signal-to-noise ratio in a frequency band, the higher the likelihood of a larger rate.

Rate Encoding Module. While the generic 8 kbps EVRC supports Rate 1 (i.e., 8 kbps), Rate 1/2 (i.e., 4 kbps), Rate 1/4 (i.e., 2 kbps), and Rate 1/8 (i.e., 1 kbps), Service Option 3 for the EVRC utilizes only three rates, 1, $1/2$, and $1/8$. The rate encoding module attempts to characterize the excitation signal and to quantize the LSPs obtained from the model parameters consistent with the selected data rate. With Rate 1/8, this module makes no attempt to capture periodicity of the residual speech signal but rather characterizes the energy contour. With a Rate 1 or Rate 1/2, the RCELP (Relaxed Code Excited Linear Prediction) algorithm is used to match the time-warped version of the original speech residual and to determine a set of parameters that would help reconstruct the speech signal at the receiver. Furthermore, Rate 1 processing also incorporates additional parameters, such as spectral transition indicator and delay difference $(\Delta\tau)$, to achieve good performance in poor channel conditions.

The RCELP algorithm can be considered a generalization of the original Code Excited Linear Prediction (CELP) algorithm. While the conventional CELP encoder tries to match the original speech residual signal exactly, the RCELP algorithm attempts to match a time-warped version of the original speech residual signal that adheres to a simplified pitch contour. Such pitch contour is obtained by estimating the pitch delay just once in the 20 ms frame and then linearly interpolating the pitch from one frame to the next. The use of the RCELP algorithm instead of traditional CELP algorithm makes more bits available in each encoded speech packet to represent the stochastic excitation and for channel impairment protection. In summary, RCELP improves the error performance without adversely affecting the speech quality in good channel conditions.

The RCELP algorithm determines which of the predetermined vectors of the codebook (i.e., which sequence of `1`s and `-1`s) minimizes the mean-squared weighted error between the original and synthesized speech. A 20 ms speech frame is divided into three subframes. The RCELP algorithm calculates some parameters for each of the three subframes. Examples of major parameters determined by this module are the adaptive codebook gain index for each subframe, the fixed codebook shape indices for each subframe, the fixed codebook gain indices for each subframe, the delay transmission code, the spectral transition indicator, and the delay difference DDELAY.

Table 2.1: Contents of the Encoded Speech Packet

Parameter	Rate 1	Rate 1/2	Rate 1/8
Spectral Transition Indicator	1		
LSP	28	22	8
Pitch Delay τ	7	7	
Pitch Delay Difference $\Delta\tau$	5		
Adaptive Codebook Gain	9	9	
Fixed Codebook Shape	105	30	
Fixed Codebook Gain	15	12	
Frame Energy			8
Reserved	1		
Total Number of Bits per Speech Block	171	80	16

Packet Formatting Module. The Packet Formatting Module formats the encoded speech packet to convey the parameters for the selected rate.

Table 2.1 lists the number of bits corresponding to various speech encoding parameters for all three rates for Service Option 3. Observe that the parameters such as adaptive codebook gain, fixed codebook shape, and fixed codebook gain are obtained from the Rate Encoding Module.

2.3.2.2 Speech Decoder for EVRC at the Receiver Figure 2.6 gives a simplified view of the EVRC decoder at the receiver. The speech packet arrives at the speech decoder. The speech decoder also receives information about the type of the packet. Specifically, the physical layer of the air-interface protocol stack at the receiver retrieves the speech packet from the air interface, estimates whether the received speech frame is in error, and determines the rate of the received speech

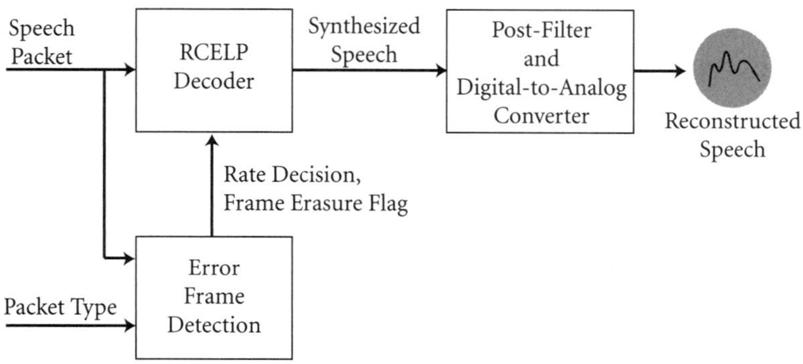

Figure 2.6: EVRC Decoder at the Receiver

packet. The packet type in Figure 2.6 contains such speech frame error and the data rate information. In addition to extracting frame error and rate information, the Frame Error Detection Module also performs a sanity check on the rate estimation of the physical layer by observing some parameters of the speech packet. The rate decision and the frame erasure (i.e., error) flag are conveyed by the Frame Error Detection Module to the RCELP decoder. The decoder utilizes the speech parameters contained in the received speech packet to reconstruct the 20 ms speech block for the estimated rate. If the frame erasure flag indicates the presence of a frame error, the frame error recovery logic is invoked. The speech synthesized by the RCELP decoder is post-filtered and converted from digital to analog to recover the original analog speech signal.

2.3.3 Adaptive Multirate Codec

2G GSM systems, 3G UMTS systems, and 4G Long-Term Evolution (LTE) systems use the AMR speech codec. This codec supports source data rates ranging from 4.75 kbps to 12.2 kbps, with a potential for changing the rate every 20 ms frame. The specific data rates are 12.2 kbps (also called GSM Enhanced Full Rate), 10.20 kbps, 7.95 kbps, 7.40 kbps, 6.70 kbps (also called PDC[23] Enhanced Full Rate), 5.90 kbps, 5.15 kbps, and 4.75 kbps. Early commercial deployments of UMTS utilized a fixed source rate of 12.2 kbps instead of variable source rates. Fixed source rates of 12.2 kbps and 5.90 kbps have been popular in later UMTS deployments.

A Question for You... What would be the main benefit and drawback of using the data rate of 4.75 kbps instead of 12.2 kbps to carry speech? Find the answer in the footnote.[24]

The AMR speech codec supports the Source Controlled Rate mode of operation, where the speech coder encodes just the background noise with a much lower rate than the normal rate used to encode active speech. In general, during a normal conversation, each person alternates speaking so that about half the time is used to talk and the other half is used to listen. The AMR coder part of the codec implemented at the transmitter is discussed in Section 2.3.3.1, while the AMR decoder part of the codec implemented at the receiver is discussed in Section 2.3.3.2

2.3.3.1 AMR Coder at the Transmitter Figure 2.7 summarizes the main components of the AMR coder. The speech is sampled at 8 kHz.

[23]Personal Digital Cellular (PDC) is a cellular system used in Japan.

[24]A lower data rate consumes less transmit power and increases capacity. However, voice quality is lower because fewer bits represent the speech.

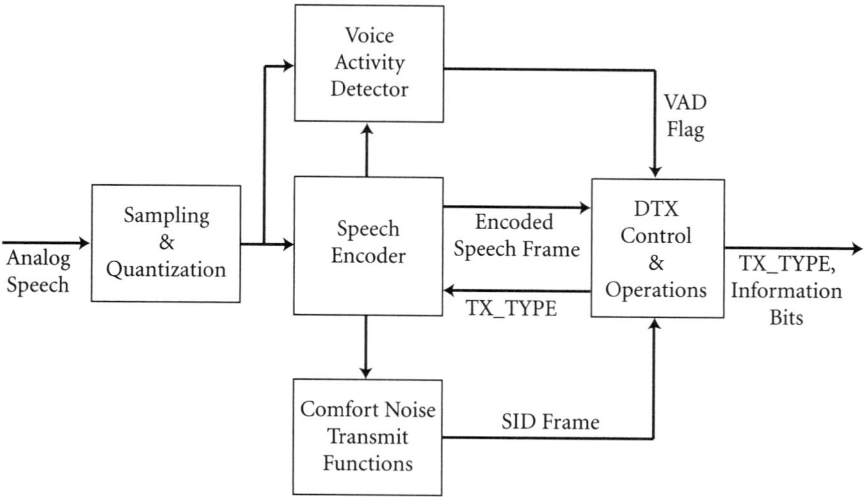

Figure 2.7: AMR Codec at the Transmitter

Do You Recall... Do you remember why the speech is sampled at 8 kHz? Refer to the footnote[25] for the answer.

The speech samples are then quantized using thirteen-bit uniform pulse coded modulation. The speech encoder represents the speech samples in the form of the ACELP (Algebraic Code Excited Linear Prediction) parameters. The main idea behind the ACELP encoder is that two excitation vectors from a fixed codebook and an adaptive codebook are added with suitable gains to construct an input vector to a short-term linear prediction synthesis filter. The optimum excitation sequence is chosen by searching through the codebooks, and the goal is to minimize the error between the original and the synthesized speech.

A block of 160 speech samples, corresponding to a 20 ms speech block with 8 kHz sampling rate, is analyzed to extract the ACELP parameters. Examples of the ACELP parameters are linear prediction filter coefficients, plus gains and indices of the adaptive and fixed codebooks. The speech encoder provides the encoded speech frame containing such parameters to the three modules: the Comfort Noise Transmit Functions Module, the VAD, and the DTX (Discontinuous Transmission) Control and Operations Module. The Comfort Noise Transmit Functions Module at the transmitter

[25]The signal needs to be sampled at the Nyquist frequency, which is twice the bandwidth. For human speech, the bandwidth is considered to be less than 4 kHz. Hence, the speech signal is sampled at $2 * 4$ kHz $= 8$ kHz.

regularly sends a Silence Descriptor (SID) frame containing noise parameters to the receiver so that the receiver can generate *comfort noise*[26] during the silence intervals.

The VAD Module analyzes the samples of the 20 ms speech block and the ACELP speech parameters to decide whether the current speech frame contains *active speech* (rather than silence). The DTX Control and Operations Module determines TX_TYPE and forwards the encoded speech frame or SID frame to a suitable entity for eventual transmission over the air. The parameter TX_TYPE conveys information such as the number of bits in a good or bad speech frame (e.g., 244 bits for 12.2 kbps AMR speech), 35 comfort noise bits (i.e., a SID frame), and NO_DATA (i.e., nothing to be transmitted). If the Source Controlled Rate operation is disabled, the AMR codec continuously generates speech frames. When the Source Controlled Rate operation is enabled, the VAD Module dictates what type of information is sent over the air for a given 20 ms speech block. For example, after the VAD Module detects eight consecutive silence frames, a SID frame containing the information on background noise is transmitted. During the remainder of the silence interval, no data is transmitted; only updated SID frames are transmitted every eighth frame. The speech encoder is apprised of TX_TYPE so it can evaluate comfort noise during the suitable 20 ms frames.

Table 2.2 summarizes the numbers of bits that represent active speech for a 12.2 kbps AMR speech. Since the total number of bits in the encoded speech frame is 244 bits for a 20 ms speech block, the data rate is 244 bits/20 ms = 12.2 kbps. In an active speech frame, the set of 244 bits is divided into three classes of bits, Class A bits, Class B bits, and Class C bits. Class A includes the bits that are most sensitive to errors; any error in these bits is likely to yield a corrupted speech frame which requires suitable error concealment. Hence, Class A bits are covered by the cyclic redundancy check and use a higher level of redundancy as part of physical layer processing of FEC. Class B bits and Class C bits contain bits that could degrade speech quality, but decoding of an erroneous speech frame is usually possible without significant impact on speech quality. Since Class B bits are more sensitive to errors than Class C bits, Class B bits undergo a higher level of FEC redundancy. Out of the total of 244 bits, 81 bits are Class A bits, 103 bits are Class B bits, and 60 bits are Class C bits. The SID frame consists of 39 bits, which include 35 comfort noise bits. All 39 bits are considered Class A bits for notational convenience.

2.3.3.2 AMR Decoder at the Receiver The decoder is relatively less complex than the encoder because the decoder relies upon the received encoded speech packet to provide all the information required to reconstruct speech. Figure 2.8 shows the block diagram of the AMR decoder. The DTX Control and Operations Module receives the encoded speech frame containing information bits, the parameter RX_TYPE, which indicates the type of the frame, such as good active speech frame or bad frame indicator for an active speech frame, and the SID frame (where

[26]If complete silence occurred during moments of no speech between the conversing parties, the callers might think the call had been dropped. To reassure the callers that the call is indeed being maintained, a background noise is generated, i.e., *comfort noise*.

Table 2.2: Number of Bits for 12.2 kbps AMR Speech

Speech Parameter	Subframe 1	Subframe 2	Subframe 3	Subframe 4	Total Number of Bits per Frame
Two LSP Sets				38	
Pitch Delay (Adaptive Codebook Index)	9	6	9	6	30
Pitch Gain (Adaptive Codebook Gain)	4	4	4	4	16
Algebraic Codebook (Positions and Signs of Pulses)	35	35	35	35	140
Fixed Codebook Gain	5	5	5	5	20
Total Number of Bits					244

applicable). If the received speech frame is good, it is passed onto the speech decoder that utilizes ACELP encoder to resynthesize the speech signal. When the received speech frame is in error, an error concealment technique is invoked by the Speech Frame Substitution Module. The speech decoder, in turn, utilizes predicted speech parameters to reconstruct speech. If the current 20 ms time window is part of the silence interval, the SID frame is used by the Comfort Noise Receive Functions Module to instruct the speech decoder to generate comfort noise. Recall that the SID frame is received every eighth frame to update the background noise parameters.

The standards and their related technical reports provide an overview of the AMR codec [51], details of ACELP coding and decoding [52], performance quantification of the AMR codec [53], and definitions of Class A, B, and C bits [54].

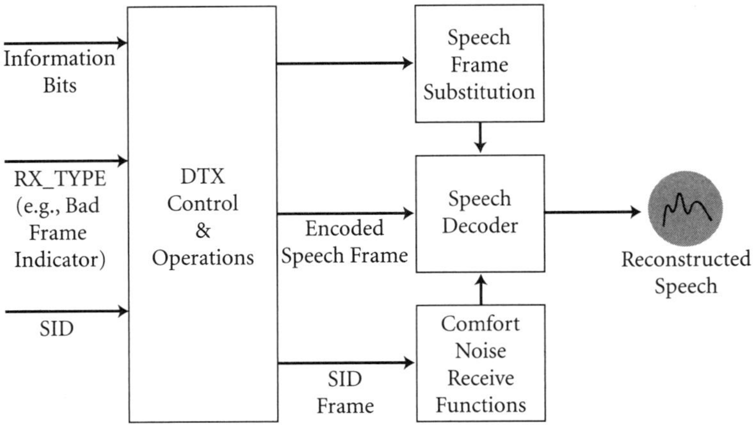

Figure 2.8: AMR Codec at the Receiver

2.4 FORWARD ERROR CORRECTION CODING AND DECODING

Information bits such as speech bits or data bits (e.g., email or web pages) pass through layers of the air-interface protocol stack and arrive at the physical layer. A cyclic redundancy check is added to the set of information bits, and FEC (also called channel coding) is carried out.

To appreciate the role FEC plays, let's first consider the case of FEC's absence. Assume that a transmitter sends out a sequence of information bits `101` as shown in Figure 2.9. The radio channel causes a fade such that the signal becomes weaker in the second bit position. Hence, the receiver misinterprets the second bit to be `1` instead of `0`. In other words, the receiver estimates the bit sequence to be `111` even though the transmitter has sent out the sequence `101`. Since the radio channel could cause the signal to fade at any time, the original bit stream must be recovered in spite of some errors. That's where FEC enters the scene.

FEC introduces redundancy such that the original bit stream can be retrieved correctly even though some bits are in error. The simplest form of FEC is simple repetition. In our example, if the transmitter intends to send the set of information bits `101`, it actually sends out `111000111`, essentially sending the same information multiple times as shown in Figure 2.9. The radio environment again causes an error in the second bit position. The decoder in our example, however, looks at three consecutive code symbols before making a decision about the transmitted bit. Since the three symbols include two `1`s and one `0`, the decoder estimates the first transmit bit to be `1`. Furthermore, three `0`s correspond to the second transmitted bit, and thus the second transmitted bit is estimated to be `0`. Finally, the last set of three `1`s means that the last transmitted bit is `1`. In summary, the decision about the transmitted bit is made by observing multiple received code symbols associated with the transmitted

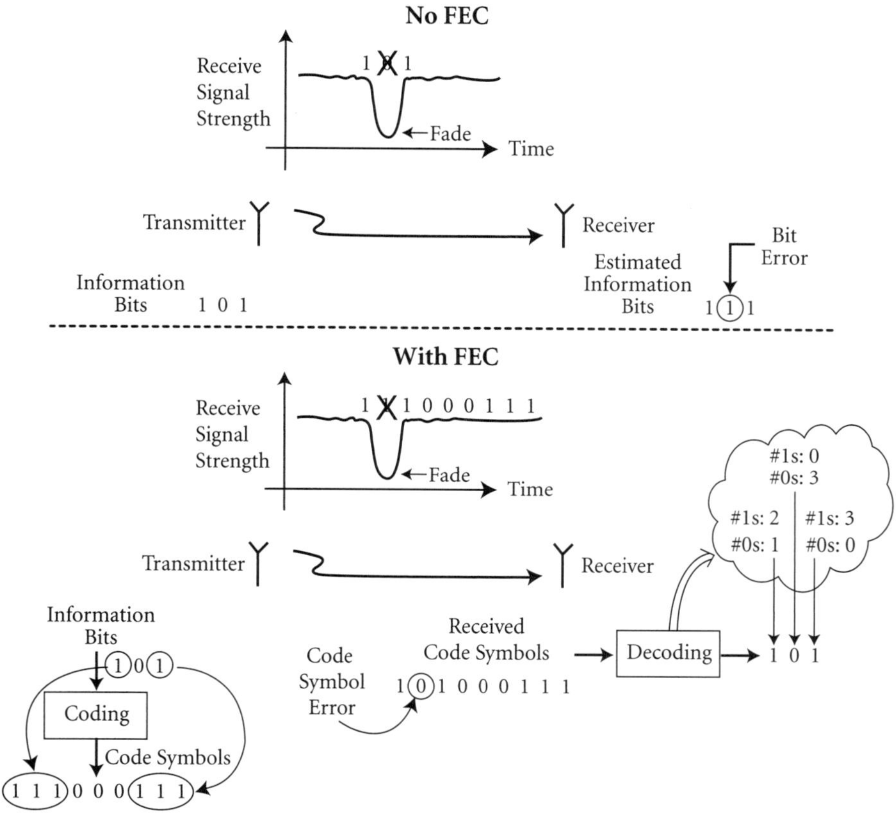

Figure 2.9: Motivation for and Basic Idea of Channel Coding

bit. Assuming that the democracy works, the majority of the received code symbols will lead us to the correct conclusion about the transmitted bits.

While the input to the FEC coding block is considered *bits*, the output of the FEC coding block is said to consist of *code symbols, coded symbols, or encoded bits*. The amount of redundancy introduced by the FEC coding is quantified by the *coding rate*, which is the ratio of the number of input bits and the number of output code symbols. In the simple example under consideration, the coding rate is $3/9 = 1/3$. In other words, we have used channel coding with a rate of $1/3$. The coding rate is an important parameter that characterizes the channel coding mechanism. While simple repetition of information bits would work as an FEC coding technique, more efficient coding techniques are available. Convolutional coding and Turbo coding are the two most popular FEC techniques used in commercial cellular networks.[27] Sections 2.4.1 and 2.4.2 discuss Convolutional coding and decoding mechanisms, and Sections 2.4.3 and 2.4.4 describe Turbo coding and decoding mechanisms. As shown in Figure 2.9, the coding operation is done at the transmitter, and the decoding operation is performed at the receiver.

2.4.1 Convolutional Coding at the Transmitter

Convolutional coding is a simple yet efficient channel coding technique. Figure 2.10 illustrates the structure of one of the Convolutional coders used in 3G UMTS-based systems [57]. In the figure, n represents a sample instant, D represents a delay element or shift register, $x(n)$ is the input bit at the current sample instant n, and $y_i(n)$ is the output at the current sample instant n. (i ranges from 0 to 2.) The state of the coder is given by the contents of the shift register at a given time instant. Before the processing of the input packet commences, all the shift registers are initialized to 0s. For every input bit appearing at time instant n, three output code symbols are generated by the Convolutional coder.

A Question for You... What is the coding rate for the Convolutional coder of Figure 2.10? See the footnote[28] to verify your answer.

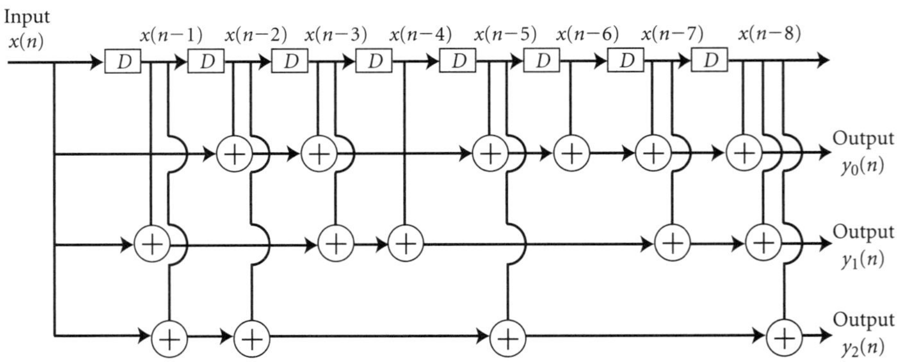

Figure 2.10: Convolutional Coder

[27] Low-density parity-check (LDPC) coding is an efficient coding technique that was invented by Robert Gallager in the 1960s. High computational complexity prevented earlier adoption of LDPC coding, but powerful modern processors make implementation of LDPC coding and decoding feasible. The interested reader is referred to Gallager's work [55, 56] for details on low-density parity-check coding.

[28] The coding rate is $1/3$, the ratio of the number of input bits to the number of output code symbols.

The output of the Convolutional coder depends, in general, on the current input bit and a certain number of previous bits. Specifically, three outputs, $y_0(n)$, $y_1(n)$, and $y_2(n)$, are calculated as follows.

$$\begin{aligned} y_0(n) = x(n) \oplus x(n-2) \oplus x(n-3) \oplus x(n-5) \oplus x(n-6) \\ \oplus\, x(n-7) \oplus x(n-8) \end{aligned} \tag{2.1}$$

$$y_1(n) = x(n) \oplus x(n-1) \oplus x(n-3) \oplus x(n-4) \oplus x(n-7) \oplus x(n-8) \tag{2.2}$$

$$y_2(n) = x(n) \oplus x(n-1) \oplus x(n-2) \oplus x(n-5) \oplus x(n-8) \tag{2.3}$$

In Eqs. (2.1)–(2.3), $x(n-k)$ is the output of the kth shift register and represents an input bit delayed by k sample instants. For example, assume that the input packet, often referred to as a code block, consists of the bits $\{A, B, C, D, E, \ldots\}$. At time instant $n = 0$, all shift registers hold 0s, and $x(0) = A$. In the next time instant $n = 1$, the input bit is B. Hence, $x(1) = B$. The first bit A has now propagated into the first register, making $x(0) = A$. All remaining shift registers have 0s. Now, the time clock again advances to $n = 2$ where the input bit is C, so $x(2) = C$. Also, at $n = 2$, the first shift register contains $x(n-1) = x(2-1) = x(1) = B$, and the second shift register now contains $x(n-2) = x(2-2) = x(0) = A$. In summary, the input bits gradually propagate through the shift registers one-by-one.

In Eqs. (2.1)–(2.3), $\oplus$ denotes the XOR operation. Recall that XOR is a mathematical operation that returns 0 when two inputs bits are the same (i.e., two 0s or two 1s). Otherwise, it returns 1. For example, when the XOR operation is performed on two bits 0 and 0, the answer is 0. However, when the XOR operation is applied to two bits 0 and 1, the answer is 1. The first output of the Convolutional coder, $y_0(n)$, is the XOR operation of the variables such as $x(n), x(n-2), x(n-3), \ldots,$ and so on. Similarly, the second output, $y_1(n)$, is the XOR operation of the variables such as $x(n)$, $x(n-1)$, $x(n-3), \ldots$, and so on. When the input stream of bits is $\{x(0), x(1), x(2), \ldots\}$, the output stream of Convolutional code symbols is $\{y_0(0), y_1(0), y_2(0), y_0(1), y_1(1), y_2(1), y_0(2), y_1(2), y_2(2), \ldots\}$.

So far, we have mentioned just one Convolutional coder parameter, the coding rate. Another important coder parameter is *constraint length.* The constraint length dictates how many previous input samples influence the code symbol. The constraint length is one plus the number of shift registers in the coder. In Figure 2.10, the constraint length is nine, as there are eight shift registers. When the constraint length is nine, $9 - 1 = 8$ previous samples influence the output. For example, $y_i(n)$ is affected by the longest delayed sample $x(n-8)$.

One way to specify the Convolutional coder output is via an explicit equation such as Eq. (2.1). Another way to specify the coder output is by defining a generator polynomial. For Figure 2.10, three generator polynomials can be specified as

$$g_0(D) = 1 + D^2 + D^3 + D^5 + D^6 + D^7 + D^8 \tag{2.4}$$

$$g_1(D) = 1 + D^1 + D^3 + D^4 + D^7 + D^8 \tag{2.5}$$

$$g_2(D) = 1 + D^1 + D^2 + D^5 + D^8. \tag{2.6}$$

The generator polynomial $g_i(D)$ is used to generate the output $y_i(n)$. In Eqs. (2.4)–(2.6), D^k represents the input sample delayed by k sample instants. Since $g_0(D)$ uses the shift registers corresponding to $k = 2, 3, 5, 6, 7,$ and 8 in addition to the current input bit (represented by the first term 1 in Eqs. (2.4)–(2.6)), $y_0(n)$ is the XOR operation of the samples $x(n)$, $x(n-2)$, $x(n-3)$, $x(n-5)$, $x(n-6)$, $x(n-7)$, and $x(n-8)$.

The generator polynomial can be specified by an explicit equation such as Eq. (2.4). Another way to specify the polynomial is to represent the connections to the shift registers via binary numbers. g_0, g_1, and g_2 can be specified in octal numbers as $G_0 = \mathbf{557}, G_1 = \mathbf{663}$, and $G_2 = \mathbf{711}$, respectively [57]. **557** in octal is `101101111` in binary, i.e., **5** in octal is `101` in binary, repeat that for the second numeral, and then **7** in octal is `111` in binary. A total of nine binary numbers, numbered 0 to 8, correspond to one current input sample and eight shift register outputs. Since the first bit and fourth bit are `0`s, the generator polynomial $g_0(D)$ contains current and all delay samples except D^1 and D^4 as in Eq. (2.4).

2.4.2 Example Convolutional Decoding at the Receiver

The most popular and de facto approach for Convolutional decoding is to use the Viterbi Algorithm (VA), which has numerous applications, such as channel equalization[29] and channel decoding. For Convolutional decoding, the VA estimates the most likely sequence of transmitted bits (i.e., input bits to the encoder) based on its knowledge of how the Convolutional encoding would have processed these bits at the transmitter. Fleming offers a very nice and simple example of how the VA can be used to implement Convolutional decoder, and his efforts have simplified our work in this section. We have borrowed that example here to explain the operation of the VA [58]. *We request your patience and undivided attention as we tackle a quite elaborate process of Viterbi decoding! (No iPhones, Droids, and Blueberries—oh, we meant Blackberries—please!)*

We will illustrate the encoding operation for a simple Convolutional coder for a small data packet first. Then, we will illustrate the concepts used by the VA for decoding. Finally, we will conclude by performing the decoding operation using the VA. Figure 2.11[30] specifies the Convolutional encoder that we would focus on [58]. In Figure 2.11 [58],

$$y_0(n) = x(n) \oplus x(n-1) \oplus x(n-2) \tag{2.7}$$

$$y_1(n) = x(n) \oplus x(n-2). \tag{2.8}$$

[29]Channel equalization is the process of undoing the effects of the radio channel. For example, if the radio channel causes the modulation symbol to undergo a 90° phase shift, the channel equalization would help detect such a phase shift so that the phase of the transmitted modulation symbol is regained.

[30]Figure 2.11 is adapted from a figure in Fleming's work [58]. The original figure is copyrighted by Robert F. Fleming, III, and the adaptation appears here with Fleming's permission.

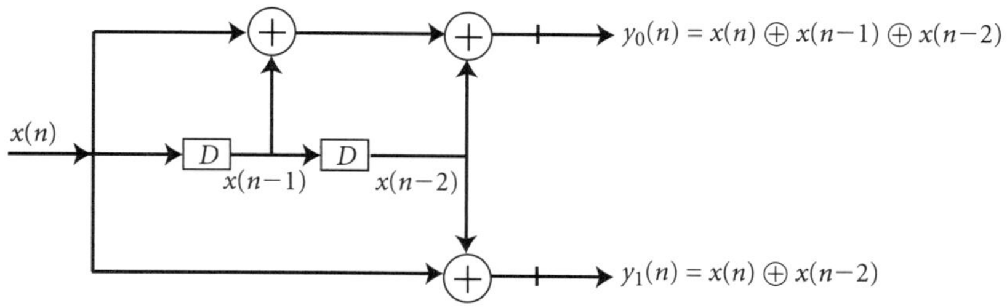

Input
Sequence: 0 1 0 1 1 1 0 0 1 0 1 0 0 0 1 (15 bits) + 0 0 (2 encoder TAIL bits)
(17 bits)

Encoder State: $\{x(n-1)\ x(n-2)\}$

Figure 2.11: Example Convolutional Encoder for Viterbi Decoding

A Question for You... How can these two equations be represented in the form of the generator polynomials? See the footnote[31] to validate your answer.

We'll use a fifteen-bit test sequence, `010111001010001`, to discuss the encoding and decoding operation. Adding *TAIL bits* to the end of the intended message sequence is a usual practice. Such TAIL bits, usually zeros, facilitate the decoding process as we will see later. The number of TAIL bits is the same as the number of shift registers. Since we have two shift registers, we will perform coding on a seventeen-bit sequence containing fifteen regular bits and two TAIL bits. To illustrate the encoder operation, we will use the concept of the state of the encoder, which is defined as the contents of the shift register. For example, if the first shift register on the left in Figure 2.12[32] holds `0` and the second shift register on the right in the figure holds `1`, the state of the encoder is said to be `01`. The operation of the encoder is fully specified using two tables: the state transition table and the output table.

Figure 2.12 summarizes the state transition table for our example encoder. Assume that the initial contents of the shift registers are AB before the input bit C is processed. Once the input bit is applied to the encoder, it moves into the first shift register, and the bit B of the second shift register is now replaced by bit A of the first shift register. If the state of the encoder is `10` at the time instant n, the application of the input bit `0` results in the transition to the state `01` at time $n + 1$ and the application of the input bit `1` results in the transition to the state `11` at time $n + 1$. Now, let's take a look at the

[31] $g_0(D) = 1 + D + D^2$ and $g_1(D) = 1 + D^2$. In octal representation, G_0 = **7** (i.e., `111`) and G_1 = **5** (i.e., `101`).

[32] The table in the bottom portion of Figure 2.12 is adapted from a table in Fleming's work [58]. The original table is copyrighted by Robert F. Fleming, III, and the adaptation appears here with Fleming's permission.

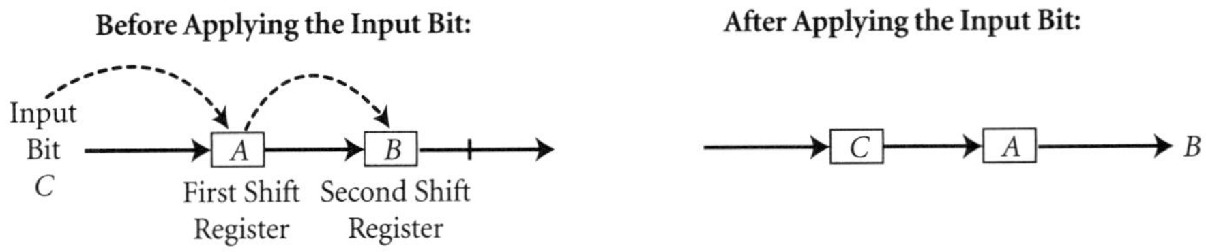

State at Time Instant n	State at Time Instant $n+1$	
	Input Bit 0	Input Bit 1
00	00	10
01	00	10
10	01	11
11	01	11

Figure 2.12: State Transition Table for Convolutional Coding

output table as summarized in Table 2.3.[33] The information bit present at the input of the encoder and the state of the encoder are adequate in determining the encoder outputs. For example, if the input bit $x(n)$ is 0 and the encoder state $x(n-1)x(n-2)$ is 01, the encoder output $y_0(n)y_1(n)$ will be 11 based on Eqs. (2.7) and (2.8).

We will use the state transition table of Figure 2.12 and the output table of Table 2.3 to construct a *trellis diagram*. The Viterbi decoder uses the trellis diagram to estimate the input bit sequence. The trellis diagram yields several pieces of information for an arbitrary[34] input sequence: the encoder states that could be valid[35] at a given time

Table 2.3: Output Table for Convolutional Coding

Current State $\{x(n-1)\ x(n-2)\}$	Output Code Symbols $\{y_0(n)\ y_1(n)\}$	
	Input Bit $x(n)=0$	Input Bit $x(n)=1$
00	00	11
01	11	00
10	10	01
11	01	10

[33] Table 2.3 is adapted from a table in Fleming's work [58]. The original table is copyrighted by Robert F. Fleming, III, and the adaptation appears here with Fleming's permission.

[34] The Viterbi decoder considers all possible state transitions and all possible input sequences to reach the conclusion about the bit stream transmitted by the transmitter.

[35] It will become apparent in the next few *minutes* (and not *hours*, we hope) that not all states are possible at all time instants even for an arbitrary input sequence.

instant; the state transitions from one time instant to the next as a function of the input bit; and the encoder output for the given combination of the input bit and the encoder state. *Here is our plan of action regarding the trellis diagram.* We will first construct an entire trellis diagram for an *arbitrary input sequence*. Such a diagram can be used to illustrate encoder and decoder operation. We will then depict the actual trellis path taken by our *test sequence* at the *transmitter*. Once we are done with the transmitter, we will summarize the steps of a Viterbi decoder followed by the receiver. We will execute the first few steps of the decoding operation by pretending *we* are a Viterbi decoder. (Yes, fasten your seat belts for a thrilling ride!) Finally, we will summarize the overall Viterbi decoder operation for our test sequence.

Figure 2.13[36] displays the trellis diagram for an arbitrary input sequence and also highlights the specific trellis path followed by our test sequence. The four encoder states are represented by tiny dots and the lines joining the two states from one time instant to the next represent state transitions. Each state transition line has an associated tuple $a : b, c$ where a is the input bit $x(n)$ that has caused the related state transition and b, c represents encoder output, $y_0(n), y_1(n)$. Let's discuss what happens when the first two inputs are processed by the encoder one after the other. We will use the state transition table and the output table to explain the development of the trellis diagram.

At time $t = 0$, the encoder is initialized with the state 00, and, hence, the trellis starts at the state 00. The input bit arriving at $t = 0$ may be 0 or 1. If it is 0, the state transition table indicates that the next state at time $t = 1$ is 00 and the

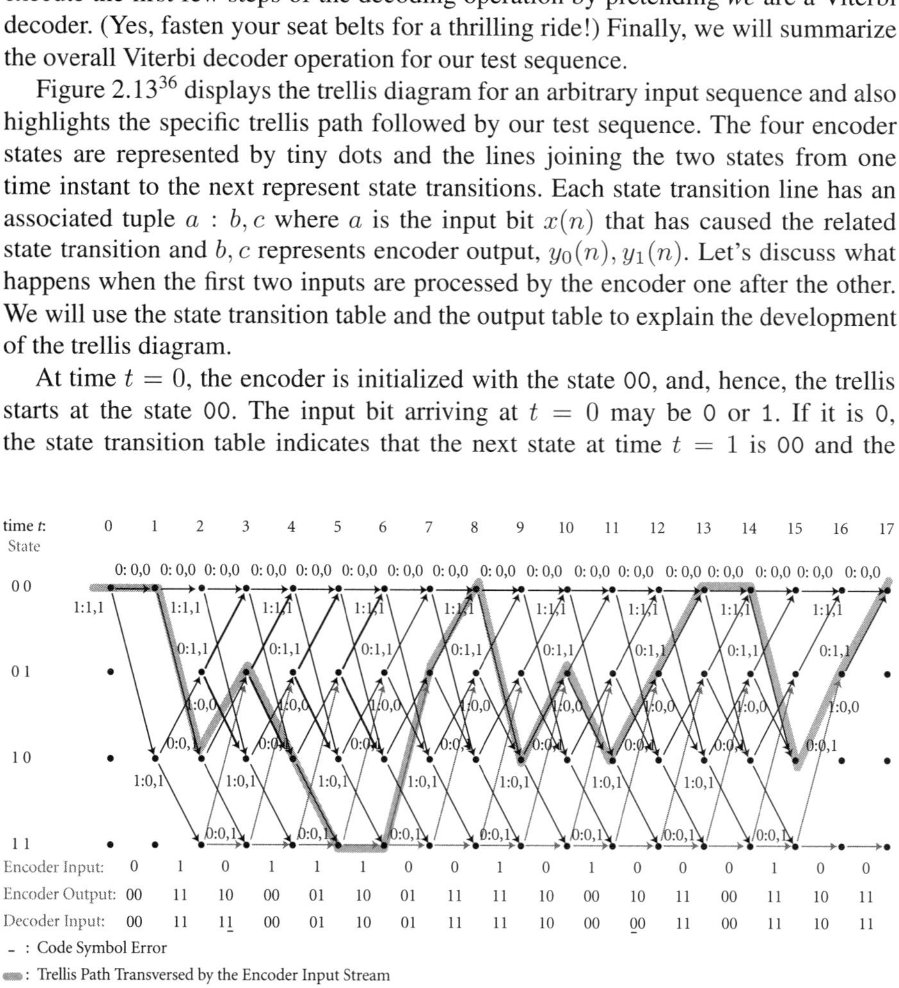

Figure 2.13: A Comprehensive Trellis Diagram for an Arbitrary Input Sequence

[36]Figure 2.13 is adapted from a figure in Fleming's work [58]. The original figure is copyrighted by Robert F. Fleming, III, and the adaptation appears here with Fleming's permission.

output table indicates that the encoder output is 00 corresponding to $x(n) = 0$ and $\{x(n-1)x(n-2)\} =$ 00. The input bit and the encoder output associated with the state transition 00 to 00 are displayed as $0 : 0, 0$ above the line that connects two 00 states at $t = 0$ and $t = 1$. If the input bit at $t = 0$ is 1, the next state at $t = 1$ is 10 from the state transition table and the resulting encoder output is 11 from the output table. At $t = 0$, there was just one valid state, 00. However, at $t = 1$, we have two potentially valid states 00 and 10 because we may reach these states based on the value of the input bit. Let's go one step further. Let's increment time t. When the second input bit is processed at $t = 1$, we need to consider potential transitions from two valid states 00 and 10. For the first valid state 00 at $t = 1$, the application of the input bit 0 again yields the transition to the same state 00 at $t = 2$ and we get the encoder output of 00. Similarly, processing of the input bit 1 while in state 00 results in the next state being 10 at $t = 2$ and encoder output being 11. When input bit 0 is processed for the second valid state 10, the input bit 0 leads to the next state of 01 and the encoder output 10. The application of the input bit 1 causes the state transition from 10 to 11 and yields the encoder output 01.

Now, let's fast-forward to $t = 15$ where all four states are valid. Since the encoder input is now the first TAIL bit 0, we need to consider only the input bit 0 while determining the next state at $t = 16$. At $t = 16$, only the second TAIL bit 0 is present at the encoder input, and two valid states, 00 and 01, are present. Observe that the encoder state at $t = 17$ is now 00, the same as the state at $t = 0$. That's the beauty of including 0s as encoder TAIL bits—we reset the encoder so that it is ready to process the next packet! The decoder will also have the same all 0s state once it is done with the decoding of the entire received packet. We now know how to create a comprehensive trellis diagram containing all valid states at all time instants for an arbitrary input sequence; possible state transitions; and the encoder output for a given combination of the state and the input bit. Tracing the path followed by our test sequence is quite straightforward. From the comprehensive trellis diagram, we just need to pick the state transitions that correspond to the actual input bit for our test sequence. For example, at $t = 0$, the first bit of the test sequence is 0, leading to the next state 00 at $t = 1$ and the encoder output 00. Similarly, the second input 1 of the test sequence is 1, which causes the state transition from 00 at $t = 1$ to 10 at $t = 2$ and the encoder output 11. The entire trellis path traversed by the seventeen-bit test sequence is highlighted in Figure 2.13.

Let's turn our attention to Viterbi decoding since we have satisfied all the prerequisites. (*Quite a few of them, right?*) Viterbi decoding at the receiver can rectify some errors present at its input. Hence, we will change two code symbols, shown as underlined in Figure 2.13 and in the text here, to mimic the error of the radio environment. The stream of code symbols received by the Viterbi decoder is 00 11 11 00 01 10 01 11 11 10 00 00 11 00 11 10 11. The Viterbi decoder attempts to estimate the encoder's trellis path by observing the received code symbols and minimizing the total cost or error metric of such a path. One way to define the cost is the Hamming distance between the set of the received symbols and the code symbols associated with a specific state transition. As an example, assume that the received code symbols are 11. Further assume that we are considering the cost of the transition from one state at $t = n$ to another state at $t = n + 1$ and that the code symbols corresponding to this transition are 10. The Hamming distance would then be number of disagreements between 11 and 10, which is 1. If the expected code symbols for the state transition were 11, the Hamming distance would be 0. If the expected code symbols for the state transition were 00, the Hamming distance would be 2. *The basic steps of the simplified Viterbi decoder are described here,* and we'll follow them just a bit later while pretending to be a Viterbi decoder. For simplicity of notation, we will represent the encoder state by its decimal representation. In other words, the encoder state 00 corresponds to the decimal number **0** and the state 11 corresponds to decimal value of **3**.

1. Initialization at $t = 0$ involves making state 00 as the starting point of the trellis path and making its cost $C(t, s) = C(0, 0) = 0$ where t and s in $C(t, s)$ represent the time index and the state, respectively. The validity of the state s at time t is specified by $V(t, s) = 1$. $V(t, s) = 0$ means that the state s at time t is invalid. Construct the entire array $V(t, s)$ based on the comprehensive trellis diagram of Figure 2.13 such that we have all of $V(0, 0), V(0, 1), \ldots, V(0, 3), V(1, 0), \ldots, V(1, 3), \ldots, V(17, 0), \ldots,$ $V(17, 3)$ defined as 0 or 1 based on the validity of the state s at time t.

2. Increment the time index by setting $t = t + 1$. At this new time instant t, find the set of valid states s_v by checking $V(t, s)$ for all values of s.

3. For each valid state s_v, find the total cost of reaching the state s_v from a predecessor state valid at time $(t - 1)$, designated s_p here, to the state s_v at the current time t. This cost is $C(t, sv) = C(t - 1, s_p) + T(s_p, s_v)$ where $C(t - 1, s_p)$ is the total cost to reach s_p at time $(t - 1)$ and $T(s_p, s_v)$ is the cost of the transition (often called branch metric) from s_p to s_v. If there are multiple paths to reach s_v (i.e., multiple s_p values can reach s_v), s_p is chosen such that the total cost to reach s_v is minimal.

4. Repeat Steps 2 and 3 until the entire sequence of code symbols is processed.

Now, the moment you've waited for! Let's become a Viterbi decoder and execute the first couple of steps. Figure 2.14[37] shows that we begin with the state 00 at $t = 0$. At $t = 1$, we have two valid states 00 and 10, which is indicated by $V(1, 0) = 1$ and $V(1, 2) = 1$, and $V(1, 1)$ and $V(1, 3)$ are both 0. First, let's consider the transition to state 00. Based on Figure 2.13, there is only one path to reach state 00 at $t = 1$, which is from the state 00 at $t = 0$ to 00 at $t = 1$. The total cost of reaching state 00 at $t = 1$ for this path is sum of the cost of reaching state 00 at $t = 0$ and the cost of transition from 00 at $t = 0$ to 00 at $t = 1$. In other words, $C(1, 0) = C(0, 0) + T(0, 0)$. $C(0, 0)$ is initialized to 0. $T(0, 0)$ is the Hamming distance between the received code symbols 00 and the code symbols 00 corresponding to the transition 00 to 00. Hence, $T(0, 0) = 0$ and $C(1, 0) = 0 + 0 = 0$. We also record the previous state, s_p,

[37]Figure 2.14 is adapted from a figure in Fleming's work [58]. The original figure is copyrighted by Robert F. Fleming, III, and the adaptation appears here with Fleming's permission.

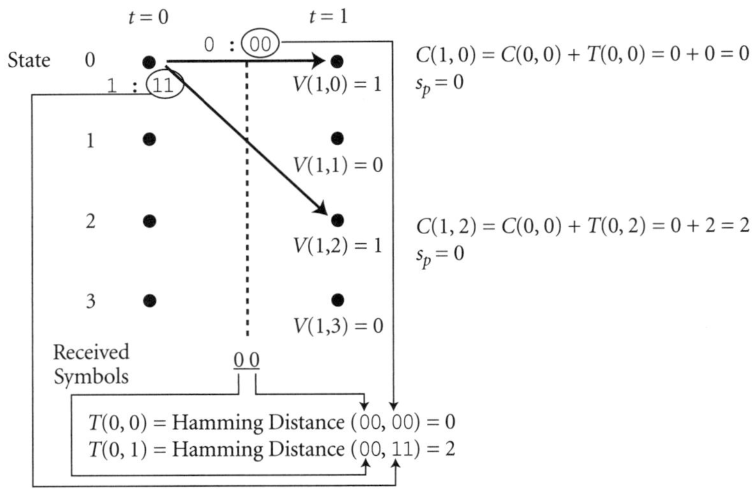

Figure 2.14: Viterbi Decoding: Processing of the First Set of Code Symbols

to be 0 (i.e., 00), corresponding to the chosen path from the previous state 00 to the current state 00. Now, let's turn our attention to the second valid state 10. There is just one way of reaching the state 10, i.e., from the state 00. The total cost of reaching the state 10 can then be expressed as $C(1,2) = C(0,0) + T(0,2)$. $T(0,2)$ is the Hamming distance between the received code symbols 00 and the code symbols 11 corresponding to the transition 00 to 10, which is 2: $C(1,2) = 0 + 2 = 2$. We also record the previous state, s_p, to be 0 (i.e., 00), corresponding to the chosen path from the previous state 00 to the current state 10.

Let's execute one more step of the Viterbi decoder as illustrated in Figure 2.15.[38] The second set of received code symbols (i.e., 11) causes the state transition from $t = 1$ to $t = 2$. All the four states are valid at $t = 2$. In other words, $V(2,0) = V(2,1) = V(2,2) = V(2,3) = 1$. We will begin with the state 00. Only one path reaches state 00 at $t = 2$, which is from the state 00. The total cost of reaching state 00 at $t = 2$ is sum of the cost of reaching state 00 at $t = 1$ and the cost of transition from 00 at $t = 1$ to 00 at $t = 2$: $C(2,0) = C(1,0) + T(0,0)$. Further, $C(1,0) = 0$, and $T(0,0)$ is the Hamming distance between the received code symbols 11 and the code symbols 00 corresponding to the transition 00 to 00. Hence, $T(0,0) = 2$ and $C(2,0) = 0 + 2 = 2$. We note the previous state, s_p, to be 0 corresponding to the path from the previous state 00 to the current state 00.

The second valid state is 01. The only way of reaching the state 01 is from the state 10. The total cost of reaching the state 01 is $C(2,1) = C(1,2) + T(2,1)$. $T(2,1)$ is the Hamming distance between the received code symbols 11 and the code symbols

[38]Figure 2.15 is adapted from a figure in Fleming's work [58]. The original figure is copyrighted by Robert F. Fleming, III, and the adaptation appears here with Fleming's permission.

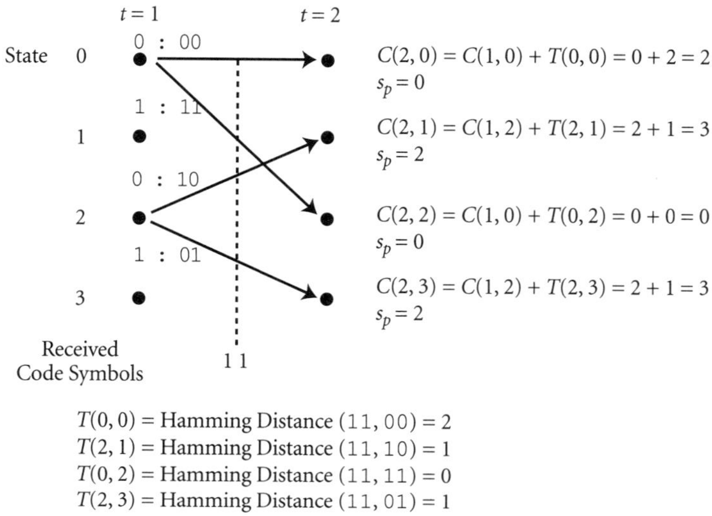

Figure 2.15: Viterbi Decoding: Processing of the Second Set of Code Symbols

10 corresponding to the transition 10 to 01, which is 1. $C(2,1) = 2 + 1 = 3$. We also observe s_p to be 2, corresponding to the chosen path from the previous state 10 to the current state 01.

10 is the third valid state. The only way to reach the state 10 is from the state 00. The total cost of reaching the state 10 is $C(2,2) = C(1,0) + T(0,2)$. $T(0,2)$ is the Hamming distance between 11 and the code symbols 11 corresponding to the transition 00 to 10, which is 0: $C(2,2) = 0 + 0 = 0$. We make s_p equal to 0 corresponding to the chosen path from the previous state 00 to the current state 10.

Finally, the last valid state is 11. The total cost of reaching the state 11 from 10 is $C(2,3) = C(1,2) + T(2,3)$. $T(2,3)$ is the Hamming distance between 11 and the code symbols 01, which is 1: $C(2,3) = 2 + 1 = 3$. Furthermore, we set s_p to 2, corresponding to the chosen path from the previous state 10 to the current state 11.

Let's summarize our findings for the entire decoding process for the seventeen-bit input sequence in Figure 2.16,[39] which summarizes the total cost of reaching a given state at time t and the related predecessor state valid at time $(t-1)$. For example, in the row corresponding to state 1 (i.e., 01), the entry (2,3) at $t = 4$ means that the total cost of reaching the state 1 is 2, and the path associated with this cost includes a transition from the predecessor state 3 at $t = 3$ to the current state 1 at $t = 4$.

Once the information shown in Figure 2.16 is available, we can start performing the decoding operation. Decoding of the Viterbi algorithm involves the *trace back* operation. Let's start the trace back operation at $t = 17$ where the current state is 0.

[39]Figure 2.16 is adapted from multiple tables in Fleming's work [58]. The original tables are copyrighted by Robert F. Fleming, III, and the adaptation appears here with Fleming's permission.

State \ Time	0	1	2	3	4	5	6	7	8	9	10	11	12	13	14	15	16	17
0	(4,0)	(0,0)	(2,0)	(3,1)	(3,0)	(3,1)	(3,1)	(4,0)	(1,1)	(3,0)	(4,0)	(3,1)	(3,0)	(2,1)	(2,0)	(4,0)	(5,0)	(2,1)
1			(3,2)	(1,2)	(2,3)	(2,3)	(3,2)	(1,3)	(4,3)	(4,2)	(1,2)	(4,3)	(2,2)	(3,3)	(4,2)	(4,2)	(2,2)	
2		(2,0)	(0,0)	(2,0)	(1,1)	(3,1)	(3,1)	(4,0)	(3,1)	(1,0)	(4,0)	(1,1)	(4,1)	(3,0)	(3,1)	(2,0)		
3			(3,2)	(1,2)	(2,3)	(1,2)	(1,3)	(3,2)	(4,3)	(4,2)	(3,2)	(4,3)	(2,2)	(3,3)	(4,2)	(4,2)		

State Transition	0→0	0→2	2→1	1→2	2→3	3→3	3→1	1→0	0→2	2→1	1→2	2→1	1→0	0→0	0→2	2→1	1→0
Estimated Input Bits	0	1	0	1	1	1	0	0	1	0	1	0	0	0	1	0	0

Figure 2.16: Summary of the Viterbi Decoding: Predicted Trellis Path and Estimated Bit Sequence

The predecessor state s_p is 1. Hence, we go from state 0 at $t = 17$ to state 1 at $t = 16$. For state 1 at $t = 16$, s_p is 2, and, hence, we go back to state 2 at $t = 15$. Eventually, we will reach state 0 at $t = 0$.

Once the full trellis path is known, we can move forward in time and observe the state transitions. As shown in Figure 2.16, from $t = 0$ to $t = 1$ we have a transition from (current) state 0 to (current) state 0. From $t = 1$ to $t = 2$, we are transitioning from (current) state 0 to (current) state 2. Similarly, from $t = 2$ to $t = 3$, the state transition is from state 2 to state 1. The final state transition occurs from state 1 at $t = 16$ to state 0 at $t = 17$. Since we already know the state transition table for our Convolutional coder, we can easily estimate the encoder input bits. For example, the state transition table shown in Figure 2.12 indicates that the encoder input bit associated with the state transition from 0 to 0 is 0 and that the encoder input bit associated with the state transition from 0 to 2 is 1. These are the first two input bits for the encoder. The estimated bit sequence is specified in Figure 2.16. We have now completed the *entire* process of Viterbi decoding! (You can now breathe a sigh of relief!)

A practical decoder does not wait to process the entire packet, which may be thousands of bits long, before tracing back; it may process few incoming received symbols and start the trace back operation to retrieve a few bits. The decoder then moves onto the next set of received symbols and traces back after processing them. Such a partial trace back operation simplifies the decoder without sacrificing too much performance. As an example of a partial trace back operation, assume that we start the trace back process at $t = 9$ in Figure 2.16 rather than at $t = 17$. We pick the state that yields the minimum cost at $t = 9$, which is state 10. The total costs of reaching the states 00, 01, 10, and 11 are 3, 4, 1, and 4, respectively. Executing the trace back operation beginning with state 10 at $t = 9$ yields the same estimates of the input bits as the full trace back process in our example.

Though we have calculated the cost as a Hamming distance, any suitable metric can be used. Such a metric may enable soft-decision Viterbi decoding as opposed to hard-decision decoding. In the hard-decision decoding discussed here, we have estimated the bit to be 0 or 1. We can use the Euclidian norm or mean-squared error along with real-valued received code symbols to execute Viterbi decoding. See Hagenauer's work [59] for more information.

2.4.3 Turbo Coding at the Transmitter

Turbo coding is a more powerful FEC coding technique than Convolutional coding. Figure 2.17 shows a generic structure of the Turbo coder that consists of two constituent Convolutional encoders, a Turbo interleaver, and a symbol puncturing and repetition module. The original bit stream is processed by one constituent encoder, and an interleaved bit stream is processed by the other encoder. The Turbo interleaver operates similarly to a regular interleaver and shuffles the order of the bits. Though a traditional interleaver rearranges bits in a systematic fashion, the Turbo interleaver should reshuffle bits so that is appears more random. Such random order results in better performance. See Section 2.5 for a closer look at interleaving. In Figure 2.17,

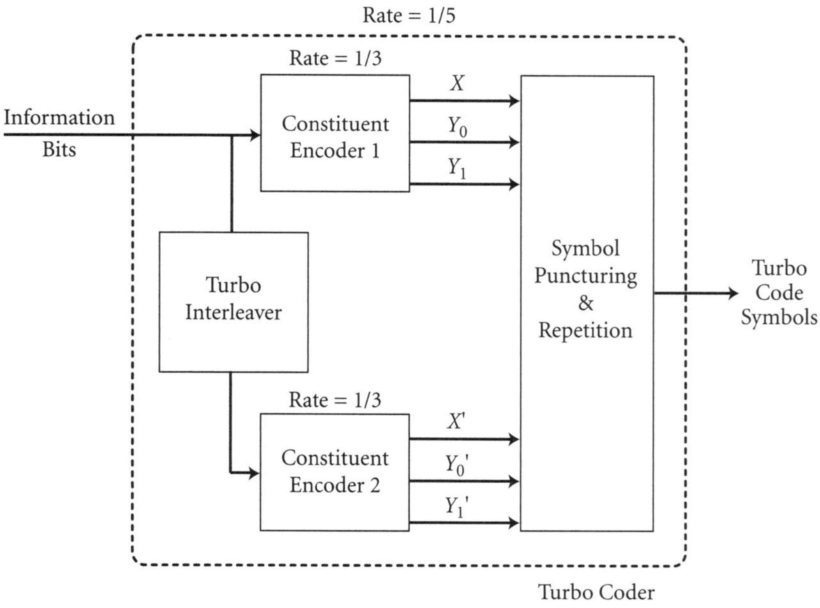

Figure 2.17: Structure of a Turbo Coder

each constituent encoder has the coding rate of $1/3$, and, hence, three streams of Convolutional code symbols appear at the output. The symbol puncturing and repetition module ensures that an appropriate number of bits exit at the output of the Turbo coders. For example, if the desired coding rate for the Turbo coder is $1/5$, the Symbol Puncturing and Repetition Module selects five of six symbols, X, Y_0, Y_1, X', Y_0', and Y_1'. Sometimes, the radio frame responsible for shipping the code symbols has more space, and then the code symbols are repeated.

The constituent encoder can be any regular Convolutional encoder. Figure 2.18 specifies the encoder used in CDMA2000-based systems. It is a rate $1/3$ recursive encoder, which includes feedback from the encoder output to the encoder input. Both constituent encoders are identical in CDMA2000-based systems. Each encoder uses the following generator polynomials.

$$n_0(D) = 1 + D + D^3 \tag{2.9}$$
$$n_1(D) = 1 + D + D^2 + D^3 \tag{2.10}$$
$$d(D) = 1 + D^2 + D^3 \tag{2.11}$$

The second output stream (i.e., the B stream) of the constituent encoder is said to have the transfer function n_0/d, and the third output stream of the constituent encoder is said to have the transfer function n_1/d. The transfer function quantifies the relationship between the output and the input of a signal processing block. The recursive nature of the encoder evident from Figure 2.18 requires a denominator in the transfer function. In the absence of the recursive nature of the encoder, only the numerator would appear in the transfer function.

One encoder takes the original bit stream as the input and yields the output streams A, B, and C, which correspond to X, Y_0, and Y_1 in Figure 2.17. Note that A is the same as the input stream, and hence, X in Figure 2.17 is the same as the original bit stream. *This is an important characteristic of a Turbo coder: the first output stream is nothing but the original input bit stream!* Such first stream code symbols are

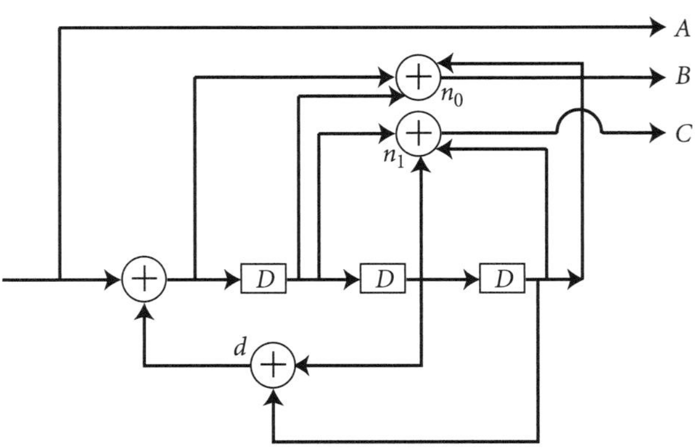

Figure 2.18: Constituent Convolutional Encoder

often called *systematic bits*. The code symbols from other streams are called *parity*, *redundancy*, or *protection* bits. The second constituent encoder takes the input bit stream processed by the Turbo interleaver and yields the output streams, A, B, and C, which are X', Y_0', and Y_1' in Figure 2.17.

2.4.4 Turbo Decoding at the Receiver

The Viterbi decoder becomes too complex when used to decode Turbo code symbols. A more prevalent decoding approach relies upon iterative decoding of Turbo code symbols using some form of the *BCJR algorithm*, named after the authors who developed a symbol-by-symbol maximum a posteriori (MAP) algorithm [60]. The specific decoder structure depends on the encoder structure. We will use the Turbo encoder and decoder shown in Figure 2.19[40] to summarize the overall decoding process.

The Turbo encoder uses two identical constituent Convolutional encoders in Figure 2.19, with an interleaver sending reshuffled input bits to the second encoder. The encoders are *Recursive Systematic* Convolutional encoders, with *recursive* meaning that the output of the encoder at sample instant k acts as an input to the encoder at the sample instant $k+1$ and *systematic* meaning that the input sequence itself is also part of the overall output sequence. For example, the Turbo encoder of rate $r = 1/2$ shown in Figure 2.19 has two output code symbols for every input bit u_k:

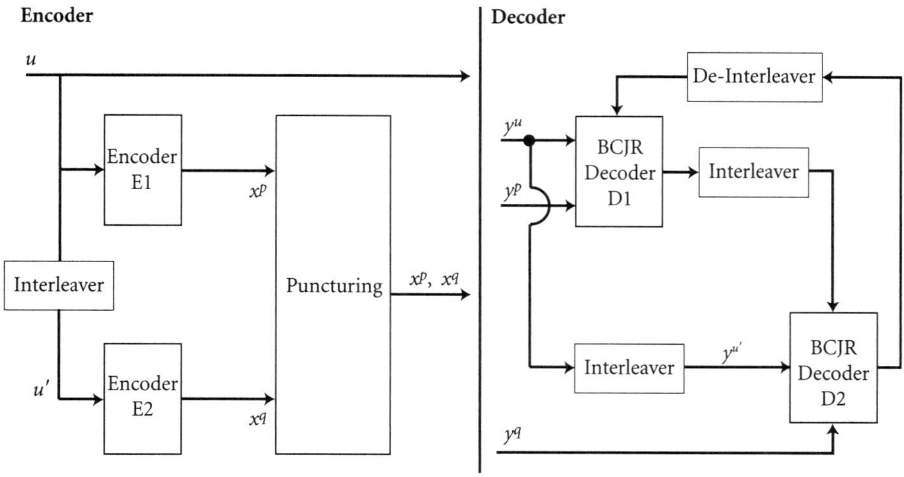

Figure 2.19: Example Encoder and Decoder Structures

[40]Figure 2.19 is adapted from Figures 12.1 and 12.6 in Valenti's work [61]. Reprinted from *Handbook of RF and Wireless Technologies*, M.C. Valenti and J. Sun, "Turbo Codes," pages 375–399, © 2004, with permission from Elsevier.

u_k and either x_k^p or x_k^q where k is the sample instant. u is the set of encoder input bits (including the encoder TAIL bits to force the final state of the encoder E1 to 0). x^p and x^q are sets of parity bits generated by the encoder E1 and the encoder E2, respectively. Puncturing selects one of the two parity bits generated by the two encoders at a given time instant k. A rate $r = 1/3$ Turbo encoder can be realized by avoiding the puncturing altogether and by choosing all of u_k, p_k, and q_k for every input u_k.

Now, let's turn our attention to the *iterative* decoder depicted in Figure 2.19. An iterative decoder goes through several iterations in an attempt to estimate the encoder input bits. The decoder inputs are y^u, y^p, and y^q and represent the noisy[41] versions of x^u, x^p, and x^q, respectively. Both decoders shown in Figure 2.19 are BCJR decoders. Since the encoder E1 processes the input stream, the first decoder D1 utilizes y^u and y^p as inputs. Furthermore, since the encoder E2 processes the interleaved input stream, the second decoder D2 utilizes interleaved y^u and y^q as inputs. Additionally, the decoders exchange *extrinsic* or *soft* information (as opposed to bits generated from hard decisions) with each other. See Valenti's work for an efficient decoder structure [61] and for a MATLAB code that implements Turbo encoding and Turbo decoding [62].

Each decoder in Figure 2.19 is an iterative parallel-concatenated convolution code decoder. Such decoder iteratively estimates the *a posteriori* probabilities $\Pr(u_k|\mathbf{y})$ where u_k is kth data bit and $\mathbf{y}$ is the codeword received at the decoder in an AWGN channel. k ranges from 1 to K, and $\mathbf{y}$ is given by

$$\mathbf{y} = \mathbf{c} + \mathbf{n} \tag{2.12}$$

where $\mathbf{c}$ is the codeword (i.e., the output of the encoder) and $\mathbf{n}$ is the noise word containing AWGN samples. The symbol-by-symbol BCJR algorithm applies the MAP rule to estimate u_k. The BCJR algorithm estimates $u_k = +1$ if it finds $P(u_k = +1|\mathbf{y})$ to be greater than $P(u_k = -1|\mathbf{y})$. Otherwise, it estimates $u_k = -1$. Such a decision $\hat{u}_k$ can be expressed compactly as

$$\hat{u}_k = \operatorname{sgn}[L(u_k)] \tag{2.13}$$

where $L(u_k)$ is the log-likelihood ratio that is expressed as (adapted from Eq (12-2) in Valenti's work [61])

$$L(u_k) = \log\left(\frac{\Pr(u_k = +1|y)}{\Pr(u_k = -1|y)}\right). \tag{2.14}$$

In practice, the log-MAP algorithm is employed to simplify the processing. See Robertson's and Viterbi's works for details on the log-MAP algorithm [63, 64].

2.4.5 What Should I Choose: Convolutional or Turbo?

We have just summarized the basic principles of the Convolutional and Turbo coding and decoding. Now, you may be wondering which coding technique should be used.

[41]We will assume that the radio channel is modeled as AWGN (Additive White Gaussian Noise), and, hence, some noise is added to the encoder-generated code symbols.

Table 2.4: Convolutional versus Turbo Coding

Perspective	Convolution Coder	Turbo Coder
Implementation	Simple	Complex
Processing Requirements	Lower	Higher
Processing Delay	Shorter	Larger
Error Performance	Order of 10^{-3}	Order of 10^{-6}
Suitability	Voice and Low Data Rate Services	Medium or Higher Rate Data Services

Table 2.4 provides a brief comparison of the two coding techniques. Since a Turbo coder includes two Convolutional coders, it is more complex and therefore demands higher processing power and longer processing delay. The most significant benefit of Turbo coding is its superior error performance compared to Convolutional coding. The larger the amount of data in the data packet, the better the Turbo coding performance would be compared to Convolutional coding. Hence, Convolutional coding is used for voice services that need low data rates and can tolerate some errors. The higher-rate applications, such as email and web-browsing, are expected to provide error-free services, and, hence, Turbo coding is used. Consider this: high-speed data-centric 3G systems such as 1xEV-DO, HSDPA, HSUPA, and HSPA+ use only Turbo coding for the data traffic channels; Convolutional coding is *not even defined* for use with the data traffic channels!

2.5 INTERLEAVING AND DE-INTERLEAVING

Let's assume that the transmitter sends a set of information bits $\{A, B, C, \ldots\}$ represented by the set of code symbols $\{A, a, \alpha, B, b, \beta, C, c, \gamma, \ldots\}$. For example, information bit A is represented by three code symbols: A, a, and α. However, as shown in Figure 2.20, the radio environment causes a fade to occur. The length and the position of the fade are such that the signal energy is quite low for the code symbols B, b, and β. Hence, the receiver errs in estimating these code symbols, resulting in three consecutive errors. If the received code symbols have such consecutive errors, the decoder cannot recover the original packet $\{A, b, C, \ldots\}$. That's the way decoder works; it does not work well with consecutive code symbol errors.

Now, we are facing a dilemma! If we give code symbols with consecutive errors to the decoder, it will be unable to rectify the errors. On the other hand, the radio environment will cause consecutive errors. The solution to this dilemma is interleaving at the transmitter and de-interleaving at the receiver. The main idea is to somehow spread out the code symbol errors at the receiver so the decoder receives no consecutive symbol errors.

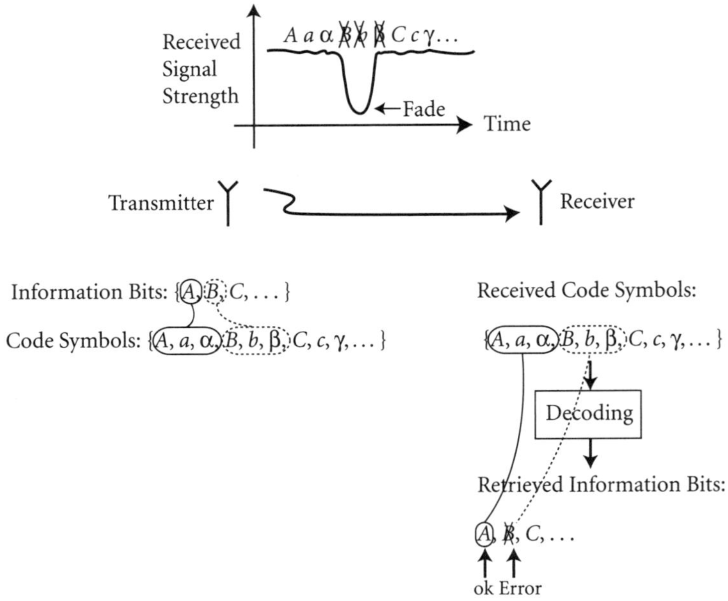

Figure 2.20: Motivation for Interleaving

Figure 2.21 illustrates the concepts of interleaving and de-interleaving. The original packet consists of the information bits $\{A, B, C, \ldots\}$, and the corresponding code symbols are $\{A, a, \alpha, B, b, \beta, C, c, \gamma, \ldots\}$. Next, the transmitter carries out interleaving. The interleaving shuffles the code symbols such that the code symbols representing a given information bit get scattered.

In our example, the code symbols are rearranged such that the interleaved code symbols follow the sequence $\{A, B, C, a, b, c, \alpha, \beta, \gamma, \ldots\}$. Just like before, the radio environment causes a fade, leading to three consecutive symbol errors (i.e., errors on a, b, and c in Figure 2.21). The receiver does the opposite of interleaving: de-interleaving. The de-interleaving process restores the original sequence of code symbols $\{A, B, C, a, b, c, \alpha, \beta, \gamma, \ldots\}$. There are still three errors on code symbols a, b, and c. However, notice that *the decoder no longer experiences consecutive code symbol errors*. Now, the decoder works its magic, and we will be able to retrieve the original packet, $\{A, B, C, \ldots\}$. In summary, interleaving at the transmitter and de-interleaving at the receiver ensure that the consecutive symbol errors caused by the radio environment are scattered at the input of the decoder, thereby facilitating the operation of the decoder.

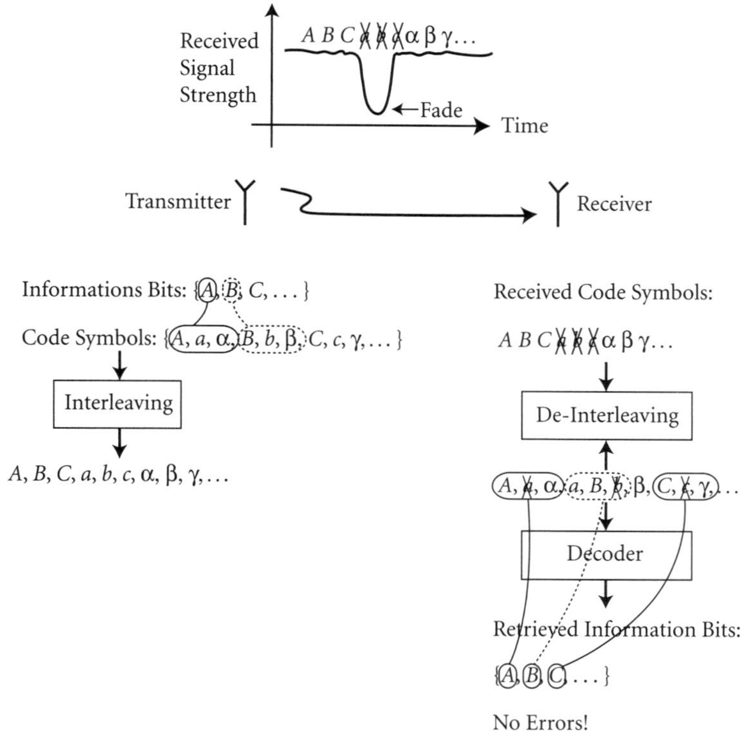

Figure 2.21: Basics of Interleaving and De-Interleaving

2.6 DIGITAL MODULATION AND DEMODULATION

Let's say that we want to send interleaved code symbols over the air. We need to represent the symbols, 1s and 0s, by some waveforms. Modulation enables us to carry the information, specifically interleaved code symbols, using sinusoidal signals. Modulation is typically carried out in two phases. The first phase can be thought of as a digital modulation phase that involves determination of the amplitude and phase of the sinusoidal signal to represent a set of code symbols. Such processing is carried out by a modulation mapper. The second phase is analog modulation in which an RF carrier signal, which is combination of a sine and cosine wave, is sent over the air. The analog modulation is nothing but frequency upconversion, which transports the baseband signal to the carrier RF frequency. The amplitude and phase of the carrier signal are the parameters determined during the first phase. The frequency upconversion is briefly discussed in Section 2.7 as part of RF processing.

2.6.1 Modulation Schemes

The modulation schemes widely used in 2G and 3G cellular systems are BPSK (Binary Phase Shift Keying), GMSK, QPSK, 8-ary Phase Shift Keying (8-PSK), and 16-ary Quadrature Amplitude Modulation (16-QAM) and 64-QAM. GSM utilizes GMSK, and BPSK and QPSK are used in 3G 1x and UMTS systems. 8-PSK is used in EDGE and 1xEV-DO. 16-QAM is used in 1xEV-DO and HSPA.

Let's take a simple example of BPSK modulation. Figure 2.22 shows a signal constellation diagram for the BPSK modulation. One BPSK modulation symbol represents one interleaved code symbol. There are two BPSK modulation symbols in the constellation diagram, s_1 and s_2. In our example, modulation symbol s_1 represents code symbol `0` and modulation symbol s_2 represents code symbol `1`. In Figure 2.22, the coordinate of s_1 is $(0, 1)$, and the coordinate of s_2 is $(0, -1)$. The I (in-phase) and Q (quadrature phase) coordinates are denoted (I, Q) such as (a, b). For s_1, $a = 0$ and $b = 1$. The modulation symbol can be represented by a complex number $a + jb$ where a is the I coordinate and b is the Q coordinate. If the I channel component maps onto a cosine wave and the Q channel component maps onto a sine wave, a modulation symbol s_i is represented by the combined wave form $(a\cos(2\pi ft) + b\sin(2\pi ft))$ where f is the RF carrier frequency and t denotes time. For example, $s_1 = \sin(2\pi ft)$ and $s_2 = -\sin(2\pi ft)$. From the amplitude and phase perspective, s_1 can be said to have the amplitude of 1 and phase of 90° (measured from positive I channel), such as $\cos(2\pi ft + 90) = \sin(2\pi ft)$. Similarly, s_2 has the amplitude of 1 and phase of 270° such as $\cos(2\pi ft + 270) = -\sin(2\pi ft)$.

Figure 2.22 is just one possible implementation of the many ways of designing a BPSK signal constellation diagram. Defining s_1 and s_2 to be on I channel [e.g.,

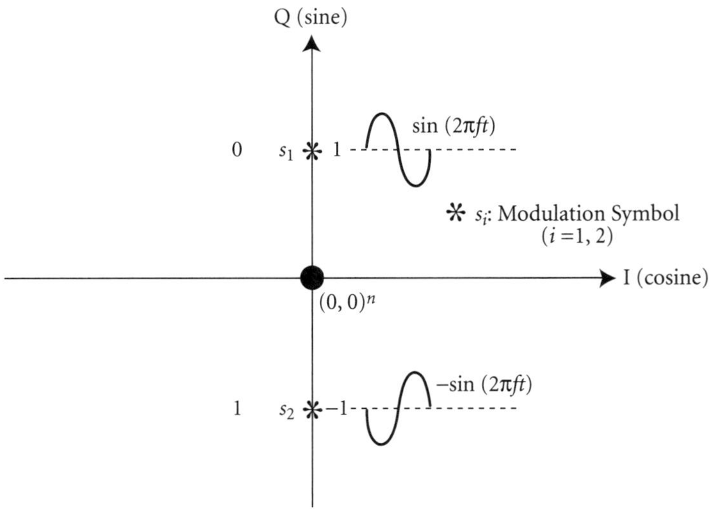

Figure 2.22: BPSK Signal Constellation Diagram

$s_1 = \cos(2\pi ft)$ and $s_2 = -\cos(2\pi ft)$] would be another way of constructing the BPSK constellation.

Gaussian Minimum Shift Keying (GMSK) used in GSM is conceptually similar to BPSK in that one modulation symbol contains one information bit or code symbol. The GMSK modulation signal of GSM is designed with the so-called BT product being 0.3 where B is the signal bandwidth and T is the bit period. The target signal bandwidth of B is ensured using a Gaussian filter with the bandwidth of 81.3 kHz, yielding $T = 3.692\,\mu s$. The bit frequency is the reciprocal of T and is $1/3.692\,\mu s = 270.833$ kHz. The GMSK modulation is a special case of Frequency Shift Keying (FSK). It is a constant-envelope modulation scheme, meaning that the amplitude of the transmitted signal remains the same regardless of which modulation symbol is transmitted. Let's look at basic FSK first, Minimum Shift Keying (MSK) next, and GMSK last.

Figure 2.23 illustrates an example of the *FSK modulation scheme*. Bit 1 is represented by one frequency (e.g., a low frequency such as f_1), and bit 0 is represented by another frequency (e.g., a higher frequency such as f_2). Based on the information bit to be transmitted, we *shift* the frequency of the signal to one value or the other, hence the name frequency *shift* keying. The receiver knows the predefined mapping between the bit value and the frequency and is able to estimate the transmitted bit pattern (more accurately, code symbol pattern) by observing the frequencies of the received signal. While FSK is simple to implement using oscillators, that sudden

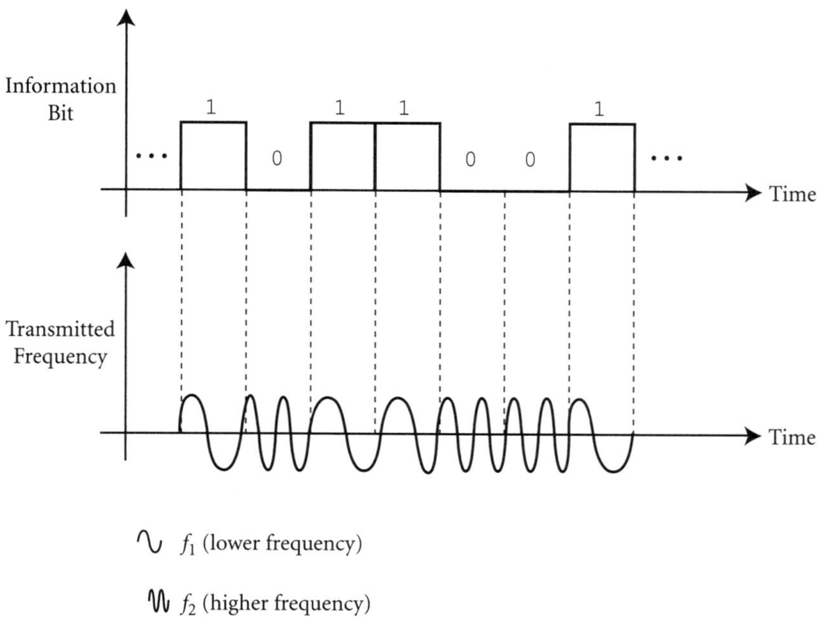

Figure 2.23: Conceptual Example of the Frequency Shift Keying Modulation

change in the signal frequency from one bit period to the next leads to relatively large spectral side lobes outside the main spectral band of the signal. Such side lobes essentially mandate larger spectrum bandwidth to carry the information signal. That's where continuous phase frequency shift keying (CPFSK) comes into the picture. CPFSK varies the signal frequency continuously, resulting in smaller spectral side lobes outside the main spectral band [65]. As we will see later, MSK is a special case of CPFSK.

Let's go one step further and discuss MSK [66]. Consider Figure 2.24 in which the information stream is 1001011000. Let $+1$ represent the information bit 1 and -1 represent the information bit 0. Now, let's prepare the groundwork for construction of an MSK signal. The information stream is divided into two sets, a set containing odd bits and a set containing even bits. If the bit period, or the modulation symbol period, is T, the duration of a given odd bit and the duration of a given even bit are both $2T$. For example, in Figure 2.24, the first bit in the information stream, bit 1, lasts for time T, but the time period of the same bit as an odd bit lasts for the time duration $2T$. Similarly, the third bit in the information stream, bit 0, lasts for time T, but the time period of the same bit as an odd bit lasts for time duration $2T$. The same value of a given odd or even bit is held for two consecutive bit periods. Since the data rate of the information stream is 270.833 kbps, the data rates of the odd bit stream and the even bit stream are half of 270.833 kbps or 135.4 kbps.

Since MSK is a variation of basic FSK, different frequencies will represent the information stream in the transmitted signal. In Figure 2.24, Waveform A is a high-frequency signal and Waveform B is a low-frequency signal, and some forms of these frequencies will be present in the final transmitted MSK-modulated signal. Waveforms A and B are our reference signals, with the structures shown in Figure 2.24 *designated as the positive-sense signals*. If Waveform A is inverted, its sense becomes negative, resulting in Waveform C. Similarly, a negative sense of Waveform B is Waveform D. In summary, we have defined Waveforms A and B as our positive-sense reference signals and Waveforms C and D as negative-sense waveforms for Waveform A and Waveform B.

Now that we have laid the foundation, let's create an MSK signal. We will follow two simple rules. First, during a given bit period of duration T, the MSK signal will be identical to one of the four waveforms, A, B, C, or D. Secondly, the specific form taken by the MSK signal during a given bit period depends on the combination of the odd bit and the even bit as summarized in the MSK table in Figure 2.24. Let's use an example. During the bit period denoted T_1, the odd bit value is $+1$ and the even bit value is -1. Such bit combination in the MSK table corresponds to Waveform B. Hence, the MSK signal is identical to Waveform B during the bit period T_1. As another example, the (odd, even) bit combination during the bit period T_2 is $(-1, -1)$, corresponding to Waveform C.

The receiver observes two consecutive bit periods to reliably estimate a given information bit. For example, consider a sliding window consisting of T_1 and T_2. Since the receiver observes Waveforms B and C during T_1 and T_2, respectively, it knows that the (odd, *even*) bit combination is $(1, -\mathit{1})$ during T_1 and $(-1, -\mathit{1})$ during T_2. Since two consecutive even bits are -1, the receiver estimates that the

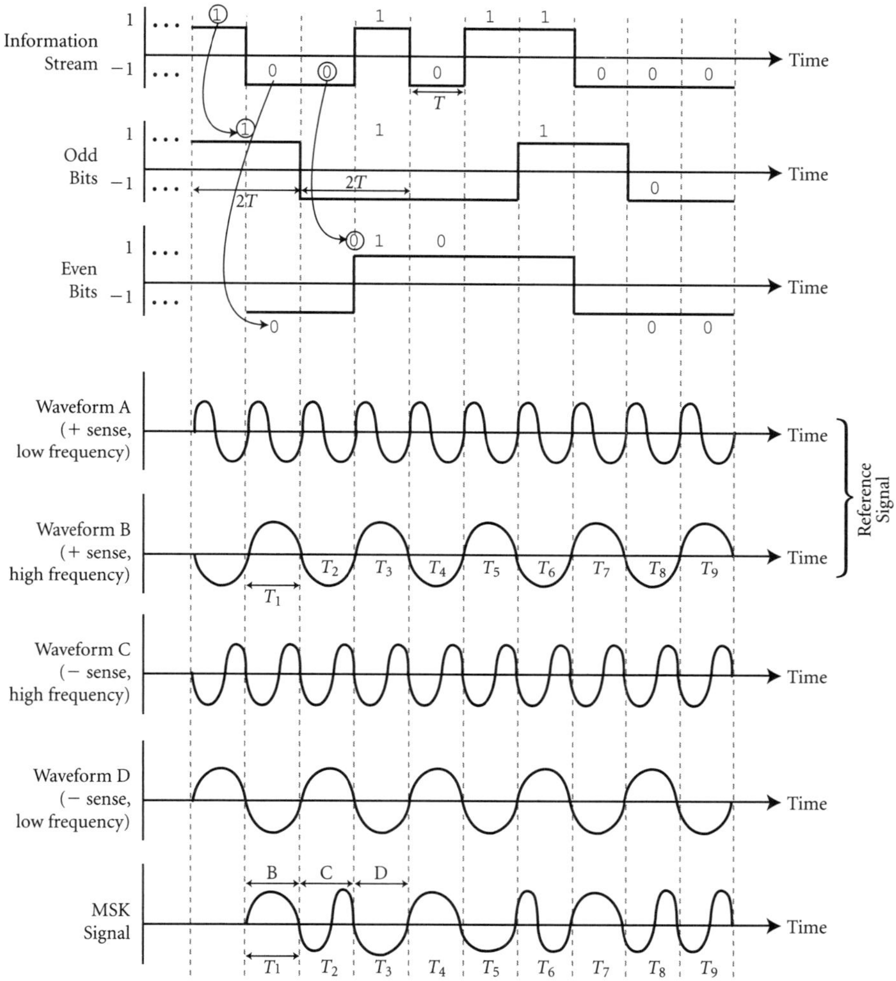

MSK Table

Odd Bit Value	Even Bit Value	MSK Signal Waveform
1	1	A
−1	1	D
1	−1	B
−1	−1	C

Figure 2.24: Example of an MSK Signal

even bit being transmitted during T_1 and T_2 must be 0 since 0 is mapped onto -1 in Figure 2.24. Similarly, by observing the waveforms during T_2 and T_3, the receiver learns that the (*odd*, even) bit combinations are $(-1, -1)$ and $(-1, 1)$, leading to the receiver concluding that the odd bit 0 must have been transmitted as the information bit. Since the receiver needs to observe two consecutive bit periods to estimate a given information bit and since the MSK signal during a given bit period (equivalent to the modulation symbol period) represents two bits (an odd bit and an even bit), MSK allows the receiver to extract one information bit during one bit period. Hence, from the perspective of representing information at the transmitter and extracting information at the receiver, MSK is like BPSK in which every modulation period contains one information bit.

Ever Wondered... Why does the word *minimum* appear in MSK? As seen in Figure 2.24, two frequencies are present in the transmitted signal. Let these two frequencies be f_1 and f_2. The separation between the two frequencies should be such that they are orthogonal to each other during the bit period. Orthogonality between the frequencies facilitates the job of the receiver as the receiver attempts to match the received waveform with possible waveforms. Two sinusoidal signals with frequencies f_1 and f_2 are said to be orthogonal over the time interval τ if

$$\int_0^{\tau} \sin(2\pi f_1 t)\sin(2\pi f_2 t)dt = 0.$$

The *minimum* frequency separation to ensure orthogonality in such case is $\Delta f = f_2 - f_1 = 1/2\tau$. This frequency separation corresponds to the modulation index h of $1/2$ for a CPFSK signal [65]. The modulation index, h, is formally defined as $h = 2f_d\tau$ where f_d is the peak frequency deviation (around the center carrier frequency) and τ is the time interval of interest where orthogonality is desired. In GSM, the value of $\tau = T = 3.69\mu s$, leading to $\Delta f = 1/2\tau = 135.4$ kHz and $fd = \Delta f/2 = 135.4/2 = 67.7$ kHz. These parameters correspond to the modulation index h of $2 * 67.7$ kHz $* 3.69\ \mu s = 0.5$. Hence, when a radio channel with carrier frequency f_c is allocated to the user, the actual frequencies carrying the information are situated $135.4/2 = 67.7$ kHz below and above f_c. In summary, since we use *minimum frequency separation* between the FSK signals carrying information bits while ensuring orthogonality, the resulting variation of the FSK is called *minimum* shift keying.

Now that we know what MSK is, we can tackle GMSK! GMSK is nothing but MSK operating on a given information stream, as shown in Figure 2.24, followed by Gaussian filtering! Each filter has a certain bandwidth that enables a range of frequencies to pass through it, and the bandwidth of such a Gaussian filter for GSM is 81.3 kHz.

The signal constellation diagrams shown in Figures 2.25–2.27 are used in 3G 1xEV-DO systems [67]. Figure 2.25 shows the signal constellation diagram of the

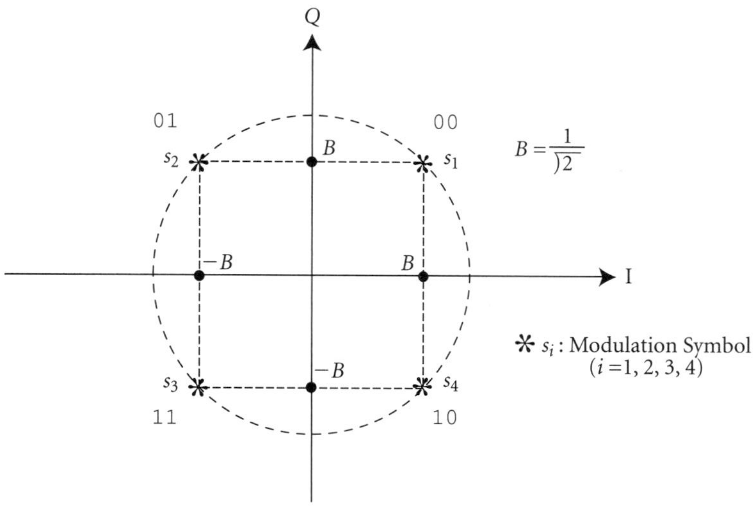

Figure 2.25: QPSK Signal Constellation Diagram

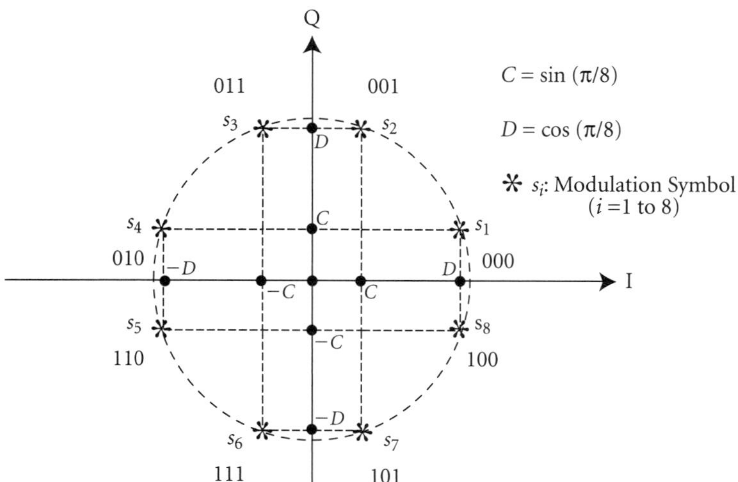

Figure 2.26: 8-PSK Signal Constellation Diagram

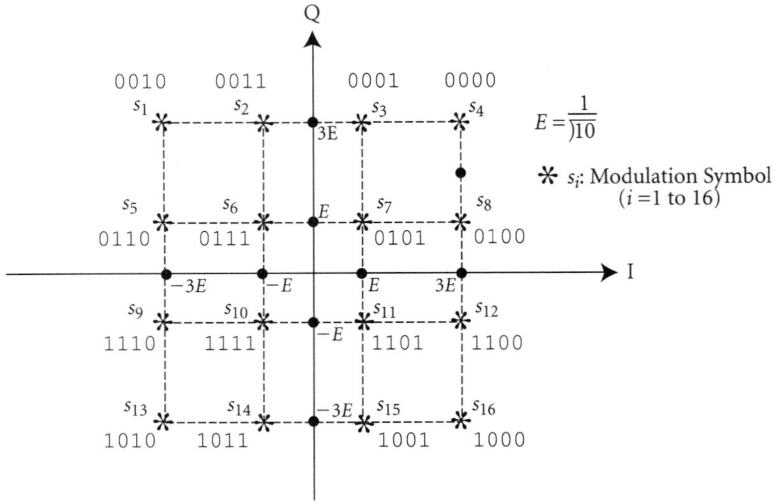

Figure 2.27: 16-QAM Signal Constellation

QPSK modulation scheme. There are four QPSK modulation symbols, and each modulation symbol represents two bits (or interleaved code symbols). In Figure 2.25, the modulation symbol s_1 represents a group of two bits 00. The symbol s_1 is specified as $B + jB$ with $B = 1/\sqrt{2}$.

An example signal constellation diagram for the 8-PSK modulation scheme is shown in Figure 2.26. Eight modulation symbols apears in the 8-PSK scheme with each modulation symbol representing a group of three interleaved code symbols. For example, the modulation symbol s_4 represents the code symbols 010 and is specified by the coordinates $(-D, C)$ where $C = \sin(\pi/8)$ and $D = \cos(\pi/8)$.

Figure 2.27 illustrates one possible 16-QAM signal constellation diagram. A 16-QAM scheme consists of sixteen modulation symbols, and each modulation symbol represents a group of four code symbols. The symbol s_8 represents 0100 and is denoted by $(3E, E)$ or $3E + jE$ where $E = 1/\sqrt{10}$.

Figure 2.28 shows an example 64-QAM signal constellation diagram used in HSPA+, which is a Third Generation Partnership Project (3GPP) Release 7 feature. Each 64-QAM modulation symbol represents six code symbols.

2.6.2 Demodulation Approaches

The modulation scheme maps one or more code symbols onto a single modulation symbol, but demodulation is a process that recovers one or more code symbols by observing the received modulation symbol. The demodulator observes the amplitude and phase of the received modulation symbol, compares it with the possible modulation symbols, and picks the modulation symbol that is closest to the received symbol. Figure 2.28 shows a simple example of the demodulator operation.

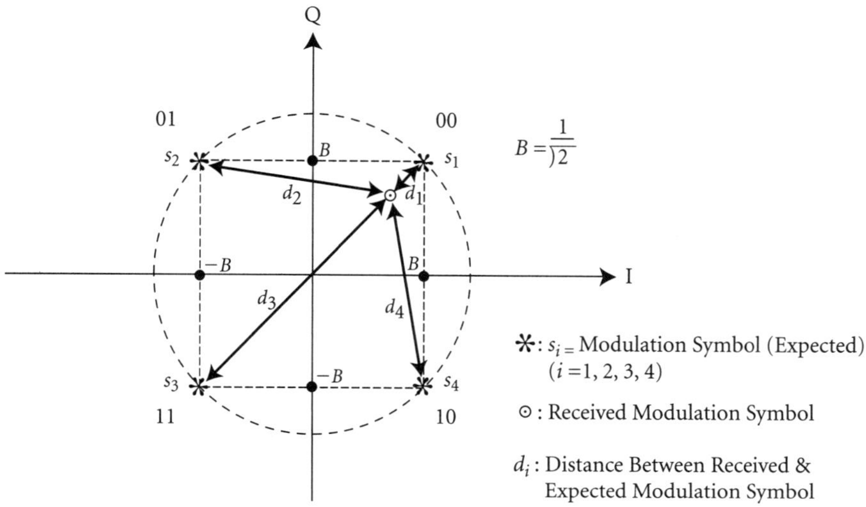

Figure 2.28: Decision-Making by the Demodulator

A Question for You... Which modulation scheme is being depicted in Figure 2.28? Hint: There are four modulation symbols with same amplitude. See the footnote[42] for the answer.

In the example here, the demodulator compares the received QPSK modulation symbol with the four expected options. Note that both transmitter and receiver know about the modulation scheme and the associated signal constellation diagram. The Euclidean distances between the received modulation symbol and all the possible modulation symbols are calculated. Let $R = x + jy$ be the received modulation symbol and $s_i = u + jv$ be an expected modulation symbol. The Euclidean distance, which is an indicator of the degree of error, between R and s_i can be calculated as

$$d_i = \sqrt{(x-u)^2 + (y-v)^2}. \tag{2.15}$$

The modulation symbol corresponding to the minimum Euclidean distance can be chosen as the estimated modulation symbol sent by the transmitter. In our example, assuming that your attention has not wandered and that Figure 2.28 does not qualify as abstract art, you will recognize that s_1 would be estimated as the transmitted modulation symbol.

Figure 2.29 compares the two main classes of demodulation techniques: non-coherent demodulation and coherent demodulation. In *non-coherent demodulation*, the demodulator relies upon the received modulation symbols on the traffic channel to

[42]It is QPSK. See Figure 2.25.

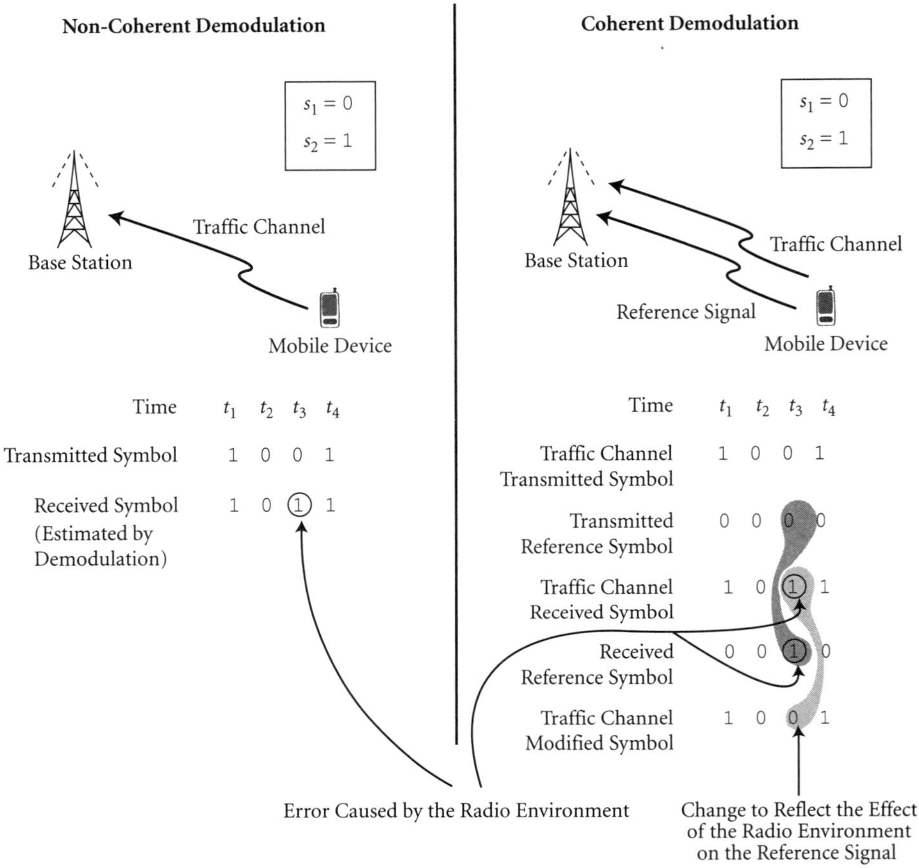

Figure 2.29: Coherent versus Non-Coherent Demodulation

estimate the modulation symbols. In the example shown here, the BPSK modulation symbols s_1 and s_2 correspond to bit 0 and 1, respectively. The mobile transmitter sends symbols s_2, s_1, s_1, and s_2 at time instants t_1, t_2, t_3, and t_4, respectively. The modulation symbol at t_3 is affected by the radio environment such that the demodulator estimates the received symbol to be s_2 rather than s_1, leading to the symbol error. *Coherent demodulation* utilizes a reference signal in addition to the regular traffic channel to aid the demodulation. Here, this reference signal could be locally generated using, for instance, a phase-lock loop or sent from the transmitter using a pilot channel or reference channel. The transmitter sends a traffic channel as well as a reference signal.

In Figure 2.29, the transmitter sends all 0s as a predefined pattern for the reference signal and sends s_2, s_1, s_1, and s_2 at time instants t_1, t_2, t_3, and t_4, respectively.

The radio environment affects the traffic channel and the reference signal in the same fashion since both pass the same radio channel. The radio environment causes errors in the received signal at time t_3 such that 0 of the reference signal becomes 1 and the 0 of the traffic channel becomes 1. The receiver expects all 0s as the reference signal and realizes that 0 has become 1 at t_3. To correct the error on the reference signal, it converts 1 back to 0. It makes same modification to the received traffic channel symbol at t_3 and converts the received symbol s_2 (i.e., 1) into symbol s_1 (i.e., 0). Such modification of the received traffic channel rectifies the error introduced by the radio environment. Now, the demodulator can correctly estimate the transmitted bits by processing the modified traffic channel symbols. The demodulator adjusts the received traffic channel symbols based on the knowledge of how the radio environment has affected the reference signal and uses the adjusted modulation symbols to estimate the transmitted modulation symbols. The error correction capability of coherent demodulation makes it a popular choice. For example, coherent demodulation with a reference signal is used in the forward link of IS-95 and in forward and reverse links of 1xRTT, 1xEV-DO, UMTS, HSPA, HSPA+, WiMAX, and LTE. The reverse link of IS-95 uses non-coherent demodulation. The coherent demodulation provides about a 2.5 to 3 dB better performance. As an example, the required reverse link E_b/N_0 to meet the 2% target error rate is 7 dB in IS-95 and 4.5 dB in 1xRTT. *In summary, the main difference between the coherent and non-coherent demodulation is the existence of a reference signal in coherent demodulation, which leads to better performance.*

2.6.3 Choosing a Modulation Scheme

Let's compare the use of two modulation schemes, BPSK and 16-QAM. Each has its benefits and drawbacks.

A Question for You... Based on the way a demodulator works (see Figure 2.29), which modulation scheme do you think would be easier for the receiver to handle—BPSK or 16-QAM? See the footnote[43] for the answer.

Now do you know *why* BPSK is easier for the receiver to handle? The Euclidean distance between two closest modulation symbols is larger in BPSK and smaller in 16-QAM. The receiver is thus more likely to make an error in 16-QAM than in BPSK when the channel modifies the transmitted modulation symbol significantly. The benefit of 16-QAM is, of course, a higher effective data rate because a single 16-QAM modulation symbol represents four bits (or code symbols), and a BPSK symbol corresponds to just one bit (or code symbol). An efficient cellular system chooses the modulation scheme based on the prevailing channel conditions. Good channel conditions will be reflected by a large signal-to-noise-plus-interference ratio. The transmitter can then use 16-QAM to increase the data rate. However, if the channel

[43] BPSK is the answer. Keep reading to discover why.

conditions are poor, the use of BPSK is preferred due to its higher reliability. Such dynamic change in the modulation scheme is called *adaptive modulation*. Systems such as 1xEV-DO, HSDPA, HSUPA, HSPA+, WiMAX, and LTE utilize adaptive modulation. Adapting both the modulation scheme and the amount of coding to the channel conditions is quite common and is called adaptive modulation and (adaptive) coding, and adaptive coding then encompasses adaptive modulation.

2.7 RF PROCESSOR FUNCTIONS AT THE TRANSMITTER AND THE RECEIVER

We discussed several baseband processing blocks, such as a speech vocoder, channel coder, and digital modulator. Once the baseband processor has done its job, the RF processor must send the signal over the air at the transmitter and receive the signal from the air at the receiver. Figure 2.30 illustrates the main RF and non-RF components of the transmitter and the receiver for the base station.

Let's focus on the transmitter first. The RF components of the transmitter for a given carrier frequency or technology (e.g., Technology X in Figure 2.30) include a frequency upconverter, RF filter, High Power Amplifier (HPA), diplexer, duplexer, jumper cable, transmission line (or feeder), and antenna. The signal from the digital baseband processor passes through a digital baseband filter that confines the bandwidth of the baseband signal. As the name suggests, the digital-to-analog converter (DAC) transforms the digital signal to the analog signal. The analog signal is then transported from the baseband to the RF frequency from which the service operator has acquired licensed spectrum. The HPA amplifies the signal so that the signal can reach the receiver reliably. A good chunk of the total cost of the base station comes from the HPA. Remember that the letter H in HPA stands for *high*—it generates a lot of heat! Air conditioning is typically required to alleviate this heat. The HPA is operated in the linear region to avoid distortion. The RF filter helps meet the spectral mask requirements of the regulatory authorities in a given country (e.g., the FCC in the U.S.). The spectral mask indicates how much power or energy is allowed at different RF frequencies. Such a spectral mask ensures that all service operators play by the rules, resulting in an interference-controlled environment that is both safe and suitable for commercial-grade services to consumers.

The transmitter chain from the baseband processor to the RF filter is typically carried out separately for different technologies. When the same antenna is used to transmit signals from different technologies or different carrier frequencies of the same technology, a diplexer is used to facilitate the signal transmission. The duplexer enables transmission and reception of signals using the same physical antenna. The signal to be transmitted passes from the bottom of the base station tower to the top of the tower via the jumper cables and the transmission line. The transmission cable is quite thick, and wherever the cable needs to be bent, a jumper cable is used. Finally, the signal energy is radiated into the air by the transmit antenna.

The receiver chain includes components such as the antenna, (optional) Tower Mounted Amplifier (TMA, also known as Tower Top Low Noise Amplifier), jumper

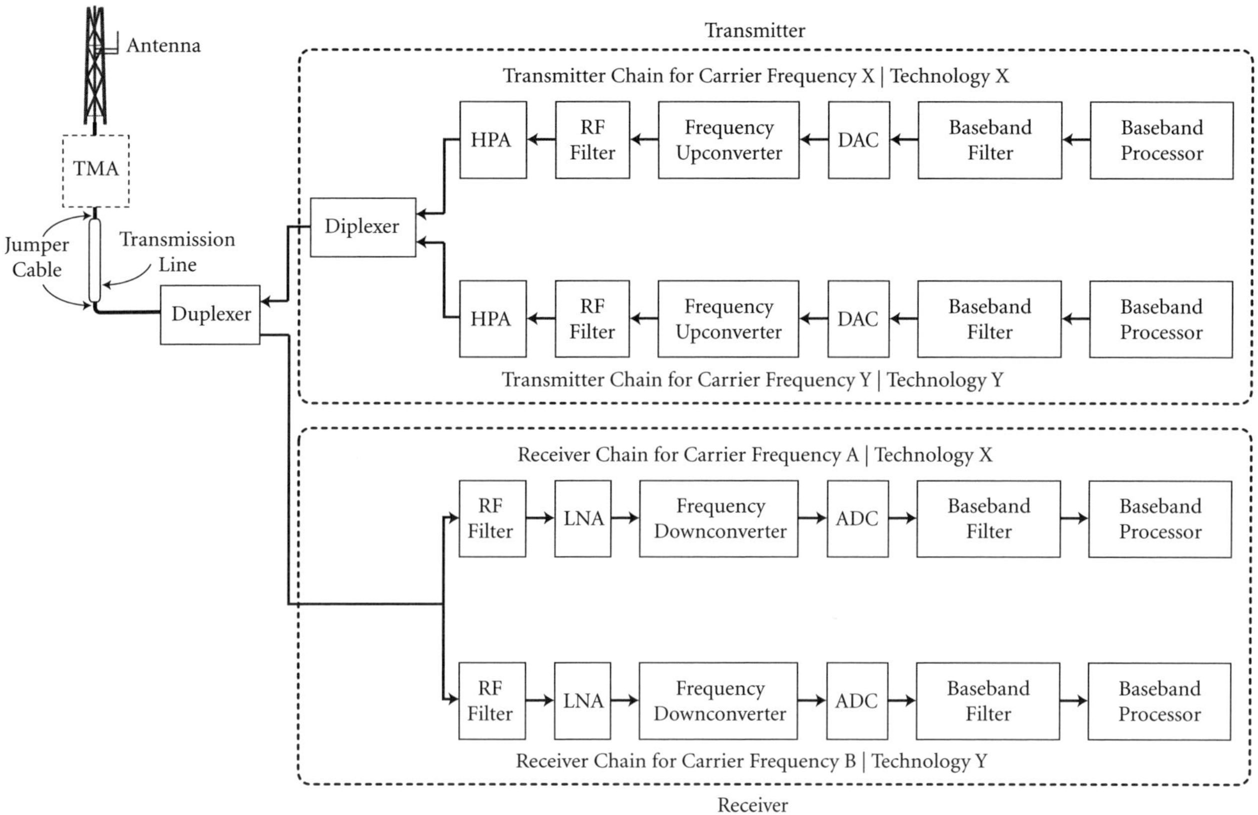

Figure 2.30: Base Station Transmitter and Receiver Functions from the RF Perspective

cable, transmission line (perhaps we should say reception line), duplexer, RF filter, Low Noise Amplifier (LNA), frequency downconverter, analog-to-digital converter (ADC), baseband filter, and, finally, baseband processor. The antenna welcomes the received signal. (It also welcomes interference; it does not discriminate!) A TMA may be used to amplify the signal (e.g., about 10 to 15 dB) at the top of the base station tower itself. Such amplification helps detect weak received signals. The TMA increases the loading on the tower and introduces reliability issues, but its benefit is overall reduction in the noise figure of the receiver. (We'll discuss the principle behind TMA soon.) The transmission line helps get the signal from the top of the tower to the bottom of the tower where the signal encounters the duplexer.

The RF filter extracts the signal with a certain bandwidth around the carrier frequency. The LNA is the power amplifier. The frequency downconverter brings the signal from the RF to the baseband. The ADC yields a digital signal that is processed by the baseband filter. Finally, the baseband processor works its magic to retrieve the information. Let's take a closer look at the TMA and antennas. The TMA effectively increases the range or cell radius. Antennas influence the radio network planning and design and the network optimization. In fact, commercial systems rely heavily on the adjustment of antenna parameters to optimize network performance. So now let's take a closer look at the TMA and the antennas.

The TMA helps increase the uplink or reverse link cell coverage by allowing the base station to detect weaker signals received from far-away mobile devices. The longest supportable cell radius is typically influenced by the link budget, and the uplink is usually the limiting link for the link budget. One of the parameters is the overall noise figure of the receiver. The higher the value of the noise figure, the more noise the receiver is adding to the received signal, which will make it harder for the receiver to detect a weak signal. Therefore, reducing the noise figure is important for enabling the receiver to detect weak signals. The TMA helps reduce the overall noise figure in the system by exploiting its low noise and high gain. According to Friis's formula for the calculation of the overall noise figure of the system containing n cascaded stage [68]

$$NF_{total} = NF_1 + \frac{NF_2 - 1}{G_1} + \frac{NF_3 - 1}{G_1 G_2} + \cdots + \frac{NF_N - 1}{G_1 G_2 \ldots G_{N-1}} \tag{2.16}$$

where NF_i is the noise figure (in normal and not decibel units) of stage i and G_i is the power gain of stage i. Eq. (2.16) indicates that if the first stage has a sufficiently high gain and low noise, its noise figure would dominate the overall noise figure of the system. The total noise floor for a BS system with a TMA can be expressed as [69]

$$NF_{BS} = NF_{short} + \frac{NF_{TMA} - 1}{G_{short}} + \frac{NF_{cable} - 1}{G_{short} G_{TMA}} + \frac{NF_{receiver} - 1}{G_{short} G_{TMA} G_{cable}}. \tag{2.17}$$

In Eq. (2.17), the overall BS receive system is considered to have four stages: a short cable between the antenna and the TMA, the TMA, the long cable from the top of the tower to the bottom of the tower, and the receiver. G_{short} is nearly 1 and NF_{short} is quite low, too, as the cable is quite short. NF_{TMA} is the noise figure of the TMA and G_{TMA} is the gain of the TMA. NF_{cable} is the noise figure of the long cable, and G_{cable}

is the gain of the cable, which is, in fact, a loss! The key observation in Eq. (2.35) is that the noise added by the cable is quite high and that placing a high-gain and low-noise TMA before the cable reduces the noise figure of the overall BS receive system as G_{TMA} will reduce the impact of the noise added by the long cable.

Antenna parameters play a critical role in the network optimization. Examples of antenna parameters that may be changed to optimize the radio network include antenna height, azimuth, beamwidth, and down-tilting. We'll discuss these parameters soon, but first let's look at the most commonly deployed antenna configuration, illustrated in Figure 2.31. The reader is encouraged to take a quick look at the antenna basics and the propagation basics explained in Chapter 3. (Of course, if the reader wants to take a *slow* look, we are fine with that!) In Figure 2.31, the base station is controlling three 120° sectorized regions called sectors or cells, i.e., α, β, and γ. Within each sector or cell, two antennas with about 10λ to 20λ separation between them are used where λ is the wavelength. One antenna is a dual transmit or receive antenna and is used to transmit signals to and receive signals from the mobiles.

The second antenna is just a receive antenna and is used to receive signals from the mobiles. The base station transmits signals to all the mobiles using just one antenna but receives signals from the mobiles using two antennas. (Remember that *antennae* are only for arthropods!) The use of multiple receive antennas represents implementation of space or spatial diversity in which independent fading on the propagation paths between the transmit antenna of the mobile and the receive antennas at the base station is exploited.

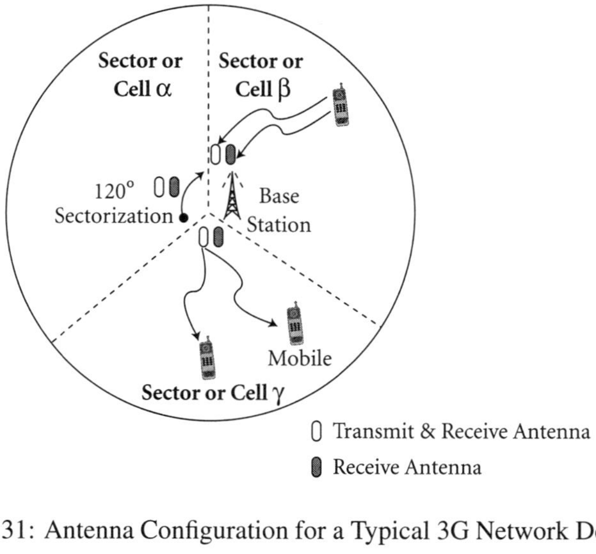

Figure 2.31: Antenna Configuration for a Typical 3G Network Deployment

Do You Recall... A disclaimer is warranted. The next question assumes that you have looked at the propagation basics covered in Section 3.4 of Chapter 3. What kind of fading would account for differences seen between the signals received on two different receive antennas: short-term fading that we model using Rayleigh or Ricean fading or long-term fading that we model using a log-normal distribution? Please see the footnote[44] to confirm your answer.

The difference in the signals received at two antennas may be as large as 20 to 30 dB depending upon the severity of fades on the antennas. Note that the fading on the antennas is independent assuming they are separated sufficiently and that the probability that two antennas experience the fade at the same time is small. Since space diversity is so useful and easily implemented at the receiver, it is widely used in commercial deployments of cellular networks. Some commercial deployments exploit polarization diversity (instead of space diversity), which houses two physically separate antennas in the same casing called a *radome* and requires no specific spacing between the antennas. These antennas transmit (and receive) signals at different polarizations, hence the name polarization diversity. Polarization refers to the direction of the electrical field of the RF signal, which is an electromagnetic wave. A typical antenna in a cellular system uses a vertical-appearing physical antenna and vertical polarization for transmission. In contrast, a satellite uses a horizontal-appearing antenna, resulting in horizontal polarization. The signal strength is maximized when the polarization at the transmit antenna matches that at the receive antenna.

As promised, let's visualize the impact of various antenna parameters on coverage. As mentioned earlier, we alter the antenna parameters to influence the coverage area where the antenna radiates. Figure 2.32 shows how a change in the antenna height changes the coverage. The taller the antenna is, the larger the coverage area. For example, an antenna height of 50 m would result in a longer cell radius than the antenna height of 30 m. See Section 3.4.1 of Chapter 3 to quantify the change in the cell radius as a function of the (base station) antenna height.

Figure 2.33 indicates the effect of changing the antenna azimuth or antenna orientation. Azimuth $= 0^\circ$ refers to the case in which the antenna radiation pattern is pointing to the North. As another example, azimuth $= 60^\circ$ means that the maximum antenna gain is achieved at a 60° angle relative to North. The simplified antenna radiation pattern shown here is for the horizontal plane.

Let's see what role the antenna beamwidth plays using Figure 2.34. For now, do not worry about the exact definition of the antenna beamwidth; we'll define it *very soon*. When the antenna beamwidth is large, the antenna is spreading energy over a wide range of angles. Since the energy is spread in a wider region, the energy cannot

[44]Short-term fading is the correct answer. If you did indeed skip ahead and read Section 3.4 of Chapter 3, then you will recall that short-term fading occurs even for little changes in time or distance, and long-term fading are slow variations in the signal as a result of environment or topology. Since two propagation paths between the transmit antenna and two receive antennas are likely to encounter the same topology (e.g., buildings along the way), long-term fading would be quite similar.

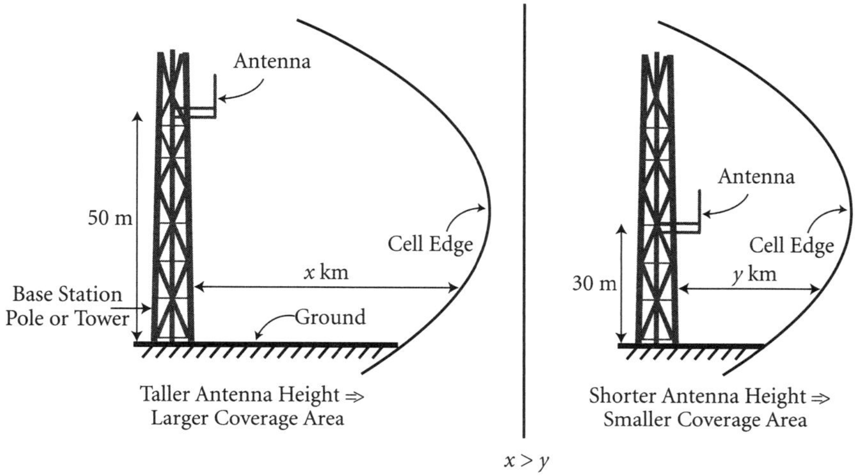

Figure 2.32: Influence of the Antenna Height on Coverage

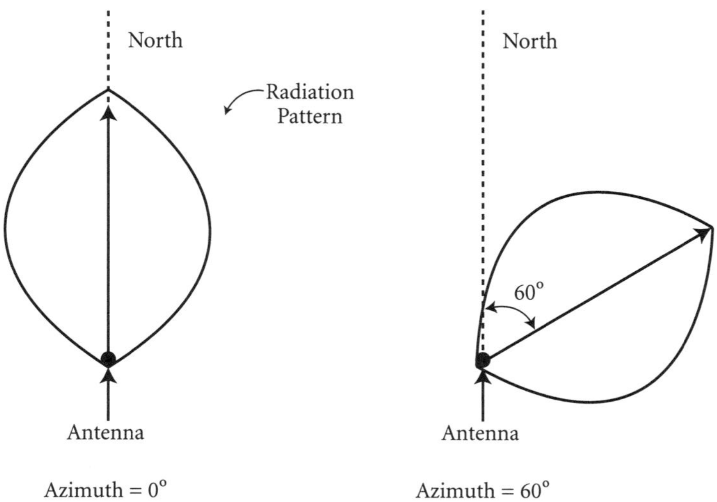

Figure 2.33: Influence of the Antenna Azimuth on Coverage

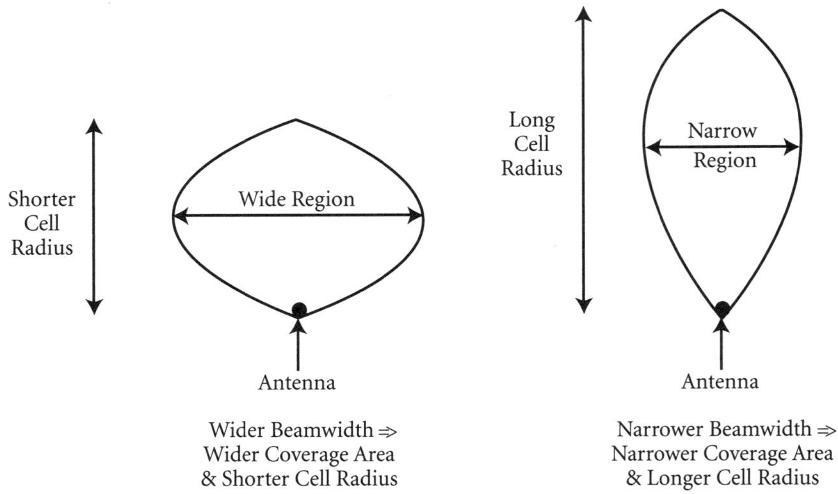

Figure 2.34: Influence of the Antenna Beamwidth on Coverage

extend far from the source, resulting in a shorter cell radius. In contrast, the narrow beamwidth means that the energy is focused over a narrow range of angles, enabling the signal to propagate farther from the base station.

The antenna gain characteristics are usually defined using a diagram such as Figure 2.35. We are illustrating a simplified antenna radiation pattern for the horizontal plane, and a similar diagram exists for the vertical plane. The antenna pattern shows the normalized antenna gain for the entire range of angles, 0° to 360°. The center of the antenna radiation pattern has 0 dB gain, and the antenna transmits (and collects) maximum energy in the center of the pattern. As we move away from the center angle, the antenna gain decreases. The antenna gain pattern quantifies such relative gain (actually, loss) as a function of the angle. Let's take a couple of examples so that we are comfortable with our interpretation of the antenna pattern. Assume that the transmitter–receiver separation is kept the same in some stationary and open-space environment so that we can ignore the impact of fading. Further assume that we are at point A corresponding to 0°, and the measured baseline reference signal level is X dBm. At point B, we are at the angle 32.5°, and the antenna pattern indicates the relative antenna gain of −3 dB because the antenna pattern crosses the −3 dB level at 32.5°. We would measure the signal strength of $X - 3$ dBm at point B. At point C, the antenna gain pattern touches the −20 dB level, and, finally, at point D, the antenna pattern points to the −30 dB antenna gain. Hence, we would measure $X - 20$ dBm and $X - 30$ dBm at points C and D, respectively.

Now that we know how to read the antenna gain diagram, such as Figure 2.35, let's define some useful antenna parameters. The antenna beamwidth is the total range (in degrees) over which the antenna gain is −3 dB or above. In Figure 2.35, the antenna gain crosses the −3 dB relative gain level at 327.5° and 32.5°, leading to the range of

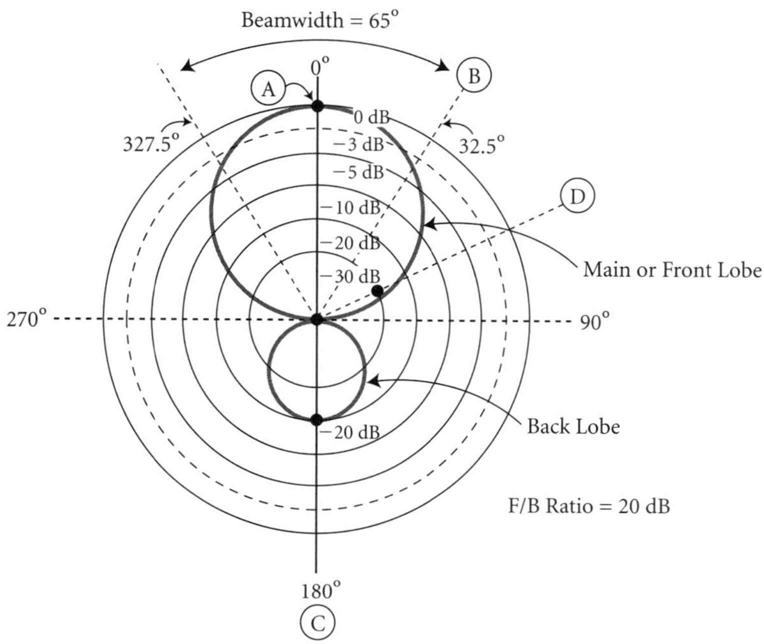

Figure 2.35: Example of an Antenna Radiation Pattern

angles of 327.5° to 32.5° where the relative antenna gain is −3 dB or above. The total width of this range is $\{(360° - 327.5° = 32.5°) + (32.5° - 0° = 32.5°) = 65°\}$. By the way, the antenna that has the beamwidth of 65° is very popular for covering a 120° sectorized region. The front-to-back ratio is the ratio of the maximum gain of the front lobe (i.e., antenna radiation area) to the maximum gain of the back lobe. In our diagram, the front lobe has the maximum (relative) gain of 0 dB and the back lobe has the maximum gain of −20 dB, leading to the front-to-back ratio of $\{0 \text{ dB} - (-20 \text{ dB}) = 20 \text{ dB}\}$. The higher the value of the ratio, the more desirable the antenna since it is spilling less energy in the undesired direction. We want to maximize energy in the front lobe and minimize energy in the back lobe. Figure 2.35 shows a simplified antenna pattern. A real antenna has side lobes in addition to front and back lobes. A good antenna spreads little energy in the back lobe and side lobes. The side lobes are physically located between the front lobe and the back lobe.

Let's now turn our attention to the last antenna parameter, the antenna down-tilt, which refers to the process of changing the antenna pattern in the vertical plane such that the effective coverage shrinks. Figure 2.36 explains this concept. When the down-tilting is absent, the antenna radiates energy over a large area, and the signal can propagate much farther from the antenna. One way to control the overspraying of energy is to down-tilt the antenna so the overspraying of energy decreases as illustrated by the physical tilting of the antenna in Figure 2.36. When you hold a

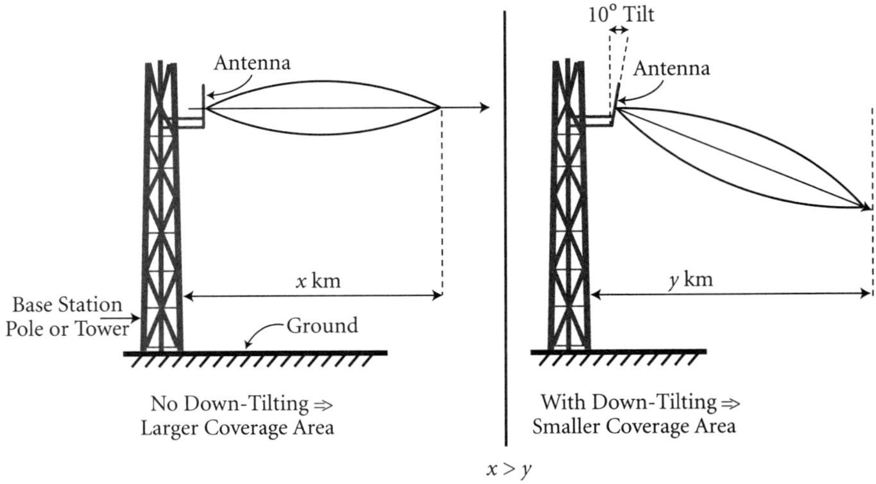

Figure 2.36: Influence of the Antenna Down-Tilting on Coverage

flashlight horizontally so it is the parallel to the ground, the light travels far away from the flashlight. However, when you tilt the flashlight toward the ground, the light is restricted to the nearby area. In the same fashion, down-tilting the antenna reduces the effective footprint of the antenna radiation.[45] When the antenna pattern is influenced by changing the physical orientation of the antenna, it is called mechanical down-tilting. A more prevalent approach is electrical down-tilting in which the phases of the electrical signals fed into the antenna are modified to shrink the coverage. Remote electrical tilt is an efficient and cost-effective way of implementing antenna down-tilting. One final note before we leave the topic of antennas: Of all the antenna parameters we have briefly discussed, antenna down-tilting is the most widely used mechanism for network optimization.

We have so far discussed a traditional BS configuration shown in Figure 2.30. Another BS configuration that is gaining popularity in certain cases utilizes the *Remote Radio Head* (RRH). The basic idea behind the RRH is that a digital signal (instead of an analog RF signal in the case of the traditional BS configuration) is sent from the bottom of the BS tower to the top of the tower where the RRH resides. The RRH carries out all the RF processing, such as frequency upconversion and power amplification. The main benefit of the RRH is (near) elimination of the cable loss altogether. Such a reduction in cable loss improves the overall noise figure and increases the link budget. Recall that the larger the link budget is, the larger the cell size would be. When the cellular network is deployed at higher frequency bands, such as 2.5 GHz, every decibel saved in the link budget is welcome since the propagation

[45]If you have ever tried to make a cell phone call from a private airplane, you will note that coverage is poor and very sporadic even at low altitudes since down-tilting reduces coverage in the air.

path loss is quite high at higher carrier frequencies! As an example, RRHs are quite prevalent in WiMAX deployments because many such deployments use higher carrier frequencies. The main drawbacks of RRHs are lower reliability and longer down-time of the cell site during RRH failures. In general, RRHs have more reliability issues, and, if something goes wrong, troubleshooting and fixing the problem take a relatively longer amount of time.

2.8 POINTS TO REMEMBER

We looked at major processing blocks for the transmitter and the receiver of a digital cellular system.

- Speech coding is typically based on linear predictive speech coding in which a block of speech over a 20 ms time period is represented by a set of filter parameters. Rather than sending quantized speech samples, these filter parameters are sent to the receiver, and, the receiver reconstructs the speech block using these parameters.

- Convolutional and Turbo coding are two main channel coding techniques, which introduce redundancy to enable information retrieval even in the presence of challenging radio environment. Convolutional coding is attractive when processing delay is of concern and some errors can be tolerated. Turbo coding is suitable when a lower error rate is the target, and, processing delay and complexity are not a concern. Convoluional coding is typically used for voice and Turbo coding is typically used for high-speed data services. The receiver can use Viterbi decoding to process Convolutional code symbols and a MAP algorithm to process Turbo code symbols.

- Digital modulation symbols represents code symbols, and the most popular modulation schemes used in 2G, 3G, and 4G systems are BPSK, GMSK, QPSK, 8-PSK, 16-QAM, and 64-QAM. One modulation symbol for these schemes represents one bit, one bit, two bits, three bits, four bits, and six bits, respectively. Since the signal constellation diagram is known to the receiver, it compares the received modulation symbol with the possible symbols and picks the symbol that yields the smallest error.

- Coherent demodulation involves the use of a reference signal, often obtained from the pilot channel, to recover the traffic channel bits. The impact of the radio environment can be determined by the coherent demodulator by comparing the received reference signal with the expected reference signal. Such channel estimation knowledge can then be used to modify the received traffic channel modulation symbols prior to their comparison with the signal constellation points.

- Spreading and scrambling are carried out for a CDMA system. Spreading increases the signal bandwidth, and scrambling facilitates identification of the source that is sending out the CDMA signal.

- In addition to the baseband processing, the transmitter involves a DAC, filter, frequency upconverter, HPA, diplexer, duplexer, jumper, transmission line, and antenna. The DAC converts a digital baseband signal into an analog baseband signal. The filter helps keep the signal energy within the intended bandwidth. The HPA amplifies the signal. The diplexer enables the sharing of the same antenna between two transmitters corresponding to two technologies or two carrier frequencies of the same technology. The duplexer allows the same antenna to be shared between the transmitter and the receiver. The thinner and flexible jumper cable can be used to connect the thicker transmission line with other components. The antenna transmits the RF signal over the air.

- Antenna parameters, such as down-tilt, beamwidth, azimuth, and height, can be modified to optimize the network performance.

- The receiver includes the receive antenna, TMA, transmission cable, jumper, duplexer, LNA, filter, frequency downconverter, and ADC as well as the baseband processor. The receive antenna collects energy from the air. The TMA amplifies the received signal while reducing the overall noise figure of the system compared to the case in which the TMA is absent; the TMA increases the range. The transmission line brings the received signal to the bottom of the base station tower. The LNA amplifies the received signal. The frequency downconverter transports the RF signal to the baseband. The ADC converts the analog signal to a digital signal for further processing by the baseband processor.

CHAPTER 3

RADIO FREQUENCY PROPAGATION

The cellular radio signal encounters a dynamic propagation environment as it travels from the transmitter to the receiver. This chapter discusses various aspects of the radio propagation at the cellular radio frequencies. The signals in cellular systems are one type of *radio frequency* (RF) waves. RF waves are introduced first. Free-space propagation is described as a starting point. Cellular propagation is more complex than free-space propagation. Major mechanisms involved in cellular propagation, such as reflection, diffraction, scattering, and absorption, are described. While predicting the signal strength at a given distance away from the transmitter, three major phenomena are considered: distance-based path loss, large-scale fading, and small-scale fading. These phenomena are extensively discussed, along with analytical or mathematical formulations that model these factors. Various statistics that quantify the nature of the propagation environment are specified. Examples are also provided of the areas in which the knowledge of the RF propagation is applied.

3.1 RADIO FREQUENCY WAVES

An electromagnetic wave consists of an electric component and a magnetic component as shown in Figure 3.1. An example of an electromagnetic wave is the radio

Cellular Communications, First Edition. Nishith D. Tripathi and Jeffrey H. Reed

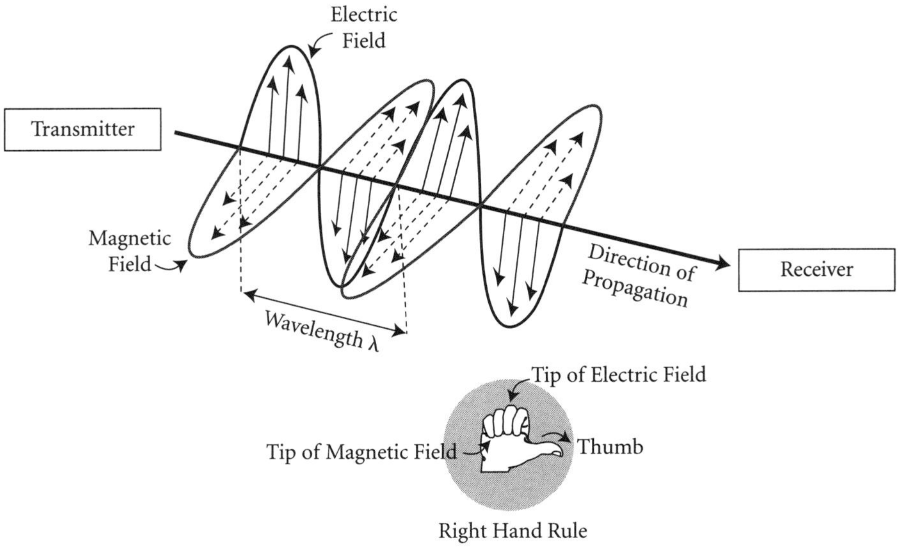

Figure 3.1: Anatomy of an Electromagnetic Wave

signal used in cellular communications between the base station (BS) and the mobile station (MS).

The electric and magnetic fields are at a right angle to each other. The distance between successive peaks or valleys in the waveform is equal to the wavelength of the electromagnetic wave. Given the directions of the electrical-field lines and magnetic-field lines, how can one figure out the direction in which the wave is propagating? The answer is the hitch-hiking gesture! Technically speaking, it's called the Right Hand Rule. Make a fist using the right hand with the thumb out. The fingers should be curling from the tip of electric-field lines to the tip of magnetic-field lines. The direction of the thumb represents the direction of the propagation of the electromagnetic wave. The electromagnetic frequency spectrum ranges from 0 to ∞ Hz. The range from 3 kHz to 300 GHz is referred to as the RF range. The focus of this chapter is on the radio frequency propagation, with particular focus on RF propagation in cellular communications. The signal in a cellular communication system travels at the speed of light. The wavelength of the signal is related to the carrier frequency of the signal according to the following formula.

$$c = f_c \lambda \tag{3.1}$$

where c is the speed of light (3×10^8 m/s), f_c is the carrier frequency in Hertz, and λ is the wavelength in meters. At 900 MHz, the wavelength is 0.33 m. As the carrier frequency increases, wavelength decreases. At 1800 MHz, the wavelength is halved (i.e., $0.33 \text{ m}/2 = 0.165$ m) compared to the wavelength at 900 MHz.

Radio waves travel or propagate in different ways at different frequencies [70]. The very-low-frequency (3 to 30 kHz) and low-frequency (30 to 300 kHz) waves follow the curvature of the earth. Such waves are called *ground waves*. Since the earth is not a perfect electrical conductor, the ground waves attenuate as they follow the earth's surface. The very-low-frequency and low-frequency radio waves are used with ships and submarines for military communications. The high-frequency (3 to 30 MHz) waves rely upon skywave propagation in which the radio waves are refracted in the ionosphere. Amateur radios and commercial marine and aircraft communications utilize skywave propagation. The very-high-frequency (30 to 300 MHz), ultra-high-frequency (300 to 3000 MHz), and super-high-frequency (3 to 30 GHz) waves undergo line-of-sight (LOS) propagation, an example being the propagation between the satellite and the ground receiver.[46] The radio waves in cellular communication systems fall largely in the ultra-high-frequency range in which LOS propagation plays an important role. This chapter discusses various mechanisms involved in radio propagation in cellular networks.

3.2 FREE-SPACE PROPAGATION

Consider an isotropic antenna radiating power in free space. Such an antenna spreads power uniformly in all directions, which can be visualized as a point source spreading energy over the surface of a sphere as shown in Figure 3.2 [71]. At a distance d from the isotropic antenna, the power flux, which is the power per unit area measured in Watts per square meter, is given by

$$w = \frac{p_T G_T}{4\pi d^2} \tag{3.2}$$

where p_T is the antenna input power and G_T is the antenna gain. The product $(p_T G_T)$ is called the *effective radiated power* (ERP) of the transmitter. The antenna gain G_T

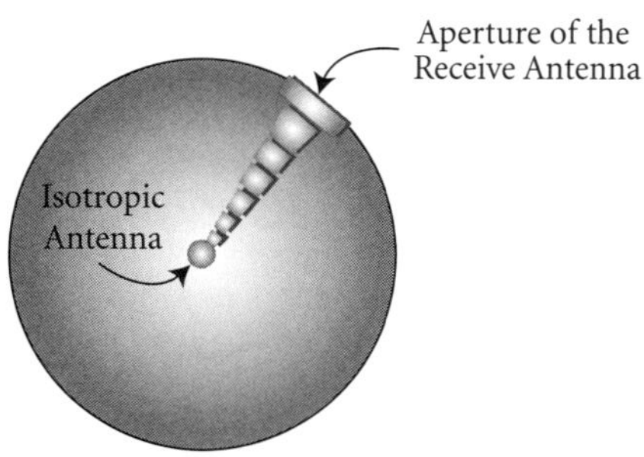

Figure 3.2: Radiation of Energy by an Isotropic Antenna

[46]Microwaves also experience LOS propagation and range from 300 MHz to 300 GHz.

is the ratio of the power flux radiated in a given direction by an antenna to the power flux radiated in the same direction by an isotropic antenna. Hence, $G_T = 1$ for an isotropic antenna. When a directional antenna directs or focuses energy in a specific direction, the power flux in that direction would be larger than the power flux of an isotropic antenna because the isotropic antenna radiates equally in all directions. Such antenna directivity results in antenna gain G_T. For example, when a BS antenna directs energy in a $120°$ region, G_T could be 18 dB or dBi.[47] When the gain of an antenna is 18 dB in a given direction, the received power at a given distance from the transmitter would be 18 dB higher than the received power at the same location if an isotropic antenna were to be used instead of a directional antenna. Figure 3.3 shows the impact of the gain of a directional antenna.

At a given location, the received power is -80 dBm in the case of a directional transmit antenna used at the BS. If an isotropic antenna were used at the BS, the received power at the same location would have been -98 dBm. The difference in power, 18 dB, represents the gain of the antenna used at the BS.

Let us offer a word of caution before you start shopping for an isotropic antenna: No such antenna exists! It is a hypothetical antenna, and in reality, all antennas have some directivity. In practice, the closest one can get to the isotropic antenna is a half-wave dipole antenna. A half-wave dipole antenna has a gain of 2.15 dB relative to an isotropic antenna. Sometimes the antenna gains are specified in the decibel gain relative to a half-wave dipole (decibel dipole, or dBd). An antenna gain of X dBi

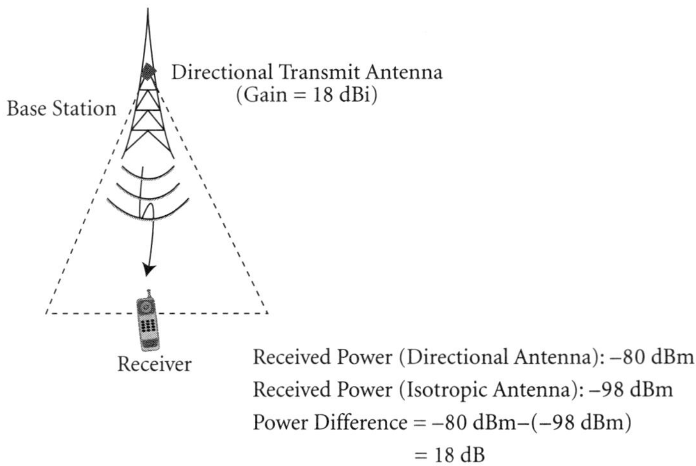

Figure 3.3: Gain of a Directional Antenna

[47] Decibels relative to isotropic (dBi) represents the power gain of the antenna in decibels relative to an isotropic antenna. In practice, the letter i in dBi may sometimes be dropped, and the antenna gain would then be represented in decibels.

corresponds to $(X - 2.15)$ dBd. In other words, antenna gains expressed in decibel dipole are 2.15 dB smaller than the antenna gains expressed in decibels relative to the isotropic antenna (decibels isotropic, or dBi).

The receive antenna at the receiver will capture some of the energy radiated by the transmit antenna depending upon its effective area, often called *aperture*. Let G_R be the receive antenna gain and A be the aperture of the receive antenna. Then, the power at the receive antenna is given by

$$p_R = wA = \left(\frac{p_T G_T}{4\pi d^2}\right)\left(\frac{G_R \lambda^2}{4\pi}\right) \tag{3.3}$$

where the receive antenna gain G_R is given by

$$G_R = \frac{4\pi A}{\lambda^2} \tag{3.4}$$

and λ is the wavelength of the radio signal, which follows the rule $c = f_c \lambda$, in which c is the speed of light and f_c is the carrier frequency of the radio signal.

Path loss between the transmitter and the receiver can be calculated as

$$L_p = \frac{p_R}{p_T} = \frac{G_T G_R \lambda^2}{16\pi^2 d^2}.$$

For an isotropic antenna, $G_T = G_R = 1$. The path loss L_p can then be expressed in decibels as

$$\begin{aligned} L_p\,(\text{dB}) &= 10\log_{10}\left(\frac{\lambda^2}{16\pi^2}\right) - 10\log_{10}(d^2) \\ &= 20\log_{10}\left(\frac{\lambda}{4\pi}\right) - 20\log_{10}(d). \end{aligned} \tag{3.5}$$

Since $c = f_c \lambda$,

$$\lambda = \frac{c}{f_c} = \frac{3 \times 10^8\ \text{m/s}}{f_c \times 10^6\ \text{Hz}} = \frac{300}{f_c}$$

where the numerical value of f_c is expressed in megahertz (e.g., 800 in case of f_c being 800 MHz). The distance between the transmitter and the receiver is often expressed in kilometers instead of meters. Then, Eq. (3.5) becomes

$$\begin{aligned} L_p &= 20\log_{10}\left(\frac{300}{4\pi f_c}\right) - 20\log_{10}(1000 \times d) \\ &= 20\log_{10}(23.87) - 20\log_{10}(f_c) - 20\log_{10}(1000) - 20\log_{10}(d). \end{aligned} \tag{3.6}$$

Simplifying Eq. (3.6) as

$$L_p = -32.44 - 20\log_{10}(f_c) - 20\log_{10}(d) \tag{3.7}$$

provides the free-space path loss in decibels, with f_c given in megahertz and d expressed in kilometers.

3.3 CELLULAR PROPAGATION MECHANISMS

Free-space path loss is typically used in satellite communications with some provision for atmospheric effects. In cellular communications, the actual path loss experienced by a cellular radio signal is usually much higher than the free-space path loss. Ray theory is widely used for analyzing radio wave propagation. Several rays can be viewed as a single entity. As this single entity or a set of rays propagates from the transmitter to the receiver, four basic mechanisms influence the overall path loss experienced by the radio signal: reflection, diffraction, scattering, and absorption or penetration.

3.3.1 Reflection

When an electromagnetic wave encounters an object that has large dimensions compared to the wave's wavelength, reflection occurs. The receiver in such a case could receive a direct LOS signal and a reflected path from the transmitter. The buildings in an urban environment, the mountains in an open environment, and the earth's surface are examples of objects that cause reflections of the RF wave. When the RF wave traveling in one type of medium (e.g., the air) impinges upon an object that represents another type of medium, part of the energy may be reflected back into the first medium, part of the energy may be absorbed by the second medium, and part of the energy may continue to flow in a wave propagating into the second medium. In the case of the second medium being a perfect dielectric, no absorption occurs. If the second medium is a perfect conductor, only reflection occurs. Figure 3.4 shows reflection due to the ground surface.

In a cellular radio environment, a propagation model that considers a direct path and a ground-reflected path, as shown in Figure 3.4, gives a more accurate prediction than the model that considers only the direct path.

$$P_r = \frac{P_t G_t G_r h_t^2 h_r^2}{d^4} \tag{3.8}$$

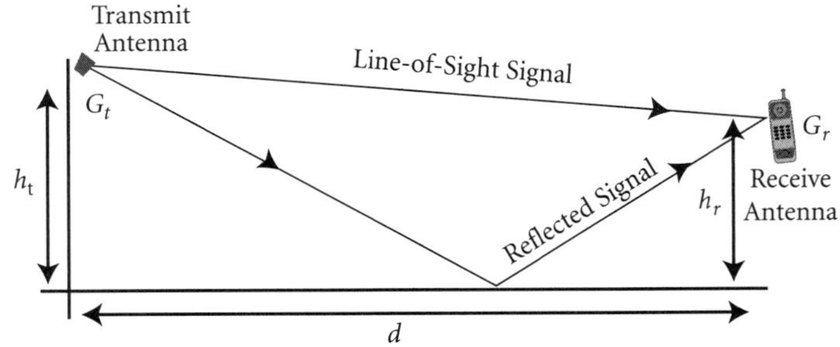

Figure 3.4: Two-Ray Ground Reflection Model

In Eq. (3.8), P_r is the received power, P_t is the transmit power, G_t is the transmit antenna gain, G_r is the receive antenna gain, h_t is the height of the transmit antenna relative to the ground, h_r is the receive antenna height relative to the ground, and d is the distance between the transmitter and the receiver. The two-ray ground reflection model specified above is sometimes used as a basis for path loss prediction in LOS conditions for microcells [72].

3.3.2 Diffraction

When an electromagnetic wave encounters an object with sharp irregularities, such as edges, it bends around the object. This effect is called *diffraction*, and it enables signal propagation in the absence of LOS and behind obstacles. The signal strength starts to decrease quickly in the shadowed region behind obstacles. According to Huygen's principle, the points on a wavefront encountering an obstacle act as sources of secondary waves that become a new wavefront, facilitating propagation in the shadowed region [40].

The concept of a Fresnel[48] zone is important in quantifying the diffraction loss. Consider a sinusoidal signal that is sent by the transmitter and that encounters an obstruction at a distance d_1 from the transmitter and d_2 from the receiver, as illustrated in Figure 3.5.

Imagine a direct path from the transmitter to the receiver encountering the obstruction. Based on Huygen's principle, a secondary wavefront emerges at the obstruction, leading to a refracted wave. Since the refracted wave has to travel a longer distance

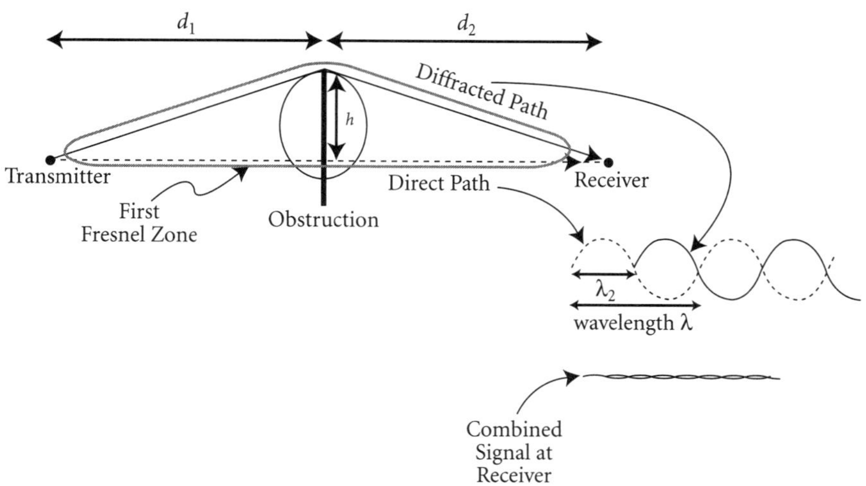

Figure 3.5: Concept of a Fresnel Zone

[48]Fresnel is pronounced "fray-NELL," although one would be tempted to pronounce it as "frez-nul!"

than the direct wave, it arrives a little late at the receiver. When the diffracted wave reaches the receiver, it may be out-of-phase relative to the direct wave. What happens when two signals are out-of-phase? In such a case, the diffracted wave cancels out the direct wave, resulting in minimal signal strength at the receiver. This is an example of destructive interference resulting from the secondary wavefront. When the diffracted wave reaches the receiver, it may be in-phase relative to the direct wave. That would be an example of constructive interference resulting from the secondary wavefront. Figure 3.5 shows the case of the first Fresnel zone in which the excess path length for the secondary wavefront is $1 * \lambda/2 = \lambda/2$. The second Fresnel zone would correspond to the case in which the excess path length for the secondary wavefront is $2 * \lambda/2 = \lambda$. Fresnel zones are concentric circles on an imaginary plane (representing an obstruction), with each circle corresponding to a secondary wavefront reaching the receiver with a phase shift related to $(n * \lambda/2)$ excess path length, where n is a non-zero integer specifying the number of the Fresnel zone.

A knife-edge diffraction model can be used to get an idea of the magnitude of the diffraction loss when a single obstruction exists in the transmitter-to-receiver path. Consider Figure 3.6 for various parameters useful for prediction of diffraction loss. For Figure 3.6, the Fresnel–Kirchoff diffraction parameter ν is defined by [40]

$$\nu = h\sqrt{\frac{2(d_1 + d_2)}{\lambda d_1 d_2}} \tag{3.9}$$

where h is the height of the obstruction relative to the line joining the transmitter and the receiver, d_1 is the distance between the transmitter and the obstruction, d_2 is the distance between the obstruction and the receiver, and λ is the wavelength.

The phase shift (in radians) experienced by a diffracted wave is

$$\phi = \frac{\pi \nu^2}{2}. \tag{3.10}$$

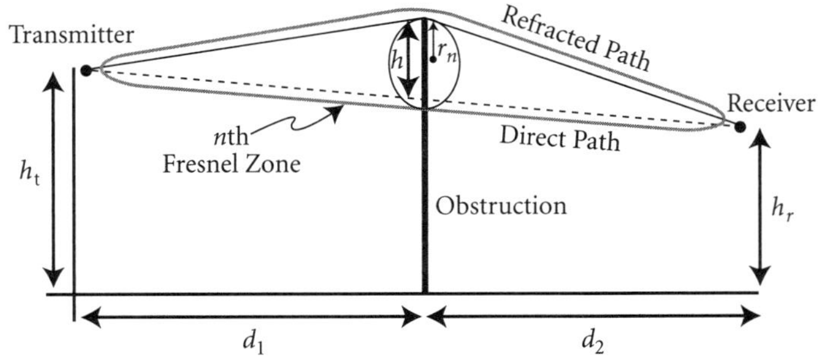

Figure 3.6: Geometry for Knife-Edge Diffraction Model

The radius of the nth Fresnel zone is given by

$$r_n = \sqrt{\frac{n\lambda d_1 d_2}{d_1 + d_2}}. \tag{3.11}$$

This estimation of the radius is applicable to long distances such that $d_1 \gg r_n$ and $d_2 \gg r_n$ [40].

When the wavefront encounters an object, part of the energy is diffracted as the obstacle blocks some Fresnel zones. Relative to the free-space, the diffraction gain (or negative loss) is given by

$$G_d\,(\text{dB}) = 20\log_{10}\left(|F(\nu)|\right) \tag{3.12}$$

where $F(\nu)$, the Fresnel integral, is

$$F(\nu) = \left(\frac{1+j}{2}\right)\int_{\nu}^{\infty} e^{((-j\pi t^2)/(2))}dt. \tag{3.13}$$

Lee [73] solves the above integral for different values of ν. Eq. (3.13) then becomes as follows.

$$G_d = 0 \text{ for } \nu \le -1 \tag{3.14}$$

$$G_d = 20\log_{10}(0.5 - 0.62\nu) \text{ for } -1 \le \nu \le 0 \tag{3.15}$$

$$G_d = 20\log_{10}(0.5e^{-0.95\nu}) \text{ for } 0 \le \nu \le 1 \tag{3.16}$$

$$G_d = 20\log_{10}\left(0.4 - \sqrt{0.1184 - (0.38 - 0.1\nu)^2}\right) \text{ for } 1 \le \nu \le 2.4 \tag{3.17}$$

$$G_d = 20\log_{10}\left(\frac{0.225}{\nu}\right) \text{ for } \nu > 2.4 \tag{3.18}$$

Eqs. (3.14)–(3.18) can be used to estimate the diffraction loss (in decibels) due to the presence of an obstacle in the transmitter-to-receiver path. The diffraction loss is minimal if the obstacle does not block a large portion of the first Fresnel zone. In microwave link designs, an attempt is made to design for a 55% clearance of the first Fresnel zone [40]. In other words, 55% of the radius (e.g., r_1 in Figure 3.6 with $n = 1$ for the first Fresnel zone) must be clear of the obstruction. Such a clearance would help avoid excessive diffraction loss. When multiple obstructions occur in the transmitter-to-receiver propagation path, each object can be represented by a knife-edge. A series of such knife-edges can be replaced by a single equivalent knife-edge, and a single knife-edge diffraction model can be used to estimate the diffraction loss [74].

3.3.3 Scattering

When an electromagnetic wave comes across objects that are smaller than its wavelength, these objects scatter energy in many directions. This effect is called *scattering*, and it gives rise to numerous scattering waves, which are capable of providing additional energy relative to the combined effect of diffraction and reflection. In other words, the received signal power measured at the receiver could be larger than the power predicted by the diffraction and reflection phenomena because scattering provides additional energy. Examples of objects that cause scattering include tree leaves and street signs. Even while the object itself may be much larger than the wavelength (e.g., a building), the surface roughness of such an object can result in scattering rather than just reflection. The question now is how rough is "rough enough" to cause scattering? A surface is considered smooth if the ratio of minimum protuberance to maximum protuberance is less than the critical height h_c [40]. The critical height h_c is expressed by the Rayleigh criterion as

$$h_c = \frac{\lambda}{8\sin(\theta_i)} \tag{3.19}$$

where λ is the wavelength and θ_i is the angle of incidence. For a rough surface, reflection is influenced by scattering.

The path loss (equaling the receive power less the transmit power) reflecting the effect of scattering due to a distant object is given by

$$\begin{aligned} L_p = G_T &+ 20\log_{10}(\lambda) + RCS - 30\log_{10}(4\pi) \\ &- 20\log_{10}(d_T) - 20\log_{10}(d_R) \end{aligned} \tag{3.20}$$

where G_T is the transmit antenna gain in decibels isotropic, λ is the wavelength, RCS is the radar cross section of the scattering object in decibel-meters squared, d_T is the distance between the transmitter and the scattering object, and d_R is the distance between the receiver and the scattering object. RCS is defined as the ratio of the power density of the signal scattered in the direction of the receiver to the power density of the signal incident on the scattering object. Based on measurements made in some European cities, RCS ranges from 14.1 dBm^2 to 55.7 dBm^2 [40, 75].

A Word of Caution... Scattering as part of the large-scale propagation path loss is the influence of an object distant from the receiver. Scattering and reflections in the vicinity of the receiver lead to sudden and large short-term variations when the movement of the receiver is very small relative to its surrounding environment (e.g., even on the order of a wavelength). These sudden changes in received signal strength could significantly weaken the signal and will be considered later as part of small-scale fading.

3.3.4 Absorption and Penetration

While atmospheric effects such as rain cause absorption at higher frequencies (e.g., more than 10 GHz) and are an important consideration for satellite communications, the ultra-high-frequency radio signals are unaffected by atmospheric effects. Hence, rain effects are ignored in cellular system design. However, seasonal changes often lead to significant changes in vegetation, and such changes are occasionally considered for network planning and deployment. For example, wet forests could cause a loss of about 5 dB [76].

When the area to be served contains buildings, a special consideration of building penetration loss is required because the signal could experience a significant attenuation. The building penetration loss is a function of several factors, such as the construction material, type of windows, frequency, and floor.[49] A concrete wall could cause an 8 to 15 dB attenuation, and a metal wall could cause attenuation as large as 20 to 25 dB. A tinted window results in a 3 to 30 dB loss compared to a non-tinted window. As the frequency increases, the penetration loss decreases. As an example, about 1 dB of additional loss can occur when the frequency is reduced from 1800 MHz to 900 MHz. Though the exact floor-dependent attenuation is a function of BS antenna height and vertical antenna pattern, penetration loss decreases at the rate of about 2 dB per floor (starting from the ground floor) up to a certain floor level (e.g., ninth or fifteenth) and begins to increase above such floor level.

In some propagation models, attenuation, such as the building penetration loss, is often considered implicitly by *clutter classes*, which represent the environment surrounding the transceiver. Urban environment, dense-urban environment, forest, open-space rural environment, and water are examples of clutter classes.[50] One clutter class (e.g., urban environment) is selected as the base clutter class corresponding to 0 dB attenuation. Additional attenuation or gain relative to such a base clutter class is specified for other clutter classes. For example, relative to the base clutter class of urban environment, the clutter classes of dense-urban and forest may have attenuation of 2 to 5 dB and the clutter class of open-space rural environment and water may have 10 dB gain.

3.4 PREDICTION OF RECEIVED SIGNAL STRENGTH

The transmitted signal reaches the receiver via the various propagation mechanisms discussed in Section 3.3. One approach to predict the received signal strength is the use of the analytical models presented in Section 3.3. At a given location, one or more propagation mechanisms may be dominant. Another prediction approach is the use of an empirical model based on measured data in a given environment. Such an empirical model automatically incorporates the effects of all propagation mechanisms. *This section describes the empirical propagation path loss models widely used in the industry for simulation-based analysis, network planning, network deployment, and network optimization.*

Assume that the BS is transmitting a signal to the MS at a constant power and that the MS is moving away from the BS. A graph of the received signal strength measured at the MS and averaged over a short time or distance (e.g., a few wavelengths) could

[49] See Rappaport's work [40] to find additional details on the numerical values mentioned here.
[50] See the Atoll reference guide [77] for additional details.

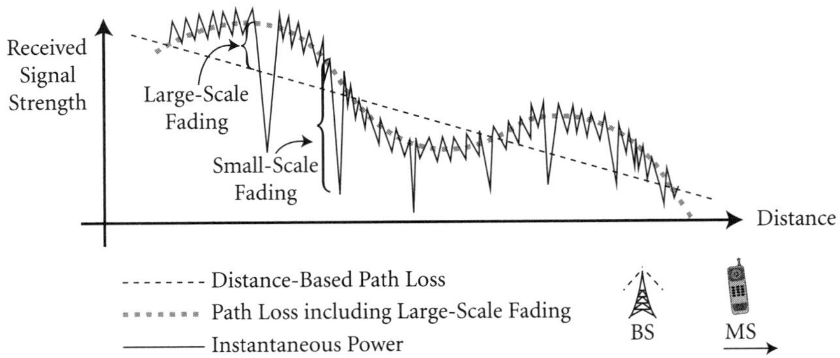

Figure 3.7: Constituents of the Received Signal Strength

look like Figure 3.7. The received signal strength at a given *time instant* can be viewed as a sum of three components: distance-based path loss, large-scale fading loss, and small-scale fading loss. The distance-based path loss considers the influence of factors such as distance between the transmitter and the receiver, frequency of the RF signal, antenna heights, and type of environment (e.g., urban or rural) on the path loss. The distance-based path loss monotonically increases as the distance between the transmitter and the receiver increases. The large-scale fading loss is an adjustment to the distance-based path loss; it accounts for differences in the average signal strength at different locations with the same transmitter-to-receiver separation. The large-scale fading, often termed *shadow fading*, is obtained when instantaneous signal strength is averaged over longer time periods or distances (e.g., 5λ to 30λ) [40]. The large-scale fading can be viewed as variations in average signal strength around the distance-based path loss. The small-scale fading loss reflects sudden changes in the signal strength as the receiver moves a small distance (e.g., on the order of the wavelength or even a fraction of the wavelength). The variations in instantaneous signal strength around the combined distance-based path loss and the large-scale fading loss represent small-scale fading.

We will discuss quite a bit about mathematical modeling of the cellular radio environment in the remainder of this chapter. What can you do with such propagation knowledge? Figure 3.8 lists example areas where the propagation knowledge becomes very valuable.

Assume that you are tasked with deploying a cellular network in a given geographical area. You would need to decide the cell size so that you calculate how many cells or BSs you would need to buy from an infrastructure vendor. The cell size is easily determined based on the cell radius. The cell radius is estimated by doing the *link budget analysis*. The link budget essentially gives the propagation path loss in decibels and depends on the technology chosen to serve a given area. Assume that the link budget is 135 dB for some radio access technology. Now, 135 dB path loss is plugged in a suitable propagation path loss model such as the COST 231–Hata model

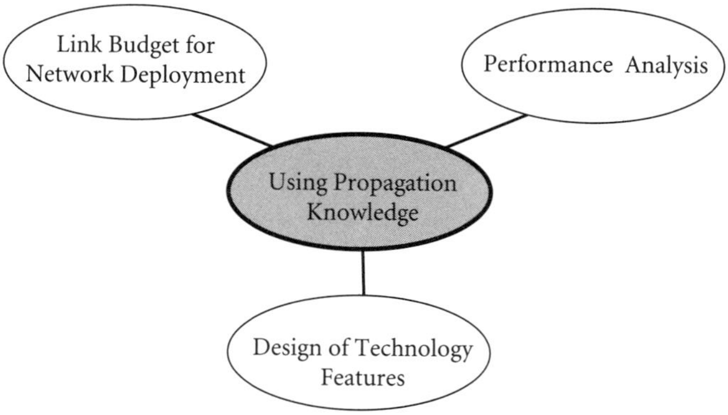

Figure 3.8: Exploiting the Power of the Propagation Knowledge

to determine the propagation distance. This distance is nothing but the cell radius! In summary, the propagation path loss models help estimate the cell radius when the network design is carried out for deployment.

Let's assume that we have chosen a specific radio access technology for cellular network deployment. How many subscribers can we support in a given sector? How much throughput can be supported at a given location? To answer such questions, simulation-based analysis and analytical modeling using a commercial network planning tool, such as Atoll and Planet EV, are carried out before any networks are laid out. Such analysis utilizes the propagation models including the path loss models and shadow fading and short-term fading models to predict how many users can be supported in a sector. Such information is valuable in determining the resources (e.g., amount of spectrum) required to offer cellular services to subscribers.[51]

When a new radio access technology is being standardized, relevant propagation channel models are used to evaluate performance of various mechanisms so that a decision on which mechanism should be included in or excluded from the standard can be made. Furthermore, some mechanisms are useful in channel conditions of one type, while other mechanisms are attractive in other types of channel conditions. For example, when the fades are very short, which is the case for terminals moving at high speeds, interleaving and channel coding minimize bit errors. In the case of slowly changing channels, power control and/or adaptive modulation and coding prove quite beneficial.

[51]Providing indoor coverage inside some large buildings can be quite challenging and may necessitate deployment of special systems, such as a distributed antenna system (DAS). See Tolstrup's work [78] for further details on indoor radio network planning.

3.4.1 Distance-Based Path Loss

Several empirical models useful in predicting distance-based path loss are discussed below. These empirical propagation models and their variations are widely used in the cellular industry for commercial deployments.

3.4.1.1 Hata's Model Hata's propagation model is the basis for several widely used propagation models in the cellular industry.[52] The main attraction of Hata's model is its simplicity with reasonable accuracy, and its main drawback is its constraints on the ranges of some parameters. Hata derived empirical formulas for propagation path loss based on Okumura's report containing graphs such as median field strength versus distance [80, 81]. Hata's basic model includes path loss for an urban environment and provides correction factors for other environments, such as suburban and open areas. Caution should be exercised while using Hata's model because it is valid only for specific cases. Hata's model makes the following assumptions: path loss is between isotropic antennas and the terrain is quasi-smooth and regular.

The median propagation path loss in an urban area is given by

$$\begin{aligned} L_{p(urban)} &= 69.55 + 26.16 \log_{10} f_c - 13.82 \log_{10} h_b \\ &\quad - a(h_m) + (44.9 - 6.55 \log_{10} h_b) \log_{10} R \end{aligned} \tag{3.21}$$

where L_p is the propagation path loss in decibels, f_c is the carrier frequency in megahertz in the range of 150–1500 MHz, h_b is the BS effective antenna height in meters in the range of 30–200 m, R is the distance between the transmitter and the receiver in kilometers in the range of 1–20 km, and $a(h_m)$ is the correction factor for h_m with h_m being the MS antenna height in meters in the range of 1–10 m. In Eq. (3.21), $a(h_m)$ is 0 dB for $h_m = 1.5$ m. If the path loss at 900 MHz at a distance of 5 km is to be estimated, the numerical values of f_c and R are 900 and 5, respectively, in Eq. (3.21). For a *medium-small city*, $a(h_m)$ is given by

$$a(h_m) = (1.1 \log_{10} f_c - 0.7) h_m - (1.56 \log_{10} f_c - 0.8) \tag{3.22}$$

for f_c in 150–1500 MHz range. For a *large city*, $a(h_m)$ is given by

$$\begin{aligned} a(h_m) &= 8.29(\log_{10}(1.54 h_m))^2 - 1.10 \text{ for } f_c \leq 200 \text{ MHz} \\ a(h_m) &= 3.2(\log_{10}(11.75 h_m))^2 - 4.97 \text{ for } f_c \geq 400 \text{ MHz}. \end{aligned} \tag{3.23}$$

Eqs. (3.21)–(3.23) can be used to estimate the path loss in an urban environment containing large, medium, or small cities for the parameter ranges specified above.

The path loss in a suburban environment is given by

$$L_{p(suburban)} = L_{p(urban)} - K_r \text{ with } K_r = 2\left[\log_{10}(f_c/28)\right]^2 + 5.4. \tag{3.24}$$

In Eq. (3.24), K_r is the correction factor for the suburban environment relative to the urban environment and f_c is the carrier frequency expressed in megahertz.

[52] Stanford University Interim (SUI) model and ECC-33 model are some of the newer empirical models. The WiMAX Forum uses the SUI model [79].

The path loss for open areas is given by

$$L_{p(open)} = L_{p(urban)} - Q_r \quad \text{with } Q_r = 4.78\left[\log_{10}(f_c)\right]^2 - 18.33\log_{10}(f_c) + 40.94. \tag{3.25}$$

In Eq. (3.25), Q_r is the correction factor for the open areas relative to the urban environment and f_c is the carrier frequency expressed in megahertz.

3.4.1.2 COST 231–Hata Model Hata's basic model is valid in the frequency range of 150–1500 MHz. European COST 231 extended the validity of Hata's model to higher frequencies by analyzing Okumura's propagation graphs in the upper frequency band. Such a model is frequently referred to as the COST–Hata Model.

For the urban environment, Eq. (3.21) for the propagation path loss is modified as

$$\begin{aligned} L_p &= 46.3 + 33.9\log_{10}(f_c) - 13.82\log_{10}(h_b) - a(h_m) \\ &\quad + (44.9 - 6.55\log(h_b))\log(R) + C_m \end{aligned} \tag{3.26}$$

where L_p is the propagation path loss in decibels, f_c is the carrier frequency in megahertz in the range of 1500–2000 MHz, h_b is the BS effective antenna height in meters in the range of 30–200 m, R is the distance between the transmitter and the receiver in kilometers in the range of 1–20 km, $a(h_m)$ is the correction factor for h_m, and h_m is the MS antenna height in meters in the range of 1–10 m. $a(h_m)$ is given by Eq. (3.22), and C_m is specified as

$$C_m = \begin{cases} 0 \text{ dB (medium-sized cities \& suburbs; medium tree density)} \\ 3 \text{ dB (metropolitan centers)} \end{cases}. \tag{3.27}$$

The COST–Hata model is valid for small and large macrocells in which the BS antenna heights are above rooftop levels in the vicinity of the BS. This model is widely used in the industry for cell coverage area prediction and for cellular system performance analysis. This model is unsuitable for microcells.

A modified version of COST 231–Hata model is used by the Third Generation Partnership Project 2 (3GPP2) for performance analysis of CDMA2000 (Code-Division Multiple Access 2000) and its successor technologies. With $h_b = 32$ m, $h_m = 1.5$ m, and $f_c = 1.9$ GHz, the path loss is specified as[53]

$$L_p = 28.6 + 35\log_{10}(R). \tag{3.28}$$

Eq. (3.28) gives a 3 dB smaller path loss compared to the basic COST 231–Hata model.

3.4.1.3 ITU-R M.1225 Vehicular Path Loss Model (Macrocell Environment) The model specified here is suitable for an environment that consists of

[53]Please note that R in Eq. (3.28) is expressed in meters (not kilometers). This model is used for distances greater than 35 m.

large cells, high transmit power, and BS antenna heights above the average rooftop level. It is applicable to urban and suburban environments. This model is based on the COST 231–Walfisch–Ikegami Model and on Xia's work [82].

$$L_p = L_{fs} + L_{rts} + L_{msd} \tag{3.29}$$

where L_{fs} is the free-space path loss, L_{rts} is the diffraction loss from rooftop to the street, and L_{msd} is the loss due to multiple screen diffraction due to rows of buildings. L_{fs} and L_{rts} are independent of the BS antenna height, and L_{msd} depends on the BS antenna height relative to the building heights.

The free-space path loss is given by

$$L_{fs} = -10\log_{10}\left[\left(\frac{\lambda}{4\pi R}\right)^2\right] \tag{3.30}$$

where λ is the wavelength of the RF signal, 2 is the free-space path loss exponent, and R is the distance between the transmitter and the receiver.

The diffraction loss is expressed as

$$L_{rts} = -10\log_{10}\left[\left(\frac{\lambda}{2\pi^2 r}\right)\left(\frac{1}{\theta} - \frac{1}{2\pi + \theta}\right)^2\right] \tag{3.31}$$

where λ is the wavelength of the RF signal and r and θ are quantified to be

$$\theta = \tan^{-1}\left(\frac{|\Delta h_m|}{x}\right) \tag{3.32}$$

and

$$r = \sqrt{(\Delta h_m) + x^2}. \tag{3.33}$$

Δh_m is the difference between the mean building height and the MS antenna height ($\Delta h_m = h_{roof} - h_m$) and x is the horizontal distance between the MS and the diffracting edges.

The multiple screen diffraction loss in the case of a BS antenna height above the rooftop level is

$$L_{msd} = -10\log_{10}\left[\left(2.35^2\left(\Delta h_b\sqrt{d/\lambda}\right)^{1.8}\right) \Big/ \left(R^{2(1-4\times10^{-3}\Delta h_b)}\right)\right] \tag{3.34}$$

where Δh_b is the BS antenna height measured from the average rooftop ranging from 0 to 50 m ($\Delta h_b = h_b - h_{roof}$) and d is the average separation between rows of buildings.

In typical urban and suburban environments, $\Delta h_m = 10.5$ m, $x = 1.5$ m, and $d = 80$ m. Then, Eq. (3.28) is simplified as

$$\begin{aligned} L_p = {} & \left[40(1 - 4\times10^{-3}\Delta h_b)\right]\log_{10}(R) \\ & - 18\log_{10}(\Delta h_b) + 21\log_{10}(f_c) + 80. \end{aligned} \tag{3.35}$$

Since the above-specified model uses the COST 231–Walfisch–Ikegami model, it is valid for f_c ranging from 800 MHz to 2000 MHz, h_b ranging from 4 to 50 m, h_m ranging from 1 to 3 m, and R ranging from 20 m to 5 km.

3.4.1.4 ITU-R M.1225 Outdoor-to-Indoor and Pedestrian Path Loss Model (Microcell Environment) The model specified here is suitable for an environment that consists of small cells, low transmit power, outdoor locations of the BS, and low BS antenna heights. It is applicable to urban and suburban environments and is inspired by the COST 231–Walfisch–Ikegami Model and Xia's and Mac's work [83, 84].

As in Eq. (3.28), the overall path loss is given by

$$L_p = L_{fs} + L_{rts} + L_{msd}. \tag{3.36}$$

The free-space path loss is given by Eq. (3.30) and the diffraction loss from rooftop to street is given by Eq. (3.31).

In the case of the BS height being close to the mean rooftop level, the multiple screen diffraction loss is given by

$$L_{msd} = -10\log_{10}\left[\left(\frac{d}{R}\right)^2\right]. \tag{3.37}$$

For typical urban and suburban environments, $\Delta h_m = 10.5$ m, $x = 15$ m, and $d = 80$ m. Then, the overall path loss becomes

$$L_p = 40\log_{10}(R) + 30\log_{10}(f_c) + 49. \tag{3.38}$$

3.4.1.5 ITU-R M.1225 Indoor Path Loss Model (Picocell Environment) This model is suitable for an indoor environment that is characterized by small cells, low transmit power, and indoor locations of the BS and the mobile devices. This model is inspired by the COST 231 indoor propagation model.

The overall path loss is given by

$$L_p = L_{fs} + L_c + \sum k_{wi}L_{wi} + n^{((n+2)/(n+1))-b} \times L_f. \tag{3.39}$$

Here, L_{fs} is the free-space path loss, L_c is constant loss, k_{wi} is the number of penetrated walls of type i, L_{wi} is the loss for wall type i, L_f is the loss between adjacent floors, n is the number of penetrated floors in the path, and b is an empirical parameter. For an average indoor office environment, n is 3 or 4, L_f is 18.3 dB, L_{w1} is 3.4 dB for thin internal walls ($<$ 10 cm) made of plasterboard and with windows, L_{w2} is 6.9 dB for thick internal walls ($>$ 10 cm) made of concrete or bricks, and $b = 0.46$. For these assumed average values, the overall path loss for an indoor office environment can be simplified as

$$L_p = 37 + 30\log_{10}(R) + 18.3n\left(\frac{n+2}{n+1} - 0.46\right). \tag{3.40}$$

Table 3.1: Scope of the Propagation Models

Propagation Model for Path Loss Prediction	Parameter Constraints			
	Carrier Frequency (MHz)	Base Station Antenna Height (m)	Cell Radius	Mobile Station Antenna Height (m)
Hata	150–1500	30–200	1 km–20 km	1-10
COST231-Hata	1500–2000	30–200	1 km–20 km	1-10
ITU-R M.1225 (Macrocell Environment)	800–2000	4–50	20 m–5 km	1–3
ITU-R M.1225 (Microcell Environment)	Any	Any	Any (Microcell)	Any
ITU-R M.1225 (Picocell Environment)	Any	Any (Indoor)	Any (Indoor)	Any (Indoor)

3.4.1.6 *Summary of Applicability of Empirical Propagation Models*

Different propagation models have different constraints. Table 3.1 summarizes the ranges of major parameters where the models are valid.

3.4.2 Large-Scale Fading

Large-scale fading is another important factor that influences the average signal strength at a given location. The concept of large-scale fading and mathematical structures that model large-scale fading are summarized below.

3.4.2.1 *Large-Scale Fading: Why Does It Occur?* Let's say we have two MSs at different locations in a cell, but each one is at the same distance from the BS. Do you expect similar signal strengths or different signal strengths at these MSs when the omnidirectional BS transmits the same transmit power to these two MSs? Experiments indicate that two different receiver locations with identical transmitter-to-receiver separation usually have different average received signal strengths. An example is shown in Figure 3.9 for receivers MS1, MS2, and MS3. The differences in the signal strengths for identical transmitter-to-receiver separation can be attributed to the differences in the environments experienced by the propagation wave as it travels from the transmitter to the receiver.

The signal strength difference is a random variable around the distance-based mean signal strength. In other words,

$$L_p\,(\mathrm{dB}) = \overline{L_p}\,(\mathrm{dB}) + X \tag{3.41}$$

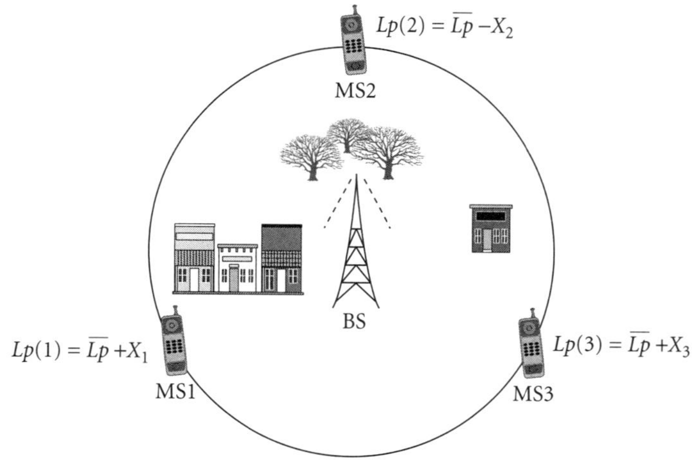

Figure 3.9: An Example of Large-Scale Fading

where L_p is the path loss at a location, $\overline{L_p}$ is the average distance-based path loss (e.g., such as the one predicted by the Hata model), and X is the random variable representing the variations in the received signal strength around the distance-based mean signal strength. X follows Gaussian distribution with a mean of 0 dB and a standard deviation of σ dB [85–87]. For example, if we observe the differences in signal strengths at various locations with the same transmitter-to-receiver separation and analyze the statistics of X, it would appear to conform to the Gaussian distribution. Since the random variable X follows the normal (i.e., Gaussian) distribution, the associated signal strength variations are referred to as log-normal fading. Furthermore, since the log-normal fading is a result of the surrounding environment, such as shadows of buildings, it is also known as shadow fading or log-normal shadowing.

3.4.2.2 Correlated Shadow Fading

When we want to predict the average signal strength at a given location, we can use a propagation path loss model, such as the COST 231–Hata model, to predict the distance-based path loss and add to it a random value of the shadow fading (which could be positive or negative). In practice, two types of correlations exist in the shadow fading. Consider an MS traveling away from BS1 and toward BS2 in Figure 3.10.

Let $S_1(t_1)$ and $S_1(t_2)$ be the strength of the signal received from BS1 at time instant t_1 and t_2. In practice, the shadow fading components of $S_1(t_1)$ and $S_1(t_2)$ are correlated [88]. Now, let $S_2(t_1)$ and $S_2(t_2)$ be the strength of the signal received from BS2 at time instant t_1 and t_2. Even the signals arriving at the MS from two different BSs may be correlated to a certain extent.[54] For example, the shadow fading

[54]The correlation between the signals reaching the MS from two different BSs depends on how similar or different these propagation paths are. In simulation-based analysis of cellular systems, 50% correlation is often assumed for such propagation paths.

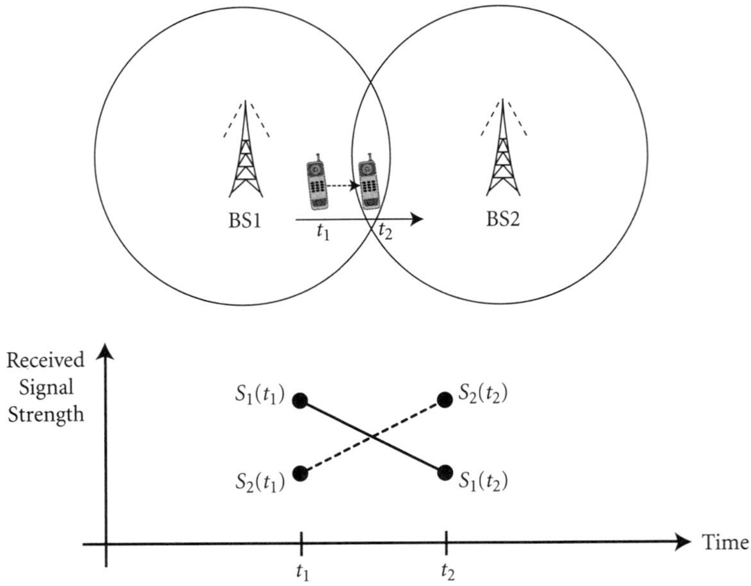

Figure 3.10: Correlated Shadow Fading

constituents of $S_1(t_1)$ and $S_2(t_1)$ are also correlated. An exponential autocorrelation function has been found to be adequate for large and medium cells while trying to model the log-normal fading between the MS and the BS as a function of time or distance [88]. The correlation at a distance of 100 m was found to be 0.82 with $\sigma = 7.5$ dB for a suburban environment in macrocells. The correlation of 0.3 was measured for the distance of 10 m for microcells. The basic exponential autocorrelation model is more accurate for the large macrocells and less accurate for microcells. A modified autocorrelation function was derived and reported to be accurate for short distances as well [89]. Graziosi and Santucci [90] integrate Gudmundson's exponential correlation [88] with Graziano's inter-BS cross-correlation [85] for a more realistic modeling of the shadow fading. In summary, correlated shadow fading would more accurately model the shadow fading components of $S_1(t_1)$ and $S_1(t_2)$ and $S_1(t_1)$ and $S_2(t_1)$.

3.4.2.3 Modeling of Correlated Shadow Fading

The 3GPP2 has suggested the following approach to model the correlated log-normal shadow fading between different BSs and a given MS at a given location and time [91]. Let X_i represent the shadow fading between the BS i and the MS. X_i follows Gaussian distribution with 0 mean and σ standard deviation. X_i is generated as

$$X_i = aZ + bZ_i \tag{3.42}$$

where Z is common across all BSs , Z_i is specific to the BS i, and the parameters a and b control the shadow fading cross-correlation between the BS i and any other BS.

Both Z and Z_i are Gaussian random variables with 0 mean and σ standard deviation. The parameters a and b are constrained by

$$a^2 + b^2 = 1.0. \tag{3.43}$$

For 50% correlation between any two given BSs, $a^2 = b^2 = 0.5$. The recommended value of σ is 8.9 dB.

The use of Gudmundson's exponential autocorrelation model [88] is also recommended for adjacent shadow fading values [92, 93]. Specifically, the normalized autocorrelation function is

$$R(\Delta x) = e^{((-|\Delta x|)/(d_{cor}))\log_e(2)} \tag{3.44}$$

where Δx is the distance difference between adjacent locations and d_{cor} is the environment-dependent decorrelation length. d_{cor} of 20 m and 5 m are recommended for the vehicular environment (as well as outdoor-to-indoor) and pedestrian environments, respectively.

3.4.3 Small-Scale Fading

Small-scale fading is characterized by sudden fluctuations in the instantaneous received signal strength over short distances or short time intervals. When the receiver moves by a short distance, perhaps just one wavelength, the received signal could change by as much as 20 to 30 dB! For such short distance travel, both the distance-based path loss and the large-scale fading are essentially the same. So, what changes the signal strength so quickly and by such a large amount? Several factors influence the variations in signal strength. Two main factors are multipath signals and mobility in the radio environment.

At a given point in space, the receiver experiences an onslaught of numerous waves that are reflected and scattered from the objects that lie between the transmitter and the receiver. These RF signals add with one another at the receive antenna, sometimes constructively but usually destructively.[55] Since these waves travel different distances and arrive from different angles, they have different relative phase shifts. For example, if one radio wave has the phase of 0 radians, another might have a phase of 0.1 radians. At a given location, when the signals add destructively, the average signal strength drops significantly, causing a deep fade. Have you ever had an experience where you had a poor-quality signal on your cell phone at one location and, just a short distance away, suddenly your call quality got much better? *That* is an example of short-term fading! Different locations in a small local area experience different types of combining of multipath signals, and signal strength measurements could be quite different at these locations. As the MS travels in an environment rich with multipath signals, it experiences significant changes in the instantaneous power. When the signal strength measurements are averaged over longer times or distances, the average signal becomes reasonably stable due to the distance-based path loss and large-scale fading.

[55] Multipaths completely cancelling out one another is a rare event.

Mobility in the radio environment refers to the relative motion between the transmitter and the receiver and the movement of the objects in the radio channel for a fixed transmitter-to-receive separation. Such movements introduce a change in the frequency of each multipath signal of the received signal. This frequency shift is called the *Doppler shift*. When the MS velocity is much higher than its surrounding environment, the speed of the environment can be ignored. However, if the MS is stationary, the effect of the speed of the surrounding environment (e.g., people or vehicles passing by) dominates. Multipath signals with different Doppler shifts add up, constructively or destructively, at the receive antenna, resulting in rapid fluctuations in the received signal strength.

The impact of multipath signals and movement on the received signal is analyzed in detail next.

3.4.3.1 *Analysis of the Influence of Multipath Signals* Imagine an MS in an urban environment with an enviable skyline. (Pick your favorite city!) Figure 3.11 shows an example scenario without worrying about doing injustice to such a skyline.

The BS transmits a signal to the MS. The MS receives three copies of the same signal. P_1 is the direct LOS path, and P_2 and P_3 are reflected paths. P_1 arrives first at the MS because it is traveling on the shortest path. However, P_2 and P_3 travel longer distances and arrive somewhat later.

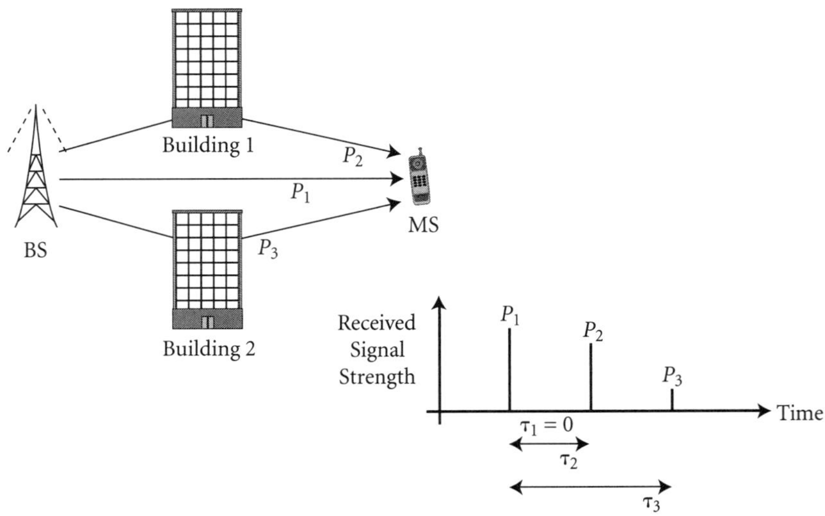

Figure 3.11: Multipath-Rich Radio Environment

Do You Recall...... Do you remember how fast these waves propagate? See the footnote[56] for the answer.

Statistics of *excess delay* are useful in quantifying time dispersion introduced by the multipath propagation channel. Excess delay refers to the propagation delay difference of a multipath signal relative to the first arriving path. In Figure 3.11, the excess delays of P_1, P_2, and P_3, are τ_1, τ_2, and τ_3, respectively. The graph of power levels versus excess delay for the detectable multipath signals is called *multipath power delay profile*. Since all excess delays use the arrival time of the first multipath as the timing reference, τ_1 is zero. Using the multipath power delay profile, useful statistics of the excess delay can be calculated as shown here [40].

The *mean excess delay* is

$$\bar{\tau} = \frac{\sum_{k=1}^{M} P(\tau_k)\tau_k}{\sum_{k=1}^{M} P(\tau_k)}. \tag{3.45}$$

In Eq. (3.45), τ_k is the excess delay for multipath k, $P(\tau_k)$ is the relative power level of multipath k, and M is the number of (usable) multipath signals. The strongest multipath signal's power level can be normalized to 1.0, and power levels of other multipath signals can be expressed relative to the strongest multipath signal. For example, if the second strongest multipath signal is 3 dB below the strongest, $P(\tau_2) = 0.5$ as $10 * \log 10(0.5) = -3$ dB. It is assumed that $P(\tau_1) = 1.0$.

The *root mean square (rms) delay spread* σ_τ is

$$\sigma_\tau = \sqrt{\overline{\tau^2} - (\bar{\tau})^2} \tag{3.46}$$

with $\bar{\tau}$ as specified in Eq. (3.45) and $\overline{\tau^2}$ specified as

$$\overline{\tau^2} = \frac{\sum_{k=1}^{M} P(\tau_k)\tau_k^2}{\sum_{k=1}^{M} P(\tau_k)}. \tag{3.47}$$

Another useful quantity is *excess delay spread* or maximum excess delay of the power delay profile. It is the excess delay beyond which the multipath power drops X dB below the maximum multipath power.

It is important to note that a single power delay profile is a result of numerous measurements (specifically, impulse response measurements) being averaged over

[56]The RF signal travels at the speed of light!

time or space in a local area. The power delay profile from one local area can help quantify the statistics specified in Eqs. (3.45)–(3.47). In practice, measurements are made in many local areas so that the statistics calculated above represent a large area [94].

Multipath signals lead to time dispersion of the transmitted signal as they travel different distances between the transmitter and the receiver. Essentially, the time period during which the modulation symbol is received is longer than the duration of the transmitted modulation symbol. As a result of the multipath time delay spread, frequency-selective fading may occur depending upon the baseband bandwidth of the transmitted signal in relation to the coherence bandwidth of the radio channel.

The coherence bandwidth specifies the channel bandwidth (i.e., the range of frequencies) over which the channel will let the transmitted signals pass with equal gain change and linear phase.

When a frequency correlation of 0.9 or greater is used to define channel similarity, the coherence bandwidth is approximated as

$$B_c = \frac{1}{50\sigma_\tau} \tag{3.48}$$

where σ_τ is the rms delay spread.

When a frequency correlation of 0.5 or greater is used to define channel similarity, the coherence bandwidth is approximated as

$$B_c = \frac{1}{5\sigma_\tau}. \tag{3.49}$$

(See Lee's work [95] for additional details on the coherence bandwidth.) The coherence bandwidth plays an important role in deciding which type of small-scale fading would be experienced by the transmitted signal. When the bandwidth of the transmitted signal is smaller than the coherence bandwidth of the radio channel, *flat fading* occurs. In other words, all frequencies in the transmitted signal get affected similarly by the radio channel. In the frequency domain, the mobile radio channel is represented by a constant gain and linear phase response within the coherence bandwidth as seen in Figure 3.12.[57] In the time domain, multipaths will cause variations in the amplitude, making the radio channel gain a time-dependent variable. At time instant t_1, the channel has gain G_1, while at time instant t_2, the channel has gain G_2. This results in different amplitudes of the received signal at time instants t_1 and t_2. Hence, a flat-fading channel is often referred to as an amplitude-varying channel. Furthermore, since the bandwidth of the transmitted signal is narrow compared to the coherence bandwidth, it is known as a *narrowband channel*.

Note that the receiver in the flat-fading channel cannot differentiate multipath signals because the symbol duration is relatively longer compared to the multipath delay spread. Subsequently, the receiver obtains a combined signal at the antenna. A flat-fading channel often causes deep fades. In summary, a flat-fading radio channel

[57] A flat-fading channel can be viewed as a Finite Impulse Response (FIR) filter with a single tap.

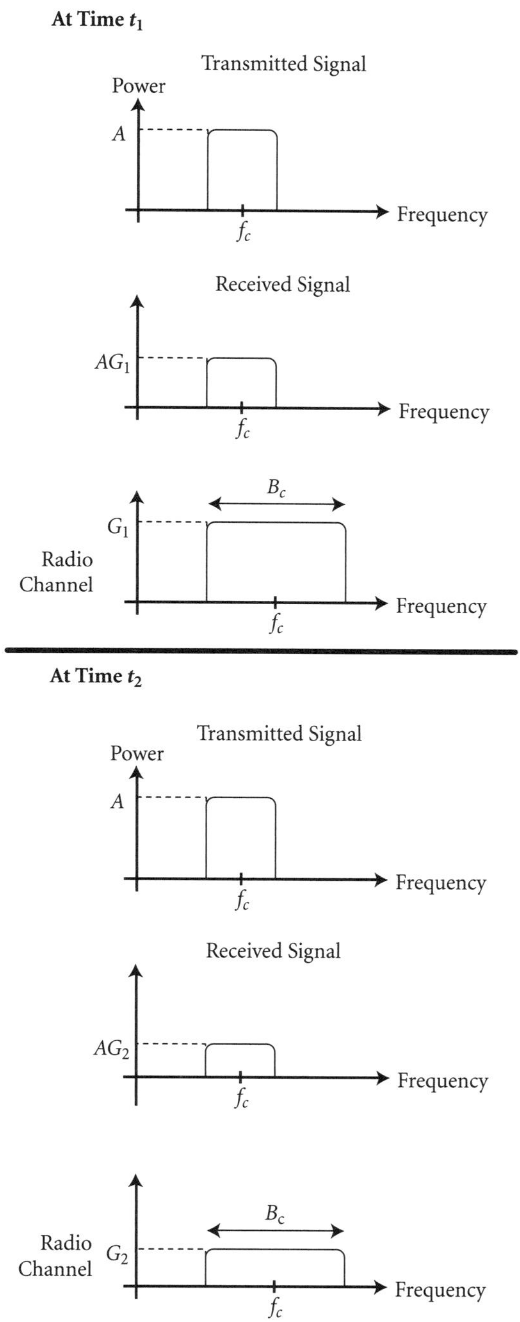

Figure 3.12: An Example of Flat Fading

causes fluctuations in the amplitude (and, hence, power) of the received signal while preserving the spectral characteristics (i.e., frequencies). The time-varying fluctuations in the amplitude of the received signal are usually modeled as a Rayleigh distribution in a vehicular environment. This is why small-scale fading is often called Rayleigh fading.

When the bandwidth of the transmitted signal is larger than the coherence bandwidth of the radio channel, *frequency-selective fading* occurs. In other words, different frequencies in the transmitted signal are affected by the radio channel differently as shown in Figure 3.13. In a frequency-selective fading channel, the symbol duration of the transmitted signal is similar to or even shorter than the delay spread of the channel. Hence, there could be a significant overlap between the current symbol appearing on the shortest multipath signal and the previous symbol arriving on a delayed multipath signal, causing significant *inter-symbol interference*. One method to counteract inter-symbol interference is the implementation of an equalizer at the receiver. This approach is used in cellular systems based on Global Systems for Mobile Communications (GSM). In CDMA-based systems, the transmitted signal bandwidth is very large (e.g., 1.25 MHz or 5 MHz). In such systems, the RAKE receiver resolves the multipath signals and combines energy from multipaths to obtain

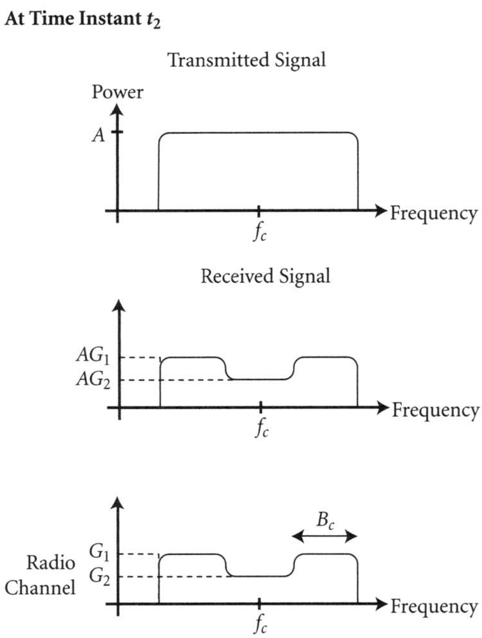

Figure 3.13: An Example of Frequency-Selective Fading

a more powerful signal.[58] A frequency-selective fading channel is more difficult to model than a flat-fading channel. While one Rayleigh fading distribution suffices for a flat-fading channel, a two-ray Rayleigh fading distribution or measured impulse responses are required for a frequency-selective fading channel.

A channel can be classified as a flat-fading channel or a frequency-selective fading channel by comparing the transmission signal bandwidth with the coherence bandwidth of the channel. In terms of the symbol time and the rms delay spread of the channel, another rule of thumb considers a channel to be a flat-fading channel if the symbol time equals or exceeds $10\sigma_\tau$. Otherwise, the channel is considered a frequency-selective fading channel [40].

3.4.3.2 Analysis of the Influence of Mobility As the MS moves in the multipath environment, each multipath signal between the BS and the MS experiences the Doppler shift (a change in frequency). Consider the MS moving toward the BS as in Figure 3.14. The Doppler shift is given by

$$f_D = \frac{v \cos\theta}{\lambda} \tag{3.50}$$

where v is the mobile speed, θ is the angle between the line connecting the BS and the MS and the arrow representing the direction of MS velocity (or movement), and λ is the wavelength. The signal is transmitted at the frequency f_c, and the frequency of the signal at the receiver is $f_c' = f_c + f_D$. When the MS is moving toward the BS, θ is less than $90°$ and f_D is positive. This will result in a higher received frequency than carrier frequency. However, what would be the impact on the received frequency if the MS is moving away from the BS? Draw the line representing the MS velocity (i.e., movement) and find the angle. You will see that is greater than $90°$. This will

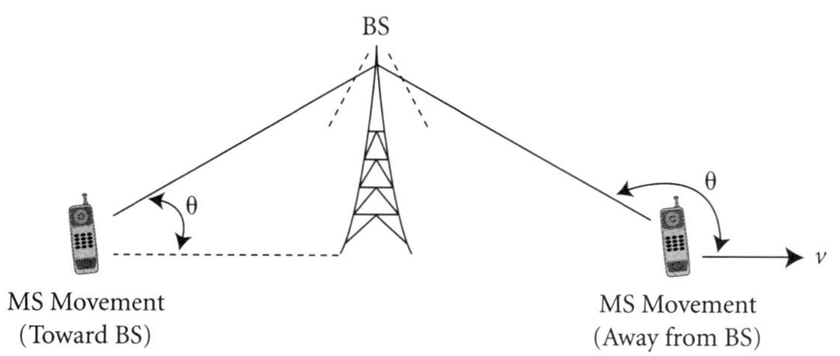

Figure 3.14: Influence of User Mobility

[58]Receive diversity is another method widely used in commercial cellular networks to address the short-term fading and uses multiple receive antennas to receive signals from the transmitter. Different propagation paths between the transmitter and the multiple receive antennas will likely experience different fading intensities. See Chapter 15 for more details on receive diversity.

make f_D negative in Eq. (3.50), and the received frequency is now less than the carrier frequency!

As the MS moves around in a cellular system, both v and θ change. This will lead to variations in the Doppler shift. The resulting new range of frequencies forms the *Doppler spectrum*. The *maximum Doppler shift* f_m occurs when θ is $0°$ and is expressed as

$$f_m = \frac{v}{\lambda}. \tag{3.51}$$

If the RF carrier is f_c, the Doppler spectrum ranges from $f_c - f_m$ to $f_c + f_m$ with f_c in the center of the spectrum. As an example, when the mobile speed is 65 mph or 29.0576 m/s, the maximum Doppler shift is 82.33 Hz in case of 850 MHz. The wavelength at 850 MHz is 0.35 m. At 20 mph, which could be a school-zone speed limit, the maximum Doppler shift is 25.33 Hz at the carrier frequency of 850 MHz. We have given examples of these two speeds only for illustration purposes; we are not endorsing the use of cell phones in school zones! Some U.S. states impose a penalty for using a cell phone in a school zone, so consider the safety of children first and foremost.

The variations in the frequency of the received signal are quantified by the *Doppler spread*, which is defined as the range of frequencies for which the Doppler spectrum is non-zero (i.e., containing detectable energy). In a multipath-rich environment, each multipath experiences a Doppler shift that depends on the relative mobile speed and the angle between the mobile velocity and the angle of arrival of the multipath wave according to Eq. (3.50). The broadening of the frequency spectrum due to mobility can be viewed as frequency dispersion just like the broadening of the symbol duration is time dispersion due to the multipath delay spread. The radio channel can be characterized as a slow-fading channel or a fast-fading channel depending upon the relationship between the transmitted symbol duration and coherence time of the radio channel.

Coherence time indicates how fast the radio channel is changing. The radio channel (i.e., the channel impulse response) remains similar during the time period of coherence time. In other words, the signal received at two time instants inside a time window of coherence time length are likely to be highly correlated and would thus be affected by the radio channel in a similar fashion (e.g., attenuation by X dB). When the target correlation between the signals received at two time instants is 0.5 or above, coherence time is given by

$$T_c = \frac{9}{16\pi f_m} = \frac{0.179}{f_m} \tag{3.52}$$

where f_m is the maximum Doppler shift [96]. A more popular rule of thumb defines coherence time as

$$T_c = \sqrt{\frac{9}{16\pi f_m^2}} = \frac{0.423}{f_m} \tag{3.53}$$

where f_m is the maximum Doppler shift [40].

Let's use the coherence time to characterize the type of small-scale fading. Figure 3.15 shows examples of slow and fast fading. Slow fading occurs when the

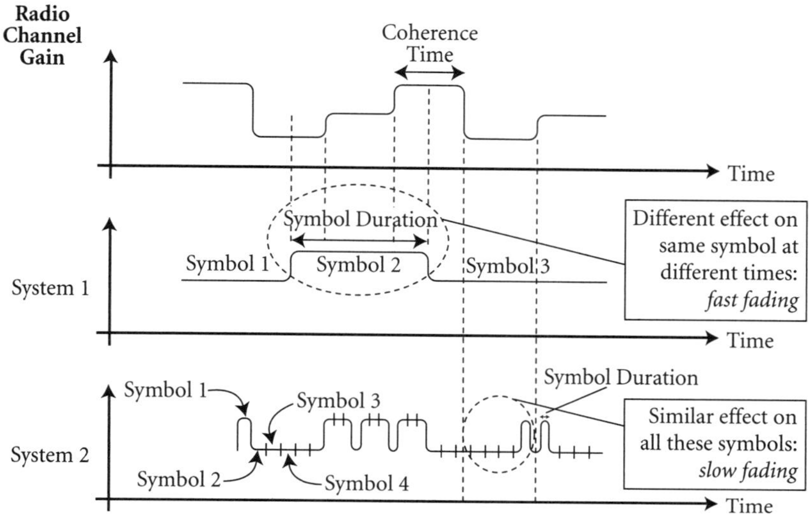

Figure 3.15: Slow and Fast Fading

coherence time of the radio channel exceeds the duration of the transmitted symbol. The radio channel is changing slowly relative to the rate of the transmission rate of the symbols. In this case, many transmitted symbols experience similar channel conditions. When the coherence time of the radio channel is smaller than the duration of the transmitted symbol, fast fading occurs. The symbol is distorted in the time domain and frequency dispersion occurs. Every transmitted symbol could experience different channel conditions.

The top part of Figure 3.15 shows a simplistic view of how the radio channel gain is changing as a function of time. The gain of the channel during the coherence time is similar (i.e., the constant gain in Figure 3.15). In the case of System 1, the modulation symbol periods are longer than the coherence time of the channel. Hence, the radio channel would affect the same transmitted symbol differently at different time instants. Thus, the transmitted symbols in System 1 experience a fast-fading radio channel. In the case of System 2, the symbol duration is smaller than the coherence time, and the radio channel would have a similar impact on several consecutive symbols. The transmitted symbols in System 2 encounter a slow-fading radio channel. *In summary, coherence time is an indicator of how fast or slow the radio channel characteristics are changing.*

Challenge! Show that even for an MS moving at a high speed (e.g., 70 mph), the MS experiences a slow-fading channel. Make the following assumptions.

- Operating frequency of the cellular system: 1.9 GHz

- Voice bit rate: 10 kbps
- Modulation: Binary Phase Shift Keying (BPSK)
- Channel coding: Absent (i.e., no coding)

See the footnote[59] for the calculations.

3.4.3.3 Representation of the Radio Channel from the Small-Scale Fading Perspective Modeling of the small-scale fading of the propagation channel requires determination of the multipath structure (e.g., the relative strengths of multipaths and the delay differences between any two multipaths) and superimposing the impact of mobility on each multipath. These two aspects of channel modeling can be integrated into a discrete wide-sense stationary uncorrelated scattering channel model that represents the radio channel by a channel impulse response model. Consider Figure 3.16 in which the transmission of a single impulse results in multipath signals at the receiver. The structure of the received signal, meaning multipaths with

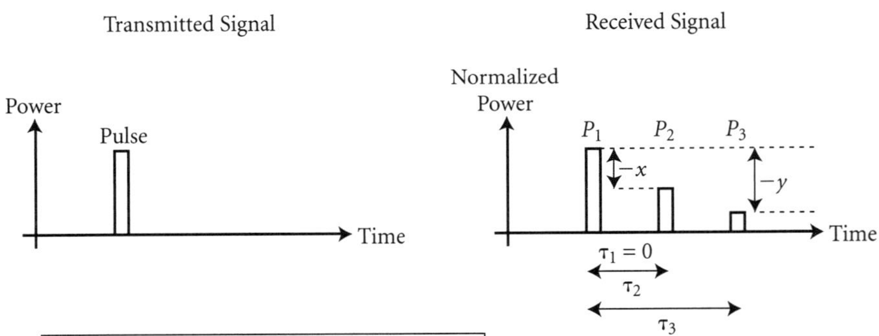

Multipath	Relative Delay	Relative Strength (dB)
P_1	0	0
P_2	τ_2	$-x$
P_3	τ_3	$-y$

Figure 3.16: Channel Impulse Response Model

[59]At 1.9 GHz and 70 mph, the maximum Doppler shift is 198.19 Hz. Using Eq. (3.53), the coherence time is 0.002 s. Since the data rate is 10 kbps and BPSK modulation is used, the bit period and, equivalently, the code symbol period and the modulation symbol period are 0.0001 s. Since the coherence time is larger than the symbol period, the MS experiences a slow-fading channel.

the observed relative strengths and relative delay differences, represents the impulse response of the radio channel.

In Figure 3.16, three detectable multipaths are observed at the receiver for a given transmitted signal. The first multipath happens to be the strongest multipath. The power of the second and third multipath are expressed relative to the strongest multipath. Multipath P_2 is X dB below the strongest multipath, and multipath P_3 is Y dB below the strongest multipath. The propagation delay difference between the first arriving multipath P_1 and the second multipath P_2 is τ_2 and the delay difference between P_1 and P_2 is τ_3. The table in Figure 3.16 summarizes the multipath structure.

Do You Recall... Do you remember the name of the graph of multipath power levels versus multipath delay? See the footnote[60] for the answer.

Let N be the number of multipaths with p_n being the strength of the nth multipath and τ_n the relative delay of the nth multipath compared to the first multipath. When each multipath is assumed to be a zero-mean complex Gaussian time variant process, the received signal $w(t)$ can be expressed as [92]

$$w(t) = \sum_{n=1}^{N} \sqrt{p_n} g_n(t) z(t - \tau_n) \tag{3.54}$$

where $z(.)$ is the transmitted signal and $g_n(t)$ is the complex Gaussian process. The power spectrum of g_n in the frequency domain is the Doppler spectrum of the nth multipath. In general, since each multipath has a different angle relative to the mobile velocity (τ in Eq. (3.50)), each multipath experiences a different Doppler shift. The Doppler spectrum could be classic or flat. The *classic Doppler spectrum* is suitable for an outdoor environment and is specified for a given multipath as [92]

$$P(\nu) = \frac{1}{\pi\sqrt{(V/\lambda)^2 - \nu^2}} \text{ with } |\nu| < \left(\frac{V}{\lambda}\right) \tag{3.55}$$

where V is the mobile velocity, λ is the wavelength, and ν is the Doppler frequency shift in Hertz (at baseband).[61] Such classic spectrum is identical to Clark's and Jakes' widely cited work [97,98]. It is used to model narrowband channels. The *flat Doppler spectrum* for a given multipath is specified as [92]

$$P(\nu) = \frac{\lambda}{(2V)} \text{ with } |\nu| < \left(\frac{V}{\lambda}\right) \tag{3.56}$$

where V is the mobile velocity and λ is the wavelength. In summary, Eq. (3.54) represents the received signal by the sum of the delayed copies of the transmitted

[60] The graph is called the multipath power delay profile.

[61] We have here preserved the naming convention of the standard [92]. Here V is the same as v in Eq. (3.51).

signal, with each copy carrying certain weight relative to the strongest multipath and each copy specified by the delay relative to the first arriving multipath component.

A Question for You... Would you expect the Doppler shifts on different multipaths to be the same or different? Why? Focus on θ in Eq. (3.50). See the footnote[62] for the answer.

Now, how can we determine the multipath structure for a given radio environment? The basic idea is to extract the channel impulse response by *sounding the channel* using a test signal. Figure 3.16 illustrates one possible channel sounding approach: send a pulse from the transmitter and observe the received signal at the receiver. This is a *direct RF pulse* method where short pulses of width T_{bb} are periodically transmitted and then detected at the receiver, which utilizes a bandpass filter of bandwidth $(2/T_{bb})$. Such a system can resolve multipath signals that are at least T_{bb} apart from each other. Hence, if a high resolution is desired, T_{bb} has to be shorter, leading to higher frequency of the transmitted signal. The primary advantage of the RF pulse method is simplicity. (See Rappaport's work [94, 99] for an application of this method to derive the multipath delay profile.) The disadvantages of this method are susceptibility to noise and interference and collapse of the approach if the first arriving multipath cannot be detected reliably (e.g., due to a fade).

Another channel sounding technique relies upon the *spread-spectrum principle*. In this technique, a pseudo-noise (PN) sequence (a set of 1s and 0s) is transmitted. Each bit of the PN sequence is referred to as a *chip*. At the receiver, a sliding correlator is used to detect the PN sequence. Whenever the correlator finds a local peak in the correlation between the received PN sequence and the expected PN sequence, it represents the existence of a multipath signal at that time instance. (Refer to Cox for details of the spread-spectrum channel sounding technique [100].) In practice, this technique can resolve multipaths that are at least one chip apart. The main advantage of this technique is its tolerance to interference and noise, and a drawback is its longer processing delay due to more processing requirements.

Yet another technique to extract the channel impulse response is *frequency-domain channel sounding*. A known signal is transmitted and a frequency sweeper is used to scan discrete frequencies present in the received signal. The measured frequency response (e.g., amplitude at a given frequency) is fed into the inverse discrete Fourier transform function to yield the time-domain channel impulse response. The benefit of this method is that it yields both amplitude and phase information for the multipath signals. However, it requires precise synchronization between the transmitter and the receiver. Hence, it is suitable for only indoor applications, in which the transmitter and receiver can be physically connected to the same source for synchronization. (See Pahlavan's work [101] for additional details of this method.)

[62] In general, since each multipath has a different angle relative to the mobile velocity, i.e., θ in Eq. (3.50), each multipath experiences a different Doppler shift.

In summary, channel sounding techniques based on the use of the RF pulse, spread-spectrum principle, and frequency-domain processing can be used to extract the channel impulse response. Such response defines the multipath structure of the small-scale fading radio channel in a given environment.

Channel sounding techniques utilize measured data to extract the channel impulse response. To predict the performance of a commercial cellular system, real-world measurements are quite valuable. When simulation-based analysis is required, a set of multipath structures that mimics the real-world environment is frequently used when the actual measurements are unavailable. Various multipath structures applicable to different scenarios or *test environments* are summarized in Tables 3.2–3.5. These structures are defined by the International Telecommunication Union (ITU) [93] and represent the expected multipath structure in a given radio environment by a tapped-delay-line filter. Figure 3.17 illustrates how the multipath delay structure can be represented by the tapped-delay-line filter. The filter tap delays correspond to time differences in multipath arrivals, while the filter tap gains correspond to the relative strengths of the multipaths.

In the ITU models below, for a given environment, Channel A corresponds to the low delay spread case and Channel B corresponds to the median delay spread. In

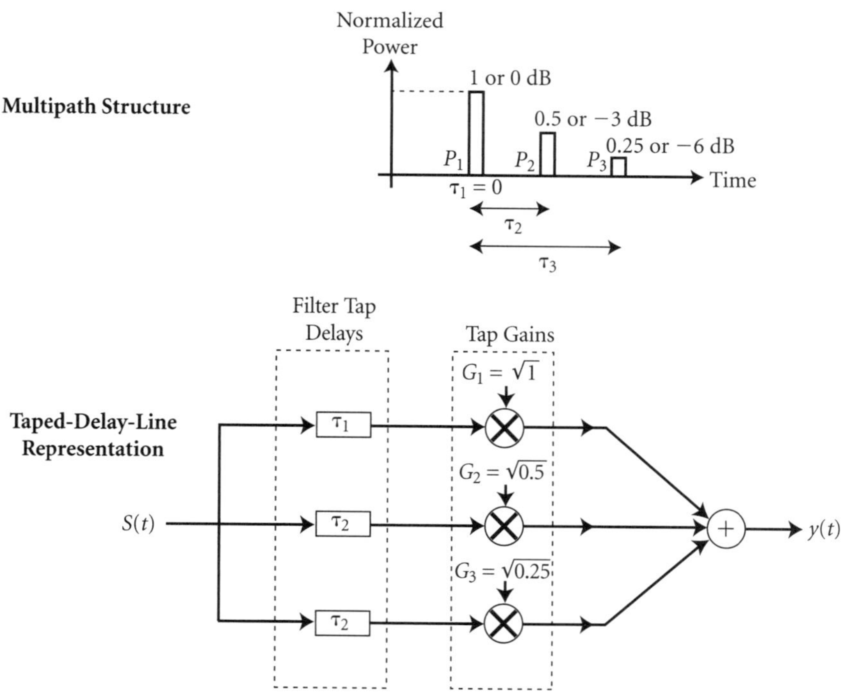

Figure 3.17: Representation of the Multipath Structure via a Tapped-Delay-Line Filter

Table 3.2: High-Level Characterization of the Radio Channel

Environment	Channel A (Low Delay Spread)		Channel B (Median Delay Spread)	
	σ_τ (ns)	P (%)	σ_τ (ns)	P (%)
Indoor Office	35	50	100	45
Pedestrian and Outdoor-to-Indoor	45	40	750	55
Vehicular	370	40	4000	55

Table 3.3: Multipath Structure for an Indoor Office Environment

Filter Tap	Channel A (Low Delay Spread)		Channel B (Median Delay Spread)		Type of Doppler Spectrum
	Delay Difference (ns)	Relative Power (dB)	Delay Difference (ns)	Relative Power (dB)	
1	0	0	0	0	Flat
2	50	-3.0	100	-3.6	
3	110	-10.0	200	-7.2	
4	170	-18.0	300	-10.8	
5	290	-26.0	500	-18.0	
6	310	-32.0	700	-25.2	

Table 3.4: Multipath Structure for Pedestrian and Outdoor-to-Indoor Environments

Filter Tap	Channel A (Low Delay Spread)		Channel B (Median Delay Spread)		Type of Doppler Spectrum
	Delay Difference (ns)	Relative Power (dB)	Delay Difference (ns)	Relative Power (dB)	
1	0	0	0	0	Classic
2	110	-9.7	200	-0.9	
3	190	-19.2	800	-4.9	
4	410	-22.8	1200	-8.0	
5	-	-	2300	-7.8	
6	-	-	3700	-23.9	

Table 3.5: Multipath Structure for Vehicular Environment

Filter Tap	Channel A (Low Delay Spread)		Channel B (Median Delay Spread)		Type of Doppler Spectrum
	Delay Difference (ns)	Relative Power (dB)	Delay Difference (ns)	Relative Power (dB)	
1	0	0	0	-2.5	Flat
2	310	-1.0	300	-0	
3	710	-9.0	8900	-12.8	
4	1090	-10.0	12900	-10.0	
5	1730	-15.0	17100	-25.2	
6	2510	-20.0	20000	-16.0	

practice, large delay spreads occur infrequently compared to low delay spread and median delay spread cases. In Table 3.2, conditions of Channel A and Channel B occur 95% of the time, and large delay spreads occur the rest of the time (i.e., 5%). For example, in the case of the vehicular environment, 40% of the time Channel A is experienced by the receiver and 55% of the time Channel B is experienced by the receiver. The rms delay spread in such a case is 370 ns for Channel A and 4000 ns for Channel B. The multipath structures for the three types of environments are summarized in Tables 3.3–3.5.

3.4.3.4 Modeling of the Radio Channel for Small-Scale Fading According to the central limit theorem, the sum of a large number of independent random processes produces a Gaussian or normal distribution provided that the random variables have similar ranges of magnitudes. The probability density function of the Gaussian distribution is given by

$$f(x) = \frac{1}{\sqrt{2\pi\sigma}} e^{((-(x-\mu)^2)/(2\sigma^2))} \tag{3.57}$$

where μ and σ are mean and variance of the Gaussian process, respectively.

Let a_c and a_s be independent Gaussian random processes with 0 mean and σ variance. The envelope r of the sum of two Gaussian random processes follows Rayleigh distribution and is given by[63]

$$r = \sqrt{a_c^2 + a_s^2}. \tag{3.58}$$

The probability density function for the Rayleigh distribution is

$$p(r) = \frac{r}{\sigma^2} e^{((-r^2)/(2\sigma^2))}. \tag{3.59}$$

The Rayleigh distribution is widely used in cellular systems to model the envelope of a multipath component. From the perspective of the cellular radio environment, σ^2 in Eq. (3.59) is the time average power of the received signal that is superimposed on the distance-based path loss and large-scale fading. Recall that the short-term fading represents short-term variations in the signal strength around the *local mean* signal strength in a given area. Samples of the received signal are usually taken at spatial separations on the order of a fraction of the wavelength (e.g., $\lambda/4$). The local mean signal strength can be easily found by taking an average over a long time period (or, equivalently, many wavelengths, such as 5λ to 40λ).

The mean of the Rayleigh distribution is

$$\mu_{\text{Rayleigh}} = \sigma\sqrt{\frac{\pi}{2}}, \tag{3.60}$$

and the variance of the Rayleigh distribution is

$$\sigma^2_{\text{Rayleigh}} = \sigma^2\left(2 - \frac{\pi}{2}\right). \tag{3.61}$$

[63] a_c and a_s are amplitudes of orthogonal sinusoidal signals, i.e., sine and cosine, and Eq. (3.58) is the envelope of the combined signal.

$\sigma^2_{\text{Rayleigh}}$ is the AC power in the signal envelope and is often referred to as non-line-of-sight power.

The median of the Rayleigh distribution is

$$\text{Median}_{\text{Rayleigh}} = 1.779\sigma. \tag{3.62}$$

When a dominant non-fading component occurs in the received signal, the envelope becomes

$$r = \sqrt{A_c + a_c^2 + a_s^2}. \tag{3.63}$$

The Ricean distribution can be expressed as

$$p(r) = \frac{r}{b_0} e^{((-(r^2+A_c^2))/(2b_0))} I_0\left(\frac{rA_c}{b_0}\right) \tag{3.64}$$

where b_0 is σ^2, A_c is the peak amplitude of the dominant or DC signal, and $I_0(.)$ is the modified Bessel function of the 0th order and of the first kind. The modified Bessel function is given by [102]

$$I_0(a) = \frac{1}{2\pi} \int_0^{2\pi} e^{(a\cos(u))} du. \tag{3.65}$$

The Ricean parameter K is defined to be

$$K = \frac{A_c^2}{2b_0} = \frac{A_c^2}{2\sigma^2}. \tag{3.66}$$

K is often expressed in decibels as

$$K\ (\text{dB}) = 10 \log_{10}\left(\frac{A_c^2}{2\sigma^2}\right). \tag{3.67}$$

As is evident from Eq. (3.67), the smaller the value of A_c becomes, the lower the K factor becomes. Now, what happens when $A_c = 0$? Compare Eq. (3.59) with Eq. (3.64). The Ricean distribution turns into the Rayleigh distribution! Intuitively, it makes sense. If no dominant signal exists, we are left with just the scattered time-varying AC power, which is represented by the Rayleigh distribution. The Rayleigh fading can thus be considered a special case of the Ricean fading with $K = 0$ or $-\infty$ dB. A negative K value means that the DC signal power is smaller than the AC power, corresponding to poor radio channel conditions as the AC power would be reflected in widely varying received signal amplitudes. A large positive value of K in decibels means that the DC power is much stronger than the AC power and that the signal amplitude variations can be expected to be smaller. For your visual pleasure, we have plotted in Figure 3.18 the Gaussian, Rayleigh, and Ricean distributions based on Eqs. (3.57), (3.59), and (3.64). (See Loo's and Rappaport's work [40, 103] for additional details on this topic.)

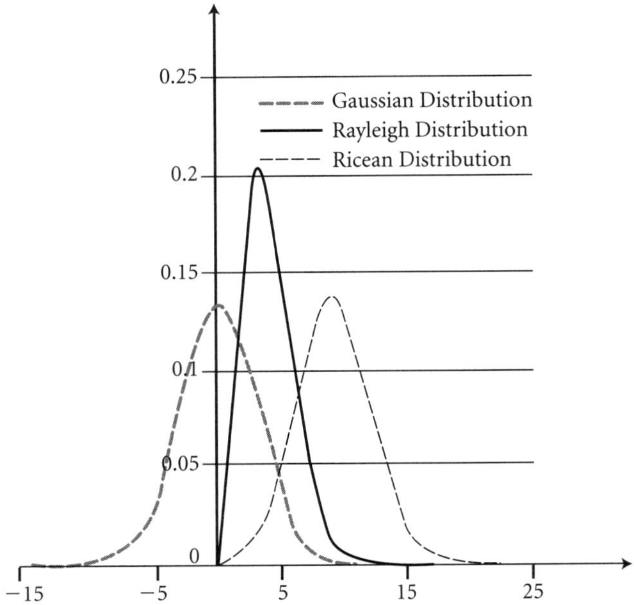

Figure 3.18: Probability Density Functions for Gaussian, Rayleigh, and Ricean Distributions

In summary, the envelope of the received signal or the envelope of the arriving multipath component follows Rayleigh or Ricean distribution. Recall that each component also undergoes a Doppler effect. The influence of the Doppler effect on the spectrum of the received signal was analyzed by Gans [104], and the result is reproduced here.

The power spectral density of the received signal is given by

$$s(f) = \frac{A\{p(\alpha)G(\alpha) + p(-\alpha)G(-\alpha)\}}{f_m\sqrt{1-((f-f_c)/(f_m))^2}} \tag{3.68}$$

where f_c is the center carrier frequency, f_m is the maximum Doppler shift, f is the frequency ranging from $f_c - f_m$ to $f_c + f_m$, A is the average received power relative to an isotropic antenna, α is the angle of arrival (same as θ in Figure 3.14), $G(.)$ is the azimuthal gain at the receive antenna, $p(.)$ is the probability density function indicating how the received power is distributed as a function of λ, and $S(f)$ is zero outside the frequency range of f.

If the receive antenna is a vertical $\lambda/4$ antenna and α is uniformly distributed from 1 to 2π (i.e., $p(.) = 1/(2\pi)$), Eq. (3.68) becomes [40]

$$S(f) = \frac{1.5}{\pi f_m\sqrt{1-((f-f_c)/(f_m))^2}}. \tag{3.69}$$

While carrying out a simulation-based analysis of a communication system, it is often important to model small-scale fading. Smith developed a simulation code that incorporates both the Rayleigh fading effect and Doppler effect to simulate small-scale fading [40, 105]. Recall that two independent Gaussian processes can be used to generate Rayleigh fading distribution and Gans' spectral shaping can be used to mimic the Doppler effect. Figure 3.19 illustrates the block diagram of Smith's Rayleigh fading simulator.

Each Gaussian process undergoes spectral shaping to simulate the Doppler effect. The shaped baseband frequency spectrum is upconverted to f_c. Two frequency up-converted Gaussian signals are added to obtain a Rayleigh-faded signal. This signal can be interpreted to represent one multipath component. As an example, if there are three resolvable multipaths, each of the three multipaths can be modeled in a similar fashion by repeating the same process.

Figure 3.20 illustrates the flowchart for generating the Rayleigh fading samples influenced by the Doppler spectrum using the block diagram shown in Figure 3.19 [40, 105]. The following inputs are required: the maximum Doppler shift f_m; N as an indicator of the frequency spacing between adjacent frequencies of a Doppler spectrum. The flowchart generates N time samples of a Rayleigh-faded signal for one multipath component.

The basic steps for generating Rayleigh fading samples are as follows.

1. Obtain the input parameters f_m and N. The frequency spacing or resolution of the Doppler spectrum becomes $\Delta f = (2f_m)/(N-1)$. If higher resolution is required, choose a larger value of N. N is typically chosen as an integer that is a power of 2 (e.g., 1024) so that Inverse Fast Fourier Transform (IFFT) can be performed efficiently. The time period of the Rayleigh-faded duration would be $T = 1/\Delta f$. In other words, N samples of a Rayleigh-faded signal would be generated for the time period of T.

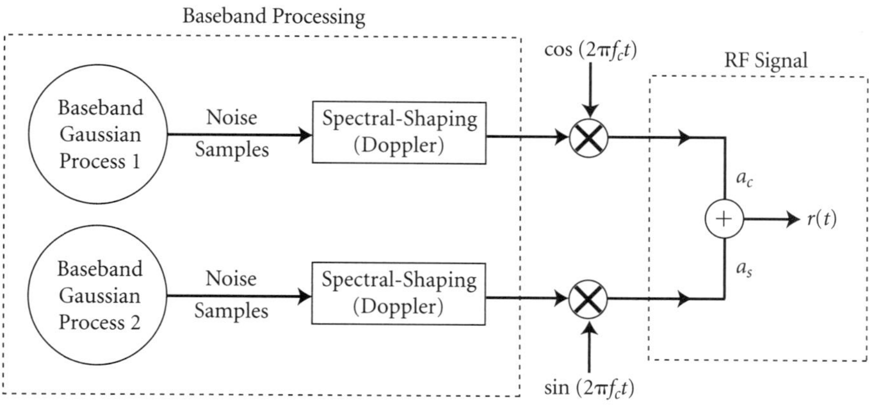

Figure 3.19: Structure of Smith's Rayleigh Fading Simulator

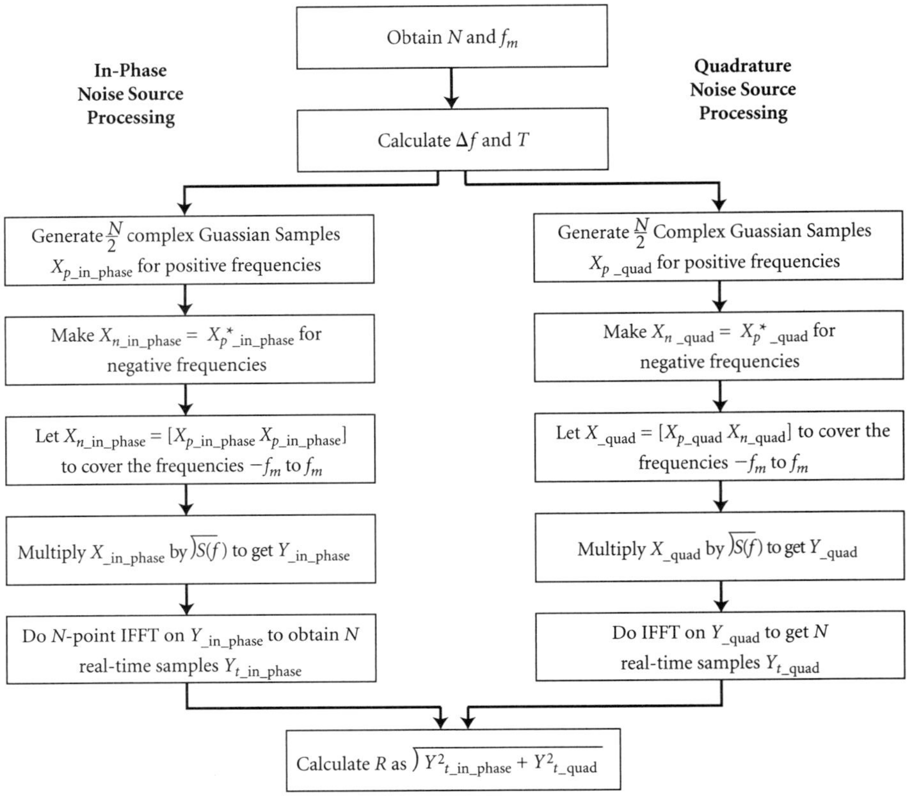

Figure 3.20: Basic Steps for Generating Rayleigh Fading Samples

2. Now, two Gaussian random processes are required to generate N samples. One Gaussian process represents an in-phase noise source, and the other process represents a quadrature phase Gaussian noise source. Execute Steps 3–7 for the two Gaussian processes with i equal to in-phase for the in-phase noise source and i equal to quad for the quadrature noise source.

3. Generate $N/2$ *complex* Gaussian samples X_{p_i}, which is a sequence representing amplitudes at positive frequencies.

4. Derive $(N/2)$ complex Gaussian samples X_{n_i}, which is the complex conjugate of X_{p_i}.

5. Determine a sequence X_i of N complex numbers by concatenating X_n and X_p. X_i now covers the entire baseband frequency range from $-f_m$ to f_m.

6. To mimic the Doppler effect, multiply X_i by $\sqrt{S(f)}$ and get Y_i.

7. Do an N-point IFFT on Y_i that consists of N complex numbers. Let the result be Y_{t_i}, a sequence of N real numbers.

Calculate

$$R = \sqrt{Y_{t\_\text{in\_phase}}^2 + Y_{t\_\text{quad}}^2}.$$

R is a sequence of N samples representing a Rayleigh faded signal that reflects the Doppler effect. *Mission accomplished!*

Jakes' model, a similar approach, is also widely used to model Rayleigh fading and is based on summing sinusoids [98, 106]. Assume that a kth multipath (i.e., kth waveform) arrives at the receiver. Each multipath or waveform is represented by sum of M equal-strength rays (e.g., $M = 10$ rays). Let n denote a ray that models one sinusoidal signal. In other words, n ranges from 1 to M. The ray n experiences the Doppler shift of

$$f_n = f_d \cos \alpha_n \tag{3.70}$$

where α_n is the uniformly distributed angle of arrival of the scattered ray (usually set to 0) and could be expressed as

$$\alpha_n = \frac{2\pi n}{M}. \tag{3.71}$$

Rayleigh fading on the kth waveform (i.e., kth multipath) as a function of time t is given by

$$\begin{aligned} R(t,k) = 2\sqrt{2} \Bigg[& \left\{ \sum_{n=1}^{M} (\cos\beta_n + j\sin\beta_n)\cos\left(2\pi f_n t + \theta_{n,k}\right) \right\} \\ & + \left\{ \frac{1}{\sqrt{2}} (\cos\alpha + j\sin\alpha)\cos\left(2\pi f_d t\right) \right\} \Bigg] \end{aligned} \tag{3.72}$$

where α is phase (usually set to 0),

$$\beta_n = \frac{\pi n}{M+1}, \tag{3.73}$$

and

$$\theta_{n,k} = \beta_n + \frac{2\pi(k-1)}{M+1}. \tag{3.74}$$

β_n is chosen to ensure zero correlation between the real and imaginary parts of $R(t)$. According to Jakes, $\theta_{n,k}$ in Eq. (3.74) provides uncorrelated waveforms. However, nearly zero cross-correlation between waveform j and waveform k is achieved only when $\theta_{n,j} - \theta_{n,k} = i\pi + \pi/2$ for some integer i [107]. Furthermore, Jakes' model is deterministic.

A modified Jakes' model utilizes Walsh–Hadamard sequences to ensure zero cross-correlation among the multipath waveforms [108]. The model parameters are chosen to be

$$\alpha_n = \frac{2\pi(n-0.5)}{M+1} \tag{3.75}$$

and

$$\beta_n = \frac{\pi n}{M}. \tag{3.76}$$

The modified Jakes' model is then given by

$$R(t,k) = \sqrt{\frac{2}{M}} \sum_{n=1}^{M} A_k(n)(\cos\beta_n + j\sin\beta_n)\cos(2\pi f_n t + \theta_{n,k}) \tag{3.77}$$

where $A_k(n)$ is the kth Walsh–Hadamard sequence in n with the sequence comprising $(+1\text{s})$ and (-1s). The Walsh–Hadamard sequences are orthogonal to one another and have zero cross-correlation among them. $\theta_{n,k}$ can be initialized as random numbers.

3.4.3.5 Useful Statistics for Small-Scale Fading

Several statistics quantify the small-scale fading experienced by the transmitted signal. (See Rappaport's work [40] for additional details.) For a Raleigh fading distribution, the *average fade duration* is given by

$$\tau(R) = \frac{e^{\rho^2} - 1}{\rho f_m \sqrt{2\pi}} \tag{3.78}$$

where ρ is defined as R/R_{rms} with R as the signal level defining the fade level and R_{rms} as the rms value of the signal and f_m is the maximum Doppler shift. Expressing Eq. (3.78) using the speed of the mobile terminal (or, more accurately, the speed of the mobile terminal relative to the nearby environment),

$$\tau(R) = \frac{\lambda(e^{\rho^2} - 1)}{\rho v \sqrt{2\pi}}. \tag{3.79}$$

In Eq. (3.79), λ is the wavelength and v is the (relative) speed of the mobile terminal. As Eq. (3.79) suggests, the higher the speed of the mobile terminal, the shorter the average fade duration is. Similarly, the lower the speed of the mobile terminal, the longer the fades are. Hence, in vehicular environments, fades are expected to be short, and in a fixed wireless or pedestrian environment, fades are expected to be long.

Example: Impact of Mobile Speed on the Fade Duration If a fade is defined as the signal level, which is 10 dB below the R_{rms}, $20\log_{10}(\rho) = -10$ dB, leading to $\rho = 0.32$. The average fade duration in the case of 13.9 m/s (or 50 km/hr or 31 mph) and $\lambda = 0.33$ m (corresponding to f_c of 900 MHz) would be 3.19 ms.

The *level crossing rate* (LCR) is defined as the number of times a given signal level is exceeded in the positive direction per unit time. Consider Figure 3.21 in which the signal level R is exceeded three times during the observation window of 0.5 second. Then, the LCR associated with the threshold level R is $3/0.5 = 6$ per second. If the signal level R is considered a threshold level below which the signal is considered to

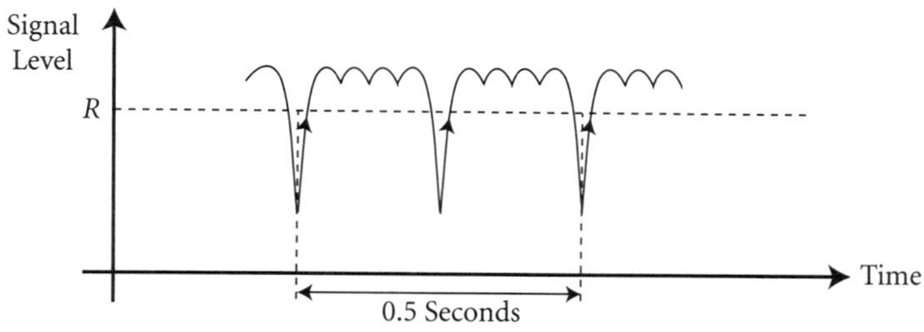

Figure 3.21: Level Crossings of a Rayleigh-Faded Signal

be in a fade, the LCR for the signal level R indicates how frequently the fade occurs. For a Rayleigh fading distribution, the LCR is given by

$$LCR = \sqrt{2\pi} f_m \rho e^{-\rho^2} = \sqrt{2\pi} \frac{v}{\lambda} \rho e^{-\rho^2}. \tag{3.80}$$

The statistics on the average fade duration and the LCR can be used in the simulation-based analysis to evaluate the expected performance of a cellular network in a given radio environment.

3.5 POINTS TO REMEMBER

Let's summarize some of the key points elaborated on in this chapter.

- The cellular radio environment is quite dynamic.
- Cellular propagation is more complex than free-space propagation. The basic mechanisms related to cellular propagation are reflection, diffraction, scattering, and absorption and penetration.
- The received signal strength at a location can be predicted using three major phenomena: distance-based path loss, large-scale fading, and small-scale fading. The distance-based path loss considers the transmitter-to-receiver separation, BS and MS antenna heights, carrier frequency, and the type of environment, such as rural and urban environments.
- A variety of models exist for estimating the distance-based path loss. These models are suitable for specific scenarios (e.g., macrocells versus microcells and 900 MHz versus 1900 MHz). Examples of such propagation path loss models are the COST 231–Hata and ITU models.
- Large-scale fading considers the variations in signal strength for the same transmitter-to-receiver separation. These variations consider the effects of the environment clutter (e.g., the presence of buildings and trees) on the received signal strength.

- Large-scale fading, also known as shadow fading, is modeled via log-normal distribution.
- The small-scale fading models signal strength variations over short distances or short time intervals due to multipaths and user movement relative to the surrounding environment. Small-scale fading could be frequency-selective fading or flat fading based on the coherence bandwidth of the radio channel being larger than or smaller than the signal bandwidth. Small-scale fading could be classified as fast fading or slow fading depending upon the coherence time of the channel being shorter or longer than the symbol duration. Small-scale fading is often modeled by Rayleigh fading or Ricean fading.
- A Rayleigh-faded signal can be constructed using Gaussian processes and the sum of sinusoids.
- Various statistics quantify the short-term fades. Examples of such statistics include average fade duration and level crossing rate.
- Propagation knowledge is quite valuable in predicting system performance and estimating the cell size.

CHAPTER 4

IP AND ASSOCIATED TECHNOLOGIES FOR A CELLULAR SYSTEM

The Internet is a network of the networks. The Internet was a result of research opportunities funded by the U.S. government. The creation of the Internet began as a project within the U.S. Advanced Research Projects Agency. Vinton Cerf and Robert Kahn are the two people most often credited as being the fathers of the Internet [109].[64]

This chapter provides an overview of Internet Protocol (IP), which is at the heart of the Internet and related technologies from the cellular system perspective. A short introduction to IP is given first. Each layer of the IP stack is discussed. The IP packets contain information such as web pages and email, and the path these packets travel is influenced by routing protocols, so major routing protocols are summarized. The host that is the final destination of an IP packet usually resides on some physical network, such as an Ethernet network. Interworking between IP and Ethernet is explained. Within a cellular network, the transport of information is implemented using mechanisms such as Asynchronous Transfer Mode (ATM) and Multi-Protocol Label Switching (MPLS), and such transport is also described. As the industry

[64] If you love history, other sources can tell you more about *exactly* how the Internet came to exist [110–115].

transitions from circuit-switched voice to Voice over IP (VoIP), understanding the roles of relevant protocols is quite important, and, therefore, the basic operation of such protocols is also covered. The chapter primarily focuses on IP version 4 (IPv4), but a brief introduction to IP version 6 (IPv6) is also given.

4.1 WHY INTERNET PROTOCOL? (OR, PERHAPS, WHY *NOT* INTERNET PROTOCOL?)

The cellular business has changed. The early cellular business of first-generation and second-generation systems focused on voice services. However, the service operator now faces declining voice revenues and growing data revenues. The advent of the Internet has changed the cellular industry landscape forever. Third-generation (3G) and emerging fourth-generation (4G) technologies pay a lot of attention to providing IP-based services using an IP-based, packet-switched core network.

Something to Think About. . . Our colleague, Dr. Best, likes to give this example to make people realize the impact of the Internet. Google, a company that primarily relies on the Internet for its revenue, was born in 1998. In less than a dozen years, Google reached market capitalization of more than $100 billion, and dictionaries now include *google* as a verb!

To get a feel for what the wireless Internet involves, consider Figure 4.1 in which a user is reading news posted on a website, news.com, using a 3G mobile device. Recall from Chapter 1 that the cellular network has a radio access network and a core network.

For the user to read news, the mobile device and the news.com web server need to exchange IP packets. Each IP packet includes the source IP address and the destination IP address. In a simple example of packet exchange, the application layer of the mobile device sends a *Hypertext Transfer Protocol (HTTP) request* message to the news.com web server, and the web server replies with an HTTP response message containing the news.com homepage. These messages are carried inside IP packets. The web browser window on the mobile device then displays the homepage on the screen.

Let's trace the path traversed by an IP packet from the mobile device all the way to the web server. Assume that the mobile device has established a packet data session and has received an IP address from a gateway in the wireless packet core network. An application layer protocol in the mobile's protocol stack creates a message. Since we are talking about web-browsing, the suitable application layer protocol is HTTP. The HTTP forwards a message, such as an HTTP request message, to the lower-layer protocol, such as Transmission Control Protocol (TCP), which adds its header to the HTTP message and forwards it to the lower IP layer. The IP layer adds its own header to the TCP message. (The source and destination IP addresses that we mentioned earlier are examples of the IP header contents.) With the addition of the

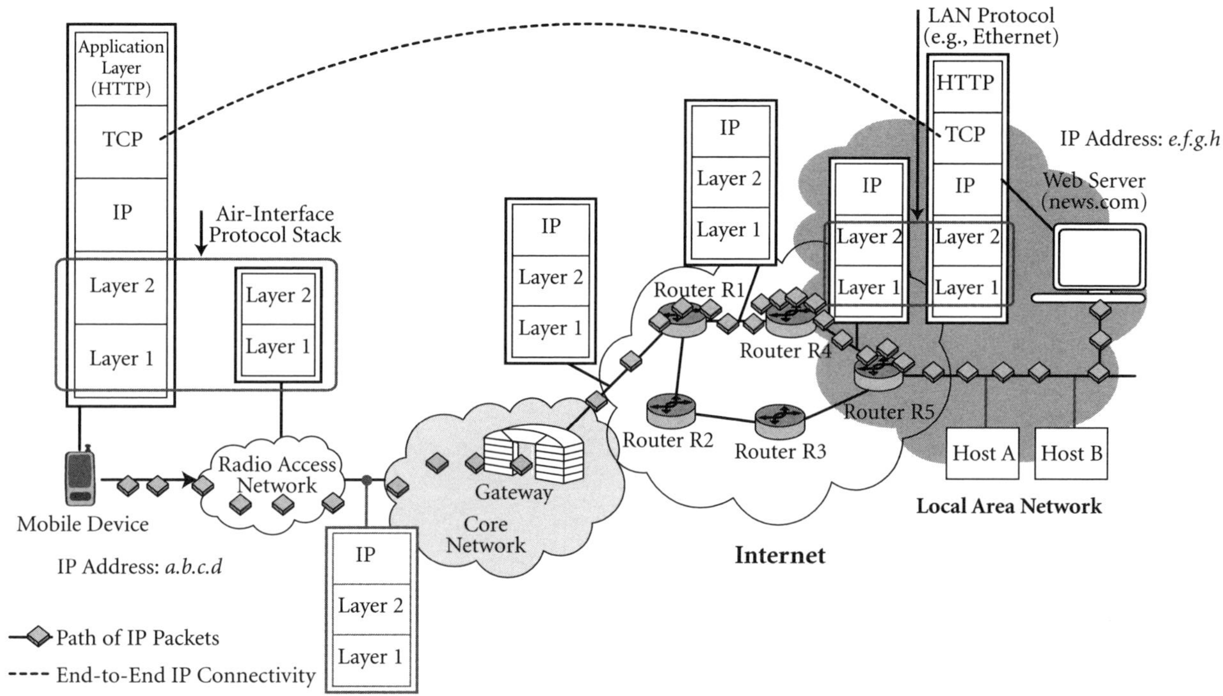

Figure 4.1: Accessing the Internet via a Cellular Network

IP header, we now have what is called an IP packet (i.e., datagram). It is this IP packet that travels (generally) unchanged between the user's mobile device and the web server. The IP layer forwards the IP packet to layer 2, and layer 2 adds its header and forwards the message to layer 1. Since we are discussing a cellular system, layer 1 and layer 2 are defined by the air-interface protocol of the cellular technology, such as a 3G Universal Mobile Telecommunication System (UMTS) system or a 3G 1xEvolution–Data-Optimized (1xEV-DO) system. Layer 1 and layer 2 of the mobile device and the radio access network enable the packet to go from the mobile device to the base station controller or the radio network controller that interfaces with the core network.

Do You Recall... In Chapter 1, we mentioned two types of core networks. Do you remember which type of core network is relevant to Internet access? See the footnote[65] to verify your answer.

The IP packet travels on the radio-core interface to reach a gateway in the wireless packet core network. This is the gateway that has allocated the IP address *a.b.c.d* to the mobile device and that interfaces with the public and global Internet. In our example, we are showing just a tiny portion of the Internet,[66] which has five routers, R1 through R5. The gateway observes the destination IP address, which is *e.f.g.h*, and decides to forward it to R1 because R1 is considered to be in the path that leads to the destination. R1 forwards the IP packet to R4, which in turn forwards the IP packet to R5. R5 finds the destination address to be on the same physical network and, therefore, determines that it need not forward the packet to any other router. A local area network (LAN)[67] is an example of a physical network, and Ethernet is the layer 2 protocol that operates a LAN. Layer 1 in a LAN could be just some physical coaxial cable.

R5 knows the physical address [e.g., a medium access control (MAC) address] of the computer that acts as the news.com web server and that has the IP address *e.f.g.h*. Hence, R5 uses the Ethernet protocol and the MAC address of the web server to transport the IP packet of the mobile device to the computer hosting the web server. The Ethernet protocol receives the Ethernet frame that includes the Ethernet header. The Ethernet header is stripped off, and the original IP packet is forwarded to the IP layer. The IP layer strips off the IP header and forwards the resulting segment to the TCP layer. Finally, the TCP layer removes the TCP header and forwards the packet to the HTTP application at the application layer.

As seen from Figure 4.1, the two endpoints of the communication link—the mobile device and the web server—have five layers in the Internet Protocol stack: layer 1,

[65]The packet-switched core network interfaces with the Internet. (The other type of core network, the circuit-switched core network, interfaces with the public-switched telephone network for voice calls.)

[66]Technically, even the mobile device, the gateway, and the web server are considered part of the Internet.

[67]LAN is pronounced as a word rather than as individual letters, such that it rythmes with "can."

layer 2, IP as layer 3, TCP as layer 4, and the application layer as layer 5. (Section 4.2 takes a closer look at all the five layers of the Internet Protocol stack.) This stack is different from the seven-layer Open-System Interconnect model. The application layer is considered layer 7 in the Open-System Interconnect model. In Figure 4.1, the packets passed through R1, R4, and R5. Each router needs to choose the next router (called *next hop*) for an incoming IP packet in an attempt to move the packet in the right direction. The routing protocols determine the next hop for an arriving packet. (Section 4.3 discusses major routing protocols and provides basics of the packet forwarding.) When the packet reaches the router that is part of the same LAN as the host with the destination IP address, LAN forwarding is used. (Section 4.3 also describes cooperation between IP and Ethernet in the LAN and summarizes how the IP packets are carried from one host to another in the LAN setting.) In Figure 4.1, we illustrated how the IP packets from the mobile device pass through the radio access network and reach the gateway residing in the core network. Section 4.4 describes typical approaches for transporting the IP packets within the radio access network and core network. The industry is transitioning from circuit-switched voice to VoIP due to the expected advantages of cost reduction and efficiency. Section 4.5 addresses various protocols that form the basis for VoIP. Section 4.6 deals with several IP-based mechanisms frequently encountered in the operation of a cellular system. The primary focus of this chapter is on IPv4, but IPv6 is becoming important and relevant, so Section 4.7 summarizes the main characteristics of IPv6.

4.2 PROTOCOL STACK FOR THE INTERNET

The TCP/IP suite or the IP stack consists of five distinct layers. Figure 4.2 provides a glimpse of such a protocol stack. Many sources consider the IP suite as having four layers, with the lowest layer being the link layer. Some sources break the link layer into data link layer (sometimes referred to as network interface layer) and physical layer [116]. We will use the latter approach.

The top-most layer is the application layer, which interfaces with the user through programs such as Outlook and Internet Explorer. Example protocols at the application layer include HTTP for web-browsing, Simple Mail Transfer Protocol (SMTP) for email, and File Transfer Protocol (FTP) for uploading and downloading files. The transport layer provides an end-to-end connectivity between two endpoints or ports of the communication session. TCP, Stream Control Transmission Protocol (SCTP), and User Datagram Protocol (UDP) are examples of layer 4 protocols. Layer 3 is the internetwork layer, which implements packet forwarding to carry information from source to destination. Example protocols residing at layer 3 are IP and Internet Control Message Protocol (ICMP). Layer 2 is the data link layer, which provides an interface to the physical hardware or physical medium. Popular layer 2 protocols are Ethernet, ATM, and cellular air interface. Layer 1 is the physical layer that is responsible for actual transfer of information on some physical medium. Examples of layer 1 implementation are the cellular air interface, such as UMTS and 1xEV-DO, and coaxial cable. Finally, comparing the IP stack to the famous Open System Interconnect

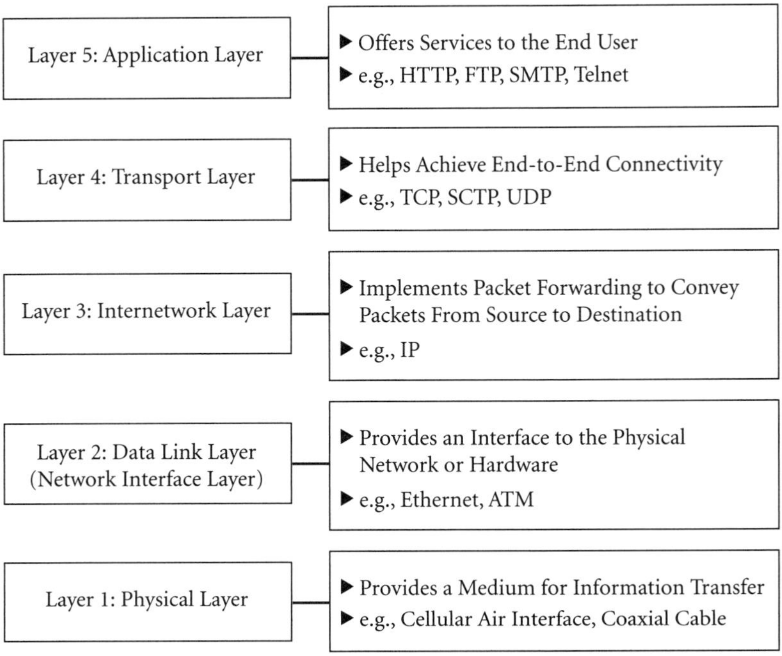

Figure 4.2: TCP/IP Suite (IP Stack)

seven-layer model, the three upper layers of the Open System Interconnect model—application layer, presentation layer, and session layer—combine to become the application layer of the IP stack.

Now that we have a basic understanding of various layers, let's get the basic understanding of various protocols that operate at different layers of the stack. See Parziale et al. [117] for high-level overviews of various application layer protocols. The transport layer protocols of TCP, SCTP, and UDP are described in Section 4.2.1. Various aspects of IP are discussed in Section 4.2.2. Section 4.2.3 summarizes the link layer protocols of Ethernet and ATM.

4.2.1 Overview of the Transport Layer (Layer 4) Protocols

The concepts of a port and a socket are quite important in the IP stack. Pretend that you are a transport layer! You just received a packet from the IP layer sitting right below you. How will you, as a transport layer, figure out to which application layer protocol you should be forwarding the packet? The answer lies in the concept of *ports*. The port number specifies the application layer protocol associated with the packet. The port is a sixteen-bit number used by the host-to-host protocol (e.g., transport layer protocol, such as TCP or UDP) to identify to which higher-layer protocol (e.g., HTTP) or application program (e.g., an Internet Explorer web browser) it should deliver the incoming messages [117]. Three types of ports exist: *well-known*, *registered*, and *dynamic/private*. The Internet Assigned Number Authority assigns and controls the well-known and registered ports.

The port number that falls in the range 0 to 1023 is called a *well-known* port. The port number within the range 1024 to 49151 is called a *registered port*. Finally, the *dynamic/private* ports are the ports from 49152 through 65535. Well-known ports are used by standard servers. Examples of well-known ports are 20 for data connection and 21 for control connection for FTP, 23 for Telnet, 25 for SMTP, 53 for DNS (Domain Name System), and 80 for HTTP or www. Defining the well-known ports helps clients easily find servers. A given well-known port number may be available to both TCP and UDP. Registered ports provide a standardized approach to facilitate interaction between two products. For example, 1102 is a registered port for Adobe servers. Dynamic/private ports are dynamic in nature, and clients use them to identify their side of the connection with the servers. Since the client is initiating the connection, it can inform the server what ephemeral port number it is using. The server then uses such a dynamic port number to contact the client. The host on which the client process is running allocates the dynamic port number to the client. To ensure that two applications running on the same host (e.g., two web browser windows) get different dynamic port numbers, the TCP/IP suite software can be used for port allocation.

The connection between two processes is specified using the concept of *sockets*. The socket provides a full-duplex byte stream connection to the communicating processes. The Berkeley socket application programming interface is a de facto standard for TCP/IP socket implementation. The socket address takes the form of the three-tuple <protocol, local IP address, local port>. For example, <tcp, 192.168.2.3, 8080> is a valid socket address [117]. The connection between two processes is specified by an association. The association merges the socket addresses for two processes to construct a five-tuple of the form <protocol, local IP address, local-port, foreign IP address, foreign-port>. For example, the five-tuple <tcp, 192.168.14.100, 1500, 192.168.10, 80> is an example of an association where TCP is used as the transport layer protocol for a TCP connection between two hosts with one client host having the IP address 192.168.14.234 and using an ephemeral port 1500 and the server host having the IP address 192.168.44 and well-known port 80 (associated with an HTTP or www server). And, of course, a server is capable of managing connections through a single well-known port.

The processes of multiplexing and demultiplexing utilize ports to enable the transport layer to support multiple application layer protocols and multiple instances or processes of a given application layer protocol simultaneously [118]. Multiplexing is done at the source, and demultiplexing is done at the destination. Assume that we have two applications, A and B, running between two hosts, H1 and H2, and that packets are being sent from H1 to H2. Multiplexing at H1's transport layer could use source port SP-A and destination port DP-A to identify A and source port SP-B and destination port DP-B to identify B. Demultiplexing at H2's transport layer would use SP-A, DP-A, SP-B, and DP-B to deliver a received packet to the correct application.

4.2.1.1 Transmission Control Protocol Overview The main goal of TCP is to provide an end-to-end reliable connection between the ports (or hosts with one process running on one host).[68] TCP at the sending process receives data from the application layer, distributes data bytes into TCP segments, and forwards data to the lower-layer protocol, IP, for transmission. IP of the sending process then delivers the IP packets to the IP of the receiving process, which in turn forwards the packet to the receiving TCP. Finally, the application layer of the receiving process obtains the data.

Let's briefly summarize the overall operation of TCP, which is a byte-stream–based protocol. A TCP segment is a set of bytes. Each transmitted byte has an associated sequence number in TCP. The TCP header includes the sequence number of the first data byte (e.g., 100) of the accompanying TCP segment. The sequence number serves several purposes. The receiver TCP can rearrange potentially out-of-order TCP segments. It can also discard duplicate TCP segments. The receiver TCP must respond to each received segment and its response includes a positive acknowledgment (ACK) for the last contiguously received segment. The receiver TCP also conveys the amount of data it can accommodate without experiencing a buffer overflow. Such window mechanism is a form of flow control because it enables control of the amount of data being sent from the sender to the receiver. The start of the window corresponds to the first data byte of the packet that has successfully reached the receiver. The end of the window corresponds to the first data byte of the packet that the sender can send without overflowing the receiver buffer. In other words, the window size is expressed in terms of the number of bytes. The window size quantifies the maximum number of unacknowledged bytes that the sender can transmit. As the TCP segments are acknowledged to have been received, the TCP sender slides the window forward so that more TCP segments can be sent. The initial window size is determined by the receiver during the TCP connection establishment. The ACK from the receiver indicates the window size that the receiver can handle. If the sender TCP receives no ACK from the TCP of the receiving process within a certain timeout interval, it retransmits the unacknowledged data bytes.

In the rest of this section on TCP, we will first summarize TCP's main characteristics. The structure of a TCP segment will be specified next. We will then take a closer look at the establishment and teardown of a TCP connection. Finally, two important TCP mechanisms, congestion control being the first and fast retransmission and recovery being the second, will be explained.

TCP Characteristics. TCP is connection-oriented, meaning that two endpoints have a specific connection between them, and TCP port numbers identify this connection. TCP provides reliable communications by retransmitting a TCP segment if necessary. TCP implements flow control so that the receiver is not overwhelmed by an onslaught of TCP segments. TCP performs congestion control by adjusting the

[68] The technical documentation provides additional details [119].

amount of data it sends into the network, thereby alleviating network congestion. TCP provides an end-to-end communications service to the endpoints.

Structure of a TCP Segment. Figure 4.3 highlights selected fields that appear in a TCP segment.[69] The sixteen-bit port numbers specified in the TCP segment identify the associated application, with the source port being the segment sender's port and the destination port being the segment receiver's port. The thirty-two–bit sequence number specifies the sequence number of the first data byte of the segment. When the synchronize (SYN) control bit is set, the sequence number is referred to as the initial sequence number and is typically a random number. The use of a random number rather than a fixed number, such as 0, helps distinguish two different TCP connections (e.g., old and new) between the same hosts and the same port numbers. The first data byte in such a case is *initial sequence number* + 1. When the ACK control bit is set, the thirty-two–bit acknowledgment number specifies the next sequence number that the receiver is expecting. By setting the final (FIN) control bit, the sender tells the receiver that it has no more data to send and that it wants to close the TCP connection.

The sixteen-bit Window field specifies the total number of data bytes that the segment sender will gladly accept, and the first data byte is indicated by the acknowledgment number. The sixteen-bit Checksum field helps detect a potential error in the received segment and is calculated using a *pseudo-IP header* (which contains fields for information such as source and destination IP addresses, protocol, and TCP length) and the TCP segment.[70]

An optional Options field extends the flexibility of TCP. We'll look at two options, the maximum segment size and selective acknowledgment (SACK). As the name suggests, the maximum segment size is indeed the maximum size of the TCP segment that the segment sender can receive. This option is used during the establishment of a TCP connection when the SYN control bit is set. The SACK option enables the data receiver to indicate to the data sender which blocks of TCP segments have been received successfully. In this case, the data sender will send only the missing TCP segments.

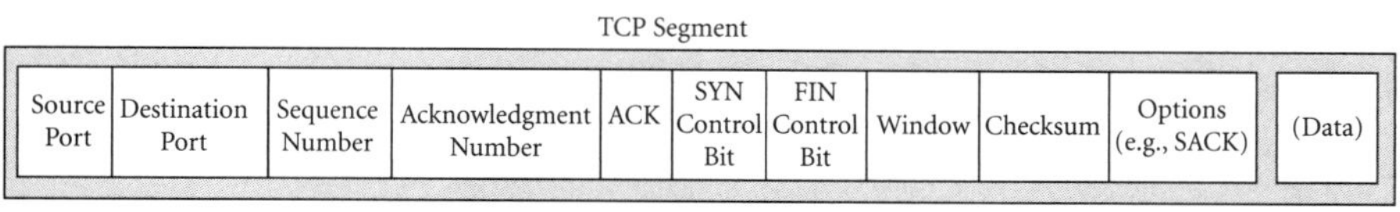

Figure 4.3: A Selective View of a TCP Segment

[69]Other sources offer the format of a full-fledged TCP segment [117, 119].

[70]For the purpose of the checksum calculation, the Checksum field of the TCP segment is assumed to be 0.

A block may encompass one or more TCP segments, and information on up to four blocks can be included in a given TCP segment. *The minimum length required by a TCP header (i.e., excluding the data portion and options) is forty bytes.*

TCP Connection Establishment and Teardown. Figure 4.4 summarizes how a TCP connection is established using a famous three-way handshake. Host A initiates the establishment of a TCP connection by sending a TCP segment that includes an SYN control bit set to 1 and the sequence number set to some initial value, such as SN_A. Host B responds with a TCP segment that also sets the SYN control bit to 1 and includes an initial sequence number SN_B. Host B also informs Host A that it expects the next TCP segment containing data from Host A to start with the sequence number $(SN_A + 1)$. Finally, Host A sends a TCP segment that includes the sequence number of the segment and the acknowledgment number indicating the expected sequence number for the future data segment from Host B.

We talked about TCP connection establishment and data transfer. Now let's summarize how a TCP connection is closed. A host wishing to close the TCP connection sends a TCP segment with the FIN control bit set to 1 to indicate that it wants to send no more data.

Do You Recall... Is the TCP connection is simplex or duplex? The footnote[71] gives the answer.

Transmission of such a TCP segment closes the TCP connection in one direction, i.e., from sender of this segment to the receiver. The other end of the communication link sends any remaining data and then sends a TCP segment with FIN = 1. The connection is now considered closed since FIN = 1 for both directions of the communication link.

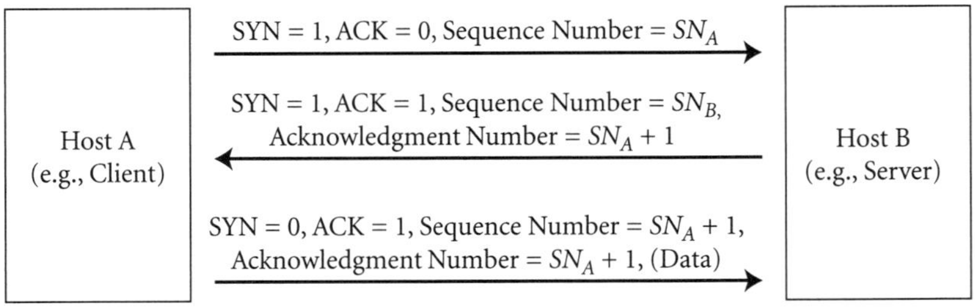

Figure 4.4: TCP Connection Establishment via a Three-Way Handshake

[71] It is duplex.

Now that we know how a TCP connection is established, let's discuss some features of TCP using the data transfer scenario depicted in Figure 4.5. We are sending data from a TCP sender to a TCP receiver (e.g., uploading a file from a client to a server using FTP as an application layer protocol). Such basic operation is enhanced by several features that will be discussed later in this section.

Assume that the maximum segment size is 1500 bytes and the window is 4550 bytes. In other words, the maximum number of bytes that a segment will contain is 1500, and the buffer at the receiver can receive a total of 4550 bytes at a given time instant. From the sender's perspective, the left edge of the initial window is byte number 1000 and the right edge of the window is byte number 5549, assuming the initial

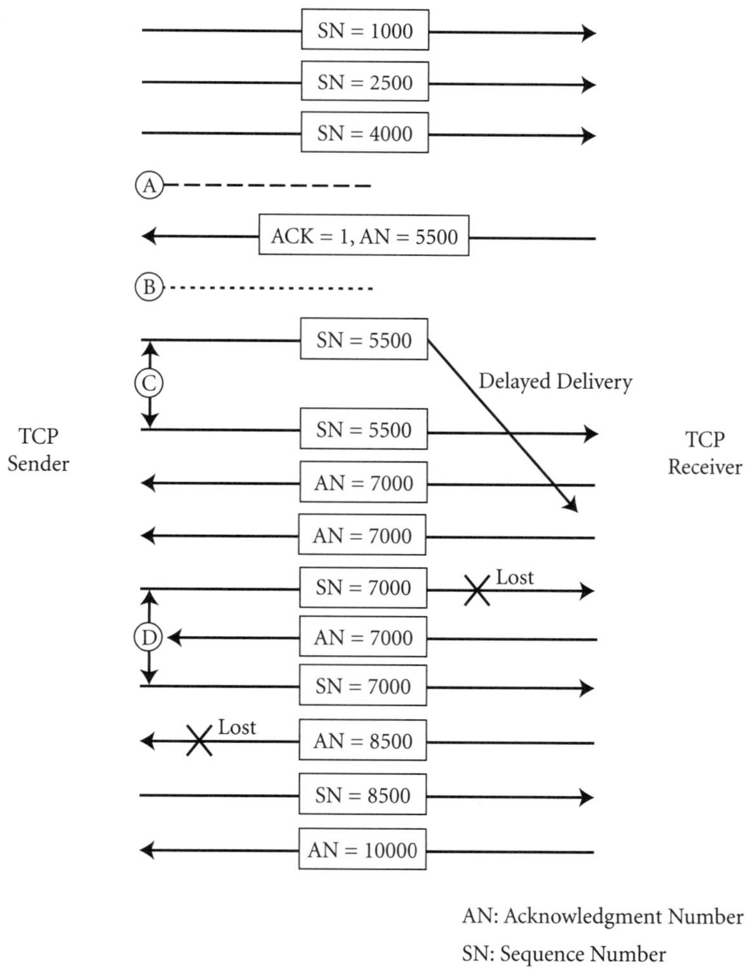

Figure 4.5: Basic Data Transfer using TCP

sequence number is 999. The TCP sender in Figure 4.5 attempts to maximize the throughput by sending three TCP segments in sequence with the sequence numbers being 1000, 2500, and 4000. The first segment contains the data bytes numbered from 1000 to 2499. The second TCP segment includes the data bytes 2500 through 3999. The third segment has the data bytes 4000 to 5549. Then the TCP sender can send no more segments because it has reached the window size advertized by the receiver (event A in Figure 4.5).[72] The TCP sender receives an ACK from the receiver with the acknowledgment number 5500. The TCP sender knows that all three segments have been successfully received.

Once an acknowledgment to a segment is received, the sender can slide the window and send more data. In our example, the sender slides the window such that the new window starts at byte 5500 and ends at byte 10000 (Event B in Figure 4.5). The TCP sender goes ahead and transmits a TCP segment with the sequence number 5500; however, the path from sender to receiver is congested, delaying the delivery of the TCP segment. When the sender gets no acknowledgment from the receiver within a certain time interval,[73] a timeout occurs for the TCP segment numbered 5500 (event C in Figure 4.5). The TCP sender retransmits the segment numbered 5500, and the receiver responds with an ACK with the acknowledgment number 7000. The delayed TCP segment with sequence number 5500 now arrives, and the receiver sends a duplicate ACK with the acknowledgment number 7000. The TCP sender now sends a new segment with sequence number 7000. This TCP segment with the sequence number 7000 gets lost somewhere in the Internet,[74] and a timeout occurs for this segment (event D in Figure 4.5). The retransmission of this TCP segment is received successfully by the receiver, and the receiver sends an ACK with the acknowledgment number 8500. However, this ACK gets lost along the way. Since the window[75] is still open, the sender sends another TCP segment with the sequence number 8500. The receiver responds with an ACK that includes the acknowledgment number 10000 as it has received all the data bytes up to byte 9999.

Let's summarize key points from Figure 4.5. The TCP sender ensures that it does not overflow the receiver buffer by keeping track of the window. The window size is the maximum number of unacknowledged bytes that the sender can transmit. The receiver acknowledges the reception of each TCP segment by sending an ACK with the acknowledgment number that specifies the next expected data byte. Reception of an ACK corresponding to the left edge of the window results in the sliding of the Window because all the consecutive bytes so far have been acknowledged by the receiver. When the sender receives no ACK corresponding to a TCP segment, it retransmits the TCP segment.

[72]Events A, B, C, and D represent simply our own nomenclature for the purpose of illustration and do not appear in the TCP documentation!

[73]The sender dynamically determines such a timeout value based on the estimated round-trip time delay estimated using ACKs received in response to the TCP segments.

[74]The Internet is a big place, and the packet may be lost! A busy router may discard or drop a packet.

[75]The window at this time ranges from byte 7000 to byte 11549.

Congestion Control. Even though the window size is a form of flow control to prevent buffer overflow at the receiver, we still need a flow control to minimize congestion in the network. Remember that the network is a common resource for all, and we need to use it wisely with respect for others!

Figure 4.6 illustrates the operation of a congestion control mechanism for a TCP version called TCP Reno [118, 120, 121]. A new window called the *congestion window*, cwnd, is defined for congestion control. The sender maintains cwnd and sends only cwnd bytes during a given transmission round. In practice, the sender considers the constraints due to cwnd and due to the receive window to determine the number of bytes to be sent. In other words, the sender uses the minimum of cwnd and the receiver window as the maximum amount of unacknowledged bytes it can send. In our example in Figure 4.6, we assume that the receive window is so large that cwnd effectively dictates the amount of data sent by the sender. This will isolate the effect of the congestion control mechanism and enable us to discuss the congestion control aspect of the TCP.

A typical TCP sender does not inject multiple TCP segments to fully fill up the window at the beginning of the connection. Rather, TCP carries out the *slow-start* mechanism at the beginning of the connection. The slow-start mechanism involves the initial transmission of just one segment and exponential growth of the number of transmitted segments thereafter in response to received ACKs. The sender continues to operate in the slow-start mode as long as cwnd is less than the slow-start threshold, ssthresh. The initial value of ssthresh upon connection establishment is set to a large value (i.e., 65535 bytes). In Figure 4.6, the sender starts with cwnd of one segment and, hence, sends just one segment in the first transmission round. After it gets an ACK, the sender doubles the number of segments to $1 \times 2 = 2$. Reception of additional ACKs results in the doubling of the transmission to four, then eight, then sixteen segments.

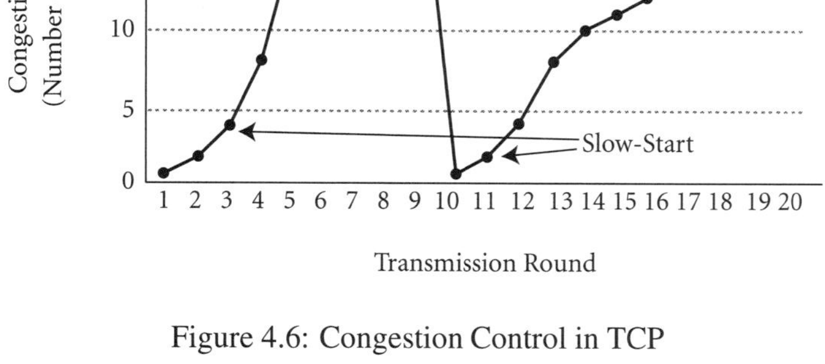

Figure 4.6: Congestion Control in TCP

Once the sender transmits sixteen segments, it receives a duplicate ACK, which is a potential indicator of network congestion. The sender becomes conservative, exits the slow-start phase, and enters the congestion avoidance phase in which the sender linearly (rather than exponentially) increases cwnd. In Figure 4.6, the sender increases cwnd linearly from sixteen to twenty during the transmission rounds five to nine. After cwnd reaches twenty segments, the sender experiences a timeout for a transmitted segment, which is an indicator of network congestion. The sender takes an aggressive corrective action by setting ssthresh to half the value of cwnd, reducing cwnd to 1, and entering the slow-start phase. In our example, ssthresh is set to $cwnd/2 = 20/2 = 10$. The slow-start phase remains in effect until transmission round thirteen since cwnd stays below the ssthresh of 10. Beginning with transmission round fourteen, at which cwnd is set to ssthresh, the sender again enters the congestion avoidance phase.

As seen from Figure 4.6, TCP responds differently to duplicate ACKs and timeouts. TCP aggressively reduces cwnd in response to a timeout and conservatively reduces cwnd in response to a duplicate ACK. The rationale behind such behavior is based on the nature of the network congestion. A timeout results after a sufficiently long time period has elapsed for the corresponding ACK to arrive, and hence a timeout is a strong indicator of severe network congestion. Reception of duplicate ACKs means that at least some TCP segments are getting through the network, indicating that network congestion may be minor. Figure 4.6 shows an exponential growth in cwnd during the slow-start, but the actual cwnd growth may be slower because a TCP receiver may implement a delayed ACK, in which it does not immediately send an ACK but waits for another TCP segment to arrive. If it gets another in-sequence TCP segment within a certain time period, it sends just a single ACK acknowledging the reception of two TCP segments. However, if the receiver gets no other TCP segment soon or if it gets an out-of-sequence TCP segment, it does send an ACK in response to the earlier TCP segment.

Do You Recall... Do you recall when a TCP sender retransmits a TCP segment? See the footnote[76] to validate your answer.

Fast Retransmission and Fast Recovery. Fast retransmit does not wait for the TCP timeout; it retransmits a TCP segment when three duplicate ACKs, i.e., a total of four ACKs with the same acknowledgment number, are received for a given TCP segment. Such fast retransmission reduces the overall delay and improves throughput. TCP waits for a small number of duplicate ACKs before fast retransmission because a duplicate ACK may be caused by a lost TCP segment or the out-of-sequence arrival of the TCP segments at the receiver. If the duplicate ACKs are indeed due to re-ordering of segments, a couple of duplicate ACKs should provide adequate time for the receiver to re-order the segments. However, if several (e.g., three) consecutive duplicate ACKs are received by the sender, the missing TCP segment has most

[76] When a suitable timer at the sender expires, the sender transmits the relevant segment.

likely been lost. In that case, TCP would then retransmit a (potentially) missing TCP segment without waiting for the timeout associated with such a segment. Once fast retransmission occurs, TCP behaves as if a timeout had occurred. Hence, TCP may initiate the slow-start mechanism, setting ssthresh to cwnd/2 and making cwnd one maximum segment size.

The idea behind the fast recovery mechanism is to avoid the slow-start mechanism by properly adjusting cwnd. In fast recovery, TCP stays in the congestion avoidance phase, sets ssthresh to cwnd/2 (but not less than two segments), and sets cwnd to ssthresh plus three times the segment size. Note that such a cwnd setting reflects that three TCP segments (corresponding to three duplicate ACKs) beyond the missing segment are not contributing to network congestion because they have already entered the receiver buffer. Additionally, the sender increments cwnd by one segment for every additional duplicate ACK received. The fast recovery mechanism improves throughput, especially when the network congestion is moderate [117].

4.2.1.2 Stream Control Transmission Protocol Overview SCTP is suitable for efficient and reliable delivery of signaling messages. Although SCTP exploits several features of TCP, it overcomes certain shortcomings of TCP (e.g., the inability of the window to slide due to unacknowledged bytes of an early segment and message splitting across TCP segments). We'll discuss the basics of SCTP here.[77] Figure 4.7 summarizes key features of SCTP.

Like TCP, SCTP supports full-duplex transmission. Furthermore, just like TCP, SCTP is connection-oriented whereby two endpoints of the communication link establish a logical link or *association*[78] between one another and preserve it for the duration of the data transfer. SCTP provides reliability by delivering the packets

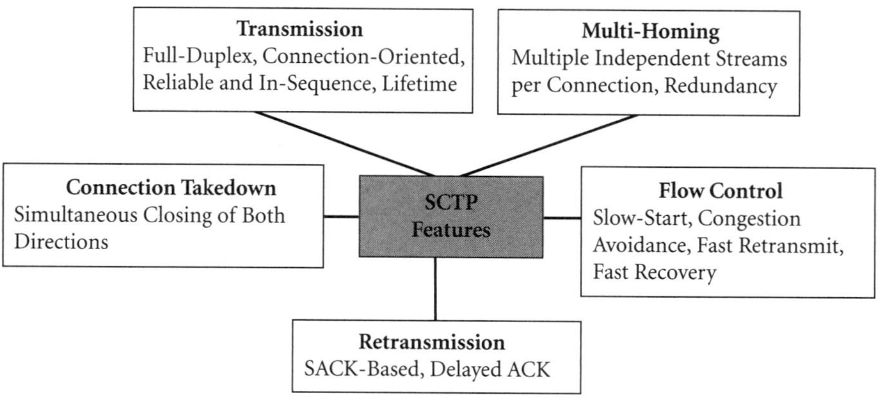

Figure 4.7: Salient Features of SCTP

[77] See the technical documentation for SCTP details [122, 123].

[78] A *connection* in SCTP is formally termed *association*; we'll use these terms interchangeably.

without error and in sequence. The basic TCP transmission is byte-oriented with no specific framing of bytes. On the contrary, SCTP transmission is message-oriented with framing of individual messages. SCTP also allows the application layer to specify a lifetime for data to be transmitted. In such a case, if data has not been transmitted within the specified lifetime, it is discarded. The lifetime concept makes the delivery of time-sensitive signaling messages more efficient and relevant.

An SCTP connection supports multiple streams with independent in-sequence delivery of stream packets. Such independent control of individual streams ensures that the packet delivery for different streams is impervious to problems with just one stream. In the case of a message loss for a stream, the receiver delivers messages to the unaffected streams and buffers the messages for the affected stream. The buffered messages will be delivered in sequence after a successful retransmission. SCTP is quite attractive for telephony signaling in which one logical connection will be able to carry signaling messages for multiple users but message loss for one user does not affect delivery of messages for other users.

Multi-homing is the ability of an endpoint to support multiple IP addresses for a given SCTP connection and enables the sharing of a single connection among multiple hosts. Multi-homing can also be exploited to provide redundancy in the presence of network failures [117]. The redundancy aspect of SCTP involves maintenance of a counter that tracks the number of unsuccessful retransmissions to a specific destination address. When the counter exceeds a threshold, the SCTP starts using an alternate IP address to send data chunks [123]. The redundancy mechanism also involves periodic transmission of *heartbeat chunks*[79] to non-primary alternate addresses and maintenance of a counter that tracks the number of missing ACKs to the heartbeats. When such a counter exceeds a threshold, the associated destination IP address is considered inactive. The heartbeats are periodically sent to even an inactive address, and the reception of an ACK to the heartbeat makes the inactive address an active address. The list of IP addresses is exchanged between the endpoints during the initiation of the SCTP association. Just one port number is used for all IP addresses at an endpoint.

Just like TCP, SCTP also implements flow control to address network congestion. SCTP has no need to re-invent the wheel; it reuses the TCP mechanisms of slow-start, congestion avoidance, and fast recovery and fast retransmit. The main difference in TCP and SCTP implementations of these mechanisms is the need for SCTP endpoints to manage the conversion between the bytes and the transmission sequence numbers. A single congestion control mechanism applies to all streams within an SCTP connection, thereby using less overhead than would be required if individual streams had independent congestion control.

The SCTP endpoints exchange information in the form of data chunks or messages. SACK is optional in the basic TCP but is mandatory for SCTP. As with TCP, a timeout or the reception of a SACK results in retransmission. Upon receiving data chunks, the receiver responds with SACK chunks using a delayed ACK approach with one SACK

[79] A heartbeat chunk is a low-overhead message that ensures that the destination address is alive and well.

for every other chunk received. A SACK is immediately sent if a gap is detected. The SACK chunks specify both the cumulative and non-cumulative transmission sequence numbers. The sender can then retransmit missing chunks. The sender retransmits the data chunk upon receiving the fourth consecutive SACK specifying the missing chunk. Such an approach reduces unnecessary retransmissions when the receiver acquires chunks that are out of sequence due to normal causes, i.e., those unrelated to congestion. SCTP enhances the overall TCP retransmission mechanism by updating the retransmission timeout to reflect the prevailing round-trip delay, resulting in an exponential back-off as the rate of message loss increases.

Closing an SCTP connection, called *takedown*, involves a three-message exchange, and this process is different from the TCP connection close. These messages are SHUTDOWN, SHUTDOWN ACK, and SHUTDOWN COMPLETE. TCP can have one direction of the connection closed and the other direction open, but the SCTP connection close allows no new data transfer once the takedown procedure has begun.

Figure 4.8 shows selected fields of an SCTP packet. The SCTP packet consists of a common header and one or more chunks. A chunk may have SCTP control information or user data. These chunks could be different messages. The source and destination port numbers in the header enable SCTP to support different SCTP associations for a given pair of the (source, destination) IP addresses. A thirty-two–bit Verification Tag protects against insertion of a malicious packet into the SCTP association. A thirty-two–bit Checksum field helps detect errors in the SCTP packet. Within each chunk, fields such as Chunk Type, Chunk Flags, Chunk Length, and Chunk Value are included. The Chunk Type field indicates if it is a control chunk or data chunk. For example, Chunk Type 0 corresponds to payload data, 1 corresponds to initiation of the SCTP connection, and 3 corresponds to SACK. The Chunk Flags field facilitates specific tasks, such as segmentation at the sender and reassembly at the receiver as well as supports retransmission via inclusion of Transmission Sequence Number, Stream ID, Stream Sequence Number, and a Payload Protocol Identifier fields. A port number can be used to implicitly identify the application layer protocol, but the Payload Protocol Identifier field is still present for future flexibility. The application layer running on top of SCTP controls the bundling of multiple data and control chunks into a single message, increasing the efficiency of transport. The total length of a chunk, including the Chunk Type, Chunk Flags, Chunk Length, and Chunk Value fields is a multiple of four bytes. The Chunk Value field contains the

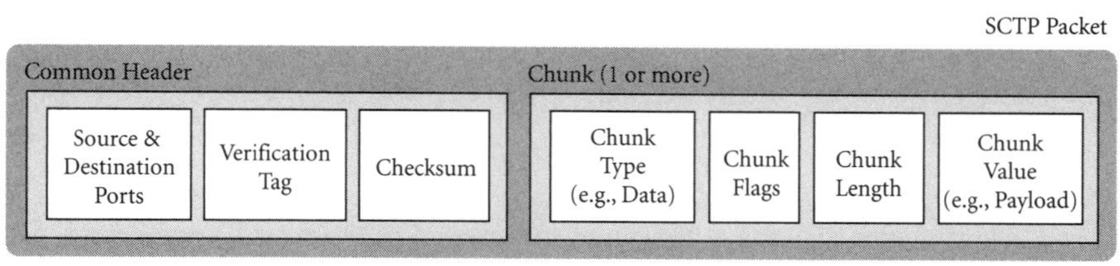

Figure 4.8: Inside an SCTP Packet: A Selected View

actual information to be transferred (e.g., an SCTP control message, such as SACK, and the user data).

4.2.1.3 User Datagram Protocol Overview Let's take a brief look at another popular protocol, UDP. [80] In general, UDP is preferred over TCP when only a small amount of data is to be transferred (e.g., signaling messages) or when some loss of data is tolerable (e.g., multimedia streaming) or for applications that are intolerant to delay. If reliability is desired, the application layer running on top of UDP would have to provide that reliability since UDP itself offers none.

Figure 4.9 shows the format of a UDP datagram, which has an eight-byte header followed by data octets (i.e., bytes). The Source Port specifies the port number of the sending process in the source host, and the Destination Port indicates the port number of the destination host to which the datagram is heading. (Perhaps we should replace "is" with "should be" since UDP offers no guarantee of delivery!) The Length field specifies the total number of octets required by the entire datagram, which includes both header and actual data. The sixteen-bit Checksum field helps detect an error in the received datagram and is calculated using the same procedure TCP uses. Checksum is calculated over the pseudo-IP header, the UDP header, and the UDP data.

Do You Recall. . . Considering our TCP discussion, what could be in the pseudo-IP header? See the footnote[81] for the answer.

As seen from Figure 4.9, the beauty of UDP is its simplicity. When one is conscious about overhead and when an occasional loss of the datagram is tolerable, UDP offers a better option than TCP. The TCP header is at least twenty bytes long, and UDP header consumes just eight bytes. VoIP in a cellular system exploits UDP to achieve

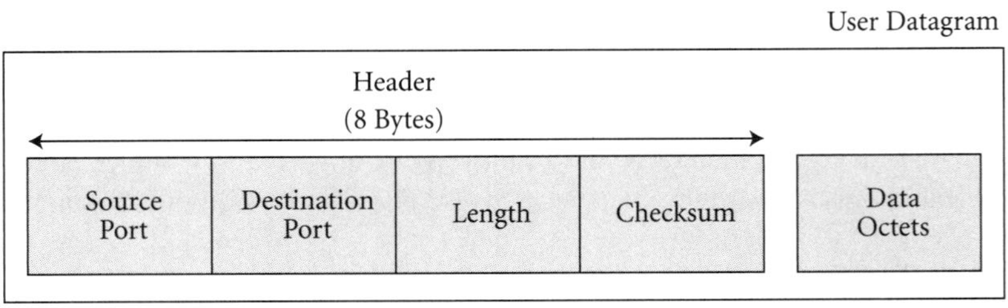

Figure 4.9: Inside a UDP User Datagram

[80]The interested reader is referred to the documentation [124] and Parziale et al.'s tutorial [117] for additional information on UDP. Please do not expect too many details even in the documentation. UDP follows the philosophy of minimalism; the documentation describing UDP is just three pages long, and, of those three pages, the references occupy almost half a page!

[81]Example contents of the pseudo-IP header are the source IP address, the destination IP address, and the protocol field.

efficiency over the air interface and inside the infrastructure network. DNS is yet another application that uses UDP.

4.2.2 Overview of the Internetwork Layer (Layer 3) Protocols

IP is the most widely used protocol at layer 3. Examples of layer 3 (i.e., internetworking layer) protocols that help IP are ICMP, Address Resolution Protocol (ARP), and Dynamic Host Configuration Protocol (DHCP). IP is the heart of the Internet we know today. For example, when you browse the web, your device exchanges information in the form of IP packets with a web server.

Figure 4.10 highlights the main roles IP plays that make it quite popular. When we want to send information from a source host to a destination host via the Internet, we can use ATM at layer 2 or IP at layer 3 to transport the information. IP as a transport mechanism results in automatic discovery of the path or route from the source to the destination. Additionally, IP enables the network to adapt to the prevailing network conditions and to determine alternative routes from the source host to the destination host if the packets cannot travel via the initial route. Furthermore, a large chunk of the configuration is automatic.

For example, the hosts can obtain their IP addresses automatically (rather than via manual configuration) through some mechanism such as DHCP and Mobile IP. Just like the post office delivers mail based on the destination address, IP delivers the IP packet to a destination host using the IP address of that host. IP provides *scalability* to the network whereby network usage growth can be easily addressed by adding a few nodes and doing some relatively simple configuration (e.g., increasing the size of the pool of IP addresses).

Widespread use of IP-based nodes such as routers and the standardized nature of IP components reduce deployment cost of IP networks. A service provider can buy

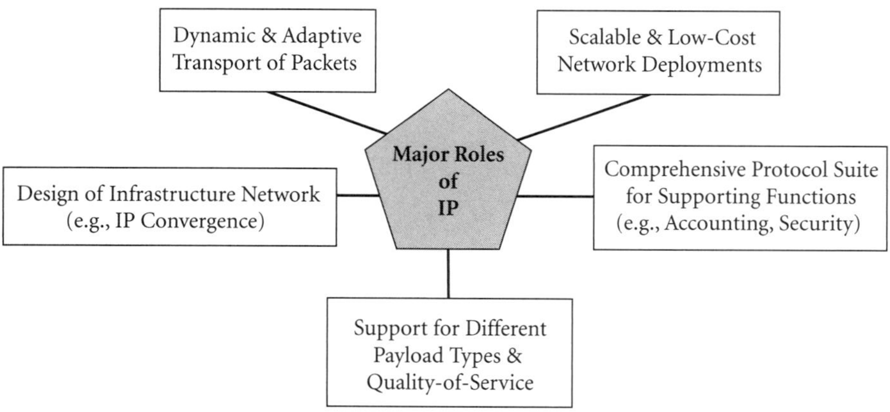

Figure 4.10: The Many Roles of IP: A Versatile Actor!

IP components from different vendors and still maintain smooth interaction among them. Numerous protocols defined by the Internet Engineering Task Force support IP, making it easier for a service operator to manage operational aspects, such as accounting and security. For example, RADIUS (Remote Authentication Dial-In User Service) can be used to implement accounting and IPsec (IP Security) protocols can be used to provide security. IP supports the transport of different payload types, such as web page, email, voice, and video using suitable quality-of-service (QoS) mechanisms, such as DiffServ (Differentiated Services), where necessary.

IP encourages innovation in services. For example, in the next-generation 4G cellular technologies, IP Multimedia Subsystem is emerging as a service delivery platform that facilitates the introduction of new IP-based services. IP plays an important role in defining the architecture of the next-generation cellular infrastructure. IP, through IP Convergence or Fixed Mobile Convergence, facilitates the design of a common unified core network that can interface with any type of access network (wireline or wireless) to provide the user with a seamless communication experience across a variety of access technologies.

Something to Think About... IP as a transport mechanism helps carry an IP packet from the source to the destination. IP provides a best-effort delivery service and offers no guarantee that the packet will indeed reach the destination. However, as we are all aware, IP does indeed work and we are usually able to use applications such as web-browsing and email. Dr. Tripathi's colleague, Ramki Rajagopalan, often says, "IP does not make any promises, but it delivers. On the other hand, politicians make a lot of promises but..." We'll let you complete the statement! On a side note, if the reliable delivery of packets is indeed required, an upper-layer protocol such as TCP comes to the rescue.

4.2.2.1 IP Addresses of All Kinds! A host is identified by an IP address at the IP layer. There are two different types of IP addresses: *static* and *dynamic*. When the IP address allocated to a device remains the same and cannot be allocated to this device or another device, it is called a *static* IP address. In contrast, if the IP address is allocated to a device for only a certain time and can then be re-allocated to another device, it is called a *dynamic* IP address. In commercial cellular networks, dynamic IP addresses are more common. The cellular network temporarily allocates a dynamic IP address to the mobile device for a limited time period as part of the call setup. The IP address may be an IPv4 or an IPv6 address, with IPv4 having a quasi-monopoly in commercial cellular networks as of early 2009. An IPv4 address consists of thirty-two bits, and an IPv6 address consists of 128 bits. We will primarily focus on IPv4 here but will certainly give you a glimpse of IPv6 in Section 4.7.

The IP address may be private or public. A private IP address can be used to communicate only with other hosts within a given private network (e.g., within some company or university). A public IP address enables a host to communicate with hosts outside the private network (e.g., public Internet). For example, when you want to read news from www.cnn.com from your workplace or university, you need a public

IP address because Internet routers understand only public (not private) IP addresses. Public IP addresses are globally unique. Are you ready for a simple exercise with IP? Go to your operating system's command prompt. If you are using Windows, type `ipconfig`. If you are using a Unix-like system, type `ifconfig`. You may see something like Figure 4.11. Since the author's laptop has both LAN and wireless LAN connectivity, two IP addresses, 10.25.208.94 and 10.25.208.93, are allocated to the laptop. Private IP addresses are in one of the following address ranges: $10.x.x.x$; $192.168.x.x$; 172.16.0.0 to 172.31.255.255. All other IP addresses are public.

Let's take a closer look at the format of the IP address. The IP address has two parts: the network part and the host part. For example, in the IP address 128.10.20.30, 128.10 denotes the network number and 20.30 is the host number. An IPv4 address contains thirty-two bits or four bytes. The integers 128, 10, 20, and 30 in the IP address are decimal representation of the set of eight bits. The decimal value for each integer thus ranges from 0 to $2^8 - 1 = 255$. The network number is administered by one of five regional internet registries, and the host number is controlled within the entity that owns the network number. The five regional internet registries are the American Registry for Internet Numbers, the Reseaux IP Europeans, the Asia Pacific Network Information Centre, Regional Registry for Internet Number Resources for Africa, and Internet Address Registry for Latin America and the Carribean. Figure 4.12 specifies structures of the IP addresses for various classes.

A Class A address allows seven settable bits for the network number and twenty-four settable bits for the host number. This yields $2^7 - 2 = 126$ Class A networks, and each network is capable of supporting $2^{24} - 2 = 16,777,214$ hosts, resulting in $126 * 16777214 = 2,113,928,964$ (i.e., more than two billion) IP addresses. A Class B address allows fourteen settable bits for the network number and sixteen settable bits

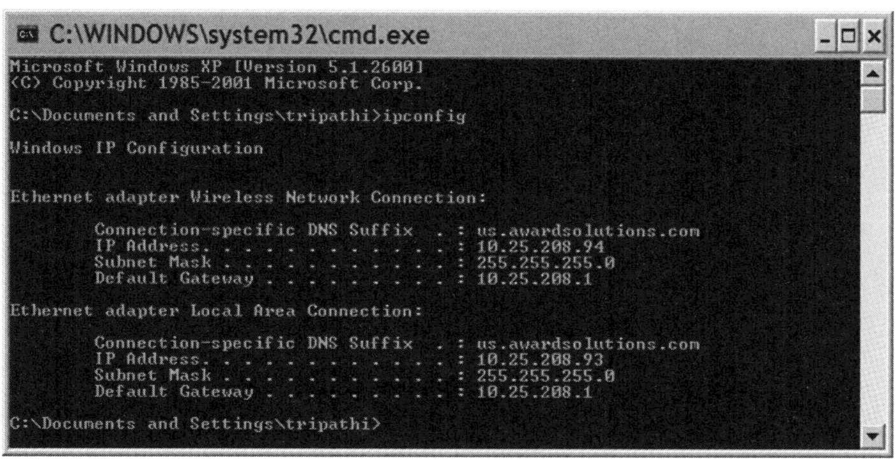

Figure 4.11: What is Your IP Address Now?

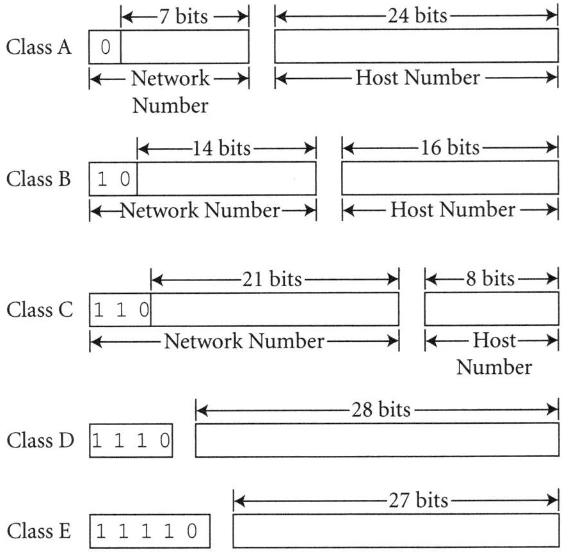

Figure 4.12: Classes of IP Addresses

for the host number. A total of $2^{14} - 2 = 16,382$ Class B networks can be defined, each with $2^{16} - 2 = 65,534$ hosts, resulting in $16,382 * 65,534 = 1,073,577,988$ hosts (i.e., more than one billion hosts). A Class C address involves twenty-one settable bits for the network number and eight settable bits for the host number, leading to 2,097,150 networks, 254 hosts per network, and a total of 532,676,100 (i.e., over half a billion) hosts. Class A, Class B, and Class C were intended for large enterprises, medium enterprises, and small businesses, respectively. A Class D address is reserved for multicasting (i.e., sending an IP packet to a set of hosts). Finally, a Class E address is reserved for future and experimental use.

All IP addresses must follow a set of rules.

- An address with all 0s as the host number portion indicates "this host." When a router on the network is acting as a gateway to the outside world and receives an IP packet, it compares the network number contained in the IP packet to the IP address of the network (with 0s as the host number) to determine if the IP packet belongs to a host on the network.

- All 1s as the host number, such as the last sixteen bits in the IP address 128.10.255.255, imply a broadcast IP address, and all the hosts of the relevant network, such as the network number 128.10, receive such an IP packet.

- All 1s as the settable bits in the network number, such as 255.255, corresponds to all the networks that use a sixteen-bit network number. Such a setting is used to create a broadcast address.

Assume that a company, CellNet, has received a Class B network number 128.10 from a regional internet registry. CellNet can then allocate the host numbers 0.1 through 255.254. The combinations for all 0s and all 1s as the host number are not allowed as mentioned previously. The full IP address for a host of CellNet would be in the range from 128.10.0.1 to 128.10.255.254.

Based on our discussion of classes of IP addresses, we can say that about 3.5 billion general-purpose (i.e., Class A, B, and C) unique IP addresses are available in IPv4. That may sound a lot of IP addresses, but, actually, we are in danger of running out of IP addresses. Such a problem is called *IP address exhaustion* [117]. One easy way to solve the problem is to define a longer IP address. IPv6 uses that approach and defines a 128-bit IP address. One simple calculation, 2^{128}, shows the total number of IP addresses to be $3.40 * 10^{38}$. There are about 50,000 IPv6 addresses per square meter of land on Earth [117]! Another way is to use private IP addresses within a private network (e.g., within a company or university). Three ranges of IP addresses are reserved for private IP addresses [117]: 10.0.0.0, 172.16.0.0 through 172.31.0.0, and 192.168.0.0 through 192.168.255.0.

A Question for You... Which class would the address 10.0.0.0 belong to? See the footnote[82] for the answer.

Imagine an organization is using private IP addresses. In such a case, we can allocate the IP addresses 192.168.0.1 to 192.168.255.254 to the hosts within the network. These hosts can then talk to one another using the private IP addresses. Other organizations can also use identical IP addresses within their private networks as well, reducing the need for public IP addresses that are globally unique. A private IP address is like a phone extension number that you can dial on your desk phone to talk to your colleague, who is also using a desk phone. In this case, we need not have separate dedicated phone numbers for all individuals in the company. Now, if someone outside the company wants to call an individual on the desk phone, we need a regular phone number. Similarly, if a host with a private IP network wants to communicate with a host outside the private network, such as a web server, a public IP address is required. *Note that the routers in the Internet cannot route IP packets with private IP addresses.* The mapping between the private (or internal) and public (or external) IP addresses is carried out by Network Address Translation (NAT). Figure 4.13 illustrates how a host with a private IP address can exchange IP packets with the outside world using traditional NAT and Network Address Port Translation (NAPT).[83]

Traditional NAT allocates an external IP address from its pool of addresses to a host only when the host wants to communicate with the external network. In these cases, a limited number of public IP addresses are adequate. In Figure 4.13, when a host with the IP address 192.168.0.1 sends an IP packet to the web server with

[82] 10.0.0.0 is a Class A address because the first bit in the network number is 0.

[83] See the documentation for details on NAT and NAPT [125, 126].

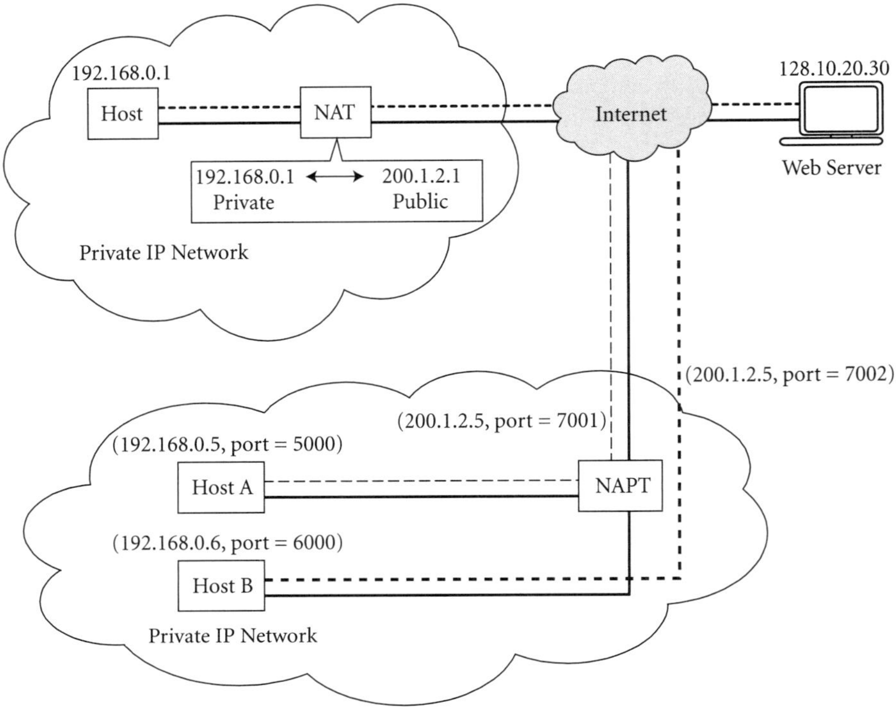

Figure 4.13: Use of Private IP Addresses

the public destination IP address 128.10.20.30, NAT translates the source IP address 192.168.0.1 to 200.1.2.1, which is one of the public IP addresses in its pool. When the external host replies with an IP packet that has the destination IP address 200.1.2.1, the IP packet is sent to NAT, which then replaces the public destination IP address with the internal IP address 192.168.0.1 and forwards the packet toward the initial host. In addition to the substitution of the IP addresses, NAT must also complete other tasks, such as adjusting some fields in the IP packet, such as the Checksum field in the IP header [117]. When the private–public association is unused for a configurable time period (e.g., fifteen minutes), the public IP address is returned to the NAT pool [117]. NAT can be used to control which hosts can use NAT and thus communicate with the outside world. If only a few hosts simultaneously communicate with the outside world, basic NAT may be adequate. However, when numerous hosts need to simultaneously communicate with an external network, NAPT becomes quite attractive.

NAPT maps a private IP address and a private port to the assigned public IP address and the assigned port. In Figure 4.13, NAPT translates the pair (Private IP Address 192.168.0.5, private port 5000) into the pair (Public IP Address 200.1.2.5, assigned port 7001) and the pair (Private IP Address 192.168.0.6, private port 6000) into the pair (Public IP Address 200.1.2.5, assigned port 7002). Like NAT, NAPT must also process some fields of an IP packet (e.g., the Checksum field of the IP header). NAT and NAPT help alleviate the IP address exhaustion problem, but they have certain drawbacks. NAT and NAPT are computationally intensive because each IP packet requires NAT or NAPT lookup and modifications. Furthermore, for applications that have IP information buried in the application data,[84] the NAT and NAPT implementations are more complex than the simplified NAT and NAPT implementations that we just discussed. It is more difficult for the outside world to send an unsolicited packet to a host in the private network.

The allocation of a pool of IP addresses to an organization can be made more efficient when we are no longer constrained to the classes of IP addresses. Classless inter-domain routing (CIDR) indeed utilizes such approach. The CIDR defines the IP address with the structure $a.b.c.d/n$ where n is the number of most significant bits forming the network portion of the IP address. For example, 100.128.0.0/11 implies that the network is identified by the first eleven bits of the IP address 100.128.0.0, which are `0110 0100 100` with the first eight bits, `0110 0100`, corresponding to the decimal number `100` and the last three bits, `100`, representing three most significant bits of `1000 0000` corresponding to the decimal number 128. The remaining $32 - 11 = 21$ least significant bits are used to separate hosts within the network. A total of $2^{21} - 2 = 2,097,150$ hosts can be allocated distinct IP addresses within the network. The host IP address in such a network ranges from 100.128.0.1 to 100.159.255.254. Note that all `0`s and all `1`s in the host portion of the IP address are unavailable for allocation to a host. In addition to alleviating the IPv4 address exhaustion problem, CIDR also simplifies the routing tables for the routers as a single entry for the entire network suffices. The routers only consider the network portion of the IP address, the *network prefix*, to make the packet forwarding decision. We will discuss CIDR further in Section 4.2.2.3.

4.2.2.2 Inside an IP Header IP transfers information in the units of IP datagrams. An IP datagram includes an IP header and upper-layer data (e.g., TCP or UDP).[85] IP delivers datagrams[86] in a *best-effort* and connectionless fashion. *Best-effort* here means that IP packets may be lost, arrive out of sequence, or be duplicated [117]. IP was designed to keep the network simple and the end devices smarter and complex. The U.S. Department of Defense's design aimed for a network that would be operational even if parts of the country were destroyed. The Internet is indeed such a network, and IP is a critical component of the Internet. IP supports fragmentation and re-assembly of datagrams, with the maximum datagram size being 65,535 octets. All IP hosts are required to support a datagram size of at least 576

[84]SIP-based (where SIP represents Session Initiation Protocol) Voice-over-IP applications have messages that contain IP addresses.

[85]Details of IP are available in many other sources [117, 127–133].

[86]We will use the terms IP datagrams and IP packets interchangeably, recognizing that IP datagram is the formal name.

octets. IP provides best-effort delivery and includes no feedback or acknowledgment mechanism.

Figure 4.14 shows the fields of an IP datagram.[87] The minimum size of the IP header is twenty octets. The Version field indicates the version of the IP, such as 4 for IPv4. The IP Header Length field specifies the length of the header portion of the datagram.

The Type of Service octet includes three fields: Precedence, Type of Service, and Must Be Zero. The Type of Service octet indicates the QoS requested for the datagram. Early implementations largely ignored the Type of Service octet field, but today it is widely used and its definition has been updated [128].[88] The Precedence field indicates the requested priority level for the datagram (e.g., `000` meaning routine and `001` meaning priority). The Type of Service field specifies how the network should make tradeoffs between throughput, delay, reliability, and cost (e.g., `0000` meaning normal service and `1000` meaning minimize delay). The Must Be Zero field is reserved for experimentation and future use.

The Total Length field specifies the total length of the IP datagram. The Identification field facilitates fragmentation because it includes a number that is the same in each fragment of the fragmented datagram. The Flags field contains some control flags related to fragmentation (e.g., to allow or disallow fragmentation and to indicate the last fragment). The Fragment Offset field, as the name suggests, supports

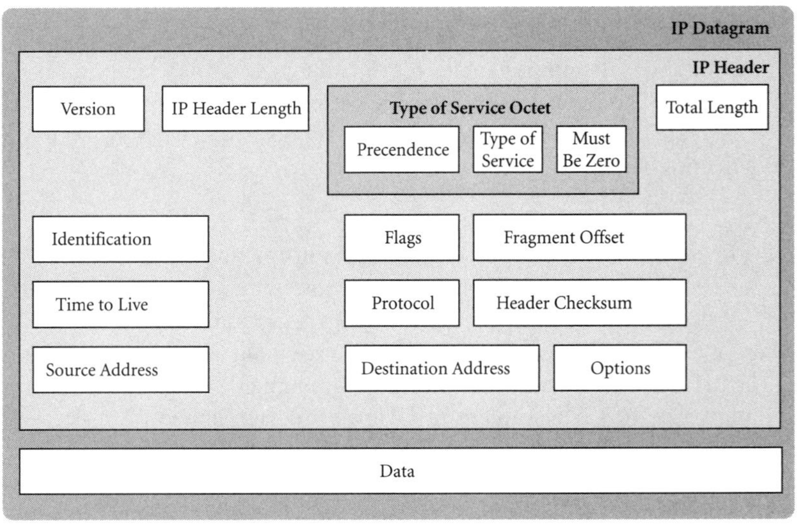

Figure 4.14: Format of an IP Datagram

[87] See the documentation for the full-fledged format of an IP datagram [127, 128].

[88] Newer implementations insert DSCP (Differentiated Services Code Point) in the Type of Service field.

fragmentation and specifies the number of sixty-four–bit data segments (i.e., excluding the header) contained in earlier fragments.

The Time to Live (TTL) field was intended to indicate the time (in seconds) that the IP datagram is allowed to travel. In practice, this field has become the number of remaining hops for which the IP datagram is allowed to travel. In other words, each router reduces the value of TTL by one, and, when it reaches zero, the IP datagram is discarded. The main idea behind the TTL field is to prevent the wandering of an IP datagram in a loop. The Protocol field specifies the upper-layer protocol to which IP must deliver the data portion of the IP datagram (e.g., the field value of 6 for TCP and 17 for UDP). The Header Checksum field is, of course, a checksum for the IP header. The Source Address and Destination Address fields are thirty-two–bit IP addresses of the source host and the destination host for the IP datagram. The Options field contains zero or more options. For example, an option called Record Route helps determine an end-to-end path from the sender to the receiver. If an option is indeed included in the header, the Padding field, containing 0s, ensures that the overall header length reaches the next thirty-two–bit boundary. Finally, the Data field contains upper-layer protocol information, such as a TCP segment or a UDP datagram.

4.2.2.3 Classless Inter-Domain Routing and IP Subnetting As mentioned in Section 4.2.2.1, addresses in IPv4 are quite precious, and we should use them wisely and efficiently. That's where CIDR enters the scene. CIDR helps us move away from the constraint of allocating class-based address blocks (Class A, B, or C). Recall that a Class B network supports 65,534 hosts and a Class C network supports 254 hosts. For most companies and institutions, a Class C network address is inadequate, and a Class B network is wasteful. We need something that is between Class B and Class C to efficiently use the available IP address space. Imagine a small company that has just ten hosts. Assigning a Class C network address to such a company will result in $254 - 10 = 244$ unused IP addresses. What a waste of precious IPv4 addresses! CIDR is that *something* we need between Class B and Class C. The Internet registries no longer use IP address classes to allocate IP address ranges; rather, they use CIDR.

Simply put, the IP address of a subnet or a subnetwork does not fall into any of the traditional classes of IP address but is able to support the required number of hosts efficiently. The main idea behind CIDR is to steal some bits from the host number of the IP address to construct a subnet. Figure 4.15 illustrates the concept of subnets. Assume that an IP service provider, SpeedNet, owns a Class B address 128.128.0.0. Three customers, BigCo, MidCo, and SmallCo, approach SpeedNet for Internet connectivity. BigCo has 15,000 employees, MidCo has 3,200 employees, and SmallCo has just nine employees. All need public IP addresses for all employees. SpeedNet designs three subnets, one for each company. The size of the subnet should be such that it can support the required number of hosts. Each subnet is identified by two parameters: the subnet number and the prefix. The bits specifying the subnet number include all the bits of the parent network number bits and some bits stolen

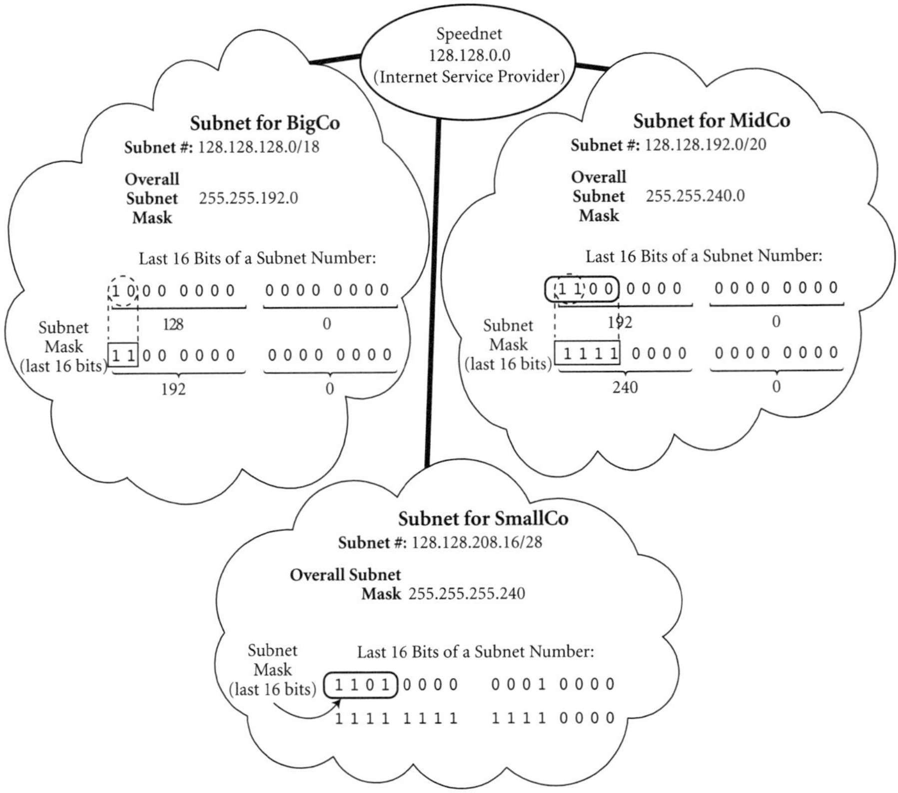

Figure 4.15: Example of CIDR and Subnets

from the host number part of the IP address. The subnet number has all 0s in the host portion of the IP address.

Now, let's discuss the subnets for each company. The subnet for BigCo must have at least fourteen bits for the host number because thirteen bits to represent a host number can address only approximately 8k hosts. Since we need fourteen bits for hosts, the total number of bits that define the subnet number is $32 - 14 = 18$. Since sixteen bits of the network number already determine the first sixteen bits of the subnetwork number in our example, only $18 - 16 = 2$ additional bits need to be selected. These two bits are 10 in Figure 4.15. The subnet number, including prefix, is then 128.128.128.0/18 where the prefix value of 18 is the total number of bits specifying the actual subnet number. The parameter called *overall subnet mask* is derived by using 1 in the bit positions that define the subnet number. In BigCo's subnet, the subnet mask has eighteen consecutive 1s in the eighteen most significant bits, leading to the subnet mask of 255.255.192.0. Either the prefix or the subnet mask is needed to determine how many bits of the IP address are used to calculate the subnet number.

Now, let's move on to the subnet for MidCo. The MidCo subnet needs twelve bits to represent 3200 hosts, and $32 - 12 = 20$ bits for the subnet number. Since the network number already specifies the first sixteen bits, we need to determine just an additional $20 - 16 = 4$ bits to fully specify the subnet number. The only constraint in specifying these four bits is that we should specify an IP address space that is distinct from other already-defined subnets. Since the subnet for BigCo uses `10` for bit positions 17 and 18,[89] we must use another two-bit combination for these bit positions. Therefore, the combination `11` is chosen for bit positions 17 and 18. The remaining two bit positions, 19 and 20, could be anything, and `00` is chosen for those positions. The subnet number then becomes 128.128.192.0/20, and the associated subnet mask is 255.255.240.0.

Finally, the subnet for SmallCo needs just four bits for the host number to represent nine hosts. Again, we need to ensure that the IP address space is unique for this third subnet (relative to the previous two subnets). Since the MidCo subnet IP space is different from BigCo's, using MidCo's subnet number as the starting point is easier. The bit positions 17, 18, 19, and 20 for the SmallCo subnet need to be different from the `1100` used for MidCo. Choosing `1101` as four such bits yields the subnet number 128.128.208.16/28 and the subnet mask of 255.255.255.240.

In the absence of CIDR, the internet service provider (ISP) would need a Class B block of IP addresses but has no need for that many IP addresses. CIDR allows the ISP to obtain a smaller block of IP addresses, yielding the savings in IP address space.

4.2.2.4 Internet Control Message Protocol Overview

ICMP formally exists at the internetworking layer (like IP) but acts as if it were a protocol at a layer higher than IP.[90] ICMP sends messages using IP. The main purpose behind ICMP is to report errors associated with the processing of IP datagrams and to facilitate diagnostic testing of IP routes. Let's take a look at some examples of how ICMP is used. An application called *ping*[91] is useful to measure the round-trip time for a packet. Ping is also used to check the reachability of a host. For example, if you type "ping google.com" at your command prompt, it will provide the round-trip time between your computer and Google's server. Ping uses the ICMP messages Echo and Echo Reply.

A Word of Caution... A *ping of death* is a type of attack on a computer in which a ping with a large message is sent to a computer [135]. The computer being attacked

[89] For a thirty-two–bit IPv4 address, the most significant bit is considered bit position 1 and the least significant bit is considered bit position 32 in our nomenclature.

[90] See [117, 129, 134] for additional details.

[91] The term *ping* is borrowed from the sonar operation that locates an underwater object and stands for Packet Internet Groper [117].

could crash due to buffer overflow. Earlier computer systems could not handle a ping packet larger than the maximum IP packet size (i.e., 65,535 bytes), causing a crash. Beginning in the late 1990s, however, most systems can handle such an attack. A different type of attack called *ping flooding* has become widespread in recent years and involves sending the target computer so much ping traffic that regular traffic does not reach the computer system.

Another application, tracert (i.e., traceroute), uses ICMP to find the path the IP datagrams take from the source to the destination. In traceroute, the sender first sends a UDP/IP datagram with a TTL of 1 to the destination host. The first router decreases the TTL by one to $1 - 1 = 0$. Since TTL is 0, the router discards the IP datagram and sends an ICMP Time Exceeded message to the source. The sender now learns the IP address of the first hop in the (source, destination) path. The sender then sends an IP datagram with TTL = 2, which results in the identification of the second hop. The sender keeps increasing TTL values, and, eventually the UDP/IP datagram will reach its destination.

Upon receiving the UDP/IP datagram, the destination finds a port number outside the standard-defined range. The destination will then respond with an ICMP Destination Unreachable message, the header of which will contain the code associated with "port unreachable." The sender now knows that the UDP/IP datagram has arrived at the destination, and, in this process, the sender learns the entire (source, destination) path!

One more application of the ICMP is to enable an IP router to automatically discover other neighboring IP routers. A host or a router can use two ICMP messages, Router Advertisement and Router Solicitation, for this purpose. IP routers can, based on their configuration, periodically (e.g., every ten minutes) send out an ICMP Router Advertisement message on relevant subnets to advertise their IP addresses using a multicast address, 224.0.0.1, or the broadcast address, 255.255.255.255. Routers also respond to the ICMP Router Solicitation messages sent by hosts (e.g., three solicitation messages at three-second intervals) [117]. The hosts use the multicast address, 224.0.0.2, or the broadcast address, 255.255.255.255, to send solicitation messages. The hosts may decide to just wait for advertisements rather than sending solicitation messages.

4.2.3 Overview of the Link Layer (Layer 2) Protocols

IP routers forward layer 3 IP datagrams to the destination network in which the help of the lower layers (i.e., layers 2 and 1) is sought to deliver the IP packet to the destination host. Let's consider two widely used layer 2 protocols, Ethernet (or IEEE 802.3) in Section 4.2.3.1 and ATM in Section 4.2.3.2. In general, Ethernet (or IEEE 802.3) is typically used for small LANs, such as offices and homes,[92] and

[92] An emerging technology called Carrier Ethernet uses the basic concept of Ethernet but can function over longer distances.

the ATM is typically used to carry information over long distances (e.g., one city to another).

4.2.3.1 Overview of Ethernet or IEEE 802.3 and Associated Physical Layer

A LAN uses one of two frame formats, the basic Ethernet standard format or IEEE 802.3 standard format, with earlier systems using the old Ethernet standard and new systems typically using IEEE 802.3.[93] The basic Ethernet standard was specified by Digital Equipment Corporation, Intel Corporation, and Xerox Corporation and is often referred to as DIX Ethernet.[94]

Figure 4.16 shows two frame structures: the DIX frame (also known as Ethernet Version 2 or the Ethernet II frame) and the IEEE 802.3 MAC frame, containing IEEE 802.2 headers. First, we will discuss the format of the basic Ethernet frame, and then we'll contrast it with the IEEE 802.3 frame format.

The Preamble field is a bit pattern that helps the LAN hosts with synchronization. The Destination Address field specifies which host(s) should receive the frame, and the Source Address field specifies the host that is sending the frame. The Type field indicates the frame type of the upper-layer protocol to allow different protocols to operate on the same physical medium, such as the coaxial cable. The Type field is set to a value greater than 1536. In the basic implementation of the Ethernet in LANs, the Ethernet II frame is most common [137]. As the name suggests, the Data field contains upper-layer data bytes, which can contain 46 to 1500 bytes. If the length of the Data field is less than forty-six bytes, the Data field is extended by adding a Pad to make the total length of the Data field forty-six bytes. Finally, the Frame Check Sequence field includes a thirty-two–bit cyclic redundancy check to facilitate error detection at the receiver.

Let's briefly highlight the differences in the basic Ethernet and IEEE 802.3 frame formats. The eight-byte preamble of the Ethernet is divided into a seven-byte Preamble field (i.e., seven occurrences of the pattern `10101010`) and a one-byte Start-of-Frame Delimiter (SFD) field in IEEE 802.3. Just like the Preamble field, the SFD field is also an alternating pattern of `1`s and `0`s but ends with two consecutive `1`s (i.e., `10101011`), pointing to the end of the first eight bytes of the frame. In IEEE 802.3, the Length field replaces the Type field. The Length field specifies the number of bytes in the Data field and is set to a value less than or equal to 1500. Furthermore, a typical 802.3 frame includes two extra headers in the first few bytes of the Data field of the basic Ethernet: the IEEE 802.2 Link Service Access Point header and the 802.2 Subnetwork Access Protocol header. The Link Service Access Point header is a three-byte header including a Destination Service Access Point field, a Source Service Access Point field, and a Control field. The Subnetwork Access Protocol header is an extension to the Link Service Access Point header, and its existence is indicated by using *source service access point* = 170 and *destination and source service access point* = 170 in the Link Service Access Point header. The Subnetwork Access Protocol header uses zero as the Organization Code

[93] See the documentation for additional information on Ethernet or IEEE 802.3 [117, 136].

[94] DIX is an acronym: D for Digital, I for Intel, and X for Xerox.

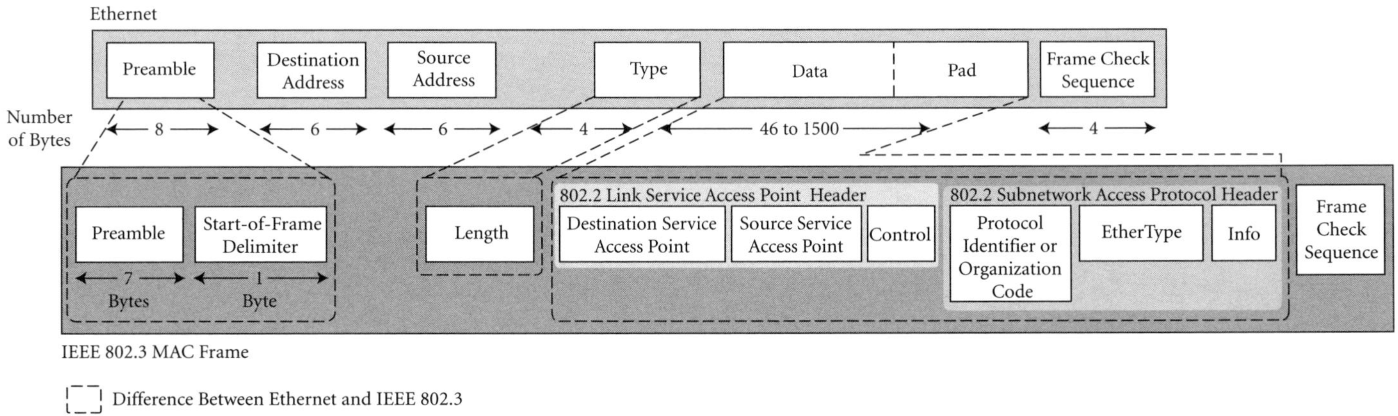

Figure 4.16: Frame Formats for Ethernet and IEEE 802.3

field and uses the values 2048, 2054, and 32821 in the EtherType field to specify the upper-layer protocols of IP, ARP, and Reverse ARP, respectively. The Protocol Identifier field indicates the EtherType field value for the protocol running over the Subnetwork Access Protocol when the Organization Code field is zero. If the Organization Code field is non-zero, the Protocol Identifier contains the value assigned by such an organization to the protocol running over the Subnetwork Access Protocol. Finally, the Info field contains the upper-layer data, and the Frame Check Sequence facilitates error detection at the receiver. The same values are used in the Type field of the basic Ethernet frame. The maximum number of bytes carried by an Ethernet frame is 1526 bytes. Since there are twenty-six header bytes, the frame can carry up to 1500 bytes of upper-layer data. IEEE 802.3 carries 1500 bytes for 10 Mbps networks but more bytes for higher transmission speeds.

Now, let's move on to the basic transmission approach in the Ethernet protocol. The IEEE 802.3 MAC operates in half-duplex (i.e., *either* transmit or receive at one time) or full-duplex mode (i.e., *both* transmit and receive *simultaneously*), with mandatory support for half-duplex operation. The half-duplex operation uses the CSMA/CD[95] protocol. Each Ethernet station (i.e., host) determines on its own whether or not it should send a frame. Each station continuously observes the physical medium to detect periods of inactivity. Once a station detects an idle medium, it can start transmission of the Ethernet frame. If two or more stations transmit at the same time, a collision occurs, corrupting the frame. As soon as a station detects a collision, it ceases the transmission of the frame, waits for a random back-off period, and reattempts the transmission later. In contrast, the full-duplex operation involves no collisions or retransmissions.

Let's take a quick look at the evolution of the basic Ethernet protocol. The basic Ethernet protocol works at 10 Mbps; *Fast Ethernet* operates at 100 Mbps. The Gigabit Ethernet protocol defined by IEEE 802.3z works at 1 Gbps. The 10 Gigabit Ethernet protocol, 10GbE, defined by the IEEE 802.3-2005 standard, supports 10 Gbps.

There is renewed interest in the Ethernet protocol. The concept of Metro Ethernet or Carrier Ethernet is gaining popularity and involves the use of the Ethernet principles for a Metropolitan Area Network (e.g., a city or even wider area) [138, 139]. The Metro Ethernet Forum has defined the Metro Ethernet as a network with five specific attributes that distinguish it from the traditional Ethernet: standardized services, scalability, reliability, service management, and QoS [140]. These service attributes are used to create a variety of services. Three basic service types are Ethernet Line Service, which provides point-to-point connectivity between two user network interfaces, Ethernet LAN Service, which provides multipoint-to-multipoint connectivity, and Ethernet Tree Srvice, which provides rooted point-to-multipoint connectivity. Metro Ethernet is typically used to interconnect enterprise LANs and to connect the subscribers and businesses to a wide-area network such as the Internet. Metro Ethernet utilizes uses layer 2 or Ethernet switches to interconnect its stations. In contrast to traditional routing where the Ethernet frame headers change, the Ethernet frame in

[95]CSMA represents Carrier Sense Multiple Access, and CD represents Collision Detect.

Metro Ethernet remains intact as it travels from the source to the destination. MPLS is not mandatory for Metro Ethernet; Metro Ethernet is transport-agnostic. Large service providers typically use MPLS in conjunction with Metro Ethernet.

Since Ethernet is a layer 2 protocol, the Ethernet device implements just the bottom two layers of the Internet Protocol stack. This device is typically implemented as a network interface card on the Ethernet station. Examples of the physical media the Ethernet device uses include a coaxial cable, twisted-pair cable, and optical fiber cable.

4.2.3.2 Asynchronous Transfer Mode Overview ATM is widely used in cellular networks to quickly transport information over long distances and on the backhaul between the base station and the base station controller, all while meeting a target QoS. Cellular operators are gradually replacing ATM by the IP/MPLS approach to transport information over long distances. We present here a brief overview of ATM, summarizing its main features, defining an ATM cell, and discussing the implementation of QoS.

ATM transports information in units of *cells*. (We will take a closer look at the structure of a cell in just a moment in Figure 4.17.) The basic function of ATM is to switch ATM cells quickly. When information is sent over long distances, it passes through several ATM switches. Short ATM headers, fixed-size cells, and hardware-based (as opposed to software-based) switching all contribute to enable fast ATM switching. ATM supports segmentation and re-assembly, in which a large upper-layer packet (e.g., an IP packet) can be divided into several smaller ATM cells at the sender and re-assembled at the receiver. Think of ATM as a moving train with the train compartments representing ATM cells. The train keeps moving, and the train compartment may be full or empty depending upon whether there is something to transport or not. Similarly, whenever we have something to transport, we fill up the ATM cell; otherwise, the cell is empty. Since the ATM cells transfer information based only on demand (not periodically), such a transfer mode is considered asynchronous.

ATM needs a connection setup between the two endpoints of the communications link before data transfer can begin. Hence, ATM is considered connection-oriented, which introduces initial connection setup delay and requires maintenance of the connection state. During such a connection setup, suitable resources are reserved to guarantee QoS. Since pure ATM just transports information, we need some layer above ATM to adapt traffic to ATM to provide QoS. ATM adaptation layers (AALs) are defined to facilitate QoS control. ATM is more suitable for low-rate real-time traffic than for high-rate non-real-time traffic due to the overhead associated with small cell sizes. Some of the main drawbacks of ATM are that it requires manual configurations of the connections among the ATM nodes and that it is not easily scalable. The addition of new ATM nodes or switches requires significant manual configuration to interconnect with various existing nodes.

Let's discuss the structure of an ATM cell, shown in Figure 4.17. An ATM cell consists of two parts: a five-octet header and a forty-eight–octet payload. The payload remains intact between two endpoints, and the cell header is updated by each ATM switch along the way. The cell header facilitates fast switching. The User Network

Interface is the interface between the endpoint (e.g., a host) and the ATM network, and the Network Node Interface is the interface between two points in the ATM network, such as between two ATM switches. As seen in Figure 4.17, the majority of cell headers are the same for User Network Interface and Network Node Interface. The Generic Flow Control field exists only for User Network Interface to facilitate sharing the same ATM connection among multiple terminals or hosts. The Virtual Path Identifier (VPI) and Virtual Channel Identifier (VCI) fields are used by ATM switches to route ATM cells, with VPI providing coarse or high-level routing and VCI providing fine or low-level of routing. (We'll discuss VPI and VCI further later in this section.)

The Payload Type field specifies the type of data in the cell and is primarily used to identify special-purpose cells, such as maintenance cells, cells indicating the network congestion, and the last cell of a multi-cell message. The Cell Loss Priority field indicates if the cell can be discarded in favor of another cell if a cell, unfortunately, must be sacrificed!

The Header Error Check field is a checksum that helps detect error in the cell header, excluding the Generic Flow Control field. Another potential use of the Header Error Check field is to identify or verify the boundary of the cell or header. The Header Error Check field match for the preceding four octets would point to the correct identification of the header boundary. In such a case, ATM need not rely upon the underlying transmission equipment to detect cell boundaries.

Ever Wondered... Why do we have fifty-three bytes for the ATM cell? Nope, we were not searching for some good prime number! The reason for fifty-three bytes is more political and less technical! It is a classic example of give-and-take compromises made during standardization. The American National Standards Institute in the U.S. wanted a larger cell size of sixty-four bytes because that was considered a good compromise between the smaller size preferred for voice due to lower latency and the larger size preferred for data due to lower overhead. However, the European

User Network Interface Header

Byte Number			
1	Generic Flow Control	VPI	
2	VPI	VCI	
3	VCI	VCI	
4	VCI	Payload Type	Cell Loss Priority
5	Header Error Check		

Network Node Interface Header

Byte Number			
1	VPI		
2	VPI	VCI	
3	VCI	VCI	
4	VCI	Payload Type	Cell Loss Priority
5	Header Error Check		

Figure 4.17: Inside an ATM Cell

Telecommunications Standards Institute wanted a smaller cell size of thirty-two bytes to minimize the need for echo cancellation for voice. The compromise was the payload size of $(64 + 32)/2 = 48$ bytes. Just like we have to pay sales tax on the items we buy, we need to pay cell tax to use ATM! A 10% cell tax was considered reasonable, leading to five bytes (i.e., 10% of 48 bytes is 5 bytes) of header. That's how the overall ATM cell size of $48 + 5 = 53$ bytes was chosen [141].

Figure 4.18 illustrates the concepts of VPI and VCI via an example of switching. A virtual path identified by VPI carries one or more virtual channels identified by VCIs. A physical connection, such as an optical fiber, between two ATM switches carries one or more virtual paths. A set of VPIs and VCIs between two endpoints define the overall ATM connection between two endpoints. The term VPI:VCI is used to indicate that both VPI and VCI are being considered together. An ATM switch may use a VCI or VPI field of the ATM cell for cell switching. A switch that uses only VCI for switching is called a *virtual channel switch* or *virtual channel cross-connect*. Similarly, a switch using only the VPI field for switching is called a *virtual path switch* or *virtual path cross-connect*. To implement switching, the ATM switch has a translation table that includes a mapping between the incoming switch port and incoming VPI:VCI and the outgoing switch port and outgoing VPI:VCI. When an ATM cell arrives on a specific incoming port, the switch tries to match the cell header contents with the relevant incoming VPI:VCI entry in the translation table. When a match is found, the switch will replace the relevant VPI and VCI field based on the outgoing VPI:VCI and send out the updated ATM cell on the associated outgoing

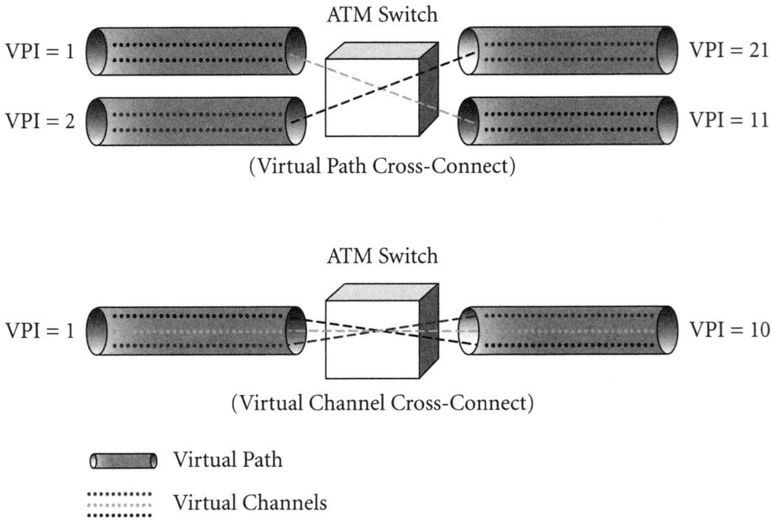

Figure 4.18: ATM Switching, Virtual Paths, and Virtual Channels

switch port. The ATM switch's update of such an ATM cell header facilitates cell switching for the next switch in the source-to-destination path.

ATM offers advantages such as such as fast switching and suitability for real-time services. ATM also supports QoS while transporting information via cells.

Do You Recall... What layer exists on the top of ATM to facilitate implementation of QoS? See the footnote[96] for the answer!

Several AALs have been defined to cater to different types of services or traffic. The International Telecommunications Union has defined four service classes, A, B, C, and D, to quantify service characteristics for suitable QoS implementation. Table 4.1 defines the characteristics of the International Telecommunications Union traffic classes and also provides a mapping between the traffic class and the type of AAL.

The earlier intention was to define a separate AAL protocol for each of the International Telecommunications Union service or traffic classes. Hence, A, B, C, and D traffic classes are associated with AAL1, AAL2, AAL3, and AAL4, respectively, However, AAL3 and AAL4 were found to be very similar and were merged into a single protocol called AAL3/4, which allows multiplexing of several channels onto the same connection. ATM is widely used to carry IP traffic, and AAL3/4 was deemed too complex for such purpose. Hence, AAL5 was designed as a simpler alternative to AAL3/4 but without the multiplexing capability. Traffic Class A targets a connection-oriented service with strict timing requirements and a fixed data rate. Examples of Class A services are uncompressed voice (e.g., 64 kbps pulse coded modulation voice) and uncompressed video. Class B also targets a connection-oriented service with strict timing control but allows a variable data rate. Compressed voice and video are examples of Class B services. Both Class C and Class D are for non-real-time services with variable data rates; Class C focuses on connection-oriented mode and Class D on connectionless mode. The connection-oriented mode requires the establishment of an end-to-end connection between two communicating entities. TCP/IP applications, such as web-browsing and email, fall under Class C, and UDP/IP applications, such as instant messaging, fall under Class D.

Two other AALs are AAL Type X and Signaling AAL. Type X AAL can be used to access the raw cells of the ATM layer, and Signaling AAL is designed, as the name suggests, to carry signaling messages [142]. Signaling AAL is a variation of AAL5 but is more complex than other AALs. Multiple AAL bearers can share the same (VPI, VCI) combination using a suitable AAL header. For example, different voice calls can use different channel IDs associated with their AAL2 bearers, and all such calls can share the same ATM connection [143].

AALs were designed with International Telecommunications Union traffic classes in mind, but a more common discussion these days is the ATM QoS from the perspective of the International Telecommunications Union traffic classes, such as available

[96] It is ATM Adaptation Layer (AAL)!

Table 4.1: International Telecommunications Union Traffic Classes and AALs

Traffic Characteristics	Traffic Classes			
	A	B	C	D
Data Rate	Constant	Variable	Variable	Variable
Timing	Real-Time	Real-Time	Non-Real-Time	Non-Real-Time
Connection Mode	Connection-Oriented	Connection-Oriented	Connection-Oriented	Connectionless
Service	Uncompressed	Compressed	TCP/IP Application	UDP/IP Application
Example	Voice and Video	Voice and Video	(e.g., Web-Browsing)	(e.g., Instant Messaging)
	ATM Adaptation Layers			
	AAL1	AAL2	AAL3/4 AAL5	AAL3/4 AAL5

bit rate, constant bit rate, non-real-time variable bit rate, real-time variable bit rate, and unspecified bit rate [144]. The data rate or bandwidth is guaranteed by constant bit rate, real-time variable bit rate, and non-real-time variable bit rate, is optional for available bit rate, and is impossible with unspecified bit rate. Constant bit rate and real-time variable bit rate are suitable for real-time services. The AAL information is provided in the payload portion of the ATM cell and is negotiated during the connection setup. The ATM Forum has specified a Generic Rate-Control Algorithm to enforce the QoS [144]. According to the Generic Rate-Control Algorithm, the ATM switch places the incoming ATM cells into the input queue as long as the queue is not full. If the queue is full, the incoming ATM cell is dropped. The switch maintains a pool of tokens. If the pool of tokens is not empty and if a cell is waiting in the input queue, one token is removed from the pool and the cell is transmitted. The QoS negotiated as part of the connection setup influences the rate at which the tokens enter the pool and the size of the input queue.

4.3 ROUTING AND FORWARDING OF IP PACKETS

Let's take a quick look at how IP and Ethernet work together in the LAN. Upon receiving an IP packet, the IP router or IP gateway observes the destination IP address and determines the packet's next hop. The process of sending the received IP packet to a suitable next hop is called (IP) *forwarding* and involves matching the destination IP address with the entries of the *routing table*, which has a mapping between the destinations and the next hops. The process of determining the entries of the routing table is called (IP) *routing*. In summary, routing helps construct the entries in the routing table, and the router uses the routing table to forward an incoming packet to a suitable next hop such that the IP packet (ideally) travels toward its final destination. We'll first look at three major routing protocols used in the industry and then explain forwarding via an example.

The routing protocols of interest to us are Routing Information Protocol (RIP), Open Shortest Path First (OSPF), and Border Gateway Protocol (BGP). Before moving forward, we must define an *autonomous system* (AS), which is a network controlled and managed by a single administrative entity, such as a corporation, university, or ISP. A company can manage all the IP routers in the company and implement various policies (e.g., which routing algorithm to use for routing within the autonomous system). The Internet can be considered a network of numerous such autonomous systems. An autonomous system is also called a *routing domain*; the autonomous system can designate some of its routers as external routers or external gateways. The routing protocol that allows the routers to exchange information *within* an AS is called an Interior Gateway Protocol; OSPF and RIP are examples of such a protocol. In contrast, an Exterior Gateway Protocol enables the routers to exchange information *between* ASs; the BGP is an example of such a protocol. We'll first discuss RIP and OSPF and then turn our attention to the BGP.[97] OSPF runs directly over IP and has

[97]The reader interested in additional details is referred to the documentation [117, 145–147].

its own reliable transmission mechanism. RIP runs over UDP, and BGP runs over TCP [148].

RIP is suitable for small ASs and is a *distance vector* protocol. Let's summarize the basics of a distance vector protocol. According to the distance vector routing, each router in the local network, such as an AS, keeps track of its distance (or cost) to known destinations. The routes or paths corresponding to shorter distances or smaller costs are preferred to those corresponding to longer distances or larger costs. From a source to a destination, the preferred path has the smallest cost. The distance vector table maintained in each router is dynamically updated to reflect the prevailing network conditions, such as network congestion. A given router advertises its own table to its neighbors, and such advertisements help ensure that the information stored in the distance vector table is current. The router calculates its distance to various other routers by processing such advertisements. The advantages of the distance vector algorithm are simplicity of implementation and easy debugging and troubleshooting. The routing behavior of a router is somewhat unpredictable when the distance vector tables have not yet stabilized. The periodic advertisements, though critical in estimating costs, represent overhead and consume link capacities.

To support distance vector routing, the RIP specifies two types of packets: request packets and response packets. A router sends a request packet to its neighbors to obtain their partial- or full-distance vector tables. The response packet advertises the local distance vector table and is sent periodically (i.e., every thirty seconds) and in response to a received request packet. The router, upon receiving a response packet, updates its distance vector table if a new lower-cost route to a destination is found. RIP sends and receives packets using UDP port 520.

As mentioned earlier, incorrect routes may exist before a steady-state is reached. Routing loops, which cause the distance from a source to a destination to continually increase, may also exist, and such a problem is referred to as *counting to infinity*. The RIP limits the maximum number of hops to sixteen. Enhancements to the basic distance vector algorithm, such as split horizon with poison reverse and triggered updates, have been made in new RIP algorithms to minimize the counting to infinity problem.[98] Two versions of RIP exist, the basic RIP or RIP Version 1 (RIP-1) [145] and RIP Version 2 (RIP-2) [149]. When the version number is unspecified, RIP refers to RIP-1.

The OSPF protocol utilizes the concept of *link state routing* and aims to overcome some drawbacks of distance vector routing such as suitability for small size ASs only. Let's have a glimpse of how link state routing operates. A link state specifies an interface on a router (e.g., IP address and subnet mask) and the cost of the link with the neighboring router. Each router maintains a *link state database*, also called a *topology database*, that stores such link states. Routers send out link state advertisements (LSAs) that include the lists of the direct links with their immediate neighbors (i.e., those with no other routers in-between) and the costs of the links. Each router then builds the network topology database for the network, which would be the same for all

[98] See the technical documentation for additional details [145].

routers in the network in steady-state. Once the topology database is built, the router determines the most desirable route, which would be the path of least cost, to each destination network. OSPF possesses several features lacking in RIP, which make OSPF a popular routing protocol in large networks. Examples of OSPF's advantages are equal-cost load balancing,[99] logical partitioning of the network,[100] support for authentication, faster convergence time, and support for classless inter-domain routing and Variable-Length Subnet Masks[101] [117].

OSPF relies upon the shortest-path-first algorithm to implement link state routing. Specifically, the shortest-path-first algorithm considers the topology database to build a tree-representation of the network; the router executing the shortest-path-first algorithm is the root of the tree. The shortest-path-first algorithm constructs shortest paths defining the preferred routes to each destination network within an AS. All routers have the same link state database, but each router has a unique tree due to its overall position in the network.

OSPF divides the network into several logical areas called *OSPF areas*. Each OSPF area, identified by a thirty-two–bit area ID, includes a logical collection of networks and routers. Each OSPF area may be defined based on geographic or administrative boundaries. All routers within an area have the same link state database describing the routing devices and links within the area. The routers in one OSPF area know nothing of the topologies outside their own OSPF area, but they do know the routes to other OSPF areas. The concept of OSPF areas makes the link state database more manageable, reduces the signaling load of link state advertisements, and saves processing power at the routers [117].

Figure 4.19 illustrates the OSPF areas and introduces several specific types of routers. An OSPF network contains at least one area called *area 0* or the *backbone area*. The goal of the backbone area is to physically connect all other OSPF areas. Routing information flows between the backbone area and the OSPF areas. Three classes of routers are defined: intra-area, area border, and AS boundary routers. An intra-area router resides completely inside an OSPF area and keeps track of the topology database for the OSPF area. The area border router connects to the backbone area and one or more other OSPF areas. The area border routers maintain awareness of the topology databases for the OSPF areas with which they are connected. Finally, the autonomous system boundary router is at the periphery of the overall OSPF network. The autonomous system boundary router is a gateway of the OSPF network and exchanges routing information with external networks. The autonomous system boundary router also announces external link state advertisements. When neighboring routers exchange link state information to synchronize their topology databases, they are forming *adjacency*, i.e., the adjacent OSPF routers exchange link state advertisements. Once a router creates or updates a link state advertisement, it forwards

[99]When multiple paths with equal cost are available, load balancing across those paths increases the utilization efficiency of network resources.

[100]The network can be divided into logical parts called *OSPF Areas* to make the announcements more efficient, e.g., reducing the propagation of outage information during adverse conditions.

[101]Such support allows the network administrator flexibility in allocating IP addresses.

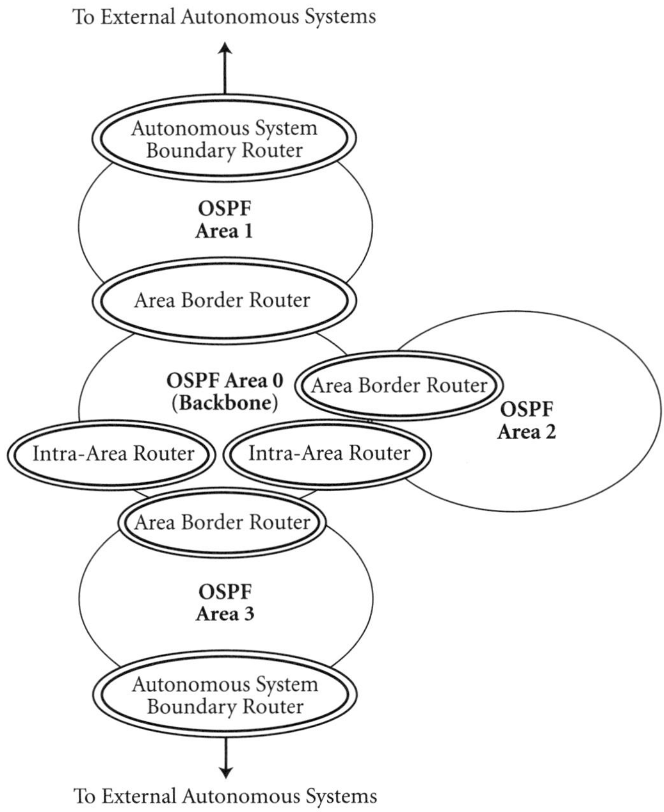

Figure 4.19: OSPF Areas and Classes of Routers

the link state advertisement to each adjacent device. Each device or router receiving the link state advertisement sends it to its neighbors. This link state advertisement forwarding process is referred to as *reliable flooding* and enables the OSPF to derive the optimal paths from the source to the destination. The OSPF packets are transmitted in the form of IP datagrams with 89 in the Protocol Identifier field and 0 in the Type of Service field of the IP header.

Now that we are familiar with the routing protocols for *intra*-AS routing, let's understand how BGP can be used to implement *inter*-AS routing. BGP utilizes *path vector routing*, which works similarly to distance vector routing in that a router advertises the destinations it can reach. Distance vector routing advertises the *distance* to reach a destination, and path vector routing specifies the *entire path* to the destination in the form of vectors. The path is specified as a sequence of domains and confederations from a source to a destination. For example, from the source source.com to the destination destination.com, the overall path may be specified as source.com, abc.com, def.com, ghi.com, . . . , xyz.com, destination.com. When a router can identify multiple paths from itself to a target domain, it can choose the path that has the fewest domains.[102]

Path vector routing is simpler to implement than link state routing. Link state routing works as intended when the routing domains implement homogeneous policies for route selection (e.g., they have the same definition of cost), but the route selection policy used by one routing domain may be unknown to other routing domains. The path vector protocol operates with no implicit assumption on the route selection policies. Routing loops can easily be avoided because the entire path is explicitly advertised. Since routing information is provided to the adjacent domains after the routes have been determined, only the routing information related to the routes a domain selects is propagated, thereby reducing the amount of routing information.

Let's discuss BGP Version 4, but we will simply call it BGP.[103] Figure 4.20 introduces several BGP terms. A router that supports BGP is referred to as a *BGP speaker*. Two neighboring routers that are also BGP speakers are called *internal BGP neighbors* if both belong to the same AS and *external BGP neighbors* if they belong to a different AS. A given pair of BGP neighbors uses a TCP session via TCP port 179 to exchange routing information, and such a TCP session is maintained through *keep alive* messages. A BGP router uses an OPEN BGP message to establish an

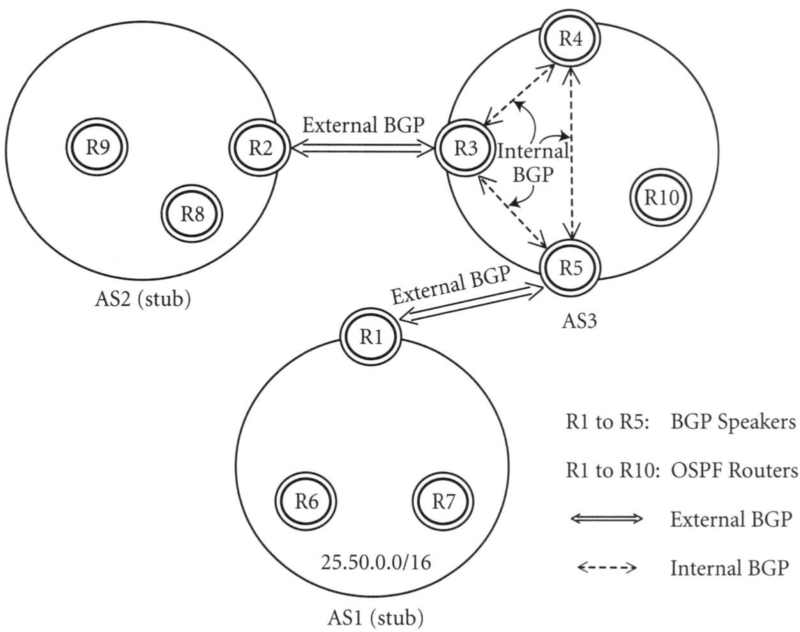

Figure 4.20: BGP Terminology

[102]The router can also avoid any path that includes its competitors!

[103]Details of BGP can be found in the documentation [147].

association and a session with its peer BGP router. When two speakers create a BGP session between them, whole routing tables containing AS paths to destinations are exchanged. The updates to the routing tables are exchanged dynamically thereafter as incremental updates. The traffic originating or terminating within the AS is called *local traffic*; non-local traffic is called *transit traffic*. BGP defines three ASs: *stub*, *multi-homed*, and *transit*. A stub AS connects to just one external AS and carries only local traffic. A multi-homed AS connects to two or more other ASs and can be configured to omit any transit traffic. A transit AS, like a multi-homed AS, connects to two or more ASs but carries local *and* transit traffic.[104]

In Figure 4.20, AS1 and AS2 are stub ASs, and AS3 is an altruistic transit AS assuming that it allows transit traffic. To advertise routes, BGP uses Network Layer Reachability Information, which is specified as a series of networks represented by the tuple <*length*, *prefix*> [117]. For example, the tuple <16, 100.200.0.0> represents the network 100.200.0.0/16 where the first sixteen bits of the prefix 100.200.0.0 represent the network number. The internal BGP routers have fully meshed sessions among them as shown for AS3. Let's look at an example of how BGP propagates routing information. The network 25.50.0.0/16, hereafter referred to as the network in our example, resides in AS1. The BGP speaker R1 advertises the existence of the network to the BGP speaker R5 using external BGP neighbors. R5 in AS3 uses internal BGP to advertise the network to R3 and R4, which may not be physically adjacent to R5 but are edge routers in the AS. R3 in AS3 uses external BGP neighbors to inform R2 of AS2 about the network. BGP is adequate to advertise routing information across ASs. Both BGP and intra-AS routing protocols (e.g., RIP and OSPF) are used to propagate routing information through an AS. When a BGP speaker receives routing information from its internal BGP neighbor, it uses external BGP neighbors to convey the routing information to its external BGP neighbors but does not forward the information to its other internal BGP neighbors.

We discussed three routing protocols (i.e., OSPF, RIP, and BGP) that help create routing tables. The routing table essentially indicates the next hop for a given destination. Figure 4.21 shows the immediate neighborhood of the router R1 as well as selected entries of a routing table used by R1.

R1 has built the table using the routing protocols we just discussed. When an IP datagram arrives at the router, the router determines the subnet number that provides the longest match with the destination IP address present in the IP header of the incoming IP datagram. For your convenience, we have included a column titled "Bits for Subnet Number" in Figure 4.21.[105] Assume that packet A with the destination IP address 130.20.192.3 arrives at R1. R1 will compare the destination address with the first N bits of the subnet number for each entry in the table (where N corresponds to the actual number of bits that represent the subnet for a given row in the table). In our example, R1 will forward the packet to R4 because the first eighteen bits of the IP

[104] After all, we need at least some ASs to be unselfish! Otherwise, how will packets travel from one part of the planet to another?

[105] Perhaps we should have created a homework assignment for you here rather than giving away this information!

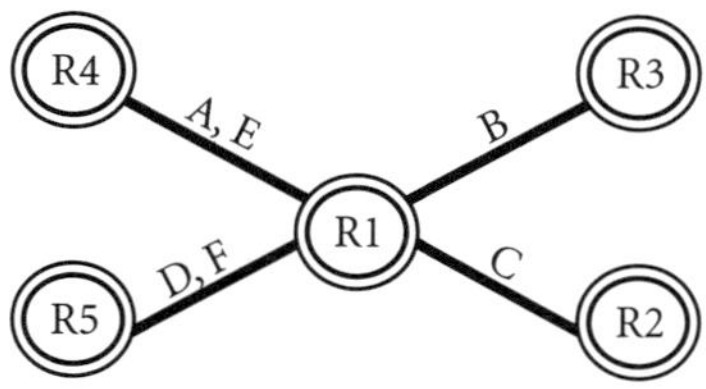

Incoming Packet at R1	Destination IP Address
A	130.20.192.3
B	130.20.96.1
C	130.20.112.4
D	130.20.128.0
E	130.20.193.6
F	128.1.2.3

Routing Table for R1 (Example Entries)

Subnet Number	Subnet Mask	Bits for Subnet Number	Next Hop
130.20.112.0	255.255.240.0	20	R2
130.20.96.0	255.255.224.0	19	R3
130.20.192.0	255.255.192.0	18	R4
130.20.0.0	255.255.0.0	16	R5
Default			R5

Figure 4.21: Forwarding by an IP Router Using a Routing Table

address of the incoming packet match the subnet number associated with R4. Please note that the concept of *longest match* is used. If we observe just the first sixteen bits of the IP address of the incoming packet, we will find a match with both R4 and R5. However, we want to match as many bits as possible for the subnet number and the initial portion of the IP address of the incoming packet. Hence, we would forward the IP packet to R4 (and not R5).

In addition to the entries appearing in Figure 4.21, some entries in the routing table specify the subnets corresponding to the LANs to which the router belongs. These entries have no IP router as the next hop, but a layer 2 Interface identification is associated with such LAN entries. For example, the router R1 may be physically connected to two LANs corresponding to the subnets 130.20.

Assume that an IP packet arrives at a router, and the routing table indicates no next hop. The host with the destination IP address is on the same physical network (e.g., a LAN) as the router, i.e., no more IP forwarding! The IP router would then place the packet on the LAN using the help of a layer 2 LAN protocol.

A Question for You. . . Can you name some layer 2 protocols? See the footnote[106] for the answer!

Layer 2 utilizes a forty-eight–bit physical hardware address, also known as an Ethernet or MAC address, of the destination so the packet can be sent to the correct host on the LAN. We need some mechanism that can map the destination IP address to the MAC address, and that mechanism is ARP [117, 150]. If the router does not know the mapping between the destination IP address, it sends an *ARP Request* message containing the destination IP address as well as its own IP address and the MAC address in a broadcast fashion to all the hosts on the physical network. Since all hosts are reading the ARP Request message, they can update their database with (IP address, hardware address) mapping for the sending host. The host that owns the destination IP address responds only to the sending host (*not* to *all* hosts) with an *ARP Reply* message that includes its MAC address and the IP address. The ARP requestor stores the mapping between the IP address and the MAC address in an *ARP cache* for future use. The host must send out an ARP Request message only when the ARP cache has no mapping between the IP and MAC address.

The ARP works as described here with Ethernet or IEEE 802. With ATM, however, each ATM host registers with an ARP server to find mappings between MAC addresses and IP addresses. Reverse ARP accepts a physical address as an input to yield the associated IP address [151].

4.4 TRANSPORT OF INFORMATION WITHIN THE NETWORK INFRASTRUCTURE

We will turn our attention to the application of some technologies to the cellular system. Specifically, we will discuss how Ethernet, ATM, and IP are used in a typical physical implementation of the cellular network. The network architecture of a typical 3G cellular system consists of a radio access network and two core networks (a circuit-switched core network and a packet-switched core network). The radio access network includes base station controllers (also known as radio network controllers or RNCs) and base stations, with one base station controller typically controlling several hundred base stations. The base station communicates with the mobile device over-the-air. The base station controller uses the help of the circuit-switched core network components, such as the mobile switching center, to support voice calls, and the packet-switched core network components, such as some gateways to provide Internet access. Figure 4.22 illustrates a typical physical implementation of the network infrastructure. Different service providers have their own variation of the example architecture depicted in Figure 4.22.

The base station includes various components (such as the transceiver, remote electrical tilt, and battery backup units) and a cell-site gateway. Such components

[106] Ethernet or IEEE 802 and ATM are examples of layer 2 protocols.

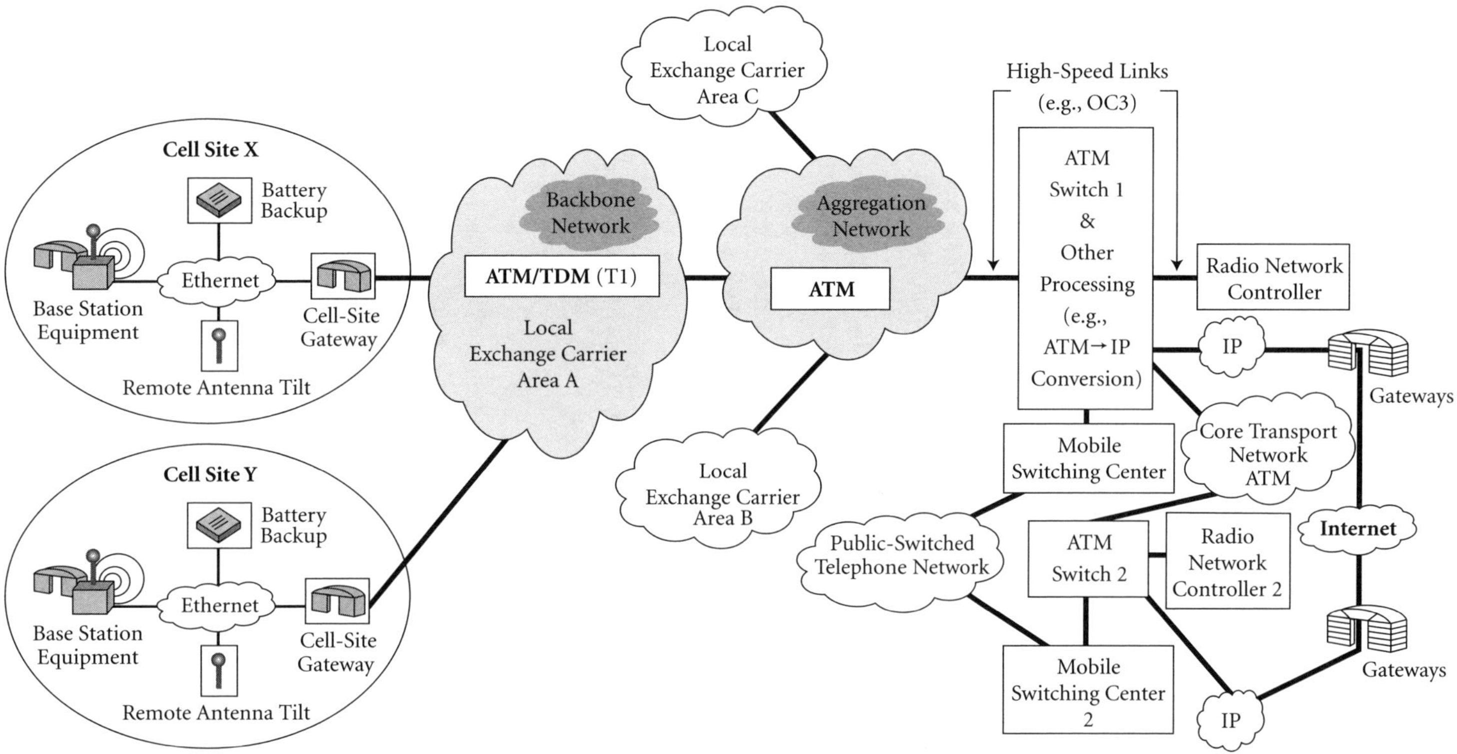

Figure 4.22: Traditional Transport in a 3G Cellular Infrastructure Network

reside on the same physical LAN. The cell-site gateway connects to an ATM/TDM (where TDM is Time Division Multiplexing) network, which usually consists of several T1 links. The TDM network is owned by local exchange carriers. The TDM networks feed into an aggregation network such that links with speeds higher than those offered by competitive local exchange carriers' networks can be used to forward information to the processor that implements ATM switch functions and other functions such as protocol conversion. The processor interfaces with the radio network controller, the mobile switching center, an ATM-based core transport network, and an IP-based transport network. The ATM-based core transport network helps transport information between the MSCs and the gateways, and the IP-based network enables the radio network controller to communicate with gateways to access the Internet.

The ATM-based transport has been quite popular for the physical network architecture of Figure 4.22 but is gradually being replaced by IP/MPLS-based transport as illustrated in Figure 4.23. The emerging transport approach replaces ATM by Metro Ethernet and/or IP/MPLS in the core network and the aggregation and backhaul networks. The interface between the base station and the radio network controller is referred to as the *backhaul*. The core network transport upgrade typically precedes the backhaul and aggregation network transport upgrades. Let's discuss the emerging technology of MPLS, which gets its name from its ability to support any network and internetworking layer protocols (even though IP is the basic one) and its use of labels to switch packets quickly [152].

MPLS represents a marriage between the best qualities of IP and ATM. Specifically, MPLS relies upon an ATM-type philosophy for fast switching of packets. ATM switching is implemented using hardware and, hence, is fast.

A Question for You... What kind of IDs does the ATM use to switch packets? Hint: The names of such IDs start with the letter V. See the footnote[107] to confirm your answer!

Furthermore, MPLS exploits IP's dynamic and scalable nature (e.g., routing protocols) to circumvent some of the manual configuration typically required for ATM. MPLS supports traffic engineering, i.e., traffic must be routed via specific paths to ensure an even utilization of various links. Such optimal link utilization helps prevent over-provisioning of link capacities and thus reduces cost. MPLS also enables implementation of IP-based QoS approaches, such as Integrated Services and Differentiated Services, to provide better user-perceived experience. MPLS facilitates IP convergence by enabling a unified common core network to support a variety of services and to interface with dissimilar networks such as ATM and frame relay networks. For example, if an enterprise wants to connect two remote offices via ATM or Ethernet, the core network of an MPLS service provider will accept layer 2 packets from one office, construct a layer 2 tunnel between the two offices to mimic a layer 2 virtual private network, and provide layer 2 packets to the other office. If another

[107] These IDs are VPI and VCI.

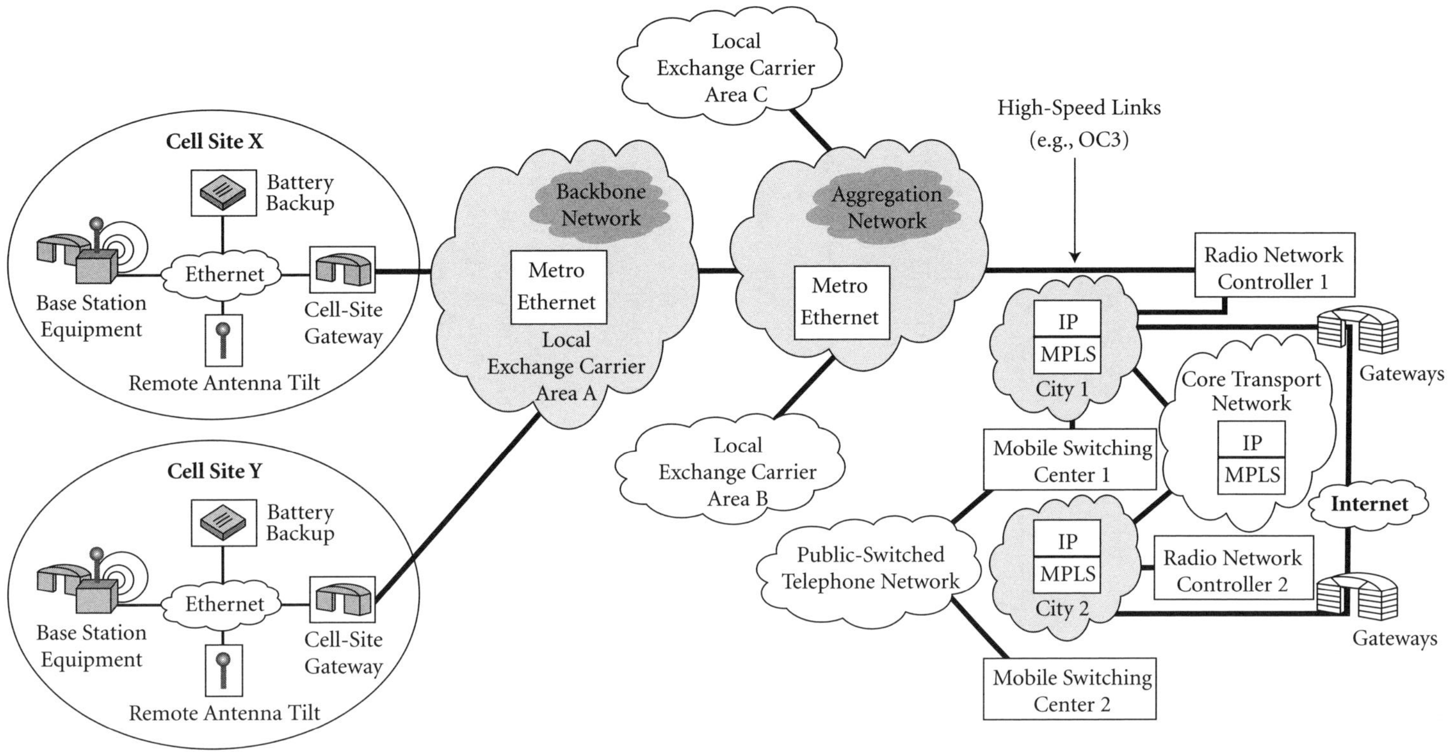

Figure 4.23: Emerging Transport Structure in the Cellular Infrastructure Network

enterprise has similar requirements, the MPLS service provider will reuse its core network elements to create two different layer 2 virtual private networks, one for each enterprise. Similarly, MPLS allows a layer 3 virtual private network whereby it accepts layer 3 packets (i.e., IP packets) at one end of the tunnel and yields layer 3 packets at the other end of the tunnel, providing layer 3 connectivity between two remote hosts.

MPLS is often considered a layer 2.5 protocol because it sits between layer 3 and layer 2. Figure 4.24 illustrates how packets are forwarded from one host to another through an IP/MPLS network. In Figure 4.24, Host A (with IP address $a.b.c.d$) wishes to send an IP packet to Host B (with IP address $e.f.g.h$). Host A sends the packet to its outgoing router R1, which Enterprise X owns. Because R1 interfaces with an MPLS service provider's MPLS network, it is called a *Customer-Edge* router. From the MPLS service provider's perspective, router R2 is on the edge of the provider's network because it interfaces the Customer-Edge router. Hence, R2 is called a (service) *Provider-Edge* router. The Provider-Edge router is also called a *Label-Edge* router because it is on the edge of the provider's network and is the router that starts using labels. Routers R3 through R7 are entirely inside the (service) provider's network and are referred to as *Provider* routers or *Label-Switching Routers*. In Figure 4.24, R8 is another Provider-Edge router and R9 is another Customer-Edge router. When an IP packet arrives at the Provider-Edge or Label-Edge router, it finds the mapping between the destination IP address and a previously established label from its label-forwarding information base.[108] (We'll summarize the process of determining labels in the next paragraph; for now, assume that each router has built a suitable label-forwarding information base.) R2 gives the packet the label number (e.g., 21 in our example) and forwards the packet to R3 based on its label-forwarding information base. R3 swaps label 21 with label number 31 and forwards the packet to R4. R4 replaces label 31 with label 41 and forwards the packet to R8. R8 finds from its label-forwarding information base that the incoming label is associated with the next hop, and hence no label-swapping is needed. R8 forwards the IP packet to the Customer-Edge router R9, which delivers it to the host using LAN forwarding.

MPLS supports various ways of constructing an label-forwarding information base. Two example approaches are the new Label Distribution Protocol or the existing Reservation Protocol with MPLS extensions. The specific structure of an MPLS label depends on the layer that sits just below MPLS. If the ATM layer is just below the MPLS, the (VPI, VCI) field is used as the label. If the Ethernet layer is just below MPLS, a *shim header* is introduced between the traditional IP header and the traditional Ethernet header and used to place a label.

4.5 VOICE-OVER-IP–RELATED PROTOCOLS

A traditional voice call in a cellular system involves the circuit-switched core framework establishing a dedicated circuit, and a relatively high data rate is often required

[108] The label-forwarding information base replaces routing tables.

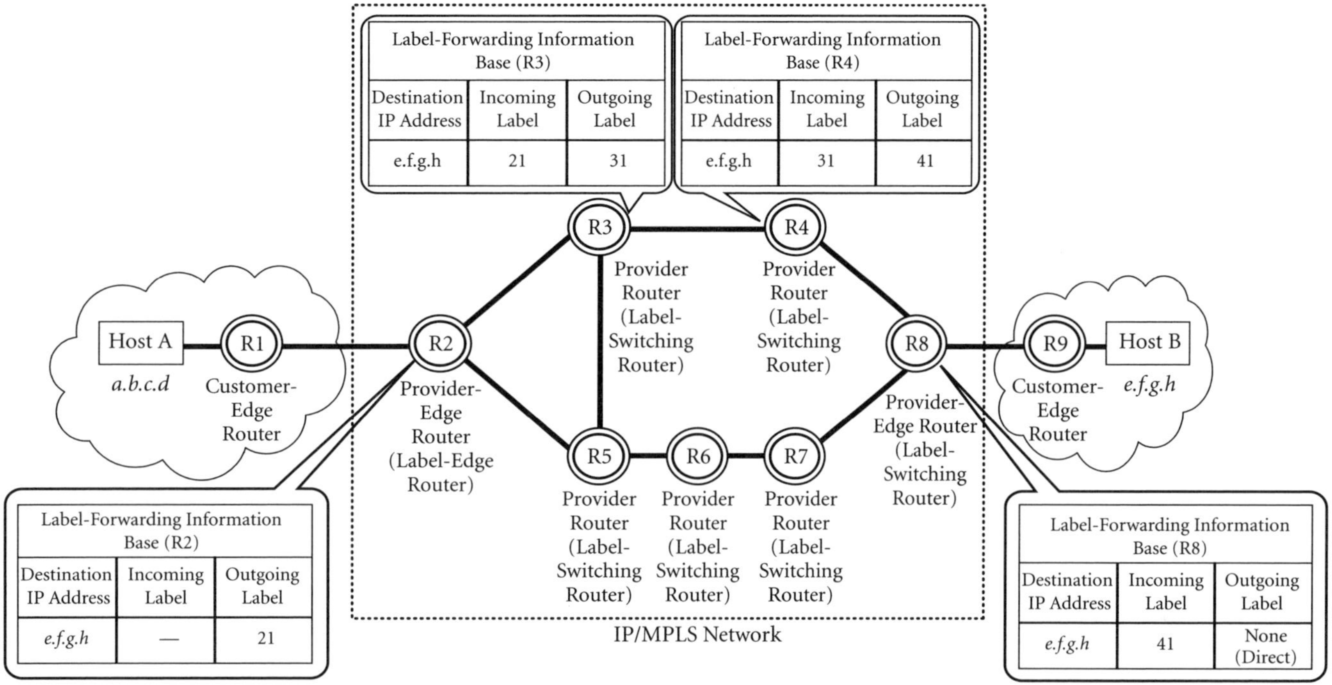

Figure 4.24: Packet Forwarding using MPLS

on some of the links through which the speech frames are passing. To set up a call, VoIP relies upon a less expensive and more scalable packet-switched core network. Furthermore, relatively low data rates can be adequate for VoIP due to efficient speech vocoders, and statistical multiplexing of VoIP packets during the silence interval can improve transport efficiency. Session Initiation Protocol is the main VoIP-related signaling protocol and gets help from three protocols: Session Description Protocol (SDP), Real-Time Transport Protocol (RTP), and Real-Time Transport Control Protocol (RTCP). We will look at all four protocols, starting with Session Initiation Protocol (SIP).[109]

4.5.1 Session Initiation Protocol

Session Initiation Protocol is an application layer control protocol that enables the creation, modification, and termination of multimedia sessions with one or more participants. A VoIP call and a multimedia conference are examples of multimedia sessions. SIP can run over any transport layer protocol, but UDP is typically used with the default port of 5060. The endpoints of a session such as participants are called *user agents*. A user agent that makes a SIP request is called a *User Agent Client*, and a user agent that responds to a SIP request is called a *User Agent Server*. Of course, the same device may be a User Agent Client at one time and a User Agent Server at another time. Other SIP entities include the proxy server, redirect server, registrar, and location service. A proxy server interprets a request message from a User Agent Client and forwards it (after possible modification of the request message) to an entity closer to the destination user. The proxy server may perform functions such as accounting as well as authentication and authorization for a user. A redirect server is a User Agent Server that directs a User Agent Client to some other entity by sending 3xx responses. A registrar accepts users' REGISTER requests and updates the location service with the information it has received. The proxy, location, and registrar servers are logical entities and may be implemented in the same physical device.

Figure 4.25 summarizes how SIP can be used to set up and terminate a session [153]. SIP is a text-based protocol, i.e., the message contents are coded as text. Jerry wants to make a VoIP call to his friend George. Jerry's *softphone*[110] will construct an INVITE request message to initiate the establishment of a SIP session. A SIP message consists of a start line specifying the message type, one or more header fields, an empty line indicating the end of the header fields, and an optional message body. The header fields, among other things, identify both the source user and the destination user. Each SIP user is identified by a SIP identity, i.e., a Uniform Resource Identifier (URI).[111] In this case, Jerry's URI is sip:jerry@abc.com and George's URI

[109] SIP is pronounced as the word *sip*, rather than as the individual letters.

[110] A *softphone* can be thought of as a PC or laptop that has been transformed into a *phone* by the SIP application *soft*ware [153].

[111] When you type a website name in the address portion of the web browser, you are typing a URI, which is also called a www address. An example of URI is the string http://www.cnn.com.

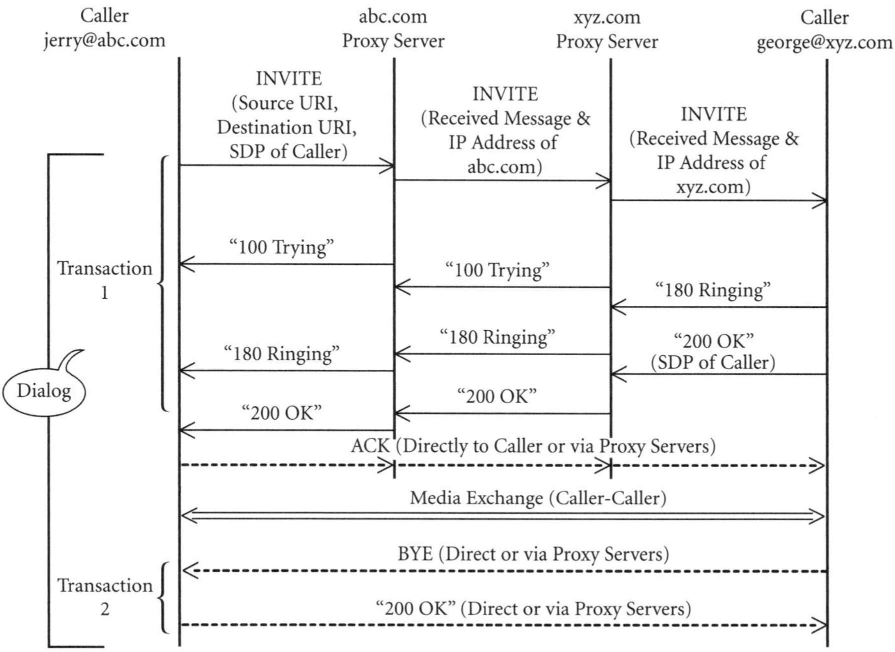

Figure 4.25: Session Setup and Termination using SIP

is sip:george@xyz.com. Within the URIs, abc.com and xyz.com are the domain names of the SIP service providers serving Jerry and George, respectively.

SIP itself provides no description of the session (e.g., the type of media and codec). Another protocol, such as SDP, is used for the session description, which resides in the message body of the INVITE request message. The header in the INVITE request message indicates SDP as the content type, and SDP in the message body would actually describe the media stream(s). The INVITE request message carries SDP just like an email message carries an attachment and an HTTP message carries a web page [153].

The INVITE request message goes from Jerry's softphone to the abc.com SIP proxy server. Jerry's phone could have obtained the IP address of abc.com proxy server via configuration or via DHCP. The abc.com proxy server sends a SIP response to Jerry upon receiving the INVITE request message. A SIP response includes a status line as the start line, which specifies a three-digit status code and an associated description. In our example, the abc.com proxy server sends the status code 100 and phrase "Trying" to acknowledge the reception of the INVITE request message and to inform the user that the proxy server is trying to forward the INVITE request message toward the destination.

The abc.com proxy server finds the IP address the of xyz.com proxy server using DNS,[112] adds a header to the original INVITE request message to specify its IP address, and forwards the message to xyz.com. Now, the xyz.com SIP proxy server acknowledges the reception of the INVITE request message with a "100 Trying" response. The xyz.com proxy server contacts a location service to retrieve George's current IP address. We'll see later that the location service learns via registration about a SIP user's current IP address. The xyz.com proxy server inserts its own header in the INVITE request message to specify its IP address and forwards the updated message to George.

George's SIP phone, which could be a regular phone with SIP functionality, rings to alert George. George's SIP phone sends a "180 Ringing" response to inform its proxy server about alerting the user. This response finds its way back to Jerry through the use of IP addresses available in the header fields in the SIP response. Jerry's softphone would then play an audio ringtone or display some message.[113] George does not want to jeopardize his friendship with Jerry and hence answers the call! George's phone sends a "200 OK" response, and this response finds its way to Jerry just like the "180 Ringing" response. The message body of this response describes the session George prefers. Recall that the INVITE request message contained the session description that Jerry preferred. To complete the session establishment, Jerry's softphone sends an acknowledgment (i.e., ACK) request message. Now, the media start flowing *directly between* the SIP-capable devices; note that the actual media do not traverse through proxy servers. Once a party decides to terminate the call (hopefully, amicably), a BYE request message is sent, and a "200 OK" message terminates the session.

The final request–response exchange may occur directly between the devices or via proxy servers. When the service operator wants to control the call (e.g., for billing purpose), proxy servers would be involved in the final message exchanges between the end devices. Such proxy server involvement can be realized using a special header called Record-Route. (See Chapter 14 on IP Multimedia Subsystem for additional details.) During the session, any endpoint can modify the session characteristics by sending an INVITE (often referred to as a Re-INVITE) request message again. When the other party sends a "200 OK" response to the new INVITE request message, the new INVITE request message supersedes the previous session characteristics.

The location service uses *registration* to learn about the current location[114] of a SIP user. Authentication is typically performed during registration. A SIP device sends a REGISTER message to the registrar server at initialization and again periodically. This message includes the association between the SIP URI and a specific IP address

[112]We will discuss DNS in Section 4.6.3.

[113]Ringback in SIP can be achieved using a local ringback or remote ringback. In local ringback, the user agent client receives the "180 Ringing" response and plays a ringback tone, which could be a default tone or a tone specified by the Alert-Info header. In remote ringback, the user agent server encodes the ringback tone as media packets and sends them to the UAC. For carrier networks, remote ringback is more common.

[114]From the perspective of SIP, *location* means a routable address in the network.

where the SIP user can be contacted.[115] SIP supports a challenge-based mechanism for authentication similar to the one used in HTTP [154]. Two user agents can authenticate each other, or a proxy server can authenticate the User Agent Client.

The Registrar Server responds to the registration request by sending a "401 Unauthorized" response. This response includes a header called WWW-Authenticate that provides numerous parameters, such as a random string called *nonce*, another string called *opaque*, the *realm* (e.g., Jerry@abc.com or George@xyz.com), the name of the authentication algorithm (e.g., Message Digest 5 or MD5 being the default), the QOP-Options field indicating the quality of protection (e.g., the value "auth" meaning authentication and the value "auth-int" meaning authentication with integrity protection), and the digest value generated by the authentication algorithm using various parameters, such as nonce, URI, username, and password. The client does not process the string opaque but simply returns the string unaltered in the Authorization header of a subsequent request message. The server uses nonce for authentication and for imposing time validity.

Upon receiving the "401 Unauthorized" response, the SIP user then goes through its own MD5 calculations using stored parameters (such as username and password) to calculate a 128-bit response to the random challenge nonce. The SIP user re-submits the REGISTER request with the Authorization header that contains the cnonce and response. Cnonce helps avoid plaintext attacks, facilitates mutual authentication, and provides some level of message integrity protection. If the Registrar Server finds the response to be valid based on its earlier MD5 calculations, it sends a "200 OK" response to imply successful authentication. SIP registration can be conducted securely by first constructing a secure encrypted tunnel between the SIP user and the server. Upon successful authentication, the Registrar Server stores this URI and IP Address association or binding in the location service database.

4.5.2 Session Description Protocol

SDP describes the multimedia session in sufficient detail to enable users to participate in the session [155]. SDP can be used by a variety of protocols, such as SIP and HTTP. The session description consists of a *session-level description* applicable to the entire session (and all the media streams within the session), *a time-level description* specifying the session times, and zero or more *media-level descriptions* applicable to a specific media stream. Let's understand the session description using an example shown in Figure 4.26.

The session-level description begins with the v= line. The v= field indicates the version of the SDP, which is 0. The o= or Origin field provides the following:

[115] SIP separates the concept of identity and address. The REGISTER message tells the network the user's current routable address. Separating identity and address enables *follow-me* services in which calls follow the user from home to office to car—wherever the user has registered. Furthermore, the user can register multiple identities, such as user@home, user@work, and user@school, to a single address. The user can also register the same identity at multiple addresses; in such a case, the system can either simultaneously ring at multiple locations or ring each location sequentially.

	v=0
	o=mhandley 2890844526 2890842807 IN IP4 126.16.64.4
	s=SDP Seminar
Session-level Description	i=A Seminar on the session description protocol
	u=http://www.cs.ucl.ac.uk/staff/M.Handley/sdp.03.ps
	e=mjh@isi.edu (Mark Handley)
	c=IN IP4 224.2.17.12/127
Time-level Description	t=2873397496 2873404696
	a=recvonly
	m=audio 49170 RTP/AVP 0
	m=video 51372 RTP/AVP 31
Media-level Description	m=application 32416 udp wb
	a=orient:portrait

Figure 4.26: Session Description Example
SOURCE: M. Handley and V. Jacobson, "SDP: Session Description Protocol," [155]. © The Internet Society (1998). Used with permission.

- username (e.g., mhandley),
- session ID typically specified as the Network Time Protocol timestamp (e.g., 2890844526),
- version number of the announcement or message containing the SDP, typically specified as Network Time Protocol timestamp (e.g., 2890842807),
- type of network (e.g., IN IP4 for Internet IPv4), and
- the IP address (e.g., 126.16.64.4).

The s= field mentions the session name (e.g., SDP Seminar). The i= field provides information about the session (e.g., A Seminar on the session description protocol). The i= fields are labels for media streams and hence are quite useful when one session includes multiple distinct media streams of the same media type (e.g., one whiteboard for the presentation slides and one whiteboard for questions [155]). The optional u= field specifies a URI that points to additional information about the session (e.g., `http://www.cs.ucl.ac.uk/staff/M.Handley/sdp.03.ps`). The e= field provides the email address. The p= field may be substituted for the e= field to provide the phone number. The c= or Connection field provides the network type, the address type, and the combination connection address and TTL field (e.g., IN IP4 224.2.17.12/127 for IPv4) corresponding to the data source, relay, or sink. /TTL is present only for a multicast IP address (e.g., a conference call) and is absent for a unicast IP address. The b= field, required bandwidth in kilobits per second, is an example of optional fields within the session-level description.

The time-level description begins with the t= line that has the format t=start time stop time, with the time expressed as Network Time Protocol seconds. The use of 0s as start and stop times would indicate a permanent session. Multiple pairs of start and stop times for the session may be specified. A regularly recurring session would specify repetition-related information.

The a= or Attribute field may be present at the session-level and/or media-level. A *property* attribute has a binary value as a session property (e.g., a=recvonly to indicate that the message sender would receive only the media stream), and a value attribute can assume one of several values as a media property (e.g., a=orient:portrait).

The media-level description begins with the m= line and includes the following:

- media type (e.g., audio and video),
- transport port to which the media stream will be sent,
- transport protocol (e.g., UDP and RTP/AVP[116] over UDP), and
- media formats (e.g., the RTP/Audio-Video Profile Payload Type 0 corresponding to the μ-law-companded pulse coded modulation audio with the sampling rate of 8 kHz).

Furthermore, the a=rtpmap field can be used to define the media coding for the given RTP Payload Type. For example, in a=rtpmap:98 L16/16000/2, the subfield 98 denotes the Payload Type, the subfield L16 is name of the encoding format, the subfield 16000 is the clock rate, and the optional subfield 2 is part of the list of additional encoding parameters, which in audio streams is the number of audio channels.

Though the i=, u=, and e= parameters shown in Figure 4.26 are valid SDP parameters, they are not commonly used in SIP. (Furthermore, t= is a required SDP parameter but is not very useful for SIP because signaling is used to indicate the start and end time of the session). t= is usually set to 0 0 or 287339746 0 to indicate the current start time; the end time is typically unspecified. Additionally, the c= line usually matches the IP address in the o= line.

4.5.3 Real-Time Transport Protocol and Real-Time Transport Control Protocol

Now, let's discuss the protocols that are instrumental in supporting VoIP media transmission: RTP and RTCP [156]. RTP implements end-to-end network transport functions to facilitate delivery of real-time media, such as audio and video, in a unicast or multicast fashion. Helping RTP, RTCP is a control protocol with features such as (minimal) QoS monitoring and control and participant identification via out-of-band signaling.

[116]RTP represents Real-Time Transport Protocol, and AVP represents audio-video profile.

Since different IP packets carrying media may be delayed by different times and since the IP packets may arrive out-of-order at the receiver, RTP specifies adequate information for the receiver to faithfully reconstruct the original media. An RTP packet includes a header and a payload; the twelve bytes of the fixed header are mandatory. The header can be extended by indicating such via an extension bit in the fixed header. The payload would contain the actual data, such as voice bits or video bits. The mandatory portion of the fixed header includes fields such as Payload Type, Sequence Number, Timestamp, and Synchronization Source.[117] The Payload Type field specifies the format of the RTP payload (e.g., Payload Type 0 for the μ-law-companded pulse coded modulation audio with the sampling rate of 8 kHz).

The documentation defines the default mappings for RTP audio and video [157]. The initial value of the sixteen-bit Sequence Number field is chosen randomly and is subsequently incremented by one for each new RTP data packet sent. The receiver can use the Sequence Number field to detect packet loss and to restore the original packet sequence. A thirty-two–bit Timestamp field corresponds to the sampling instant of the first byte in the RTP data packet. The clock frequency for the purpose of sampling is a function of the Payload Type field (e.g., 8 kHz for Payload Type 0). Just like the Sequence Number field, the initial value of the Timestamp field is also chosen randomly. The Timestamp field is included to enable the receiver to present the media to the user in a synchronized fashion. The Timestamp field helps minimize jitter (i.e., variations in delay). The Synchronization Source field indicates the source of RTP packets. RTP packets from a microphone and camera, both originating from the same source (e.g., the same host), have the same thirty-two–bit Synchronization Source field so the receiver can group the packets for playback.

RTCP periodically transmits control packets using the same delivery mechanism as the data packets. RTCP performs four different functions. The first and primary function is to provide feedback via sender reports and receiver reports to the sender regarding the quality of the data distribution, and such feedback may enable the sender to carry out adaptive encoding of RTP data packets (e.g., the addition of more redundancy to reduce the number of packet errors). The second function of the RTCP is to carry a persistent transport-level identifier for an RTP source called the *canonical name*, which keeps track of the participant if a program must be restarted or if two participants in a multicast happen to choose the same Synchronization Source field. The Synchronization Source field must change if two participants in a multicast happen to choose the same Synchronization Source field. The third function of RTCP is to control the rate of control packets. Since a given participant sends RTCP packets to all other participants, the control traffic could grow significantly as the number of session participants increases. In such situations, the RTCP dynamically adjusts the interval between RTCP packet transmissions. The fourth RTCP function is optional and transports session control information (e.g., display of the participant identification on the user interface).

[117] See the documentation for the full header details [156].

Let's talk a bit more about sender reports and receiver reports. Active senders transmit the sender reports to convey transmission and reception. Non-active senders send receiver reports containing reception statistics. Reception statistics are sent in sender reports and receiver reports as frequently as possible to provide good resolution of the statistics. The reports can be useful not only for senders and receivers but also for third parties that can quantify the performance of the network by just observing RTCP packets and thus avoid processing RTP data packets. Examples of performance metrics that can be calculated using the information in RTCP packets are packet loss rate, average payload data rate, average packet rate, and jitter [156].

4.6 A POTPOURRI OF IP-CENTRIC PROTOCOLS

Several miscellaneous protocols and mechanisms play an important role in the operation of cellular packet data networks. These protocols and mechanisms are DHCP, Mobile IP, DNS, RADIUS, Diameter,[118] Integrated Services, Differentiated Services, and Generic Routing Encapsulation (GRE), and IPsec. Let's take a look at each one.

4.6.1 Dynamic Host Configuration Protocol

DHCP provides two basic functions: allocating an IP address to a host and providing configuration information to a host (e.g., the subnet mask and the IP address of a default router). DHCP is intended for hosts, not routers. DHCP is based on the Bootstrap Protocol that enables a host with a minimal IP stack to obtain configuration information. DHCP messages have a structure similar to that for Bootstrap Protocol messages.

Three mechanisms are defined by DHCP to allocate an IP address. In the *automatic allocation* approach, DHCP allocates a permanent, static, or fixed IP address to a client. DHCP can be used by a system administrator to manually configure the IP address for a host in the *manual allocation* approach. Finally, DHCP temporarily allocates to a client an IP address from a pool of available addresses in the *dynamic allocation* approach. The dynamic allocation approach is attractive when IP addresses are scarce because each IP address can be reused for a different host when its original host is idle. The message exchange between the host and the server can extend the lease time if some data activity is still occurring.

Let's summarize how a DHCP server allocates an IP address to a DHCP client (i.e., a host). The DHCP client uses the UDP port 68 and the DHCP server uses the UDP port 67. The DHCP server has a pool of available IP addresses. A DHCP client broadcasts a DHCPDISCOVER message in the local physical subnet, and this message contains the hardware address such as MAC address of the host. One or more

[118]Diameter is not an acronym! The basic RADIUS protocol was enhanced, and, to show that the enhancement is bigger and better than RADIUS, the resulting protocol was given the name Diameter. So, the next time someone asks you what Diameter is, just say, "It is twice as good as RADIUS!" If you like this little joke, the credit goes to our colleague, Chris Reece. If you don't like it, well, what can we say, blame Chris Reece! Just kidding! The authors will take the blame for including this engineering humor.

DHCP servers may respond to such a message by sending to the client's specific hardware address a DHCPOFFER message containing the IP address the server selected for the client from a pool of IP addresses. The client accepts one of the offers and broadcasts the DHCPREQUEST message containing the selected IP address. The relevant DHCP server responds with a DHCPACK message that specifies additional configuration parameters, such as the lease time for the IP address. By observing the DHCPREQUEST message, other DHCP servers recognize that the client has chosen the IP address offered by another DHCP server. Hence, the IP addresses these servers had offered are returned to their pool for future allocation to another client. We have provided a glimpse of DHCP here; refer to the documentation [158] to dive deeper into DHCP.

4.6.2 Mobile IP

Mobile IP or IP mobility is quite critical for cellular networks to prevent the user from losing ongoing packet data sessions while moving. Using simple IP jeopardizes the end-to-end connectivity between the device and the application servers as illustrated in Figure 4.27.

The user has established a web-browsing session with a web server, such as www.cnn.com. The user is in the area or subnet controlled by a local Gateway X, and therefore Gateway X has assigned an IP address 130.200.5.6 to the user device. The

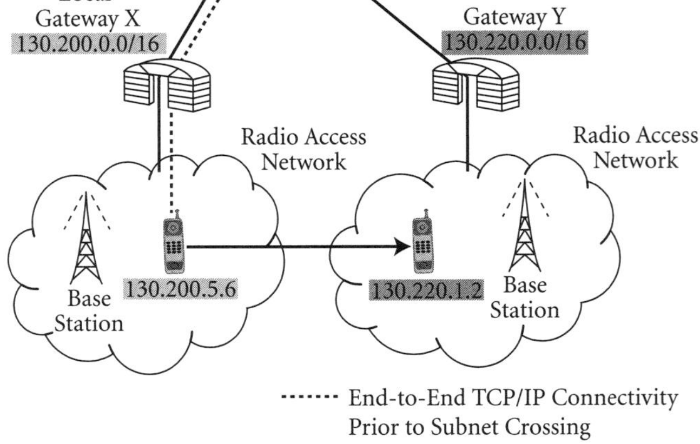

Figure 4.27: Shortcomings of Simple IP

HTTP requests from the device pass through the radio access network, Gateway X, and various routers in the Internet to reach the web server. The HTTP web server sends web pages in the form of IP datagrams that travel through the Internet routers, Gateway X, and the radio access network, eventually arriving at the device. When the user moves to an area controlled by another radio access network and another local gateway, Gateway Y, the end-to-end HTTP/TCP/IP session between the web server and the device is lost. Two radio access networks belong to two different subnets in our example. The packets for the device can reach just Gateway X due to packet forwarding based on the IP address 130.200.5.6 but cannot be sent over-the-air to the device; the user is no longer reachable in the previous radio access network. For the HTTP/TCP/IP session to remain valid across the subnets, the device's IP address must be maintained and routed to the new geographical area. Note that packets containing the IP address that local Gateway X allocated to the device (i.e., 130.200.5.6) cannot be routed to the new subnet controlled by Gateway Y. In summary, the use of simple IP cannot guarantee seamless mobility and data session continuity when the users are moving around in cellular systems. That's where the mobile IP comes to our rescue!

Figure 4.28 shows the same mobility scenario with one of several possible solutions. In Figure 4.28, we have introduced a new gateway designated as the home gateway,[119] which allocates a *home address* (HoA) to the device. The device will use this HoA when it wants to communicate with the outside world, such as a web server.

As part of the IP address allocation mechanism, the home gateway also knows which local gateway is taking care of the mobile device and the *care of address* (CoA) that the local gateway has allocated to the mobile device. Let's say that the mobile device is initially in the area Gateway X controls. The packets destined for the mobile device originate at the web server, pass through the Internet routers, and reach the home gateway. (The packets first reach the home gateway, not the local gateway, because packet forwarding is based on the HoA, which the home gateway owns and allocates.) The home gateway looks at its database and finds that Gateway X is the correct gateway and that the associated CoA is 130.200.5.6. Since Gateway X owns this CoA, the packets travel through the cellular service provider's private IP network to go from the home gateway to Gateway X. The packets will then reach the mobile device via the radio access network.

Now, let's see what happens when the mobile device moves from one geographical area (e.g., a city) to another. The mobile device, the new local gateway, Gateway Y, and the home gateway communicate with one another such that Gateway Y allocates a new CoA (130.220.1.2) to the mobile device and the home gateway re-registers the mobile device to reflect the change in the local gateway and the CoA for the mobile device. The packets from the web server continue to reach the home gateway. Since the home gateway has updated its database to reflect the new location of the mobile device, it forwards the packets to Gateway Y based on new CoA. Gateway Y forwards the packets to the radio access network, and the packets eventually reach the device.

[119]Home gateway is our nomenclature for the purpose of explanation and is not a standard term. In the Internet Engineering Task Force world, the home gateway is the home agent.

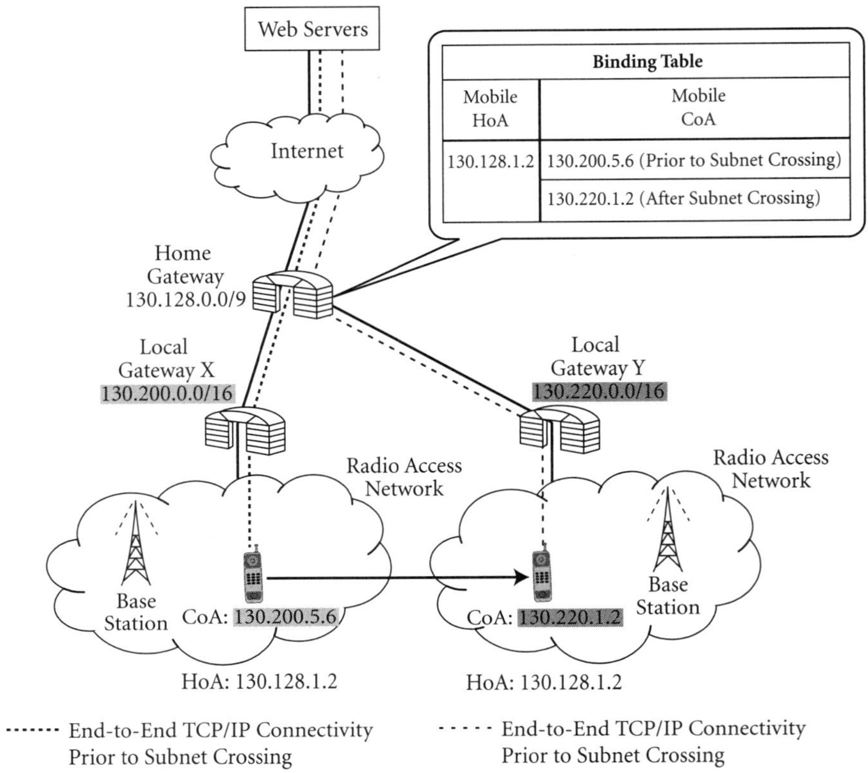

Figure 4.28: Basic Operation of Mobile IP

In summary, the preservation of the IP address (i.e., HoA), the anchoring of mobility at the home gateway, and the maintenance of the home gateway database with information regarding the local gateway for a given device facilitate seamless IP mobility even when the users are moving in a cellular system.

Something to Think About. . . Mobile IP operates like a post office. Professor Jones usually gets his mail delivered to his home address in Blacksburg, Virginia. He goes on a sabbatical in Washington, D.C., to work on a government-sponsored science project. He would like to get his mail delivered to his hotel address in D.C. He contacts the post office and provides the hotel address. Now, upon receiving mail addressed to Prof. Jones's home address, the post office forwards the mail to the D.C. hotel address. Think of the post office as a home gateway and the home and hotel as local gateways. Just as the post office knows where to direct the mail, the home gateway's database indicates which local gateway should receive the mobile's IP packets to maintain an end-to-end packet data session!

Let's briefly discuss two main flavors of Mobile IP: Client Mobile IP and Proxy Mobile IP. Their main differences are related to the device's role; in Client Mobile IP, the device itself understands and works with Mobile IP messages defined in the documentation [159], but in Proxy Mobile IP, the device has no role to play. Figure 4.28 is directly applicable to Client Mobile IP. We'll discuss one typical Client Mobile IP scenario widely used in commercial 3G 1xRTT and 1xEV-DO deployments. Two mobility agents are defined for Client Mobile IP: Foreign Agent (FA) and Home Agent (HA). The local gateway performs the FA functions, and the home gateway performs the HA functions. Specifically, the network sends an Agent Advertisement message[120] using a special extension to the regular Router Advertisement message. The mobile node then acquires the CoA from the Agent Advertisement message and requests an HoA via a Registration Request message. This CoA is called the *foreign agent care-of address*,[121] where the CoA is an IP address of the FA itself. Such a CoA can be shared among multiple mobile nodes, thereby alleviating pressure in the limited IPv4 address space. We'll focus on this type of CoA here. The FA forwards the Registration Request message to the HA, which responds with a Registration Reply message containing the HoA for the mobile node. The Registration Reply message goes from the HA to the FA to the mobile node. As a result of registration, a Mobile IP tunnel between the HA and the FA is established and carries an encapsulated IP datagram from the HA to the FA.

See Figure 4.29 to visualize this Mobile IP tunnel. Assume that a server sends IP datagram X to the mobile node. The routers forward this datagram to the HA because the destination IP address is the IP address of the mobile node that the HA *owns* (and allocates). The HA recognizes the IP address that it has allocated to the mobile node and uses the Mobile IP tunnel to forward the IP datagram to the FA that is currently in charge of the mobile node. More specifically, the HA puts IP datagram X, which becomes an inner IP datagram, in another *outer* IP datagram. The outer IP datagram has the HA address as the source IP address and the FA address as the destination address. The basic idea behind these addresses is that the IP network between the HA and the FA can easily forward the IP datagrams for the mobile node irrespective of the IP address of the mobile node itself. When the mobile node moves from one FA to another FA, its IP address remains the same and the mobile IP tunnel between the new FA and the HA would carry the mobile node's IP datagrams. Though Figure 4.29 shows a packet moving through the tunnel from the HA to the FA and then from the FA to the mobile node, the packet originating from the mobile node to the HA traverses the reverse path.

The IP header of the outer IP packet uses the HA IP address and the FA's address (i.e., the mobile node's CoA) as the source and destination addresses, respectively,

[120]This message is sent usually autonomously but sometimes in response to an Agent Solicitation message (which is a special extension to the basic Router Solicitation message). Recall that both Router Solicitation and Router Advertisement messages are defined by ICMP.

[121]Another kind of CoA is called a *co-located care-of address*, which the mobile node obtains via means such as DHCP. When such a CoA is used, the Mobile IP tunnel runs between the HA and the mobile node, and the mobile node itself decapsulates the IP datagrams.

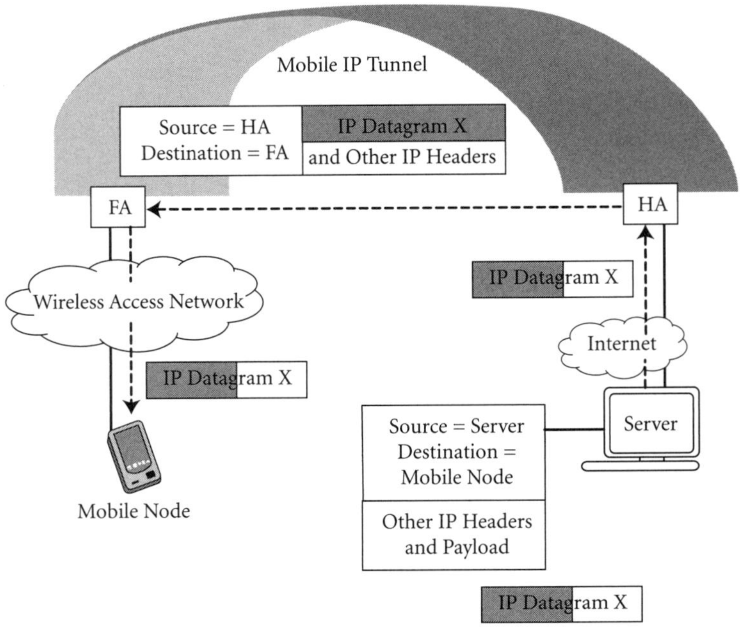

Figure 4.29: Mobile IP Tunnel

for the outer IP packet. The original IP packet is intact inside the outer IP packet. The FA decapsulates the IP packet and delivers the inner original IP packet to the mobile node. Now, the mobile node learns the HoA and the time period for which the registration is valid (i.e., the lifetime of the registration). The registration can be renewed prior to expiration. Furthermore, the mobile node observes the Agent Advertisement message, and upon detecting a change in the FA area, it performs (re)registration using the same Registration Request and Reply messages.

The Registration Request and Registration Reply messages could include authentication extensions (e.g., the Mobile-Home Authentication field) to facilitate Mobile IP authentication. The Mobile-Home Authentication field includes parameters such as the Security Parameter Index and Authenticator. The Security Parameter Index specifies the authentication algorithm and the secret type (e.g., a shared key or a public and private key pair). The Hash-Based Message Authentication Code-Message Digest 5 is the default authentication algorithm. The Authenticator field contains the output of the authentication algorithm (e.g., 128-bit Hash-Based Message Authentication Code) calculated using inputs such as the Registration Request and Reply message bits and the secret type.

IP datagrams from the outside world to the mobile node pass through the HA, but the IP datagrams originating from the mobile node may go directly to the Internet via the FA. Service operators, however, often want traffic originating from mobile nodes to pass through the HA (e.g., for data usage statistics) in a process called *reverse tunneling*. The mobile node requests reverse tunneling by setting the reverse tunneling bit (or the T field) to 1 in the Registration Request message.

Now that we know how Client Mobile IP works, let's spend few moments on the basics of Proxy Mobile IP. Using Proxy Mobile IP with IPv6 (referred to as PMIPv6) within the network is the trend in emerging 4G cellular networks; of course, the network could allocate an IPv4 or IPv6 HoA to the mobile node based on the capabilities of the mobile node and the network operator policies. Proxy Mobile IP is often considered *network-based mobility* because the network is responsible for managing IP mobility [160]. The mobile node is uninvolved in any IP mobility signaling; the signaling required to provide seamless IP mobility to the mobile node is done entirely within the network. Two main mobility entities are the Local Mobility Anchor (LMA) and the Mobile Access Gateway (MAG) as shown in Figure 4.30. The LMA is like a home agent; it is the anchor for the IP address or network prefix[122] and hides the mobility of the mobile node from the outside world. The MAG interfaces with the mobile node's access network to track the mobile node's movements. The MAG, on behalf of the mobile node, informs the LMA about the mobile node's location. A bidirectional (Proxy Mobile IP) tunnel is established between the LMA and the MAG to support IP mobility. As the user moves, the tunnel between the LMA

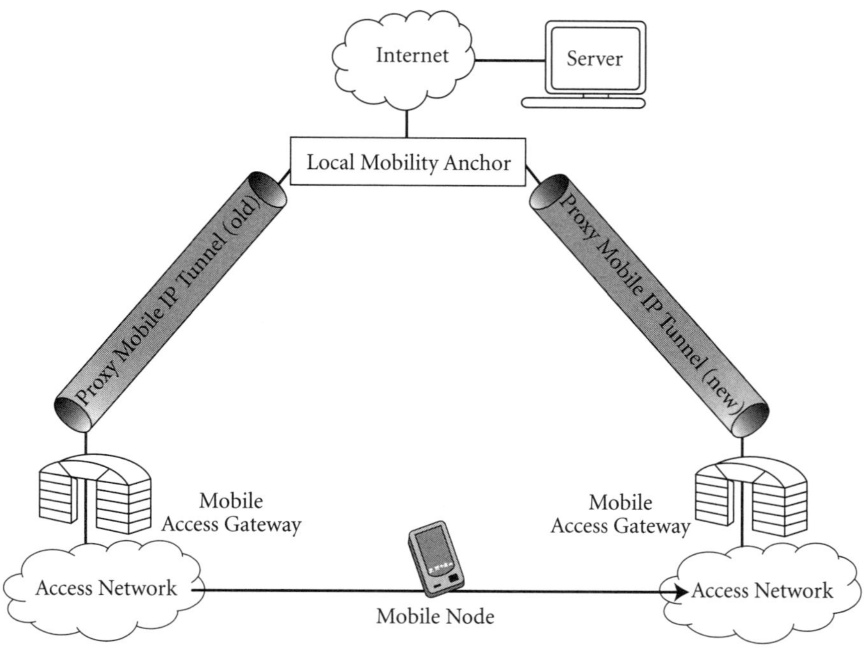

Figure 4.30: Seamless Mobility via Proxy Mobile IP

[122]The network prefix is an IPv6 term and refers to part of the overall IPv6 address.

and the previous MAG is replaced by a new tunnel between the LMA and the new MAG.

Let's take a quick look at basic signaling messages involved in the implementation of PMIPv6. Assume that the mobile network attaches to the network, perhaps after power-on. The mobile node sends a Router Solicitation message that reaches the MAG via the access network. Note that the MAG acts like a default router for the mobile node. The MAG sends a Proxy Binding Update message to inform the LMA about the mobile node's existence. The LMA responds with a Proxy Binding Acknowledgment message and sets up its endpoint of the bidirectional Proxy Mobile IP tunnel. The MAG then sets up its endpoint of the tunnel. The MAG sends the Router Advertisement messages in response to the original Router Solicitation message. The mobile node then obtains its IP address based on the configuration information present in the Router Advertisement Message. Now that the tunnel is set up and the mobile node has a valid IP address, IP packets can flow between the mobile node and the outside world.

Now assume that the mobile node changes its point of attachment. The previous MAG informs the LMA about the mobile node's detachment. The LMA removes the binding with the previous LMA and waits for the Proxy Binding Update message from the new MAG. Once the new MAG sends this message to the LMA and the LMA responds with a Proxy Binding Acknowledgment message, a new Proxy Mobile IP tunnel is established. The new MAG then provides configuration parameters to the mobile node in the Router Advertisement message; the IP address of the mobile node (i.e., HoA) remains intact.

3G UMTS systems use no Internet Engineering Task Force compliant Proxy Mobile IP, but they do implement *network-based mobility* in a UMTS-specific fashion. A gateway called a Gateway General Packet Radio Service (GPRS) Support Node allocates an IP address to the mobile device, and a GPRS Tunneling Protocol tunnel is established between the Gateway GPRS Support Node and a gateway called Serving GPRS Support Node (SGSN). As the user moves, the Gateway GPRS Support Node side of the tunnel endpoint remains intact, but the SGSN side of the tunnel endpoint changes with a change in the SGSN. An SGSN is assigned to a specific geographical area and other SGSNs to other areas. The SGSN can be thought of as a local gateway, and the Gateway GPRS Support Node can be thought of as a home gateway in Figure 4.28; however, no CoA is given to the mobile node. Hence, depending upon of the mobile node's physical location, the IP packets for the mobile node travel through the GPRS Tunneling Protocol tunnel between the Gateway GPRS Support Node and a suitable SGSN.

4.6.3 Domain Name System

One of the DNS's important functions is to provide the mapping between a symbolic name (of a server) and the IP address. Furthermore, through the use of the Naming Authority Pointer Record, the DNS can locate special servers (e.g., SIP proxy servers and mail servers) and some IMS nodes within a specified domain. When we want to read news from the CNN website, we need not remember the IP address of the CNN website; we just type in www.cnn.com in the web browser window. The DNS provides the mapping between the symbolic name www.cnn.com and the IP address of the web server responsible for cnn.com. The two basic types of messages—queries and replies—are defined to implement the DNS. The query includes a question section, and the replies and forwarded queries have additional sections, such as the answer section. The question section contains the symbolic name, and the answer section includes the relevant IP address. The messages are sent as UDP datagrams or TCP packets. Let's take a look at the operation of the DNS and introduce some relevant terminology.[123]

A domain name that is complete (e.g., www.cnn.com. with the trailing dot formally required but typically unnecessary) is called a *fully qualified domain name* or an *absolute domain name*. www.cnn would be an *incomplete* domain name, and the DNS may make it an FDQN by appending a suffix, such as .com. The domain names follow a hierarchical tree structure with the *root* as the root of such tree. The top-level domain names are referred to as *generic top-level domains*. Examples of generic top-level domains are com for commercial organizations, edu for educational organizations, and gov for U.S. government agencies.

The mapping between the symbol name and the IP address is retrieved, stored, and later provided upon request by *DNS name servers*. When the symbolic name www.cnn.com is entered in the Internet Explorer window, a program called a *name resolver* is executed and communicates with the name server of the local domain where the host is located. The name server helps the host obtain the required IP address. Consider Figure 4.31, which depicts two basic approaches—an iterative approach or mode and a recursive approach or mode, with the support for the iterative approach being mandatory—of finding the IP address associated with a symbolic name.[124] First, we will discuss the iterative approach and then we will look at the recursive approach.

Host A is part of the abc.edu domain and needs to find the IP address of www.xyz.com. Host A's name resolver program sends a query to its abc.edu name server. The name server fails to find the required IP address and hence forwards the query to the root name server. Numerous countries have contributed to the set of global root servers. The root server replies with the IP address of the .com name server. The abc.edu name server now sends a query to the .com name server, which replies with the IP address of the xyz.com name server. The abc.edu name server contacts the xyz.com name server, which returns the IP address of the www.xyz.com web server. Now, the abc.edu name server is able to provide to Host A the required IP address from its name server. In summary, the iterative approach keeps the local name server in charge, and this name server iteratively exchanges query and reply messages with other name servers to retrieve the required IP address.

According to the recursive approach, the local abc.edu name server sends the query to the root server, and the root server forwards the query to the .com name server.

[123] See the documentation [117, 161, 162] to dig deeper into the DNS.

[124] A flag bit in the query message indicates the type of query preferred by the client, and a flag bit in the response message specifies the server's support of the recursive queries.

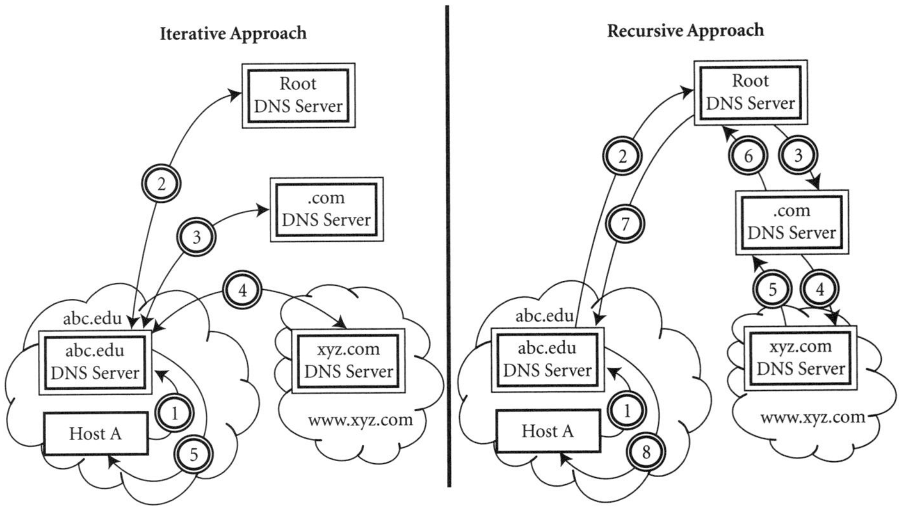

Figure 4.31: Operation of the DNS: Iterative and Recursive Approaches

The .com name server forwards the query to the xyz.com name server. The xyz.com name server knows the exact IP address of the www.xyz.com web server and replies to .com with the relevant IP address. The .com name server sends the reply to the root server, which replies to the abc.edu name server. Now, Host A receives the IP address from its name server. In summary, the recursive approach involves successive queries from one name server to the next until the name server with knowledge of the required IP address is found. Once such a name server is found, the replies follow the reverse path to eventually reach the original host requesting the IP address.

4.6.4 RADIUS and Diameter

RADIUS is used by various cellular networks, such as CDMA2000 1xRTT and 1xEV-DO. RADIUS facilitates authentication, authorization, and accounting. First, we will discuss the use of the basic RADIUS protocol for authentication and authorization, and then we will summarize how it is used for accounting.

RADIUS follows the client/server model [163]. A network access server acts as a RADIUS client and forwards information about the user to a suitable RADIUS server. Examples of a network access server are a wireless gateway and ISP gateway. The RADIUS server authenticates the user and sends relevant configuration information to the client so that the client can provide services to the user. The RADIUS server supports numerous authentication mechanisms, such as Password Authentication Protocol and Challenge Handshake Authentication Protocol. The transactions between the RADIUS client and server are also authenticated via a shared secret. The user passwords passing between the client and the server are encrypted for additional security.

In support of authentication, the client constructs an Access-Request message that includes information such as the user name, an encrypted user password, and the port the user is accessing. When the RADIUS server receives the request, it first validates the client itself based on a shared secret. Once the validity of the client is confirmed, the server tries to match the received user name and password with its database entries. The server can also evaluate whether the user is authorized to access certain ports associated with specific services or applications. The server may also throw a challenge (e.g., a random number) to the user by constructing an Access-Challenge response. The client uses the information in the Access-Challenge response to authenticate the user using some other protocol, such as Point-to-Point Protocol, Mobile IP, and SIP. The authorized user can correctly calculate the response based on an algorithm and a secret key. Upon obtaining the user's response to the challenge, the client re-sends the Access-Request message to the server, which contains the encrypted response to the challenge instead of the user password. Finally, the server could respond to the new Access-Request message by sending an Access-Accept message to the client, which would eventually reach the user. This message contains configuration information for the user, such as an IP address for simple IP and a desired compression technique. If the client gets no response from the server within a certain time period, the server can re-send the request multiple times and can also forward requests to another server.

A RADIUS packet includes fields, such as Code, Length, Authenticator, and Attributes.[125] The Code field indicates the type of packet (e.g., a setting of 1 indicates the Access-Request message and a setting of 2 the Access-Accept message). The Length field specifies the total length of the packet excluding the padding. The Authenticator field is used by the client to validate the reply from the RADIUS server. In the Access-Request message, this field contains a random number. RADIUS uses UDP as the transport layer protocol due to RADIUS' simplicity of implementation, inherent reliability, and lower overhead. Just one RADIUS packet is inserted in the UDP Data field, and UDP port = 1812 is used in the request messages. The reply exchanges the source and destination ports.

Now, let's see how RADIUS can be used for accounting [164]. Similar to RADIUS authentication, RADIUS accounting follows a client/server model. A network access server acts like a RADIUS client, conveying the user accounting information to a specific RADIUS accounting server. The RADIUS accounting server informs the client about the successful reception of the request. When service delivery begins, the client constructs an Accounting Start packet that specifies the type of service and the associated user. The client sends such a packet to the server, and the server responds with an acknowledgment. Once the service delivery is over, the client generates an Accounting Stop packet that specifies the type of service and (optionally) service statistics, such as elapsed time and the number of input and output octets, or the

[125]See the documentation [163] for full details of the packet format.

number of input and output packets. The client sends such a packet to the server, and the server acknowledges the reception of the packet. The RADIUS packet format for accounting is similar to that for authentication. The code value of 4 corresponds to the Accounting-Request message, and the code value of 5 corresponds to the Accounting-Response message.

Diameter enhances the basic functionality of the RADIUS protocol [165, 166]. However, Diameter is not backward-compatible with RADIUS. Although RADIUS uses UDP, Diameter uses TCP or SCTP to provide reliability. Diameter supports security mechanisms such as IP security, where payload can be encrypted and the authenticity of the sender can be validated. Diameter typically follows a client-server protocol, but some server-initiated messages are also allowed. Diameter enables dynamic discovery of peers. Reliability is further enhanced using features such as application layer acknowledgments, error notification, and failover methods. Compared to RADIUS, Diameter can be easily extended by defining new commands and attributes. RADIUS has been widely used in 3G cellular standards (e.g., 1xEvolution–Data-Optimized), but emerging 4G cellular standards, such as Long-Term Evolution, use Diameter.

4.6.5 IP Quality-of-Service Implementation: Integrated Services and Differentiated Services

The lack of QoS support in IP was one of the primary reasons for the selection of ATM instead of IP to transport information within the second- and third-generation cellular infrastructure network. The basic IP provides best effort service with no guarantees of bandwidth, delay, or packet loss rate. Now, however, IP is indeed capable of controlling QoS, thanks to two basic approaches: Integrated Services (IntServ) and Differentiated Services (DiffServ).[126] Let's explore how IntServ and DiffServ provide QoS in an IP network.[127]

4.6.5.1 Integrated Services Model In the IntServ model, resources required to meet a target QoS for a given *flow* (e.g., video streaming) are reserved throughout the sender-to-receiver path through signaling prior to actual data transfer [167]. The IP routers must remember the flow parameters (e.g., QoS parameters) to implement QoS-compliant forwarding of IP packets. Before discussing QoS implementation, let's get familiar with the relevant terminology. A flow consists of a stream of related IP datagrams from a unique sender to a unique receiver [117]. A flow is associated with a *flow descriptor*, which consists of a *filter specification* or *filterspec* and a *flow specification* or *flowspec*. The filterspec indicates the sender's IP address and source port. The flowspec is divided into two parameters, *traffic specification* and *service request specification*. The traffic specification outlines the traffic characteristics of the

[126]Integrated Services and Differentiated Services are also sometimes abbreviated as IS and DS, respectively.

[127]The industry trend is to rely upon an architecture called Policy and Charging Control to specify and enforce QoS for a given application. See Chapter 14 for more details.

requested service in the form of two token bucket filter parameters: token rate r (e.g., X bytes per sec) and the bucket capacity b (e.g., Y bytes). The token bucket essentially acts as a flow control mechanism; tokens are added to the bucket periodically at rate r and the router lets an IP packet pass through only if the packet length is less than or equal to the number of tokens in the buffer. The amount of data passing through a router would not exceed $rT + b$ during the time period of T seconds. The service request specification is the requested QoS and can be quantified by parameters such as bandwidth, maximum packet delay, and maximum packet loss rate.

The IP router uses components such as admission control, packet classifier, and packet scheduler to control the traffic passing through it. The admission control component determines if the QoS request for a new flow should be accepted based on the availability of routing resources. Once the admission control has agreed to serve a flow, flow-related information is given to the packet classifier and packet scheduler components. The packet classifier associates an incoming packet to a certain service class based on the source and destination IP addresses and port numbers. The packet scheduler forwards the received IP packets based on the service class to ensure that the required QoS is achieved.

The IntServ framework utilizes the Resource Reservation Protocol [168] to reserve resources required for the target QoS. Figure 4.32 illustrates how Reservation Protocol is used to support the IntServ model.

Two major Reservation Protocol messages play an important role in the resource reservation, Path and Resv. In the IntServ model, the IP packets of a given flow travel the same path. The Path message is sent from the sender host to the receiver host via a set of routers, and the usual IP routing algorithms determine which routers will be part of a given path.

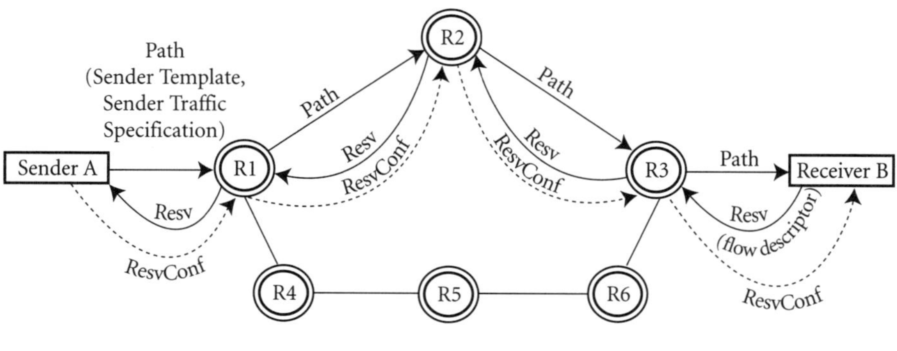

Figure 4.32: Integrated Services: Resource Reservation using Reservation Protocol

A Question for You... What are the examples of the routing algorithms? See the footnote[128] to verify your answer!

Sender A sends the Reservation Protocol Path message to router R1, which stores the IP address of the previous hop, replaces that address with its own IP address, and forwards the modified message to router R2. The Path message includes the *Sender Template* field, which specifies the sender's IP address (and optionally the sender's UDP/TCP port), and the *Sender Traffic Specification* field, which specifies the traffic characteristics of the flow the sender will generate. The Sender Template field helps identify the packets of the specific flow or sender. The Sender Traffic Specification field helps avoid over-reservation, leading to better utilization of network resources. Router R2 stores the IP address of R1 and forwards the modified Path message to router R3. The Path message finally reaches Receiver B. By this time, all routers know that the sender will generate a flow with certain traffic characteristics. In a typical IntServ scenario, the receiver would now send a Resv (i.e., Reservation) message containing the flow descriptor to request a specific QoS. Recall that the flow descriptor includes filterspec and flowspec. The Resv message follows the reverse path from the receiver to the sender as shown in Figure 4.32. When a Resv message arrives at the router, it executes admission control and configures the packet classifier with the filterspec information and the packet scheduler with the flowspec information. The receiver could request a reservation confirmation message, and the sender would then send the ResvConf message upon reception of the Resv message from its immediate router. IntServ is suitable for small networks and thus is not scalable. The need for explicit signaling for QoS establishment is another drawback of IntServ.

4.6.5.2 Differentiated Services Model Unlike the Integrated Services model, the Differentiated Services model needs no QoS-related signaling messages [169]. In the Differentiated Services model, QoS is differentiated based on service types and not individual service flows. For example, all voice service flows may be aggregated into one service type, and all video flows may be aggregated into another service type. The DiffServ model involves *classification* and *conditioning* of traffic and association of traffic with the *behavior aggregate*. The behavior aggregate is identified by a Differentiated Services Code Point (DSCP), which is a set of packets with the same DSCP that traverses a link in a given direction. Once the IP packet is marked with a certain DSCP, the packet is forwarded based on the per-hop behavior (PHB) associated with the DSCP. The mapping between the DSCP and the PHB may follow Internet Engineering Task Force recommendations [170] or may be locally customized by a network administrator [170]. The services experience different QoS depending upon traffic conditioning and PHB and DSCP-based packet forwarding. A PHB forwards the packets of a given DiffServ behavior aggregate in a certain way. For example, a PHB may guarantee a behavior aggregate a minimal bandwidth allocation of $X\%$ of an outgoing link during a certain time window [169]. The PHB can be quantified in

[128] RIP, OSPF, and BGP are examples of IP routing protocols.

numerous ways (e.g., buffer and bandwidth priority relative to other PHBs and delay and packet loss or drop rate for a PHB).

Traffic *classification* essentially steers the packet that matches a predefined rule to a certain element of the traffic conditioner for further processing. A behavior aggregate classifier uses DSCP matching. A multi-field classifier matches parameters, such as source IP address, destination IP address, DiffServ field, protocol ID, source port, and destination port numbers.

A traffic *conditioner* may include elements such as *meter*, *marker*, *shaper*, and *dropper*. A *meter* compares the traffic stream with a *traffic profile* and decides whether a packet is in-profile or out-of-profile. A *traffic profile* quantifies the temporal properties of the traffic stream. The traffic profile may look like "For DSCP $= X$, use token-bucket (r, b)" [169]. A packet arriving at a time when the number of tokens is insufficient is classified as an out-of-profile packet. A *marker* marks the DiffServ field of the packet with a specific DSCP. For example, the marker may mark the packet with one of the DSCPs to associate the packet with a specific PHB in the supported or configured PHB group by considering its traffic profile determined by the meter. A *shaper* delays packets and a *dropper* (as the name reveals) drops packets so that the outgoing traffic conforms to the requirements of the target traffic profile. Allowing in-profile packets to enter the DiffServ domain without further conditioning and changing the DSCP are two examples of *traffic conditioning*.

The DiffServ field in the IP header consists of eight bits with six bits specifying the DSCP and two bits currently unused (CU). The DSCP space of sixty-four possible values is divided into three pools for allocation and management. Pool 1 consists of thirty-two values recommended by the standard. Pool 2 has sixteen DSCPs and is reserved for experimental or local use. Finally, Pool 3 consists of sixteen DSCPs that are initially available for experimental or local use but that may be standardized if Pool 1 is exhausted. Eight recommended DSCPs are specified [170]. The DSCPs that are backward-compatible with the IP Precedence field defined in IPv4 are called class selector DSCPs, and they use only the first three of six bits.

We'll summarize here the interpretation of DSCPs.[129] The DiffServ standard uses the same three IP precedence bits for the priority setting and describes the use of the next three bits in the DSCP [171]. For the first three bits, corresponding to the original IP precedence field, DiffServ reserves two bit patterns, `111` and `110`, for the link layer and routing protocols and uses the remaining six bit patterns to define four *assured forwarding* classes, one *expedited forwarding* class, and one *default forwarding* class [171,172]. The *assured forwarding PHB* guarantees a certain amount of bandwidth and provides extra bandwidth if available [173]. The four assured forwarding classes are specified by the three-bit patterns `100`, `011`, `010`, and `001`. For each assured forwarding class, three packet drop probabilities correspond to the bit patterns `010`, `110`, and `100`. In other words, a total of twelve different levels of service corresponding to twelve DSCPs are possible with assured forwarding. The *expedited forwarding PHB* is suitable for a low-loss, low-latency, low-jitter,

[129]See the documentation [171] for additional details.

and assured bandwidth [174]. DSCP = 101110 is recommended for the expedited forwarding PHB. The *default forwarding PHB* provides best-effort service and is associated with the default DSCP = 000000. In summary, a total of fourteen DSCPs are standardized, with twelve for assured forwarding and one each for expedited forwarding and default forwarding.

Before we leave the topic of DiffServ, let's take a quick look at the DiffServ architecture. A DiffServ domain consists of boundary nodes and interior nodes where a boundary node connects one DiffServ domain to another DiffServ or non-DiffServ domain and an interior node connects to another node within the same DiffServ domain. Both types of nodes implement same PHB for a given DSCP. Furthermore, the boundary nodes may carry out traffic conditioning specified by a traffic conditioning agreement between their DiffServ domain and a neighboring domain. A host in a DiffServ domain can act like a boundary node for its own application traffic. One or more contiguous DiffServ domains constitute a DiffServ Region. To implement QoS for the services that span across the domains of the DiffServ Region, the same service provisioning policy and the PHB-DSCP mappings could be used. Another option is to establish a Service Level Agreement, which is a service contract between the customer and the service provider that specifies the type of forwarding service a customer should receive. The Service Level Agreement may also include the definition of a complete or partial traffic conditioning agreement.

DiffServ is much more widely used than IntServ in wirleess networks due to its distinct advantages. First of all, DiffServ needs no explicit QoS signaling. It is scalable and suitable for large networks. Furthermore, it requires minimal configuration.

4.6.6 IP Tunneling, GTP, and GRE

The IP router forwards an incoming IP packet toward its destination based on the packet's destination IP address. In a cellular network, the mobile device may get an IP address in Area A, which Gateway X controls, and then move away from that area into Area B, which Gateway Y controls. Assume that Gateways X and Y are responsible for allocating IP addresses to the mobile devices. Since these gateways own these IP addresses, the IP packets from the Internet can certainly find their way to the original gateway and, hence, Area A, but the mobile device is no longer in Area A where it received its IP address. The IP packets now need to be sent to Area B, i.e., where the mobile device actually is. Since the mobile device's IP address is associated with Area A, we need some mechanism that helps us forward the IP packets from Area A to Area B. IP tunneling, or IP encapsulation, is the mechanism that will come to our rescue! Figure 4.33 illustrates the scenario for IP tunneling.

The mobile device receives its IP address, 100.100.10.20, in Area A from Gateway X. Initially, the mobile device receives its packets through Gateways X and X1. The mobile device moves to Area B where Gateway Y is in charge of allocating IP addresses. To maintain connectivity between the mobile device and the server, the mobile device's IP address remains unchanged. Furthermore, an IP tunnel is established between Gateway X and Gateway Y2. This IP tunnel is created by encapsulating the original IP packet inside an outer IP packet. The IP packet traveling from

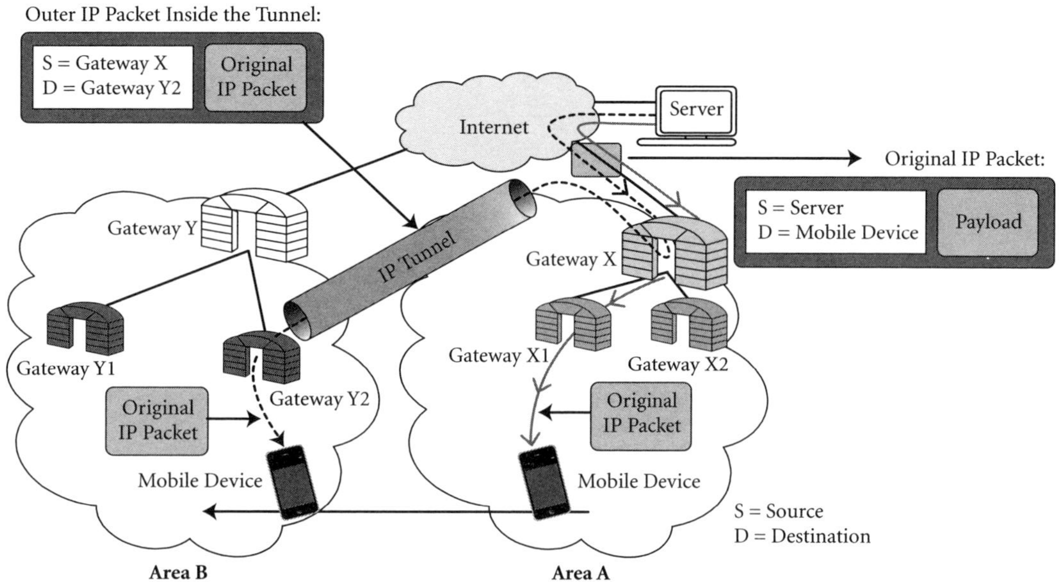

Figure 4.33: IP Tunneling Scenario

Gateway X to Gateway Y2 uses Gateway X's IP address as the source IP address and Gateway Y2's IP address as the destination IP address. Since the outer IP packet's destination IP address belongs to Area B, routers can correctly forward the outer IP packets to Gateway Y2. The original IP packet for the mobile device remains intact during its journey from Gateway X to Gateway Y2 as it rides the outer IP packet.

Two IP tunneling protocols widely used in commercial cellular networks are GRE and GPRS Tunneling Protocol (GTP).[130] GRE is used in 1xRTT and 1xEV-DO networks, and GTP is used in UMTS networks.[131] Figure 4.34 summarizes the basic idea behind GRE and GTP.

Assume that we have an IP tunnel between two gateways, Gateway G1 and Gateway G2, and that we need to send packets for two mobile devices from Gateway G1 to Gateway G2. A mobile device may have more than one IP flows, such as one flow for web-browsing and one flow for video streaming. Furthermore, the gateways may be responsible for hundreds of thousands of mobile devices during a given time period. Since all the IP packets traveling from Gateway G1 to Gateway G2 would have Gateway G1's IP address as its source IP address and Gateway G2's IP address as its destination IP address in the outer IP packet, we need to separate multiple IP flows for a single mobile device and multiple IP flows for multiple mobile devices. The GRE protocol uses GRE keys to separate IP flows, with one key per IP flow. Similarly, the GTP uses tunnel endpoint identifiers (TEIDs) to separate IP flows. Both GRE keys and TEIDs consist of four octets (i.e., four bytes or thirty-two bits), enabling 2^{32} IP flows between a given pair of gateways. A given GRE or GTP tunnel also uses sequence numbers to facilitate in-order delivery of IP packets for a given IP flow. We have discussed the flow of IP packets from Gateway G1 to Gateway G2, and similar concepts apply to the flow of IP packets from Gateway G2 to Gateway G1.

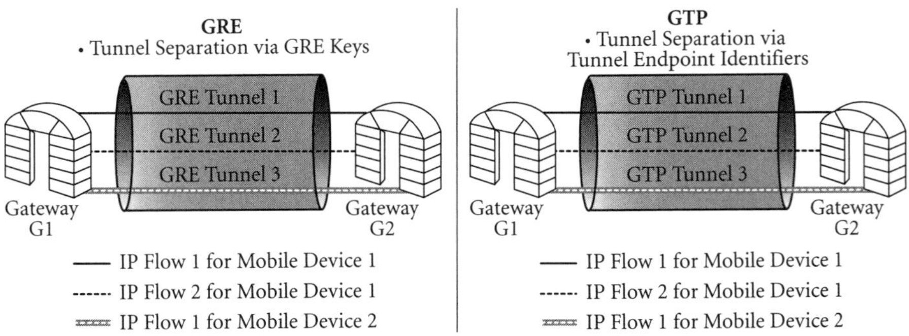

Figure 4.34: IP Tunneling Protocols: GRE and GTP

[130] See Dommety's work [175] for GRE details, and see the standard [176] for GTP details.

[131] 1xRTT, 1xEV-DO, and UMTS refer to 1x Radio Transmission Technology, 1x Evolution–Data-Optimized, and Universal Mobile Telecommunication System, respectively.

For example, another GRE key or GTP TEID would be used for an IP flow from Gateway G2 to Gateway G1.

4.6.7 IPsec

IPsec, or IP security, is becoming very important because emerging cellular technologies rely heavily on IP. Two security protocols that form the core of the IPsec suite are the Authentication Header (AH) protocol and the Encapsulating Security Payload (ESP) protocol [177].

The AH protocol helps the receiver authenticate the sender. The sender constructs the authentication header by considering the content of the entire IP packet and inserts such a header in the original IP packet. The receiver can ensure the integrity of the received IP packet by observing the authentication header. If the contents of the IP packet are modified (e.g., by a rogue entity) between the sender and the receiver, the received authentication header no longer matches the expected authentication header at the receiver. The receiver can then discard the modified IP packet. The sender or the source and the receiver or the destination need to establish a unidirectional (i.e., from source to destination) security association (SA) between them to decide upon a specific hashing algorithm and a key to be used to generate the authentication header. Only the source and the destination know about such an algorithm and key.

The AH protocol ensures the validity of the source and the integrity of data, but it provides no privacy. In other words, someone could indeed observe the content of the IP datagram. If you are interested in protecting or hiding the content, the ESP protocol is the one you need. The ESP protocol actually encrypts the original IP datagram so that the original content is hidden from any entity between the source and the destination. The ESP protocol uses IP encapsulation in which the encrypted original IP packet is placed inside an outer IP packet. The ESP header part of the outer IP packet and helps the destination decrypt the inner original IP datagram. Similar to the case of the AH protocol, the establishment of an SA is a prerequisite for the ESP protocol.

4.6.8 Robust Header Compression

An IP packet includes the IP header and the payload. The minimum and typical size of the IP header is twenty bytes. Furthermore, the typical IP packet length is about 1500 bytes for applications such as web-browsing and email in which the amount of data to be sent is reasonably large. Since the payload is much larger than the header in most cases, the negligible amount of header is usually of no concern. However, for some applications, such as VoIP, the payload could be quite small, resulting in the header consuming a significant percentage of the IP packet. Transmission of such VoIP packets becomes inefficient if we use a full-fledged IP header. IP header compression rectifies such a situation by reducing the size of the IP header, making the transmission of small payloads efficient. Robust header compression is a popular IP header compression mechanism in commercial cellular standards such as Long-Term Evolution (LTE).

Do You Recall. . . Even a simple question can be difficult to answer in a long chapter. Do you remember the minimum size of the IP header? The footnote has the answer.[132]

Robust header compression is applicable to the combination of RTP, UDP, and IP, which is the case for a VoIP application. The overall overhead for a VoIP packet is forty bytes with twenty bytes of an IPv4 header, eight bytes of a UDP header, and twelve bytes of an RTP header. In contrast, the size of the payload would be about thirty-one bytes in case of full-rate 12.2 kbps adaptive multirate speech codec that generates a speech frame every 20 ms. Robust header compression reduces the size of the total RTP, UDP, and IP header from forty bytes to just one to three bytes [178].

IP header compression is feasible because significant redundancy occurs between the IP headers being sent from one entity to the other. For example, the source and destination IP addresses for all the IP packets associated with a given application of the mobile device would be the same. Hence, during the initial stage of packet transfer, the full-fledged IP header is conveyed. Then, the compressed IP header can indicate to the receiver that the contents of the previously sent full IP header are valid for use. The compressed header thus eliminates the need to include the full-fledged IP header. Relevant information from past packets is stored in a context. A robust header compression algorithm ensures consistency of the context at the sender or compressor and the receiver or decompressor. A compression profile specifies which and how header fields are compressed.

The robust header compression scheme is operated in one of three modes at any given instant: Unidirectional mode (U-mode), Bidirectional Optimistic mode (O-mode), and Bidirectional Reliable mode (R-mode) [179]. The U-mode involves packet transmission from compressor to decompressor only. This mode is suitable for links where a return path from decompressor to compressor is either unavailable or undesirable.

The O-mode is similar to the U-mode in many respects, but a feedback channel sends error recovery requests and (optionally) acknowledgments of significant context updates from the decompressor to the compressor. The O-mode attempts to maximize compression efficiency and minimize usage of the feedback channel. Hence, the O-mode omits periodic refreshes. The R-mode is quite different from the other two modes. The R-mode extensively uses the feedback channel and strives to prevent loss of context synchronization between compressor and decompressor. Feedback includes the acknowledgment of context updates. The R-mode attempts to maximize robustness against loss propagation even under header loss or error burst conditions.

4.7 ESSENCE OF IPv6

As the Internet continues its phenomenal growth and as non-PC IP devices such as electronic point-of-sale terminals and web-enabled television receivers become

[132]The minimum size of the IP header is twenty bytes.

popular, we find IPv4 to be less and less capable of handling the address requirements. IP version 6, or IPv6, is the enhanced version of IPv4 [180]. One of the main motivating factors behind IPv6 is "significant" expansion of the IP address space. How much is "significant?" Well, there can be about fifty thousand IPv6 addresses per square meter of land on Mother Earth [117]! IPv4 uses a thirty-two–bit IP address, but IPv6 uses a 128-bit IP address. An IPv6 address is specified as eight groups of hexadecimal numbers separated by colons such as 1234:0000:0000:5678:9ABC:DEF0:1234:5678. Leading zeroes in a group can be removed as exemplified by 1234:0:0:5678:9ABC:DEF0:1234:5678. Additionally, one occurrence of consecutive groups of all zeroes can be replaced by a double colon such as 1234::5678:9ABC:DEF0:1234:5678.

There are different types of IPv6 addresses, such as a *unicast* address. A unicast address is an identifier assigned to a single interface (e.g., a specific host). One type of a unicast address is an IPv4-compatible address used when IPv6 traffic must be tunneled across existing IPv4 networks. In automatic tunneling, the endpoints of the tunnel are hosts. In contrast, configured tunneling has routers as the tunnel endpoints. *An IPv6 node could be a host or a router.* Constructing IPv4-compatible addresses is quite easy; just append ninety-six zeroes in front of a valid thirty-two–bit IPv4 address! Another type of unicast address is an IPv4-mapped address that has the format of ::FFFF:<IPv4_address>; this address is useful when an IPv6 host needs to communicate with an IPv4 host [117]. Either a dual-stack (i.e., IPv6 and IPv4) host or dual-stack router is required to carry out IP header translations. When an IPv6 host sends data to an IPv4 host that has the IPv4 address *a.b.c.d*, the destination IP address of ::FFFF:0a0b:0c0d is specified in the IPv6 header.

Finally, let's get introduced to the IPv6 address format called the *global unicast address format* [181, 182], which will be a dominant format for an IPv6 host [117]. The global unicast address format has three parts: a global routing prefix, subnet ID, and interface ID. The global routing prefix has a hierarchical structure and is managed by regional internet registries and internet service providers. The subnet ID identifies a subnet within the site (e.g., a university or an enterprise) and is managed by site administrators. The interface ID points to a specific interface on a link (e.g., some IPv6 host). A typical unicast address includes sixty-four–bit long interface IDs. An IPv6 node typically obtains its IP address via *autoconfiguration*. In stateful address autoconfiguration, the node uses DHCP to acquire an IP address. However, in stateless address autoconfiguration, the node builds a global unicast IP address based on its forty-eight–bit MAC address and the prefix received from a Router Advertisement message [183]. Even when stateless address autoconfiguration is used, DHCP still remains relevant because the node uses DHCP to obtain configuration options (e.g., DNS servers and SIP servers) based on the directive received in the Router Advertisement message [117, 184].

IPv6 is more than just the number of IP addresses. Let's summarize other benefits of IPv6 compared to IPv4 [117]. The use of a prefix in the globally unique and hierarchical addressing scheme reduces the size of routing tables in the backbone. Autoconfiguration of network interfaces is supported via ICMP and DHCP. Different classes of service provide differential QoS treatment to different types of data traffic. Specifically, an eight-bit Traffic Class field includes a six-bit DSCP and a two-bit Explicit Congestion Notification, which facilitates active queuing management at a router instead of brute force dropping of a packet. Multicast routing is enhanced in IPv6. Support for IP Security is mandatory in IPv6. Two headers are available: Authentication Header and Encapsulated Security Payload. The Authentication Header carries the information relevant to authentication, and the Encapsulated Security Payload includes the encrypted data. Various mechanisms are defined to ensure interworking between IPv4 and IPv6. For example, as mentioned earlier, an IPv6 packet can be tunneled to pass through an IPv4 network. One way to ensure smooth interworking between a new IPv6 node and existing IPv4 systems is to implement a dual-stack IPv6/IPv4. Such a dual-stack node is allocated two independent IPv6 and IPv4 addresses and is thus capable of sending and receiving IPv6 packets *and* IPv4 datagrams. A 6-to-4 router is another key element for IPv4–IPv6 interworking. Such a router sits at the border of IPv4 and IPv6 networks and translates between IPv4 and IPv6 headers. An IPv6 source node dynamically determines the maximum transmission unit supported by all links along a path and sends the packets that are smaller than the path's maximum transmission unit, thereby avoiding IP fragmentation caused by routers. IPv6 requires each link to be capable of supporting at least 1280 bytes of the maximum transmission unit.

Let's take a quick peek at the format of an IPv6 packet [117]. An IPv4 has a variable header length with twenty bytes being the minimum length, but IPv6 has a fixed forty-byte header. The IPv6 header consists of eight bytes of control information, a sixteen-byte source IP address, and a sixteen-byte destination IP address. The reduction in control information, from twelve bytes in IPv4 to eight bytes in IPv6, helps reduce the processing time in a router. The Control field includes the IP Version Number (which is set to six, of course), Traffic Class, Flow Label, Payload Length, Next Header, Hop Limit, and Data field. A *flow* is a sequence of related packets (e.g., VoIP packets or video-streaming packets) between the source and the destination with specific packet handling instructions for the routers (e.g., minimizing latency). The packet handling instructions can be given using an extension header or a separate protocol such as Reservation Protocol. The Next Header field indicates the type of header that immediately follows the basic IP header. It could specify an upper-layer protocol such as TCP or an IP option header indicative of the presence of extension headers. The extension header helps convey optional information, such as the Authentication Header and Encapsulated Security Payload. The lengths of jumbo payload sizes in excess of 65535 bytes can also be specified using an extension header. The Hop Limit field now truly represents the number of remaining hops through which the packet is allowed to pass.

Ever Wondered... We have talked a lot about IPv4 and some about IPv6. Have you ever wondered about IPv5? Well, IPv5 does exist, on paper at least. The Internet Stream Protocol was defined in 1979 and later revised. This later revision is considered

IPv5 since it uses version number 5 [185], and it was intended to provide QoS in an IP network.

4.8 POINTS TO REMEMBER

- IP plays an essential role in existing and emerging cellular networks.
- Five layers of the Internet Protocol stack are the application layer, transport layer, internetwork layer, link layer (or layer 2), and physical layer (or layer 1).
- The user's applications reside at the application layer. For example, HTTP is the application layer protocol for web-browsing and SMTP is the application layer protocol for email.
- Three main transport layer protocols are TCP, UDP, and SCTP (SCTP being a more efficient version of TCP). TCP provides connection-oriented and reliable end-to-end packet delivery through means such congestion control and retransmissions. UDP is a connectionless, low-overhead, and low-latency transport protocol.
- IP is the main internetwork layer protocol and is based on the philosophy of "simple network, complex or intelligent endpoints." IP is dynamic and scalable, and much of its dynamic nature stems from its routing protocols, such as RIP, OSPF, and BGP. RIP and OSPF are suitable for intra-AS routing, and BGP is suitable for inter-AS routing. Once the routing tables are developed using the routing protocols, the router uses these tables to forward a packet to the correct next hop based on the destination IP address. Different classes of IP addresses exist, and subnetting and private IP addresses enable efficient use of the existing IPv4 address space.
- Within a cellular network, the transport of information is implemented using mechanisms such as ATM and IP/MPLS. Numerous auxiliary protocols and mechanisms, such as DNS, DHCP, RADIUS, Diameter, Mobile IP, and Differentiated Services, facilitate the implementation of functions such as symbolic name mapping to an IP address, dynamic IP address allocation, authentication and accounting, preservation of an end-to-end session between two endpoints even in the presence of user mobility, and maintenance of QoS.
- The protocols, such as SIP, SDP, RTP, and RTCP, play a crucial role in the implementation of VoIP.
- IPv6 overcomes several of IPv4's shortcomings and offers significantly larger address space, better QoS, increased security, and support for user mobility.

PART II

2G, 2.5G, AND 3G CELLULAR TECHNOLOGIES

CHAPTER 5

OVERVIEW OF GSM

This chapter presents an overview of Global System for Mobile Communications (GSM), a second-generation (2G) digital cellular system. A brief summary of the salient features of GSM is followed by a discussion of an end-to-end system architecture. The GSM core network forms the circuit-switched core network of a 3G Universal Mobile Telecommunication System (UMTS) system. The GSM air interface is discussed next. Various radio channels used for over-the-air communications are described along with the structure of the radio frame. Power-on initialization and GSM network acquisition for the mobile device are summarized. The steps of a voice call setup are illustrated. Operations such as handover[133] and power control are discussed. Finally, the evolutionary path of GSM is outlined.

[133]Handover and handoff are just two terms for the same operation; no difference exists between the two.

Cellular Communications, First Edition. Nishith D. Tripathi and Jeffrey H. Reed

5.1 INTRODUCTION

GSM is a 2G digital cellular system. The number of GSM subscribers was about 4 billion at the end of the second quarter of 2010, with the total number of mobile subscribers being about 4.9 billion, representing more than 80% of the mobile market [186, 187]. What an enviable market share!

5.1.1 The Origin of GSM and GSM Milestones in Two Minutes

GSM was initially referred to as Groupe Speciale Mobile, which was later changed to Global System for Mobile Communication. If you are curious about how GSM was born, your curiosity will be satisfied in the next two minutes. Let's say that you start driving from Dallas, Texas. You will have to drive for several hours to get out of Texas. (As they say, *everything* is big in Texas.) But, if you are in Europe, you will be crossing into another *country* if you drive for several hours. If you take a ride in a high-speed train, you can get out of a country even faster! Prior to the arrival of GSM, several analog cellular systems existed in Europe. When people traveled from one country to a neighboring country (which is quite common in Europe), they could not roam seamlessly due to the dissimilar analog systems that different countries used. The Confederation of European Posts and Telecommunications formed GSM to design a pan-European standard. A single cellular standard deployed throughout Europe would enable seamless mobility, allowing people to roam across countries.

Table 5.1 lists selected GSM milestones.[134] The latest and greatest happenings in the GSM and GSM evolution arena can be found on the GSM Association's website [188]. The GSM Association is a global trade association representing over 700 GSM operators across more than 200 countries [189].

Table 5.1: GSM Milestones

Year	Event
1982	The Confederation of European Posts and Telecommunications forms GSM to design a pan-European mobile technology.
1986	European governments decide to reserve the 900 MHz spectrum band for GSM.
1987	The basic parameters of the GSM standard are agreed upon, and the Memorandum of Understanding is signed. The GSM Association is founded.
1991	Radiolinja makes the first GSM call in Finland.
1994	GSM Phase 2 circuit-switched data and fax services are launched.

[134] See `www.gsmworld.com` [188] for additional events in the history of GSM.

Do You Recall... Do you recall examples of 1G analog celular systems in Europe? See the footnote[135] for the answer.

5.1.2 Highlights of a GSM System

A GSM system offers several salient features, such as support for voice and circuit-switched data services, availability of a convenient Subscriber Identity Module (SIM), ability to offer numerous supplementary services, and use of the time-division multiple access (TDMA) technique [186, 187].

GSM supports voice and circuit-switched data services, with a focus on voice. A circuit-switched data service requires the reservation of resources throughout the call even when no data transfer occurs during some time intervals. GSM allows data rates of up to 14.4 kbps, supporting basic circuit-switched services.

The GSM system exploits a SIM card to enable users to personalize services beyond a single physical mobile device. The SIM card stores information about the subscriber, such as subscriber identity and a personal telephone directory. This SIM card can be inserted into any suitable mobile device. Hence, the user can experience the same services on different physical devices. Furthermore, the SIM card is programmed with security parameters, such as the identity of the subscriber, so the user can be authenticated and the air-interface communications can be secured. Specifically, suitable authentication and ciphering algorithms and related keys are programmed into the SIM card. Additionally, the SIM card also stores information about the cellular network (e.g., location area identity) as the subscriber moves around in a geographical area.

GSM and its evolutionary forms tend to emphasize the services aspect of the cellular network and offer a comprehensive set of supplementary services, such as call forwarding and caller identification. The user can place an emergency call without knowing the local emergency phone number (which can change from one country to another). Furthermore, international dialing is facilitated by dialing + before the country-specific code so that the user need not remember whether to first dial 0 or 00 to make an international call.

GSM uses the TDMA technique to allow multiple mobile devices to use the system simultaneously. The GSM radio interface utilizes 200 kHz radio channels.[136] Each channel can be shared among eight users via eight time-slots, with one user using the channel during one time-slot. GSM is also deployed at various frequency bands. It typically operates in the 900 MHz and 1.8 GHz bands in Europe and the 1.9 GHz and

[135]Examples of 1G cellular systems in Europe were ETACS, NMT, C-Netz, Radiocom2000, and RTMI (Radio Telefono Mobile Integrato).

[136]The GSM channel is actually 271 kHz wide, but the channels are spaced 200 kHz apart. Their ability to be spaced this way is due to the use of a modulation scheme called GMSK (Gaussian Minimum Shift Keying). We will use the term *200 kHz channels* for the sake of convenience.

850 MHz bands in the U.S. The 850 MHz band is also used for GSM in Australia, Canada, and several South American countries.

The original and main frequency bands of GSM were the standard or Primary GSM 900 Band (P-GSM), Extended GSM 900 Band (E-GSM), and Digital Cellular System 1800 Band (DCS 1800). The P-GSM system uses the 890–915 MHz band in the uplink and the 935–960 MHz band for the downlink. The E-GSM system includes the standard GSM 900 band and uses the 880–915 MHz band for the uplink and the 925–960 MHz band for the downlink. The DCS 1800 system operates in the 1710–1785 MHz band for the uplink and in the 1805–1880 MHz band for the downlink. The term GSM 900 denotes any GSM system that operates in any 900 MHz band.

Operators may deploy networks operating at multiple frequency bands, and, in such cases, multi-band mobile terminals give users additional coverage flexibility. In some cases, a nationwide operator has one frequency band in one market and another frequency band in another market. A multi-band mobile will then work seamlessly across the markets.

GSM is deployed in more than 210 countries, and it facilitates international roaming. Such a capability provides GSM subscribers with seamless and same number connectivity around the globe. The main benefits of GSM compared to 1G cellular systems are improved voice quality, higher voice capacity, support for roaming, and air-interface security. Air-interface security mechanisms include *ciphering* (i.e., encryption) of voice and signaling messages.

The 2G GSM standard has evolved to 2.5G GPRS (General Packet Radio Service) and EDGE (Enhanced Data Rates for GSM Evolution), which are discussed in Chapter 6. GPRS enhances GSM's air-interface protocol stack and introduces a new packet-switched core network to enable packet-switched services. EDGE offers additional radio network enhancements to improve packet data performance.

5.2 GSM SYSTEM ARCHITECTURE

Figure 5.1 illustrates the high-level architecture of a GSM PLMN (Public Land Mobile Network). A GSM-based system or network consists of the radio access network called the BSS (Base Station System) and the core network called the NSS (Network Switching Subsystem).[137] The primary difference between the mobile equipment (ME) and the mobile station (MS) is that the MS is the ME (i.e., the device) *and* the SIM (i.e., the card or the chip containing the user's subscriber identity and the information about personalized services).

One BSS consists of one Base Station Controller (BSC), one TRAU (Transcoder Rate Adaptation Unit), and multiple Base Transceiver Stations (BTSs). Two BSSs are shown in Figure 5.1.[138] The MS communicates with the BTS using the TDMA-

[137] Other works [190] provide more details on the BSS and the NSS.

[138] Although the TRAU is logically in the BSS, it is usually physically close to (often in the same room with) the Mobile Switching Center (MSC) to perserve backhaul resources on the A interface. Even the BSCs are typically in the same room with the MSC as well. Also, in the figure, VLR represents the Visitor Location Center.

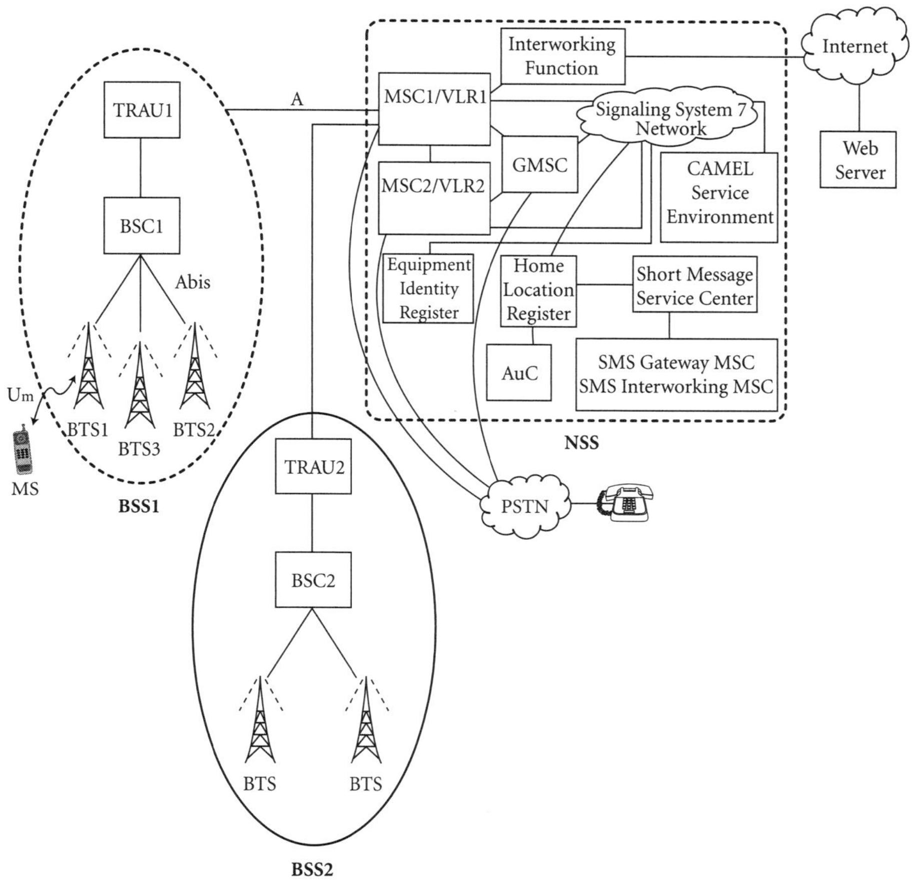

Figure 5.1: GSM PLMN Architecture

based air interface called the *Um interface*. Formally speaking, one BTS controls one cell, which may be an omnidirectional cell or a sector (e.g., a 120° region). In practice, one physical BTS cabinet controls multiple cells (e.g., three cells in 120° sectorization). One BSC controls hundreds of BTSs depending upon the BSC capacity. The interface between the BTS and the BSC is called the Abis interface. Though any physical medium can be used, T1/E1 links are typically used on the Abis interface. Sometimes microwave frequencies are used as a physical medium to implement a wireless Abis interface. A BSC connects to a TRAU, which performs data rate conversion between the BSC and the NSS. Section 5.2.1 discusses the BSS elements in more detail. The NSS entities essentially participate in functions such as authentication, call routing, billing, and short message service (SMS). Several core network entities use Signaling System 7 to exchange signaling messages.

Let's see how speech travels from a GSM MS to a landline phone. Speech enjoys a flight from the MS to the BTS over the Um interface. Then it traverses the Abis interface from the BTS to the BSC. The BSC sends the speech to the TRAU. The speech data rate is 16 kbps (called *full rate*) or 8 kbps (called *half rate*) between the MS and the TRAU. These nominal data rates for speech correspond to the original codecs a standard supported. [Support for adaptive multirate (AMR) speech coding, in which the data rate for speech ranges from 4.75 kbps to 12.2 kbps, was added later.[139] The AMR codec has been quite popular in commercial GSM deployments.] The TRAU now converts the low-rate speech into a high-rate, 64 kbps, Pulse Coded Modulation signal and passes the 64 kbps speech to the Mobile Switching Center (MSC). The MSC interfaces with the PSTN (Public-Switched Telephone Network). The MSC conveys the speech to an Inter-Exchange Carrier, which, in turn, forwards the speech signal to a Central Office switch. The Central Office switch directly delivers the speech to the destination landline phone or uses other Inter-Exchange Carriers to eventually reach the intended landline phone. The GSM NSS's functional elements depicted in Figure 5.1, with some upgrades, are reused in the 3G evolution of GSM called UMTS (see Chapter 9). Section 5.2.2 describes the roles played by the NSS components.

5.2.1 Base Station System

As shown earlier in Figure 5.1, a BSS consists of a BSC, a TRAU, and BTSs. The functions of each component and interface require explanation.

The BTS communicates with the MSs via the TDMA-based air interface. The BTS implements the TDMA physical layer for over-the-air communications with MSs by using one or more TRXs (i.e., transceivers). One TRX supports one 200-kHz radio channel and eight time-slots of a TDMA frame, essentially providing eight basic radio channels. One TRX can serve eight full-rate, circuit-switched users on a GSM Traffic Channel (TCH) carrier or seven full-rate users on a Broadcast Control Channel (BCCH) carrier because one time-slot on the BCCH carrier is used to transmit control information to the MSs. More details on the transmitter and receiver operations including the radio frame structure are given in Section 5.4.

Simultaneous transmission and reception between the MS and the BTS become possible due to Frequency-Division Duplexing in which different frequency bands are used for the uplink and downlink. Furthermore, transmission and reception at the MS are disjointed in time, leading to simpler and lower-cost MS design. The GSM standard (and its future evolutions) formally defines a BTS as a network component that serves *one* cell [190]. A BTS has one or more TRXs. One cell could be using omnidirectional antennas, and, in that case, one *physical* BTS (*physical* here means something you can touch) corresponds to one *logical* BTS that controls just one cell. If 120° sectorization is used, one physical BTS houses three logical BTSs, with each BTS serving one cell and each cell covering a 120° region.

[139] See Chapter 2 for AMR details.

The BSC configures and controls the BTSs. One BSC typically controls hundreds of BTSs. Several radio resource (RR) management algorithms, such as call admission control and handover algorithms, are implemented at the BSC. Most of the BSS intelligence resides at the BSC since the BSS architecture is highly centralized. The motivation behind such a centralized architecture is that the BSC has a more comprehensive view of the prevailing resource utilization at several BTSs and can make better decisions about how to manage precious radio resources on airwaves. Remember that while Mother Nature allows us to breathe the air without any fees, governments charge billions for airwaves!

In general, if we give the BTS more control, it may be unable to make the best decisions because it knows only about the resources being used in its own cell. For example, when the MS is in the neighborhood of multiple cells, the BSC can ask the MS to talk to the best possible BTS based on prevailing radio channel conditions and the current resource utilization in various cells. Examples of other RR-related functions include MS power control, cell selection and reselection, and MS measurement reporting. The MS and the BSC exchange air-interface–related messages via the BTS. Certain messages, such as those related to services and user authentication, are exchanged between the MS and the NSS. The BSC transparently lets these messages pass through. The BSC and MS cipher traffic and signaling messages to provide air-interface security.

The TRAU is often referred to as a transcoder. Consider this scenario to discover the motivation behind the TRAU. Let's say that we are using a GSM MS to talk to our friend Bob, who is using a landline phone. Bob's speech signal travels through the switches of the PSTN and arrives at the MSC. The MSC receives from the PSTN a 64 kbps Pulse Coded Modulation signal carrying Bob's speech.

A Question for You... Will using a 64 kbps radio channel to carry voice be considered efficient? Just answer "yes" or "no." You have an at least 50% chance of being correct! The footnote[140] has the answer.

The use of a 64 kbps pipe to carry just one voice call is quite expensive. We need to use the precious radio resources more efficiently. Now, let the TRAU enter the scene. The TRAU receives the 64 kbps Pulse Coded Modulation signal from the MSC and converts it into a lower rate, such as 12.2 kbps full-rate signal using the AMR speech codec. Now, we will consume only a 12.2 kbps pipe over the air interface to carry voice rather than a 64 kbps pipe! We are now increasing the efficiency by a factor of $64\text{ kbps}/12.2\text{ kbps} = 5.2$. Use of a lower data rate on the air interface reduces overall interference and results in lower transmit power. Reduction in the transmit power and interference, in turn, enables us to exploit a more aggressive frequency reuse, thereby increasing the capacity of a GSM system. In summary, the TRAU plays a significant role in making a successful business case of offering voice services!

[140]No, it will be considered too expensive from the perspective of resource utilization.

Such TRAU operation is often called *speech compression* because we are compressing a 64 kbps speech signal into a 12.2 kbps signal. The TRAU can also carry more users' speech traffic between the BSC and the BTS because it occupies only 12.2 kbps of bandwidth between the BTS and the BSC for one voice call in contrast to 64 kbps of bandwidth. The TRAU is typically located near the MSC to maximize the cost savings. The TRAU also carries out rate adaption in support of circuit-switched data calls and fax. Since the MSC expects and accepts just a 64 kbps data rate, the TRAU adapts the radio interface data rates to 64 kbps for the MSC. The Interworking Function (IWF) is connected to the MSC and is used for fax and data calls.

The Abis interface allows the BTS and the BSC to communicate with each other. The BTS manages the radio resources of the cell based on the directives received from the BSC and sent on the Abis interface. Messages (between the MS and the BSC and between the MS and the NSS) and user traffic, such as speech bits, travel through the Abis interface.[141]

Something to Think About. . . The curious word *Abis* can mean *A-prime* or A'. The French phrase *À la fin du concert, le groupe a joué deux bis,* means "At the end of the concert, the group played two encores." This suggests that *bis* means *repeat* or *another*, which we often intend when we apply a prime mark to a variable. The *A* interface is seen *again* on the other side of the BSC in the *Abis* interface.[142]

In principle, specification of the Abis interface enables us to connect BTSs from different manufacturers. However, in practice, the BTSs and the BSC from the same manufacturer usually communicate with one another. The basic digital transmission rates supported by this interface are 1.536 Mbps for a T1 link (and 2.048 Mbps for an E1 link) and 64 kbps, with lower sub-rates also supported for rate-adapted speech. Two types of communication channels exist over the Abis interface: the stand-alone dedicated channel, carrying traffic such as speech or data, and the signaling channel, carrying both MS–BSC signaling and BTS–BSC signaling. The BSC utilizes the signaling channel to direct the BTS to set up a dedicated radio link with the MS and to establish an Abis bearer to carry the user-specific speech traffic. The established Abis bearer for the MS represents the stand-alone dedicated channel.

On one end of the A interface is the BSC, and on the other end is the MSC. This interface carries user traffic (e.g., speech) and three types of signaling messages between the BSS and the MSC. The types of signaling messages include Direct Transfer Application Part messages, BSS Management Application Part messages, and BSS Operation and Maintenance Application Part messages [192]. The Direct Transfer of Application Port and BSS Management Application Part messages together are considered BSSAP (BSS Application Part) messages [193]. The Direct Transfer of Application Port is used to transfer call control and mobility management messages

[141] See the standard [191] for additional details on the Abis interface.

[142] Thanks to Dr. Malcolm Oliphant, one of the reviewers for this chapter, for providing the material for this sidebar.

between the MS and the MSC. Authentication messages are examples of Direct Transfer of Application Port messages. The NSS, not the BSS, is responsible for authenticating the user. Hence, such messages between the MS and the NSS transparently flow through the BSS. The Direct Transfer of Application Port messages are *direct transfer* messages; the BSS does not interpret them. Most radio interface messages are transferred across this interface with Direct Transfer of Application Port messages.

In contrast to the Direct Transfer of Application Port, the BSS Management Application Part supports the procedures that require interpretation and processing of the messages traveling over the A interface. For example, assignment is a procedure that ensures that a suitable dedicated radio resource be allocated or re-allocated to an MS. During the call setup, the MSC analyzes the call control information received from the MS and sends an ASSIGNMENT REQUEST message to the BSS using the A interface. This message includes details of the required resources (e.g., channel rate and channel type). The BSS can then perform admission control and accept the request if resources are available.

5.2.2 Core Network

As mentioned earlier, the NSS entities can be reused in systems evolving from GSM (e.g., UMTS). Such potential for reuse reduces the service operator's cost in the long run. See Figure 5.1 to refresh your memory about the network architecture. For example, a new piece of NSS equipment could support all the protocols for the relevant GSM and UMTS interfaces, and the service provider then would need to maintain only a single piece of the NSS equipment.

The Home Location Register (HLR) is a central database that stores information about mobile subscribers. One large nationwide service operator could have one logical HLR with multiple physical HLRs geographically distributed in the country, such as one in each of the eastern, western, northern, and southern parts of a country. These HLRs contain the same information and update each other quickly over a high-speed backbone network. The exact number of physical HLRs depends upon several factors, such as the number of subscribers, capacity of the equipment, and target delays. The type of information stored in the HLR includes user subscription information (e.g., subscribed voice and data services), service restrictions (e.g., roaming limitations), supplementary services, user location to enable routing of calls to an MSC in which the MS has registered [e.g., MS roaming number, VLR (Visitor Location Register) number, MSC number, and local MS identity], International Mobile Subscriber Identity (IMSI),[143] and Mobile Station ISDN (Integrated Services Digital Network) number(s).

The VLR tracks the MS in a given geographical area once the MS registers with the VLR. The MSC in charge of the area transfers to the VLR the identity of the location area where the MS is situated. The VLR and the HLR exchange

[143] IMSI is usually pronounced as "eem-zee."

messages for proper call routing. Examples of information the VLR stores to handle the calls for the MS include IMSI, mobile station ISDN, mobile station roaming number, temporary mobile subscriber identity (TMSI),[144] location area identity, and supplementary services.

The Authentication Center (AuC) Center stores subscriber data, such as a shared secret, to enable authentication of the subscriber and ciphering (i.e., encryption) of the radio path between the MS and the BSC. The AuC is usually integrated into the HLR. See Section 5.5 for details of the authentication process carried out as part of call setup.

The Equipment Identity Register stores International Mobile Equipment Identities for the MSs. An MS is classified as *white listed*, *grey listed*, *black listed*, or *unknown*. For example, a black-listed MS may be entirely blocked from using any services, but a grey-listed MS may be allowed some basic services. The Equipment Identity Register helps minimize fraud by preventing the use of lost or stolen MSs. If a subscriber loses a cell phone, the subscriber informs the service provider, and the relevant International Mobile Equipment Identity is now properly classified in the Equipment Identity Register to prevent unauthorized use of the MS. The Equipment Identity Register may or may not exist in a commercial system depending upon the service provider's preference.

The MSC builds a bridge between the PLMN and the fixed network, such as the PSTN. It carries out tasks necessary to handle circuit-switched services for the MSs. The MSC is an exchange that performs switching and signaling functions for the MS in the location areas the MSC serves. The MSC participates in location update and handover procedures. More specifically, the MSC contains a layer 3 entity called *mobility management* (MM) that knows whether or not the MS is reachable. If the MS is reachable, the MSC knows the MS's location area so that the network can send a paging message to the MS.

When a network cannot communicate with the HLR of a target GSM PLMN to deliver a call, the call is routed to the GMSC. The GMSC interrogates the HLR and then routes the call to the MSC with which the MS has registered. For example, when someone using a landline phone tries to call a GSM MS, the PSTN contacts the GSM PLMN's GMSC so that the call can be routed to an appropriate geographical area that the MSC services. The service operator can configure one or more MSCs to perform GMSC functions.

The IWF implements the functions that provide interworking between the PLMN and the fixed networks such as ISDN, PSTN, and Packet Data Networks. The IWF converts protocols between the PLMN protocols and the fixed network protocols.

Two types of MSCs are defined to support the SMS. The SMS Gateway MSC resides between the short message service center and the PLMN that delivers the SMS from the short message service center to the MSs. For example, when the short

[144]The purpose of TMSI (usually pronounced as "tem-zee") is to provide anonymity, thereby increasing the radio link security and privacy. Use of the TMSI minimizes the use of IMSI. One of the few times that the IMSI is sent over the radio link is during the execution of a location update procedure. Using TMSI makes it difficult for anyone monitoring the radio link to determine to whom the device belongs.

message service center intends to send an SMS to an MS, it contacts the SMS Gateway MSC. The SMS Gateway MSC determines a suitable GSM PLMN that can deliver the SMS to the MS. The SMS Interworking MSC resides between the GSM PLMN and the short message service center to allow the SMSs from the MSs to be submitted to the short message service center. In other words, there is an SMS Gateway MSC or an SMS Interworking MSC between the MS and the short message service center for a given SMS. The MS communicates with the short message service center via the SMS Gateway MSC for an incoming SMS and with the short message service center via the SMS Interworking MSC for an outgoing SMS. A service operator may designate an MSC to act as an SMS Gateway MSC and/or an SMS Interworking MSC.

The CAMEL Service Environment is a logical entity that consistently offers operator-specific services regardless of the serving mobile network [194]. Specifically, the subscriber receives operator-specific services even while roaming outside the home PLMN. The services offered as part of the CAMEL feature are considered separate from the supplementary services (e.g., call forwarding and call hold). The CAMEL Service Environment provides Intelligent Network services to users via proper provisioning. Examples of such services are prepaid services and toll-free dialing. For example, when a prepaid subscriber makes or receives a call, the MSC contacts the CAMEL Service Environment, which, in turn, evaluates the current balance for the subscriber and admits or blocks the call. Furthermore, the CAMEL Service Environment can announce the current balance to the subscriber and can forward the call to a suitable entity so that the subscriber can increase the available credit.

5.3 AIR-INTERFACE PROTOCOL STACK BETWEEN MS AND BSS

The MS and the BSS communicate with each other over the air interface using the TDMA multiple access technique and a suite of air-interface protocols. Section 5.3.1 provides a high-level view of how the MS and the BSS communicate. Section 5.3.2 discusses details of the air-interface protocol stack. This protocol stack formulates operating principles that enable the MS and the BSS to communicate over the wireless interface. Detailed time-slot and frame structures are described in Section 5.4.

5.3.1 High-Level View of MS–BSS Communications

GSM exploits the concepts of both TDMA and FDMA (see Chapter 1 for basics of various multiple access techniques). In a pure TDMA system, only one radio frequency channel exists; different users use the same radio channel but during different time intervals. In an FDMA system, different users are allocated different radio frequency channels. Consider Figure 5.2 in which a BTS is serving sixteen users simultaneously in the downlink using two 200-kHz TCH carriers.

The BTS uses two frequency channels, f_1 and f_2, in a cell. Sixteen MSs are also present in the cell. The BTS has allocated f_1 to eight users and f_2 to the remaining eight users. Focus on the radio channel f_1. The BTS first sends information to MS1

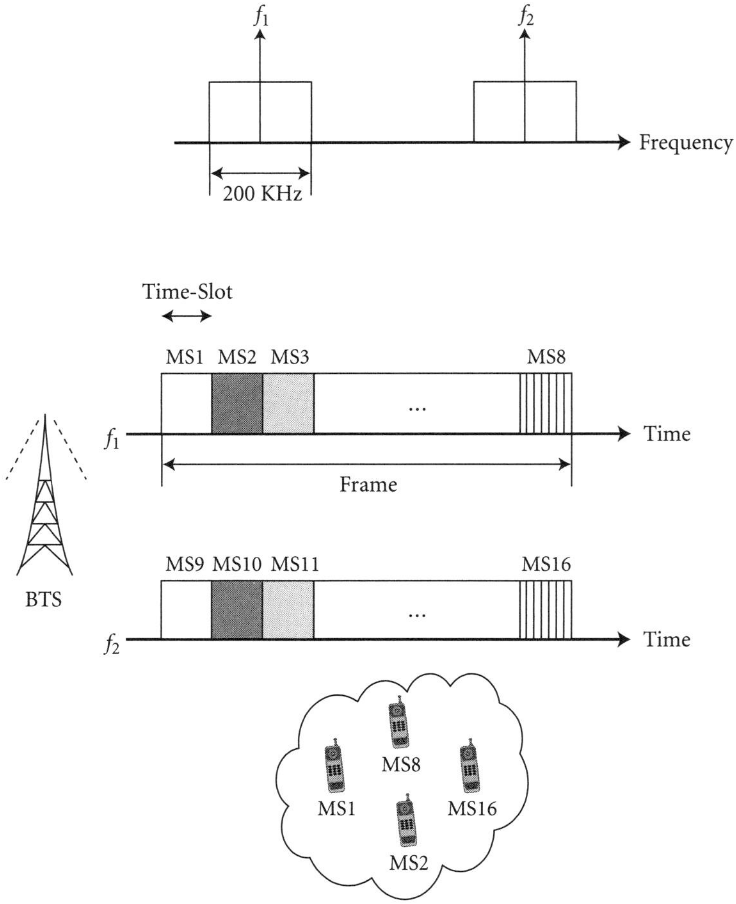

Figure 5.2: Multiple Access in a GSM Radio Network

in one time-slot on f_1. Then it sends information to another MS, MS2, in the second time-slot on f_1. The BTS shares the same frequency among eight users by sending information to different users during different time-slots. Eight time-slots together constitute a frame, formally called a *TDMA frame*. While MS1 through MS8 are getting their information on f_1, eight more users, MS9 through MS16, are getting their information during eight time-slots on frequency f_2. In summary, the GSM network uses multiple radio channels and multiple time-slots to simultaneously serve multiple users.[145]

[145]Commercial GSM networks exploit frequency hopping in which a given MS uses different Absolute Radio Frequency Channel Numbers (ARFCNs) at different times during the call.

5.3.2 Air-Interface Protocol Stack: A Brief Overview

Figure 5.3 depicts the protocol stack that the MS and the BSS use to communicate over the radio or air interface. This protocol stack is motivated by the ISDN protocol stack. The air interface, also called the *Um interface*, consists of three layers. Layer 1 is a physical layer, which is responsible for over-the-air transmission of traffic and signaling messages. The data link layer is layer 2 and consists of the Link Access Protocol for the mobile D (LAPDm) channel. (No, LAPD here does not stand for Los Angeles Police Department!) The LAPDm conveys information between layer 3 entities (i.e., MS and BSS) across the GSM PLMN radio interface using the mobile D or Dm channel. The Dm channel refers to the collection of all the signaling channels required in the GSM system. Layer 3 includes entities such as Connection Management, Mobility Management, and Radio Resource. The Connection Management entity coordinates call establishment and release. The Mobility Management entity facilitates authentication and location update. The Radio Resource entity interfaces with layer 2 and coordinates over-the-air transmissions and receptions. Details of the three layers of the Um interface are discussed in Sections 5.3.2.1 through 5.3.2.3.

5.3.2.1 Layer 1: Physical Layer The physical layer or L1 (i.e., layer 1) transfers information over the radio interface. Figure 5.4 illustrates the basic physical layer processing. The BTS and the MS carry out such processing individually in every cell. Recall that a cell could be omnidirectional, covering 360°, or sectorized, covering, for example, 120°.

The physical layer receives a data stream from layer 2, the data link layer. Figure 5.4 shows a simplified view of how the physical layer processes information and eventually transmits it over the air [195]. However, keep in mind that different types of radio channels undergo somewhat different processing; some aspects of these channels will be discussed in Section 5.4.

Let's go through the processing blocks of Figure 5.4 one-by-one. The information bits first encounter the block coding processing block. The information bits could be speech bits, data bits, or a signaling message. In the case of speech, the set of

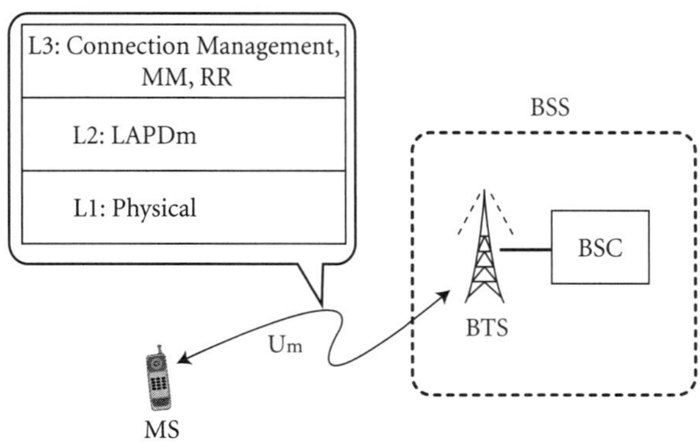

Figure 5.3: Air-Interface Protocol Stack

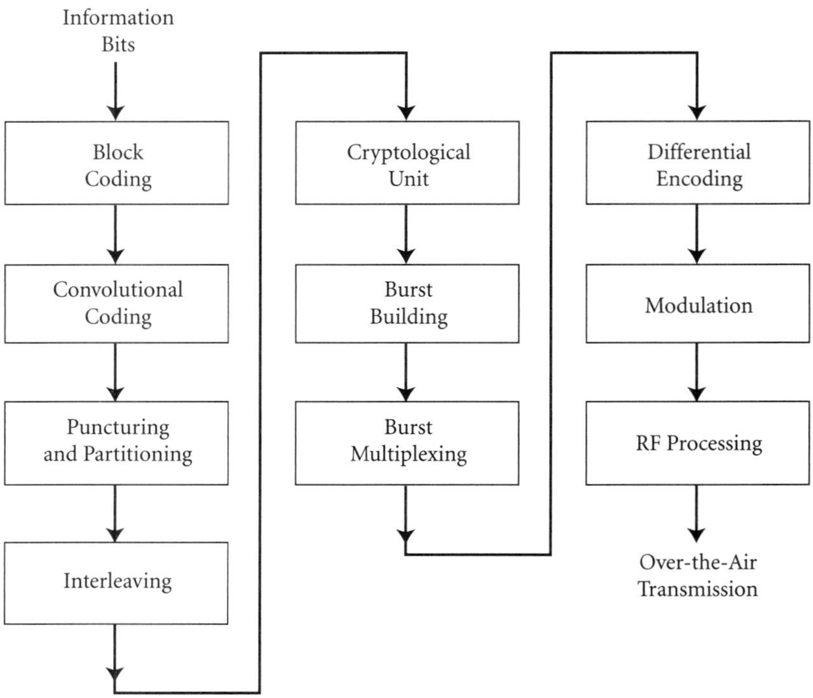

Figure 5.4: Simplified Physical Layer Processing at the Transmitter

information bits includes the entire speech frame. For example, when enhanced full-rate speech is used, one speech frame contains 20 ms worth of speech bits at the rate of 12.2 kbps, yielding 244 speech bits.

Let's see how many bits we have for half-rate speech. Well, half-rate speech contains 112 bits. (You divided 244 by 2 and were expecting this number to be $244/2 = 122$. Right? It is really 112 bits and *not* 122 bits based on the way the speech codec bits are constructed and accommodated by the physical layer!)

The typical number of bits in a signaling message is 184. Block coding processes the information bits to yield bits that include original information bits, some parity bits, and some repetition bits. The output of the block coding includes all the input bits and is called *systematic coding*. Block coding has two main goals: to introduce parity bits for important bits at the transmitter so that the receiver can detect the presence of errors in the received bits and take suitable actions and to provide redundancy to those bits that would not undergo subsequent channel coding (i.e., convolutional coding here).

Two types of block codes are used, cyclic code and fire code. Full-rate and half-rate speech use the cyclic code, and signaling messages typically use the fire code. In the case of enhanced full-rate speech, block coding (also called preliminary channel coding) adds sixteen redundancy bits to the original set of 244 speech bits. The overall block of 260 bits appearing at the output of the block coding includes 182 bits of Class 1 and 78 bits of Class 2. The Class 1 bits are further divided into Class 1a and Class 1b. *Definitions of multiple classes enable us to minimize the impact of bit errors on the perceived speech quality*. During the subsequent forward error correction stage (summarized in the next paragraph), Class 1a bits are protected by both a cyclic code and the convolutional code, Class 1b bits are protected just by the convolutional code, and Class 2 bits are unprotected. The reason for different levels of protection for different classes of speech bits is to ensure good speech quality while minimizing redundancy because some speech bits are more sensitive to errors and could degrade the voice quality significantly if they are in error.

Out of these sixteen redundancy bits the block coding introduces, eight Cyclic Redundancy Check or parity bits protect the sixty-five most important Class 1a bits and Class 1b bits and eight repetition bits correspond to a set of four bits that are important but unprotected by the subsequent forward error correction coding. The block coding yields 260 bits for full-rate speech, 121 bits for half-rate speech, and (typically) 228 bits for a signaling message. See other works [196] for additional details on the block coding.

As mentioned earlier, the forward error correction coding involves either the combination of the cyclic code and the convolutional code or just the convolution code. The convolutional coding is used due to its simplicity of implementation and adequate error performance for voice and low-rate data services. The cyclic code adds very little redundancy and typically covers the most important bits, such as Class 1a bits for the enhanced full-rate speech speech. In the case of enhanced full-rate speech, Class 1a bits (all fifty of them) undergo a cyclic code to yield three additional parity bits that are added to the original fifty Class 1a bits. Fifty Class 1a information bits, 132 Class 1b bits, and three Class 1 parity bits are re-ordered. The re-ordered set of 185 bits ($50 + 132 + 3$) is appended by four tail bits, defining a total of 189 Class 1 bits as input to the convolutional coding.

As discussed in Chapter 2, the main goal of the channel coding, also termed forward error correction coding, is to add redundancy so that the receiver can recover the original set of information bits even in the presence of errors caused by the not-so-kind radio environment.[146] The supported coding rate is $1/2$ for speech and signaling messages and variable for circuit-switched data. Details of a generic convolutional coder can be found in Chapter 2.

A Question for You... Can you guess what might be the constraint length of the convolutional coder when four tail bits are used in the input to the coder? The footnote[147] has the answer.

[146]The cyclic redundancy check accompanying the frame helps detect the presence of one or more errors at the receiver. In such an error scenario, the receiver may ask for retransmission. In a voice frame, though, the receiver may use a previously received correct speech frame to recreate the speech instead of using a current erroneous frame.

[147]Constraint Length $K = 5$. The number of tail bits is equal to the number of shift registers so that the decoder at the receiver has all 0s in the shift registers after processing the received coded bits. This will prepare the decoder to receive the next set of coded bits. The constraint length K is one plus the number of shift registers. Hence, $K = 1 + 4 = 5$.

The convolutional coder with $K = 5$ and rate $= 1/2$ is used for 189 Class 1 bits of the enhanced full-rate speech and the coder provides $189 * 2 = 378$ bits. Add 78 Class 2 bits (which do not undergo convolutional coding), and we will get a total of $378 + 78 = 456$ bits at the completion of convolutional coding. Two generator polynomials for rate $1/2$ convolutional coder are specified as

$$G_0 = 1 + D^3 + D^4 \tag{5.1}$$

$$G_1 = 1 + D + D^3 + D^4. \tag{5.2}$$

The number of coded bits at the conclusion of the forward error correction process is 456 for full-rate speech, most signaling messages, and data traffic and is 228 for half-rate speech.

The encoded bits are re-ordered and partitioned into eight or four blocks. For example, full-rate speech encoded bits are partitioned into eight blocks, and half-rate speech encoded bits are partitioned into four blocks. One data block always carries fifty-seven coded bits. Re-ordering is done such that one re-ordered data block contains data bits from two consecutive unordered blocks, with the higher-numbered unordered block contributing even numbered data bits and the lower-numbered unordered block contributing odd numbered bits. The re-ordered blocks are then interleaved. Such interleaving is block-diagonal for speech and some signaling messages and block-rectangular for other signaling messages. Within each burst containing fifty-seven interleaved bits, two bits are appended. One of the bits indicates the stealing of even-numbered bits by a signaling message, and the other bit indicates the stealing of odd-numbered bits by a signaling message. For each traffic channel block not stolen for signaling, those two bits are set to `0`. These bits are set to `1` to indicate stealing of the bits. In full-rate speech, the signaling message being sent on the full-rate Fast Associated Control Channel (FACCH) steals one full frame of data. The FACCH's primary use is to carry a handover command for a circuit-switched voice call. Section 5.4.2 describes all the radio channels including the FACCH.[148] See other works [196] for additional interleaving details. *In essence, the interleaving is primarily intended to shuffle the bits to achieve time diversity and resultant gain.*

Once interleaving is carried out, the *burst* (i.e., the specific set of bits transmitted during a time-slot) is encrypted based on the agreement between the MS and the GSM network. For example, encrypted full-rate speech bits received from the cryptological unit are transmitted in a *normal burst*. The concept of a burst enables the MSs to stay synchronized with the network and reduce overall interference. The five types of bursts are described in Section 5.4. The GSM Association, not the GSM standard, defines the specifics of the encryption mechanisms.

The encrypted burst now encounters a differential encoder. Before the bits of the burst enter the differential encoder, the encoder is initialized to a state as if a continuous stream of `1`s had been transmitted in the last time-slot. Let d_i be the input bit to the differential encoder, with d_i assuming the value of `0` or `1`. The output of the

[148]The FACCH's presence is indicated by the reversed state of the stealing flags on either side or both sides of a training sequence that is present in the time-slots.

differential encoder is given by

$$d_i = d_i \oplus d_{i-1} \tag{5.3}$$

where $\oplus$ denotes the XOR operation or modulo 2 addition. The output of the differential encoder is mapped to $+1$ or -1 by using the relationship

$$\alpha_i = 1 - 2d_i. \tag{5.4}$$

The modulating data value α_i represents input to the GMSK (Gaussian Minimum Shift Keying) modulator with BT $= 0.3$ [195]. See Chapter 2 for details on GMSK. A pulse-shaping filter with a 3 dB bandwidth and α_i influence the phase of the signal being transmitted. The modulating bit rate is $1/T = 1625/6 = 270.833$ kbps. See other works [197] for additional details on filtering and modulation.

A Funny Detail... If you search the GSM modulation technical report [197] for the term *GMSK*, you will not find it. Do not blame the search tool! Even though the report describes GMSK implementation, it never uses the term. If your mind will not be at ease until you actually see *GMSK* in the standard, see the technical report on the physical layer on the radio path [195].

The actual RF signal being transmitted from the antenna, $x(t)$, has the form

$$x(t) = (2E_c/T)^{0.5} \cos\left(2\pi f_0 t + \varphi(t) + \varphi 0\right) \tag{5.5}$$

where E_c is the energy per modulating bit, T is the modulating bit period (which is $(6/1625)$ ms or about 3.69 microseconds), f_0 is the center carrier frequency (e.g., 900 MHz), and $\varphi(t)$ is a time-varying phase dependent upon α_i. In Eq. (5.5), φ_0 is a *random* phase that is *constant* during one burst. Talk about randomness!

Something to Think About... We have discussed the transmitter operations in this section so far. The GSM standard has no receiver chain specifications. Only the receiver performance requirements are specified. You can find a wealth of MS performance requirements in the standard [198]. Hence, the receiver manufacturers are not constrained to one specific implementation; they can build proprietary high-performance receivers that use advanced signal processing techniques. Conceptually, the receiver carries out the reverse processing. It performs RF processing to bring the signal from RF to the baseband. It carries out demodulation, differential decoding, burst demultiplexing, burst recovery (i.e., the undoing of the transmitter's burst building), deciphering, de-interleaving, and channel decoding. Since the radio environment can adversely affect the transmitted signal, the receiver design incorporates advanced features, such as adaptive equalization, to achieve good performance. See Laurent's representation of GMSK as an approximately linear modulation; such characterization facilitates equalization [199].

In addition to implementing the processing blocks identified in Figure 5.4, the physical layer also participates in several upper-layer (e.g., the radio link layer or layer 2 and the RR entity at layer 3) functions: power control, synchronization, handover measurements, and cell selection and reselection [195].

The power-control mechanisms adjust the output transmit power level of the MS (and, optionally, of the BTS) to ensure that the desired quality is achieved with the minimum possible transmit power. The power control at the MS also increases the MS's battery life. The power control at the MS and (optionally) the BTS reduces the amount of interference in the system. The transmit power can be changed with a step size of 2 dB. See Section 5.9 for details of the power-control operation.

The receiver must be properly time-aligned and frequency-aligned to minimize interference issues and to optimally extract information from the received signal. Section 5.5 deals with power-on synchronization procedures.

The physical layer at the MS first observes the prevailing radio channel conditions. The MS then sends measurement reports to the BSS, which makes a handover decision if another BTS can provide better quality to the MS. See Section 5.7 for handover details. The requirements are specified in a technical report on radio subsystem link control [200].

When the MS is powered-on, it looks for the BTS that can provide the highest-quality communications. Such a process is called *cell selection*. When the MS is in the Idle Mode, i.e., its power is on but it has no dedicated radio resources, it periodically searches for the best cell. As the MS travels from one cell to the next in a given geographical region, it keeps looking for a better cell to lock onto and reads some common control channels from such a cell. Observation of these channels gives the MS important information about the system and enables the MS to receive a PAGE message.

5.3.2.2 LAPDm The LAPDm framing protocol helps carry signaling messages between two layer 3 entities on the two ends of the air interface, i.e., the MS and the BSS. We will discuss in Section 5.4 various radio channels that can carry signaling messages over the air. The LAPDm supports signaling messages over various radio channels (e.g., the paging channel and broadcast control channel).

The LAPDm performs several major functions. It allows one or more data link connections on a Dm channel and distinguishes among these connections by a data link connection identifier. It enables layer 3 messages to be passed transparently between layer 3 entities via *unacknowledged* or *acknowledged* mode operation. In the *unacknowledged* mode operation, the layer 3 information is transmitted in Unnumbered Information frames, and the LAPDm does not acknowledge the Unnumbered Information frames. Furthermore, flow control mechanisms and error recovery mechanisms are undefined. The LAPDm, through the unacknowledged operation, carries the layer 3 messages on the BCCH and the Paging and Access Grant Channel (PAGCH), i.e., PCH (Paging Channel) + AGCH (Access Grant Channel). In the *acknowledged* mode operation, layer 3 information is transmitted in numbered information frames that the receiving data link layer acknowledges. Additionally, error recovery procedures based on retransmission of unacknowledged frames are specified. If the data link layer cannot correct the errors in the frames, it issues a report to the layer 3 entity. Flow control procedures are also defined for this operation. The channels, such as Stand-Alone Dedicated Control Channel (SDCCH), Slow Associated Control Channel (SACCH), and FACCH, support both the acknowledged mode operation and the unacknowledged mode operation.

The LAPDm also implements sequence control so that the order of frames across a data link connection is maintained. It detects operational errors on a data link and notifies the layer 3 entity of any unrecoverable errors. Note that this protocol just reports the errors; it is the responsibility of a layer 3 entity to recover from such errors. The LAPDm also participates in contention resolution after the MS has made an access request on the random access channel. For additional details on the LAPDm, refer to the GSM technical reports [201, 202].

5.3.2.3 Layer 3 Entities Three main sublayers residing within layer 3 are the Radio Resource (RR) sublayer, Mobility Management (MM) sublayer, and Connection Management (CM) sublayer.

The main purpose of the RR sublayer is to establish access for radio communications between the MS and the BSS. Examples of functions this sublayer performs are coordinating the transmissions and receptions and maintaining the connection between the MS and the network. When the connection is no longer needed, the RR sublayer coordinates the release of the connection between the MS and the network. The RR sublayer also makes the initial request for a dedicated radio channel and executes the ciphering procedure to guarantee confidentiality of communications. The RR sublayer at the MS informs the BSS about the quality of radio channel conditions for surrounding cells by sending MEASUREMENT RESULT messages. These messages are used in the handover algorithm as discussed in Section 5.7.

The main goal of the MM sublayer is to support the mobility of MSs. The MM sublayer facilitates authentication of the MS by working with the MSC/VLR. When the MS powers up, the MM sublayer also carries out an IMSI attach operation whereby the MS registers its IMSI with the NSS. This informs the NSS of the existence of the MS (for a possible page). Furthermore, a temporary identity called TMSI is allocated to the MS so that TMSI is used instead of IMSI, hiding the real identity for security over the open air interface. Such a feature of hiding the original MS identity is called *anonymity*. Authentication typically follows the IMSI attach. When the MS powers down, this sublayer performs the IMSI detach operation to convey unavailability of the MS.

In addition, when the MS goes from one location area to another location area, the MM sublayer performs a location update by sending the Location Update Request. This sublayer at the BSS responds with the LOCATION UPDATE ACCEPT message. The location area (LA) is usually a group of multiple cells. The network associates the MS with one LA at any instant. As long as the network knows the LA in which the MS resides, it can send a page to the MS in the correct geographic area for an incoming call. If the LA consists of only one cell, the MS would need to tell the network where it is every time the MS performs cell reselection. The purpose of grouping multiple cells into an LA is to keep the amount of air-interface signalling to a practical minimum by requiring the MS to update its location when the MS travels to another LA. TMSI is often re-allocated during the location update process because the NSS (more specifically, the VLR) assigns the TMSI, and, when the MS moves from an area that, for example, VLR 1 serves into another area that VLR 2 serves, VLR 2 must allocate a new TMSI to the MS. The TMSI that VLR 1 assigned to the MS is no longer valid since the MS is no longer in the area that VLR 1 serves.

The MM sublayer in the BSS could ask the MS for identification using the IDENTITY REQUEST message. The MM sublayer in the MS then responds with the IDENTITY RESPONSE message, providing its identity.

Something to Think About... An alias is also referred to as a pen name or pseudonym [203]. Pen names have been used in place of original names throughout human history. Publius is one of most famous aliases used in American history. Alexander Hamilton, James Madison, and John Jay used this pen name in honor of Roman consul Publius Valerius Publicola while they were authoring *The Federalist Papers*, a series of 85 articles advocating the ratification of the United States Constitution. In addition, Mark Twain, often referred to as the father of American literature, was a pen name for Samuel Langhorne Clemens.

TMSI is like an alias. Instead of using the original identity of IMSI, we use a pseudo-identity. Hence, whenever we want to have communications between the MS and the GSM network and we need to identify the MS, we use a TMSI instead of an IMSI. Minimizing the use of IMSIs over the air gives "bad guys" very little opportunity to steal the original MS identity.

The CM sublayer resides above the RR and MM sublayers. It consists of three independent entities: Call Control, SMS Support, and Supplementary Services Support. The main function of the CM sublayer is to coordinate call setup and call release and to facilitate SMS and supplementary services. The CM sublayer uses the services of the RR and MM sublayers to carry out its functions.

The CM sublayer performs other functions as well. The Call Control entity of the CM sublayer coordinates MS-initiated call establishment, MS-terminated call establishment, and call release. For example, when the MS wants to initiate a call, the Call Control entity of the CM sublayer in the MS generates a CM SERVICE REQUEST message. If the MSC accepts the request, it responds with the CM SERVICE ACCEPT message. After the CM connection is established, the MM connection is established.

In addition, the SMS and Call Control entities of the CM sublayer coordinate the necessary functions for SMS messaging and Supplementary Services Support messaging between the peer SMS and Supplementary Services Support applications residing at the MS and a suitable NSS entity.

5.4 RADIO INTERFACE: FRAME STRUCTURE AND CHANNELS

In a GSM system, the 200-kHz–wide radio frequency channels and time-slots are our precious radio resources because GSM is based on FDMA and TDMA. Section 5.4.1 provides a detailed look at the frequency channels and timing structure. Various radio channels are defined in GSM to support a variety of functions. These channels are discussed in Section 5.4.2. A time-slot contains a burst, and Section 5.4.3 summarizes the five types of bursts defined in GSM.

5.4.1 Frequency and Time Structures

Consider a BTS communicating with an MS in a cell as shown in Figure 5.5. The BSS has allocated time-slot 0, uplink (UL) frequency f_1, and downlink (DL) frequency f_2

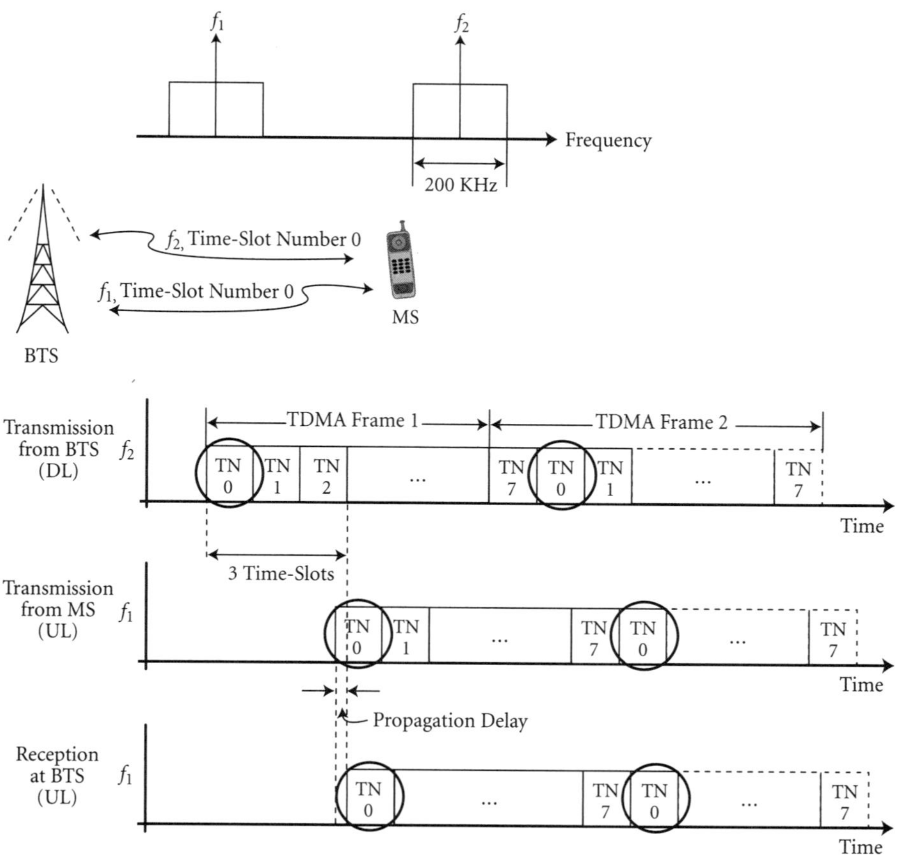

Figure 5.5: Two-Way Communications Between the MS and the BTS

to MS1. The BTS sends information (e.g., full-rate speech) to MS1 using time-slot 0. A TDMA frame contains eight time-slots, numbered 0 to 7. The time-slot number (TN) is specified to the MS during call setup. In our example, $TN = 0$. The same TN is used for both the DL and the UL for a given MS. In other words, the MS receives information from the BTS during $TN = 0$ and sends information to the BTS during $TN = 0$. Since we are using Frequency-Division Duplexing, one ARFCN, f_1, is used in one direction, and another ARFCN, f_2, is used in the other direction. The difference between the DL transmission of the time-slot and the UL reception of the same time-slot as observed by the BTS in a cell is three time-slots. Note that the BTS observes the time difference here. The MS needs to transmit the assigned time-slot at a suitable time instant such that when the BTS receives such a time-slot, the time-slot difference between the DL transmission and the UL reception is three time-slots. We need to account for the propagation delay, which is variable for different MSs. The process of adjusting the MS's transmission timing advance is called *adaptive frame alignment* or *timing advance* [204]. The gap of three time-slots between the DL and the UL simplifies the design of the transmitter and the receiver because only a transmit or receive operation must be carried out during a given time-slot. This gap gives the transceiver (i.e., transmitter + receiver) adequate time to switch from transmit operation to receive operation and vice versa.

5.4.1.1 Time Resources The GSM standard defines several time structures as illustrated in Figure 5.6. The time-slot is the shortest time unit that can carry information (such as a signaling message) or user traffic (such as speech). The length of the time-slot is $(3/5200)$ seconds or 577 μs. Eight time-slots, $TN = 0$ to $TN = 7$, constitute one TDMA frame. The length of the TDMA frame is $8*577\mu s = 4.615$ ms. One multiframe consists of twenty-six or fifty-one TDMA frames. Two types of multiframes are defined. The multiframe with fifty-one TDMA frames is used for broadcast and common control channels. The multiframe with twenty-six TDMA frames is used for traffic and associated control channels. A superframe combines either twenty-six multiframes with fifty-one TDMA frames per multiframe *or* fifty-one multiframes with twenty-six TDMA frames per multiframe. Essentially, one superframe contains a total of $26 * 51 = 1326$ TDMA frames. A combination of 2048 superframes yields a hyperframe, which is the longest time structure defined in GSM. One hyperframe corresponds to $2048 * 1326 = 2,715,648$ frames. In fact, each TDMA frame is given a frame number (FN) that ranges from 0 to $FN\_MAX - 1$ where $FN\_MAX = 26 * 51 * 2048 = 2,715,648$. In other words, the FN is reused in every hyperframe. Now, why do we have all of these time units defined? Table 5.2 summarizes the purpose behind each time structure.

The time-slot carries user traffic or a signaling message. The TDMA frame accommodates up to eight users, with one user per time-slot. The GSM standard defines two types of channels, logical and physical (see Section 5.4.2). The multiframe helps locate the physical channel that carries the logical channel of interest. The superframe always contains 1326 frames and supports two types of multiframes. Hence, it provides a unified structure that facilitates multiplexing of traffic multiframes and

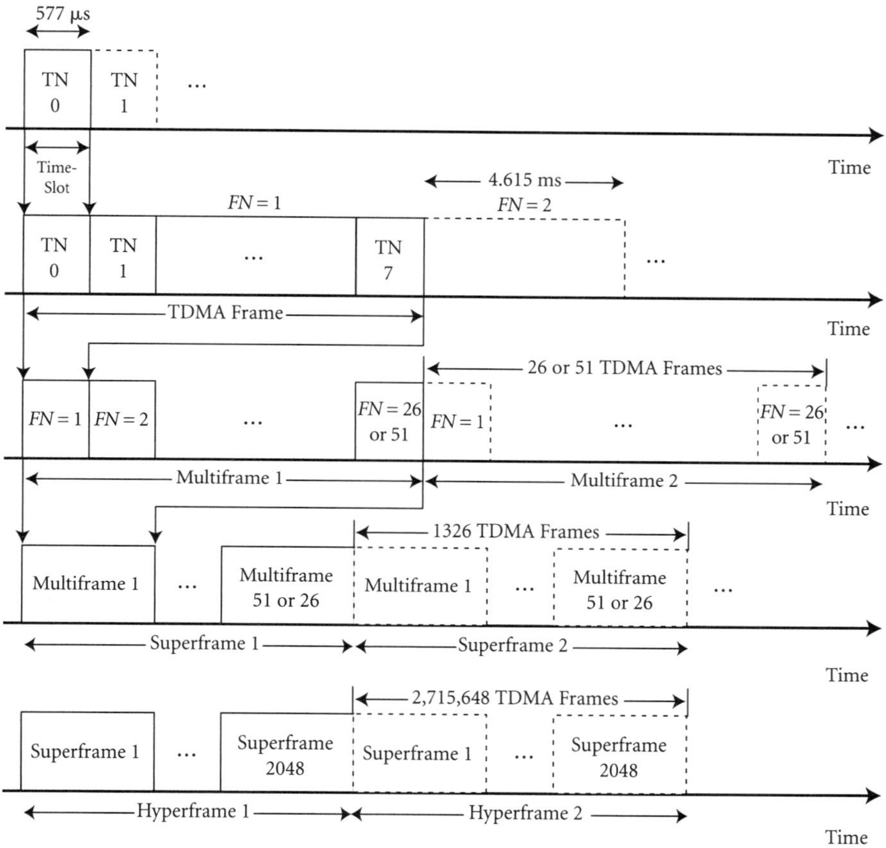

Figure 5.6: Timing Structures for the Radio Interface

Table 5.2: Importance of Time Structures

Time Structure	Purpose
Time-Slot	Carries information (traffic or signaling)
TDMA Frame	Allows multiple access by sharing eight time-slots among the users
Multiframe	Identifies the location of the physical channel that carries a specific logical channel
Superframe	Provides a unified structure for signaling messages and user traffic (e.g., speech) by encompassing traffic multiframe and broadcast multiframe
Hyperframe	Facilitates encryption

common control multiframes. Finally, the hyperframe number is used as an input parameter for the encryption algorithm [205].[149]

5.4.1.2 Frequency Resources

Now we know the time structure for the GSM air interface. One TDMA frame consists of eight slots, and eight users, each occupying a time-slot, can be supported per frame. A 200-kHz–wide ARFCN is thus shared via the time-slots of a TDMA frame. If we have two such ARFCNs, we can support up to sixteen users, as shown in the simplified example of Figure 5.5, when each MS is assigned a full-rate channel. The BSS allocates the ARFCNs, such as f_1 and f_2 in Figure 5.5, to the MS by specifying the ARFCN. For example, $ARFCN = 10$ for P-GSM 900 corresponds to the uplink ARFCN at $890 + 0.2 * n = 890 + 0.2 * 10 = 892$ MHz and the downlink ARFCN of $uplink\ ARFCN + 45 = 892 + 45 = 937$ MHz where n is the ARFCN number [198].

A service operator may have several ARFCNs per cell based on a deployment strategy and available spectrum. See Section 5.10 for GSM deployment considerations. Each cell is allocated a subset of ARFCNs available in the operator's spectrum, and this subset is called *cell allocation*. The MS is allocated a subset of cell allocation, which is termed *mobile allocation*. The standard requires one of the cell allocation's ARFCNs to carry synchronization information and the BCCH. Such an ARFCN is often referred to as the BCCH carrier. The time-slot $TN = 0$ in the TDMA frame for the BCCH carrier carries several control channels, including the BCCH; the remaining time-slots of this TDMA frame could be allocated to the MSs to carry user traffic, such as speech.

5.4.1.3 Frequency Hopping

GSM allows frequency hopping whereby the MS utilizes different ARFCNs from one time-slot to the next. Frequency hopping implementation is optional for the network and mandatory for the MS. Frequency hopping is defined such that MSs communicating with a given BS avoid using the same ARFCN at a given instant and that MSs communicating with different BSs have random collisions of ARFCNs. The standard defines an algorithm that considers several input parameters, such as the TN, the mobile allocation, and a given mobile allocation index offset to map the TDMA frame number to a specific ARFCN [206]. In other words, one-to-one mapping occurs between the ARFCN and the FN for a given MS in a given cell.

A Question for You... What are the advantages and disadvantages of frequency hopping? See the footnote[150] for the answer.

[149] The GSM air link uses the Vernam cipher for its ciphering. This cipher's encryption code (or cipher mask) continuously varies but is synchronized over the radio link by a hyperframe counter in the BTS and the MS. This encryption scheme yields a changing 116-bit mask for each normal burst.

[150] The main benefits of frequency hopping are frequency diversity and interference diversity. Frequency diversity ensures that the MS is not stuck on an ARFCN that is experiencing a fade. Similarly, interference diversity reduces the probability that the MS is on an ARFCN that has high interference. The drawback is increased complexity.

5.4.2 Air-Interface Channels

Two main types of radio channels are defined, a logical channel and a physical channel. The logical channel specifies the type of information being conveyed, and the physical channel actually carries one or more logical channels over the air. For example, a speech traffic channel is a logical channel. A specific time-slot (e.g., $TN = 2$) in a number of TDMA frames represents a physical channel, and such a physical channel carries the logical channel (such as the speech traffic channel) over the air. A given physical channel uses the same TN in every TDMA frame. The basic physical channel thus corresponds to a specific TN in a sequence of TDMA frames. Recall that the FN ranges from 0 to $FN\_MAX - 1$. Section 5.4.2.1 discusses the logical channels, and Section 5.4.2.2 describes the mapping of the logical channels onto physical channels. Refer to the standard [206] for details about the logical and physical channels.

5.4.2.1 Logical Channels To begin our discussion of logical channels, we must first explain some terms. Figure 5.7 illustrates these terms. All MSs in a cell's coverage area listen to *broadcast channels.* In other words, the information being carried on broadcast channels is not customized for one MS; the same information is valid for all the MSs. In Figure 5.7, all the MSs, MS1 through MS5, get the same information, such as the values of some system parameters.

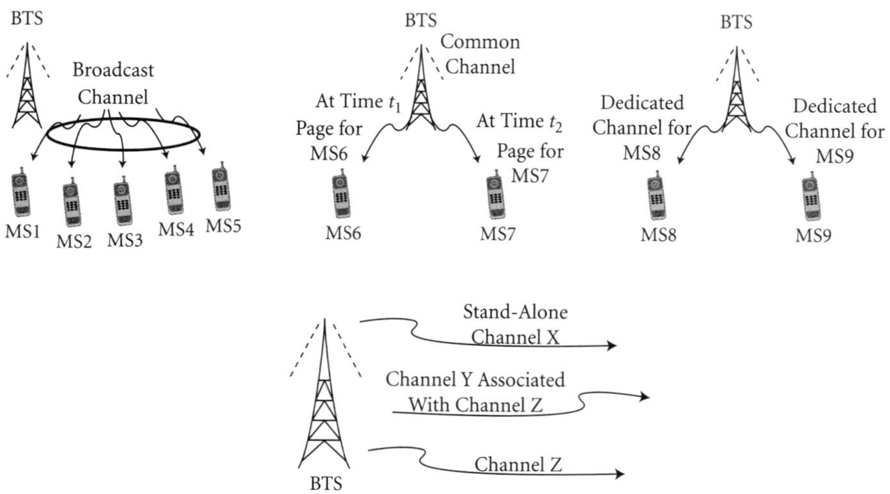

Figure 5.7: Logical Channel Nomenclature

Multiple MSs can share some *common channels*. However, at a given time and on a given ARFCN, messages specific to one MS travel on a given common channel. In other words, the common channels sometimes carry one MS's messages and sometimes another MS's messages. In Figure 5.7, a common channel sends a PAGE message to MS6 at time instant t_1. The same common channel sends a PAGE message to another MS, MS7, at another time instant t_2; the same channel is used for both MSs.

Dedicated Channels are customized for individual MSs. For example, one dedicated channel is used by only one MS, and another dedicated channel is for another MS. Figure 5.7 shows two dedicated channels for two different MSs.

A *stand-alone channel* can stand on its own feet; it requires no support from any other channel. An *associated channel* needs another channel for its existence. In Figure 5.7, Channel X is a stand-alone channel, and Channel Y is an associated channel. Channel Y is associated with Channel Z; without Channel Z, Channel Y cannot exist. Channel Z could exist on its own. (Thank God this is not a chicken-and-egg problem! If Channel Z does not exist, Channel Y will certainly not exist.)

Table 5.3 lists the logical channels the GSM standard defines along with their purpose. Let's briefly touch upon these channels.

There are two main types of logical channels, traffic channels and control channels. The TCH contains user traffic. Two types of traffic channels are speech traffic channels and data traffic channels.[151] The speech traffic channel conveys user speech, and the data traffic channel carries user data (e.g., email). The TCH is also classified according to the rate, which could be full rate or half rate. The full-rate TCH and half-rate TCH correspond to the gross bit rates of 22.8 kbps and 11.4 kbps, respectively. As an example, TCH/FS (Traffic Channel/Full-Rate Speech) is a TCH that carries speech at the full rate. TCH/F9.6 (Traffic Channel/Full-Rate 9.6 kbps) is a TCH that carries 9.6 kbps data using the full-rate channel. An example list of traffic channels includes TCH/FS, TCH/HS, TCH/F9.6, TCH/F4.8, TCH/H4.8, TCH/H2.4, and TCH/F2.4.[152] In the channel name, H to the right of the slash denotes half rate and F denotes full rate.

The four categories of the control channel are the broadcast control channel, common control channel, dedicated control channel, and cell broadcast control channel. The Frequency Correction Channel (FCCH) enables the MS to apply frequency correction. The Synchronization Channel (SCH) specifies the Base Station Identity Code (BSIC) that helps the MS identify the BTS. The SCH also carries a reduced frame number in the form of three parameters, T1, T2, and T3, which are used by the MS to calculate the current TDMA frame number. The BCCH specifies numerous BTS-specific parameters, such as the configuration of common control channels. For example, the BCCH specifies the number of common control channels and indicates how the CCCH (Common Control Channel) and SDCCH are used in a cell. The PCH is used to send a PAGE message to the MS. The MS uses the Random Access Channel

[151] These are circuit-swiched channels in GSM. GPRS and EDGE support packet-switched channels.
[152] Additional types of traffic channels have been introduced in later versions of the standard.

Table 5.3: Logical Channels and Their Purpose

Logical Channel Category	Type of Logical Channel	Logical Channel	Purpose
Traffic	Dedicated	Speech Traffic Channel	Conveys user speech
		Data Traffic Channel	Carries user data
Control	Broadcast	FCCH	Facilitates frequency correction at the MS
		SCH	Helps frame synchronization and BTS identification
		BCCH	Transports BTS-specific parameters, such as configuration of common control channels
	Common Control	PCH	Sends a PAGE message to the MS
		RACH	Carries the MS's request for a dedicated control channel (SDCCH)
		AGCH	Allocates SDCCH or TCH to the MS
	Dedicated Control	SACCH/TF	Carries (at a slow rate) control messages associated with a full-rate TCH
		FACCH/F	Carries (at a fast rate) control messages associated with a full-rate TCH
		SACCH/TH	Carries (at a slow rate) control messages associated with a half-rate TCH
		FACCH/H	Carries (at a fast rate) control messages associated with a half-rate TCH
		SDCCH/8	Carries (at a slow rate) stand-alone control messages
		SACCH/C8	Carries (at a slow rate) control messages associated with SDCCH/8
		SDCCH/4	Acts as a combined SDCCH and CCCH
		SACCH/C4	Carries (at a slow rate) control messages associated with SDCCH/4
	Cell Broadcast	CBCH	Broadcasts short messages (e.g., advertisements) in a cell

(RACH) to make a request for an SDCCH. The AGCH allocates the SDCCH or TCH to the MS.

The SACCH/TF (Slow Associated Control Channel/Traffic Channel Full Rate) is a dedicated control channel that carries at a slow rate the control messages that are associated with a full-rate TCH. The FACCH/F (Fast Associated Control Channel/Full Rate) carries at a fast rate the control messages that are associated with a full-rate TCH. The SACCH/TH (Slow Associated Control Channel/Traffic Channel Half Rate) carries at a slow rate the control messages associated with a half-rate TCH. The FACCH/H (Fast Associated Control Channel/Half Rate) carries at a fast rate the control messages associated with a half-rate TCH. The SDCCH/8 (SDCCH/8 kbps) carries stand-alone control messages. The SACCH/C8 (Slow Associated Control Channel/Control Channel 8 kbps) carries the control messages associated with SDCCH/8. The SDCCH/4 (SDCCH/4 kbps) acts as a combined SDCCH and CCCH. The SACCH/4 (Stand-Alone Associated Control Channel/4 kbps) carries at a slow rate the control messages associated with SDCCH/4. Lastly, the Cell Broadcast Channel can be used to broadcast short point-to-point messages (e.g., advertisements) in a cell. These messages are sent to multiple users in a cell and are different from the user's SMS messages.

5.4.2.2 Mapping Between the Logical and Physical Channels The mapping between one or more logical channels onto a physical channel is discussed in the specifications [195, 206]. *The multiframe structure defines the mapping between the logical channel and the physical channel.* The twenty-fix–TDMA frame multiframe is 120 ms long and carries TCH (and SACCH/T) and FACCH. The 51-TDMA frame multiframe is 235.4 ms long and carries the BCCH, CCCH (i.e., a combination of the AGCH, PCH, and RACH), and SDCCH (and the SACCH/C).[153] An example of the logical-to-physical channel mapping in the case of a twenty-six–frame multiframe is shown in Figure 5.8.

We will see in Section 5.6 that the MS is assigned a TCH during the call setup. Two channels are associated with the TCH, the FACCH and the SACCH. The MS uses these associated channels to report measurements to facilitate operations such as handover and power control. The SACCH, the FACCH, and the TCH use the same physical channel. In other words, the SACCH and the FACCH are born along with the TCH, and all use exactly the same physical channel. As shown in Figure 5.8, the SACCH is transmitted in the twelfth TDMA frame of a multiframe. The twenty-fifth TDMA frame is an idle frame, which the MS uses to synchronize with adjacent cells and to make measurements of GSM or other radio access technologies. The TCH and the FACCH may use the remaining twenty-four frames. Let's ignore the FACCH for now.

The set of twenty-four TDMA frames carries full-rate speech. One full-rate speech block equivalent to a 20 ms speech frame consumes four bursts but is interleaved over eight bursts to achieve an enhanced time diversity benefit from interleaving. Six

[153]Figure 5.11 in Section 5.5 gives an example of how a fifty-one–frame multiframe maps one or more logical channels onto physical channels.

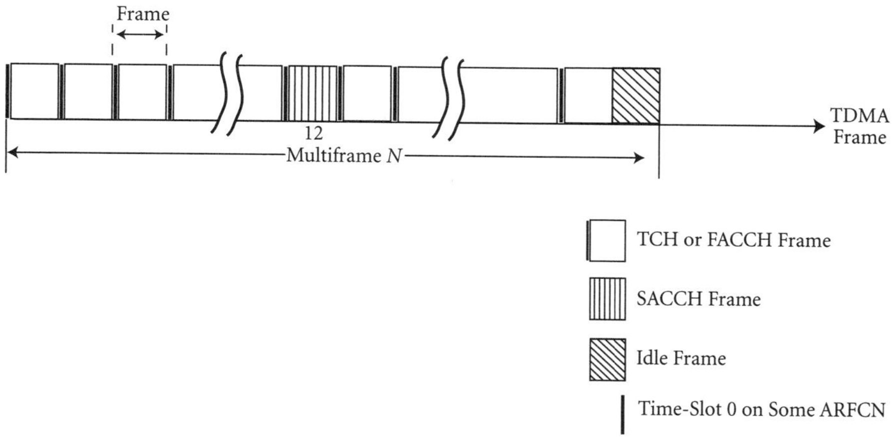

Figure 5.8: Logical-to-Physical Channel Mapping for a Traffic Multiframe

blocks of full-rate speech can be transmitted in one multiframe. Hence, three blocks of full-rate speech can be transmitted in one multiframe. When a message is to be sent on the SACCH, such a message needs a total of four TDMA frames. Since one multiframe allocates only one frame to the SACCH, the duration of four multiframes (i.e., $26 * 4 = 104$ TDMA frames or 480 ms) is needed to send a message on the SACCH. No wonder it is called a *slow* associated control channel! Delay-tolerant messages take a ride on the SACCH. For time-sensitive messages, we need a faster-moving vehicle, and that's where the FACCH comes into the picture. As shown in Figure 5.8, the FACCH has no reserved allocation during a multiframe; it just preempts the TCH to send a message quickly. In other words, instead of sending speech on the TCH during a given TDMA frame, the FACCH sends a signaling message. Since such preemption of the speech occurs once in a while, the user's perception of speech quality degrades only insignificantly. When the FACCH replaces or steals speech bits, the stealing flags convey such a replacement of speech bits (see Section 5.3.2.1).

5.4.3 Bursts: An Inside Look

We have seen that a time-slot is the basic time structure that carries information such as speech or a signaling message. Now let's see what is inside such a time-slot. Remember that the physical contents of the time-slot are referred to as a *burst* [195]. A stream of bits modulates the RF carrier (e.g., an ARFCN at 900 MHz) to form a burst during the time-slot. For example, in the case of full-rate speech, the bits the cryptological unit encrypts are placed into a burst. The time-slot carrying the burst is said to consist of 156.25 bit periods. Five types of bursts are defined in which each burst type uses the burst length of 156.25 bit periods differently [195,206]. Figure 5.9 illustrates the structure of these bursts.

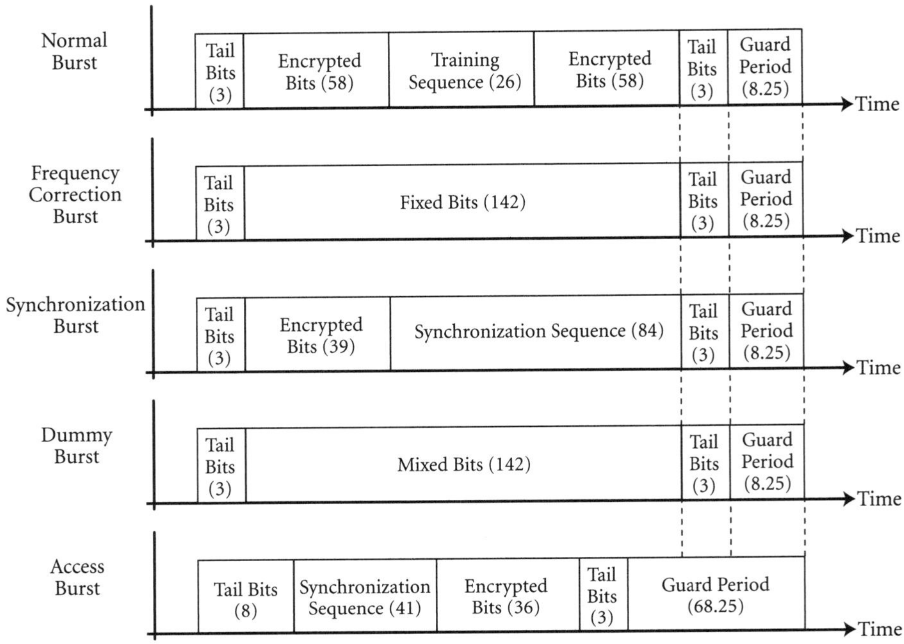

Figure 5.9: Bursts on a GSM Air Interface

Do You Recall... Do you recall the approximate length of the time-slot? See the footnote[154] for the answer.

The first type of burst, the normal burst, carries information on traffic and control channels excluding the RACH. It contains 116 encrypted bits in two separate blocks. The initial part of the normal burst includes three tail bits. (By the way, should they not be called *head bits* instead of *tail bits* since they are at the *beginning* of the normal burst?) The training sequence facilitates channel estimation and channel equalization. The tail-end of the normal burst contains three tail bits and a guard period that lasts for the duration of 8.25 bit periods (equivalent to 30.46 μs).

The second type, the frequency correction burst, facilitates the MS's frequency synchronization. It is equivalent to an unmodulated carrier, shifted in frequency. It contains tails bits and the guard period just like the normal burst. This burst is broadcast along with the BCCH. The sequence of the frequency correction bursts is nothing but the FCCH.

[154]The length of the time-slot is (3/5200) seconds or 577 μs.

The synchronization burst helps the MS with time synchronization. It identifies the BTS to the MS and enables the MS to extract the TDMA frame number. It is broadcast along with the frequency correction burst. The sequence of the synchronization bursts is the SCH.

The fourth type is the dummy burst, and it contains a predefined set of 142 bits. Since the dummy burst is filled with a special sequence of bits, the MS differentiates it from a burst carrying information or from the frequency correction burst. If the dummy burst were not filled with some bit pattern, it would have appeared like an unmodulated carrier of the frequency correction burst, thereby confusing MSs searching for the frequency correction burst. In a GSM system, the BTS must transmit a burst in every time-slot of every TDMA frame on the BCCH carrier frequency so that MSs can make suitable power measurements of the BCCH. Hence, the dummy burst is transmitted on the BCCH carrier during the unoccupied time-slots.

Finally, the MS uses the access burst to access the network. The access burst includes a longer guard period (252 μs instead of 30.46 μs) than the other bursts. This longer guard period allows the MS to access the system at up to 35 km away from the BTS without causing excessive interference to other users and without significantly increasing the access delay. This burst is also used after a handover. See the access mechanism specifications [207] for cell radii longer than 35 km.

5.5 NETWORK ACQUISITION

After the MS is powered on, it attempts to acquire the network and enters an Idle Mode. Such network acquisition and Idle Mode operations are discussed in this section. Further details of the Idle Mode activities can be found in the specifications [204,208].

After the GSM MS's power is turned on, the MS looks for a GSM system in its neighborhood. Figure 5.10 summarizes the basic steps of cell identification and synchronization. The MS looks for the FCCH on the BCCH carrier frequency. Once the MS finds a suitable BCCH carrier frequency, it has achieved frequency synchronization and coarse timing synchronization. Then, the MS proceeds to the next stage, which is acquiring the timing. The MS synchronizes to the SCH's synchronization word to achieve fine (as opposed to coarse) timing synchronization. The MS also extracts the current radio frame number and some system identification parameters by decoding the message the SCH is carrying. Finally, it processes the messages received on the BCCH to learn important information about the system with which it is communicating. Let's discuss these various stages in more detail.

The MS uses the FCCH, the SCH, and the BCCH to synchronize with the GSM system and to obtain various system parameters. Let's see where these channels are located. Figure 5.11 specifies the location of these channels [195]. This can be considered the mapping of the logical channels, such as the FCCH, the SCH, the BCCH, and the CCCH, onto the physical channels. Recall that a physical channel is the time-slot number, TN, in a sequence of TDMA frames. The multiframe with fifty-one TDMA frames for the BCCH carrier is shown in Figure 5.11 in which $TN = 0$ carries the logical channels of interest. For example, within a multiframe, the TDMA

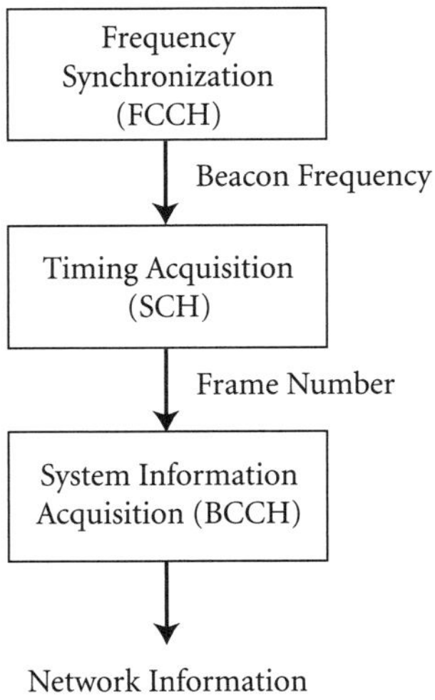

Figure 5.10: Cell Search and Synchronization

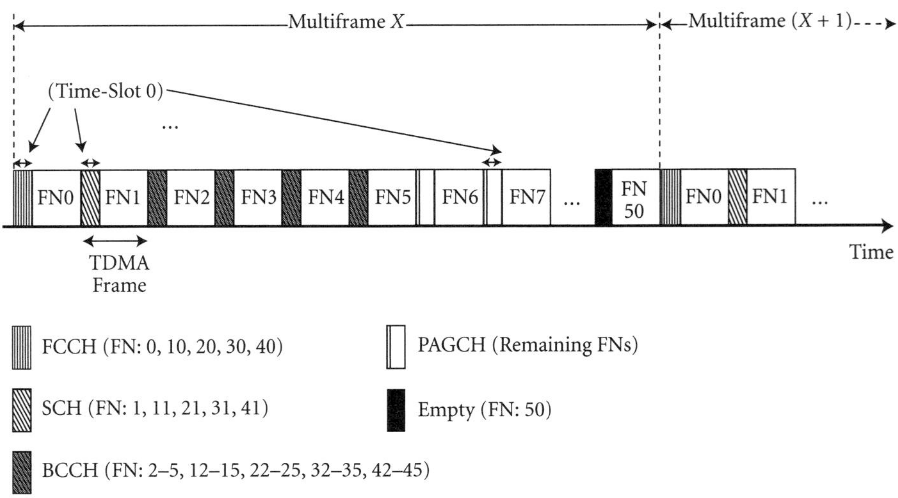

Figure 5.11: Multiframe of the Downlink BCCH Carrier Frequency for Signaling

frames with $FN = 0, 10, 20, 30$, and 40 contain the FCCH. The TDMA frames that have $FN = 1, 11, 21, 31$, and 41 carry the SCH. A set of four consecutive TDMA frames, numbered $FN = 2$ through 5, 12 through 15, 22 though 25, 32 through 35, and 42 through 45 carry the BCCH. The last TDMA frame, corresponding to $FN = 50$, in a multiframe is empty. The remaining TDMA frames in the multiframe can be used to carry the downlink CCCH, which consists of the PCH and the AGCH. The combination of the PCH and the AGCH is called the PAGCH.

Now that we know where to find the FCCH, the SCH, and the BCCH, let's see in detail how the MS carries out the network acquisition. The MS first looks for the BCCH carrier frequency, also known as the *beacon frequency*. The set of beacon frequencies belonging to the home PLMN is typically programmed into the SIM. Hence, the MS looks for these ARFCNs first. If it finds none of the stored home PLMN ARFCNs, it searches for all ARFCNs in the spectrum band it supports. For example, the MS may need to search 173 ARFCNs in the GSM 900 MHz band, 373 ARFCNs in the DCS 1800 band, and/or 298 ARFCNs in the PCS 1900 band. The MS could also store the last known set of BCCH carriers and may prioritize scanning of the ARFCNs based on its previous observations of the messages received from the GSM radio network.

While on a given ARFCN in search of the BCCH carrier, the MS specifically looks for the FCCH. After successfully finding a match with the fixed pattern of 142 bits,[155] the MS knows that it has achieved frequency synchronization and has detected a valid BCCH carrier or beacon frequency. Within a multiframe, the time difference between the successive FNs carrying the FCCH is ten frames (e.g., $10 - 0 = 10$ and $20 - 10 = 10$). However, the time difference between the TDMA frame carrying the FCCH in one multiframe and the TDMA frame carrying the FCCH in the next multiframe is twelve (i.e., the time difference between $FN = 40$ in one multiframe and $FN = 0$ in the next multiframe). Such a twelve-frame time difference helps the MS determine the zeroth TDMA frame within the multiframe.

Do You Recall... Do you remember the type of burst that is associated with the FCCH? (Hint: there are five types of bursts.) See the footnote[156] for the answer.

Since the MS has identified the FCCH and since the multiframe structure is as shown in Figure 5.11, the MS knows where to find the SCH. Recall that the synchronization burst type contains the BTS ID and the timing information. More specifically, the synchronization burst includes the six-bit BSIC that identifies the BTS and the nineteen-bit RFN (Reduced Frame Number), consisting of parameters T1, T2, and T3. The BSIC consists of a three-bit network color code (i.e., a shortened PLMN number) and a three-bit BTS color code (i.e., a shortened BTS ID). The RFN enables the MS to determine the current frame number. Only one TRX in a BTS transmits the BCCH carrier frequency, and this carrier cannot hop.

[155]This pattern is just a series of 0s, leading to the transmission of a pure sinusoidal signal.
[156]It is the Frequency Correction burst.

Now that the MS knows the TDMA frame number, it proceeds to read the BCCH on the BCCH carrier. The BCCH carries information about the system by periodically broadcasting system information messages. For example, the BCCH periodically sends SYSTEM INFORMATION TYPE 2, SYSTEM INFORMATION TYPE 3, and SYSTEM INFORMATION TYPE 4 messages [209]. The MS uses such information to decide whether and how to access the system that it is monitoring. For example, if the BCCH indicates that the current cell belongs to the Barred Cell category, the MS will avoid accessing such a cell and will look for a candidate cell that it can access.

To determine how to access the system, the BCCH informs the MSs in a cell about the parameters to use while using the common RACH in the uplink. For example, the BCCH sends a message called SYSTEM INFORMATION TYPE 1, which carries, among other parameters, Cell Channel Description and RACH Control Parameter information. The Cell Channel Description specifies the ARFCNs that are part of the Cell Allocation. The RACH control parameter indicates whether the cell is barred for access and specifies the number of retransmissions allowed on the RACH during the access.

5.5.1 Beyond Radio Network Acquisition: MS–Core-Network Communications

Once the MS has acquired the GSM radio network, it has achieved frequency and timing synchronization and has learned about the GSM system. Even if the MS is not initiating a call, it still needs to inform the network of its location so that the network can send a page or an SMS to the MS for an MS-terminated call or SMS. This process of contacting the network to give an update on the MS location is called *location updating*. The MS attaches its IMSI with the network so that the network knows that this IMSI is alive and kicking! Formally, the MS performs *IMSI attach* upon power-up when the LAI (location area identity) stored in SIM and the LAI received on the BCCH are the same; otherwise it performs *normal location updating*.[157] By observing the SYSTEM INFORMATION TYPE 3 message, the MS learns whether the IMSI attach is required. First, we'll take a high-level view of the location updating procedure, and then Sections 5.5.1.1 through 5.5.1.7 will provide a detailed look at the procedure. The opposite of IMSI attach is *IMSI detach* in which the MS informs the network about its unavailability (e.g., due to power-off).

Another case of IMSI detach is *implied detach* in which the mobile is paged in the last known LA but the MS doesn't respond to the paging message (even after the maximum number of paging messages are sent). In such a case, the network silently detaches the mobile and sends the mobile no more paging messages until the MS performs a location area update or re-attaches to the network. The *implied detach*

[157] Three types of location updates are defined. A Type 1 location update is a normal update in which the MS moves from one location area to another or the MS is powered up in a new location area. IMSI attach is a Type 2 location update. Finally, a Type 3 location update is periodic; the MS periodically performs the location update procedure. The periodicity of such updates is expressed in units of decihours and ranges from 0 (i.e., no periodic update) to 255 decihours (i.e., 25.5 hours).

usually occurs when the mobile is sent a page while it is out of the radio network coverage. This is often the cause of that familiar phenomenon in which someone calls you but your phone doesn't ring and the caller gets routed directly to your voicemail.

Figure 5.12 summarizes the main activities of the MS after network acquisition. See the specifications [209] for details of the relevant signaling messages.

The MS makes a request for a dedicated radio channel by sending a CHANNEL REQUEST message on the RACH to the network. The BSS allocates a dedicated radio channel to the MS so that the rest of the signaling messages can be reliably exchanged between the BSS and the MS. This allocation is conveyed to the MS via the CCCH. The MS now informs the core network of its existence by sending a LOCATION UPDATING REQUEST message.

The core network initiates the authentication procedure to ensure validity of the MS. The core network sends a random number to the MS via an AUTHENTICATION REQUEST message. The MS executes the authentication algorithm using the received random number and its secret key as inputs and generates a response. The MS sends this response to the core network through an AUTHENTICATION RESPONSE message. The core network validates the MS.

The security aspects are considered next via exchange of CIPHER MODE COMMAND and CIPHER MODE COMPLETE messages. The network instructs the MS whether to start ciphering and which ciphering algorithm to use. The core network now responds to the location update request from the MS by sending a LOCATION UPDATING ACCEPT message with a (possibly) new TMSI for the MS and

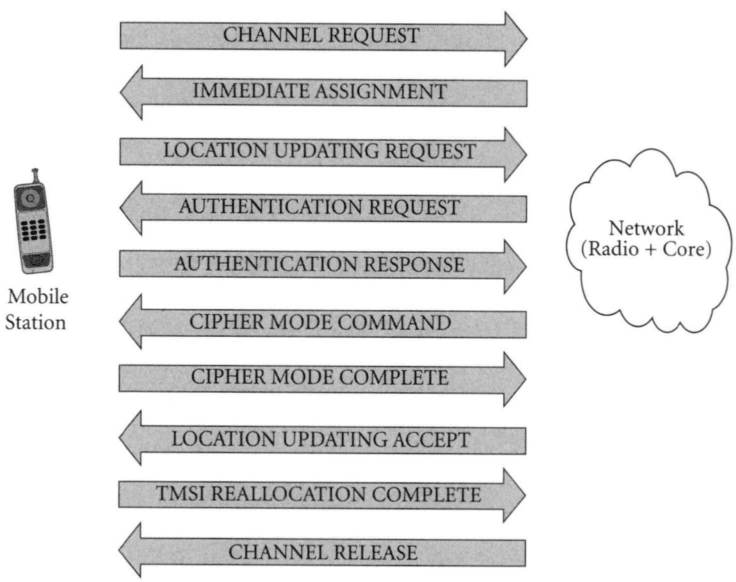

Figure 5.12: Overview of IMSI Attach and Location Updating Procedure

the associated LAI. The MS typically uses TMSI instead of IMSI for the sake of confidentiality. Now let's discuss the steps shown in Figure 5.12 in more detail.

5.5.1.1 CHANNEL REQUEST Message: Establishing the First Contact

After completing the power-on network acquisition, the MS executes a location updating procedure. The MS needs to establish a radio connection with the BTS to perform the IMSI attach. Since the MS has no dedicated radio connection yet, it needs to use some common channel in the uplink to contact the BTS, which is the RACH in the UL. The use of the RACH is influenced by the MS's access class.

Each MS is arbitrarily provisioned with one of the ten access classes, and this access class is stored on the SIM. An access class is denoted by a number ranging from 0 to 9. Furthermore, an MS may belong to one of five special access classes, numbered 11 through 15, and the special access class number is also stored on the SIM card [210]. These special classes are reserved for PLMN staff, emergency services, public utilities (such as water and gas suppliers), security services, and PLMN use. An additional access class, 10, is used by the BCCH to influence the ability of an MS to make emergency calls. For example, if an MS has been assigned an access class from 0 to 9 or if the MS has no IMSI, it will fail to access the system if the network bars access class 10 as indicated by the BCCH. If an MS is allocated to one of the five special access classes, it fails to access the system if both access class 10 and the MS's special class are barred. The BCCH periodically sends system information messages that specify the list of authorized access classes and authorized special access classes. The MS learns from the BCCH if its access class is allowed in a cell at a given time.

Assuming that the access class of the MS is allowed to use the cell, the MS makes an access attempt in the cell. Specifically, the RR entity of the MS initiates the access process by scheduling transmission of a CHANNEL REQUEST message on the RACH. Consider Figure 5.13[158] in which the MS sends the first CHANNEL REQUEST message on the RACH.

The CHANNEL REQUEST message contains an establishment cause (e.g., speech or data call origination and location updating) and a uniformly distributed random number called *random reference*. Since our MS is in poor channel conditions, the BTS failed to decode the message due to the weaker received signal strength. Because the MS has received no feedback from the BSS, it sends another CHANNEL REQUEST message. The number of slots between the two successive CHANNEL REQUEST messages is a uniformly distributed random number between S and $S + \textit{Tx-integer} - 1$ where S depends on the CCCH configuration parameters and *Tx-integer*.

Let's say that the second RACH transmission is also unsuccessful. Now, the question becomes how many times should the MS retry? The MS sends a maximum of (*max retrans* + 1) CHANNEL REQUEST messages on the RACH. The parameters *Tx-integer* and *max retrans* are RACH Control Parameters, which are part of the BCCH messages, such as SYSTEM INFORMATION TYPE 1, SYSTEM

[158] Tx represents "transmit."

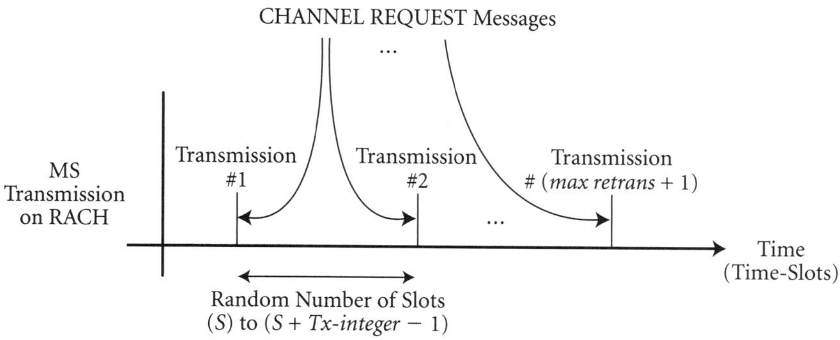

Figure 5.13: Transmissions on the RACH

INFORMATION TYPE 2, SYSTEM INFORMATION TYPE 2BIS, SYSTEM INFORMATION TYPE 3, and SYSTEM INFORMATION TYPE 4 messages.

After sending the CHANNEL REQUEST message, the MS listens to a suitable downlink CCCH time-slot corresponding to its CCCH group. If the MS gets no response from the BSS after sending a total of (*max retrans* + 1) CHANNEL REQUEST messages, it aborts the access attempt.

5.5.1.2 Eureka! Reception of the BSS Response at the MS

After the BSS successfully decodes the CHANNEL REQUEST message from the MS, it allocates a dedicated channel to the MS by sending one of the two types of immediate assignment messages. An IMMEDIATE ASSIGNMENT message contains allocation for just one MS, and an IMMEDIATE ASSIGNMENT EXTENDED message contains assignments for two MSs at the same time. The dedicated channel being assigned to the MS could be an SDCCH or TCH with the channel mode set to "signaling only." The assignment messages include information such as the description of the allocated dedicated channel (e.g., cell allocation, time-slot number, and hopping sequence number), the random reference number of the CHANNEL REQUEST message, the frame number when the CHANNEL REQUEST message was received, the initial timing advance, the mobile allocation (i.e., the ARFCN(s) allocated from the cell allocation), and (optionally) a starting time indication.

A Question for You... Why does the BSS include the random reference number of the CHANNEL REQUEST message in the assignment message? See the footnote[159] for the answer.

[159] If multiple MSs have used the RACH, the BTS needs to identify to which MS it is directing the CHANNEL REQUEST message. Hence, the BTS includes such a number. Recall that the MS includes a random number when it uses the RACH.

After the MS receives the assignment message from the BSS, it starts using the allocated channel with the channel mode set to "signaling only." Such a dedicated radio link between the MS and the BSS represents the main signaling link that will be used to exchange future signaling messages. *The RR connection is said to be established now.*

If the BSS has no radio resources for the MS, it sends an IMMEDIATE ASSIGNMENT REJECT message to the MS on the same CCCH time-slot on which the CHANNEL REQUEST message was received. This reject message contains the random reference and a wait indication containing a value for a timer T3122. The MS can make no new access attempts in the same cell until T3122 expires.

5.5.1.3 MS to the Network:* Cogito, Ergo Sum *(Translation from Latin:* I Think, Therefore I Am*) The MS informs the network of its existence by sending a LOCATION UPDATING REQUEST message to the GSM network. This message includes information such as the MS identity, LAI, MS Classmark, Location Updating Type, and Ciphering Key Sequence Number. The MS identity is typically TMSI or IMSI. It may also be International Mobile Equipment Identifier and International Mobile Equipment Identifier *with* its software version number. The MCC (Mobile Country Code), MNC (Mobile Network Code), and LAC (Location Area Code) specify the LAI. The service provider assigns an LAC to a given LA within its PLMN(s). The MS classmark describes MS characteristics, such as its power class, which defines the MS's maximum transmit power. The Location Updating Type gives the reason for this message, which may be IMSI attach, normal location area updating, or periodic location updating. In our example scenario, it would be IMSI attach.

The Ciphering Key Sequence Number is the currently stored value at the MS, which will enable the network to determine the value of the ciphering key, Kc, without invoking the authentication procedure. If the network does not want to authenticate the MS, it can skip the authentication procedure without worrying about security aspects because the MS already has the required ciphering information. The network can *optionally* choose to initiate the classmark interrogation procedure to inquire about the MS's encryption capabilities and the identification procedure if it is unable to get the IMSI based on the TMSI and LAI (Location Area Identifier) that the MS uses as identification.

5.5.1.4 Authentication Procedure in GSM The network typically carries out the authentication procedure after receiving the LOCATION UPDATING REQUEST message from the MS. Figure 5.14 illustrates the authentication apparatus in GSM and indicates a typical sequence of steps.

The core network initiates the authentication procedure. The VLR determines the HLR of the MS's Home PLMN based on its TMSI/IMSI. The VLR queries the HLR by sending a SEND AUTHENTICATION INFO message. Specifically, the HLR/AuC executes an authentication algorithm, an A3 algorithm, to generate an authentication using a shared secret called Ki. This triplet consists of a 128-bit random number called RAND, a 32-bit Signed Response (SRES), and a ciphering key called Kc. Different service operators can implement different A3 algorithms. The HLR/AuC forwards the RAND, SRES, and Kc to the MSC/VLR. The MSC/VLR forwards

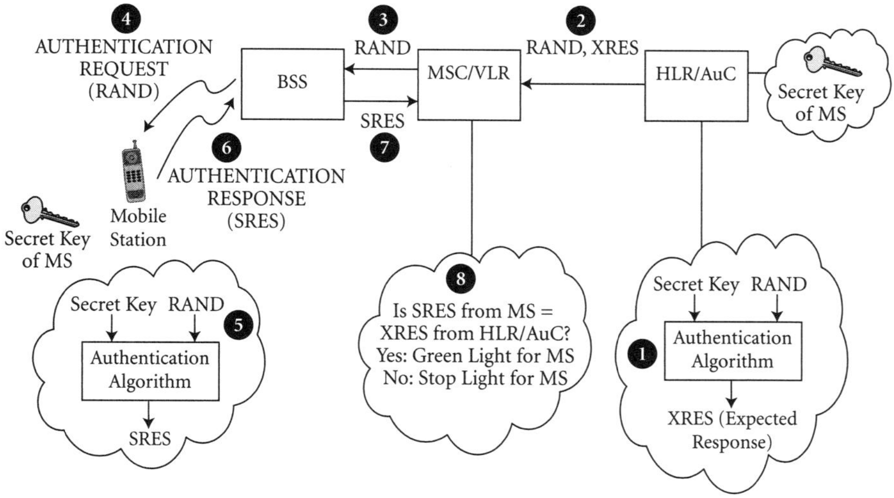

Figure 5.14: Authentication Apparatus in GSM

to the BSC the RAND and the locally generated, ciphering key sequence number. The BSC sends the MS the AUTHENTICATION REQUEST message containing the RAND and the ciphering key sequence number. The MS processes the challenge and responds with the AUTHENTICATION RESPONSE message containing the signed response, SRES. The MS calculates a new ciphering key based on the challenge information and stores it along with the ciphering key sequence number on the SIM. The BSC forwards the received SRES to the MSC/VLR, and the MSC/VLR compares that SRES with XRES. If both values match, the MS is successfully authenticated; the MS is a law-abiding citizen!

Observe that only two entities know the secret key (i.e., shared secret Ki) for authentication—the MS and the HLR. Furthermore, note that the security parameters for authentication and ciphering are related. In other words, the MS calculates both the SRES and the ciphering key based on the RAND and the shared secret. The importance of the ciphering key sequence number is that the MS indicates such sequence numbers in its first message to the network (e.g., LOCATION UPDATING REQUEST, CM SERVICE REQUEST, PAGING RESPONSE, CM RE-ESTABLISHMENT REQUEST), and this allows ciphering to begin on an RR connection without carrying out an explicit authentication procedure.

5.5.1.5 Securing the Air Interface: Ciphering Mode Settings The network may initiate the ciphering mode setting procedure, especially when a new TMSI is to be allocated. The network sends to the MS the CIPHER MODE COMMAND message, which specifies whether the MS must use ciphering and which ciphering algorithm to use. The MS responds with the CIPHER MODE COMPLETE message, indicating to the network that the MS has started the ciphering and deciphering mechanism.

As mentioned during the authentication discussion, a ciphering key, Kc, is generated during the authentication procedure. Kc will then be used to cipher or encrypt all data, regardless of the type of information. Specifically, a 114-bit ciphering sequence[160] is constructed by executing the ciphering algorithm, an A5 algorithm, with two inputs: a twenty-two–bit value based on the frame number and a sixty-four–bit Kc. The A5 algorithm generates two distinct 114-bit ciphering sequences for the uplink and the downlink for each TDMA frame.

A Question for You. . . Why is the ciphering sequence changed rather than kept constant? See the footnote[161] for the answer.

Recall that the user information is sent in bursts of 114 bits (and a set of two bits classifying the 114-bit burst as a traffic or signaling burst). At the receiver, the deciphering occurs by performing an XOR operation on the received encrypted sequence using the same ciphering sequence.

An Exercise for You. . . What would be the result of ciphering when the input bit sequence is `1001` and the ciphering sequence is `0111`? See the footnote[162] for the answer.

5.5.1.6 Good News for the MS. . . Location Updating Successful!

Now the network sends the LOCATION UPDATING ACCEPT message to the MS. If the identity confidentiality service is active, the allocated TMSI and the LAI are included in the LOCATION UPDATING ACCEPT message. The MS stores the received TMSI in the SIM. The MS responds with TMSI REALLOCATION COMPLETE message.

5.5.1.7 Cleaning Up. . . Connection Release

After the location updating procedure is completed, the RR connection is released based on either network-triggered messaging or time-based expiration. The MSC/VLR informs the BSS that the location updating has been completed by sending a Clear Command. Now, the BSS releases the RR connection. Such network-triggered connection release involves the BSS sending the CHANNEL RELEASE message to the MS. This message includes the cause for the release (e.g., "normal event" in our example).

[160] The cipher mask itself is 116 bits long (114 bits of the ciphering sequence and two bits of stealing flags) for each time-slot. The FACCH is said to steal bits from other, less important channels, and such stealing is done in halves of a slot. Hence, two stealing flags tell the receiver which of the fifty-seven bits in a time-slot the FACCH has preempted.

[161] Changing the ciphering sequence makes it harder for anyone to guess the sequence, leading to tighter security of the GSM air interface.

[162] Ciphering is like an XOR operation. Hence, the ciphered sequence is a bit-by-bit XOR operation of `1001` and `0111`, which is `1110`.

5.6 VOICE CALL SETUP

Let's discuss how to set up a voice call in GSM. MS-originated call setup is discussed in Section 5.6.1, and MS-terminated call setup and its associated paging procedure are summarized in Section 5.6.2.

5.6.1 MS-Originated Call Setup

After you dial the number on your cell phone and before you start talking, your cell phone is quite busy. Your cell phone exchanges numerous signaling messages with the BSS and the core network. Such a signaling exchange tends to be quite comprehensive in GSM and GSM-evolved systems, such as UMTS, compared to other technologies, such as CDMA2000 1x.

Figure 5.15 summarizes the main steps of the call setup. Prior to the call setup, the MS is in the Idle Mode, periodically observing the BCCH messages. As is evident from Figure 5.15, the voice call setup involves several messages we already encountered during the discussion of the IMSI attach and location updating.

- The MS uses the RACH to send the CHANNEL REQUEST message on the RACH to make a request for a dedicated radio channel. The establishment cause would specify speech call origination. The BSS allocates dedicated radio resources to the MS via the IMMEDIATE ASSIGNMENT message. Note that these radio resources help create an RR connection between the MS and the BSS. Furthermore, such a dedicated radio channel carries signaling messages reliably. No speech traffic can yet flow through this radio channel because it is purely a signaling-only channel.

- The MS makes a request for service by sending a CM SERVICE REQUEST message to the core network. Remember the core network (*not* the radio network) is responsible for service authorization.

- The core network and the MS exchange AUTHENTICATION REQUEST and AUTHENTICATION RESPONSE messages as discussed earlier.

- The air-interface security is initiated by preparing for ciphering. As discussed earlier in Section 5.5, the network instructs the MS whether to start ciphering and which ciphering algorithm to use via the CIPHER MODE COMMAND message. The MS responds with the CIPHER MODE COMPLETE message to inform the network that it has started the ciphering and deciphering process.

- The MS now sends a SETUP message indicating the phone number that it is trying to reach. The core network replies with a CALL PROCEEDING message to indicate to the MS that the progress is being made.

- The dedicated radio connection that the MS has been using is for signaling only. Another radio channel capable of carrying speech traffic is required. Hence, the BSS establishes a dedicated radio channel for the MS via the ASSIGNMENT

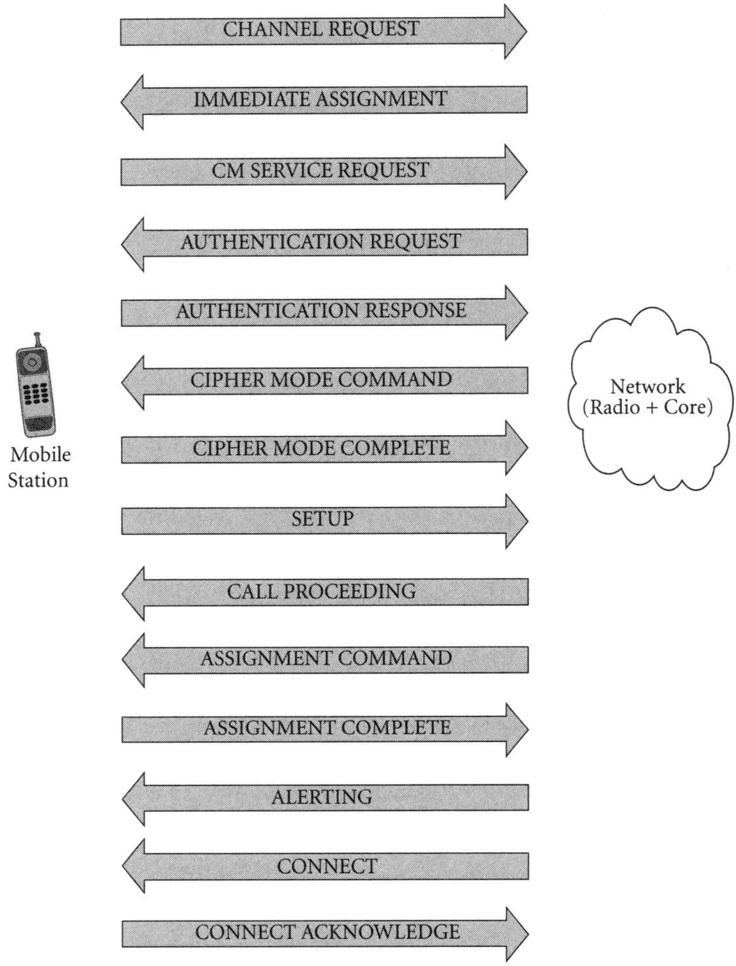

Figure 5.15: Voice Call Setup

COMMAND message. This radio channel would carry speech traffic at the end of call setup. The MS responds with the ASSIGNMENT COMPLETE message.

- The core network sends an ALERTING message to inform the MS that the phone is ringing at the called party. When the called party picks up the phone, the PSTN sends an ANSWER message to the core network (i.e., MSC/VLR), which, in turn, sends a CONNECT message to the MS. The MS answers with a CONNECT ACKNOWLEDGE message.

Do You Recall... Which two variables does the MSC/VLR compare to authenticate the MS? (Hint: The name of one variable starts with X and the name of the other variable starts with S). See the footnote[163] for the answer.

Now that the call is up, the MS can exchange speech traffic with the other end. At the end of the call, the MS or network transmits a DISCONNECT message. The recipient of this message sends out a RELEASE message. The acknowledgment to the RELEASE message is a RELEASE COMPLETE message.

5.6.2 MS-Terminated Call Setup and Paging Procedure

In MS-terminated (or, equivalently, network-originated) call setup, the network sends a PAGE message to the MS. Recall that the MS performs an IMSI Attach or location updating procedure to convey its location area to the core network. When the MSC receives a call or an SMS for the MS, it could request that all BSCs in the location area page the MS. Specifically, the MSC sends the page request to the BSC with the IMSI and TMSI of the MS. The BSC uses the IMSI to figure out the paging group and paging channel. As a general rule, the TMSI is used to send a PAGE message whenever possible. If the TMSI cannot be used, the IMSI is used. For example, if the MS had not registered in a given MSC/VLR and a call was routed to such an MSC/VLR, the page would contain the IMSI since the MSC/VLR under consideration has not yet allocated the TMSI to the MS. Paging involves the broadcast of the TMSI or IMSI on the PCH of all cells in the location area of the MS.

As you may remember, time-slot 0 on the BCCH carrier is reserved for the BCCH. When the BCCH is unused, up to nine paging channels can be configured on the BCCH carrier frequency. The PCH sends a PAGING message using a set of four TDMA frames. There are three types of paging requests, Type 1, Type 2, and Type 3. Type 1 carries two identities of any type (e.g., two TMSIs or two IMSIs), Type 2 carries two TMSIs plus one identity of any type (which could be TMSI or IMSI), and Type 3 carries four TMSIs.

The process of saving battery power by observing the PCH intermittently (as opposed to continuously) is called the *Discontinuous Reception* (DRX) mode of operation. The DRX operation exploits the concept of paging groups to give the MS an opportunity to go to sleep and save battery power when its paging group is not being paged. The drawback of the DRX operation is a slight increase in the paging delay, essentially increasing the call setup delay. Since the overall call setup delay is on the order of seconds (e.g., five to ten seconds), such an increase in paging delay is considered insignificant.

Based on the BCCH's observation, the MS knows to which paging group it belongs. Hence, the MS periodically looks at the PCH for a potential message. Different MSs are paged during different time intervals corresponding to different paging groups.

[163]The MSC/VLR compares the XRES received from the HLR and the SRES received from the MS.

The BSS sends the PAGING REQUEST message first. The MS and the network exchange CHANNEL REQUEST and IMMEDIATE ASSIGNMENT messages to establish a dedicated radio channel. The MS sends the PAGING RESPONSE message using the dedicated signaling channel.

Other steps (e.g., authentication and ciphering mode setting) are quite similar to those discussed for the MS-originated call setup. See the specifications [209] for additional details on the mobile-terminated call setup.

5.7 MOBILITY MANAGEMENT IN IDLE MODE

After the MS is powered on, it enters the Idle Mode. If the MS has a dedicated radio channel with the BSS, it is in the Active or Connected Mode. As the MS travels in a given region, its mobility is managed differently in the idle and active states. This section discusses the Idle Mode mobility, and Section 5.8 describes the Active Mode mobility. Additional details of mobility management can be found in the specifications [208, 209]. Let's take a look at the three main Idle Mode tasks: PLMN selection, cell selection and reselection, and location updating.

The MS usually works in its home PLMN, but it may work in a visited PLMN if the home PLMN is inaccessible (e.g., due to lack of radio coverage of the home PLMN). The Automatic mode considers a list of PLMNs in priority order, and the highest-priority and available PLMN is selected. In the manual mode, the MS gives the user a list of available PLMNs, and the user selects a PLMN. The MS's periodic observation of the BCCH and the CCCH is called *camping on the cell*, and thus the MS attempts to camp on a cell associated with the selected PLMN.

The MS attempts to find a suitable cell in the selected PLMN and observes the BCCH and the CCCH. From the paging perspective, the MS listens to the paging subchannel for the paging group to which the MS belongs. The paging group is a function of the IMSI [211]. The network is required to send valid layer 3 messages continuously on all paging subchannels. As discussed in Section 5.5, the MS performs an IMSI attach or location updating procedure. Camping on the cell enables the MS to receive system information from the PLMN, to make a call by using the RACH in the cell, and to receive calls by observing the CCCH/PCH in the cell. If the MS can find no suitable cell to camp on in the selected PLMN, it camps on a cell irrespective of the PLMN and enters a *limited service* state in which it can make only emergency calls.

The MS adheres to several rules while selecting a cell to camp on. For example, the cell should belong to the selected PLMN and should not be barred by the network. Furthermore, the cell should not be part of an LA that is a member of the list of forbidden LAs for roaming. Additionally, the propagation path loss between the MS and BTS must be below a threshold the PLMN operator sets. When all these constraints are met, the MS has found a set of suitable cells. Then the MS picks from this set the cell that yields the best received signal strength. This process of looking for a cell to camp on after power-on network acquisition is called *cell selection*. The MS uses a *path loss criterion* parameter called C1 to determine whether a cell is suitable to camp on. C1 is a function of the following four parameters: the average received signal strength; RXLEV_ACCESS_MIN (which corresponds to the minimum received signal strength that the BTS could detect during the MS's access attempt); MS_TXPWR_MAX_CCH (which is the maximum transmit power that an MS may use during the access attempt); and the MS's maximum transmit power. RXLEV_ACCESS_MIN and MS_TXPWR_MAX_CCH are broadcast on the BCCH. The value of C1 indicates whether the MS can overcome the path loss and successfully access the BTS in the absence of uplink interference. See the specifications [200] for details on C1 calculation.

Once the MS is camping on the selected cell, it periodically looks for a better cell to camp on, and this process is called *cell reselection* as discussed earlier. The network facilitates the MS's cell search operation by advertising a list of BCCH carriers on the BCCH. The MS searches thirty ARFCNs for GSM 900 and forty for DCS 1800 [208]. When the MS performs cell reselection, it switches to the new cell's BCCH and learns about the new cell (e.g., ARFCNs of the neighboring cell) by observing the BCCH messages.

GSM supports cell prioritization whereby MSs are encouraged to prefer some cells to others. For example, the operator may assign lower priority to umbrella cells than to microcells due to the microcells' higher capacity. During the process of cell selection, a cell with a lower-priority indication is selected only if no suitable normal-priority cell is found. During cell reselection, cell prioritization is achieved by (optionally) broadcasting specific parameters that influence the value of a parameter called C2. Specifically, C2 depends on the following parameters: Cell Reselect Offset, Penalty Time, and Temporary Offset. C2 is a per-cell positive or negative offset to encourage or discourage MSs to reselect that cell (e.g., to avoid ping-ponging between the cells). See the specifications [200] for additional details.

5.8 MOBILITY MANAGEMENT IN ACTIVE OR CONNECTED MODE

When the MS has a dedicated radio connection (e.g., RR connection), it may be on the SDCCH or the TCH. Handover is defined as a process of changing the MS's radio channel from one channel to another. Generally (but not always), the old radio channel belongs to one cell and the new radio channel belongs to another cell. Deteriorating signal quality on the current channel usually triggers handover to another channel or cell to maintain the call. Such handover is frequently referred to as *rescue handover* and comprises more than 90% of handovers. Another type of handover is *load balancing* handover in which the network transfers some users from a suitable cell with heavy traffic to a suitable cell with less traffic; the goal is to accommodate more MSs in the cell that is experiencing heavy traffic.

We'll start with a brief overview of handover, and then we'll discuss handover in detail. The entities that could participate in the overall handover process are the MS, BSS, and MSC. Figure 5.16 summarizes the overall handover process.

The first stage in handover is evaluation of the prevailing radio channel conditions. The MS makes measurements of the downlink channel conditions of its serving

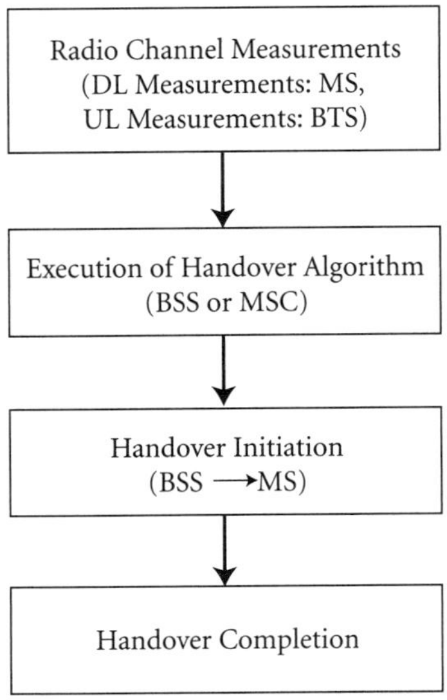

Figure 5.16: Overview of the Handover Process

cell and surrounding cells (e.g., signal strength). The MS reports these downlink measurements to the BSS. The BSS quantifies the uplink performance for the MS and observes the amount of interference on its idle traffic channels. The BSS follows a handover algorithm to make a handover decision. In some scenarios, such as handover across MSCs, the MSC may execute a handover algorithm. For such inter-MSC handover, the BSC and MSC need to communicate with each other. See the specification [193] for details of handover signaling between the BSS and the MSC.

The handover algorithm or strategy is beyond the scope of the specifications, and infrastructure vendors can implement different proprietary algorithms. Once the network makes a handover decision, it initiates the handover process and notifies the MS of the handover decision. Handover is completed when the MS tears down the communications with the current cell and establishes a new radio channel with the new cell. Such handover is a *break-before-make* handover (also called *hard handover*) because the previous connection is broken first and then a new connection is made. The GSM handover is *mobile-assisted hard handover* (MAHO) in which the MS provides measurements to the BSS and thus assists the network in making a handover decision. The various stages of handover are discussed in Sections 5.8.1 through 5.8.3.

5.8.1 What are the Channel Conditions?

The MS quantifies the downlink channel conditions, and the BTS quantifies the uplink channel conditions. The channel conditions information is used in the handover algorithm. Let's examine the radio channel measurements in detail.

5.8.1.1 Determining Which Cells Should be Monitored The MS takes measurements of the BCCH carriers in surrounding cells. Since a given BCCH carrier frequency may be reused in reasonably close cells, we need a way of identifying which cell's BCCH carrier the MS is measuring. For example, consider Figure 5.17 in which the cluster size of four (i.e., a frequency reuse factor of $1/4$) is used along with omnidirectional cells. In other words, the same BCCH carrier is used in every fourth cell. The BCCH carrier f_1 is used in cells 1, 5, 9, and 13. When the MS measures the quality of f_1, it may be measuring the BCCH carrier of cell 1, 5, 9, or 13, but the BSS needs to know *which* cell's BCCH the MS is measuring. That's where the concept of BSIC comes into the picture. The BSIC identifies the cell uniquely in a given geographical area. The BSIC consists of six bits with the PLMN part of the BSIC called the *PLMN color code*. We can uniquely identify 2^6 or sixty-four unique cells. In a set of sixty-four cells, the BSIC will always be unique. The BSICs will be reused in another set of 64 cells. For example, $BSIC = 1$ is used in Cell 1 and Cell 13. Hence, when MS5 is able to identify the BSIC, it is (for all practical purposes) measuring f_1 corresponding to $BSIC = 1$ in Cell 1 because $BSIC = 1$ in Cell 13 will be too weak for MS5 to detect. In other words, the reuse distance for the BSIC is so large compared to the reuse distance of the BCCH carrier frequency that the cell's identity is ambiguous.

The MS demodulates the surrounding cells' BCCH carriers and identifies the BSIC. The MS uses at least four idle frames per SACCH multiframe to decode the BSICs. One SACCH multiframe corresponds to 104 TDMA frames (i.e., 480 ms) for a TCH and 102 TDMA frames (i.e., 470.8 ms) for an SDCCH. Since the MS is searching for the BSIC during these frames, they are called *search frames*. The MS demodulates the SCH of each surrounding cell and decodes the BSIC as frequently as possible, the minimum requirement being once every ten seconds. For a multi-band MS, the network specifies to the MS in the parameter Multi-Band Reporting how many cells for each frequency band should be included in the measurement report. The idle frame occurs every twenty-sixth frame, and the FCCH and SCH structure is

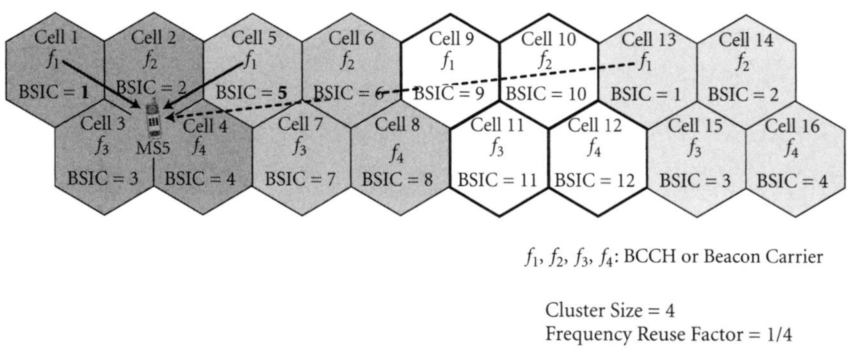

Figure 5.17: Motivation for the BSIC: A Simplified Example

repeated every fifty-two frames. Hence, if the MS keeps observing the idle frame, it is guaranteed to eventually find the FCCH and SCH.

5.8.1.2 Determining What Measurements Should be Collected The MS makes measurements over the duration of one SACCH period and sends a measurement report on the SACCH. Two quantities are measured: received signal strength and received signal quality.

First, we'll discuss signal strength. The MS and the BTS measure the received signal level at the receiver input in the range of −110 dBm to −48 dBm. The specific parameter being reported is called RXLEV, which is the average of the received signal level measurements (in decibels relative to one milliwatt) within the period of one SACCH multiframe. The MS considers the physical channels associated with the BCCH carrier (and the allocated TCH/SDCCH) for RXLEV, and the BSS makes RXLEV measurements on the physical channels associated with the uplink TCH or SDCCH allocated to the MS and SACCH. Of course, idle time-slots are ignored for the measurements. The measured received signal level is mapped onto a six-bit RXLEV value. For example, RXLEV = 0 corresponds to the signal level less than −110 dBm, and RXLEV = 63 corresponds to the signal level greater than −48 dBm.

Now let's discuss signal quality. The MS and the BTS quantify the received signal quality in the form of a bit error rate (BER) before channel decoding. The BER is calculated over the reporting period of one SACCH multiframe. The estimated BER is mapped onto a three-bit quantity called RXQUAL. For example, RXQUAL_0 indicates that the average BER is less than 0.2%. RXQUAL_1 corresponds to an average BER greater than 0.2% but less than 0.4% [200]. RXQUAL_7 corresponds to a BER greater than 12.8%.

5.8.1.3 Determining When to Report Measurements The measurement reporting method varies depending upon whether the MS has a TCH or just an SDCCH. The MS with a TCH reports the measurements during specific TDMA frames during the 104-TDMA-frame (i.e., 480 ms) reporting period. For example, assume that the MS has been allocated time-slot 0 and a full-rate TCH. This MS reports the measurements on the SACCH during those TDMA frames when the result of $TDMA\ Frame\ Number \mod 104$ is 12, 38, 64, or 90. The specific measurements that the MS with a TCH reports are

- RXLEV for the BCCH carrier of the six cells with the highest RXLEV,
- RXLEV_FULL and RXQUAL_FULL for the full set of TCH and SACCH TDMA frames,[164] and
- RXLEV_SUB and RXQUAL_SUB for the subset of four SACCH frames and the SID (Silence Descriptor) TDMA frames or layer 2 fill frames.

[164]The full set of TDMA frames is one hundred frames (104 frames with four idle frames excluded) for a full-rate TCH and fifty-two frames for a half-rate TCH.

The MS with an SDCCH reports measurements during the 102-TDMA-frame (470.8 ms) reporting period. The specific measurements that the MS with an SDCCH reports are

- RXLEV for the BCCH carrier of the six cells with the highest RXLEV,
- RXLEV and RXQUAL for the full set of twelve frames (eight SDCCH frames and four SACCH frames), and
- RXLEV_SUB and RXQUAL_SUB (which are kept the same as RXLEV_FULL and RXQUAL_FULL because DTX is unused on the SDCCH).

The MS sends a MEASUREMENT REPORT message on the SACCH to report these measurement results. Just like the MS, the BTS also calculates RXLEV_FULL, RXLEV_SUB, RXQUAL_FULL, and RXQUAL_SUB based on these measurements.

5.8.2 Making a Handover Decision: Who is in Charge Around Here?

A handover algorithm or strategy makes a handover decision based on the handover measurements and some other considerations. Section 5.8.2.1 describes the principles of a handover algorithm, and Section 5.8.2.2 summarizes various types of handover that may arise at the end of the execution of the handover algorithm.

5.8.2.1 Handover Algorithm As mentioned earlier, the standard offers no specific handover algorithm mandate but does offer an example of the overall algorithm [200]. Potential inputs to the handover algorithm are the handover measurements the MS reports (e.g., downlink RXLEV, downlink RXQUAL, and RXLEV of neighboring cells) and measurements the BTS determines, e.g., uplink RXLEV, uplink RXQUAL, distance between the MS and the BTS (a variable called MS_BTS $Distance$), and amount of interference level in unallocated time-slots. The BTS can estimate the MS-to-BTS distance based on the adaptive frame alignment mechanism. The BTS averages the measurements, e.g., non-weighted average, weighted average, or modified median that excludes outliers (i.e., too high and too low values), based on instructions from the O&M (Operation and Maintenance). The the O&M controls the averaging process through parameters *Hreqave* and *Hreqt*. *Hreqave* is the time period over which an average is calculated, and *Hreqt* is the number of averaged results. Figure 5.18[165] highlights potential handover criteria that a handover algorithm implemented at the BSC can use. Descriptions of the nomenclature used in conjunction with the handover algorithm follow.

- RXLEV_NCELL(n): Averaged RXLEV for neighboring cell n.
- RXLEV_DL: Averaged RXLEV for the currently serving cell.
- PWR_C_D: Difference between the maximum DL power allowed in the cell and the actual DL power.

[165] Several parameters in the figure are undefined in the standard and are thus undefined here as well.

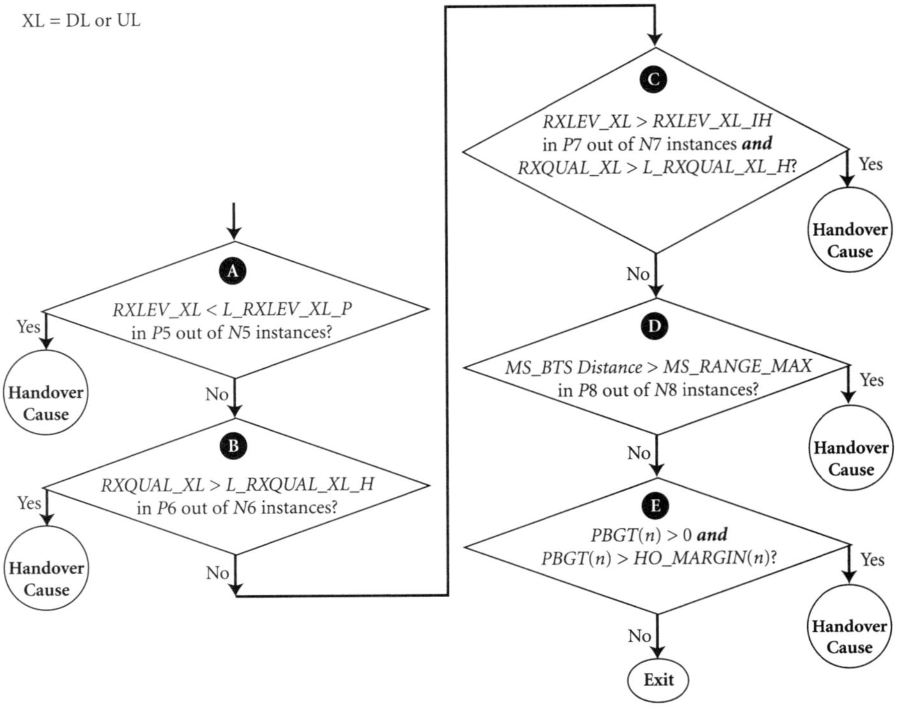

Figure 5.18: Handover Criteria for a Handover Algorithm at the BSC

- MS_TXPWR_MAX: Maximum transmit power that an MS can use on a TCH in the currently serving cell.
- MS_TXPWR_MAX(n): Maximum transmit power that an MS can use on a traffic channel in the neighboring cell n.
- P: Maximum transmit power capability of the MS.
- PBGT(n): Power budget for the neighboring cell n.
- MS_RANGE_MAX is the maximum allowed distance between the BTS and the MS.

The handover algorithm checks for several conditions, and, if one of the conditions is satisfied, a handover decision is made and a suitable target cell is selected. These conditions are briefly discussed below, followed by an algorithm for target selection. *A word of caution: the conditions are named A through E for convenience of our discussion only. Do not come after the authors when you cannot find such nomenclature in the standard!* (The standard just lists various conditions without giving any names.) Both UL and DL channel conditions are evaluated where applicable.

Condition A. If the RXLEV for the MS in the current cell is smaller than a threshold, the MS is likely to be away from the BTS. Handover can be initiated if such a condition is met in $P5$ of $N5$ instances. Recall that new averaged measurements become available every SACCH time period, and each such time period is one instance. $N5$ can thus be viewed as an observation window. If the majority of the observations indicates a low RXLEV, handover is probably a good idea. The parameters $P5$ and $N5$ could be ten and twelve, respectively. The example settings for the parameters P^* and N^* (e.g., $N = 5$) are suggested in the specifications [200].

Condition B. If RXQUAL for the MS exceeds a threshold, the error rate is high. Handover in such a case could yield better quality communications with another cell. $P6$ and $N6$ could be five and seven, respectively.

Condition C. Let's say that RXLEV is high, but RXQUAL is also high. An intra-cell handover to a radio channel with lower interference would improve the signal quality. $P7$ and $N7$ could be set to ten and twelve, respectively.

A Question For You. . . What can we say about the channel conditions when both RXLEV and RXQUAL are high? Pick your answer from these options: (a) the MS is far away from the serving BTS and interference is low, (b) the MS is far away from the serving BTS and interference is high, or (c) the MS is close to the serving BTS and interference is high. See the footnote[166] for the answer.

Condition D. If the MS is too far away from the BTS, it is exceeding the planned cell boundaries. Such a scenario would cause more interference because the co-channel distance would decrease. Hence, a distance-based handover decision can be made to circumvent such a scenario. $P8$ and $N8$ may be set to eight and ten, respectively.

Condition E. Sometimes the MS has reasonably good signal quality with the serving BTS, as indicated by good RXLEV and RXQUAL values. However, choosing another BTS over the current BTS may enable the MS to conserve battery power. Perhaps a new BTS would enable the MS to have reasonably good signal quality with much lower transmit power. Such a transmit power-based handover decision is made by calculating the power budget for the neighboring cells.

$$\begin{aligned} PBGT(n) =& \text{Min}(MS\_TXPWR\_MAX, P) - RXLEV\_DL - (PWR\_C\_D) \\ & - (\text{Min}(MS\_TXPWR\_MAX(n), P) - RXLEV\_NCELL(n). \end{aligned}$$

While the algorithm discussed here focuses on radio-environment–related criteria, other criteria, such as resource utilization and load balancing, can also be considered.

[166] The answer is (c) because a high RXLEV means that the MS is close to the BTS and a high RXQUAL points to a high error rate, potentially due to high interference.

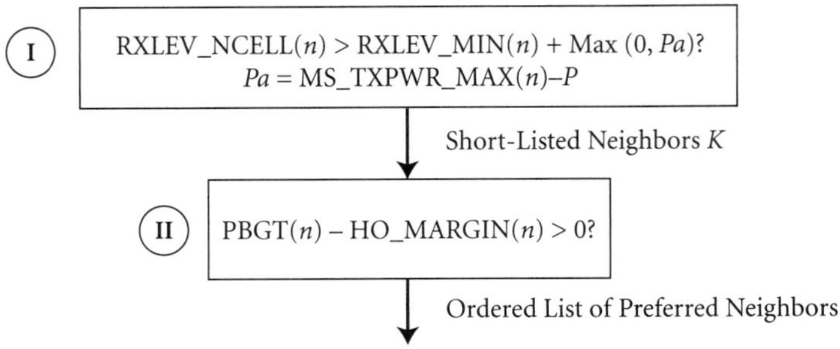

Figure 5.19: Selection of the Handover Target Cell

Now that we've discussed Conditions A–E and a handover decision is made, we need to choose a good target cell for handover. Figure 5.19 shows a potential algorithm for selecting a target cell for handover. (This algorithm is called the BSS Decision Algorithm in the specifications [209].)

The target cell is selected in two stages. In Stage I, the DL signal strength for each neighboring cell n is compared with a threshold (i.e., RX_LEV_MIN) that defines the minimum acceptable signal level. The neighbors with adequate signal strength pass the signal strength test and are now in the short-list of the target cells. Remember that we need just one cell for handover in GSM. The short-listed neighbors are tested using the power budget criterion. The neighbors passing the second-stage test are neighbors that can provide good signal strength to the MS and that enable the MS to communicate with low transmit power. Stage II yields an ordered list of neighbors, with the first neighbor corresponding to the highest power budget and the last neighbor in the list corresponding to the lowest power budget. Assuming that a higher-ranking cell has adequate radio resources to support the MS, such a cell becomes the new serving cell. The new serving cell allocates to the MS the radio channel that belongs to the lowest interference category. The available radio channels are divided into five interference categories. The interference limits for different categories, 0 to $X5$, are configurable via an O&M command [193]. Such a strategy ensures good communication quality and helps minimize call drops.

5.8.2.2 Types of Handover Depending upon where the target cell is, different types of handover occur. The left side of Figure 5.20 illustrates the two main types of handover: *inter-cell handover* and *intra-cell handover* [200].

Inter-cell handover occurs from the serving cell to a neighboring cell. As the MS moves away from the currently serving cell and toward a neighboring cell, the channel conditions are likely to deteriorate for the current cell and to improve for the neighboring cell. These changing conditions will be reflected in a low RXLEV and/or RXQUAL for the current cell and a higher RXLEV for the neighboring cell. Furthermore, the MS might be able to communicate with the neighboring cell using lower transmit power. The handover during the call setup itself is also supported in GSM. For example, the initial cell that has allocated a DCCH to the MS may find

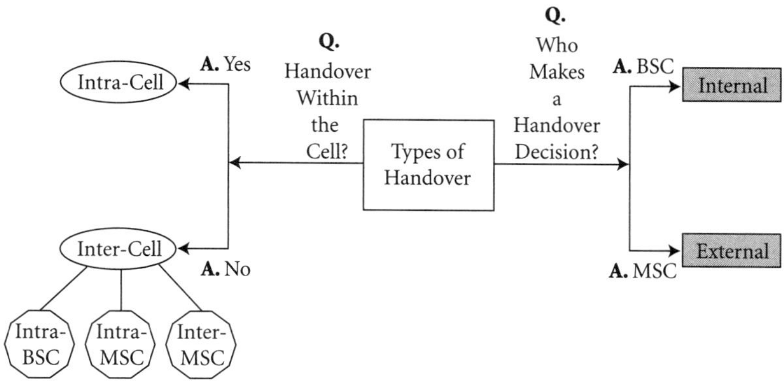

Figure 5.20: Types of Handover

out later that it lacks adequate resources to establish a TCH for the MS. In such a case, inter-cell handover helps establish a TCH in the new cell, thereby avoiding call blocking in the initial cell. Inter-cell handover is also supported between the cells operating at different frequency bands.

Intra-cell handover occurs from one ARFCN or time-slot to another ARFCN or time-slot in the serving cell. Consider a scenario in which the currently serving cell's RXQUAL is low but its RXLEV is high. Such measurements tell us that the ARFCN and time-slot combination that the MS is currently using is experiencing a higher degree of interference even though the MS is reasonably close to the serving cell's BTS. An intra-cell handover from the current ARFCN and time-slot combination to another is likely to improve the quality of communications if the new ARFCN and time-slot combination corresponds to a lower amount of interference. Just like inter-cell handover across multiple frequency bands, intra-cell handover across multiple frequency bands is also supported.

An inter-cell handover occurs in a variety of scenarios. An intra-BSC handover occurs between two cells of the same BSC. An intra-MSC handover is the handover between two cells that belong to the same MSC. An inter-MSC handover involves handover between cells that are associated with different MSCs. Furthermore, an inter-cell handover could be internal or external. When the handover algorithm at the BSC makes the handover decision, this is called an *internal handover*. However, when the MSC makes a handover decision, such handover is an *external handover*. If the handover algorithm at the BSC can make no handover decision, which may occur when the BSC does not control the handover target cell, the BSC notifies the MSC and the MSC performs an external handover. If the MSC makes a handover decision, it instructs the target BSC to allocate the necessary radio resources to accept the handover. In an internal handover, the BSC itself allocates the resources without any MSC participation.

If an internal handover is required, the MSC need not be involved. However, if an external handover is required, the BSC sends a HANDOVER REQUIRED message containing the preferred list of target cells to the MSC. The MSC works with the target BSC to get resources for the MS. For an inter-MSC handover, the currently serving MSC (called an *anchor MSC*) sends an MAP Prepare Handover Request message to the target MSC. The target MSC, in turn, sends the HANDOVER REQUEST to the target BSC. The target BSC allocates the resources to the MS and specifies these resources in a HANDOVER REQUEST ACKNOWLEDGE message. This message reaches the currently serving BSC, which, in turn, sends a HANDOVER COMMAND message to the MS.

When an MSC needs to execute a handover algorithm (e.g., in inter-MSC handover), it considers the handover criteria in this order of priority: RXQUAL, RXLEV, DISTANCE, and PBGT. Once the network makes a handover decision, it *initiates* a handover. *Once the handover algorithm is executed, handover may be initiated if circumstances warrant it.* The network initiates the handover procedure by sending a HANDOVER COMMAND message to the MS on the main DCCH. Example parameters of this message include the Frequency List, Frequency Short-List, and Cell Channel Description. The MS then releases the link layer connections, disconnects the physical channels, and initiates activation of the new data links and physical channels. Example contents of the HANDOVER COMMAND message are [209]

- characteristics of the new radio channels (e.g., TCH and Associated Control Channels),
- characteristics of the new cell (e.g., a frequency list to support slow frequency hopping),
- a power command specifying the power level for the MS to use for the initial power on the new channel(s),
- an indication of which of the four physical channel establishment procedures is to be used,
- a handover reference, and
- (optionally) a timing advance to be used on the new cell and a cipher mode setting (with the ciphering key kept the same even after handover).

A successful handover requires establishment of new physical channels. *Four procedures to establish physical channels during handover are defined in the specifications: finely synchronized, non-synchronized, pseudo-synchronized, and pre-synchronized* [204,209]. The MS is required to support three of them, and the support for the pseudo-synchronization procedure is optional at the MS. The MS's classmark characteristics indicate whether the MS supports the pseudo-synchronized handover. Recall that the network controls which procedure the MS is to use.

In the finely synchronized cell procedure, the old cell and the new cell are time-synchronized. This procedure is suitable when the timing advance (TA) with the new cell is within range or if the new cell accepts an out-of-range timing advance. In this procedure, the MS sends the HANDOVER ACCESS message four times in four successive layer 1 frames on the main DCCH in an access burst after switching to the new channels. The contents are reduced to just the handover reference. The only purpose of the handover reference is to identify the MS's access. The network could make the transmission of these four non-ciphered messages optional by specifying so in the HANDOVER COMMAND message. The MS then activates the channels in both sending mode and receiving mode and connects the channels if required. Applicable ciphering is also started.

The non-synchronized cell procedure involves an old cell and a new cell that are not time-synchronized. After switching to the new channels, the MS repeatedly sends the unciphered HANDOVER ACCESS message in consecutive physical layer frames on the main DCCH (and optionally on the SACCH). This message is considered part of an access burst. As in a finely synchronized cell, this message is quite simple: it contains just the handover reference. The MS now activates the new channels in receiving mode only and prepares to receive on these channels. Applicable deciphering is also initiated. The network sends a ciphered PHYSICAL INFORMATION message containing various physical layer parameters on the main DCCH to facilitate the MS's transmission of the access bursts. The BSS repeats the PHYSICAL INFORMATION message if necessary. Upon receiving the PHYSICAL INFORMATION message, the MS ceases transmission of the access bursts and now activates the physical channels in both sending and receiving modes.

The pseudo-synchronized cell procedure applies when the BSS knows the timing difference between the old cell and the new cell. Hence, the MS can easily compute the timing advance (TA) to be used with the new cell using the real-time difference specified in the unciphered HANDOVER COMMAND message. After switching to the new channels, the MS sends a HANDOVER ACCESS message in the random mode in four successive slots on the main DCCH. This message contains the handover reference and does not conform to the basic format. As in the synchronized cell procedure, the network could make the transmission of these four messages optional. The MS now activates the channels in both sending and receiving modes. Applicable ciphering is also started.

For the pre-synchronized cell procedure, the MS, after switching to the new channels, sends an unciphered HANDOVER ACCESS message as an access burst in four successive slots on the main DCCH. Again, this message contains the handover reference. Additionally, the network can make the transmission of these four messages optional. The MS activates the channel in both sending and receiving modes. Now the timing advance the MS uses for the new cell is the value specified in the HANDOVER COMMAND message. If no such value is specified, the MS uses the default value for pre-synchronized handover. Applicable ciphering is also started.

5.8.3 Follow the Leader: Handover Completion

Once the MS has established all the layer connections, it responds with the HANDOVER COMPLETE message on the main DCCH, specifying a "normal event" as the cause. Now the network knows that the MS has successfully established the main signaling link. Based upon the request in the HANDOVER COMMAND message, the MS may include the observed time difference between two cells (the old cell and the new cell).

5.9 POWER CONTROL

GSM specifies a power-control mechanism to control the MS's transmit power and to enable the BTS to change the BTS transmit power. In other words, uplink power control is fully defined, and downlink power control is left to vendor implementation. The overall goal of the power-control apparatus is to enable the achievement of a target error rate while offering the desired quality-of-service (QoS) and while transmitting at a reasonable transmit power. The power-control mechanism aims to achieve the minimum power level that will sufficiently meet the error rate requirements.

A Question For You... What might be the benefits of using the lowest possible transmit power at the MS? See the footnote[167] for the answer.

Let's start with a brief overview of uplink power control. Section 5.9.1 discusses power control in more detail, and Section 5.9.2 provides more information on power reduction.

Figure 5.21 provides a high-level view of the basic closed-loop power-control operation for the uplink. A closed-loop power mechanism involves the receiver sending explicit power-control commands to the transmitter so that the transmitter

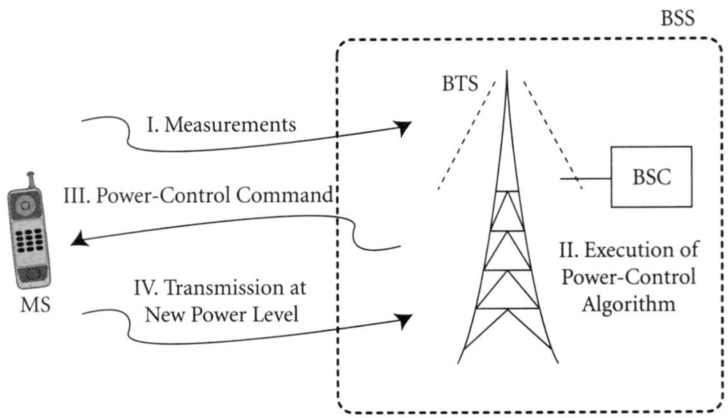

Figure 5.21: An Overview of the Closed-Loop Uplink Power Control

[167] The main benefit of uplink power control in GSM is increased MS battery life due to minimal power consumption. Another benefit for GSM is lower interference.

can adjust its transmit power. The power-control operation is summarized in four basic steps.

- Step I. The BTS makes measurements to quantify the uplink channel conditions in RXLEV and RXQUAL.
- Step II. The power-control algorithm implemented at the BSC processes the averaged RXLEV and RXQUAL and decides what transmit power level the MS must use to maintain good quality.
- Step III. The BSS conveys to the MS the specific power level that the MS should use.
- Step IV. The MS increases or decreases its transmit power so that the power level the BSS specified is reached.

Let's discuss the power-control stages in more detail.

5.9.1 A Detailed Look at Power Control

In Step I, as mentioned during the handover discussion, the BTS makes several measurements, such as RXLEV and RXQUAL. Furthermore, the MS determines the downlink measurements for the currently serving cell and the neighboring cells and reports such measurements to the BSS via the SACCH.

Step II involves the execution of the power-control algorithm. Figure 5.22 illustrates the potential algorithm that the BSS can implement to issue power-control commands to the MS and to determine how much power the BTS should transmit to a given MS [209]. The BSS compares the averaged measurements with the relevant thresholds every SACCH multiframe to determine if certain conditions are met.

- Condition A: Weak Signal Strength. If the RXLEV is below a threshold in $P1$ out of $N1$ instances, the signal strength is weak. Hence, the transmit power needs to be increased to improve the quality of the communication link. Example settings for $P1$ and $N1$ could be ten and twelve, respectively [209].
- Condition B: Strong Signal Strength. If RXLEV is usually greater than a threshold, the signal strength is quite good, perhaps much better than necessary for the target QoS! In such a case, the transmit power is reduced. The parameters $P2$ and $N2$ may be set to ten and twenty, respectively.
- Condition C: Poor Quality. If the achieved bit error rate (BER) is higher than a threshold, the transmit power may be inadequate under current channel conditions. Thus, the transmit power needs to be increased. Potential settings for $P3$ and $N3$ could be five and seven, respectively.
- Condition D: Much Better Quality. If the actual BER is much lower than the target BER, the transmit power being used may be excessive. In such a case,

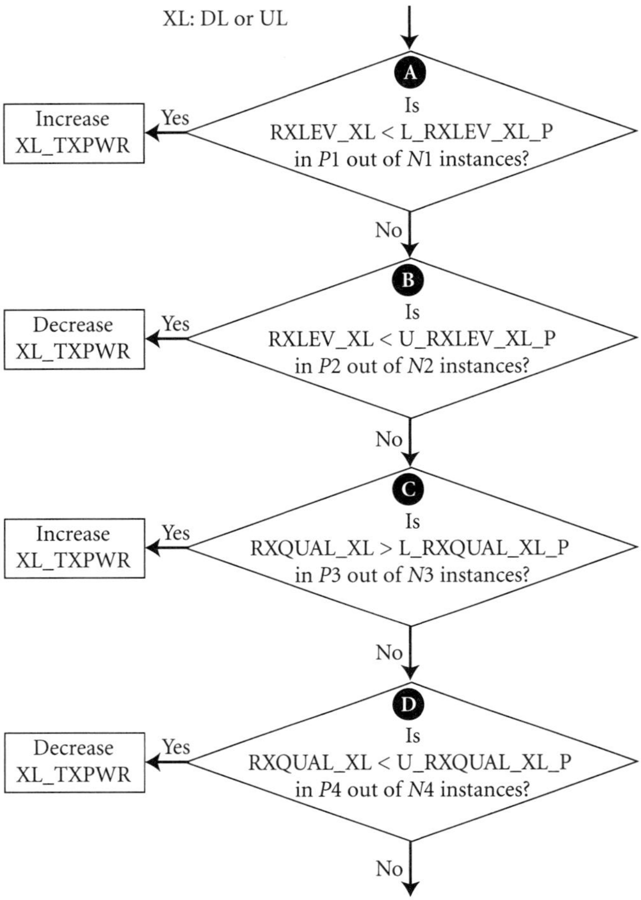

Figure 5.22: BSS Implementation of the Power-Control Algorithm

the transmit power should be reduced. The parameters $P4$ and $N4$ may be set to fifteen and eighteen, respectively.

Note that if the downlink power control is implemented in the BSS, the BTS's transmit power can change in 2-dB steps with a dynamic range (i.e., minimum to maximum) being 30 dB.

In Step III, the BSS specifies the transmit power level for the MS the use. The BSS has two choices for sending this power-control command: it can use the layer 1 header of each downlink SACCH message block [212] or of a dedicated signaling block [209].

In Step IV, the MS utilizes the most recent power-control command to transmit its bursts on channels, such as a TCH, FACCH, SACCH, or SDCCH. The MS also conveys to the BTS the specific power level that the MS used in the last burst of the previous SACCH period, and this information is included within the SACCH's layer 1 header. More specifically, the MS changes its transmit power to the required level at the rate of one 2-dB power-control step (which could be a step up or down) every thirteen TDMA frames (i.e., 60 ms). Hence, the MS may take up to fifteen steps with each 2-dB step to cover its 30-dB dynamic range. The total time the MS takes to cover the entire transmit power range would be 900 ms (i.e., 60 ms $*$ 15). The MS starts changing its transmit power at the first TDMA frame of the next reporting period. The MS continues to update its transmit power level regardless of whether actual transmission occurs.

Is closed-loop power control the only form of power control? Let's say that the MS is performing a random access procedure to access a cell. How much transmit power should the MS use? The GSM MSs and class 1 and class 2 DCS 1800 MSs use the power level MS_TXPWR_MAX_CCH defines. The class 3 DCS 1 800 MSs use the power level $MS\_TXPWR\_MAX\_CCH + POWER\ OFFSET$. Both MS_TXPWR_MAX_CCH and POWER OFFSET are broadcast on the BCCH.

5.9.2 Further Reduction in the Transmit Power Consumption: DTX

The closed-loop power control in the uplink helps minimize the transmit power, leading to increased MS battery life. An additional battery-life enhancement comes from an attractive GSM feature called *Discontinuous Transmission* (DTX). As the name implies, the transmitter transmits discontinuously, thereby reducing the average transmit power. You may be concerned about the quality of communication if we just stop transmitting, but we'll explain why you need not worry.

The concept of a speech vocoder or codec is discussed in Chapter 2. The speech vocoder generates speech bits during the active speech period and no bits during the silence interval. Hence, during the active speech period, we can perform regular transmission using suitable power levels. During the silence interval, we can refrain from transmitting a continuous stream of dummy bits and thus save battery power. Figure 5.23 illustrates the use of DTX in conjunction with the speech codec operation.

As discussed in Chapter 2, the MS and the network coordinate with each other such that comfort noise is generated at the receiver to avoid the misconception that the call has dropped. Such coordination occurs in the form of the Silence Descriptor (SID) frame. The transmitter sends an SID frame at the end of an active speech period. The SID frame contains relevant parameters to facilitate the generation of comfort noise at the receiver. The SID frames are periodically transmitted to reflect new comfort noise parameters. The receiver can observe the SID frame's signal level and signal quality to quantify any change in radio channel conditions. Though the DTX in the uplink is mandatory, the DTX implementation in the downlink is optional. Use of the DTX in a cell can be conveyed to the MS in two ways: via the BCCH and via the SACCH. The BCCH and the SACCH send the DTX parameter in the Cell Options information element. This DTX parameter instructs the MSs in the cell whether they *may* use DTX in the uplink, they *must* use DTX in the uplink, or they *must not* use DTX in the uplink [209]. Note that the DL DTX is fully under the BSS's control.

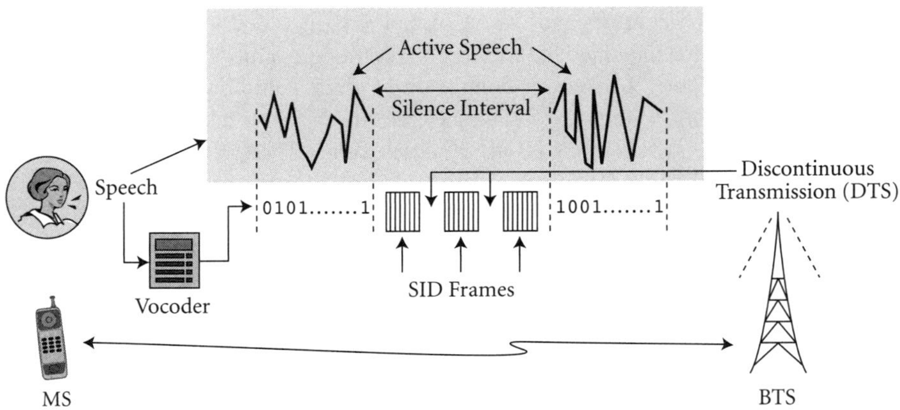

Figure 5.23: Speech Vocoder and Discontinuous Transmission

Using the DTX results in the effective data rate decreasing to 500 bps rather than the full rate of 12.2 kbps, significantly reducing power consumption and overall noise in the uplink. The full-rate speech involves one speech frame transmission every 20 ms, and the DTX operation involves less frequent frame transmission, the frequency of which depends on the type of speech codec.

5.10 DEPLOYMENT CONSIDERATIONS

GSM is a hybrid TDMA/FDMA-based system. Hence, the frequency reuse factor plays a critical role in the deployment. The SIR (Signal-to-Interference Ratio) required to guarantee a target QoS (e.g., 2% error rate) dictates the cluster size. Recall that the cluster contains the cells that cannot use the same ARFCN. Using a suitable cluster size ensures minimum co-channel distance. Such minimum separation between the cells using the same ARFCN ensures a good SIR value, and a GSM service operator's subscribers will be happy because they will be experiencing good QoS.

If the frequency reuse factor is $1/4$, what is the cluster size? What is the significance of cluster size? Recall from Chapter 1 that $Cluster\ size = 1/frequency\ reuse\ factor = 1/(1/4) = 4$. Cluster size dictates how often we can reuse the same radio channel in a given geographical area. If the cluster size is one, we can reuse the same radio channel in every single cell in a system. In Figure 5.24, each omnidirectional cell in a cluster is using a distinct BCCH carrier frequency or beacon frequency. The available number of ARFCNs can then be equally distributed among four cells in a cluster. Let's say that we are a GSM service operator with 5 MHz of spectrum for the DL and 5 MHz for the UL. Since the carrier spacing is 200 kHz, a total of $5\ \text{MHz}/200\ \text{kHz} = 25$ ARFCNs is available in each cluster. With a frequency reuse factor of $1/4$, each cell gets about six ARFCNs. One of the six ARFCNs is

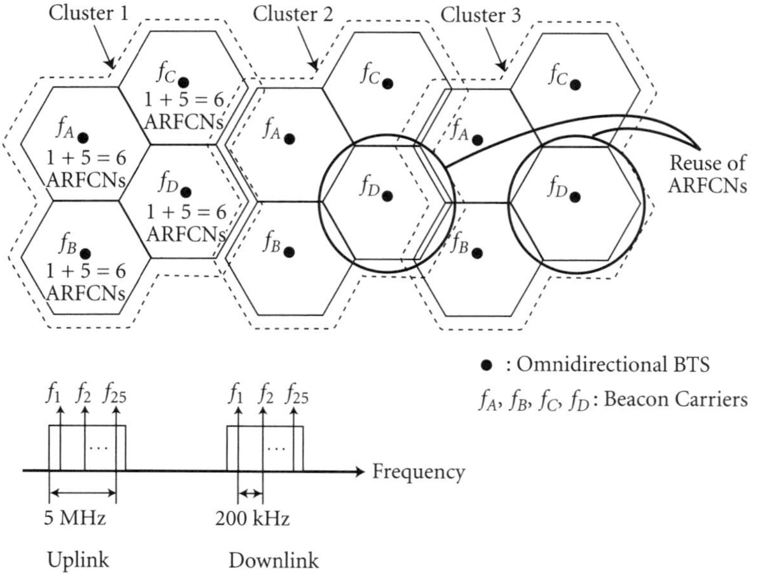

Figure 5.24: Simplified Example Deployment of Early GSM Networks

reserved for the beacon frequency, leaving five ARFCNs for the user traffic. Since each ARFCN supports eight time-slots with one user per time-slot in case of full-rate speech, each cell supports a maximum of $5 * 8 = 40$ users in this configuration. Though time-slot 0 of the BCCH carrier frequency is reserved for the BCCH, other time-slots can be used to carry user traffic. Hence, a little more than forty full-rate speech users can be supported per omnidirectional cell.

Like AMPS, initial GSM systems used the frequency reuse factor of $1/7$, and several enhancements led to the frequency reuse factor of $1/4$. If a simple channel assignment strategy is used with this frequency reuse factor for the traffic channel, we can deploy a GSM system as shown in Figure 5.24.

Can we do better than a fixed channel assignment approach? Rather that using the fixed channel assignment strategy, a dynamic channel assignment strategy can be used. Such a strategy can have fixed allocations in different cells or sectors for only BCCH carriers; users in any cell are allocated ARFCNs (let's call them *traffic ARFCNs*) from a pool of remaining ARFCNs. If we use the frequency reuse factor of $1/4$ for BCCH carriers, we have now a pool of $25 - 4 = 21$ ARFCNs. These twenty-one ARFCNs are not assigned to a specific cell in a cluster. In such a scenario, if one cell has more than forty users, we can still allocate an ARFCN to a new call. Note that such a call would have been blocked in the configuration of Figure 5.24 due to a hard capacity limit of forty users. For the dynamic channel assignment strategy under consideration, each cluster of four omnidirectional cells can support $21 * 8 = 168$ users with these users distributed anywhere (not necessarily uniformly) in the cluster.

The advantage of such a dynamic channel assignment strategy is better performance (e.g., lower call blocking probability) in the case of non-uniform distribution of user traffic. The drawback of such an approach is the potential for higher co-channel interference since two cells of two different clusters could be neighbors and may be allocated the same channel. We would then violate the sanctity of the co-channel distance with no guarantees of target QoS.

Though early GSM network deployments followed a simple strategy, such as the one shown in Figure 5.24, the strategies shown in Figure 5.25 have become quite popular in GSM deployments. Each cell in Figure 5.25 consists of three 120° sectors. Since the MSs must reliably detect the BCCH, the BCCH's frequency reuse is conservatively set to $1/12$.[168] In other words, a given BCCH carrier is used only once in a given cluster of twelve sectors. In Figure 5.25, a total of 12 BCCH carriers,

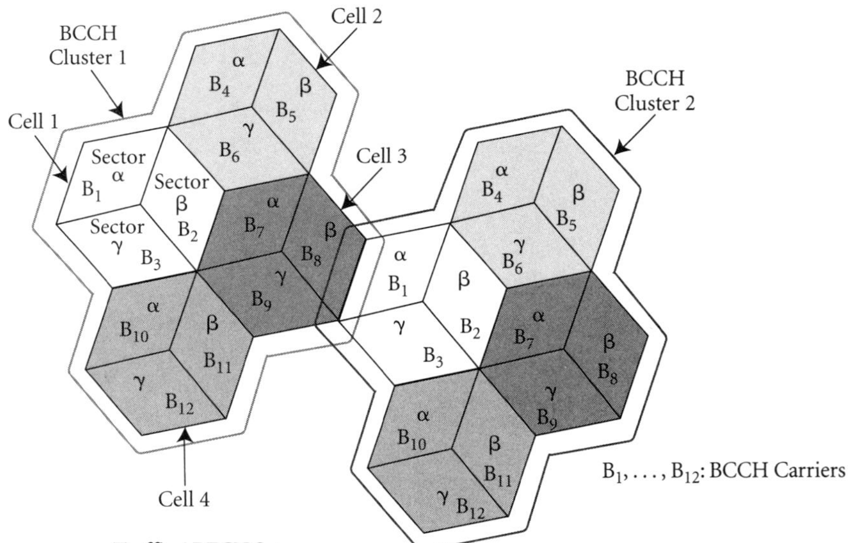

Traffic ARFCN Sets

Sector	Deployment Configuration 1	Deployment Configuration 2
α	$\{f_1, \dots, f_6\}$	$\{f_1, \dots, f_6\}$
β	$\{f_1, \dots, f_6\}$	$\{f_7, \dots, f_{12}\}$
γ	$\{f_1, \dots, f_6\}$	$\{f_{13}, \dots, f_{18}\}$

Figure 5.25: BCCH and TCH Frequency Reuse in GSM Deployments

[168]Such BCCH configuration is called $4/12$ because each BCCH cluster consists of four cells and twelve sectors. Other example BCCH configurations are $3/9$ and $5/15$.

B_1 through B_{12}, exist in each BCCH cluster. Each sector must have its own BCCH carrier.

The traffic ARFCNs are a different story. Each sector within a cell has a set of traffic ARFCNs to implement frequency hopping. Traffic ARFCNs have two prevalent deployment configurations. The first deployment configuration uses the frequency reuse of 1, i.e., all sectors have the same set of traffic ARFCNs (e.g., the same set of $\{f_1, \ldots, f_6\}$ in sector α, sector β, and sector γ in each cell). The second deployment configuration uses the frequency reuse of $1/3$, i.e., three sectors of the same cell use different sets of traffic ARFCNs, e.g., $\{f_1, \ldots, f_6\}$ for sector α, $\{f_7, \ldots, f_{12}\}$ for sector β, and $\{f_{13}, \ldots, f_{18}\}$ for sector γ.

A Question for You... What are the main advantage and the main disadvantage of the first deployment configuration compared to the second one in Figure 5.25? See the footnote[169] to compare your view with ours!

Features such as power control and frequency hopping help realize an aggressive frequency reuse as illustrated in Figure 5.25 (relative to the configuration depicted in Figure 5.24). A word of caution is warranted. In Figure 5.25, one TRX is using a given traffic ARFCN set; hence, such a TRX uses a single ARFCN from this set during a given frame based on the frequency hopping algorithm. Keep in mind that six different TRXs in sector α are not using six different ARFCNs f_1 through f_6 at a given instant when the ARFCN set is to $\{f_1, \ldots, f_6\}$ for sector α.[170] One TRX chooses just one of the ARFCNs from the set available to the TRX. *Fractional reuse* quantifies how aggressively a given ARFCN is being reused; it is the ratio of the number of TRXs in a sector to the number of ARFCNs available to the sector. For example, if a sector has just one TRX and a total of six traffic ARFCNs, the fractional reuse is $1/6$. The higher the fractional reuse number, the more aggressive the frequency reuse for the traffic ARFCNs.

To minimize co-channel interference among the sectors of the same cell and to randomize interference among sectors of different cells, suitable settings of the parameters, such as Hopping Sequence Number and mobile allocation index offset,[171] are used to implement frequency hopping. For example, different sectors in a given cell use the same Hopping Sequence Number value but different mobile allocation index offset values (e.g., $\textit{mobile allocation index offset} = 0$, 2, and 4 in sector α, sector β, and sector γ, respectively). Furthermore, two different cells use two different Hopping Sequence Number values (e.g., $\textit{Hopping Sequence Number} = 1$ in one cell and $\textit{Hopping Sequence Number} = 2$ in another cell).

[169] Since the first deployment configuration is aggressively reusing the frequencies, its main advantages are higher capacity and less need for spectrum bandwidth. Its main drawback is potential QoS degradation due to increased interference.

[170] Of course, you could have one ARFCN set for one TRX in sector α, and a different ARFCN set for another TRX in the same sector α.

[171] Hyper frame number also influences the ARFCN used at a given instant in a sector.

A final note on commercial GSM deployments: Offline and non-real-time automatic frequency planning tools are used to dynamically and periodically (e.g., every few months) adjust the ARFCN set to be used for user traffic and the ARFCN for the BCCH carrier in each sector of a cluster based on observed ARFCN usage patterns and interference patterns.

5.11 EVOLUTION OF GSM

GSM is a 2G digital cellular system with a circuit-switched core network. The main focus of GSM is on voice services and low-rate data services like SMS. In a circuit-switched system, the circuit has to be dedicated to the user throughout the call, and resources need to be reserved throughout the network. If we are surfing the Internet, we truly use the resources only when we download a web page. When we read the downloaded web page, we really utilize no resources, such as the time-slot and the ARFCN, but we hold onto these resources and thus make them unavailable to other users. That's where a packet-switched core network comes into the picture.

In a packet-switched world, the majority of critical resources are occupied only during data activity. In the absence of data activity, the resources are released and made available to other users. When the data activity resumes, the resources are quickly allocated to the user.

The first step in the GSM evolution is the introduction of a packet-switched core network and air-interface protocol stack enhancements to support packet-switched data services. This evolution is a 2.5G system called GPRS. The next step beyond GPRS is EDGE, which provides higher data rates, on the order of tens and hundreds of kilobytes per second, than the GSM data rate (14.4 kbps). Another step in GSM's evolution is 3G UMTS, which provides even higher data rates and much better spectral efficiency. We will learn about 2.5G and 3G systems in Chapters 6 and 9.

5.12 POINTS TO REMEMBER

- GSM originated as a unified cellular system that provides seamless services in different European countries. It expanded its footprint to other parts of the world and boasts about 80% of cellular market share (as of late 2011).
- GSM is a 2G digital cellular system based on the hybrid TDMA/FDMA-based air interface.
- Different users in a cluster access the system at the same time using different combinations of the ARFCN and time-slots.
- GSM's network architecture is centralized; the radio network consists of BTSs and the BSC, and the core network consists of components such as the MSC/VLR, HLR/AuC, and short message service center.
- The focus of GSM is voice services and low-rate circuit-switched data services.

- Call setup includes the establishment of a dedicated signaling connection for reliable exchange of signaling messages; authentication; ciphering; and establishment of a dedicated traffic channel.
- Handover in GSM is mobile-assisted and network-controlled hard handover in which the MS provides measurement reports to the network and the network makes a handover decision.
- Power control and DTX reduce power consumption, thereby improving the battery life and reducing interference.
- The GSM evolutionary path includes 2.5G GPRS, 2.5G EDGE, and 3G UMTS.

CHAPTER 6

GPRS AND EDGE

GPRS (General Packet Radio Service) and EDGE [Enhanced Data Rates for GSM (Global System for Mobile Communications) Evolution, sometimes also referred to as Enhanced Data Rates for Global Evolution] represent second-and-a-half–generation (2.5G) cellular networks, an evolution of second-generation (2G) GSM (Global System for Mobile Communications) networks. This chapter discusses the main concepts of GPRS and EDGE. First, a brief introduction to these 2.5G systems is given. The network architecture is illustrated at length, including discussions of new network elements that did not exist in GSM. An end-to-end packet flow is described. The air-interface protocol stack consisting of physical and Radio Link Control and Medium Access Control is discussed.

GPRS and EDGE reuse the basic timing structure of the GSM radio interface; this timing structure is briefly reviewed, followed by a discussion of various new radio channels introduced by GPRS and EDGE. The mobile station's (MS's) network acquisition after power-on is illustrated along with two GPRS-specific procedures called *GPRS attach* and *Packet Data Protocol (PDP) context activation*. GPRS and EDGE are focused on efficient packet data services; both MS-originated and MS-terminated packet transfer are summarized. Both STANDBY state and READY state mobility are considered. Deployment of 2.5G systems and the network evolution path beyond 2.5G are summarized.

6.1 INTRODUCTION

GPRS and EDGE are 2.5G digital cellular standards that represent the evolution of the 2G GSM standard. GPRS and EDGE are a natural path of network evolution for the service operators that have deployed GSM. GPRS and EDGE are applicable only to packet data services; the mobile station (MS) and supporting GSM, GPRS, and EDGE technologies would use GSM for traditional circuit-switched voice services. Let's discover the motivation for GPRS and EDGE. Then we will summarize the main features of GPRS and EDGE.

6.1.1 The Motivation for GPRS

In a 2G GSM system, users access wireless data services using a circuit-switched data approach. Consider Figure 6.1 in which a GSM user is accessing a web server to read news.[172] The user sets up a call and begins communicating with the web server. Since 2G cellular systems are digital in nature, the signal from the user's MS to the IWF (Interworking Function) is digital.

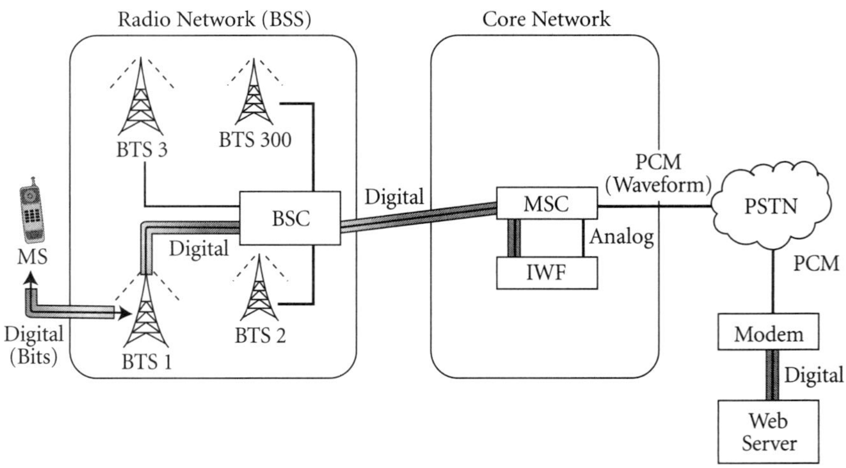

Figure 6.1: Circuit-Switched Data Approach in 2G Systems

[172]Recall from Chapter 5 on GSM that BSS represents the base station subsystem, BSC the base station controller, BTS the base transceiver station, MSC the mobile switching center, and PSTN the public-switched telephone network.

A Question for You... Do you remember what is inside the IWF and for what the IWF is used? The footnote[173] contains the answer.

A modem in the IWF maps the digital signal into the analog signal that travels through the PSTN to the web server. The modem at the web server retrieves the digital signal by processing the received analog signal received via the PSTN. The overall process is reversed for a communication from the web server to the user. Once the call is set up, a dedicated connection exists between the MS and the web server, and the resources reserved for such end-to-end connection are exclusively for the MS; no other MS can touch these resources. For example, if ARFCN f_1 and time-slot 2 are allocated to the MS under consideration, no other MS can use f_1 and time-slot 2 as long as the call exists.

Now, let's discuss what happens during the web-browsing. Figure 6.2 illustrates data activity as a function of time. Once the call is set up, just before the time instant t_1, the user clicks on a web page. The data of the web page is downloaded during the time interval from t_1 to t_2. The reserved resources, such as an ARFCN and a time-slot, are utilized very well during this interval. When the user is reading the downloaded web page during the time interval from t_2 to t_3, the reserved ARFCN and time-slot carry no data. Similarly, when the second and third web pages are downloaded during the intervals from t_3 to t_4 and t_5 to t_6, the reserved resources are again utilized for downlink data transfer. The call is released at t_7 following the termination of the web-browsing application.

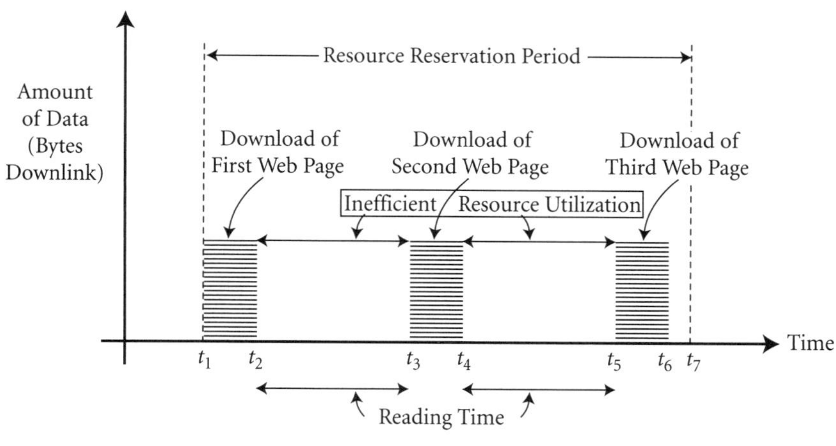

Figure 6.2: Resource Utilization during a Typical Web-Browsing Session

[173]The IWF contains a pool of modems that perform conversion between waveforms and digital signals (i.e., bits) and support digital data services, such as email and web-browsing, using a waveform-based (e.g., Pulse-Coded-Modulation–based, or PCM-based) core network, such as the PSTN.

As shown in Figure 6.2, the reserved resources are used inefficiently during the reading time periods. Even then, no other user can utilize the resources reserved for the user under consideration during the entire time period of t_1 through t_7. *GPRS exploits such data inactivity periods to share the unused resources with several users by using a distinct packet-switched core network.* When the resources are fully occupied and used for the transfer of real data, the total amount of data the network transfers during a given observation period increases. In other words, throughput, which is measured in bits per second, increases. Recall that throughput is one of the most important performance metrics for a data system. *In summary, GPRS introduces a new packet-switched core network to the basic GSM core network to efficiently support packet data services. Furthermore, GPRS increases the peak data rate of 171.6 kbps from the practical data rate of 9.6 or 14.4 kbps in GSM.* The peak rate of 171.6 kbps assumes that all eight time-slots of an ARFCN are reserved for GPRS. In commercial GPRS deployments, two time-slots have been quite common, allowing the peak rate of $171.6 * 2/8 = 42.9$ kbps.

6.1.2 The Motivation for EDGE

Let's say that you want to download an email containing photos from your friend. If you use a GSM- or GPRS-capable MS, your download speed would be slow because the data rates GSM and GPRS offer are quite low (e.g., 9.6 kbps for GSM and 42.9 kbps for GPRS). Next time, you would perhaps ask your friend to avoid sending too many photographs via email! EDGE attempts to improve this situation by supporting higher data rates (e.g., the peak rate of 473.6 kbps when eight time-slots are reserved and 236.8 kbps when four time-slots are reserved). If the MS and the network support the peak data rate of 236.8 kbps, the download speed would be faster by a factor of approximately 236.8 kbps $/42.9$ kbps $= 5.6$. *EDGE offers higher data rates than GSM and GPRS and improves the user's data services experience.*

6.1.3 GPRS in Two Minutes

GPRS offers packet data services via a packet-switched core network with data rates higher than those offered by commercial basic GSM networks. In other words, GPRS can be considered both an air-interface enhancement and a core network enhancement for GSM. GPRS specifies a new Time-Division Multiple Access (TDMA) based air interface without breaking the time-slot and radio channel structure of GSM. Hence, GPRS is fully backward-compatible with GSM, allowing the same carrier frequency (e.g., ARFCN) to be shared among GSM and GPRS users. The packet-switched core network of GPRS enables packet mobility, so users can move across geographical regions without losing an end-to-end link with an Internet entity such as a web server.

As mentioned earlier, GPRS makes the resource utilization efficient by allocating resources, such as time-slots, per *block* instead of per *call*. Such an approach can keep the resources well-utilized and support more users by exploiting the burst nature of several packet data services, such as email and web-browsing. Recall that the circuit-switched system reserves the resources as long as the call lasts.

6.1.4 EDGE in Two Minutes

EDGE is a radio network enhancement to the GPRS air interface, which increases over-the-air data rates for (usually) packet data services. EDGE can be viewed as a super-set of GPRS. EDGE consists of GPRS plus the following main radio interface enhancements: high-order modulation of 8-ary Phase Shift Keying (8-PSK), which permits faster data rates of three bits per symbol rather than one bit per symbol achievable with GPRS; Hybrid Automatic Repeat Request (H-ARQ),[174] which reduces the number of radio link retransmissions under poor radio conditions; and radio link adaptation, which exploits periods of good propagation by selecting modulation and coding schemes that have less protection and more throughput when possible [213]. EDGE makes no alteration to the core network; it relies upon the pre-existing core network, such as a GPRS core network, to connect to the packet data network (e.g., the Internet).

You may come across the term EGPRS (Enhanced GPRS), which means that a GPRS network (i.e., combined radio network and core network) has been enhanced to exploit new EDGE air interface. In other words, EGPRS refers to the overall radio and core network, and EDGE refers to the air-interface aspects of the new air interface. Consider an EGPRS network shown in Figure 6.3. EDGE is fully backward-compatible with GPRS, and GPRS is fully backward-compatible with

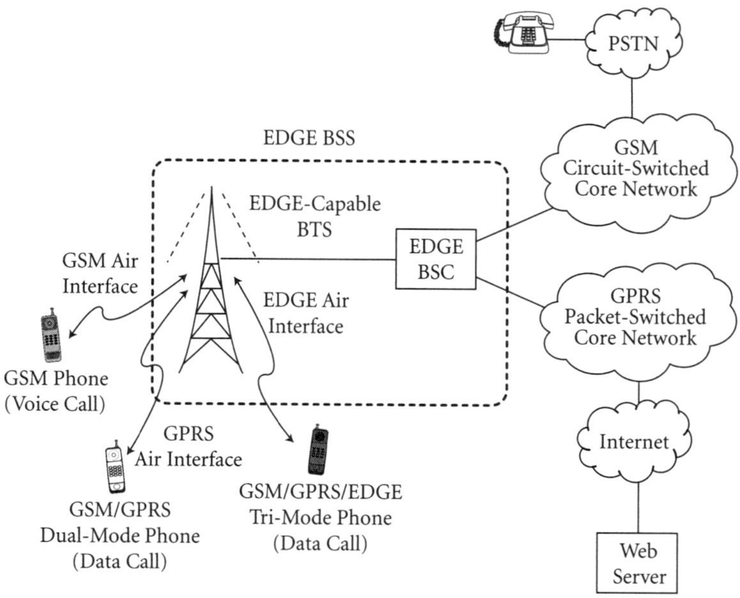

Figure 6.3: Mobile Devices in an EGPRS Network: A Simplified View

[174] H-ARQ is pronounced like the word "hark."

GSM. In Figure 6.3, the base station system (BSS) [i.e., base transceiver stations (BTSs) and the base station controller (BSC)] supports EDGE, and the core network includes the basic GSM circuit-switched core network and the packet-switched core network introduced by GPRS.

Three mobile devices are present in the cell controlled by the BTS in Figure 6.3: a GSM-only MS, a dual-mode GSM/GPRS MS, and a tri-mode GSM/GPRS/EDGE MS. When a voice call is made using the GSM-only phone, the basic GSM air-interface features are used for communications between the MS and the BTS. When a GSM/GPRS MS makes a data call, GPRS radio interface features are used to obtain better data rates than those available with the GSM air interface. Similarly, a GSM/GPRS/EDGE MS is served using the EDGE air interface for better performance in the case of a data call. In summary, whenever possible, a better air interface is used.

6.2 SYSTEM ARCHITECTURE

GPRS introduces a new packet-switched or packet-domain core network to efficiently transfer packet data and signaling. The packet data services could be non-real-time services, such as web-browsing and email, or real-time services, such as a VoIP or video call. The packet-domain core network makes optimal use of the BSS and the core network resources. Just as the GSM circuit domain (i.e., a circuit-switched network) keeps the radio network and the core network functions separate, the GPRS packet domain also maintains the separation of the radio network and the core network. Such separation enables the same packet domain to be reused with a future non-GPRS and non-EDGE radio access network, such as a 3G Universal Mobile Telecommunication System (UMTS) radio network.

Section 6.2.1 illustrates the core network architecture of the GPRS network. The components and the interfaces of the packet-domain core network are discussed in detail in Sections 6.2.2 and 6.2.3, respectively.

6.2.1 Packet-Domain Core Network Architecture: An Overview

Figure 6.4 depicts the core network architecture GPRS introduces [214]. Such a packet-domain core network is available for use with GPRS and EDGE radio access technologies and possibly future radio access technologies.

A GPRS packet-domain core network consists of two main components, the SGSN (Serving GPRS Support Node) and the GGSN (Gateway GPRS Support Node). The BSS connects to the SGSN. The SGSN works with the HLR (Home Location Register) to authenticate the MS as part of a process called the *GPRS attach*. The GGSN is a gateway to the outside world for the MS: the data packets that the MS sends leave the service operator's core network from the GGSN, and the data packets destined for the MS arrive at the GGSN from the external PDN (Packet Data Network), such as the Internet. The GGSN also allocates an Internet Protocol (IP) address, often termed

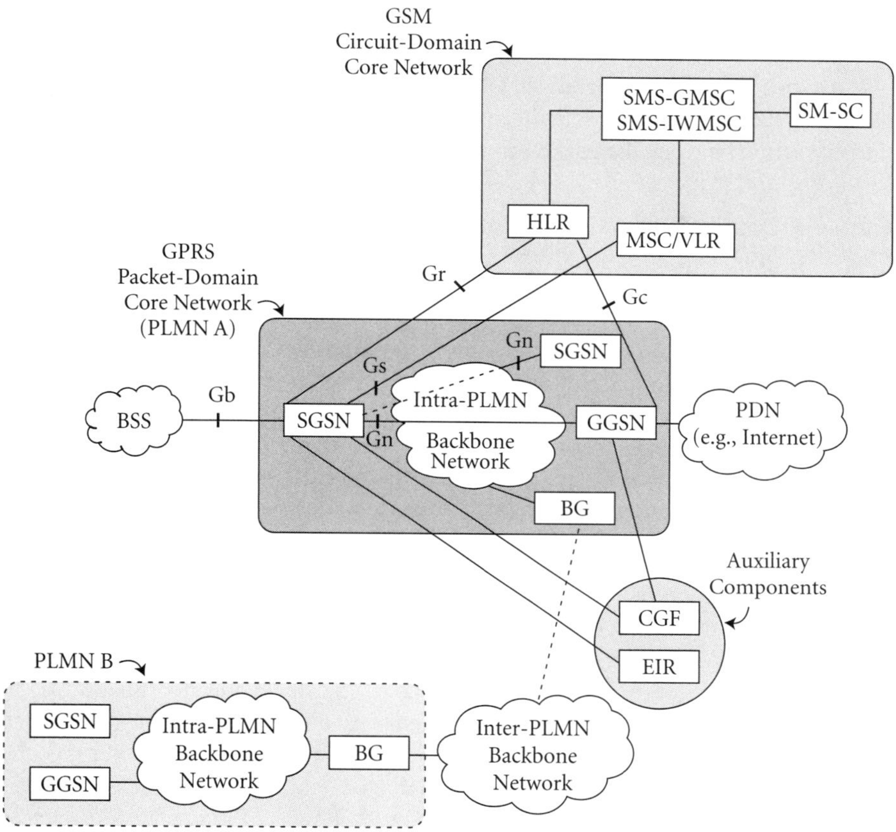

Figure 6.4: GPRS Packet-Domain Core Network

a PDP (Packet Data Protocol) address, to the MS as part of a process called *PDP context activation*.

An intra-PLMN (Public Land Mobile Network) backbone network connects SGSNs and GGSNs. A typical nationwide GPRS network consists of hundreds of geographically distributed SGSNs and dozens of GGSNs. One service operator's PLMN can be connected to another service operator's PLMN via border gateways (BGs) and an inter-PLMN backbone network. Intra-PLMN and inter-PLMN backbone networks are high-speed transport networks. The packet-domain core network communicates with the pre-existing GSM PLMN circuit-domain core network, such as the HLR, Mobile Switching Center (MSC)/Visitor Location Register (VLR), Short Message Service (SMS) Gateway MSC (GMSC) or Interworking MSC (IWMSC), Charging Gateway Function (CGF), and Equipment Identity Register (EIR). The interfaces between the packet-domain core and GSM circuit-domain core support functions such as SMS transfer, billing, and denial of service requests from lost or stolen MSs.

Chapter 5 discusses the details of GSM circuit-domain core network components. Hence, as far as the circuit-domain components are concerned, the focus here is on the interaction between the packet-domain and specific circuit-domain components.

6.2.2 A Detailed View of the Components of the Packet-Domain Core Network

The packet-domain core network makes the packet data services more efficient by avoiding the reservation of various resources in the absence of data activity. It allows the user to access the Internet and numerous IP-based services. Furthermore, the functions that the circuit-domain core network performs for a circuit-switched call are performed by the packet-domain core network for packet-switched services. Examples of these functions are authentication and call setup. The roles of various components in supporting packet-switched services are discussed in detail next.[175]

6.2.2.1 Gateway GPRS Support Node The GGSN enables the MS to communicate with a packet data network (PDN) external to the GPRS network. The Internet is an example of a PDN. Upon the MS's completion of the GPRS attach and the PDP context activation, the GGSN knows which SGSN is currently responsible for the MS. In Figure 6.5, two MSs, MS1 and MS2, have attached themselves to the GPRS packet-domain core network and have established PDP contexts. When a packet for the MS arrives at the GGSN, the GGSN tries to match the IP address of the received packet with the list of IP addresses in its database and forwards the packet to a suitable SGSN. In our example, when the IP address of the received packet is 100.20.30.40, the GGSN forwards this packet to SGSN1 because MS1 is in the area SGSN1 controls. Similarly, when the received IP packet contains MS2's IP address as the destination address, the GGSN forwards the packet to SGSN2.

The GGSN may contact the HLR via the optional Gc interface to locate the MS. The Gc interface is typically unused in commercial networks. Figure 6.5 illustrates some information stored at the GGSN, and Table 6.1 provides additional examples of such information. The GGSN maintains such information for activated PDP contexts. See the references for an exhaustive list of the parameters the GGSN stores for the MS-specific PDP context [214].

The IMSI (International Mobile Subscriber Identity) is used as the MS's basic identity. Hence, the IMSI can be used to point to any MS-specific information. The PDP type indicates the type of PDP being used (e.g., IP). The PDP Address is the address that the GGSN allocates to the MS; hence, it is usually an IP address. The Dynamic Address parameter indicates whether the assigned PDP address is static or dynamic. Dynamic IP addresses are typically used in commercial networks. The record "APN in Use" indicates which APN (Access Point Name) is being used for the given MS and for the given PDP address. For example, if the service operator

[175] Readers interested in further details can refer to the standard [214].

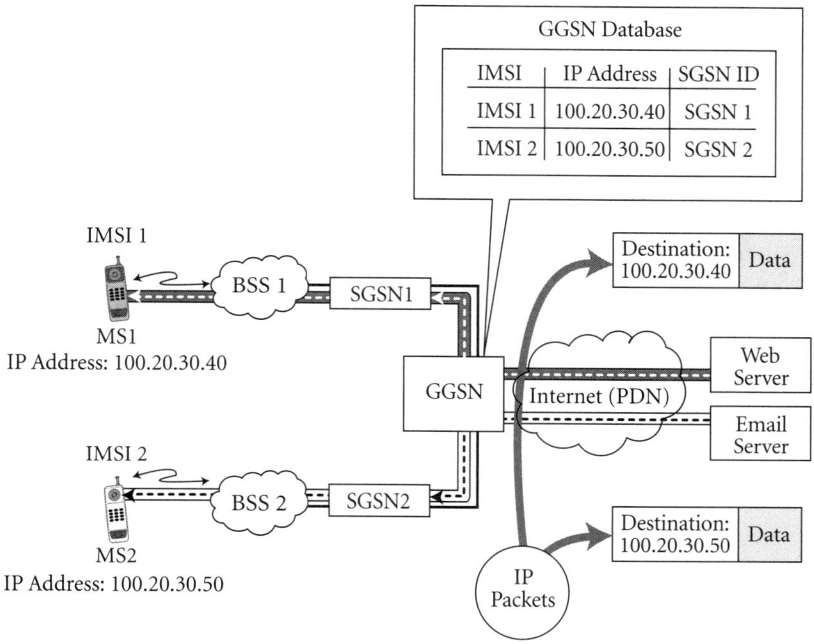

Figure 6.5: Packet Forwarding by the GGSN

Table 6.1: Example Contents Stored at the GGSN

Parameter	Example Setting
IMSI	123...9
PDP Type	IP
PDP Address	100.20.30.40
Dynamic Address	Dynamic
APN in Use	`ServiceOperator.com`
SGSN Address	20.30.40.50
TEID	10
GTP SND	100
GTP SNU	100
TFT	All Allowed
QoS Profile Negotiated	Best Effort
Charging ID	1000
Charging Characteristics	Normal

is AT&T, the APN in Use may be `att.com`. An APN is associated with a specific PDN. For example, `att.com` may be associated with the regular Internet as the PDN, and `blackberry.net` may be associated with the PDN (i.e., the email network) managed by Blackberry.

The SGSN Address is the IP address of the SGSN currently responsible for the MS. The parameters TEID (Tunnel Endpoint Identifier),[176] GTP SND (GPRS Tunneling Protocol Sequence Number Downlink), and GTP SNU (GPRS Tunneling Protocol Sequence Number Uplink) are associated with the communication link between the GGSN and the SGSN for the MS under consideration. When a packet for an MS needs to be transferred between the GGSN and the SGSN, a GTP-U (GPRS Tunneling Protocol–User Plane) tunnel needs to be established. Such a tunnel ensures the security of the packet and enables reservation of resources, such as bandwidth, to ensure a suitable QoS (quality-of-service). The TEID then identifies a link that carries packets for the MS. The parameter GTP SND specifies the GTP-U sequence number of the next downlink packet to be sent to the SGSN. Similarly, the parameter GTP SNU specifies the GTP-U sequence number of the next uplink packet to be received from the SGSN. The parameters GTP SND and GTP SNU essentially facilitate suitable ordering of the packets flowing in a given direction between the GGSN and the SGSN. The parameters TFT (Traffic Flow Template) and QoS Profile Negotiated are related to the services for the MS. The TFT specifies rules for packet filtering, which may dictate which types of service flows or applications (e.g., video streaming vs. email) are allowed or blocked. The QoS Profile Negotiated quantifies the QoS negotiated between the MS and the GGSN. Consider the term *negotiated* as a misnomer; the MS will be only *informed* of which QoS the network is giving the MS! As shown in Figure 6.4, the GGSN also connects to the CGF. In such a case, the parameters Charging ID and Charging Characteristics are important. Charging ID is an identifier for the charging records generated by the GGSN (and the SGSN). Furthermore, Charging Characteristics may indicate the method of charging the associated PDP context (e.g., normal or prepaid).

Note that the MS or the cellular network can initiate a PDP context. If the PDP context is enabled due to the cellular-network–requested PDP context activation, parameters such as IMSI, PDP Type, PDP Address, and SGSN Address are considered valid even when the PDP context is inactive and even when the MS has not yet performed the GPRS attach procedure [214].

6.2.2.2 Serving GPRS Support Node The SGSN keeps track of the MS's location via the GPRS attach and routing area update procedures, works with the HLR to authenticate the MS, and works with the BSS and GGSN to participate in the QoS implementation. The SGSN interfaces with the BSS via the Gb interface and with the GGSN via the Gn interface. Furthermore, the SGSN interfaces with the MSC/VLR and HLR for location update and SMS transfer.

The SGSN stores two contexts for the MS. During the GPRS attach (also called *packet-switched attach*) procedure, the SGSN establishes a *mobility management*

[176]See Chapter 4 for an overview of GTP and TEID.

(MM) context. This context stores information about the mobility and security for the MS. Furthermore, during the PDP context activation procedure, the SGSN establishes a PDP context with the GGSN. Such a context stores information about packet routing for the MS. Each MM context contains zero or more of the PDP contexts. For example, if the MS has performed only the GPRS attach and not the PDP context activation, only the MM context exists; no PDP context exists for such an MS. The shaded rows in Table 6.2 correspond to the PDP context parameters. You will see a lot of similarities between these parameters and the PDP context parameters stored at the GGSN (shown in Table 6.1). In summary, the SGSN stores for the MS the parameters that facilitate mobility management and PDP context management.

Table 6.2: Example Contents Stored at the SGSN

Parameter Name	Definition
IMSI	Identifies the MS and is used as a pointer to various parameters of the contexts.
MM State	The state of the MS from the perspective of mobility management state, e.g., READY.
P-TMSI	Packet Temporary Mobile Subscriber Identity, a temporary identity given to the MS by the SGSN. The SGSN assigns the P-TMSI to the MS during the PS attach procedure so that the MS can use P-TMSI instead of its original identity, IMSI.[a]
IMEI	International Mobile Equipment Identity, identifies the physical equipment of the MS.
Routing Area	Specifies the MS's current location.
VLR Number	The VLR number of the MSC/VLR currently serving the MS, can be used to track the location of the MS from the perspective of the circuit domain.
New SGSN Address	The New SGSN Address is the IP address of the new SGSN where the currently serving SGSN needs to forward the buffered packets not yet sent to the MS. This parameter is useful in the case of inter-SGSN mobility. When the MS travels from an area controlled by one SGSN to an area controlled by another SGSN, handover is required. If the previously serving SGSN forwards the not-yet-transmitted packets to the new SGSN, the new SGSN can then send these packets to the MS. Such inter-SGSN packet forwarding minimizes packet delays and packet loss rate.

Continued on next page

Table 6.2: (continued from previous page)

Parameter Name	Definition
Authentication Vectors	As mentioned earlier, the SGSN works with the HLR to authenticate the MS. The parameters associated with authentication and ciphering (e.g., the ciphering algorithm to be used) are stored in Authentication Vectors. In GPRS and EDGE systems, the SGSN secures the air interface by ciphering (i.e., encrypting) user data.
MS Radio Access Capability	The radio-access–related capabilities of the MS (e.g., supported frequency bands) are stored as MS Radio Access Capability.
MS Network Capability	The core-network–related capabilities of the MS (e.g., supporting ciphering algorithms) are stored as MS Network Capability.
DRX Parameters	DRX Parameters contain the information for the discontinuous reception (DRX) at the MS.
Subscribed Charging Characteristics	The SGSN can store charging information (e.g., normal or prepaid) in Subscribed Charging Characteristics.
PDP Type	e.g., IP
PDP Address	e.g., an IP address
APN In Use	e.g., `operator.com`, often associated with a GGSN. For example, one GGSN may be associated with one APN, and another GGSN may be associated with another APN. APN in Use helps the SGSN choose a suitable GGSN because the MS conveys APN in Use to the SGSN during the call setup.
GGSN Address in Use	Specifies the IP address of the GGSN currently used, which is useful in routing the packets received from the MS to the correct GGSN.[b]

Continued on next page

Table 6.2: (continued from previous page)

Parameter Name	Definition
TEID for Gn and Gp	Contains the Tunnel Endpoint Identifiers for the Gn and Gp interfaces. The TEID for the Gn interface between the SGSN and the GGSN is used for MS-specific information transfer. The Gn interface occurs between the SGSN and the GGSN of the same PLMN, and the Gp interface occurs between the SGSN of one PLMN and the GGSN of another PLMN; the TEID for the Gp interface is used for MS-specific information transfer between the SGSN of one PLMN and the GGSN of another PLMN.
QoS Profile Subscribed	The QoS to which the MS has subscribed.
QoS Profile Requested	The QoS the MS requested during the packet data call setup.
QoS Profile Negotiated	The QoS that the network has assigned to the MS.
Aggregate BSS QoS Profile Negotiated	While the MS can request any QoS, the network looks at the subscribed QoS and the availability of resources to determine the QoS to be granted to the MS. The MS may be running several applications (e.g., email and web-browsing). Aggregate BSS QoS Profile Negotiated specifies the cumulative BSS QoS assigned to the packet flow for the PDP context under consideration.
GTP SND	The GTP-U sequence number of the downlink packet to be sent to the MS.
GTP SNU	The GTP-U sequence number of uplink packet to be sent to the GGSN.
Send N PDU (Protocol Data Unit) Number	The SNDCP (Subnetwork Dependent Convergence Protocol) sequence number of the next downlink packet to be sent to the MS.
Receive N PDU Number	The SNDCP sequence number of the next uplink packet expected from the MS. The SNDCP sequence number is for the BSS-SGSN interface. (See the standard [215].)
PDP Context Charging Characteristics	Records the charging characteristics applicable to the PDP context under consideration, e.g., normal or prepaid.

[a] Minimizing the use of IMSI over the air interface provides additional security. GSM uses TMSI instead of IMSI whenever possible.

[b] TEID facilitates implementation of tunneling in which the original IP packet is placed inside another IP packet. Tunneling enables IP mobility such that the MS can keep using the same IP address even while it travels from one geographic area to another. See Chapter 4 for additional details.

Do You Recall... Do you recall the name of a feature in which a temporary pseudo-identity is used instead of IMSI? Do you remember the main benefit of the DRX? See the footnote[177] for the answer.

6.2.2.3 Home Location Register The HLR has a record of the MS-specific information related to packet-domain subscription (e.g., web-browsing and email services), routing (e.g., which SGSN is currently serving the MS), and the authentication (e.g., the shared secret of the MS). The SGSN can contact the HLR via the Gr interface and the GGSN can contact the HLR via the Gc interface. Table 6.3 summarizes some of the HLR contents from the perspective of the packet domain. The HLR is a huge database that stores subscriber information. The HLR plays an important role in the MS's authentication and the routing of an MS-terminated call. The HLR typically carries out Authentication Center (AuC) functions (e.g., storing an IMSI-specific shared secret and generating authentication parameters).

6.2.2.4 Mobile Switching Center/Visitor Location Register and Short Message Services Entities The SGSN can optionally be connected to the MSC/VLR for an efficient coordination between packet-switched services and circuit-switched services. The Gs interface between the SGSN and the MSC/VLR can be used to send a page to the MS using packet data resources even for a CS call. Furthermore, when the MS travels from one location area to another location area, combined GPRS and non-GPRS location update is possible if the SGSN and MSC/VLR are connected. To support such enhanced coordination, the MSC/VLR stores the SGSN number of GPRS-attached MSs that are also IMSI-attached. The interface between the SGSN and SMS GMSCs or SMS IWMSCs facilitates the SMS transmission via the SGSN. Prior to GPRS, the SMS transmission had to rely solely upon the circuit-domain core network.

6.2.2.5 Charging Gateway Function and Equipment Identity Register As we discussed earlier, the HLR, the SGSN, and the GGSN store information related to charging. The CGF collects the charging records from the SGSN and GGSN for the user. See the standard [216] for CGF details. The service operator configures the SGSN and the GGSN to collect charging per individual MS and/or per PDP context by properly setting the Subscribed Charging Characteristics and/or PDP Context Charging Characteristics in the HLR.

For a given MS, the SGSN records the charging information according to the radio access network usage (e.g., the amount of data transmitted and received), and the GGSN records the charging information according to the external packet data network usage (e.g., amount of data sent to and received from the external data network). Furthermore, the SGSN and the GGSN record the charging information

[177]The feature is called *anonymity*. The DRX saves battery power and processing power at the MS because the MS need not continuously receive the downlink information.

Table 6.3: Example Contents of the HLR Database

Parameter	Definition
IMSI	Used as the MS identifier to point to MS-specific information.
SGSN Number	The SS7 (Signaling System 7) number of the SGSN currently serving the MS. (Recall that SS7 is used for communication among the circuit-domain components.)
SGSN Address	The IP address of the SGSN currently serving this MS. (Remember that the HLR should always know the latest location of the MS.)
Subscribed Charging Characteristics	Contains information on how to charge the user (e.g., normal or prepaid). During the call setup, the HLR provides such information to the SGSN so that the SGSN can store such information in its own database.
MNRG	Specifies whether the MS is reachable via an SGSN. If the MS has not performed GPRS attach (e.g., as a result of power being OFF), no SGSN is responsible for the MS. Hence, the MS would be marked as not reachable at the HLR.
GGSN List	Includes the GSN (GPRS Support Node) number and optional IP address associated with the GGSN that is contacted when MS starts data activity and MNRG indicates that the MS is reachable via an SGSN.
PDP Type, PDP Address, Access Point Name, & PDP Context Charging Characteristics	Each IMSI may be associated with zero or more PDP contexts. As in case of the GGSN and SGSN, the HLR keeps a record of the PDP Type, PDP Address, Access Point Name, and PDP Context Charging Characteristics.
QoS Profile Subscribed	Provided to the SGSN during call setup.

based on usage of the packet-domain core network resources (e.g., the time period for which the MS has used the PDP address). The role of the EIR for the PS domain is the same as that for the circuit-switched domain; lost or stolen devices can be barred from using the services.

6.2.2.6 The MS and Its Database The GPRS or EDGE MS stores MM and PDP contexts. Table 6.4 provides example parameters the MS stores. We have already discussed these parameters in conjunction with SGSN and GGSN, and hence we will

Table 6.4: Example Contents Stored at the MS

Parameter Name
IMSI
MM State
P-TMSI
IMEI
Routing Area
Authentication Vectors
MS Radio Access Capability
MS Network Capability
PDP Type
PDP Address
APN Requested
QoS Profile Requested
QoS Profile Negotiated
Aggregate BSS QoS Profile Negotiated
GTP SND
GTP SNU
Send N PDU Number
Receive N PDU Number

not waste time (and trees) in discussing them again here. The shaded table entries correspond to the PDP context(s).

Twelve multislot classes are defined for a GPRS MS as summarized in Table 6.5 [217]. One radio frame consists of eight time-slots. Downlink Slots specifies the maximum number of downlink slots that the MS is capable of receiving. Uplink Slots defines the maximum number of uplink slots that the MS is able to use for transmission. Active Slots indicates the maximum number of slots that the MS can use for simultaneous uplink transmission and downlink reception. The higher the multislot class, the more capable the MS and the higher the peak rates the MS can support.

A GPRS MS may belong to Class A, Class B, or Class C; the class defines the MS's technology capabilities. Class A mobile devices can connect to both GPRS and GSM services simultaneously. Class B mobile phones are capable of both GPRS and GSM services; they support only one service at a time, but the mobile device can automatically switch between the two services. For example, the network and the MS suspend the GPRS services during voice calls or an SMS session and *automatically* resume them once the voice call or the SMS session has terminated. Class B devices are commonly used in commercial deployments. Class C mobile devices, like Class B devices, support either GPRS services or GSM services at a given time instant; in contrast to Class B devices, these devices require user intervention to switch *manually* between the services.

Table 6.5: Classes for GPRS

Multislot Class	Downlink Slots	Uplink Slots	Active Slots
1	1	1	2
2	2	1	3
3	2	2	3
4	3	1	4
5	2	2	4
6	3	2	4
7	3	3	4
8	4	1	5
9	3	2	5
10	4	2	5
11	4	3	5
12	4	4	5

6.2.3 Interfaces of the Packet-Domain Core Network

We just looked at the nodes of the packet-domain core network. These nodes communicate using interfaces. Let's briefly discuss the roles of certain interfaces to support packet-switched services [214].

The Gb interface exists between the BSS and the SGSN and enables their exchange of signaling messages and user traffic. A given physical resource, such as a T1 or E1 link on the Gb interface, is shared among multiple users.

Do You Recall... Where is the GSM A interface? As always, the footnote[178] has the answer.

The main difference between the GSM A interface and the Gb interface is that the A interface resources stay with the user throughout the call regardless of the activity (e.g., active speech interval or silence interval), and resources on the Gb interface are used on a need basis. The Gb interface resources are utilized for the user only during periods of data activity; these resources become available to other users in the case of data inactivity for a given user. In other words, the A interface follows a circuit-switched approach, and the Gb interface follows a packet-switched approach. Another characteristic of this interface is that GSM and GPRS signaling and user data are sent in the same user plane; no dedicated physical resources are needed for signaling. The physical layer for the Gb interface is undefined by the standard, but T1 or E1 links are widely used in practice. The Gb interface link layer is based on Frame Relay [218]. Later revisions of the standard also allow an IP-based Gb interface.

[178]The A interface exists between the BSS and the MSC and supports circuit-switched services.

The link layer involves establishment of Frame Relay permanent virtual circuits (PVCs) between the SGSN and the BSS. Multiplexing of LLC (Logical Link Control) PDUs from different users is allowed on such PVCs. Furthermore, one or more Frame Relay PVCs are established between an SGSN and a BSS to transport BSSGP (BSS GPRS Protocol) PDUs. BSSGP's main task is to provide information related to the radio interface, QoS, and routing so that the user data can flow between the BSS and the SGSN. BSSGP provides an interface between LLC frames and RLC/MAC (Radio Link Control/Medium Access Control) blocks. BSSGP also operates node management control functions for the BSS and the SGSN. Furthermore, BSSGP handles the paging requests going from the SGSN to the BSS.

The SGSN and GGSN in a given PLMN are connected via the Gn interface. However, when the SGSN and GGSN are in different PLMNs, they communicate using the Gp interface. The Gp interface is just the Gn interface plus support for security functions, and such support provides a secure inter-communication to keep the bad guys at bay! The specific security mechanism implemented depends upon the mutual agreement between the service operators that own the PLMNs. The main function of the SGSN-GGSN interface is to facilitate smooth packet flow for an MS.

The SGSN and the HLR are connected via the Gr interface. The Gr interface, based on the Mobile Application Part (MAP) and Signaling System 7 (SS7), enables the SGSN to keep the HLR informed of the latest location of the MS. Furthermore, the SGSN can retrieve the subscription profile and authentication-related parameters from the HLR.

The interface between the SGSN and the MSC/VLR is the Gs interface. This optional interface enables coordination of paging between the circuit-switched services and GPRS services for a hybrid GSM/GPRS MS. The Gs interface is not commonly used in commercial deployments.

The SMS GMSC and SMS IWMSC can be connected to the SGSN via the Gd interface to enable the SGSN to support SMS. Two SGSNs in a PLMN connect to each other via the Gn interface. This interface facilitates inter-SGSN seamless mobility by supporting handover for the MSs crossing the SGSN boundaries.

GPRS Roaming Exchange (GRX) enables secured roaming among service operators. For example, a GRX service provider provides a hub to the GPRS service operators so that two GPRS service operators can quickly connect their cellular networks (i.e., PLMNs) via a single connection to the GRX network. The beauty of such a GRX network is that it obviates the need for a GPRS service operator to individually connect many other GPRS service operators. In addition to offering seamless roaming, the GRX network also provides inter-PLMN security so that the PLMNs stay clear of the rogue elements in society! If we were to use the public Internet to interconnect two GPRS PLMNs, just like a computer virus making its way to our desktop and laptop computers, our GPRS devices would be vulnerable to such mischief.

6.2.4 End-to-End View of the Data Transfer

Figure 6.6[179] shows various layers of the protocol stacks implemented at different entities to support an end-to-end packet-switched data transfer between the MS and an external entity, such as a web server.

Assume that the MS is downloading a web page from a web server. Since the MS is involved in the web-browsing, the application layer protocol is HTTP (Hypertext Transfer Protocol) and the transport layer protocol is TCP (Transmission Control Protocol) from the perspective of the five-layer TCP/IP stack. However, according to the GSM protocol stack depicted in the standard [214], both are considered part of the application layer. Since the application layer and the transport layer are end-to-end protocols, these layers are shown as peer protocols here. The web page is put together by HTTP and passed down to the TCP layer and then to the IP layer. The IP packet containing the web page travels through the routers in the PDN, which is the Internet in our example, and reaches the GGSN. The GGSN uses the help of the GTP-U protocol and UDP to carry the packet to the SGSN via an IP backbone network.

The relay function in the SGSN relays PDP PDUs between the Gb and Gn interfaces. The SGSN has two protocols that communicate with their peer protocols in the MS: the SNDCP and the LLC. The SNDCP maps the network layer (e.g., IP) characteristics onto the characteristics of the underlying network, which is the

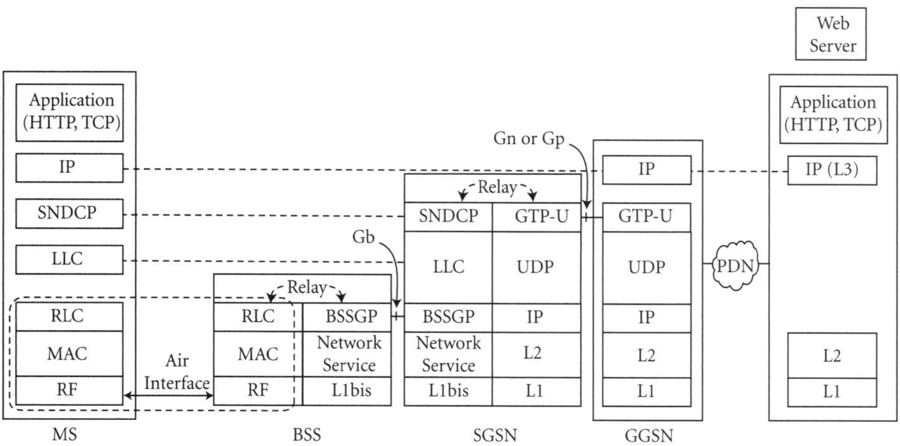

Figure 6.6: Protocol Stacks for End-to-End Packet-Switched Data Transfer

[179] L1 and L2 in Figure 6.6 represent layer 1 and layer 2. UDP is the user datagram protocol.
SOURCE: Figure 6.6 is adapted from Figure 4 in the standard [214] and is used here with permission.

GPRS or EDGE radio network in our example. Examples of functions the SNDCP performs include compression or decompression of user data (i.e., user's IP packets), segmentation of an IP packet into LLC frames, and reassembly of LLC frames into an IP packet. (See the standard [215] for details on SNDCP.) The LLC layer provides a reliable and secure logical link between the MS and the SGSN via ciphering. The LLC is independent of the lower-layer radio interface protocols so that new non-GPRS (e.g., EDGE) protocols can be supported in the future with minimal changes in the core network. (See the standard [219] for LLC details.) The BSSGP carries information about routing and QoS information between the BSS and the SGSN [220].

The Network Service layer transports the BSSGP PDUs and is based on the Frame Relay connection between the BSS and the SGSN [218]. The L1bis is the physical layer carrying information between the BSS and the SGSN and may use T1s or E1s. The relay function in the BSS relays LLC PDUs between the Um (i.e., the interface between the MS and the BSS) and Gb interfaces. The RLC/MAC layer performs two functions [221]: the RLC function increases the reliability of the radio link, and the MAC function controls access to the radio channel and performs mapping between the LLC frames and the GPRS or EDGE physical channel. Finally, the physical (PHY) layer between the MS and the BSS is referred to as GSM RF, which carries information over the air. Section 6.3 focuses on the communications between the MS and the BSS by discussing the MAC/RLC and PHY layers.

6.3 MS–BSS AIR-INTERFACE COMMUNICATIONS

Um introduces enhancements to the basic GSM air interface. A high-level view of the protocol stack for the air interface is given first, followed by details of the protocol layers, such as the physical layer and the RLC/MAC layer.

Let's focus on the radio interface between the MS and the BTS shown in Figure 6.7. The user traffic from the upper layers, such as the LLC, arrives at the RLC/MAC

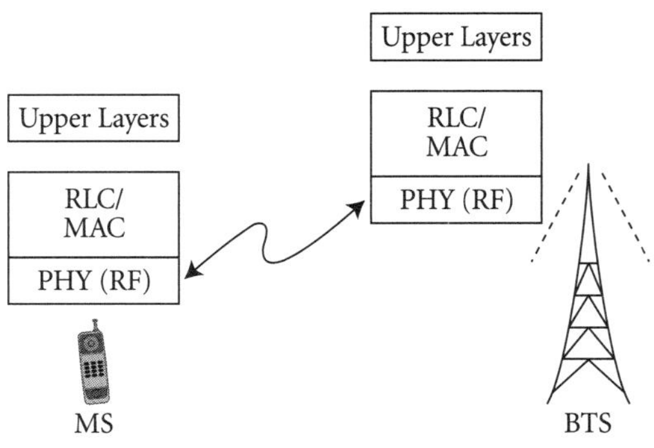

Figure 6.7: Air-Interface Protocols Between the MS and the BSS

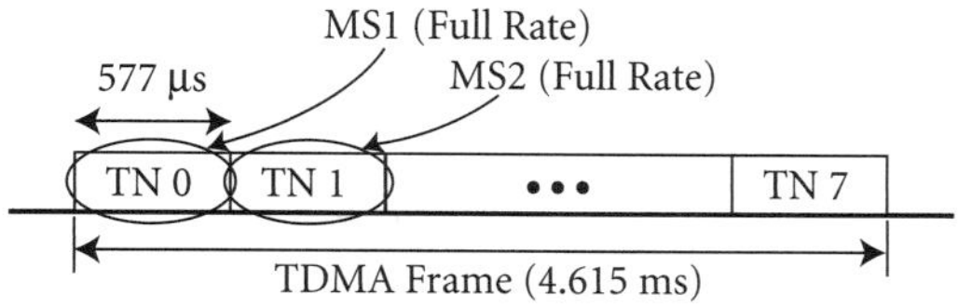

Figure 6.8: Multiple Access Using a 2G GSM Frame Structure

layer. The RLC function adds a sequence number to the traffic packets to allow the receiver to identify missing packets. If a packet is missing, the receiver requests retransmission so that the packet can then be retrieved and with no error if we're lucky! The MAC function controls access to the physical layer and takes care of the mapping between the LLC frames and the physical layer resources. A given ARFCN is utilized based on the radio frame structure shown in Figure 6.8 where the time-slot number (TN) varies from 0 to 7.

Do You Recall... What are the physical layer resources in the GSM system? See the footnote[180] for the answer.

In the basic GSM system, two MSs with full rates (MS1 and MS2 in Figure 6.8) use one time-slot each on a given ARFCN. GPRS and EDGE preserve such a frame structure and utilize the time-slots for efficient data transfer. Section 6.3.1 discusses details of the physical or RF layer, and Section 6.3.2 describes the principles of the RLC/MAC layer.

6.3.1 Physical Layer of GPRS and EDGE

The physical layer or layer 1 (also referred to as L1) helps carry information over the radio interface. Figure 6.9 illustrates the basic physical layer processing for GPRS and EDGE. The MS and BTS carry out such processing in every cell individually. The physical layer receives a data stream from layer 2 for transmission of such a stream over the air. Figure 6.9 shows a simplified view of how the physical layer processes information and eventually transmits it over the air by the physical layer [222], which is quite similar to that for the basic GSM-based system (see Chapter 5). Due to the similarities between GSM L1 processing and GPRS and EDGE L1 processing, only the new features GPRS and EDGE introduce are discussed in more detail; other basic GSM features are only briefly summarized.

Different types of radio channels undergo somewhat different processing, but generic processing still follows Figure 6.9. See the standard [222] and Section 6.4 for details on the different types of channels and the timing structure for GPRS and

[180]Since GSM is a hybrid TDMA and FDMA system, the radio resources are time-slots and RF channels.

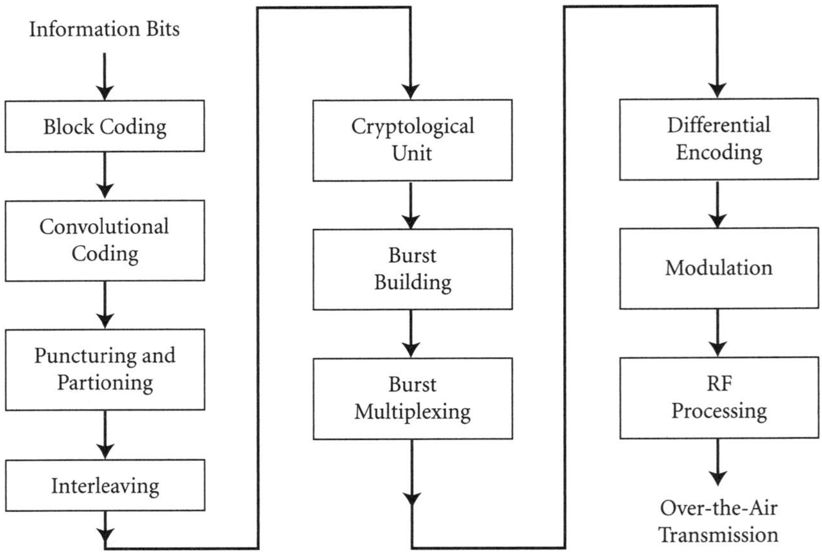

Figure 6.9: Simplified L1 Processing at the GPRS and EDGE Transmitter

EDGE. The main channel of interest to us for high-data-rate packet data services for GPRS and EDGE is the logical channel called PDTCH (Packet Data Traffic Channel), which is mapped onto the physical channel that is a combination of one or more time-slots and an ARFCN. The PDTCH contains higher-rate data packets that flow across the MS, the BSS, the packet-domain core network, and a suitable server (e.g., an email or web server).

The upper-layer data, such as RLC/MAC blocks, represent information bits. We add redundancy to these bits using convolutional coding. The coded bits or code symbols undergo puncturing and partitioning to achieve a suitable data rate at the physical layer. The selected coded bits are interleaved to facilitate the decoding operation at the receiver when the environment causes consecutive errors in code symbols. The cryptological unit then encrypts the interleaved bits or symbols.[181] The encrypted bits are then distributed among the bursts, and such bursts are multiplexed. Differential encoding[182] and modulation are carried out next. The modulation symbols now represent the information bits in some fashion. The RF processing helps transmit the modulation symbols over the air.

Let's look at Figure 6.9 in more detail. The overall coding scheme involves block coding and channel coding. The block coding facilitates error detection at the receiver, and the channel coding adds redundancy at the transmitter so the receiver

[181]Encryption mentioned here occurs at the physical layer for GSM. In GPRS and EDGE, encryption occurs at the LLC layer rather than the physical layer.

[182]Here differential encoding is applicable to the GMSK modulation.

can reliably estimate the original information bits even when the radio environment is quite hostile.

Do You Recall... Let's ensure that you are not using this book as a sleeping aid! Do you recall the types of FEC coding techniques GSM supports? The footnote[183] has the answer.

6.3.1.1 Block Coding, Convolutional Coding, and Puncturing Several data structures called *Radio Blocks* are defined for data transfer and control message transfer. The Radio Blocks arrive at the physical layer from the upper RLC/MAC layer and contain information bits. The physical layer modifies these Radio Blocks, and they are eventually transmitted over the air in a sequence of *four normal bursts*. The Radio Block structures for data transfer are different for GPRS and EGPRS, but they are the same for control message transfer for both GPRS and EDGE. See the standard [221] for additional details on Radio Blocks.

Figure 6.10 shows the Radio Block structure for data transfer in GPRS. The Radio Block arriving at the physical layer from the RLC layer consists of a MAC header, an RLC header, and one RLC data block or RLC PDU. The variable-length RLC header contains different control fields in the uplink and downlink directions. The fixed-length MAC header includes different control fields in the uplink and the downlink. The RLC Data field contains octets from one or more upper-layer (i.e., Logical Link Control) PDUs. Details of the RLC/MAC for GPRS are discussed in Section 6.3.2.

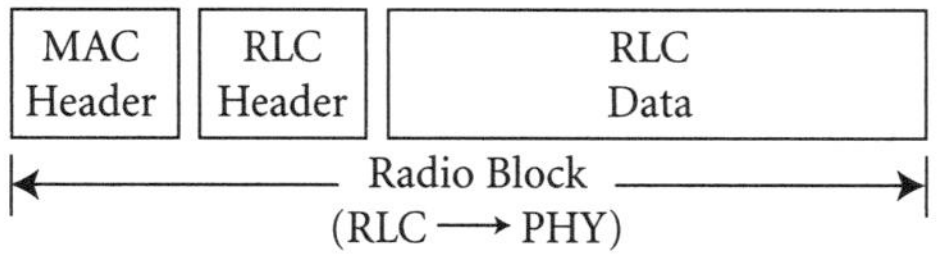

Figure 6.10: Structure of the Radio Block for the Data Transfer in GPRS

Figure 6.11 shows the Radio Block structure for data transfer with EDGE. As with the GPRS Radio Block for data transfer, the EDGE Radio Block for data transfer is

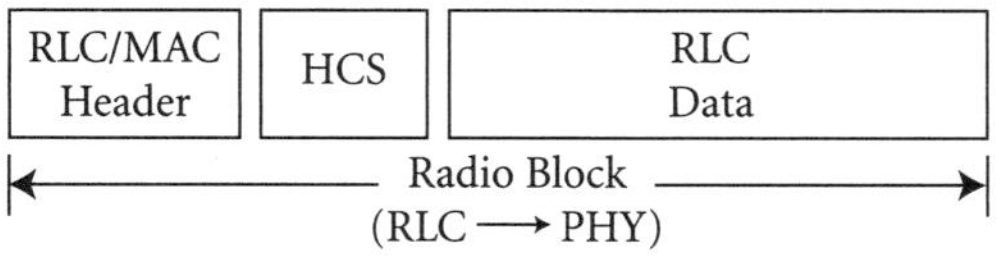

Figure 6.11: Structure of the Radio Block for the Data Transfer in EDGE

[183]GSM supports the cyclic code and the convolutional code.

carried by four normal bursts. This Radio Block consists of an RLC/MAC header, HCS (Header Check Sequence), and one or two RLC data blocks. The variable-length RLC/MAC header contains different control fields in the uplink and the downlink. The HCS enables error detection for the header. The RLC data field includes octets from one or more LLC PDUs. Details of the RLC/MAC for EDGE are discussed in Section 6.3.2.

The Radio Blocks for data transfer are processed by blocks such as block coding, channel coding, and puncturing or repetition as illustrated in Figure 6.9. Processing by such blocks is specified by one of the thirteen coding schemes, with four schemes for the GPRS radio interface and nine schemes for the EDGE radio interface [223]. In contrast to the basic GSM system, which typically supports the convolutional code with rate 1/2, GPRS and EDGE support a wide range of coding rates.

6.3.1.2 GPRS Coding Schemes: A Brief Overview Four coding schemes (CSs), CS1 to CS4, are available for the GPRS Radio Blocks carrying the RLC data blocks. CS1 is typically used for the GPRS Radio Blocks carrying the RLC/MAC control message blocks. Different CSs help achieve different data rates.

Do You Recall... Do you remember the name of the code used in the BCS sequence?[184]

The *first step* for CS1 to CS4 is the addition of a Block Check Sequence (BCS) for error detection at the receiver. The *second step* differs for different CSs.

- For CS1, the second step involves the addition of four tail bits and convolutional coding that is punctured to achieve the target coding rate.
- For CS2 and CS3, the second step involves precoding of the USF (Uplink State Flag);[185] the addition of four tail bits; and convolutional coding that is punctured to achieve the desired coding rate.
- For CS4, the second step involves precoding of the USF and the addition of four tail bits.

Table 6.6 summarizes the characteristics of the GPRS CSs. Use of CS3 and CS4 is uncommon in practice.

Let's consider the case of CS1. The USF field takes just three bits because it goes through no precoding. The BCS field occupies forty bits. The size of the Radio Block coming from the RLC layer is 181 bits. The number of tail bits is four. The total number of input bits for the convolutional coding is 3 USF bits $+$ 40 BCS bits $+$ 181 Radio Block bits $+$ 4 tail bits $=$ 228 bits. Since the base convolutional coding rate is $1/2$, the output of the convolutional coding is $228 * 2 = 456$ coded bits. Since

[184]It is called *fire code*.

[185]We will discuss USF in Sections 6.3.2 and 6.4.3.1.

Table 6.6: Characteristics of the GPRS Coding Schemes

Coding Scheme	USF	Precoded USF	BCS	Size of the Radio Block Without USF and BCS	Tail Bits	Coded Bits	Punctured Bits	Data Rate (kbps)
CS1	3	N/A[†]	40	181	4	456	0	9.05
CS2	3	6	16	268	4	588	132	13.4
CS3	3	6	16	312	4	676	220	15.6
CS4	3	12	16	428	N/A	456	N/A	21.4

[†] N/A represents *Not Applicable*.

no puncturing is performed, the base coding rate and the effective coding rate are both $1/2$. The data rate can be approximated as 181 bits /20 ms $= 9.05$ kbps. Note that this data rate calculation does include overhead from RLC and other upper layers.

As another example, let's look at CS2. The total number of input bits for the convolutional coding is 6 USF bits + 16 BCS bits + 268 Radio Block bits + 4 tail bits = 294 bits. Since the base convolutional coding rate is $1/2$, the output of the convolutional coding is $294 * 2 = 588$ coded bits. The total number of coded bits that can be transmitted is fixed at 456, and the number of bits punctured is $588 - 456 = 132$. The effective coding rate can be calculated as approximately $268/456 = 1/1.7 = 0.6 \approx 2/3$. The approximate data rate is 268 bits /20 ms $= 13.4$ kbps. The data rates mentioned in Table 6.6 assume that one time-slot on an ARFCN is allocated to the user.

6.3.1.3 A Brief Overview of EDGE Modulation and Coding Schemes

GPRS continues to use the Gaussian Minimum Shift Keying (GMSK) modulation just like GSM, but EDGE introduces an additional modulation scheme, 8-PSK (8-ary Phase Shift Keying).

Do You Recall... Let's test your memory. How many bits does an 8-PSK modulation symbol represent? You know the drill: the footnote[186] contains the answer!

One GMSK modulation symbol essentially enables retrieval of one information bit (a coded bit to be more accurate). If we do nothing else and just use 8-PSK instead of GMSK, our data rate can roughly triple; three times as much information can be transferred via 8-PSK than via GMSK during a given observation interval.

Something to Think About... If 8-PSK allows the data rate to be about three times higher than GMSK, why not use 8-PSK all the time? Though 8-PSK enables the data rate to be higher, it also increases the probability of error under given radio channel conditions. As you may know from Chapter 3, the radio environment is quite dynamic. The sayings "Only change is constant!" and "Change is inevitable except from a vending machine!" apply to the radio environment. The changing radio environment may cause signal fades, and, if we use 8-PSK, the receiver is more likely to make an error in estimating the modulation symbol that was transmitted.

Recall from Chapter 2 that the average distance between two modulation symbols in the signal constellation diagram is larger for Binary Phase Shift Keying (BPSK) (which is like GMSK) and smaller for 8-PSK. The shorter the distance, the higher the error probability. In practice, 8-PSK is used when the radio channel conditions are good so that a higher data rate can be achieved. A low-order modulation scheme such as GMSK or BPSK is used when the channel conditions are poor so that the

[186]One 8-PSK modulation symbol represents three (information) bits.

signal can be reliably detected. The mechanism of changing the modulation scheme in response to the changing channel conditions is called *adaptive modulation.*

The EDGE Radio Blocks carrying the RLC data blocks can use any of the nine modulation and coding schemes (MCSs): MCS1 to MCS9. However, CS1, defined for GPRS, is typically used for the EDGE Radio Blocks that carry RLC/MAC control blocks. GPRS uses the base convolutional coding rate of $1/2$ to derive other effective code rates, and EDGE utilizes the base convolutional coding rate of $1/3$. The MCSs are distributed among three families: A, B, and C.[187] Each family utilizes a different basic payload unit or size [224]. Families A and B allow one, two, or four payload units to be transmitted in a burst. Family C allows transmission of one or two payload units. These families enable different data rates and different effective coding rates. One RLC block is adequate to carry one or two payload units, but two RLC blocks are required to carry four payload units.

The *first step* in the EDGE coding procedure is the addition of the BCS. We just covered BCS for GPRS. The *second step* involves the addition of six tail bits for resetting the decoder at the receiver at the conclusion of the block decoding process; rate $1/3$ convolutional coding for redundancy; and suitable puncturing to achieve the target effective coding rate. The USF field is part of the RLC/MAC header and is represented by three bits. These three bits are encoded to twelve symbols (just like GPRS) in GMSK modulation and thirty-six symbols in 8-PSK modulation. Table 6.7 summarizes characteristics of the nine MCSs for EDGE [222, 225].

Do You Recall. . . What is the BCS's use? (And by BCS, we are not referring to U.S. collegiate sports' Bowl Championship Series!) The footnote[188] provides the answer!

Consider MCS1. GMSK modulation is used. The total number of input bits for the convolutional coding of the data part (*not* the header part) is 176 raw RLC information or data bits + 1 Extension Bit + 1 Final Block Indicator Bit + 12 BCS or parity bits + 6 tail bits = 196 bits. See the standard [221] for details on the Extension Bit and Final Block Indicator Bit. Since the base convolutional coding rate is $1/3$, the output of the convolutional coding is $196 * 3 = 588$ coded bits. After puncturing is performed, the total number of coded bits is reduced to 372. The effective coding rate then is $196/372 = 0.53$. The data rate can be approximated as 176 bits$/20$ ms $= 8.9$ kbps. See the standard [223] for MCS details.

[187] Note that these three families have no relationship whatsoever with the three classes of GPRS MSs that we mentioned earlier in this chapter.

[188] The BCS facilitates error detection at the receiver.

Table 6.7: Characteristics of the EDGE Modulation & Coding Schemes

MCS Scheme	Effective Coding Rate	Modulation Scheme	Family	Raw RLC Payload (Bits)	BCS	Tail Bits	Data Rate (kbps)
MCS1	0.53	GMSK	C	176	12	6	8.8
MCS2	0.66	GMSK	B	224	12	6	11.2
MCS3	0.85	GMSK	A	296	12	6	14.8
MCS4	1.0	GMSK	C	352	12	6	17.6
MCS5	0.37	8-PSK	B	448	12	6	22.4
MCS6	0.49	8-PSK	A	592	12	6	29.6
MCS7	0.76	8-PSK	B	$2 \times 448 = 896$	$2 \times 12 = 24$	$2 \times 6 = 12$	44.8
MCS8	0.92	8-PSK	A	$2 \times 554 = 1088$	$2 \times 12 = 24$	$2 \times 6 = 12$	54.4
MCS9	1.0	8-PSK	A	$2 \times 592 = 1184$	$2 \times 12 = 24$	$2 \times 6 = 12$	59.2

Note that HCS $= 8$ for all MCS schemes.

The $1/3$ convolutional coding rate uses three generator polynomials:

$$G4 = 1 + D^2 + D^3 + D^5 + D^6$$
$$G7 = 1 + D + D^2 + D^3 + D^6$$
$$G5 = 1 + D + D^4 + D^6.$$

Three output streams of the convolutional coder are specified as

$$C(3k) = u(k) + u(k-2) + u(k-3) + u(k-5) + u(k-6)$$
$$C(3k+1) = u(k) + u(k-1) + u(k-2) + u(k-3) + u(k-6)$$
$$C(3k+2) = u(k) + u(k-1) + u(k-4) + u(k-6).$$

Note that $u(k) = 0$ for $k < 0$.

6.3.1.4 Other Basic GSM Features GPRS and EDGE share other features with GSM as well.

Interleaving The USF field, data bits, and header bits are considered a single entity for the purpose of interleaving. Block rectangular interleaving is used [223].

Cryptological Unit Processing The burst is encrypted based on the agreement between the MS and the GSM network. As mentioned earlier, encryption for GPRS and EDGE is done at the higher LLC layer. The GSM Association, not the GPRS or EDGE standard, defines the specifics of the encryption mechanisms.

Burst Building and Burst Multiplexing The coded and interleaved data bits are distributed among the four normal bursts, spanning a nominal period of 20 ms.

Differential Encoding and Modulation Differential encoding is used as with GSM for the GMSK modulation. As mentioned earlier, GPRS uses the GMSK modulation just like GSM. However, EDGE can use GMSK or 8-PSK. The modulation symbol rate for GPRS and EDGE is identical to that of GSM: 270.83 ksps.

RF Processing The actual RF signal being transmitted from the antenna, $x(t)$, has the form defined by Eq. (5.5) in Chapter 5 when GMSK modulation is used. When 8-PSK modulation is used, $x(t')$ is [226]

$$x(t') = \sqrt{\frac{2E_s}{T}}\,\mathrm{Re}\left[y(t') \cdot e^{j(2\pi f_0 t' + \varphi_0)}\right] \tag{6.1}$$

where E_s is the energy per modulating symbol, T is the modulating bit period, f_0 is the center carrier frequency (e.g., 900 MHz), $\varphi(t)$ is a time-varying phase, $y(t')$ is a function of pulse-shaped rotated 8-PSK modulation symbols, and φ_0 is a *random* phase that is *constant* during one burst. To perform blind detection of the modulation scheme, the EDGE MS exploits knowledge about the structures of GMSK and 8-PSK modulation symbols.

The physical layer in GPRS and EDGE also performs functions such as power control, synchronization, measurement for handover, and cell selection and reselection, similar to the physical layer in GSM [222].

6.3.2 RLC/MAC Layer and Its Functions

As shown in Figure 6.12, the MAC/RLC layer is situated between the LLC layer and the physical layer in the radio interface protocol stack. See the standard [224, 225] for details on the RLC/MAC layer.

From the downlink transmission perspective, SNDCP (Subnetwork Dependent Convergence Protocol) sits on top of the LLC layer and segments an IP packet into one or more LLC frames. An LLC frame consists of a frame header, an information field containing user traffic, and a frame check sequence (FCS) for error detection at the LLC layer. The RLC/MAC layer segments the LLC frames into RLC data blocks. Each RLC data block consists of a block header to facilitate retransmission of erroneous RLC blocks (explained later in this section); an information field housing the LLC frame contents; and a BCS for error detection at the RLC layer. The contents of the entire LLC frame are distributed among the information fields of the RLC blocks, with the first RLC block called the *primary block*. A given RLC block is usually transmitted by the physical layer using four consecutive normal bursts; exceptions occur with EDGE MCS-7, MCS-8, and MCS-9 where two RLC blocks are transmitted using four normal bursts.

The RLC/MAC layer implements the *MAC functions* and the *RLC functions*. The *MAC function* controls access to the physical layer resources, such as radio channels, to allow multiple MSs to share these resources. In the uplink, when MSs with no dedicated radio resources attempt to access the network, the MAC function arbitrates among multiple MSs by defining collision avoidance, detection, and recovery procedures. The MAC function may allocate multiple physical channels in parallel to a given MS, enabling such an MS to experience higher data rates, and the *RLC function* allows selective retransmission of erroneous RLC data blocks to reduce the

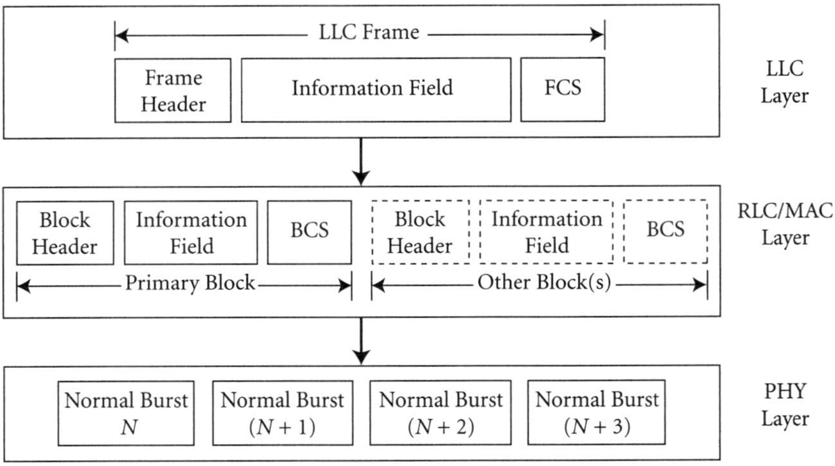

Figure 6.12: RLC/MAC Layer in the MS–BSS Protocol Stack

effective error rate the receiver experiences. Though radio channels related to packet data transmission in GPRS and EDGE are discussed in detail in Section 6.4, we'll also take an ultra-quick look at two of them here.

The MAC function at the network side controls access to the physical layer resources by multiplexing user data traffic and control signaling on both the uplink and downlink. Specifically, a downlink scheduling algorithm controls the downlink resource utilization for an ongoing packet data transfer, and an uplink scheduling algorithm that allocates physical layer resources in response to service requests from MSs controls the uplink resource utilization for an ongoing packet data transfer. Additionally, the uplink resource utilization for MS-originated channel access is governed by collision avoidance, detection, and recovery mechanisms that the MAC function defines.

Before diving into the details of how the MAC manages resources, let's discuss a key concept in MAC management: a *Temporary Block Flow* (TBF) is a physical connection that the two peer Radio Resource (RR) entities at the MS and the network use to support the unidirectional transfer of the upper-layer (i.e., LLC) PDUs. Hence, if the MS is involved in an application that involves bidirectional data transfer, the MS is allocated two TBFs, one for the uplink and one for the downlink. The TBF is allocated radio resources on one or more packet data channels (PDCHs) and helps carry the RLC/MAC data blocks that contain LLC PDUs.

As suggested by the name, the TBF is indeed temporary; it is maintained for only the duration of the data transfer and is released during the period of inactivity. The network assigns a Temporary Flow Identity (TFI) to a TBF. Since a TBF is unidirectional, the same TFI could be used in the uplink and the downlink for two unidirectional TBFs. Once the TFI is associated with a TBF in a resource assignment message prior to data transfer, the TFI is included in every RLC header of an RLC block related to a given TBF.

The physical layer resource that the MAC function controls is a PDCH. The PDCH can be dedicated to just one MS or shared among multiple MSs. In contrast to the GSM radio interface, which supports only symmetric (i.e., same data rate) dedicated channels, the GPRS radio interface allows *asymmetric* and *independent* uplink PDCH and downlink PDCH. As an example, more downlink slots and fewer uplink slots on a given ARFCN could be allocated to an MS. Furthermore, an MS can be allocated multiple PDCHs.

Figure 6.13 summarizes the medium access modes that the MAC functions support. These modes include *dynamic allocation*, *fixed allocation*, *extended dynamic allocation*, and *exclusive allocation*.

Dynamic and extended dynamic medium access modes utilize the concept of the USF, which enables multiplexing of Radio blocks from different MSs on a given PDCH as well as control of the Packet Random Access Channel (PRACH). The USF concept also allows the same PDCH to be dynamically shared among GPRS users and EDGE users. As seen from Table 6.6, the USF consists of three bits and is located at the beginning of each *downlink Radio Block*. Three USF bits enable the coding of eight different USF states that facilitate multiplexing of the uplink traffic.

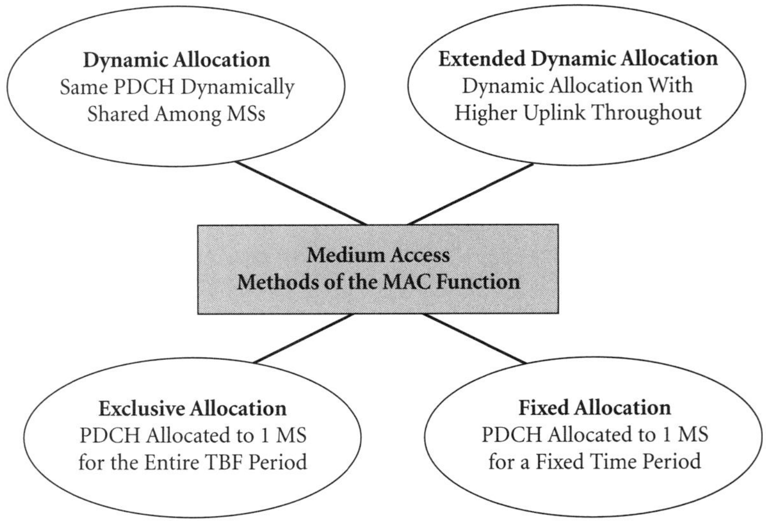

Figure 6.13: Medium Access Modes of the MAC Functions

The USF points either to the very next uplink Radio Block or the sequence of four uplink Radio Blocks starting with the very next uplink Radio Block. The control of the PRACH utilization is achieved by using a specific USF value on the Packet Common Control Channel (PCCCH); other USF values are used to *reserve* the uplink resources for the MSs. When a PDCH carries no PCCCH, all eight USF values are used to reserve the uplink resources for different MSs. The extended dynamic allocation medium access method extends the basic dynamic allocation medium access method to achieve higher uplink throughput by allowing simultaneous transmission on multiple PDCHs.

Fixed allocation refers to the scenario in which the uplink PDCH is reserved *for only one MS during a certain time period*. Furthermore, this allocation mode allows multiplexing of GPRS MSs and EDGE MSs on the same uplink PDCH.

Exclusive allocation refers to the method in which the uplink PDCH is reserved for *only one MS throughout the life of the Temporary Block Flow*. Thus, the MS can use all the uplink blocks of the PDCH in this allocation method.

GPRS and EDGE support the dynamic allocation and fixed allocation medium access modes. MSs using DTM (Dual Transfer Mode) support the exclusive allocation. The DTM allows both Dedicated Mode and Packet Transfer Mode, which enables concurrent voice and data calls.

Now that we've discussed the MAC functions, let's see how important RLC functions are for packet data services. Consider Figure 6.14 in which a user wants to use a wireless device to read news from a website. Assume no RLC function. The overall path followed by the packets includes the web server, the GGSN, the SGSN, the BSC, the BTS, and the MS. The physical layer for GSM circuit-switched

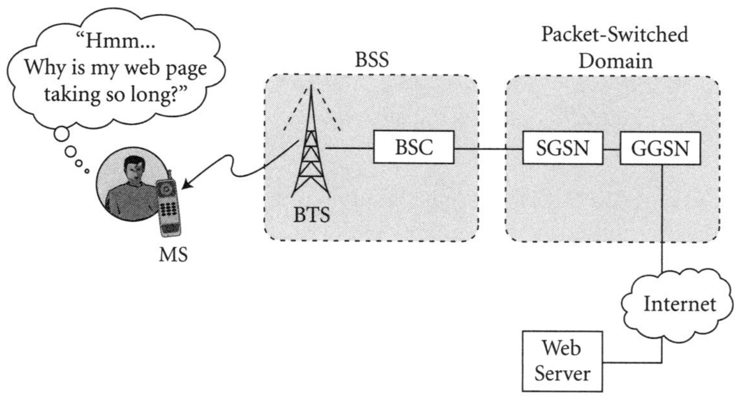

Figure 6.14: Motivation for the RLC Function

voice can tolerate a maximum of approximately 2% frame error rate, and the GPRS and EDGE RLC layer is designed to tolerate a maximum of 10% frame error rate although most GPRS and EDGE systems attempt to operate at around a 2% frame error rate. Consumers may note such higher target error rates for GPRS and EDGE at times when they are out of coverage for GSM voice but are still able to receive and send email via GPRS and EDGE.

When errors occur at the physical layer, the TCP layer initiates the congestion control mechanism, thinking that the packets have encountered network congestion. The TCP layer at the web server drastically slows down the rate at which it is injecting the packets for the MS into the Internet. The user experiences very low throughput because the packets experience delays on the order of seconds. In the extreme case, only part of the web page may be displayed. In summary, the TCP assumes an essentially error-free transmission, though it is normal for the physical layer to operate at a 2% error rate. Such conflicting behaviors of the TCP layer and the physical layer result in a poor user-perceived experience. That's where the RLC enters the scene. When an error occurs at the physical layer, the RLC function retransmits the lost RLC data blocks with delays on the order of hundreds of milliseconds. The TCP layer then encounters not a 2% error rate of the physical layer but perhaps a 0.1% error rate of the RLC/MAC layer. The RLC function is a key component of the overall radio network design that aims to offer packet data services with good QoS.

Due to the dynamic nature of the radio propagation environment and the interference environment, some errors occur at the physical layer. When the physical layer is operated at a 10% frame error rate, ten frames of a hundred consecutive physical layer frames would, in general, be in error. Furthermore, when a frame error occurs, one or more bits in such a frame may be in error.

A Question for You... Why do we not operate the physical layer at a 0.01% error rate? After all, the lower the error rate is, the better the user-perceived experience could be. The footnote[189] contains the answer!

The RLC function carries out segmentation of the LLC frames into the RLC data blocks at the transmitter and re-assembly of the RLC data blocks into the LLC frames at the receiver. The RLC function allows selective retransmission of erroneous RLC data blocks; such a process is often referred to as *backward error correction* because the corrective action is taken *after* an error has occurred. Recall that the forward error correction coding techniques, such as convolutional coding, add redundancy in anticipation of an error to enable the receiver to (we hope) retrieve the original information bits even in the presence of errors. The BCS is actually appended toward the end of the RLC block—not by the RLC/MAC layer but by the physical layer—to detect the presence of an error in the RLC data block.

The RLC function for EDGE performs one additional task to support the *Incremental Redundancy* mode in which soft values of the original erroneous RLC data block bits are combined with retransmissions of the RLC data block. Since the bit energy for the combined data block is larger than that for a single data block, the likelihood of decoding a given bit correctly is higher. The RLC/MAC function operates in an *unacknowledged mode* or an *acknowledged mode*.

A Word of Caution... Regarding the nomenclature in the rest of Section 6.3.2, the *sender* is attempting to send data traffic (i.e., IP packets in the form of LLC frames) and the *receiver* is attempting to retrieve the data traffic. However, both the sender and the receiver are capable of performing transmit and receive functions.

6.3.2.1 Acknowledged Mode for RLC/MAC Operation (GPRS)

The acknowledged RLC/MAC mode implements a selective ARQ mechanism. The RLC data blocks within one TBF are allocated numbers called Block Sequence Numbers (BSNs). Consider the sequence of events A through F shown in Figure 6.15.

- **Step A**. The sender, such as the MS or the network, transmits the RLC data blocks within a window that has a certain number of BSNs. The window size governs the maximum number of unacknowledged RLC data blocks that the sender can send. In our example, the BSNs for an existing buffer are numbered from 1 to 100, and the window size is twenty BSNs. Hence, the sender is able to send the RLC blocks with BSN ranging from one to twenty.

[189]The lower the target error rate is, the larger the required signal-to-interference ratio would be. Since the radio resources are limited, we need to minimize overall interference power relative to the signal power. This would essentially limit the number of users that can be supported at a very low target error rate. What service operator would settle for low revenue as a result of low capacity? In summary, a balancing act occurs between the QoS that users can live with and the capacity of the system.

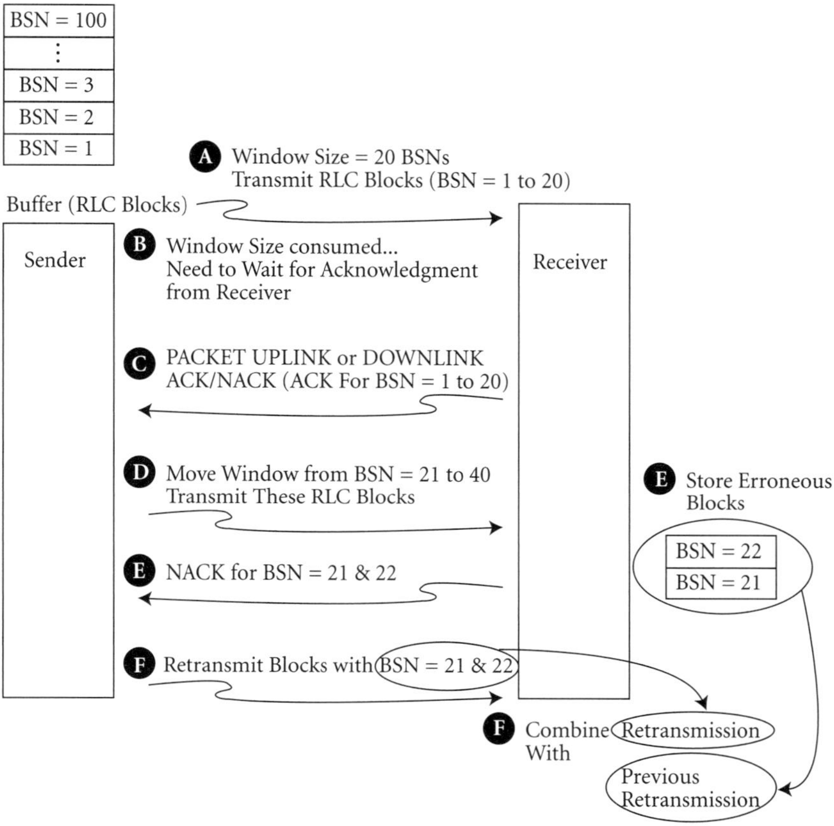

Figure 6.15: RLC ARQ Operation

- **Step B**. The sender has not yet received any feedback on the transmitted BSNs from the receiver. Hence, the sender *cannot* send an RLC block with the sequence number 21; it waits for an acknowledgment from the receiver for $BSN = 1$. The window size restricts the maximum number of RLC blocks that can remain unacknowledged by the receiver.

- **Step C**. The receiver gets some RLC blocks and responds with a message, such as PACKET UPLINK ACK/NACK or PACKET DOWNLINK ACK/NACK. This message acknowledges all the RLC data blocks correctly received up to the BSN specified in the message. In our example, the acknowledged BSN is twenty, meaning that the receiver has successfully retrieved all BSNs up to and including BSN $= 20$. When the BSS, as a receiver, receives erroneous RLC blocks, it allocates additional resources for the MS to retransmit. When the MS, as a receiver, sends the PACKET DOWNLINK ACK/NACK, it may initiate an uplink TBF. When the BSS, as a receiver, sends the PACKET UPLINK

ACK/NACK, the network could allocate uplink resources to the MS using the fixed allocation approach.

- **Step D**. The feedback from the receiver to the sender helps the sender move the beginning of the window. In the example under consideration, the sender moves the window by twenty BSNs and hence can send the RLC data blocks corresponding to the next set of BSNs such as BSN $= 21$ through BSN $= 40$.
- **Step E**. The packet containing BSN $= 21$ and BSN $= 22$ gets lost. Hence, the receiver informs the sender about the lost RLC data blocks. The receiver uses a bitmap to selectively request retransmission of only erroneously received RLC data blocks.
- **Step F**. The sender retransmits the erroneous RLC data blocks. The receiver combines the retransmitted RLC blocks with the first transmission of the RLC data blocks with BSN $= 21$ and BSN $= 22$. If the receiver can successfully retrieve these data blocks, it will send an ACK; otherwise, it sends a NACK.

6.3.2.2 Acknowledged Mode for RLC/MAC Operation for EDGE EDGE enhances the basic acknowledged mode operation of the GPRS. Furthermore, it allows an additional approach called *Type II H-ARQ* [often referred to as *Incremental Redundancy* (IR)]. The basic GPRS approach when used in EDGE is called a *selective Type I ARQ* mechanism. The overall EDGE Type I ARQ mechanism operates similar to the GPRS approach. As shown in Figure 6.16, the sender transmits blocks within a window, and the receiver sends a PACKET UPLINK ACK/NACK or PACKET DOWNLINK ACK/NACK. Though the MCS for the first transmission of an RLC block is selected to reflect the prevailing radio channel conditions, the same or a different MCS from the same family can be selected for retransmissions of the RLC block.

Figure 6.16 illustrates the operating principle of the EGPRS Type II H-ARQ mechanism. Packet P is the RLC data block containing the information bits. The convolutional coding with the base coding rate of $1/3$ yields a set of coded bits. In Step A, the receiver applies a puncturing scheme on the coded bits to construct packet P1, which represents packet P's original information bits. Since the received RLC data block is in error, the receiver sends a NACK in Step B. In response, as shown by Step C, the sender transmits packet P2 containing a set of coded bits, which are obtained by applying the puncturing scheme 2 on the initially derived coded bits. The receiver combines P2's retransmission with the first transmission of the RLC data block P1 and attempts to decode the original data block P in Step D.

The presence of a sequence number in the packet header informs the receiver of whether the transmission is a new packet transmission or a retransmission. The combining method that the receiver utilizes is implementation-specific. If the channel conditions remain poor and the sender continues to receive NACKs from the receiver, all the predefined puncturing schemes are utilized, and the sender then cycles through the set of puncturing schemes again. EDGE also supports another way of providing redundancy (i.e., coded bits) to the receiver. The first transmission could use MCS-5,

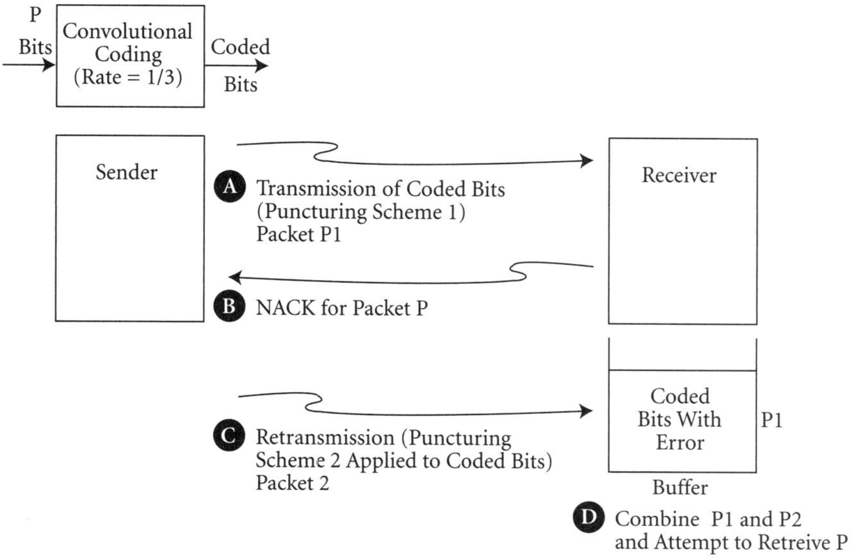

Figure 6.16: Operation of Type II H-ARQ for EDGE

and all retransmissions of the same RLC data block could then use MCS-7. The first transmission and all retransmissions could also use MCS-6 and MCS-9, respectively.

Ever Wondered... What is *hybrid* about H-ARQ? In a simple ARQ, *the same set of bits* is retransmitted. For example, if packet P consists of the information bits `1001`, then `1001` is transmitted for the first time and during subsequent retransmissions. The word *hybrid* is applicable because the specific contents being transmitted need not be (and usually are not) identical. For example, packet P with information bits `1001` may be represented by the coded bits `1010 1110 0001`. During the first transmission, we may send the first set of four bits `1010`, and the first *re*transmission may carry the second set of bits `1110`. Since *different coded bits*, as opposed to identical bits, are being sent over the air during different (re)transmissions, such an ARQ is referred to as a *hybrid* ARQ. In general, H-ARQ provides better error performance than simple ARQ.

6.3.2.3 Unacknowledged Mode of the RLC/MAC Operation As the name indicates, the receiver sends no ACK or NACK to the sender in response to the error-free or erroneous RLC blocks in this mode. Hence, no retransmissions of an RLC data block occur. The receiver may replace the lost bits of an RLC block with dummy information bits to preserve the total amount of user data. The same RLC/MAC signaling messages are used for the acknowledged mode and the unacknowledged mode; of course, the message contents are different. These messages carry control signaling, such as reports on the downlink channel quality for link adaptation and the timing advance correction for the uplink transmission.

In the unacknowledged mode, the sender transmits some radio blocks and polls (i.e., asks) the receiver to send an acknowledgment message. The receiver then responds with the PACKET UPLINK ACK/NACK or PACKET DOWNLINK ACK/NACK. In contrast to the acknowledged mode, such an acknowledgment message includes no modification in the current resource allocation. Similar to the acknowledged mode, the MS may initiate an uplink TBF using the PACKET DOWNLINK ACK/NACK, and the network may allocate uplink resources to the MS in the PACKET UPLINK ACK/NACK using the fixed allocation approach.

6.4 RADIO INTERFACE: FRAME STRUCTURE AND CHANNELS

In a GSM system, ARFCNs and time-slots are our precious radio resources since GSM utilizes FDMA and TDMA. GPRS and EDGE continue to preserve the structures of the ARFCN and the time-slot. The time structures of GSM are reused in GPRS and EDGE. Such structures are briefly reviewed first, and a discussion follows regarding the new radio channels that GPRS and EDGE introduce [222].

6.4.1 Review of Timing Structures

Recall from Chapter 5 that the time-slot is the shortest time unit that can carry information, i.e., a signaling message or user traffic such as speech. Furthermore, a frame consists of several time-slots.

Do You Recall... Do you remember how long the time-slot is and how many time-slots constitute a frame? As always, the footnote[190] has the answer.

See Figure 6.17 to refresh your memory regarding the time structures such as time-slot, frame, and multiframe. A time-slot carries a burst. GSM defines five types of bursts, with each burst type using the time-slot differently. See Chapter 5 for the structure of these bursts. GPRS and EDGE use the type of the burst called the normal burst to send packet data. The normal burst carries information on traffic and control channels and includes 116 bits (i.e., GMSK or 8-PSK modulation symbols) in two separate blocks.

[190]The length of the time-slot is $(3/5200)$ seconds or $577\ \mu s$. Eight time-slots constitute one TDMA frame. The length of the TDMA frame is $8 * 577 = 4.615$ ms.

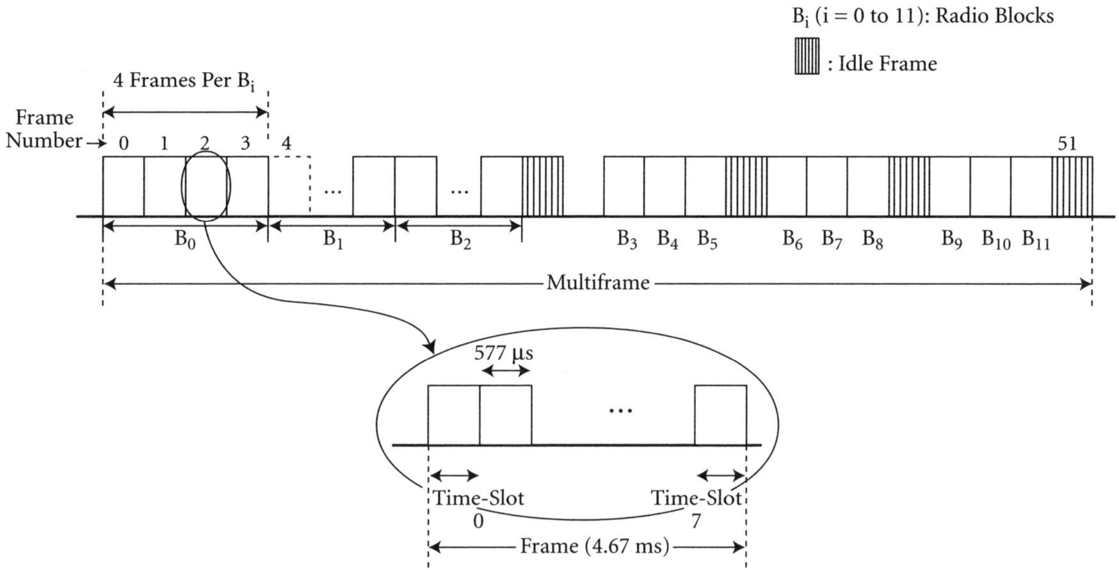

Figure 6.17: Time-Slot, Frame, and Multiframe Structures for GPRS and EDGE

6.4.2 New Logical Channels in GPRS and EDGE

As you may recall, the multiframe structure defines the mapping between the logical channel and the physical channel. GSM defines two types of multiframes: a twenty-six–frame multiframe to carry information such as broadcast information and a fifty-one–frame multiframe to carry information such as user traffic. As shown in Figure 6.17, GPRS and EDGE introduce a new type of multiframe, a fifty-two–frame multiframe, to support packet data. This new fifty-two–frame multiframe carries new logical channels to support packet data services. Figure 6.18 summarizes new logical channels of GPRS and EDGE that support the packet-switched services [227].

Do You Recall. . . What is the difference between the logical channel and the physical channel? The footnote[191] has the answer.

As in the case of GSM, two main types of logical channels exist: traffic and control. The new traffic channel is the PDTCH, which transmits user data in the packet-switched mode. One MS may be allocated multiple PDTCHs, and such a configuration is called a *multislot packet configuration*. A full-rate PDTCH using GMSK modulation transports information at an instantaneous peak rate of 22.8 kbps. When the full-rate PDTCH uses the 8-PSK modulation, the peak rate jumps to 69.6 kbps. The PDTCHs are unidirectional. In other words, only downlink PDTCH and no uplink PDTCH may exist. Such a scenario may apply to an application such as real-time video streaming.

Do You Recall. . . Why does the peak data rate increase by approximately a factor of 3 (i.e., from 22.8 kbps to 69.6 kbps)? The answer is hidden (actually, pretty obvious) in the preceding three or four sentences! Of course, the footnote[192] has the answer.

The types of control channels are broadcast control, common control, and dedicated control channels. The Packet Broadcast Control Channel (PBCCH) broadcasts parameters that the MS uses to access the network. The PBCCH also replicates the information being transmitted on the BCCH (Broadcast Control Channel) so the MSs in the GPRS-attached mode can monitor only the PBCCH. (In the GPRS-attached mode, the existence of the MS is known to the core network as a result of a previous GPRS attach.) Note that the MS may be a hybrid-mode device that supports circuit-switched services in addition to packet-switched services, and the BCCH in GSM

[191] The logical channel specifies what type of information is being carried, and the physical channel actually carries one or more logical channels over the air using the time-slot, frame, and frequency hopping structures.

[192] One GMSK modulation symbol corresponds to one bit, and one 8-PSK modulation symbol corresponds to three bits. Hence, during a given observation window, if we send information using 8-PSK, we can squeeze in three times as much information compared to using GMSK. That's why the peak rate approximately triples from 22.8 kbps to 69.6 kbps.

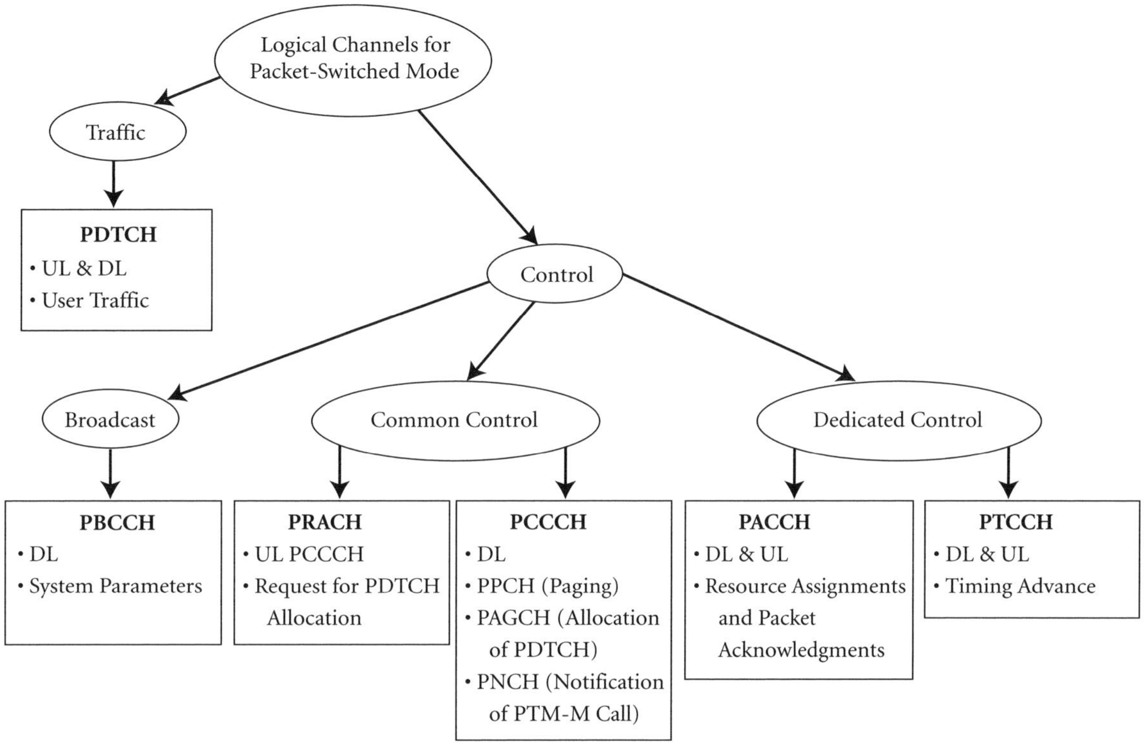

Figure 6.18: New Logical Channels for GPRS and EDGE

transmits information relevant to circuit-switched services. The BCCH indicates the existence of the PBCCH in the cell; if no PBCCH exists, the BCCH itself broadcasts information for the packet-switched operation.

Examples of the packet-operation–related parameters being sent on the PBCCH are BS_PBCCH_BLKS, specifying the number of blocks allocated to the PBCCH in the multiframe; BS_PCC_CHANS, indicating the number of physical channels carrying PCCCHs; and BS_PRACH_BLKS, providing the number of blocks reserved in a fixed manner for PRACH. Several common control channels exist: the Packet Paging Channel (PPCH), PRACH (Packet Random Access Channel), Packet Access Grant Channel (PAGCH), and Packet Notification Channel (PNCH).

The PPCH is a downlink channel used to page an MS. The PRACH is an uplink channel that an MS uses to request allocation of one or more PDTCHs for uplink and/or downlink. The PAGCH is a downlink channel the network uses to allocate one or more PDTCHs to the MS. Finally, the PNCH is a downlink channel used only by the network to inform the MS of a PTM-M (Packet-Transfer-Mode–Mobile-Terminated) call. If no PCCCH is allocated, the Common Control Channel (CCCH) carries the information related to the packet-switched operation. Furthermore, if a PCCCH is indeed allocated, it may transmit information related to the circuit-switched operation.

There are three important packet dedicated control channels: the Packet Associated Control Channel (PACCH) is bidirectional and is used for resource assignments and packet acknowledgments; the Packet Timing Advance Control Channel (PTCCH) is for the uplink (PTCCH/U); and the PTCCH is for the downlink (PTCCH/D). The MS uses the PTCCH/U to transmit random access bursts so that the network can estimate the timing advance for the MS. The network uses PTCCH/D to convey the timing advance updates for multiple MSs.

6.4.3 Association of Logical and Physical Channels

The PDCH is a physical channel that carries packet-switched logical channels. The PDCH is created using the fifty-two–frame multiframe. The full-rate PDCH consists of twelve blocks of four consecutive frames plus two idle frames and two frames carrying the PTCCH. The ordered list of blocks is defined as B0, B6, B3, B9, B1, B7, B4, B10, B2, B8, B5, and B11. The type of logical channel could change from one block to the next. In general, the block header specifies the type of the logical channel. Section 6.4.3.1 discusses the mapping of logical channels onto physical channel for the uplink, and Section 6.4.3.2 discusses the mapping of logical channels onto physical channel for the downlink.

6.4.3.1 Uplink Logical and Physical Channels Let's see which new logical and physical channels are defined for the uplink in GPRS and EDGE to support the packet data services efficiently.

Uplink PDTCH and PACCH. In the case of dynamic allocation and extended dynamic allocation medium access modes, the USF in the header of the downlink block $B(x-1)$ indicates if the MS can use the uplink block $B(x)$ on an allocated

PDCH. At the multiframe boundary, the last downlink block of the previous frame, $B(n)$ where $n = 11$ for the full-rate PDCH, indicates if the MS can use the first uplink block, $B(0)$, of the next multiframe. When the parameter USF_GRANULARITY is set to 1, the MS should use four consecutive blocks in the uplink. Of course, in cases of fixed allocation, extended dynamic allocation, or exclusive allocation, the MS need not monitor the USF on all allocated PDCHs. The MS transmits a PDTCH block or a PACCH block on any of the allocated uplink blocks the MS uses. The network instructs the MS about uplink transmission of the PACCH associated with the allocated PDTCH [225].

PTCCH. When one or more PDCHs carrying the uplink PTCCH for an MS is defined in the resource allocation message, the uplink PTCCH is mapped onto one of the time-slots that are allocated to the regular uplink PDTCH(s).

Mapping of the Uplink PRACH. The PRACH is dynamically allocated in groups of four PRACH blocks indicated by USF $=$ FREE similar to the allocation of an uplink PDTCH. Alternatively, a set of the blocks can be allocated to the PRACH in a fixed manner in which the parameter BS_PRACH_BLKS, which is broadcast on the PBCCH, specifies the number of allocated blocks. A fixed allocation is also evident from the use of USF $=$ FREE.

6.4.3.2 Downlink Logical and Physical Channels

Several new logical and physical channels are introduced in GPRS and EDGE to benefit from the new air-interface characteristics.

PDTCH and PACCH. The network specifies in resource allocation messages the PDCH from which the MS may expect its downlink PDTCH(s) or PACCH. The PDTCH and PACCH can be dynamically mapped on any blocks except those used for the PBCCH. The block header specifies the type of the logical channel. The TFI (Temporary Flow Identity) specifies the MS that owns a given PDTCH or PACCH/D.

PTCCH. The PTCCH is mapped onto resources similar to the PDTCH. The PTCCH transports signaling messages (e.g., timing advance information) for the MSs that share the PTCCH on the same PDCH.

PBCCH. The PBCCH is mapped onto just one PDCH as specified by the regular BCCH. The PBCCH is mapped on BS_PBCCH_BLKS blocks in each multiframe. The parameter BS_PBCCH_BLKS is broadcast on the PBCCH itself in block B0. The PBCCH data is packaged in the form of PACKET SYSTEM INFORMATION (PSI) messages. The network transmits certain types of PSIs in specific PBCCH blocks in specific multiframes to simplify the operation of the MS.

PCCCH. The network can dynamically map the PCCCH and its constituent logical channels, such as PAGCH, PPCH, and PNCH, and the message header specifies such mapping. The mapping is a function of the parameters that are broadcast on the PBCCH.

6.5 NETWORK ACQUISITION

After the MS is powered on, it acquires the network and enters a STANDBY state. Such network acquisition and STANDBY state operations are quite similar in GSM, GPRS, and EDGE, and they are briefly touched upon in Section 6.5.1. Further details of the STANDBY state activities can be found in the standard [204, 228]. Two new procedures, GPRS attach and PDP context activation, enable GPRS and EDGE to support packet data services efficiently. These procedures are discussed in Sections 6.5.2 and 6.5.3, respectively.

6.5.1 Cell Search and Synchronization: A Brief Review

After the GPRS or EDGE MS is powered on, it looks for a GPRS or EDGE system in its neighborhood. Figure 6.19 summarizes the basic steps of the cell identification and synchronization. The MS looks for the FCCH on the BCCH carrier frequency, also known as the *beacon frequency*. Once the MS finds a suitable BCCH carrier frequency, it has achieved frequency synchronization. The MS then observes the SCH to extract the current radio frame number. Next, the MS processes the messages received on the BCCH to learn important information about the system. These various stages are discussed in more detail in Chapter 5. The BCCH informs the MS of whether the

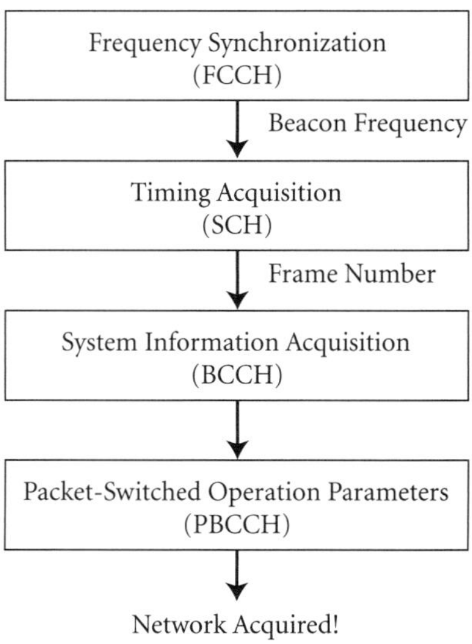

Figure 6.19: Cell Search and Synchronization for GPRS and EDGE

PBCCH exists in the cell. If the PBCCH is defined for the cell, the MS observes the PBCCH to obtain certain packet-switched parameters.

6.5.2 The MS's GPRS Attach Procedure: Making the First Contact with the Core Network

Even though the MS has acquired the network by following the major steps outlined in Figure 6.19, the network is uninformed thus far about the MS's existence. To solve this problem and to receive GPRS services, the MS performs a GPRS attach procedure by communicating with the SGSN [214, 229]. Several flavors of GPRS attach are available (e.g., combined GPRS attach and IMSI attach as well as stand-alone GPRS attach). Figure 6.20 summarizes the main stages of the combined GPRS attach and IMSI attach procedure.

The MS sends a GPRS ATTACH REQUEST message to the SGSN and provides information such as its identity, the type of attach, and its classmark. If the MS has a valid P-TMSI, it provides P-TMSI as its identity and the associated RAI (Routing Area Identity). The RAI indicates the geographical area where the MS obtained its most recent P-TMSI. A geographic area, called a routing area (RA), is assigned an RAI, which, in GPRS and EDGE, is analogous to an LAI (Location Area Identity) in GSM. An SGSN is in charge of one or more RAs. RAI consists of the PLMN ID (i.e.,

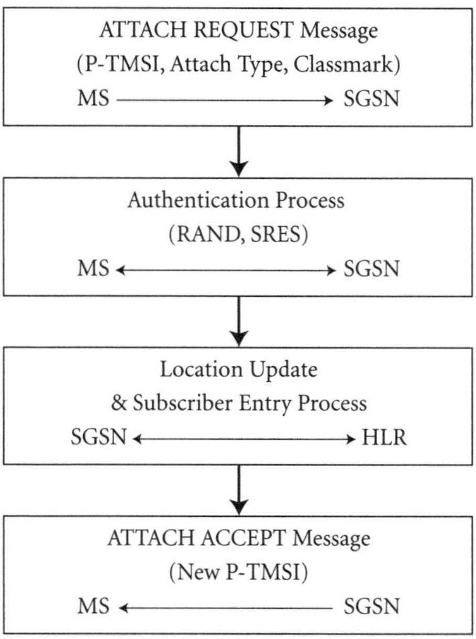

Figure 6.20: Simplified View of the Combined GPRS Attach and IMSI Attach

the combination of Mobile Country Code and Mobile Network Code) and Routing Area Code (RAC).

A cellular service provider associates different RAs with different RACs within its PLMN(s). A geographic area corresponding to one Location Area (LA) may have one or more RAs. RAI is used for STANDBY state mobility management; see Section 6.7.1 for details. If the MS has no valid P-TMSI, it provides its IMSI to the network. The type of attach may be GPRS attach only, GPRS attach while already IMSI-attached, or combined GPRS and IMSI attach. The MS classmark consists of two distinct and independent information elements: the radio access classmark containing the MS radio access capability information element and the MS network capability. The MS radio access capability information element contains the GSM radio capabilities of the MS, such as its multislot capability and power class, to facilitate the BSS's management of the radio resources for the MS. When GPRS attach occurs, the SGSN provides the MS radio access capability to the BSS using the Gb interface.

Upon receipt of the ATTACH REQUEST message, the (potentially new) SGSN may contact the old SGSN based on the P-TMSI that the MS previously received from the old SGSN. The old SGSN responds with an IDENTIFICATION RESPONSE (IMSI, Authentication Triplets) message. However, if the old SGSN fails to recognize the MS, the new SGSN obtains IMSI from the MS by exchanging IDENTITY REQUEST and IDENTITY RESPONSE messages. Authentication is mandatory if the old SGSN fails to recognize the MS.

GPRS and EDGE continue to rely upon GSM for the authentication procedure. (See Chapter 5 for authentication details.) The SGSN sends a random number, RAND, to the MS, and the MS responds with SRES. The SGSN compares the SRES with the expected response to authenticate the MS. Such GSM authentication is essentially carried out based on the information stored on the MS's SIM and Authentication Triplets in the SGSN. Since the SGSN communicates with the HLR for Authentication Triplets, the SGSN can be said to authenticate the MS with the help of the HLR. A new GSM ciphering key (Kc) agreement is established between the SGSN and the MS.

The next step is the routing area update in which the SGSN provides the SGSN number and SGSN address to inform the HLR of the MS's latest location. The HLR sends an INSERT SUBSCRIBER DATA message to the new SGSN to provide GPRS Subscription Data associated with the MS's IMSI. The MM context for the MS is said to be successfully established at this time.

The SGSN accepts the MS's GPRS ATTACH REQUEST by sending an ATTACH ACCEPT message to the MS. If the new SGSN is different from the old SGSN, the new SGSN allocates a new P-TMSI to the MS in an ATTACH ACCEPT message.

Authentication is one aspect of security, and ciphering is another aspect. Ciphering is done at the LLC layer between the MS and the SGSN. The GPRS ciphering key Kc generated during the authentication procedure is an important input to the encryption algorithm. The MS starts ciphering after sending the AUTHENTICATION

AND CIPHERING RESPONSE message, and the SGSN starts ciphering when the AUTHENTICATION AND CIPHERING RESPONSE message is received from the MS.

We looked at the GPRS attach procedure through which the MS makes its presence known to the SGSN so that it can receive GPRS services. It should come as no surprise that GPRS detach is the opposite of GPRS attach! The MS uses the GPRS detach procedure to inform the network that it is no longer interested in receiving SGSN-based services. Well, GPRS cannot be so simple, right? Different types of detach are available, such as IMSI detach, GPRS detach, and combined GPRS/IMSI detach [214]. Furthermore, the detach could be explicit or implicit. In explicit detach, the network or the MS explicitly requests detach. In implicit detach, the network detaches the MS without notifying the MS, perhaps after some configurable timer that keeps track of the MS's reachability expires or after the radio link gets disconnected due to a harsh radio environment.

6.5.3 Getting Ready for Packet Data Transfer: PDP Context Activation

Upon completion of the GPRS attach procedure, the MS is said to be in the PMM-CONNECTED (the PMM stands for Packet Mobility Management) or READY state. Furthermore, MM contexts are established in the MS and the SGSN. Next, the MS may activate *PDP contexts* as described here.

The MS's GPRS subscription includes subscription to the service of one or more PDP addresses with the most commonly used PDP being IP (Internet Protocol). Each PDP address, such as an IP address, is described by one or more PDP contexts stored in the MS, the SGSN, and the GGSN. Furthermore, each PDP context may be associated with a TFT (Traffic Flow Template), which specifies the QoS policy. It is possible to have at most one PDP context associated with a given PDP address with no TFT assigned to it. The PDP contexts of a given subscriber are associated with a single MM context for the IMSI of the subscriber. A given PDP context resides in one of two states: an ACTIVE state and an INACTIVE state.

In the INACTIVE state, the MS has no valid PDP or IP address, and hence the data transfer for the subscriber's PDP context is disabled. The PDP context includes no routing information to process PDP PDUs related to that PDP address. When the GGSN receives the mobile-terminated PDP PDUs in this state, two possible scenarios exist: the GGSN may initiate the network-requested PDP context activation procedure; or the GGSN may discard an IP packet, and an ICMP (Internet Control Message Protocol) packet error notification is returned to the sender of the IP packet. See the standard [230] for details on ICMP.

In the ACTIVE state, the MS has a PDP or IP address allocated to it, and the data transfer for such a subscriber PDP address is enabled. The PDP context for the PDP address in use is activated in the MS, SGSN, and GGSN. The PDP context now contains the routing information for transporting the PDP PDUs between the MS and the GGSN via the SGSN.

Let's see how to activate a PDP context. Figure 6.21 summarizes the PDP context activation procedure. See the standard [214] for additional details. The MS sends

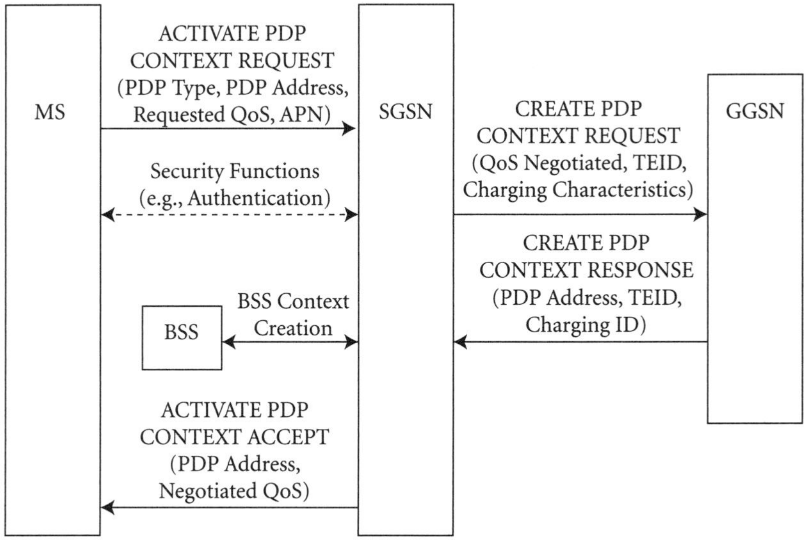

Figure 6.21: Simplified View of the MS-Initiated PDP context activation

an ACTIVATE PDP CONTEXT REQUEST message to the SGSN. Examples of the message contents include

- PDP Type (e.g., IP),
- PDP Address (e.g., request for a static versus dynamic IP address with an empty field corresponding to the request for a dynamic IP address),
- Access Point Name (e.g., to facilitate selection of a specific external network such as `operator.com`), and
- QoS Requested (e.g., peak data rate requested is set to 100 kbps).

The SGSN may optionally perform security functions, such as authentication, upon the receipt of the PDP CONTEXT REQUEST message. The SGSN may validate the ACTIVATE PDP CONTEXT REQUEST by checking received PDP Type, PDP Address, and Access Point Name against the relevant information stored in the PDP context subscription record. The SGSN may limit the MS's requested QoS to construct QoS Negotiated. The SGSN selects a suitable GGSN based on the received APN.

The SGSN sends a CREATE PDP CONTEXT REQUEST to the GGSN to provide parameters such as QoS Negotiated for the MS; TEID that the GGSN can use to send the packets of the MS; and Charging Characteristics. The GGSN creates a new entry in its PDP context database to enable routing of the packets between the SGSN and the GGSN and generates a Charging ID to start charging. The GGSN sends a CREATE PDP CONTEXT RESPONSE message to the SGSN with information such as the TEID that the SGSN can use to send the packets of the MS to the GGSN;

the PDP Address that it has allocated to the MS; and the Charging ID. The BSS and the SGSN may create a BSS Context so that the QoS the BSS can support under current resource utilization is reflected in the end-to-end QoS between the MS and the GGSN.

Finally, the SGSN sends an ACTIVATE PDP CONTEXT ACCEPT to the MS and specifies parameters, such as Negotiated QoS and the PDP Address the GGSN has allocated to the MS. The SGSN is now ready to route PDP PDUs between the GGSN and the MS and to start charging. If the MS does not like the Negotiated QoS (remember the MS actually has no negotiating power), it may deactivate the PDP context.

Ever Wondered... How about different QoS requirements for different applications at the same IP address? If the MS with one IP address is running applications with different QoS requirements (e.g., web-browsing and video streaming), providing the necessary QoS is possible by activating secondary PDP contexts.

In network-initiated PDP context activation, the GGSN informs the SGSN of the need to activate a PDP context. The SGSN, in turn, sends an ACTIVATE PDP CONTEXT REQUEST (TI, PDP Type, PDP Address, APN) message to the MS. The remaining process is similar to that of the MS-initiated PDP context activation.

As the name suggests, *PDP context deactivation* is a process that is opposite to PDP context activation. The MS or the network can initiate the PDP context deactivation. For example, if the MS has just finished some application, such as FTP (File Transfer Protocol), and no longer needs the PDP context, it can initiate the deactivation procedure. See the standard [214] for details on the PDP context deactivation procedure.

6.6 PACKET TRANSFER IN GPRS AND EDGE

Assume that the MS is in the STANDBY state and has already completed the GPRS attach and PDP context activation procedures. The MS or the network may initiate the packet transfer. Section 6.6.1 discusses the MS-originated packet transfer, and Section 6.6.2 describes the network-originated (or, equivalently, MS-terminated) packet transfer. See the standard [224] for details about these two types of packet transfer. MS-terminated packet transfer involves paging, details of which for both circuit-switched and packet-switched services are given in Section 6.6.3.

6.6.1 MS-Originated Packet Transfer

Figure 6.22 summarizes the MS-originated packet transfer. An MS initiates a packet transfer by sending a PACKET CHANNEL REQUEST message on the PRACH or RACH. The network returns a PACKET UPLINK ASSIGNMENT message on the PAGCH or AGCH, respectively. One of the two flavors of packet access method,

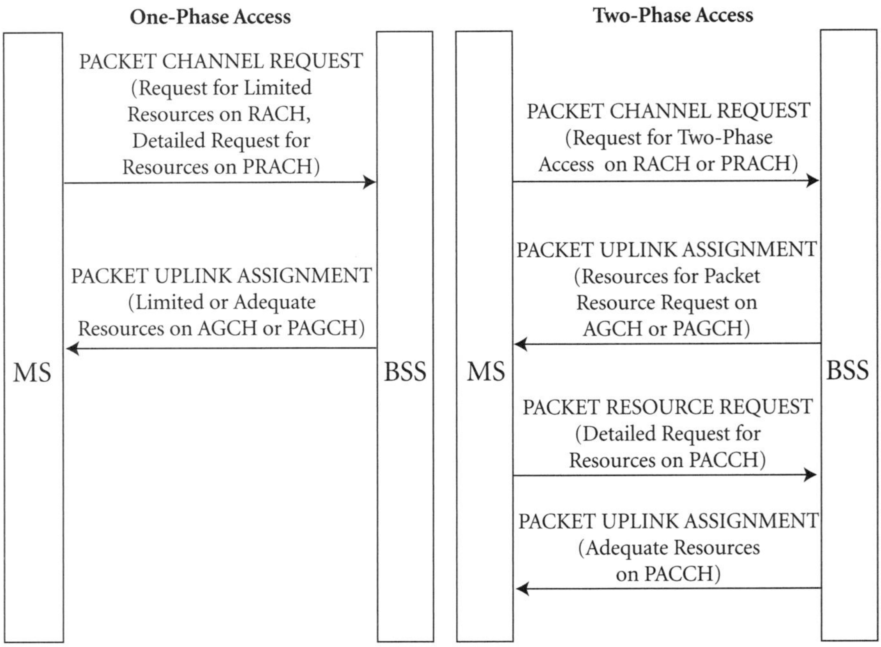

Figure 6.22: One- and Two-Phase Access for MS-Originated Packet Transfer

one- and two-phase methods, is chosen depending upon the contents of the PACKET CHANNEL REQUEST and PACKET UPLINK ASSIGNMENT messages.

The one-phase access method involves a single exchange of request and assignment messages as shown in Figure 6.22. The main idea is that the network reserves PDCH resources for the MS and allocates these resources to the MS in the PACKET UPLINK ASSIGNMENT message to enable the MS to transmit a certain number of radio blocks on the PDCH(s). PDCH resource reservation depends upon the information regarding requested resources, and the MS provides this information in the PACKET CHANNEL REQUEST message. When the MS uses the RACH to transmit a PACKET CHANNEL REQUEST, it cannot provide information about the requested resources because only two cause values are available for GPRS. Hence, the RACH usage essentially results in the request for limited resources. However, when the MS uses the PRACH, the PACKET CHANNEL REQUEST message can now contain detailed information about the requested resources (e.g., the preferred medium access method for a given TBF), and this information will enable the network to allocate adequate uplink PDCH resources to the MS.

The two-phase access method involves an exchange of two sets of request and assignment messages. The MS can initiate the two-phase access method by a suitable parameter setting in the PACKET CHANNEL REQUEST message. The network can initiate the two-phase access method by a suitable parameter setting in the PACKET

UPLINK ASSIGNMENT message. Ultimately, the network controls whether a one- or a two-phase method should be used. The network responds to the PACKET CHANNEL REQUEST message received on the RACH or PRACH with a PACKET UPLINK ASSIGNMENT. This message establishes an uplink PACCH for the MS so that the MS can now send the PACKET RESOURCE REQUEST message that contains a more detailed description of the requested resources for the uplink transfer. The network establishes an uplink PDCH that can take care of the MS's uplink data transfer requirements by sending a PACKET UPLINK ASSIGNMENT on the downlink PACCH. Let's see how the uplink resources are allocated in the PACKET UPLINK ASSIGNMENT message for different medium access approaches.

With the dynamic or extended dynamic allocation approach, the PACKET UPLINK ASSIGNMENT message includes the list of PDCHs and the corresponding USF value per PDCH. A unique TFI is allocated to the MS and is included in subsequent RLC data and control blocks of that TBF. As we discussed earlier, the MS monitors the USFs on the downlink PDCHs and transmits Radio Blocks on the uplink PDCH(s) based on the USF value reserved for the MS.

With the fixed allocation approach, the PACKET UPLINK ASSIGNMENT message includes a detailed fixed uplink resource allocation to the MS, which specifies a start frame, slot assignment, and block assignment bitmap representing the assigned blocks per time-slot. In contrast to the dynamic allocation approach, the fixed allocation excludes the USF and the MS need not monitor the downlink for the USF. An "Unused USF value" is sent on the downlink PDCH to prevent other MSs from transmitting. Of course, if the current allocation is insufficient, the MS may request additional resources using one of the assigned uplink blocks.

For the exclusive allocation approach, the PACKET UPLINK ASSIGNMENT message specifies a start frame and a slot assignment. The exclusive allocation specifies no USF or block assignment bitmap; the MS is at liberty to transmit on the uplink in the allocated slot.

6.6.2 MS-Terminated Packet Transfer

Figure 6.23 illustrates the MS-terminated, or, equivalently, network-originated packet transfer. Note that the GGSN assigns an IP address to the MS during the PDP context activation, and, hence, the IP packets from the outside world (e.g., the Internet) arrive at the GGSN when they are destined for the MS. If the GGSN has given the MS a private IP address, then the GGSN would perform NAPT (Network Address and Port Translation) so the MS and the outside world can communicate. When the network wants to initiate a packet transfer, it needs to send a page message. Specifically, the network sends one or more PACKET PAGING REQUEST messages on the downlink PPCH or PCH (Paging Channel). The MS responds by initiating an MS-originated packet transfer as discussed earlier. The MS and the network exchange request and assignment messages such that the MS has some uplink resource allocation, which enables the MS to send a PAGING RESPONSE message to the network. This response is simply one or more RLC/MAC data blocks containing an arbitrary LLC frame.

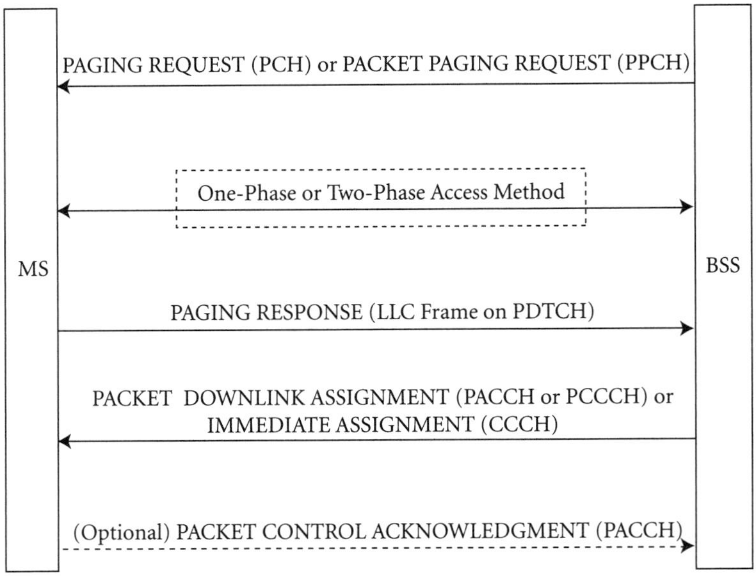

Figure 6.23: MS-Terminated Packet Transfer

The network allocates some downlink radio resources to the MS by sending a PACKET DOWNLINK ASSIGNMENT message on the PACCH if an uplink packet transfer is in progress; a PACKET DOWNLINK ASSIGNMENT message on the PCCCH if the PCCCH is allocated in the cell; or the IMMEDIATE ASSIGNMENT message on the CCCH. The assignment message specifies the PDCH(s) established for the downlink transfer.

If available, the network provides information on the Timing Advance and Power Control. If such information is unavailable, the network requests the MS to respond with a PACKET CONTROL ACKNOWLEDGMENT. The network can now commence transmission of the RLC/MAC blocks associated with a TBF on the downlink PDTCH allocated to the MS.

6.6.3 Management of Circuit-Switched and GPRS Paging

As with GSM, we need to send a paging message to the MS in MS-terminated or network-initiated call setup. Three network operation modes are defined for paging. These modes characterize how the network may coordinate paging for CS and packet-switched or GPRS services as shown in Table 6.8.

In Network Operation Mode I, the network sends a CS paging message to a GPRS-attached MS on the GPRS paging channel (which could be the PPCH or PCH) or on a GPRS traffic channel. A GPRS/EDGE MS may be in the READY or STANDBY state after the GPRS attach procedure establishes for the MS a context between the MS and the SGSN. The MS in the READY state has a dedicated radio

Table 6.8: Network Operation Modes in GPRS and EDGE

	Network Operation Mode		
	I	II	III
Where is CS Paging?	PPCH, PCH, or PDTCH	PCH	PCH
Where is GPRS Paging?	PCH or PPCH	PCH	PCH or PPCH

connection with the BSS, and the SGSN knows the MS's location at the cell level. In the READY state, the MS is *active* or *connected* to the radio and the core networks and is transmitting and receiving packets. The MS in the STANDBY mode has no dedicated radio connection with the BSS, and the SGSN knows the MS's location at the RA (Routing Area) level. In the STANDBY mode, the MS is uninvolved in packet data transfer. In the STANDBY state, the MS monitors only one paging channel for a page message. However, when the MS is allocated a PDTCH, it receives the CS paging messages on the PDTCH.

In Network Operation Mode II, the network sends a CS paging message for a GPRS-attached MS on the PCH, which is the CCCH paging channel. The MS continues to monitor the CCCH paging channel for potential CS paging even if it has been assigned a PDTCH. Network Operation Mode III involves the network sending a CS paging message for a GPRS-attached MS on the PCH and a GPRS paging message on either the PPCH or the PCH. In other words, no paging coordination occurs and the MS must monitor both paging channels. Network Operation Mode II is commonly used in practice.

Recall that the interface between the MSC/VLR and the SGSN is called the Gs interface. When the Gs interface is present, the network operates in Mode I because all MSC-originated CS paging messages go to the MS via the SGSN. Since the SGSN originates GPRS paging, the SGSN now has a view of both CS paging and GPRS paging, enabling the network to coordinate paging. However, in the absence of the Gs interface, the MSC-originated CS paging utilizes the A interface between the MSC/VLR and the BSS, leading to a lack of paging coordination for CS and GPRS paging. In the absence of the Gs interface, the network operates in Mode II or III. In Mode II, no PPCH is allocated in the cell, and in Mode III, the PPCH is available in the cell for GPRS paging.

The MS must know which Network Operation Mode the cell is using to decide which channels it needs to monitor and whether it can attach to GPRS services, non-GPRS services, or both GPRS and non-GPRS services based on its capabilities. The network informs the MS about its Network Operation Mode using the system information blocks being sent on the BCCH. All the cells in a given Routing Area (RA) implement the same Network Operation Mode for proper operation.

6.7 MOBILITY MANAGEMENT

After the MS is powered on, it enters the STANDBY state. If the MS has a dedicated radio channel with the BSS, it is said to be in the READY state. As the MS travels in a given region, its mobility is managed differently in the STANDBY and READY states. Section 6.7.1 discusses the STANDBY state mobility, and Section 6.7.2 describes the READY state mobility.

6.7.1 Routing Area Update: STANDBY State Mobility Management

There are three main STANDBY state tasks: PLMN selection, cell selection and reselection, and routing update. The PLMN selection involves finding a suitable PLMN among the PLMNs available in a given geographical area. Cell selection involves monitoring of the best cell in the selected PLMN, and cell reselection involves observation of the downlink broadcast and common control channels of a new cell as the MS moves from an old cell to a new cell. See Chapter 5 for details on PLMN selection and cell selection and reselection.

Routing area update is a new concept for GPRS and EDGE and will be discussed in detail here. Having said that, routing update is analogous to location update for the circuit-switched services in GSM. As the GSM MS goes from one location area (LA) to another, it needs to perform a location update procedure.[193] Similarly, as the GPRS/EDGE MS travels from one RA to the next, the MS performs a routing area update procedure, which is similar to the GPRS attach procedure discussed in Section 6.5.2. Furthermore, just as one MSC takes care of one or more location areas, the SGSN takes care of one or more RAs. The routing area update ensures that the network (specifically, the HLR and SGSN) knows the latest location of the MS to facilitate paging.

Recall that the MS stores an MM context when GPRS attach succeeds. The MS stores the cell identity and the RAI in the MM Context. The BSS periodically broadcasts the cell identity and the RAI in each cell. The MS compares the stored RAI with the RAI observed in the new cell to detect its entry into a new RA. When the MS detects that it has now crossed into a new RA, it performs RA Update or combined RA and LA update procedures. Other causes can initiate RA updates, such as expiration of the periodic RA update timer.

The MS sends a ROUTING AREA UPDATE REQUEST to the SGSN and includes stored P-TMSI, stored RAI, and Update Type. The parameter Update Type indicates whether it is truly an RA update or just a periodic RA update.

Upon receiving the ROUTING AREA UPDATE REQUEST message, the SGSN may carry out security functions such as authentication. The SGSN may check subscription data and may allocate a new P-TMSI. Of course, if the SGSN controlling the RAI stored by the MS and the SGSN receiving the ROUTING AREA UPDATE REQUEST are different, the new SGSN allocates a new P-TMSI to the MS. The

[193] See Chapter 5 on GSM for details on LAs.

SGSN then responds with a ROUTING AREA UPDATE ACCEPT that includes the new P-TMSI. The MS acknowledges the receipt of a new P-TMSI by sending a ROUTING AREA UPDATE COMPLETE message to the SGSN.

A combined RA and LA update occurs for Network Operation Mode I in the following cases:

- the MS enters a new RA,
- a GPRS-attached MS performs an IMSI attach,
- the MS needs to indicate new access capabilities to the network, and
- the BSS fails to resume GPRS services for a suspended MS.

The MS initiates the combined RA and LA update procedure by sending a ROUTING AREA UPDATE REQUEST with an indicator that an LA update needs to be performed, too. The SGSN forwards the LA update to the VLR.

6.7.2 Mobility Management in READY State

In the early versions of the GPRS and EDGE specifications, no GSM-type Active Mode handover was defined for an MS in the READY state. In early GPRS and EDGE systems, the packet transfer mobility control was achieved via MS-controlled cell reselection as it was in GSM Idle Mode operations [213]. When the MS in the READY state entered a new cell with a better Received Signal Strength Indication (RSSI), the MS performed cell reselection as if it were in the STANDBY state.

To facilitate cell reselection, the MS acquires cell selection and cell reselection parameters typically broadcast on the Broadcast Control Channel (BCCH). The MS evaluates the beacon frequency's RSSI and obtains selected broadcast information for the strongest neighbors. The basic idea behind cell reselection is to ensure that the MS has enough transmit power to overcome the path loss between the MS and the new cell and that the new cell is better than the current cell. Upon performing cell reselection, the MS aborts the existing TBF and the network marks all unacknowledged radio blocks in the transmit buffer as not sent [225]. The MS then executes the cell update procedure.

Let's summarize the cell update procedure with the two-phase access method.[194] The MS sends a PACKET CHANNEL REQUEST message on the PRACH with an indication for cell update. The BSS responds by sending on the PAGCH a PACKET UPLINK ASSIGNMENT message specifying the information related to the uplink PACCH. The MS provides the details of the required uplink TBF in the PACKET RESOURCE REQUEST message sent on the PACCH. The BSS now specifies the channel allocation for the uplink TBF in the PACKET UPLINK ASSIGNMENT message sent on the downlink PACCH. The MS informs the SGSN of the cell update by sending an LLC PDU. Once the uplink TBF is addressed, the network turns to the

[194] Feel free to refer to Figures 6.22 and 6.23 and Section 6.6.1 to refresh your memory.

downlink TBF setup. The network sends a PACKET DOWNLINK ASSIGNMENT message on the PCCCH. The network resumes data transfer in the new cell beginning with the last unacknowledged radio blocks; no data would be lost during such cell reselection in the acknowledged mode of operation. The uplink data transfer is similar to the downlink data transfer.

The MS-controlled cell reselection method ensures that no data are lost during the mobility across cells, but many duplicate radio blocks are often sent and the data interruption is relatively long. For example, when cell reselection interrupts the transmission of a Logical Link Control (LLC) frame after receiving only one of one hundred radio blocks, ninety-nine duplicate radio blocks would be sent in the new cell, lowering the overall data throughput and inefficiently using the radio spectrum. Depending upon the interruption time, Transmission Control Protocol (TCP) retransmissions may also be triggered, further lowering the achievable throughput.[195]

Enhancements to the MS-controlled cell reselection approach for the MS in the READY state were defined in later releases of GPRS and EDGE. These enhancements can be viewed as variations of the basic GSM handover technique. Cell Change Order (CCO) and Network Assisted Cell Change (NACC) are examples of such enhancements.

The CCO procedure can be viewed as network-controlled cell reselection. In such a case, the network initiates the cell change order procedure by sending on the PCCCH or PACCH a PACKET CELL CHANGE ORDER message to the MS. This message specifies the characteristics of the new cell, such as the combination of BSIC and the BCCH frequency. The network may utilize measurement reports from the MS to make the cell reselection decision. The CCO procedure is considered complete when the MS has received a successful response to its CHANNEL REQUEST or PACKET CHANNEL REQUEST message on the new cell. The CCO procedure could provide better performance than the MS-controlled cell reselection because the network may use other information unavailable to the MS to make the cell reselection decision.

The NACC involves two distinct procedures [225]. In the Cell Change Notification (CCN) procedure, the MS notifies the network of the cell reselection decision and delays the actual cell reselection to enable the network to respond with the neighbor cell system information. In the other procedure, the network conveys to the MS the neighbor cell's system information. A cell shall indicate the availability of the CCN procedure on the BCCH or the PBCCH using the parameter CCN_ACTIVE. The network may alternatively use messages, such as a PACKET MEASUREMENT ORDER or PACKET CELL CHANGE ORDER message, to ask an MS to apply the CCN procedure. During the CCN procedure, the MS sends a PACKET CELL CHANGE NOTIFICATION message to the network. The network then conveys via the BCCH or PBCCH the neighboring cell's information in the PACKET NEIGHBOUR CELL DATA message. The NACC significantly reduces the packet interruption time.

[195] See Chapter 4 for details on TCP retransmissions.

MS-controlled cell reselection, network-controlled cell reselection, or NACC may be in action in a given commercial network depending on the capabilities of the MS, the network equipment, and the operator preferences.

6.8 DEPLOYMENT CONSIDERATIONS AND NETWORK EVOLUTION

GPRS and EDGE are TDMA- and FDMA-based cellular systems. Hence, traditional frequency planning is required for deployments of such systems. A service operator usually has a GSM system already deployed, and an upgrade is carried out to support GPRS and EDGE. In practice, reuse of cell sites is highly desired to reduce the infrastructure costs and alleviate real-estate acquisition issues. Evolution from GSM to GPRS or EDGE is simplified due to full backward-compatibility of GPRS and EDGE with prior GSM systems. Both hardware and software upgrades are required for the BSS. A new packet-switched core network is required while evolving from GSM to GPRS or EDGE, but the circuit-switched network can remain intact.

A typical deployment approach is to reserve some time-slots (e.g., two or three) on an ARFCN for GPRS and EDGE users' packet data services. The remaining time-slots on such an ARFCN and other ARFCNs are then used for pre-existing GSM users. When additional data capacity is required, more time-slots on an ARFCN are reserved. If even higher data throughput is needed, some ARFCNs can be fully reserved for GPRS and EDGE and new ARFCNs could be added to meet GSM capacity requirements.

The Third Generation Partnership Project (3GPP) defined the evolution of EDGE, called Evolved EDGE, in Release 7 [231, 232]. Evolved EDGE's general goal is to enhance the radio network while backward-compatibly reusing the packet-switched core network. Evolved EDGE provides an enhanced user experience compared to that of basic EDGE when the user moves outside the coverage area of technologies such as UMTS and Long-Term Evolution (LTE). Evolved EDGE reduces latency from about 135 ms to 70 ms [232]. Delay-sensitive applications benefit from such a latency reduction.

The downlink dual carrier feature enables the MS to receive data on two channels instead of just one, increasing the throughput by a factor of two or more. MS receive diversity supports two antennas at the MS, improving the reliability of communications, especially near the cell edge. Using the 32-ary Quadrature Amplitude Modulation (32-QAM) high-order modulation scheme, rather than only 8-PSK, increases the data rate by a factor of $5/3$. Evolved EDGE can be quite attractive to operators when larger spectrum bandwidths are unavailable.

Where do we go from 2.5G GPRS and EDGE? The next major stepping stone in the evolution path is a 3G UMTS network. While GPRS and EDGE are based on TDMA, UMTS utilizes CDMA (Code-Division Multiple Access) as the multiple access technique. See Chapter 9 for an overview of UMTS.

6.9 POINTS TO REMEMBER

We discussed various aspects of GPRS and EDGE in this chapter. Let's summarize the main points.

- GPRS and EDGE are a TDMA-based 2.5G evolution of 2G GSM networks.
- The main GPRS enhancements are the introduction of a packet-switched core network and air-interface protocol stack changes to support packet data services.
- The packet-switched core network consists of the SGSN and GGSN. The SGSN facilitates MS location tracking by maintaining the MM context and authenticates the MS with the help of the HLR. The GGSN allocates a dynamic IP address to the MS and interfaces with the outside world to enable an MS to access the Internet.
- GPRS also makes some air-interface enhancements such as the design of RLC/MAC layer to improve reliability of the air-interface communications.
- GPRS and EDGE continue to rely upon time-slot and frame structures that GSM defines; a new fifty-two–frame multiframe is defined to carry new packet data channels.
- Several new logical radio channels, such as the PDTCH and PACCH, are introduced for packet data services.
- The MS's network acquisition after power-on is similar to that of GSM with an additional PBCCH available to broadcast GPRS- and EDGE-specific information.
- Two GPRS-specific procedures, GPRS attach and PDP context activation, play an important role in providing packet data services. The MS's GPRS attach procedure informs the network of the MS's location and gives the network an opportunity to authenticate the MS. PDP context activation is used to allocate an IP address to the MS and to establish adequate resources to provide an end-to-end QoS to the MS and the GGSN.
- The READY state mobility for GPRS and EDGE is quite different from that for GSM.
- The STANDBY state mobility management involves RA updates.
- GPRS and EDGE evolve to UMTS—a 3G cellular system.

CHAPTER 7

FUNDAMENTALS OF CDMA, WCDMA, AND IS-95

Welcome to a very efficient multiple access technique: code-division multiple access (CDMA)! IS-95 (Interim Standard-95) is a digital 2G cellular network based on CDMA. GSM (Global System for Mobile Communication) and IS-136,[196] both based on time-division multiple access (TDMA), are examples of other 2G standards. A special flavor of CDMA, *wideband CDMA* (WCDMA), is used in 3G cellular networks using Universal Mobile Telecommunication System (UMTS).

This chapter consists of two parts. The first part discusses CDMA and WCDMA as generic multiple access techniques and covers CDMA principles used in networks that are based on UMTS, IS-95, and 1x.[197] The second part describes IS-95 as the first CDMA-based cellular standard. A brief overview of CDMA and IS-95 is given first. An example of how CDMA allows multiple users to access the system simultaneously is illustrated. Various characteristics of CDMA, such as wideband transmission, universal frequency reuse, interference management, soft handoff, and power control, are discussed next. The capacity of a CDMA system is soft; an analytical approach

[196] IS-54 evolved to IS-136 and was eventually primarily replaced by GSM in commercial cellular networks.

[197] 1x is an evolution of IS-95. Chapter 8 takes a detailed look at 1x.

Cellular Communications, First Edition. Nishith D. Tripathi and Jeffrey H. Reed

to estimate CDMA capacity is narrated. As part of the IS-95 discussion, network architecture is illustrated first. The air-interface protocol stack and various radio channels that enable over-the-air communication between the mobile station and the radio network are explained. The method of generating orthogonal codes, which are critical in a CDMA-based system, is specified. The process of setting up a call in IS-95 is described. The mobile device's network acquisition after power-up as well as mobility management are addressed next. Important deployment considerations are also mentioned. IS-95A, the first version of IS-95, was widely deployed, but the second version, IS-95B, experienced a limited deployment. IS-95B features are summarized. Finally, key points about CDMA and IS-95 are highlighted.

7.1 A BRIEF INTRODUCTION TO CDMA, WCDMA, AND IS-95

CDMA's spectral efficiency[198] propelled both 3G cellular systems (UMTS and CDMA2000 1x) to adopt CDMA as the air-interface multiple access technique. 2G GSM has evolved to UMTS, and 2G IS-95 has evolved to CDMA2000 1x.

An ancient saying goes, "All roads lead to Rome." In recent times, we could say, "All major technologies lead to the U.S. military!" CDMA's roots as a spread-spectrum technique are found with the U.S. military.

A Question for You... One company played a vital role in commercializing the use of CDMA for cellular communications; that's how IS-95 was born. Can you guess the name of this company? Hint: The name starts with the letter Q. Another hint: If you had joined this company by the early 1990s, you could have become a millionaire in about seven to ten years as a result of the soaring company stock price. As always, the footnote[199] has your answer.

CDMA possesses several attractive features, which are discussed in Section 7.3. A CDMA signal occupies a wide bandwidth. The users in frequency divison multiple access (FDMA) and in hybrid TDMA/FDMA systems utilize narrow bandwidth signals, such as 30 kHz in IS-136 and 200 kHz in GSM. However, each user in a CDMA system occupies the same wide bandwidth, e.g., 1.25 MHz in IS-95 and 1xRTT[200] and 5 MHz in UMTS. Such a wideband signal enables CDMA to recover information in the presence of frequency-selective fading because only a fraction of the information would be corrupted by fading.[201] CDMA is deployed with the frequency reuse factor of 1, enabling the same spectrum to be reused in every single sector. Such frequent spectrum reuse significantly increases the system capacity. To

[198]For a voice-centric system, one way of quantifying spectral efficiency is to measure the number of voice users supported in a sector per megahertz of spectrum.

[199]Qualcomm, it is!

[200]1x Radio Transmission Technology

[201]In military applications, such a wideband signal makes it difficult for enemies to detect and jam the signal.

minimize co-channel interference, traditional hybrid TDMA/FDMA systems need to use different radio channels in different sectors that a base station serves. Soft handoff is inherent in a CDMA system; the mobile terminal simultaneously communicates with multiple sectors and/or multiple cells based on its location. Soft handoff increases the reliability of communications near sector or cell borders. A CDMA system utilizes a *RAKE* receiver, which combines the energy of multipaths received from one or more sectors. Power-control mechanisms in CDMA reduce interference, increase capacity, and improve the mobile device's battery life.

IS-95, with the brand name of cdmaOne, is the first cellular standard to implement CDMA. Two versions of IS-95 have been commercially deployed: IS-95A and IS-95B. cdmaOne refers to both IS-95A and IS-95B. IS-95A's deployment was a huge success because it offered a multi-fold capacity increase [e.g., eight to ten times the capacity of Advanced Mobile Phone System (AMPS) capacity and four to five times GSM's capacity] at the time of its initial deployment. IS-95A is much more prevalent in global deployments (e.g., Sprint and Verizon Wireless deployments in the U.S. and by other providers in Canada, China, Argentina, and Brazil), and IS-95 B is deployed in few places (e.g., SK Telecom and LG Telecom deployments in South Korea) [233]. This chapter primarily focuses on IS-95A; however, enhancements that IS-95B has introduced are summarized in Section 7.9. In this chapter, the term IS-95 generally refers to IS-95A unless explicitly stated otherwise. The CDMA Development Group offers details on IS-95 and its evolution [234].

The initial IS-95 standard was first published in July 1993, and the IS-95A revision was published in May 1995. IS-95A implements the basic CDMA features, such as handoff and power control. It uses frequency-division duplexing with 1.25 MHz in both the downlink and the uplink.[202] The main focus of IS-95A is on circuit-switched voice services, but it also supports 14.4 kbps circuit-switched data services. IS-95A was first deployed by Hutchison Telecom in Hong Kong in September 1995.

IS-95B, often considered a 2.5G CDMA-based system, supports 64 kbps packet-switched data in addition to voice services. IS-95B was first deployed in September 1999, in South Korea.

cdmaOne provides a voice capacity advantage compared to 1G AMPS. cdmaOne simplifies network planning and deployment because the same carrier frequency is used in every single sector in the system. The use of various types of codes provides enhanced privacy. Since CDMA offers a better link budget or link margin than 1G AMPS and 2G GSM, it has greater coverage and hence requires fewer cell sites. Power control and efficient variable-rate operation together increase battery life and hence talk time compared to 1G AMPS.

cdmaOne has evolved to 3G CDMA2000 1x, often called just 1x. 1x increases voice capacity beyond cdmaOne and provides higher peak data rates in excess of 150 kbps. Chapter 8 discusses 1x.

[202]The downlink is the same as the forward link and represents a communication link from the radio access network to the mobile device. The uplink is the same as the reverse link and represents a communication link from the mobile device to the radio access network.

7.2 CDMA OR WCDMA: AN EFFICIENT RESOURCE-SHARING TECHNIQUE

Let's see how CDMA enables multiple users to utilize the system simultaneously. We will discuss multiple access from the downlink perspective here. An analytical perspective on downlink capacity and uplink capacity will be presented in Section 7.3.6.

Consider Figure 7.1 in which two mobile stations (MSs), MS1 and MS2, are receiving information from a base station (BS).[203] A CDMA system utilizes *codes* to enable multiple access. In our example, we have allocated codes $[1\,1]$ and $[1\,-1]$ to MS1 and MS2, respectively. Each code in our example consists of a sequence of two bits called *chips*. In other words, we are using two-chip codes. Such codes are referred to as Walsh codes in IS-95 and 1xRTT and as Orthogonal Variable Spreading Factor (OVSF) codes in UMTS. One code can be considered one physical channel.

Our goal is to send bit 0 to MS1 and bit 1 to MS2. Let's map bit 0 to the value $+1$ and bit 1 to the value -1 to facilitate generation of the transmission signal. Let α be the amplitude of the signal for MS1 and β be the amplitude of the signal for MS2. The information bit 0 for MS1 corresponds to the amplitude $+1 * \alpha = \alpha$, and the information bit 1 for MS2 corresponds to the amplitude $-1 * \beta = -\beta$. The BS will use codes to carry the bits, multiplying the user-dependent amplitudes by the user code to generate chips per user and then adding the chips for different users. The chips for MS1 are obtained by multiplying its code $[1\,1]$ by its amplitude represented by $+\alpha$. So, the chips for MS1 are $[1\,1] * (+\alpha) = [\alpha\,\alpha]$. The chips for MS2 are $[1\,-1] * (-\beta) = [-\beta\,\beta]$. The BS now adds chips for all MSs. During the first chip period, the cumulative chip is $\{\alpha + (-\beta)\} = \alpha - \beta$. The second chip period

* For MS1: $Code = [1 \quad 1]$, $Bit = 0$ mapped to $+1$, $Amplitude = \alpha$
* For MS2: $Code = [1 \quad -1]$, $Bit = 1$ mapped to -1, $Amplitude = -\beta$
* $N_c = Number\ of\ Chips\ per\ Code = 2$

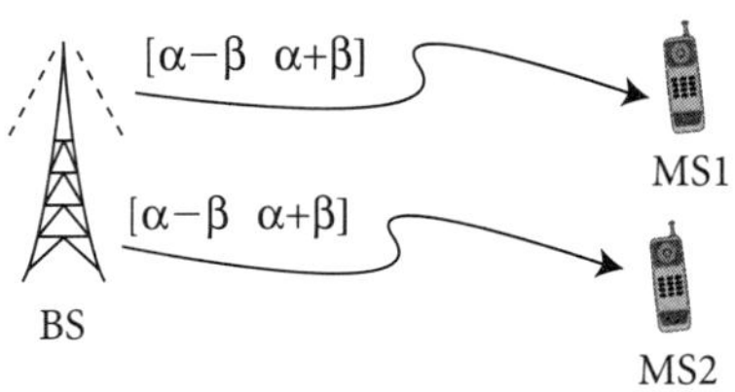

Processing at MS1

$$\frac{\Sigma\,[\alpha-\beta \quad \alpha+\beta] * [1 \quad 1]}{N_c} = \alpha$$

Processing at MS2

$$\frac{\Sigma\,[\alpha-\beta \quad \alpha+\beta] * [1 \quad -1]}{N_c} = -\beta$$

Figure 7.1: Multiple Access in a CDMA System: A Simplified View

[203] Another term for the BS is the Node B.

corresponds to the cumulative chip of $\{\alpha + \beta\}$. In summary, the BS successfully transmits two cumulative chips, $[(\alpha - \beta)\,(\alpha + \beta)]$.

In our example, both MSs receive the cumulative chips during the observation window of two chips. Each MS performs the following processing.

- Step 1: Multiply the received chips by the MS-specific code.
- Step 2: Add all the chips to calculate the final sum, and divide the final sum by the number of chips in the code.
- Step 3: Compare the final sum with some threshold to estimate the bit.

Let's pretend that we are MS1! We want to extract our information bit 0 represented by a positive value $+1$. As specified in Step 1, we will multiply the received chips, $[(\alpha - \beta)\,(\alpha + \beta)]$, by our code, $[1\,1]$, in a chip-by-chip fashion to get $[(\alpha - \beta) * 1\,(\alpha + \beta) * 1] = [(\alpha - \beta)\,(\alpha + \beta)]$. Then, as mentioned in Step 2, we will calculate the final sum as $(\alpha - \beta) + (\alpha + \beta) = 2\alpha$. Next, we will divide the final sum, 2α, by the number of chips, which is 2. We will get $2\alpha/2 = \alpha$. Such division by the number of chips normalizes the final result for proper bit estimation. Since we have only two possible values, $+1$ and -1, representing the bits, we can use the threshold value of 0 to estimate the bit. At the receiver, we can use following rule. If the final result is greater than 0 (which is our chosen threshold), the bit is estimated to be 0. Otherwise, the bit is estimated to be 1. In our example, $\alpha > 0$, and, hence, bit 0 must have been transmitted for MS1. You can perform a similar calculation and convince yourself that MS2 will be able to estimate that bit 1 was transmitted to it.

In summary, the transmitter uses codes to carry channel-specific information bits and combines all chips. The receivers can extract their own information by applying a suitable channel-specific code.

In our preceding example, we had just two users, with each user using one code. The more users we have, the more complex the signal would look. For example, for five users, the first cumulative chip sent by the BS may be $\alpha + \beta - \theta + \delta - \nu$! You may be wondering what enables users to extract their own information out of a seemingly complicated signal. The answer is orthogonality! The codes that we have used are orthogonal to each other. As long as we use orthogonal codes, users can apply their own code to extract their information from a seemingly complicated received signal.

Let's test if the codes that we used, $[1\,1]$ and $[1\ -1]$, are orthogonal. Do chip-by-chip multiplication, and add the multiplication results. If the answer is 0, we have orthogonality. $[\{1*1\} = 1$ for the first chip and $\{(1)*(-1)\} = -1$ for the second chip$]$ are two multiplication results and their addition yields $(1+(-1)) = 0$. It is this property of orthogonality that is instrumental in making CDMA work efficiently. See Section 7.5.3.1 to learn the properties of Walsh or OVSF codes and a method to construct these codes.[204] We can share the available spectrum among numerous users by allocating orthogonal codes to these users. For a two-chip code, we have two orthogonal codes. Similarly, a four-chip code has four orthogonal codes. IS-95 supports

[204] See Section 9.4.2 in Chapter 9 to learn about the process of creating OVSF codes using a code tree and the guidelines on the usage of the OVSF code tree.

64 orthogonal codes, and, hence, as long as the number of users in a sector is less than or equal to 64,[205] we can (ideally) maintain orthogonality in the sector, leading to good performance. The number of available orthogonal codes is *one of the potential capacity bottlenecks* in CDMA-based systems. We will see later that codes are not the real capacity-constraining culprits in IS-95. The codes are a capacity bottleneck under certain circumstances in CDMA2000 1xRTT and UMTS.

As is evident from Figure 7.1, one user information bit takes the form of two chips when a two-chip code is used. If the original information bit rate is X kbps, the data rate after the multiplication process becomes $(X * N_c)$ kcps (kilo chips per second) where N_c is the number of chips in the code. Since we are increasing the bandwidth of the signal, such a multiplication process is called *spreading*. The number of chips in the code is called the *spreading factor*; the bandwidth of the input signal increases by the spreading factor due to the spreading process. Figure 7.2 gives examples of spreading.

The signal bandwidth at the end of the spreading process is constrained by the available spectrum bandwidth. Commercial CDMA systems fix the value of the spreading bandwidth. For example, IS-95 and CDMA2000 1x have the spreading rate of 1.2288 Mcps or the spreading bandwidth of 1.2288 MHz, and UMTS R99 has the spreading bandwidth of 3.84 MHz.

UMTS R99 uses a suitable spreading factor for the orthogonal code to support variable packet data rates. For example, when the input bit rate to the spreading process is 30 kbps, the code with the spreading factor of 128 chips is used to yield

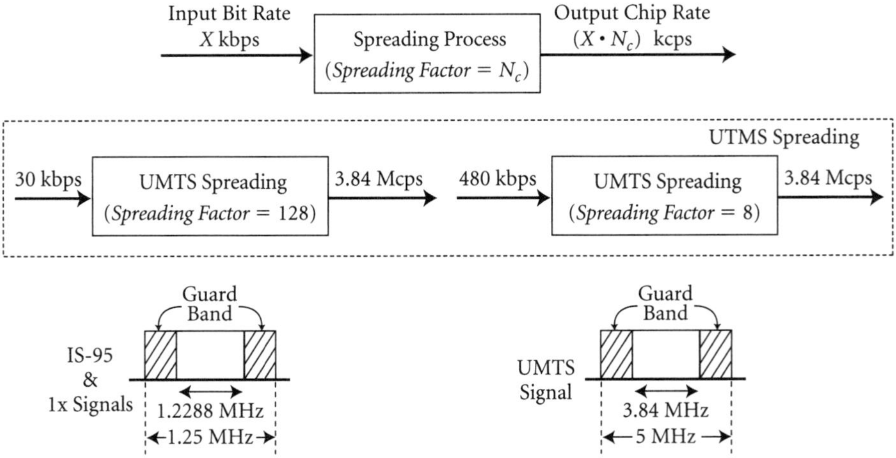

Figure 7.2: Process of Spreading

[205]The actual number of Walsh codes available for users is 61 because three of the 64 codes are used for other purposes.

30 kbps $*$ 128 $=$ 3.84 Mcps. However, when the input bit rate is 480 kbps, the code with the spreading factor of eight chips is used to yield 480 kbps $*$ 8 $=$ 3.84 Mcps. In summary, the variable spreading factor enables a CDMA-based system to support different data rates. Now you know why UMTS refers to the codes as orthogonal variable spreading factor codes![206]

When guard bands at the left and right edge of spectrum are included to minimize adjacent spectrum interference, the total required spectrum bandwidth becomes 1.25 MHz for IS-95 and CDMA2000 1x and 5 MHz for UMTS R99.[207]

7.3 CHARACTERISTICS OF CDMA

Various characteristics of CDMA, such as wideband transmission, universal frequency reuse, efficient interference management, reliable soft handoff, and power control, are discussed in this section. Such characteristics make CDMA an efficient multiple access technique.

7.3.1 Wideband Transmission

A CDMA signal occupies a wide bandwidth, such as 1.25 MHz or 5 MHz in commercial cellular systems. Figure 7.3 illustrates a snapshot of the frequency-selective fading that may be observed at some time instant.

Do You Recall... Let's refresh your memory about fading. Do you recall what frequency-selective fading is? The footnote[208] gives the answer.

If the transmitted signal occupies a narrow bandwidth (e.g., 30 kHz or 200 kHz), we might lose a significant portion of the signal energy due to frequency-selective fading. However, a CDMA signal occupies a much larger bandwidth compared to the bandwidth of the information signal that it represents; hence, most signal energy remains intact even in the presence of frequency-selective fading. In summary, the CDMA signal's wider bandwidth makes it more resilient to frequency-selective fading than narrowband signals typically used in non-CDMA systems.

[206] Though we have used the unit of bits per second for the input to the spreading process, such input can be more accurately represented as modulation symbols per second.

[207] Additional guard bands are needed between two adjacent IS-95 or 1x carrier frequencies. For example, in the 5-MHz spectrum, three (not four) IS-95 or 1x carrier frequencies can be placed. Seven IS-95 or 1x carrier frequencies can be deployed in the 10-MHz bandwidth. In UMTS, the 10-MHz bandwidth can accommodate two side-by-side 5-MHz carrier frequencies; additional guard bands are unneeded.

[208] Frequency-selective fading is said to occur when part of the transmitted signal experiences a fade and part does not. Flat fading is said to occur when the entire signal is in a fade or out of a fade.

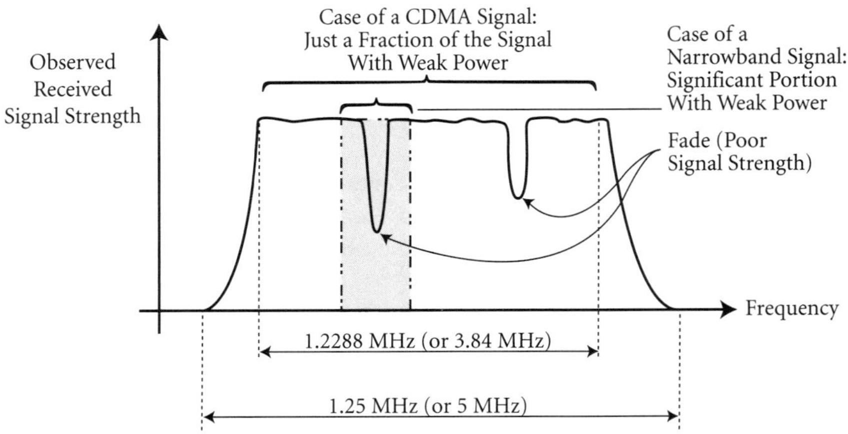

Figure 7.3: Frequency-Selective Fading and a CDMA Signal

7.3.2 Frequency Reuse and Impact of Sectorization

A CDMA system uses the frequency reuse factor of 1, meaning that the same carrier frequency is used in every single sector of the system. Figure 7.4 illustrates a typical deployment scenario for omnidirectional and sectorized systems.

Let f_1 be the downlink carrier frequency and f_2 be the uplink carrier frequency. In the case of omni cells, all cells use the same f_1 and f_2. When 120° sectorization is used, the same carrier frequencies, f_1 and f_2, are still used in all the sectors throughout the system. Contrast that with a hybrid TDMA/FDMA system in which

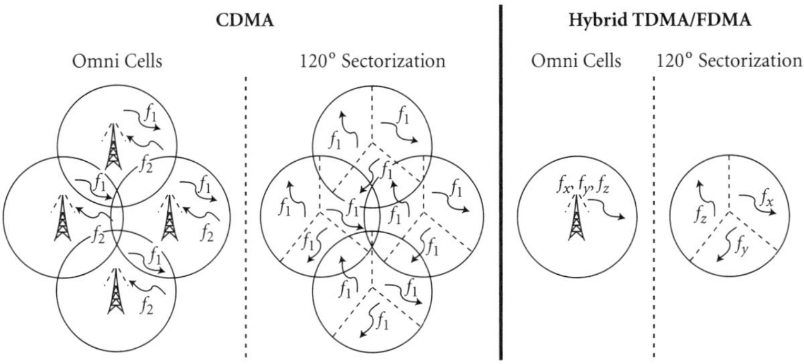

Figure 7.4: Frequency Reuse in CDMA and Hybrid TDMA/FDMA Systems

the same radio channel cannot be used continuously[209] in the sectors of a base station. For example, if a cell is configured to use three carrier frequencies, f_x, f_y, and f_z, in the downlink, when $120°$ sectorization is utilized, one sector uses f_x, another sector uses f_y, and the last remaining sector uses f_z. In other words, a given sector of the cell cannot use all three carrier frequencies. We would point out though that sectorization in hybrid TDMA/FDMA systems does decrease interference and that we can squeeze in multiple narrowband hybrid TDMA/FDMA carrier frequencies in the bandwidth consumed by one wideband CDMA carrier frequency.

We know from Chapter 1 that the more frequently we can reuse the available spectrum in a given geographical region, the higher the capacity of the system can be. In summary, sectorization does not adversely impact the frequency reuse in CDMA systems. In fact, sectorization *increases* the system capacity. For example, in a CDMA system, if an omni cell supports N users, a cell with $120°$ sectorization supports about $2.5N$ to $2.6N$ users; the cell capacity for $120°$ sectorization does not become $3N$ users due to factors such as increased handoff.

7.3.3 Interference and its Control in CDMA

In hybrid TDMA/FDMA systems, adequate spacing between the radio channels controls adjacent channel interference and separating the cells or sectors by a sufficient distance controls co-channel interference. However, interference is controlled quite differently in CDMA than in hybrid TDMA/FDMA systems. Before we discuss how to control interference, let's see where the interference is coming from. Figure 7.5 illustrates the sources of interference in a single-path environment, such as a rural environment with a flat terrain.

The nature of interference in the downlink differs from that in the uplink. Let's consider the downlink first. Within a sector, synchronous downlink transmissions to the users occur via orthogonal codes. Hence, in a single-path environment, the downlink contains no intra-sector interference.[210] Major interference in the downlink comes from other sectors because the base stations are transmitting signals to their own users. In the uplink, the signals being transmitted from different users have no orthogonality. Hence, the uplink experiences significant intra-sector and inter-sector interference. The intra-sector interference in the uplink can be as high 60% of the total uplink interference.[211] In Figure 7.5, when a BS attempts to decode MS1's signal in Sector α, the users in Sector α and the users in nearby sectors cause interference to the signal of MS1.

[209] When frequency hopping is used in conjunction with an identical set of carrier frequencies in all sectors of a GSM system, these frequencies randomly collide with one another (but not continuously). Furthermore, multiple GSM carrier frequencies exist within the wideband spectrum bandwidth occupied by one wideband CDMA carrier frequency.

[210] In IS-95 and 1xRTT systems, all the downlink radio channels use codes orthogonal to one another, avoiding intra-sector interference in a single-path environment case. However, in UMTS, some downlink channels are transmitted with no orthogonal code. Hence, UMTS' downlink has non-zero intra-sector interference.

[211] This value is based on observations of typical CDMA deployments.

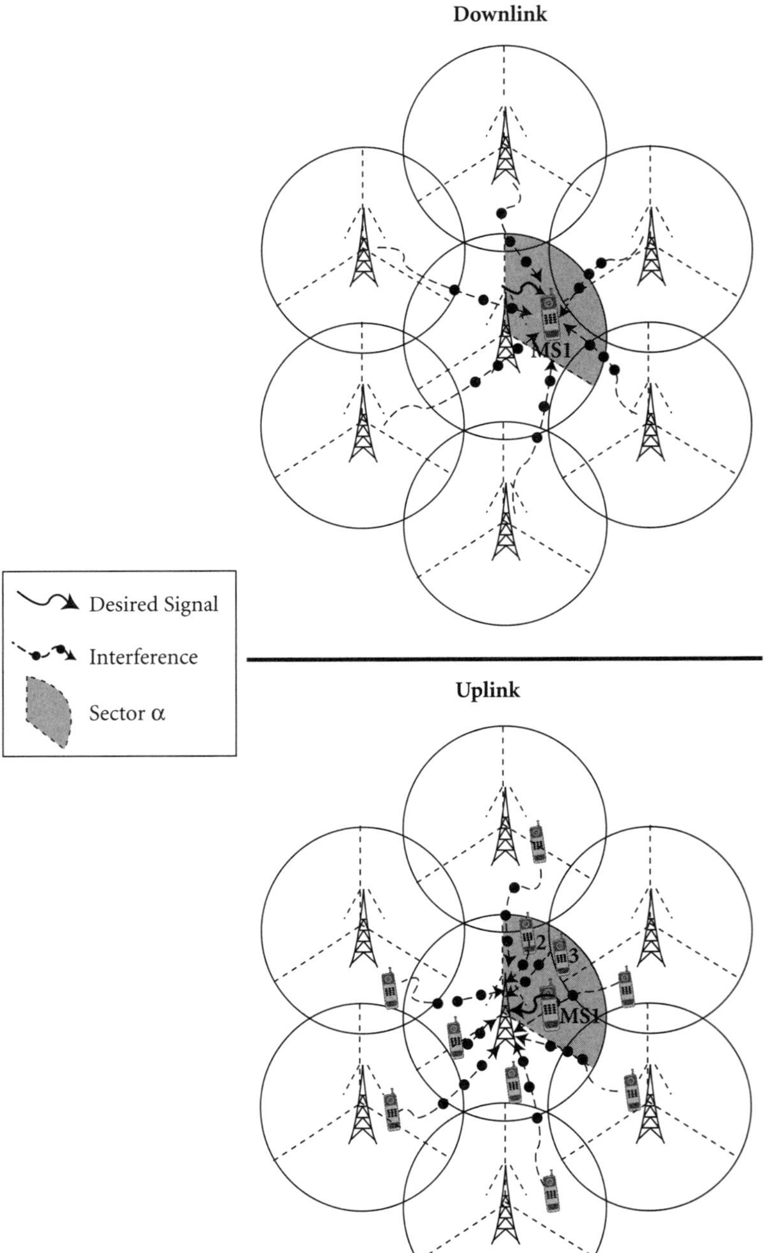

Figure 7.5: Interference in a Single-Path Environment in CDMA

In practice, an embedded sector (i.e., a sector with surrounding sectors as opposed to a border sector on the edge of cellular coverage) experiences significant interference from the first two tiers of surrounding sectors. In summary, in a single-path case, the downlink contains zero or minimal intra-sector interference, and the uplink contains significant intra-sector interference. Furthermore, out-of-sector interference is significant in the downlink and uplink.

Is it possible that the downlink contains non-negligible intra-sector interference? The answer is *yes*. In the case of a multipath-rich environment, significant intra-sector interference occurs as illustrated in Figure 7.6.

The BS sends to MS1 a signal from a transmit antenna. Part of the signal energy travels along the direct or line-of-sight path to MS1. However, part of the signal energy gets reflected from a building and reaches MS1 via a non-direct or reflected path. The BS transmits just one signal for MS1, but MS1 receives two copies of the signal. Since both paths contain the same code for MS1 and since the reflected path takes longer to reach MS1, self-interference occurs at MS1. Note that the orthogonality is guaranteed only when the orthogonal codes at the receiver are fully aligned. In our example, let $[1\,1\,-1\,-1]$ be the code allocated to MS1 in the downlink. The reflected path has a time offset of two chips, meaning that it arrives two chips later than the direct path.

Let's use the time period of four chips, which we will call *processing window* hereafter, to test for orthogonality. In one processing window, the direct path chips are $[-1\,-1\,1\,1]$, and the set of four chips in the reflected path are $[1\,1\,-1\,-1]$. Carrying out chip-by-chip multiplication and subsequent addition yields $\{(1)(-1)+(1)(-1)+(-1)(1)+(-1)(1)\} = -4$, which is not 0. Hence, no orthogonality exists between the direct and reflected paths. In summary, when a time offset occurs between the orthogonal codes, orthogonality of the codes is no longer guaranteed.

Now that we know the sources of interference in a CDMA system, let's briefly look at how to counteract such interference. Major mechanisms that facilitate interference

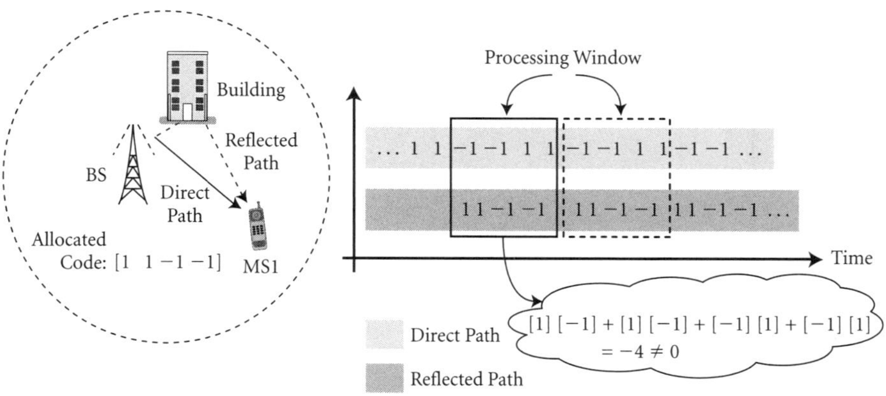

Figure 7.6: Intra-Sector Downlink Interference in a Multipath-Rich Environment

control are soft handoff or handover, power control, and RAKE receivers. Soft handoff leads to reduction in the average transmit power requirements near the cell edge, reducing average interference. Section 7.3.4 deals with soft handoff. The goal of a power-control mechanism is to help achieve a target error rate while minimizing transmit power. The lower the transmit power, the less the interference. Section 7.3.5 discusses power control in CDMA. A CDMA system exploits a special type of a receiver called a *RAKE* receiver. Let's take a look at the basic operating principle of a RAKE receiver, which is illustrated in Figure 7.7.

A RAKE receiver gets its name from its resemblance to a gardening rake; it consists of multiple fingers. Each finger maintains its lock on a specific signal (and demodulates such a signal), such as direct or reflected multipaths, and these signals are given an appropriate weight before signal combining occurs. Examples of generic signal combining techniques include equal-gain combining, selection combining, and maximal ratio combining. Equal-gain combining gives the same weight to each signal. For example, the combined signal is calculated as $s = 0.5 * s_1 + 0.5 * s_2$ where s_1 and s_2 are two input signals. Here, 0.5 represents 50% weight. Selection combining just selects one of the input signals. For example, the combined signal is calculated to be s_1 or s_2. The term *selection combining* is a misnomer because no real *combining* occurs; only selection occurs! However, the term *selection combining* has been widely used in the literature.

Finally, maximal ratio combining is our greatest interest at this time. Why? Because the RAKE receiver typically uses maximal ratio combining to combine the signals! For example, the combined signal in the case of maximal ratio combining can be calculated as $s = \alpha s_1 + \beta s_2$ where α and β are complex weights that may be a function of the signal-to-interference ratio (SIR) on a given finger. Numerical values of α and

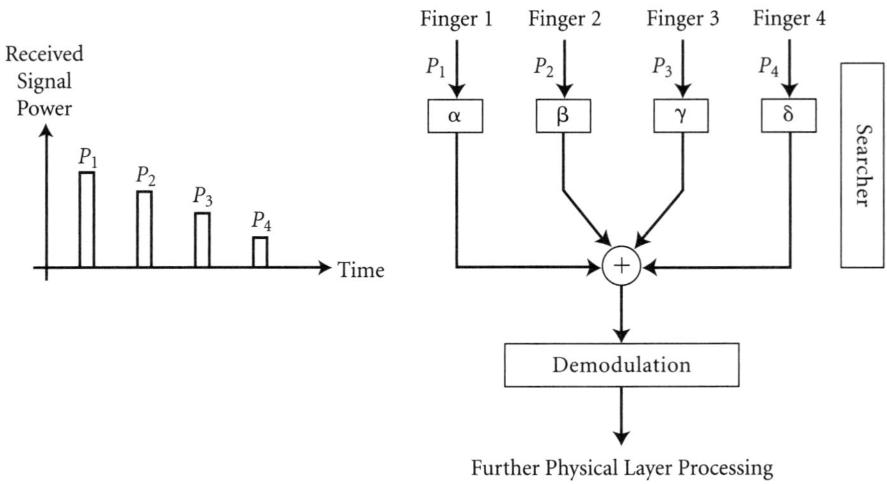

Figure 7.7: A Simplified View of the Operation of a RAKE Receiver

β may be 0.9 and 0.1, respectively, if the signal s_1 has very high SIR and the signal s_2 has low SIR. Figure 7.7 shows four fingers that contribute modulation symbols to the demodulation process. Early RAKE receivers implemented four fingers to simplify implementation and to save processing power. To enhance performance in a multipath-rich environment, six or even eight fingers became available at receivers in late 2010.

Multipath signals cause inter-symbol interference. In non-CDMA systems, such as GSM, multipath signals are enemies of the receiver; the receiver is designed to fight against multipath signals! However, in CDMA systems, we welcome multipath signals! In Figure 7.7, four multipath signals, P_1 through P_4, are present with P_1 being the first arriving multipath signal and P_4 being the last arriving multipath signal. Four fingers are able to track four paths, and the RAKE receiver exploits energy contained in these multipath signals to obtain a more reliable combined signal. The receiver dynamically determines the weights, α, β, γ, and δ, for every modulation symbol.

Of course, if a fifth multipath signal is present, we can no longer combine its energy with other multipath signals, and then the power of such a multipath signal contributes to *dispersed power*, which acts as pure interference. The RAKE receiver also has a searcher finger that continuously looks for a better multipath signal. If a stronger multipath signal is found, it replaces one of the existing multipath signals the fingers are processing. The RAKE receiver is implemented at the MS and the BS. The BS has one RAKE receiver per user in the uplink, even when multiple sectors are simultaneously communicating with a given user.

7.3.4 Handoff

As the user travels from one sector to the next, the radio channels for the user must switch to maintain the communications between the user and the cellular network. Such handoff is carried out differently for different cellular systems. By the way, you may find some folks in the communications industry who are under the impression that handoff and handover are different. *Handoff and handover are the same; there is absolutely no difference between the two*. It is simply a difference of semantics; historically, the word *handoff* has been used in conjunction with systems based on IS-95, 1x, and 1xEV-DO (1xEvolution—Data-Optimized or —Data-Only), and the word *handover* is used in discussing UMTS.

In a hybrid TDMA/FDMA system, the handoff is *hard* or *break-before-make*, meaning that the communications link between the old sector and the MS is broken first, and then the link between the new sector and the MS is established. In contrast, CDMA systems inherently support *soft* or *make-before-break* handoff in which the MS simultaneously communicates with multiple sectors, and the link between the new sector and the MS is established before the link between the old sector and the MS is (eventually) terminated. Figure 7.8 compares handoff in TDMA and CDMA systems.

The MS is initially communicating with Cell A and is moving into Cell B. As the MS moves away from Cell A and toward Cell B, the signal quality for the communication between Cell A and the MS gradually degrades and the achievable

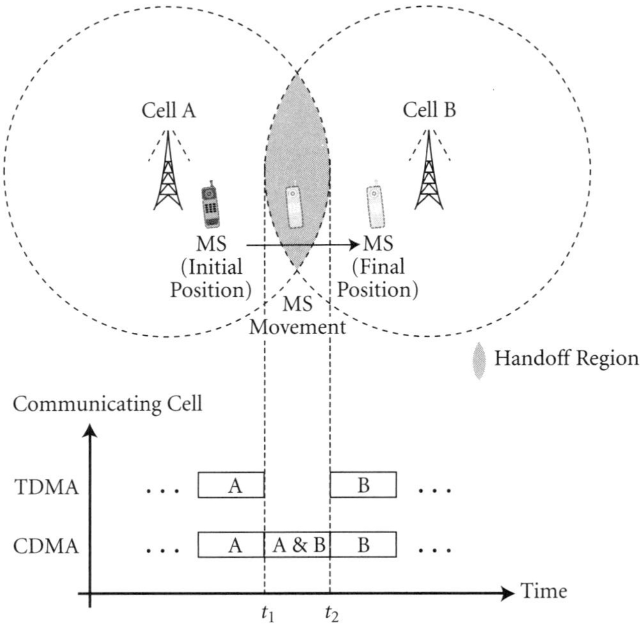

Figure 7.8: Hard Handoff versus Soft Handoff

signal quality for the communications between Cell B and the MS gradually improves. Hence, at some point, we need to execute handoff. Let's discuss the handoff scenario in TDMA systems. The MS communicates with Cell A and receives a handoff command from Cell A at time instant t_1. The radio link between Cell A and the MS is released. Communication between Cell B and the MS begins on a new radio channel at t_2. No communication between the MS and the network occurs during the time interval between t_1 and t_2.

Now, let's turn our attention to handoff for CDMA. The MS communicates with Cell A up to t_1. The MS then enters the handoff region about midway between the two cell sites. In this region, the MS communicates with both Cell A and Cell B simultaneously via two radio channels from t_1 to t_2. The MS moves farther away from Cell A, and it begins communicating with only Cell B at t_2. In summary, in hybrid TDMA/FDMA system handoff, some communications gap occurs and the MS communicates with just one sector at a given time instant. In contrast, in CDMA system handoff, no communications gap occurs and the MS communicates with one or more sectors at a given time instant. The cells or sectors communicating with the MS in a CDMA system constitute an *Active Set*. All sectors in the Active Set transmit identical information bits to the MS, and they listen to the MS for the uplink information bits it transmits. The Active Set may have up to six sectors in IS-95 and 1x.

Do You Recall... Time to test your ultra–short-term memory! What is the name of the receiver used in CDMA, and what is the benefit of such a receiver? Of course, the footnote[212] has the answer.

The main benefits of soft handoff are enhanced link reliability in both the downlink and uplink and reduced average transmit power per link. The fingers of the RAKE receiver can track multipaths of a signal being transmitted from one sector antenna as well as signals being transmitted from multiple sectors. For example, in a rural environment with predominantly one path (i.e., just one propagation path between the transmitter and the receiver), one finger of the MS's RAKE receiver can lock onto one sector and another finger can lock onto another sector when the MS is communicating with two sectors. In the uplink, if the BS serves three sectors and each sector has two receive antennas, the RAKE receiver for a given user has access to the received signals from all six antennas. A four-finger RAKE receiver then can choose the four best signals from those received from the MS at suitable sector antennas. The RAKE receiver can separate one path from another path (which could be for the same sector or a different sector) as long as these paths are at least one chip apart when they arrive at the receiver.

Different types of soft handoff are (basic) *soft handoff*, *softer handoff*, and *softer-soft handoff*. When we use the term *soft handoff*, we generally mean all these types of handoff. We will differentiate among these handoff types when required. Let's discuss these handoffs using Figure 7.9 as a basis. In Figure 7.9, BS 1 controls Cell 1, which consists of three sectors, A, B, and C. Similarly, BS 2 is in charge of Cell 2, which controls Sectors D, E, and F. Consider MS1, which has the Active Set $\{B, D\}$. When the Active Set includes multiple BSs with just one sector being contributed by each BS, the type of handoff is termed soft handoff. Since Sector B is controlled by BS 1 and Sector D is controlled by BS 2, we essentially have two BSs in the Active Set. Hence, MS1 is said to be in (basic) soft handoff.

Now, consider MS2 with the Active Set $\{A, C\}$. When the Active Set includes multiple sectors of the same BS, the type of handoff is called *softer handoff*. Since both Sector A and Sector C belong to the same BS, MS2 is in softer handoff.

Finally, when soft handoff and softer handoff are combined, the resulting handoff is *softer-soft handoff*. MS3 has the Active Set $\{B, C, F\}$. Since multiple sectors of the same BS, BS 1 in this case, are in the Active Set, it corresponds to softer handoff. However, two BSs, BS 1 and BS 2, are involved in the Active Set, corresponding to soft handoff. Since we have both softer handoff and soft handoff, the type of handoff is referred to as softer-soft handoff. Defining such handoffs allows proper resource utilization quantification. For example, compared to softer handoff, soft handoff and softer-soft handoff need larger backhaul bandwidth because the same user information is flowing between the base station controller (BSC)[213] and multiple BSs.

[212] It is the RAKE receiver. The main benefit is improved reliability due to maximal ratio combining of received signals such as multipath signals.

[213] Another term for the BSC is the radio network controller (RNC).

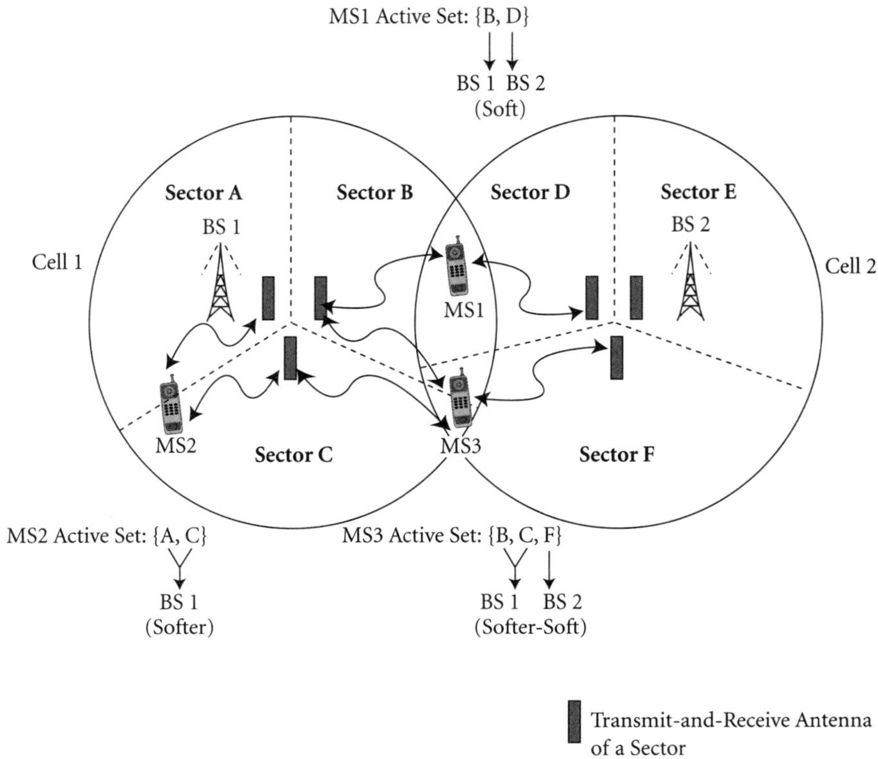

Figure 7.9: Types of Soft Handoff

Now that we know the basics of soft handoff, let's see how signals are processed in the downlink and the uplink using the most comprehensive soft handoff scenario, which is softer-soft handoff. For simplicity of discussion, we will assume a single-path environment; on a given receive antenna, one copy of the transmitted signal will be received. However, the concepts being discussed are valid even when we have more than one path. Figure 7.10 discusses signal processing in the downlink during soft handoff. Assume that the Active Set for the MS is $\{B, C, F\}$ with BS 1 and BS 2 participating in handoff.

Let's say that we want to send a frame containing voice bits to the MS. The BSC receives such a frame from the circuit-switched core network. Since two BSs, BS 1 and BS 2, are part of the Active Set because they control relevant sectors, they receive the same speech frame from the BSC. Physical layer processing is carried out at BS 1 and BS 2, and such a processed speech frame is transmitted over the air. Since BS 1 controls Sectors B and C, the same processed speech frame is sent from the transmit antenna of Sector B and the transmit antenna of Sector C. The transmit antenna of Sector F also sends the processed speech frame to the MS. The RAKE receiver at the MS now receives three signals (because we have assumed a single-path environment).

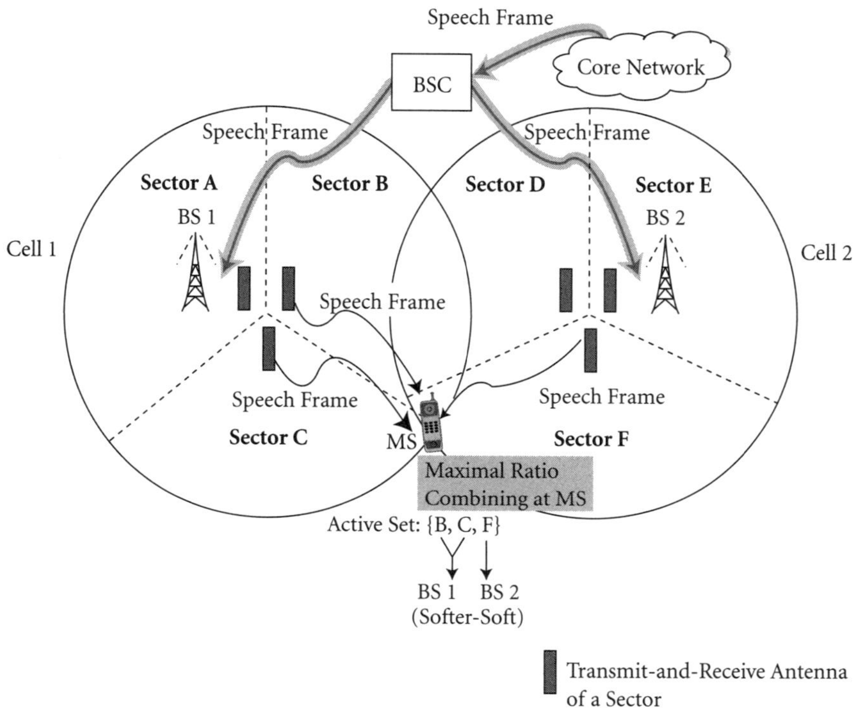

Figure 7.10: Downlink Processing During Soft Handoff

It performs maximal ratio combining of the three signals to retrieve the speech frame. In summary, during soft handoff, the sectors in the Active Set transmit the signal to the MS and the MS carries out maximal ratio combining.

Now, let's turn our attention to soft handoff processing in the uplink, as illustrated in Figure 7.11, which is more *interesting* (the authors' euphemism for *complex*!) than that of the downlink. Our goal here is to send the speech frame from the MS to the core network. The MS sends out the speech from its omnidirectional antenna. All the sectors in the Active Set listen to the MS. Since two receive antennas are typical at a sector due to space diversity, each sector now receives two signals from the MS, i.e., one signal from each antenna. Since the *base station* (and most certainly not the *sector*!) has the RAKE receiver for the MS, the receiver at BS 1 has access to a total of four signals and the receiver at BS 2 has access to two signals. Each RAKE receiver performs maximal ratio combining and attempts to recover the speech frame.

BS 1 and BS 2 send the retrieved speech frames, X and Y, respectively, to the BSC along with an indication about the frame quality. Such an indicator specifies whether the frame is good or contains some bit errors. The BSC, in our example, receives these two frames, selects one of the frames (a good one if available), and forwards the frame to the core network. In summary, during soft handoff, the sectors in the Active

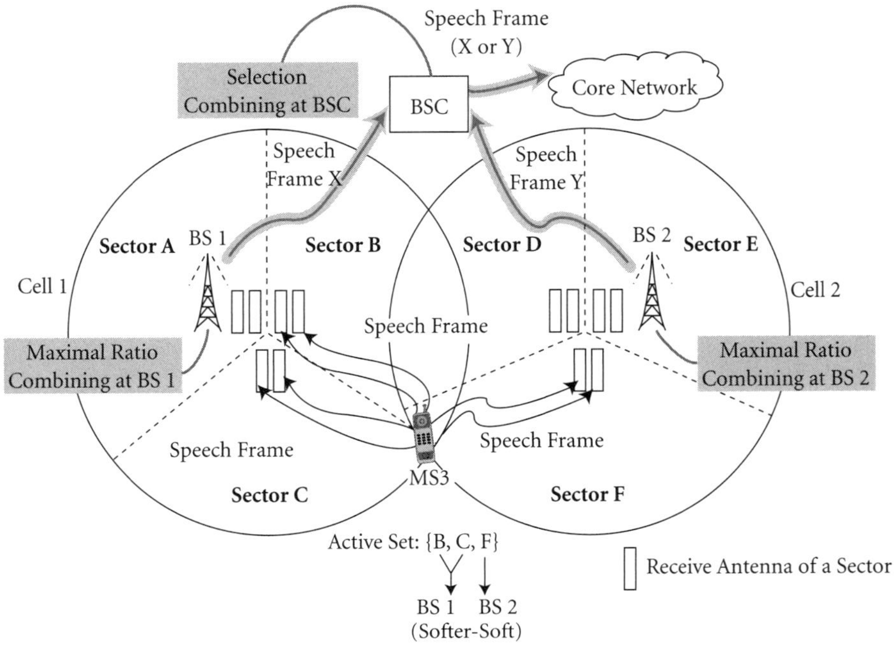

Figure 7.11: Uplink Processing During Soft Handoff

Set receive the signal from the MS, the BS performs maximal ratio combining, and the BSC carries out selection combining. The MS houses the RAKE receiver, and the BS contains CEs that house RAKE receivers. For every MS that is communicating with the BS, one CE is reserved regardless of the number of BS sectors in the Active Set. In other words, whether an MS has one BS sector or two sectors of the same BS in the Active Set, a single CE is needed. In Figure 7.11, one CE is reserved for MS3 in BS 1 and another CE is reserved for MS3 in BS 2. If a BS is communicating with ten MSs, the BS uses ten CEs.

Sectors per user (SPU) is a useful metric that quantifies how much soft handoff we have in the system. If an MS communicates with one sector one-third of the time, two sectors one-third of the time, and three sectors one-third of the time, the average SPU is $(1/3) * 1 + (1/3) * 2 + (1/3) * 3 = 2.0$. Observations of optimized commercial deployments suggest that the typical SPU value ranges from approximately 1.6 to 2.2 with a lower SPU more common for UMTS and a higher SPU more common for IS-95 and 1x systems. The larger the SPU, the more reliable the communication and the greater the degree of resource utilization. The *right* SPU amount finds an optimal balance between the capacity and the communication reliability. SPU also quantifies the radio resource utilization. For example, when $SPU = 2$, the MS, on average, requires downlink power to be allocated in two sectors and requires two Walsh codes or OVSF codes.

To enable the RAKE receiver to combine received signals properly, the BSs' transmission timings must be synchronized. As long as the BS transmissions are within a few chips, the RAKE receiver at the MS can combine the signals from different BSs. The RAKE receiver can separate two propagation paths as long as these paths are more than one chip apart. Systems based on IS-95, 1xRTT, and 1xEV-DO use GPS (Global Positioning System) satellites for synchronous downlink transmissions from BSs. UMTS-based systems typically do not use GPS and hence need the BSC's help to instruct new BSs that are added to the Active Set to adjust the BS transmission timings. The uplink transmission timing is typically relative to the DL timing and hence needs no special considerations once the DL transmissions are properly time-aligned.

Before leaving the topic of handoff, let's summarize the implications of soft and softer handoff when the Active Set contains two sectors. Since soft handoff involves multiple base stations and softer handoff involves two sectors of the same base station, spatial diversity gain is likely to be higher in soft handoff. Resource utilization, such as backhaul between the base station and the base station controller and channel elements, would be lower in softer handoff. Since the channel element performs the RAKE receiver processing for a given user, MRC is possible at the base station. The typical presence of two receive antennas per sector enables the MRC gain at the base station, but overall MRC gain may be larger for softer handoff due to the availability of more signals for combining.

7.3.5 Power Control

Power control is one of the most critical design elements of a CDMA system. *The goal of the power-control mechanism is to determine the minimum transmit power that will adequately meet the target quality-of-service.* One way of quantifying quality-of-service is the frame error rate (FER) or block error rate. For example, if speech is being carried in the form of speech frames or speech blocks, FER or block error rate is a good quality-of-service metric. Typical FER or block error rate targets for voice services are 1% to 3% in commercial deployments. The goal of the power control is then to ensure that the actual FER is as close as possible to the target FER. Several types of power-control mechanisms are available as summarized in Figure 7.12.

In an open-loop power-control approach, the transmitter on its own changes the transmit power level; the receiver offers the transmitter no instructions about how to change the power level. In contrast, a closed-loop power-control method involves explicit feedback from the receiver to the transmitter on how to change the power level. Specifically, the receiver sends power-up or power-down commands to the transmitter, and the transmitter changes its power level accordingly. The closed-loop power control is possible to implement in both the downlink and uplink and consists of two mechanisms, an inner-loop power control and an outer-loop power control. The outer-loop power control dynamically determines a suitable value of the target SIR that helps achieve the target FER and provides this value to the inner-loop power

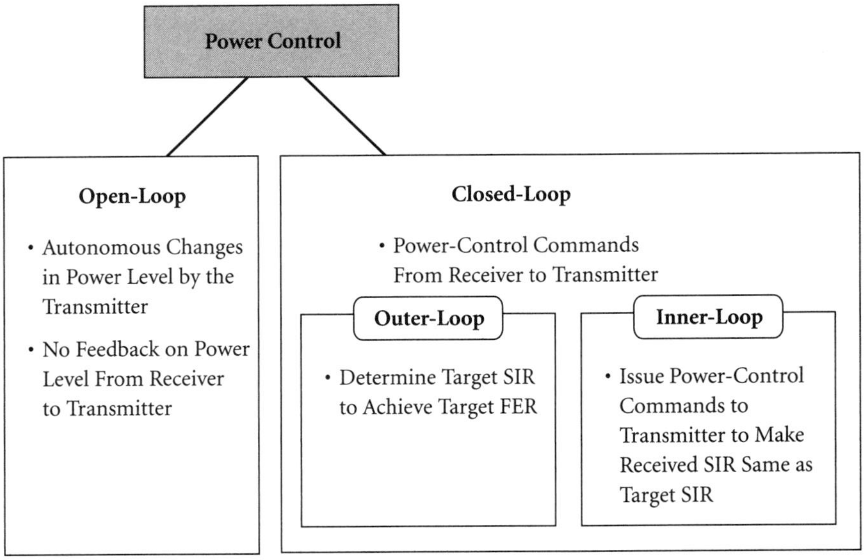

Figure 7.12: Summary of Power-Control Methods

control. The inner-loop power control issues power-up or power-down commands to make sure that the actual SIR and the target SIR are as close as possible to the target levels. Section 7.3.5.1 discusses the open-loop power control, and Section 7.3.5.2 provides an overview of the closed-loop power control.

7.3.5.1 Open-Loop Power Control As mentioned earlier, the open-loop power control involves the transmitter's voluntary power-level adjustments with no explicit feedback from the receiver. Such power control is typically used for MS transmit power control when the MS has no dedicated radio channel.[214] Specifically, the MS adjusts the transmit power of the random Access Channel, which is a common uplink channel that the MS uses to contact the radio network. Figure 7.13 illustrates how the MS adjusts its transmit power while transmitting on the random Access Channel.

The signal the MS sends on the random Access Channel is often referred to as an *access probe*. The MS transmits the first access probe at some initial power level that depends on network parameters and the estimated path loss[215] between the MS and the BS. If the MS gets no response from the BS, it assumes that its transmit power is inadequate. It then sends another access probe with a higher power level, such

[214]Sometimes the closed-loop power control makes adjustments using the open-loop power control as the baseline.

[215]To estimate the path loss, the MS uses the transmit power of a channel such as the Pilot Channel that the network advertises as well as the received power level of such a channel.

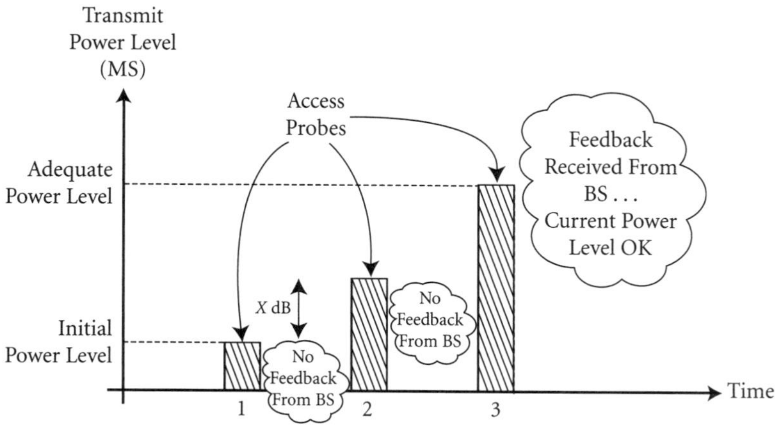

Figure 7.13: Open-Loop Power Control

as X dB above the previously used power level.[216] In our example, the MS gets no response from the BS after transmitting the second access probe. The MS then sends the third access probe at a power level that is X dB greater than the power level of the second probe.

The BS is now able to detect the probe from the MS and sends an acknowledgment to the MS. The MS determines that it no longer needs to use a larger transmit power since it has found an adequate transmit power level under current channel conditions. In summary, the MS, on its own, is changing its transmit power; the BS sends no explicit power-control commands to the MS. Since no explicit feedback is sent from the receiver (which is the BS in our example) to the transmitter (which is the MS in our case), such a power-control mechanism is called an open-loop power control.

7.3.5.2 Closed-Loop Power Control Let's see how we can implement a closed-loop power-control mechanism. A word of caution: the specific power-control algorithm is usually implementation-dependent[217] and is different for different network infrastructure vendors. We will look at a simple algorithm here.

[216] The transmit power from the MS could possibly be adequate, but the access probe transmission from one MS could collide with the access probe transmission from another MS. In such a case, the BS cannot correctly decode the access probe transmissions, resulting in the absence of an acknowledgment from the BS.

[217] For example, the exact condition under which and the exact amount by which the SIR value is changed may differ from one vendor to another vendor.

Do You Recall... Do you recall the main difference between open-loop power control and closed-loop power control? The footnote[218] has the answer.

Figure 7.14 illustrates the basic concept of the closed-loop power control in the uplink. A similar concept is applicable to the downlink as well.

We know that the closed-loop power control consists of an outer-loop power control and an inner-loop power control. In the uplink, the BSC implements the outer-loop power-control algorithm, and the inner-loop power-control algorithm is implemented at the BS.

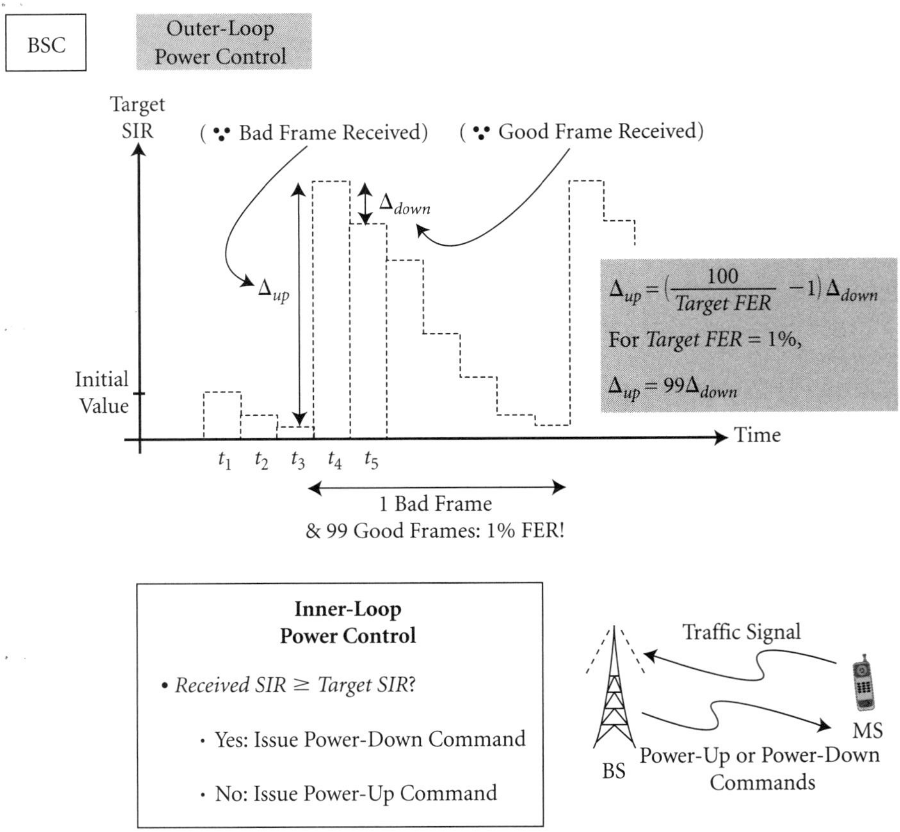

Figure 7.14: Closed-Loop Power Control for the Uplink

[218]The closed-loop power control involves explicit power-control commands from the receiver to the transmitter. The open-loop power control has no such explicit feedback.

The outer-loop power-control algorithm attempts to determine a suitable target SIR that would meet the required target FER. A simple algorithm that dynamically determines a target SIR value is shown in Figure 7.14. Assume that the radio channel conditions remain the same. For such channel conditions, some SIR value would help meet the target FER.

The algorithm follows a simple rule: When a good frame is received, it reduces the target SIR by Δ_{down}, and when a bad frame is received, the target SIR is increased by Δ_{up}. The relationship between Δ_{up} and Δ_{down} can be defined as $\Delta_{up} = ((100/(Target\ FER)) - 1)\,\Delta_{down}$. Hence, if the target FER is 1%, $\Delta_{up} = 99\Delta_{down}$. Let's see how such a simple implementation could get us pretty close to the target FER. The initial target SIR is assumed to be some value. During the time interval t_1, the received frame is free of errors. Hence, the target SIR is decreased by Δ_{down}. Similarly, during the time intervals t_2 and t_3, the target SIR is decreased. We are decreasing the target SIR in hopes that a lower SIR target may still be adequate to meet the target FER. In such a case, a lower SIR target would lead to a lower transmit power and hence lower interference without adversely affecting the target FER. The next frame received during the time interval t_4 is bad. Hence, the target SIR is increased by Δ_{up}. Since the channel conditions are assumed to be the same, we will again have an error after taking 99 down-steps. Within the observation interval of 100 frames, we will have one step-up and 99 steps-down corresponding to one bad frame and 99 good frames, resulting in 1% FER, which is our target!

The inner-loop power control periodically receives updates on the target SIR from the outer-loop power control. In the uplink, the BS measures the received SIR and compares the actual SIR with the target SIR. If the received SIR is greater than or equal to the target SIR, the signal power is more than needed for the prevailing interference. Hence, the BS sends a power-down command to the MS. However, if the received SIR is less than the target SIR, the signal power is inadequate under the prevailing interference. So, the BS sends a power-up command to the MS. If you are ever on a game show and are asked how fast the power control is, you'll know the BS sends to the MS 800 power-control commands per second in IS-95 and 1x and 1,500 power-control commands per second in UMTS.

In summary, the inner-loop power-control algorithm implemented at the BS regularly receives target SIR values from the BSC, and the BS sends power-control commands to the MS. The outer-loop power-control algorithm at the BSC calculates the required target SIR to meet the target FER.

Something to Think About... What happens to the target SIR if channel conditions change? No need to worry at all! If the channel conditions continue to improve, we would keep decreasing the target SIR until a frame error occurs. Similarly, if the channel conditions worsen, we would have more frame errors and more steps-up. In essence, the outer-loop power-control algorithm would converge to the correct target SIR that would result in an average FER close to the target FER.

We used uplink power control as an example of closed-loop power control. A similar approach can be used for the downlink power control. Of course, both the outer-loop power-control algorithm and the inner-loop power-control algorithm would need to be implemented at the MS. The outer-loop power-control algorithm attempts to find a suitable target SIR to achieve the target FER for the downlink. The inner-loop power-control algorithm sends power-up or power-down commands to the BS so that the actual downlink SIR and the target SIR are as close as possible.

Let's re-visit the uplink closed-loop power control. Since the BS is making actual SIR measurements, it is in the best position to implement the inner-loop power control because it can quickly send out power-control commands to the MS. If we were to rely upon the BSC to come up with the power-control commands, we would introduce an unacceptable delay. This delay arises from the backhaul delay between the BS and the BSC and processing delay at the BSC. The idea behind closed-loop power control is to quickly adapt to the changing channel conditions; introduction of delay would adversely affect the timeliness of closed-loop power control.

A Question for You... Why is the outer-loop power control implemented at the BSC and not at the BS? After all, the inner-loop power control is implemented at the BS. A short answer would not do justice to this question! The answer has to do with soft handoff. We will learn about the interaction between soft handoff and power control, and, along the way, we will get the answer to our question.

Consider Figure 7.15 in which the MS is in softer-soft handoff. The Active Set of the MS is $\{B, C, F\}$. BS 1 performs maximal ratio combining of the signals received from the MS and forwards the retrieved Frame P to the BSC. Similarly, BS 2 forwards Frame Q to the BSC. In our example, Frame P has no errors, but Frame Q is erroneous. The BSC performs selection combining and forwards Frame P, a good frame, to the core network. Since the received frame at the BSC is good, the outer-loop power control decreases the target SIR. The BSC sends an updated target SIR, $SIR = X$ dB, to BS 1 and BS 2. The inner-loop power-control algorithms running at individual BSs compare the measured SIR with target SIR. The measured SIR at BS 1 exceeds the target SIR, and, hence, BS 1 sends a power-down command to the MS from Sectors B and C.

A Question for You... The same power-control command is sent from the sectors of a given BS. Can you guess the resulting benefit? See the footnote[219] for the answer!

[219]The same power-control command from the sectors of a given BS enables the MS to perform maximal ratio combining of the power-control commands to reliably detect the command. Furthermore, the SIR measured by the BS reflects the signal received from multiple sectors of the same BS. The BS combines multipaths received from all the sectors and therefore sends a single PC command to reflect the combined received signal.

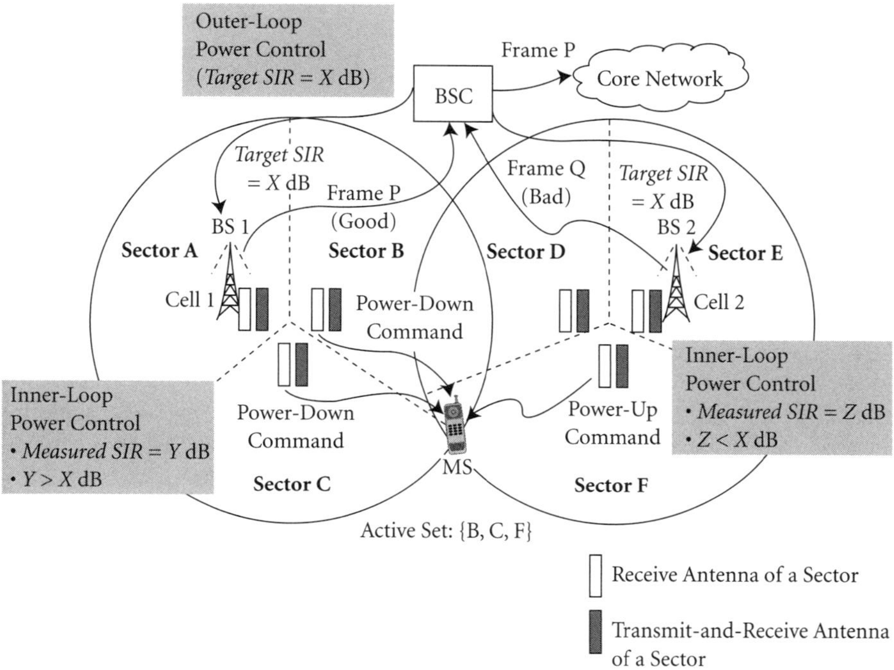

Figure 7.15: Interaction Between Soft Handoff and Power Control

The SIR measured by BS 2 is lower than the target SIR, and, hence, the inner-loop power control at BS 2 sends a power-up command to the MS. Our MS has now received two conflicting power-control commands, a power-down command from BS 1 and power-up command from BS 2. Since the goal of power control is to use the absolute minimum power to meet the target FER, the MS decreases its power if it receives one or more power-down commands. Doing so will save battery power at the MS and reduce the uplink interference without adversely impacting the quality.

Now, coming back to our original question, why is the outer-loop power control implemented at the BSC? Obviously, the BSC knows the *final* quality of the received uplink frame.[220] An individual BS knows the quality of the frame it has received but knows nothing of the frame quality at another BS. Hence, the BSC is best positioned to update the target SIR needed to ensure the target FER.

One last thought on power control before we leave the topic: The average received power level for different MSs at the BS is the same due to power control when the MSs are using the same data rate. The nature of the uplink interference is such that the total interference is the same for all MSs in such a case. Different distances from

[220]In the absence of soft handoff, one BS has the correct knowledge of the final frame quality.

the BSs result in different transmit power levels for the MSs but similar received power levels. The uplink power control plays a critical role in solving an important problem called the *near-far effect*, in which the MS *near* a BS causes significant uplink interference to the MS that is *far* from the BS. In the absence of uplink power control, the signal power received at the BS from the nearby MS could be much larger than that from the distant MS due to differences in the propagation path loss. Hence, the signal for one MS could be lost in the interference caused by another MS's signal. The uplink power control ensures that the received power levels for all MSs are similar and that no single MS is able to dominate other MSs.

7.3.6 Capacity of a CDMA System

Sector capacity can be defined as the maximum number of users that can be supported in a sector. The capacity of a CDMA system is soft (as opposed to hard); it depends on the prevailing interference. Determination of the CDMA capacity is complex or interesting; the capacity is different for the downlink and the uplink because the interference differs. Section 7.3.6.1 discusses the uplink capacity, and Section 7.3.6.2 describes an approach to calculate the downlink capacity. The downlink usually contrains the sector capacity for voice services. In other words, we usually run out of downlink capacity before we run out of uplink capacity.

7.3.6.1 Uplink Capacity Assume that the BS receives power S from each of the N MSs in a sector.

$$\frac{E_b}{N_0} = PG * \frac{S}{(N-1)S + I_{oc} + N_{th}} \tag{7.1}$$

where E_b/N_0 is the effective SIR to meet the target error rate; PG is the processing gain, which is equal to W/R, which is equal to the spreading rate over the data rate; S is received signal power; I_{oc} is out-of-cell interference; and N_{th} is thermal noise power. Let I_{ic} represent in-cell interference. Let

$$f = \frac{I_{oc}}{I_{ic}}.$$

Therefore,

$$\frac{E_b}{N_0} = \frac{S}{I_{ic} + fI_{ic} + N_{th}} = \frac{S}{(1+f)I_{ic} + N_{th}} \tag{7.2}$$

$$\frac{E_b}{N_0} = PG\frac{S}{(1+f)S(N-1) + N_{th}}. \tag{7.3}$$

If vaf is the voice activity factor, the signal experiences interference of $(S \cdot vaf)$ on average.

$$\frac{E_b}{N_0} = PG\frac{S}{(1+f)(S \cdot vaf)(N-1) + N_{th}}. \tag{7.4}$$

Let's ignore N_{th} for a moment.

$$\frac{E_b}{N_0} = PG\frac{S}{(1+f)S(N-1)vaf} \tag{7.5}$$
$$\frac{E_b}{N_0} = \frac{PG}{(1+f)(N-1)vaf}$$
$$\therefore (N-1) = \frac{PG}{((E_b)/(N_0))\,(1+f)vaf}$$
$$\therefore N = 1 + \frac{PG}{((E_b)/(N_0))\,(1+f)vaf}$$
$$F = \frac{1}{(1+f)} = \mathit{frequency\ reuse\ efficiency}$$
$$\therefore N = 1 + \frac{PG \cdot F}{((E_b)/(N_0))\,vaf} \tag{7.6}$$

N in Eq. (7.6) is called the *pole capacity*, N_{pole}.

$$N_{pole} = 1 + \frac{PG \cdot F}{((E_b)/(N_0))\,vaf} \tag{7.7}$$

In Eq. (7.7) all the signal power is used to fight interference, and no signal power is used to overcome path loss. In other words, Eq. (7.7) corresponds to a theoretical case where all MSs are at the cell site. In practice, the uplink is operated at a certain decibel level above the thermal noise floor, such as 3 dB. This is called *rise-over-thermal* noise. When *rise-over-thermal noise* $=$ 3 dB, the interference and thermal noise power are equal, and the total received power at the sector is 3 dB above thermal noise floor. The 3-dB rise-over-thermal noise corresponds to 50% loading (i.e., 50% of pole capacity).[221] The actual number of users in the uplink cannot exceed N_{pole}.

7.3.6.2 Downlink Capacity Let's calculate the downlink capacity. When the power available at the BS is a bottleneck,

$$N = \frac{100 - P_{overhead}}{P_{user} \cdot SPU} \tag{7.8}$$

where N is the number of users per sector and equals the downlink capacity; $P_{overhead}$ is the percentage of the BS amplifier power reserved for the overhead channels and soft handoff; P_{user} is the percentage of the BS amplifier power required per user to meet target FER; and SPU is the average sectors per user. Eq. (7.8) is empirically derived. It is based on observations that a certain amount of power (as a fraction of the total power) is required to overcome downlink interference and that the BS power is properly distributed among active radio links according to channel conditions on those links.

[221] *Rise-over-thermal noise* (in decibels) $= 10\log_{10}(1/(1-X))$ where X is loading and is calculated as the ratio of the target number of users to the pole capacity. When the loading X is 50% or 0.5, rise-over-thermal noise in decibels is $10\log_{10}(1/(1-0.5)) = 3$ dB.

In early IS-95 systems, $P_{overhead} = 35\%$ was used with about 25% overhead power and 10% soft handoff reserved power. As an example, $SPU = 2.0$. For a specific IS-95 configuration, called Rate Set 2, $N = 20$ and $P_{user} = 1.7\%$ based on observations of typical IS-95 deployments. In summary, for Rate Set 2, the sector capacity is about 20. The data rate for Rate Set 2 is 9.6 kbps for voice. Furthermore, $vaf = 40\%$ is reflected in $P_{user} = 1.7\%$. Eq. (7.8) assumes the downlink power as the capacity bottleneck, which, based on the experience, is valid for commercial CDMA deployments of 2G and typical 3G systems with focus on voice services. Two main radio resources in the downlink are power and Walsh or OVSF codes. The nature of downlink interference is such that the BS runs out of downlink power (while fighting downlink interference) before running out of codes.

Caution! If the maximum power of the BS in each sector is increased from X Watts (e.g., 20 Watts) to Y Watts (e.g., 45 Watts), the sector capacity generally does not increase. The nature of downlink interference is such that that the use of larger transmit power will increase interference. Each user, in turn, requires more absolute power. P_{user}, a percentage of maximum BS power, remains unchanged. Due to independence of the downlink capacity from the *absolute* transmit power of the BS in a sector, Eq. (7.8) uses percentage power levels instead of absolute power levels.

7.4 IS-95 SYSTEM ARCHITECTURE

Figure 7.16 depicts the network architecture of an IS-95–based system, which consists of the radio access network (RAN) and a circuit-switched core network. The RAN

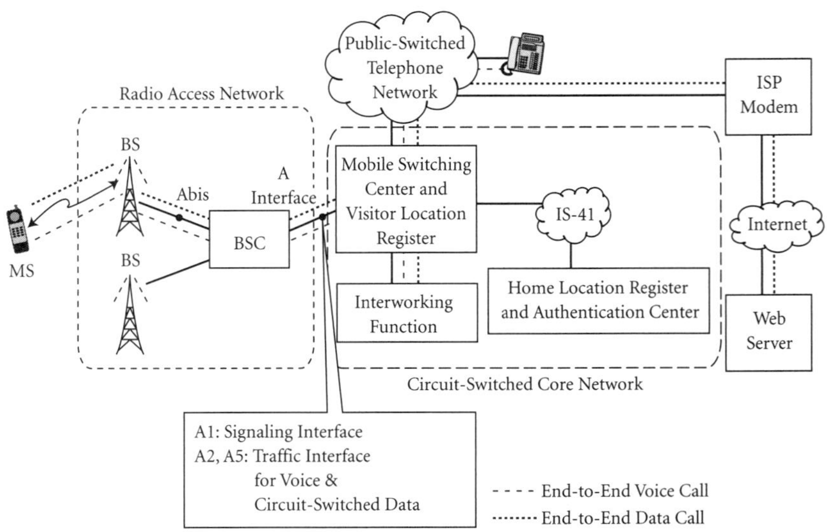

Figure 7.16: IS-95 Network Architecture

consists of BSs and the BSC. The BS communicates with the MSs over the radio interface using CDMA. The BSC typically controls and configures hundreds of BSs using the Abis interface. Most of the RAN intelligence resides at the BSC, and the BSC makes the most of the radio resource management decisions. For example, the call admission control algorithm, handoff algorithm, and uplink outer-loop power-control algorithm are implemented at the BSC. The exchange of signaling messages between the MS and the RAN more specifically occurs between the MS and the BSC. The RAN architecture of IS-95 can thus be classified as a centralized architecture.

The circuit-switched core network includes components such as the mobile switching center, home location register, authentication center, visitor location register, and interworking function. The mobile switching center switches a call between the RAN and the public-switched telephone network and performs billing functions. The home location register stores subscriber information including the last known location of the subscriber. The authentication center stores the secret key of the subscriber to facilitate user authentication. The home location register and the authentication center are usually co-located and implemented in the same physical box.

The visitor location register supports roaming and participates in the user authentication process by acting as a mediator between the MS and the home location register or authentication center. The visitor location register informs the home location register of the most recent location of the MS. The mobile switching center and the visitor location register are usually co-located and are in the same physical cabinet(s). The interworking function consists of a pool of modems to support a circuit-switched data call. From the functional perspective, the circuit-switched core network in IS-95 and the circuit-switched core network in GSM perform similar tasks. Hence, the reader is referred to Chapter 5 on GSM for details on the circuit-switched core network components.

The IS-95 standard itself focuses on the air interface between the MS and the RAN. The interface between the BSC and the mobile switching center is the A interface, with A1, A2, and A5 taking care of signaling, voice traffic, and circuit-switched data traffic, respectively. The A interface is defined in IS-634. The circuit-switched core network components, such as the visitor location register and the home location register, communicate using the IS-41 signaling standard.

7.5 IS-95 MS–RADIO-NETWORK COMMUNICATIONS: AN OVERVIEW

IS-95 uses frequency-division duplexing for simultaneous transmission and reception. Section 7.5.1 discusses duplexing for IS-95. The MS and the RAN follow the rules defined by an air-interface protocol stack. Section 7.5.2 provides a glimpse of this protocol stack, and protocol details are given in Sections 7.5.4 through Section 7.5.6. Various radio channels are discussed in Section 7.5.3.

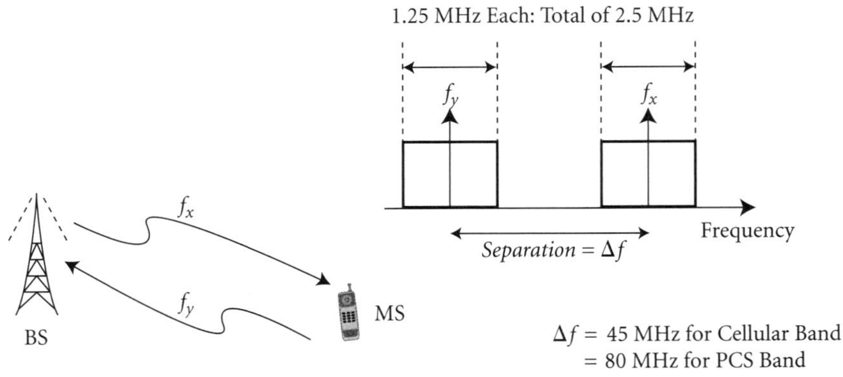

Figure 7.17: Duplexing and Frequency Bands for IS-95

7.5.1 Duplexing: Simultaneous Downlink and Uplink Transmissions

The IS-95 standard defines two frequency bands for operation, the cellular band and the PCS band. Figure 7.17 illustrates how these bands are utilized in the downlink and the uplink.

Since IS-95 is a CDMA-based technology, a wide spectrum bandwidth is required. A total of 2.5 MHz is required for operation, with 1.25 MHz in each direction. In Figure 7.17, the uplink carrier frequency is f_y and the downlink carrier frequency is f_x, with $f_x > f_y$. The separation between the downlink carrier frequency and the uplink carrier frequency is specified to be 45 MHz for the cellular band and 80 MHz for the PCS band.

Do You Recall... We hope that you know why $f_x > f_y$. In the rare event that you can't figure out why, see Chapter 3 on propagation for a detailed answer. A lazy person's answer is in the footnote.[222] Let's see who dares to look at the footnote!

7.5.2 IS-95 Protocol Stack Overview

Figure 7.18 shows the air-interface protocol stack that IS-95 defines. Recall that IS-95 is an air-interface standard that focuses on MS–RAN communications. IS-95 itself has nothing to do with communication between the RAN and circuit-switched core network; IS-634 handles such communication.

[222]The higher the frequency is, the larger the propagation path loss. Since the MS has much less power than the BS, we want to help the MS by whatever fraction of a decibel. Hence, we use a lower frequency for the MS transmission.

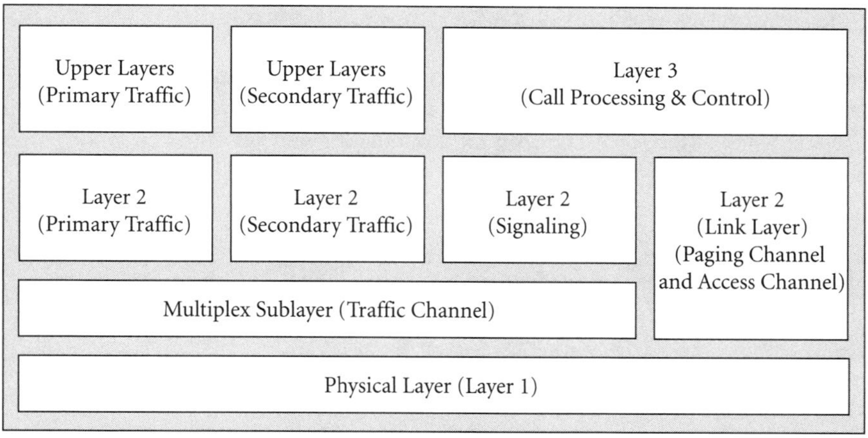

Figure 7.18: Air-Interface Protocol Stack

The bottom-most layer is layer 1, frequently referred to as the physical layer. The physical layer incorporates various processing modules to facilitate CDMA-based over-the-air communications. Examples of such modules include framing, forward error correction coding, modulation, and spreading. Sandwiched between layer 1 and layer 2 is a multiplex sublayer that enables user traffic (e.g., voice) and user-specific signaling messages to share a physical radio channel. In other words, a given physical layer frame for a user can carry both user traffic and a user-specific signaling message on a *traffic channel*. We will provide an overview of all IS-95's radio channels in Section 7.5.3. At this time, it would suffice to know that the traffic channel is a channel dedicated to a given user to transport user-specific information, such as speech bits.

From the perspective of user data, protocol layering above the multiplex sublayer depends on *service options*. For example, the upper layer for Service Option 3 would contain an enhanced variable-rate speech vocoder that generates speech bits at 8 kbps [49]. These speech bits then constitute layer 2 for primary traffic. Secondary traffic at layer 2 may be some form of circuit-switched data, such as fax or short message service. From the perspective of signaling messages, two higher layers are defined, layer 2 and layer 3. The layer 2 Signaling Protocol is responsible for reliable delivery of layer 3 signaling messages between the MS and the BS through mechanisms such as message retransmission and duplicate detection.

The layer 3 Signaling Protocol constructs (at the transmitter) and interprets (at the receiver) the call processing and call control messages, such as those related to call setup, handoff, and power control. The messages are sent on the dedicated traffic channel or some common channel, such as the paging channel in the downlink and the Access Channel in the uplink.

7.5.3 Downlink and Uplink Radio Channels of IS-95

We will now discuss the radio channels defined in IS-95. Both downlink channels and uplink channels will be described. The BS in a sector transmits the downlink channels using orthogonal codes in an attempt to avoid interference among these channels.

Do You Recall... Do you remember the name of these orthogonal codes in IS-95? See the footnote[223] for the answer!

The process of generating the Walsh codes is summarized in Section 7.5.3.1. An overview of the downlink and uplink radio channels is given in Section 7.5.3.2.

7.5.3.1 Generation of Orthogonal Codes in IS-95 Let's learn how to construct orthogonal Walsh codes. Before we get into the nitty-gritty details of Walsh code construction, let's take care of some nomenclature. A Walsh code is denoted W_i^N where N is the number of chips (or bits) in the code (which is the same as the spreading factor of the code) and i is the code number at the associated spreading factor (SF). The codes at a given SF are numbered from 0 to $SF - 1$. The number of orthogonal codes at spreading factor SF is SF itself. For example, eight orthogonal Walsh codes are present at $SF = 8$ and they are denoted W_0^8 through W_7^8.

When $SF = 8$, for example, the bandwidth of the signal is increased by a factor of eight as each digital modulation symbol of the signal is multiplied by or undergoes the XOR operation with eight chips of the code. If the code is expressed as consisting of 1s and 0s, the input signal is assumed to consist of 1s and 0s, and we carry out the XOR operation. However, if the input signal is expressed in the form of $+1$s and -1s, the code can be expressed in the form of $+1$ and -1s, and a simple multiplication takes place for spreading.

The basic way of constructing Walsh codes is to derive a suitable Walsh-Hadamard matrix and pick rows of such a matrix as Walsh codes. A Walsh-Hadamard matrix with the dimensions $(2k \times 2k)$ is derived from the Walsh-Hadamard matrix of dimension $(k \times k)$ as shown below.[224]

$$W_{(2k\times 2k)} = \begin{bmatrix} W_{(k\times k)} & W_{(k\times k)} \\ W_{(k\times k)} & \overline{W}_{(k\times k)} \end{bmatrix} \tag{7.9}$$

where $\overline{W}_{(k\times k)}$ is the complement of $W_{(k\times k)}$. The complement of a matrix is obtained by replacing 0s of the original matrix by 1s and by replacing 1s of the original matrix by 0s.

[223] The orthogonal codes in IS-95 (and its successor technologies such as CDMA2000 1xRTT) are called Walsh codes.

[224] Just so we understand one another, the first number, m, in the matrix dimensions $(m \times n)$ specifies the number of rows in the matrix and the second number, n, in the matrix dimensions $(m \times n)$ specifies the number of columns in the matrix.

The baseline Walsh-Hadamard matrix is $W_{(1\times1)} = [1]$ with dimension (1×1). Let's construct a (2×2) Walsh-Hadamard matrix using Eq. (7.9).

$$W_{(2\times2)} = \begin{bmatrix} W_{(1\times1)} & W_{(1\times1)} \\ W_{(1\times1)} & \overline{W_{(1\times1)}} \end{bmatrix} = \begin{bmatrix} 1 & 1 \\ 1 & 0 \end{bmatrix} \tag{7.10}$$

Let's go one step further.

$$\begin{aligned} W_{(4\times4)} &= \begin{bmatrix} W_{(2\times2)} & W_{(2\times2)} \\ W_{(2\times2)} & \overline{W_{(2\times2)}} \end{bmatrix} \\ &= \begin{bmatrix} \begin{bmatrix} 1 & 1 \\ 1 & 0 \end{bmatrix} & \begin{bmatrix} 1 & 1 \\ 1 & 0 \end{bmatrix} \\ \begin{bmatrix} 1 & 1 \\ 1 & 0 \end{bmatrix} & \begin{bmatrix} 0 & 0 \\ 0 & 1 \end{bmatrix} \end{bmatrix} \\ &= \begin{bmatrix} 1 & 1 & 1 & 1 \\ 1 & 0 & 1 & 0 \\ 1 & 1 & 0 & 0 \\ 1 & 0 & 0 & 1 \end{bmatrix} \end{aligned}$$

The Walsh-Hadamard matrix, $W(4 \times 4)$, contains four orthogonal codes, which are rows of the matrix.

$$\begin{aligned} W_0^4 &= [1 \quad 1 \quad 1 \quad 1] \\ W_1^4 &= [1 \quad 0 \quad 1 \quad 0] \\ W_2^4 &= [1 \quad 1 \quad 0 \quad 0] \\ W_3^4 &= [1 \quad 0 \quad 0 \quad 1] \end{aligned}$$

Let's see whether the Walsh codes are orthogonal. Let's perform a chip-by-chip XOR operation between two codes and add the resulting four chips using binary summation. For W_0^4 and W_1^4, a chip-by-chip XOR operation yields

$$[1 \text{ XOR } 1 \quad 1 \text{ XOR } 0 \quad 1 \text{ XOR } 1 \quad 1 \text{ XOR } 0] = [0 \quad 1 \quad 0 \quad 1].$$

Let's map bit 0 to $+1$ and bit 1 to -1. The result is $[1 \ -1 \ 1 \ -1]$. Adding all the chips gives $(1) + (-1) + (1) + (-1) = 0$. Hence, the codes W_0^4 and W_1^4 are orthogonal. Another way to show orthogonality is explained in Section 7.2.

Ever Wondered... Why are they called *Walsh Codes*? The Walsh codes are named after an American mathematician, Joseph Leonard Walsh, who wrote a paper titled "A Closed Set of Normal Orthogonal Functions" [235,236]. As you know by now, the orthogonality of Walsh codes enables CDMA to be very efficient. Though the term *Walsh codes* is widely used in the industry, other terms, such as Walsh-Hadamard codes and Hadamard codes or sequences, are also used.

7.5.3.2 Overview of Radio Channels Table 7.1 summarizes the four downlink channels defined in IS-95: the *Pilot* Channel, *Sync* Channel, *Paging* Channel, and *Forward Traffic* Channel.

The *Pilot* Channel consists of all 0s, and this sequence of 0s is spread by W_0^{64}. The standard refers to W_0^{64} as Walsh function W_0. Since all the Walsh codes being used in IS-95 have the spreading factor of 64, the standard does not use 64 while specifying a Walsh function (Walsh code). We will see in the chapters on CDMA2000 1xRTT (Chapter 8) and UMTS R99 (Chapter 9) that variable spreading factors are being used to support variable data rates. So, let's start getting familiar with the nomenclature!

After power-on, the MS looks for a specific sequence of chips of the Pilot Channel called a PN (pseudo-noise) sequence to acquire the radio network. The PN sequence is essentially the Pilot chips that result from the spreading operation and that a PN code scrambles. One PN code with different chip offsets is used by different sectors. Let's take a simple example. If the chips of the base PN sequence with no chip offset is 100011**0**, the PN sequence with an offset of one chip is **0**100011 because the first chip 1 moves to the right by one chip along with other chips. The last chip **0** of the base sequence wraps around and becomes the first chip in the new chip sequence with a one-chip offset.

In practice, 512 PN offsets are available, numbered from 0 to 511, with each sector allocated one of the PN offsets. A PN sequence with a PN offset of N, with N ranging from 0 to 511, has an offset of $64 * N$ chips relative to the base PN sequence with PN offset set to 0. *The use of the different PN offsets in different sectors enables the MS to identify different sectors. Furthermore, the use of the PN sequence helps the MS with the timing synchronization with one-chip timing resolution.* The MS observes

Table 7.1: Downlink Channels and Their Use

Channel	Main Use	Channel Identification
Pilot	-Sector identification via a PN offset -Timing reference for synchronization -Downlink channel quality measurements for handoff -Reference for coherent demodulation	W_0^{64}
Sync	System time synchronization and determination of the PN offset during network acquisition	W_{32}^{64}
Paging	-Broadcast messages for system information -MS-specific messages, such as a page	W_1^{64} to W_7^{64} available
Forward Traffic	-MS-specific user data traffic -MS-specific signaling messages	W_8^{64} to W_{63}^{64}

the pilot channels of surrounding sectors to evaluate the prevailing downlink channel conditions in such sectors. Such MS measurements facilitate handoff since better sectors than the currently serving ones are encountered.

The Pilot Channel also aids coherent detection or demodulation when the MS is attempting to estimate the information on its Forward Traffic Channel. The Pilot Channel acts as a reference signal in such an instance, and the MS estimates the channel conditions by observing the Pilot Channel and applies such knowledge while demodulating the traffic channel.

The *Sync* Channel plays a role during power-on network acquisition and is sent using W_{32}^{64}. The MS receives a message on the Sync Channel that provides information such as the system time and the specific PN offset of the sector the MS has locked onto.

The *Paging* Channel transmits broadcast messages (e.g., a message containing the list of neighboring sectors) that are heard by all MSs that are in the Idle Mode. The Paging Channel, as the name suggests, sends a page to a specific MS for a mobile-terminated (network-originated) call setup. The standard allows a maximum of seven Paging Channels that can be transmitted on W_1^{64} to W_7^{64}, with just one Paging Channel being a common implementation.

Finally, the *Forward Traffic* Channel carries the user traffic such as speech and the user-specific signaling messages, such as a message asking the MS to perform handoff. The Forward Traffic Channel can use any of the Walsh codes in the range of W_8^{64} to W_{63}^{64}. The Forward Traffic Channel can also use the Paging Channel's unused Walsh codes. The BS combines the downlink channels for a given sector as shown in a simplified diagram in Figure 7.19.

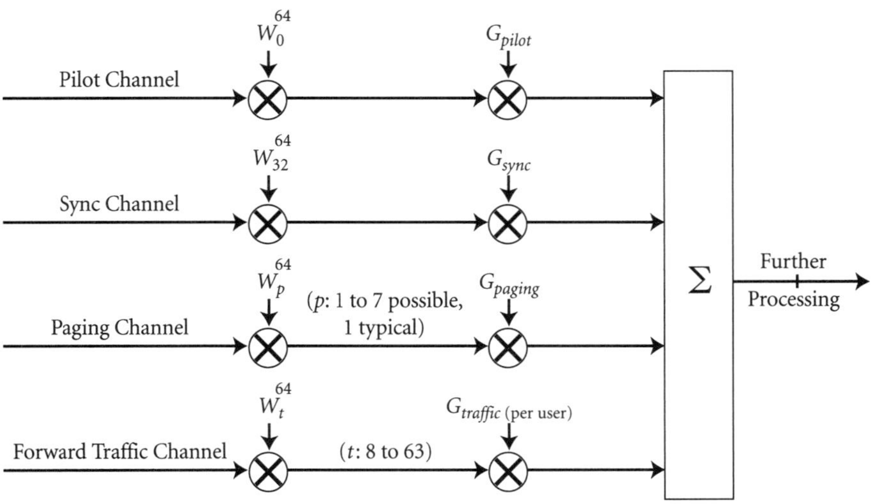

Figure 7.19: Combining of the Downlink Channels at the Base Station

The Pilot Channel, Sync Channel, one or more Paging Channels, and zero or more Forward Traffic Channels are spread using suitable Walsh codes. The Pilot Channel, Sync Channel, and Paging Channel together are often referred to as overhead channels because they carry no user traffic. The chips of the channels are multiplied by some digital gain or a weight factor (i.e., G_{pilot}, G_{sync}, G_{paging}, and $G_{traffic}$) because power is managed differently for different channels.

The power level is proportional to the square of the digital gain. Think of the digital gain as one of the important factors that influence the amplitude of the sinusoidal signal that will eventually be transmitted over the air. Typically, the gains of overhead channels G_{pilot}, G_{sync}, and G_{paging} are fixed, essentially leading to about 25% of the total high-power amplifier power being used for overhead channels in a sector. When about 10% of the power is reserved for handoff, a total of 25% to 35% of the power becomes overhead power, resulting in 65% to 75% of the power being left for traffic channels.

$G_{traffic}$ varies dynamically for each user because power control modifies it to reflect prevailing channel conditions. The chips with different power levels are added together and undergo further processing, which involves steps such as sector-specific scrambling using a PN sequence with a specific PN offset, quadrature spreading, and frequency upconversion. See Section 7.5.4 for an overview of the physical layer processing.

The MS transmits just two channels in the uplink, the Access Channel (ACH) and the Reverse Traffic Channel (RTC), which are summarized in Table 7.2. The ACH is a common uplink channel that an MS uses when the radio network has allocated it no dedicated radio channel. The MS uses the ACH to send messages, such as an Origination message to initiate an MS-originated call setup and a Page Response message to establish an MS-terminated or network-originated call setup. For each Paging Channel defined for a sector, up to thirty-two ACHs can be used. In practice, one ACH is commonly used. The RTC is a dedicated radio channel that the MS uses

Table 7.2: Uplink Channels and Their Use

Channel	**Use**	Channel Identification
Access Channel	-To contact the radio network in the absence of a dedicated radio channel -Example messages: (Call) Origination, Page Response	ACH-Specific Long-Code Mask
Reverse Traffic Channel	-To communicate with the radio network in a dedicated fashion -To carry user traffic, such as speech, and MS-specific messages, such as a Handoff Completion message	RTC-specific Long-Code Mask

to exchange MS-specific user traffic, such as speech and signaling messages (e.g., a Handoff Completion message that is the last message in the sequence of handoff-related signaling exchange between the MS and the radio network).[225] The ACH and the RTC are distinguished based on the channel-specific long-code masks.

7.5.4 Physical Layer (Layer 1) Processing

Different channels undergo somewhat different physical layer processing. We will use the Forward Traffic Channel (FTC) as an example to illustrate typical physical layer processing in IS-95 downlink. Figure 7.20 shows a simplified block diagram of the physical layer processing for the FTC for the *Rate Set 1* configuration, which supports four variable rates: 9.6 kbps, 4.8 kbps, 2.4 kbps, and 1.2 kbps. These rates correspond to a specific speech codec configuration (i.e., 8k Enhanced Variable-Rate Codec) that generates speech bits according to voice activity. The data rate of 9.6 kbps corresponds to *full-rate speech* that represents the *speech on* period, and the data rate of 1.2 kbps corresponds to *eighth-rate* speech that represents silence. The remaining two data rates, 4.8 kbps and 2.4 kbps, are transient data rates representing transitions between the *speech on* period and silence interval.

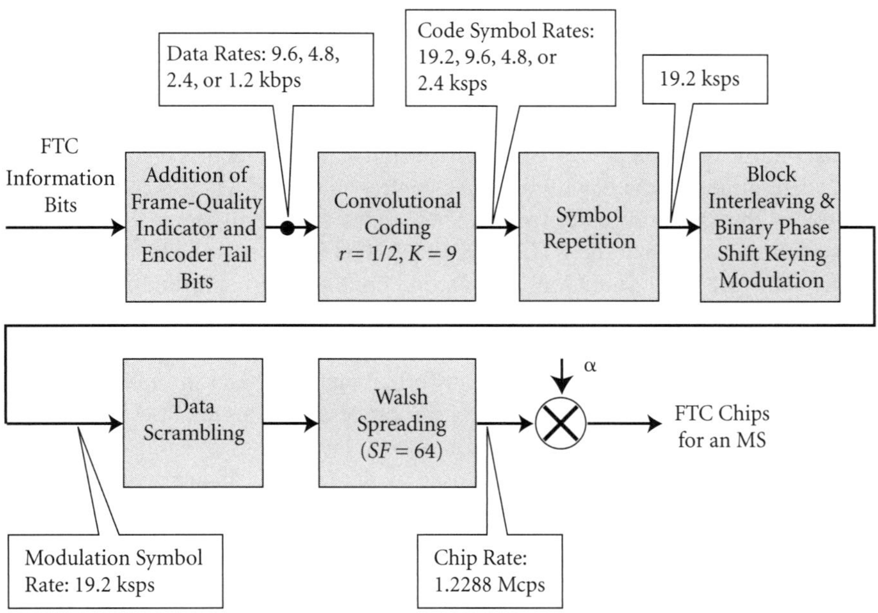

Figure 7.20: Physical Layer Processing for the Forward Traffic Channel

[225]The ACH and the RTC are distinguished based on the channel-specific long-code masks. The long-code mask, which is essentially a specific bit pattern, and the channel's PN chips undergo an XOR operation.

For a given MS involved in a voice call, the FTC carries information bits, which are speech bits from the speech codec, in 20 ms speech frames. The physical layer first adds a frame-quality indicator[226] that enables error detection at the receiver as shown in Figure 7.20. Encoder tail bits are also added, leading to the data rate ranging from 9.6 kbps to 1.2 kbps.

Do You Recall... What is the use of adding encoder tail bits? The footnote[227] contains the answer!

Convolutional coding with a coding rate of $1/2$ and constraint length $K = 9$ is used to add redundancy as an insurance policy in anticipation of an adverse radio environment. The coding mechanism doubles the data rate. Symbol repetition is carried for the encoder output data rates other than 19.2 ksps (kilo code symbols per second) so that the input rate to the block interleaver is fixed at 19.2 ksps. Block interleaving provides some degree of time diversity. More importantly, it breaks up error bursts to enable the decoder at the receiver to recover errors.[228] Binary phase shift keying modulation is used for reliability, and one interleaved code symbol corresponds to one binary phase shift keying modulation symbol. The modulation symbol rate is thus 19.2 ksps (kilo modulation symbols per second). The modulation symbols undergo user-specific scrambling. The code for such scrambling is derived by using the user-specific mask on a long PN code with a period of $(2^{42} - 1)$ chips.

The scrambled modulation symbols are spread to the bandwidth of 1.2288 MHz via the user-specific Walsh code of $SF = 64$ since 19.2 ksps $* \; 64 = 1.2288$ Mcps. The user traffic chips are allocated a user-specific gain, specified as α in Figure 7.20, that is determined by the downlink power-control algorithm. The BS combines such user traffic chips with the chips from other channels (shown in Figure 7.21).

The basic data rate of the FTC in Figure 7.20 is 9.6 kbps, which is the full rate for the Rate Set 1 configuration. The second configuration specified in IS-95 is a Rate Set 2 configuration that has the full rate of 14.4 kbps. The Rate Set 2 FTC configuration has the data rate of 28.8 ksps after symbol repetition and 19.2 ksps at the input of the block interleaver. The transformation from 28.8 ksps into 19.2 ksps is achieved via puncturing. Other physical layer processing steps are the same as those described for the Rate Set 1 configuration.

The BS combines FTCs of different users with the Pilot Channel, Sync Channel, and Paging Channel as shown in Figure 7.21. The power-weighted chips of all downlink channels at the rate of 1.2288 Mcps are combined and the same set of combined chips are sent to both I and Q channels. The pilot PN-offset–based I-channel and Q-channel PN sequences are used to scramble the combined signal. In

[226]The frame-quality indicator is a CRC (cyclic redundancy check) that helps the receiver perform error detection and blind rate detection.

[227]When the packet is processed by the decoder of the receiver, the state of the decoder is reset to all 0s due to addition of the encoder tail bits at the transmitter. Such resetting prepares the decoder to process the next packet and results in simplified decoder implementation.

[228]See Chapter 2 for interleaving details.

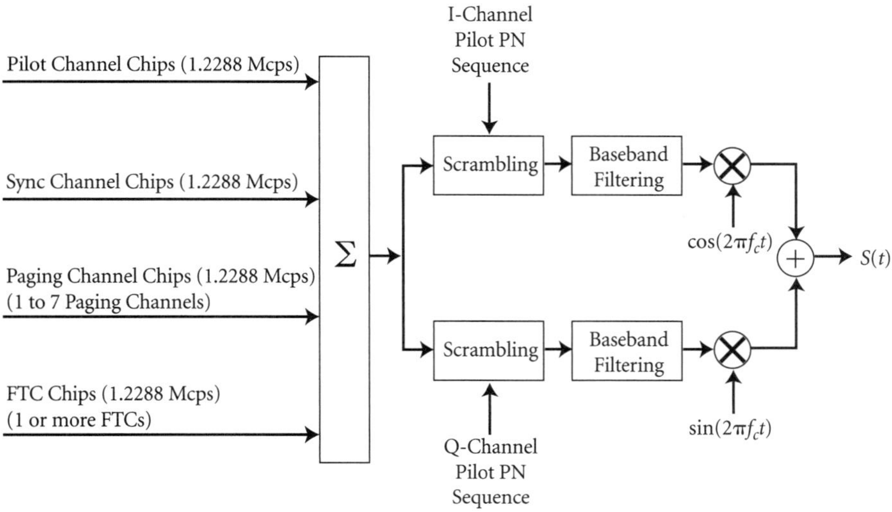

Figure 7.21: Combining of Downlink Radio Channels at the Base Station

other words, different sectors transmit the signals that are scrambled differently as a result of the use of sector-specific PN offsets. Since the PN code being used for sector-specific scrambling has a period of $(2^{15} = 32768)$ chips, it is referred to as a short PN code. Since scrambling differs between the I and Q channels, the signal constellation diagram has the appearance of a QPSK modulation; however, the modulation scheme is binary phase shift keying because one modulation symbol represents just one interleaved code symbol. The scrambled baseband signal undergoes baseband filtering to ensure that most signal energy stays within the 1.25 MHz system bandwidth. Frequency upconversion or analog modulation takes place using sinusoidal signals at the carrier frequency f_c and the sum of the sine and cosine signals is transmitted over the antenna.

We discussed downlink physical layer processing in which the BS combines various channels in a CDMA fashion. Let's see what happens in the uplink. We will use Rate Set 1 configuration of the RTC as an example in Figure 7.22.

The RTC carries information bits in 20 ms frames. As with the FTC, the physical layer adds frame-quality indicator and encoder tail bits, leading to the data rate ranging from 9.6 kbps to 1.2 kbps. The FTC uses rate $1/2$ convolutional coding, but the RTC uses $coding\ rate = 1/3$ and constraint length $K = 9$. The data rate is tripled by the coding mechanism, yielding the rate ranging from 28.8 ksps to 3.6 ksps. Symbol repetition results in the constant rate of 28.8 ksps. Block interleaving provides interleaved code symbols to a 64-ary orthogonal modulator. The modulator arranges the interleaved symbols in groups of six bits to construct modulation symbols and maps the resulting modulation symbol onto a Walsh function.

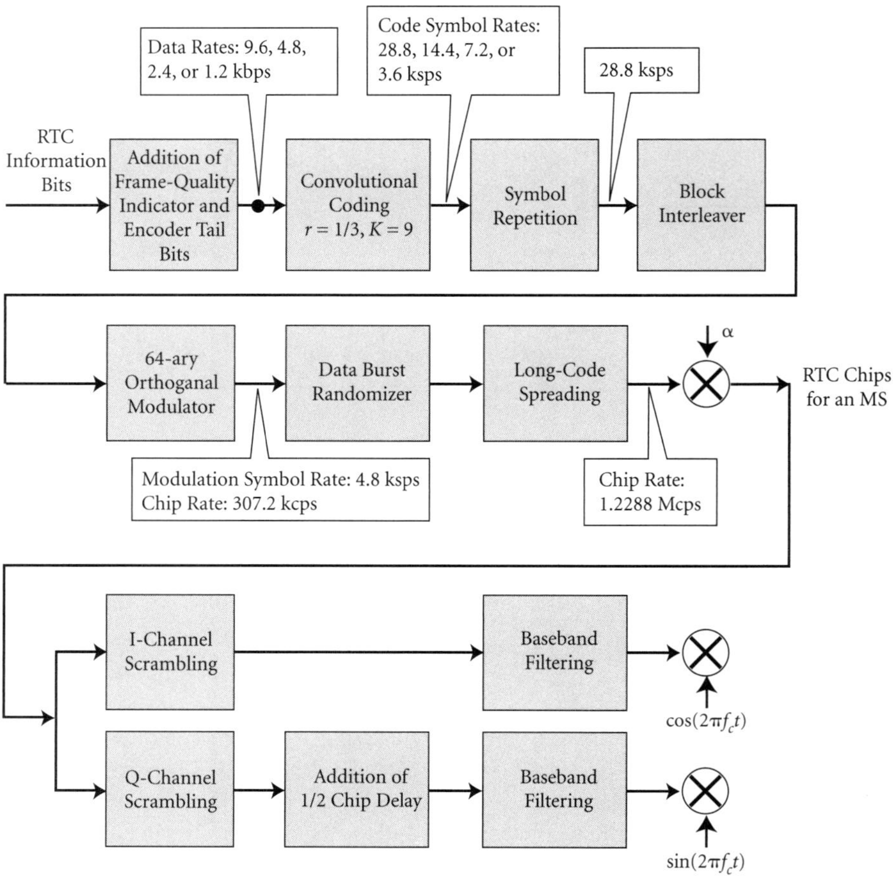

Figure 7.22: Physical Layer Processing of the Reverse Traffic Channel

For example, the set of six interleaver output symbols $\{0, 0, 0, 1, 0, 1\}$ corresponds to the integer 5, which is decimal representation of the binary string `000101`. Since each 64-ary modulation symbol represents six bits, the modulation symbol rate is 28.8 ksps$/6 = 4.8$ ksps. Since each 64-ary modulation symbol is mapped onto a 64-chip Walsh function (e.g., W_1^{64} in our example), the chip rate is $(28.8$ ksps$/6) * 64 = 307.2$ ksps. The next processing block of the data burst randomizer randomizes data using a user-specific long PN code.

The user-specific long PN code is also used to spread the signal by a factor of four to yield 307.2 ksps $* \ 4 = 1.2288$ Mcps. The use of a user-specific long code uniquely identifies the RTC of a user and provides some level of privacy or security. The power-control algorithm gives the RTC chips suitable weight and sends them to the I and Q channels. Zero-offset short PN codes are used to generate scrambling PN sequences for the I and Q channels. Such a scrambling process is often called quadrature spreading as both I and Q channels are used to carry the same chips coming out of the long-code–based spreading. The concept of such quadrature spreading is also used in the downlink as discussed earlier. A 1/2 chip delay is added to the Q channel scrambled chips to minimize the peak-to-average-power ratio of the transmitted signal. The processing blocks, such as baseband filtering and frequency upconversion, are similar to those for the FTC.

The Rate Set 2 configuration differs from the Rate Set 1 configuration primarily in following respects: the data rate ranges from 14.4 kbps to 1.8 kbps for Rate Set 2; convolutional coding with a rate of 1/2 is used for Rate Set 2; and an erasure indicator bit is sent along with the information bits in a 20 ms frame in Rate Set 2 to inform the network about the quality of the previously received downlink frame (whereas Rate Set 1 relies upon explicit messages for such quality reports). Rate Set 1 is more popular than Rate Set 2 due to its higher voice capacity at a reasonably good voice quality, which is attributable to the Enhanced Variable-Rate Codec.

7.5.5 Multiplex Sublayer (Between Layer 1 and Layer 2): An Overview

Figure 7.23 summarizes the main concepts for the multiplex sublayer. The multiplex sublayer utilizes mechanisms, such as dim-and-burst and blank-and-burst, to share a

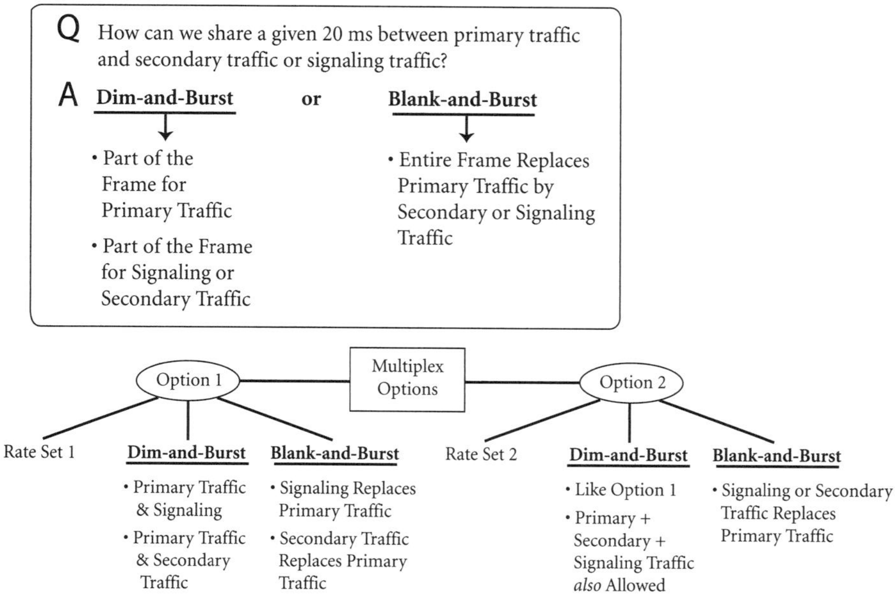

Figure 7.23: Summary of Multiplex Options

given 20 ms frame among primary traffic, secondary traffic, and/or signaling traffic. The dim-and-burst mechanism involves replacing the primary traffic during part of the frame. In other words, some portion of the frame carries primary traffic, and the remaining portion of the frame carries secondary and/or signaling traffic. The blank-and-burst mechanism blanks out the primary traffic completely by replacing the primary traffic with secondary or signaling traffic.

Two multiplex options are defined, Multiplex Option 1 and Multiplex Option 2. Multiplex Option 1 is for Rate Set 1 configuration, and Multiplex Option 2 is for Rate Set 2 configuration. Multiplex Option 2 allows *all* primary traffic, secondary traffic, and signaling traffic to share the same 20 ms frame because the higher data rate of Rate Set 2 allows more bits to be squeezed into the frame. The Multiplex Option 1 uses the dim-and-burst mechanism to have *either* primary traffic and secondary traffic *or* primary traffic and signaling. Multiplex Option 1 uses the blank-and-burst mechanism to replace the primary traffic with secondary traffic *or* with signaling traffic.

Multiplex Option 2 uses the dim-and-burst mechanism to send secondary traffic and/or signaling traffic along with the primary traffic in a given frame. Furthermore, as with Multiplex Option 1, Multiplex Option 2 uses the blank-and-burst mechanism to wipe out primary traffic and to send secondary traffic *or* signaling traffic instead. In summary, the multiplex options define various methods of sending primary traffic, secondary traffic, and signaling traffic using a 20 ms frame. In general, time-critical signaling messages, such as those related to handoff, can benefit from the blank-and-burst approach, and non-time-critical secondary traffic and signaling traffic can be sent using the dim-and-burst approach.

7.5.6 Layer 2 Versus Layer 3

Let's take a brief look at the functions layer 2 and layer 3 perform. Figure 7.24 contrasts these two layers using an example of a signaling message. Layer 2 is often referred to as the *link layer*, and layer 3 is called the *control process layer* [237]. Within a signaling message (e.g., the Origination message), some fields are defined to be layer 2 fields and the remaining fields are said to be layer 3 fields.

The early part of the message contains layer 2 fields. The main goal of layer 2 is to enable the reliable exchange of signaling messages. Figure 7.24 shows two examples of layer 2 fields, MSG_SEQ and ACK_REQ, for the Origination message that the MS

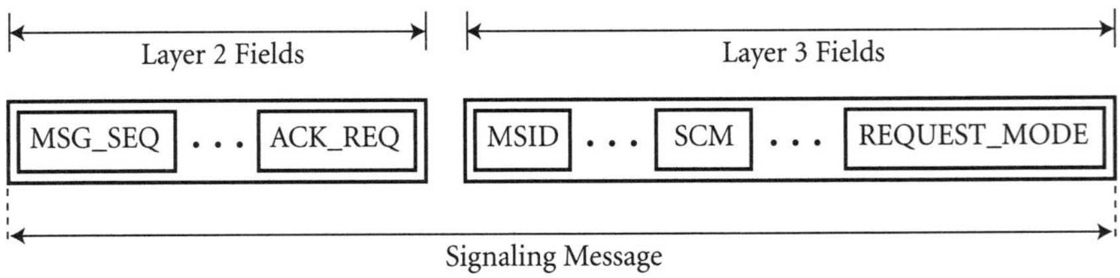

Figure 7.24: Layer 2 Versus Layer 3 Functions

transmits to the network. The MSG_SEQ field is the message's message sequence number, and the ACK_REQ field specifies whether an acknowledgment is required from the message's recipient. If MSG_SEQ is 5 and ACK_REQ is 1, the BS includes $ACK\_SEQ = 5$ (i.e., an acknowledgment of the message sequence number 5) in its response to the MS.

As far as layer 3 is concerned, everything except the layer 2 fields is a layer 3 field. Hence, the rest of the signaling message is a layer 3 message. In Figure 7.24, examples of layer 3 fields are MSID, SCM, and REQUEST_MODE. MSID is the identity of the MS, such as the electronic serial number or international mobile subscriber identity. SCM is the station class mark, which specifies characteristics of the MS, such as CDMA-only or dual-mode.[229]

In practice, the MS and the BSC usually implement layer 3 functionalities, and some layer 2 functionalities do reside at the BS as far as the infrastructure is concerned. For example, the BS (not the BSC) sends a layer 2 acknowledgment upon reception of the Origination message.

7.6 IS-95 CALL SETUP

Now that we are familiar with radio channels and the basic functionalities of all three layers, let's discuss call setup in IS-95! We will consider an MS-originated call setup in which the MS initiates the process. Figure 7.25 highlights the basic steps of such call setup.

When the MS wants to initiate a call, it must contact the network. However, since the network has yet to allocate to the MS a dedicated radio channel, the MS must use the common ACH. The MS sends an access probe on the Access Channel (as explained in Section 7.3.5.1 during the discussion of the open-loop power control). Each ACH is uniquely identified by a long code in a sector. Two ACHs in two adjacent sectors essentially have different long codes since the long-code masks used in generating the long codes depend upon a sixteen-bit identity of the BS and the pilot PN offset being used in the sector.

Once the BS detects the access probe from the MS, it responds with a layer 2 acknowledgment on the common Paging Channel (PCH). The MS now knows that it has found an adequate transmit power level such that the BS can detect its transmissions. The BS now starts transmitting null traffic channel data on the FTC to help the MS acquire the FTC in near future.[230] The network reserves the radio resources for the MS and allocates dedicated radio channels to the MS by transmitting a Channel Assignment message on the PCH. For example, the network allocates a specific Walsh code to the MS for the FTC. The MS learns that it has been allocated dedicated channels in the downlink and the uplink, and, hence, it starts transmitting a traffic channel preamble on the RTC. It can now process the null traffic channel data being

[229] A dual-mode phone supports AMPS and CDMA.

[230] Since the physical layer processing is different for the Pilot Channel and the FTC, we need to ensure that the MS can indeed decode the FTC and not just the Pilot Channel.

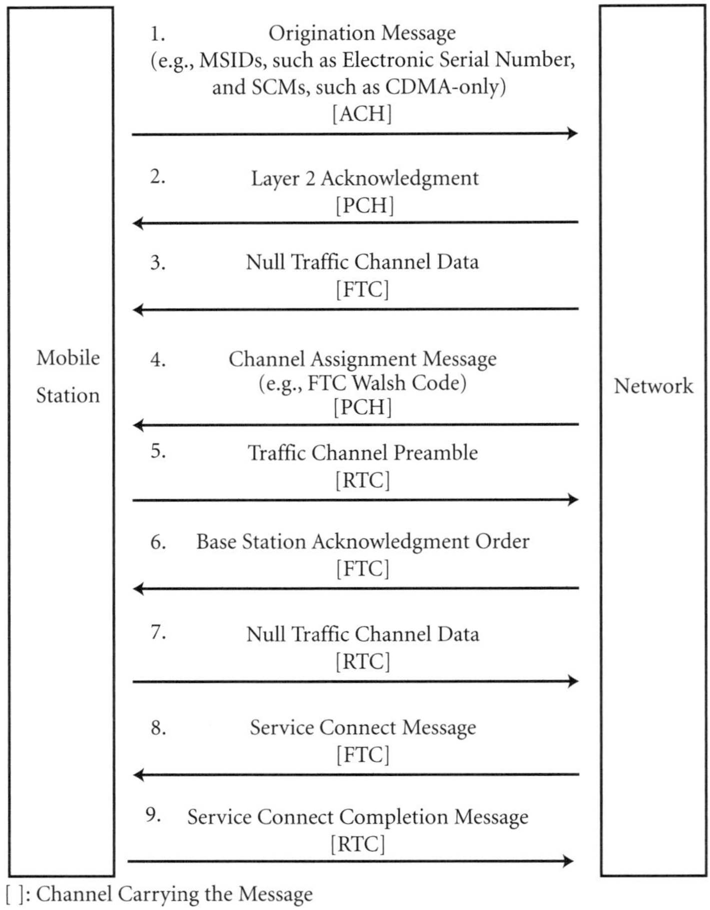

Figure 7.25: MS-Originated Call Setup

sent on the FTC and thus acquire the FTC. This preamble enables the BS to acquire the MS's RTC.

The BS now sends a Base Station Acknowledgment Order message on the FTC to acknowledge successful detection of the RTC preamble. Since the BS has already detected the RTC preamble, the MS starts transmitting null traffic channel data just like the BS. Finally, a Service Connect message from the BS and a Service Connect Completion message from the MS complete the call setup. Conversation can now begin, and real speech frames from the speech codec replace the null traffic channel data being sent on the FTC and the RTC.

7.7 IS-95 MOBILITY MANAGEMENT

When the MS is powered on, it searches for an IS-95 system and acquires the network. Such network acquisition is discussed in Section 7.7.1. After completing the network acquisition, the MS enters the Idle Mode. Section 7.7.2 provides a glimpse of Idle Mode activities of the MS. Once an MS-originated or MS-terminated call setup occurs, the MS is in the Connected Mode,[231] and mobility management becomes quite *interesting*. Section 7.7.3 illustrates handoff in the Connected Mode.

7.7.1 Power-On Network Acquisition—One-Way Traffic!

Upon power-up, the MS enters the Initialization State and attempts to acquire the network as summarized in Figure 7.26. Note that the processing shown in Figure 7.26 is a one-way street; the network has no idea the MS is doing all this hard work! The MS vendor-specific system selection algorithm determines the system to use. For

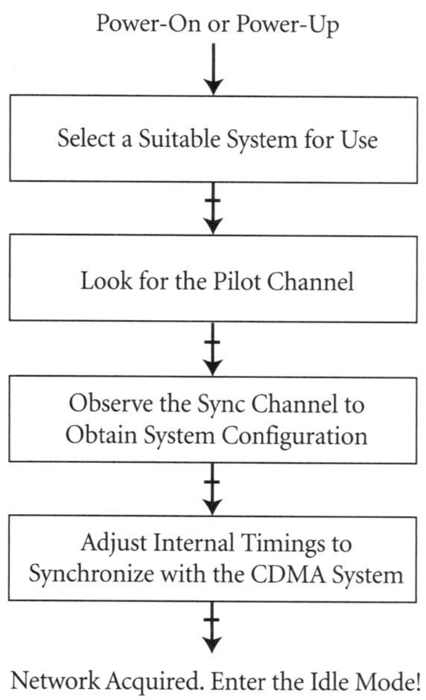

Figure 7.26: Power-Up Initialization and Network Acquisition

[231]We use the terms *Connected Mode* and *Traffic Mode* interchangeably to refer to the case in which the MS has the traffic channel established with the radio network. When the MS has no traffic channel with the radio network, we will consider the MS to be in the *Idle Mode*.

example, a preferred roaming list of CDMA channels (i.e., carrier frequencies) is used to select the system for communications.[232]

Once a CDMA system is selected, the MS looks for the Pilot Channel. Recall that different sectors use only one PN sequence with different offsets. The MS looks for such code and acquires the Pilot Channel. The MS achieves PN chip-level timing synchronization at this time. Next, the MS proceeds to demodulate the Sync Channel to extract the Sync Channel message. This message contains important system configuration. A system, such as a cellular or PCS system, consists of many networks and is identified by an *SID* (System Identification). A network contains a group of BSs, and an *NID* identifies a network within the system. A system may span a city, a metropolitan area, or one or more counties.

The combination of the SID and NID, called the (SID, NID) *pair*, uniquely identifies the geographical coverage area of a service provider. The concept of the SID and NID facilitates operations such as registration and roaming. For example, when an MS goes from an area associated with $SID = X$ to an area associated with $SID = Y$, it is said to be a *SID-roamer.*

Pilot PN Offset is the PN offset of the Pilot Channel the MS has just locked onto. *CDMA System Time* refers to the common time across all sectors of a CDMA system; the BSs use the Global Positioning System consisting of U.S. government satellites to obtain the CDMA system time. *Long-Code State* refers to the state of the shift registers used to generate a long code. The MS now adjusts its system time and long-code timing to synchronize with those of the CDMA system using Pilot PN Offset, Long-Code State, and CDMA System Time. The MS has now officially acquired the network and enters the Idle Mode.

7.7.2 MS to Network: "Here I Am"—Registration and Other Idle Mode Activities

When the MS is in the Idle Mode, it is not really relaxing! It still does something although not continuously! Specifically, the MS carries out functions such as monitoring the Paging Channel (Section 7.7.2.1), carrying out registration (Section 7.7.2.2), and participating in idle handoff (Section 7.7.2.3). Let's take a brief look at these Idle Mode activities.

7.7.2.1 Monitoring of the Paging Channel Let's first understand some basic terms used in conjunction with the PCH. The PCH is divided into 80 ms paging slots, with 2048 being the total number of slots. The CDMA System Time is used to calculate the specific paging slot number. A group of a certain number of consecutive paging slots is termed a *paging slot cycle*. In our example in Figure 7.27, one paging slot cycle consists of sixteen consecutive paging slots.

An MS operating in a *slotted mode* is allocated a paging slot number within the paging slot cycle. In our example, the MS is allocated paging slot number 4. Hence,

[232]If the MS is outside its home service area, the preferred roaming list helps the MS select a roaming partner of the home service provider.

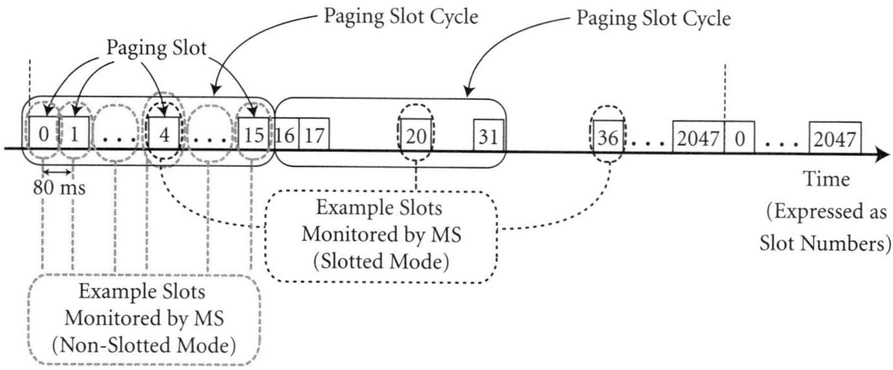

Figure 7.27: Monitoring of the Paging Channel

the MS monitors the PCH during paging slot 4, paging slot 20 (i.e., $4 + 16 = 20$), paging slot 36 (i.e., $20 + 16 = 36$), and so on. In other words, the MS looks at the Paging Channel once per paging slot cycle. The MS itself actually randomly calculates the specific paging slot number allocated to the MS using a hash function that uses the MS's international mobile subscriber identity as one of the input parameters. The minimum paging slot cycle duration is 1.28 seconds, and the paging slot cycle duration is specified in units of 1.28 seconds (e.g., 1.28, $1.28 * 2$, $1.28 * 4$, and so on).

The MS operating in the *non-slotted mode* continuously monitors all the slots as shown in Figure 7.27. The MS may switch between the non-slotted mode and the slotted mode. For example, an MS operating in the slotted mode leaves the slotted mode and enters the non-slotted mode during idle handoff. Now that we know the basic difference between two operating modes, a question is in order.

A Question for You... What are the benefits and drawbacks of the slotted mode compared to the non-slotted mode? The footnote[233] contains the answer!

Examples of the messages that the Paging Channel transmits during a given paging slot are the System Parameters message specifying various configuration parameters (e.g., the maximum paging slot cycle length and registration parameters) and the Neighbor List message specifying the list of surrounding sectors. And, of course,

[233] The slotted mode helps the MS save some battery power since the MS can conserve power when it is not monitoring the PCH. The drawback is that the delay to send a message to the MS increases since the MS is not monitoring the PCH all the time. By the way, the actual length of the paging slot cycle used by the MS is the minimum of these two values: its preferred slot cycle index and the maximum slot cycle index specified in the System Parameters message. An MS in the slotted mode observes the Paging Channel for one or two slots per slot cycle. The length of the slot cycle, T, is expressed in units of 1.28 seconds and is given by $T = 2i$ where i is the slot cycle index.

how can we forget the page message itself? The PCH carries a page message for the MS for MS-terminated calls in a Data Burst message or a General Page message. The MS knows whether the page is directed to it by finding a match between the address in the PCH message and its international mobile subscriber identity, temporary mobile subscriber identity, or electronic serial number. Once the MS finds its page message, it responds with a Page Response message and exits the Idle Mode to set up the call.

7.7.2.2 Registration—Oh, So Many Kinds! Let's summarize what has happened so far in the Idle Mode after power-up. The MS has monitored the PCH and obtained important system configuration parameters. So far the MS was only downloading information and digesting it. Now, it is time to make the first contact with the network. So far, after power-on, the MS has yet to inform the network of its existence. Until the MS does so, it will receive no page message from the network! The process of contacting the network to notify it of the MS's location and capabilities is formally called *registration*. After power-up, the MS registers with the network. However, several kinds of registration are possible, and Figure 7.28 lists various types.

The MS on its own handles *autonomous registration*. For example, after power-up and before power-down, the MS performs registration to let the network know of its availability or unavailability for a page message. If the MS has traveled a long distance since its last registration, it performs distance-based registration. If a long time has elapsed since its last registration, the MS conducts timer-based registration. If the MS crosses from one paging zone to another, it completes zone-based registration.

An *implicit registration* occurs when the network successfully receives the Origination message or Page Response message from the MS because the network can easily infer the MS's location. In *ordered registration*, the BS requests that the MS register; perhaps the network has not heard from the MS for a long time (timer-based

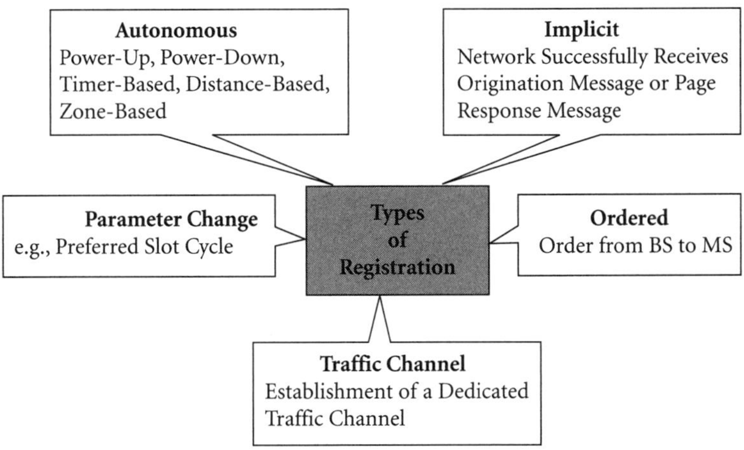

Figure 7.28: Registration Methods

registration may have been disabled) and the network wants to make sure that the MS does exist!

When a traffic channel is successfully established and the network has the MS's necessary information, *traffic channel registration* is said to occur. When the MS modifies some parameters, such as the preferred paging slot cycle and SCM, *parameter-change registration* occurs. Registration typically involves the MS sending a Registration message and the network responding with a Registration Accepted Order message. The network may also initiate authentication procedure upon receiving a Registration message from the MS.

7.7.2.3 Idle Handoff The MS in the Idle Mode monitors the PCH of the best sector in a slotted or non-slotted mode. If the MS travels into another sector, the PCH from the previous sector would be received with poor quality. Hence, the MS ceases to monitor the PCH of the old sector and starts monitoring the PCH of the new sector. This switching is called *idle handoff* and ensures that the MS is listening to the PCH of the sector that can provide the best downlink quality. To hunt for a better sector, the MS needs to know which sectors are in the neighborhood.

Each sector sends out messages, such as the Neighbor List message, which provides a list of neighboring sectors. Specifically, pilot PN offsets of the sectors are specified. The MS then evaluates the quality of the Pilot channels coming from different sectors to determine which sector is capable of providing the best quality downlink. For example, if an MS travels from Sector α to Sector β and finds the Pilot Channel of Sector β provides better quality than that of Sector α, it stops monitoring Sector α's PCH and starts monitoring Sector β's.

7.7.3 Handoff in Connected or Traffic Mode—Now We Are Getting Really Busy!

Mobility management in the Idle Mode is quite straightforward—the MS autonomously evaluates downlink channel conditions and monitors the PCH of the best sector. The network is uninvolved in the idle handoff process. However, when the MS has dedicated traffic channels, the FTC and the RTC, handoff becomes quite complex. Both the MS and the network must do quite a bit of processing for handoff in the Connected Mode. Such a Connected Mode is referred to as the *traffic channel state* in the standard.

The handoff concepts discussed next are quite important and are similar to basic concepts used in CDMA2000 1xRTT and UMTS. Before moving forward, let's make sure that you are fully awake. The wake-up alarm that we have for you is a question (or two)!

A Question for You... What is soft handoff and what are its benefits? See the footnote[234] for the answer!

Let's discuss the overall process of handoff using a scenario illustrated in Figure 7.29. The MS is traveling from BS1 to BS2. Both BS1 and BS2 control omnidirectional cells, and, hence, each BS controls just one sector. BS1 is in charge of the sector that uses the pilot PN offset 100, and BS2 controls the sector using the pilot PN offset 200. The MS is initially communicating with the sector using PN offset 100. In other words, at time instant t_1, the Active Set is $\{\text{PN100}\}$. The MS is measuring strengths of the pilot channel. The pilot strength is quantified as pilot E_c/I_0 where E_c is the energy per chip and I_0 is the power spectral density in Watts per Hertz. The pilot E_c/I_0 can be viewed as a ratio of the received pilot channel power and the total received power (both in Watts or mW). The total received power, of course, includes the pilot power, too. So, E_c/I_0 is always less than 1. In decibels, E_c/I_0 is always negative (e.g., $E_c/I_0 = -10$ dB).

At t_2, E_c/I_0 of the Pilot Channel with PN offset 200, called $(E_c/I_0)_{200}$ in Figure 7.29, exceeds a pilot add or detection threshold, T_ADD. A typical value of T_ADD is -14 dB. Since PN200 is deemed to be strong, the MS sends a Pilot Strength Measurement message containing $(E_c/I_0)_{200}$ (e.g., -12 dB). The BSC receives the Pilot Strength Measurement message and executes a vendor-specific proprietary soft handoff algorithm. In our example, this soft handoff algorithm makes a decision to add PN200 into the Active Set. So, our new Active Set at the BSC becomes $\{\text{PN100}, \text{PN200}\}$. The BSC sends the MS an Extended Handoff Direction message at t_3 to convey the soft handoff decision. The Extended Handoff Direction message allocates dedicated radio resources, such as a Walsh code, to the MS in the new sector with PN200.

The MS updates its Active Set from $\{\text{PN100}\}$ to $\{\text{PN100}, \text{PN200}\}$. It starts using the super-duper RAKE receiver with the paths received from the sectors using PN100 and PN200 and thereby reaps soft handoff's benefits. The MS responds with a Handoff Completion message, indicating successful completion of soft handoff. As the MS continues its journey away from BS1 and toward BS2, $(E_c/I_0)_{100}$ continues to degrade.

At t_5, $(E_c/I_0)_{100}$ drops below a pilot drop threshold, T_DROP, prompting the MS to start a drop timer. A typical value of T_DROP is -16 dB. If $(E_c/I_0)_{100}$ continues to stay below T_DROP for more than T_TDROP (e.g., four seconds), PN100 is considered too weak to be part of the Active Set. Such a condition is met at t_6. The MS sends a Pilot Strength Measurement message at t_7 to report the weak sector, PN100, to the BSC. The BSC executes the soft handoff algorithm and decides to remove PN100 from the Active Set. The BSC sends an Extended Handoff Direction message to the MS at t_8.

[234] In soft handoff, the MS simultaneously communicates with multiple sectors or cells. The main benefit is enhanced reliability of communications in the handoff region.

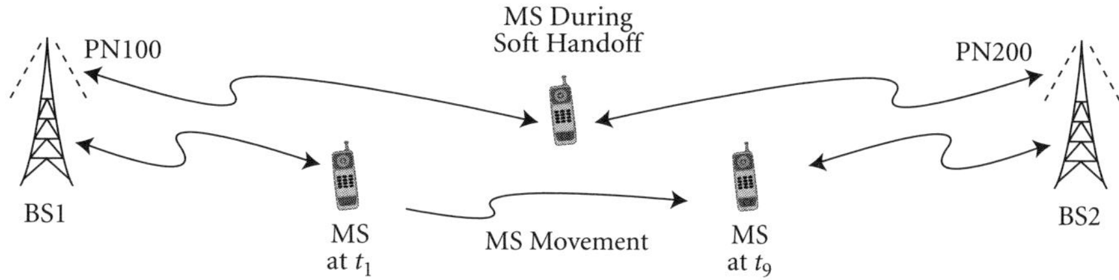

Time	Event
t_1	MS ⟷ PN100 (Communications)
t_2	MS ⟶ Network (Pilot Strength Measurement Message)
t_3	Network ⟶ MS (Extended Handoff Direction Message)
t_4	MS ⟶ Network (Handoff Completion Message)
t_5	Drop Timer Started
t_6	Drop Timer Expired
t_7–t_9	Like t_2–t_4

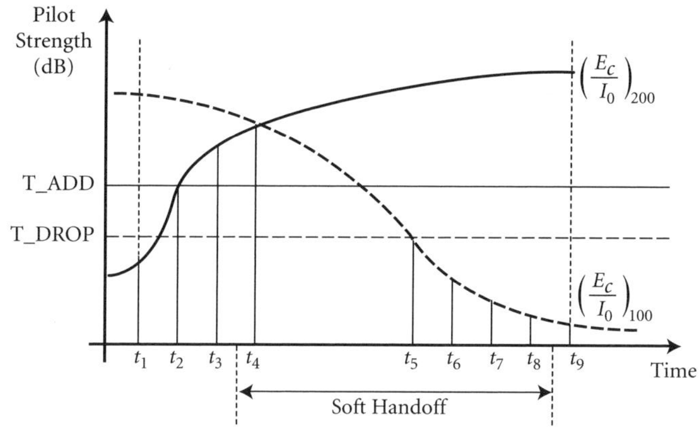

Figure 7.29: Soft Handoff Scenario

The MS acknowledges the receipt of the Extended Handoff Direction message and responds with a Handoff Completion message. The Active Set is just {PN200} now. In summary, we want to have strong sectors in the Active Set so that the maximal ratio combining at the RAKE receiver gives an excellent overall signal-to-interference ratio, and we want to keep weak sectors out of the Active Set to minimize resource utilization.

The scenario described here is one possible set of signaling messages. The IS-95 standard defines numerous messages to provide additional flexibility. For example, instead of sending an Extended Handoff Direction message, the BSC may send a General Handoff Direction message. And, in case you are curious, the thresholds, T_ADD, T_DROP, and T_TDROP, are specified in the System Parameters message.

Ever Wondered. . . What are Candidate Set, Neighbor Set, and Remaining Set? In addition to the Active Set, the MS maintains a Candidate Set, Neighbor Set, and Remaining Set. As the name suggests, the Candidate Set includes the pilots that are candidates for the Active Set but that are not yet in the Active Set. In Figure 7.29, the MS places PN200 into the Candidate Set at t_2 because PN200 is a strong contender to be part of the Active Set. After the MS receives the Extended Handoff Direction message from the network, the MS moves PN200 from the Candidate Set into the Active Set.

The Neighbor Set consists of the pilots the MS is evaluating for a potential handoff and excludes Active Set and Candidate Set pilots. The MS receives the Neighbor Set from the network, and the Neighbor Set usually includes the surrounding sectors. Finally, all other pilots that the MS detects but that are not in Active, Candidate, or Neighbor Set constitute the Remaining Set. Such pilot classification helps the MS prioritize the pilot search and pilot evaluation.

For example, Active Set pilots are evaluated more frequently than the Neighbor Set pilots. Furthermore, this classification influences measurement reports. For example, if a Candidate Set pilot exceeds the pilot strength of an Active Set pilot by $(\text{T\_COMP} * 0.5)$ dB, the MS sends a Pilot Strength Measurement message to the network.[235] Such a report screams to the network that it would be useful to make the reported Candidate Set pilot a member of the Active Set because it is better than even an Active Set pilot!

7.8 IS-95 POWER CONTROL

Earlier we discussed power-control mechanisms, open-loop power control and closed-loop power control. The generic closed-loop power control includes an outer-loop power control and an inner-loop power control. As you may recall, the goal of a power-control mechanism is to find the minimum transmit power level sufficient in meeting

[235]The parameter T_COMP is one of the parameters that influence the time instant when the MS sends a Pilot Strength Measurement message.

the target quality-of-service, such as 2% FER. The uplink power control is described in Section 7.8.1 and the downlink power control is described in Section 7.8.2.

A Question for You... What are the benefits of power control? The footnote[236] has the answer!

7.8.1 Uplink Power Control

The uplink power-control mechanism controls the transmit power of the MS. We have two channels in the uplink, the ACH and the RTC. The ACH uses the open-loop power control whereby the MS autonomously changes its transmit power so that the BS can hear the MS's access probe transmissions. The RTC uses the combination of the open-loop power control and the closed-loop power control. (See Section 7.3.5 if you want to refresh your memory on these power-control mechanisms.)

Figure 7.30 summarizes the main steps for the uplink power control of the RTC. The transmit power for the RTC can change from one power control group (PCG) to the next. A 20 ms frame is divided into sixteen PCGs, 0 through 15, as shown in Figure 7.30. The transmit power level of the RTC remains the same during the PCG

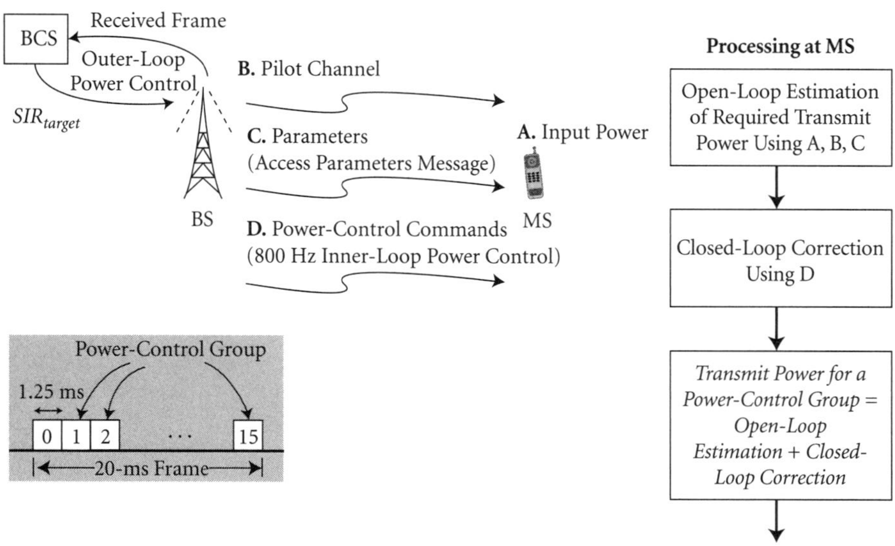

Figure 7.30: Power Control of the Reverse Traffic Channel

[236] The main benefits of power control are elimination of the near-far problem, increased MS battery life, lower interference, higher capacity, and ability to meet quality-of-service requirements.

that has the length of $20 \text{ ms}/16 = 1.25$ ms. The MS determines the transmit power level for a given PCG by combining the open-loop estimation and the closed-loop correction.

The MS carries out the open-loop estimation of the required power level by considering the total mean input power received at the antenna connector, the pilot E_c/I_0, the parameters received via the Access Parameters message, the band-class–specific power offset,[237] and the cumulative corrections made in the power levels during the transmission on the ACH.[238] The MS then combines the open-loop estimation with the closed-loop correction that is based on the power-control command received from the BS. The uplink power control in IS-95 is considered fast—the MS receives 800 power-control commands per second, one command every PCG. (That's why the IS-95 uplink is said to be using 800 Hz power control.)

Recall from Section 7.3.5 that the outer-loop power-control algorithm at the BSC specifies the SIR target to the BS to achieve a target FER, and the inner-loop power-control algorithm of the BS issues power-up or power-down commands to the MS to keep the actual SIR in the uplink and SIR target as close as possible. The BS sends the power-control commands to the MS in the form of power-control bits, one power-control bit per PCG.[239] A power-control bit of 0 means the power-up command and a power-control bit of 1 the power-down command.

The amount by which the power is increased or decreased, which is simply the closed-loop correction, is the power-control step size. The power-control step size can assume one of the three values, 1.0 dB, 0.5 dB, and 0.25 dB, with 1.0 dB being popular.[240] Different MSs receive their customized (*customized!*) power-control bits on their dedicated FTCs.

In summary, the MS sums the open-loop estimation that primarily reflects path loss and the closed-loop correction that reflects power-control commands and thereby calculates its required transmit power level per PCG.

7.8.2 Downlink Power Control

The design of the IS-95 standard (incorrectly) assumed that the uplink was the capacity bottleneck, and, hence, a lot of effort was focused on making the uplink efficient. The uplink power control in IS-95 is quite elaborate as we just described. The downlink power control pales in comparison. In fact, enhancement in the downlink

[237]Band Class 0 corresponds to the 800 MHz cellular band, and Band Class 1 corresponds to the 1.8 to 2.0 GHz PCS band. The band-class–specific power offset accounts for different propagation path losses in different frequency bands.

[238]See the standard [237] for details of the specific formula the MS uses.

[239]The power-control bit replaces two scrambled modulation symbols for Rate Set 1 and one scrambled modulation symbol for Rate Set 2 on the FTC. The Data Scrambling block in Figure 7.20 yields scrambled modulation symbols. The power-control bit is said to be punctured onto the FTC. The specific modulation symbol the power-control bit replaces is a function of the user-specific long code.

[240]The network specifies in the Power Control message the specific power-control step size the MS will use. Furthermore, if the BS measures the uplink SIR during PCG N, the BS sends the associated power-control bit during the PCG $(N + 2)$.

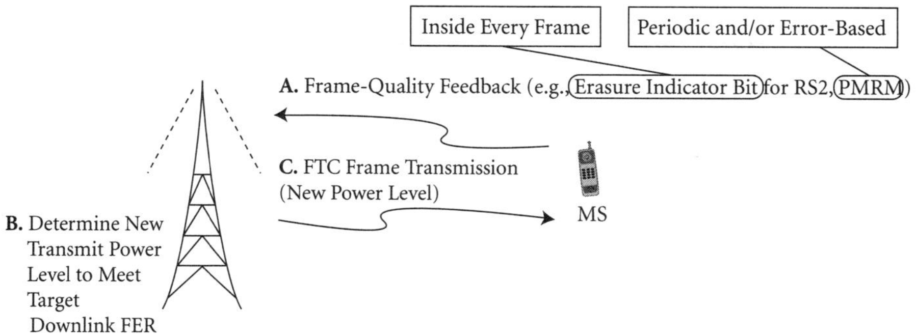

Figure 7.31: Power Control of the Forward Traffic Channel

power-control mechanism is one of the key design enhancements of IS-95's next incarnation, CDMA2000. The FTC's power control in IS-95 is considered slow—the transmit power can change as fast as once every 20 ms. This power change amounts to 50 Hz power control. Compare that to fast uplink power control, which operates at 800 Hz!

The Pilot Channel, Sync Channel, and Paging Channel are overhead channels and are transmitted at fixed power levels. Together, these overhead channels may consume about 25% of the BS's high-power amplifier's power in a given sector on a given carrier frequency. The FTCs are individually power-controlled for the MSs because the FTC chips of a given user are transmitted using user-specific power gain. Let's use Figure 7.31 as a basis for exploring how the FTC's transmit power level is controlled in IS-95.

The MS reports the quality of the received FTC frames to the BS. In Rate Set 2 configuration, the MS includes an erasure indicator bit in the RTC frame. This bit indicates to the BS if the previous downlink frame the MS received was erased (i.e., in error). Another frame-quality feedback mechanism is an explicit Power Measurement Report message containing information on the frame error rate. For example, the Power Measurement Report message may include the total number of frames observed during some time period and the number of erroneous frames during the same time period. Such a report enables the BS to easily calculate the FER.

The question is when should the MS send Power Measurement Report messages? The standard is quite flexible in that respect; the network can configure the MS for periodic Power Measurement Report message reporting and/or event-based Power Measurement Report message reporting. With periodic reporting, the BS can, for example, ask the MS to send the Power Measurement Report message every 100 frames. In error-based reporting, the MS sends a Power Measurement Report message if it observes a certain number of frame errors (e.g., two or three frame errors) since the last Power Measurement Report message.

Once the BS receives the frame-quality feedback, it makes a decision to adjust the transmit power level to be used in the next 20 ms frame in an attempt to achieve the target downlink FER. For example, if the BS learns that a frame error has occurred, the BS may increase the transmit power level by a large amount Δ_{up}, or if no frame error has occurred, the BS may decrease the transmit power level by a small amount Δ_{down}. The relationship between Δ_{up} and Δ_{down} can be suitably defined by the network to achieve the target FER (e.g., like the target SIR determination approach discussed in Section 7.3.5).

Comparing the downlink power control with the uplink power control, you will observe that no inner-loop power control is present in the downlink; the MS sends no power-control commands to the BS. The BS autonomously adjusts the FTC transmit power. The downlink power control can still be considered closed-loop (as opposed to open-loop) since explicit feedback from the MS to the BS occurs regarding frame quality.

7.9 IS-95: DEPLOYMENT, EVOLUTION, AND ENHANCEMENTS

IS-95 is the first cellular standard based on CDMA, and CDMA's efficiency led to IS-95's successful deployment. Since IS-95 is based on CDMA, it requires no elaborate frequency planning. The same 1.25 MHz spectrum is used in all sectors of the system. In other words, the same 1.25 MHz spectrum bandwidth is used in every sector of an omni cell, 120° sectorized cell, or even 60° sectorized cell. Different sectors are allocated different pilot PN offsets, and when we run out of 512 PN offsets, we reuse them. In practice, not all 512 PN offsets are used. Typically, pilot increments of four, two, and three are popular. If we use pilot increments of four, we are using the Pilot PN offsets $0, 0 + 4 = 4, 4 + 4 = 8, 8 + 4 = 12$, and so on.

Once the system is deployed, it can handle subscriber growth to a certain extent, and beyond that additional steps are required to maintain target grade of service.[241] For example, the sector capacity in early deployments was about twelve users for Rate Set 2 and twenty users for Rate Set 1, and Rate Set 1 was more popular. Later IS-95 deployments saw the sector capacity increase to about twenty-four. If we consistently see higher call blocking, we need to increase the offered capacity. In CDMA-based systems, a typical approach is to deploy multiple 1.25 MHz carrier frequencies. Handoff from one carrier frequency to the other is hard (we do not mean *difficult*; we mean *not soft*!).

IS-95A was enhanced to IS-95B, which is the topic of discussion in the very next section. The 2G CDMA-based IS-95 standard evolved to 3G CDMA2000. Furthermore, CDMA2000 evolved to 1xEV-DO. We will discuss these 3G standards in Chapters 8, 10, and 12.

As mentioned earlier, IS-95A was deployed widely, and IS-95B had limited deployment. However, IS-95B did provide a strong foundation for CDMA2000;

[241] Call block probability is an example of a performance metric that quantifies grade of service.

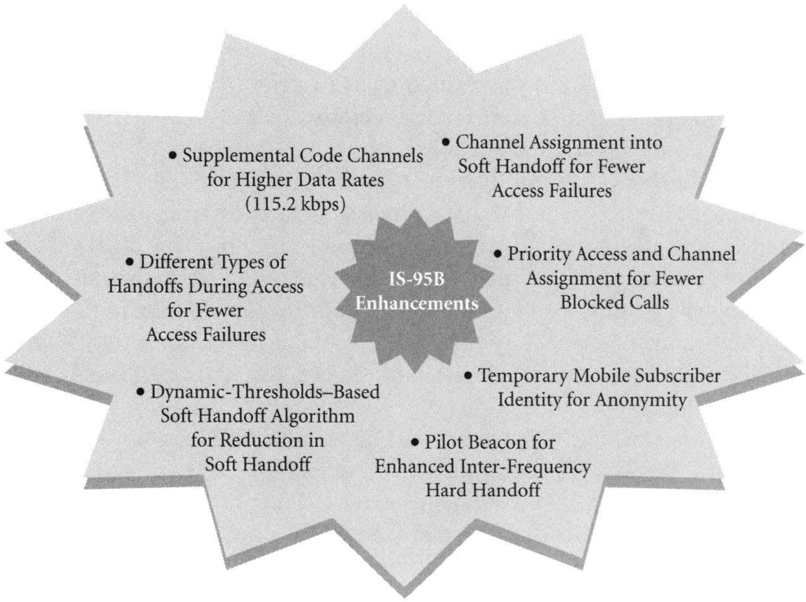

Figure 7.32: Summary of IS-95B Enhancements

commercial deployments of CDMA2000 exploit many IS-95B features. Figure 7.32 lists the main enhancements IS-95B introduced relative to IS-95A.

IS-95A allows the maximum data rate of 14.4 kbps when Rate Set 2 is used. IS-95B defines supplemental code channels. When a user is allocated multiple supplemental code channels (up to eight), the data rate increases. The peak supportable data rate for a given user is $14.4 \text{ kbps} * 8 = 115.2 \text{ kbps}$.

In IS-95A–based systems, the MS in the handoff region first establishes communications with just one sector, and additional sectors are added to the Active Set later. Such an approach introduces delay in adding strong sectors, and these sectors may cause significant interference until they become part of the Active Set. IS-95 B allows multiple sectors to be part of the Active Set when the traffic channel is established. Such a channel-assignment-into-soft-handoff feature improves the access performance.

The priority access and channel assignment features enable a high-priority call to be queued if the resources are unavailable. When resources are freed up, the BS asks the MS to re-originate. If implemented, such an approach reduces the number of blocked calls.

IS-95B supports a temporary mobile subscriber identity to provide anonymity since the use of the MS's original identity, the international mobile subscriber identity, is minimized.

When multiple CDMA carrier frequencies are deployed, hard handoff across the carrier frequencies is required to avoid call drop. Traditionally, such hard handoff in IS-95 failed quite often. The feature called *pilot beacon* significantly improves hard handoff performance, increasing the success rate to beyond 90%. The basic idea is that a pilot beacon signal is transmitted from a carrier frequency that supports no traffic. The MS detects and measures such a pilot beacon and reports it to the network. The network can now make a hard handoff decision.

Soft handoff improves reliability in the handoff region, but its downside is resource consumption at multiple sectors. In an attempt to reduce the amount of soft handoff, a dynamic-thresholds–based algorithm is defined in IS-95B. This algorithm delays the addition of a new pilot into the Active Set when current Active Set members can provide adequate quality. Furthermore, this algorithm aggressively removes a pilot from the Active Set if the pilot fails to significantly contribute to the overall quality. In essence, the amount of time for which a pilot stays in the Active Set is now reduced, leading to a reduction in soft handoff.

In IS-95A systems, the MS sticks to just one sector for a long time to access the network. However, IS-95B allows the MS to quickly try accessing another sector if the current sector is found to be unresponsive. Such handoff during the access stage increases the likelihood of successful access. As mentioned earlier, many IS-95B enhancements, such as channel assignment into soft handoff and handoff during access, are used in commercial 3G CDMA2000 networks.

7.10 POINTS TO REMEMBER

- IS-95 is the first commercial cellular standard that exploits CDMA.
- CDMA possesses various characteristics that make it a very efficient multiple access technique.
- Spread-spectrum signal transmission enables CDMA to provide a huge processing gain that ultimately results in higher capacity.
- The RAKE receiver helps mitigate multipath interference and is able to combine energy of multipaths instead of fighting against multipaths.
- CDMA supports soft handoff in which the MS communicates with multiple sectors simultaneously. The right amount of soft handoff provides a good balance between reliability and capacity.
- Power control in CDMA helps minimize interference, improves battery life of the MS, and increases the capacity, and enables achievement of the FER.
- An IS-95 system includes a radio network consisting of BSCs and BSs and a circuit-switched core network consisting of components such as a mobile switching center or visitor location register, home location register or authentication center, and interworking function.
- The protocol stack comprises layer 1 for CDMA processing, the multiplex sublayer to multiplex user traffic and signaling, the link layer for reliability, and layer 3 for control signaling.

- The downlink radio channels include the Pilot Channel, Sync Channel, Paging Channel, and Forward Traffic Channel.
- The uplink radio channels are the Access Channel and Reverse Traffic Channel.
- The call setup procedure involves the use of common Access Channel and Paging Channel during the early stage and establishes the dedicated FTC and the RTC for communications.
- The MS provides measurements of the Pilot Channels of surrounding sectors so the network can make a handoff decision and to facilitate soft handoff.
- The uplink implements fast 800 Hz power control, and the downlink allows a slow 50 Hz power control.
- CDMA2000 represents the 3G evolution of 2G IS-95.

CHAPTER 8

CDMA2000 1x

CDMA2000 is a digital third-generation (3G) cellular technology based on Code-Division Multiple Access (CDMA). The CDMA Development Group provides the latest information surrounding CDMA2000 and its companion technologies [238]. CDMA2000 and Universal Mobile Telecommunication System (UMTS) are the two 3G families of technologies in existence. Second-generation Global System for Mobile Communications and second-and-a-half-generation General Packet Radio Service and Enhanced Data Rates for Global Evolution have evolved to 3G UMTS, and second-generation IS-95 has evolved to 3G CDMA2000.

This chapter provides an overview of CDMA2000 1x, a radio transmission technology (RTT) that is a CDMA-based (no surprise here given the name!) 3G cellular system. A brief history of CDMA2000 1x, often called CDMA2000 1xRTT or just 1x, is given. The system architecture of CDMA20001x is illustrated. The communication between the mobile station (MS) and the network is described by discussing the air-interface protocol stack and the radio channels defined for the forward and reverse links. The call setup procedure is illustrated. The way in which data transfer occurs in the forward link and the reverse link is described. Operations such as handoff and power control are discussed. Deployment and evolution of CDMA2000 1x are briefly discussed.

Cellular Communications, First Edition. Nishith D. Tripathi and Jeffrey H. Reed

8.1 A BRIEF OVERVIEW OF CDMA2000

You may be feeling some discomfort because we have already used three distinct terms: CDMA2000 1xRTT, CDMA2000 1x, and CDMA2000. Let's make you very comfortable; learning is much more fun when comfortable! The International Telecommunication Union challenged the cellular industry to develop a 3G standard. The Third Generation Partnership Project offered UMTS as the evolution of Global-System-for-Mobile-Communications–based systems, and the Third Generation Partnership Project 2[242] developed CDMA2000 specifications to provide an evolution path to IS-95 systems. The initial release of CDMA2000 was IS-2000 (Interim Standard 2000).

Two RTTs, 1xRTT and 3xRTT, were defined as part of IS-2000. 1xRTT uses the chip rate of 1.2288 Mcps (mega chips per second) and the spectrum bandwidth of 1.25 MHz for each of the forward and reverse links. 1.2288 Mcps is 1.2288 Mcps $\times$ 1, hence the name 1xRTT. Similarly, 3xRTT uses the chip rate of 3 $\times$ 1.2288 Mcps channels for each of the forward and reverse links. 3xRTT needs 5 MHz of spectrum for deployment with some guard bands already part of the 5 MHz bandwidth. However, the industry paid no attention to the development and deployment of 3xRTT products. Commercial deployments use CDMA2000 1xRTT, which uses 1.25 MHz channels just like IS-95, allowing 1xRTT to be fully backward-compatible with IS-95. CDMA2000 1xRTT is so elegantly designed that the same 1.25 MHz spectrum can be shared transparently and seamlessly among IS-95 MSs and CDMA2000 1xRTT MSs.

This chapter primarily focuses on 1x; unless specified otherwise, we will be discussing CDMA2000 1x. Furthermore, we will refer to CDMA2000 1x simply as 1x.[243] Additionally, when we mention IS-95, we mean IS-95A; references to IS-95B will be explicitly clear.

The International Telecommunication Union approved 1x as an IMT-2000 standard (where IMT stands for International Mobile Telecommunications) in November 1999. The world's first 3G commercial system was a CDMA2000 1x system that was launched by SK Telecom in South Korea in October 2000 [238]. 1x has evolved to 1xEV-DO Revision 0.[244] 1xEV-DO Revision 0 has evolved to 1xEV-DO Revision A. Chapter 10 discusses 1xEV-DO Revision 0, and Chapter 12 deals with 1xEV-DO Revision A.

Let's summarize the main enhancements of 1x relative to IS-95. First and foremost, 1x supports more voice users than IS-95. For example, IS-95 supports about twenty users in a 1.25-MHz bandwidth, and 1x supports about thirty-five users in the same

[242] The Third Generation Partnership Project and the Third Generation Partnership Project 2 are two separate organizations and both are still in existence.

[243] Though the authors most certainly want to reduce their amount of typing, the term 1x's widespread use in the industry is the real reason for this convention.

[244] What does 1xEV-DO stand for? Well, the early literature defined 1xEV-DO to be 1xEvolution–Data-Only. However, the term was later redefined to be 1xEvolution–Data-Optimized, which is certainly much more appealing than the original definition.

bandwidth. 1x is also more efficient than UMTS in voice capacity. UMTS R99 supports about sixty users per sector per carrier frequency in the 5-MHz bandwidth. Three 1x carrier frequencies fit in a 5 MHz spectrum, leading to the voice capacity of $35 * 3 = 105$ users. Inter-carrier frequency load balancing can further increase the 1x voice capacity by about 10%. In summary, 1x is about twice as efficient for voice capacity as UMTS R99 when a 12.2 kbps full-rate speech codec is used for UMTS. The voice capacity numbers mentioned here assume 9.6 kbps Enhanced Variable-Rate Codec (EVRC) speech for 1x and 12.2 kbps AMR speech for UMTS R99, which is the case for deployments as of late 2010.

1x includes two distinct types of core network, a circuit-switched core network and a packet-switched core network, whereas IS-95 has only a circuit-switched core network. The packet-switched core network enables 1x to utilize valuable resources efficiently.

The highest data rate that a user could experience in practice is about 153.6 kbps or 153.6 kbps $+ 9.6$ kbps $= 163.2$ kbps depending upon the system configuration. A theoretical peak rate of 307.2 kbps is defined in the standard, but early commercial products shied away from supporting 307.2 kbps. New radio channels called *Supplemental channels* make such high data rates possible.

Slow 50 Hz power control in IS-95 forward link (FL) is replaced by fast 800 Hz power control in 1x, and such fast power control is a key contributing factor to the capacity increase of 1x relative to IS-95. Several radio configurations corresponding to different channel coding rates and data rates are supported in 1x. *Turbo coding*, a channel coding scheme more powerful than convolutional coding, is used to provide large data packets with a good level of protection against radio channel dynamics. From another perspective, turbo coding yields power savings compared to convolutional coding at a given error rate for large packets. The reverse link (RL) in 1x introduces the pilot channel to support coherent demodulation. The RL pilot channel increases the link budget by about 2.5 dB and increases theoretical pole capacity of the RL.

Two sublayers, the Radio Link Protocol (RLP) sublayer and the Medium Access Control (MAC) sublayer, are defined at the link layer in 1x. The RLP sublayer improves the reliability of communications for packet data services through retransmissions. The MAC sublayer allows multiplexing of multiple information streams. The RLP sublayer functionality is new in 1x, and the MAC sublayer functionality is similar to IS-95's multiplex sublayer functionality.

The operations, such as system access and paging, are enhanced to reduce the access failure rate and to increase battery life, respectively. In addition to the basic soft handoff algorithm of IS-95, a dynamic-thresholds–based soft handoff algorithm is available for use in 1x. This new algorithm attempts to aggressively reduce the amount of soft handoff to lower the overall resource utilization. Although the 1x standard specifies support of advanced antenna techniques, such as orthogonal transmit diversity and beamforming, commercial deployment of these techniques has gained no traction.

8.2 CDMA2000 1x NETWORK ARCHITECTURE

A CDMA2000 1x network includes the radio network and the core network. The core network includes both the circuit-switched core network and the packet-switched core network. Various features of the radio network are discussed throughout the chapter. The circuit-switched core network in 1x is similar to that in an IS-95 network. However, the packet-switched core network is a new addition in a 1x system; an IS-based system has no packet-switched core network. Section 8.2.1 provides an overview of the overall network architecture, and Section 8.2.2 discusses the packet-switched core network in more detail.

8.2.1 Overview of the Network Architecture

Figure 8.1 depicts the network architecture of a 1x system. A 1x system consists of the centralized radio access network and two core networks, the circuit-switched core network and the packet-switched core network. The radio access network architecture includes an all-powerful base station controller (BSC), controlling hundreds of base

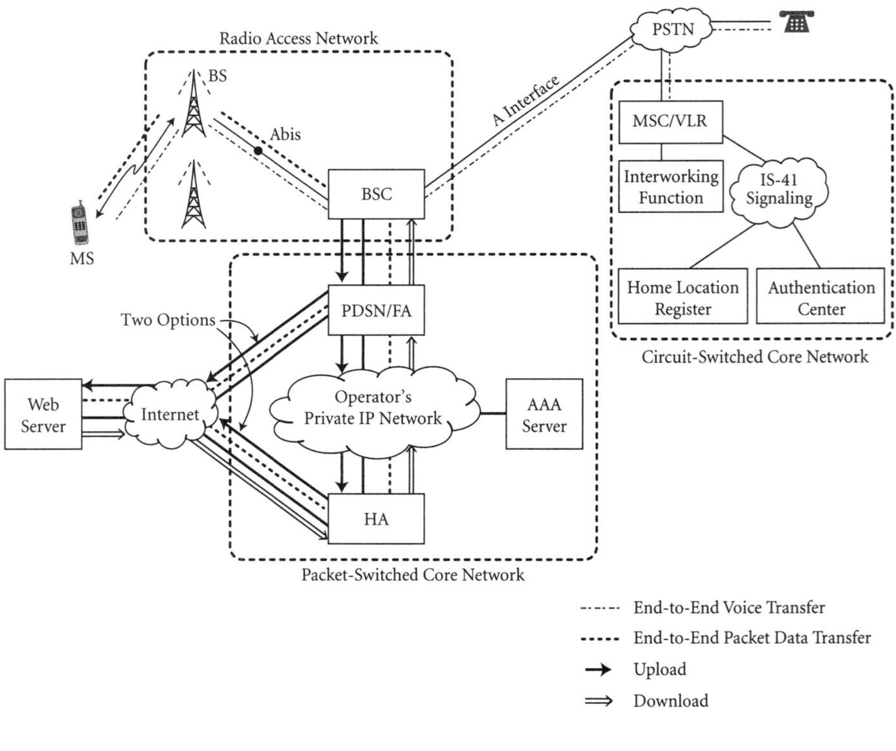

Figure 8.1: 1x Network Architecture

stations (BSs). The BS communicates with the MSs over the radio interface using CDMA. As in IS-95, most of the radio access network intelligence resides at the BSC, and the BSC makes most radio resource management decisions. This is the perfect place to ask you a question!

A Question for You... Which radio access network element executes the handoff algorithm? Why? The footnote[245] has your answer.

Examples of radio resource management algorithms implemented at the BSC include the call admission control algorithm, handoff algorithm, and RL outer-loop power-control algorithm. The radio access network architecture of 1x is thus centralized like IS-95.

A Word of Caution... The 1x standard refers to the entire infrastructure with the term *base station*, and, in practice, only some of the infrastructure functions are actually carried out by the physical base station. We will use the term *base station* to refer to the physical piece of equipment that includes suitable hardware and software to communicate wirelessly with the MS.

The circuit-switched core network architecture in 1x is the same as that in IS-95; it includes components such as the mobile switching center (MSC), home location register, authentication center, visitor location register (VLR), and interworking function. The end-to-end flow of voice between an MS and a landline phone involves the BS, the BSC, the MSC, and the PSTN as shown in Figure 8.1. The general functions these circuit-switched core network elements perform are the same in Global System for Mobile Communications, IS-95, and 1x. (Chapter 5 offers additional information about these circuit-switched core network elements.)

Let's quickly summarize main functions of the circuit-switched core network nodes. The MSC switches a call between the radio access network and the public-switched telephone network (PSTN). The home location register stores the subscriber database information including the last-known location of the subscriber. The authentication center stores the secret key of the subscriber to facilitate user authentication. The home location register and the authentication center are usually co-located and implemented in the same physical box. The VLR acts as a mediator between the MS and the home location register and authentication center during the user authentication process. The VLR informs the home location register of the MS's most recent location. The MSC and the VLR are usually co-located and are in the same physical cabinet(s). The interworking function consists of a pool of modems to support a

[245]The BSC implements the handoff algorithm because it processes the measurement messages from the MS. Thus the BSC has a global view of the prevailing channel conditions between the MS and a given sector. The BS itself has only a limited, local view of some aspects of the channel conditions (e.g., only the RL quality measurements).

circuit-switched data call. The IS-95 standard itself focuses on the air interface between the MS and the radio access network. The interface between the BSC and the MSC is the A interface, with the A1, A2, and A5 interfaces taking care of signaling, voice traffic, and circuit-switched data traffic, respectively. The A interface is defined in IS-634. The circuit-switched core network components, such as the VLR and the home location register, communicate using the IS-41 signaling standard.

CDMA2000 introduces a new type of core network, a packet-switched core network, to support packet data services, as depicted in Figure 8.1. The typical nodes of the packet-switched core network include the authentication, authorization, and accounting (AAA) server, packet data serving node (PDSN), foreign agent (FA), and home agent (HA). The AAA server participates in the user *authentication* process during the packet data call setup. The AAA server's *authorization* function ensures that the user is accessing only the services to which the user has subscribed. The AAA server's *accounting* function obtains information on the data transfer that the user has carried out and charges the user according to the subscribed service plans or agreements.

The PDSN maintains a logical link, a Point-to-Point Protocol (PPP) session, with the MS when the MS is in a *dormant* mode, i.e., a mode in which the MS has no active packet data transfer but the PDSN has stored relevant parameters to facilitate fast transition from the dormant mode to the Active Mode. The MS transfers data in the Active Mode.

If simple Internet Protocol (IP) is supported, the PDSN allocates an IP address to the MS. In Mobile IP (MIP), two additional entities exist, an HA and an FA. MIP is widely used in commercial 1x deployments and is more prevalent than simple IP. One HA is typically associated with tens or hundreds of FAs. A nationwide cellular network may have tens of HAs geographically distributed, and the AAA server associates an MS with a suitable HA during the call setup. The FA allocates a temporary care-of address to the MS, and the HA allocates the home IP address to the MS. When the MS constructs an IP packet to communicate with the outside world, such as during a web-browsing session, it uses the home address as its source IP address. The FA and the HA are also involved in MIP authentication during the packet data call setup. The PDSN usually implements the FA functionality, and, hence, the PDSN/FA is typically observed as a single physical entity. The FA and the HA are the MIP components defined in the standard as part of the MIP framework [159] and are discussed in Chapter 4.

A detailed view of the 1x network architecture, including the PDSN's and AAA server's functions, are described in Section 8.2.2. The end-to-end packet flow from an external entity, such as a web server, to the MS traverses the HA, the PDSN/FA, the BSC, and the BS. Two options are available for the packet flow from the MS to the web server. When reverse tunneling is used, the packets flow from the MS to the web server via the BS, the BSC, the PDSN/FA, the HA, and the Internet. In the absence of reverse tunneling, the packets flow from the MS to the web server via the BS, the BSC, the PDSN/FA, and the Internet. See Chapter 4 for more information on reverse tunneling. See the standard [239] for additional details of the core network.

8.2.2 A Detailed View of the Packet-Switched Core Network

The packet-switched core network consists of components such as the PDSN, the FA, the HA, and the AAA server, as shown in Figure 8.2. The radio network communicates with the packet network using the radio-packet (R-P) interface. Specifically, the R-P session exists between the packet control function (PCF) and the PDSN. The PCF also keeps track of the MS's packet data session's status (e.g., active or dormant) and buffers the FL packets received from the PDSN when required. The PCF is usually integrated into the BSC. In other words, a physical BSC cabinet houses both the BSC functions and the PCF functions, making the interface between the BSC and the PCF internal to the physical BSC.

The communication between the BSC and the PCF occurs via A8 and A9 interfaces. The A8 interface carries user traffic (e.g., web page or email), and the A9 interface carries signaling messages. The PCF sends to the BSC a request for the radio resources required or desired for a given MS in the active state. Note that the BSC is the main piece of equipment that buffers user data for the FL and RL for an active packet data session. However, when radio resources are inadequate or unavailable (e.g., during the dormant to active transition) to support the packet flow from the PDSN, buffering is required at the PCF. The PCF also supports accounting by providing the PDSN with information regarding the MS's air-interface usage.

The PDSN routes the packets between the radio network and the packet-switched core network. When a PPP session exists, the user-specific R-P session also exists. One BSC can be connected to multiple PDSNs, and, in such a case, the BSC may select the PDSN based on the MS's International Mobile Subscriber Identity (IMSI). The user-specific R-P session is established at the beginning of a packet data session.

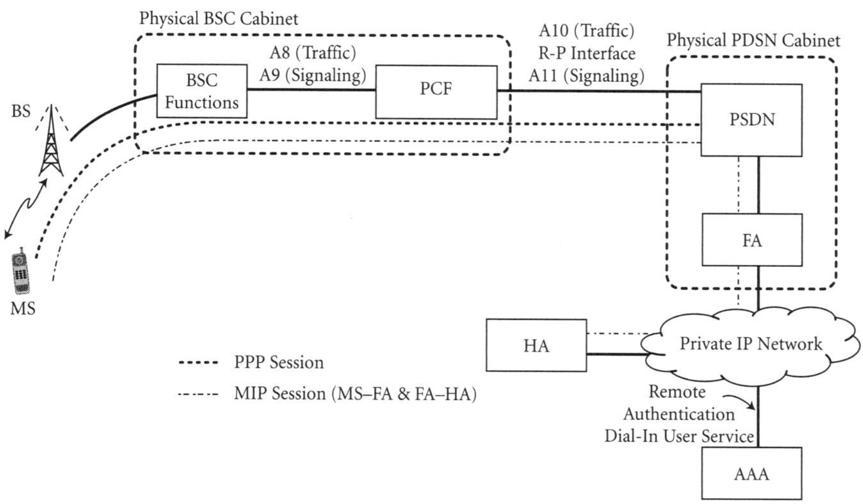

Figure 8.2: A Closer Look at the Packet-Switched Core Network

The PDSN stores the association between the IP address allocated to the MS (or the MS's Network Access Identifier) and the identity of the R-P link so it can associate the packets for the MS with a relevant R-P link. (See Chapter 4 for Network Access Identifier details.)

The PPP connection exists when the MS is in the *active* or *dormant* state from the perspective of the packet data session. The *active* state involves data transfer between the MS and the PDSN, and the *dormant* state involves idle time (no actual data transfer). The R-P session is preserved during the dormant state so that the data transfer can be resumed quickly upon a transition from the dormant to active state. The R-P session is released when a packet data session terminates. Usually, several hours of data inactivity results in the termination of the PPP connection and hence the R-P session.

The A10 interface on the R-P link is simply the *bearer path* carrying data traffic bits, and the A11 interface on the R-P link is the *signaling path* carrying signaling messages. User Datagram Protocol over IP is used for the signaling messages on the A11 interface. The PPP frames containing user traffic data are encapsulated in the Generic Routing Encapsulation (GRE) tunnels on the A10 interface since GRE over IP is used for the bearer path. (See Chapter 4 for basic information about Generic Routing Encapsulation.) The A11 interface helps establish the R-P session by exchanging messages, such as Registration Request and Registration Reply. The PCF can initiate the release of the R-P session by using $Lifetime = 0$ in the Registration Request message. The PDSN can also initiate the release of the R-P session using the Registration Update message.

The PDSN also facilitates security implementation using an ingress filter that blocks suspicious IP packets from entering the network. The PDSN contains a Remote Authentication Dial-In User Service (RADIUS) client that communicates with the AAA server using the RADIUS protocol to support authentication, authorization, and accounting. From the perspective of accounting, the PDSN incorporates the airlink records received from the PCF to facilitate construction of the User Data Record that helps with billing. The PDSN participates in the quality-of-service (QoS) implementation by downloading the QoS profile from the AAA server and then dynamically marking the IP packets with DiffServ code points.

The AAA server stores the subscriber profile and takes care of authentication, authorization, and accounting functions in the packet-switched core network. In addition to the PDSNs, the AAA server also connects to AAA clients (such as the PSDN) as well as remote AAA servers (e.g., a home or visited AAA server). The AAA server utilizes the RADIUS protocol over User Datagram Protocol over IP to communicate with the PDSN, more specifically, with the RADIUS client portion of the PDSN. The RADIUS client passes the user-specific information to the RADIUS server. For example, during the packet data session setup, the PDSN client sends an Access Request message to the AAA server. This message includes information such as the username, password, and port identity. To provide additional security between the PDSN and the AAA server, the password is hidden using the Rivest-Shamir-Adleman message digest algorithm (e.g., Message Digest version 5). The AAA server may issue a challenge to the user by sending an Access Challenge response.

Once the AAA server receives the Access Request message, it usually responds with an Access Accept or an Access Reject message. The Access Accept message contains configuration parameters for the user such as the desired compression technique.

If simple IP is used, the PDSN usually allocates an IP address to the MS during the establishment of a PPP link, and the MS uses such an IP address for communicating with the outside world (e.g., the Internet). Another way of allocating an IP address in simple IP is the use of a Dynamic Host Configuration Protocol (DHCP) server. When MIP is used, the HA assigns an IP address called the *home address* to the MS and the FA allocates a temporary care-of address to the MS. As the MS moves from one FA to another FA, its care-of address changes but its home address remains the same.[246] In MIP, two MIP tunnels exist: one between the MS and the FA and one between the FA and the HA. Since the PDSN usually implements the FA functionality, the PDSN is involved in both tunnels.

The PDSN/FA communicates with the HA using a secure tunnel established using IP Security (IPsec). (See Chapter 4 for details on simple IP, MIP, and IPsec.) The PDSN forwards the IP packets received from the MS to the Internet or the HA. In reverse tunneling, which is the typical scenario, the PDSN forwards the packets from the MS to the HA. The packets arriving from the Internet and destined for the MS first encounter the HA, and the HA forwards the IP packets to the PDSN/FA. The PDSN/FA, in turn, forwards the packets to the PCF. As an FA, the PDSN takes care of agent advertisements[247] and keeps track of care-of address and home address mapping.

8.3 MS–RADIO-ACCESS-NETWORK COMMUNICATIONS

An air-interface protocol stack defines how the MS and the radio access network communicate with each other. Section 8.3.1 provides a glimpse of this protocol stack. The FL channels, the RL radio channels, and the radio configurations that facilitate different data rates on the forward and reverse links are briefly discussed in Section 8.3.2. Processing done at layer 1, layer 2, and layer 3 (and other upper layers) of the protocol stack is discussed in Sections 8.3.3 through 8.3.5.

8.3.1 Overview of Air-Interface Protocol Stack

Just like IS-95, 1x uses frequency-division duplexing as the duplexing mechanism. Various frequency bands, such as cellular and personal communications service, are supported. Figure 8.3 shows the air-interface protocol stack used for MS–network communications. The bottom-most layer of the protocol stack uses a given frequency band that the operator has deployed in a given geographic region (which could even be an entire country!).

[246] The home address enables the MS to maintain connectivity with the IP networks, such as the Internet, and the care-of-address is used to forward the IP packets to the MS's current location.

[247] An agent advertisement conveys a care-of-address for an MS.

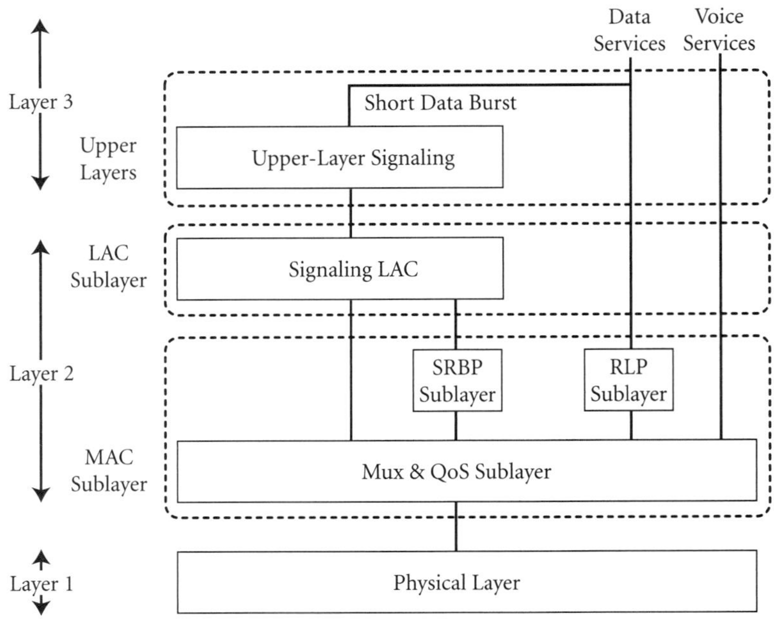

Figure 8.3: Air-Interface Protocol Stack

A Question for You. . . What is the main benefit of duplexing? See the footnote[248] for the answer.

The bottom-most layer, layer 1, is the physical layer that encompasses several processing blocks to exploit CDMA for wireless communications. Examples of the physical layer processing blocks are forward error correction coding, modulation, and spreading.

1x introduces a new layer 2, the MAC sublayer, which itself consists of three sublayers: the multiplexing (Mux) and QoS sublayer, the Signaling Radio Burst Protocol (SRBP) sublayer, and the RLP sublayer. The MAC sublayer's Mux and QoS sublayer allows user traffic (e.g., voice) and user-specific signaling messages to share a physical radio channel. The Mux and QoS sublayer also enforces the negotiated QoS levels via suitable prioritization of access requests. The SRBP sublayer controls how the common channels, such as the Paging channel, Access channel (ACH), and Synchronization (Sync) channel, are used for signaling. The MAC's sublayer's

[248] Duplexing enables simultaneous communications on the FL and the RL. In other words, it allows a user to talk and listen at the same time! Two main types of duplexing are frequency-division duplexing and time-division duplexing; see Chapter 1 for more details on duplexing.

RLP sublayer is new in 1x and enables reliable transmission over the radio link by retransmitting erroneous packets.

The Link Access Control (LAC) sublayer is the upper-most sublayer of layer 2 and interfaces with the MAC sublayer's Mux and QoS sublayer and with the upper-layer signaling entity. The LAC sublayer defines a data link protocol that enables correct delivery of signaling messages that layer 3 generates. The upper-layer signaling entity takes care of delivery of signaling messages and short data bursts (SDBs).

8.3.2 Radio Channels and Radio Configurations of 1x

Let's discuss both the FL and RL radio channels defined in 1x. The BS transmits the FL channels in a sector using orthogonal codes to avoid interference among these channels.

Do You Recall... Do you remember the name of these orthogonal codes in IS-95? See the footnote[249] for the answer.

See Chapter 7 to refresh your memory regarding generation of the orthogonal codes. An overview of the FL and RL radio channels is given in Sections 8.3.2.1 and 8.3.2.2, and radio configurations are the topic of focus in Section 8.3.2.3.

8.3.2.1 Overview of Forward Link Radio Channels Table 8.1 summarizes the FL channels defined in 1x. Three basic overhead channels of IS-95, Pilot, Sync, and Paging, are also used in commercial 1x deployments. Let's make sure that you still remember these channels.

Do You Recall... Do you recall the main uses of the Pilot, Sync, and Paging channels? The footnote[250] summarizes the answer.

The Pilot channel of IS-95 is now called the Forward Pilot channel to distinguish it from the Reverse Link Pilot channel. The Forward Pilot channel uses the Walsh code W_0^{64} for both IS-95 and 1x. Just like IS-95, 1x also has a total of 512 pilot pseudo-noise (PN) offsets, and these PN offsets are used to generate a PN sequence to carry out source-based (i.e., sector-based) scrambling. The Sync channel carries Sync channel messages that provide synchronization information and network information, such as the system identification, network identification, Pilot PN, system time, and CDMA channel number.

The Paging channel transmits mobile-station–specific messages, such as a General Page message, as well as system overhead information. The rate of the paging

[249]These orthogonal codes are called Walsh codes.

[250]The Pilot channel is used for synchronization, coherent demodulation, and handoff measurements. The Sync channel is used for synchronization and determination of the pseudo-noise offset. The Paging channel is for paging the MS for a network-initiated call.

Table 8.1: Forward Link Channels and Their Use

Channel	Main Use	Channel Identification	Comments
Forward Pilot	Same as IS-95	W_0^{64}	Also exists in IS-95
Sync	Same as IS-95	W_{32}^{64}	Also exists in IS-95
Paging	Same as IS-95	W_1^{64} to W_7^{64} available	Also exists in IS-95
Forward Fundamental	MS-specific user data traffic, MS-specific signaling message	Generally any unused code at spreading factor of 64 and 128	Called Forward Traffic channel in IS-95
Auxiliary Pilot	Beamforming		New in 1x; unused in typical 1x deployments
Transmit Diversity Pilot	Orthogonal Transmit Diversity	W_{16}^{128}	New in 1x; unused in typical 1x deployments
Auxiliary Transmit Diversity Pilot	Orthogonal Transmit Diversity in conjunction with Auxiliary Pilot channel	Any	New in 1x; unused in typical 1x deployments
Broadcast Control	Overhead information	Any	New in 1x; unused in typical 1x deployments
Forward Common Control	MS-specific messages	Any	New in 1x; unused in typical 1x deployments
Quick Paging	Increased battery standby time	W_{80}^{128}, W_{48}^{128}, W_{112}^{128}	New in 1x; unused in typical 1x deployments
Common Assignment	Allocation of Reverse Common Control channel	Any	New in 1x; unused in typical 1x deployments
Common Power Control	Power control of Reverse Common Control channel	Any	New in 1x; unused in typical 1x deployments
Forward Supplemental Code	Low-rate data services	Any	Also exists in IS-95B for Radio Configuration 1 and 2;[a] unused in typical 1x deployments
Forward Dedicated Control	MS-specific dedicated signaling messages	Any	New in 1x; unused in typical 1x deployments
Forward Supplemental	Medium and high-rate data services	Any	New in 1x : Radio Configurations 3, 4, & 5 only

[a] A Radio Configuration is associated with a specific set of data rates.

channel—full-rate 9.6 kbps or half-rate 4.8 kbps—is specified by the Sync channel message. The BS uses the Paging channel to send messages to the MS if the MS has no dedicated channels.

The Forward Traffic channel in IS-95 is called the Forward Fundamental channel (F-FCH) in 1x. The Forward Traffic channel carries user traffic, such as speech bits and data bits, and user-specific signaling messages, such as a Measurement Report message.[251] In addition to the basic Forward Pilot channel, three more pilot channels are introduced in 1x relative to IS-95: the Auxiliary Pilot channel, the Transmit Diversity Pilot channel, and the Auxiliary Transmit Diversity Pilot channel. These additional pilot channels are unused in commercial 1x deployments. These channels support transmit diversity and beamforming, which are *optional* features in 1x.

The Auxiliary Pilot channels are transmitted from auxiliary antennas[252] and used as reference signals to support beamforming. Orthogonal transmit diversity is supported in the FL to improve performance. Orthogonal transmit diversity involves transmission of code symbols from multiple antennas using orthogonal codes. To support coherent demodulation, reference signals are required from multiple antennas. The Transmit Diversity Pilot channel and Auxiliary Transmit Diversity Pilot channel provide such reference signals for an auxiliary antenna. When orthogonal transmit diversity is supported in a given sector, the Transmit Diversity Pilot channel is transmitted continuously. Orthogonal transmit diversity is uncommonly used in commercial deployments.

The Broadcast Control channel carries overhead messages, and the Paging channel carries both overhead messages and MS-specific messages. The Forward Common Control channel carries MS-specific messages, such as a page message. In an IS-95 system, the MS in the idle state needs to process potentially large messages on the Paging channel while looking for its page message. The Quick Paging channel relieves the MS of such intensive work; it informs the MS via a short message of whether a potential page message awaits on the Paging channel or the Forward Common Control channel. The Quick Paging channel increases the standby time of the MS battery.

A Question for You. . . What do you think is the relationship among 1x's Broadcast Control channel, 1x's Forward Common Control channel, and IS-95's Paging channel? See the footnote[253] to verify your answer.

[251] The Measurement Report message contains the measurements the MS made to quantify the prevailing radio channel conditions and thus facilitate handoff.

[252] A typical 1x sector utilizes a single transmit antenna in a commercial network. An auxiliary antenna is an additional transmit antenna that helps transmit signals such as the Auxiliary Pilot channel.

[253] IS-95's Paging channel can be replaced by two 1x channels, the Broadcast Control channel and the Forward Common Control channel! The benefits would be savings in the processing requirements at the MS and improved MS battery life. The drawback would be a reduction in capacity since more overhead power results in less power available for traffic channels.

The Common Assignment channel assigns a specific Reverse Common Control channel to the MS to minimize the probability of collision. The MS temporarily uses the assigned Reverse Common Control channel instead of the regular Access channel to send a message so that the access performance is enhanced. The Common Power Control channel carries time-multiplexed power-control commands for the power control of multiple Reverse Common Control channels and Enhanced Access channels. Such power control aims to reduce RL interference. The Common Assignment channel, Reverse Common Control channel, Common Power Control channel, and Enhanced Access channel are uncommon in typical deployments.

The Forward Supplemental Code channel is same as in IS-95B; it enables low-rate data services to provide backward-compatibility to IS-95B. The Forward Dedicated Control channel is a new channel in 1x that can carry MS-specific signaling messages at the rate of 9.6 kbps or 14.4 kbps during the call. The Forward Supplemental channel (F-SCH) is a new 1x channel that supports high-rate data services; typical 1x commercial deployments support 153.6 kbps F-SCH.

The term *Forward Traffic channel* is often used to refer to the Forward Dedicated Control channel, the F-FCH, the F-SCH, or the Forward Supplemental Code channel. *The BS combines various FL channels just like in IS-95 (as discussed in Chapter 7).*

8.3.2.2 Overview of Reverse Link Radio Channels The MS may transmit the RL channels as summarized in Table 8.2. All of the RL channels defined in IS-95 are available for use in 1x to preserve backward-compatibility. Furthermore, new RL channels are introduced in 1x.

Do You Recall... The Access channel and the Reverse Fundamental channel (R-FCH) in 1x are same as those in IS-95. Do you recall how these channels are used? See the footnote[254] to confirm your answer.

The BS uses the Reverse Pilot channel[255] to coherently demodulate the R-FCH. Furthermore, the FL power-control commands also travel on the Reverse Pilot channel. As the name suggests, the Enhanced Access channel is an access channel with some enhancements relative to the basic access channel of IS-95. Examples of such enhancements include a different channel structure to enable the MS to request a Reverse Common Control channel and to facilitate easier detection of the MS transmission due to the use of the Reverse Pilot channel as an Enhanced Access Channel Preamble, to support for higher rates such as 9.6 kbps, 19.2 kbps, and 38.4 kbps rather than only the 4.8 kbps basic Access channel rate; and to support for power control. Such enhancements reduce the possibility of collision, reduce interference, and send more data within a given time interval. The BS can temporarily allocate a Reverse

[254]The MS uses the Access channel to send messages to the network in the absence of a dedicated radio channel. The R-FCH (i.e., Reverse Traffic channel in IS-95) carries MS-specific signaling messages and user traffic during the call.

[255]The Reverse Pilot channel provides a reference signal to the BS so that the BS can estimate the reverse link channel conditions and implement coherent demodulation of the Reverse Traffic channel.

Table 8.2: Reverse Link Channels and Their Use

Channel	Use	Channel Identification	Comments
Access	Same as IS-95	ACH-specific long-code mask	Also exists in IS-95; provides full backward-compatibility with IS-95
Reverse Fundamental	Same as IS-95	R-FCH-specific long-code mask	Also exists in IS-95; Reverse Traffic channel
Reverse Pilot	Coherent demodulation of the R-FCH and power-control commands for the FL		New in 1x; represents the most significant reverse link enhancement of 1x relative to IS-95
Enhanced Access	Higher-data-rate access channel		New in 1x; unused in typical 1x deployments
Reverse Common Control	Collision-free access		New in 1x; unused in typical 1x deployments
Reverse Dedicated Control	MS-specific signaling		New in 1x; unused in typical 1x deployments
Reverse Supplemental Code	MS-specific traffic (low data rates)		Radio Configuration 1 and 2 only; provides full backward-compatibility with IS-95
Reverse Supplemental	MS-specific traffic (high data rates)		Radio Configuration 3 and 4 only

Common Control channel to the MS during access to virtually eliminate collisions. In such a case, the MS immediately switches from the Enhanced Access channel to the Reverse Common Control channel to send a message to the network.

Just like its FL counterpart, the Reverse Dedicated Control channel is a new 1x channel that can carry signaling messages at the rate of 9.6 kbps or 14.4 kbps during the call. The Reverse Supplemental Code channel is an IS-95B channel useful for low-data-rate services for IS-95B MSs. The Reverse Supplemental channel, just like the F-SCH, carries medium- to high-speed data for 1x MSs, with a practical peak data rate of 153.6 kbps. Typical 1x commercial deployments utilize two new (relative to IS-95) RL channels: the Reverse Pilot channel and the Reverse Supplemental channel.

The term *Reverse Traffic channel* in 1x refers to the Reverse Dedicated Control channel, Reverse Fundamental channel, Reverse Supplemental channel, or Reverse Supplemental Code channel. Six radio configurations for the Reverse Traffic channel are defined in 1x and are covered in Section 8.3.2.3.

8.3.2.3 Overview of Radio Configurations The radio configurations (RCs) define characteristics of the Forward Traffic channel and the Reverse Traffic channel. Table 8.3 summarizes the FL RCs for 1x.

The first two RCs, RC1 and RC2, provide backward-compatibility of 1x with IS-95 since they are only Rate Set 1 and Rate Set 2, respectively. Four data rates for RC1 and RC2 correspond to different rates of the speech frames the speech vocoder generates for a given 20 ms speech block. RC3 and RC4 are widely used in commercial deployments. They support higher data rates through the use of the F-SCH and the 8k EVRC speech vocoder.[256] Furthermore, the higher-order modulation scheme of quadrature phase shift keying (QPSK) is used in RC3 through RC9 [rather than RC1's and RC2's use of binary phase shift keying (BPSK)], which makes the system more efficient.

The network configures a 1x MS to operate in RC3 or RC4 during the call setup. When the lack of FL power is found to constrain the system performance, RC3 is usually preferred. However, when the number of available Walsh codes is the limiting factor, RC4 is preferable. RC3 uses the Walsh codes at a spreading factor of 64, and RC4 uses the Walsh codes at a spreading factor of 128. Hence, RC3 has fewer Walsh codes than RC4. The benefit of RC3 is more channel coding gain and hence less required transmit power per radio link. The practical peak rate is limited to 153.6 kbps. RC5 uses a 13 kbps speech codec and supports higher data rates via F-SCH. Since RC3 and RC4 exploit the 8k EVRC vocoder and since the 8k EVRC vocoder enables higher voice capacity compared to a 13 kbps speech codec at a reasonably good speech quality, RC5 is less attractive to a service operator and is typically unused in commercial deployments.

Now, let's turn our attention to the RL RCs. Table 8.4 summarizes the RL RCs for 1x. As is the case with the FL, RC1 and RC2 provide backward-compatibility with IS-95. RC3 and RC4 are intended to be used with an 8k EVRC and a 13k speech

[256] See Chapter 2 for details on the 8k EVRC speech vocoder.

Table 8.3: Forward Link Radio Configurations

Radio Configuration	Data Rates	Channel Coding Rate	Modulation Scheme	Comments
1	1.2, 2.4, 4.8, and 9.6 kbps	1/2	BPSK	Same as Rate Set 1 in IS-95; enables 1x to be fully backward-compatible with IS-95
2	1.8, 3.6, 7.2, and 14.4 kbps	1/2	BPSK	Same as Rate Set 2 in IS-95; enables 1x to be fully backward-compatible with IS-95
3	1.2, 1.35, 1.5, 2.4, 2.7, 4.8, 9.6, 19.2, 38.4, 76.8, and 153.6† kbps	1/4	QPSK	New in 1x; supports higher data rates; typically used in commercial deployments; suitable when FL power is the performance bottleneck
4	1.2, 1.35, 1.5, 2.4, 2.7, 4.8, 9.6, 19.2, 38.4, 76.8, and 153.6† kbps, and 307.2 kbps	1/2	QPSK	New in 1x; supports higher data rates; typically used in commercial deployments; suitable when Walsh codes are the performance bottleneck
5	1.8, 3.6, 7.2, 14.4, 28.8, 57.6, 115.2, and 230.4 kbps	1/4	QPSK	New in 1x; supports higher data rates; typically unused in commercial deployments
6, 7, 8, and 9	From 1.2 kbps to 1.0368 Mbps	1/2, 1/3, 1/4, and 1/6	QPSK	New in 1x; only for Spreading Rate 3 (and unavailable in 1xRTT) to support higher data rates

† 153.6 is the highest rate in typical commercial deployments.

Table 8.4: Reverse Link Radio Configurations

Radio Configuration	Data Rates	Channel Coding Rate	Modulation Scheme	Comments
1	1.2, 2.4, 4.8, and 9.6 kbps	1/3	64-ary orthogonal modulation	Same as Rate Set 1 in IS-95; enables 1x to be fully backward-compatible with IS-95
2	1.8, 3.6, 7.2, and 14.4 kbps	1/2	64-ary orthogonal modulation	Same as Rate Set 2 in IS-95; enables 1x to be fully backward-compatible with IS-95
3	1.2, 1.35, 1.5, 2.4, 2.7, 4.8, 9.6, 19.2, 38.4, 76.8, 153.6† kbps, and 307.2 kbps	1/4 and 1/2	BPSK	New in 1x; enables coherent demodulation; supports higher data rates; widely used in commercial deployments
4	1.8, 3.6, 7.2, 14.4, 28.8, 57.6, 115.2, and 230.4 kbps	1/4	BPSK	New in 1x; enables coherent demodulation; supports higher data rates
5 and 6	from 1.8 to 230.4 kbps	1/2, 1/3, and 1/4	BPSK	New in 3x; only for Spreading Rate 3 (and unavailable in 1xRTT) to support higher data rates

† 153.6 is the highest rate in typical commercial deployments.

codec, respectively. Furthermore, RC3 and RC4 support higher data rates through the Reverse Supplemental channel (R-SCH). RC3 is widely used in commercial deployments due to the efficiency of the 8k EVRC vocoder. The practical peak rate is limited to 153.6 kbps. The new RCs in 1x use BPSK modulation rather than RC1's and RC2's 64-ary orthogonal modulation. Furthermore, the new RCs utilize a Reverse Pilot channel to facilitate coherent detection of the Reverse Traffic channel. The use of coherent demodulation in the RL is one of the important enhancements of 1x relative to IS-95. Such enhancement becomes more useful for higher data rates. RC5 and RC6 are included here just for the sake of completion; they are unused in commercial 1x deployments because they belong to 3xRTT, not 1xRTT!

8.3.3 Overview of Physical Layer (Layer 1) Processing

Different radio channels and radio configurations involve somewhat different physical layer processing. We will focus here on the F-FCH and F-SCH processing for RC3 and RC4 when 20 ms frames are used. Figure 8.4 shows a simplified block diagram of the physical layer processing for the F-FCH and the F-SCH. Information bits from the MAC sublayer arrive at the physical layer. The number of these bits ranges from 16 to 3048 for RC3 and 16 to 6120 for RC4. A variable number of Frame Quality Indicator (FQI) bits and eight encoder tail bits are added to achieve a data rate ranging

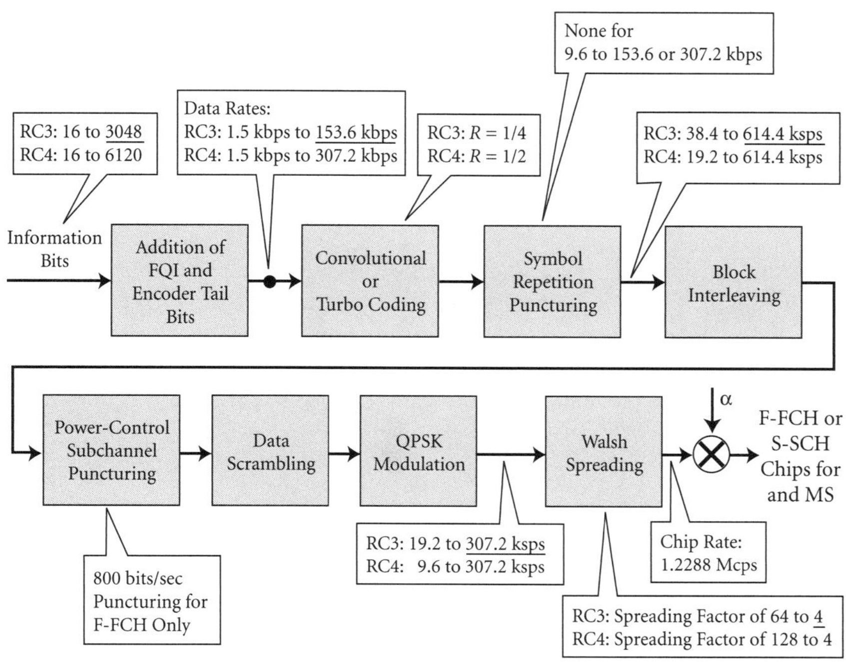

Figure 8.4: Physical Layer Processing for the F-FCH and F-SCH

from 1.5 kbps to 153.6 kbps for RC3 and from 1.5 kbps to 307.2 kbps for RC4. The FQI bits are the cyclic redundancy check bits. With RC3 and a 153.6 kbps data rate, sixteen FQI bits are added, resulting in $3048 + 16 + 8 = 3072$ bits in a 20 ms frame. Observe that $3072 \text{ bits}/20 \text{ ms} = 153.6$ kbps. As another example, a total of 192 bits per 20 ms frame are obtained by combining 172 information bits, twelve FQI bits, and eight encoder tail bits to achieve the rate of 9.6 kbps.

Two forward error correction coding schemes are supported, convolutional coding with $K = 9$ and turbo coding. *Convolutional coding is mandatory, and turbo coding is available for use when the number of information bits is 360 or more.* Support for the high-data-rate Supplemental channel (SCH) and turbo coding is a significant air-interface enhancement in 1xRTT compared to IS-95. The forward error correction coding rate R is $1/4$ for RC3 and $1/2$ for RC4. Symbol repetition or puncturing achieves data rates of 38.4 ksps[257] to 614.4 ksps for RC3 and 19.2 ksps to 614.4 ksps for RC4. Block interleaving rearranges the order of the symbols for transmission. The interleaved code symbols of the F-FCH experience power-control subchannel puncturing at the rate of 800 bits per second. The interleaved symbols are scrambled using a long-code mask that is a function of the electronic serial number (ESN).

The use of QPSK modulation reduces the rate to the range of 19.2 ksps to 307.2 ksps for RC3 and 9.6 ksps to 307.2 ksps for RC4. The modulation symbols are spread to the chip rate of 1.2288 Mcps through the use of Walsh codes, with the spreading factors ranging from 64 to 4 for RC3 and from 128 to 4 for RC4. For example, the modulation symbol rate of 307.2 ksps and spreading factor of 4 gives the chip rate of $307.2 \text{ ksps} * 4 = 1.2288$ Mcps. The resulting F-FCH or F-SCH chips are allocated a suitable digital gain to reflect the influence of the FL power control. The chips from various FL channels are combined and scrambled as shown in Chapter 7.

The physical layer processing for the RL in 1x is different from that in IS-95 in the following main respects: the addition of a pilot channel in 1x, the use of BPSK modulation instead of 64-ary modulation, and the addition of a power-control subchannel to support fast FL power control. A review question is in order!

Do You Recall... In IS-95, the network sends RL power-control commands to the MS, and the MS increases or decreases its power based on the power-control commands received at 800 Hz. How many times per second does the MS send power-control commands to the network so that the FL power for the MS can be adjusted? See the footnote[258] for the answer.

Figure 8.5 summarizes the physical layer processing for the R-FCH, R-SCH, Reverse Pilot channel, and Reverse Dedicated Control channel (R-DCCH) in RC3. RC3 in the RL is widely used in commercial deployments.

[257] kilo code symbols per second

[258] Sorry! It was a trick question! The MS sends *no* power-control commands to the network in IS-95. The BS, with guidance from the BSC, autonomously adjusts the FL power for the MS based on the FL error information the MS sends.

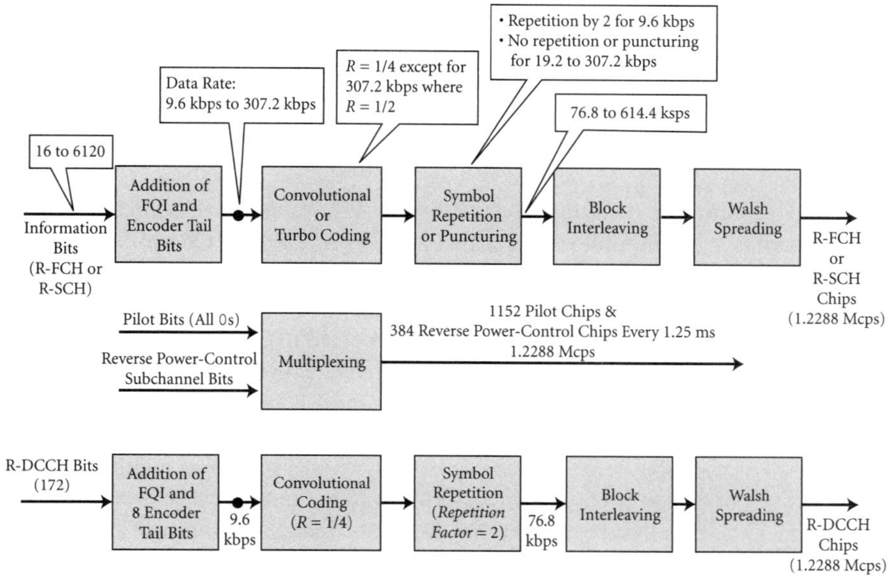

Figure 8.5: Physical Layer Processing of the R-FCH, R-SCH, Reverse Pilot Channel, and R-DCCH for RC3

Let's discuss the processing of the R-FCH and the R-SCH first. The information bits from the MAC sublayer arrive at the physical layer for the R-FCH and the R-SCH, with the number of bits ranging from 16 to 6120. Once the FQI bits and the encoder tail bits are added, the data rate ranges from 9.6 kbps to 307.2 kbps. The number of FQI bits is variable; for example, the FQI bits are twelve bits for 9.6 kbps and sixteen bits for the rate ranging from 19.2 kbps to 307.2 kbps. The number of encoder tail bits is eight. All eight bits are used as true encoder tail bits for convolutional coding. However, in turbo coding, the first two of the eight bits are reserved and the rest of the bits are internally generated. The maximum data rates are 9.6 kbps and 307.2 kbps for the R-FCH and the R-SCH, respectively, in RC3. However, commercial deployments use only the highest R-SCH data rate of 153.6 kbps.

Both convolutional coding and turbo coding are supported, with turbo coding available for an R-SCH with a data rate of 19.2 kbps or higher. The coding rate used is $R = 1/4$ for all rates except 307.2 kbps where $R = 1/2$ is used. Convolutional coding uses the constraint length $K = 9$. The code symbols are repeated and/or punctured to achieve rates that range from 76.8 kbps to 614.4 kbps. For example, 9.6 kbps R-FCH is repeated by a factor of two to reach the data rate of 9.6 kbps $* 4 * 2 = 76.8$ ksps) where 4 is the inverse of the coding rate $R = 1/4$.

Block interleaving rearranges the order of symbols for transmission. More specifically, the Walsh covers of length 2, 4, or 8 are used along with suitable repetition of these Walsh covers by the factor of 1, 2, 4, or 8 to reach the magic number of 1.2288 Mcps. For example, the 153.6 kbps R-SCH is spread by a Walsh cover of two chips, $[+1\ +1]$ (which is referred to as $(++)$ in the standard [240]) with a repetition factor of one since the interleaver output rate is 614.4 ksps for 153.6 kbps R-SCH. As another example, the 38.4 kbps R-SCH corresponds to 153.6 ksps at the output of the interleaver, which is spread by the Walsh cover $(++)$ and repeated by a factor of four, essentially leading to the spreading factor of $2 * 4 = 8$ and yielding 153.6 ksps $* 8 = 1.2288$ Mcps.

Now, let's turn our attention to the Reverse Pilot channel. The Reverse Pilot channel carries time-multiplexed pilot bits and Reverse Power Control subchannel bits. Such multiplexing is performed such that the Reverse Pilot channel carries 1536 chips in each Power Control Group with 1152 pilot signal chips that are all 0s and 384 reverse power-control subchannel chips that represent one power-control bit. The length of the Power Control Group is 1.25 ms. Since the power-control bits represent one power-control command every 1.25 ms for the FL of the MS, the FL power control is said to operate at $1/1.25$ ms $= 800$ Hz. The existence of fast FL power control is a key contributor to the increase in capacity of 1x systems relative to IS-95 systems. The power control is termed fast compared to the slow 50 Hz power control possible in IS-95. The voice capacity in the sector increases from twenty users to thirty-five users from IS-95 to 1x, thanks in large part to fast FL power control. Other factors contributing to the capacity increase in 1x are the optimization of the power levels of the FL overhead channels (e.g., pilot channel) and the optimization of handoff; both are based on experience with IS-95 deployments.

Finally, let's summarize the R-DCCH processing. The physical layer receives 172 bits of the R-DCCH. With the addition of twelve FQI bits and eight encoder tail bits, the total number of bits becomes 192 bits over a 20 ms frame, resulting in the data rate of 9.6 kbps. The set of 192 bits is protected using convolutional coding with a coding rate of $1/4$ and constraint length $K = 9$. Symbol repetition by a factor two yields the code symbol rate of 76.8 ksps. The repeated symbols undergo block interleaving. The interleaved code symbols are spread by a Walsh code of length sixteen chips to reach 1.2288 Mcps.

Now that we know how to generate chips for various radio channels, let's see how to combine these channels. The radio channels have some gain relative to the reverse pilot channel. The weighted chips from various radio channels are combined for the I-channel and the Q-channel. Specifically, the Reverse Pilot channel, R-DCCH, and the R-SCH2 contribute to the I-channel chips, and the R-FCH and R-SCH1 contribute to the Q-channel chips. *Two R-SCHs are shown in Figure 8.6*[259] *for the sake of completion only; the commercial deployments utilize only R-SCH1.* R-SCH2 is used only when two R-SCHs are used in parallel. Furthermore, R-SCH1 uses the Walsh cover W_1^2 (i.e., $(+-)$) for 153.6 kbps and 307.2 kbps only. R-SCH1 uses the Walsh cover W_2^4 (i.e., $(++--)$) for all other lower rates.

The I-channel chips and the Q-channel chips of the Reverse Traffic channel shown in Figure 8.6 are multiplied by a complex spreading sequence to achieve MS-specific scrambling. The in-phase spreading sequence is constructed using the I-channel

[259] Tx represents "transmit."

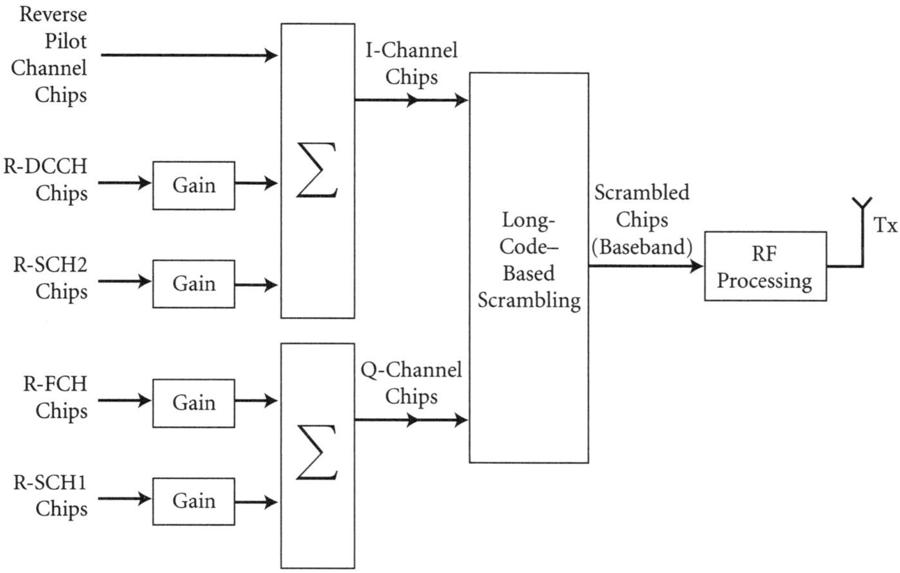

Figure 8.6: Combining of Radio Channels in the Reverse Link

PN sequence and the I-channel long-code sequence. The quadrature-phase spreading sequence is a function of the Walsh function W_1^2 (i.e., $(+\,-)$), I-channel PN sequence, I-channel long-code sequence, Q-channel PN sequence, and Q-channel long-code sequence. The I-channel long-code sequence and the Q-channel long-code sequence are generated using the ESN-based public long-code mask with a long-code generator. The I-channel PN sequence and Q-channel PN sequence are periodic with a period of 2^{15} chips. The scrambled PN chips go through baseband filtering and frequency upconversion for over-the-air RF transmission. Once a dedicated radio channel is set up and the MS is doing packet data transfer, typical channels present simultaneously include the R-FCH, Reverse Pilot channel, and R-SCH1. R-DCCH is a substitute for R-SCH and is uncommonly used.

8.3.4 Overview of Link Layer (Layer 2) Processing

Various functions of layer 2 are discussed below. Specifically, Section 8.3.4.1 discusses the MAC sublayer, which carries out SRBP, RLP, and Mux and QoS delivery functions, and Section 8.3.4.2 summarizes the LAC sublayer, which performs signaling functions. Sections 8.3.4.3 and 8.3.4.4 summarize the RLP functions and the Mux and QoS functions of the MAC sublayer. Section 8.3.4.5 provides a brief overview of the LAC functions.

8.3.4.1 Layer 2 and Its Interfaces with Layer 1 and Upper Layers

We just discussed layer 1 or the physical layer. Let's look the next higher-up layer, layer 2. Figure 8.7 shows how layer 2 interfaces with layer 1 and layer 3 or other upper layers.

The LAC sublayer and the MAC sublayer are layer 2's two main sublayers. The LAC sublayer consists of a signaling block that interfaces with layer 3 signaling. Layer 3 signaling messages pass through the signaling LAC, and the LAC adds some message-dependent headers, such as the identity of the MS. See Section 8.4 on call setup for examples of LAC headers.

Let's summarize the information units that are exchanged between the MAC sublayer and other sublayers. The information unit between the signaling LAC and the MAC sublayer's Mux and QoS sublayer of MAC and between the signaling LAC and the MAC sublayer's SRBP is an LAC protocol data unit. The information unit

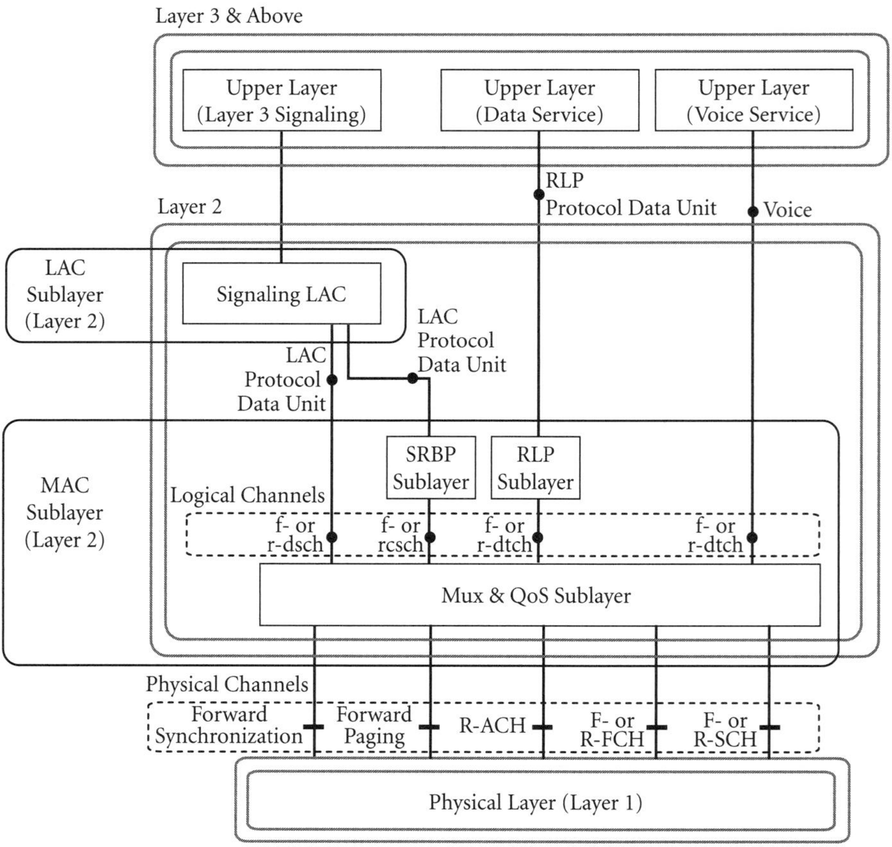

Figure 8.7: Layer 2 Architecture: A Closer Look

between the upper layer providing data services and the MAC sublayer's RLP is an RLP protocol data unit.

Finally, voice frames are exchanged between the upper layer providing voice service and the MAC sublayer's Mux and QoS sublayer. The Mux and QoS sublayer carries out a mapping between logical channels and physical channels. Such mapping could be a default specified in the standard [241], or the network could customize such a mapping for the MS. The logical channels enable layer 2 to convey signaling or traffic information to layer 1. The names of the logical channels are specified with lowercase letters, and the physical channels are specified as capital letters. Three types of logical channels are f- or r-csch (forward or reverse common signaling channel), f- or r-dsch (forward or reverse dedicated signaling channel), and f- or r-dtch (forward or reverse dedicated traffic channel). The signaling messages from the f-csch are typically carried by Forward Synchronization channel and Forward Paging channel. The r-csch messages can be transported by the R-ACH (Reverse Access channel). The f-dsch maps to F-FCH, and r-dsch maps to R-FCH (Reverse Fundamental channel) for MS-specific signaling messages. The f-dtch typically maps onto F-FCH and F-SCH, and r-dtch typically maps onto R-FCH and R-SCH.

8.3.4.2 Medium Access Control Sublayer and Signaling Radio Burst Protocol Functions The MAC sublayer takes care of the SRBP functions. The SRBP is used for transmission and reception of f- or r-csch signaling messages.

Do You Recall... Do you remember the uses of the Access channel? The footnote[260] has the answer.

The SRBP manages the operation of Sync, Paging, and Access channels so these channels can carry intended signaling messages. For example, the SRBP decides whether an access probe needs to be sent on the Access channel during an access attempt. See Section 8.6.2 for details on the access operation in 1x. In addition to the signaling messages, the SRBP is also responsible for transport of the short data burst that contains bearer data (e.g., short message service) generated by the short data burst teleservice.

8.3.4.3 Medium Access Control Sublayer and Radio Link Protocol Functions The MAC sublayer implements the RLP functions for packet data traffic. The RLP used in 1x is based on Negative Acknowledgments (NAKs) whereby the receiver informs the transmitter which data packets have yet to be received. Recall that the upper layers, such as Transmission Control Protocol (TCP), expect lower layers to provide essentially error-free transmission of packets.

[260]The MS uses the Access channel to contact the network when it has no dedicated channel.

A Question for You... Pretend for a moment that there is no RLP. What impact would a harsh radio environment have on user-perceived throughput—very little, some, or significant? The footnote[261] has the answer.

The RLP alleviates radio environment problems by retransmitting packets so that the effective error rate the TCP sees is low. 1x uses RLP Type 3. RLP Types 1 and 2 were defined for IS-95B. (Curious readers can refer to the standard [242] for details on the RLP.) RLP in this chapter refers to RLP Type 3 unless explicitly stated otherwise.

RLP works in units of octets (i.e., bytes). The higher-layer packets (e.g., IP packets) and the RLP frames have no direct and predefined relationship. One RLP frame may contain multiple small IP packets, or multiple RLP frames may carry one large IP packet. Figure 8.8 provides an example of how the RLP in 1x operates. Note that the standard is quite flexible, and Figure 8.8 represents one possible implementation. The sender sends RLP data frames with sequence numbers A, B, and C. The goal is to make an attempt to deliver these frames error-free from the sender to the receiver. (A, B, and C are used only for the sake of illustration; in practice, a twelve-bit sequence number identifies a given RLP data frame.)

The radio environment causes the RLP frame with sequence number B to be in error. When the receiver correctly receives the frames with sequence numbers A and C, it realizes that it is missing an RLP frame that has a sequence number B. The receiver sends a NAK to the sender identifying the missing data frame. The NAK is sent as part of a NAK control frame that can request the retransmission of one or more data frames. For example, the receiver can request retransmission of one or more groups of the data frames where each group contains consecutive octets that have been missing.

Where the RLP frames containing the octets 0 through 99, 200 through 499, and 800 through 999 have been correctly received, the receiver may request retransmission of the RLP frames containing the octets 100 through 199, 500 through 799, and 1000 through 1199. Note that the use of twelve bits allows the RLP frame sequence number to be in the range 0 to $2^{12} - 1 = 4095$. In Figure 8.8, the receiver is missing just one frame containing the sequence number B and sends a NAK for just one frame. The sender retransmits the missing frame. The RLP receiver's transmission of one NAK and the RLP transmitter's one retransmission constitute one NAK round.

In Figure 8.8, the RLP receiver still has no error-free RLP frame with sequence number B. Now it transmits two NAKs, and the RLP transmitter responds with two identical RLP frames. At the end of the second NAK round, the receiver still has no good RLP frame with sequence number B. The third NAK round consists of the RLP receiver sending three NAKs and the RLP transmitter sending three copies of the

[261] The user-perceived throughput could be quite poor. The TCP now experiences the physical layer errors caused by the harsh environment. Errors at the physical layer would mean that TCP would miss some packets and would initiate the congestion control mechanism, drastically reducing the amount of traffic that could be sent on the air interface. See Chapter 4 for TCP details.

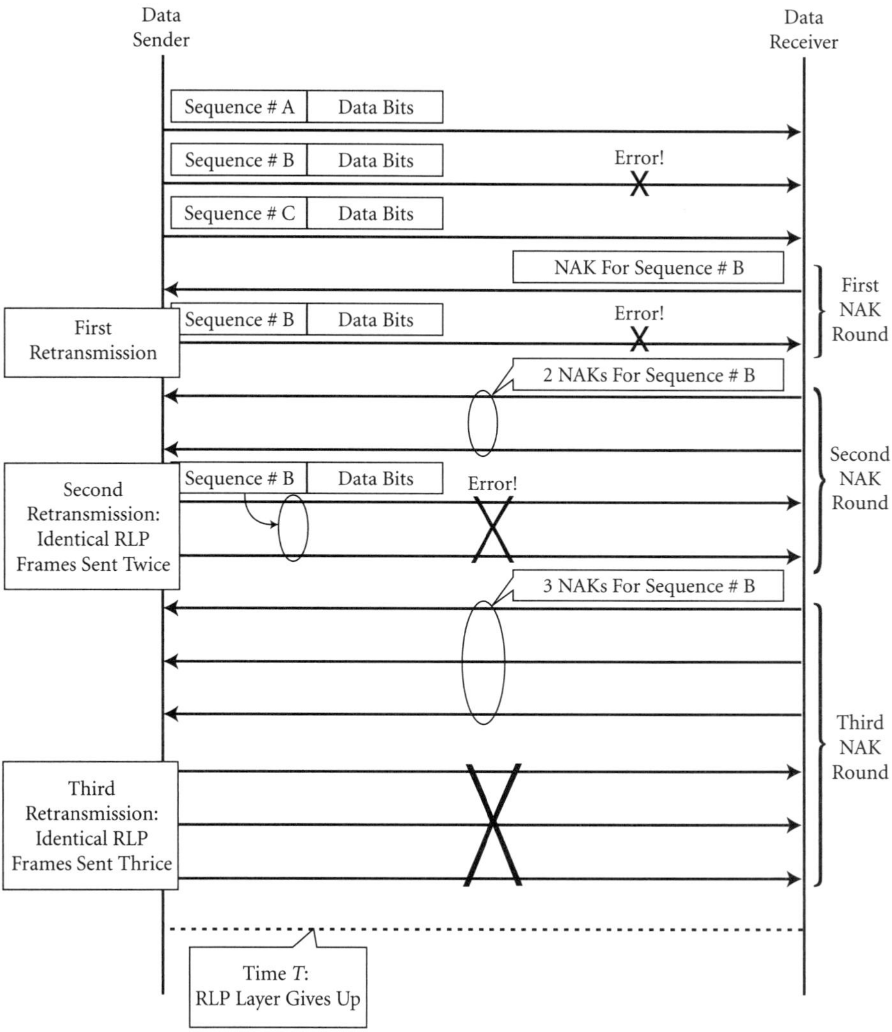

Figure 8.8: Example Implementation of RLP

RLP frame with sequence number B. If the receiver still has no error-free RLP frame, the RLP sublayer gives up. The upper layers must then determine how to recover from the error. For example, the TCP typically retransmits a lost packet resulting from erroneous packets at the lower layers. The RLP retransmission scheme shown in Figure 8.8 is called an RLP "1-2-3 scheme" where the receiver sends one NAK in the first round, two NAKs in the second round, and three NAKs in the third round. The MS is required to support at least three NAK rounds.

The network and the MS determine the RLP parameters during the call setup or service negotiation by exchanging the parameter RLP_BLOB (where BLOB represents the block of bits) with the MS. The MS sends RLP_BLOB to the network and informs the network of the maximum number of NAK rounds it can support for the Forward Traffic channel and the Reverse Traffic channel. The network configures the MS with the NAK parameters (e.g., number of NAK rounds and number of NAK requests per NAK round) by including them in RLP_BLOB.

Now that we understand the basic RLP operation, let's look at some RLP characteristics. These characteristics influence the overall RLP functions.

Physical Channels for Transmission of Radio Link Protocol Frames. A *fundicated data block* is one that can be sent on a Fundamental channel or a Dedicated Control channel. The data blocks are typically send on the Fundamental channel in commercial deployments. When an RLP frame is given to the multiplex sublayer for transmission in a fundicated data block, it is called a *fundicated RLP frame*. Furthermore, an RLP frame to be carried in a supplemental channel data block is called a *supplemental RLP frame*. The RLP provides the frames at the rate the multiplex sublayer dictates.

Priority of Frames. The RLP classifies the RLP frames into three priority classes: RLP control frames (highest priority), retransmitted data frames, and new data frames (lowest priority). Furthermore, identical retransmitted data frames are sent in *consecutive* 20 ms intervals, rather than in the *same* 20 ms interval, to exploit time diversity. In general, when control or data frames are unavailable, an idle frame or a fill frame is sent as a fundicated RLP frame and a blank RLP frame is sent as a supplemental RLP frame.

Support for Encryption. RLP supports the transport of frames in an encrypted mode and a non-encrypted mode, with the non-encrypted mode being the de facto implementation. The network controls the use of encryption as part of service negotiation during the call setup.

8.3.4.4 Medium Access Control Sublayer and Multiplexing and Quality-of-Service Delivery Functions The MAC sublayer's Mux and QoS sublayer multiplexes or combines information from the upper-layer sources, such as layer 3 signaling, voice service, and data service, to prepare service data units for physical layer transmission. Relative service priority is considered while multiplexing the upper-layer information. However, 1x commercial deployments utilize the QoS delivery sublayer very little. For 1x-related systems, true QoS control is lacking until 1xEV-DO Rev. A.

The Multiplex sublayer operates in Mode A or Mode B. Mode A works with the physical layer RC1 or RC2 and hence is applicable to IS-95 systems, and Mode B works with RC3 and higher. In other words, Mode A ensures full 1x backward-compatibility with IS-95, and Mode B allows 1x to achieve better performance than IS-95 by using CDMA2000 physical layer features (e.g., a supplemental channel).

As mentioned earlier, the Mux sublayer combines the upper-layer information to prepare a physical layer service data units. The rules that govern how to combine information are defined by a Multiplex Option. In essence, the Multiplex Option specifies the available data rates for the physical channel, such as the FCH or SCH. Multiplex Option 0x01 for the FCH corresponds to Rate Set 1 operation, and Multiplex Option 0x02 for the FCH corresponds to Rate Set 2 operation. Figure 8.9 defines the nomenclature used in conjunction with the bit combining operation of the MAC sublayer's multiplex sublayer.

The multiplex sublayer places the information bits received from the signaling entity, voice service, or data service into a data block. It then multiplexes one or more data blocks into a multiplexing protocol data unit (MuxPDU) and combines one or more MuxPDUs into a physical layer service data units. There are four types of MuxPDU. MuxPDU Type 1 is defined for a 9.6 kbps (and lower subrates such as $9.6/2 = 4.8$ kbps) channel, and MuxPDU Type 2 is defined for a 14.4 kbps (and lower subrates) channel. The FCH, the Dedicated Control channel (DCCH), and Rate 1x SCH can use MuxPDU Type 1 and Type 2. For Rate Set 1, Rate 1x SCH corresponds to the base rate of 9.6 kbps, and, for Rate Set 3, Rate 1x SCH corresponds to the base rate of 19.2 kbps. MuxPDU Type 3 supports the data transmission on an SCH with a rate 2x or larger (e.g., 19.2 kbps or larger for Rate Set 1). MuxPDU Type 4 carries a 5 ms data block. Since typical deployments use the frame length of 20 ms, MuxPDU Type 4 is uncommonly encountered in practice.

A physical layer FCH and DCCH service data unit contains just one MuxPDU of Type 1 or 2, with the MuxPDU containing a header and one or more data blocks. The header for MuxPDU Type 1 and 2 describes how the data blocks are multiplexed

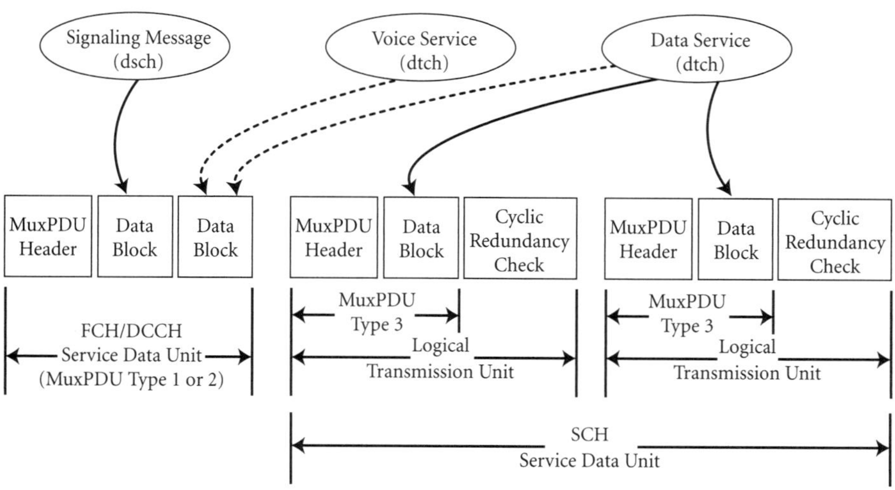

Figure 8.9: Bit Combining at the Multiplex Sublayer of MAC

together. For example, the header for MuxPDU Type 1 contains $MixedMode = 0$ to indicate that the data block carries 171 bits of primary traffic.

A physical layer SCH service data unit includes one or more MuxPDUs. A Logical Transmission Unit is defined for a 2x or higher-rate SCH and consists of one MuxPDU Type 3 and the related sixteen-bit cyclic redundancy check. The header for an MuxPDU Type 3 uses the parameter sr_id (service reference identity) to specify the logical channel associated with the MuxPDU.

The multiplex option associated with the SCH defines the number of MuxPDUs in the service data units and the size of the data block in each MuxPDU. Two block sizes, single-size and double-size, are defined for MuxPDU Type 3. With Rate Set 1, the single-size data block has 170 bits, and the double-size data block has 346 bits. For Rate Set 2, the single-size and double-size data blocks have 266 bits and 538 bits, respectively. The multiplex option associated with the SCH specifies the number of MuxPDUs in the service data units and the size of the data block in each MuxPDU. For example, when the MS is configured with the SCH multiplex option set to 0x921, it means that the MS should operate using Rate Set 1 with the highest possible rate being 153.6 kbps and up to eight MuxPDUs of Type 3 can contain double-size data blocks. When the scheduler dynamically allocates SCHs, the multiplex option *in use* changes depending upon the rate of the allocated SCH during a given time period.

8.3.4.5 Link Access Control Sublayer for Signaling: An Overview *The main purpose of the LAC sublayer is to facilitate delivery of signaling messages across the air interface between the MS and the network.* Figure 8.10 depicts the architecture

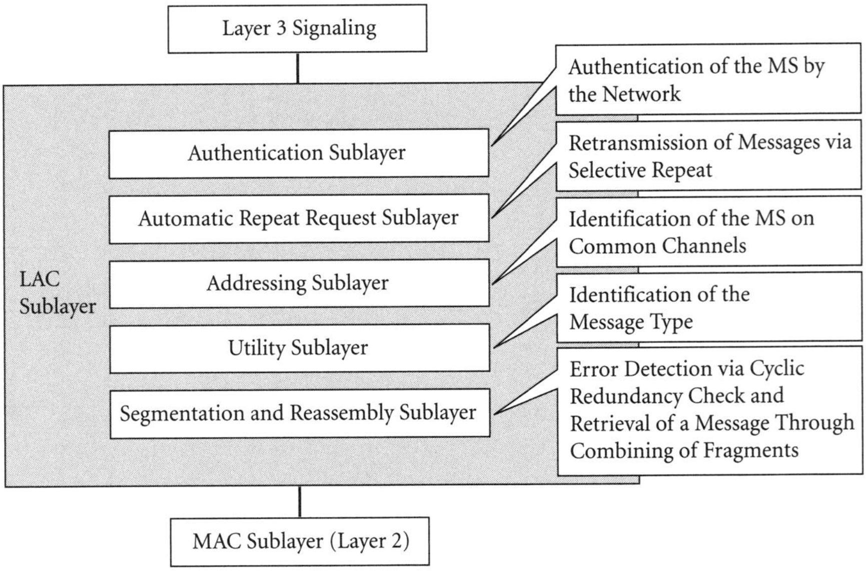

Figure 8.10: LAC Architecture and its Major Functions

of the LAC sublayer. The LAC sublayer works on the logical channels, f- or r-csch and f- or r-dsch, that deal with signaling messages. The LAC sublayer is uninvolved with the logical channels carrying traffic (i.e., f- or r-dtch). Five sublayers defined for the LAC sublayer are Authentication, Automatic Repeat Request, Addressing, Utility, and Segmentation and Reassembly.

The Authentication sublayer enables the network to confirm the identity of the MS, but the MS cannot authenticate the network. The Authentication sublayer includes information such as a random challenge, the identity of the MS (such as an IMSI) and, in the LAC header of a suitable message (e.g., Origination or Registration message), the associated authentication response from the authentication algorithm to the network. The network executes the authentication algorithm and compares the generated response with the response specified by the authentication sublayer in the received message. See Section 8.6.1 and the standards [243–245] for additional details on authentication.

The Automatic Repeat Request sublayer implements a selective repeat mechanism to increase the probability of the successful delivery of a message. Assume that Message X needs to be delivered correctly due to its importance. In such a case, Message X can be retransmitted if it is thought to have been lost in transit. The main idea behind the selective repeat mechanism is to include a sequence number in Message X and request the receiver to send a positive acknowledgment upon reception of Message X. If Message X's sender receives no positive acknowledgment within a predefined time period, the sender repeats Message X. Retransmission of Message X thus increases the chances that the receiver can correctly retrieve Message X.

When the receiver successfully decodes Message X, it responds with Message Y that includes Message X's sequence number as the Acknowledgment. Upon decoding Message Y, Message X's sender realizes that it no longer needs to retransmit Message X. The Automatic Repeat Request sublayer also keeps track of statistics that quantify the performance of various channels. For example, it calculates the value of parameter ACC_8 that indicates the number of unsuccessful access attempts. If the message is encrypted on f- or r-dsch, the Automatic Repeat Request sublayer provides the relevant information.

When the MS has no dedicated radio link, it relies on the common channels for communication with the BS. Since multiple MSs are using the common channels, we need some way of identifying the MS that is the target recipient of a message. The Addressing sublayer takes care of such MS identification by including the addressing fields as part of an LAC header. Examples of the MS identities that the addressing sublayer can use include IMSI, ESN, and Temporary Mobile Station Identity. Various formats of IMSI have been defined including a Mobile-Identification-Number–based IMSI.

The Utility sublayer specifies the type of the message and provides a report on the radio environment. For example, $message\ type$ = 000100 corresponds to an Origination message, and $message\ type$ = 000001 corresponds to a Registration message on the r-csch. As another example, $message\ type$ = 00000101 represents Pilot Strength Measurement message for the r-dsch. The Utility sublayer enables the MS to report the measured pilot strengths (i.e., E_c/I_0) of strong pilots in the neighborhood as part of a radio environment report. The Segmentation and Reassembly sublayer allows the receiver to put together the entire message by combining the fragments of the message and supports error detection on the message by using a thirty-bit cyclic redundancy check.

We briefly discussed five sublayers of the LAC. Keep in mind, however, that not all logical channels are processed at all five sublayers. For example, the Addressing sublayer is uninvolved with the f- or r-dsch because the dedicated logical channels no longer need to identify the MS. However, we do need to identify the MS from all other MSs when a common logical channel, f- or r-csch, is used. Hence, the Addressing sublayer steps in and includes the MS's identity in the messages carried over the f- or r-csch.

8.3.5 A Brief Overview of Upper-Layer Processing

Conceptually, three types of entities, a signaling entity, a voice service entity, and a data service entity, constitute the upper layers. The signaling entity is a layer 3 signaling entity that is responsible for the construction of layer 3 signaling messages (e.g., an Origination message) at the transmitting side and the interpretation of the signaling message at the receiving side. The voice service entity includes the speech vocoder, such as an 8k EVRC vocoder. The data service entity provides data traffic in chunks of IP packets.

Short data burst is a service in which the signaling entity and data service entity work together. The layer 3 or upper-layer signaling entity receives the short data burst containing the data burst service data units from the short data burst teleservice that interfaces with the data traffic services. Example attributes of a short data burst being sent to an MS are the data block, the length of the data block, the service option number, the Service Reference Identifier, and the MS's address. The receiver short data burst teleservice uses the specific Service Reference Identifier to direct other attributes of the short data burst to the corresponding data service instance. The receiver uses the service option number to provide the data block and its length to the appropriate data service instance link layer entity. The upper-layer signaling entity transports the short data burst via a Data Burst message over the air. Specifically, the BURST_TYPE field of the Data Burst message is set to "Short Data Burst" and the CHARi fields are set to the data burst itself. A layer 2 acknowledgment is required for the Data Burst messages.

8.4 1x CALL SETUP

The overall call flow for setting up a voice call in 1x is similar to that in IS-95. Section 8.4.1 provides an overview of the voice call setup in 1xRTT, and details of the call setup steps are discussed in Section 8.4.2. Section 8.4.3 discusses how a data session is set up. MS-terminated call setup is summarized in Section 8.4.4.

8.4.1 Voice Call Setup in 1x: An Overview

Figure 8.11 summarizes the main steps of the MS-initiated voice call setup in 1xRTT for an MS-originated call. The MS sends the Origination message on the Access channel as the first step in the call setup process. The network responds with the Extended Channel Assignment message on the Paging channel. Observe that no dedicated radio channel yet exists between the MS and the BS, and hence both the MS and the BS are using the common radio channels for communication. The Extended Channel Assignment message establishes the dedicated radio channels between the MS and the BS. For example, this message includes the allocated FL Walsh code for the MS in one or more sectors.

The MS starts sending the traffic channel preamble on the allocated dedicated Reverse Traffic channel. An Order message containing Base Station Acknowledgment Order is sent to the MS on the Forward Traffic channel to acknowledge that the BS has successfully detected the preamble. Until the real information bits (e.g., speech bits) become available and the call is fully set up, the MS sends null traffic channel data on the Reverse Traffic channel. The network sends the Service Connect message to specify to the MS which service option should be used for FL and RL transmissions. The MS acknowledges the reception of this message by sending the Service Connect Completion message. The call is now established, and the MS and the BS communicate with each other using the agreed-upon service option.

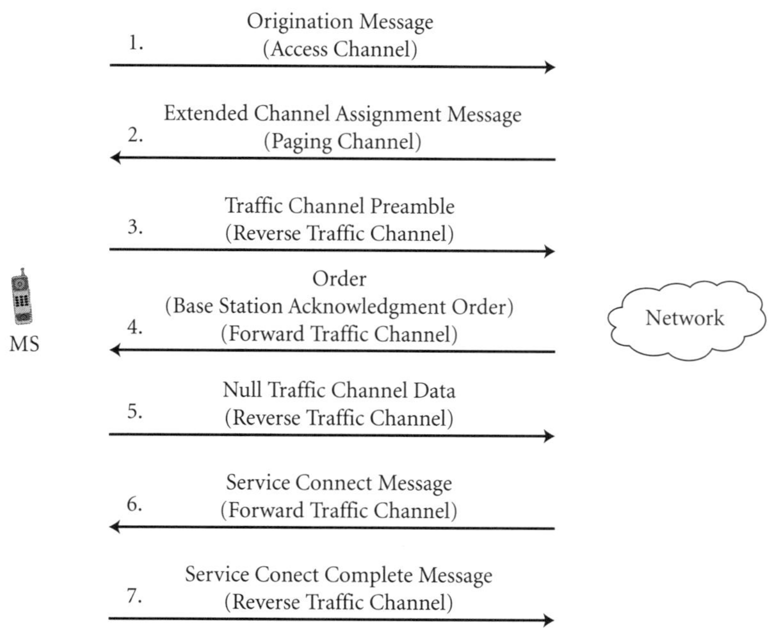

Figure 8.11: Example Call Flow for an MS-Initiated Call Setup

8.4.2 Voice Call Setup in 1x: A Closer Look

Let's dig a little deeper into the call setup process in 1x. Examples of the message fields are given in the discussions below; see the standard [246] for the entire set of message fields. The MS sends an Origination message, and the network responds with an Extended Channel Assignment message. As shown in Figure 8.12, the MS-initiated call setup involves the transmission of the Origination message from the MS and the transmission of the Extended Channel Assignment message from the network.

The MS identifies itself and provides its capabilities in the Origination message. Layer 2 fields include identities of the MS, such as the ESN and IMSI, and reports about the radio environment, such as pilot strengths of strong pilots the MS has observed. If the MS prefers the default service option, which is Service Option 0, the SPECIAL_SERVICE field is set to 0. If the MS prefers a non-default service option, it sets the SPECIAL_SERVICE field to 1 and indicates its preferred service option. Table 8.5 summarizes some of the main service options; see the standard [49] for the comprehensive list of service options. Commercial deployments typically use Service Option 3 for voice services and 33 for packet data services. The service option in the Origination message informs the network of whether the MS is making a voice call or a data call.

Since typical 1x deployments implement Release 0 of CDMA2000 1xRTT, the mobile protocol revision number parameter, MOB_P_REV, is set to 6. The MS provides all digits of the phone number of the dialed party in the Origination message

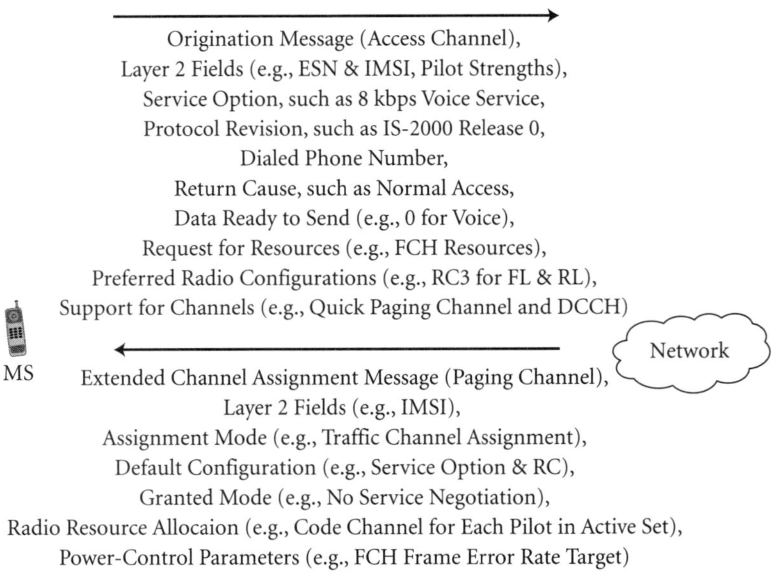

Figure 8.12: Origination and Channel Assignment Messages

Table 8.5: Example Service Options for 1xRTT

Service Option Number	Type of Service
1	Basic Variable-Rate Voice Service (8 kbps)
3	Enhanced Variable-Rate Voice Service (8 kbps)
5	Group 3 Facsimile (9.6 kbps)
6	Short Message Services (Rate Set 1)
7	Packet Data Service: Internet and International Organization for Standardization Protocol Stack
13	Group 3 Facsimile (14.4 kbps)
14	Short Message Services (Rate Set 2)
17	High-Rate Voice Service at 13 kbps
22	High-Speed Packet Data Service: Internet and International Organization for Standardization Protocol Stack (Rate Set 1 for Forward and Reverse Links)
33	144 kbps Packet Data Service: Internet and International Organization for Standardization Protocol Stack

itself or places additional digits in the Origination Continuation message. The MS specifies the reason for sending the Origination message as RETURN_CAUSE. For example, RETURN_CAUSE may be registration or normal access for a call initiation or registration. If the MS is making a data call and has some data to be sent, the Data Ready to Send field is set to 1. The Data Ready to Send field is set to 0 for a voice call. The MS makes a request for resources by specifying which radio channel it prefers in the CH_IND field. Examples of channels include FCH, DCCH, or both FCH and DCCH. Typical deployments involve a request for the FCH. The MS also conveys its preferred radio configurations for the forward and reverse links through the parameters FOR_RC_PREF and REV_RC_PREF. RC3 and RC4 are typically used in commercial deployments. To provide flexibility of configuration and to enhance performance, the MS informs the network if it supports channels such as the Quick Paging channel and DCCH.

Upon successfully detecting the Origination message, the network executes the call admission control algorithm. In particular, it evaluates the availability of radio resources, such as the availability of the FL power and the FL Walsh codes as well as the rise-over-thermal noise at one or more sectors. Assuming that the network finds some resources for the MS, it reserves these resources for the MS and sends an Extended Channel Assignment message to the MS specifying the call configuration and reserved resources. Since the Extended Channel Assignment message is sent on the Paging channel, the identity of the MS, such as its IMSI, is included in the layer 2 fields. The network informs the MS of the assignment mode, ASSIGN_MODE, that the MS should use. For example, an ASSIGN_MODE of 0 corresponds to the traffic channel assignment for the 1x system, and an ASSIGN_MODE of 2 tells the MS to acquire an analog system.

The network provides a default configuration to the MS specifying the multiplex option and a radio configuration to use before the MS receives a Service Connect message. The network allocates resources, such as Walsh codes, for each Active Set member. Note that the network can place the MS into soft handoff right from the beginning by allocating the MS a CODE_CHAN (i.e., a code channel specified by a Walsh code) in each sector. The MS is also instructed to use a specific FRAME_OFFSET that determines the timing of its FL and RL traffic channels relative to the system time. Different MSs are usually allocated different frame offsets. Such a frame offset allocation approach facilitates backhaul management between the BS and the BSC as well as the processing of packets (especially speech packets) since the transmission of packets and the processing load within the network are not concentrated.

The network dictates the timing of service negotiation by setting the value of GRANTED_MODE. For example, when GRANTED_MODE is 2, the MS cannot start the service negotiation until the network has had a chance to send a Service Connect message. Since 1xRTT implements fast power control in the FL, relevant parameters, such as target FL frame error rate (FER) and the E_b/N_t settings for the F-FCH outer-loop power control, are also specified in the Service Connect message.

Figure 8.13 summarizes initial communications between the MS and the network after the network has set up a dedicated radio link with the MS. Once the MS receives the Extended Channel Assignment message, it starts using the dedicated traffic channels in the forward and reverse links. Specifically, the MS sends the traffic channel preamble on the Reverse Traffic channel. This preamble consists of the Reverse Pilot channel that transmits all 0s as data with the Walsh code W_0^{32} spreading. Furthermore, the Reverse Traffic channel is scrambled using the MS-specific long code. The network acknowledges the successful detection of the Reverse Traffic channel by sending an Order message that includes the Base Station Acknowledgment Order. The MS begins to transmit null traffic data consisting of all 1s on the R-FCH that is part of the Reverse Traffic channel.

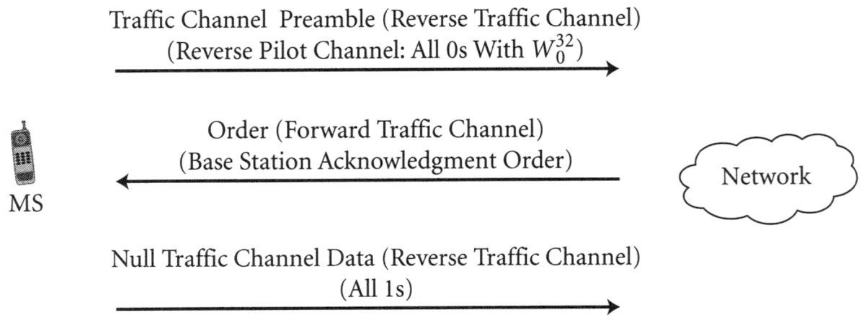

Figure 8.13: Initial MS–Network Communications on Dedicated Traffic Channels

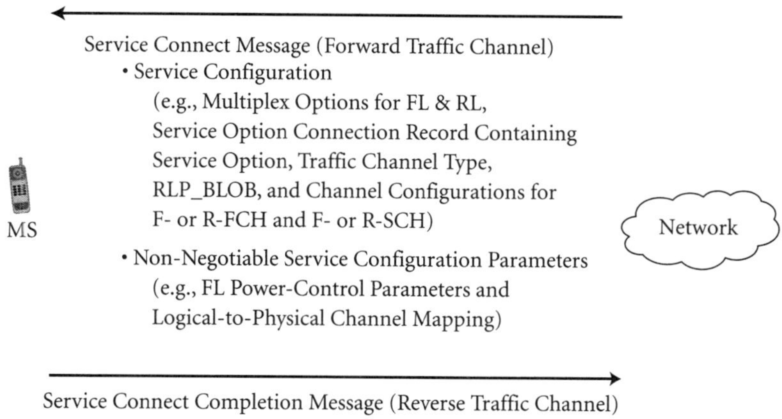

Figure 8.14: Closing the Deal in Call Setup: Service Connect

Figure 8.14 shows the final stage in the call setup, which is the exchange of service connect messages. The basic idea behind the service connection is to configure the MS with various parameters. The network sends two main types of parameters: basic service configuration parameters and non-negotiable service configuration parameters. Examples of basic service configuration parameters include multiplex options for forward and reverse links (e.g., Multiplex Option 1) and a Service Connection Record, specifying the service option, the type of traffic carried by the traffic channel (e.g., primary traffic on F-FCH and R-FCH), the RLP_BLOB, specifying RLP parameters for data services, and the configurations of FL and RL FCHs and SCHs, such as the RCs of the channels. Examples of non-negotiable service configuration parameters are those related to fast FL power control (e.g., F-FCH target FER and E_b/N_t setpoints) and mapping between the logical and physical channels. A default logical-physical channel mapping can, of course, be used. *The MS or the BS can use the Service Request message to propose a new or different service configuration.* Such a proposal is made after the Service Connect message and Service Connect Completion message have been exchanged.

8.4.3 How About Data Call Setup in 1x?

The data call setup and the voice call setup are almost identical with the two following main differences. First, the Origination message differs. The service option points to a data service instead of a voice service. For example, the Origination message includes $Service\ Option = 33$ for a 144 kbps Packet Data Service instead of a service option associated with a voice service (e.g., $Service\ Option = 3$ for 8 kbps voice). The Data Ready to Send field in the Origination message is set to 1 in case the MS has some data that is ready to be shipped!

Secondly, the Service Connect message changes. Since the requested service option is now for a packet data service, the Service Connect message reflects such a service option. Furthermore, the network needs to configure the MS to process future F-SCH and R-SCH assignments for high-rate data transfer. Hence, the network specifies parameters such as Radio Configuration (e.g., RC3 for both F-SCH and R-SCH), multiplex option (e.g., 0x921 corresponding to 16x rate and a maximum of eight double-size MuxPDU Type 3 for F-SCH and 0x811 corresponding to 8x rate and a maximum of four double-size MuxPDU Type 3 for R-SCH), and coding (e.g., convolutional coding only or frame-size–dependent convolutional or turbo coding for both F-SCH and R-SCH). Once the data call is set up, the scheduler at the BSC dynamically allocates the F-SCH and R-SCH by sending messages, such as the Extended Supplemental Channel Assignment message, to the MS; see Section 8.5 for additional details.

8.4.4 What About MS-Terminated Call Setup?

Let's contrast an MS-initiated call setup with an MS-terminated or network-initiated call setup. Figure 8.15 summarizes the first two messages involved in the MS-terminated call setup. The remaining call setup messages are the same as those for the MS-initiated call setup.

The MS in the idle state periodically wakes up (e.g., once every 5.12 seconds in a typical commercial deployment) and looks for a page message on the Paging channel. As shown in Figure 8.15, the network sends a General Page message on the Paging channel, and multiple MSs process this message because they are monitoring the common Paging channel. Since the Paging channel is a common channel, we need to identify our target MS. In a CDMA2000 system, an IMSI usually identifies an MS. The IMSI's maximum length is fifteen numerical characters. The IMSI's first three

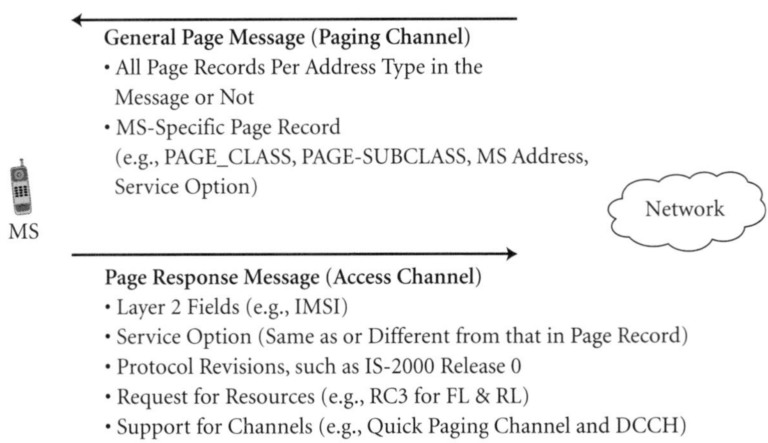

Figure 8.15: Initial Phase of an MS-Terminated Call Setup

digits are the Mobile Country Code, and the remaining digits are called the National Mobile Station Identity. Furthermore, the National Mobile Station Identity consists of the Mobile Network Code and the Mobile Station Identification Number.

An IMSI that is fifteen digits in length is called a class 0 IMSI, and an IMSI that has a length of less than fifteen digits is referred to as a class 1 IMSI.[262] As Figure 8.15 indicates, the overall General Page message includes layer-3–generated common fields and MS-specific page message fields (e.g., the service option for the new call) as well as layer-2–generated fields (e.g., an MS address inserted by the LAC's addressing sublayer). The layer 3 common fields indicate if the message contains all page records for a given type of MS address (e.g., class 0 IMSI or class 1 IMSI). If only a few MSs are being paged, one message is adequate in sending page records for those MSs.

For a General Page message, the LAC's addressing sublayer includes the parameters PAGE_CLASS and PAGE_SUBCLASS to identify the type of the page record included in the message. For example, a page record with $PAGE\_CLASS = 0$ is used to page the MSs that have previously registered with a class 0 IMSI. Within $PAGE\_CLASS = 0$, PAGE_SUBCLASS indicates a specific format of the MS address. For example, $PAGE\_SUBCLASS = 0$ means that IMSI_S has been included, and $PAGE\_SUBCLASS = 1$ means that IMSI_S and IMSI_11_12 have been included. IMSI_S is a ten-digit number based on IMSI, and IMSI_11_12 is a two-digit number based on IMSI.

The MS replies with a Page Response message in which it identifies itself and provides its capabilities. Many of the message fields are the same as those in the Origination message. Layer 2 fields include the identity of the MS, such as its IMSI, and the radio environment report. If the MS accepts the service option specified in the General Page message, it includes that service option in the Page Response message. Otherwise, it can suggest another service option. The mobile protocol revision number, MOB_P_REV, is set to 6 to imply the use of Release 0 of CDMA2000 1xRTT. The MS makes a request for resources by specifying which radio channel it prefers as CH_IND field (e.g., FCH). The MS indicates its preferred radio configurations (e.g., RC3) for the forward and reverse links. The MS also tells the network if it supports channels such as the Quick Paging channel and DCCH.

The remainder of signaling for the MS-terminated call is essentially same as that for an MS-initiated call. For example, the next message in the call setup sequence is the Extended Channel Assignment message, which the network transmits and which allocates dedicated resources to the MS.

[262] See the standard [246] for additional details on IMSI.

8.5 OVER-THE-AIR TRANSMISSION OF TRAFFIC FOR FORWARD AND REVERSE LINKS

1xRTT supports voice and packet data services on the same carrier frequency, which complicates radio resource management. Resources need to be managed efficiently while meeting individual QoS needs of voice and packet data services.

Do You Recall. . . What are examples of significant radio resources for the FL and RL in a CDMA-based system? As always, see the footnote[263] for your answer.

The over-the-air transmission of speech in 1x is similar to that in IS-95. The network chooses the RC (e.g., RC3 versus RC4) for the MS and allocates a suitable Walsh code for the forward link. Compared to IS-95, 1x has more opportunities for optimization while allocating radio resources for a voice call. For example, RC3 has better overall coding gain than RC4 and hence requires lower transmit power. However, for a given data rate, RC3 consumes larger Walsh code space, limiting the number of available Walsh codes at a given spreading factor. The network can use such knowledge to configure the MSs to operate in RC3 or RC4, depending upon the expected capacity or throughput bottleneck. In other words, if the FL power is becoming the capacity bottleneck, RC3 can be configured for new calls, and, if the sector is running out of Walsh codes, RC4 can be configured for new calls. The transmission of data packets is new in 1x compared to IS-95. We will discuss FL and RL transmission of data traffic in Sections 8.5.1 and 8.5.2, respectively.

8.5.1 Forward Link Packet Data Transmission

The 1x system uses the F-SCH to provide a higher data rate to the MS. As discussed earlier, the network configures the MS to operate using a given SCH multiplex option and RC during the call setup. Once such configuration is complete, the scheduler, typically implemented at the BSC, dynamically allocates the F-SCH to an MS. In general, the data traffic can be sent on just the F-FCH, both the F-FCH and F-SCH, or just the F-SCH. Note that the F-SCH can carry only data traffic; it carries no signaling messages. Hence, if only the F-SCH is used, layer 3 signaling messages go over the F-FCH. Furthermore, if only the F-FCH is used, both signaling messages and data traffic are carried over the F-FCH. Section 8.5.1.1 provides a high-level view of FL data transmission, and Section 8.5.1.2 provides a more detailed view.

8.5.1.1 Forward Link Packet Data Transmission: A High-Level View

Figure 8.16 illustrates typical steps of the FL packet data transmission. Let's summarize the main steps of the FL data transmission first. The network estimates the

[263] The FL resources are power at the sector (typically normalized to maximum high-power amplifier power, such as a certain percentage of total power) and Walsh codes. The RL resource is the interference margin (e.g., 3 to 5 dB rise-over-thermal noise) that is used in the link budget design.

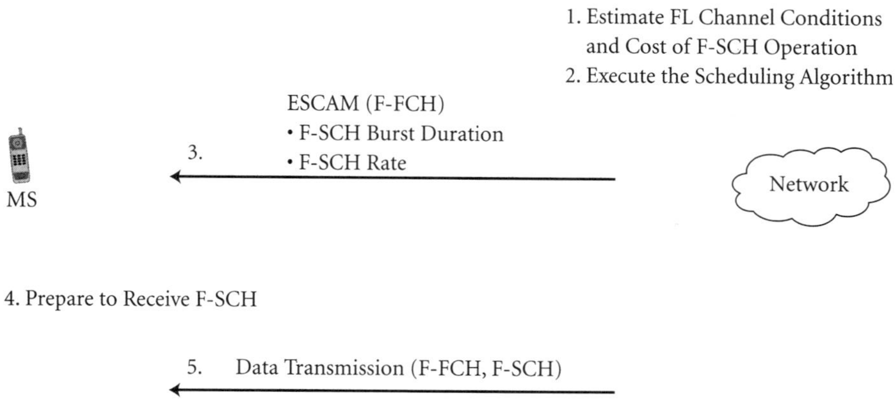

Figure 8.16: Overall Steps of the FL Data Transmission

RF power and the Walsh code resource requirements for various F-SCH rates. A proprietary scheduling algorithm considers such resource requirements for the MS along with other pieces of information, such as available resources, amount of data in the buffer, and resource needs for other MSs, to determine the F-SCH rate for the MS. The network informs the MS about the allocated F-SCH parameters such as the duration (e.g., 256 frames with each frame being 20 ms long) and rate (e.g., 76.8 kbps). Such resource allocation is sent on the F-FCH via messages, such as the Extended Supplemental Channel Assignment message (ESCAM). Upon receiving the ESCAM, the MS prepares to receive the F-SCH at the time specified in the ESCAM. The BS starts transmitting data traffic on the F-FCH and F-SCH. Steps one through five are repeated for an active data session that requires radio resources for data transfer.

8.5.1.2 Forward Link Packet Data Transmission: A Closer Look Let's take a closer look at the five main steps shown in Figure 8.16. The network estimates the prevailing average channel conditions for the MS to estimate the radio resource requirements or cost of supporting an F-SCH. Potential pieces of information that the MS can use for such an estimate is the average F-FCH power being used for the MS and reported pilot E_c/I_0, with the F-FCH power being more reflective of the most recent channel conditions. An explicit real-time feedback on the FL channel conditions (e.g., signal-to-interference ratio) from the MS is absent in 1x systems. Hence, the network needs to estimate the FL channel conditions on its own; the BS and the BSC work with each other for such an estimation.

Assume that an MS is allocated an average of $X\%$ power on the F-FCH for a given radio channel environment. A 19.2 kbps F-SCH may require $X\%$ power and an F-SCH of 38.4 kbps may require $2X\%$ power for the MS under prevailing channel conditions. Such a simplistic estimate assumes the following: the F-FCH is operated at 1% to 2% FER; the F-SCH is operated at 5% FER; and the signal-to-interference ratio required is the same to meet F-FCH and F-SCH FERs. Elaborate simulations, lab tests, and field tests can be used to develop a more accurate and sophisticated approach to estimate the required F-SCH from the observed F-FCH power.

Another aspect of the F-SCH cost is the Walsh code space that it consumes. Calculation of such cost is quite straightforward. For a given RC, the 19.2 kbps F-SCH occupies twice as much space as the 9.6 kbps F-FCH because the required spreading factor is halved when the data rate is doubled. Hence, if we keep track of the number of Walsh codes consumed at a spreading factor (i.e., Walsh length) of 64 for RC3, the 9.6 kbps F-FCH consumes one Walsh code and the 19.2 kbps F-SCH consumes one Walsh code at spreading factor of 32, or, equivalently, two Walsh codes at a spreading factor of 64. In other words, when an MS is allocated one F-FCH and one F-SCH, it occupies the Walsh space that spans three Walsh codes at spreading factor 64. Furthermore, the Walsh space for RC3 and the Walsh space for RC4 have a direct relationship.

Do You Recall... Do you remember the spreading factors used by the F-FCH in RC3 and RC4? The footnote[264] has your answer.

The total number of Walsh codes at a spreading factor of 128 is twice the total number of Walsh codes at a spreading factor of 64. The Walsh cost of an F-FCH for RC3 can thus be considered twice the Walsh cost of an F-FCH for RC4. In summary, the network needs to estimate the cost of supporting an F-SCH based on the estimated power cost and the estimated Walsh code usage for various F-SCH rates for a given RC.

Implementation-dependent, vendor-specific scheduling uses the estimated F-SCH cost along with numerous pieces of information, such as the amount of data in the buffer, the history of allocated resources to the MS, available resources, and resource requirements for other MSs, to determine the rate and the duration of the F-SCH. The F-SCH could be allocated indefinitely (i.e., until re-allocation) or for a finite duration selected from the set {1, 2, 3, 4. 5, 6, 7, 8, 16, 32, 64, 96, 128, 256} where each number represents the number of frames. Recall that each frame is 20 ms long. Since the signaling message, such as ESCAM itself, consumes a 20 ms frame, F-SCH allocations of very short duration (e.g., one or two frames) are inefficient from the overhead perspective.

Note that the network can override the allocation by sending a new ESCAM, even if the start time of the new allocation falls within the effective time period of the previous allocation. In such a case, the old allocation remains valid until the start time of the new allocation. Depending upon the implementation, the MS may need about three to four frames to process the received ESCAM and to prepare to retrieve the packet from the F-SCH. Hence, the scheduler should carefully specify a suitable

[264] RC3 uses the spreading factor of 64, and RC4 uses the spreading factor of 128.

start time for the allocation. Due to long delays, true channel-sensitive scheduling is challenging.

Upon receiving the ESCAM, the MS prepares to decode the F-SCH. As mentioned earlier, the network may send packets to the MS on both the F-FCH and the F-SCH. The packets that the MS's physical layer correctly decodes are forwarded to layer 2, and the RLP at the MS makes a request for the FL retransmission if and when necessary.

Figure 8.17 summarizes the design considerations for optimizing the FL data transmission involving the F-SCH. The network may allocate to the MS an F-SCH for a finite or an infinite duration. The finite duration enables the network to dynamically allocate resources to existing data users in a sector since frequent ESCAMs are sent to the MS. The infinite duration saves the overhead of ESCAM signaling since the allocation remains valid for a long time, but the radio environment may change significantly, requiring adjustments in the allocated resources. Of course, a new ESCAM can always be sent to switch from the infinite duration to the finite duration and vice versa.

The finite duration of the F-SCH allocation enables the scheduler to respond to a change in long-term channel conditions and to dynamic utilization of resources, such as FL power. Of course, signaling load is higher with finite duration F-SCH allocation. The power-control–related parameters, such as the F-SCH target FER, mode of power control, and initial E_b/N_t setpoint, must be tuned properly to optimize the utilization of FL power. Since the signaling messages are sent on the F-FCH, it is operated at a low FER, such as 1% or 2%. However, the F-SCH carries just data traffic. Furthermore, RLP retransmissions can reduce the effective packet error rate that upper layers, such as TCP, experience. Hence, we can afford to run the F-SCH at a higher FER.

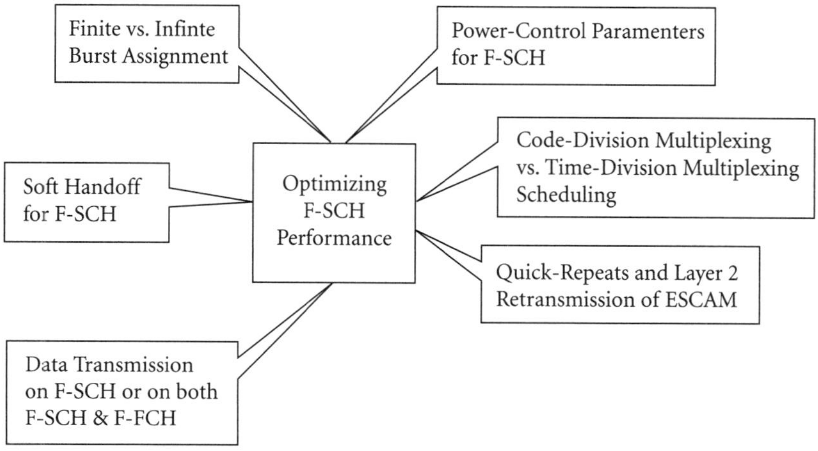

Figure 8.17: Optimization of the F-SCH Performance

A Question for You. . . What benefits may come from using a higher FER for a given rate F-SCH—savings in power or savings in Walsh code space? The footnote[265] discusses the answer.

The mode of power control influences how the F-SCH is power-controlled. For example, the 800 Hz power control can be shared between the F-FCH and the F-SCH, with 400 Hz power control for the F-FCH and 400 Hz power control for the F-SCH. In such a case, the F-SCH has its own target FER and outer-loop power-control E_b/N_t setpoint. Long F-SCH duration or infinite F-SCH duration provides adequate time for the power-control loop to reach a steady state and thereby use the power more efficiently.

The F-SCH can be shared in a time-division multiplexing or a code-division multiplexing fashion. When a given F-SCH (e.g., a specific Walsh code) is allocated to different users during different time periods, it is called *time-division multiplexing scheduling*. However, when multiple F-SCHs (i.e., different Walsh codes) are used for simultaneous transmission to different users during a given time interval, *code-division multiplexing scheduling* occurs. The scheduling algorithm must decide between pure time-division multiplexing scheduling, pure code-division multiplexing scheduling, or combined time-division multiplexing and code-division multiplexing scheduling. In general, time-division multiplexing scheduling is simpler, and code-division multiplexing scheduling can provide better resource utilization and/or user experience.

Another important consideration for F-SCH optimization is the reliability of ESCAM detection. If the MS incorrectly decodes the ESCAM, it misses the resource allocation, and the data packet sent on the F-SCH is lost. Example approaches to improve ESCAM detection reliability are layer 2 retransmission of the ESCAM and a quick-repeat of the ESCAM transmission at layer 3. A layer 2 retransmission means that the MAC sublayer is handling the retransmission. A layer 3 quick-repeat means that the same message is sent multiple times to exploit time diversity at the expense of increased signaling load. Data can be sent on only the F-SCH or on both F-FCH and F-SCH, with the latter resulting in a somewhat higher throughput. The F-SCH can be placed into soft handoff or can be allocated to the MS in one sector only. When soft handoff is used, the Active Set for the F-SCH must be a subset of the F-FCH Active Set and the maximum number of pilots in the F-SCH Active Set should not be too high. Switching from one-way handoff to two-way handoff offers significant benefits, but the incremental gain offered by switching from two-way to three-way and beyond is minor.

[265]The higher the operating FER is, the lower the required signal-to-interference ratio becomes. This will require lower transmit power for a given rate of the F-SCH. The amount of Walsh code space is determined by the rate of the F-SCH and hence remains the same when the data rate is unaltered. Such a reduction in the required transmit power enables the sector to support more users and/or a higher throughput.

8.5.2 Reverse Link Packet Data Transmission

The RL transmission can be carried out using the R-SCH and the R-FCH. Figure 8.18 summarizes the main steps of the RL packet data transmission.

When the MS has adequate data to send, it sends a Supplemental Channel Request message to request an R-SCH. So, how *adequate* is adequate? An implementation-specific algorithm makes that determination. If the R-FCH cannot send the buffer data within a certain time period, it should request an R-SCH to reduce the average packet delay. The Supplemental Channel Request message includes pilot E_c/I_0s of the Active Set and Neighbor Set pilots so that the network has an idea of the prevailing radio environment the MS is experiencing. The Supplemental Channel Request message also includes the requested R-SCH parameters as part of the SCRM_REQ_BLOB field that is within the REQ_BLOB field. The main parameters that are included in the SCRM_REQ_BLOB field are the preferred rate (e.g., as little as 9.6 kbps or as high as 153.6 kbps) and duration (e.g., 256 frames or infinite). Another parameter called RES_DATA specifies how much data will be transmitted (e.g., one-fourth or three-fourths of the data in the buffer) at the requested rate for the requested duration. This will help the network decide the duration of the R-SCH allocation to the MS.

The network considers available resources, such as the rise-over-thermal noise and the R-SCH parameters the MS requested, to determine the rate and duration for the R-SCH. The network must estimate the cost of supporting a given R-SCH rate. For example, compared to a 38.4 kbps R-SCH, a 76.8 kbps R-SCH will require almost twice as much transmit power, leading to higher rise-over-thermal noise. An extensive set of simulation results, lab tests, and field tests can be used to fine-tune the algorithm that estimates the increase in rise-over-thermal noise as a result of the

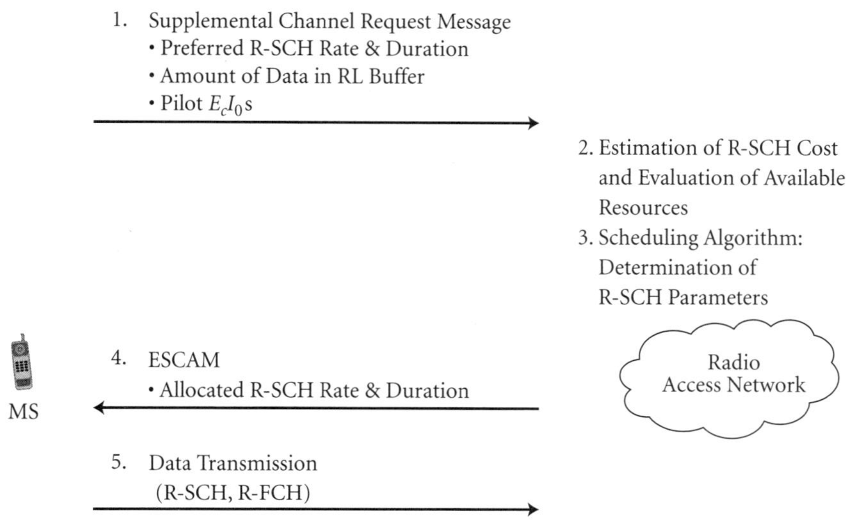

Figure 8.18: Reverse Link Packet Data Transmission

use of higher-rate R-SCHs. Sarkar's work [247] offers a discussion of various issues related to the RL's design and optimization.

The network conveys the R-SCH allocation parameters to the MS via an ESCAM. Recall that the ESCAM conveys both FL and RL resource allocations. Upon receiving the ESCAM containing the R-SCH specifications, the MS starts transmitting the R-SCH at the specified rate for the specified duration. The allowed data rates and allocation periods for the R-SCH are same as those for the F-SCH. Once the data transfer has started, the MS can, of course, send a new Supplemental Channel Request message to reflect the changes in the amount of data in the RL buffer. Discontinuous transmission is allowed on the R-SCH. The network allows a certain length of R-SCH discontinuous transmission operation, and the MS can indicate its desire for a longer discontinuous transmission by omitting a REQ_BLOB field in the case of a finite or infinite R-SCH duration.

8.6 MOBILITY MANAGEMENT

Upon power-on, the MS searches for a 1x system and acquires the network. Such network acquisition in 1x is similar to that in IS-95. (See Section 7.7 in Chapter 7 for details on network acquisition.) The MS looks for a Pilot channel on the selected carrier frequency to achieve timing synchronization, and, after finding the Pilot channel, it demodulates the Sync channel to retrieve the Sync channel message. Example contents of this message are the system identification, the network identification, the CDMA system time, and the Pilot PN offset of the sector.

To learn about the system, the MS then retrieves overhead messages being transmitted on the Paging channel. The MS then carries out registration and enters the Idle Mode. In an MS-terminated call, the MS's registration tells the network that the MS exists and that a General Page message can be sent from a set of suitable sectors. Section 8.6.1 takes a detailed look at registration and authentication. Since the MS has no dedicated channel immediately after power-on, it must use the Access channel to establish the first contact with the network. Section 8.6.2 discusses how the MS uses the Access channel. Once registration has been completed, the MS enters the Idle Mode. A 1x MS and an IS-95 MS carry out similar Idle Mode activities. (See Section 7.7 in Chapter 7 for details on Idle Mode activities.) In the Idle Mode, the MS searches for a sector that provides the highest E_c/I_0. In the strongest sector, the MS looks for a General Page message and any updates in the system information reflected in modified overhead messages (e.g., a System Parameters message).

Idle handoff occurs when the MS stops monitoring one sector for FL messages and starts monitoring another sector.[266] The MS may re-register if required. Furthermore, when the MS wants to set up a call, it sends an Origination message on the Access channel. If the MS finds a page message directed to it, it replies with a Page Response

[266]The MS carries out idle handoff with no network involvement. If the network needs to send a page to the MS, it does so by transmitting a page in multiple sectors of a paging zone. If the MS travels from one paging zone to another during the idle handoff, the MS informs the network about its existence in the new paging zone so that the network can send the page in the correct paging zone.

message on the Access channel. After an MS-originated or MS-terminated call setup concludes, the MS enters the Connected Mode. Section 8.6.3 discusses handoff in the Connected Mode and highlights 1x's handoff enhancements.

8.6.1 Registration and Authentication

Numerous types of registration exist, such as power-up and power-down registration; see Section 7.7.2 in Chapter 7 for a summary of the types of registration. The MS processes the System Parameters message to learn about the registration-related parameters. Figure 8.19 offers examples of the registration parameters.

The System Parameters message here shows selected registration parameters; the entire message contains several other pieces of information, such as handoff-related parameters. The sector includes the Pilot PN offset to identify itself. Both a system identification and a network identification identify the operator's cellular system. Recall that one system identification may encompass multiple network identifications. The registration zone number to which the sector belongs is also specified. The MS stores the REG_ZONE field and the corresponding (system identification, network identification) pair in its memory and retains the information for at least forty-eight hours, even after power-off. The network enables or disables different types of registration to control which type of registration the MS should perform. Typical commercial deployments support power-up, power-down, and timer-based registrations. Distance-based registrations are usually unsupported. The network also

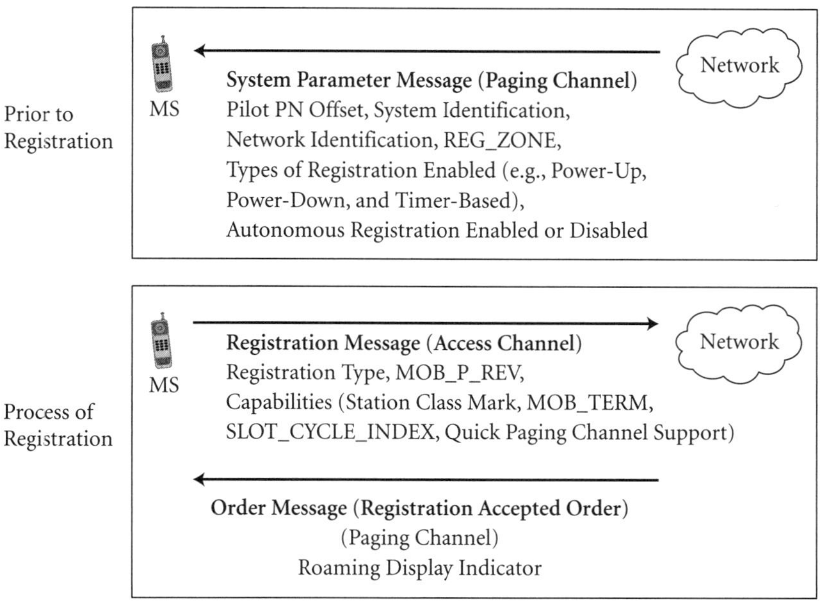

Figure 8.19: Overview of Registration

dictates whether non-roaming and roaming MSs can perform autonomous registrations.

Upon digesting the System Parameters message, the MS knows the events that would trigger registration. The MS performs registration by sending a Registration message on the Access channel. This message informs the network about the MS's existence to facilitate paging and conveys some of the MS's capabilities. The specific event (e.g., power-up) that resulted in this message is indicated as the registration type. The parameter $MOB\_P\_REV = 6$ indicates that the MS supports the CDMA2000 Rev. 0. The Station Class Mark specifies MS capabilities, such as its band class and its ability to work in a CDMA-only mode or a dual mode (i.e., CDMA and analog).

The parameter MOB_TERM indicates if the MS will support MS-terminated calls while roaming. The MS also indicates if it supports a Quick Paging channel; if so, the network can use a Quick Paging channel for the MS to help increase the MS's battery life. The MS that is operating in the slotted mode sets the value of SLOT_CYCLE_INDEX to its preferred slot cycle. For example, let $I = SLOT\_CYCLE\_INDEX = 2$. The slot cycle is then given by

$$1.28 * 2^{SLOT\_CYCLE\_INDEX} = 1.28 * 4 = 5.12 \text{ seconds.}$$

In a 1xRTT system, the network uses authentication to confirm the identity of the MS [245]. Note that this authentication is a one-way street; the MS cannot validate the network, but the network can validate the MS. The authentication occurs when the MS and the network have stored the same *shared secret data*. See the standards for the specifics of the algorithm itself [243] and for the inputs and outputs of the authentication algorithm [244]. There are two types of authentication procedures, a global challenge procedure and a unique challenge procedure. In a global challenge procedure, the MSC directs the radio network to inform all the MSs via a single common message to execute the authentication procedure at a suitable time. In contrast, in the unique challenge procedure, the MSC directs the radio network to send an Authentication Request message to a specific MS. The global challenge procedure is widely used in commercial deployments.

A Question for You... Can you think of the main benefit of the global challenge compared to the unique challenge? The footnote[267] has the answer.

Let's discuss authentication based on the global challenge procedure as summarized in Figure 8.20. The radio network follows the MSC's directive and sets the AUTH field of the Access Parameters message to 1. This value indicates to the MSs in a sector that they should execute the authentication procedure and include an authentication response to the network when they transmit certain messages on

[267]The global challenge procedure reduces the signaling load on the FL because just one Authentication Request message is sent for multiple MSs. Transmission of MS-specific Authentication Request messages will result in numerous FL messages, increasing the FL interference. Such an increase will reduce the capacity.

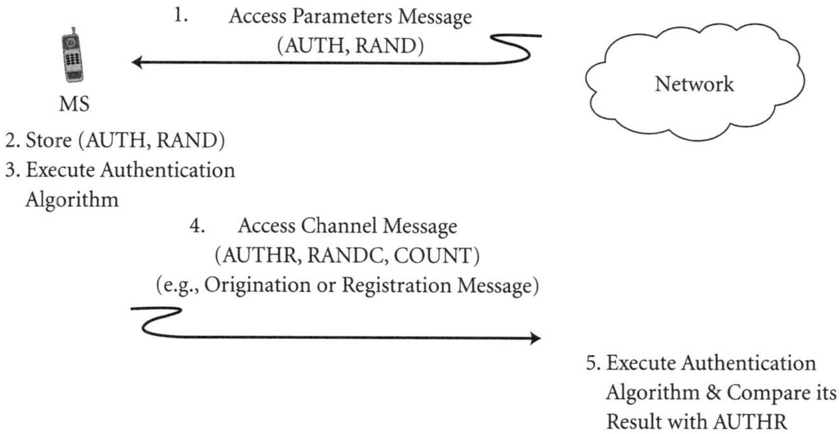

Figure 8.20: Overview of Authentication

the access channels. Examples of these messages include the Origination message, Page Response message, Registration message, and Data Burst message. The Access Parameters message also includes a random number RAND.

The MS in a sector stores the parameters in its memory and then executes the authentication algorithm called Cellular Authentication and Voice Encryption. The inputs and outputs of the Cellular Authentication and Voice Encryption algorithm are summarized in Figure 8.21. The inputs to the Cellular Authentication and Voice Encryption algorithm are RAND, ESN, AUTH_Data, and SSD_A. AUTH_Data varies depending upon the message, and its examples include IMSI_S1 and dialed digits. For

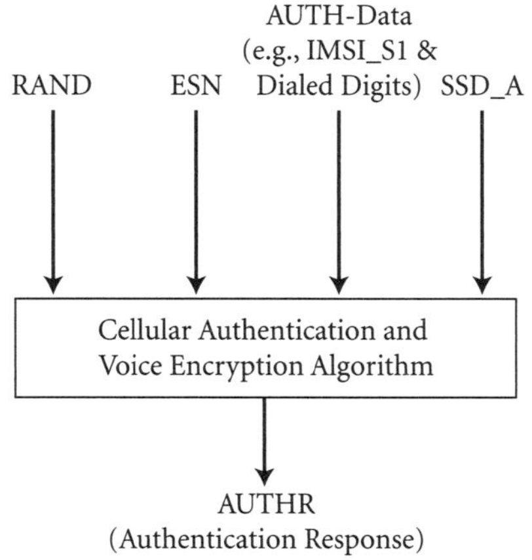

Figure 8.21: Inputs and Outputs of the Authentication Algorithm

example, in the Origination message, AUTH_Data is set to the dialed digits. SSD_A is the shared secret data for authentication.

The output of the Cellular Authentication and Voice Encryption algorithm is the authentication response called AUTHR. When the MS sends a message on the Access channel, it includes authentication-related parameters, such as AUTHR, RANDC, and COUNT. RANDC is the eight most significant bits of the RAND variable that the MS obtained from the Access Parameters message and used to calculate AUTHR. COUNT is the MS-specific call count and provides an additional layer of security. COUNT is often referred to as the call history parameter. The core network compares AUTHR with the expected authentication response.

A Question for You. . . Which core network elements get involved in authenticating the MS? The footnote[268] has the answer.

More specifically, the VLR compares the RANDC of the received message with the most significant eight bits of stored RAND. Assuming a match, the network calculates AUTHR just like the MS; it uses stored SSD_A and IMSI and compares the calculated value of AUTHR with the AUTHR received from the MS. If the AUTHRs match, authentication is considered successful, and the MS is all set to receive a service to which it has subscribed.

8.6.2 Using the Access Channel

When the MS has no dedicated radio channel, it uses the Access channel in the RL to contact the network. Once the MS has a dedicated channel allocated to it, it no longer uses the Access channel. 1xRTT makes some enhancements to the basic access operation carried out in IS-95. The network controls the use of the Access channel by specifying parameters such as PWR_STEP, MAX_REQ_SEQ, and MAX_RSP_SEQ. These parameters are discussed in Section 8.6.2.1. Once the MS reads the Access Parameters message from the Paging channel and learns about these parameters, it knows how to contact the network using the Access channel. 1xRTT enhances the overall access performance compared to IS-95A. Example enhancements include support for various types of handoff during the access process and support for *Channel Assignment into Soft Handoff*. While Channel Assignment into Soft Handoff was briefly introduced in Chapter 7, it is also discussed here along with other R-ACH enhancements. Section 8.6.2.1 describes the overall access operation, and Section 8.6.2.2 discusses access enhancements of 1xRTT relative to IS-95A.

8.6.2.1 Using the Access Channel: The Basics for 1xRTT The MS sends an access probe on the R-ACH. The access probe consists of two parts, an R-ACH preamble of length $(PAM\_SZ + 1)$ frames and a message capsule that may occupy

[268]The VLR (usually part of the MSC) compares the MS response with the expected response it obtained from the home location register.

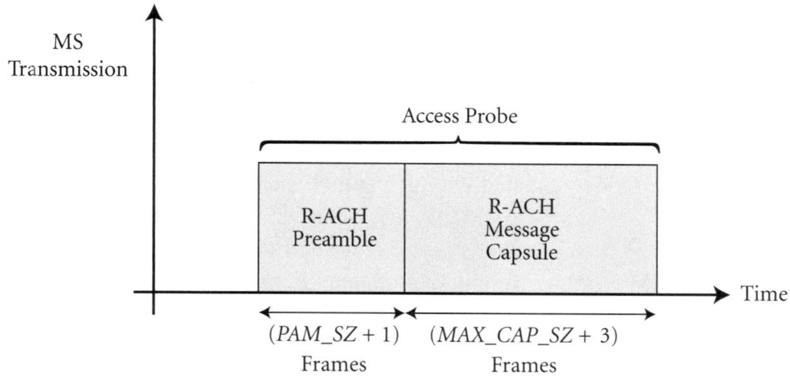

Figure 8.22: Anatomy of an Access Probe

up to $(MAX\_CAP\_SZ + 3)$ frames, as shown in Figure 8.22. The message capsule contains the message, such as an Origination message and Registration message.

Let's consider some typical numerical values for an access probe. For example, when $PAM\_SZ = 2$, the preamble lasts for $(PAM\_SZ + 1) = (2 + 1) = 3$ R-ACH frames. Furthermore, when MAX_CAP_SZ is set to 3, the maximum number of R-ACH frames that the MS can use for the message capsule is $(MAX\_CAP\_SZ + 3) = (3 + 3) = 6$. The MS may use fewer than $(MAX\_CAP\_SZ + 3)$ frames if the message size is small. An R-ACH slot is said to occupy the entire time period $((PAM\_SZ + 1) + (MAX\_CAP\_SZ + 3))$ frames.

Each R-ACH frame in an access probe carries ninety-six information bits since the R-ACH data rate is 4.8 kbps and the R-ACH frame length is 20 ms. Since we are using CDMA as the multiple access technique, we need to process these bits to reach the chip rate. The chip rate in 1x is the same as that in IS-95.

Do You Recall... Do you recall the chip rate for IS-95? The footnote[269] provides the answer.

The set of information bits passes through coding, repetition, and interleaving, eventually leading to a 28.8 ksps (kilo code symbols per second) signal. This set consists of ninety-six zeros for a preamble frame and ninety-six bits with eighty-eight information bits plus eight encoder tail bits for each message capsule frame. A 64-ary orthogonal modulator spreads the 28.8 ksps signal into a 307.2 kcps (kilo chips per second) signal.

A set of six code symbols is replaced by one modulation symbol, leading to $28.8\ \text{ksps}/6 = 4.8$ ksps (kilo modulation symbols per second). Furthermore, each

[269]The chip rate in IS-95 and 1x is 1.2288 Mcps. With a suitable guard band on either side of the signal, the overall required bandwidth becomes 1.25 MHz for a given carrier frequency.

such modulation symbol is spread by a sixty-four–chip Walsh function, yielding 4.8 ksps $* \ 64 = 307.2$ kcps. Finally, a long-code generator further spreads the 307.2 kcps signal into a 1.2288 Mcps signal. This long-code generator generates a PN sequence that is a function of an R-ACH–specific (in contrast to other channels, such as a traffic channel) long-code mask, the number of the R-ACH,[270] the code channel number for the associated Paging channel, the current base station's identity (i.e., BASE_ID), and the sector's PILOT_PN. The MS can obtain the BASE_ID and Pilot_PN from the System Parameters message. The 1.2288 Mcps signal is scrambled using an I-channel PN sequence and a Q-channel PN sequence and given a suitable power level. The I-channel and Q-channel PN sequences are generated from two distinct polynomials of length $(2^{15} - 1)$. The baseband signal is then shipped to the RF frequency and sent over the air.

Let's examine the overall access procedure as illustrated in Figure 8.23. A total of $(1 + NUM\_STEP)$ access probes constitute one access probe sequence. Multiple access probe sequences form an access sub-attempt, and several access sub-attempts comprise an access attempt.

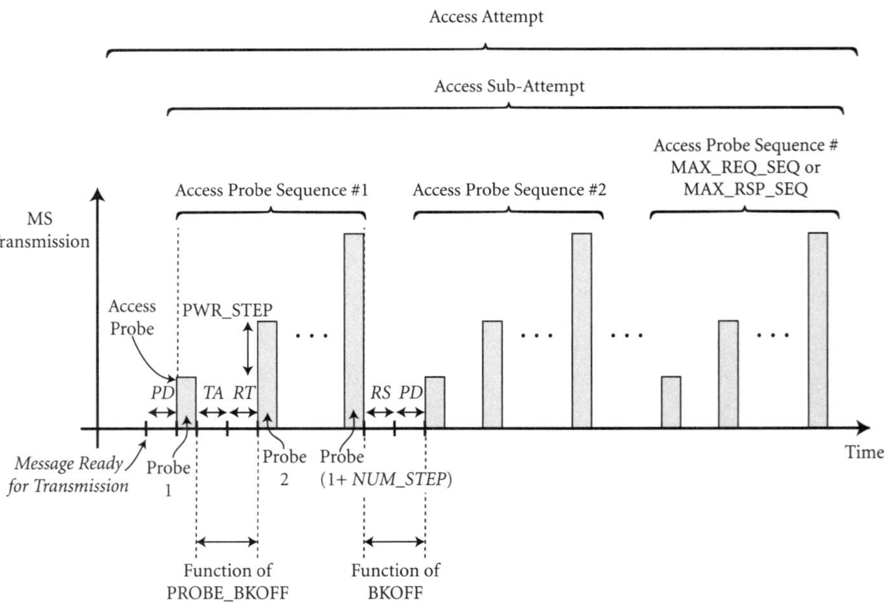

Figure 8.23: Overall Access Attempt

[270]Thirty-two R-ACHs could be present for a given Paging channel, and multiple Paging channels can exist in a sector. In practice, one R-ACH and one Paging channel are commonly encountered.

The maximum number of access probe sequences depends on the type of the message. MAX_REQ_SEQ is the maximum number of access probe sequences for an Access channel request (e.g., Origination message), and MAX_RSP_SEQ is the maximum number of access probe sequences for an Access channel response (e.g., Page Response message). The time interval between two consecutive access probes is random and its randomness depends on various parameters. In Figure 8.23, such a time interval includes components such as PD, TA, RT, and RS. Such a non-constant time interval minimizes the probability of collision on the R-ACH and hence increases the probability of a successful access.

PD is persistence delay that depends on the outcome of a persistence test. PD exists before each access probe sequence. This test involves the generation of a random number from 0 to 1 and the comparison of such a number with a parameter P. The parameter P can be conveniently expressed as [245]

$$P = 2^{-\alpha} 2^{-\beta} \tag{8.1}$$

where the persistence parameters α and β depend upon the access overload class and the type of the message (or a message other than a Registration message). Sixteen access overload classes are defined. For example, the access overload class 10 is for test MSs and class 11 is for emergency MSs. As an example, the parameter α would be $PSIST(12)$ for the access overload class 12 and β would be REG_PSIST for the Registration message where $PSIST(12)$ and REG_PSIST are defined in the Access Parameters message.

In practice, P usually turns out to be 1 based on the persistence parameters specified in the Access Parameters message. In other words, persistence is not commonly used in practice but can be an effective way of improving the access performance when the R-ACH is heavily utilized. The persistent delay continues until the slot-by-slot persistence test passes! Note that no persistence test is needed for sending a response message: $PD = 0$ in such a case. After sending an access probe, the MS waits for an acknowledgment from the BS for the duration given by

$$TA = (2 + ACC\_TMO) \times 80 \text{ ms} \tag{8.2}$$

where ACC_TMO is the access channel timeout specified in the Access Parameters message. If the MS receives no acknowledgment order from the BS, it continues the access probe transmission as shown in Figure 8.23.

TA varies from 160 ms to 1360 ms. RT is the random back-off between two access probes and can be anywhere between 0 and $(1 + PROBE\_BKOFF)$ slots. RS is the random sequence back-off between two access probe sequences and varies from 0 to $(1 + BKOFF)$ slots. While having just one R-ACH per Forward Paging channel is common, multiple R-ACHs may be associated with the current Forward Paging channel. In such a case, the R-ACH used for each access probe sequence (or even each access probe within a sequence!) is chosen pseudo-randomly.

Assume that the MS has overcome all the randomness and even passed the persistence test. We need to take care of just one more thing, and then the MS can transmit an access probe. At the beginning of an R-ACH slot, the MS introduces a time delay of *RN* PN chips relative to the start time of an R-ACH frame (or slot). Such a transmit timing adjustment results in the delay of the direct-sequence spreading long code and delay of the quadrature spreading I and Q Pilot PN sequences. It effectively allows the BS to detect distant MSs and to distinguish among transmissions from multiple MSs in the same Access channel slot, especially when such MSs are at a similar distance from the BS [240]. The parameter PROBE_PN_RAN facilitates implementation of such time randomization. More specifically, the MS delays its transmission from system time by *RN* PN chips where *RN* is a number between 0 and $(2^{PROBE\_PN\_RAN} - 1)$ PN chips.

Now, let's talk about the transmit power level of a given access probe. The access probe in each access probe sequence is transmitted at a power level determined by the physical layer relative to the nominal open-loop power level as specified by [240]

$$\begin{aligned} mean\ output\ power\ (\text{dBm}) = &- mean\ input\ power\ (\text{dBm}) \\ &+ offset\ power + interference\ correction \\ &+ (NOM\_PWR - 16 \times NOM\_PWR\_EXT) \\ &+ INIT\_PWR + (PWR\_LVL \times PWR\_STEP). \end{aligned} \tag{8.3}$$

In Eq. (8.3), mean input power is the received power of the Pilot channel. Offset power is defined to consider the influence of the frequency band (e.g., $offset\ power = -73$ for Band Class 0 and -76 for Band Class 1). The parameter interference correction is calculated as $\min\big(\max(-7 - ECIO, 0), 7\big)$ with $ECIO$[271] being the E_c/I_0 of the strongest Active Set pilot. PWR_LVL indicates the access probe number in an access probe sequence, and the power level increment between two successive access probes is PWR_STEP dB (e.g., 4 dB). The adjustment $(NOM\_PWR - 16 \times NOM\_PWR\_EXT)$ enables the MS to reflect different operating environments. INIT_PWR is intended to compensate for occasionally decorrelated or asymmetric path losses between the FL and the RL. Typical commercial deployments set NOM_PWR and INIT_PWR to 0 dB.

8.6.2.2 Access Enhancements in 1xRTT IS-95 uses a simple access operation. However, different types of handoff are supported in 1xRTT to enhance the access performance. An *access entry handoff* occurs when the MS receives a Page message from one BS but sends a Page Response message to another BS. An *access probe handoff* occurs when the MS stops transmitting to a specific BS the access probes of an access attempt and starts transmitting them to a different BS. In such a case, the access probes being transmitted to a given BS constitute an access sub-attempt as illustrated in Figure 8.23. An access sub-attempt involves transmission of one or more access probe sequences. An access attempt consists of a single access sub-attempt in the absence of an access probe handoff, which is the case in Figure 8.23.

[271] $ECIO$ is the variable name used in the standard. (Yes, the standard uses a letter O rather than a zero, and we're using the term as it appears in the standard.)

An access handoff is essentially a handoff of the Paging channel where the MS receives a layer 2 acknowledgment to an Origination or Page Response message from one BS but receives the Channel Assignment message from another BS. The network controls the use of the access entry handoff, access handoff, and access probe handoff by setting the parameters ACC_ENT_HO_ORDER, ACCESS_HO, and ACCESS_PROBE_HO, respectively, in the Extended System Parameters message. When a relevant parameter is set to 1, the network permits the MSs to exploit that type of access handoff. The access entry handoff and access handoff are commonly used in commercial deployments. The network also specifies which neighbor sectors can be involved in the relevant type of handoff. Of course, not all neighboring sectors can participate due to the limitations arising from the geographical layout of sectors and MS transmit power!

Among different types of access handoffs, the access entry handoff and access handoff are used in some commercial deployments. Access probe handoffs are uncommonly used. Some deployments may use no access handoff to simplify the MS's and network's operation. In a typical case, the BSC sends a Traffic Channel Assignment message to the MS allocating a dedicated radio channel to the MS in a sector. If the MS is in the handoff region, the network sends an Extended Handoff Direction message a little later to place the MS in soft handoff. If the time interval between the Traffic Channel Assignment message and the Extended Handoff Direction message is long, the MS may experience a radio link failure due to significant interference coming from strong cells or from sectors that are not yet in soft handoff with the MS.

A feature called *Channel Assignment into Soft Handoff* addresses this issue by allowing multiple sectors in the Active Set right from the beginning. The layer 2 Radio Environment Report field in the message (e.g., an Origination message) from the MS includes pilot E_c/I_0s. The BSC then decides to set up radio resources in multiple suitable sectors and sends a message, such as an Extended Channel Assignment message. Such a message specifies the resources (e.g., CODE_CHAN or code channel index to convey a specific Walsh code for the F-FCH in a sector) reserved for the MS in multiple Active Set sectors. Now, the MS benefits from soft handoff instead of getting overwhelmed by interference!

8.6.3 Handoff in Connected Mode: 1x Enhancements

The mobility management in the idle state is quite straightforward: the MS autonomously evaluates FL channel conditions and monitors the Paging channel of the best sector. The network is not actively involved in the idle handoff process. However, when the MS has dedicated traffic channels—the Forward Traffic channel and the Reverse Traffic channel—handoff becomes quite complex. Both the MS and the network must do quite a bit of processing for handoff in the Connected Mode. Such a Connected Mode is referred to as the *traffic channel state* in the standard. Before moving forward, let's make sure that you are fully awake. The wake-up alarm that we have for you is a question!

Do You Recall... What is soft handoff and what are its benefits? See the footnote[272] for the answer.

We already know the basic handoff algorithm defined in IS-95A (see Chapter 7 for details). The basic handoff algorithm involves thresholds, such as T_ADD, T_DROP, and T_TDROP. Pilots above T_ADD are considered worthy of an Active Set membership, and the Active Set pilots that stay below T_DROP for T_TDROP seconds are considered weak and hence removed from the Active Set. Typical values of T_ADD, T_DROP, and T_TDROP are -13 or -14 dB, -15 or -16 dB, and four seconds, respectively. The MS learns about these parameters by observing the System Parameters message. Though some commercial 1x deployments indeed use such a basic algorithm for simplicity and ease of optimization, a more advanced or complex algorithm facilitates control of the resource utilization to a certain extent.

The new soft handoff algorithm was introduced in IS-95B and briefly summarized in Chapter 7. Let's contrast the new dynamic-thresholds–based algorithm with the static-thresholds–based basic algorithm via a handoff scenario. The main goal of the new algorithm is to reduce the amount of soft handoff by judiciously choosing the add and drop thresholds. Though soft handoff has several benefits, excessive soft handoff uses more resources, such as more OVSF or Walsh codes, more cumulative power from all the sectors or cells, and increased bandwidth between the BS and the BSC. Hence, the ideal amount of handoff will balance capacity, resource utilization, and reliability. The enhanced 1x handoff algorithm helps achieve such balance. First, let's summarize how the IS-95A soft handoff algorithm works, and then we will compare that algorithm with the new, enhanced algorithm.

Consider Figure 8.24. The MS is traveling from a sector with *Pilot PN offset* = 100 (PN100) to a sector with *Pilot PN offset* = 200 (PN200). As it travels, the pilot strength of PN100 gradually decreases and the pilot strength of Pilot PN200 increases. The initial Active Set consists of {PN100}. At time instant t_1, the MS finds PN200 to be greater than T_ADD and sends a Pilot Strength Measurement message. The BSC executes a soft handover algorithm and decides to add PN200 to the Active Set, making the new Active Set {PN100, PN200}. The BSC instructs the MS to start using the new Active Set by sending a message such as an Extended Handoff Direction message. The MS begins using both PN100 and PN200 and responds with a Handoff Completion message to acknowledge the receipt of the Extended Handoff Direction message. After a while, when PN100 drops below T_DROP at t_3, the MS starts a drop timer. When PN100 stays below T_DROP for T_TDROP, the MS knows that PN200 is indeed weak.

The MS then sends Pilot Strength Measurement message at t_4. The BSC removes the weak PN100 from the Active Set and asks the MS to do the same via an Extended Handoff Direction message. The MS now makes the Active Set {PN200} and replies

[272]Soft handoff involves the MS simultaneously communicating with multiple sectors or cells. The main benefits are enhanced reliability of communications in the handoff region, better capacity, and lower transmit power requirements.

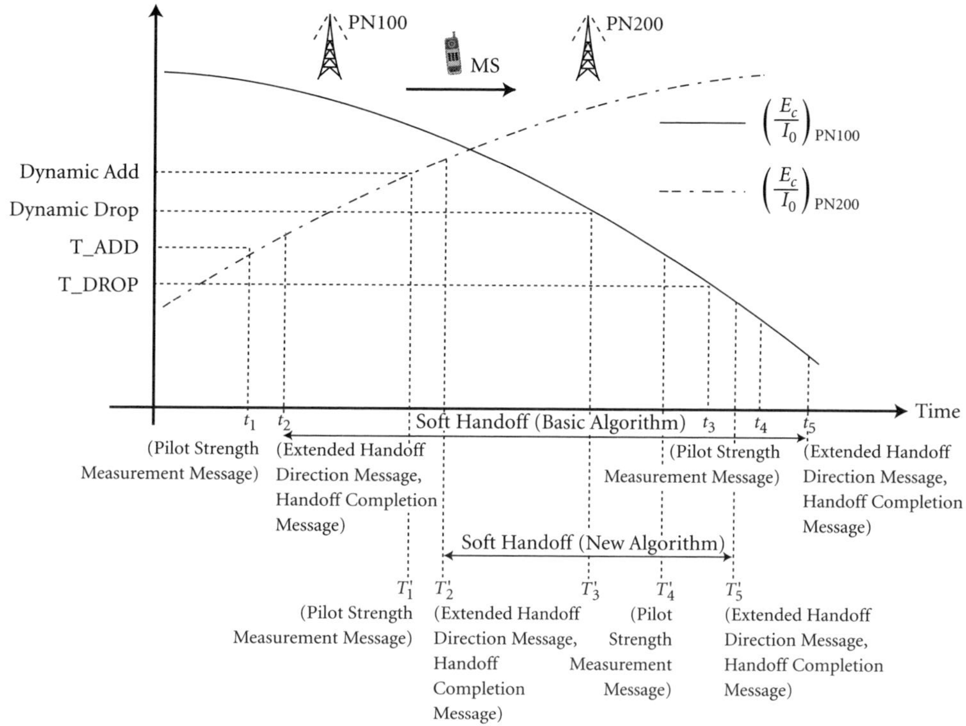

Figure 8.24: Soft Handoff: Basic Versus New Soft Handoff Algorithms

with an Handoff Completion message. In summary, the MS stays in soft handoff during the time interval t_2 to t_5. A word of caution: Though Figure 8.24 shows the same time instant for Extended Handoff Direction message and Handoff Completion message, some non-zero time interval does exist between Extended Handoff Direction message and Handoff Completion message.

Now is the time for the new soft handoff algorithm to show us some action! Recall that the goal of the new algorithm is to reduce the amount of soft handoff while limiting the potential loss in reliability. Hence, the new algorithm tries to delay the addition of a new pilot into the Active Set if the existing Active Set provides a reasonably good combined E_c/I_0 and the addition of a new pilot would only insignificantly improve the combined E_c/I_0. Similarly, the new algorithm accelerates the removal of an Active Set pilot if the combined E_c/I_0 without such a pilot remains sufficiently strong. The MS keeps track of two dynamic thresholds, an add threshold and a drop threshold, to facilitate the addition and removal of pilots into and from the Active Set, respectively. See the standard [240] to see how the MS calculates these dynamic thresholds on an ongoing basis.

The MS finds PN200 to be greater than a dynamic add threshold at time T_1' and sends a Pilot Strength Measurement message. Observe that the dynamic add threshold is greater than T_ADD and the difference between T_ADD and the dynamic add threshold depends on the strength of the existing Active Set pilots. The BSC makes {PN100, PN200} a new Active Set, and the BSC and the MS exchange the Extended Handoff Direction message and the Handoff Completion message to update the Active Set. As the MS continues its journey away from PN100, PN100 drops below a dynamic drop threshold at time T_3'. When PN100 stays below the dynamic drop threshold for the duration of T_TDROP, the MS sends a Pilot Strength Measurement message at T_4'. The BSC then removes PN100 from the Active Set and exchanges the Extended Handoff Direction message and the Handoff Completion message with the MS. The MS stays in soft handoff from T_2' to T_5'. As seen from Figure 8.24, the enhanced soft handoff algorithm reduces the time during which the MS is in soft handoff compared to the basic algorithm. In summary, the use of dynamic thresholds enables the new handoff algorithm to reduce the degree of soft handoff.

The dynamic add and drop thresholds are functions of parameters such as SOFT_SLOPE (e.g., 16), ADD_INTERCEPT (e.g., 4), and DROP_INTERCEPT (e.g., 0). These parameters are broadcast in the Extended System Parameters message. (Other works [246] offer formulas that use these parameters to calculate dynamic add and drop thresholds.) The dynamic add threshold increases as the combined Active Set pilot strength increases, making it more difficult for a new, relatively weak pilot to enter the Active Set. Furthermore, the dynamic drop threshold also increases as the combined Active Set pilot strength (excluding the weak Active Set pilot(s)) increases, making it easier for a relatively weak Active Set pilot to be removed from the Active Set.

Note that the MS uses the basic or dynamic thresholds to determine when to send a Pilot Strength Measurement message; the BSC actually makes the handoff decision and constructs the Active Set. The BSC usually follows the MS's handoff suggestion but could override such a suggestion by considering factors such as resource availability and independent evaluation of the need to add or drop a pilot to or from the Active Set.

8.7 1x POWER CONTROL

We discussed power-control mechanisms—open-loop power control and closed-loop power control—in Chapter 7. The MS uses the open-loop power-control mechanism while using the R-ACH in a 1x system. Once the dedicated traffic channels are established in the FL and the RL, the closed-loop power control is used. A generic closed-loop power-control mechanism includes an outer-loop power control and an inner-loop power control. As you may recall, the goal of a closed-loop power-control mechanism is to find the minimum transmit power level that is sufficient in meeting the target QoS, such as 2% FER.

Do You Recall... What are the benefits of power control? The footnote[273] has the answer.

The RL power control for the R-FCH is described in Section 8.7.1, and the FL power control for the F-FCH is described in Section 8.7.2.

8.7.1 Reverse Link Power Control for the Reverse Fundamental Channel

The RL closed-loop power control for the R-FCH is conceptually quite similar to that for the Reverse Traffic channel of IS-95. (See Chapter 7 if you want to refresh your memory on the RL power control in IS-95.) The MS in IS-95 uses the open-loop estimation for the Reverse Traffic channel as the baseline and modifies it to reflect the power-control commands received at the 1500-Hz rate. In contrast, the MS in 1x uses the Reverse Pilot channel as the baseline channel and maintains a certain power offset between the transmit power of other channels, such as R-FCH and R-SCH, and the transmit power of the Reverse Pilot channel. See the standard [240] for formulas to determine a suitable power offset.

Recall that the Reverse Pilot channel is considered part of the Reverse Traffic channel. After a dedicated Reverse Traffic channel is allocated to the MS, the MS starts transmitting the Reverse Pilot channel. The *initial* transmit power of the Reverse Pilot channel is a function of the received power and various parameters, including access-related parameters such as NOM_PWR, NOM_PWR_EXT, INIT_PWR, PWR_LVL, and PWR_STEP. The basic idea is that the MS must overcome the path loss between the MS and the BS and that the access operation helps determine an approximate transmit power level as a good starting point. After the MS receives the first power-

[273]The main benefits of power control are increased MS battery life, lower interference, higher capacity, and ability to meet QoS requirements.

control bit from the network, the associated closed-loop power-control correction is added to the open-loop power for the Reverse Pilot channel.

When the MS transmits on the Reverse Fundamental channel and the Reverse Supplemental channel, the power levels of these channels are transmitted with specific offsets relative to the Reverse Pilot channel. These offsets are functions of various parameters, such as Nominal_Attribute_Gain, RLGAIN_TRAFFIC_PILOT, and RLGAIN_SCH_PILOT; the standard [240] offers the full definition of these offsets. As an example, consider Nominal_Attribute_Gain, which is a function of three parameters, Data Rate, Frame Duration, and Type of Coding. For example, Nominal_Attribute_Gain is 30 for a 9.6 kbps data rate, a 20-ms frame, convolutional coding, and a target FER of 1%. As another example, Nominal_Attribute_Gain is 76 for a 153.6 kbps data rate, a 20-, 40-, or 80-ms frame length, turbo coding, and a target FER of 5%. RLGAIN_TRAFFIC_PILOT is a correction factor that specifies the gain adjustment of the Reverse Traffic channel relative to the Reverse Pilot channel. The parameter RLGAIN_TRAFFIC_PILOT can be adjusted to achieve the target FER on the R-FCH since it influences the transmit power of the R-FCH (and the R-SCH).

Note that the R-FCH FER may very well be close to the target FER because the standard specifies default settings of various relevant parameters. However, if the combined operation of the outer-loop power control and the inner-loop power control consistently yields a higher FER than the target FER, RLGAIN_TRAFFIC_PILOT may be increased. Similarly, the parameter RLGAIN_SCH_PILOT can be modified to bring the actual R-SCH FER closer to the target R-SCH FER. The network can configure the parameter RLGAIN_TRAFFIC_PILOT for all MSs via an Extended System Parameters message and the parameter RLGAIN_SCH_PILOT per MS via an Extended Supplemental Channel Assignment message. The RLGAIN_TRAFFIC_PILOT parameter can also be specified per MS via a General Handoff Direction message.

When the 1x MS receives the power-up or power-down command from the BS, it adjusts the transmit power of its Reverse Pilot channel by PWR_CNTL_STEP dB, which can be 0.25 dB, 0.5 dB, or 1.0 dB [240]. In general, the MS supports 0.5 dB and 1 dB step sizes. The MS may support a 0.25 dB step size.

8.7.2 Forward Link Power Control for the Forward Fundamental Channel

As mentioned earlier, fast power control on the FL is the one of the main contributors to the higher capacity of 1x compared to IS-95. Another contributor is the increased number of Walsh codes in RC4. In the FL, the MS implements the outer-loop power control with the initial value and range of the E_b/N_t setpoint (i.e., the target signal-to-interference ratio) for the network-specified F-FCH and F-SCH. Specifically, the network configures the MS with FPC_FCH_INIT_SETPT (where FPC represents forward power control; e.g., 7 dB) as the initial E_b/N_t and FPC_FCH_MIN_SETPT (e.g., 3 dB) and FPC_FCH_MAX_SETPT (e.g., 10 dB) as the minimum and maximum E_b/N_t for the F-FCH outer-loop power control. The MS attempts to meet the target FER of the F-FCH, specified as FPC_FCH_FER while staying within the specified E_b/N_t limits.

The network specifies the F-FCH power-control–related parameters to the MS in messages, such as the Service Connect message, the Extended Channel Assignment message, and the Power Control message. Similarly, the network configures the MS with FPC_SCH_FER, FPC_SCH_INIT_SETPT, FPC_SCH_MIN_SETPT, and FPC_SCH_MAX_SETPT via an Extended Supplemental Channel Assignment message to facilitate the F-SCH outer-loop power control. The network can also configure the MS to send an outer-loop report message in case the MS exceeds a threshold (called FPC_SETPT_THRESH_SCH) by setting FPC_THRESH_SCH_INCL to 1. If the MS does send such a report, the network could update the E_b/N_t thresholds so that the MS can meet its obligation of achieving the target FER!

Two RL power-control subchannels are defined to support the fast inner-loop power control for the FL. The Primary Reverse Power Control subchannel supports power control of the F-FCH or the Forward-DCCH, and the Secondary Reverse Power Control subchannel supports power control of the F-SCH. To implement the inner-loop power control for the FL, the MS sets the power-control bit to 0 when the actual E_b/N_t is less than the corresponding setpoint. Otherwise, the power-control bit is set to 1. Specific Power Control Groups will carry the power-control bits for the F-FCH and the power-control bits for the F-SCH based on FPC_MODE.

Four FPC modes are supported for the inner-loop power control, $FPC\_MODE = 0, 1, 2,$ and 3. Table 8.6 summarizes the main characteristics of the FPC modes. These modes essentially specify how the Reverse Power Control subchannel is shared between the Primary Reverse Control subchannel and the Secondary Reverse Control subchannel and what information is sent on these subchannels. The Primary Reverse Control subchannel carries power-control commands to influence the transmit power

Table 8.6: Characteristics of Forward Power-Control Modes

FPC_MODE	Characteristics
0	800-Hz power control of an F-FCH via Primary Reverse Power Control subchannel; Absence of Secondary Reverse Power Control subchannel; Indirect power control of an F-SCH via relative power offsets
1	400-Hz power control of an F-FCH via a Primary Reverse Power Control subchannel or of an F-SCH via a Secondary Reverse Power Control subchannel
2	200-Hz power control of an F-FCH via a Primary Reverse Power Control subchannel or 600-Hz power control of an F-SCH via a Secondary Reverse Power Control subchannel
3	50-Hz power control of F-FCH via Primary Reverse Power Control subchannel; Absence of Secondary Reverse Power Control subchannel; Same erasure indicator bit in all Power Control Groups

of the F-FCH, and the Secondary Reverse Control subchannel carries power-control commands to influence the transmit power of the F-SCH.

Do You Recall... Do you remember the use of a power-control group from Chapter 7? See the footnote[274] for the answer.

$FPC\_MODE = 0$ enables the fastest power control for the F-FCH. This mode is adequate if only an F-FCH is allocated to the MS. If an F-SCH is also allocated to the MS with $FPC\_MODE = 0$, the network determines a suitable power offset between the F-FCH and the F-SCH to achieve a target FER for the F-SCH. $FPC\_MODE = 2$ and 3 provide direct control over the F-SCH FER as separate outer- and inner-loops are defined for the F-SCH. $FPC\_MODE = 1$ has equally fast F-FCH and F-SCH inner-loop power control, and $FPC\_MODE = 2$ has faster F-SCH power control. $FPC\_MODE = 2$ and 3 have two parallel power-control mechanisms running, one for the F-FCH and another for the F-SCH. Finally, $FPC\_MODE = 3$ is simply IS-95 FL power control! The erasure indicator bit is replicated in each of the sixteen Power Control Groups.

8.8 1x DEPLOYMENT AND EVOLUTION

1x was the first 3G cellular system deployed in the world. Recall that UMTS is the *other* 3G cellular system! Capacity of a 1x system is quite high, with about thirty-five voice users per sector supported on a 1.25-MHz FL when an 8-kbps EVRC vocoder is used. UMTS faced quite a few hiccups during its initial launch (e.g., lack of mobile device availability, higher cost, smoother migration path, and a challenging business and market environment), but 1x was almost an instant success. Since 1x is a CDMA-based system like IS-95, it requires no elaborate frequency planning. The same 1.25 MHz spectrum is used in all sectors of the system. Deployment and optimization issues are quite similar for IS-95 and 1x, with 1x providing better potential for optimization through means such as the tuning of parameters related to the enhanced handoff algorithm and the fast FL power control.

The basic CDMA optimization techniques, such as coverage control via antenna parameters, are valid for IS-95 and 1x as well as UMTS. On the other hand, support for packet data services on the same carrier frequency as voice services makes the radio resource management in 1x more complex since voice and data have different QoS requirements and consume different amounts of resources. As in IS-95, once the system is deployed, it can manage the subscriber growth to a certain extent, and multiple 1.25 MHz carrier frequencies are deployed to maintain good voice capacity and/or data throughput levels. It is common to have two to three 1.25 MHz carriers in a given market.

[274] A power-control bit carrying a power-up or power-down command is sent in a Power Control Group. A Power Control Group facilitates fast power control. Since a Power Control Group occupies a 1.25-ms time period, the power-control frequency is $1/1.25$ ms $= 800$ Hz.

Where do we go beyond 1x? Well, there *were* two evolutionary alternatives defined by the Third Generation Partnership Project 2: 1xEV-DV (1xEvolution—Data-and-Voice) and 1xEV-DO. However, simplicity, efficiency, and early availability of 1xEV-DO (perhaps coupled with Qualcomm's convincing marketing department?) made 1xEV-DO the sole survivor. 1x evolved to 1xEV-DO Revision 0 and then to 1xEV-DO Revision A. *1x and 1xEV-DO are not backward-compatible; they are companion technologies.* 1xEV-DO requires a separate 1.25 MHz spectrum in the FL and in the RL and cannot share the same spectrum bandwidth with 1x. If we have no or few packet data users in a given region at a given time, the 1.25 MHz spectrum of 1xEV-DO would be sitting idle. Remember that service operators have spent a lot of money on acquiring the spectrum. (Now you know why we mentioned Qualcomm's marketing department!) Technically speaking, 1xEV-DO is quite an elegant and efficient packet data system, and Chapter 10 will introduce you to 1xEV-DO Rev. 0. Eventually, support of Voice over IP will remove 1xEV-DO's main technical drawback.

8.9 POINTS TO REMEMBER

- A 1xRTT system is a CDMA-based 3G cellular system and is an evolution of IS-95. It consists of a centralized or hierarchical radio access network and two types of core networks. The radio access network consists of BSCs and BSs. 1x can reuse the circuit-switched core network that IS-95 uses. Relative to IS-95, a packet-switched core network is new in 1x. The main components of a packet-switched core network are PDSN, FA, HA, and AAA Server.

- The air-interface protocol stack between the MS and the radio access network includes the physical layer at layer 1, the MAC and LAC sublayers at layer 2, and a signaling entity or upper traffic layers at layer 3 and above. The physical layer or layer 1 carries out tasks such as forward error correction coding, modulation, and spreading. The MAC sublayer at layer 2 consists of the Mux and QoS sublayer, the SRBP sublayer, and the RLP sublayer. The Mux and QoS sublayer enables user traffic and user-specific signaling messages to share a physical radio channel. The SRBP controls the Paging channel, Access channel, and Synchronization channel. The RLP sublayer enables reliable transmission over the radio link by retransmitting erroneous packets. The LAC sublayer is the upper-most sublayer of layer 2 and defines a data link protocol that enables correct delivery of signaling messages generated by layer 3. The upper-layer signaling entity takes care of delivery of signaling messages and SDBs.

- The call setup process in IS-95 and 1x are similar. However, the call may be a voice call or a data call. A suitable service option is used to specify the type of call.

- The F-SCH and R-SCH are two new radio channels in 1x to support data transmission at rates up to 153.6 kbps.

- Introduction of the Reverse Pilot channel makes the RL coherent, just like the FL of IS-95 and 1x.
- The enhanced soft handoff algorithm exploits dynamic thresholds to reduce the amount of handoff while minimizing the potential reduction in reliability.
- Fast power control in the FL is a major contributor to the higher capacity of 1x relative to IS-95.
- Reverse link power control in 1x is similar to that in IS-95, with an additional support for the R-SCH power control.
- 1xEV-DO is the natural path of evolution from CDMA2000.

CHAPTER 9

UNIVERSAL MOBILE TELECOMMUNICATION SYSTEM RELEASE 99

Benefitting from an enviable footprint of legacy Global System for Mobile Communication (GSM), General Packet Radio Service (GPRS), and Enhanced Data Rates for GSM Evolution (or Enhanced Data Rates for Global Evolution, i.e., EDGE) networks, Universal Mobile Telecommunication System (UMTS) is a third-generation (3G) cellular system. This chapter provides an overview of UMTS Release 99 with a focus on the frequency-division duplexing (FDD) mode. First, a brief overview of UMTS is given, followed by the description of an end-to-end system architecture that includes a radio access network and a core network. The air-interface protocol stack implemented in the mobile device and the radio network is discussed. Various radio channels used for over-the-air communications are described. Cell search, a process whereby the mobile device finds the existence of a UMTS network in its neighborhood, is summarized. Voice call setup and data session setup are illustrated. UMTS' implementation of handover, power control, and quality-of-service (QoS) are also discussed. Finally, UMTS' evolutionary path is depicted.

Cellular Communications, First Edition. Nishith D. Tripathi and Jeffrey H. Reed

9.1 UNIVERSAL MOBILE TELECOMMUNICATION SYSTEM RELEASE 99: EXECUTIVE SUMMARY

UMTS and CDMA2000 are two examples of 3G cellular systems. UMTS represents the evolution of GSM, GPRS, and EDGE systems, which are time-division-multiple-access–based second-generation (2G) and second-and-a-half-generation (2.5G) systems. Release 99, abbreviated as R99, is the first version of the UMTS standard deployed in commercial networks. The Third Generation Partnership Project standardized UMTS. Let's summarize the salient features of UMTS R99.

The air interface benefits from the higher spectral efficiency of code-division multiple access (CDMA).[275] For example, UMTS voice capacity could be approximately more than two to three times the GSM voice capacity when a full-rate speech vocoder is used for GSM and UMTS systems [248].

The air interface allows two duplexing modes, FDD and time-division duplexing (TDD). The FDD mode requires a total of 10 MHz of spectrum with the downlink (DL) and the uplink (UL) each using 5 MHz of spectrum. The TDD mode allows the same 5 MHz of spectrum to be shared between the uplink and the downlink via slots. The FDD mode is much more prevalent than the TDD mode in commercial deployments, primarily due to simplified design, deployment, and optimization. The 3.84 Mcps chip rate helps keep the signal within the 5 MHz of reserved channel bandwidth. We'll focus on FDD in this chapter.

UMTS supports traditional circuit-switched voice and packet-switched data services on the same carrier frequency and within the same 10 MHz of spectrum bandwidth. Furthermore, concurrent voice and data services are also supported such that a user can send a picture, video, or email while talking.

Higher data rates and spectral efficiency enable UMTS to offer multimedia services. The theoretical peak user data rate in R99 is 2 Mbps though commercial R99 deployments have a lower supported peak data rate, such as 384 kbps. UMTS supports various types of applications by classifying them into one of four QoS or traffic classes: conversational, streaming, interactive, and background.

The UMTS architecture includes a new CDMA-based radio network, but the core network components are identical to the GSM circuit-switched core network components and GPRS packet-switched core network components. These core network components support new interfaces and protocols. In practice, an operator often deploys a separate new core network for UMTS that could eventually be shared between 2G and 2.5G radio networks and the UMTS radio network.

UMTS aims for open interfaces between the components of the network so that a service provider can mix-and-match products from different vendors. In practice,

[275] In this book, we use the term *CDMA* to refer to a multiple access technique that uses spread-spectrum principles. *CDMA* is often used in the industry to refer to IS-95 (and sometimes CDMA2000 1xRTT). Consider Chapter 7 as the prerequisite for this chapter because the CDMA basics discussed in Chapter 7 will come in handy as you read about the CDMA-based UMTS in this chapter. The air-interface UMTS is also called *wideband CDMA* or WCDMA.

however, the radio network in a given geographic region is from one vendor, and the core network components could be from several vendors.

As of early 2012, the Third Generation Partnership Project (3GPP) Releases 4, 5, 6, 7, 8, 9, and 10 have been specified by the standard. The 3GPP website, `www.3gpp.org`, offers the latest information on UMTS.

9.2 UMTS NETWORK ARCHITECTURE

Figure 9.1 illustrates the network architecture of the UMTS-based Public Land Mobile Network (PLMN) [249,250]. One PLMN covers a large geographical region. For

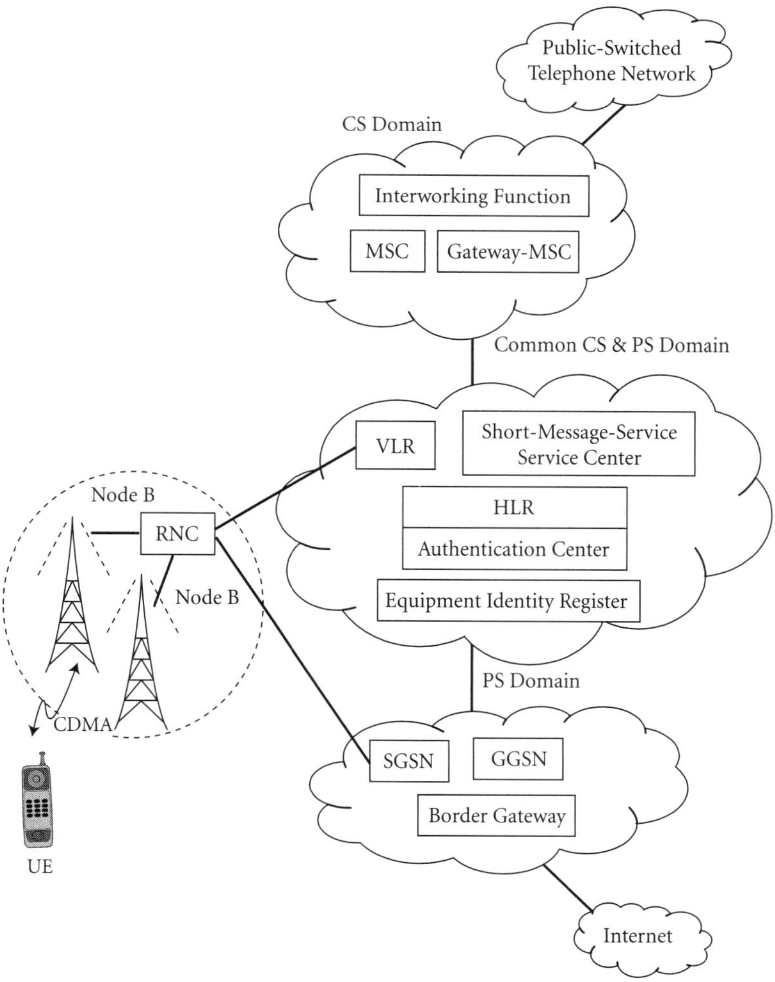

Figure 9.1: Typical Network Architecture of a UMTS-Based PLMN

example, one service provider's PLMN may cover the entire country. Furthermore, one service provider may have multiple PLMNs in the country. The PLMN is identified by parameters such as the Mobile Country Code and Mobile Network Code [251]. A UMTS-capable mobile device is called the User Equipment (UE) and contains a UMTS Subscriber Identity Module and Mobile Equipment. A UMTS-based network consists of the access network and the core network.

The access network is formally called the Universal Terrestrial Radio Access Network (UTRAN), which consists of one or more radio network subsystems. The radio network subsystem provides a new wideband-CDMA air interface to the UEs. The radio network subsystem includes Node Bs and one radio network controller (RNC). The core network consists of the circuit-switched (CS) domain and packet-switched (PS) domain. The core network entities include the entities common to the CS and PS domains; the entities specific to the CS domain; and the entities specific to the PS domain [249]. The common CS- and PS-domain entities include the Home Location Register (HLR), Visitor Location Register (VLR), Authentication Center, Equipment Identity Register, and short-message-service–related Mobile Switching Centers (MSCs).

The CS-domain entities include the MSC, Gateway-MSC, and Interworking Function. The PS-domain entities are the 3G Serving GPRS Support Node (SGSN), Gateway GPRS Support Node (GGSN), and Border Gateway. As shown in Figure 9.1, the basic core network entities in UMTS are the same as in the GSM core network and the GPRS core network. See Chapters 5 and 6 for details of their functions. Various components in the UMTS network communicate via a transport newtwork based on Asynchronous Transfer Mode (ATM). For example, the MSC sends a speech block to the RNC using ATM. The industry trend is to replace ATM with Internet Protocol (IP) as the transport mechanism. We show the MSC in Figure 9.1 because we are discussing R99 in this chapter, but several UMTS operators have started using a Release 4 CS-domain architecture. Release 4 decomposes MSC into an MSC server and a media gateway to increase efficiency and to reduce operating expenses.

A Short Detour... UMTS R99 includes an MSC in the circuit-switched core network. The MSC in R99 is responsible for signaling and user traffic, such as speech. Hence, when the service operator wants to determine the number of MSCs required to cover a certain number of subscribers in a given geographic area, both the MSC's signaling capacity and traffic capacity need to be considered. Depending upon the signaling and traffic loads, either the MSC's signaling capacity or traffic capacity may be underutilized in a given area. Furthermore, for a long-distance voice call (e.g., between Dallas and Chicago), a relatively high-rate transport channel, such as a 64 kbps DS0 between the MSC near the call-originating UE and the MSC near the call-terminating UE, would be needed to carry voice traffic. These issues of capacity and long-distance trunked channels lead to expensive over-provisioning of the MSCs. Release 4 addresses these issues by breaking down the monolithic or unified MSC into two components, an MSC-Server and a Media Gateway. The MSC-Server is responsible for only signaling, and the Media Gateway is responsible for only user traffic. In a given area, the service operator can now independently provision the number of MSC-Servers and the number of Media Gateways to match the signaling capacity and the traffic capacity. Furthermore, a low-rate [e.g., about 12.2 kbps for full-rate adaptive multirate (AMR) speech] trunked channel is needed between the distant Media Gateways. The terminating Media Gateway would perform the conversion between the low-rate speech and the high-rate (e.g., 64 kbps) speech suitable for the Public-Switched Telephone Network. In summary, the Release 4 architecture enables significant cost-savings for the operator by breaking a monolithic MSC into an MSC-Server and a Media Gateway.

Section 9.2.1 describes the radio network subsystem elements and the associated interfaces. Section 9.2.2 introduces the protocol stacks used for various network interfaces.

9.2.1 Radio Network Subsystem

Figure 9.2 illustrates the UTRAN's components and interfaces. The UTRAN consists of several radio network subsystems with each radio network subsystem made up of Node Bs and an RNC. The air interface between the UE and the Node B is called Uu. Iub is the interface between the Node B and the RNC. The Iur interface between two

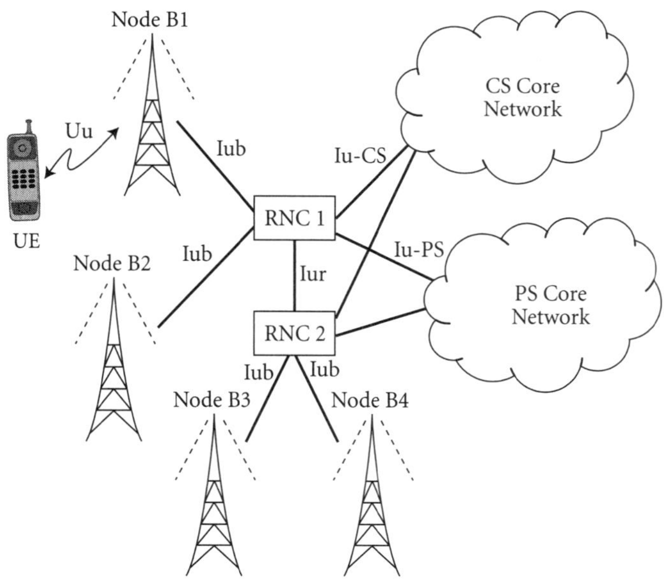

Figure 9.2: UMTS Radio Access Network Architecture

RNCs facilitates seamless handover as the UE travels from one RNC area into another RNC area. The RNC communicates with the core network via the Iu interface. More specifically, Iu-CS is the interface between the RNC and the MSC, and Iu-PS is the interface between the RNC and the SGSN. The UTRAN components are discussed here, and interfaces and associated protocol stacks are discussed in Section 9.2.2.

The Node B communicates with the UEs via the CDMA-based air interface. The Node B implements the CDMA physical layer for over-the-air communications with UEs. Two air-interface modes are supported, FDD and TDD. The chip rate is 3.84 Mcps. The Node Bs in a commercial deployment typically transmit asynchronously because the standard contains no mandate for the use of a common timing reference, such as a global positioning system. In other words, two Node Bs start transmitting radio frames at different time instants. The standard, however, also allows for the use of a global positioning system for a synchronous operation. When a UE is in soft handover, the RNC controls transmission on dedicated channels from different Node Bs individually such that the signals arrive at approximately the same time, enabling the UE to combine the signals. One Node B controls several cells. For example, if 120° sectorization[276] is used, that Node B controls three cells, each cell covering a 120° region. If omnidirectional antennas are used at a Node B, that Node B controls just one cell.

Ever Wondered... So what exactly does the *terrestrial* in UTRAN mean? Let's take a short detour before addressing that question. Remember the rovers, Spirit and Opportunity, that NASA sent to Mars in 2004? We received beautiful pictures of the Mars landscape for several years, even though the expected lifespan of these rovers was just three months [252]. A new NASA rover, Curiosity, also landed on Mars in August 2012; it started sending pictures quite soon after landing. When humans go to Mars and deploy UMTS, the access network will be called the Universal *Extraterrestrial* Radio Access Network! That was just a joke; *terrestrial* isn't really used to separate it from futuristic *extraterrestrial!* Nevertheless, a UMTS-based satellite system was indeed envisioned. To separate the traditional cellular UMTS from the satellite UMTS, the term *terrestrial* was introduced.

The RNC configures and controls the Node Bs. One RNC can control hundreds of Node Bs depending upon the RNC processing capabilities and the availability of backhaul bandwidth. Several radio resource management algorithms are implemented at the RNC. For example, algorithms related to call admission control, handover, and data rate assignment are implemented at the RNC. Most of the UTRAN intelligence resides at the RNC. Many such algorithms are vendor-specific and proprietary. The UE and the RNC exchange air-interface–related messages via the Node B. Certain messages, such as those related to services and user authentication, are exchanged between the UE and the core network. The RNC transparently lets these messages pass through without processing them; such RNC functionality is called *direct transfer*.

[276] See Chapter 1 for the basics of sectorization.

The RNC that controls the logical resources of the Node Bs is called the *controlling RNC* [250]. In Figure 9.2, RNC2 is the controlling RNC for Node B3, and RNC1 is the controlling RNC for Node B1. The RNC encrypts signaling messages (and traffic, if desired) to provide air-interface security.

9.2.2 Protocol Stacks: From User Equipment to the Edge of the Network

Figure 9.3 shows the protocol stacks for the control plane and user plane on various interfaces on the network [253–256]. The Control Plane is used to exchange signaling messages between two network nodes, and the User Plane carries user traffic, such as voice and data, as well as user-network signaling messages (e.g., handover-related messages). ATM is used as a transport mechanism on the Iub, Iu, and Iur interfaces. See Chapter 4 for details on ATM and ATM adaptation layers.

The control plane of Iub carries signaling messages between the Node B and the RNC. The user plane of Iub carries messages between the UE and the RNC and between the UE and the core network; user traffic, such as speech bits and packet data; and common messages, such as broadcast and paging messages. The Node B Application Part is the Iub control plane protocol in charge of Iub. The Node B Application Part runs over the Signaling ATM Adaptation layer, and the ATM is below the Signaling ATM Adaptation layer. The Signaling ATM Adaptation layer helps carry signaling traffic over the ATM.

The RNC configures Node Bs using the Node B Application Part. The Node B provides the RNC via the Node B Application Part with information on the resource utilization. For example, the Node B indicates to the RNC the total amount of power being transmitted in the downlink in the cells it is controlling. The RNC sends information about the system (e.g., PLMN identification and cell selection and reselection thresholds) to the Node B through the Node B Application Part. The RNC utilizes the Node B Application Part to direct the Node B to set up a dedicated radio link with the UE and to establish an Iub bearer[277] to carry the user-specific traffic. The established Iub bearer now represents the user plane for the UE. Q2630.2 protocol is used to set up and manage ATM Adaptation layer 2 bearers that are part of the Iub user plane. ATM Adaptation layer 2 is used on the Iub user plane to meet stringent timing requirements due to factors such as support for circuit-switched voice and soft handover.

The Radio Access Network Application Part manages the Iu control plane. The Radio Access Network Application Part helps establish and modify radio access bearers that are end-to-end links between the UE and the core network. These bearers facilitate QoS implementation. The SGSN or MSC requests radio resources from the RNC using the Radio Access Network Application Part. The Radio Access Network Application Part runs on top of the Signaling Connection Control Protocol, which provides UE-specific logical connections. A Signaling ATM Adaptation layer or IP-

[277] A bearer is a logical connection that carries information, such as user traffic.

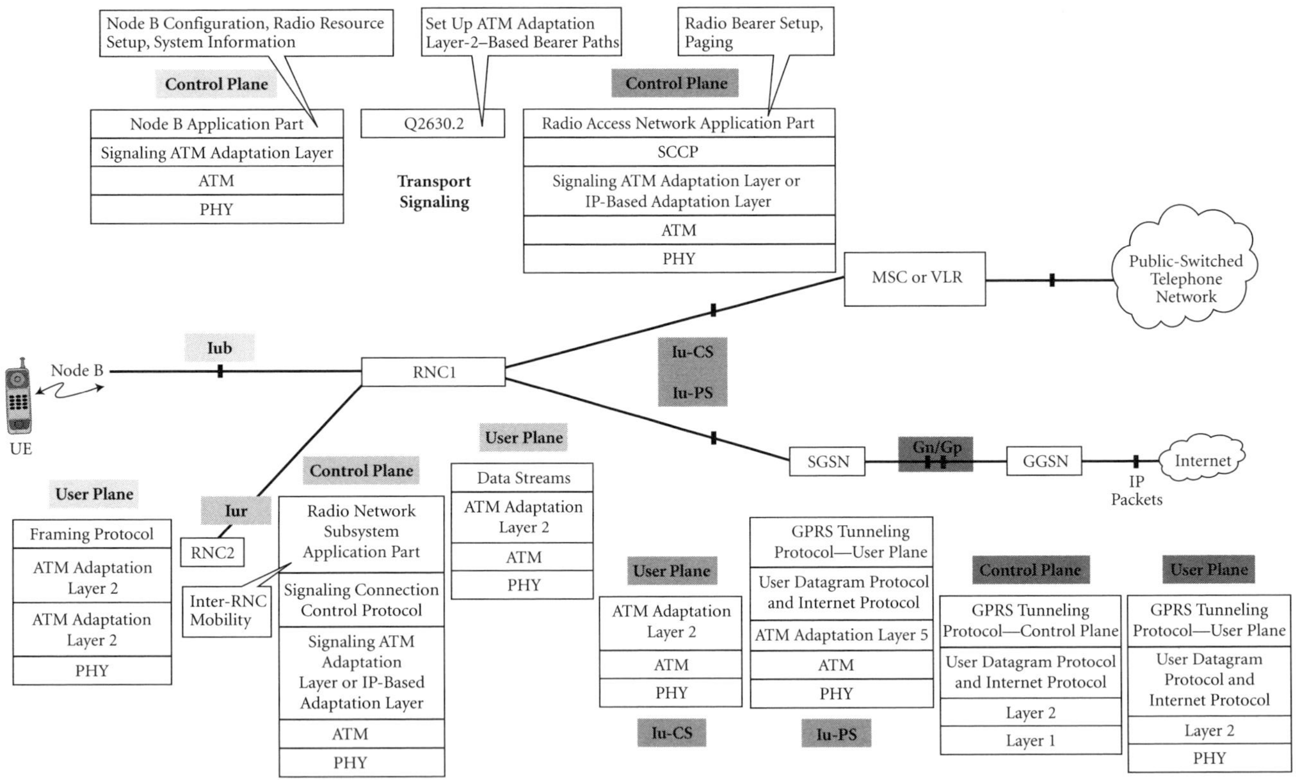

Figure 9.3: Protocol Stacks for the Control Plane and User Plane

based adaptation layer can be used to adapt signaling traffic for transport over the ATM. The Signaling ATM Adaptation layer is available for Iu-CS and Iu-PS, and an IP-based adaptation layer is available for the Iu-PS interface. The Iu user plane transports voice between the RNC and the MSC as well as IP packets between the RNC and the SGSN. The ATM Adaptation layer 2 is used on the Iu-CS interface to adapt real-time, variable-rate (e.g., due to active speech and silence intervals), and connection-oriented traffic for transport over the ATM. The Iu-PS user plane uses the ATM Adaptation layer 5 to adapt variable-rate and connectionless IP traffic for transport over the ATM.

The control plane on the Iur interface carries signaling between two RNCs, and the user plane carries user traffic, such as speech or packet data. When two RNCs are simultaneously supporting the UE (e.g., during soft handover across RNCs), the RNC that provides the UE the logical link with the core network is called the *serving RNC*, and the other RNC is called the *drift RNC*. In other words, the serving RNC terminates the Iu interface for the UE, and the controlling RNC supports the serving RNC by providing radio resources in the Node Bs that it controls [250]. The Radio Network Subsystem Application Part is in charge of the Iur control plane and facilitates soft handover implementation across RNCs by setting up resources in the drift RNC. The Iur user plane carries traffic between two RNCs for DL transmission to the UE and for UL reception from the UE. The Iur user plane transports voice and data streams using the help of the ATM Adaptation layer 2 over the ATM.

The SGSN and GGSN exchange signaling messages via the GPRS Tunneling Protocol–Control Plane protocol and IP traffic packets via the GPRS Tunneling Protocol–User Plane protocol. The GPRS Tunneling Protocol–Control Plane protocol helps establish GPRS Tunneling Protocol–User Plane tunnels that provide QoS on the Gn or Gp interface. The GGSN provides the SGSN the IP address to be assigned to the UE using the GPRS Tunneling Protocol–Control Plane protocol. Any layer 1 or physical (PHY) layer and any layer 2 can be used on the Gn or Gp interface. The GGSN exchanges the UE's IP packets of the UE with external networks, such as the Internet. In the CS domain, the MSC or VLR exchanges Integrated Services Digital Network User Part signaling with the public-switched telephone network, and the voice frames are carried over digital-signal-0 trunks.

9.3 RADIO INTERFACE PROTOCOL STACK BETWEEN USER EQUIPMENT AND UNIVERSAL TERRESTRIAL RADIO ACCESS NETWORK

UMTS has defined an air-interface protocol stack that delivers efficiency and flexibility. We will discuss the air-interface protocol stack implemented at the UE, the Node B, and the RNC. Section 9.3.1 provides an overview of the protocol stack, and details are presented in Sections 9.3.2 through 9.3.7.

9.3.1 Radio Interface Protocol Stack: A Brief Overview

Figure 9.4 depicts the protocol stack that the UE and the UTRAN use to communicate over the radio or air interface [257]. The radio interface consists of three layers: the

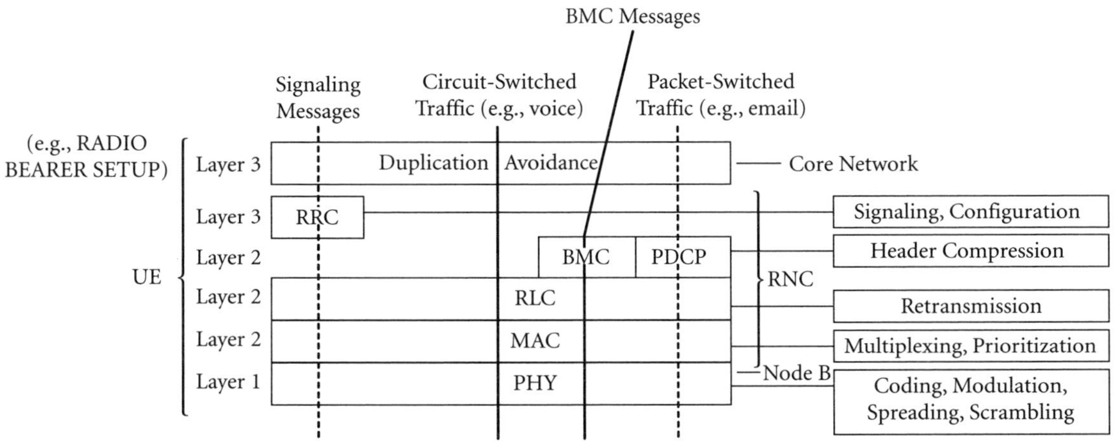

Figure 9.4: Radio Interface Protocol Stack between the UE and UTRAN

physical layer as layer 1, the data link layer as layer 2, and the network layer as layer 3. Layer 2 is split into the sublayers of medium access control (MAC), radio link control (RLC), Packet Data Convergence Protocol (PDCP), and broadcast-multicast control (BMC). Layer 3 is split into the upper sublayer of Duplication Avoidance and the lower sublayer of Radio Resource Control (RRC). The Duplication Avoidance sublayer terminates in the core network, and the RRC sublayer resides in the UTRAN.

Signaling messages pass through the Duplication Avoidance, RRC, RLC, and MAC sublayers and the PHY layer. The Duplication Avoidance sublayer helps prevent loss of data in the case of serving radio network subsystem relocation[278] [258] but is uncommonly used. The RRC layer constructs air-interface–related signaling messages and configures lower layers of the protocol stack.

User traffic for circuit-switched services, such as speech, passes through the Duplication Avoidance, RLC, and MAC sublayers and the PHY layer. Packet-switched traffic, such as web-browsing or email packets, passes through the Duplication Avoidance, PDCP, RLC, and MAC sublayers and the PHY layer. The BMC messages pass through the Duplication Avoidance, BMC, RLC, and MAC sublayers and the PHY layer. The UE implements all the layers. In the UTRAN, primarily the Node B (with a bit of help from the RNC) implements the PHY layer, and the RNC implements the MAC, RLC, RRC, and PDCP sublayers. The BMC sublayer is uncommonly used in commercial deployments.

The PHY layer transmits and receives information over the air. The MAC sublayer multiplexes different types of information and determines the instantaneous data rate. The RLC sublayer can retransmit in case of an error at the PHY layer, thereby increasing the reliability. The RRC sublayer configures the PHY layer and the MAC and RLC sublayers and creates signaling messages for purposes such as call setup and handover. The PDCP sublayer performs header compression to reduce transmission of overhead information over the air.

The standard defines three types of radio channels: logical, transport, and physical. Numerous channels (oh, yes, new acronyms!) exist within each of the three types of channels. Why so many channels? Well, it is UMTS, and simplicity and UMTS do not go together! These three types of channels have distinct purposes as summarized in Figure 9.5. A *logical channel* specifies what type of information is being transferred. A *transport channel* defines how and with what characteristics the information is transmitted. A *physical channel* actually carries information over the air. Logical channels exist between the RLC and MAC sublayers. Transport channels exist between the MAC sublayer and the PHY layer. Physical channels exist within the PHY layer. When a broadcast message conveying information about the system is to be sent, a logical channel called a Broadcast Control channel is used. When email for a user is to be carried, a logical channel called a Dedicated Traffic channel (DTCH) is used. Applications can also be viewed as logical channels. In other words, web-browsing would correspond to one DTCH, and email corresponds to another DTCH.

[278]Hold that thought! We'll discuss serving radio network subsystem relocation in Section 9.8.

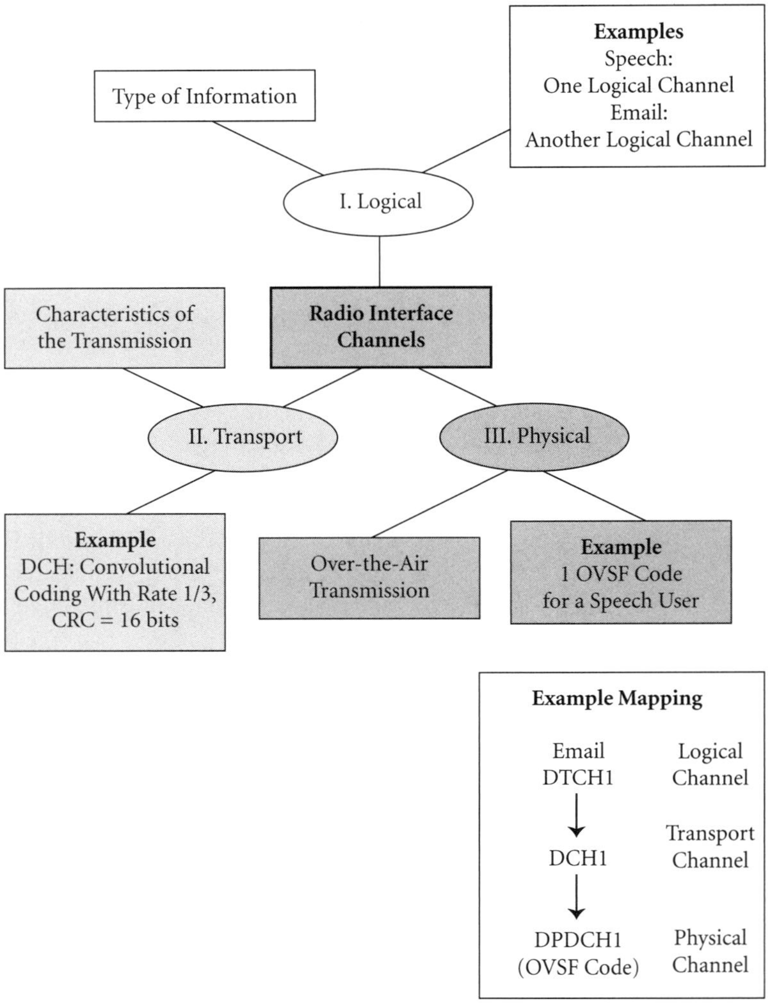

Figure 9.5: Types of Radio Interface Channels

A transport channel called Dedicated channel (DCH) may use convolutional coding with the coding rate of 1/3 along with a cyclic redundancy check (CRC) of sixteen bits. Another DCH may use turbo coding with rate 1/3. A physical channel carries information over the air. For example, speech for a user can be carried in the downlink on a Dedicated Physical Data channel (DPDCH), which is represented by an OVSF (orthogonal variable spreading factor) code[279] in a specific cell.

[279] See Section 9.4.2 to learn about OVSF codes.

Usually, a logical channel is associated with a transport channel, and a transport channel is associated with a physical channel [257]. Some channels exist as physical channels only; they have no mapping with logical or transport channels. Section 9.4 discusses the physical channels and identifies associated logical and transport channels where applicable.

Now that we know a little about logical, transport, and physical channels, let's see where they exist in the radio interface protocol stack. At the transmitter, the RLC sublayer creates logical channels and passes them to the MAC sublayer. The MAC sublayer maps logical channels onto transport channels. The PHY layer converts transport channels into physical channels and sends these physical channels over the air. At the receiver, reverse-processing is carried out. Section 9.4 provides details of the radio channels.

9.3.2 Physical Layer

The PHY layer or layer 1 transfers information over the radio interface. Figure 9.6 illustrates the basic physical layer processing for the downlink. The UE carries out such processing as does the Node B for every cell.

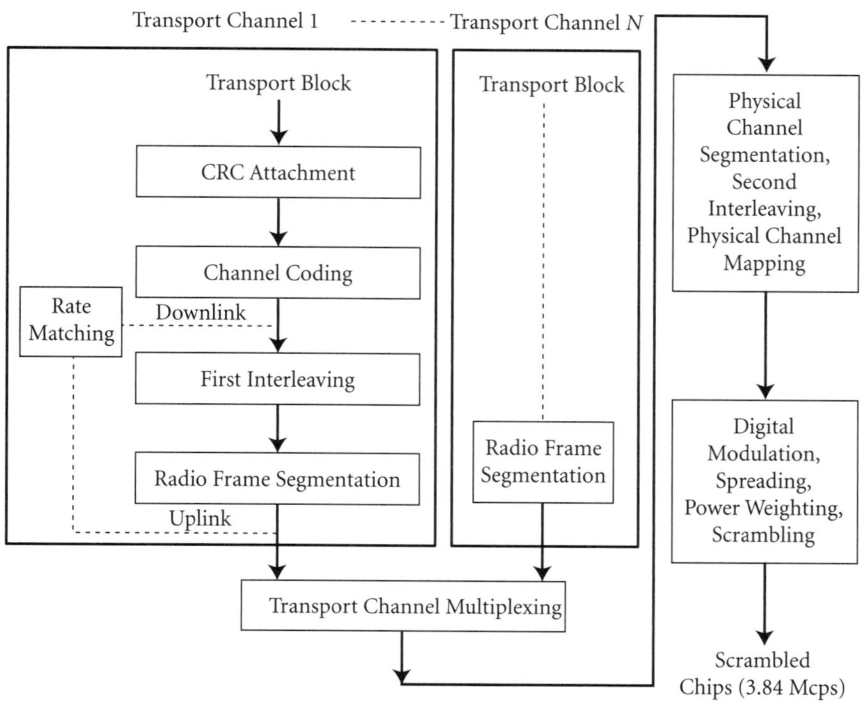

Figure 9.6: Simplified Physical Layer Processing

The physical layer receives a data stream from the MAC layer in units of transport blocks every transmit time interval (TTI). The TTI could be 10, 20, 40, or 80 ms. For example, signaling messages often use a 40 ms TTI, AMR speech uses a 20 ms TTI, and packet data services typically use a 10 ms TTI. Figure 9.6 shows a simplified view of how a transmitter creates a physical channel in the uplink and the downlink [57]. Multiple physical channels are then combined together for over-the-air transmission as explained in Sections 9.4.13 and 9.4.14.

A CRC is attached to a given transport block depending upon the transport channel. For example, in AMR speech,[280] a twelve-bit CRC is attached to the transport block carrying Class A speech bits, and no CRC is attached to the transport block carrying Class B or Class C speech bits [259]. The CRC helps detect presence of errors in a transport block.

More than one transport block could be transmitted in a given TTI. Hence, the transport blocks along with CRCs are serially concatenated to prepare a set of input bits to the channel coding scheme. Two channel coding schemes are supported, convolutional coding and turbo coding. The maximum number of input bits to the convolutional and turbo coding schemes is 504 bits and 5,114 bits, respectively. Hence, if the actual number of input bits exceeds the maximum limits, the bits are essentially equally divided into multiple code blocks, and each code block passes through the channel coding scheme to generate code symbols. For example, if the total number of input bits to convolutional coding is 600 (which is greater than the maximum of 504), two code blocks with 300 bits each are constructed, and each code block undergoes convolutional coding. The supported coding rates are 1/2 and 1/3 for convolutional coding and 1/3 for turbo coding.[281] The convolutional coding scheme has the constraint length $K = 9$ with eight shift or delay registers.

The turbo coding scheme consists of two parallel concatenated convolutional codes with two eight-state constituent encoders and a turbo code internal interleaver. Each constituent encoder has three shift registers and yields one output. Since the coding rate is 1/3, the turbo coding scheme provides three output streams; the first is the same as the input bits, and the second and third streams are outputs from the first and second constituent encoders. The number of output bits, or code symbols, is calculated as $Y = (1/\mathit{coding\ rate}) * (X + N)$ where X is the number of input bits to the channel coding scheme, the coding rate is 1/2 or 1/3, and $(1/\mathit{coding\ rate}) * N$ is the effective number of tail bits. We need a question here to make sure that you still remember the channel coding and decoding basics!

Do You Recall... What is the purpose of tail bits?[282]

[280] See Chapter 2 for the details of the AMR speech codec.

[281] See Chapter 2 for coding details.

[282] The tail bits reset the decoder at the receiver to the all-zero state to prepare for the next data arrival. Inclusion of tail bits simplifies the decoding operation.

N is 8 for convolutional coding and 4 for turbo coding. In the case of multiple code blocks, after each code block passes through the channel coding scheme, the output code symbols are serially concatenated.

The next stage is first interleaving in which a block interleaver[283] with inter-column permutations is used. The input data is placed into a matrix column-wise with the number of columns being one, two, four, or eight for 10, 20, 40, or 80 ms TTI, respectively. The number of rows in the matrix is the number of input bits divided by the number of columns. The output of the interleaver consists of the data from permuted columns instead of regular columns. For example, if the TTI is 40 ms, the number of columns is four. The output sequence in such a case consists of picking the bits from columns zero, two, one, and three, successively. *In the case of 10 ms and 20 ms TTI, no interleaving occurs as no column permutations are done.* In summary, the first interleaving is primarily intended to shuffle the bits for longer TTIs to achieve higher interleaving gain.

The radio frame segmentation segments the code symbols into the proper number of blocks, and each block is to be transmitted during one radio frame. Recall that a radio frame is 10 ms long, which is the minimum TTI during which information is transferred. However, some information (e.g., speech and signaling messages) is transferred using a TTI longer than 10 ms. The output code symbols need to be distributed among the radio frames. In the case of 12.2 kbps full-rate speech, the convolutional code symbols are distributed between two 10 ms radio frames since a 20 ms TTI is used.

Rate matching is a process that repeats or punctures bits so that the total bit rate after transport channel multiplexing is the same as the channel bit rate of the assigned dedicated physical channel. Rate matching yields a constant bit rate for over-the-air transmission. The rate matching is carried out after radio frame segmentation in the uplink and immediately after channel coding in the downlink. A parameter called the rate matching attribute is used to control the amount of repetition and/or puncturing.

The transport channel multiplexing unit serially concatenates the radio frames from multiple channels and creates a Coded Composite Transport channel. One radio frame from each transport channel is submitted to this unit every 10 ms. For example, in 12.2 kbps full-rate AMR speech, three transport channels carrying Class A, Class B, and Class C speech bits provide code symbols to this unit during a 10 ms interval. This unit then combines all these code symbols to create a Coded Composite Transport channel for speech. In the next 10 ms interval, another set of code symbols would constitute a Coded Composite Transport channel. Since a speech block is carried during a 20 ms TTI, two consecutive 10 ms radio frames carrying the Coded Composite Transport channels are required.

Physical channel segmentation may occur next, in which the Coded Composite Transport channel bits are now equally divided into P sets where P is the number of physical channels. Though the R99 supports $P > 1$, commercial R99 systems typically use one physical channel (i.e., one OVSF code). For each physical channel,

[283] See Chapter 2 for the basics of interleaving.

second interleaving is carried out to obtain interleaving gain within a 10 ms radio frame. The second interleaver utilizes inter-column permutation with thirty columns in the matrix. The input bits are placed row-wise, and the output of the interleaver consists of getting the data column-wise.

The interleaved code symbols are further processed quite differently in the downlink and the uplink. Details of such processing are available in Section 9.4. In the downlink, a physical channel utilizes quadrature phase shift keying (QPSK) modulation yielding I- and Q-channels, and the modulation symbols are spread using an OVSF code to reach 3.84 Mcps on the I-channel and Q-channel. A suitable power weight is given to each physical channel, and all the weighted channels are combined together. The combined signal is then scrambled using the downlink scrambling code of the cell. In the uplink, the data channel is sent on the I-channel and the control channel is sent on the Q-channel. On a given I- or Q-channel, the interleaved symbols pass through BPSK modulation and are spread using an OVSF code. Since both I- and Q-channels carry information simultaneously, the overall modulation of the transmitted signal resembles QPSK. The spread signal on an I- or Q-channel is weighted by a gain and combined together. The combined signal is scrambled by a UE-specific scrambling code. The 3.84 Mcps signal needs a 5 MHz spectrum including the guard bands. The process of rate matching discussed earlier ensures that the final bandwidth after spreading the signal by an OVSF code is 3.84 Mcps. Once the baseband signal is constructed, as illustrated in Figure 9.6, the RF processor takes over. It upconverts the filtered baseband signal to the center of the operator's spectrum, such as 850 MHz and 1900 MHz, and radiates the RF signal into space via an antenna. Similarly, the signal received by one or more antennas needs to be filtered, downconverted, and amplified prior to the baseband processing at the receiver.

Example: A Simplified View of Reaching 3.84 Mcps in the Downlink Assume that the raw downlink data rate is 384 kbps. A turbo coding rate of 1/3 yields 384 kbps$/(1/3) = 1152$ ksps (kilo code symbols per second). Rate matching reduces the data rate to 960 ksps via puncturing. The use of QPSK modulation further reduces the data rate to 960 ksps$/2 = 480$ ksps (kilo modulation symbols per second). Finally, spreading by a factor of eight leads to 480 ksps $* 8 = 3.84$ Mcps. See the standard [259] for the physical layer details regarding how a packet is formed for the raw data rate of 384 kbps.

The physical layer also performs several additional functions. If no information is to be transferred on a dedicated channel, *discontinuous transmission* (DTX) is used to save power and to reduce interference. DTX is supported in both the uplink and downlink. If data is unavailable, it is not transmitted. However, control information, such as a power control command, is always transmitted as long as a dedicated radio connection exists. Since CDMA is used on the air interface, both the Node B and the UE have RAKE receivers. Recall that the RAKE receiver utilizes the maximal ratio combining technique to combine multiple received signals. The RNC performs selection combining of the frames received from the Node Bs in the UE's Active Set.

Section 9.8 discusses handover. A 1500 Hz closed-loop power control is used in both uplink and downlink. See Section 9.9 for power control details. The physical layer makes measurements of important quantities such as signal-to-interference ratio, error rate, and transmit power and reports them to the upper layers so that the upper layers can carry out functions such as call admission control and data rate control.

9.3.3 Medium Access Control Sublayer

The MAC sublayer manages the physical layer resources based on the directives from the RRC layer. There are two main entities in the MAC layer, the MAC-common/shared[284] entity and the MAC-dedicated entity [260]. The MAC-common/shared entity is responsible for common transport channels, and the MAC-dedicated entity controls access to the dedicated transport channels. Assume that the network needs to page a UE. In such a scenario, the MAC-common/shared entity maps the Paging Control channel, which is a common logical channel, onto a common transport channel, such as the Forward Access channel (FACH) or Paging channel (PCH). If the UE location is known at the cell level, the Paging Control channel is mapped onto an FACH, and if the UE is to be paged in a set of cells, the Paging Control channel is mapped onto the PCH. Now, assume that the network wants to send an email packet to a UE. The MAC-dedicated entity maps a dedicated logical channel called the DTCH onto the DCH, which is a dedicated transport channel. The UE has one MAC-common/shared entity and one MAC-dedicated entity. The UTRAN has one MAC-common/shared entity and multiple MAC-dedicated entities with one MAC-dedicated entity per UE.

Figure 9.7 summarizes the main functions that the MAC sublayer performs. The MAC sublayer maps logical channels coming from the RLC sublayer onto transport

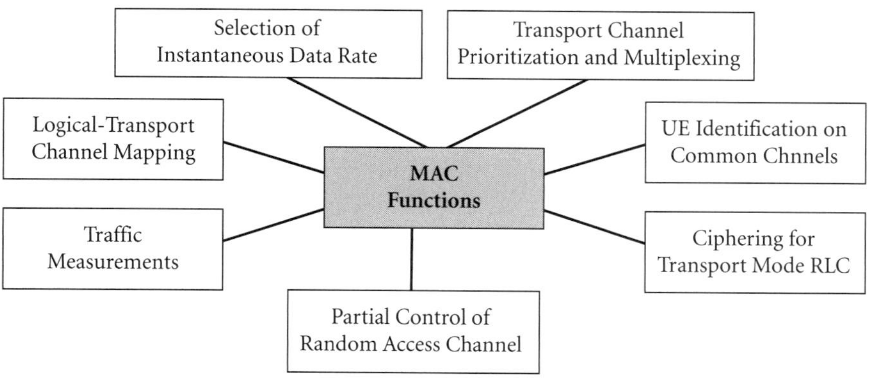

Figure 9.7: Examples of Functions the MAC Sublayer Performs

[284]The MAC-common/shared entity is referred to as the MAC-c/sh in the standard.

channels. For example, a DTCH is a logical channel that is mapped onto the DCH, which is a transport channel.

The MAC sublayer selects an instantaneous data rate within the limit set by the RRC sublayer. The RRC sublayer limits the maximum data rate for a given radio connection initially via the call setup as well as (possibly) dynamically via explicit RRC signaling messages. The transport format combination indicator (TFCI) hints at the exact data rate chosen by the MAC sublayer at the transmitter. For example, assume that the UE has been assigned a 384 kbps data rate channel in the downlink during the call setup, but little data is present in the buffer at the RNC. In such a case, the MAC layer can choose a 64 kbps instantaneous data rate and indicate this choice via a TFCI. When a UE has multiple transport channels, the MAC sublayer *prioritizes a transport channel* by choosing a TFCI that maximizes the transmission of that transport channel's bits during a given TTI.

The MAC sublayer multiplexes various transport channels into a transport block. One or more transport blocks can be transmitted in a TTI. Identities of the logical channels and the amount of data for these logical channels are included in the MAC header.

When a UE-specific message is to be sent on a common channel in the uplink or the downlink, the MAC sublayer places the UE identity inside the packet. For example, when a UE with no dedicated radio channel sends a short packet (considered to be the contents of the DTCH logical channel) on the Physical Random Access channel (PRACH), the MAC header includes the UE's identity, such as its Cell Radio Network Temporary Identifier [257]. Similarly, in the downlink, when a dedicated logical channel is to be transmitted to the UE on a common channel, such as the Secondary Common Control Physical channel (S-CCPCH), the UE's identity must be included in the MAC header [257]. When the UE performs cell update or UTRAN Registration Area (URA) update procedures, which help the network decide to which area the page should be sent, the UE includes the UTRAN Radio Network Temporary Identity in the header while sending relevant messages. When the Transparent Mode RLC is used, the MAC sublayer performs *ciphering* (i.e., encryption). (Three modes of RLC operation are discussed in Section 9.3.4.) For example, 12.2 kbps AMR speech is transmitted using a Transparent Mode RLC, and the MAC sublayer ciphers speech bits.

The UE's MAC sublayer determines several parameters, such as the Access Service Class, to use a common uplink access channel called the random access channel. Section 9.4 discusses the use of the random access channel in detail. The MAC makes measurements of traffic volume and provides the number of bytes available in the RLC sublayer to the RRC layer.

9.3.4 Radio Link Control Sublayer

The main function of the RLC sublayer is to provide reliable data transfer. A selected list of RLC functions is given in Figure 9.8. When the RLC sublayer of the transmitter receives a large packet from the upper layer, it segments the packet into smaller RLC PDUs (protocol data units). An RLC PDU consists of an upper-layer payload [also

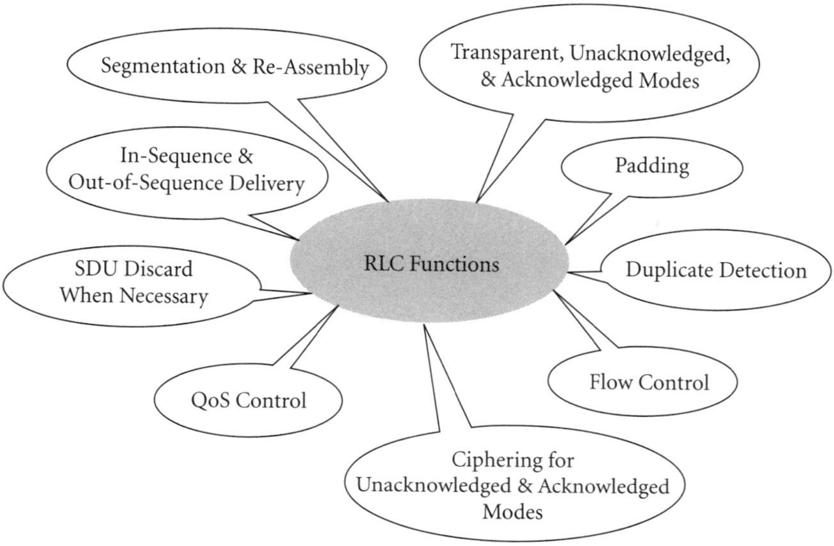

Figure 9.8: Functions of the RLC Sublayer

called an RLC service data unit (SDU)] and an RLC header. The RLC sublayer of the receiver conducts the reverse-processing and re-assembles the smaller RLC PDUs to reconstruct the original large packet for the upper layer. For example, with a 384 kbps packet-switched service such as web-browsing, the RLC sublayer at the RNC breaks down the IP packet into RLC PDUs, each with the payload size of 320 bits and sixteen bits of RLC header. This will lead to $320 + 16 = 336$ bits of RLC PDU.

The RLC layer supports three modes: the *Transparent Mode*, *Unacknowledged Mode*, and *Acknowledged Mode*. In the Transparent Mode, the RLC sublayer transparently lets the upper-layer data pass thorough it and adds no RLC header to such data. In this mode, the RLC may perform segmentation and re-assembly. This mode is suitable for delay sensitive applications that can tolerate some errors. A 12.2 kbps AMR speech service utilizes the Transparent Mode RLC. Some signaling messages, such as the RRC CONNECTION REQUEST message,[285] are transferred using the Transparent Mode RLC.

In the Unacknowledged Mode, the transmitter's RLC sublayer adds an RLC header to the RLC SDU so that the receiver's RLC sublayer can use the RLC sequence number from the RLC header to detect erroneous data. Some signaling messages, such as the RRC CONNECTION SETUP message, are transferred using the Unacknowledged Mode RLC.

The Acknowledged Mode involves the addition of an RLC header at the transmitter, the detection of errors at the receiver, the sharing of feedback from the receiver to

[285] Section 9.6 discusses these messages as part of call setup.

the transmitter about missing RLC PDUs, and the transmitter retransmitting missing RLC PDUs. The RLC entity at the receiver delivers only error-free SDUs to the upper layer. The Acknowledged Mode RLC entity is suitable for delay-tolerant services with stringent error-free transmission requirements. Numerous packet-switched services, such as web-browsing and email, use the Acknowledged Mode RLC. Some signaling messages, such as the MEASUREMENT CONTROL message,[286] are transferred using the Acknowledged Mode RLC. The modes of RLC operation are described in Sections 9.3.4.1 to 9.3.4.3.

When the bits to be transmitted fill less than the entire RLC PDU, padding bits are added. These padding bits could be any set of bits (i.e., not necessarily all 0s). The RLC sublayer detects duplicate RLC PDUs and delivers them only once to the upper layer. The receiver's RLC layer controls the rate at which the transmitter's RLC layer sends information. The RLC sublayer can encrypt the RLC SDU for Unacknowledged and Acknowledged Modes to prevent unauthorized access to the information. The RRC layer configures various parameters that control the RLC retransmission process. Hence, the RLC layer plays a key role in achieving QoS for a given service. The benefits of RLC-aided QoS control are higher throughput and shorter latency compared to cases in which no RLC sublayer exists. The transmitter's RLC sublayer discards the SDU when the maximum number of retransmissions is exhausted. Furthermore, corrupted SDUs are discarded; only error-free SDUs are delivered to the upper layer. Error detection is performed through an RLC sequence number check. The RLC sublayer supports both in-sequence and out-of-sequence delivery mechanisms.

9.3.4.1 Transparent Mode The Transport Mode allows information to be quickly and efficiently transferred because no RLC overhead is added. The transmitting RLC entity receives information—a message or traffic—from the upper layer in the units of RLC SDUs. The length of an RLC SDU is a multiple of valid Transparent Mode PDU sizes. One RLC PDU is equivalent to one RLC SDU since no RLC header is added to the RLC SDU. One TTI carries one or more Transparent Mode PDUs of the same size. Segmentation may be performed if the RRC sublayer configures the segmentation and if one RLC SDU cannot fit into one Transparent Mode PDU. The Transparent Mode PDUs are submitted to the MAC sublayer on a logical channel.

The receiving RLC entity obtains RLC PDUs from the MAC sublayer through a logical channel. When the Transparent Mode RLC PDU is received with a CRC error, it can be delivered to the upper layer or discarded by the RLC based upon whether the RRC layer has set the "Delivery of Erroneous SDUs" parameter to "yes" or "no."

9.3.4.2 Unacknowledged Mode Figure 9.9 illustrates the operation carried out by the transmitting and receiving RLC entities. The transmitting RLC entity may segment the RLC SDU received from the upper layer if the RLC SDU cannot fit into an Unacknowledged Mode PDU. Similarly, multiple RLC SDUs can be

[286]Section 9.8 provides details of this message while describing handover.

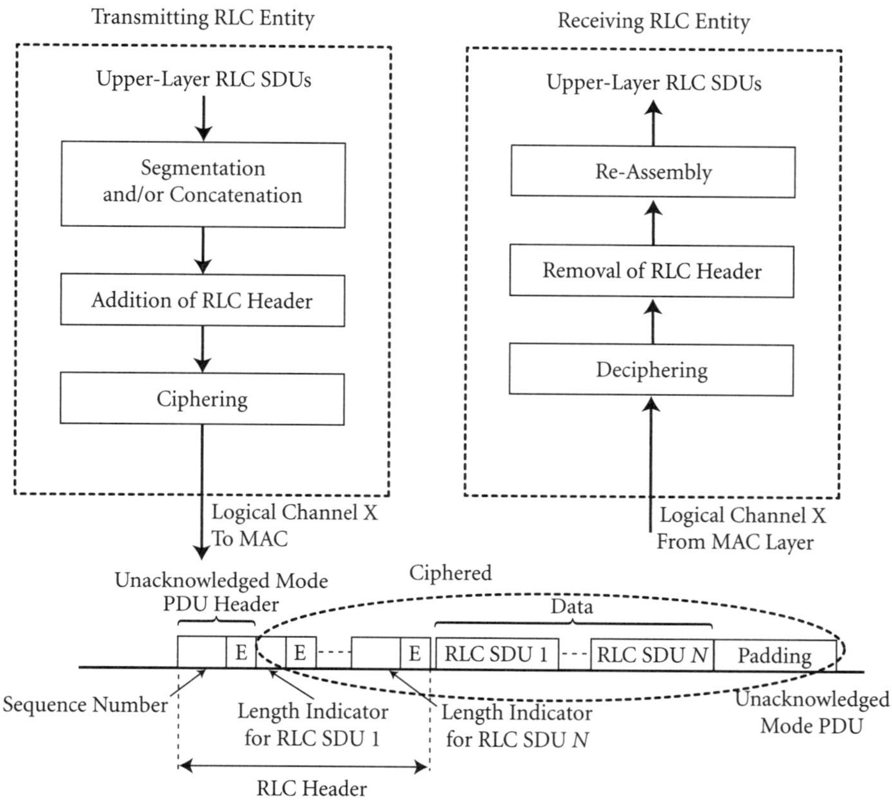

Figure 9.9: Unacknowledged Mode RLC Operation

concatenated to create a Unacknowledged Mode PDU. A suitable number of padding bits may be added to ensure a valid overall length of the Unacknowledged Mode PDU. Specific contents of the Unacknowledged Mode PDU are shown in Figure 9.9. The Unacknowledged Mode PDU header includes fields such as the sequence number, extension (E) bit, and length indicators. The sequence number is a seven-bit field that helps detect a missing PDU and is used for re-assembly. The E bit indicates whether the next field is data (i.e., an RLC SDU) or is a length indicator and E bit. The length indicator is a seven-bit or fifteen-bit field that indicates the number of octets in each RLC SDU and is used to define the boundaries between th RLC SDUs within an Unacknowledged Mode PDU. Some length indicators have special purposes. For example, all 1s in the length indicator specifies the existence of padding bits in the Unacknowledged Mode PDU. When the receiver detects all 1s in the length indicator field, the receiver knows that the rest of the bits in the Unacknowledged Mode PDU are padding bits. If the RRC configures the ciphering, the Unacknowledged Mode PDU excluding the Unacknowledged Mode PDU header is ciphered as highlighted in Figure 9.9. The Unacknowledged Mode PDU is submitted to the MAC sublayer over a logical channel.

The receiving RLC entity obtains the Unacknowledged Mode PDU over a logical channel and performs deciphering. The RLC header is then removed, and the RLC SDUs are re-assembled based on segmentation or concatenation performed at the transmitter.

9.3.4.3 Acknowledged Mode The Acknowledged Mode involves the retransmission of missing PDUs. Figure 9.10 shows the overall operation that the entities in the Acknowledged Mode carry out. The transmitting side segments and/or concatenates the radio link set SDUs received from the upper layer into the Acknowledged Mode PDUs of a fixed length. The RLC header includes the D/C bit, sequence number, P (polling) bit, HE (header extension) bits, length indicator, and E bit as shown in Figure 9.10. The functions of the sequence number, length indicator, and E bit fields are the same for the Acknowledged Mode PDU and the Unacknowledged Mode PDU. The D/C bit indicates if the PDU is a data PDU or a control PDU. A data PDU contains RLC SDUs from the upper layer, and a control PDU carries control information generated by the RLC layer. For example, STATUS PDU is a control PDU the receiver sends to the transmitter to inform it of which PDUs are missing at the receiver. Based on such information, the transmitter can now retransmit the missing PDUs. The transmitting entity uses the P bit to request that the receiving entity send a STATUS PDU. The HE is a two-bit field that indicates if the next set of bits contains data or the combined length indicator and E bit.

The RLC header is placed at the front of data to create an Acknowledged Mode PDU. One or more Acknowledged Mode PDUs containing new RLC SDUs are sent to the multiplexing unit and the retransmission buffer. The multiplexing unit concatenates these Acknowledged Mode PDUs with possibly the PDU(s) to be retransmitted. Now, some RLC header contents may be updated. For example, the padding bits may be replaced by a piggybacked STATUS PDU, and the length indicator would specify the existence of the STATUS PDU. A polling bit may also be set. The Acknowledged Mode PDU, excluding the header, is ciphered. Ciphered Acknowledged Mode PDUs are submitted to the MAC layer over one or two logical channels. When two logical channels are used, one channel is used for data PDUs and the other for control PDUs.

The receiving RLC entity receives ciphered PDUs from the lower layer over a configured logical channel and deciphers them. The deciphered PDUs are delivered to the buffer. The receiver acknowledges the successful reception of the PDUs or requests retransmission of the PDUs by sending one or more STATUS PDUs to the transmitter. The RLC header is removed next. The RLC SDUs are now re-assembled and delivered to the upper layer.

The following RLC characteristics illustrate the flexibility of the overall RLC operation.

- The receiver sends a STATUS PDU periodically or upon the detection of a missing PDU.

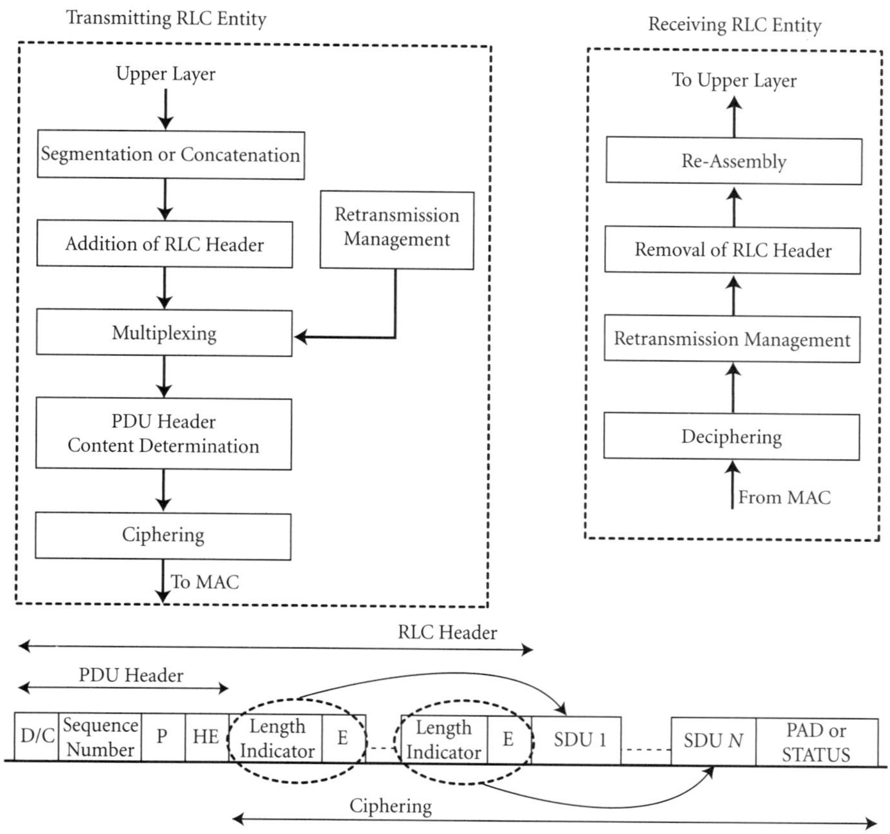

Figure 9.10: Acknowledged Mode RLC Operation

- The RRC layer configures the maximum number of retransmissions by setting the parameter MaxDAT. The number of transmissions for the same Acknowledged Mode PDU equals $MaxDAT - 1$ [261].

- The sender can poll the receiver to obtain a status report using various triggers, such as the last PDU in the transmit buffer and the expiration of timers. The sender can discard the RLC PDUs from its buffer using several approaches, such as timer-based discard or MaxDAT-based discard.

- When the RLC PDU is received with a CRC error, it is discarded in the Unacknowledged Mode and Acknowledged Mode.

- The receiver can store only a limited number of packets, and the starting point of such a receiving window is the oldest expected (i.e., missing) Acknowledged Mode PDU that has the lowest RLC sequence number. When the sender is unable to successfully deliver an Acknowledged Mode PDU for a period of time or for a number of transmissions, it asks the receiver to update the receiving window. This operation enables the receiver to continue getting new packets rather than infinitely waiting for the missing PDU. In this case, patience is *not* a virtue!

9.3.5 Packet Data Convergence Protocol Sublayer

The PDCP sublayer enables efficient use of the radio resources by minimizing the upper-layer header information. The PDCP layer typically accepts an IP packet from the upper layer.

Do You Recall... What does the PDCP layer do to reduce overhead with IP packets?[287]

Let's summarize the functions the PDCP layer carries out. To implement header compression, the transmitter's PDCP layer compresses Transmission Control Protocol and IP or Real-Time Transport Protocol, User Datagram Protocol and Internet Protocol headers to minimize the number of overhead bits associated with these upper layers. The receiver's PDCP layer decompresses these headers to retrieve full-fledged original headers. The PDCP layer receives data from the packet-switched core network and transfers data to the RLC sublayer in the network-to-UE information transfer. Similarly, with UE-to-network information transfer, the PDCP sublayer passes data from the upper layer to the RLC sublayer for eventual transfer to the core network. The PDCP sublayer maintains the PDCP sequence numbers to support lossless serving radio network subsystem relocation as the UE moves from one RNC to another RNC. Commercial networks typically omit lossless serving radio network subsystem relocation. See Section 9.8 for more details on handover.

Figure 9.11 summarizes the PDCP sublayer's operation. The transmitting side adds a PDCP header to the received upper-layer data. If the PDCP is to be bypassed completely, no PDCP header is added. The PDCP header, when present, consists of two fields, PDU type and packet identifier, or three fields, PDU type, packet identifier, and sequence number. The three-bit PDU type field essentially indicates whether the PDCP PDU contains a sequence number. The five-bit packet identifier field indicates the header compression and packet type. For example, *packet identifier* = 0 indicates that the header compression is absent, *packet identifier* = 1 indicates that a full header has been included, and *Packet Identifier* = 2 means that the packet includes a compressed Transmission Control Protocol header. When lossless serving radio network subsystem relocation is supported, the PDCP header includes a sixteen-bit

[287]The PDCP sublayer carries out header compression to reduce overhead. With short IP packets, such a reduction can be quite significant because the size of the overhead can be comparable to the payload itself.

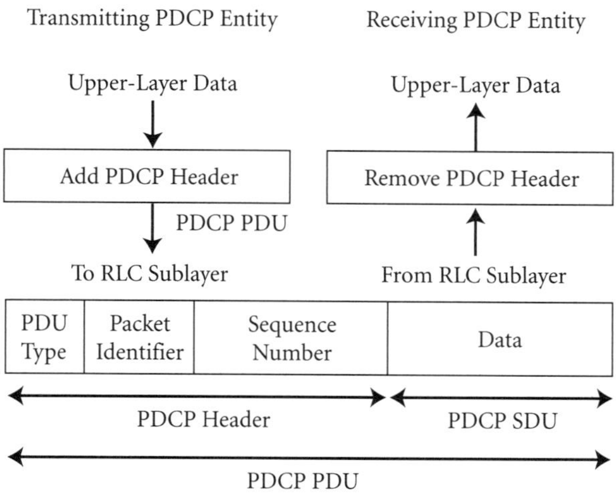

Figure 9.11: Operation of the PDCP Sublayer

PDCP SDU sequence number. If both UE and UTRAN support lossless serving radio network subsystem relocation, the UTRAN can configure the UE to implement it. During the RRC connection setup process, the UE informs the UTRAN about its capabilities, and the UTRAN then learns about the UE's lossless serving radio network subsystem relocation capability. Note that lossless serving radio network subsystem relocation is allowed only when the RLC is configured for in-sequence delivery and Acknowledged Mode. During the lossless serving radio network subsystem relocation process, the UE and the source or current RNC exchange PDCP sequence numbers of the transmitted but unconfirmed uplink and downlink PDUs. Furthermore, the source RNC sends the target RNC relevant PDCP SDUs and PDCP sequence numbers. Hence, the packet loss is avoided when the serving radio network subsystem relocation occurs. The data field could contain an uncompressed PDCP SDU, a header-compressed PDCP SDU, or header-compression–related control information. The only supported header compression protocol is specified in Internet Engineering Task Force RFC 2507 [262]. Once the PDCP PDU is constructed, it is submitted to the lower RLC sublayer.

The receiving entity obtains the PDCP PDU from the RLC sublayer and removes the PDCP header. It processes the data field of the PDCP PDU based on the associated header. For example, if the packet identifier field indicates a compressed header, the receiving entity decompresses the header contained in the data field. The processed PDCP SDU is then transferred to the upper layer, such as the IP layer.

9.3.6 Broadcast-Multicast Control Sublayer

The main task of the BMC sublayer is to support the existing Cell Broadcast functionality of GSM and GPRS systems. A Cell Broadcast service transfers short text messages to many users in a given area. The BMC sublayer in the RNC receives the Cell Broadcast messages and their scheduling information from the Cell Broadcast Center and stores them. This layer in the RNC also determines the required data rate for transmission of the received Cell Broadcast messages and requests the RRC sublayer for appropriate resources. Furthermore, it creates and transmits the schedule messages and transmits the Cell Broadcast messages based on the schedule. The BMC sublayer in the UE processes the schedule messages and works with the RRC sublayer so that the RRC sublayer can configure the lower layers to receive the Cell Broadcast messages through discontinuous reception. The UE's BMC sublayer delivers error-free Cell Broadcast messages to the upper layer. As mentioned earlier, typical commercial deployments omit the BMC sublayer. Now you know why we dedicated so little space to the BMC sublayer!

9.3.7 Radio Resource Control Layer

The RRC layer controls lower layers and is responsible for layer 3 signaling messages exchanged between the UE and the RNC. Example functions of the RRC layer are summarized in Figure 9.12.

The RRC layer broadcasts system information, cell-specific information, and paging messages to the UEs. Certain information such as system information is usually repeated periodically. The contents the RRC broadcasts are generated by the core network or the RNC.

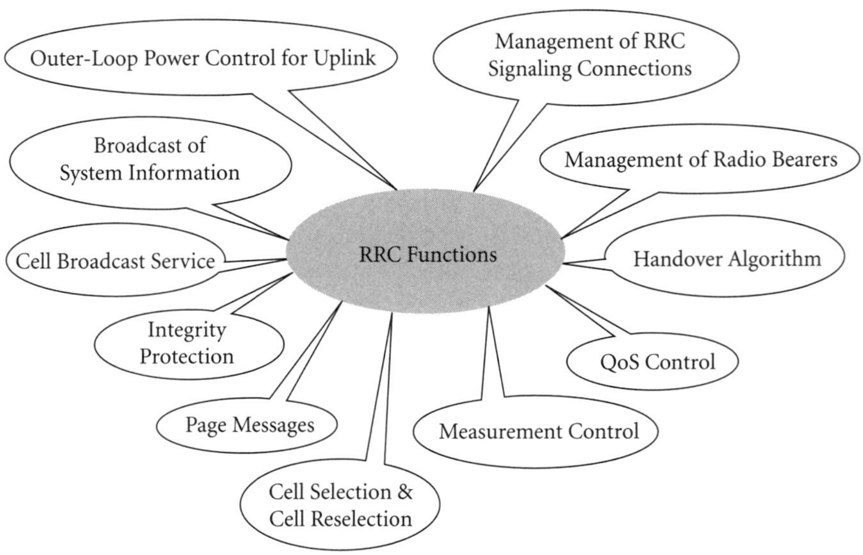

Figure 9.12: RRC Layer Functions

The RRC layer at the RNC determines the target signal-to-interference ratio for the uplink power control and conveys it to the Node B so that the Node B can execute inner-loop power control. Recall that outer-loop power control and inner-loop power control are two components of closed-loop power control.

Do You Recall. . . Do you remember the main goal of closed-loop power control?[288]

The RRC layer at the RNC also determines the target error rate so that the UE can execute outer-loop power control for the downlink. See Section 9.9 for power control details.

This layer establishes, maintains, and tears down the RRC signaling connection between the UE and the RNC. This signaling connection enables the UE and the RNC to reliably exchange signaling messages to set up the call and maintain a dedicated radio connection. The call admission control algorithm is executed to determine if resources, such as OVSF codes and transmit power for the DL, are available to support the RRC connection.

A radio bearer can be viewed as a dedicated link between the UE and the RNC, which includes a dedicated radio connection between the UE and the Node B and a UE-specific Iub bearer between the Node B and the RNC. The RRC layer allocates appropriate resources to radio bearers to ensure QoS for a given UE. The call admission control and scheduling algorithm play an important role in establishing and reconfiguring radio bearers. Multiple radio bearers are established to support services with different QoS requirements for a given UE.

The RRC layer in the RNC implements the handover algorithm that determines the Active Set in case of intra-frequency handover and makes inter–radio-access-technology handover decisions based on the measurements the UE reports. The RRC layer also participates in the cell selection and location update procedures. See Section 9.8 for mobility details.

The RRC layer is in charge of controlling the requested QoS from the perspective of radio resources. For example, during the call setup, the core network requests from the RNC for a specific QoS, such as 384 kbps maximum data rate in the downlink. If the RNC has adequate resources, such as power and OVSF codes, its call admission control algorithm will set up a radio bearer to achieve 384 kbps. In summary, a radio bearer implements QoS. The RNC also configures the number of retransmissions at the RLC layer to enforce QoS metrics, such as residual error rate.

The RRC layer at the RNC controls both the measurements that the UE will make as well as how the UE reports measurements to the RNC. For example, the RRC layer can configure the UE to make pilot strength measurements of a set of cells and report these measurements when certain events occur. Section 9.8 discusses such various events.

[288]The goal is to transmit minimum power while achieving a target error rate.

When the UE is in the Idle Mode, it performs its initial cell selection after power-on and continues performing cell reselection as it crosses cell borders. In the Idle Mode, the UE periodically wakes up and checks for a possible paging message.

The RRC layer in the RNC broadcasts paging messages and associated notifications to UEs. The RRC layer adds a Message Authentication Code called MAC-I to the RRC signaling messages to implement integrity protection, which ensures authenticity of the sender. The RRC layer at the RNC allocates radio resources for transmission of the Cell Broadcast Service messages. The UE's RRC layer configures the lower layers for discontinuous reception of these messages.

Caution! Our intention is not to drop you into a maze! We will highlight major aspects of Figure 9.13 next. The figure illustrates various states of the RRC layer. Certain activities are allowed for the UE, and specific channels are used in these states. The UE with an RRC connection with the UTRAN is said to be in RRC Connected Mode. In contrast, a UE without an RRC connection is in the Idle Mode. In the Idle Mode, the UTRAN knows nothing about the UE; the core network knows where to page the UE.[289] Four RRC states are defined for a UE in the RRC Connected Mode: CELL_DCH, CELL_FACH, CELL_PCH, and URA_PCH. The first word, e.g., "CELL" in CELL_DCH, indicates the granularity with which the UTRAN knows about the location of the UE. The second word, "DCH" in CELL_DCH, indicates

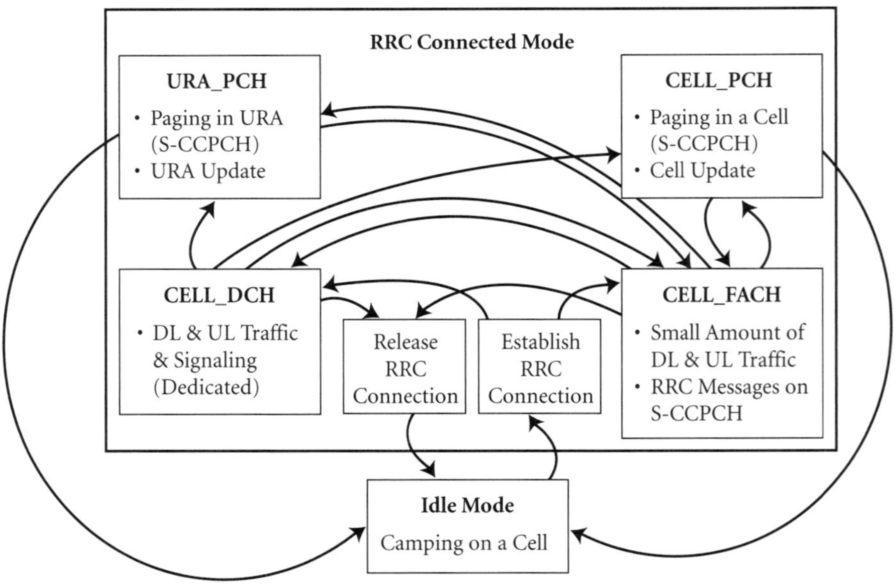

Figure 9.13: RRC States

[289]We will discuss paging in the Idle Mode and in CELL_PCH and URA_PCH states in Section 9.8.

the specific transport channel used for communication with the UE. The transitions allowed from one RRC state to another and from an RRC state to the Idle Mode are also depicted in Figure 9.13.

When an RRC connection between the UE and the UTRAN is established, the UTRAN specifies to the UE the RRC state in which the UE should operate. Typically, the UTRAN asks the UE to operate in the CELL_DCH state upon establishment of an RRC connection. The UE can exchange speech and data and dedicated signaling messages in the UL and the DL in CELL_DCH state. The CELL_FACH state enables the UE to receive RRC messages on the FACH. Though a small amount of data transfer is allowed in the CELL_FACH state, the CELL_DCH is commonly used in practice even for the transfer of short messages. In the CELL_PCH and URA_PCH states, the UE looks for a paging indication[290] and (if needed) the S-CCPCH carrying the PCH to retrieve a paging message. When the UE goes from one cell to another, it performs a cell update in the CELL_PCH state, and when it goes from one URA to another URA in the URA_PCH state, it performs a URA update. Commercial deployments commonly omit the CELL_PCH and URA_PCH states. No data transfer is allowed in the CELL_PCH and URA_PCH states, and the UE observes only the DL PCH.

9.4 RADIO INTERFACE BETWEEN THE UE AND THE UTRAN

We'll focus on the radio interface here. More specifically, we will consider radio frame structure, OVSF codes, and radio channels. Section 9.4.1 provides a brief overview of this topic. Since UMTS is based on CDMA, it uses OVSF codes over the air interface. These codes are discussed in Section 9.4.2. Sections 9.4.3 through 9.4.14 take a closer look at all the radio channels; fasten your seatbelt because we have plenty of channels to cover!

9.4.1 Overview of Frame Structure and Radio Channels

The radio interface in a UMTS system utilizes 10 ms radio frames. Figure 9.14 illustrates the structure of a radio frame. Each radio frame consists of fifteen slots, and each slot is $10\ \text{ms}/15 = 0.667$ ms long. Since the UMTS air interface is based on CDMA, chips are transmitted during a frame. The UMTS chips rate is 3.84 Mcps (i.e., 3.84×10^6 chips per second), which corresponds to 2560 chips during a slot. Transmission of user traffic and signaling messages occurs in transmit time intervals TTIs.

Do You Recall... Do you remember the TTI for a speech block?[291]

[290]We will discuss various physical channels, including a paging indication channel and S-CCPCH, in Section 9.4.

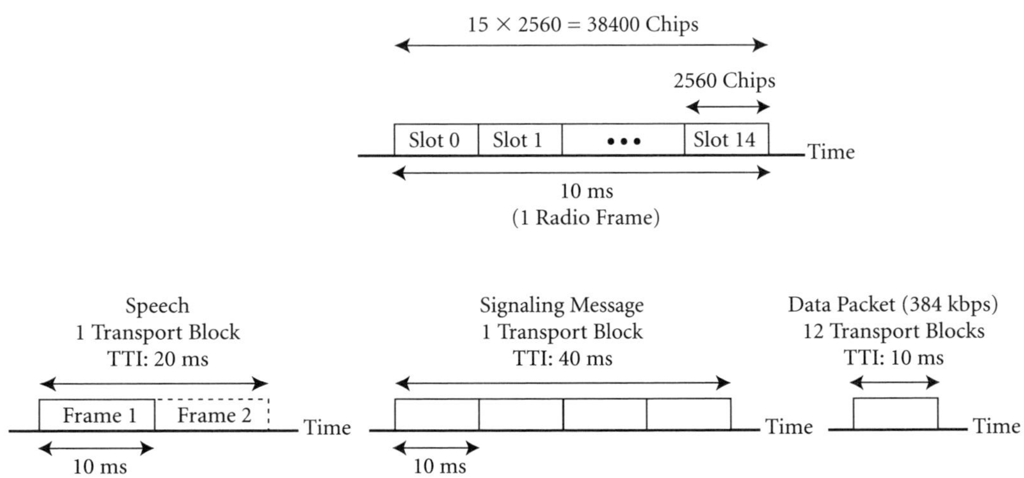

Figure 9.14: Frame Structure and Frame Contents

A signaling message is usually transmitted using a 40 ms TTI. If a packet is transmitted at a high rate (e.g., 384 kbps), multiple transport blocks are transmitted in a single 10 ms TTI. The user traffic or signaling message passes through the protocol stack (i.e., the RRC or PDCP, RLC, MAC, and PHY layers) and gets mapped onto a logical channel, a transport channel, and finally a physical channel along the way.

In addition to user traffic and signaling messages, some fast physical layer signaling indicators (e.g., a page indicator, discussed later in this section) and physical signals (e.g., a preamble sent on a Physical Random Access channel, also discussed later in this section) are also mapped onto some physical channels and transmitted during a frame. Furthermore, some channels, such as a Primary Common Pilot channel, are transmitted only as physical channels and have no association with transport or logical channels.

A radio channel can be common for all users or dedicated to a specific user. Figure 9.15 illustrates how the UE uses various radio channels upon power-up. We'll take a closer look at power-up synchronization in Section 9.5.

The UE looks for the Primary Synchronization channel (P-SCH) after power-up and learns of the slot boundaries at the end of the synchronization with the P-SCH. The P-SCH is a physical channel with no association with logical or transport channels. It facilitates the UE's cell search operation.

The UE next looks for the Secondary Synchronization channel (S-SCH) that also exists only as a physical channel. The S-SCH helps the UE identify the radio frame boundaries. Additionally, once the UE synchronizes with the S-SCH, the UE

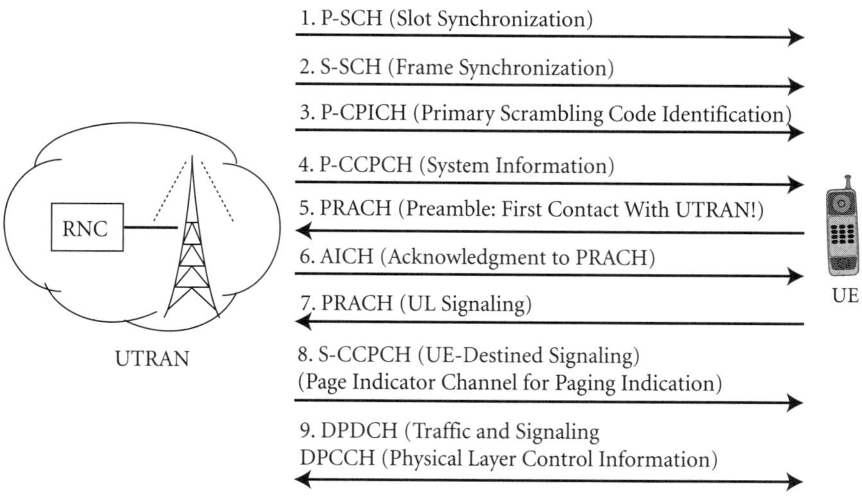

Figure 9.15: Radio Channels the UE Encounters upon Power-Up

[291] A speech block equivalent to one transport block is transmitted during the 20 ms TTI, meaning two consecutive radio frames are required for one speech transport block.

automatically learns the identity of the group to which the primary scrambling code belongs. For example, assume that the UE is in a cell that uses the primary scrambling code 1, which belongs to the primary scrambling code group 0. Upon synchronization with the S-SCH, the UE knows that it is inside a cell that belongs to group 0.

Since the UE already knows the group number of the primary scrambling code, it tries eight possible primary scrambling codes associated with the identified group number to descramble the Primary Common Pilot channel (P-CPICH) and finds a match with one of them. The P-CPICH channel also exists purely as a physical channel. It consists of all 0s spread by the OVSF code $C_{ch,256,0}$ and scrambled by the primary scrambling code. The UTRAN can configure the UE to use the P-CPICH for channel estimation and coherent demodulation. The UE measures the strength of the P-CPICH and reports the measurements to the UTRAN so the UTRAN can execute handover.

After achieving synchronization with the downlink, the UE is eager to learn all about the network. (We certainly hope that you are just like the UE—eager to learn!) The Primary Common Control Physical channel (P-CCPCH) conveys system-specific and cell-specific information for all UEs. For example, the P-CCPCH provides the identity of the PLMN as well as radio channel information, such as the UMTS Absolute Radio Frequency Channel Number (ARFCN), which specifies the carrier frequency. This channel provides a timing reference for all the other downlink and uplink physical channels. The system frame number that is transmitted on the P-CCPCH identifies a specific radio frame. The P-CCPCH is associated with the transport channel called Broadcast channel and the logical channel called the Broadcast Control channel.

Note that the UE has been observing the downlink after power-up but the network has no idea thus far about the UE's existence! So, the UE attempts to contact the network to announce its presence and to enable delivery of a paging message to the UE if required. The UE uses the PRACH to send a preamble to the network. The PRACH is an uplink physical channel corresponding to the transport channel called the Random Access channel (RACH). The Common Control channel is a logical channel that can be associated with the RACH. When a UE has no dedicated radio channel and needs to send information to the network, it utilizes the PRACH.

The Node B is always looking for a preamble on the PRACH, and as soon as it detects one, it informs the UE using the Acquisition Indicator channel (AICH) that it has successfully detected the preamble. The AICH is a downlink physical channel with no mapping with logical or transport channels.

In a UMTS system, the UE attempts to access the system in two stages. First, the UE sends a preamble on the PRACH, and, if an AICH successfully acknowledges the preamble, the UE sends an RRC CONNECTION REQUEST message on the PRACH. The Node B replies with an RRC CONNECTION SETUP message on the S-CCPCH. The S-CCPCH is associated with the transport channel of the FACH (Forward Access channel) or the PCH. The FACH is used for sending UE-specific messages on a common channel and for cell-specific paging when the UE location is known at the cell level. If the UE location is unknown at the cell level, the paging message is sent to a larger area containing multiple cells, using the PCH as the transport channel in such scenario.

When a paging message is to be sent on the S-CCPCH, a physical channel called the Page Indicator channel (PICH) makes such paging more efficient. The PICH is a downlink physical channel that indicates to a UE whether a paging message (for this or some other UE) would be sent on the S-CCPCH after a specific time period. Since a page indicator is quite short (as few as two bits), the UE requires little processing power to look at the PICH. If no page indicator is present, the UE need not look for potentially large paging messages on the S-CCPCH, leading to savings in processing power and thereby increasing the battery life.

The Node B typically allocates dedicated resources using the RRC CONNECTION SETUP message. From now on, the UE can use dedicated physical channels in the uplink and the downlink. These channels are DPDCH and Dedicated Physical Control channel (DPCCH). The DPDCH has a mapping with a transport channel called the dedicated channel and a (possible) logical channel called the DTCH. This channel carries user traffic, such as speech or data packets, and user-specific signaling messages. The DPCCH is only a physical channel and conveys the control information generated at the physical layer. For example, it specifies the data rate of the accompanying DPDCH via the TFCI; the power control adjustments via the TPC (transmit power control) command; and the dedicated pilot bits for channel estimation and coherent demodulation.

For your convenience, Table 9.1 summarizes common radio channels and their basic uses, and Table 9.2 summarizes the dedicated radio channels. Details of the channels are discussed in the Sections 9.4.3 to 9.4.13 and can also be found in the standard [257, 263]. The standard is quite flexible (consider this a euphemism for complex!) in terms of the mapping among logical, transport, and physical channels. The mappings shown here are representative only; refer to the standard [257] for all possible channel mappings.

9.4.2 Orthogonal Variable Spreading Factor Codes

The channelization codes that separate one physical channel from another are referred to as OVSF codes in the standard. The OVSF codes are generated using a code tree shown in Figure 9.16 [264].

The channelization code helps spread the signal to the chip rate of 3.84 Mcps and is formally specified as $C_{ch,SF,n}$ where ch refers to channelization, SF stands for spreading factor, and n is the code number. For example, $C_{ch,4,2}$ refers to the channelization code $[1 \quad -1 \quad 1 \quad -1]$, which is the code number 2 (out of the set $\{0,\ 1,\ 2,\ \text{and}\ 3\}$) at $SF = 4$. When an input signal with bandwidth X Hz is multiplied by $C_{ch,4,2}$, the resulting signal has the bandwidth of $4X$ Hz, which would be 3.84 MHz in the case of UMTS. The channel's required data rate generally dictates the SF.

Let's discuss how to construct OVSF codes by starting with the code $C_{ch,1,0} = [1]$. Two branches emerge from this code, the upper branch and a lower branch. Both branches include $C_{ch,1,0}$ as the first part of the new code. The upper branch then

Table 9.1: Common Radio Channels

Physical Channel	Transport Channel	Logical Channel	Direction	Main Use
P-SCH	–	–	DL	Cell search: slot synchronization
S-SCH	–	–	DL	Cell search: frame synchronization and primary scrambling code *group* identification
P-CPICH	–	–	DL	Cell search: primary scrambling code identification, coherent demodulation, handover
P-CCPCH	Broadcast channel	Broadcast Control channel	–	Broadcast of system information, timing reference
PICH	–	–	DL	To indicate presence of a paging message on S-CCPCH
S-CCPCH	PCH or FACH	Paging Control channel, DTCM, Dedicated Control channel, Common Traffic channel, Common Control channel	DL	To send a page or other messages to a UE
PRACH	RACH	Common Control channel, CDDH, DTCH	UL	To send a message to UTRAN in the absence of a dedicated channel
AICH	–	–	DL	To acknowledge successful detection of the PRACH

Table 9.2: Dedicated Radio Channels

Physical Channel	Transport Channel	Logical Channel	Direction	Main Use
DPDCH	DCH	DTCH, Dedicated Control channel	UL, DL	Dedicated user traffic or signaling messages
DPCCH	–	–	UL, DL	Data rate indication, power-control commands, dedicated pilot

repeats the code $C_{ch,1,0}$, creating a new code $[C_{ch,1,0} \quad C_{ch,1,0}] = [1 \quad 1]$. The lower branch complements $C_{ch,1,0}$ by multiplying with -1 to create the second part of the code. The lower branch constructs the code $[C_{ch,1,0} - 1 * C_{ch,1,0}] = [1 \quad -1]$. We have generated two codes, $C_{ch,2,0}$ and $C_{ch,2,1}$. For a given spreading factor, the code is numbered from 0 to $SF - 1$.

Let's do one more set of calculations. Let's create two codes from $C_{ch,2,0}$. The upper branch yields the code $[C_{ch,2,0} \quad C_{ch,2,0}] = [1 \quad 1 \quad 1 \quad 1]$ because we are simply

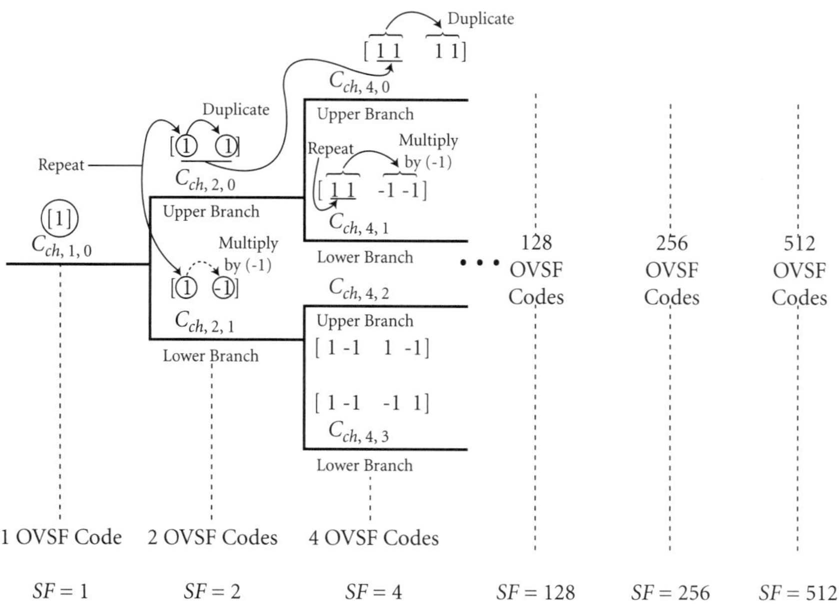

Figure 9.16: Constructing an OVSF Code Tree

repeating the code $C_{ch,2,0}$. The lower branch emanating from $C_{ch,2,0}$ gives the code $[C_{ch,2,0} - 1 * C_{ch,2,0}] = [1 \quad 1 \quad -1 \quad -1]$. You can confirm your understanding by generating $C_{ch,4,2}$ and $C_{ch,4,3}$ from $C_{ch,2,1}$; that's your homework assignment!

If you are curious about why the channelization codes are called OVSF codes, this is your paragraph! Let's follow the "divide and conquer" method and divide the term "orthogonal variable spreading factor" into "orthogonal" and "variable spreading factor." The channelization codes are *orthogonal* to one another. To confirm orthogonality, consider $C_{ch,4,2}$ and $C_{ch,4,3}$. Multiply individual chips of the codes and add all resulting chips. In other words, $[1 \quad -1 \quad 1 \quad -1] * [1 \quad -1 \quad -1 \quad 1] = [(1*1) \quad (-1*-1) \quad (1*-1) \quad (-1*1)] = [1 \quad 1 \quad -1 \quad -1]$. Adding the resulting chips yields $1+1-1-1=0$!

Do You Recall... What is the benefit of orthogonality of these codes?[292]

Now, let's focus on variable spreading factor. The chip rate must be 3.84 Mcps, not one chip more and not one chip less. Hence, if the data rate is high, the spreading factor is small, and, if the data rate is small, the spreading factor is high. For example, when the modulation symbol rate entering the spreading block is 480 ksps, $SF = 8$ is used, resulting in 480 ksps $* 8 = 3.84$ Mcps. On the other hand, if the modulation symbol rate is 30 ksps, $SF = 128$ is used to achieve 3.84 Mcps. In summary, the variable spreading factor enables us to support different data rates. Furthermore, a higher SF provides larger processing gain and requires less transmit power to meet a target error rate at the receiver, increasing the capacity. (See Chapter 7 for more details on spreading and processing gain.) Observe that in Figure 9.16 the number of bits or chips in a code is the spreading factor of the code and the number of orthogonal codes is the same as the spreading factor. While the standard defines the largest spreading factor to be 512, commercial deployments typically go up to only $SF = 256$. In practice, the data rate of 384 kbps uses an OVSF code at $SF = 8$ and the full-rate AMR codec rate of 12.2 kbps uses an OVSF code at $SF = 128$ in the downlink.

Before we leave the topic of the OVSF code tree, we would like to mention some rules on the use of the OVSF code tree. To maintain orthogonality among the OVSF codes of different spreading factors, we observe the following rules: all the parent codes of an allocated code become unavailable, and all the children codes of an allocated code become unavailable as well. A parent code is the code that helps create a child code. For example, in Figure 9.16, $C_{ch,4,2}$ and $C_{ch,4,3}$ are the children codes generated from the parent code $C_{ch,2,1}$. If we were to allocate $C_{ch,4,2}$ to a physical channel, all the parent codes such as $C_{ch,2,1}$ and $C_{ch,1,0}$ can no longer be allocated because $C_{ch,2,1}$ is the parent code of $C_{ch,4,2}$ and $C_{ch,1,0}$ is the parent code of $C_{ch,2,1}$. On the other hand, if we were to allocate $C_{ch,2,0}$ to a physical channel, all the children codes such as $C_{ch,4,0}$, $C_{ch,4,1}$, $C_{ch,8,0}$, $C_{ch,8,1}$, $C_{ch,4,1}$, $C_{ch,8,1}$, $C_{ch,8,2}$,

[292]Orthogonality of codes minimizes interference (ideally, 0) among the codes, making CDMA operational and more efficient.

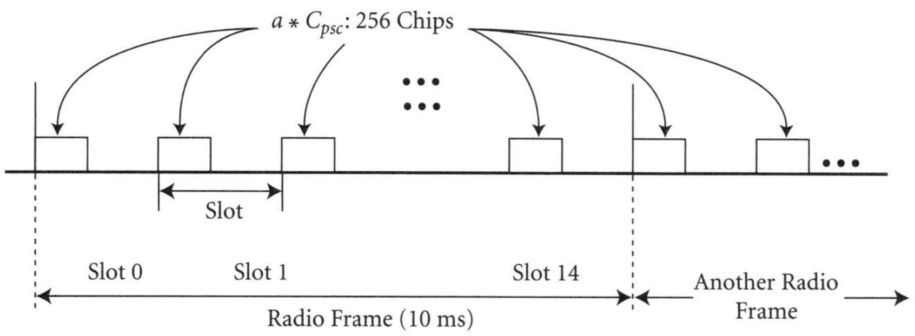

Figure 9.17: Primary Synchronization Channel

$C_{ch,16,0}$, $C_{ch,16,1}$, $C_{ch,16,1}$ and so on, become unavailable. Observe that $C_{ch,4,0}$ and $C_{ch,4,1}$ are children codes of $C_{ch,2,0}$ and that $C_{ch,8,0}$ and $C_{ch,8,1}$ are children codes of $C_{ch,4,0}$.

9.4.3 Primary Synchronization Channel

The primary synchronization channel consists of the first 256 chips in every slot of a radio frame as shown in Figure 9.17. These 256 chips are $a * C_{psc}$ where a is $+1$ if P-CCPCH (yes—your eyes are just fine; it is indeed P-CCPCH, not P-SCH) is utilizing the space time transmit diversity technique and -1 if P-CCPCH is utilizing a single antenna rather than the space time transmit diversity technique and C_{psc} is a 256-chip primary synchronization code. C_{psc} is constructed using a Golay sequence and has good autocorrelation properties. C_{psc} is defined in the standard [264]. Typical commercial deployments omit space time transmit diversity, and $-1 * C_{psc}$ is transmitted during every slot. Every cell in a UMTS system around the globe utilizes the same C_{psc}. The P-SCH is used for slot synchronization. The P-SCH chips are transmitted over the air in a cell with no cell-specific scrambling.

9.4.4 Secondary Synchronization Channel

The S-SCH transmits $a * C_{ssc,i}$ during the first 256 chips of a slot as shown in Figure 9.18, where a is $+1$ when P-CCPCH utilizes space time transmit diversity and -1 when P-CCPCH omits space time transmit diversity and $C_{ssc,i}$ is the secondary synchronization code (SSC) number i. The SSC consists of 256 chips generated using Hadamard sequences and a Golay sequence [264].

There are sixteen SSCs, $C_{ssc,1}$ to $C_{ssc,16}$. Sixty-four secondary synchronization sequences are generated using the sixteen SSCs with each sequence containing fifteen SSCs. For example, the first (or 0th) secondary synchronization sequence consists of following fifteen SSCs: $C_{ssc,1}$, $C_{ssc,1}$, $C_{ssc,2}$, $C_{ssc,8}$, $C_{ssc,9}$, $C_{ssc,10}$, $C_{ssc,15}$, $C_{ssc,8}$, $C_{ssc,10}$, $C_{ssc,16}$, $C_{ssc,2}$, $C_{ssc,7}$, $C_{ssc,15}$, $C_{ssc,7}$, and $C_{ssc,16}$ [264]. One SSC from the secondary synchronization sequence is used in one slot, with all fifteen SSCs of the sequence used in the fifteen slots of the entire frame. Every frame repeats the same pattern of SSCs. The UE utilizes the S-SCH to achieve frame synchronization and to identify the primary scrambling code group. Just like the P-SCH, the S-SCH chips are transmitted over the air in a cell with no cell-specific scrambling.

9.4.5 Primary Common Pilot Channel

Of the two types of common pilot channels, Primary Common Pilot channel and Secondary Common Pilot channel, in practice, only P-CPICH is typically used. The P-CPICH is broadcast over the entire cell, and the Secondary Common Pilot channel, if present, is transmitted over part of a cell or the entire cell. Figure 9.19 illustrates how the P-CPICH chips are constructed. In a system without transmit diversity, which is a typical deployment scenario, the P-CPICH consists of twenty pilot bits, which

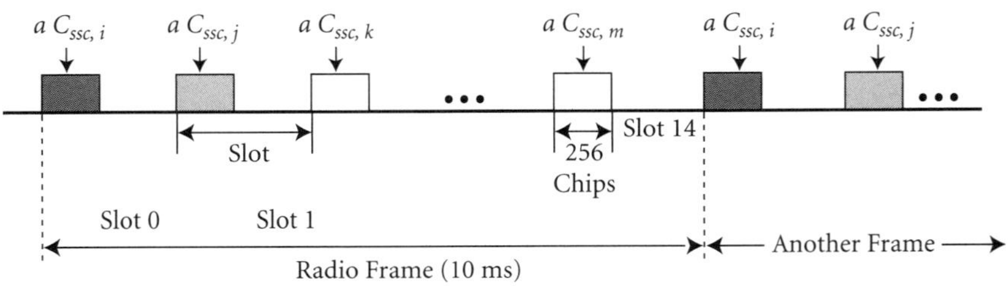

Figure 9.18: Secondary Synchronization Channel

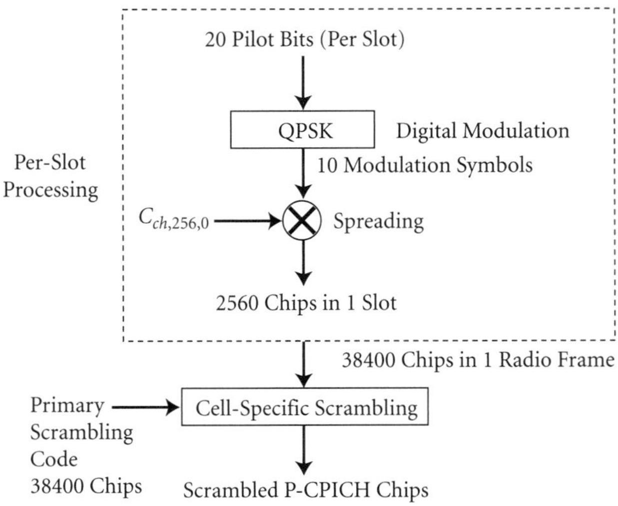

Figure 9.19: Constructing the P-CPICH

are all 0s, in a slot. These twenty 0s pass through the QPSK modulation block, yielding ten QPSK modulation symbols. These symbols are then spread using the OVSF code $C_{ch,256,0}$, resulting in $10 * 256 = 2560$ chips in a slot. Since a radio frame has fifteen slots, a total of $2560 * 15 = 38400$ chips are generated. Spreading by a factor of 256 helps achieve the chip rate of 38400 chips$/10$ ms $= 3.84$ Mcps. These chips are scrambled by a cell-specific scrambling code that also consists of 38400 chips. After all, scrambling is an XOR operation! The RNC allocates the 512 primary scrambling codes to different cells. If a Node B controls three cells as a result of 120° sectorization, the RNC allocates three primary scrambling codes to the Node B, one per cell.

The P-CPICH and P-CCPCH transmission timings are identical. In other words, the first chip of the P-CPICH and the first chip of the P-CCPCH are transmitted at the same time. The P-CPICH provides a reference signal that the UE can use as a phase reference for numerous downlink physical channels (e.g., P-CCPCH and S-CCPCH). By default, the P-CPICH is also a phase reference for the UE-specific DPDCH and DPCCH. The UTRAN can configure the UE to use a specific phase reference (e.g., P-CPICH or DPCCH) in an RRC message (e.g., RRC CONNECTION SETUP).

9.4.6 Primary Common Control Physical Channel

This downlink physical channel carries the Broadcast channel that transports information about the system in the form of a Master Information Block and System Information Blocks (SIBs). (See the standard [265] for details about SIBs.) Figure 9.20 shows how the P-CCPCH chips are created. A set of eighteen P-CCPCH bits pass through modulation every slot, yielding nine modulation symbols. A fixed

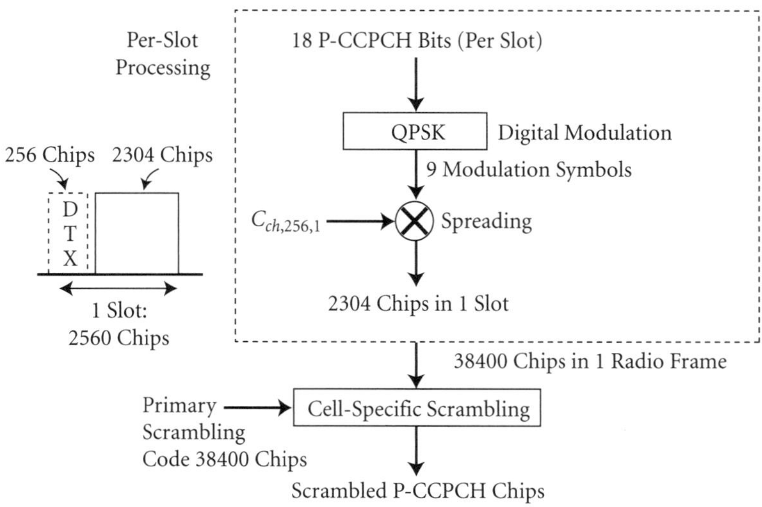

Figure 9.20: Creation of the P-CCPCH

OVSF code, $C_{ch,256,1}$, is used to spread the signal, leading to $9 * 256 = 2304$ chips. No P-CCPCH is transmitted during the first 256 chips of the slot. The overall slot still corresponds to 256 chips that are not transmitted and are shown as DTX in Figure 9.20 as well as 2304 P-CCPCH bits, with a total of $256 + 2304 = 2560$ chips. The first 256 chips of the slot are used to transmit the P-SCH and the S-SCH instead of the P-CCPCH.

9.4.7 Page Indicator Channel

The PICH is a downlink physical channel that carries paging indicators, which indicate the presence or absence of a paging message on the associated S-CCPCH. While the standard allows multiple S-CCPCHs in a cell, one S-CCPCH is quite common. The UE selects a S-CCPCH based on its International Mobile Subscriber Identity [266]. Figure 9.21 shows the preparation of the PICH.

The PICH consists of a total of 300 bits during a 10 ms radio frame. The last twelve bits are not transmitted, leading to a total of $300 - 12 = 288$ bits to represent paging indicators. These twelve bits are separate from the PICH, they keep the PICH structure a little simpler, and they are reserved for possible future use [263]. The number of paging indicators, Np is 18, 36, 72, or 144. If $Np = 144$, each paging indicator is represented by two bits so that $144 * 2 = 288$ bits are transmitted during a 10 ms frame. In such a case, $\{1, 1\}$ represents the paging indicator of 1 and $\{0, 0\}$ represents the paging indicator of 0. Similarly, if $Np = 18$, each paging indicator is represented by sixteen bits. A smaller value of Np (e.g., 18) is quite common in practice to achieve higher reliability of the PICH detection at the expense of a slight

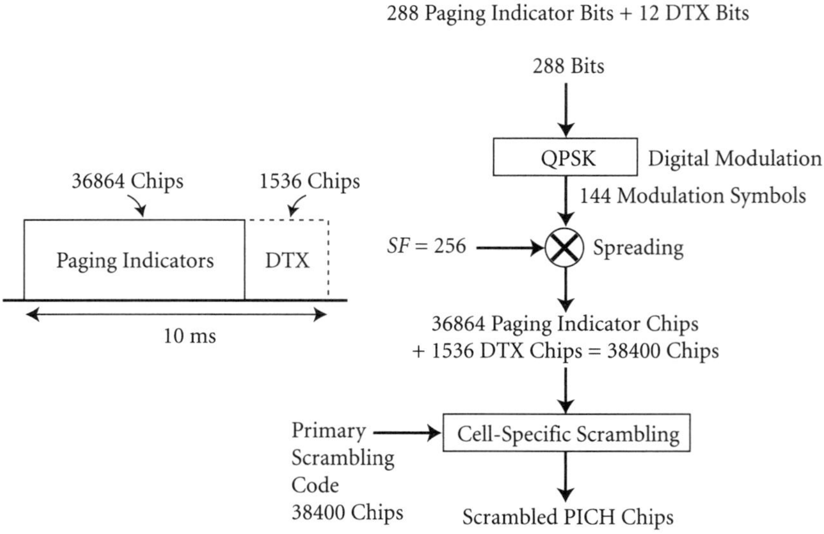

Figure 9.21: Format of the PICH

increase in the overall paging delay in a heavily loaded systems. The value of Np is conveyed to the UEs in the PICH info field present in SIB Type 5 [265], and each UE calculates which Np paging indicator is applicable to it [266]. Assume that there are 144 paging indicators and that the UE calculations result in the paging indicator number 100. When the value of the paging indicator 100 is set to 1, this UE reads the appropriate radio frame carrying the S-CCPCH to look for a paging message potentially directed to itself. If the value of the paging indicator 100 is set to 0, the UE does not process the S-CCPCH.

As depicted in Figure 9.21, a set of 288 paging indicator bits passes through QPSK modulation and gets converted into 144 modulation symbols. Spreading by a factor of 256 yields $144 * 256 = 36864$ chips. The exact OVSF code is undefined in the standard, and the RNC can assign any suitable OVSF code at $SF = 256$ to the PICH. The PICH info field present in SIB Types 5 and 6 transmitted on the P-CCPCH informs the UEs in a cell which OVSF code carries the PICH [265]. Together with 1536 DTX chips, a total of 38400 chips are available for a 10 ms radio frame.

A UE in the Idle Mode uses the discontinuous reception to conserve its battery power. SIB Type 5 transmitted on the P-CCPCH includes numerous parameters that govern the UE operation in the Idle Mode. The UE monitors its paging indicator in the PICH frame that corresponds to the system frame number given by the paging occasion, and such monitoring occurs once per discontinuous reception cycle. The UE calculates the paging occasion based on the International Mobile Subscriber Identity, the number of S-CCPCHs in the cell, the system frame number, the discontinuous reception cycle length, and the frame offset [266]. The discontinuous reception cycle length is calculated as 2^k frames where k is the core network domain-specific discontinuous reception cycle length coefficient specified in the SIB Type 1. For example, if $k = 6$, the UE looks for its paging indicator once every $2^6 = 64$ frames or 640 ms. The frame offset is the timing difference between the PICH and the P-CCPCH and is 7680 chips earlier than the offset of the S-CCPCH specified in SIB Type 5. Within a PICH frame, the UE calculates the specific paging indicator number relevant to itself as a function of the International Mobile Subscriber Identity and Np [266].

9.4.8 Secondary—Common Control Physical Channel

The S-CCPCH carries the transport channels FACH and PCH. Figure 9.22 summarizes the overall operations in creating a S-CCPCH. The S-CCPCH may or may not include a TFCI. When TFCI bits are absent, they become DTX bits. The S-CCPCH bits pass through QPSK modulation and spread by a variable spreading factor to yield 3.84 Mcps. $SF = 128$ and $SF = 64$ are widely used.

Once the UE finds its paging indicator set to 1, it utilizes the information received in SIB Type 5 to find the OVSF code of the S-CCPCH that is potentially carrying its paging message. Both the spreading factor and the code number are specified in SIB Type 5 because the S-CCPCH has no fixed spreading factor. The timing offset between the P-CCPCH and the S-CCPCH and presence or absence of a TFCI are also specified. The PICH frame is transmitted 7680 chips earlier than the S-CCPCH

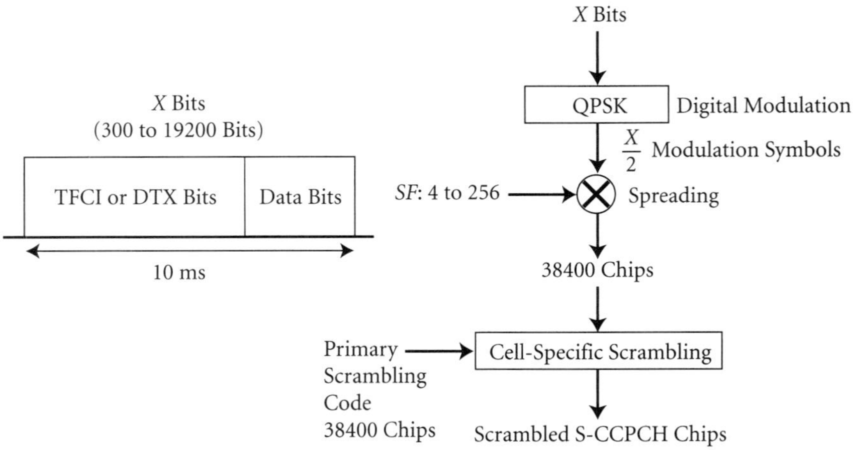

Figure 9.22: S-CCPCH Structure

frame potentially carrying the paging message for the UE. Such timing offset gives the UE adequate time to prepare for an incoming paging message.

9.4.9 Physical Random Access Channel

The PRACH carries the transport channel RACH. The PRACH is a common uplink channel that the UE uses when it has no dedicated uplink channel for transmission. Figure 9.23[293] shows the overall timing structure for the PRACH. The access operation utilizes the concept of access slots in which one access slot consists of two normal slots.

Do You Recall... How long is a slot?[294]

One access slot consists of $2*2560 = 5120$ chips. There are fifteen access slots that span two 10 ms frames. The P-CCPCH and AICH transmission times are identical. Hence, these channels are received at the UE at the same time. The PRACH is divided into fifteen access slots where the PRACH access slots are ahead of the AICH access slots by 7680 or 12800 chips for AICH_Transmission_Timing parameter is set to 0 or 1, respectively. This parameter is specified in SIB Type 5. A total of twelve RACH subchannels are defined, with every access slot having an association with one subchannel during a given 20 ms time interval. For example, during a radio frame when system frame number modulo $8 = 0$, access slot 0 belongs to subchannel 0, access slot 1 belongs to subchannel 1, and so on. See the standard [267] for a table of

[293] Rx represents "receive," and Tx represents "transmit."

[294] A slot is $10/15$ ms long.

mapping between the access slot number and the subchannel number as a function of the system frame number.

Let's discuss the overall access operation summarized in Figure 9.24. The UE's MAC layer decides whether to start the PRACH transmission in the current TTI based on the *persistence* value P_i, which is the transmission probability for the Access Service Class i. (P_i is not the mathematical π representing the numerical value 3.14!) P_i is 1 for Access Service Class 0, and the UE calculates P_i based on the parameters specified in SIB Type 5, Type 6, and/or Type 7 [265]. The RRC selects the Access Service Class for sending an RRC CONNECTION REQUEST message; the MAC layer makes the selection in other cases.

If the PRACH transmission is allowed in the current TTI based on the persistence test, the physical layer carries out the actual PRACH transmission as outlined in the next paragraph. The persistence test is successful if $R \leq P_i$ where R is a random number the UE generates and $0 \leq R < 1$. If the persistence check fails, another check is carried out in the next TTI. The persistence checks are carried out every TTI until transmission is permitted.

Something to Think About... "Persistence is the twin sister of excellence. One is a matter of quality; the other, a matter of time" [268]. It is just be a matter of time before the UE's persistence pays off and the UE can start transmitting on the PRACH!

Once the persistence test succeeds, the UE selects an access slot to begin the PRACH transmission. The PRACH transmission involves one or more preambles and a message part and could span numerous access slots. The UE selects an access slot randomly among the access slots available to the UE to start the PRACH transmission.

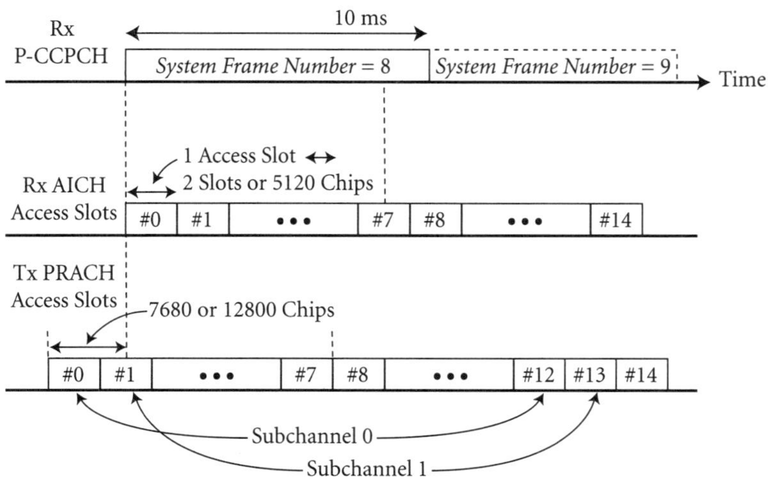

Figure 9.23: Timing Relationships Among the P-CCPCH, AICH, and PRACH

The UE is configured with the Access Class, and the UE obtains the mapping between the Access Class and the Access Service Class from SIB Type 5. Furthermore, the association between an Access Service Class and a set of subchannels is defined in SIB Type 5, which also specifies the entire set of subchannel numbers available in a given cell [265]. By default, all the subchannels are available for a given Access Service Class.

The Access Class ranges from zero to fifteen, and the Access Service Class ranges from zero to seven. $AccessServiceClass = 0$ corresponds to the highest priority. In practice, defining just one Access Service Class is quite common, and all the subchannels are made available to this Access Service Class. However, if different treatment is to be given to different UEs, the standard allows the Access Service Classes to configure different Access Classes for different UEs and specify different numbers of subchannels for different Access Service Classes, thereby prioritizing access for some UEs and gaining some flexibility. If a UE is a member of multiple Access Classes, it selects the Access Service Class corresponding to the highest Access Class. Once the UE learns which subchannels it can use, it determines the access slots associated with these subchannels based on the mapping defined in the standard [267]. Then, it randomly picks one such access slot to initiate the PRACH transmission.

The UE then randomly selects one of the available signatures that are specified in SIB Type 5. The signature influences the exact bit pattern sent as a preamble. The

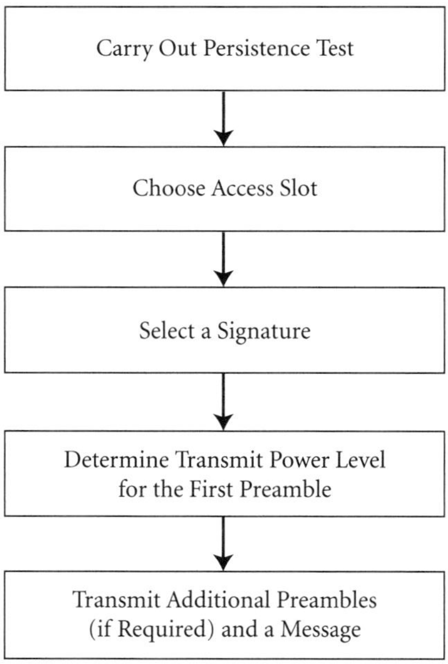

Figure 9.24: Overall Access Operation

UE calculates the transmit power of the first preamble, Preamble_Initial_Power, using Eq. (9.1) [265].

$$\begin{aligned} \textit{Preamble\_Initial\_Power} =& \textit{Primary CPICH Tx Power} - \textit{CPICH\_RSCP} \\ & + \textit{UL Interference} + \textit{Constant Value} \end{aligned} \tag{9.1}$$

where RSCP is the received signal code power. The parameters Primary CPICH Tx Power (where CPICH represents the Common Pilot channel), UL Interference, and Constant Value are specified in SIB Type 5, Type 6, and SIB Type 7, and the UE measures CPICH_RSCP. Eq. (9.1) considers the uplink interference and the path loss between the Node B and the UE and introduces Constant Value to provide additional flexibility. Now, the UE transmits the first preamble using the access slot, signature, and Preamble_Initial_Power.

Figure 9.25 depicts the UE's overall PRACH transmission and the interaction between the UE and the network. The Node B sends a positive acknowledgment on the AICH in an appropriate access slot after it detects the preamble from the UE. However, if the UE's transmit power is inadequate, the Node B may be unable to reliably detect the preamble and the AICH contains no acknowledgment for the UE. In such a case, the UE transmits another preamble with a higher transmit power, a new randomly selected signature, and a new randomly selected access slot. The difference between successive access preambles is the Power Ramp Step.[295] The UE transmits a maximum of Preamble Retrans Max preambles during a preamble ramping cycle such as the one shown in Figure 9.25. If the UE fails in one cycle, it retries in another cycle. The UE tries a maximum of M_{max} preamble ramping cycles. Once the UE receives a positive acknowledgment from the Node B, it sends the message part at the time instant shown in Figure 9.25 at a power level $Pp - m$ higher than the power of the last transmitted preamble. If AICH_Transmission_Timing is set to 0, the message part is sent 15360 chips after the last successfully transmitted preamble.

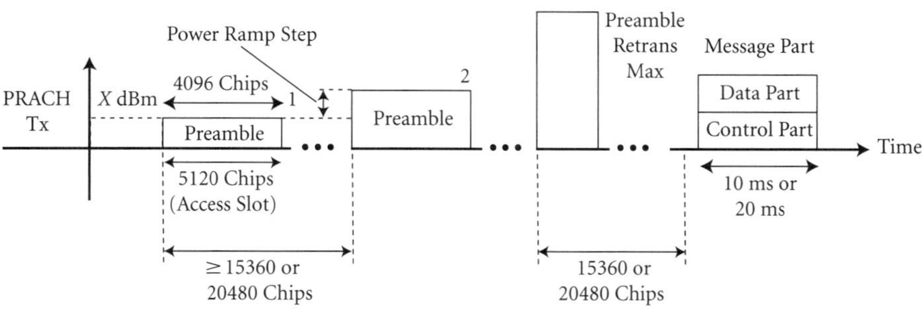

Figure 9.25: Transmission on the PRACH

[295]The parameters Power Ramp Step, Preamble Retrans Max, M_{max}, $Pp - m$, and AICH_Transmission_Timing are defined in SIB Type 5.

However, if AICH_Transmission_Timing is set to 1, the message part is sent 20480 chips after the last successfully transmitted preamble. If the UE receives a negative acknowledgment on the AICH, the UE's MAC layer performs a random back-off and a persistence check to explore the possibility of the PRACH transmission during another TTI [260].

Now that we know the overall PRACH transmission, let's take a closer look at the structure of the preamble and the message part. The preamble consists of a complex-valued code of 4096 chips. This code is constructed using a preamble scrambling code and a signature sequence [264]. There are 8192 preamble scrambling codes, divided into 512 groups with each group containing sixteen codes. The preamble scrambling code used by the UE belongs to the group number that is associated with the downlink primary scrambling code of the cell the UE is accessing. Within the preamble scrambling code group, the exact code number (from 0–15) that the UE can use in a cell is specified in SIB Type 5. The signature sequence is obtained by repeating a sixteen-chip signature 256 times. The UE randomly selects such a sixteen-chip signature during each preamble transmission from a total of sixteen Hadamard codes defined in the standard [264].

The message part is transmitted in a 10 ms or 20 ms TTI and includes the *control part* and the *data part* as shown in Figure 9.26. The data and control parts are transmitted in parallel. The data part consists of $10 * 2^k$ bits, which are spread by a variable spreading factor. The data part carries the transport channel RACH.

The control part includes eight pilot bits and two TFCI bits, which are spread by a fixed spreading factor of 256. The TFCI indirectly specifies the data rate of the PRACH. A PRACH scrambling code scrambles the chips from the data and control parts. A total of 8192 PRACH scrambling codes are defined. As evident from Figure 9.26, the control part effectively experiences BPSK modulation and the data part experiences BPSK modulation. However, since they are transmitted in parallel over the I-channel and Q-channel, the signal constellation of the transmitted signal appears like a QPSK constellation.

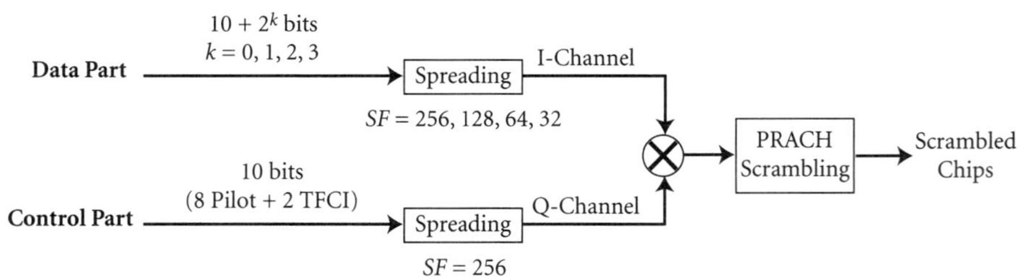

Figure 9.26: PRACH Message Part Creation

9.4.10 Acquisition Indicator Channel

When a Node B successfully decodes an access preamble on the PRACH from a UE, it sends a positive acknowledgment on the AICH. The timing difference between the PRACH transmission instant and the acquisition indicator (AI) receive instant as observed at the UE is 7680 or 12800 chips depending on whether the AICH_Transmission_Timing is 0 or 1 as mentioned earlier [263]. Figure 9.27 illustrates the structure of the AICH.

Recall that there are sixteen sixteen-chip signatures, and the UE selects one of them during an access slot for the preamble. When a UE uses a signature number s, the Node B also uses the same signature number s on the AICH to send an acknowledgment to the UE. A positive acknowledgment is $AIs = +1$ and a negative acknowledgment is $AIs = -1$. A positive acknowledgment gives the UE the green light to transmit an actual message on the PRACH.

A Question For You... How do you think the Node B might use a negative acknowledgment?[296]

The signature sequence used on the AICH consists of thirty-two real-valued symbols (i.e., $\{-1, 1\}$). The AIs for different UEs is multiplied by the corresponding

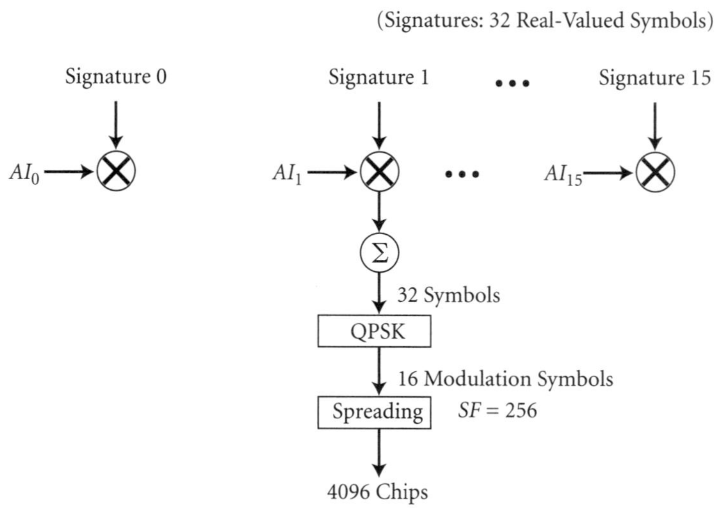

Figure 9.27: Structure of the AICH

[296] If numerous UEs are trying to use the PRACH at the same time, the Node B can separate these UEs in time to reduce interference on the PRACH and thereby improve PRACH detection reliability (especially the message on the PRACH).

sixteen signatures and summed together. The combined thirty-two real-valued symbols pass through QPSK modulation to create sixteen modulation symbols, which are spread by $SF = 256$ to generate 4096 chips. The OVSF code number of the AICH is specified in SIB Type 5. During the access slot on the AICH, the first 4096 chips carry AIs, and the remaining $5012 - 4096 = 1024$ chips are not transmitted.

9.4.11 Uplink Dedicated Physical Data Channel and Dedicated Physical Control Channel

When a UE is assigned dedicated radio resources for the UL, it no longer transmits on the PRACH; it starts using the uplink DPDCH and DPCCH. The DPDCH carries the transport channel DCH that conveys user or bearer traffic, such as speech bits or data packet bits. The DPCCH carries layer 1 control information. Figure 9.28 shows the construction of the DPDCH and the DPCCH.

The DPDCH carries $10 * 2^k$ modulation symbols to be transmitted in the each of the fifteen slots of a 10 ms radio frame where k ranges from 0 to 6. These symbols are generated as part of the physical layer processing illustrated in Figure 9.6 of Section 9.3.2. More specifically, these symbols are available after the transport block(s) has passed through stages such as channel coding, second-stage interleaving, physical channel mapping, and finally BPSK modulation. These symbols are spread using a variable spreading factor ranging from 256 (for $k = 0$) to 4 (for $k = 6$) to yield 2560 chips for the slot. When the UE uses one UL DPDCH, the spreading factor SF is specified to the UE during the call setup, and the UE determines the exact DPDCH OVSF code to be $C_{ch,SF,((SF)/4)}$. A 38400-chip UE-specific scrambling code scrambles the chips resulting from the spreading operation. The UTRAN assigns the number of the uplink scrambling code to the UE during the call setup.

The DPCCH consists of ten bits per slot. These ten bits are distributed among the pilot bits, TFCI bits, feedback indicator bits, and TPC bits. The slot format

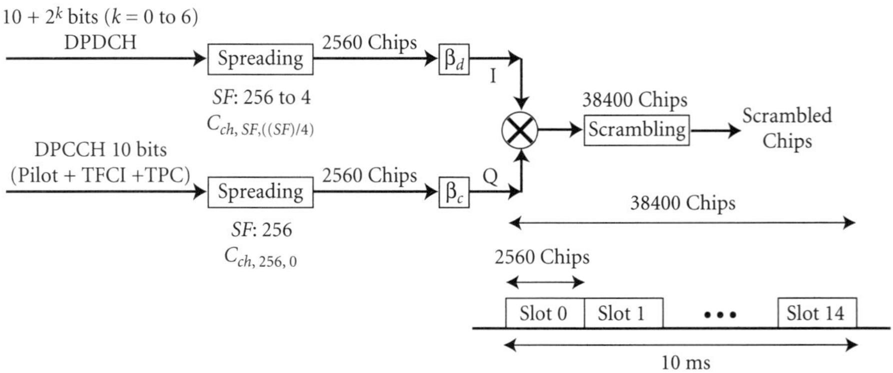

Figure 9.28: Creation of the Uplink DPDCH and DPCCH

specified during the call setup dictates the exact distribution of these ten bits. The feedback indicator bits are usually absent in R99 commercial systems because these bits support transmit diversity, which is typically omitted. The pilot bits represent a dedicated pilot channel to facilitate channel estimation for coherent demodulation at the Node B. The TFCI bits indicate how many bits from which transport channel are being carried on the DPDCH. The pilot bits consist of different patterns in each of the fifteen slots. For example, when eight bits are configured for pilot and two bits are configured for TPC in the case of 12.2 kbps AMR speech, the eight pilots bits are `1 1 1 1 1 1 1 1` during the zeroth slot and `1 0 1 0 1 1 1 0` during the first slot [263]. When a power-up command is to be issued, the TPC command is 1, which is represented by two TPC bits `1 1`. A power-down command corresponds to the TPC command 0, represented by the TPC bits `0 0`. The set of ten DPCCH bits are spread using the predefined OVSF code $C_{ch,256,0}$, leading to 2560 chips in a slot.

As seen from Figure 9.28, the DPCCH and DPDCH each experience BPSK modulation as the code symbols are directly passed to the spreading block instead of a modulation block. However, since the DPCCH and DPDCH are transmitted in parallel on I-channels and Q-channels, the combined signal modulation constellation diagram appears like a QPSK constellation diagram. Furthermore, the DPCCH and the DPDCH are transmitted at the power levels influenced by the parameters β_c and β_d, which are configured during the call setup [265, 267]. Observe that the DPDCH and the DPCCH are code-multiplexed in the UL.

9.4.12 Downlink Dedicated Physical Channel and Dedicated Physical Control Channel

Figure 9.29 illustrates the structure of the downlink Dedicated Physical channel (DPCH) that consists of the DPDCH and the DPCCH. The DPDCH carries the DCH transport channel(s), and the DPCCH carries layer 1 control information such as TPC, TFCI, and pilot bits. The data bits shown in Figure 9.29 are code symbols of the transport channel. The UTRAN controls the distribution of the DPCCH bits among the TPC, TFCI, and pilot bits via configuration of the slot format during the call setup. We just covered the use of these DPCCH bits for the UL. So, this is a fair question.

A Question For You... What are the uses of TPC, TFCI, and pilot bits?[297]

We would like to highlight one point. The DPCCH pilot bits are *dedicated* pilot bits for a *specific user*. The P-CPICH, in contrast, is a pilot channel common to *all* UEs. The UTRAN indicates to the UE during the call setup if the UE can use common and/or dedicated pilot bits for channel estimation and coherent demodulation at the

[297] TPC bits carry power control commands. The TFCI indicates how many bits from which transport channels are being carried, thereby indicating the data rate. Finally, the pilot bits are used for channel estimation and coherent demodulation.

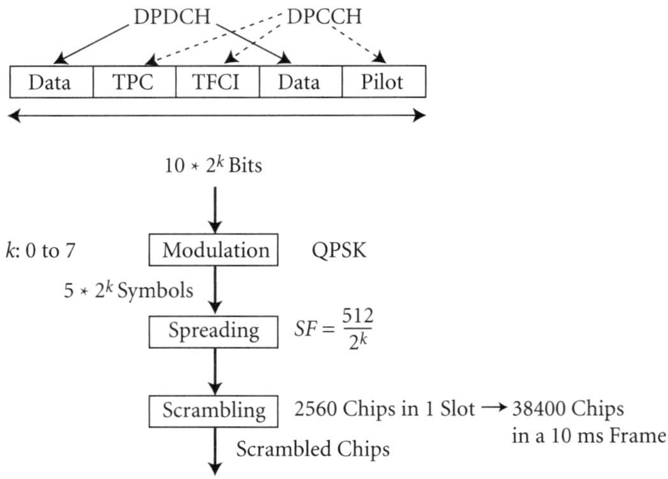

Figure 9.29: Structure of the Downlink Dedicated Physical Channel

UE. The dedicated pilot bits are specific patterns of bits (i.e., not all dedicated pilot bits are 0s) for each slot of a radio frame [263]. A total of $(10*2^k)$ bits or code symbols of the DPCH are available per slot, with k ranging from 0 to 7. These code symbols pass through QPSK modulation[298] to yield $(5*2^k)$ modulation symbols per slot. The modulation symbols are spread by a UE-specific OVSF code that has $SF = 512/2^k$, generating $(5*2^k)*(512/2^k) = 2560$ chips per slot. The UTRAN assigns an OVSF code (i.e., both the spreading factor and the code number) to the UE for the DPCH during the call setup. These chips are scrambled per radio frame using the cell-specific primary scrambling code. Observe that DPDCH and DPCCH are time-multiplexed in the DL. Though the standard allows multiple downlink DPDCHs—equivalent to multiple OVSF codes—to be assigned to the same UE, just one DPDCH is assigned to the UE in typical R99 commercial deployments.

9.4.13 Combining of the Downlink Channels in a Cell

Figure 9.30 illustrates how a Node B combines downlink channels in a cell. The only modulation scheme used in the downlink is QPSK. The P-CPICH carries pilot bits and uses the OVSF code $C_{ch,256,0}$ to achieve the chip rate of 3.84 Mcps. The P-CCPCH carries the Broadcast channel code symbols and uses the OVSF code $C_{ch,256,1}$ for spreading. Such fixed OVSF code facilitates the detection of the P-CCPCH when the UEs are in the process of acquiring the information about the UMTS network. Recall that the P-CCPCH is not transmitted during the first 256 chips of each slot.

[298]The modulation scheme to be used for a given channel is fixed in UMTS R99; it is not changed dynamically.

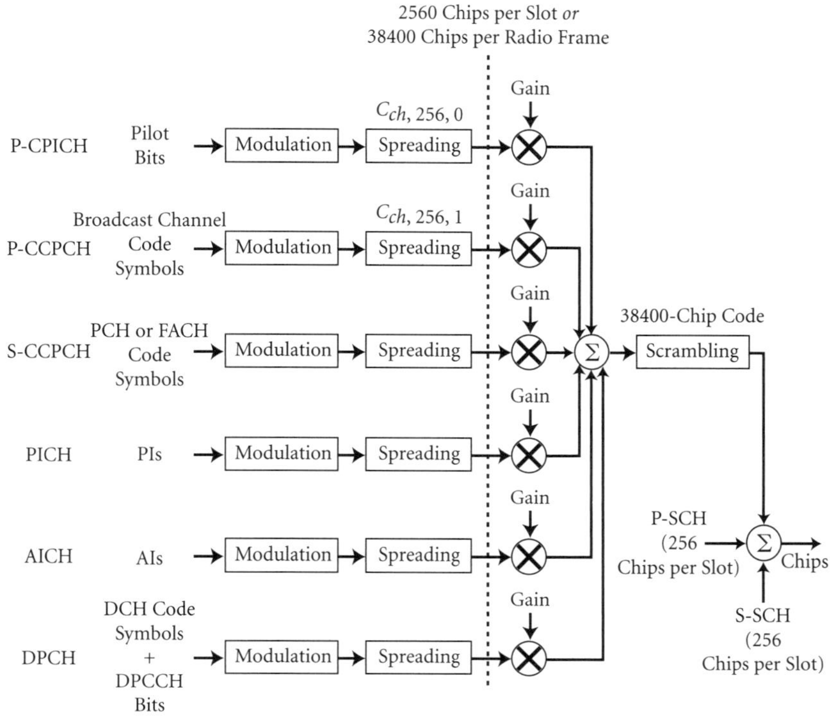

Figure 9.30: Combining of Downlink Channels in a Cell

The S-CCPCH carries the PCH and FACH code symbols and uses a variable spreading factor. The PICH and AICH carry the page indicators and acquisition indicators, respectively, and use a fixed $SF = 256$. The DPCH carries the DCH transport channel(s) on the DPDCH and the layer 1 control information on the DPCCH and uses a variable spreading factor. Though Figure 9.30 shows just one DPCH for simplicity, multiple DPCHs are required to support multiple UEs, with one DPCH per UE.

The chips from the P-CPICH, P-CCPCH, S-CCPCH, PICH, AICH, and DPCH are weighted by individual channel-specific gains so that these channels can be transmitted at different power levels. A cell-specific primary scrambling code now scrambles the power-weighted chips, 38400 chips per frame. The power-weighted chips from P-SCH and S-SCH, which are transmitted during the first 256 chips of every slot, are summed up with the combined scrambled chips from other physical channels. The combined chips now represent the overall cumulative baseband signal that occupies 3.84 MHz of bandwidth. With the guard bands, the overall required baseband spectrum becomes 5 MHz.

The downlink channels follow certain time constraints as defined in Figure 9.31. Note that these are the transmission timings as observed at the Node B. The P-CCPCH

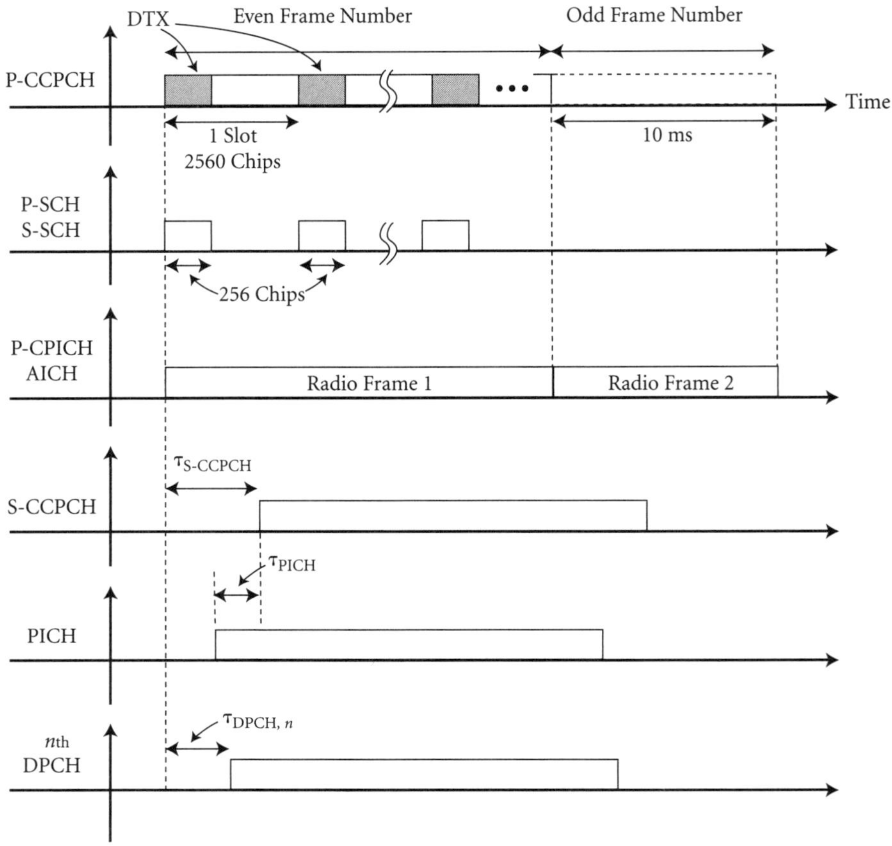

Figure 9.31: Transmission Timing Relationships among Downlink Channels in a Cell

is used as the timing reference. The transmission timings of all other physical channels are derived relative to the P-CCPCH. Every P-CCPCH radio frame is identified by a system frame number, which the UE calculates based on the SFNprime parameter that is broadcast in the SYSTEM INFORMATION message on the P-CCPCH [265]. See Chapter 13 on High-Speed Uplink Packet Access for details regarding the system frame number.

The P-SCH and S-SCH transmission timings coincide with the P-CCPCH slot timings. However, the P-CCPCH does not transmit the first 256 chips in every slot, and the P-SCH and S-SCH transmit only during the first 256 chips of a slot.

The P-CPICH transmission timing is identical to the P-CCPCH transmission timing. The AICH transmission and P-CCPCH transmission also occur at the same time; however, one AICH frame consists of two radio frames because the fifteen access slots of the AICH span over two radio frames.

The S-CCPCH is transmitted $\tau_{\text{S-CCPCH}}$ chips later than the P-CCPCH. The PICH precedes the S-CCPCH by $\tau_{\text{PICH}} = 7680$ chips. A DPCH for a UE is transmitted τ_{DPCH} chips later than the P-CCPCH. τ_{DPCH} can be different for different UEs. $\tau_{\text{S-CCPCH}}$ and τ_{DPCH} assume one of the following values: 0, $1 * 256 = 256$, $2 * 256 = 512, \ldots$, $199 * 256 = 50944$ chips.

9.4.14 Transmission of Uplink Channels by User Equipment

A UE transmits on a common channel, the PRACH, or on dedicated channels, the DPDCH and DPCCH. The UE transmits on the uplink DPCCH 1024 chips after receiving the beginning of the downlink DPCH frame from the received signal's first arriving multipath component [263]. The DPCCH and the DPDCH are transmitted at the same time.

The dedicated uplink channels from a UE undergo UE-specific scrambling using an uplink scrambling code. The twenty-four–bit representation of the uplink scrambling code enables a total of 2^{24} codes. Two long uplink scrambling code sequences are generated via position-wise modulo 2 sum of 38400 chip segments of two binary m-sequences created using two generator polynomials of degree 25. The twenty-four–bits representing the number of the uplink scrambling code become initial conditions (i.e., the states of shift registers) for one of the generator polynomials.

9.5 CELL SEARCH AND SYNCHRONIZATION

The UE looks for a UMTS network upon power-up and synchronizes with a strong cell. We take a closer look at such a cell search process here. Before diving into this process, we will summarize the overall operation in Section 9.5.1. The details will be explained in Section 9.5.2. Section 9.5.3 summarizes the communications between the UE and the core network after synchronization is completed.

9.5.1 Overview of Power-Up Synchronization and Cell Search

After the UE's power is turned on, the UE looks for a UMTS cell, identifies the radio frame boundaries, and learns the primary scrambling code of the cell. Many channels, including the P-CCPCH that carries the system information, are scrambled using the primary scrambling code. Hence, once the UE learns the primary scrambling code of the cell, it can read the P-CCPCH and several other radio channels. Figure 9.32 summarizes the basic steps of the cell identification and the synchronization.

The UE looks for a UMTS signal at a carrier frequency in the frequency band programmed in the UMTS Subscriber Identity Module. The UE first achieves slot synchronization and then achieves frame synchronization.

A Question For You... Do you remember which physical channels the UE uses to achieve slot and frame synchronization?[299]

[299]The UE utilizes the P-SCH to determine the start and end of the slot and the S-SCH to find the boundaries of the 10 ms radio frame.

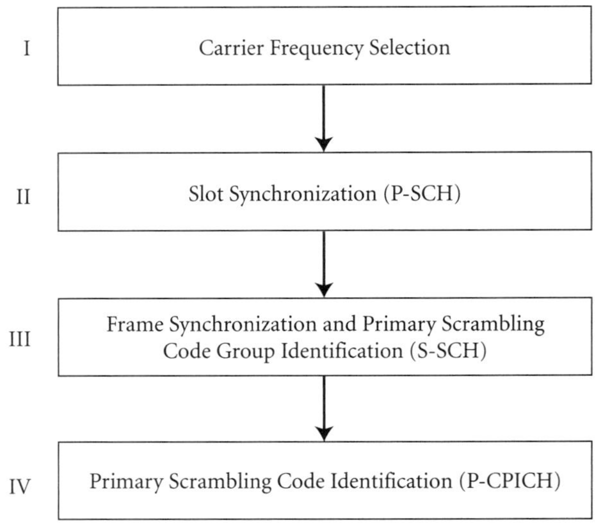

Figure 9.32: Cell Search and Synchronization

Once the UE achieves frame synchronization, it also knows the identity of the code group to which the primary scrambling code belongs. Recall that 512 primary scrambling codes are distributed among sixty-four code groups with each group having eight primary scrambling codes.

Next, the UE attempts to descramble the P-CPICH using eight possible primary scrambling code options for the code group. One of the eight options enables the UE to correctly retrieve bits on the P-CPICH, which is the primary scrambling code being used in that cell. Since the UE has determined the primary scrambling code, it can descramble broadcast information from the P-CCPCH and obtain useful information about the system and the cell, such as identities of the neighboring cells. Details of these steps are discussed below. *To enable the UE to observe the contents of the P-SCH and the S-SCH, neither of these channels is scrambled.*

9.5.2 Power-Up Synchronization and Cell Search: A Closer Look

A UE's UMTS Subscriber Identity Module card contains a list of prioritized frequency bands. The UE selects a carrier frequency within a frequency band and looks for a UMTS signal. For example, the UE may look at a 1950 MHz center carrier frequency corresponding to a certain UMTS ARFCN [269].

The UE searches for the primary synchronization code, C_{psc}, in the 5 MHz bandwidth centered at the selected carrier frequency selected. The primary synchronization code is transmitted on the P-SCH in the beginning of every slot. The UE stores C_{psc} and tries to match the chips received from the strongest cell with $a * C_{psc}$ where $a = -1$ in the absence of space time transmit diversity. Since each UMTS cell utilizes the same C_{psc}, the UE can easily achieve slot synchronization with the cell that provides the best downlink signal.

Once the UE achieves slot synchronization, it uses the S-SCH to try to find frame boundaries. The S-SCH transmits $a * C_{ssc,i}$ during the first 256 chips of a slot. Recall that sixty-four secondary synchronization sequences are generated using the sixteen SSCs with each sequence containing fifteen SSCs. Furthermore, one secondary synchronization sequence consists of a set of fifteen SSCs with one SSC per slot. Every frame repeats the same secondary synchronization sequence. The UE looks for the predefined sixty-four sequences and finds a match with one of them, thereby achieving frame synchronization. One secondary synchronization sequence is associated with one primary scrambling code group. Hence, upon frame synchronization, the UE learns the identity of the primary scrambling code group to which the current cell belongs based on the number of the secondary synchronization sequence.

For example, assume that a UE is in a cell that uses the Primary Scrambling Code 0. The primary scrambling code 0 belongs to the primary scrambling code group 0. Hence, the secondary synchronization channel utilizes the 0th secondary synchronization sequence. Since the UE is attempting to correlate the received secondary synchronization sequence with all the sixty-four sequences, it finds a match with the 0th sequence. Now, the UE knows it is in a cell that is using the primary scrambling code and that this code belongs to the primary scrambling code group 0. *In summary, the UE utilizes the S-SCH to achieve frame synchronization and to implicitly identify the primary scrambling code group.*

The UE now attempts to find the primary scrambling code of the cell where it is currently located using the P-CPICH. The same primary scrambling code is repeatedly used in every 10 ms radio frame in a given cell. Now let's look at how a primary scrambling code is created and how it is associated with a primary scrambling code group [264].

A primary scrambling code sequence is obtained by combining two real sequences into a complex sequence so that the both I-channel and Q-channels can be scrambled. 38400 chips for each of these two real sequences are obtained by performing a modulo 2 sum of two binary m-sequences that are generated using two generator polynomials of degree 18. The specific set of 38400 chips selected as a primary scrambling code depends upon the number of the Gold code sequence n. The generator polynomials of degree 18 can be used to create $2^{18} - 1 = 262143$ Gold code sequences ranging from sequence number $n = 0$ to $n = 262142$; however, only some Gold code sequences are used to construct the primary scrambling codes.

Table 9.3 summarizes the association between the primary scrambling code, Gold code sequence, and scrambling code group. A primary scrambling code i is associated with the Gold code sequence number $n = 16 * i$ where the primary scrambling code i ranges from 0 to 511. For example, the primary scrambling code $i = 0$ is created from the Gold code sequence $n = 0$, and the primary scrambling code $i = 511$

Table 9.3: Relationships among Primary Scrambling Code, Gold Code Sequence, and Scrambling Code Group

Primary Scrambling Code (i)	Gold Code Sequence Number (n)	Scrambling Code Group (j)
0	0	0
1	16	0
2	32	0
$\vdots$	$\vdots$	$\vdots$
7	112	0
8	128	1
9	144	1
$\vdots$	$\vdots$	$\vdots$
15	$\vdots$	1
$\vdots$	$\vdots$	$\vdots$
511	8176	63

is created from the Gold code sequence $n = 8176$. Furthermore, a given primary scrambling code i belongs to a specific scrambling code group j. For example, the scrambling code group j consists of the primary scrambling codes associated with the Gold code sequence numbers $n = (16 * 8 * j + 16 * k)$ where j ranges from 0 to 63 and k ranges from 0 to 7. For example, the scrambling code group 0 consists of the eight primary scrambling codes $0, 1, 2, \ldots, 7$ associated with the Gold code sequence numbers $n = 0, 16, 32, \ldots, 112$. Similarly, the scrambling code group 1 consists of the primary scrambling codes $8, 9, \ldots, 15$.

Consider a UE located in Cell X that uses the scrambling code group 0. Cell X spreads the pilot bits, which are all 0s, using the OVSF code $C_{ch,256,0}$ to generate pilot chips, which in turn are scrambled using the primary scrambling code 0. The UE receives scrambled pilot chips and attempts to descramble the received chips using all eight possible primary scrambling codes $i = 0, 1, 2, \ldots, 7$ associated with the scrambling code group $j = 0$ as well as to despread the signal using $C_{256,0}$. The UE would correctly receive all 0s as pilot bits when it descrambles the received chips with primary scrambling code 0.

9.5.3 Beyond Synchronization: User Equipment and Core Network Communications

Figure 9.33 illustrates typical operations the UE performs after achieving DL synchronization [270]. The UE reads the system information being broadcast on the

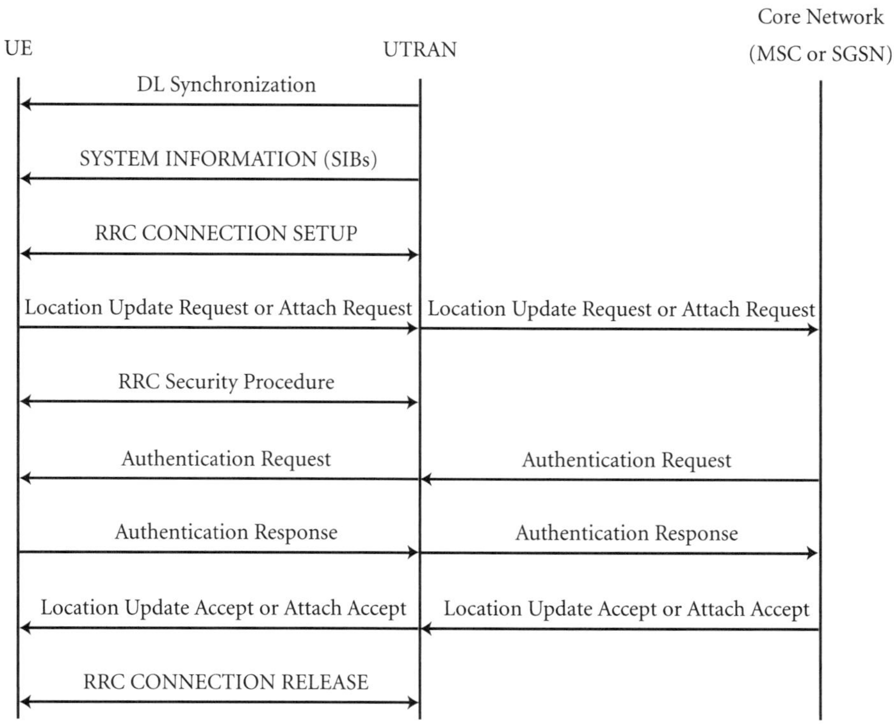

Figure 9.33: The UE's System Information Retrieval and Power-Up Registration

Broadcast Control channel or Broadcast channel or P-CCPCH. Such information is sent in a SYSTEM INFORMATION message in the form of SIBs [265].

The UE and the UTRAN establish the RRC signaling connection. The UE sends an RRC CONNECTION REQUEST message specifying the establishment cause to be "registration." The UTRAN responds with an RRC CONNECTION SETUP message and typically allocates dedicated resources to the UE for the reliable exchange of future signaling messages. The UE completes the process by sending an RRC CONNECTION SETUP COMPLETE message.

The UE sends a Location Update Request or an Attach Request to the MSC or VLR or to the SGSN. The UTRAN secures the air interface by sending a SECURITY MODE COMMAND message, and the UE replies with a SECURITY MODE COMPLETE message. Next, authentication may be carried out. Though authentication may be omitted for some call setups (to reduce call setup delay), power-up registration usually involves authentication. The MSC or SGSN sends an Authentication Request message to the UE, and the UE provides an Authentication Response message to the core network. Assuming that the authentication is successful, the core network sends a Location Update Accept or Attach Accept message to the UE.

Since the registration is now over, the RRC connection is no longer needed. The RRC connection is released, and the UE enters the Idle Mode. In the Idle Mode, the UE relaxes for a while and periodically wakes up to look for a potential paging message.

9.6 VOICE CALL SETUP

We will discuss here how a voice call is set up. Let's highlight major prerequisites for a call. The UE has already completed registration, and the UE is in the idle state. Hence, the core network knows about the UE, but the UTRAN has no link with the UE. Section 9.6.1 gives an overview of the overall call setup, and Section 9.6.2 takes a closer look at this process.

9.6.1 Overview of Voice Call Setup and Teardown

UMTS defines quite a comprehensive call setup procedure. Several seconds (e.g., often exceeding five seconds) pass between the dialing of digits and the actual talking in a heavily-loaded network. Hence, have patience while making a call; as the saying goes, patience is a virtue!

Figure 9.34 summarizes the typical steps involved during the voice call setup and teardown. In Step 1, the UE and the UTRAN establish the RRC signaling connection with each other. The UTRAN typically assigns a dedicated signaling radio connection to the UE so the signaling messages can be reliably transferred.

The UE makes a service request for a voice call in Step 2. The UTRAN directly transfers the request to the core network, which is a circuit-switched domain in this case. The CS domain and the UE now authenticate each other since UMTS supports mutual authentication. As mentioned earlier, to reduce signaling delay, authentication is omitted for some calls from a UE in an implementation-specific manner.

The UTRAN in Step 3 turns on encryption and integrity protection so the UE is configured to secure the air interface. The UE provides the phone number of the called party to the CS domain in Step 4 using the secured air interface. The UTRAN in Step 5 establishes radio bearers that can carry actual speech traffic once the call is fully set up. In Step 6, the CS domain helps ring the phone of the called party, and, when the called party answers the call, the UE receives the acknowledgment from the CS domain. As the dedicated radio bearers and various links in the network have been set up, the UE exchanges speech blocks with the called party in Step 7. Typical R99 commercial deployments use 12.2 kbps full-rate AMR speech for voice communications. Chapter 2 tells you all about AMR speech.

When the call is over, the UE and the core network exchange disconnect messages in Step 8. The radio resources are released in Step 9. The dedicated RRC signaling connection between the UE and the UTRAN also disappears, and the UTRAN no longer recognizes the UE. The UE enters the Idle Mode. If a new voice call is to be made, a new RRC connection would be required.

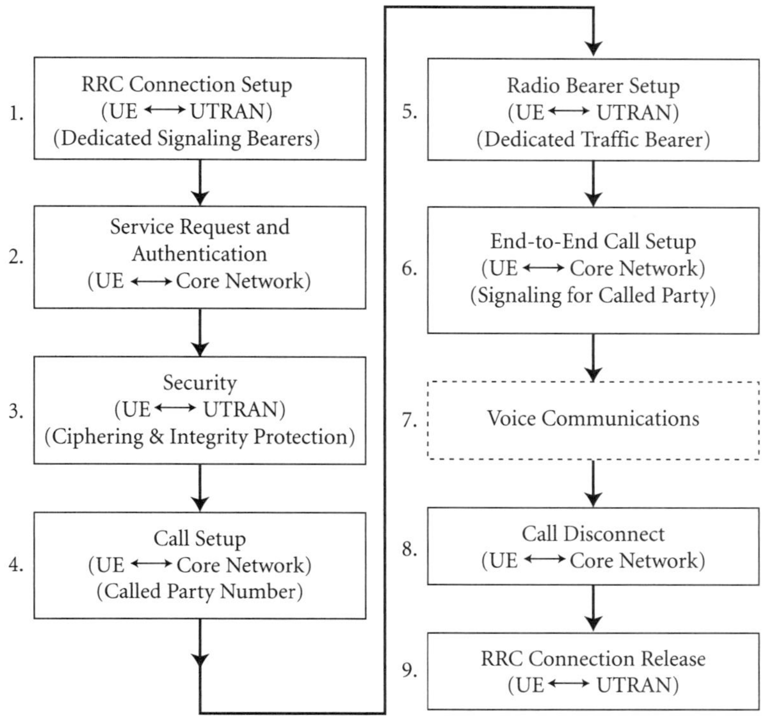

Figure 9.34: High-Level View of Voice Call Setup and Teardown

9.6.2 Detailed Voice Call Setup

Sections 9.6.2.1–9.6.2.7 present a simplified view of the detailed call setup and teardown steps and are based on typical commercial deployments; the UMTS standard has quite comprehensive and flexible message structures, so other methods of voice call setup and teardown are available. See the standard [265] for details of the RRC messages mentioned in Sections 9.6.2.1–9.6.2.7.

9.6.2.1 Radio Resource Control Connection Setup When the UE's power is on but the UE has been uninvolved in voice and data communications in the recent past, the UE operates in the Idle Mode. The UTRAN knows nothing of the existence of the UE; the CS domain knows the UE's location. Hence, the first step in a voice call setup is the establishment of an RRC signaling connection whereby the UE informs the UTRAN about its presence and its intent to exchange some messages. Figure 9.35 summarizes the messages involved in the RRC connection setup.

The UE sends an RRC CONNECTION REQUEST message to the UTRAN using the UL Common Control channel or RACH or PRACH: the Common Control channel is the logical channel, the RACH is the transport channel, and the PRACH

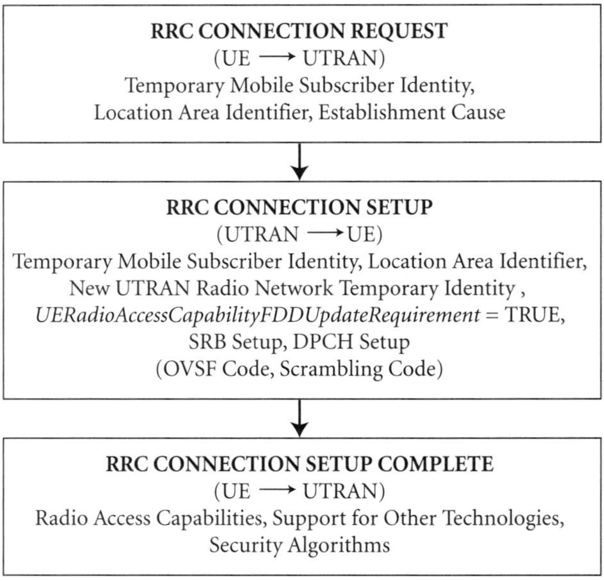

Figure 9.35: RRC Connection Setup

is the physical channel.[300] Since the PRACH is a common channel, the UE provides its Temporary Mobile Subscriber Identity to identify itself and the Location Area Identifier where it obtained its Temporary Mobile Subscriber Identity. The Location Area Identifier consists of a PLMN identification that the Mobile Country Code and Mobile Network Code specify as well as the Location Area Code. The UE also informs the UTRAN the cause for establishing the RRC connection, which can be "originating a conversational call." The UTRAN also receives the pilot E_c/N_0s of the strongest cells. Such reported measurement enables the UTRAN to decide the Active Set for the UE.

The UTRAN responds with an RRC CONNECTION SETUP message to the UE on the DL Common Control channel or FACH or S-CCPCH. Since the P-CCPCH is a common DL channel, the UE for which this message is destined is identified by the Temporary Mobile Subscriber Identity and Location Area Identifier combination that the UE reported earlier. The UE is now assigned a new identity called the UTRAN Radio Network Temporary Identity by the UTRAN. The UTRAN Radio Network Temporary Identity consists of the identity of the serving RNC and the UE identity within the serving RNC. The UTRAN also establishes signaling radio bearers (SRBs) that help convey messages between the UE and the UTRAN and between

[300] Such logical/transport/physical channel convention is used in the remainder of the chapter where appropriate.

the UE and the core network.[301] Four SRBs are typically established: SRB1 with an Unacknowledged Mode RLC for RRC messages; SRB2 with an Acknowledged Mode RLC for RRC messages; SRB3 with an Acknowledged Mode RLC for high-priority Non-Access Stratum (NAS[302]) messages; and SRB4 with an Acknowledged Mode RLC for low-priority NAS messages. Four dedicated logical channels represent four SRBs, and they are usually mapped onto one dedicated transport channel, the DCH.

The DL and UL DPCH characteristics specify the dedicated radio connection between the UE and the UTRAN. The DL DPCH involves an OVSF code in a cell with a specific DL primary scrambling code. On the other hand, the UL DPCH is characterized by an UL scrambling code and an UL spreading factor for the UL OVSF code. In the DL, the UE is assigned a specific OVSF code (e.g., $C_{ch,128,5}$). However, in the UL, the UTRAN provides just the spreading factor of the OVSF code since $C_{ch,SF,SF/4}$ uniquely defines the exact OVSF code as discussed in Section 9.4. An example configuration is a 10 ms TTI for the transport channel carrying SRBs, $SF = 64$ for the uplink DPCH, and $C_{ch,128,5}$ for the downlink DPCH. The UTRAN usually asks the UE to provide its capabilities by setting the *UERadioAccessCapabilityFDDUpdateRequirement* variable to TRUE.

The UE acknowledges the reception of the RRC CONNECTION SETUP message by sending an RRC CONNECTION SETUP COMPLETE message on the UL Dedicated Control channel, DCH, or DPDCH. The UE informs the UTRAN of its capabilities, such as support for other technologies (such as GSM), support for specific ciphering and integrity protection algorithms, support for frequency bands (e.g., 1900 MHz), its power class, and its UL and DL compressed mode[303] capabilities. Four UE power classes are defined. The maximum UE transmit power is 33 dBm $=$ 2 W, 27 dBm $=$ 500 mW, 24 dBm or 250 mW, and 21 dBm or 125 mW, for the power classes 1, 2, 3, and 4, respectively [269]. Class 4 UEs are widely used.

9.6.2.2 Service Request and Authentication

The UE initiates a call management service request to the core network using the dedicated RRC connection as shown in Figure 9.36. Specifically, the UE sends an (RRC) INITIAL DIRECT TRANSFER message containing a NAS Service Request message. This is an example of a *direct transfer* message, meaning the UTRAN does not process or act upon the message but directly transfers it to the core network. The UE mentions the service type, such as "mobile originating call establishment," and provides its Temporary Mobile Subscriber Identity in the message.

The core network may authenticate the UE before agreeing to provide the service. Within the core network, the MSC or VLR communicates with the HLR or Authentication Center to obtain a 5-tuple: a random number as well as four random-number–

[301] The UE and the core network exchange messages using Non-Access Stratum signaling in which the radio access or Access Stratum does not process the contents of such Non-Access Stratum signaling messages but simply passes them along.

[302] NAS is pronounced "nahs," not as the individual letters.

[303] The compressed mode allows a UE to temporarily leave the UMTS air interface and search for other technologies and other UMTS carrier frequencies. The compressed mode facilitates handover and is discussed in Section 9.8.

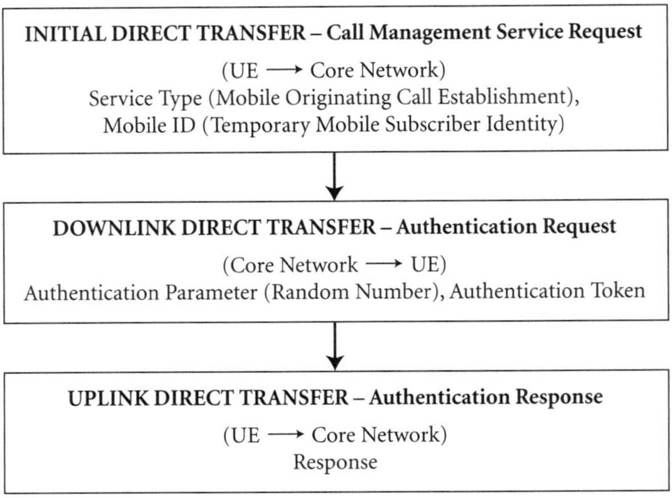

Figure 9.36: Service Request and Authentication

dependent and secret-key–dependent parameters—ciphering key (CK), integrity key (IK), expected response, and authentication token. These random-number–dependent and secret-key–dependent parameters are determined by several algorithms defined in the standard. The secret key is known only to the HLR or Authentication Center and the UE.

The MSC now sends a NAS Authentication Request message to the UE, which the RNC encapsulates in an (RRC) DOWNLINK DIRECT TRANSFER message. This message includes the random number and authentication token. The UE algorithms also generate random-number–dependent and secret-key–dependent parameters. If the authentication token generated at the UE matches the received authentication token, the UE considers the network to be a valid one. The UE now sends an (RRC) UPLINK DIRECT TRANSFER message containing a NAS Authentication Response message. This message includes a response to the authentication challenge from the core network. The MSC or VLR now compares the response received from the UE with the expected response, and, if they match, the UE is said to be authenticated. Note that the UMTS standard implements *mutual authentication* in which the UE authenticates the network and the network authenticates the UE.

9.6.2.3 Air-Interface Security and Call Setup Though using CDMA makes the UMTS air interface secure, integrity protection and ciphering[304] provide additional security. The UTRAN sends a SECURITY MODE COMMAND message to the UE to provide information on ciphering and integrity protection as shown in

[304]Integrity protection enables the receiver to validate authenticity of a message sender, and ciphering means encryption.

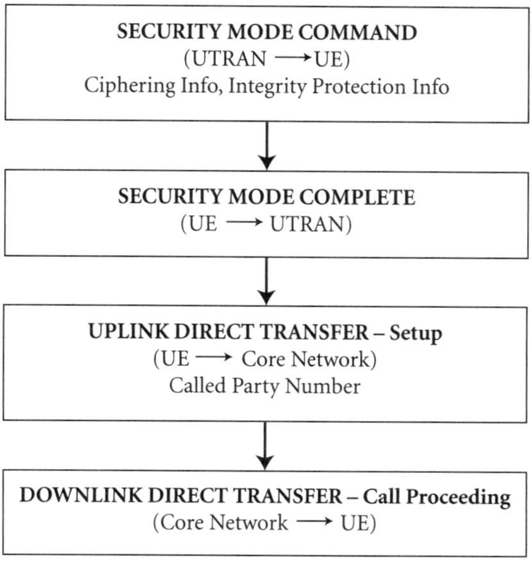

Figure 9.37: Over-the-Air Security and Call Setup

Figure 9.37. Examples of ciphering and integrity protection parameters include supported ciphering algorithms, such as UEA0 (meaning no encryption) and UEA1; a supported integrity protection algorithm, such as UIA1; and commands for the initiation of integrity protection and/or ciphering with specific algorithms, such as UIA1 and UEA1. The UE responds with a SECURITY MODE COMPLETE message as an acknowledgment. Next, the UE sends an UPLINK DIRECT TRANSFER message containing a NAS Setup message to the core network and provides the phone number of the called party over the secure radio interface.

Do You Recall... Do you think that a CDMA expert can sniff the air interface and obtain the called party number?[305]

Of course, government agencies, such as the Federal Bureau of Investigation and Central Intelligence Agency, could lawfully intercept the number using the core network entities, such as the MSC or VLR, SGSN, and GGSN. The core network responds by sending a NAS Call Proceeding message, which the RNC sends to the UE in an (RRC) DOWNLINK DIRECT TRANSFER message.

[305] Ciphering, started in the previous step, secures the air interface. Even a CDMA expert with advanced tools cannot steal the called party number!

9.6.2.4 Radio Bearer Setup The core network now sends an Radio Access Network Application Part Radio Access Bearer (RAB) Assignment Request message to the UTRAN to request establishment of an RAB. The RAB includes a radio bearer and the Iu bearer, with the radio bearer comprising the UE–Node-B air-interface channel and an Iub bearer. In summary, an RAB is an end-to-end link between the UE and the core network that ensures a specific QoS (e.g., 12.2 kbps data rate for the full-rate AMR speech).

The RNC executes an implementation-specific call admission control algorithm to evaluate the availability of downlink resources, such as the OVSF codes, and the downlink power and the uplink resources, such as the uplink rise-over-thermal noise, to see if the RAB with a specific QoS request can be met. If adequate resources are available to support the RAB, the RNC instructs the Node B to reserve resources for the UE via a Node B Application Part RADIO LINK SETUP message on the Iub interface [271] and the RNC allocates resources to the UE via an (RRC) RADIO BEARER SETUP message over the air interface as shown in Figure 9.38.

In AMR speech, three radio bearers are established, each representing one class of AMR bits.[306] Furthermore, information about the UL and DL transport and physical channels is given similar to the information provided during the RRC CONNECTION SETUP message. An example configuration is a dedicated transport channel to carry signaling messages via four SRBs over a 40 ms TTI and three dedicated logical and transport channels to carry three classes of AMR bits over a 20 ms TTI. The transport channel corresponding to the Class A bits supports transport block sizes of 0, 39, and 81 bits and its target block error rate for the downlink is specified (e.g., 2%). Convolutional coding with the rate of 1/3 and a twelve-bit CRC size are used for this transport channel. The transport channel associated with Class B bits supports transport block sizes of 0 and 103 bits with the rate of 1/3 convolution coding but with no CRC bits. The transport channel related to Class C bits allows the transport block sizes of 0 and 60 bits with the rate of 1/2 convolutional coding but with no CRC bits.

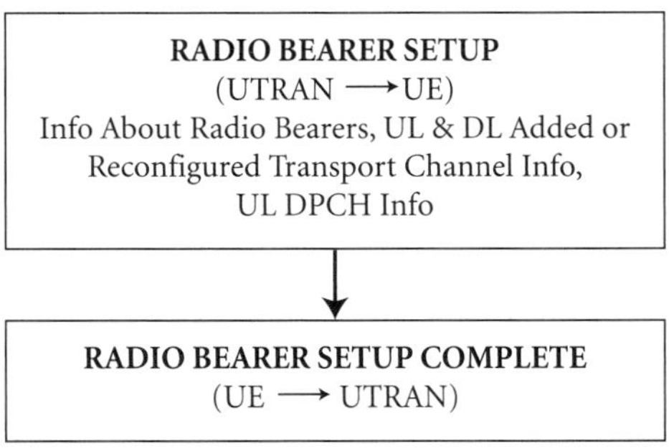

Figure 9.38: Radio Bearer Setup

[306]The AMR speech consists of Class A, B, and C bits. See Chapter 2 for details.

A Question For You. . . Can you determine the combination of Class A, B, and C bits needed to achieve a 12.2 kbps data rate?[307]

The DPCH information, such as a long scrambling code and spreading factor (e.g., 64), for the uplink can also be specified. The UE sends the (RRC) RADIO BEARER SETUP COMPLETE message in response to the (RRC) RADIO BEARER SETUP message.

9.6.2.5 End-to-End Call Setup The UE and the core network exchange a series of NAS messages to complete the remainder of the call setup as shown in Figure 9.39. The core network sends a NAS Alert message to the UE, which the RNC actually places in an (RRC) DOWNLINK DIRECT TRANSFER message. This message indicates that the called party's phone is ringing. When the called party picks up the phone to answer the call, the core network sends a NAS Connect message. The UE now responds with an (RRC) UPLINK DIRECT TRANSFER message containing a NAS Connect Acknowledge message to complete the end-to-end setup of the call.

9.6.2.6 Voice Communications Since the call is set up, the two devices exchange speech. The radio interface utilizes the dedicated radio channels that the resources specify, such as the OVSF codes and the scrambling codes. The Node B and the RNC utilize the Iub bearer and the RNC and the MSC utilize the Iu bearer to carry dedicated speech traffic. The speech traffic is carried as AMR speech within the UMTS radio and core network. The transcoder rate adaptation unit that is connected

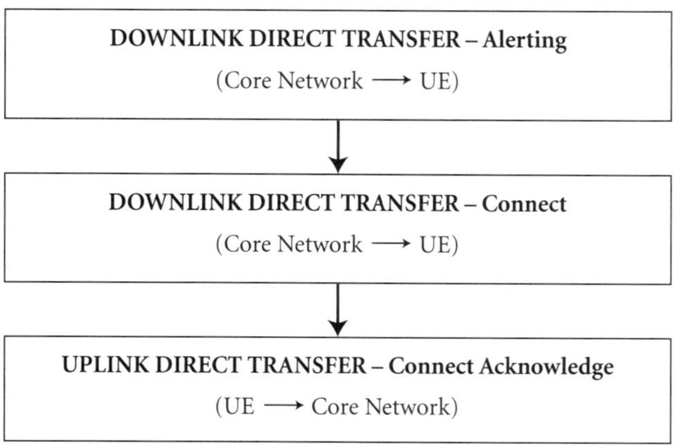

Figure 9.39: End-to-End Setup

[307] The sum of 81 Class A, 103 Class B, and 60 Class C bits gives 244 bits for a 20 ms TTI, leading to $244/20 = 12.2$ kbps! Now you know why we call it a 12.2 kbps RAB!

to the MSC translates between the AMR speech and 64 kbps PCM speech for a landline phone. The transcoder rate adaptation unit near the MSC (usually part of the MSC) allows a 12.2 kbps bandwidth (rather than a 64 kbps bandwidth) to be used on the Iu interface. The traditional landline network, such as the public-switched telephone network, operates with 64 kbps PCM speech.

9.6.2.7 Call Disconnect and RRC Connection Release The call release involves the message exchange between the UE and the core network as shown in Figure 9.40. When the UE ends the call, it sends an (RRC) UPLINK DIRECT TRANSFER message containing a NAS Disconnect message. The network responds with a NAS Release message inside an (RRC) DOWNLINK DIRECT TRANSFER message. Finally, the UE sends an (RRC) UPLINK DIRECT TRANSFER message with a NAS Release Complete message. Now, the dedicated core network resources are no longer available to carry the speech traffic. The UE, however, still has the radio connection with the UTRAN until this time.

As the call is being disconnected, no radio resources are needed. Hence, as shown in Figure 9.40, the UTRAN sends an (RRC) CONNECTION RELEASE message to the UE with the release cause set to “normal event.” The UE responds with an (RRC) CONNECTION RELEASE COMPLETE message.

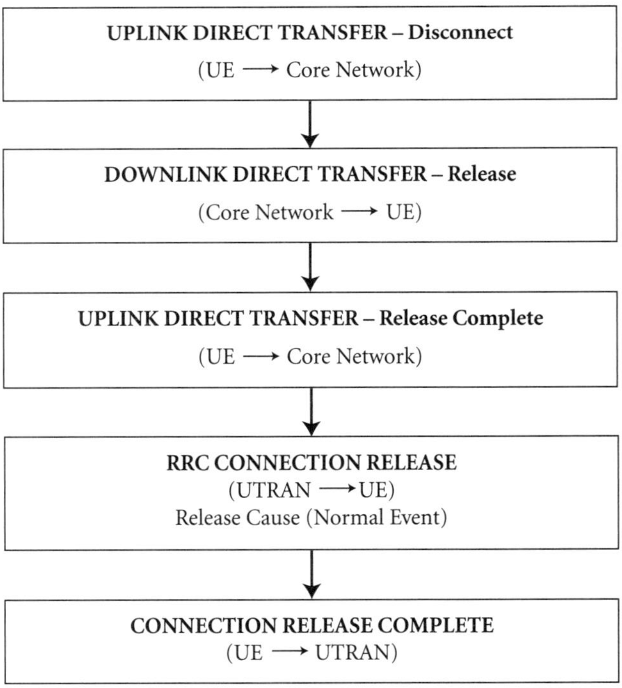

Figure 9.40: Call and RRC Connection Release

9.7 DATA CALL SETUP

We have some good news for you! Most of the steps required for a data session setup are similar to those for a voice call setup! Section 9.7.1 provides a high-level view of the data call setup, and Section 9.7.2 digs deeper into relevant details.

9.7.1 Overview of the Data Session Setup

Figure 9.41 provides a high-level view of how data session setup and teardown occur in typical commercial deployments. One major difference between voice and data call setup is the Packet Data Protocol (PDP) context activation, which provides a dynamic IP address to the UE for packet data transfer.

The first step is the establishment of the RRC signaling connection between the UE and the UTRAN. The UTRAN assigns a *low-rate* dedicated signaling radio connection to the UE so that the signaling messages can be reliably transferred. The UE next makes a service request for a data session. The UTRAN transfers the request to the packet-switched domain, and authentication may occur. Time for a review question that has a binary logic (no fuzzy logic!) answer.

Do You Recall... In UMTS, do we have one-way authentication in which the network authenticates the UE or we have mutual authentication in which the UE and the network authenticate each other?[308]

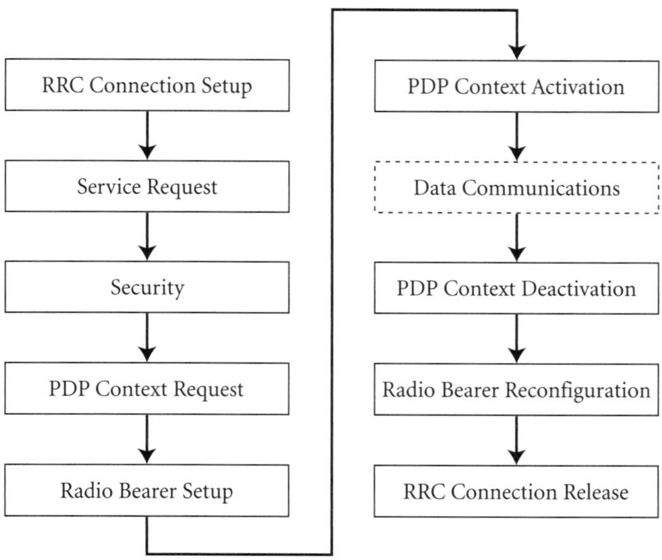

Figure 9.41: Data Call Setup and Release Overview

[308] UMTS has mutual authentication.

The PS domain authenticates the UE, and the UTRAN configures the UE to secure the air interface by activating integrity protection and ciphering. The UE sends a PDP context request message to the core network, requesting an IP address and desired QoS. The UTRAN now establishes radio bearers that can carry packet data traffic at a potentially higher data rate.

The PS domain sends the PDP context information to the UE by assigning a dynamic IP address and by specifying the negotiated QoS. The word *negotiated* is in the standard but is actually a misnomer; the UE must accept whatever QoS the core network has determined! Since the dedicated radio bearers and various links in the network are now configured, the UE exchanges IP packets with the other end, such as a web server.

When the data session is over, the PDP context is removed. The UE no longer possesses an IP address. The Dedicated radio bearers capable of carrying high-rate packets are now replaced by low-rate dedicated resources to enable transfer of signaling messages. If the data inactivity persists, the dedicated RRC signaling connection between the UE and the UTRAN also disappears, and the UTRAN no longer recognizes the UE. The UE then enters the Idle Mode.

9.7.2 Detailed Data Session Setup

The call setup stages summarized in Section 9.7.1 are discussed in detail below. The similarities and differences between voice and data call setups are highlighted where applicable.

9.7.2.1 Radio Resource Control Connection Setup and Security Activation The message exchange that establishes an RRC signaling connection between the UE and the UTRAN is similar for both voice and data calls. The UE sends an RRC CONNECTION REQUEST to the UTRAN and includes the initial UE identity and P-CPICH E_c/N_0s of strong cells. This message includes a Packet Temporary Mobile Subscriber Identity allocated by an SGSN (instead of a Temporary Mobile Subscriber Identity allocated by an MSC or VLR) as a UE identity and an establishment cause, such as “originating high-priority signaling” or “originating background call” instead of “originating a conversational call.”

The UTRAN responds with an RRC CONNECTION SETUP message. This message contains a Packet Temporary Mobile Subscriber Identity to identify the UE on the common S-CCPCH and also assigns a new UTRAN Radio Network Temporary Identity to the UE. Furthermore, the *UERadioCapabilityFDDUpdateRequirement* flag is set to TRUE to obtain the UE’s capabilities. Four dedicated SRBs are usually set up to carry RRC and NAS messages. Transport channels and physical channels for the UL and the DL are configured. For example, a 40 ms TTI can be set as the TTI to send messages using the SRBs, and a dedicated uplink radio channel can be specified via assignment of a long uplink scrambling code and one DPDCH (e.g., a spreading factor of 64). Similarly, a dedicated DL radio connection is specified by an OVSF code at a given spreading factor (e.g., $SF = 128$) in one or more Active Set cells.

The UE replies with the RRC CONNECTION SETUP COMPLETE message on the assigned dedicated radio channel in the uplink. The UTRAN learns about various UE capabilities to make appropriate decisions on radio bearer setups and/or reconfigurations in the future.

The UE next sends an uplink (RRC) INITIAL DIRECT TRANSFER message containing a NAS Service Request message and specifies PS as the core network domain and provides its Packet Temporary Mobile Subscriber Identity. The PS domain and UE may now exchange authentication messages. After such optional authentication, the UTRAN sends an SECURITY MODE COMMAND message to the UE to configure the UE to use specific algorithms such as UEA1 for ciphering and UIA1 for integrity protection. The UE responds with an SECURITY MODE COMPLETE message.

9.7.2.2 Packet Data Protocol Context Request and Radio Bearer Setup

The UE makes a request for a PDP context by sending an (RRC) UPLINK DIRECT TRANSFER message containing a NAS Activate PDP Context Request message. Two major types of requests are the requested QoS specified by parameters, such as the traffic class[309] (e.g., background), maximum bit rates for the UL and the DL, and guaranteed bit rates for the UL and the DL as well as the requested PDP address, such as an Internet-Engineering-Task-Force–based IP version 4 address, and the desired access point name, such as “xyz.com” for a user who has subscribed to services of an operator “xyz.”

The PS domain, specifically the 3G SGSN, sends an Radio Access Network Application Part RAB Assignment Request message to the RNC via the Iu interface to determine if the UTRAN can meet the requested QoS. The RNC executes a proprietary algorithm and evaluates the request against the available radio resources. If adequate resources are available, the UTRAN sends a (RRC) RADIO BEARER SETUP message to the UE to allocate dedicated radio resources to the UE for potentially high-rate data transfer. The contents of this message are similar for voice and data calls with the main differences being the use of turbo coding instead of convolutional coding, 10 ms TTI for data traffic instead of 20 ms TTI for voice, and OVSF codes’ lower spreading factor for higher data rates. The UE responds with an (RRC) RADIO BEARER SETUP COMPLETE message.

9.7.2.3 Packet Data Protocol Context Activation and Data Communications The PS domain responds to the UE’s request for activation of the PDP context by sending a DOWNLINK DIRECT TRANSFER message containing a NAS Activate PDP Context Accept message. This message includes the negotiated QoS and the Internet Engineering Task Force address, such as a public or private IP version 4 address “a.b.c.d” dynamically assigned to the UE. Since the UE has dedicated

[309] Section 9.10 talks about traffic or QoS classes.

resources within the UMTS network and an IP address, it can exchange packets with another entity, such as an application server.

9.7.2.4 Packet Data Protocol Context Deactivation and Release of Radio Bearers and Radio Resource Control Connection When the data application no longer requires resources (e.g., a file transfer has been completed) or no data transfer activity occurs for a certain time period, the PDP context is deactivated. Figure 9.42 summarizes the signaling messages for one such scenario. In Figure 9.42,

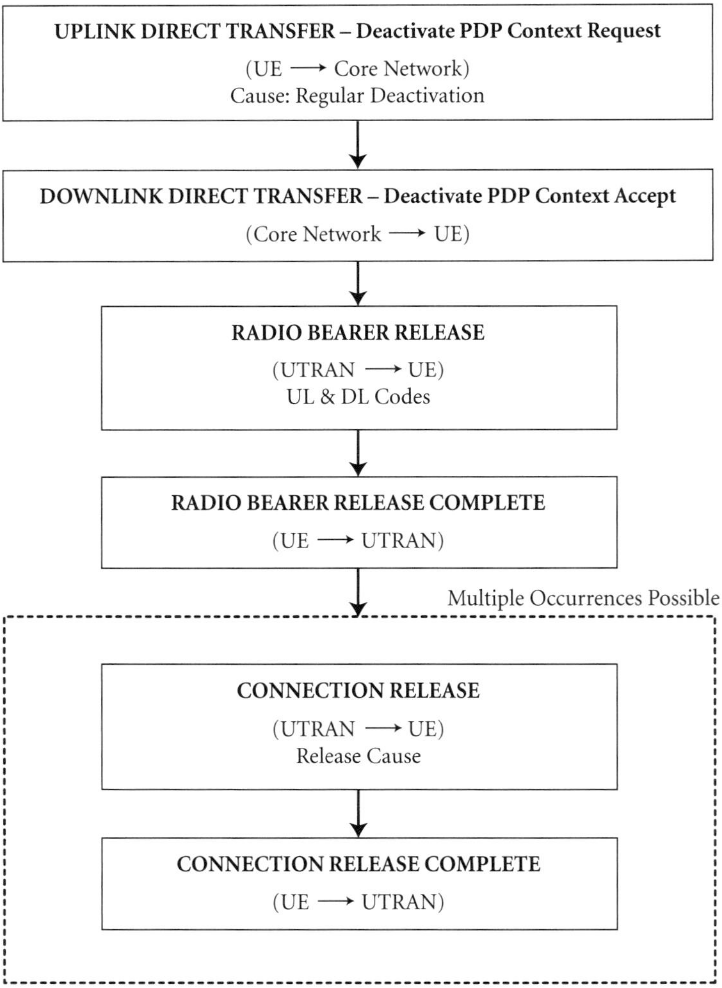

Figure 9.42: PDP Context Deactivation and Release of Radio Bearer and RRC Connection

the RRC signaling messages are exchanged between the UE and the RNC, and the NAS messages are exchanged between the UE and the SGSN.

The UE sends an (RRC) UPLINK DIRECT TRANSFER containing a NAS Deactivate PDP Context Request message with the cause for such request (e.g., regular deactivation). The PS-domain core network accepts the UE's request and replies with an (RRC) DOWNLINK DIRECT TRANSFER with a NAS Deactivate PDP Context Accept message. At this time, the core network resources reserved for the UE (e.g., the Iu traffic bearer) and the assigned IP address are released. Now the UTRAN starts the process of releasing some radio resources by sending an (RRC) RADIO BEARER RELEASE to the UE. The UE responds with an (RRC) RADIO BEARER RELEASE COMPLETE. Dedicated bearer resources capable of carrying user data traffic are released, and only the SRBs are preserved to enable the exchange of signaling messages.

An implementation-specific UTRAN algorithm can release the RRC connection based on the absence of data activity for a certain time period or on some other criteria. The UTRAN does so by sending an (RRC) CONNECTION RELEASE message to the UE with a possible cause, which could be "normal event" or "unspecified." The UE replies with an (RRC) CONNECTION RELEASE COMPLETE message. Such message exchange could be repeated to increase reliability.

9.8 MOBILITY MANAGEMENT

When the UE's power has been on for a while, it is in the Idle or Connected Mode. In the Idle Mode, the UE is known only to the core network and is recognized by NAS identities, such as the International Mobile Subscriber Identity, Temporary Mobile Subscriber Identity, and Packet Temporary Mobile Subscriber Identity. If the UTRAN needs to communicate with the UE in the Idle Mode, it needs to send a paging message in a relatively large geographical area[310] consisting of numerous cells. The UE in the Idle Mode looks for a paging message.

Do You Recall... Do you remember the name of the physical channel that carries an actual paging message and the channel that makes paging more efficient?[311]

When an RRC connection between the UE and the UTRAN is established, the UE enters the Connected Mode. In the Connected Mode, the UE location could be known at two levels of granularity: the cell level and the URA level. A URA consists of a set of cells served by an RNC via Node Bs. Mobility management in the Idle and Connected Modes is discussed next. Sections 9.8.1 and 9.8.2 provide a brief overview of the Idle Mode and Connected Mode mobility, respectively. Section 9.8.3 takes a closer look at the Connected Mode intra-frequency UMTS mobility. Inter-

[310]Such an area is called the *location area* for CS paging and the *routing area* for PS paging.
[311]The S-CCPCH carries the actual paging message, and the PICH makes paging efficient.

frequency intra-UMTS mobility and inter-RAT (radio access technology) mobility are briefly discussed in Section 9.8.4. Finally, the Compressed Mode, which facilitates inter-frequency and inter-RAT measurements, is described in Section 9.8.5.

9.8.1 Idle Mode Mobility: An Overview

When the UE's power is turned on, it remains in the Idle Mode until it transmits an RRC CONNECTION REQUEST message to the UTRAN. Furthermore, whenever the RRC connection between the UE and the RNC is released normally or abnormally, the UE enters the Idle Mode. The UE in the Idle Mode waits for initiation of communications from its own side or the core network side. For example, when the user dials a phone number on the UE, the UE initiates the process of RRC connection establishment and exits the Idle Mode. An example of core-network–initiated communications is transmission of a paging message to the UE, which makes the UE exit the Idle Mode. To enable the UE to be in the best possible radio environment while exiting from the Idle Mode and to increase UE's battery life, efficient Idle Mode management is required.

The downlink physical/transport channels that play an important role in the Idle Mode operation are P-CCPCH/Broadcast channel, PICH/Not-Applicable, and S-CCPCH/PCH or S-CCPCH/FACH. The P-CCPCH periodically transmits useful system information in units of SIBs. If the UE is moving around in a UMTS coverage area, it will continue to find reasonably strong UMTS cells. Assume that the UE is inside Cell C3 and that it is receiving a high CPICH E_c/N_0 from Cell C3. The UE periodically wakes up (e.g., every 1.28 seconds corresponding to 2^7 frames if the discontinuous reception cycle coefficient is 7,) and evaluates the presence or absence of a Paging message by observing page notification indicators on the PICH. If the paging notification indicator on the PICH points to the presence of a paging message, the UE now looks at the S-CCPCH that carries an (RRC) PAGING TYPE 1[312] message. The paging message includes the UE identity, such as a Temporary Mobile Subscriber Identity or Packet Temporary Mobile Subscriber Identity.

If the UE finds no paging message for itself, it can go to sleep to conserve battery power. (We certainly hope that you are not conserving your brain power while reading the book; we would like you to be continuously using your brain power!) When the UE periodically wakes up, it also looks for other neighboring cells to determine if another cell can provide a better quality signal in the downlink. The UE utilizes SIB 11 to find a list of neighboring cells [265]. If the UE finds a stronger cell, it starts observing the downlink channels, such as P-CCPCH, PICH, and S-CCPCH, from the new cell. The process by which the UE stops monitoring the current cell and starts monitoring a new cell is called *cell reselection*. Though the UE autonomously performs cell reselection, the network controls the process by advertising several thresholds in SIBs.

[312] A PAGING TYPE 1 message is used for a UE in the Idle Mode (i.e., without an RRC connection). A PAGING TYPE 2 message is used for PS-domain paging when a UE is already connected to the CS domain and for CS-domain paging when a UE is already connected to the PS domain.

Idle Mode mobility may involve a change in the RAT. Assume that a hybrid UE supports two radio access technologies, UMTS and GSM. As the UE leaves the UMTS coverage area, the downlink quality of the UMTS cell degrades. If the UMTS quality drops below a threshold, the UE looks for GSM cells in the neighborhood. If a GSM or GPRS cell can provide a better quality signal than the UMTS cell, the UE leaves the UMTS air interface and starts monitoring the GSM cell's broadcast channel. SIB 11 provides a list of GSM cells and various inter-RAT cell selection or reselection thresholds. In a GSM cell, the broadcast channel transmits information about the neighboring UMTS cells so that the UE can look for them when necessary. In other words, the UE can go back and forth between UMTS and GSM seamlessly.

9.8.2 Connected Mode Mobility: An Overview of Basic Concepts

When the UE is in the Connected Mode (i.e., with an RRC connection) but is actively processing no signaling messages or traffic with the UTRAN, it has no radio bearers.[313] In such cases, the UE location may be known at the cell level or the URA level. Hence, when a page arrives for the UE, the UTRAN knows in which cells the paging message should be sent. When cell-level location granularity is supported, the UE sends a CELL UPDATE message to the UTRAN so that the UTRAN knows in which cell the UE is located. Similarly, in URA-level location granularity, the UE sends a URA UPDATE message when it enters a new URA. Typical commercial deployments omit cell updates to support paging due to its signaling load. URA-level mobility is sometimes used. The most common approach involves paging in a location area controlled by an MSC or a Routing Area controlled by an SGSN when the UE is in the Idle Mode (not Connected Mode). A location area has one or more routing areas and is identified by a location area code. A routing area can have one or more URAs and is identified by a routing area code.

When the UE in the Connected Mode is actively exchanging speech or packet data traffic with the UTRAN, it has dedicated radio channels in the downlink and the uplink and an RAB exists between the UE and the core network. In such a case, the UE could simultaneously communicate with multiple cells. We will focus on mobility management in such cases in the remainder of this section.

In a traditional CDMA-based[314] system, the UE can simultaneously communicate with multiple cells as shown in Figure 9.43. Assume that the UE is involved in a voice call and is in a region where the P-CPICHs from cells 2, 3, and 4 are strong. The Node B1 controls cells 2 and 3, and the Node B2 controls cell 4. The Active Set contains cells 2, 3, and 4. When a cell is a member of the Active Set, it transmits a dedicated physical channel to the UE and receives a dedicated physical channel from the UE. Since Node Bs work *asynchronously*, the RNC provides timing guidance to

[313]Such a situation exists, for example, when the UE has completed data transfer, such as the downloading of a web page, and the inacitivity times that control the transition from the Connected Mode to the Idle Mode have not expired.

[314]CDMA here is a multiple access technique and does not imply any cellular standards.

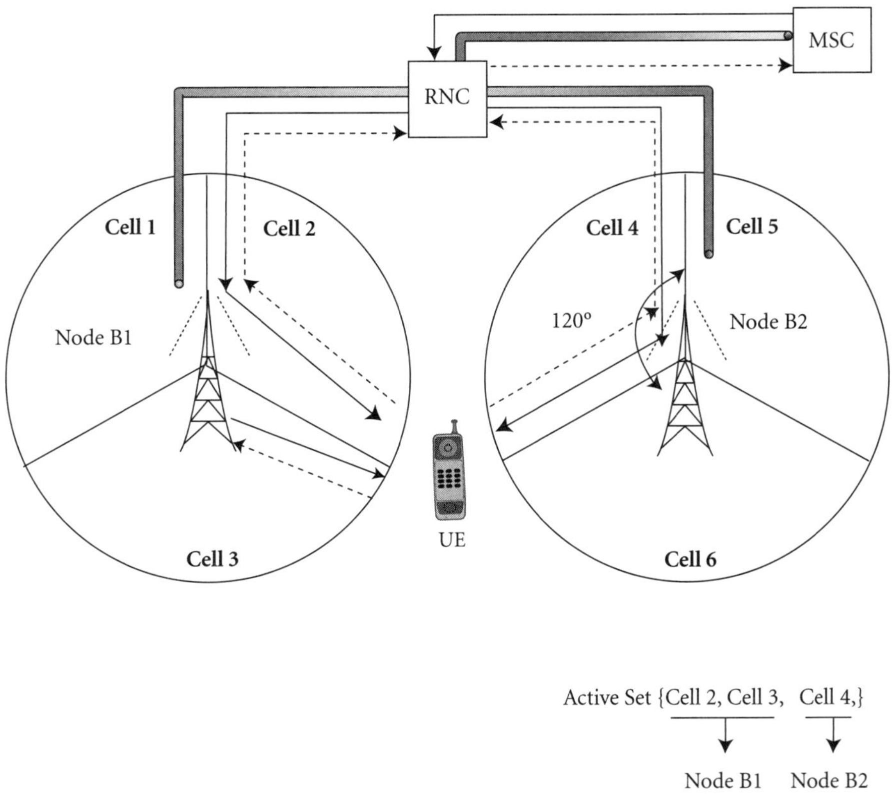

Figure 9.43: Soft Handover in a UMTS System

the Node Bs in the Active Set so that signals can be properly combined at the UE's RAKE receiver.

Let's consider the downlink operation first. The RNC receives a speech block from the CS-domain component such as the MSC or transcoder rate adaptation unit. Since Node B1 and Node B2 are in the UE's Active Set, the RNC forwards the speech block to these two Node Bs. Each cell in the Active Set transmits the DPDCH to the UE. The UE's RAKE receiver combines the received DPDCHs to retrieve the speech block.

Do You Recall... Which combining technique does the RAKE receiver utilize?[315] Hint: The name of the technique starts with an M.

[315]This technique is maximal ratio combining.

Now let's turn our attention to the uplink operation. The UE sends a speech block on the uplink DPDCH. The RAKE receiver reserved for the UE at the Node B1 performs maximal ratio combining on the uplink DPDCH received in cells 2 and 3 and forwards the reconstructed speech block, speech block 1, to the RNC. Similarly, the RAKE receiver at the Node B2 receives the uplink DPDCH in cell 4 and forwards the recreated speech block, speech block 2, to the RNC.

The RNC performs selection combining on the two speech blocks, speech block 1 and speech block 2, and forwards an error-free speech block, if available, to the MSC. For example, if speech block 1 is error-free and speech block 2 has an error, the RNC sends speech block 1 to the MSC. When multiple Node Bs are in the Active Set with one cell per Node B, such handover is called *soft handover*. See Chapter 7 for generic details on soft handover. Handover and handoff are the same. When the Active Set contains the cells from just Node B, we have *softer handover*. The combination of soft handover (i.e., multiple Node Bs) and softer handover (i.e., multiple cells but one Node B) is called *softer-soft handover*. The main reason for differentiating between soft and softer handover is that suitable resource provisioning can be carried out. For example, only one channel element is required at a Node B whether just one cell of the Node B is present in the Active Set or more cells are present. In other words, if a UE's Active Set is {cell 2, cell 3}, only one channel element[316] is required at the Node B1. However, if a UE's Active Set is {cell 2, cell 4}, two channel elements are required, one at Node B1 and another at Node B2. The term soft handover is often used in a generic sense and, in such a case, refers to any type of soft handover—soft, softer, and softer-soft. While the example given here is for a speech call, the same downlink and uplink operations occur for a data call.

A Bit of Cellular History... Have you ever wondered why the Node Bs do not use a global positioning system for synchronous transmissions in typical UMTS deployments? In the movie "Independence Day," what did the aliens attack first? Satellites. Hence, a system that relies on satellites for operation would stop working in such a case, but UMTS would continue to operate well! Of course, the ability to work in the presence of an alien attack was beyond the requirements for UMTS! Let's delve into UMTS history a little. Early standardization work on UMTS included significant contributions from European companies. Who owns the global positioning system satellites? The U.S. government! Though the U.S. government happily allows companies to use the global positioning system, it can revoke such privileges in emergencies, such as war. The European companies were apprehensive and decided to design a network that did not rely upon the global positioning system. We wonder if Qualcomm's CDMA patents played any role in this decision! A global positioning system receiver also adds to the cost of the overall Node B equipment. Other CDMA-

[316] A channel element is a hardware resource within the baseband chipset and does physical layer processing, including implementation of a RAKE receiver, demodulation, and decoding.

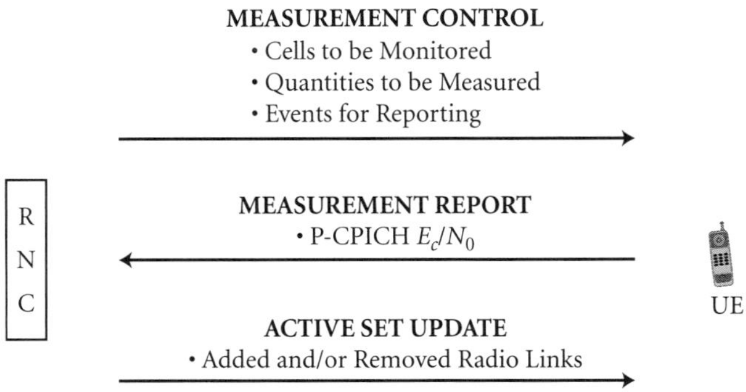

Figure 9.44: Overview of Handover Processing in a UMTS System

based standards, such as IS-95 and 1x (both widely deployed in the U.S. and some other parts of the world), do use the global positioning system.

Because a UE with dedicated radio channels travels in a region, it encounters varying propagation channel conditions, and the Active Set must reflect the propagation environment the UE experiences. Figure 9.44 illustrates the basic handover processing in a UMTS system.

The overall process of handover can be viewed as a three-stage process. During Stage 1, the UTRAN configures the UE to make measurements of the surrounding cells using a MEASUREMENT CONTROL message. This message specifies which cells the UE should monitor; which quantities the UE should measure (e.g., P-CPICH E_c/N_0); and when the UE should report the measurements. In Stage 2, the UE makes suitable measurements and sends measurement reports to the RNC via a MEASUREMENT REPORT message. During Stage 3, the soft handover algorithm implemented at the RNC processes the received measurement reports to determine the Active Set. The RNC informs the UE to update the Active Set by sending an ACTIVE SET UPDATE message. Now the UE and the cells in the updated Active Set start communicating on dedicated radio channels. Let's next look at the details of these three stages in Section 9.8.3.

9.8.3 Stages of Intra-Frequency Handover: A Closer Look

Brace yourself! We are ready to dive into the details of how intra-frequency handover occurs in UMTS.

9.8.3.1 User Equipment Configuration Recall that the RNC sends a MEASUREMENT CONTROL message to the UE to specify which cells to monitor; which measurements to make; and how to report the measurements. A total of thirty-two cells, including the current cells in the Active Set, can be specified for measurements [265].

A Question for You... What do you think may be the benefits and drawbacks of asking UEs to monitor too few or too many cells?[317]

The P-CPICH primary scrambling code identifies the cell. The P-CPICH E_c/N_0 and P-CPICH RSCP are typically measured quantities. The P-CPICH E_c/N_0 is the ratio of the energy per chip for the P-CPICH and the total energy measured in the bandwidth of the pulse-shaping filter at the receiver [272]. Since power is the ratio of energy to time, the P-CPICH E_c/N_0 can also be calculated as the ratio of the received P-CPICH power to the total received power in the channel bandwidth during a given time period. The P-CPICH RSCP is the total power measured on the P-CPICH after despreading. Recall that P-CPICH is transmitted on $C_{256,0}$ [265, 272].

The RNC configures the UE to operate in one of two basic reporting modes, *periodic* and *event-based*. The periodic reporting mode involves periodic reporting of the configured measurements at a configured interval. The event-based reporting mode, as the name implies, involves reporting due to occurrences of specific events. The main benefit of the periodic reporting mode is that the latest measurements become available on a regular basis, and the drawback is the potentially unnecessary higher signaling load. For example, if the UE is stationary and the radio environment is fairly stable, frequent periodic reports are no longer of high value. The main benefit of the event-based reporting mode is the reduced signaling load, and frequent reporting is feasible when radio conditions change. Event-based reporting is widely used in commercial deployments. The measurements are processed (i.e., averaged or filtered) according to the parameters specified in the MEASUREMENT CONTROL message.

9.8.3.2 Measurement Reporting The UE conveys the measurements to the RNC by sending a MEASUREMENT REPORT message. Periodic reporting mode is controlled by setting the amount of reporting, which is the number of reports to be sent (e.g., $1, 2, \ldots, \infty$), as well as the reporting interval, which dictates how frequently the reports are sent (e.g., every 250 ms, 500 ms, ..., 64000 ms). The event-based reporting mode is controlled by setting the event identity, which identifies a specific event, and the relevant parameters that modify the basic reporting mechanism. More than one event can be configured, and once an event occurs, periodic reporting is possible as long as the same conditions persist. Let's take a look at these parameters along with various types of events.

[317] If the monitored list is too long, the UE may be so busy making measurements that we may get delayed reports of a cell that has a rapidly increasing P-CPICH E_c/N_0. Higher interference may lead to a call drop, poor call quality, or excessive resource utilization. If the list is too small, the UE may miss detection of a cell altogether, leading to similar problems. Determination of the proper size of the monitored list is an important area of RF optimization in a CDMA system.

The basic measurement quantities that can be configured to evaluate the occurrence of an intra-frequency event are CPICH E_c/N_0, CPICH RSCP, and path loss. The path loss is calculated as P-CPICH Transmit Power minus P-CPICH RSCP where the value of the transmit power of the P-CPICH for a given cell is specified in an SIB Type 5. Table 9.4 summarizes the intra-frequency reporting events. The "Availability of Periodic Reporting" indicates if the measurements can be reported periodically once the event occurs and the same conditions that triggered the event persist.

Event 1A occurs when a P-CPICH enters the reporting range, and the UE sends a MEASUREMENT REPORT message if the reporting deactivations threshold is greater or equal to the number of the Active Set cells or is zero. If the value of the reporting interval is zero, just one report is sent; otherwise, multiple reports are sent based on the settings of the amount of reporting and reporting interval parameters for this event. Figure 9.45 illustrates the occurrence of Event 1A.

Assume that the initial Active Set contains the cells 1, 2, and 3. The P-CPICH in cell 1 provides the best E_c/N_0 and R_{1A} is the reporting range constant (e.g., 4 dB). The reporting range then becomes a range of E_c/N_0 values between P-CPICH1 and 4 dB. At time instant t_1, P-CPICH from cell 4 comes within R_{1A} dB of the P-CPICH1. Hence, event 1A is triggered, and the UE reports the measurements, such as P-CPICH E_c/N_0s of the current Active Set cells and of cell 4. Assume that the periodic reporting is configured for event 1A. As long as the P-CPICH stays within the reporting range, the UE continues to send reports (e.g., at time instants t_2, t_3, t_4, and t_5).

When a P-CPICH leaves the reporting range, event 1B is triggered, and the UE reports the measurements. As shown in Figure 9.45, P-CPICH4 leaves the reporting range at time instant t_5. Hence, the UE sends a MEASUREMENT REPORT message at t_5. Periodic reporting for event 1B is unavailable.

When a cell that is not currently a member of the UE's Active Set becomes stronger than an Active Set cell, event 1C occurs. Figure 9.46 depicts the occurrence

Table 9.4: Intra-Frequency Measurement Reporting Events

Event Identification	Event Description	Availability of Periodic Reporting
1A	P-CPICH enters the reporting range	Yes
1B	P-CPICH leaves the reporting range	No
1C	A non-active P-CPICH becomes better than an active P-CPICH	Yes
1D	Change of the best cell	No
1E	P-CPICH becomes better than an absolute threshold	No
1F	P-CPICH becomes worse than an absolute threshold	No

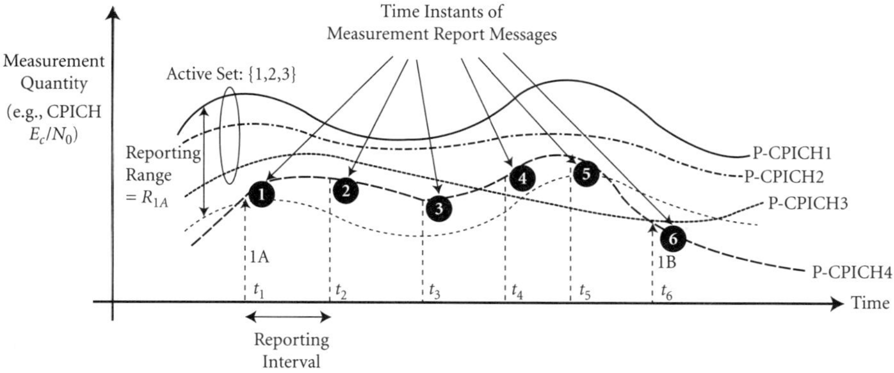

Figure 9.45: Examples of Events 1A and 1B

of event 1C. Assume that the current Active Set is $\{$cell 1, cell 2$\}$. At time instant t_1, E_c/N_0 of P-CPICH3 becomes better than E_c/N_0 of P-CPICH2. Since event 1C is triggered, the UE sends a MEASUREMENT REPORT message containing the E_c/N_0s of the Active Set cells 1 and 2 and E_c/N_0 of cell 3. The RNC may decide to update the Active Set, and, in Figure 9.46, the RNC instructs the UE to use the new Active Set containing the cells 1 and 3; the weaker cell 2 is replaced by the stronger cell 3. Periodic reporting is supported for event 1C. If the RNC fails to add cell 3 to the Active Set, the UE would continue sending periodic reports.

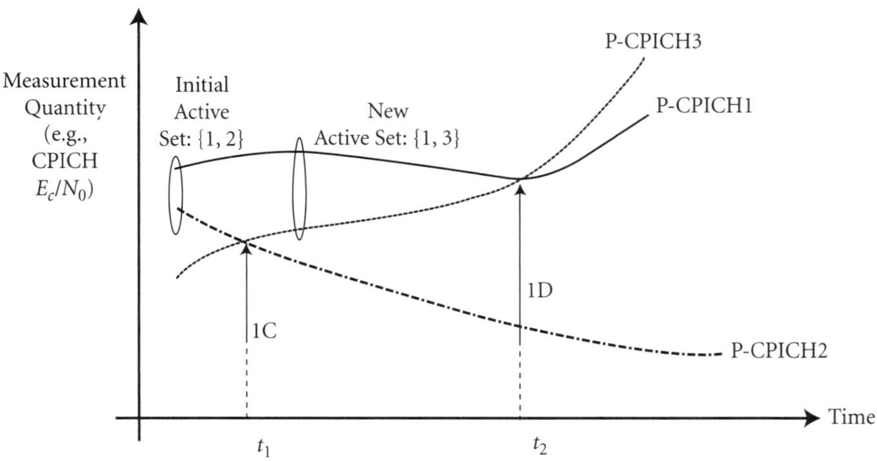

Figure 9.46: Examples of Events 1C and 1D

When the best cell changes, event 1D occurs. For example, P-CPICH from cell 1 is the best cell until time instant t_2. At t_2, the E_c/N_0 of P-CPICH3 exceeds that of P-CPICH1. Since event 1D is triggered, the UE sends a MEASUREMENT REPORT message. No periodic reporting is available for this event.

Event 1E occurs when a P-CPICH becomes larger than an absolute threshold. Figure 9.47 illustrates event 1E. The initial Active Set is {cell 1, cell 2}. At time instant t_1, P-CPICH3 E_c/N_0 exceeds a threshold, and the UE reports the measurements to the RNC. The RNC may decide to add cell 3 to the Active Set, leading to three members in the new Active Set—cell 1, cell 2, and cell 3. Periodic reporting is unavailable for this event.

Event 1F occurs when a P-CPICH becomes smaller than an absolute threshold. As illustrated in Figure 9.47, P-CPICH2, which is a member of the Active Set, drops below an absolute threshold at time instant t_2 and triggers the UE to report the measurements to the RNC. The RNC may remove cell 2 from the Active Set. The new Active Set contains cells 1 and 3. No periodic reporting is applicable to this event.

The UE sends the MEASUREMENT REPORT messages to the RNC using the RLC Unacknowledged Mode or RLC Acknowledged Mode according to the configuration dictated by the MEASUREMENT CONTROL message from the RNC. The basic event-based reporting mechanism discussed above can be modified using parameters such as hysteresis, time-to-trigger, and cell individual offset. Furthermore, a cell can be forbidden to affect the reporting range for the events 1A and 1B. For example, if the service operator knows that the quality of the measurements for a given cell is unreliable due to wide fluctuations, such a cell can be forbidden to affect the reporting range. Figure 9.48 illustrates the influence of hysteresis and time-to-trigger.

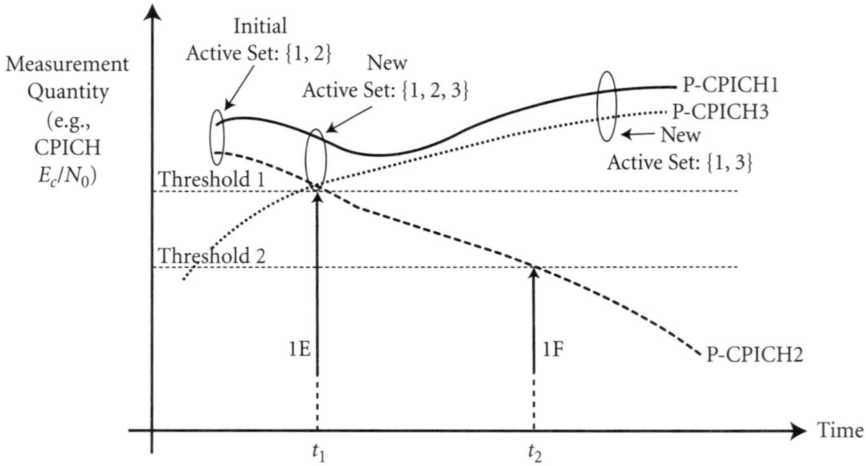

Figure 9.47: Examples of Events 1E and 1F

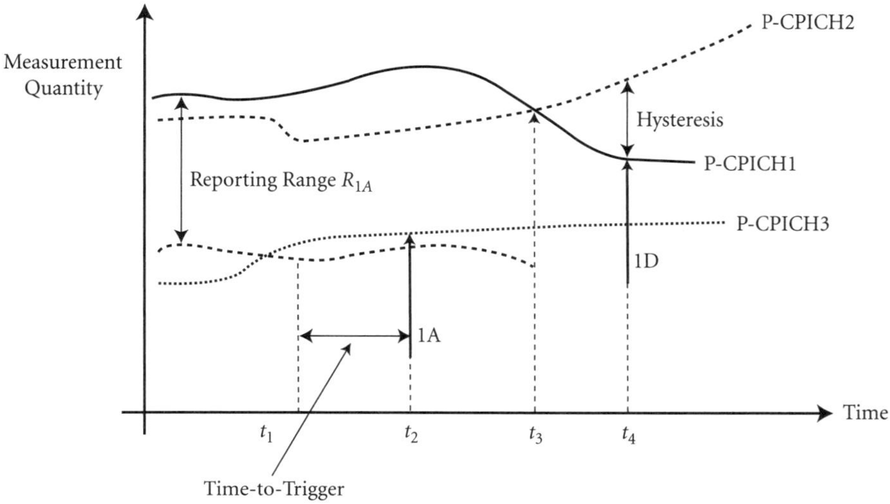

Figure 9.48: Influence of Hysteresis and Time-To-Trigger

Hysteresis (e.g., about 4 dB) and time-to-trigger (e.g., about 320 ms to 640 ms) are also used to reduce the frequency of reports.

The initial Active Set is {cell 1, cell 2}. P-CPICH3 enters the reporting range at t_1; however, the UE delays in sending a report due to the existence of the time-to-trigger parameter. Since P-CPICH3 stays within the reporting range for the duration of time-to-trigger, event 1A occurs at t_2. The RNC may now make the new Active Set {cell 1, cell 2, cell 3}. P-CPICH2 becomes stronger than P-CPICH1 at t_3; however, as a result of non-zero hysteresis, the UE delays sending the report. When P-CPICH2 exceeds P-CPICH1 by the hysteresis amount, event 1D occurs. The hysteresis and time-to-trigger parameters help limit the measurement signaling load.

Cell individual offsets could be positive, negative, or zero and can be used to give more or less importance to the reporting of specific pilots. Though the hysteresis parameter applies to all pilots, all individual offsets are applicable to specific pilots. When a positive offset is used for a given P-CPICH, the UE sends a report as if the P-CPICH were x dB better than it actually is. Such a positive offset can be used for a P-CPICH if such pilot is known to become stronger suddenly in a given area (e.g., around a street corner or at the top of a hill). Cell individual offsets are typically unused in commercial deployments.

9.8.3.3 Handover Execution The RNC receives the MEASUREMENT REPORT messages from the UE and executes an implementation-specific and vendor-proprietary handover algorithm. The handover algorithm determines the Active Set. In general, cells associated with strong P-CPICH E_c/N_0s should be made members of the Active Set, and cells with weak P-CPICH E_c/N_0s should be excluded from the Active Set. Three types of sets exist: the Active Set, the Monitored Set, and the Detected Set. The UE actively exchanges signaling and traffic with Active Set cells. The UE needs to measure the cells specified in the MEASUREMENT CONTROL message. Such cells, when absent from the Active Set, constitute a Monitored Set.

The cells that the UE detects but that are absent from the Active Set and the Monitored Set are part of a Detected Set. For example, when event 1A is triggered, the RNC could add to the Active Set the new cell with the P-CPICH entering the reporting range. If an active P-CPICH leaves the reporting range, the RNC may remove such a cell from the Active Set. The handover algorithm may attempt to aggressively reduce the degree of handover to increase capacity but at the potential risk of reducing reliability. The right degree of handover maximizes the capacity while maintaining a target QoS (e.g., 2% error rate).

9.8.4 Intra-UMTS Inter-Frequency and Inter-RAT Handover

Initial UMTS deployments utilize a single carrier frequency in a given area. When the subscriber growth demands higher capacity, multiple carrier frequencies are deployed. In such a case, the UE must be able to evaluate the quality of different carrier frequencies. When another carrier frequency can provide better quality than the current frequency, inter-frequency handover is performed. The UE calculates a frequency quality estimate using the measurement quantity of P-CPICH E_c/N_0 or P-CPICH RSCP to determine whether an inter-frequency measurement event has occurred. The UE can use P-CPICH E_c/N_0, P-CPICH RSCP, and downlink path loss to update the virtual Active Set. The frequency quality estimate for a given carrier frequency is the best E_c/N_0 or the weighted sum of the best E_c/N_0 (or RSCP) and the cumulative E_c/N_0 (or RSCP) of the true or virtual Active Set pilots. Recall that the current frequency has a real Active Set but an unused frequency has no real Active Set. The UE maintains a virtual Active Set for the unused frequency.

Table 9.5 summarizes the measurement reporting events for the inter-frequency intra-UMTS handover. The RNC sends to the UE a MEASUREMENT CONTROL message to inform the UE about which inter-frequency events would trigger a MEA-

Table 9.5: Inter-Frequency Reporting Events

Event Identification	Event Description
2a	Change of the best frequency
2b	Quality of the used frequency less than a threshold and quality of an unused frequency greater than a threshold
2c	Quality of an unused frequency greater than a threshold
2d	Quality of the current frequency less than a threshold
2e	Quality of an unused frequency less than a threshold
2f	Quality of the current frequency greater than a threshold

SUREMENT REPORT message. Both hysteresis and time-to-trigger parameters are available for configuration; however, periodic reporting is unavailable. Examples of reported measurements are cell identity, P-CPICH E_c/N_0, downlink path loss, and Universal Terrestrial Radio Access carrier received signal strength indicator.

A service operator usually deploys a new technology in a few markets first to start earning revenue from early adopters[318] and expands the footprint gradually based on the expected customer growth. Some geographical areas have UMTS coverage, and other areas would have only 2G and/or 2.5G coverage. When a hybrid UMTS/GSM/GPRS UE leaves the UMTS coverage and enters the GSM and/or GPRS coverage area, an inter-RAT handover takes place. The UE estimates the quality of the current UMTS frequency as the best E_c/N_0 or the weighted sum of the best E_c/N_0 (or RSCP) and the cumulative E_c/N_0 (or RSCP) of the Active Set pilots. The frequency quality estimate for GSM is GSM carrier received signal strength indicator. The UE reports quantities, such as the GSM carrier received signal strength indicator and observed time difference, to the GSM cell. The Broadcast Control channel ARFCN and base station identity code combination specify the GSM cell to be measured. Table 9.6 summarizes the inter-RAT reporting events.

9.8.5 Compressed Mode

A handover from the current UMTS frequency to another UMTS frequency or a GSM frequency can be made blindly or with the help of the measurements resulting from the reporting events. In blind handover, if the current frequency provides inadequate quality in the UMTS–GSM border region, the RNC could ask the UE to make a handover to another UMTS frequency or a GSM frequency without knowing the potential quality at such a UMTS or GSM frequency.

However, a better handover decision can be made when measurements of other UMTS or GSM frequencies are available. The measurement events discussed earlier provide relevant measurements to the RNC for a more informed handover decision. Since a typical UE design includes just one RF chain, the UE cannot simultaneously

Table 9.6: Inter-RAT Reporting Events

Event Identification	Event Description
3a	Quality of the current UTRAN frequency below a threshold *and* quality of another system above a threshold
3b	Quality of another system below a threshold
3c	Quality of another system above a threshold
3d	Change of the best cell in another system

[318]Early technology adopters are users who want the newest, cool mobile device and will spend a fortune for the latest technology. Other users will wait years for that technology to become available on free phones!

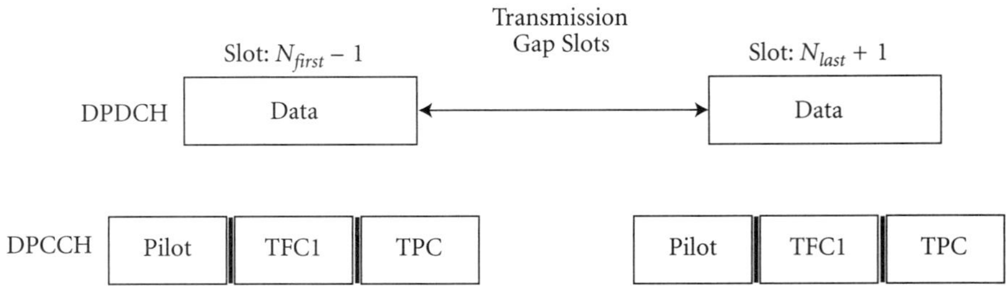

Figure 9.49: Uplink Compressed Mode Frame Structure

monitor multiple carrier frequencies. *The compressed mode enables the UE to temporarily leave the current UMTS carrier frequency and make measurements of other frequencies or technologies.* Once the measurements are made, the UE evaluates the occurrence of a measurement reporting event. The compressed mode creates a transmission gap as shown in Figure 9.49 for the uplink and Figure 9.50 for the downlink.

To account for potentially deteriorating radio channel conditions and to maintain the error rate [e.g., 2% block error rate (BLER)] in spite of reduced processing gain, the transmit power is increased before entering the compressed mode and after exiting the compressed mode. Recall that the transmission of more bits in less time increases the data rate and decreases the CDMA processing gain. The transmission gap length slots from N_{first} to N_{last} are not used for data transmission because the UE uses these slots for inter-frequency measurements.

As shown in Figure 9.49, the UL DPDCH and DPCCH transmit nothing during the transmission gap length slots. However, the downlink compressed frame structure is somewhat different in that some transmission does occur during the transmission gap length slots, as shown in Figure 9.50. The downlink compressed frame could be Type A or Type B; Type A gives the UE more time for measurements by maximizing the transmission gap length, and Type B provides better power control. Observe the transmission of the pilot field in the last slot inside the transmission gap for Type A and B frames and the transmission of the TPC field in the first slot of the transmission gap for Type B.

The transmission gap shown in Figure 9.49 and Figure 9.50 is part of the transmission gap pattern sequence illustrated in Figure 9.51. Each transmission gap sequence consists of a number of transmission gap patterns equal to the transmission gap pattern repetition count. The sequence starts in the radio frame numbered Transmission Gap Connection Frame Number. Two transmission gap patterns are defined, Pattern 1 and Pattern 2. These two transmission gap patterns alternate to create an overall sequence. Each transmission gap pattern starts at the slot numbered *transmission gap starting slot number* in a radio frame. One or two transmission gaps exist in a transmission gap pattern with the first gap of length transmission gap length 1 and the second gap

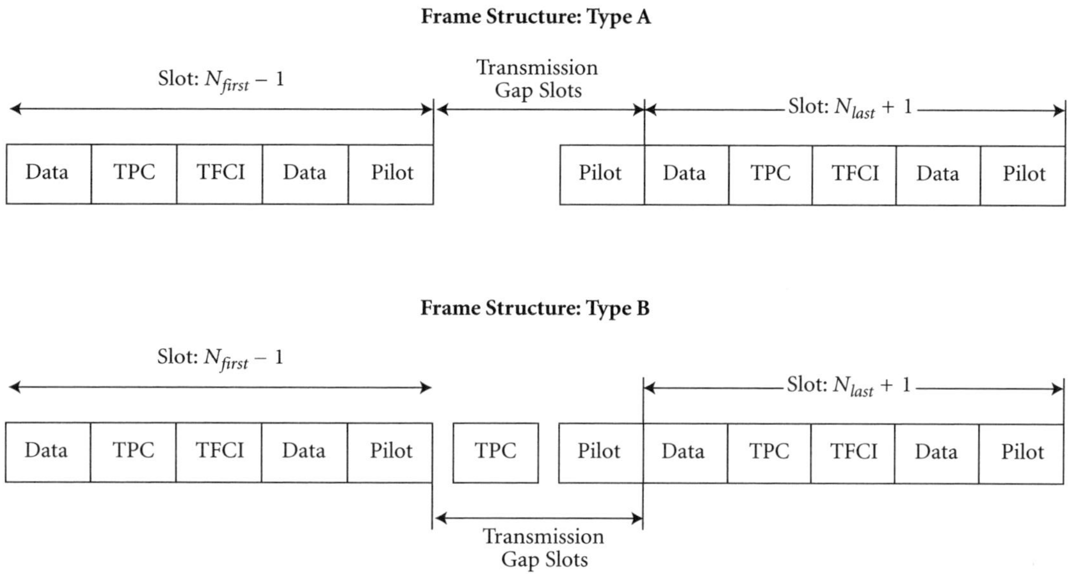

Figure 9.50: Downlink Compressed Mode Frame Structures

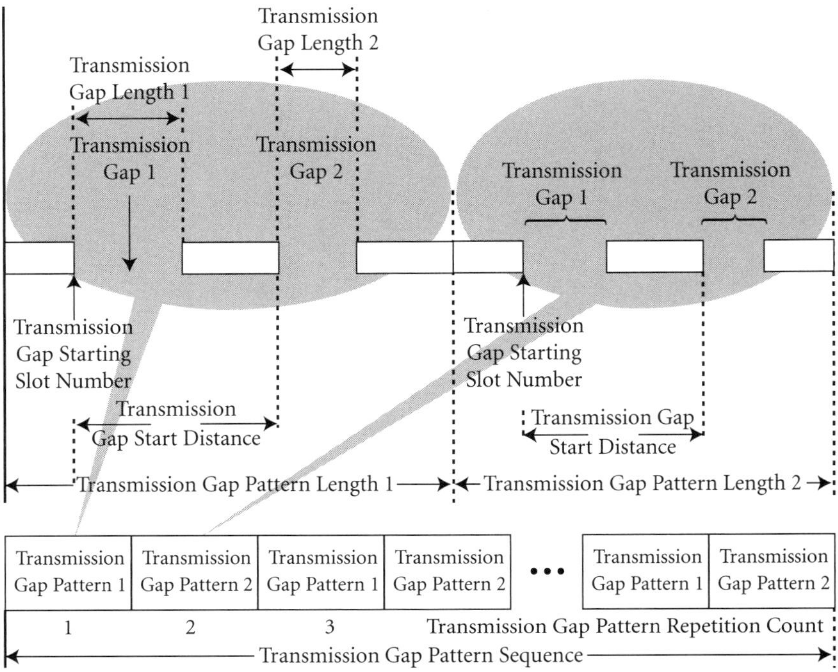

Figure 9.51: Structure of the Transmission Gap Pattern Sequence

of length transmission gap length 2, and the time difference between the start instants of the two gaps is the transmission gap start distance.

Three methods create the transmission gap in the compressed mode: puncturing, reducing the spreading factor by 2, and higher-layer scheduling [57]. In the first method, rate matching is used to create the transmission gap. In the second method, the number of information bits is the same even when the transmission gap is created because the spreading factor is reduced by 2. In the third method, the upper layer delivers fewer information bits to the physical layer and a subset of the TFCIs are then used to create the transmission gap. A single-frame method occurs when the compressed mode introduces the transmission gap length slots in a single frame. The double-frame method occurs when the transmission gap spans two frames. The number of allowed transmission gap length slots is 3, 4, 5, 7, 10, or 14 (for either the single-frame or double-frame method). A single frame can have a transmission gap of no more than seven slots.

Let's briefly summarize how compressed mode may be used for a typical UMTS to GSM handover. UTRAN configures the UE for event 2d by sending a MEASUREMENT CONTROL message in a UMTS cell near a GSM border. When event 2d occurs, the UTRAN prepares for a potential handover to GSM by configuring the UE with compressed mode parameters [e.g., gap pattern, method (such as reduction in spreading factor by 2), and use of a Type A frame] via a (RRC) RADIO BEARER RECONFIGURATION message. The UE is then configured via a MEASUREMENT CONTROL message for event 3a, and the compressed mode is actually activated. When event 3a occurs, the UE sends a MEASUREMENT REPORT message containing the identity of the inter-RAT base station identity code. The UTRAN may decide to move the UE to GSM. In such a case, the UTRAN sends a HANDOVER FROM UTRAN COMMAND message to the UE and deactivates the compressed mode and measurement reporting by indicating "release" in the MEASUREMENT CONTROL message. The reporting events of 2d and 2f and P-CPICH E_c/N_0 and P-CPICH RSCP as the frequency quality estimate are commonly used in commercial deployments.

9.9 POWER CONTROL

The power control refers to the mechanism that dynamically determines the power of the transmitter on a given physical radio channel. The UMTS supports two basic power-control methods: open-loop power control and closed-loop power control.

In an open-loop power-control mechanism, the transmitter autonomously changes its transmit power without explicit feedback from the receiver. The UL PRACH utilizes this mechanism. When the UE transmits a preamble on the PRACH, it determines the initial transmit power for the first preamble and increases its power for successive preambles unless it receives an acknowledgment from the Node B on the DL AICH. Since the UE autonomously changes its transmit power, such an operation is considered an open-loop power control. Once the UE receives an acknowledgment on the AICH, it sends a message part consisting of the control part and the data part on the PRACH. The message part is transmitted with the power offset of P_{p-m} dB relative to the last transmitted preamble. The control part is transmitted using the gain factor β_c, and the data part is transmitted using the gain factor β_d. The UE learns about these gain factors via SIB Type 5.

The closed-loop power control involves explicit feedback in the form of power-up or power-down commands from the receiver to the transmitter. The transmitter then increases or decreases the power based on such feedback.

Do You Recall... Do you remember the primary goal of closed-loop power control?[319]

The closed-loop power control utilizes two mechanisms, outer-loop power control and inner-loop power control, to achieve the target error rate. The outer-loop power control determines the target value of the signal-to-interference ratio, SIR_{target}, which can help achieve the target BLER under current channel conditions. The inner-

[319]The overall goal of closed-loop power control is to transmit the minimum power adequate to achieve a target error rate, such as 2% BLER. Such an approach will reduce the overall interference and increase the efficient utilization of power, the UE battery life, and the capacity.

loop power control ensures that the estimated value of the signal-to-interference ratio, SIR_{est}, and SIR_{target} are as close as possible. For example, under given channel conditions, if $SIR = 10$ dB leads to the achieved BLER of 2%, the outer-loop power control would, over a time period, result in a SIR_{target} of about 10 dB, and the inner-loop power control will issue power-up or power-down commands to ensure that SIR_{est} is close to SIR_{target}. The closed-loop power control is used for DPCCH and DPDCH in the UL and the DL. Section 9.9.1 discusses the uplink closed-loop power control, and Section 9.9.2 describes the downlink closed-loop power control [267].

9.9.1 Uplink Closed-Loop Power Control

The uplink closed-loop power control influences the transmit power of the UE. Figure 9.52 illustrates the uplink power control for a UE in soft handover; the situation is less complex if the UE is communicating with just one cell of a Node B or if the UE is communicating with multiple cells of just one Node B.

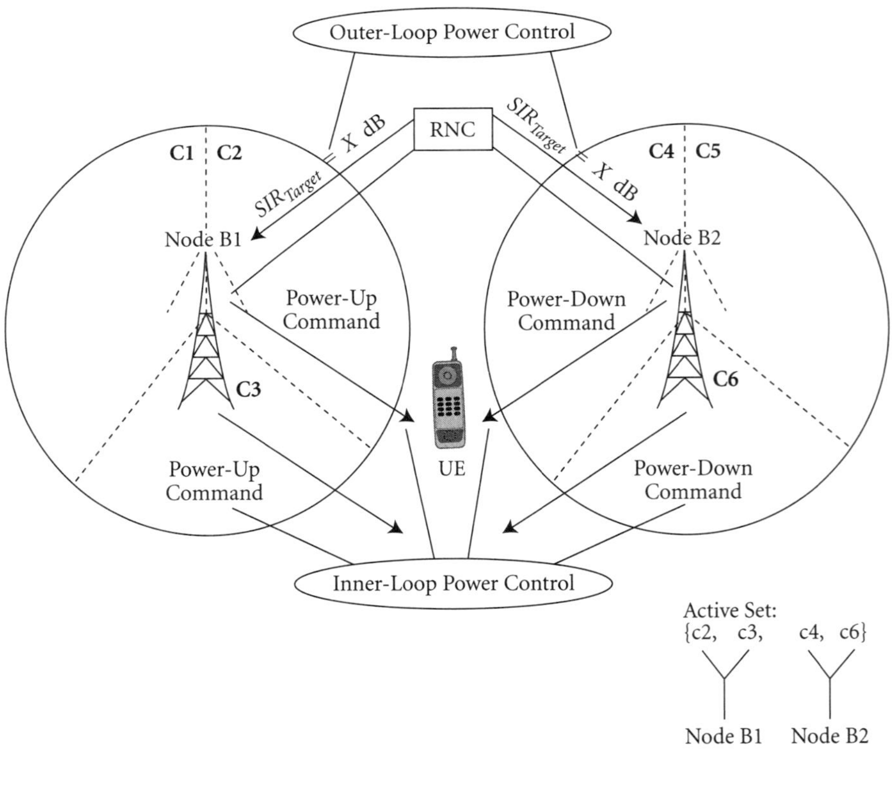

Figure 9.52: Scenario for Uplink Closed-Loop Power Control

The UE is in soft handover with the Active Set containing $\{c2, c3, c4, c6\}$. Cells c2 and c3 are controlled by Node B1, and cells c4 and c6 are controlled by Node B2. The RNC is responsible for the uplink outer-loop power control and hence implements an algorithm that determines a suitable value of the SIR_{target}. The RNC provides SIR_{target} to the Node Bs in the Active Set of a given UE. The Node B is responsible for the uplink inner-loop power control. The Node B estimates SIR_{est} based on the signals received from its own cells that are in the UE's Active Set, compares SIR_{est} with SIR_{target}, and issues power-up or power-down commands to the UE. The same power-up or power-down command is sent in multiple cells from a given Node B so that the UE can soft-combine these commands. The UE knows that certain cells contain identical power-control commands based on the "TPC combination index" parameter present in RRC signaling messages (e.g., the ACTIVE SET UPDATE message). If the TPC combination index parameter is 0, 0, 1, and 1 for the cells c2, c3, c4, and c6, respectively, the UE soft-combines the power-control commands for c2 and c3 and soft-combines the power-control commands received from c4 and c6. The cells belonging to one Node B constitute a radio link set (RLS). In Figure 9.52, RLS1 consists of the cells c2 and c3, and RLS2 includes the cells c4 and c6. The UE considers the combined power-control commands from one or more RLSs to change its transmit power. Section 9.9.1.1 describes the uplink outer-loop power control, and Section 9.9.1.2 deals with the uplink inner-loop power control.

9.9.1.1 Uplink Outer-Loop Power Control An outer-loop power-control algorithm is proprietary and implementation-specific. However, a simple algorithm shown in Figure 9.53 can be considered a baseline algorithm. Better performance could be achieved by modifying this simple algorithm.

The implementation-specific outer-loop power-control algorithm could adhere to the following guidelines to achieve a target BLER.

- Decrease SIR_{target} by Δ_{down} dB when a good (i.e., error-free) transport block is received. A transport block is considered error-free if it has a CRC match.
- Increase SIR_{target} by Δ_{up} dB when a bad block (i.e., a block with an error) is received. A transport block is erroneous when it has no CRC match.
- The relationship between Δ_{up} and Δ_{down} can be defined as

$$\Delta_{up} = \big((100/T) - 1\big) * \Delta_{down} \tag{9.2}$$

  where Δ_{up} is the SIR_{target} step-up size in decibels, T is the target BLER in percentage, and Δ_{down} is the SIR_{target} step-down size in decibels. Once a suitable value of Δ_{down} is selected (e.g., 0.01 or 0.02), Δ_{up} is given by Eq. (9.2). A large value of Δ_{down} enables faster response to the changing channel conditions; however, the drawback is that the average transmit power used may be somewhat higher. The operating range for SIR_{target}, between the maximum and minimum values in Figure 9.53, should be wide enough to meet the target BLER. For example, if the maximum SIR_{target} is inadequate, the achieved BLER may be higher than the target BLER because the maximum

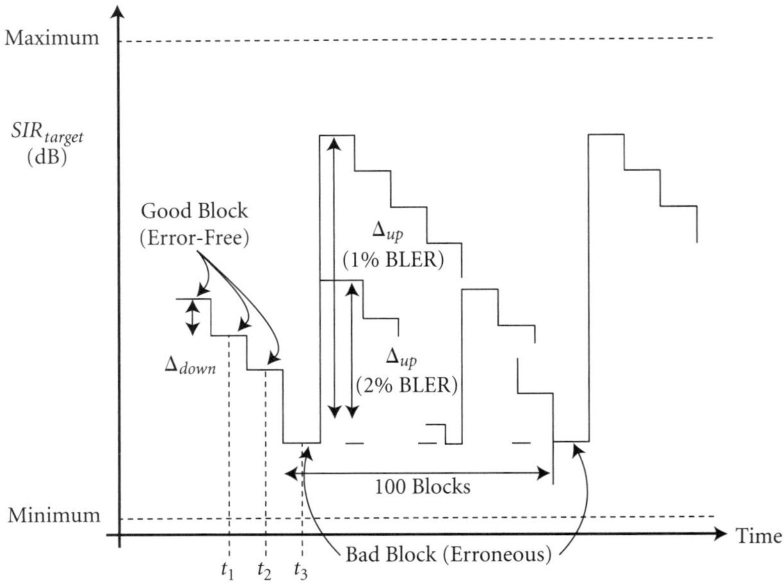

Figure 9.53: Outer-Loop Power-Control Mechanism

used transmit power would saturate. Assume that a 10 dB signal-to-interference ratio is required to achieve 2% BLER under certain channel conditions. If the maximum SIR_{target} is set to a small number, such as 7 dB, the 10 dB signal-to-interference ratio cannot be met, causing a higher than 2% BLER. Similarly, if the minimum value of SIR_{target} is very small, reaching the steady-state target BLER may take longer when channel conditions suddenly degrade from excellent to poor. If the minimum value of SIR_{target} is too high, the average transmit power used would be larger than the required level.

Assume that the target BLER is $T = 1\%$. Eq. (9.1) yields $\Delta_{up} = 99 * \Delta_{down}$. In Figure 9.53, the received block is good at time instants t_1 and t_2. Hence, SIR_{target} is reduced by a small amount Δ_{down}. However, at time instant t_3, the block is erroneous. Now, SIR_{target} is increased by a large amount, $\Delta_{up} = 99 * \Delta_{down}$. If the channel conditions remain the same, many consecutive good blocks would be received and SIR_{target} would gradually decrease until another block would be in error. If a time period of one hundred blocks is observed, one error would be found, leading to 1% BLER. Similarly, if the target BLER is $T = 2\%$, Eq. (9.1) gives $\Delta_{up} = 49 * \Delta_{down}$. SIR_{target} is increased by Δ_{up} and is reduced by Δ_{down} for every good block. Within an observation window of fifty blocks, one error would exist, and within a window of one hundred blocks, two errors would be observed. Hence, the target BLER of 2% would be achieved.

Figure 9.53 assumes that the channel conditions remain the same, and, hence, the error occurs at the same SIR_{target} value. However, in practice, the channel conditions are quite dynamic, and the algorithm would continue to adjust itself to the new channel conditions. For example, if the channel conditions become better, the algorithm would encounter an error at a lower SIR_{target} value than the SIR_{target} point shown in Figure 9.53. On the other hand, if the channel becomes worse, the algorithm would encounter an error at a higher SIR_{target} value than the SIR_{target} value in Figure 9.53. In practice, a very small reduction (e.g., 0.01 or 0.02 dB) in signal-to-interference ratio is impossible because transmit power cannot be changed with such resolution. Hence, SIR_{target} is updated less frequently (e.g., every few frames rather than every frame).

A Question for You... Why is outer-loop power control implemented at the RNC instead of at the Node B?[320]

9.9.1.2 Uplink Inner-Loop Power Control

Recall that the job of the inner-loop power control is to keep SIR_{est} as close as possible to SIR_{target}. The Node B utilizes the most recent SIR_{target} received from the RNC. The Node B estimates the SIR_{est} of the received uplink DPCH and determines a single power-control command, power-up or power-down, for *every slot* using the following guidelines.

- If $SIR_{est} \geq SIR_{target}$, the TPC command, TPC_cmd, is set to 0, which corresponds to a power-down command.
- If $SIR_{est} < SIR_{target}$, TPC_cmd is set to 1, which corresponds to a power-up command.

The TPC_cmd is sent on the DL DPCCH. In Figure 9.52, the Node B1 sends the same $TPC\_cmd = 1$ (i.e., power-up) to the UE in cells c2 and c3. Similarly, the Node B2 sends $TPC\_cmd = 0$ to the UE in cells c4 and c6.

The UE soft-combines the TPC_cmd received on two separate DL DPCCHs from the cells c2 and c3 to derive a single TPC_cmd from RLS1 as 1. Similarly, the UE derives another TPC_cmd from RLS2 as 0. In essence, the UE determines one TPC_cmd per RLS in a slot. Now, the UE needs to process all the TPC_cmd values for all RLSs to calculate the final value of TPC_cmd. This final value will dictate the change to the transmit power. Two algorithms, Algorithm 1 and Algorithm 2, are defined in the standard to specify how the UE should process the TPC commands received from RLSs [267] (see Figure 9.54). Algorithm 1 is commonly used in commercial deployments.

Let's discuss Algorithm 1 first. *This algorithm enables the UE to change transmit power every slot in response to the received power-control command.* In other words,

[320]Recall that the RNC knows the final quality of the received uplink block in a soft handover scenario. That's why the outer-loop power-control algorithm for the uplink is implemented at the RNC. If the outer-loop power control were to be implemented at the Node B, two different Node Bs could be using widely different SIR_{target} values based on the quality of the blocks they have individually received from the UE. This would lead to suboptimal performance.

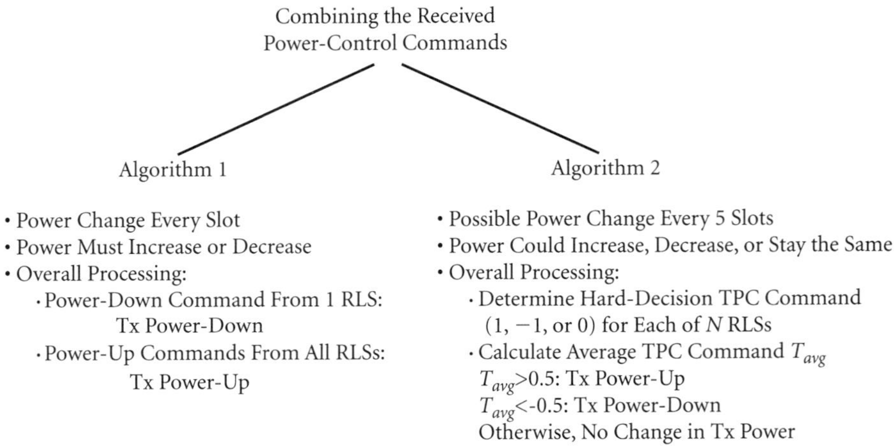

Figure 9.54: Outer-Loop Power-Control Mechanism

this algorithm enables 1500 Hz power control. When one of the RLSs specifies a power-down TPC_cmd, the UE decreases the power of the UL DPCCH by tpcStepSize dB. When all the RLSs specify the power-up TPC_cmd, the UE increases the power of the UL DPCCH by tpcStepSize dB. The parameter tpcStepSize can be 1 or 2 dB. Algorithm 1 with tpcStepSize = 1 dB is widely used in commercial deployments. RRC messages, such as RRC CONNECTION SETUP, specify the algorithm to be used and the value of tpcStepSize as part of PowerControlAlgorithm parameter. Recall that the UL DPCCH carries the pilot bits. The UL DPDCH is transmitted at a predefined power offset relative to the UL DPCCH. Such a power offset is signaled to the UE for all TFCs or calculated by the UE for the TFCs based on the gain factors for a reference TFC [267]. Specifically, the gain factors β_c and β_d dictate the relative power offset between the UL DPCCH and DPDCH. Different gain factors are required to maintain a target UL BLER for different UL DPDCH data rates.

Now, let's focus on Algorithm 2. The UE does receive power-control commands at 1500 Hz. However, the UE decides whether to change its transmit power only every five slots. *This algorithm effectively reduces the power-control frequency to* $1500/5 = 300$ *Hz*. Let's discuss how the UE carries out the overall processing.

The UE makes a hard power-up or power-down decision every slot for each RLS. For example, if the power-control command is found to be a power-up command after the UE soft-combines the received TPC_cmd values from multiple cells of a given RLS, the hard decision for such an RLS would be 1. If the power-control command were power-down, the hard decision for the RLS would be 0.

For each RLS, the UE evaluates the hard decisions made for five consecutive slots. If all five hard decisions for an RLS are 1, the temporary TPC_cmd, TPC_temp, for such an RLS is considered to be 1. If all five hard decisions for an RLS are 0, the TPC_temp is considered to be -1. In all other scenarios, TPC_temp is considered to be 0.

The UE now calculates the average TPC_temp for all the RLSs. For example, if one RLS yields $TPC\_temp = -1$ and another RLS yields $TPC\_temp = -1$, the average TPC_temp is considered to be $((-1)+(-1))/2 = -1$. The final TPC_cmd is set to 1 if the average TPC_temp is greater than 0.5 and -1 if the average TPC_temp is less than -0.5. In all other cases, the final TPC_cmd is set to 0. If the final TPC_cmd is 1, the UE increases the power of the UL DPCCH by tpcStepSize dB. If the final TPC_cmd is -1, the UE decreases the power of the UL DPCCH by tpcStepSize dB. If the final TPC_cmd is 0, the UE makes no change to the power of the UL DPCCH. The UL DPDCH maintains its power offset relative to the UL DPCCH based on the gain factors β_c and β_d.

9.9.2 Downlink Closed-Loop Power Control

The downlink closed-loop power control directs the transmit power on the DPDCH and DPCCH. Both outer-loop and inner-loop power-control mechanisms are implemented at the UE. Figure 9.55 illustrates the basic concept of the downlink closed-loop power control.

The UE's Active Set is {c2, c3, c4, c6}. The UE implements the downlink outer-loop power-control algorithm, which operates similarly to the uplink outer-loop power-control algorithm implemented at the RNC. The UE receives the DL target BLER (e.g., 2%) from the RNC through RRC signaling [e.g., the (RRC) RADIO BEARER SETUP message] and determines a suitable SIR_{target} to meet such a target error rate. The UE's downlink inner-loop power-control algorithm operates similarly to the uplink inner-loop power-control algorithm implemented at the Node B.

Recall that the inner-loop power-control algorithm attempts to keep SIR_{est} as close as possible to SIR_{target}. The UE estimates SIR_{est} as the ratio of the RSCP and interference signal code power where RSCP is the received signal code power for the DL DPCH and interference signal code power is non-orthogonal interference signal code power [267]. Since the UE is in soft handover in Figure 9.55, SIR_{est} should reflect all the downlink signals received from the Active Set cells. The UE issues power-up or power-down commands to the UTRAN via TPC bits on its UL DPCCH. Each Node B individually interprets the received power-control commands after soft-combining and increases or decreases the transmit power on the DL DPCCH in each cell that it controls by $(\Delta_{TPC} + P_{bal})$ dB.

Though Δ_{TPC} could be 0.5, 1, 1.5. or 2 dB, UTRAN must support $\Delta_{TPC} = 1$ dB; other Δ_{TPC} values are optional. P_{bal} is a correction factor for balancing the radio link power toward a common reference. The main reason for including P_{bal} is to ensure optimal performance for the maximal ratio combining at the UE. When the received power levels of the DPCHs from different cells are widely different, the gain of the maximal ratio combining decreases. Hence, to preserve good combining performance at the UE's RAKE receiver, P_{bal} is added to Δ_{TPC} as an additional

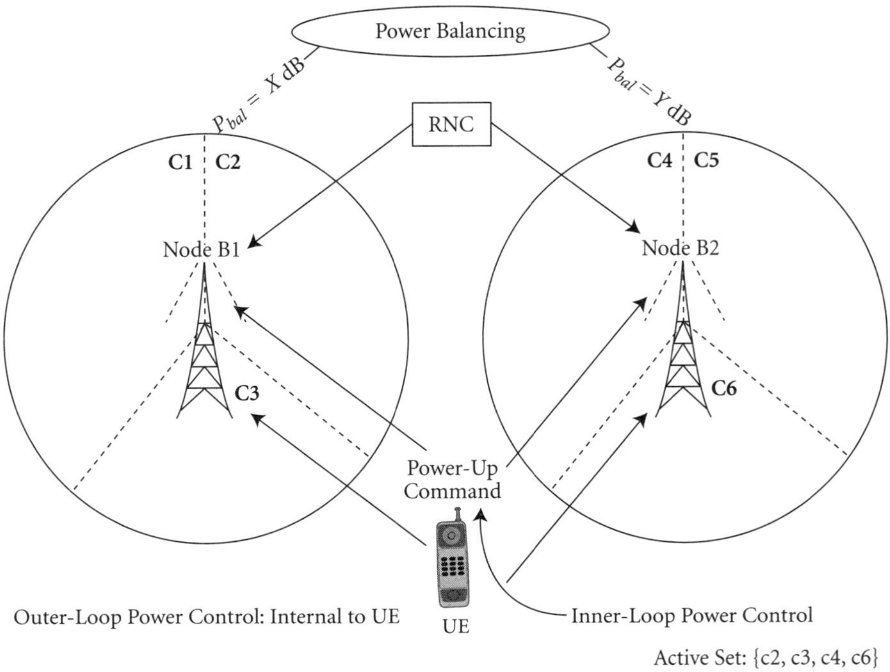

Figure 9.55: Scenario for Downlink Closed-Loop Power Control

correction factor. The RNC periodically provides an updated value of P_{bal} to the Node Bs in the UE's Active Set.

9.10 QUALITY-OF-SERVICE IN THE UNIVERSAL MOBILE TELECOMMUNICATION SYSTEM

To support a variety of services and applications, the UMTS defines four QoS classes, also called traffic classes: *conversational*, *streaming*, *interactive*, and *background* [273].

The conversational class is characterized by the preservation of time variation between the information entities of the stream and the stringent and low delay. Examples of services that could fall into this class include traditional voice calls or telephony speech, VoIP, and video conferencing calls. If the information entities, such as speech blocks, experience a long delay, human perception of the service would be quite poor. Furthermore, the variations in the arrival instants between any two successive information blocks need to be similar and minimal.

The streaming class is characterized by preservation of the time variation between the information entities of the stream. In other words, delay variations between successive information blocks should be limited. Some delay transfer is tolerable compared to the conversational class. Examples of services for this class are real-time audio streams and real-time video streams. One method to minimize delay variations is to buffer information at the device before playing the received audio- and/or video-streaming file on the device.

The characteristics of the interactive class include a request–response pattern and the preservation of the payload content. Examples of applications for this category include web-browsing, human or automatic database retrieval, server access, and polling and recording for measurements. An error-free or a low-error-rate response is expected within a reasonable time period.

The background class is characterized by a destination that has no expectation to receive the data within a certain time period and the preservation of payload content. This is the least delay-sensitive QoS class. Background download of emails falls into this QoS category. The data is expected to be error-free.

A given UMTS bearer service (e.g., a data session) is associated with one of these QoS classes and a set of QoS attributes. Examples of QoS attributes are maximum and guaranteed bit rate, error rate, packet size, transfer delay, and traffic priority. The QoS classes are primarily intended to support a variety of packet data services. When a UE requests a PDP context as part of data session setup, it specifies the QoS class and associated parameters (e.g., 384 kbps maximum bit rate). The SGSN works with the RNC to determine if such a QoS can be entertained. The PDP context activation response from the SGSN to the UE includes the granted QoS.

9.11 EVOLUTION OF THE UNIVERSAL MOBILE TELECOMMUNICATION SYSTEM

We have discussed Release 99 in this chapter, which is the first version of the UMTS standard. Figure 9.56 outlines the evolutionary path of the UMTS standard via releases of the standard.

Release 4 introduces a third air-interface mode, 1.28 Mcps TDD mode, in addition to the 3.84 Mcps FDD and TDD modes of R99. The main motivation behind this mode was to address the Chinese cellular market. An MSC server and media gateway are specified as a replacement for the MSC. These components enable a low-bandwidth transport of long-distance circuit-switched calls.

A Bit of Cellular History... The standard's authors and numerous cellular companies had hoped for a rapid deployment of Time-Division Synchronous Code-Division Multiple Access. Several companies started investing in R&D efforts on the design of Time-Division Synchronous Code-Division Multiple Access to get a share of the lucrative Chinese market. After all, developed countries had been facing subscriber saturation, while developing nations, such as China and India, still had a huge untapped cellular market. These companies waited (and waited and waited) for the Chinese government to announce the awarding of the spectrum licenses. During this

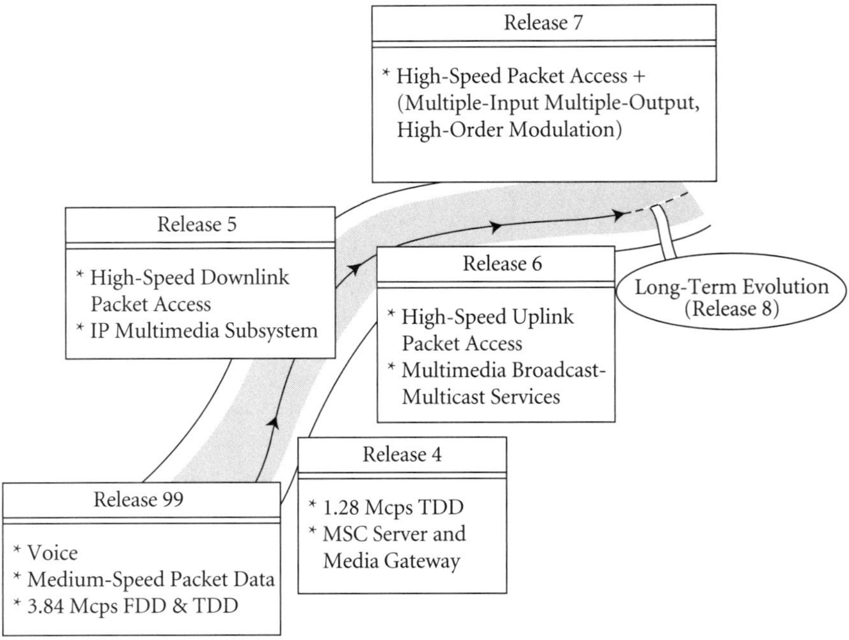

Figure 9.56: Evolutionary Path of UMTS

long wait, some exited the UMTS business and some even went bankrupt! Finally, the Chinese government awarded a Time-Division Synchronous Code-Division Multiple Access license to China Mobile in 2009 [274]. China Mobile had 1.327 million Time-Division Synchronous Code-Division Multiple Access subscribers as of August 2009. Time-Division Synchronous Code-Division Multiple Access promises to be a lucrative market in China [275].

Release 5 makes significant changes in the air interface to support the High-Speed Downlink Packet Access feature. High-Speed Downlink Packet Access allows a peak user data rate of 14 Mbps. See Chapter 11 for more on High-Speed Downlink Packet Access. Several elements of the IP multimedia subsystem are also specified. The IP multimedia subsystem allows a low-cost VoIP and multimedia call instead of a high-cost traditional circuit-switched call. Furthermore, new services can be offered easily. See Chapter 14 for IP multimedia subsystem details.

Examples of Release 6 features include High-Speed Uplink Packet Access and Multimedia Broadcast-Multicast Services. High-Speed Uplink Packet Access supports up to 5.7 Mbps peak user data rate in the uplink. Multimedia Broadcast-Multicast Services allows the efficient use of radio resources while transmitting contents such as games and movies to multiple users simultaneously. See Chapter 13 for details on High-Speed Uplink Packet Access. Multimedia Broadcast-Multicast Services had remained unused (more accurately, it hasn't seen the light of day!) as of late 2010.

One of the main features of Release 7 is High-Speed Packet Access + (HSPA+), which involves enhancements to High-Speed Downlink Packet Access and High-Speed Uplink Packet Access. Multiple-Input Multiple-Output enables transmission and reception over multiple antennas. High-Order Modulation (e.g., 64-QAM in HSPA+ compared to 16-QAM in High-Speed Downlink Packet Access) is also exploited. Multiple-Input Multiple-Output and high-order modulation significantly enhance capacity and throughput. HSPA+ also improves VoIP capacity and increases the UE battery life. See Appendix D for a summary of HSPA+ features.

Long-Term Evolution, introduced in Release 8, uses an advanced and highly-efficient multiple access technique called Orthogonal Frequency-Division Multiple Access along with multiple antenna techniques. Long-Term Evolution supports a peak user data rate of about 300 Mbps in the DL and 75 Mbps in the UL. See Chapter 15 for an overview of LTE.

9.12 POINTS TO REMEMBER

- Release 99 is the first version of the UMTS. It is one of the 3G cellular systems and is the most dominant 3G system in the world.
- The main benefits of UMTS R99 compared to its 2G (i.e., GSM) and 2.5G (i.e., GPRS and EDGE) predecessors are higher voice capacity, support for higher peak and average data rates, support for QoS, common ATM-based transport, and open interfaces.
- The UMTS network architecture utilizes a CDMA-based air interface to achieve higher spectrum efficiency and reuses GSM- and GPRS-based core networks to minimize costs. An FDD-based R99 system requires a total of 10 MHz radio spectrum. Since the air interface is based on CDMA, traditional CDMA features such as soft handover and efficient power control are essential features of UMTS.
- The handover within a UMTS network and across UMTS and other radio technologies are made possible by multi-mode UEs and the provisioning of suitable handover events.
- The air-interface protocol stack includes PHY, MAC, RLC, RRC, and PDCP layers.
- Call setup procedures for voice and data calls are quite similar and include basic steps such as RRC signaling connection, authentication, and radio bearer setup.
- The evolutionary path of UMTS enables a UMTS operator to continue to evolve the network to efficiently provide better services to more users.

CHAPTER 10

1xEVOLUTION–DATA-OPTIMIZED REVISION 0

The Third Generation Partnership Project 2's third-generation technology, CDMA2000 1xRTT (where CDMA is code division multiple access and which is referred to as 1xRTT or 1x hereafter), fails to meet third-generation requirements, such as a 2 Mbps data rate, but 1xEvolution–Data-Only or –Data-Optimized (1xEV-DO) Revision 0 (Rev. 0) does meet those requirements. 1xEV-DO is also referred to as High-Rate Packet Data. Its primary focus is on the forward link air interface compared to 1x. Though it lacks backward-compatibility with CDMA2000 1xRTT, it is a *de facto* evolution of 1x. The main benefit of 1xEV-DO Rev. 0 relative to 1x is higher data rates (e.g., a peak data rate of 2.4576 Mbps) in the forward link.

In this chapter, the unique features of 1xEV-DO are highlighted first. The network architecture is then introduced, followed by an explanation of the basic call setup. The air-interface protocol stack and forward link (FL) and reverse link (RL) channels are described. The forward link and reverse link data transmission are discussed in detail. Mobility management in 1xEV-DO is illustrated. Optimization of 1xEV-DO is considered, and its link budget is analyzed. Finally, interworking between 1xEV-DO and 1x is summarized.

Cellular Communications, First Edition. Nishith D. Tripathi and Jeffrey H. Reed

10.1 1xEV-DO: EXECUTIVE SUMMARY

A 1xEV-DO system is a spectrally efficient packet data system [276]. It is primarily based on Qualcomm's proposed High Data Rate and has been standardized as IS-856. This chapter discusses Rev. 0 of IS-856, and Chapter 12 discusses an enhanced version, Rev. A. IS-856 is essentially an air-interface standard that provides specifications for the wireless link between the user terminal, also called the access terminal (AT), and the radio network, also called the access network (AN). IS-856 is companion to IS-2000 and is not backward-compatible with IS-95 or IS-2000. An IS-856 system requires dedicated frequency bands (1.25 MHz in the forward link and 1.25 MHz in the reverse link) that must be separate from IS-2000 (and/or IS-95) frequency bands.

A Bit of Cellular History... When the Third Generation Partnership Project 2 was busy working on 1xEvolution–Data-and-Voice (1xEV-DV) as an evolution of 1x in the late 1990s, Qualcomm was working on a proprietary data technology—High Data Rate. Verizon Wireless was shopping for a high-speed wireless access technology to establish a performance advantage over competitors, and Qualcomm proposed High Data Rate. Qualcomm submitted High Data Rate to the Third Generation Partnership Project 2 for standardization, and other companies (e.g., Lucent Technologies, Motorola, and Nortel) also made some contributions to the standard. Thus, 1xEV-DO was formally created in the early 2000s. 1xEV-DO is a packet-data–only system, and, if no packet data user is present, 1.25 MHz FL spectrum and 1.25 MHz RL spectrum would be sitting idle. In contrast, 1xEV-DV can simultaneously support voice and data on the same carrier frequency. In addition to good engineers (and good attorneys, some would add!), Qualcomm must have had a remarkable sales team because they successfully sold 1xEV-DO's efficiency to Verizon! 1xEV-DV was predicted to be available in some chipset, the delivery date of which gradually slipped from one version to the next. Sprint waited for 1xEV-DV chipsets for a couple of years and finally ran out of patience in the highly competitive U.S. cellular market. Sprint also chose 1xEV-DO as their evolutionary path from 1x. Since two major CDMA and 1x carriers in the U.S. had chosen 1xEV-DO, 1xEV-DV was quickly abandoned. Countless hours and millions of dollars spent on 1xEV-DV standardization and design yielded not a single commercial network. What an impact business decisions and time-to-market have on the success or failure of a technology!

Let's summarize the salient features of IS-856 Rev. 0. It provides peak user data rates of 2.4576 Mbps in the forward link and 153.6 kbps in the reverse link. It is a packet data system that provides Internet Protocol (IP) connectivity. It is designed for packet-switched data services (not circuit-switched voice or data services) and its interface with circuit-switched core network elements (e.g., a mobile switching center) is undefined in the standard. Applications that require asymmetric bandwidth in the forward and reverse links can be better served. An example of such services is web-browsing, in which a high data rate is desirable in the forward link and a low rate is adequate in the reverse link.

Compared to 1x and Universal Mobile Telecommunication System (UMTS) R99 systems, 1xEV-DO offers superior user data rates and throughput (user and sector throughput) in the forward link. 1xEV-DO is simple yet achieves enviable performance. An equivalent competing technology is High-Speed Downlink Packet Access (HSDPA), which is a Third Generation Partnership Project (3GPP) Release 5 feature. See Appendix D for a comparison between 1xEV-DO Rev. 0 and HSDPA.

Superior packet data performance is primarily related to the air-interface enhancements, such as fast channel feedback from the AT to the AN; maximal utilization of radio resources; a fast, channel-sensitive, and user-discriminating scheduler; high-order modulation schemes; and a physical layer packet acknowledgment, known as Hybrid Automatic Repeat Request.[321]

To maximize throughput, data rate control, rather than power control, is exploited in the forward link traffic channel. The forward link channel is transmitted at full power or at a predetermined fixed-level power. In contrast, the reverse link is power-controlled, just as in IS-95 and CDMA2000 1xRTT systems.

The forward link traffic channel is shared among the active users through time-division multiplexing (TDM) instead of code-division multiplexing (CDM), which is used in IS-95 and 1x systems. The AT selects instantaneous data rates in the FL and RL. In 1x and UMTS, the radio access network selects the FL and RL data rates.

10.2 OVERVIEW OF CALL SETUP AND DATA TRANSMISSION

Data transmission in a 1xEV-DO system requires successful establishment of a logical link called a *session* between the AT and the radio network controller (RNC) as well as a radio link called a *connection* between the AT and the base station (BS). Figure 10.1 shows a simplified view of the call setup, and Figures 10.2 and 10.3 summarize the main stages for FL and RL data transmission. Detailed call setup is described in Section 10.8, while forward link data transmission and reverse link data transmission are explained in Sections 10.9 and 10.10, respectively.

As part of the call setup, a session between the AT and the AN (more specifically, the RNC) is established. The AT is assigned a temporary identity called Unicast Access Terminal Identifier (UATI). A dedicated radio connection between the AT and the BS is established for reliable session negotiation. A MACIndex[322] allocated per sector in the Active Set identifies the AT's dedicated radio connection with the BS. Several session parameters may also be negotiated between the AT and the RNC during session configuration. Next, the Point-to-Point Protocol (PPP) session between the AT and the packet data serving node (PDSN) is set up, and the AT is assigned an IP address through simple IP or mobile IP procedures, with mobile IP being more common in commercial deployments.

[321] See Chapter 6 on GPRS and EDGE for an example of how Hybrid Automatic Repeat Request works.
[322] MAC here represents Medium Access Control.

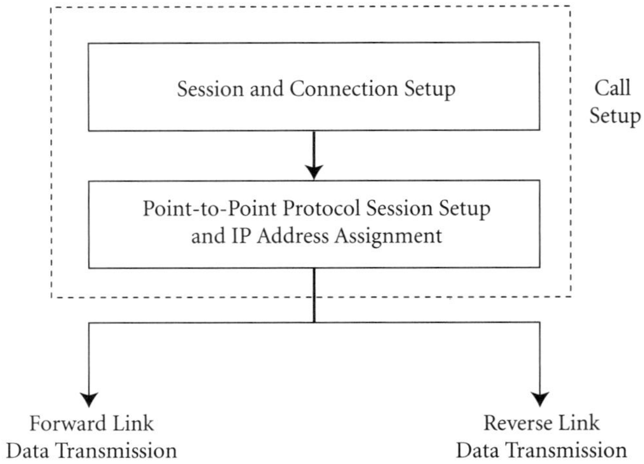

Figure 10.1: Simplified Call Setup

Once the session is set up and the AT has an IP address, data transfer can begin in the FL and RL. Figure 10.2 summarizes the basic steps of FL data transmission. The ATs observe the FL pilot channels to estimate the channel conditions prevailing in their individual Active Sets. Since there is no soft or softer handoff in the FL, the AT must pick the best sector to receive a FL packet. The AT then asks the chosen sector to send FL data at a certain data rate based on the FL channel conditions between

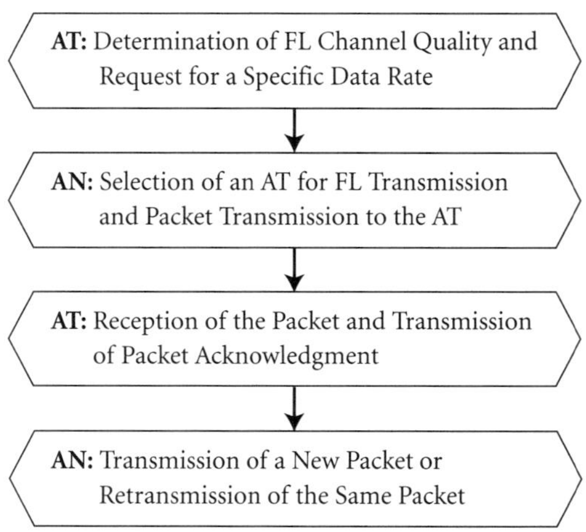

Figure 10.2: Overview of Over-the-Air FL Data Transmission

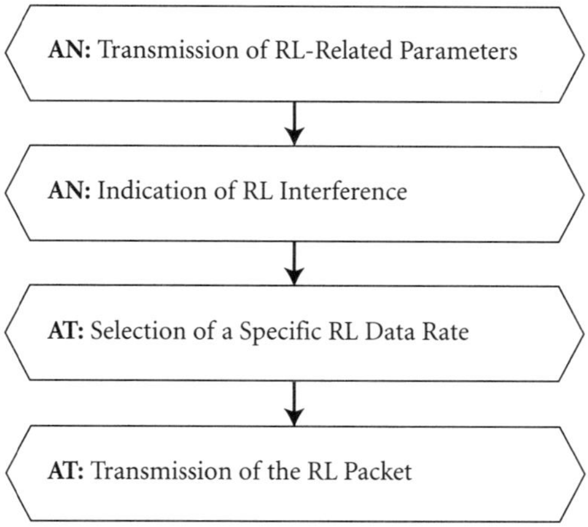

Figure 10.3: Overview of Over-the-Air RL Data Transmission

the sector and the AT. The AT targets a packet error rate of 1%[323] for the requested FL data rate. The scheduler at the BS runs every 1.67 ms slot and selects one (*and no more than one!*) AT for the transmission in the next FL slot.[324] The BS scheduler cannot upgrade or downgrade the instantaneous rate the AT requests.

Of course, the scheduler influences the average throughput the ATs experience by choosing an AT for transmission during a given slot. The ATs in a sector look for their MACIndex values in the early part of the slot. The AT that finds a match with its MACIndex attempts to retrieve the packet at the data rate requested earlier. If the AT decodes the packet correctly, it sends a positive acknowledgment to the BS; otherwise, it sends a negative acknowledgment to the BS. The BS can transmit a new packet to the AT in case of a positive acknowledgment. With a negative acknowledgment, the BS retransmits the packet if less than the maximum number of data-rate–dependent retransmissions has been reached. Once the limit on retransmissions is reached, the physical layer gives up and lets the upper-layer protocol, the Radio Link Protocol (RLP), handle the RLP retransmission.

Let's briefly discuss the overall transmission in the RL as summarized in Figure 10.3. Just as it did with the FL, the AT decides upon the instantaneous data rate for the RL as well. Observe that the AT is quite authoritative in 1xEV-DO; it determines instantaneous data rates for both the FL and RL!

Default values for certain parameters are given in the standard. The AN may send updated values for these parameters to the AT to override the default values as part of session configuration. These parameters influence the data rate the AT has chosen. We do want to make sure that the AT causes no excessive interference in the RL. The

[323]The 1% target error rate reflects the first transmission of a new packet and a number of data-rate–dependent retransmissions the physical layer carries out.

[324]The FL slot is 1.67 ms long.

sector indicates on the FL Reverse Activity channel whether it is heavily loaded. In such a case, the AT would try to reduce its RL data rate. The AT considers several parameters along with the interference indication to determine an instantaneous data rate for a given 26.67 ms RL frame. The AT transmits a packet at the selected data rate and indicates this rate to the AN so that the AN need not perform a blind detection of the RL data rate. The RL uses traditional soft and softer handoff, and all the sectors in the AT's Active Set attempt to decode the RL packet. Hybrid Automatic Repeat Request is unsupported in the RL, and, hence, no physical layer retransmissions of a packet occur in the RL.

10.3 NETWORK ARCHITECTURE

Figure 10.4 shows the functional architecture of a 1xEV-DO system. Recall that 1xEV-DO Rev. 0 is designed for packet data services only. Hence, we should see no circuit-switched core network in the diagram here!

The AT communicates with the BSs using an air-interface protocol stack. The BS is sometimes referred to as an access point. The BS performs physical layer functions, such as channel coding, modulation, and frequency up or down translation.

The RNC or access network controller controls multiple BSs. The RNC performs functions similar to those of a base station controller in a 1x system. Examples of RNC functions include determination of the Active Set and outer-loop power control for the reverse link. The packet control function (PCF) is typically integrated with the RNC and provides an interface between the RNC and the packet-switched core network. The interfaces between the RNC and the PCF are A8 and A9; A8 carries the

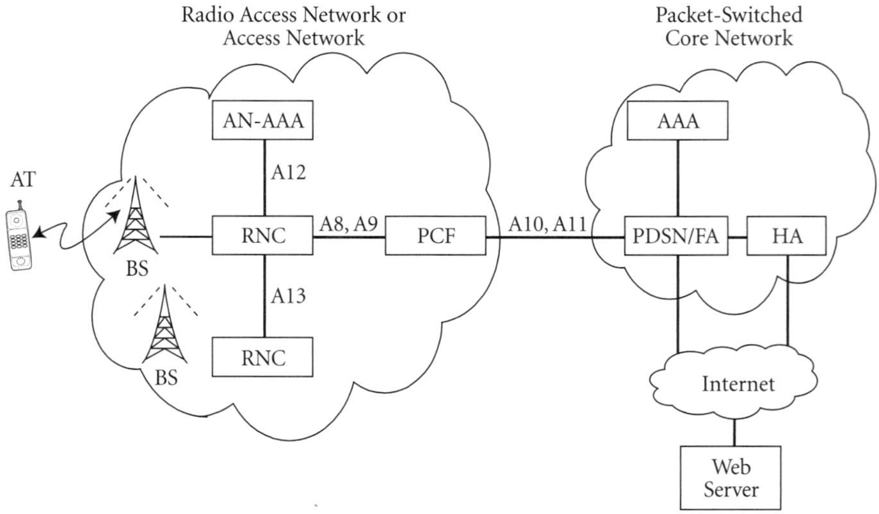

Figure 10.4: Architecture of a 1xEV-DO System

user traffic and A9 carries signaling. The PCF utilizes A10 for user traffic exchange with the PDSN and A11 for signaling exchange with the PDSN.

Two optional interfaces are A12 [between the RNC and the access-network–authentication-authorization-and-accounting (AN-AAA)] and A13 (between two RNCs). Since no 1xEV-DO system is connected to a circuit-switched core network, the home location register cannot perform device authentication. Hence, a new (optional) element, AN-AAA, has been introduced in 1xEV-DO so that device (or radio access) authentication can be performed. The AN-AAA may use the network access identifier for device authentication. If the AN-AAA is absent, the RNC may carry out such authentication. The A13 interface facilitates inter-RNC (also called inter-AN or inter–radio-access-network) handoff by expediting call setup in the new RNC. No traffic-carrying interface between the two RNCs was originally defined in the standard. However, a proprietary interface that carries user packets between the RNCs is implemented by some infrastructure vendors and is thus used in some commercial deployments. The PDSN is a packet-switched core network element that routes packets between the radio access network and the external data network, such as the Internet.

The logical link between the AT and the PDSN is a PPP session. The PPP link and the A10 and A11 interfaces enable the PDSN to route the user packets properly. When simple IP is supported, the PDSN assigns an IP address to the AT. When mobile IP is supported, the home agent (HA) assigns an IP address to the AT. Two components—a foreign agent (FA) and an HA—exist in mobile IP only. The FA is usually integrated with the PDSN. The AAA server takes care of authentication (e.g., via username and password exchange), authorization (i.e., verification of the subscription services for a given user), and accounting (e.g., billing based on the amount of data, if applicable). An application server (e.g., web or email server) is external to the 1xEV-DO system and represents an entity with which the AT is exchanging IP packets.

10.4 1xEV-DO PROTOCOL STACK: AN OVERVIEW

IS-856 defines a seven-layer protocol stack for the air interface between the AT and the AN. Note that this protocol stack is different from the Open System Interconnect seven-layer model and is applicable only to the air interface in 1xEV-DO.[325] Let's highlight the main functions of the protocol stack depicted in Figure 10.5.

The AT implements all the layers, and the AN implementation of the layers is distributed between the BS and the RNC. Though the standard fails to specify layers for the BS and the RNC separately, physical, Medium Access Control (MAC), and security layers are usually in the BS, and session, stream, and application layers are in the RNC. Part of the connection layer would be implemented at the BS, and part of the connection layer would be implemented at the RNC.

[325] The air-interface protocol stack of 1xEV-DO maps to the bottom two layers of the Open System Interconnect model.

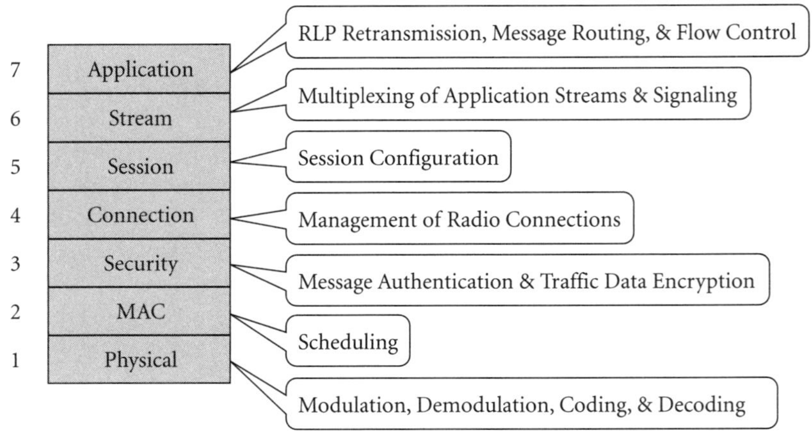

Figure 10.5: 1xEV-DO Air-Interface Protocol Stack

The physical layer carries out processing, such as coding and modulation at the transmitter and demodulation and decoding at the receiver. These functions are usually implemented in the baseband chipsets at the BS and the AT. For example, a channel card at the BS has numerous channel elements for FL and RL processing. One channel element performs functions for one AT. The MAC layer controls access to the physical layer resources through mechanisms such as user scheduling and data rate control.

The security layer performs message authentication and data encryption. In typical commercial deployments, security is usually turned off (i.e., no encryption occurs) to reduce the processing delay associated with encryption and decryption.

The connection layer is responsible for establishing, maintaining, and terminating the connection, which is the dedicated radio connection between the AT and the BS. The RNC conveys a MACIndex to the AT to identify the dedicated radio connection in the forward link. The AT utilizes a UATI-based scrambling code in the reverse link to signify the dedicated radio connection. Connections are normally released in the absence of data transfer and established in the blink of an eye (oh yes, less than half a second!) when needed. Such fast connection setups in 1xEV-DO are a significant advantage compared to 1x and competing HSPDA systems.

The session layer plays an important role in configuration of various protocols and parameters at each layer in the protocol stack. A session is the logical link between the AT and the RNC. Every AT that has a session (often referred to as a 1xEV-DO session) with the RNC has a UATI. UATI is a temporary identity the RNC gives to the AT. Once a session is established, it lasts for several hours. (Fifty-four hours is the default value in the standard.) During a session, connections come and go numerous times. In fact, the pre-existence of a session is one of the key contributors to 1xEV-DO's super-fast connection establishment and re-establishment.

The stream layer allows multiplexing of streams so that a single physical layer packet can carry all the streams at the same time. Examples of streams are a packet data stream for a web-browsing application and a signaling stream.

The application layer performs tasks such as RLP retransmission, message routing, and flow control. RLP in 1xEV-DO is much simpler than that in 1x as discussed next.[326] When a physical layer packet is in error, the RLP sends one retransmission based on the negative acknowledgment (NAK). If the error still exists after the RLP retransmission, the application layer gives up and lets the upper layers beyond the IS-856 protocol stack (e.g., Transmission Control Protocol) recover the lost packets. When a protocol layer (e.g., the MAC layer) in one entity (e.g., the AN) wants to send a message to the peer layer in another entity (e.g., an AT), it seeks help from the application layer for message routing. Flow control prevents one entity from sending excessive amounts of data that could result in buffer overflow at the receiver.

Something to Think About. . . Sure, complaints are made about 1xEV-DO. Spectrum sits idle in the absence of packet data users. Qualcomm collects royalties, and we imagine that they, like any business in a capitalist society, would accept any form of payment—credit card, check, PayPal, or whatever! The standard appears to be written by software engineers with expertise in object-oriented design! Nevertheless, 1xEV-DO's efficiency is enviable. The features we really like are the fast handoff between two sectors, the fast connection setup (or, equivalently, Idle State to Connected State transition in less than half a second), an elegant Hybrid Automatic Repeat Request design in the FL, a fighting chance for an AT to decode a packet even at the highest data rate of 2.4576 Mbps in the FL (compared to praying for a miracle to decode a packet at the highest 14 Mbps data rate in competing HSDPA!), and the existence of default parameters for protocol stack layers.

10.5 INTRODUCTION TO RADIO CHANNELS AND TIMING STRUCTURE

Now that we are familiar with the air-interface protocol stack, let's catch a glimpse of the functions of various radio channels and timing structures. Figure 10.6 summarizes the FL and RL radio channels and their main functions.

The Pilot, MAC, Control, and Traffic channels are the four main FL channels. The Pilot channel has a similar function in 1xEV-DO as it does in other technologies; it facilitates channel estimation and timing synchronization in the FL. The AT makes measurements of the Pilot channel to aid handoff and to quantify the prevailing FL channel conditions in the form of Data Rate Control (DRC) value. The MAC channel includes three channels: the Reverse Activity channel, Reverse Power Control (RPC) channel, and Data Rate Control Lock (DRCLock) channel. The Reverse Activity

[326] See Chapter 8 for RLP details in 1x.

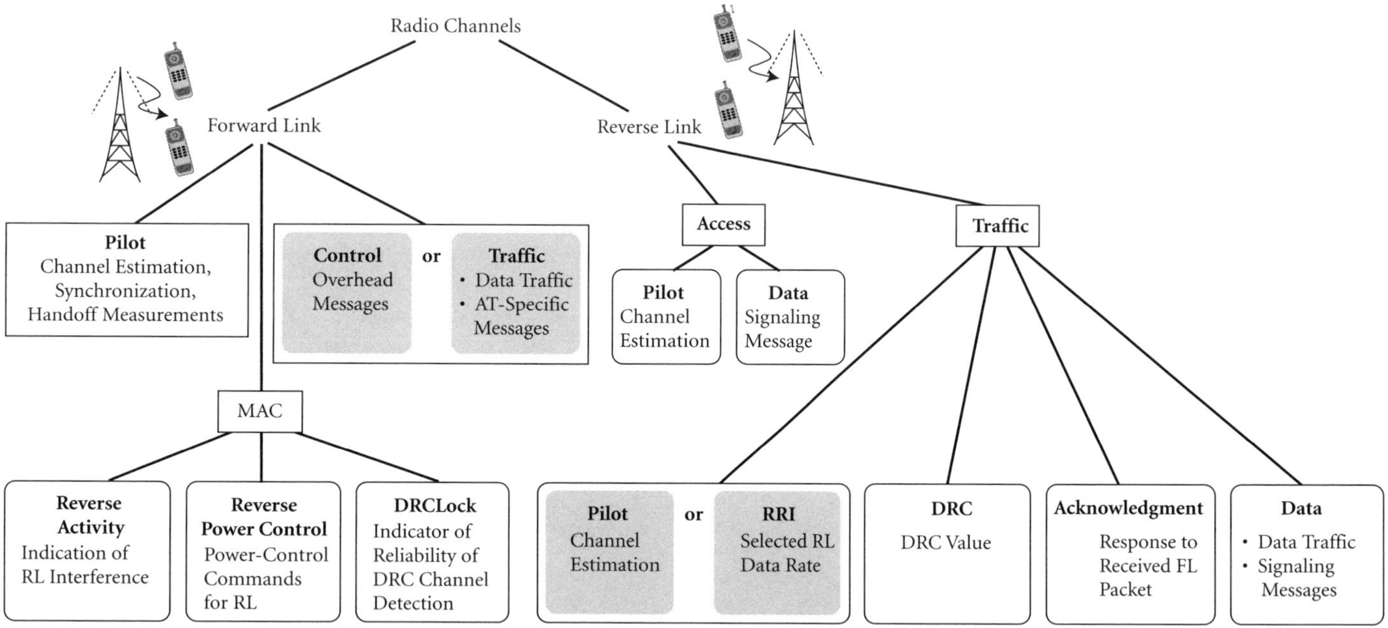

Figure 10.6: 1xEV-DO Air-Interface Channels

channel informs ATs of whether the sector is experiencing high interference in the RL. The RPC channel carries power-control commands to control ATs' transmit power levels. The DRCLock channel informs the AT of whether the BS can reliably detect the AT's DRC values, which specify the FL data rates the AT has requested. The Control channel transports overhead messages to provide information about the network. The Traffic channel carries AT-specific traffic data or signaling messages.

In the RL, the AT transmits either an Access channel or a Reverse Traffic channel at a given time. The Access channel is used when the AT has no dedicated radio connection with the AN and includes a Pilot channel for channel estimation and a Data channel to carry signaling messages. The Reverse Traffic channel (RTC) consists of a Pilot channel, Reverse Rate Indicator (RRI) channel, DRC channel, Acknowledgment channel, and Data channel. The RTC's Pilot channel helps the BS with RL channel estimation. The RRI channel indicates the Data channel's rate. The DRC channel indicates the FL data rate that the AT requested. The Acknowledgment channel informs the BS if the AT has correctly received a FL packet. The Data channel can carry user traffic or AT-specific signaling messages.

The forward link Pilot, MAC, and Control (or Traffic) channels are transmitted using TDM. The minimum transmission unit in the FL is a 1.67[327] ms long slot. In other words, when the AN sends a packet to an AT, the entire packet may be contained in a single slot. As the 1xEV-DO system maintains the over-the-air chip rate of a 1x system, the number of chips transmitted in every 1.67 ms slot is 1.2288 Mcps $*$ 1.67 ms $=$ 2048 chips. These chips are distributed among the FL channels as shown in Figure 10.7.

A slot could be active or idle. For an active slot, all 2048 chips are transmitted. However, only Pilot and MAC channels are transmitted in the idle slot. In other words, regardless of whether users exist in a sector, Pilot and MAC channels are

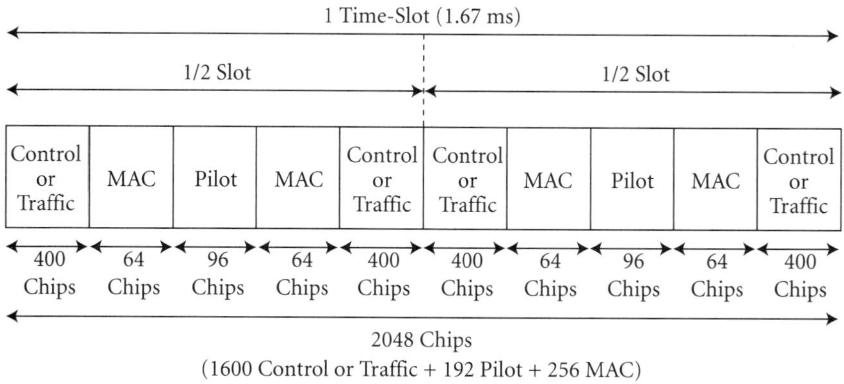

Figure 10.7: Forward Link Slot Structure

[327]The exact length of the slot is $1.\bar{6}$ ms.

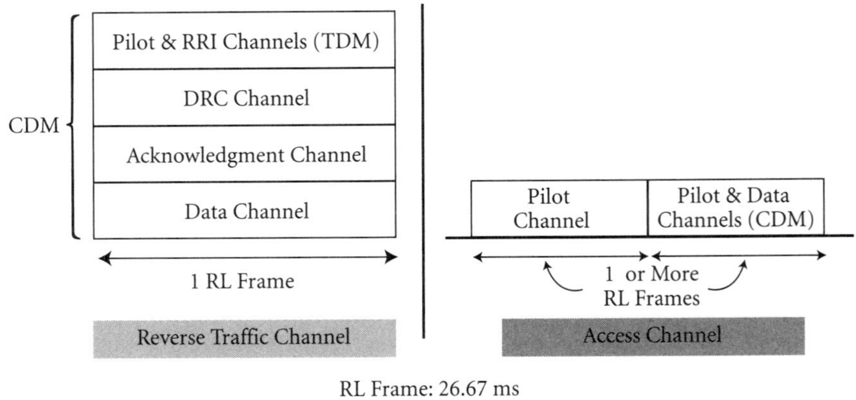

Figure 10.8: Reverse Link Timing Structure

always transmitted. In a given active slot, *either* the Control channel *or* the Traffic channel (*but not both*) is transmitted. A preamble at the front of the slot indicates whether a Control channel or a Traffic channel appears in the slot. As seen from Figure 10.7, 1600 of 2048 chips are available for a Control channel or a Traffic channel, and the remaining 448 chips are overhead chips.

Figure 10.8 shows the timing structure of the reverse link. The FL works in units of 1.67 ms slots, and the RL works in units of 26.67 ms frames. CDM is used to combine the Pilot or RRI channel, DRC channel, Acknowledgment channel, and Data channel into the RTC.[328] The Pilot and RRI channels are time-division-multiplexed. When the AT uses the Access channel, the Pilot channel occupies one or more frames as do the code-division-multiplexed Pilot and Data channels.

10.6 A CLOSER LOOK AT FORWARD LINK RADIO CHANNELS

We have provided a high-level discusion of various FL and RL channels, so now let's get into the details of these channels. We will discuss FL channels here and the RL channels in Section 10.7. The FL channels to be discussed here include the Pilot channel, MAC channel, Control channel, and Forward Traffic channel. The MAC channel includes the Reverse Activity channel, RPC channel, and DRCLock channel.

[328] CDM implies that various channels are separated via their own codes and are transmitted simultaneously.

10.6.1 Pilot Channel

The pilot channel transmits 0s on the in-phase (I) channel. Though the structure of the Pilot channel in 1xEV-DO is quite different from that in 1x, its functions are quite similar in 1x and 1xEV-DO.

Do You Recall... Do you remember the functions of the Pilot channel?[329]

The AT estimates the channel conditions by comparing the expected sequence of pilot bits with the received stream of demodulated pilot bits. Assuming the environment similarly affects both pilot bits and traffic bits, the AT can make suitable corrections to the received traffic bits. The AT also utilizes the Pilot channel to estimate carrier-to-interference (C/I) ratio so that it can request a supportable data rate under current channel conditions. Since the pilot channel is transmitted at fixed times, the AT can synchronize with the AN. Once the AT has synchronized with the Pilot channel, it can easily determine the positions of all other channels (e.g., Control and MAC channels).

The AT reports the pilot to the RNC so that the RNC can execute the soft handoff algorithm. The Sync message that the AN transmitted on the Control channel provides the AT with the exact pseudo-noise (PN) offset of the pilot the AT is currently monitoring.

10.6.2 Medium Access Control Channel

CDM is used inside the MAC channel to carry multiple channels at the same time as illustrated in Figure 10.9. During a given 1.67 ms slot, the MAC channel carries the Reverse Activity channel and *either* the RPC channel *or* the DRCLock channel *but not both.*

The Reverse Activity channel carries the reverse activity bit (RAB) to convey the presence or absence of high RL interference. The BS sets the RAB for each of its sectors individually. The algorithm the BS uses to determine the value of the RAB is unspecified in the standard and thus is implementation-specific. One possible implementation of the algorithm could set the RAB to 1 when the measured rise-over-thermal noise exceeds a configurable threshold (e.g., 3 to 6 dB). The BS must evaluate the impact that its threshold value choice will have on cell-edge reliability. Furthermore, such a threshold's interaction with the link budget design should be considered because the link budget incorporates the interference or loading margin as one of its parameters. The RAB indicates to all the ATs in the sector the status of the reverse link activity (i.e., loading or interference). When the RAB is set to 1 for a sector, the ATs interpret that setting as an indication of high RL interference and

[329]The Pilot channel is used for channel estimation (to implement coherent demodulation), data rate selection, time synchronization, sector identification via observation of the pilot pseudo-noise offset, and handoff.

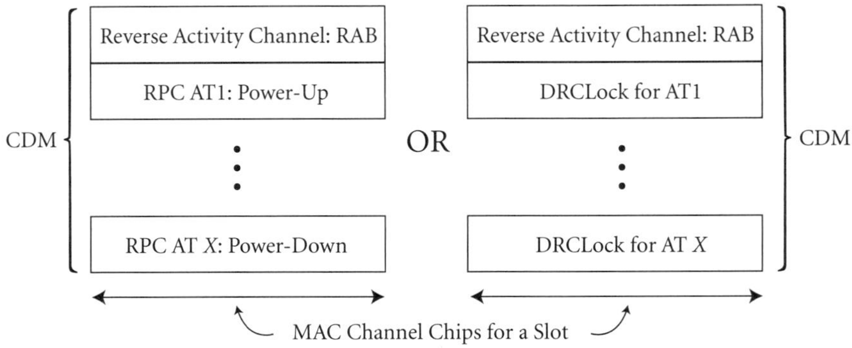

Figure 10.9: Components of a MAC Channel

consider the sector to be heavily loaded. When the RAB is 0, the sector is considered to be lightly loaded.

The RPC channel carries power-control commands for ATs individually so that each AT can increase or decrease its transmit power to facilitate achieving the reverse link target frame error rate. Individual power-control commands are possible due to the use of AT-specific Walsh codes (or, equivalently, a MACIndex). During a given slot, one power-control command is sent for an AT. Hence, an approximate power-control frequency, which is the number of power-control commands per second, is $1/1.67$ ms $=$ 600 Hz. We will soon calculate the exact power-control frequency by considering the impact of the DRCLock channel.

The DRCLock channel replaces the RPC channel once every eight or sixteen slots. The existence of the DRCLock channel helps avoid low throughput when the BS is unable to reliably detect the DRC values from the AT. Consider Figure 10.10 where two ATs, AT1 and AT2, have the Active Sets $\{\gamma\}$ and $\{\beta, \gamma\}$, respectively.

Assume that the ATs find Sector γ to be the best sector from the perspective of pilot E_c/I_0. Hence, both ATs request packets from Sector γ. After some time, perhaps due to AT2 moving away from the cell site and to relatively high reverse link interference, Sector γ is unable to reliably detect the DRC values AT2 is reporting. The BS may use the received pilot E_c/I_0 and/or the received DRC channel E_c/I_0 to determine if the DRC detection is reliable. Since the BS cannot reliably detect DRC values, it will schedule no packets for AT2, leading to a very low throughput for AT2. That's where the DRCLock channel comes into the picture. When the BS inaccurately estimates the AT's DRC values, it sets the DRCLock bit to 0. The AT monitors this bit on the DRCLock channel and knows that it is unlikely to receive data from Sector γ. Now, AT2 looks at the DRCLock bit of another sector in the Active Set, which is Sector β in our example. AT2 now chooses Sector β so it can achieve reasonably good throughput in Sector β. In the absence of the DRCLock channel, AT2 would have no choice but to keep asking for data from Sector γ. Another possible use of the DRCLock bit is to turn away the ATs to another sector when a sector has inadequate

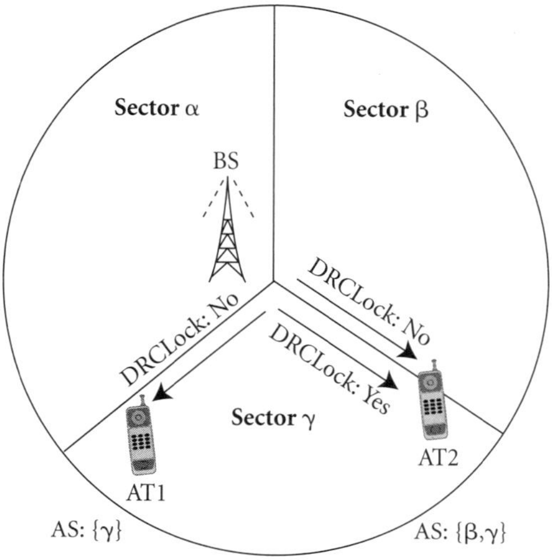

Figure 10.10: Motivation for the DRCLock Channel

resources (e.g., slots) to support additional ATs. Such use of the DRCLock bit can be considered a load-balancing mechanism.

Now, we are ready to calculate the exact power-control frequency. Within a time window of DRCLockPeriod consecutive slots, one slot transmits the DRCLock bit instead of the power-control command. Just like the power-control command, the DRCLock bits are also specified for the AT. The effective power-control frequency becomes $\{600 \text{ Hz} * ((DRCLockPeriod\ - 1)/DRCLockPeriod)\} = 562.5$ Hz for the DRCLockPeriod of sixteen slots, which is the default value in the standard.

Table 10.1 summarizes the MACIndex values available for the MAC channels. Recall that each AT with a dedicated radio connection is assigned a unique MACIndex in a given sector during the connection setup. The Reverse Activity channel is common for all the users in the sector and is transmitted during the MAC channel part of the slot using a specific Walsh cover corresponding to MACIndex 4. The AT-specific RPC channel and DRCLock channel use a Walsh cover that depends on the MACIndex allocated to the AT. The mapping between a MACIndex *i* and a 64-ary Walsh cover W

Table 10.1: Mapping Between the MACIndex and the Associated MAC Channel

MACIndex	MAC Channel
0, 1, 2, and 3	Unused
4	Reverse Activity Channel
5 to 63	RPC Channel and DRCLock Channel

is defined as follows.

$$W_{1/2}^{64} \qquad \text{for } i = 0, 2, \ldots, 62 \tag{10.1}$$

and

$$W_{((i-1)/2))+32}^{64} \qquad \text{for } i = 1, 3, \ldots, 63. \tag{10.2}$$

The MAC channel associated with an even-numbered MACIndex is assigned the I modulation and that associated with an odd-numbered MACIndex is assigned the quadrature-phase (Q) modulation.

10.6.3 Control Channel

The Control channel provides information about the system to the ATs in the sector and is transmitted at a 38.4 kbps or 76.8 kbps data rate. $MACIndex = 2$ in the preamble indicates a 76.8 kbps data rate, and $MACIndex = 3$ indicates a 38.4 kbps data rate. The Control channel carries common overhead messages such as Sync, QuickConfig, SectorParameters, and AccessParameters as well as Page messages. Multiple messages, such as QuickConfig, SectorParameters, and AccessParameters, can be multiplexed together. Figure 10.11 lists selected messages transmitted on the Control channel.

The Sync message provides information such as the system time and pilot PN offset. The QuickConfig message tells the ATs whether they need to look at detailed

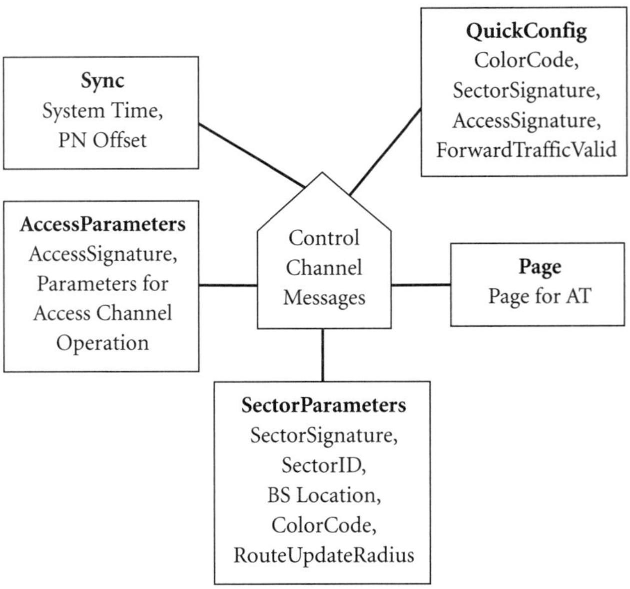

Figure 10.11: Example Control Channel Messages

SectorParameters and AccessParameters messages. The SectorParameters message conveys detailed information about the sector (e.g., the sector identity and neighbor list), and the AccessParameters message instructs ATs how to use the Access channel. The Page message, of course, carries a page for an AT. Let's take a look at some details of these Control channel messages.

The Sync message provides the air-interface protocol version,[330] pilot PN offset of the sector transmitting the message in units of sixty-four PN chips, and the system time.[331] The SectorParameters message provides information about the sector, such as the identity of the sector, sector location, route[332] update radius, neighbor list, and sector signature. The parameter SectorID is a 128-bit address of the sector, which consists of the identity of the subnet (or the RNC identity since one subnet is typically associated with one RNC) and the identity of the sector within the subnet. The subnet identity usually occupies 104 bits, and the remaining twenty-four bits represent the sector within the subnet. The SubnetMask message field specifies how many bits represent the subnet identity.

The location of the sector can be specified in latitude and longitude, but typical deployments avoid revealing sector locations[333] and set these parameters to 0! The value of the RouteUpdateRadius parameter is set to a non-zero value if the AN requires Idle State ATs to perform distance-based route updates. Such route updates could potentially be used to facilitate efficient paging. When the distance between the sector when the AT last sent a RouteUpdate message and the current sector exceeds a non-zero RouteUpdateRadius parameter, the AT sends a RouteUpdate message to the AN.

Information about the neighbors, such as pilot PN offset and search window size for the available 1xEV-DO channels, is specified. The SectorSignature parameter corresponds to all the parameters of the SectorParameters message. Two different values of the SectorSignature parameter refer to two SectorParameters messages with different contents. Hence, when values of some parameters need to be changed, the value of the SectorSignature parameter is changed (e.g., from 0 to 1).

As mentioned earlier, the AccessParameters message instructs the ATs how to use the Access channel. Examples of parameters the message carries are AccessCycle-Duration, AccessSignature, OpenLoopAdjust, ProbeInitialAdjust, ProbeNumStep, PowerSetup, PreambleLength, CapsuleLengthMax, and APersistence. Overwhelmed by this list? No worries, we will explain their use in Section 10.7.1! These parameters, along with other default parameters defined in the standard, influence how the ATs use the common Access channel.

[330]Both maximum and minimum revision are set to 0x01.

[331]The actual value specified corresponds to the system time 160 ms after the start of the Control channel cycle where such a cycle carries the Sync message and where this value is expressed in units of 26.67 ms. A Control channel cycle consists of 256 slots, and we'll discuss that later in this section.

[332]A route is a path between the AT and the AN. The concept of a route is primarily used to maintain connectivity between the AT and the AN through the AT's reporting of long-term channel conditions.

[333]Competition is fierce among providers, and they generally like to keep these locations private so their competitors cannot easily obtain this information.

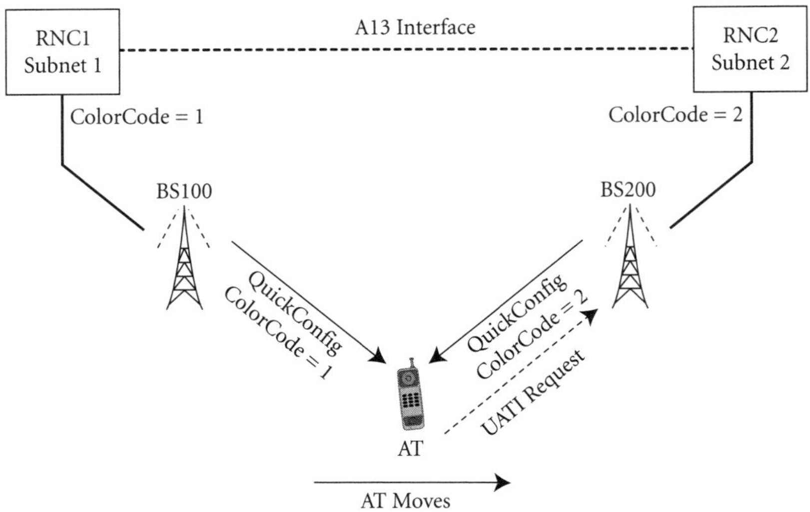

Figure 10.12: ColorCode in a QuickConfig Message

The QuickConfig message helps realize significant savings in battery power consumption and processing requirements at the AT by indicating a potential change in the contents of the overhead messages. It also facilitates an inter-RNC handoff. We'll focus on selected parameters of this message here: ColorCode, SectorSignature, AccessSignature, RPCCount, and ForwardTrafficValid. [334] ColorCode is the shortened identity of the subnet to which the sector belongs. Figure 10.12 explains the use of ColorCode. A ColorCode identifies each subnet in Figure 10.12.

Do You Recall. . . What is the relationship between a subnet and an RNC?[335]

As soon as the ColorCode seen on the Control channel of the strongest sector changes, the AT knows that it may have crossed a subnet and equivalently an RNC boundary.[336] The AT then initiates establishment of a new session with the new RNC.

Do You Recall. . . Can the AT keep its old UATI for a session with the new RNC?[337]

[334]See the standard [277] for the entire list of parameters of the QuickConfig message.

[335]One RNC is usually associated with one subnet, and, hence, a change in the subnet is an indication of the change in the RNC service area.

[336]The AT needs to read the subnet identity from the SectorParameters message to determine if the subnet has actually changed. See Rezahfar [278] for more details on the allocation of color codes and subnet identifies.

[337]No, the AT cannot keep the old UATI because part of the UATI includes the identity of the RNC, and the new RNC would give a new UATI to the AT.

Specifically, the AT requests a new UATI from the new RNC upon detecting a change in ColorCode. The session and the associated UATI are valid only within one subnet or RNC. When the A13 interface is supported, RNC2 can obtain from RNC1 all the session parameters for the AT's previous session with RNC1, avoiding the session configuration process and saving air-interface resources.

SectorSignature and AccessSignature correspond to the most recent SectorParameters and AccessParameters messages, respectively. If the AT has already stored the parameters associated with these signatures based on observations of previous SectorParameters and AccessParameters messages, the AT need not process these long, detailed messages. In other words, significant savings in processing requirements are realized since the AT just compares these signatures with previously stored signatures.

RPCCount is the maximum number of connected (i.e., with a MACIndex) ATs that this sector supports. The nth ForwardTrafficValid parameter with a value of 1 indicates that the ForwardTrafficChannel associated with MACIndex $64 - n$ is valid. A ForwardTrafficValid parameter of 0 in the nth field indicates that the related MACIndex is unassigned or the AN has released the connection for the AT with MACIndex $64 - n$. For example, assume that only two ATs exist in a sector, AT1 with $MACIndex = 63$ and AT2 with a $MACIndex = 62$. Hence, the first $n = 1$ occurrence of the ForwardTrafficValid parameter is 1 because $64 - 1 = 63$ is the MACIndex assigned to AT1. Similarly, the second occurrence of the ForwardTrafficValid parameter is also 1 because $64 - 2 = 62$ is the MACIndex allocated to AT2. If the value of the RPCCount parameter is 48, the remaining $48 - 2 = 46$ occurrences of the ForwardTrafficValid parameter are 0s because no other MACIndex is allocated to an AT in the sector. If an AT finds the value of the ForwardTrafficValid parameter to be 0 for its MACIndex, it assumes that the connection has been released or dropped and enters the Idle State.

The AN sends a Page message on the Control channel to instruct the AT to request a connection. We discussed several messages that are sent on the Control channel. These messages are contained within capsules. Two types of capsules, synchronous and asynchronous, are sent on the Control channel. The transmission of a synchronous capsule begins at predefined times, which corresponds to a Control channel cycle of 256 slots. An asynchronous capsule can be transmitted at any time. Figure 10.13[338] contrasts these two types of capsules.

The messages that need to be sent periodically are carried inside synchronous capsules. For example, since a QuickConfig message must be sent every Control channel cycle, it is in every synchronous capsule. Some messages need to be sent immediately, without waiting for the synchronous capsule; such messages are sent in the first available slot.

[338]Tx represents "transmit."

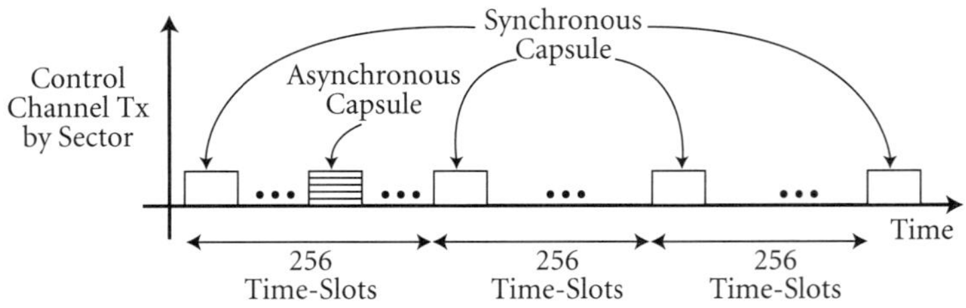

Figure 10.13: Synchronous versus Asynchronous Capsules

10.6.4 Forward Traffic Channel

The Forward Traffic channel carries a packet for an AT at the AT's requested rate. The AT specifies the requested rate on the RL DRC channel in the form of a DRC value. There are thirteen possible DRC values, 0 to 12. The DRC value of 0 corresponds to 0 kbps (null rate), 1 corresponds to 38.4 kbps, and 12 corresponds to 2.4576 Mbps.[339] The AN sends a packet to the AT via multislot transmission as shown in Figure 10.14 where N is the number of slots reserved for a user packet. The preamble identifies the

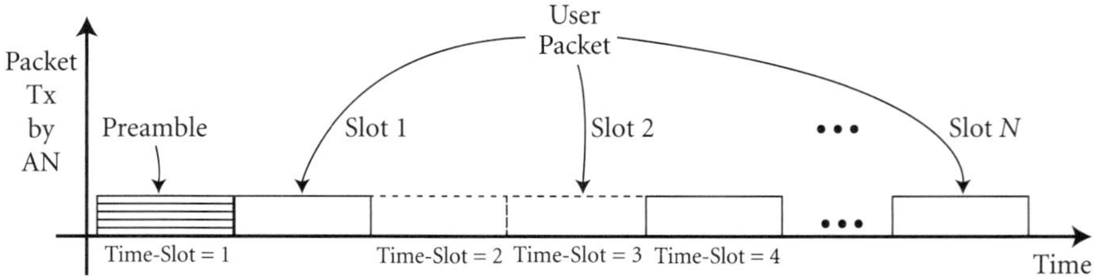

Figure 10.14: Multislot Transmission in the Forward Link

[339] 38.4 kbps is the minimum non-zero data rate in the FL, and 2.4576 Mbps is the highest data rate in the FL.

Table 10.2: Use of MACIndex in the FL Preamble

MACIndex	Preamble Use
0, 1, and 4	Unused
2	76.8 kbps Control Channel
3	38.4 kbps Control Channel
5 to 63	Forward Traffic Channel

AT(s) that should decode the packet. Table 10.2 summarizes the use of the MACIndex inside the preamble.

A given packet containing information bits arriving from the MAC layer undergoes physical layer processing to create traffic chips. These chips are sent to the selected AT in up to N slots where N depends on the DRC value. Figure 10.15 shows a

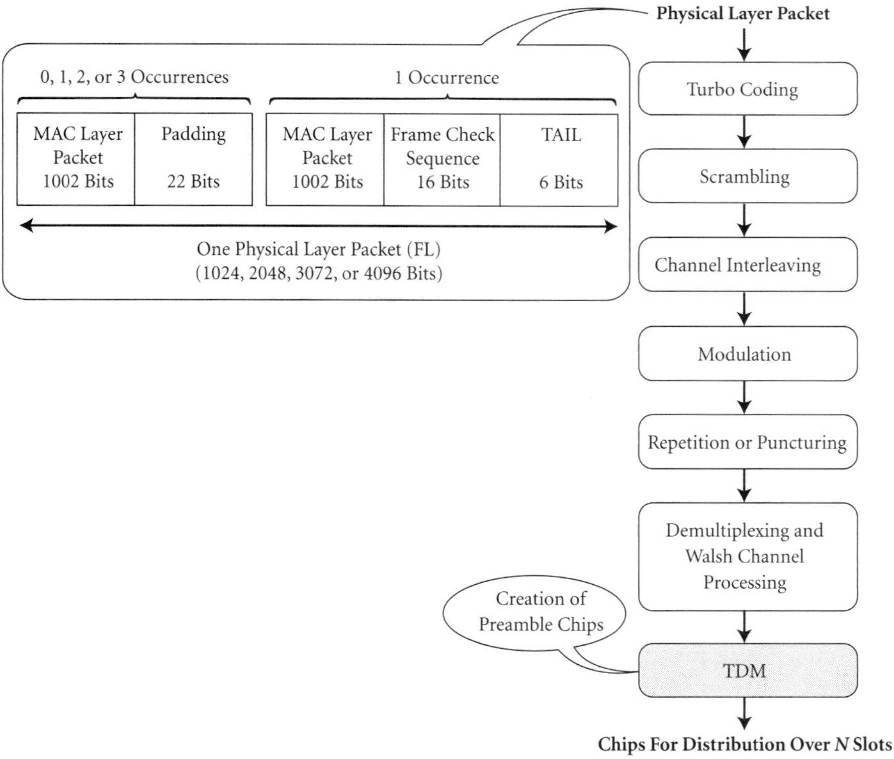

Figure 10.15: Simplified Processing for the Forward Traffic Channel

simplified block diagram of the physical layer processing, and Table 10.3 summarizes the physical layer parameters [277].

The physical layer packet for the Forward Traffic channel is encoded by the turbo coder at the rate of 1/3 or 1/5 (see Table 10.3). The turbo encoder consists of two systematic and recursive convolutional encoders in parallel along with the turbo interleaver that precedes the second convolutional encoder [277].[340]

The output of the turbo encoder is scrambled to randomize data. A seventeen-tap linear feedback shift register is used. Scrambling is a function of the MACIndex assigned to the AT and the DRC value the AT requested. Every turbo code symbol and the bit of the scrambling sequence undergo the XOR operation to create a scrambled turbo code symbol.

The scrambled turbo code symbols are interleaved through two stages of symbol re-ordering to ensure that the entire first stream of scrambled turbo code symbols remains in the initial part of the information transmitted over the air. The re-ordered symbols are permuted separately for the first stream and other streams.

The channel interleaver output is given to the modulator that generates an in-phase stream and a quadrature stream of modulated values. Three modulation schemes, quadrature phase shift keying (QPSK), 8-PSK (8-ary phase shift keying), and 16-QAM (where QAM is quadrature amplitude modulation), are used for the Forward Traffic channel. A simple but important review question is in order...

Do You Recall... How many code symbols are represented by one QPSK, 8-PSK, and 16-QAM modulation symbol?[341]

The modulator provides a specific number of modulation symbols based on the physical layer packet size, the code rate of turbo encoder, and the modulation scheme. In a 1xEV-DO system, no true CDMA-type spreading of modulation symbols exists in the Forward Traffic channel. So, one modulation symbol corresponds to one chip. The modulator, in effect, provides a certain number of traffic chips. The multislot transmission requires that an appropriate number of traffic chips be available to fill up the slots corresponding to a given DRC value. As seen from Figure 10.7, a slot carries 1600 traffic chips. The number of traffic chips required for the multislot transmission of a physical layer packet is $X = (N * 1600 - \textit{Preamble Chips})$ where N is the number of slots. Hence, if the modulator fails to provide X chips, sequence repetition is required. On the other hand, if the modulator provides Y chips, where $Y > X$, symbol puncturing is performed. Repetition or puncturing of the modulation symbols (i.e., chips) is carried out to fill up the slots with the traffic chips.

The in-phase stream is demultiplexed into sixteen parallel streams with each stream at 76.8 ksps. The quadrature stream is also demultiplexed into sixteen streams. Each

[340] See Chapter 2 for a description of a typical turbo encoder and decoder.

[341] A QPSK modulation symbol represents two code symbols. An 8-PSK modulation symbol represents three code symbols, and a 16-QAM modulation symbol represents four code symbols. See Chapter 2 for details of modulation schemes.

Table 10.3: Physical Layer Parameters for the Forward Traffic Channel

DRC Value	Nominal Data Rate (kbps)	Packet Length (Slots)	Preamble (Chips)	Packet Size (Bits)	Turbo Code Rate	Modulation Scheme	Repetition or Puncturing Factor
1	38.4	16	1024	1024	1/5	QPSK	9.60
2	76.8	8	512	1024	1/5	QPSK	4.80
3	153.6	4	256	1024	1/5	QPSK	2.40
4	307.2	2	128	1024	1/5	QPSK	1.20
5	307.2	4	128	2048	1/3	QPSK	1.00
6	614.4	1	64	1024	1/3	QPSK	2.04
7	614.4	2	64	2048	1/3	QPSK	1.02
8	921.6	2	64	3072	1/3	8-PSK	0.50
9	1228.8	1	64	2048	1/3	QPSK	1.02
10	1228.8	2	64	4096	1/3	16-QAM	0.50
11	1843.2	1	64	3072	1/3	8-PSK	1.02
12	2457.6	1	64	4096	1/3	16-QAM	0.50

individual stream at the rate of 76.8 ksps is assigned to one of sixteen Walsh channels. This is called orthogonal spreading, in which individual streams at 76.8 kbps are spread by a factor of sixteen to yield 76.8 kbps $*$ 16 $=$ 1.2288 Mcps. Each Walsh channel's power is scaled by a gain of $1/\sqrt{16}$ where the gain is normalized to a unity reference corresponding to unmodulated binary phase shift keying (BPSK) modulation symbols transmitted at full power. The scaled Walsh chips associated with sixteen Walsh channels are summed on a chip-by-chip basis. The Walsh chip rate is 1.2288 Mcps. Such demultiplexing and stream-based spreading operations facilitate decoding at the receiver but yields no processing gain associated with traditional CDMA spreading.

In the Traffic or Control channel portion of the first slot of a multislot transmission, a preamble is included to identify the AT(s) that should decode the packet. The preamble consists of all 0s, and such a sequence is covered by a thirty-two–chip bi-orthogonal sequence. The preamble sequence is repeated a specific number of times depending on the DRC value as implied in Table 10.3. For example, a DRC value of 12 involves a sixty-four–chip preamble. Hence, the thirty-two–bit preamble sequence is transmitted twice. As another example, a DRC value of 1 has a 1024-chip preamble, and the thirty-two–bit preamble sequence is transmitted thirty-two times. The mapping between the MACIndex i and the bi-orthogonal sequence W is given by

$$W^{32}_{i/2} \qquad \text{with } i = 0, 2, \ldots, 62 \tag{10.3}$$

and

$$W^{32}_{((i-1)/2))} \qquad \text{with } i = 1, 3, \ldots, 63. \tag{10.4}$$

Ever Wondered... If you have heard of an orthogonal sequence but never a bi-orthogonal sequence, you are not alone! Examples of two two-chip orthogonal codes are $[1, 1]$ and $[1, -1]$. Note that sum of the chip-by-chip multiplication is zero. In other words, if we multiply the first chips and add that product to the product of multiplying the second chips, we end up with zero: $(1 * 1) + (1 * -1) = 1 - 1 = 0$. Two orthogonal codes and four bi-orthogonal codes exist at the code length of 2. The code length here is the number of chips in the code.

There are four bi-orthogonal codes of code length 1: $[1, 1]$, $[-1, -1]$, $[1, -1]$, and $[-1, 1]$. Two bi-orthogonal codes of the same code length constitute a pair where one code is a complement of another. For example, $[1, 1]$ and $[-1, -1]$ are complements to each other as $[-1, -1] = (-1) * [1, 1]$. One code from a pair and another code from a different pair are orthogonal to each other. For example, $[1, 1]$ and $[1, -1]$ are orthogonal.

Orthogonal codes are useful when the same code will be used to indicate two different actions. For example, to send a power-up command, the orthogonal code $[1, 1]$ can be used, and to send a power-down command, the complement of the orthogonal code $[1, 1]$, which is $(-1) * [1, 1] = [-1, -1]$, can be used.

Bi-orthogonal codes are used when only one type of action is adequate for the code. For example, a bi-orthogonal code $[1, 1]$ can identify one AT, and another bi-orthogonal code $[-1, -1]$ can identify another AT. The concept of bi-orthogonality doubles the number of codes available at a given code length. For example, only two orthogonal codes but four bi-orthogonal codes of code length 2 are present.

The time-division-multiplexed preamble and Control or Traffic channel chips are quadrature-spread using a pilot PN sequence. The pilot PN sequence period is 26.67 ms,[342] leading to seventy-five repetitions every two seconds.

Let's discuss how the BS generates an Forward Traffic channel packet. Assume the AT selects Sector 1 and sends a DRC value of 3 corresponding to the nominal data rate of 153.6 kbps. The nominal data rate is defined as the ratio of the packet size to the packet length. As seen from Table 10.3, the packet size is 1024 bits and the packet length is four slots for the DRC value of 3. The nominal DRC rate is therefore $1024 \text{ bits}/(4 * 1.67 \text{ ms}) = 153.6$ kbps. Assume that the BS scheduler for Sector1 chooses to send a packet to the AT. The BS creates the traffic chips for the AT by using the parameters specified for $DRC = 3$ in Table 10.3. Figure 10.16 illustrates a simplified view of such packet formation.

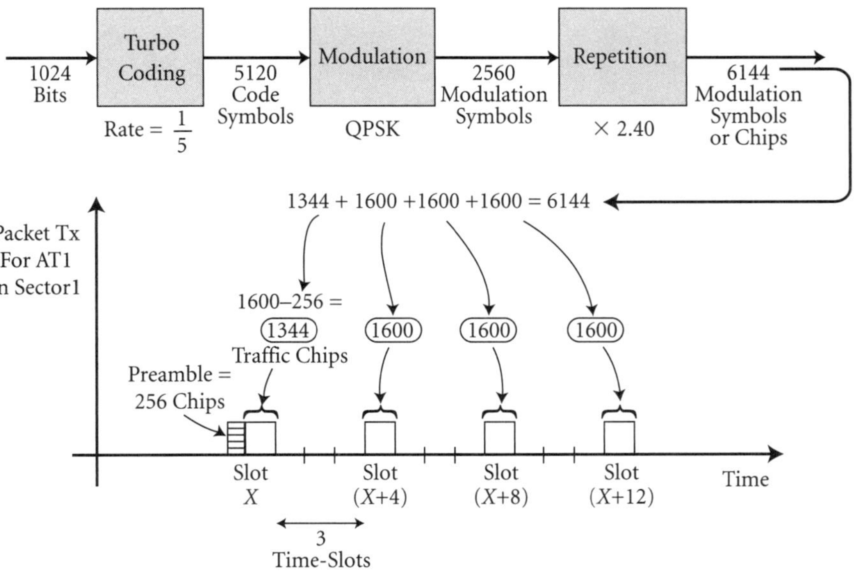

Figure 10.16: Packet Formation at the Base Station

[342]The exact amount is $26.\bar{6}$ ms, and the standard uses the term 26.66 . . . ms.

The BS takes 1024 bits from the AT's buffer and passes it though turbo coding. Since the coding rate is 1/5, the number of turbo code symbols generated is $1024 * (1/code\ rate) = (1024 * 5) = 5120$. Since the modulation scheme is QPSK for a DRC value of 3, two turbo code symbols are represented by one QPSK modulation symbol. These modulation symbols are repeated 2.40 times, creating a total of $2560 * 2.40 = 6144$ modulation symbols or traffic chips. These traffic chips are sequentially distributed among the four slots. Recall that a slot contains traffic chips as well as pilot and MAC chips; hence, only 1600 traffic chips can be transmitted in a given slot. Since the first slot must identify the AT through a preamble, not all 1600 chips are available to transmit traffic. The preamble length is 256 chips, leaving $1600 - 256 = 1344$ traffic chips for the first slot. The remaining three slots carry 1600 traffic chips each.

Only the first slot in a multislot transmission of a packet contains the preamble sequence; the remaining slots need no preamble sequence because the gap between two successive slots carrying a packet is defined to be three. Hence, once the AT detects a packet in slot X based on the preamble, it expects the packet transmission to continue at $X + 4$, $X + 8$, and so on (if applicable). Since a multislot transmission involves transmission in every fourth slot, such transmission is also called *four-slot interlacing*. The three-slot gap enables the AT to process the received traffic chips and send an Acknowledgment (ACK) or NAK to the BS. If the BS receives an ACK, it avoids the transmission of the remaining chips, and if the BS receives a NAK, it continues the transmission of the traffic chips up to the packet-length slots. The gap facilitates the implementation of Hybrid Automatic Repeat Request. While slots such as X and $X + 4$ are transmitting a given packet, the intervening slots (i.e., slots $X + 1$, $X + 2$, and $X + 3$) can be used to send different packets to the same or a different AT. Such continuous transmission from the BS increases sector throughput and user throughput.

When the set of code symbols vary from one transmission to the next during Hybrid Automatic Repeat Request retransmissions, such a Hybrid Automatic Repeat Request is referred to as *Incremental Redundancy*. However, when identical code symbols are used during all Hybrid Automatic Repeat Request (re)transmissions, such a Hybrid Automatic Repeat Request is called *Chase combining*. Incremental Redundancy has the potential for better error correction performance, and Chase combining reduces the buffer space requirements [279, 280]. Transmission in Chase combining does include the original information bits and some redundancy bits.

10.6.5 Combining the Forward Link Channels

Now that we have covered all the forward link channels, let's see how the AN combines all these channels to create a composite forward link signal for transmission. Recall that the FL transmission occurs in the units of 1.67 ms slots at the rate of 1.2288 Mcps. Figure 10.17 illustrates how the AN combines the FL channels. A binary 0 is represented by $+1$ and a binary 1 is represented by -1.

Since the FL utilizes the TDM approach, chips corresponding to the Control or Traffic channel packet (including the preamble for the first transmission), Pilot channel, and MAC channel are time-division-multiplexed on the I channel and Q channel.

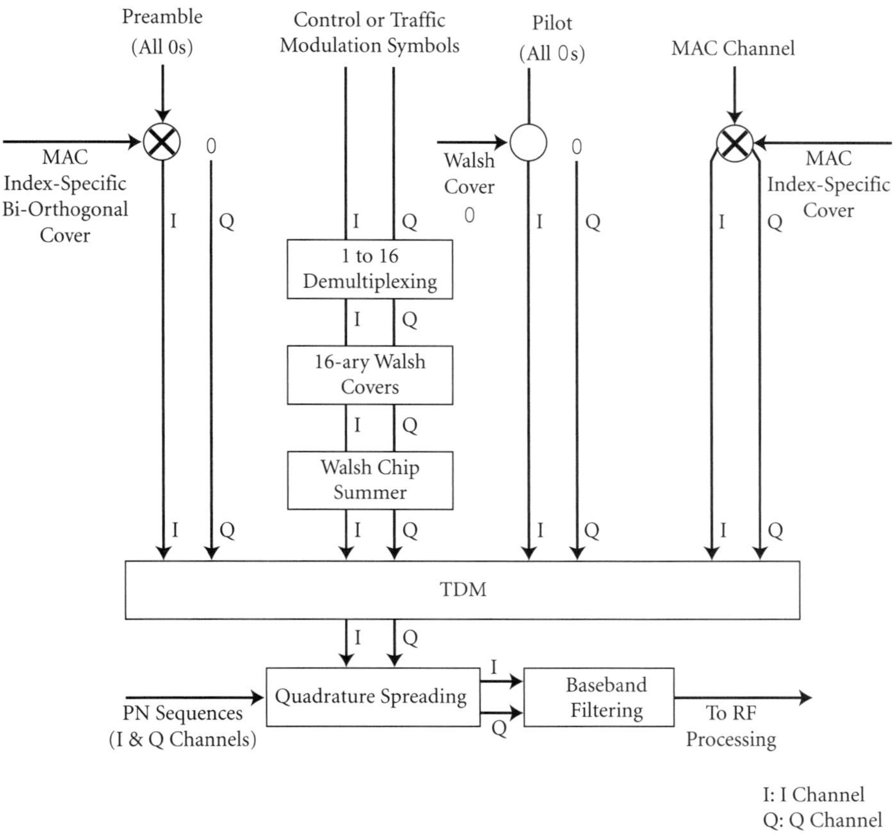

Figure 10.17: Combining of the Forward Link Channels

The preamble consists of all binary 0s,[343] and the bi-orthogonal cover associated with a suitable MACIndex covers the preamble.

Do You Recall. . . What is the significance of the MACIndex for the preamble portion of the slot?[344]

The preamble is transmitted on only the I channel with 0s (i.e., no transmission) on the Q channel. The Control channel message or the Forward Traffic channel data or message undergoes physical layer processing to yield modulation symbols.

[343]These binary 0s are mapped onto +1.

[344]The MACIndex identifies the intended recipient(s) of the Control or Traffic packet.

Since the modulation symbols are complex, both the I channel and Q channel contain useful information. The modulation symbols undergo Walsh processing whereby the stream of modulation symbols on a given I channel or Q channel are demultiplexed into sixteen streams, each stream is covered by a 16-ary Walsh cover, and the resultant chips are summed. Note that no actual spreading of the signal occurs because the signal bandwidth before and after the Walsh processing is the same. Such Walsh processing can be considered a form of coding to facilitate decoding and detection at the AT.

The Pilot channel is all binary 0s mapped onto $+1$ and is covered by Walsh Cover 0 (which is all 1s). The Pilot channel chips go over only the I channel, and hence the Q channel transmits nothing during the Pilot channel burst. Since the preamble and the Pilot channel are using only the I channel, they are using BPSK modulation, improving the reliabilty that ATs will detect them. The MAC channel involves code-multiplexing the Reverse Activity channel with either an RPC channel or a DRCLock channel during a given slot. For the MAC channel, the I channel is used for even-numbered MACIndex values, and the Q channel is used for odd-numbered MACIndex values. Both I channel chips and Q channel chips are time-division-multiplexed. The quadrature spreading occurs next where two streams, I' and Q', are constructed.

$$I' = IPN_I - QPN_Q \tag{10.5}$$

$$Q' = IPN_Q + QPN_I \tag{10.6}$$

The PN sequence generators create the PN_I and PN_Q pilot PN sequences. The Pilot channel identifies the PN offset, which is expressed in units of sixty-four chips. The PN offset specifies how much the pilot PN sequence lags behind the zero-offset pilot PN sequence. The I' and Q' sequences pass through baseband filtering and are frequency upconverted to the carrier frequency at which the network has been deployed (e.g., 1.9 GHz). Specifically, for frequency upconversion, the I' sequence is multiplied by $\cos(2\pi f_c t)$ and the Q' sequence is multiplied by $\sin(2\pi f_c t)$.

10.7 A CLOSER LOOK AT REVERSE LINK CHANNELS

Now that we have closely examined FL channels, let's turn our attention to the RL channels. The RL contains two basic channels, the Access channel and the RTC. The Access channel consists of Pilot and Data channels, and the RTC includes Pilot, Data, DRC, and Acknowledgment channels. Details of all the reverse link channels are discussed in Sections 10.7.1 through 10.7.4. Section 10.7.5 summarizes how various RL channels are combined.

10.7.1 Access Channel

Figure 10.18 illustrates how the AT utilizes the Access channel. The AT carries out an access attempt that consists of a number of probe sequences. Each probe sequence, in turn, includes a series of access probes.

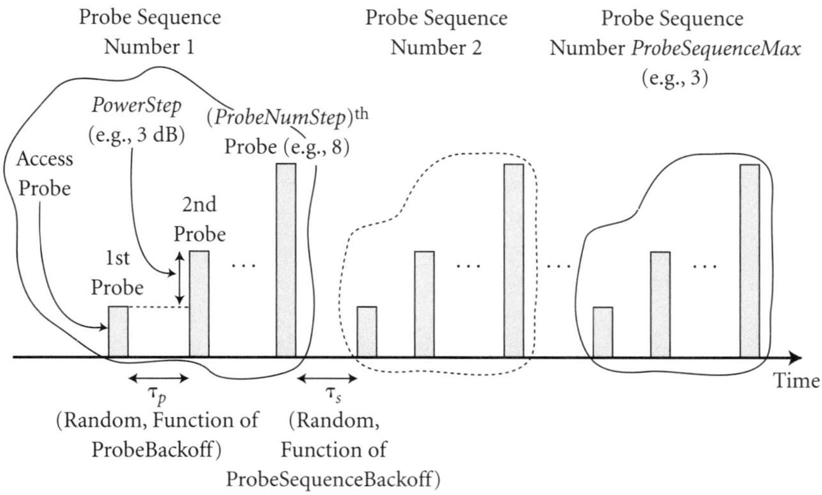

Figure 10.18: Access Attempt

The AT sends an access probe on the Access channel and waits for an ACAck message[345] from the BS. If the AT gets no feedback from the BS, it assumes that the BS failed to detect the probe. Hence, the AT increases the power level of the probe by *PowerStep*[346] and retries after waiting for a random time, τ_p, that is a function of ProbeBackoff. It sends a maximum of *ProbeNumStep* probes in a given probe sequence. *ProbeSequenceMax* is the maximum number of probe sequences that the AT utilizes during an access attempt. The time between two successive probe sequences is random and a function of ProbeSequenceBackoff. If the AT still hears nothing back from the BS after attempting all the probes and all the probe sequences, the access attempt is considered to have failed. Unlike 1x, the AT just makes an access attempt on one sector. In other words, access handoff is unsupported. If numerous access failures occur in a given area, radio frequency (RF) optimization and adjustment of some access parameters (e.g., increasing ProbeNumStep from 4 to 8 and increasing PowerStep from 2 dB to 3 dB) could be performed to reduce access failure rates to below 2%.

The AT calculates the transmit power of the first access probe in an access attempt as

$$\begin{aligned} X_0 = -&Mean\ Rx\ Power\ (\text{dBm}) + OpenLoopAdjust \\ &- ProbeInitialAdjust \end{aligned} \tag{10.7}$$

[345]Though the standard omits any explicit expansion of the acronym ACAck, it represents the Access Channel Acknowledgment.

[346]*PowerStep* is expressed in units of 0.5 dB. So, $PowerStep = 4$ means a step size of $4 * 0.5 = 2$ dB.

where *Mean Rx Power* is estimated throughout the transmission of each probe and is the average received power level at the AT.[347] The AN configures *OpenLoopAdjust* and *Probe Initial Adjust* (e.g., −86 dB and 0 dB, respectively) to allow the AT to transmit adequate power such that the BS can detect the access probe.

The AccessCycleDuration parameter governs the time instant at which the AT can start transmitting an access probe. This parameter dictates how frequently an access probe can potentially be sent. An AT can start the access probe transmission at slot T such that $T \bmod AccessCycleDuration = 0$ where T is the system time in slots. For example, it could be thirty-two slots.

Figure 10.19 shows details of an access probe. In an access probe, first PreambleLength frames contain just the pilot bits, and the remaining CapsuleLengthMax frames contain both pilot bits and messages, such as RouteUpdate and ConnectionRequest. The Pilot channel helps the BS detect the probe and also enables coherent demodulation at the BS when the BS attempts to demodulate the Data channel. The Access channel is identified by a distinct Access channel long code, which the AT constructs using a PN sequence generator that utilizes parameters such as AccessCycleNumber, SectorID, and ColorCode.

AccessCycleNumber is defined as $SystemTime \bmod 256$ where $SystemTime$ is the CDMA system time (expressed in slots) corresponding to the slot in which the preamble for the access probe is sent. The AT identifies itself by including its UATI in a message in the Data channel portion of the Access channel. If the AT has no UATI, it creates a Random Access Terminal Identifier and includes it in the message. When the AT transmits on the Access channel, it sends a RouteUpdate message along with a desired message, such as a ConnectionRequest or a UATIRequest message.

The operation of the Access channel is governed by two types of parameters, those that the AccessParameters message defines and those that are configurable parame-

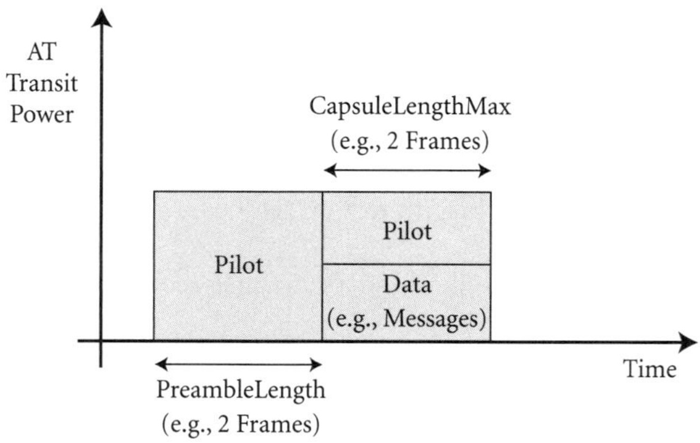

Figure 10.19: Anatomy of an Access Probe

[347]Rx represents "receive."

ters defined in the standard but absent in the AccessParameters message. Examples of configurable parameters include ProbeSequenceMax, ProbeBackoff, and ProbeSequenceBackoff. Examples of parameters present in an AccessParameters message are AccessCycleDuration, AccessSignature, OpenLoopAdjust, ProbeInitialAdjust, ProbeNumStep, PowerStep, PreambleLength, CapsuleLengthMax, and APersistence. Let's briefly discuss some of these parameters we have not yet considered.

AccessSignature is associated with all the parameters included in the AccessParameters message. When the AN wishes to change one or more parameters, it changes the value of AccessSignature. AccessCycleDuration specifies the length of the access cycle that governs when an access probe can be transmitted, and APersistence actually determines whether the access probe can indeed be sent or whether the AT needs to wait for another slot. Four classes of ATs are defined with each class potentially using a different value of APersistence. APersistence influences how quickly the AT can start transmission of an access probe. An AT must carry out a persistence test before it can start transmission of the first access probe in a given probe sequence. The AT generates a random number between 0 and 1, and if this number is less than APersistence or if the number of persistence tests exceeds $4/APersistence$, the persistence test is considered successful and the AT starts transmitting the access probe. If the persistence test fails, the AT re-executes the test in the next access channel cycle, the length of which is specified by AccessCycleDuration.

After the AT transmits the first probe in a given probe sequence, the time interval between two successive access probes is governed by the ProbeBackoff parameter. Furthermore, the ProbeSequenceBackoff parameter and the persistence test for the first probe govern the time interval between two consecutive probe sequences. The specific class the AT will use is beyond the scope of the standard with class 0 being the default for non-emergency ATs. A service provider can configure the AT with a specific class.

10.7.2 Pilot Channel and Data Channel

The RL Pilot channel facilitates coherent demodulation at the BS. It also serves as a baseline channel for RL power control with transmit power levels of other components of the RTC (e.g., the Acknowledgment channel, DRC channel, and Data channel) defined as offsets to the Pilot channel. For example, DRCChannelGain is the power offset of the DRC channel relative to the Pilot channel. Similarly, ACKChannelGain is the power offset of the AckChannel relative to the Pilot channel. DRCChannelGain (e.g., −3 dB) and AckChannelGain (e.g., 3 dB) are specified in the TrafficChannelAssignment message when a dedicated radio connection between the AT and the sector(s) is set up. The relative gain of the Data channel is a configurable attribute that depends on the data rate. The RL data rate varies from 9.6 kbps to 153.6 kbps. The default gains of the Data channel relative to the Pilot channel are specified in the standard and summarized in Table 10.4 [277]. The default values of the parameters DataOffset* (with $* = Nom$, $9k6$, and so on) are 0s. These Data channel gains are part of the PowerParameters attribute. The AN can, if desired, provide AT-specific gains during session configuration. Time for a review question...

Table 10.4: Gains of Data Channel

Data Rate (kbps)	Gain Relative to the Pilot Channel
9.6	DataOffsetNom + DataOffset9k6 + 3.75
19.2	DataOffsetNom + DataOffset19k2 + 6.75
38.4	DataOffsetNom + DataOffset38k4 + 9.75
76.8	DataOffsetNom + DataOffset76k8 + 13.75
153.6	DataOffsetNom + DataOffset153k6 + 18.50

Note: DataOffsetNom and DataOffset_k_ are defined in the PowerParameters attribute.

Do You Recall... Why does the overall required gain of the Data channel increase as the data rate increases?[348]

Let's take a closer look at the physical layer processing carried out for the Data channel as depicted in Figure 10.20. The turbo coding rate is $1/2$ for 153.6 kbps

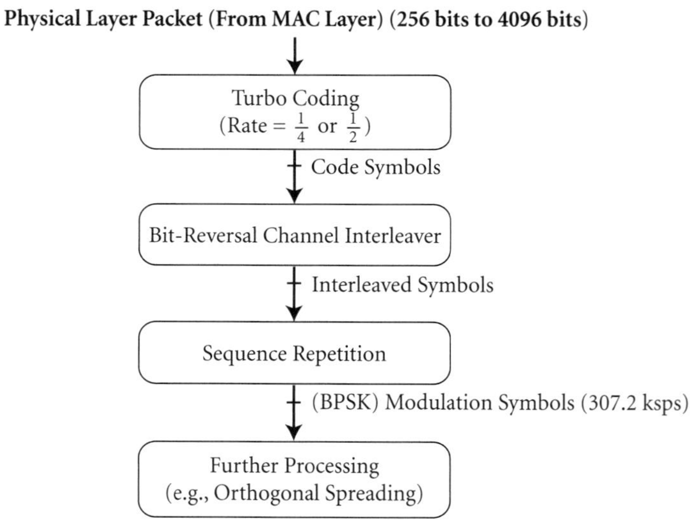

Figure 10.20: Physical Layer Processing of the Data Channel

[348]Remember that the RL in 1xEV-DO is based on traditional CDMA. As the data rate increases, the processing gain decreases. Recall that processing gain is the ratio of the spreading bandwidth to the data rate. A certain signal-to-interference-plus-noise ratio is required to achieve a target packet error rate. Signal-to-interference ratio is proportional to processing gain. We need to compensate for the reduction in processing gain at a high rate by increasing the transmit power so that a similar signal-to-interference ratio can still be maintained regardless of the data rates.

and 1/4 for all other data rates. The turbo encoder first removes the six bits of the TAIL field in the physical layer packet received from the MAC layer and encodes the remaining bits. The turbo encoder then appends internally generated TAIL bits. The use of such specific TAIL bits improves the decoder operation at the receiver. The turbo encoder utilizes two systematic and recursive convolutional encoders with a turbo interleaver preceding the second convolutional encoder.[349] The sequence of binary turbo code symbols is interleaved using a bit-reversal channel interleaver.

Ever Wondered... Never heard about a bit-reversal channel interleaver before? No problem, there is always the first time for everything in life! Assume that the sequence of code symbols is stored in an array A where the array index i ranges from 0 to $2^L - 1$. For example, when we have a total of 1024 code symbols, i ranges from 0 to 1023 and L is 10 since $2^L = 2^{10} = 1024$. The ith interleaved symbol is retrieved from the array element $A[i]$ such that $A[i] = Bit\_Reversal(i, L)$. If i is expressed in the binary form as $i = \{b_{L-1}\, b_{L-2} \ldots b_1\, b_0\}$, $Bit\_Reversal(i, L)$ would yield $A[i] = \{b_0\, b_1 \ldots b_{L-2}\, b_{L-1}\}$ [277]. Let's take a numerical example. The seventh symbol entering the interleaver corresponds to $i = 6$. The seventh interleaved symbol coming out of the interleaver would be $A[i] = A[6]$. Since 6 is expressed as a ten-bit number $\{0\ 0\ 0\ 0\ 0\ 0\ 0\ 1\ 1\ 0\}$, its bit-reversed form is $A[6] = \{0\ 1\ 1\ 0\ 0\ 0\ 0\ 0\ 0\ 0\}$.

The interleaved code symbols pass through the processing block of sequence repetition. The goal of sequence repetition is to ensure a fixed symbol rate of 307.2 ksps at the input of the (digital) modulator. This rate corresponds to 8192 symbols for transmission during a 26.67 ms frame. Hence, this block repeats the interleaved code symbols a suitable number of times. Since BPSK is used as the modulation scheme, the symbols coming out of the sequence repetition block are, in effect, BPSK modulation symbols. The BPSK modulation symbols of the Data channel undergo traditional CDMA spreading and scrambling, formally referred to as *quadrature spreading*. (Hold that thought! We'll get back to this in Figure 10.24 in Section 10.7.5.)

10.7.3 Data Rate Control Channel

When the AT is in the Connected State (i.e., with dedicated FL and RL channels), it continuously transmits DRC values on the DRC channel. As shown in Table 10.3, thirteen DRC values are supported, 0 corresponding to null rate, 1 corresponding to the lowest FL data rate of 38.4 kbps, and 12 corresponding to the highest FL data rate of 2.4576 Mbps. The AT estimates the FL C/I ratio of the selected serving sector and determines the DRC value (or, equivalently, the data rate) that it can support at 1% packet error rate. In other words, for the channel conditions leading to a specific C/I ratio, if the AT requests a DRC value of 5, the AT is guaranteeing the AN that no more

[349] See Chapter 2 for additional details on the turbo encoder and decoder.

than one of one hundred packets at $DRC = 5$ would be in error under these channel conditions. Important parameters that influence the operation of the DRC channel are DRCChannelGain, DRCLength, and DRCGating. Let's discuss their roles next. DRCChannelGain is the DRC channel's power level relative to the RTC. A higher DRCChannelGain increases the BS's DRC channel detection reliability but increases the overall interference in the RL.

DRCLength is the number of times the AT must report the same DRC value. It affects the processing requirements at the AT and the AN, the channel adaptation, and the reliability of DRC channel detection at the BS. The standard supports four values of DRCLength: one, two, four, and eight slots. Figure 10.21 illustrates the implications of setting DRCLength to two and four slots.

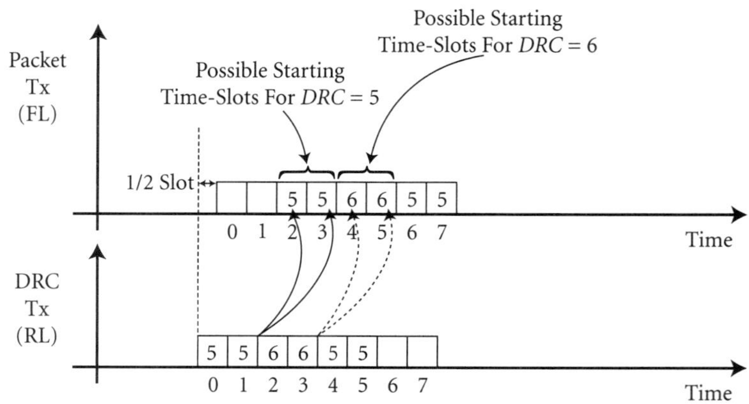

(a) $DRCLength = 2$

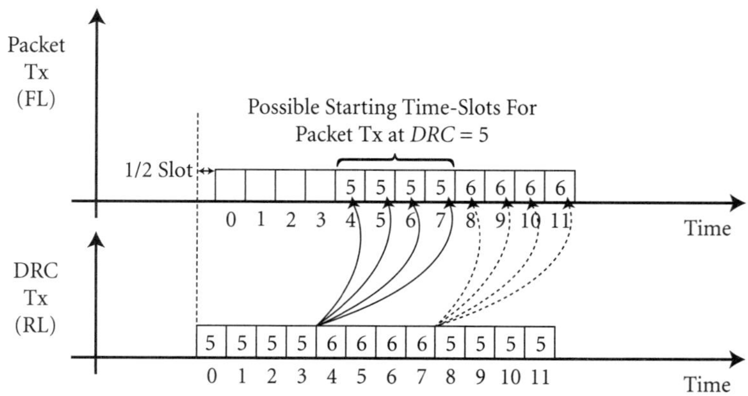

(b) $DRCLength = 4$

Figure 10.21: Impact of DRCLength Settings

The RL slot is 1/2 slot ahead of the corresponding FL slot. In Figure 10.21a, DRCLength of 2 slots is used. Just prior to RL slot 0, the AT measures the latest C/I ratio of the serving sector and estimates the supportable DRC value to be 5. Since DRCLength is 2, the AT transmits $DRC = 5$ twice. The BS combines the received DRC values and can *start* transmission of a packet at $DRC = 5$ in the FL slots 2 or 3. Again, just prior to the RL slot 2, the AT measures the FL C/I ratio and estimates the supportable DRC value to be 6 and transmits $DRC = 6$ twice.

Let's contrast Figure 10.21a with Figure 10.21b. When DRCLength is 4, the AT sends the same DRC value four times. The BS can now start a packet transmission in any of the four FL slots upon reception of all DRC values.

Let's analyze impact of DRCLength values. At small DRCLength values, the AT must estimate the most recent channel conditions frequently as (potentially) new DRC values need to be sent more frequently.

The BS scheduler makes scheduling decisions in every available slot. At large DRC values, the BS scheduler has time to make a scheduling decision. For example, in Figure 10.21a, the BS has approximately two slots to make a scheduling decision and to start transmission to an AT, and it has about four slots in Figure 10.21b.

When the DRCLength value is small, the AT can send a new DRC value as soon as the channel conditions change. However, when the DRCLength value is large, even if the channel conditions change, the AT has to continue the transmission of the previous DRC value DRCLength times. Thus, a smaller DRCLength value keeps the DRC value more reflective of the prevailing channel conditions.

When the DRCLength value is large, the BS can combine multiple received values to estimate the DRC value the AT is transmitting. Hence, a large DRCLength value provides better DRC channel detection reliability.

When DRCGating is 1, transmission of the DRC value is gated off (i.e., no transmission occurs) in $DRCLength - 1$ slots when $DRCLength > 1$. Figure 10.22 illustrates two cases, one with DRC gating and the other without DRC gating. In the absence of DRC gating, the AT continuously transmits DRC values. However, when DRC gating is turned on, the AT transmits the DRC value only once based on the C/I ratio and turns off the transmission of the DRC values in $DRCLength - 1 = 4 - 1 = 3$ slots. The benefits of DRC gating are reduced RL interference and increased battery life because the average transmit power from the AT is decreased. The disadvantage of the DRC gating is lower DRC channel detection reliability at the BS because the BS no longer has multiple DRC channel receptions to combine.

10.7.4 Acknowledgment Channel

To every transmission directed to the AT, the AT responds with a positive acknowledgment or negative acknowledgment. Some DRC values involve multislot transmissions in which a physical layer packet is transmitted over multiple slots. For example, as shown in Table 10.3, the DRC value of 2 involves an eight-slot transmission. Figure 10.23 shows the timing of the packet transmission and the associated ACK or NAK.

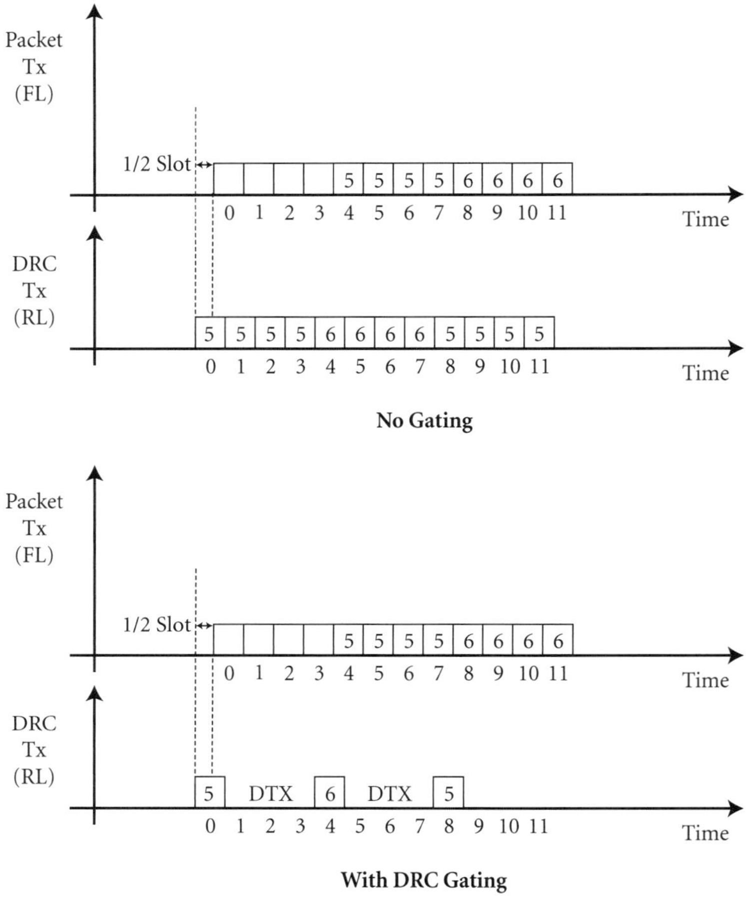

Figure 10.22: Impact of DRC Gating

In the example shown in Figure 10.23, the BS scheduler for the sector decides to transmit a packet to the AT at $DRC = 2$ based on the AT's transmission of $DRC = 2$ on the DRC channel. Recall that the BS can neither upgrade nor downgrade the data rate. If the BS chooses to send a packet to the AT, the data rate of the packet must be the same as the date rate associated with the DRC value the AT has requested. Since $DRC = 2$ requires eight slots, the BS reserves eight slots to transmit the packet, say packet P, to the AT. These slots are four slots apart, with a gap of three intervening slots between the two consecutive slots carrying the packet. Such a transmission approach is called *four-slot interlacing*.

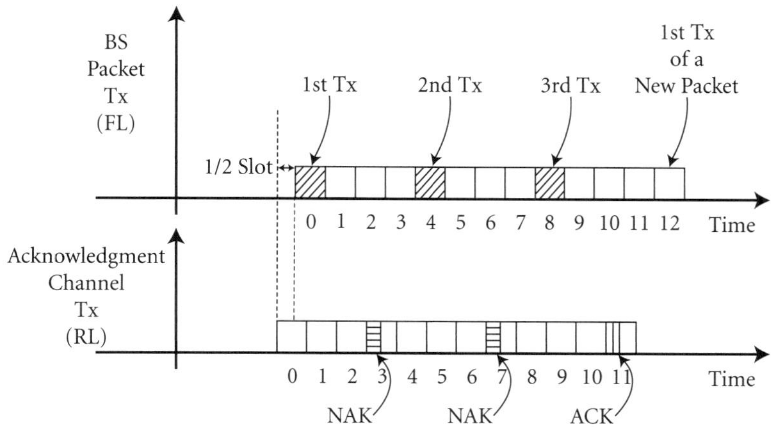

Figure 10.23: Operation of the Acknowledgment Channel

The BS transmits some traffic chips in slot 0. The AT attempts to retrieve the original information bits[350] from the received traffic chips. In this example, the AT is unable to correctly decode packet P and hence sends a NAK in RL slot 3. The length of the actual ACK or NAK is 1/2 slot. Observe the time difference between the reception of a packet and the transmission of an ACK or NAK. The BS receives this NAK approximately[351] $1/2$ slot before FL slot 4. The BS continues to transmit packet P's traffic chips in slot 4. The AT now combines the previously received traffic chips from the FL slot 0 with the newly received traffic chips in FL slot 4. In our example, the AT still fails to decode the original packet bits and hence sends another NAK in RL slot 7. The BS next utilizes the previously reserved FL slot 8 for the third transmission. The AT combines the traffic chips from all the three transmissions of packet P. The AT's persistence pays off! The AT succeeds in retrieving all of packet P's original bits and sends an ACK to the BS.[352] The BS stops transmission of packet P's remaining chips since the AT has already decoded the packet correctly. The BS now releases the remaining $8 - 3 = 5$ slots, which can be used to transmit a new packet, say packet Q, to the same or a different AT.

10.7.5 Combining of the Reverse Link Channels

Figure 10.24 illustrates how the AT combines various reverse link channels to construct the RTC. Since the reverse link is based on CDM, the RL channels are com-

[350] This example scenario corresponds to the packet containing 1024 bits for $DRC = 2$.

[351] We say approximately because the reception is affected by the actual propagation delay.

[352] The AT assumes the packet is decoded correctly if the sixteen-bit frame check sequence passes. The frame check sequence bits are part of the physical layer packet. For example, if the physical layer packet size is 1024 bits, 1002 bits represent the MAC layer packet, six bits are TAIL bits for the encoder, and sixteen bits are frame check sequence bits. See Figure 10.15 for the format of the FL physical layer packet.

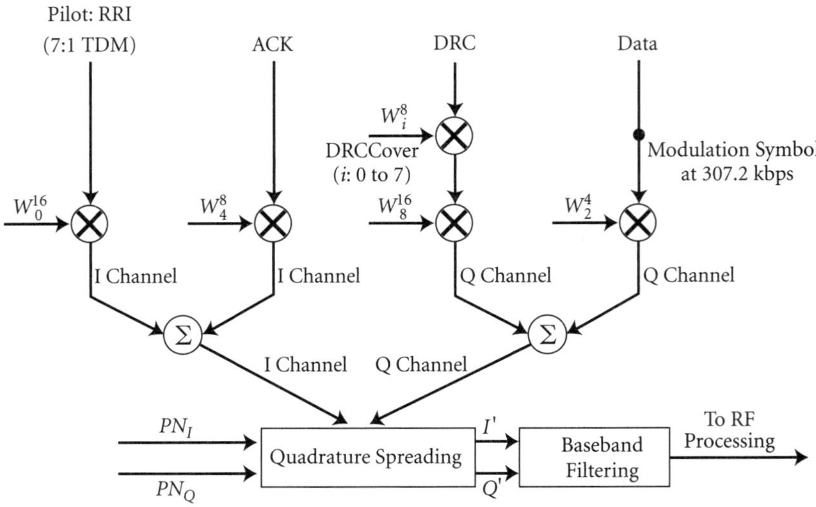

Figure 10.24: Combining of the Reverse Link Channels to Form the Reverse Traffic Channel

bined through Walsh functions (also known as Walsh covers). The Pilot or RRI and Acknowledgment channels are considered part of the I channel, and the DRC channel and Data channel are part of the Q channel. The Pilot and RRI channels are time-division-multiplexed according to a 7:1 ratio and multiplied by W_0^{16}. The Acknowledgment channel is multiplied by W_4^8. The Walsh chips from the Pilot or RRI channel and Acknowledgment channel are combined together and sent to the quadrature spreading block.

The DRC value is multiplied by DRCCover W_i^8 where i ranges from 0 to 7 and then multiplied by W_8^{16}. DRCCover identifies the sector from which the AT is requesting FL transmission. DRCCover W_0^8 is called *null DRCCover* and indicates that the AT is requesting no FL transmission from any sector. The modulation symbols from the Data channel at the rate of 307.2 ksps are multiplied by W_2^4. The Walsh chips from the DRC channel and Data channel are added and sent to the quadrature spreading block as part of the Q channel. The quadrature spreading block utilizes the PN sequences, PN_I and PN_Q, with I channel Walsh chips containing Pilot or RRI and Acknowledgment channels and Q channel Walsh chips containing DRC channel and Data channel. PN_I and PN_Q are derived from the short PN sequences and user-specific long code PN sequences. After quadrature spreading, the baseband filtering is carried out to keep signal energy within the intended bandwidth (i.e., 1.25 MHz). RF processing ships the baseband signal to the RF carrier frequency, amplifies the RF signal, and radiates the signal energy into the air via a transmit antenna.

10.8 CALL SETUP: A CLOSER LOOK

Typical steps of the 1xEV-DO call setup are highlighted in Figure 10.25. After the AT is turned on, it initializes and synchronizes with the system. A 1xEV-DO session between the AT and the AN (the RNC, to be more specific) is then established. As part of session setup, four major steps are executed. These steps include the assignment of a UATI to the AT, the establishment of a dedicated radio connection between the AT and the AN (more specifically, the BS), the session configuration, and the key exchange. Upon session setup, a logical link exists between the AT and the AN.

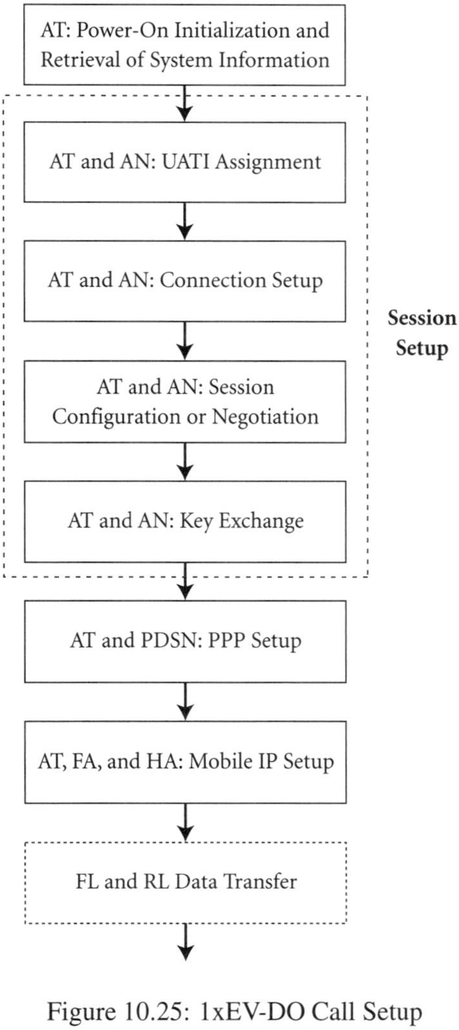

Figure 10.25: 1xEV-DO Call Setup

A link between the AT and the packet data core network is next established via a PPP session setup. Specifically, the RNC/PCF and PDSN exchange A11 registration messages to establish an A10 connection that can carry traffic [281, 282]. Usually, an IP address is assigned to the AT using client mobile IP procedures; the AT, FA, and HA participate in client mobile IP. The basic steps of the call setup are now over, and the AT can send and receive IP packets.

Let's briefly discuss how the AT initializes and retrieves system information. When the AT's power is turned on, it enters the Initialization State, in which it acquires the 1xEV-DO system. The AT chooses a specific carrier frequency based on the preferred roaming list programmed in the AT. If the AT is a hybrid 1xEV-DO and CDMA2000 1xRTT terminal, it chooses a 1xEV-DO system or a 1x system based on a proprietary implementation-specific system selection algorithm. Such a system selection algorithm could consider the pilot E_c/I_0 values of 1xEV-DO and 1x systems to choose a better system at a given time and location. On the 1xEV-DO carrier frequency, the AT attempts to find the Pilot channel, and since the pilot bits are all 0s and since only one PN code is present, the AT can synchronize more easily with the pilot channel. Once the AT locks onto the Pilot channel, it knows the timings of all the other FL channels. Now it waits for the Sync message to appear on the Control channel. Recall that the Control channel utilizes a specific MACIndex (e.g., 2 or 3). The Sync message provides the SystemTime and the PN Offset of the Pilot channel with which the AT has just synchronized. Next, the AT extracts information about the system by observing the overhead messages, such as the QuickConfig, SectorParameters, and AccessParameters messages.

Let's take a closer look at the 1xEV-DO call setup. Section 10.8.1 discusses session setup, and PPP establishment is described in Section 10.8.2. Mobile IP is summarized in Section 10.8.3.

10.8.1 Session Setup

As highlighted in Figure 10.25, the session setup involves four basic steps, UATI assignment, connection setup, session configuration, and key exchange. The AT sends a UATIRequest message to the AN and includes a random number called the Random Access Terminal Identifier as its temporary identity. The AT sends the UATIRequest message along with a RouteUpdate message on the Access channel. The BS replies to the AT with the ACAck message to indicate the successful detection of the access probe. The RNC in the AN responds to the AT with a UATIAssignment, which includes the AT's Random Access Terminal Identifier to identify the AT and the newly assigned 128-bit UATI. The AT then sends UATIComplete message, which is the final message in the UATI assignment process and acknowledges the AT's receipt of the UATI. The UATI may be specified as a set of 128 bits or as a combination of an eight-bit ColorCode to indicate the subnet identification (usually equivalent to RNC identification) and twenty-four–bit UATI024. The AT then creates the full 128-bit UATI by associating the ColorCode with the 104-bit subnet identification based on a previously received SectorParameters message and by using the UATI024 as the remaining twenty-four bits.

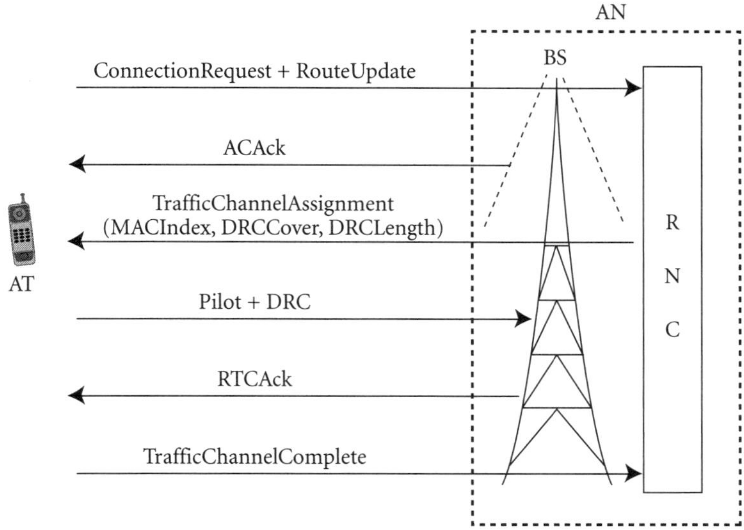

Figure 10.26: AT-Initiated Normal Connection Setup

Now a dedicated radio connection between the AT and the BS is established so that the next stage of session setup, session configuration, can be executed reliably. A connection can be set up three ways: the *AT-initiated normal setup*, the *AN-initiated normal setup*, and the *AN-initiated fast connect*. Figure 10.26 illustrates the AT-initiated normal setup, and the other two types of connection setups are discussed later in this section. The AT sends to the AN a ConnectionRequest message and a RouteUpdate message, both within an access probe. The RequestReason field of the ConnectionRequest message indicates "AT-initiated." The RouteUpdate message gives the RNC the E_c/I_0s of the strong pilots that the AT is observing. The BS responds with an ACAck message to acknowledge the successful detection of the access probe. The RNC obtains ConnectionRequest and RouteUpdate messages from the BS and sends a TrafficChannelAssignment message to the AT. The AN can directly place the AT into soft and/or softer handoff if the AT has reported multiple strong pilots. The TrafficChannelAssignment message includes DRCChannelGain, ACKChannelGain, and DRCLength, which are common to all the pilots, and MACIndex and DRCCover, which are assigned to the AT separately for each pilot in the Active Set.

Do You Recall... Do you recall the significance of DRCCover?[353]

[353]DRCCover identifies the sector from which the AT is requesting a FL packet.

After the AT receives the TrafficChannelAssignment message, it starts transmitting two channels, the Pilot and DRC channels, on the RTC. The BS sends an RTCAck message[354] to the AT to acknowledge the reception of the RTC. Now the AT sends a TrafficChannelComplete message, which indicates a successful Connection Setup. The absence of a TrafficChannelComplete message implies access failure or connection setup failure. Any type of failure beyond the TrafficChannelComplete message is interpreted as a call drop or a connection drop. Let's now briefly contrast the AT-initiated connection setup with the AN-initiated normal connection setup and the AN-initiated fast connect.

In the AN-initiated connection setup, the AN sends a Page message to the AT on the Control channel in a synchronous capsule. The AT identity, such as the UATI, is included in the MAC layer header of the packet carrying the Page message. In response to the Page message, the AT sends a ConnectionRequest message to the AN. Not surprisingly, the Request Reason field in the ConnectionRequest message indicates "AN-initiated!" The subsequent steps of the connection setup (i.e., from ACAck to TrafficChannelComplete) are identical to those illustrated in Figure 10.26.

The AN-initiated fast connect is faster (as the name suggests!) than the AN-initiated normal setup because the Page and ConnectionRequest messages are skipped and the AN directly sends a TrafficChannelAssignment message to the AT. When the AN is sure that the AT is continuously monitoring the FL Control channel, it directly sends the TrafficChannelAssignment message instead of the usual Page message. Assume that the AT has sent a ConnectionClose message to the AN including SuspendEnable set to 1 and the SuspendTime that specifies the time until which the AT would continuously monitor the FL Control channel transmissions. The AN can now feel comfortable using fast connect.

Session configuration is the third step in the session setup process. Session configuration allows the AN and the AT to determine parameter settings of the protocols and to override any default parameter settings defined in the standard. The messages involved in session configuration are displayed in Figure 10.27. Either the AT or the AN can initiate the process of session configuration or negotiation. The AN can start the process by sending a ConfigurationStart message, and the AT can start the process by sending a ConfigurationRequest message. The AT first leads the process of session configuration and sends a ConfigurationRequest message, and the AN responds with a ConfigurationResponse message. A series of such ConfigurationRequest and Response messages is exchanged based on what the AT wants to negotiate. In general, the AT and the AN negotiate the protocols at the seven layers of the 1xEV-DO protocol stack and the parameters for the negotiated protocols. Once the AT is done with the session negotiation, it sends a ConfigurationComplete message to indicate that it has completed its session negotiation. Next, if the AN wants to negotiate, it starts sending ConfigurationRequest messages, and the AT responds with Configuration-Response messages. When the AN is done with the negotiation process, it sends

[354]RTCAck is understood to represent the Reverse Traffic Channel Acknowledgment though the standard omits a formal expansion of the acronym.

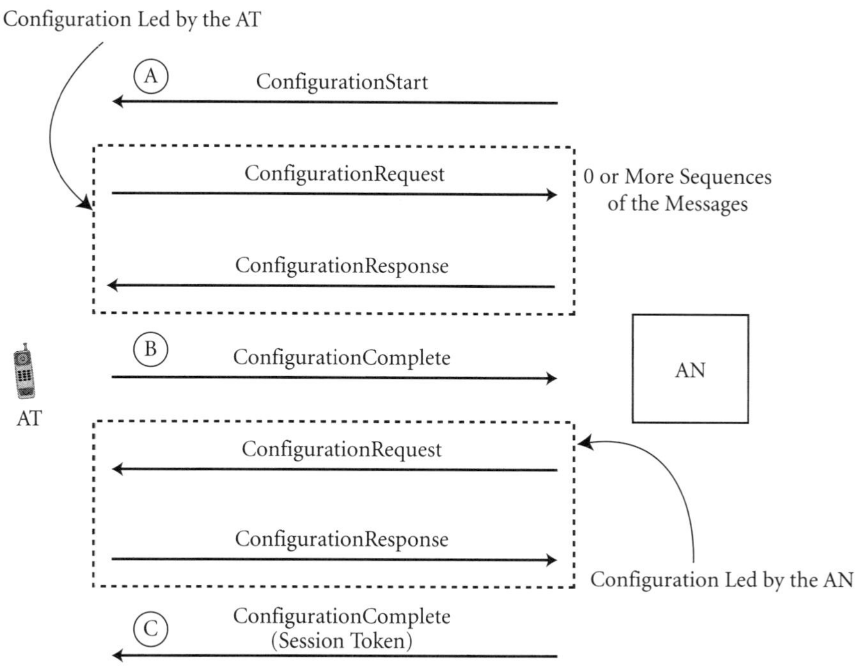

Figure 10.27: Session Configuration and Negotiation

a ConfigurationComplete message. This message indicates the end of the overall session negotiation process.

Examples of attributes or parameters that can be negotiated include transition probabilities[355] and PreferredControlChannelCycle.[356] Note that the standard, IS-856, provides default values of numerous attributes and default protocols. Hence, if both the AT and AN intend to use the default protocols and parameters, minimal session negotiation occurs. At the very least, the stream protocol should be negotiated to set up a packet data stream to support packet data applications (e.g., email or web-browsing) as the default configuration sets up the signaling stream only.

The key exchange is the fourth and the final step in the session setup. During this stage, certain information is exchanged between the AT and the AN to facilitate both authentication and encryption of messages and traffic. Typical commercial systems skip the key exchange stage in favor of reduced latency. Note that encryption and decryption add processing delays at the transmitter and receiver.

[355] See Section 10.10.1 for details on transition probabilities.

[356] This parameter facilitates interworking between 1xEV-DO and 1xRTT. When the page monitoring instants for 1xEV-DO and 1xRTT coincide, the AT sets this parameter to a suitable value that enables the AN to decide when to send a Page message to the AT.

At the end of session setup, the connection is usually closed. If the AT or AN require data transfer, a new radio connection is established. The negotiated session remains valid for a long time; the default value for the session timer, $T_{SMPClose}$, is fifty-four hours. The AT and the AN store the negotiated session attributes with the SessionToken as a reference to those attributes. During the time period corresponding to the session timer, the AT sends three SessionKeepAlive messages. For example, with the fifty-four–hour session timer, the AN sends a total of three SessionKeepAlive messages at $54/3 = 18$ hour intervals. If the AT fails to respond to all three Session-KeepAlive messages, the session ends and the AT loses the UATI and the negotiated session attributes. Of course, when the time is reset for an active AT as soon as the AT responds to the SessionKeepAlive message. During the session's validity time period, radio connections are established when data transfer is required and closed when data activity is absent for an implementation-dependent time period.

10.8.2 Point-to-Point Protocol Setup

The next step after the session setup is the establishment of a logical link between the AT and the PDSN. The RNC sends an A11 Registration Request to the PDSN to establish a radio-packet connection with the PDSN. More specifically, an A10 connection for best-effort service is established. Now, the AT and the PDSN communicate with each other via a three-stage process. In the first stage of Link Control Protocol, PPP link parameters are configured and a specific authentication protocol is selected when a sample IP is used. In mobile IP, no authentication protocol is chosen during this stage because mobile IP has its own authentication process. In the second stage of authentication, Challenge Handshake Authentication Protocol or Password Authentication Protocol can be used. Typical commercial deployments carry out no PPP authentication at this time, but client mobile IP authentication is carried out later. In the final stage of Internet Protocol Control Protocol, the PDSN assigns an IP address to the AT in the case of simple IP. However, in case of mobile IP, no IP address is assigned to the AT at this stage. The PPP session between the AT and the PDSN stays valid for the duration of the PPP timer (e.g., twenty-four hours). The service provider can configure a suitable value for the PPP timer. If no data activity occurs for the duration of the PPP timer, the PPP session is terminated.

10.8.3 Mobile Internet Protocol Setup

Mobile IP enables the system to support the AT's mobility across PDSNs while maintaining the end-to-end link (i.e., the Transmission Control Protocol or User Datagram Protocol session) between the AT and a server, such as the web server. Upon completion of the PPP setup, the PDSN/FA initiates mobile IP. Figure 10.28 depicts the basic aspects of IP address allocation and authentication using mobile IP; the exact implementation depends on the service provider's policies.

The PDSN sends an AgentAdvertisement message to the AT to offer a care-of-address and to provide an FA challenge. The AT constructs a Registration Request message and includes its care-of-address along with authentication parameters,

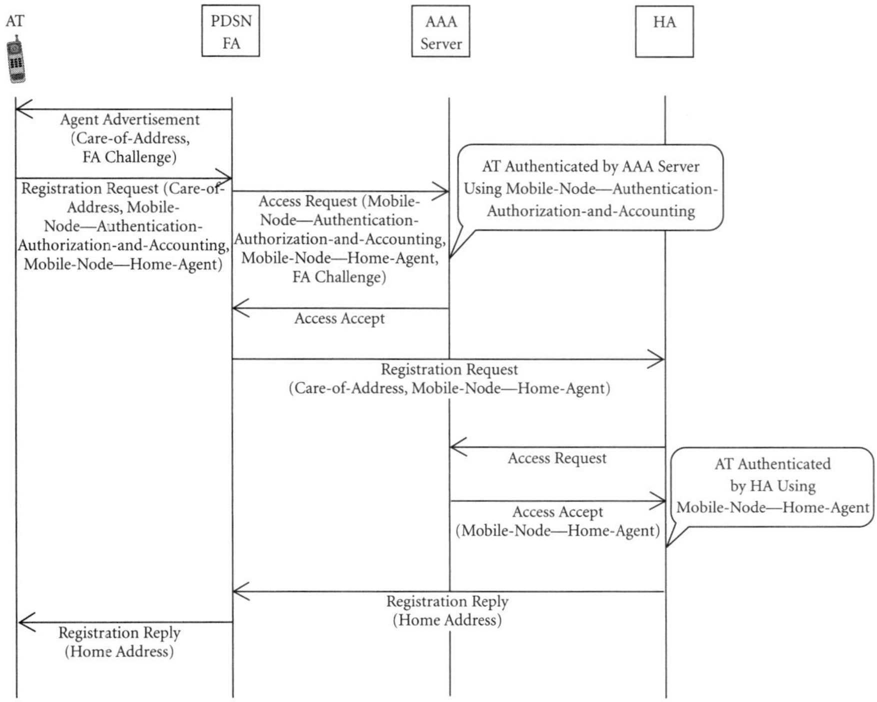

Figure 10.28: IP Address Allocation and AT Authentication using Mobile IP

mobile-node–authentication-authorization-and-accounting and mobile-node–home-agent. The PDSN/FA consults with the AAA server to validate the AT's authenticity by providing the FA challenge and the corresponding responses (i.e., mobile-node–authentication-authorization-and-accounting and mobile-node–home-agent) in an Access Request message. The AAA server compares mobile-node–authentication-authorization-and-accounting with the expected response for the FA challenge to authenticate the AT. Successful AAA authentication results in an Access Accept message.

The PDSN/FA now forwards the Registration Request to an HA. The HA keeps a record of the care-of-address to contact the FA for this AT. The HA exchanges Access Request and Access Response messages with the AAA server so that it can authenticate the AT using mobile-node–home-agent. Once the HA has authenticated the AT, it replies with a Registration Reply to the FA, which contains the newly assigned dynamic IP address called home address for the AT. [In Figure 10.28, Agent Advertisement is an Internet Control Message Protocol message, Registration Request and Registration Reply messages are mobile IP messages, and Access Request and Access Accept messages are Remote Authentication Dial-In User Service (RADIUS) messages. Refer to Chapter 4 to learn about Internet Control Message Protocol, mobile IP, and RADIUS.] Now all the requirements for data transfer are met, and the AT and the AN can engage in FL and RL data transfer.

10.9 FORWARD LINK DATA TRANSMISSION: A DETAILED VIEW

Figure 10.29 identifies the major stages involved in the forward link data transmission. In Stage 1, a connection is established, and a session is configured. The AN establishes a dedicated traffic channel with the AT in all Active Set sectors. The AT is assigned a MACIndex in each sector of the Active Set.

Stage 2 involves the AT transmitting the DRC channel. In the Connected State, the AT continuously sends a DRC value in every slot when DRC gating is off.[357] The DRCCover covers the DRC value.

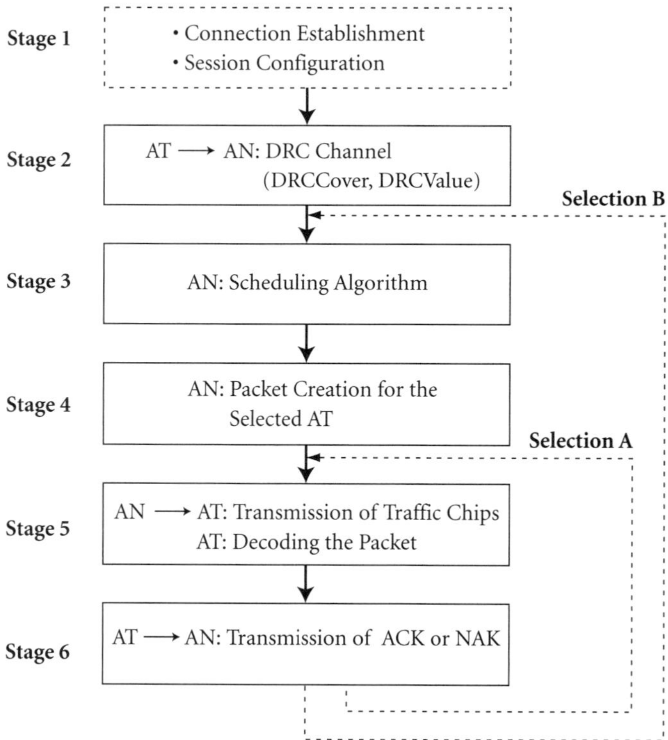

Figure 10.29: Stages of FL Data Transmission

[357] In general, DRC gating is not recommended due to the reduced reliability of DRC detection at the BS.

Do You Recall... Do you remember the significance of DRC value and DRC-Cover?[358]

Each sector in the AT's Active Set has a separate DRCCover. For example, a TrafficChannelAssignment message may specify three sectors in the Active Set as $\{PNoffset\ 100, PNoffset\ 104, PNoffset\ 200\}$ and may allocate DRCCover of 1, 2, and 3 for these sectors.

The DRCLength parameter dictates the number of slots during which the AT must send the same DRC value. The AT should estimate new channel conditions at least every DRCLength slots so that it can request a data rate that reflects the most recent channel conditions. All the sectors in the Active Set listen to the DRCCover, and the selected sector may decide to send a packet to the AT. Assume that the AT finds the sector with $Pilot\ PN\ offset = 104$ to be the best sector (i.e., with the strongest C/I ratio) and that it estimates the supportable data rate to be 921.6 kbps. Then the AT sends a DRC value of 8 with the DRCCover of 2 on the DRC channel in four consecutive slots if $DRCLength = 4$.

The BS in the AN executes the scheduling algorithm separately for each sector in Stage 3. From all the connected ATs that are sending DRC values, the scheduling algorithm selects one AT for the FL packet transmission. The IS-856 standard defines no specific algorithm; however, the proportional fairness algorithm and its variants are widely used [283]. This algorithm aims to increase the sector throughput while providing some degree of fairness.

Something to Think About... A scheduling algorithm is an implementation-specific proprietary algorithm that can be a good product differentiator for infrastructure vendors. The scheduling algorithm's decisions significantly affect packet data performance. A best-rate scheduler attempts to maximize revenue (capitalism at work!) by choosing a user who can support the highest data rate. A round-robin scheduler takes turns (socialism at its best!) to give every user a chance. A proportional-fair algorithm attempts to achieve a balance between these two philosophies (hybrid socialism and capitalism?!) by maximizing throughput and providing some degree of fairness. Specifically, it chooses a user who has the highest ratio of requested or supportable data rate to average data rate in the recent past. Users who have been treated equally in the recent past would have similar average data rates. The scheduler would then choose a user who can support the highest data rate, and such action increases throughput. However, if a user has been unlucky enough to request high data rates, that user would have a lower average past data rate. The ratio of the requested to past data rates would climb for such a user, and the scheduler would then choose the user

[358]The DRC value is the data rate the AT requested; this is the data rate that the AT can reliably receive at 1% packet error rate. DRCCover specifies the sector that the AT selected to receive forward link packets.

even when channel conditions prevent the user from supporting a high data rate. Such scheduler action ensures a degree of fairness.

Potential scheduler inputs are the requested data rate (i.e., the DRC rate), the average data rate in the recent past, the quality-of-service (QoS), the history of ACKs and/or NAKs, the amount of data in the buffer, and the priority of data versus signaling messages. The basic scheduler output is the selected AT for the next slot. Note that the inputs to the scheduler, such as average data rate in the recent past, may need to be updated in every slot. Once the scheduler chooses a specific AT for a given slot, it reserves a specific number of slots corresponding to that AT's requested DRC value. For example, if the scheduler chooses an AT that had requested the DRC value of 2 corresponding to the nominal data rate of 76.8 kbps, the scheduler reserves eight slots for the packet transmission.[359]

The BS in Stage 4 creates the traffic chips for the multislot transmission to the selected AT as discussed in Section 10.6.4. The physical layer packet size that the BS has chosen depends on the requested DRC value (e.g., 1024 bits for $DRC = 2$).

In Stage 5, the BS starts transmission of the packet in the first reserved slot. The ATs look for their packets in every slot. A review question is in order...

Do You Recall... How does the AT know that the current slot contains *its* data (not a different AT's data)?[360]

If the AT has requested that Sector Y send data, it monitors the FL slot of Sector Y and attempts to match the preamble with its MACIndex. Only the AT the BS has selected for the FL transmission in a given slot finds a MACIndex match with the preamble; all other ATs ignore the rest of the traffic chips contained in that slot. After receiving all the traffic chips, the AT tries to retrieve the original physical layer packet (e.g., a set of 1024 bits). *For all the DRC values, the target AT may be able to successfully decode the entire physical layer packet in just the first slot even for a multislot transmission scenario. In other words, the multislot transmission includes redundancy such that the first slot is adequate to recover the original information bits; the remaining slots provide redundant information.* The AT performs soft-combining of the received transmissions corresponding to the same physical layer packet because it stores all the received slots until it decodes the packet correctly or until transmission in all the slots has occurred.

If the AT successfully recovers the information bits of a physical layer packet, it sends an ACK to the AN on the Acknowledgment channel in Stage 6. If the AT is unable to retrieve the original information bits, it sends a NAK. If the AT receives

[359]See Table 10.3 for a mapping between the DRC value and the required number of nominal slots and for packet sizes for different DRC values.

[360]The preamble present in the early part of the slot conveys the MACIndex of the specific AT to which the data is being sent.

no packet in a given slot, it transmits no ACK or NAK. Hence, the transmission on the Acknowledgment channel is discontinuous. The Acknowledgment channel is transmitted with a power offset ACKChannelGain relative to the Pilot channel, and the AN specifies ACKChannelGain to the AT in the TrafficChannelAssignment message. ACKChannelGain should be large enough to facilitate its detection at the BS and small enough to create minimal interference in the RL (e.g., 4 dB).

10.10 REVERSE LINK DATA TRANSMISSION: A DETAILED VIEW

Figure 10.30 illustrates the overall data transmission for the reverse link. The AT determines the reverse link data rate. The AN can influence the process of data rate selection by specifying suitable parameters during session configuration. The BS provides to all the ATs in a sector fast feedback about prevailing reverse link loading. The AT executes an algorithm to determine the data rate based on the information the AN provides. The AT transmits data on the Data channel, which is part of the RTC. The RRI channel, which is time-multiplexed with the RL Pilot channel, specifies the rate at which the AT transmits a packet. Let's take a closer look at each stage of the RL data transmission in Sections 10.10.1 through 10.10.4.

10.10.1 Stage 1: Configuration of the Access Terminal with Reverse Link Parameters

The AN can influence the AT's reverse link data determination process via two main configurable parameters, *RateLimit* and *transition probabilities*. RateLimit is the maximum data rate that the AT can transmit in the RL. RateLimit is usually set to the maximum RL data rate of 153.6 kbps in typical commercial deployments.

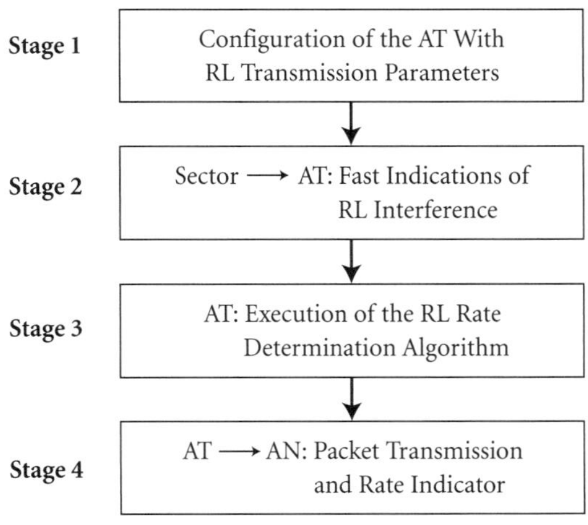

Figure 10.30: Reverse Link Data Transmission

The AN can use either of the two messages, UnicastReverseRateLimit or Broadcast-ReverseRateLimit, to convey RateLimit to the AT. The UnicastReverseRateLimit message is used to specify RateLimit for just one AT, and the BroadcastReverseRateLimit message can specify RateLimit per AT for multiple ATs. The first entry of RateLimit corresponds to the AT that has been assigned the MACIndex $64 - 1 = 63$. Similarly, the second entry of RateLimit corresponds to the AT with the MACIndex $64 - 2 = 62$, and so on. The parameter RPCCount in the BroadcastReverseRateLimit message gives the total number of ATs for which RateLimit is being specified.

In a 1xEV-DO system, the RL data rate changes are constrained to be gradual to minimize wide and fast variations in the RL loading or interference. The *transition probability* defines the probability of a transition from one rate to the next higher or lower rate. Transition probabilities place an additional upper limit on the RL data rate that the AT can transmit during a given frame. Such an upper limit is either the same as the data rate the AT used in the previous frame, called *CurrentRate*, or one step higher or lower than CurrentRate. Possible RL data rates are 9.6 kbps, 19.2 kbps, 38.4 kbps, 76.8 kbps, and 153.6 kbps.

Assume that CurrentRate is 38.4 kbps. Then the upper limit on the data rate in the next frame could be one level higher, 76.8 kbps; one level lower, 19.2 kbps; or, the same as CurrentRate, 38.4 kbps. As an example, Transition 038k4_076k8 defines the probability with which the AT can increase its data rate from 38.4 kbps to 76.8 kbps. Similarly, Transition 076k8_038k4 is the probability that the AT will decrease its data rate from 76.8 kbps to 38.4 kbps. Table 10.5 summarizes the default values of the transition probabilities specified in the standard. The AT or AN could propose different values for the transition probabilities during session configuration.[361]

Table 10.5: Transition Probabilities for Reverse Link Data Rate

Transition Probability	Default Value
Transition009k6_019k2	$48/255$
Transition019k2_038k4	$16/255$
Transition038k4_076k8	$8/255$
Transition076k8_153k6	$8/255$
Transition153k6_076k8	$255/255$
Transition076k8_038k4	$32/255$
Transition038k4_019k2	$16/255$
Transition019k2_009k6	$16/255$

[361] The default transition probabilities are generally considered to be quite conservative and are typically modified during the session configuration to enable higher RL throughput.

10.10.2 Stage 2: Fast Indications of Reverse Link Interference

The BS evaluates the amount of RL interference in each sector and sends a sector-specific RAB on the Reverse Activity channel. A RAB value of 1 corresponds to a heavily loaded sector. The algorithm that determines whether the RAB value is 0 or 1 is undefined in the standard is implementation-specific. Potential inputs to such an algorithm include the measured rise-over-thermal noise floor, the target interference margin or loading margin, the number of connected ATs, and the RL loading in the surrounding sectors. In a simple implementation, the algorithm could set RAB to 1 if the measured rise-over-thermal noise exceeds a predefined threshold. Such a threshold should be close to the loading or interference margin (e.g., 3 to 6 dB), which is one of the inputs to the RL link budget. To achieve a certain cell-edge reliability target, the RAB determination algorithm should use a threshold that is lower than the target loading margin so that a preventive action can be taken before the actual amount of interference exceeds the predetermined target.

From a sector, a given RAB value is transmitted in RABLength consecutive slots as part of the forward link MAC channel. Recall that the MAC channel is transmitted in four segments of a slot with 64 chips in each segment. The RABLength assumes one of the four values—eight, sixteen, thirty-two, or sixty-four slots. A smaller value enables a faster response to changes in RL interference, and a larger value improves the reliability of RAB detection. The RABLength is specified for each sector in the Active Set in the TrafficChannelAssignment message.

10.10.3 Stage 3: Execution of the Reverse Link Data Determination Algorithm

Each AT executes an algorithm to determine the RL data rate. This algorithm follows the constraints defined in the standard. In a 1xEV-DO Rev. 0 system, the RL data rate can be changed as fast as every 26.67 ms frame.

Do You Recall. . . For the sake of comparison, do you remember how fast the data rate can be changed in the FL?[362]

Figure 10.31 mentions the basic inputs to the algorithm: transition probabilities, CurrentRateLimit, CurrentRate, CombinedBusyBit, and constraints related to power and data. Transition probabilities and CurrentRateLimit are the parameters that the AN can configure during session configuration; their default values are defined in IS-856. Transition probabilities are specified as part of the RateParameters configurable attribute, and CurrentRateLimit is RateLimit conveyed to the AT by the AN via a UnicastReverseRateLimit or BroadcastReverseRateLimit message. If the AT never receives a RateLimit from the AN, the default RateLimit is assumed to be 9.6 kbps.

[362]In the forward link, the data rate could potentially change as fast as every 1.67 ms slot.

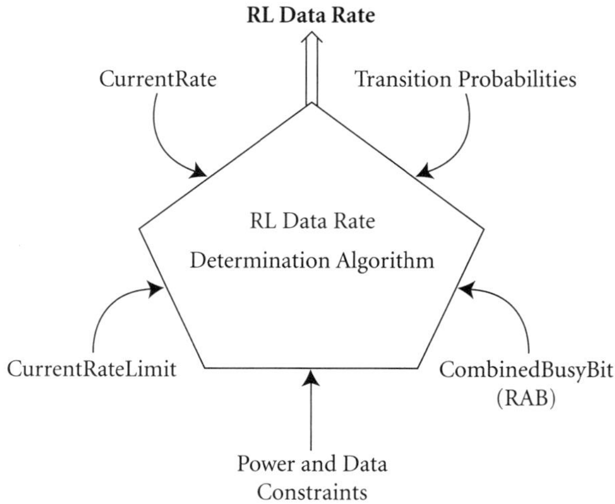

Figure 10.31: Reverse Link Data Determination Algorithm

CurrentRate is the data rate that the AT has transmitted in the previous frame.[363] If the AT has not transmitted anything in the previous frame, CurrentRate is 0 kbps.

The AT considers the RABs received from all the Active Set sectors and performs an OR operation on the RABs to calculate the CombinedBusyBit. Assume that Sector X is lightly loaded and transmitting an RAB of 0 and that Sector Y is heavily loaded and transmitting an RAB of 1. CombinedBusyBit is then obtained by applying the OR operation to $RAB = 0$ and $RAB = 1$, which results in a value of 1. If all sectors in the Active Set transmit $RAB = 0$, the CombinedBusyBit is 0. The CombinedBusyBit dicates the direction of the change in the data rate relative to the CurrentRate. For example, the CombinedBusyBit of 0 tells the AT to try to increase its data rate because the overall loading or interference is low. A CombinedBusyBit value of 1 corresponds to high interference in the RL, and the AT tries to reduce the data rate to reduce the RL interference. Even if only one sector in the Active Set has high interference, the OR operation will yield a value of 1 for the CombinedBusyBit, and the AT attempts to reduce the data rate to alleviate the interference problem in such a sector.

The AT also considers power and data constraints. Specifically, the AT checks the available transmit power for the RTC's Data channel and the amount of data to be transmitted. Recall that the RTC consists of Pilot or RRI, Acknowledgment, DRC, and Data channels. Hence, the total AT transmit power must be shared among all these channels. If the AT is near the cell edge, it is already consuming a significant amount of power and may have inadequate power to support higher data rates. If the

[363] "The previous frame" means the frame immediately preceding the frame for which the AT is in the process of determining the rate.

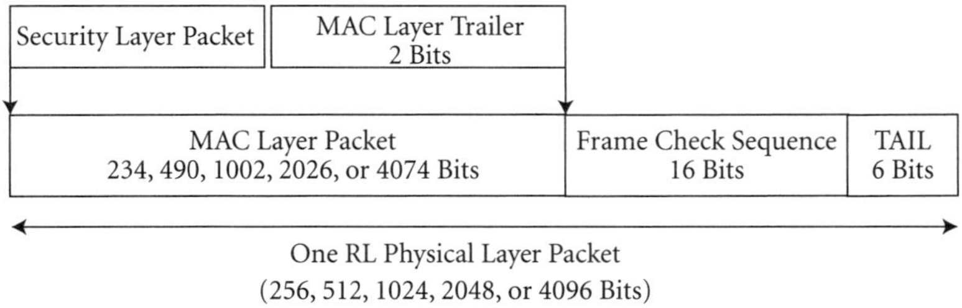

Data Rate (kbps)	Supportable Security Layer Payload Size (Bits)
0	0
9.6	1 to 232
19.2	233 to 488
38.4	489 to 1000
76.8	1001 to 2024
153.6	2025 to 4072

Figure 10.32: Format of the Reverse Link Physical Layer Packet

AT can send the available data in one frame at a low data rate, it cannot use a higher data rate. Figure 10.32 summarizes the supportable payloads for various RL data rates. The payload in Figure 10.32 is the size of the security layer packet. The MAC layer packet includes the security layer packet and a two-bit MAC layer trailer.[364] Two additional fields, frame check sequence and TAIL bits, are included in the RL physical layer packet as shown in Figure 10.32.

IS-856 provides guidelines the AT can use while determining the RL data rate. The overall algorithm is summarized in Figure 10.33. The AT executes this algorithm at the beginning of every frame in the Connected State. The steps of this algorithm are as follows.

1. Set the CurrentRate to the data rate transmitted in the previous frame. If the AT transmitted no data in the last frame, the CurrentRate is set to 0.

2. Keep a record of the latest RateLimit received from the AN and make it CurrentRateLimit.

3. Process RABs received from the Active Set sectors to determine Combined-BusyBit.

4. Use CombinedBusyBit and CurrentRate to determine the potential value of the next data rate R. For example, if CombinedBusyBit is 0, associated with low

[364]The MAC layer trailer consists of the one-bit ConnectionLayerFormat and one-bit MACLayerFormat. The ConnectionLayerFormat indicates if it is a Format A or Format B connection layer packet. The MACLayerFormat indicates if the MAC layer packet indeed contains a valid payload.

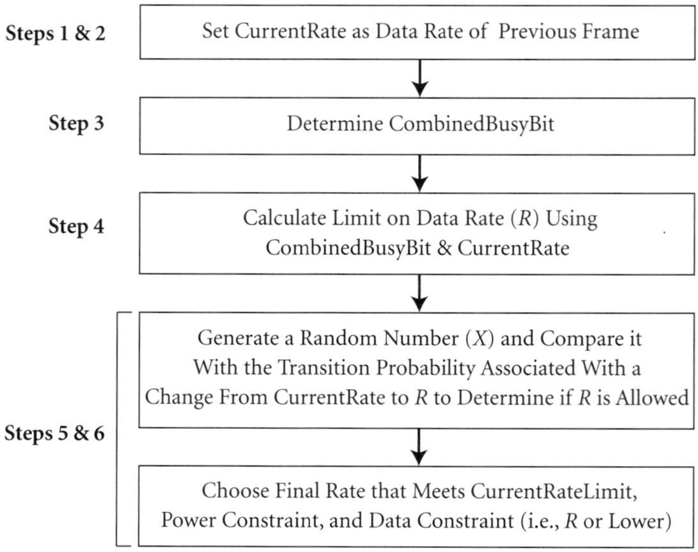

Figure 10.33: The AT's RL Data Rate Selection

interference, and CurrentRate is 38.4 kbps, the next potential data rate R is one level higher than 38.4 kbps, which is 76.8 kbps.

5. Generate a random number, X, between 0 and 1 (e.g., 0.4).

6. Compare X with the transition probability corresponding to the rate transition from CurrentRate to R. In other words, if CurrentRate is 38.4 kbps and R is 76.8 kbps, the relevant probability is $TransitionCurrentRate\_R$. If $X < TransitionCurrentRate\_R$, the AT can increase the data rate to R in the next frame as long as $R \leq CurrentRateLimit$ and the AT has sufficient transmit power and enough data. If $X \geq TransitionCurrentRate\_R$, the AT's maximum rate stays at CurrentRate.

Let's take a numerical example. Assume that the AT's CurrentRate is 38.4 kbps and CombinedBusyBit is 0. The relevant transition probability is Transition38k4_076k8. The AT generates $X = 0.4$. If $0.4 < Transition38k4\_076k8$, the AT can increase its data rate to 76.8 kbps if it has enough power and adequate data to support 76.8 kbps. If $0.4 > Transition38k4\_076k8$, the AT keeps its rate at 38.4 kbps if it has enough transmit power and adequate amount of data for the 38.4 kbps data rate.

10.10.4 Stage 4: The Access Terminal's Data Transmission

The AT starts transmitting a packet at the data rate determined in Stage 3. The physical layer parameters corresponding to allowed RL data rates are summarized in

Table 10.6: Physical Layer Parameters for Reverse Link Traffic Channel

Data Rate (kbps)	Physical Layer Packet Size (Bits)	Coding Rate	Repetition Factor	Spreading Factor
9.6	256	1/4	8	4
19.2	512	1/4	4	4
38.4	1024	1/4	2	4
76.8	2048	1/4	1	4
153.6	4096	1/2	1	4

Table 10.6. For example, a 38.4 kbps data rate uses the physical layer packet size of 1024 bits.

A Question for You... Can you do a simple calculation to define the exact relationship between 38.4 kbps and 1024 bits?[365]

The packet size of 1024 bits undergoes turbo coding at rate 1/4, yielding $1024/(1/4) = 4096$ turbo code symbols. A repetition factor of 2 results in $4096 * 2 = 8192$ code symbols. Since BPSK is used in the RL for the Data channel, 8192 BPSK modulation symbols are available. Spreading by a factor of 4 leads to the target chip rate of $8192 * 4 = 32768$ chips. The chip rate becomes 32768 chips$/26.67$ ms $=$ 1.2288 Mcps, the same as that in IS-95 and 1x!

The AT specifies the Data channel's selected rate in RRI form on the RRI channel. Recall that the RRI channel and the Data channel are code-division-multiplexed. The RRI channel obviates the BS's need for blind rate detection of the Data channel. The instantaneous transmit power of the Data channel depends on the RL Pilot channel power and the power offsets specified in Table 10.4. The RL fast power control may be as fast as 562.5 Hz as discussed in Section 10.6.2.

The sector must estimate the rise-over-thermal noise so that the RAB can be set properly. Since the noise floor can vary over time, a mechanism that allows the sector to measure the noise floor is useful. That's where the RL silence interval enters the scene. While the sector is measuring the noise floor, no ATs should be transmitting on the reverse link. If ATs are transmitting, the noise floor would include some system loading (i.e., interference). Hence, one approach is to measure the received power when the system is very lightly loaded, such as during the night. Another approach is to force ATs to periodically turn off their transmissions and then measure the received power at the base station antenna, and this approach is supported in 1xEV-DO through the silence interval. Figure 10.34 illustrates the basic idea of the reverse link silence

[365] We hope that you found the question to be easy! Packet size divided by frame duration equals data rate: 1024 bits$/26.67$ ms $= 38.4$ kbps.

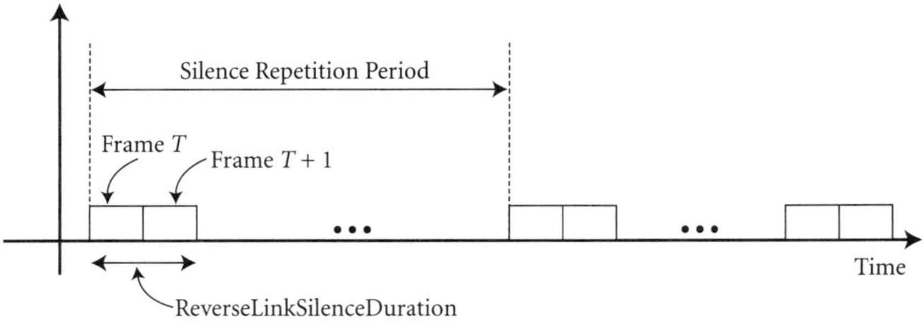

Figure 10.34: Reverse Link Silence Interval

interval.

The ATs stop their transmissions for the duration of ReverseLinkSilenceDuration frames. The silence interval starts at time T such that following condition is satisfied.

$$T \bmod \left(2048 \times 2^{ReverseLinkSilencePeriod} - 1\right) = 0 \tag{10.8}$$

where T is the CDMA system time expressed in frames and *ReverseLinkSilence Period* dictates how frequently the silence interval is repeated. Both ReverseLink-SilenceDuration and ReverseLinkSilencePeriod are two-bit fields and hence can assume the values of 0, 1, 2, or 3. If the AN wishes to omit the reverse link silence interval mechanism, it can simply set the ReverseLinkSilenceDuration to 0. These parameters are conveyed to the ATs in the SectorParameters message.

10.11 MOBILITY MANAGEMENT IN 1xEV-DO

Rather than re-inventing the wheel, 1xEV-DO relies upon the same basic mobility management principles as those used in 1x. An AT in the Idle Mode has no dedicated radio connection with the AN. Such an AT follows the paging cycle and periodically wakes up to look for a Page message in the chosen sector. Furthermore, the Idle Mode AT continues to search for a sector with the strongest FL quality that may be quantified by the pilot E_c/I_0. When the AT wants to initiate data transfer or when it receives a Page message from the AN, the AT establishes a radio connection with the AN and exits the Idle Mode.

Mobility of an AT with a dedicated radio connection with the AN is managed via the traditional soft handoff principles of CDMA. Figure 10.35 illustrates the basic steps of the handoff for a Connected Mode AT. In Step 1, the AT quantifies E_c/I_0 by observing the Pilot channels transmitted by the sectors that are part of the current Active Set and the neighbor list [277]. The current Active Set is determined based on the most recent TrafficChannelAssignment message the AT receives. The AT can generally use the neighbor list present in the SectorParameters message. The AN can provide the AT-specific neighbor list by sending a NeighborList message to the AT.

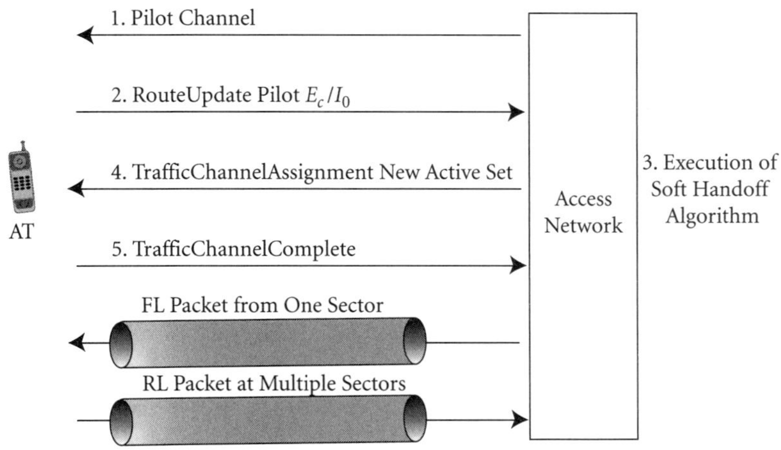

Figure 10.35: Main Steps of Handoff in 1xEV-DO

1xEV-DO E_c/I_0 is typically higher than 1x E_c/I_0 due to 1xEV-DO's full-power transmission of the Pilot channel. The Pilot channel in 1x is allocated part of the total FL power (e.g., 15% to 20%) in a given sector. In Step 2, the AT evaluates the prevailing radio environment and sends a RouteUpdate message containing pilot PN offsets and their E_c/I_0s.

Do you Recall... Do you remember the name of the IS-95 or 1x message that carries pilot E_c/I_0 measurements the mobile device made?[366]

1xEV-DO supports the basic static thresholds-based algorithm and the enhanced dynamic thresholds-based algorithm for reporting measurements to facilitate handoff.[367] A session-configurable parameter, DynamicThresholds, is used to choose between these two algorithms. A DynamicThresholds setting of 0 implies the basic algorithm, and a DynamicThresholds setting of 1 implies the dynamic thresholds-based algorithm. The default value of DynamicThresholds is 0. With the basic algorithm, the AT sends a RouteUpdate message to the AN when a neighbor pilot exceeds a threshold PilotAdd. As another example, when an Active Set pilot drops and stays below PilotDrop for PilotDropTimer seconds, the AT sends a RouteUpdate message.[368]

[366]The Pilot Strength Measurement message in IS-95 and 1x includes the pilot E_c/I_0 measurements the mobile device made.

[367]See Chapter 7 on CDMA and IS-95 for details of the basic static thresholds-based handoff algorithm. Chapter 8 on 1x describes the enhanced dynamic thresholds-based handoff algorithm.

[368]The parameters PilotAdd, PilotDrop, and PilotDropTimer in 1xEV-DO are equivalent to T_ADD, T_DROP, and T_TDROP in 1x, respectively.

In Step 3, the RNC executes a soft handoff algorithm and makes a handoff decision based on the pilot measurements received in the RouteUpdate message. The RNC may add a sector to the Active Set, drop a sector from the Active set, and/or replace an existing sector by a new sector. The soft handoff algorithm in 1xEV-DO, just like in IS-95 and 1x, is implementation-specific.

In Step 4, the RNC specifies the updated Active Set in the TrafficChannel-Assignment message. In Step 5, the AT acknowledges the reception of the Traffic-ChannelAssignment message by sending a TrafficChannelComplete message.

The AT and the AN start using the new Active Set now. More specifically, the AT can choose any one of the Active Set sectors to receive a FL packet. In the RL, all the sectors in the Active Set listen to the packet the AT sends. The BS performs maximal ratio combining of the signals received from the AT on the antennas of the sectors it controls. The RNC does selection combining of the packets received from multiple BSs in the AT's Active Set. See Chapter 7 for details of the packet processing in soft handoff.

10.12 A GLIMPSE OF 1xEV-DO OPTIMIZATION

A 1xEV-DO system is usually deployed as an overlay to a 1x system, which provides traditional circuit-switched voice services. A 1xEV-DO system (preferred because of higher throughput) or a 1x system can provide packet data services. A typical initial deployment scenario involves an area with 1x and 1xEV-DO systems surrounded by an area with only 1xRTT systems. Since the same cell sites are normally shared between 1xRTT and 1xEV-DO systems, the impact of such constraints on the performance of a 1xEV-DO system should be evaluated. Figure 10.36 illustrates an example scenario of 1xEV-DO deployment.

As shown in Figure 10.36, all the cells, C1 through C13, support CDMA2000 1xRTT systems in a given geographical area. However, 1xEV-DO is deployed only in the center of the region, in cells C1 through C4. The 1xEV-DO footprint is gradually expanded based on subscriber demand, market predictions, and business strategies. Service operators often showcase their technological prowess by displaying third-generation coverage maps on their websites. Various aspects of 1xEV-DO optimization are briefly touched upon in Sections 10.12.1 and 10.12.2.

The end-to-end system optimization involves AN optimization, packet-switched core network configuration, and Transmission Control Protocol and IP optimization. The AN optimization includes RF optimization and optimal radio network parameter settings. Example areas of packet-switched core network configuration include PPP compression and PPP timer setting at the PDSN. Examples of Transmission Control Protocol and IP optimization include IP header compression and Transmission Control Protocol parameter optimization (e.g., increasing maximum transmission unit size and Transmission Control Protocol window size for higher throughput).

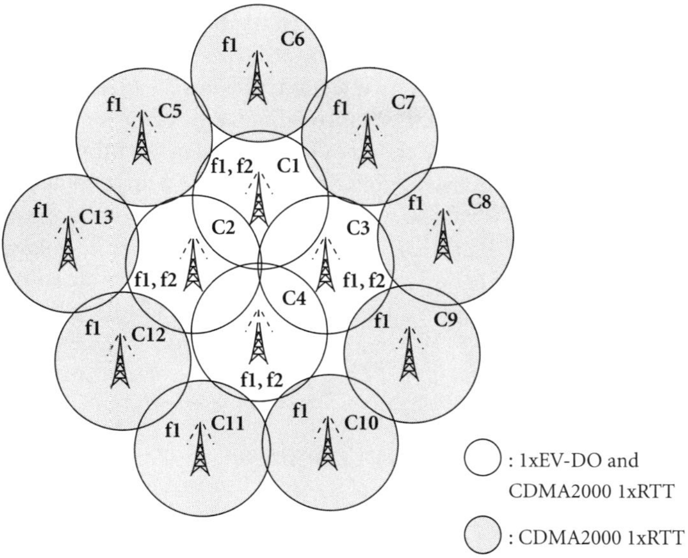

Figure 10.36: 1xEV-DO Deployment Scenario

10.12.1 RF Optimization

Generic RF optimization techniques traditionally applied to IS-95 and CDMA2000 can also be utilized to optimize the RF environment in a 1xEV-DO system. Examples of such techniques include RF coverage optimization through changes in the antenna type (e.g., 65° beamwidth instead of 90° beamwidth) and antenna parameters (e.g., downtilt, height, and orientation); neighbor list optimization per sector-neighbor pair; and active and neighbor pilot search window optimization. The neighbor list should be short enough to enable a fast search and long enough to include all important neighbors. The AT obtains the neighbor list from the SectorParameters message. When the AT is in the Connected State, the RNC can send to the AT a dedicated NeighborList message to convey the AT-specific composite neighbor list.

The size of the active pilot search window, SearchWindowActive, should be small enough to facilitate a fast search but large enough to help detect usuable multipaths of a given pilot. The delay spread statistics in a given RF environment can be used to determine the optimal SearchWindowActive. The size of the neighbor pilot search window should be such that it helps detect a neighbor pilot in a given sector. The delay difference statistics per sector-neighbor pair are useful to determine the SearchWindowNeighbor. A given sector could have as many as twenty neighbors. The IS-856 standard, unlike 1x, allows the specification of a SearchWindowNeighbor *per neighbor* rather than just one common SearchWindowNeighbor for all the neighbors.

The value of SearchWindowNeighbor can be specified in the SectorParameters and NeighborList messages.

Since 1xEV-DO Rev. 0 is a packet data system intended for non-real-time services, the main performance metric is throughput. User throughput and sector throughput are important performance metrics. Achievable user throughput can be estimated by plotting the DRC values the AT has requested (not scheduled) along a drive route and deriving a histogram of requested DRC values. After the RF environment is optimized, the requested DRC values should be higher than before. Other secondary performance metrics are access failure rate or connection setup failure rate, connection-drop rate, physical layer packet error rate, frequency of RLP NAKs, and handoff distribution including intra-BS, inter-BS, inter-RNC, and inter-technology (e.g., 1x and 1xEV-DO) handoff.

10.12.2 Parameter and Configuration Optimization

In addition to RF optimization, various configurable parameters can be tuned to enhance the overall system performance. The connection setup failure rate can be reduced by tuning parameters such as OpenLoopAdjust, ProbeInitialAdjust, ProbeSequenceMax, ProbeNumStep, PowerStep, and AccessCycleDuration. For example, if the currently used ProbeNumStep is 2 and PowerStep is 2 dB, and if numerous connection setup failures are observed in a given area that has been RF-optimized, ProbeNumStep can be increased to 8 and PowerStep can be increased to 4 dB. The side effects of such changes should also be evaluated since the average RL interference due to these changes would be somewhat higher than before. Appropriate handoff parameters and algorithms can help minimize connection drops. Examples of relevant parameters include PilotAdd, PilotDrop, PilotDropTimer, AddIntercept, DropIntercept, and SoftSlope. A choice between a traditional fixed-thresholds–based algorithm (i.e., PilotAdd and PilotDrop) and a dynamic-thresholds–based algorithm (i.e., AddIntercept, DropIntercept, and SoftSlope) must also be made.[369] In general, the fixed-thresholds–based algorithm yields higher reliability, and the dynamic-thresholds–based algorithm reduces the degree of handoff.

The FL throughput can be increased by ensuring that the RL channels, such as the DRC channel and the Acknowledgment channel, are reliably detected at the BS. Examples of relevant tunable parameters are DRCChannelGain, DRCLength, DRCGating, and ACKChannelGain. Larger values of DRCChannelGain, DRCLength, and ACKChannelGain increase the reliability of detection of the DRC channel and the Acknowledgment channel at the expense of higher RL interference and somewhat lower RL throughput due to the availability of less transmit power for the RL Data channel. The RL throughput can be influenced by tuning the RL power-control thresholds and transition probabilities. Aggressive transition probabilities could potentially increase the RL throughput at the expense of higher average RL interference. The impact of a hybrid 1x and 1xEV-DO mode on the 1xEV-DO performance can be

[369]See Chapter 8 on 1x for a discussion of fixed-thresholds–based and dynamic-thresholds–based handoff algorithms.

minimized by using a longer 1x paging cycle of 5.12 seconds instead of a shorter paging cycle of 2.56 seconds. Of course, such action will increase 1x paging delay.

10.13 1xRTT AND 1xEV-DO INTERWORKING

When a hybrid AT supporting 1x and 1xEV-DO is powered on, a proprietary system selection algorithm implemented in the baseband chipset considers the preferred roaming list of the carrier frequencies and evaluates the suitability of 1xRTT and/or 1xEV-DO. When the hybrid AT is in the Idle State on a 1xEV-DO system, it has to periodically monitor the control channel on a 1xEV-DO carrier frequency and the Paging channel on a 1x carrier frequency. For example, the AT observes the Control channel once every 5.12 seconds, the fundamental monitoring cycle defined in the standard. The hybrid AT periodically jumps from the 1xEV-DO frequency to the 1x frequency to look for a possible Page message. The 1x paging cycle is typically set to 2.56 or 5.12 sec. Hence, the hybrid AT jumps to the 1x carrier frequency every 2.56 or 5.12 sec. When the AT is in the Connected State, it still needs to periodically jump to the 1x carrier frequency for a possible Page message. To avoid the loss of a packet that the BS could transmit to the AT on the 1xEV-DO carrier frequency while the AT is on the 1x carrier frequency, the AT sends null DRCCover to the serving sector before jumping to the 1x frequency.[370]

Do You Recall. . . Do you remember the significance of DRCCover?[371]

After the hybrid AT returns to the 1xEV-DO carrier frequency, it resumes sending DRC values with the DRCCover corresponding to the selected sector, and the selected sector can resume the data transmission to the hybrid AT. The overall FL and RL throughputs of hybrid AT decrease because it is absent from the 1xEV-DO air interface any time it jumps to the 1x carrier frequency.

When the hybrid AT is in the 1xEV-DO and 1x coverage area, it periodically monitors 1xEV-DO Control channel and 1x Paging channel. When the hybrid AT in the 1xEV-DO Idle State leaves the 1xEV-DO coverage area, an implementation-specific system selection algorithm monitors just the 1x system and could look for a 1xEV-DO system with longer time intervals than the usual 5.12 seconds.

When the hybrid AT in the 1xEV-DO Connected State leaves the 1xEV-DO coverage area, the connection is closed, and the AT enters the 1xEV-DO Idle State. The hybrid AT now finds a 1x system and establishes a link with the 1x system. If mobile IP is supported, the end-to-end packet data session between the AT and the web server is preserved although the air interface has changed from 1xEV-DO

[370]We have discussed here low-cost ATs that have a single radio tuned to either 1xEV-DO or 1x at a given instant. ATs may also support different levels of simultaneous 1x and 1xEV-DO operations such as simultaneous 1x and DO 1xEV-DO Receive or simultaneous 1x and 1xEV-DO Receive and Transmit.

[371]DRCCover identifies the sector from which the AT is requesting FL data. When DRCCover is null, the serving sector schedules no packets to this hybrid AT.

to 1x. Even the PDSN/FA might be different. The hybrid AT can thus resume the packet data session. When the AT is active on the 1x system, it cannot monitor the 1xEV-DO system. However, when the AT enters the Dormant State on the 1x system, an implementation-specific system selection algorithm could look for a 1xEV-DO system. When the AT reselects the 1xEV-DO system, the 1xEV-DO session between the AT and the RNC may already be valid because of its prior establishment.

10.14 1xEV-DO LINK BUDGET

The link budget effectively determines the cell radius or cell size by specifying the maximum tolerable path loss. Various parameters go into the link budget equation, and, the output of the equation is the path loss in decibels. This path loss can be used in a propagation path loss formula such as a COST231-Hata model (see Chapter 3), to yield the estimated cell radius (e.g., in kilometers). Since the base station has a high-power amplifier capable of transmitting a large amount of power (e.g., 20 W) and the AT has limited power (e.g., 200 mW), the link budget is usually limited by the reverse link. In other words, the link budget for the reverse link determines the cell radius, and the forward link budget can then be used to determine the supportable data rates and/or capacity for that cell size. Black and Wu [284] offer additional details on the link budget approach and specific values of the link budget parameters used here.

The maximum tolerable path loss, PL_{max}, is given by

$$PL_{max}\,(\text{dB}) = EIRP\,(\text{dBm}) + Gains\,(\text{dB}) - Losses\,(\text{dB}) - ReceiverSensitivity\,(\text{dBm}). \tag{10.9}$$

In Eq. (10.9), $EIRP$ is the effective isotropic radiated power of the AT, $Gains$ are all the gains the signal experiences between the AT and the BS antenna connector, $Losses$ are all the losses that the signal encounters between the AT and the BS, and $ReceiverSensitivity$ represents the minimum signal level that the BS must be able to detect at a given QoS. Intuitively,

$$\textit{Received Signal Power} = \textit{Effective Transmit Power} + \textit{Gains} - \textit{Propagation Path Loss} - \textit{Other Losses}.$$

Hence,

$$\textit{Propagation Path Loss} = \textit{Effective Transmit Power} + \textit{Gains} - \textit{Other Losses} - \textit{Received Signal Power}.$$

That's how Eq. (10.9) is derived. Figure 10.37 summarizes the basic steps for calculating the link budget Eq. (10.9) represents. Let's take a look at the steps of this calculation.

The receiver sensitivity is the minimum signal power level that the BS must be able to detect at the base station antenna connector. This power level should be adequate

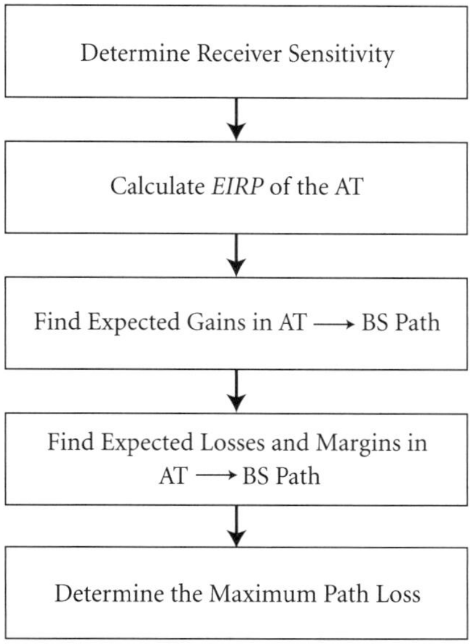

Figure 10.37: Steps for Deriving the Link Budget

to yield the target error rate under expected channel conditions. For a CDMA-based reverse link, E_b/N_0 is specified as

$$\frac{E_b}{N_0} = (PG)\left(\frac{ReceiverSensitivity}{EffectiveNoisePower * LM}\right) \tag{10.10}$$

where E_b/N_0[372] is the signal-to-noise ratio required to meet a target QoS (e.g., 2% packet error rate), PG is the processing gain and is the ratio of the spreading bandwidth (i.e., 1.2288 MHz) to the data rate (e.g., 9.6 kbps), $EffectiveNoisePower$ includes the total thermal noise power (in mW in Eq. (10.10)) in the 1.2288 MHz bandwidth and the noise figure of the BS receiver, and LM is the loading margin, which reflects how much reverse link interference is provisioned relative to the thermal noise floor. $ReceiverSensitivity$ can now be expressed as

$$\begin{aligned} ReceiverSensitivity\ (\text{dBm}) = {} & ThermalNoisePower\ (\text{dBm}) \\ & + NoiseFigure\ (\text{dB}) \\ & + \left(\frac{E_b}{N_0}\right)(\text{dB}) - PG\ (\text{dB}) + LM\ (\text{dB}). \end{aligned} \tag{10.11}$$

Table 10.7 summarizes example values for the parameters of Eq. (10.11). *Receiver*

[372] E_b/N_0 is more formally called the ratio of energy per bit to noise-plus-interference power spectral density.

Table 10.7: Settings of Selected Parameters for RL Link Budget

Parameter	Value
Processing Gain	$(1.2288 \times 10^6$ chips/sec$)/(9600$ bits/sec$) = 128$
	$10 * \log_{10}(128) = 21$ dB
Noise Figure	5 dB
Overall Thermal Noise Power	-113.2 dBm $+$ $Noise\ Figure$
$E_b/N_0(No\ Load)$	6.6 dB
Loading Margin	3 dB

$Sensitivity$ is then $-113.2+5+6.6-21+3 = -119.6$ dBm. In Table 10.7, -113.2 is kTB expressed in decibels relative to one milliwatt where k is Boltzmann's constant, T is temperature in kelvin, and B is the noise bandwidth of 1.2288 MHz.

$EIRP$ is the effective power the AT transmits. The total power that the AT radiates is the maximum transmit power of the AT plus the antenna gain less the losses at the AT. The maximum transmit power of the AT is 200 mW or $10 * \log_{10}(200 \text{ mW}) =$ 23 dBm. Since the AT uses an omnidirectional antenna, the antenna gain is 0 dBi. The letter i in dBi represents *isotropic*. If the AT is a hand-held device, a body loss of 3 dB is typically assumed. $EIRP$ can now be calculated as 23 dBm $+ 0 - 3 =$ 20 dBm.

As the signal travels from the AT to the BS connector, it experiences gains, such as BS receive antenna and soft handoff gain. The BS receive antenna gain of 17 dBi is one example value. Since the RL supports regular soft handoff, a soft handoff gain of 4.1 dB is assumed. The total gains become $17 + 4.1 = 21.1$ dB.

As the signal propagates from the AT to the BS, it experiences various losses in addition to the basic propagation path loss. Furthermore, additional margins are required to ensure good performance under a variety of channel conditions. Table 10.8 summarizes various losses and margins.

Since the RF signal typically travels between the bottom of the BS tower and the top of the tower via a cable, cable loss is introduced. Three decibels is a widely used cable loss value. The log-normal fade margin reflects the shadow fading margin due to the environment or topology (e.g., rural vs. suburban) and cell-edge reliability. For example, for shadow fading modeled as a 0 dB mean and 8 dB standard deviation

Table 10.8: Losses and Margins

Parameter	Value
BS Cable Loss	3.0 dB
Log-Normal Fade Margin	10.3 dB
Differential Fade Margin	2.1 dB
Building Penetration Loss	10.0 dB

with 90% cell-edge reliability yields 10.3 dB.[373] A soft handoff gain of 4.1 dB assumes equally balanced paths. A differential fade margin of 2.1 dB accounts for some imbalance between these two paths and reduces the effective soft handoff gain. In-building coverage requires provisioning additional margin and is assumed to be 10 dB here. The total of losses and margins is $3 + 10.3 + 2.1 + 10 = 25.4$ dB.

Because all the required quantities are known, the overall link budget can be determined using Eq. (10.9). For the parameters assumed above, the maximum tolerable path loss is 20 dBm + 21.1 dB − 25.4 dB − (−119.6 dBm) = 135.3 dB. This path loss can be plugged into any suitable propagation path loss model to find the actual cell radius. Such a path loss model considers factors such as the carrier frequency, BS antenna height, AT antenna height, and clutter type (e.g., rural versus suburban) to calculate the distance between the BS and the AT.

10.15 POINTS TO REMEMBER

- 1xEV-DO Rev. 0 or IS-856 is a packet data system that has evolved from CDMA2000 1x.
- An IS-856–based system consists of a radio access network that has BSs and RNCs and a packet-switched core network that has a PDSN, FA, HA, AAA Server, and AN-AAA.
- The forward link utilizes TDM of various channels, such as the Pilot channel, MAC channel, and Control or Traffic channel.
- The forward link has no power control because the BS transmits full power during an active slot. The forward link also has no soft or softer handoff for the Traffic channel.
- The reverse link is based on traditional CDMA and thus exploits fast power control and traditional soft and softer handoff.
- The peak user data rate is 2.4576 Mbps in the forward link and 153.6 kbps in the reverse link.
- Major advantages of 1xEV-DO Rev. 0 compared to 1x include faster call setup, higher peak and average user throughput and sector throughput in the forward link, and lower latency.
- The next revision of 1xEV-DO Rev. 0 is Rev. A in which significant attention is paid to the RL. Chapter 12 discusses 1xEV-DO Rev. A.

[373] In Microsoft Excel, the formula $\text{norminv}(0.9, 0.8)$ yields 10.3 where 0.9 means a probability of 0.9 or 90%, 0 is the mean, and 8 dB is the standard deviation.

CHAPTER 11

HIGH-SPEED DOWNLINK PACKET ACCESS

High-Speed Downlink Packet Access (HSDPA) is a Third Generation Partnership Project (3GPP) Release 5 feature.[374] The main focus of the HSDPA is to provide much higher downlink throughput to users than Universal Mobile Telecommunication System (UMTS) Release 99 (R99) offers. This chapter provides an overview of HSDPA. Unique features and characteristics of HSDPA are summarized first. Data transmission in HSDPA is briefly discussed next. Three new HSDPA channels are introduced in HSDPA. The characteristics of these channels and their roles in HSDPA data transmission are described. A closer look at the data transmission in the downlink follows. Call-setup–related changes required to support HSDPA are summarized. HSDPA relies upon efficient management of resources, such as downlink power and orthogonal variable spreading factor (OVSF) codes, and the explanation of such resource management is followed by a discussion on the impact of mobility on the HSDPA operation. The architecture of the network is illustrated with a focus on the radio interface protocol stack and interactions between the Node B and the radio

[374]Other major Release 5 features include Internet Protocol (IP) Multimedia Subsystem and IP-based Universal Terestrial Radio Access Network [285]. Chapter 14 discusses IP Multimedia Subsystems in detail.

network controller (RNC). Characteristics of different categories of HSDPA User Equipment (UE) are also specified.

11.1 OVERVIEW OF THIRD GENERATION PARTNERSHIP PROJECT RELEASE 5

Prior to Release 5 (such as in Release 99), the Universal Terestrial Radio Access Network (UTRAN) components communicate with each other via an asynchronous-transfer-mode–based network. The IP UTRAN defines the IP-based architecture such that UTRAN components can communicate using IP instead of asynchronous transfer mode. Scalability and lower cost are the expected benefits of an IP-based UTRAN compared to an asynchronous-transfer-mode–based UTRAN.

3GPP Release 5 was completed in March 2002 with HSDPA as an important feature. Initial commercial deployments of HSDPA systems occurred in late 2005 [286]. Specifically, Cingular Wireless (now AT&T) deployed the world's first HSDPA system [287]. HSDPA provides a higher peak and average user throughput and cell throughput in the downlink compared to Release 99 UMTS systems. HSDPA makes *no increase* to uplink data rates, circuit-switched data rates, and voice capacity; HSDPA *significantly increases* downlink data rates for packet data services. The HSDPA system is designed with the assumption that the radio link utilization would be asymmetric, i.e., the downlink data rate requirements are much higher than the uplink data rate requirements. Hence, the HSDPA feature is most suitable for background and interactive traffic classes that correspond to non-real-time yet possibly high data rate services or applications, such as web-browsing, email downloads, and FTP downloads. Since the HSDPA system offers no increase in uplink data rates, the user would continue to utilize Release 99 data rates in the uplink.

HSDPA is primarily an air-interface solution. The fundamental factors that contribute to superior downlink packet data performance are related to the air interface. In particular, most of the changes necessary to evolve from Release 99 to Release 5 are at the physical layer and the medium access control (MAC) layer.

Do You Recall... Do you remember the theoretical peak data rate in the downlink for Release 99?[375]

The *peak user data rate* is 14 Mbps compared to 2 Mbps in the downlink. The average *user throughput* and *cell throughput* in the downlink are expected to be higher than the Release 99 throughput values via maximum utilization of radio resources, including downlink power and OVSF codes; fast feedback about the downlink channel conditions from the UE to the Node B; the Node B's channel-sensitive and fast

[375]UMTS R99 supports a peak data rate of 2 Mbps. Commercial UMTS R99 deployments supported 384 kbps.

scheduling; adaptive channel coding; adaptive and higher-order modulation; and the UE's physical layer acknowledgment of received packets.

A Funny Detail... You may be wondering, "How likely is it that a user would experience 14 Mbps in the downlink?" To answer this question, we need to take a detour to Hollywood! You may have seen the movie "Dumb and Dumber." Our hero, Jim Carrey, asks a beautiful woman, "What is the chance that a guy like me can get a girl like you? Is it one in 10? One in 100?" The woman interrupts and says, "It is like one in a million." Jim Carrey smiles and replies, "Ah, so you are telling me that there *is* a chance!" Well, for a user, there is a chance of being able to get 14 Mbps though that chance may be one in a million. Many stars would have to align for this to happen. For example, the user would be the sole user in the cell, essentially consuming all radio resources. The interference would have to be minimal. The propagation path loss would have to be quite low, forcing the user to be close to the cell site. And, of course, we cannot have much severe fading, so our user has to be stationary. Very little channel coding is available to support a large packet size corresponding to 14 Mbps. Hence, we need the conditions mentioned here. In summary, 14 Mbps is a good marketing number, but a user is unlikely to frequently experience such a data rate in practice.

The HSDPA system is *fully backward-compatible* with an R99 system. Both HSDPA and R99 systems can share the same 5 MHz bandwidth spectrum in the uplink and the downlink. A UMTS R99 UE can use the HSDPA-capable cell by utilizing R99 radio channels. Similarly, an HSDPA-capable UE utilizes R99 radio channels in a UMTS R99 cell and HSDPA radio channels in an HSDPA-capable cell.

HSDPA enables smooth and incremental migration with no need for upgrading all R99 cells to the HSDPA simultaneously. Packet data demands in markets can drive cell upgrades.

11.2 A BIRD'S-EYE VIEW OF HIGH-SPEED DOWNLINK PACKET ACCESS DATA TRANSMISION

The main focus of HSDPA is high-speed, efficient data transmission. In this section, we will summarize how packet transfer occurs in the downlink. You will observe that the downlink transfer is quite different from that in UMTS R99. The uplink transfer is the same in HSDPA and UMTS 99.

In an HSDPA system, the UE could receive data on a high-speed downlink channel once the data session establishment includes HSDPA channel configuration. As shown in Figure 11.1, three new channels are defined for HSDPA: a high-speed data channel in the downlink called HS-DSCH/HS-PDSCH in which HS-DSCH is a transport channel and stands for High-Speed–Downlink Shared channel and HS-

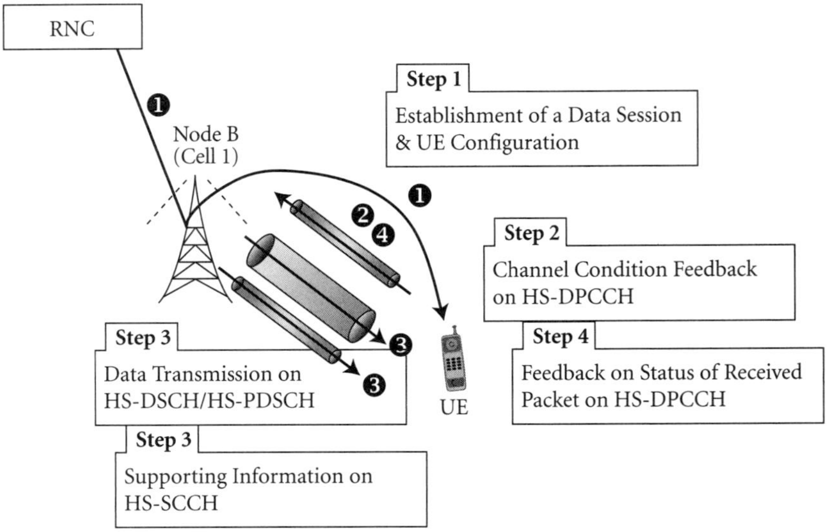

Figure 11.1: Downlink Data Transmission in HSDPA: A High-Level View

PDSCH (High-Speed–Physical Downlink Shared channel) is physical channel;[376] a low-speed control channel in the downlink called HS-SCCH (High-Speed–Shared Control channel for HS-DSCH); and a low-speed control channel in the uplink called HS-DPCCH which stands for the High-Speed–Dedicated Physical Control channel.

An HSDPA-capable cell has all Release 99 channels in addition to these new channels. The high-speed downlink data channel primarily carries the traffic packets. Figure 11.1 summarizes the basic steps for high-speed downlink data transmission.

In Step 1, a data session is established for the UE. The basic packet data call (or session) setup process for HSDPA remains the same as in Release 99. The UE undergoes procedures such as power-on utilization, system acquisition, radio resource control (RRC) connection setup, authentication, security, and Packet Data Protocol (PDP) context activation. The HSDPA-specific configuration setup is related to the new HSDPA channels. The RNC specifies to each UE which downlink control channels to monitor. In an HSDPA cell, numerous downlink control channels could be configured and simultaneously transmitted. One UE should be able to monitor up to four downlink control channels.

During an active packet data session, the UE periodically reports channel conditions to the Node B in the form of a Channel Quality Indicator (CQI) in Step 2. The fastest reporting frequency is once every 2 ms, and the slowest reporting frequency is once every 160 ms.

[376] HS-DSCH is mapped onto one or more HS-PDSCHs.

In Step 3, the Node B incorporates a scheduling algorithm that processes several inputs including channel condition reports from multiple UEs in a given cell. When the scheduler selects specific UEs, suitable HS-DSCHs/HS-PDSCHs carry the actual traffic packets. The associated HS-SCCHs identify the selected UEs and also carry the information the selected UEs require to obtain data from HS-DSCHs.

The UE that the HS-SCCH specifies attempts to decode the packet from the high-speed channel in Step 4. If the UE decodes the packet correctly, it sends a positive acknowledgment (ACK) to the Node B on the HS-DPCCH. If the UE fails to decode the packet correctly, it sends a negative acknowledgment (NACK) to the Node B.

Steps 2 through 4 are repeated as long as the UE remains in an HSDPA cell. When the scheduler selects a specific UE, it sends a new or a retransmission packet. A new packet primarily includes systematic (i.e., raw or original) bits, and the retransmission packet includes systematic or parity (i.e, protection) bits. When a retransmission packet is received, the UE combines it with the previously received packet(s) to decode the original packet. The traffic packet on the HS-DSCH is sent at a 2 ms interval (called a subframe), and the data rate can be changed every 2 ms for a given UE. In Release 99 systems, a change in data rate is achieved through layer 3 RRC signaling and, hence, incurs a delay of hundreds of milliseconds or even a few seconds.

The uplink data transmission in the HSDPA system remains the same as in a Release 99 system. The RNC configures a dedicated uplink channel. The UE transmits data on the uplink channel based on the constraints the RNC imposes. The HSDPA system focuses on superior data rates on the downlink only; therefore, the uplink data transmission mechanism for a Release 99 system remains unchanged in an HSDPA system.

11.3 HIGH-SPEED DOWNLINK PACKET ACCESS CHANNELS AND THEIR USE: A DEEP DIVE

We will give these three new channels, HS-PDSCH, HS-SCCH, and HS-DPCCH, their fair share of recognition! These channels are pivotal in making HSDPA efficient for downlink packet transmission.

11.3.1 High-Speed–Physical Downlink Shared Channel

The HS-DSCH is a downlink channel that carries user data at high rates. Several rules govern the operation of the HS-DSCH and the HS-PDSCH [288].

- An HS-DSCH is mapped onto one or more HS-PDSCHs.
- One HS-PDSCH corresponds to one physical channel at the spreading factor (SF) of sixteen. HSDPA no longer relies upon a variable SF of an OVSF code to achieve different data rates; instead, HSDPA utilizes one or more OVSF codes at $SF = 16$.

Do You Recall... How many orthogonal codes exist at $SF = 16$? Why does the standard restrict the number of usable codes for the HS-DSCH to fifteen?[377]

- One UE can receive data on only one HS-DSCH at a given time instant.
- Multiple HS-DSCHs and associated HS-PDSCHs may be actively transmitting data to multiple UEs at a given time in a cell.
- The HS-PDSCH is transmitted in units of subframes. One subframe occupies the transmit time interval (TTI) of 2 ms.
- The HS-DSCH carries exactly one transport block (yes, just one, no more and no less!) in one TTI. The contents of the transport block are transmitted on the HS-PDSCH(s) associated with the HS-DSCH.
- The HS-DSCH is not power-controlled and the UE issues no commands to change the HS-DSCH/HS-PDSCH transmit power. The Node B manages the transmit power of the HS-DSCH. The transmit power on the HS-PDSCH is typically constant during a 2 ms subframe but can change or remain constant from one subframe to the next.
- The three HSDPA channels offer neither soft nor softer handover. In other words, HSDPA operation occurs between the UE and only one cell at a given time instant.

Now that we understand the physical layer processing for the HS-PDSCH, we should look at the HS-DSCH processing in more detail as shown in Figure 11.2 [289, 290]. During a 2 ms TTI, one transport block is transmitted. The transport block contains the application layer packet, which in turn contains raw information bits and numerous headers. One transport block corresponds to one MAC-hs protocol data unit (PDU). See Section 11.8 for MAC-hs details.

The standard specifies the allowable transport block size (TBS) [291], the minimum being 137 bits and the maximum 27952 bits. Peak data rate is usually defined to be the ratio of the TBS to the TTI. Hence, the highest peak data rate is $27952 \text{ bits}/2 \text{ ms} = 13.976$ Mbps, just a little less than 14 Mbps.

A twenty-four–bit cyclic redundancy check (CRC) is attached to the transport block to facilitate error detection. The CRC parity bits are generated using a cyclic generator polynomial. Bits are scrambled to randomize distribution of 0s and 1s as in case of UMTS R99. A turbo coding of rate 1/3 is used. No convolutional coding is used. See Chapter 2 for basics of turbo coding.

[377]The total number of orthogonal codes at a given SF is the same as the SF itself. Hence, sixteen OVSF codes are available at $SF = 16$. Since $C_{256,0}$ is reserved for the P-CPICH and $C_{256,1}$ is reserved for the P-CCPCH, one OVSF code at $SF = 16$ is unusable for the HS-PDSCH. The maximum number of OVSF codes available for HSDPA at $SF = 16$ is thus $16 - 1 = 15$.

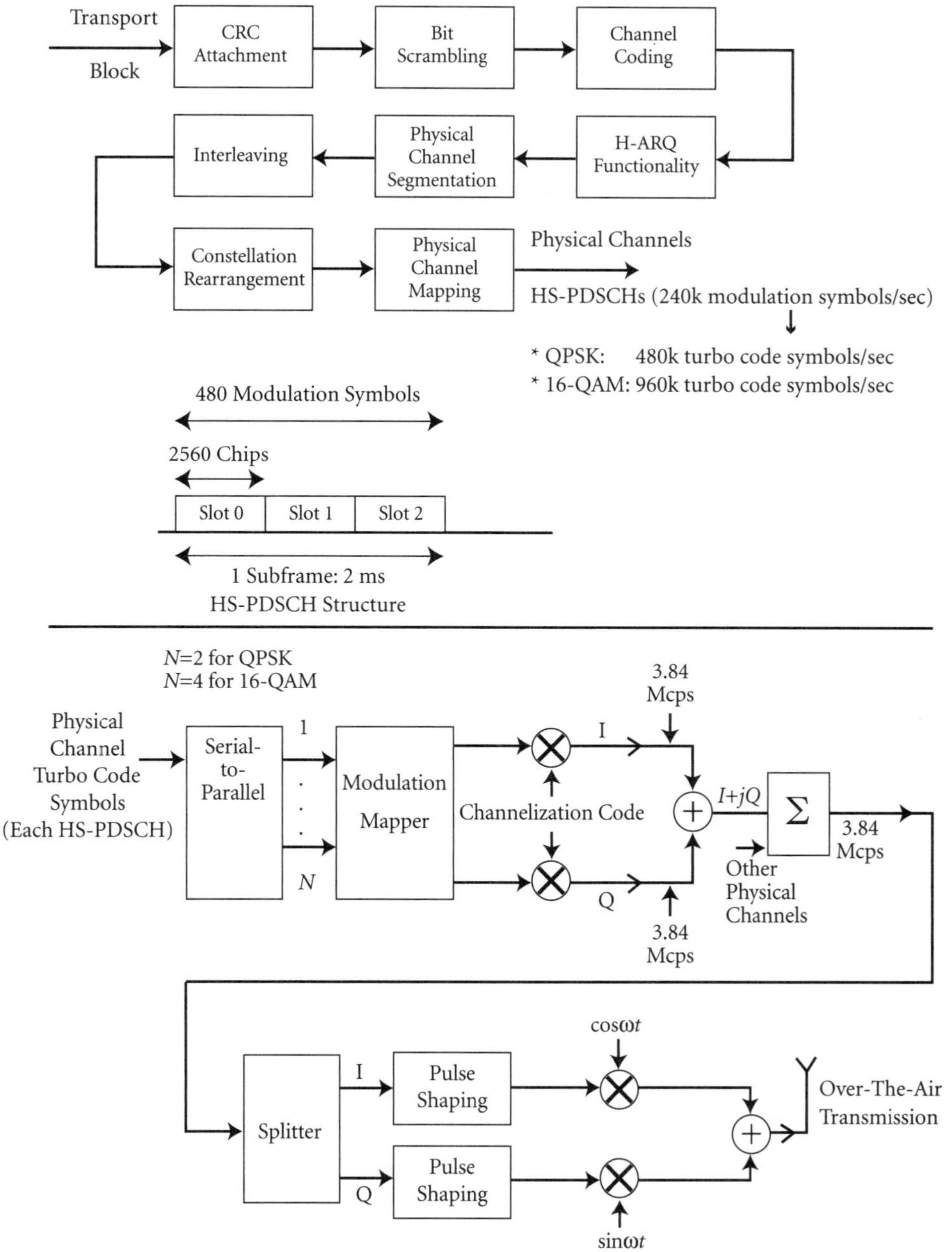

Figure 11.2: Physical Layer Processing of the HS-DSCH

HSDPA hybrid automatic repeat request (H-ARQ) functionality[378] selects a set of turbo code symbols based on the redundancy version parameters, s and r. When the value of the parameter s is 1, it indicates that the packet is self-decodable (i.e., the packet contains all the systematic bits), and $s = 0$ means that the packet is not self-decodable. The parameter r dictates how the parity bits are selected from the available turbo code symbols. Two possible redundancy versions (i.e., sets of parity bits) are available for 16-QAM (quadrature amplitude modulation) and four for quadrature phase shift keying (QPSK). When $s = 1$, systematic bits from turbo coding are prioritized. Similarly, when $s = 0$, non-systematic bits are prioritized.

A redundancy version essentially dictates what turbo code symbols are selected while forming a packet for transmission. Assume that a rate 1/3 turbo coder has processed a 1000-bit packet to yield 3000 code symbols and that we have adequate resources to transmit 1500 code symbols. Let's first consider the case of $s = 1$ and $r = 0$. Since $s = 1$, systematic bits are prioritized and all 1000 systematic bits are selected for transmission. Since we still have space for $1500 - 1000 = 500$ code symbols, $500/2 = 250$ parity 1 bits and 250 parity 2 bits are also chosen to create a packet. Now, let's consider the case of $s = 0$ and $r = 1$. Since $s = 0$, parity bits are prioritized. The packet is first filled up with parity bits and, if space is left, systematic bits are included. Since the packet can contain 1500 code symbols, $1500/2 = 750$ parity 1 bits and 750 parity 2 bits are selected to construct a packet. See the standard [290] for additional details on the redundancy versions.

Physical channel segmentation divides the code symbols equally among the physical channels being transmitted to a given UE. Interleaving is performed for each physical channel separately. The interleaver has thirty-two rows and thirty columns when QPSK modulation is used. Two such identical interleavers are used in 16-QAM modulation.

Signal constellation rearrangement is carried out only for 16-QAM modulation. A parameter called *constellation version parameter* b specifies how the 16-QAM signal constellation is modified. One basic constellation arrangement and one alternative arrangement are available. The goal of such constellation rearrangement is to ensure that the average bit error probability is similar for all bit positions in a modulation symbol. The group of four code symbols undergoes re-ordering and/or complementing, effectively changing the symbols that are transmitted over the air.

Physical channel mapping does exactly what it implies—maps the code symbols onto the physical channels. One physical channel, HS-PDSCH, corresponds to one OVSF code at $SF = 16$.

Do You Recall... Do you remember how many OVSF codes are available for HS-DSCH and the SF of these OVSF codes?[379]

[378] See Chapter 6 on GPRS and EDGE for an example of how H-ARQ works.

[379] Fifteen OVSF codes are available for the HS-DSCH, and the SF is sixteen for these codes.

The code symbols are serial-to-parallel converted to facilitate modulation. The I (in-phase) and Q (quadrature) channels for modulation symbols are multiplied (i.e., spread) by the channelization code for each physical channel. Each physical channel, HS-PDSCH, has its own channelization code at the fixed SF of sixteen. The chips from all physical channels, including overhead channels, such as P-CPICH and P-CCPCH, are summed separately for I and Q channels. Both I and Q channel chips are modulated at the 3.84 Mcps chip rate.

In summary, most of the physical layer processing for the high-speed downlink data channel is similar to that for UMTS R99. HSDPA's support for H-ARQ is the main difference between the UMTS R99 processing of the data channel and the HS-PDSCH.

11.3.2 High-Speed–Shared Control Channel

The HS-SCCH is a downlink physical channel, and, as such, its main purpose is to help UEs retrieve packets from the HS-DSCH. The UEs monitor one or more HS-SCCHs being transmitted in an HSDPA cell. Figure 11.3 illustrates how the HS-SCCH is utilized.

The RNC configures the UEs to inform them of which HS-SCCHs to monitor. In Figure 11.3, the RNC asks UE1 and UE2 to observe HS-SCCH1 and HS-SCCH2. HS-SCCH1 indicates that the data for UE1 is being carried on HS-DSCH1, and HS-SCCH2 indicates that HS-DSCH2 carries the data for UE2. Now the UEs can retrieve their packets from the appropriate HS-DSCHs.

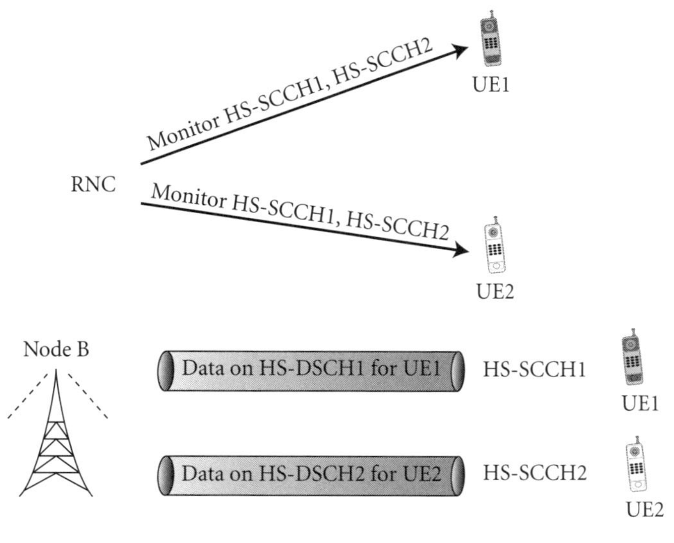

Figure 11.3: Using the HS-SCCH

Several major rules control the operation of the HS-SCCH and ensure smooth interworking between the UE and the Node B.

- The HS-SCCH identifies the UE that should collect data from the HS-DSCH. The HS-SCCH also defines several parameters characterizing the HS-PDSCHs and the packet transmission and reception.
- One HS-SCCH is associated with exactly one HS-DSCH (and, hence, one or more related HS-PDSCHs).
- A UE monitors a set of HS-SCCHs the RNC conveys to it through messages, such as the RADIO BEARER CONFIGURATION message.
- A UE is required to monitor a maximum of four HS-SCCHs. However, the Node B in a cell can transmit any number of HS-SCCHs.
- Just like HS-PDSCH, the HS-SCCH is also transmitted in 2 ms subframes.
- The maximum number of UEs that can simultaneously receive data on the HS-DSCHs during a given 2 ms time period is the number of HS-SCCHs being transmitted in the cell.
- The HS-SCCH is transmitted at the rate of 60 kbps[380] using a fixed SF of 128.
- Since the HS-SCCH is transmitted to a UE from only one cell, no soft or softer handover is available for the HS-SCCH.
- The transmit power of the HS-SCCH is not standardized. A network infrastructure vendor may implement a closed-loop power control scheme that considers power control commands UEs issue (that correspond to Release 99 downlink dedicated channels, such as the Dedicated Physical Data channel).
- When the UE is sent data on the HS-DSCH in consecutive 2 ms TTIs, the same HS-SCCH is used to convey the HS-DSCH information.

Figure 11.4 shows the two types of information, *TFRI* (transport format resource indicator) and *H-ARQ*, that the HS-SCCH carries. *TFRI* specifies the HS-DSCH transport format, and *H-ARQ* defines the implementation of the H-ARQ operation. The HS-SCCH implicitly carries the UE's identity called H-RNTI (HS-DSCH Radio Network Temporary Identifier). More specifically, the CRC is first calculated for the TFRI and H-ARQ contents. The UE's H-RNTI then masks the calculated CRC. The H-RNTI is the identity the RNC gives to the UE when the HSDPA-capable UE has HS-DSCH allocation in a cell. In other words, every UE that is asked to monitor an HS-SCCH for possible data reception on the HS-DSCH is assigned an

[380]This is actually 60 ksps (kilo symbols per second, with symbols being code symbols). The use of QPSK modulation would lead to 30 k modulation symbols per second, and $SF = 128$ would yield the required 3.84 Mcps chip rate.

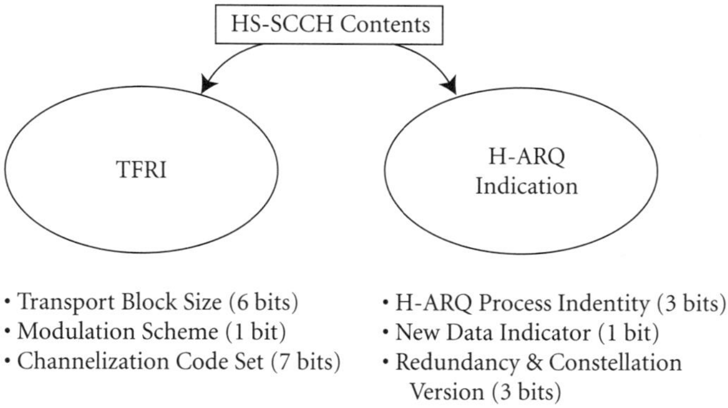

Figure 11.4: Inside the HS-SCCH

H-RNTI. Each transport block is associated with a modulation scheme and a number of channelization codes.

The parameters shown in Figure 11.4 are discussed next. The TBS being carried on the HS-DSCH is conveyed to the UE using a six-bit field. For a retransmission, TBS is set to `111111`. The actual parameter being signaled in this field is the value called k_i. When k_i is added to a parameter called $k_{0,i}$, k_t is obtained. $k_{0,i}$ is a function of the modulation scheme and the number of channelization codes.[381] For example, for the modulation scheme of QPSK and one channelization code, $k_{0,i}$ is 1. In 16-QAM modulation and with fifteen channelization codes, $k_{0,i}$ is 192. Two approaches are available to determine the mapping between k_t and the TBS value referred to $L(k_t)$: the use of a table defined in Annex A of the standard [291] (and adapted in very abbreviated form in Table 11.1) or the use of Eq. (11.1).

$$\begin{aligned} L(k_t) &= 125 + 12k_t, && \text{if } k_t < 40 \\ L(k_t) &= \left\lfloor 296 \left(\frac{2085}{2048}\right)^{k_t} \right\rfloor, && \text{otherwise} \end{aligned} \tag{11.1}$$

Example: TBS Specification
Goal: Convey $TBS = 27952$ bits to the UE.

Processing at the Node B:

1. $TBS = L(k_t) = 27952$ bits.

[381] See Table 9.2.3.1 in the standard [291].

2. Find k_t using the formula or Table 11.1.

$$k_t = 254$$

3. Find $k_{o,i}$ using Table 11.1. For 16-QAM modulation and fifteen channelization codes, $k_{o,i} = 192$.

4. Find k_i.

$$\begin{aligned} k_t =& k_i + k_{o,i} \\ \therefore k_i =& k_t - k_{o,i} = 254 - 192 = 62 \end{aligned}$$

5. Send $k_i = 62$ to the UE on the HS-SCCH using a six-bit representation `111110`.

Processing at the UE:

1. Receive $k_i = 62$ on the HS-SCCH.

2. Using the modulation scheme of 16-QAM and the fifteen channelization codes being sent on the HS-SCCH, find $k_{o,i}$ to be 192 based on Table 11.1.

3. Calculate k_t as $k_t = k_i + k_{o,i} = 62 + 192 = 254$.

4. Find $TBS = L(k_t) = L(254)$ to be 27952 bits using Table 11.1 or a formula.

The one-bit *Modulation Scheme* field specifies the modulation scheme used on the related HS-DSCH (0 for QPSK and 1 for 16-QAM). A seven-bit *Channelization Code Set* field specifies the channelization codes used to spread the HS-PDSCH(s) of the HS-DSCH. The Channelization Code Set field consists of two parts, the initial three bits to specify the code group indicator and the last four bits to specify the code offset indicator.

$$Code\ Group\ Indicator = \min(P-1,\ 15-P) \tag{11.2}$$

$$Code\ Offset\ Indicator = \left| O - 1 - \left\lfloor \frac{P}{8} \right\rfloor 15 \right| \tag{11.3}$$

where P is the number of multi-codes (each with $SF = 16$) and O is the code offset such that $C_{ch,16,O}$ to $C_{ch,16,O+P-1}$ (where ch is channelization code) defines all the OVSF codes used for the HS-PDSCHs.

A three-bit *H-ARQ Process Identifier* field specifies the identity of an H-ARQ process that is handling the packet being carried on the HS-DSCH. One H-ARQ process is in charge of a given packet's transmission or retransmission. When more than one H-ARQ process is defined for a given UE, multiple packets can be transmitted in consecutive 2 ms TTIs to the same UE.

Table 11.1: A Brief Glance at the Mapping Between k_t and $L(k_t)$
SOURCE: This table is a vastly abridged version of Annex A of 3GPP TS 25.321 v.5.14.0 © 2008. 3GPP™ TSs and TRs are the property of ARIB, ATIS, CCSA, ETSI, TTA and TTC who jointly own the copyright in them. They are subject to further modifications and are therefore provided to you "as is" for information purposes only. Further use is strictly prohibited.

Index	TBS (Bits)
1	137
⋮	⋮
42	627
⋮	⋮
81	1262
⋮	⋮
119	2524
⋮	⋮
254	27952

Let's discuss the concept of an H-ARQ process as illustrated in Figure 11.5.[382] Assume that one H-ARQ process is available at the Node B for a UE under consideration. This H-ARQ process wants to send packets P1 and P2 to the UE. The Node B transmits an HS-SCCH specifying the control information for the UE so that the UE can retrieve the packet from the HS-DSCH associated with the HS-SCCH. The H-ARQ process sends packet P1 two time-slots after the transmission of the HS-SCCH. Such an offset between the HS-SCCH and the HS-DSCH/HS-PDSCH(s) facilitates demodulation of the HS-PDSCH(s) at the UE. After experiencing the delay corresponding to the propagation time, packet P1 reaches the UE. The UE responds with an ACK or NACK on the HS-PDCCH 7.5 slots after receiving packet P1. After incurring the propagation time delay, the HS-DPCCH is received at the Node B. The H-ARQ process considers the HS-DPCCH to decide whether to transmit a new packet P2 or retransmit packet P1. In our example, the H-ARQ process transmits a new packet P2 because the HS-DPCCH carries an ACK. The Node B has a time window, referred to as scheduling time in Figure 11.5, to process the HS-DPCCH, choose one or more UEs for the next TTI, and determine the structure of the packet (e.g., modulation scheme and TBS) to be sent on the HS-DSCH.

From the figure, the time between the *beginning* of packet P1 transmission and the *beginning* of packet P2 transmission is $\Delta T = 2 * propagation\ time + 15.5\,TS + scheduling\ time \approx 15.5\,TS + scheduling\ time$ assuming negligible propagation

[382]In the figure, Tx means "transmit" and Rx represents "receive."

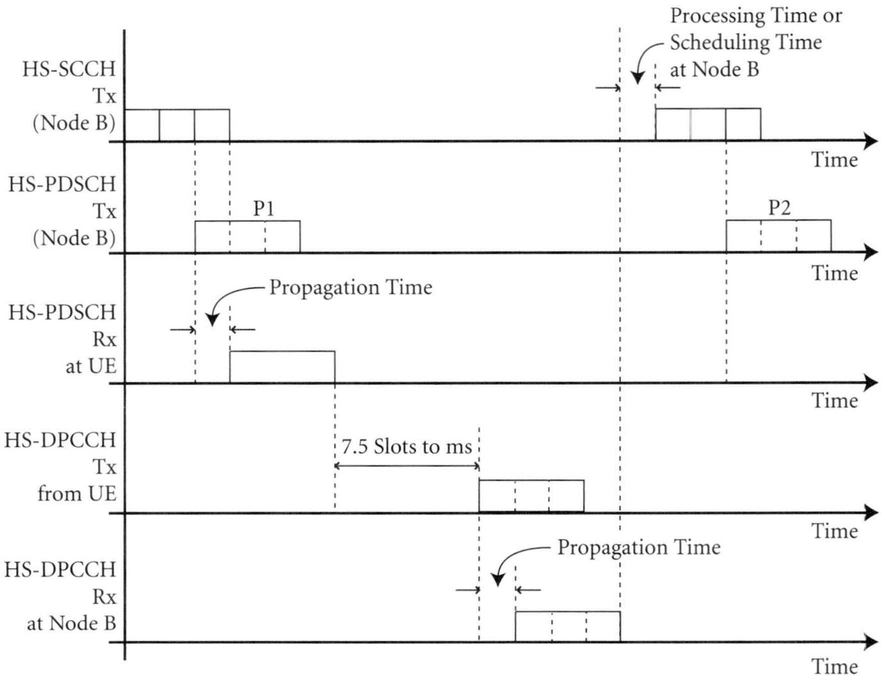

Figure 11.5: Achieving Continuous Transmission to the UE

delay and where TS represents the time-slot. Since each packet requires a 2 ms subframe, the number of H-ARQ processes required to fill ΔT is $\Delta T/2$ ms $= (15.5\,TS + \textit{scheduling time})/(3\,TS) = 5.167 + \textit{scheduling time}/3 \approx 6 + \textit{scheduling time}/3$. If the processing can be completed within 6 TSs, $6 + 6/3 = 6 + 2 = 8$ processes can remove the idle time between successive packet transmissions to the same UE, leading to continuous data transmission. From another perspective, three bits enable eight H-ARQ processes and, hence, can fill the time period of $8 * 3 = 24$ time-slots or $8 * 2 = 16$ ms. As long as the scheduling time is $24 - 15.5 = 8.5$ time-slots or shorter, continuous packet transmission to a UE can be achieved.

A one-bit field, *New Data Indicator* specifies whether the data on the HS-DSCH represents a new packet or a retransmission packet. The bit is specific to a given H-ARQ process. When a new packet, MAC-hs (high-speed) PDU to be specific, is transmitted, the New Data Indicator field is incremented by one. However, if MAC-hs PDU is retransmitted, the New Data Indicator field is not incremented.

A three-bit *Redundancy and Constellation Version* (RV) field is used to specify the redundancy version parameters r and s and constellation version parameter b. The RV parameters are required for both QPSK and 16-QAM modulation, and the constellation version parameter is used only in 16-QAM. The RV parameters define how turbo code symbols are selected in the packet formation. The constellation version parameter defines if and how the 16-QAM signal constellation has been modified for the packet transmission.

Do You Recall... Do you remember the use of the signal constellation re-arrangement?[383]

Let's discuss the physical layer processing for the HS-SCCH, which is illustrated in Figure 11.6.[384] The HS-SCCH is divided into two parts so that physical layer processing can be done differently and separately for each part. Separation between the two parts and the time offset between the HS-SCCH and the HS-DSCH enables the UE to start despreading and demodulating the associated HS-DSCH/HS-PDSCH(s). Only soft code symbols need to be stored in the UE memory for further processing, such as decoding. In the absence of such an approach, the UE would need to store a significant number of chips, requiring larger buffer space at the UE.

The first part includes Channelization Code Set field and Modulation Scheme field, and the second part includes all the other parameters, such as TBS, H-ARQ Process Identifier, RV, and New Data Indicator. In the processing of the first part, eight bits are added to the end of the code block that consists of eight Channelization Code Set field and Modulation Scheme field bits. Sixteen bits are provided as input to rate 1/3 convolutional coding, and hence forty-eight coded bits (or code symbols) are generated. Rate matching punctures these code symbols so that forty symbols are available. These forty symbols are masked using H-RNTI and mapped onto the HS-SCCH's first time-slot.

For the processing of the second part, a total of twenty-one bits containing Channelization Code Set, Modulation Scheme, TBS, H-ARQ Process Identifier, RV, and New Data Indicator are used to calculate CRC. The H-RNTI masks the sixteen-bit CRC, and the CRC is multiplexed with thirteen bits constituted by TBS, H-ARQ Process Identifier, RV, and New Data Indicator. The set of twenty-nine bits and eight tail bits are given to rate 1/3 convolutional coding to create $\left((29+8)*3\right)=111$ code symbols. Rate matching punctures these symbols to generate eighty symbols that are mapped onto the second and third time-slots of the HS-SCCH subframe.

In the spreading and modulation block, the code symbols are serial-to-parallel converted and passed through a modulation mapper that converts the groups of two code symbols into QPSK modulation symbols. The modulation symbol rate is 30 ksps because the code symbol rate is 60 ksps.[385] The modulation symbols are spread

[383] The average bit error probability is different for different bits for 16-QAM. Hence, the standard has defined the parameter b, which specifies how the signal constellation rearrangement has been performed. The signal constellation rearrangement attempts to ensure the same average bit error probability for all the bits of the modulation symbols.

[384] See the standard [290] for additional details.

[385] A total of forty code symbols in the first slot and a total of eighty code symbols in the second and third slot lead to 120 code symbols in three slots of a subframe. The code symbol rate is thus 120 code symbols$/2$ ms $=60$ ksps. Since QPSK modulation is used, two code symbols are represented by a single modulation symbol, leading to the modulation symbol rate of 30 ksps.

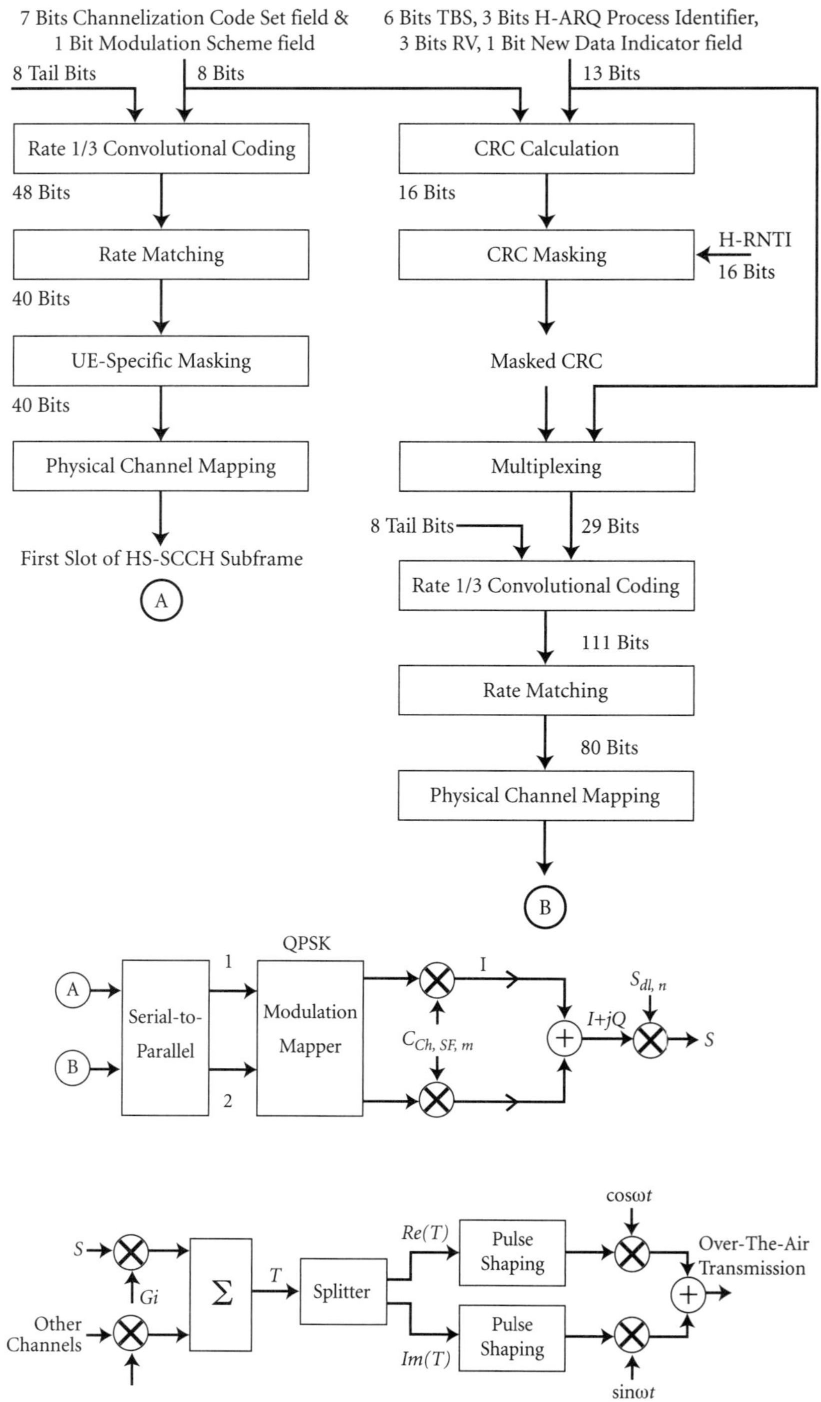

Figure 11.6: Physical Layer Processing of the HS-SCCH

using $C_{ch,SF,m}$[386] to reach the 3.84 Mcps chip rate 30 ksps $*$ 128 $=$ 3.84 Mcps. A suitable gain is applied to the HS-SCCH as determined by an implementation-specific power control algorithm. The Node B combines the HS-SCCH chips with other radio channels' chips and scrambles the combined chips by a cell-specific primary scrambling code $S_{dl,n}$ where dl refers to the downlink and n refers to the code number.

11.3.3 High-Speed–Dedicated Physical Control Channel

The HS-DPCCH is an uplink physical channel, and its main purpose is to support the downlink HS-DSCH. The HS-DSCH informs the Node B about the current downlink channel conditions. It also conveys to the Node B if the packet recently received on the HS-DSCH has been successfully decoded. Let's look at the basic rules governing the use of the HS-DPCCH.

Only one HS-DPCCH is available per UE. The UE transmits the HS-DPCCH only when the RNC configures the UE to receive data on the HS-DSCH.

The UE estimates the current downlink channel conditions and sends this estimate to the Node B in the form of the CQI, which is an integer with the range 0 to 30. $CQI = 0$ tells the Node B to refrain from sending data to the UE. The CQI value of 1 corresponds to poor channel conditions, and the CQI value of 30 corresponds to excellent channel conditions. Thus, the larger the value of CQI, the better the channel conditions.

If the UE receives a packet on the HS-DSCH, it sends on the HS-DPCCH an ACK if the packet was decoded successfully or a NACK if the packet was decoded unsuccessfully. If the UE receives no packet, no ACK or NACK is sent. In other words, transmission of an ACK or NACK is discontinuous.

The UE transmits the CQI according to the RNC's instructions. In general, the CQI transmission is continuous as long as the UE is configured to receive data on the HS-DSCH. Additional information on CQI is given later in this section.

The HS-DPCCH can exist only if the uplink DPCCH exists because the HS-DPCCH is power-controlled using the uplink DPCCH as the reference.

Repeated transmission of an ACK or NACK is allowed. The RNC sends the repetition factor of an ACK or NACK in an RRC message.[387] The UE repeats the ACK or NACK transmission in consecutive $(N\_acknack\_transmit - 1)$ subframes as shown in Figure 11.7. The Node B sends no packets to the UE in the subframes corresponding to the HS-DPCCH frames that carry repeated ACK or NACK. The UE makes no attempt to decode packets during relevant HS-DSCH subframes.

The CQI specifies the quality of the current downlink channel conditions. A given CQI value is associated with specific settings of parameters such as the TBS, number of HS-PDSCH codes, modulation scheme, and reference power adjustment [292].

[386]Recall that ch represents channelization and SF represents spreading factor (which is 128 for the HS-SCCH). m is the code number at $SF = 128$ and ranges from 0 to 127.

[387]The UMTS standard is so flexible that the same parameter can be conveyed to the UE using several RRC messages.

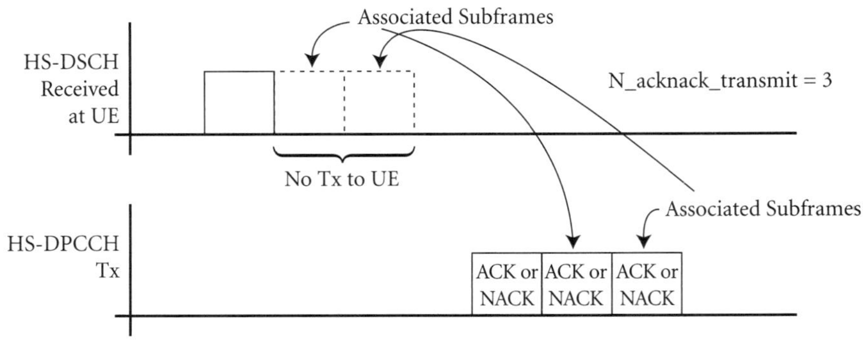

Figure 11.7: Influence of ACK or NACK Repetition Factor

If the Node B transmits a packet using the parameter settings associated with the reported CQI, the transport block error probability is 10% or less. Hence, CQI can be considered a way of guaranteeing a certain quality-of-service. A fixed value of such a target block error rate helps the UE vendor design a suitable (and proprietary) CQI estimation algorithm.

Let's dig deeper into the CQI details. The signal-to-interference ratio (SIR) has direct influence over the block error rate requirement under current channel conditions. The UE estimates the prevailing downlink interference. Furthermore, it assumes a specific signal power to calculate the estimated SIR according to the following formula.

$$P_{HS-PDSCH} = P_{P-CPICH} + \Gamma + \Delta \tag{11.4}$$

Here, $P_{P-CPICH}$ is the received P-CPICH power, Γ is the measurement power offset, and Δ is the reference power adjustment. Γ is conveyed to the UE through an RRC signaling message (e.g., RADIO BEARER SETUP or RADIO BEARER RECONFIGURATION message). Δ, associated with a given value of CQI, is defined in various CQI tables [292]. Replacing P-CPICH with S-CPICH is possible by indicating such a preference to the UE through an RRC message.

The standard has defined CQI mapping tables for all categories of the UE. A CQI mapping table specifies the parameter settings for TBS, number of HS-PDSCH codes, modulation scheme, and Δ for each value of CQI (e.g., 0 to 30). Since a total of 31 values need to be represented, the CQI requires a five-bit field. The Node B scheduler utilizes the CQI to assign a data rate for a given UE because the TBS, modulation scheme, and number of HS-PDSCH codes effectively define both the supportable data rate and how the packet should be transmitted. The CQI report indirectly provides the current SIR to the Node B, and the Node B dynamically determines the TBS, modulation scheme, and number of HS-PDSCH codes based on available resources. The assigned data rate could thus be higher or lower than the data rate directly associated with the CQI value.

Example: Simplified Data Rate or TBS Determination
Assume that Δ and Γ in Eq. (11.4) are 0. The UE assumes that the Node B would allocate same power level to the P-CPICH and the HS-PDSCH. Further assume that the Node B is actually transmitting 10% of the total transmit power on the P-CPICH and that the UE has sent $CQI = 10$ corresponding to the TBS of 1262 bits based on the estimated SIR and the CQI table associated with UEs of categories 1 to 6. Three possibilities arise. If the Node B indeed wants to allocate 10% power to the HS-PDSCH (the same as that for the P-CPICH), the Node B can choose $TBS = 1262$ bits associated with $index = 81$ using the TBS table in the standard [291] (selected entries of which are recreated in Table 11.1). However, if the Node B wants to use 20% power, it can effectively double the data rate such that the TBS of approximately $1262 * 2 = 2524$ bits can be supported. The index value of 119 in the TBS table in the standard [291] corresponds to 2492 bits, which is close to 2524 bits. The Node B then chooses $TBS = 2492$ bits. A more conservative number (i.e., a smaller number) for the TBS would increase the chances of meeting the target block error rate of 10%. Similarly, if the Node B wants to allocate just 5% power to the HS-PDSCH, the supportable TBS would be approximately $1262/2 = 631$ bits. The Node B can now choose $TBS = 627$ bits, which corresponds to $index = 42$. Our simple calculations here assume that the same SIR is required at different data rates to meet the target block error rate. In such a case, doubling the data rate requires twice as much power.

As channel conditions become gradually better, CQI value increases, leading to a higher supportable data rate. Furthermore, multiple CQI values are associated with the highest TBS for each UE category through a change in Δ. The value of Δ becomes increasingly negative as the CQI value increases, signifying that, as channel conditions improve, a lower signal power is required to meet a given SIR target corresponding to 10% block error rate. Note that the UTRAN configures the UE with the CQI reporting periodicity. The fastest rate for CQI reports is one report every 2 ms (one subframe), and the slowest rate for CQI reports is one report every 160 ms (or eighty subframes). The CQI can also be repeated based on the configured value of N_cqi_transmit. The CQI feedback can also be turned off altogether. Of course, we discourage turning off the CQI feedback since it is an important input to the scheduling algorithm!

Now that we know the parameters the HS-DPCCH carries, let's see how the physical layer processing is carried out for the HS-DPCCH. Figure 11.8 shows the structure of the HS-DPCCH. Note that the HS-DPCCH is transmitted in units of 2 ms subframes.

The ACK or NACK associated with the H-ARQ functionality is carried in the first time-slot, and CQI is carried in the second and third time-slots. If the UE wants to send an ACK, all the H-ARQ bits are `1`. If the UE sends a NACK, all ten H-ARQ bits are `0`. A group of five bits can represent the CQI values from 0 to 30. However, for better protection, these five bits are processed through channel coding to generate twenty bits. Hence, twenty CQI bits are transmitted over the air instead of five CQI bits.

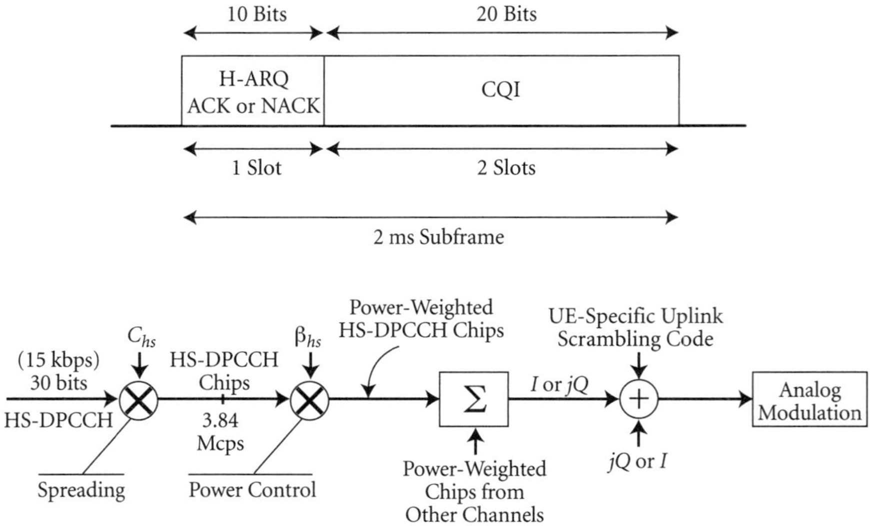

Figure 11.8: HS-DPCCH: Structure and Physical Layer Processing

Figure 11.8 illustrates the physical layer processing of the HS-DPCCH. Since thirty bits are transmitted on the HS-DPCCH during a 2 ms period, the channel data rate is $30/2$ ms $=$ 15 kbps. The BPSK modulation symbol rate is 15 ksps. The HS-DPCCH is spread to 3.84 Mcps using its own channelization (or OVSF) code at $SF = 256$. Its gain β_{hs} depends on the parameters Δ_{ACK}, Δ_{NACK}, and Δ_{CQI}, which are conveyed to the UE via an RRC message.

In Figure 11.8,

$$\beta_{hs} = \beta_c \, 10^{((\Delta_{HS-DPCCH})/20)} \tag{11.5}$$

where $\Delta_{HS-DPCCH}$ equals Δ_{ACK} when ACK is carried, Δ_{NACK} when NACK is carried, and Δ_{CQI} when CQI is carried and where β_c is the signaled or computed gain factor for the R99 DPCCH. The HS-DPCCH may be on the I branch or the Q branch depending upon the number of Dedicated Physical Data channels. The spread HS-DPCCH is added to other spread channels and multiplied by an uplink scrambling code. The overall uplink modulation appears like QPSK because symbols are transmitted on both I and Q channels simultaneously, and the effective modulation for the HS-DPCCH alone (or any other channel) is BPSK. The channelization code C_{hs} depends on $N_{max-dpdch}$ (i.e., the maximum number of Dedicated Physical Data channels) as shown in Table 11.2 [293]. As an example, in $C_{ch,256,64}$, ch refers to channelization, 256 indicates the SF, and 64 specifies the code index.

Table 11.2: Channelization Code for the HS-DPCCH

$N_{max-depth}$	C_{hs}
1	$C_{ch,256,64}$
2, 4, or 6	$C_{ch,256,1}$
3 or 5	$C_{ch,256,32}$

11.4 DATA TRANSMISSION PROCESS

Figure 11.9 characterizes the downlink data transmission process in an HSDPA system. In Step 1, UEs in a cell send CQI reports to the Node B using HS-DPCCHs. The Node B scheduler, in Step 2, considers the received CQI reports along with various factors, such as the available amount of resources and a scheduling strategy, to choose one or more UEs for transmission in the next TTI. In Step 3, the Node B identifies the UEs and specifies relevant information to these UEs on HS-SCCHs so the selected UEs can process the HS-DSCHs. The UEs that decode their packets on the HS-DSCHs send H-ARQ ACKs or NACKs on the HS-DPCCHs in Step 4. Steps 1 through 4 continue to be repeated. Let's further analyze each step shown in Figure 11.9 in Sections 11.4.1 through 11.4.4.

11.4.1 Step 1: Reporting of Channel Conditions

Each UE that the RNC[388] configures to operate using HSDPA provides CQI on the HS-DPCCH.

Do You Recall... Do you remember the significance of CQI?[389]

The RNC configures how frequently the UE should send the CQI report and the CQI repetition factor. The fastest CQI feedback cycle is once every 2 ms, and the slowest once every 160 ms. The advantage of fast CQI feedback is that the Node B can perform channel sensitive scheduling and potentially exploit some of the short-term fading-related channel variations (especially at low mobile speeds). The disadvantage of fast CQI feedback is the UE's increased computational requirements. An advantage of slow CQI feedback is lighter UE processing load, and the disadvantage is potentially lower packet data throughput. The UE estimates the SIR and predicts the data rate that it can receive on the HS-DSCH at a 10% or lower block error rate.

The higher the SIR value, the larger the data rate that can be supported at 10% (or lower) block error rate for a given amount of HS-DSCH power. A UE vendor would

[388]To perform such a configuration, the RNC sends an RRC message, such as a RADIO BEARER SETUP message.

[389]CQI quantifies the prevailing channel conditions in the downlink.

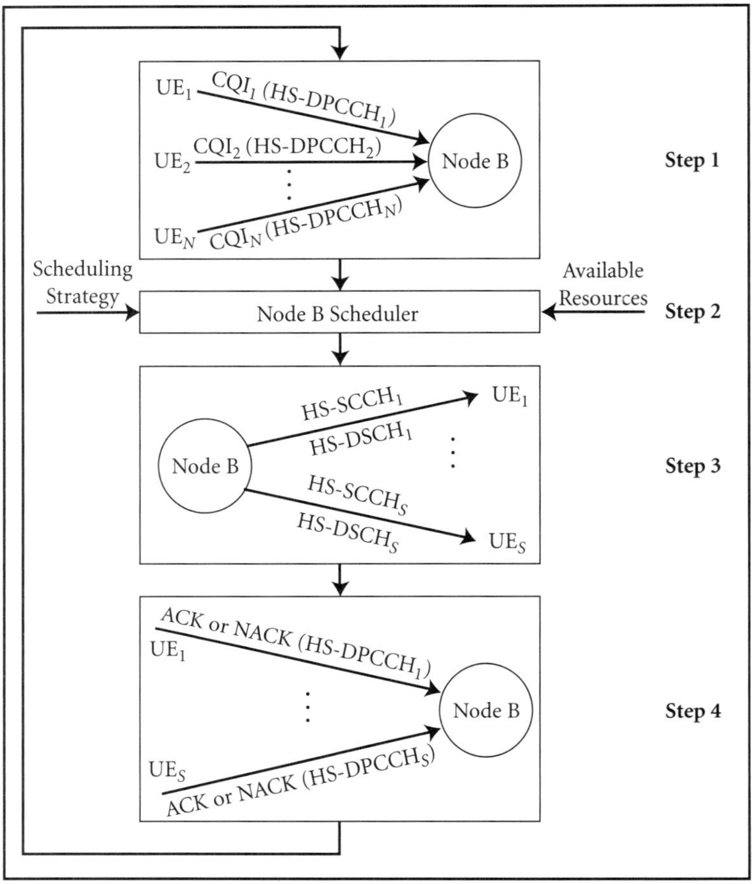

Figure 11.9: Downlink Data Transmission Steps in HSDPA

need to implement a table (see Table 11.3) that maps the SIR onto a supportable data rate (or, equivalently, CQI for a given set of parameters, including modulation scheme and number of codes). Once such table is designed into the UE, and the UE dynamically estimates SIR and utilizes this table to report CQI to the Node B.

11.4.2 Step 2: Scheduling of Users and User Packets

In an HSDPA system, the selection of users and their data rates for the downlink transmission is done at the Node B. In a UMTS R99 system, such processing is carried out at the RNC. The scheduler is non-standardized, and network vendors can implement a proprietary algorithm, which is an important product differentiator. We will first discuss how a scheduler can choose UEs for transmission. We will then summarize how the Node B can construct packets for the selected UEs.

Table 11.3: Example SIR and CQI Mapping

SIR (Fixed Signal Power)	Supportable Data Rate	CQI
3 dB	68.5 kbps	1
⋮	⋮	⋮
9 dB	325 kbps	7
⋮	⋮	⋮
19 dB	3.5 Mbps	30

11.4.2.1 Step 2A: User Selection In an HSDPA system, high-speed data transmission on the HS-DSCH occurs in units of 2 ms subframes. One or more users can be selected for transmission in a given 2 ms subframe, leading to two types of scenarios—Time-Division Multiplexing (TDM) and Code-Division Multiplexing (CDM)—as shown in Figure 11.10. In TDM scheduling, the HS-DSCH transmits a

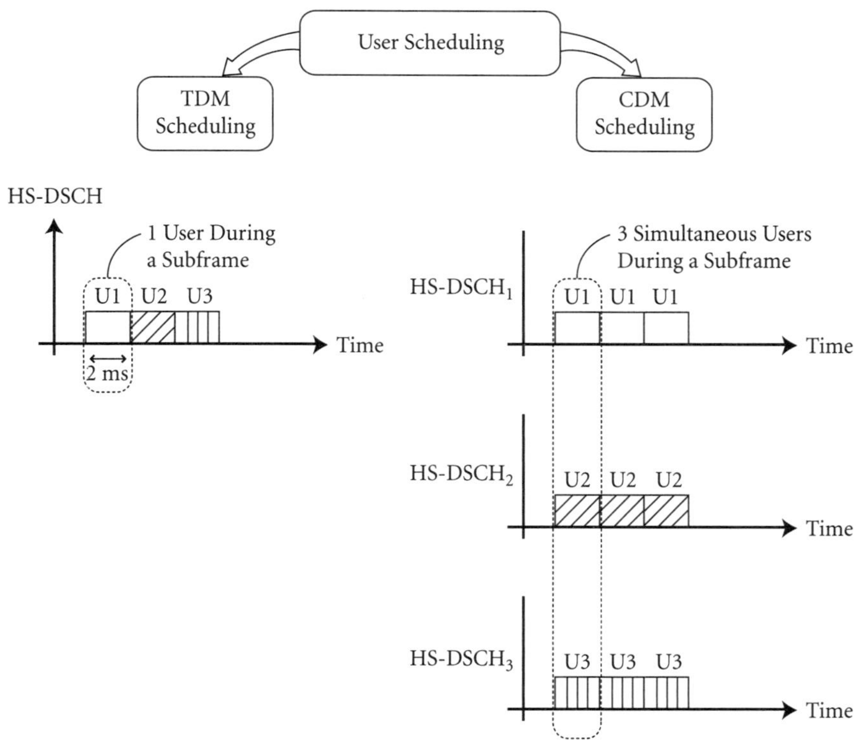

Figure 11.10: Scheduling of Users

Table 11.4: Comparison of TDM and CDM Scheduling

Feature	TDM Scheduling	CDM Scheduling
Implementation Complexity	Low	High
Flexibility	Less	More
Spectral Efficiency	Good	Better
OVSF Code Requirements	Minimum	Higher
Peak Rate Per User	Higher	Lower

packet to only one user during the subframe. In CDM scheduling, the HS-DSCHs transmit packets to more than one user simultaneously during a given 2 ms subframe. Table 11.4 summarizes benefits and drawbacks of TDM and CDM scheduling.

Implementation Complexity. The complexity of the user scheduler in TDM scheduling is low compared to CDM scheduling. The TDM scheduler needs to choose just one user. In CDM scheduling, the scheduler must decide how many users to select and which specific user to select. Furthermore, all the available resources can be assigned to just one user in TDM scheduling, but the available resources need to be properly distributed among the selected users in CDM scheduling.

A Question For You. . . Assume that three UEs are available in a cell. The Node B wants to simultaneously send packets to two UEs on HS-DSCHs during a 2 ms subframe. How many HS-SCCHs would be needed in the cell? Why?[390]

When TDM scheduling is used, only one HS-DSCH and one HS-SCCH are managed in a cell. However, CDM scheduling requires management of multiple HS-DSCHs and associated HS-SCCHs.

Flexibility. A CDM scheduler offers more flexibility than a TDM scheduler; a CDM scheduler can choose one or more users for each data transmission and can allocate different amounts of resources to different users.

Spectral Efficiency. The CDM scheduler can provide better spectral efficiency than the TDM scheduler; the CDM scheduler can optimally utilize resources by allocating available resources disproportionately among users based on the amounts of data available for transmission. With the TDM scheduler, all the available resources may be ineffectively utilized if very little data is available for transmission to users.

[390]Two HS-SCCHs are needed because one HS-SCCH can specify information about only one HS-DSCH and one UE.

OVSF Code Requirements. The TDM scheduling requires only one HS-SCCH and thus one 128-chip OVSF code. However, CDM scheduling requires multiple 128-chip OVSF codes because multiple HS-SCCHs are needed; CDM scheduling may eventually reduce the number of OVSF codes available for the HS-DSCH when the number of simultaneously scheduled UEs is very high.

Achievable User Peak Data Rate. Since TDM scheduling enables allocation of all available radio resources, such as downlink power and OVSF codes, to a single user, higher peak data rates can be easily achieved. However, when resources are distributed among multiple users in CDM scheduling, each user gets a limited amount of resources, reducing the possibility of achieving high peak data rates.

Figure 11.11 mentions potential inputs and outputs of a scheduler that selects users for data transmission in a 2 ms subframe. The design of a scheduler is a challenging task because numerous considerations are involved. As shown in Figure 11.11, the scheduler considers factors such as feedback received from UEs, available data for UEs, QoS and priorities of subscribers and their applications, and resource availability. See Chapter 10 on 1xEV-DO Rev. 0 for a brief introduction to different types of scheduling algorithms.

UE Feedback. The CQIs that the UEs report are key to channel-sensitive scheduling. The scheduler can assign higher or lower data rates based on the prevailing channel conditions for UEs. For example, when the DL channel conditions are good for a UE (reflected by high CQI values), the UE can be assigned a higher data rate. Another UE feedback that is potentially useful is the ACKs or NACKs received at

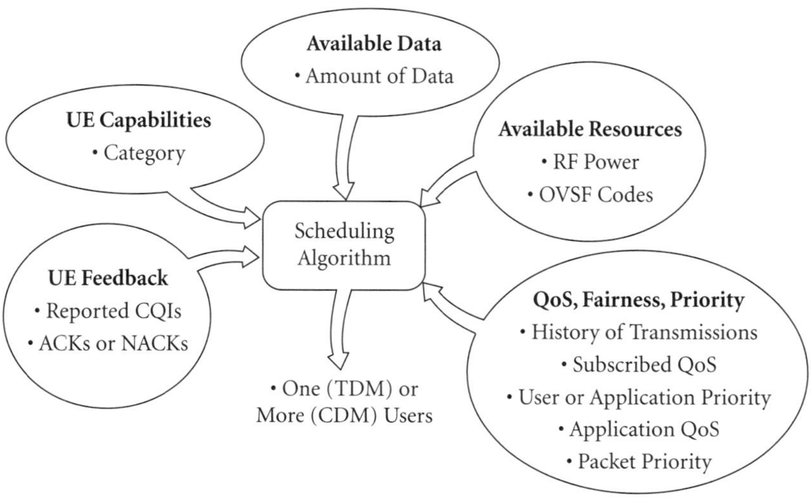

Figure 11.11: Scheduling Algorithm: Inputs and Outputs

the Node B. If the UE sends too many NACKs while reporting higher CQIs (corresponding to good channel conditions), the CQI determination algorithm in the UE may be working incorrectly. In such cases, the Node B can be more conservative when assigning data rates to such UEs.

UE Capabilities. Twelve categories of UEs exist in the perspective of HS-DSCH operation in Release 5.[391] Some categories can support a maximum of five OVSF codes at $SF = 16$, and some categories can support all fifteen codes simultaneously. Some categories cannot continuously receive data, and others can. The scheduler considers such UE capabilities.

Available Data. Air-interface resources, such as RF power and OVSF codes, are scarce and hence should be utilized efficiently. If little data is available for a UE, the scheduler could choose a lower data rate even when the UE is reporting high CQI values.

QoS, Fairness, and Priority. A good scheduling algorithm attempts to provide some degree of fairness. For example, if some users are underserved (e.g., a relatively small amount of data has recently been transmitted to these users), those users are given higher priority even though they may have relatively poor channel conditions. Users may subscribe for varying QoS levels, such as Platinum or Gold, and such QoS subscription is another parameter a scheduler can utilize. Application-specific packet priority or retransmission packet priority could also exist.

Available Resources. The available resources, such as RF power and OVSF codes, may limit the number of users that can be supported during the next 2 ms subframe. In fact, if the cell already has multiple voice calls or non-HDSPA data calls, fewer resources are available for HSDPA.

The scheduler chooses one or more users for transmission based on TDM or CDM scheduling. Examples of scheduling algorithms that select user(s) include the Round Robin Scheduler, Best Effort Scheduler, and Proportional Fairness Scheduler algorithms. See Chapter 10 for a brief introduction to these scheduling algorithms.

11.4.2.2 Step 2B: Contents of User Packets After the scheduler selects one or more users for transmission in the next 2 ms frame, the user-packet contents are determined. Specifically, the available resources, such as OVSF codes and RF power, are distributed among the selected users. The UE capability, amount of data in the Node B buffer, reported CQI, history of ACK or NACKs, and new transmission or retransmission status are considered to determine the following parameters for a packet to be transmitted: TBS, modulation scheme, redundancy version and signal constellation, number of HS-PDSCHs, and H-ARQ process identity. When a packet from a H-ARQ process is retransmitted, the TBS remains the same as in the first

[391] See Section 11.8.2 for a discussion of UE categories.

transmission, but other parameters, including modulation scheme, redundancy version, signal constellation, and number of HS-PDSCHs, can be changed. For example, the scheduler could prioritize non-systematic bits while selecting turbo code symbols, use QPSK modulation, and use more HS-PDSCHs to retransmit the same transport block. The use of low-order modulation, such as QPSK, increases the probability that the UE could successfully decode the packet for the given channel conditions.

Assume that the scheduler has chosen UE1 for transmission in the next 2 ms subframe and that UE1 (category 10) has reported a CQI of ten associated with an assumed HS-DSCH power equal to the P-CPICH power.[392] Further assume that the Node B is transmitting 10% power on the P-CPICH. The CQI of ten corresponds to the data rate of X kbps when the HS-DSCH power is 10%. If the Node B allocates exactly 10% power to UE1, the Node B would then choose the TBS such that the data rate would be X kbps. If the Node B has 20% power available for UE1, the Node B would upgrade the data rate by some factor (such as $2 * X$ kbps where 2 is the ratio of 20% available power to the UE's assumption of 10%). On the other hand, if the Node B has just 5% power available for UE1, the Node B would downgrade the data rate by some fraction such as $0.5 * X$ kbps where 0.5 is again the ratio of the available power to the power the UE assumes. The Node B determines the degree of the data rate upgrade or downgrade. When an HS-DSCH is transmitted to a given UE using multiple HS-PDSCHs, the modulation symbols and the HS-DSCH power are equally distributed among the HS-PDSCHs. However, the Node B has the flexibility of assigning different amounts of HS-DSCH power to different UEs.

11.4.3 Step 3: Packet Transmission from the Node B and Packet Reception at UEs

As part of the call setup procedure,[393] the RNC configures several important parameters via a message transmission, such as the RADIO BEARER SETUP message to the UE. Examples of such parameters include the identity of the cell for the HS-DSCH transmission, set of HS-SCCHs to be monitored in the cell, CQI feedback cycle, measurement power offset (Γ), and power offsets for the HS-DPCCH power control. All the UEs monitor the HS-SCCHs the RNC assigns to them. The scheduler can choose any of the HS-SCCHs assigned to the UE to indicate the availability of that UE's data on the HS-DSCH. From one subframe to the next, a given HS-SCCH can be associated with a different UE and a different HS-DSCH. Two examples of data transmission scenarios are discussed here, one corresponding to *TDM scheduling* and one to *CDM scheduling*.

Let's discuss TDM scheduling first. Consider Figure 11.12. Three UEs are present in a cell, and all UEs are monitoring the same HS-SCCH as the RNC instructs. Since only one UE receives data during a 2 ms subframe in TDM scheduling, one HS-SCCH is adequate. The scheduler first chooses UE1 based on its scheduling strategy. The

[392] See Eq. (11.4) on page 668.

[393] See Section 11.5 for call setup details.

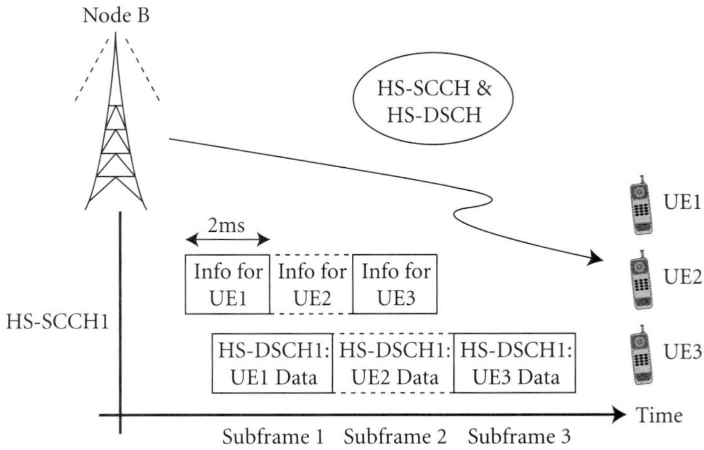

Figure 11.12: TDM-Scheduling–Based Transmission

Node B first transmits HS-SCCH1 with the H-RNTI of UE1. The first time-slot of the HS-SCCH consists of the identity of the HS-PDSCH codes and the modulation scheme with these fields masked by the H-RNTI of UE1. When UE1 correlates its H-RNTI with the first two time-slots of HS-SCCH1, it learns about the OVSF codes and the modulation scheme. UE1 now starts despreading and demodulating HS-DSCH1 while continuing to decode the last time-slot of HS-SCCH1. Once the entire HS-SCCH is decoded and the CRC (that is masked with H-RNTI) is correct, UE1 has all the information necessary to decode the entire packet. UE2 and UE3 would find no match with the first time-slot of HS-SCCH1 because this time-slot is carrying the H-RNTI for only UE1. If an error occurs in the CRC for UE1, UE1 aborts the decoding of the HS-DSCH1. In the next 2 ms interval, the scheduler chooses UE2, and the HS-SCCH1 contents are now masked with UE2's H-RNTI. HS-DSCH1 would now contain a packet for UE2. UE1, UE2, and UE3 are again monitoring HS-SCCH1, and only UE2 would find a match with HS-SCCH1. UE2 would then decode its packet from HS-DSCH1. Similarly, UE3 receives its packet in the next 2 ms subframe from HS-DSCH1 using the information on HS-SCCH1.

Let's turn our attention now to CDM Scheduling. Figure 11.13 depicts how the Node B transmits packets to multiple UEs simultaneously during a 2 ms subframe. Assume that the UTRAN has decided to allow a maximum of three simultaneous users in a 2 ms subframe. Hence, three HS-SCCHs are required in a cell. The RNC could instruct all five UEs in the cell to monitor one or more of the three HS-SCCHs. In the example shown in Figure 11.13, the RNC has configured all the UEs to monitor all three channels, HS-SCCH1, HS-SCCH2, and HS-SCCH3. The scheduler selects UE1, UE2, and UE3 for transmission. The Node B transmits HS-SCCH1 with UE1's H-RNTI1, HS-SCCH2 with UE2's H-RNTI2, and HS-SCCH3 with UE3's H-RNTI3. UE1, UE2, and UE3 retrieve their packets from HS-DSCH1, HS-DSCH2 and HS-

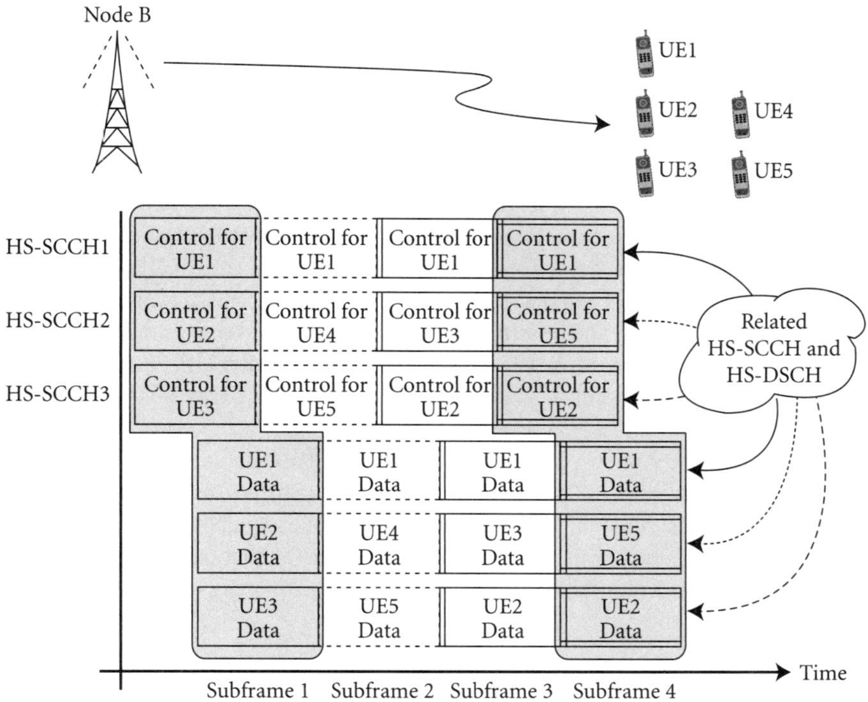

Figure 11.13: CDM-Scheduling–Based Transmission

DSCH3, respectively. UE4 and UE5 find no match between any HS-SCCH and their own HRNTIs.

In the second 2 ms subframe (i.e., Subframe 2), the scheduler has chosen UE1, UE4, and UE5, and these UEs obtain control information from HS-SCCH1, HS-SCCH2, and HS-SCCH3 to collect their packets from HS-DSCH1, HS-DSCH2, and HS-DSCH3, respectively. As is evident from Figure 11.13, one HS-SCCH is associated with one UE during a 2 ms interval. Such association can change from one subframe to the next. An exception to this case is the scenario in which data is sent to the UE in successive (i.e., consecutive) subframes. Since HS-SCCH1 is used for UE1 in the first subframe, the same HS-SCCH1 is also used for UE1 in the second subframe because data is sent to UE1 during the first and the second (consecutive) subframe.

If no data is sent in two consecutive subframes to the same UE, different HS-SCCHs can be used for a given UE. For example, UE2 receives data in the first subframe using control information from HS-SCCH2. No data is sent to UE2 in the second subframe. Therefore, in the third subframe, UE2 can be given control information on a different HS-SCCH (in this case, HS-SCCH3 rather than HS-SCCH2).

11.4.4 Step 4: UE Response to a Received Packet

A UE that receives a packet sends an ACK if it successfully decodes the packet. In the case of packet error, it sends a NACK to the Node B on the HS-DPCCH. A UE sends no ACK or NACK if it receives no packet. The UE sends an ACK or NACK approximately 5 ms or 7.5 time-slots after completely receiving the packet from the HS-DSCH as shown in Figure 11.14. The UE utilizes techniques such as Chase combining, partial Incremental Redundancy, or full Incremental Redundancy based on how the received packet was formed for retransmission (as explained later in this section). A new packet received at the UE typically contains all systematic bits (i.e., original information bits) and possibly some parity bits based on the space remaining in the packet. When a retransmission packet is received at the UE, the packet could have any redundancy version (i.e., any set of systematic and parity bits). Consider an example shown in Figure 11.14.

The Node B sends a new packet P1 containing all 1000 original bits, 250 parity 1 bits, and 250 parity 2 bits. The Node B is prioritizing the systematic bits. The UE incorrectly retrieves the 1000 original bits and hence sends a NACK to the Node B. The UE stores 1500 bits in its soft channel bits buffer. The Node B now retransmits the packet with some redundancy version, possibly prioritizing non-systematic bits.

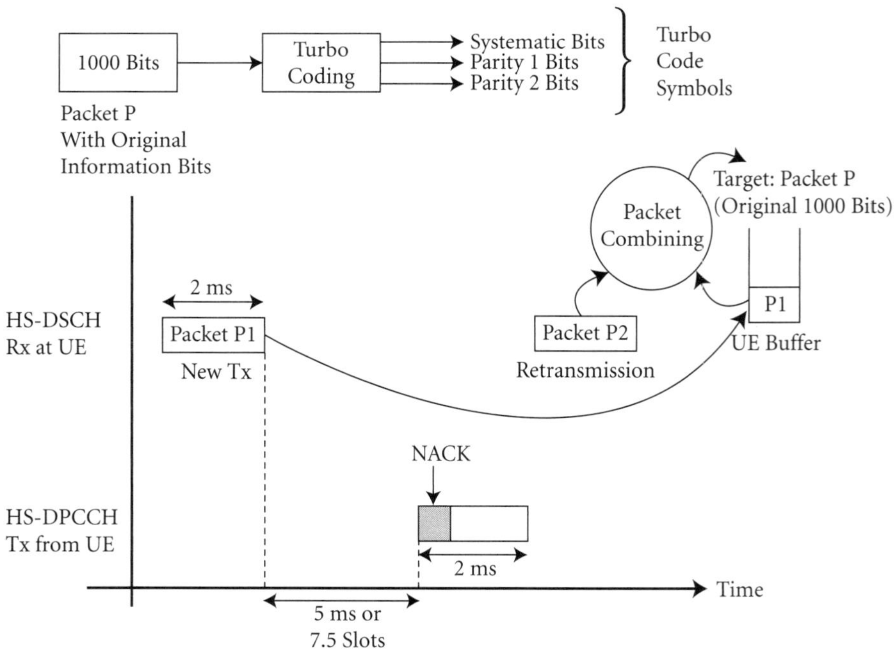

Figure 11.14: Response to a Packet Received at the UE

When the UE receives packet P2, it combines P2 with P1, which is already stored in the buffer. If the UE succeeds in collecting 1000 original bits without an error, it sends an ACK; otherwise, it sends a NACK. In summary, the UE combines all transmissions of a given packet in an attempt to decode the original information bits.

If the Node B sends an identical packet for every retransmission, the UE combines the packets through Chase combining. For example, the UE can combine retransmission packets using a weighted sum of SIRs. In such a case, relatively low buffer memory is required because the UE could just store the final result of the combining. In the case of full Incremental Redundancy, the Node B transmits an entirely different packet (e.g., with all parity bits but no systematic bits) for a retransmission. In such a case, the UE must store all transmissions of the packet in the buffer, increasing the buffer requirements. In general, Chase combining is simple to implement, requires less memory, and may be adequate under good channel conditions. Full Incremental Redundancy performs better (i.e., lower error rate or higher supportable data rate at a given SIR) than Chase combining under adverse channel conditions but requires more buffer space. The UE sends ten `1`s when it wants to send an ACK and ten `0`s when it is responding with a NACK.

11.5 HIGH-SPEED DOWNLINK PACKET ACCESS CALL SETUP

The call setup procedure in an HSDPA system is similar to that in a Release 99 system.[394] The HSDPA system utilizes existing Release 99 messages and adds HSDPA-specific information in those messages. Figure 11.15 offers a simplified view of the HSDPA call setup, focusing on messages that are important for HSDPA. Various call setup steps are summarized here.

The UE sends an RRC CONNECTION REQUEST message to establish a signaling connection with the RNC. The RNC replies with an RRC CONNECTION SETUP message to the UE and asks the UE to provide information on the UE's capabilities by setting the parameter ue-RadioCapabilityFDDUpdateRequirement to TRUE (where FDD is frequency-division duplexing). The UE responds with an RRC CONNECTION SETUP COMPLETE message and indicates its HSDPA capabilities by specifying its HS-DSCH physical layer category.

The UE and 3G-SGSN establish a signaling connection using the RNC's direct transfer functionality. More specifically, the UE sends an RRC INITIAL DIRECT TRANSFER message containing a Non-Access Stratum (NAS) Service Request message that may include information about one or more PDP contexts[395] if the UE already has PDP contexts. The UE requests the desired QoS and an IP address (PDP address) from the 3G-SGSN. The 3G-SGSN authenticates the UE using the help of a home location register and authentication center (which are typically contained within the same piece of equipment). Security is also enacted by activating encryption algorithms for the traffic packet and signaling messages. Once the security is in place,

[394]See Chapter 9 on UMTS Release 99 for a detailed discussion of the call setup in UMTS.

[395]The Network Service Access Point Identifier (NSAPI) identifies each PDP context.

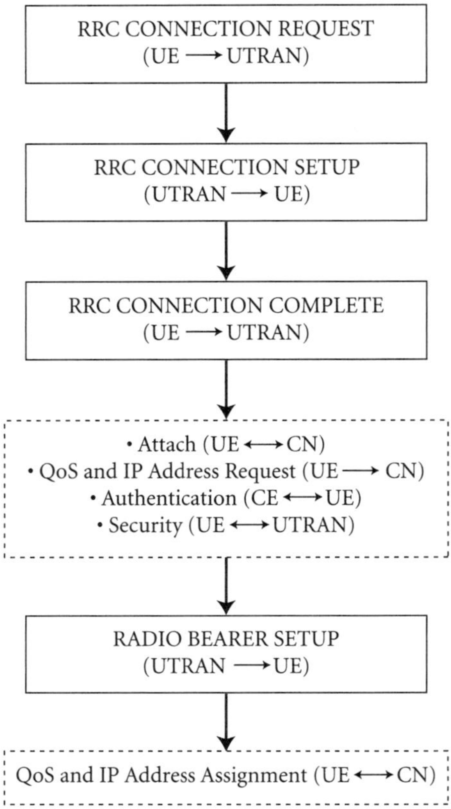

Figure 11.15: Simplified HSDPA Call Setup

the UE requests the PDP context by sending an RRC UPLINK DIRECT TRANSFER message containing an NAS Activate PDP Context Request message if the UE has no PDP context. As part of such PDP context activation, the UE requests the desired QoS and an IP address (PDP address) from the 3G-SGSN. The packet-switched domain, specifically the 3G-SGSN, sends a Radio Access Network Application Part Radio Access Bearer Assignment Request message to the RNC to determine if the UTRAN can entertain the requested QoS.

Assuming that the UTRAN has adequate radio resources for the UE, the RNC sends a RADIO BEARER SETUP message to the UE to configure the HSDPA parameters. Examples of the parameters include the identification of the HS-DSCH serving cell (as part of the downlink information for each radio link); new H-RNTI; HS-SCCH channelization code(s); downlink scrambling code for HS-SCCH(s) and HS-DSCH; measurement power offset (Γ), CQI feedback cycle, CQI repetition factor, and Δ CQI power offset (as part of the measurement feedback information); and HS-DPCCH power offsets (Δ ACK, Δ NACK) and ACK or NACK repetition factor (as part of uplink Dedicated Physical channel power control information). The SGSN responds to the UE's request for activation of the PDP context by sending an NAS Activate PDP Context Accept message, which is encapsulated in an RRC DOWNLINK DIRECT TRANSFER message. This message includes the QoS level the SGSN has allocated and the PDP address the GGSN has allocated.

Now the UE starts monitoring the set of configured HS-SCCHs. It also starts sending CQI reports based on the configured values of the CQI feedback parameters. The Node B can start sending packets to the UE on the HS-DSCH.

11.6 RESOURCE MANAGEMENT OF THE HIGH-SPEED DOWNLINK PACKET ACCESS CHANNELS

Efficient radio resource management is key to superior packet data performance. HSDPA manages the downlink radio resources more effectively than UMTS R99.

Do You Recall... What are important radio resources for HSDPA? Hint: HSDPA utilizes the CDMA-based air interface.[396]

The management of OVSF codes and power is discussed next, followed by a brief summary of how the Node B and the RNC coordinate radio resource management in HSDPA.

The HS-DSCH is a transport channel that is mapped onto one or more HS-PDSCHs, with each HS-PDSCH corresponding to one OVSF code at $SF = 16$. An HSDPA system is fully backward-compatible with a Release 99 system and hence shares the OVSF code space with a Release 99 system when both systems share the same frequency spectrum (that is, the same 5 MHz uplink bandwidth and the same 5 MHz downlink bandwidth). Furthermore, even in the absence of Release 99 UEs, an HSDPA cell must transmit Release 99 overhead channels, such as P-CPICH and P-CCPCH, because HSDPA UEs also utilize these overhead channels. Figure 11.16 shows a partial OVSF code tree.

As shown in Figure 11.16, the codes $C_{ch,16,1}$, through $C_{ch,16,15}$ can potentially be used for HSDPA UEs. Since $C_{ch,256,0}$ and $C_{ch,256,1}$ are reserved for P-CPICH and P-CCPCH, $C_{ch,16,0}$ is unavailable to the HS-DSCH to maintain orthogonality.

Recall that one HS-SCCH at $SF = 128$ is required for one HS-DSCH that could use one or more of the fifteen OVSF codes, $C_{ch,16,1}$ through $C_{ch,16,15}$. One code at $SF = 16$ is equivalent to eight OVSF codes at $SF = 128$. In other words, $C_{ch,16,0}$ is equivalent to $C_{ch,128,0}$ through $C_{ch,128,7}$. Assume that only one UE exists in a cell, and this UE is HSDPA-capable. In such a case, the UTRAN can ask the UE to monitor the HS-SCCH at $C_{ch,128,7}$ and the UE could receive data on all the codes $C_{ch,16,1}$ through $C_{ch,16,15}$. If one HS-PDSCH is used for data transmission to one UE during

[396] Two major radio resources to be managed in the downlink are RF power and OVSF codes. On a side note, managing the UE transmit power and the total interference at the Node B receiver is important in the uplink.

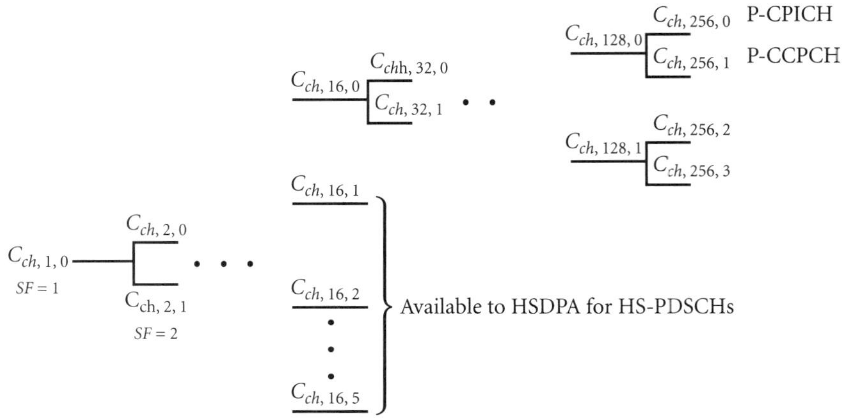

Figure 11.16: OVSF Code Tree Utilization in the HSDPA Downlink

a 2 ms TTI, fifteen UEs can simultaneously receive data on fifteen HS-PDSCHs if control information on fifteen HS-SCCHs at $SF = 128$ can be transmitted. However, when fifteen HS-SCCHs are transmitted, one more code, in addition to $C_{ch,16,0}$, at $SF = 16$ becomes unavailable. Hence, a maximum of fourteen UEs can perhaps receive data simultaneously during a 2 ms TTI with each UE utilizing one HS-SCCH at $SF = 128$ to retrieve data from one HS-PDSCH at $SF = 16$. Thus, the maximum number of UEs that can simultaneously receive data on the HS-DSCH seems to be fourteen. We are not done yet! Note that we need to send R99 Dedicated Physical channel that carries the Dedicated Physical Data channel and DPCCH. Such Dedicated Physical channels for different users will occupy some codes, perhaps at $SF = 256$.

Do You Recall... Do you remember the use of the Dedicated Physical Data channel and the DPCCH transmitted in the downlink?[397]

We will then see one more OVSF code at $SF = 16$ being blocked. Hence, in Release 5 HSDPA, the maximum number of simultaneous UEs that can receive packets simultaneously during a 2 ms subframe is thirteen. In general, numerous UEs could exist in a cell, with each UE receiving data during potentially non-consecutive TTIs.

As Release 99 voice users and packet data users arrive in a cell, fewer and fewer OVSF codes at $SF = 128$ and $SF = 16$ remain available for HSDPA. The HSDPA throughput is maximal when all fifteen OVSF codes at $SF = 16$ are available. The

[397]The Dedicated Physical Data channel is used to carry traffic and signaling messages. The DPCCH carries power control commands, the transport format combination indicator (to convey how the accompanying packet is formed), and the dedicated pilot channel.

OVSF codes should be assigned to users such that the maximum number of contiguous codes remain available in the OVSF code tree. Such a strategy ensures that HSDPA can achieve higher data rates. When the OVSF codes are shared between voice users and packet data users, the codes can be fully shared or some codes can be reserved for some service to guarantee minimum capacity (e.g., a certain Erlang capacity for voice). The reservation of codes for a service provides minimum performance guarantees for that service while denying other services access to the reserved codes. A good OVSF code allocation strategy strikes a configurable balance between the minimum performance guarantees and maximum achievable voice and packet data performance.

An HSDPA system can fully utilize all the available power, maximizing the achievable data throughput. The more power available to HSDPA, the higher the likelihood of superior throughput. If numerous voice and data users are present, the power available to HSDPA users may be little, and high data rates, such as 14 Mbps, are unattainable. Just like OVSF codes, power must be optimally shared between the voice and data users to maximize the achievable capacity and throughput. Power reservations for voice can guarantee a minimum Erlang capacity while possibly limiting achievable HSDPA packet data throughput.

When a system supports UMTS R99 users and HSDPA users in the same bandwidth, radio resource management becomes challenging. The radio resource management for R99 users is performed at the RNC, and radio resource management for HSDPA users must be performed dynamically at the Node B. Importantly, the use of radio resources is well-coordinated between the Node B and the RNC. If such coordination is managed ineffectively, the radio resources may be non-optimally utilized, leading to poor system performance. In general, the RNC specifies a certain amount of radio resources, and the Node B dynamically manages these resources for HSDPA users. Frequent exchange of radio-resource-management–related information (e.g., the available amount of downlink power) between the Node B and the RNC is desirable for optimum performance.

11.7 MOBILITY MANAGEMENT IN HIGH-SPEED DOWNLINK PACKET ACCESS

Let's discuss how user mobility is managed in HSDPA when the UE is in the Connected Mode. Time for a review question before diving into mobility management...

Do You Recall... Does HSDPA support soft and/or softer handover—yes or no? You have a 50% chance of getting it right! Of course, we hope that you have been paying attention and need not randomly pick the answer![398]

[398]HSDPA does not support soft or softer handover. The UE communicates with a single cell at a given instant.

In an HSDPA system, the HS-DSCH and the HS-SCCH are transmitted to a given UE from only one cell. Similarly, the Node B in just one cell processes the HS-DPCCH the UE transmits. When the UE goes from one cell to another, the HSDPA operation for the UE is discontinued in the old cell and initiated in the new cell. The HSDPA system utilizes Release 99 handover mechanisms without adding new messages. Figure 11.17 depicts three stages of the handover management in an HSDPA system.

During Stage 1, the RNC carries out measurement configuration for the UE. Specifically, the RNC sends a MEASUREMENT CONTROL message to the UE. In this message, the RNC informs the UE of the pilots to be monitored, the measurements to report, and the measurement reporting method. For example, the RNC may instruct the UE to monitor three cells with the pilots corresponding to primary scrambling codes 100, 104, and 108, and to report the P-CPICH E_c/I_0 and P-CPICH received signal code power to the RNC. Two methods, periodic and event-based, are available for measurement reporting. In the periodic method, the RNC specifies to the UE how frequently the measurements should be reported (e.g., every 0.5 seconds). In the event-based method, the RNC asks the UE to send measurements when a reporting event occurs. Event-based reporting is more common in commercial deployments. The UMTS R99 defines numerous events for measurement reporting. Among those events, event 1D (i.e., change of the best cell) can be exploited in an HSDPA system. See Chapter 9 for details on various UMTS R99 measurement reporting events.

In Stage 2, the UE sends measurement reports according to the configuration the RNC has selected. For example, when event 1D is configured, the UE sends a

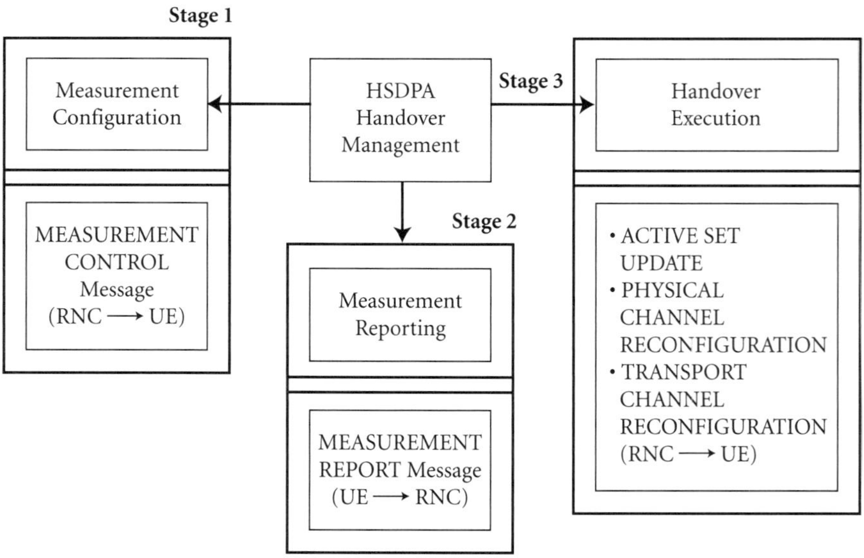

Figure 11.17: Handover Management in HSDPA

MEASUREMENT REPORT message as soon as it detects a P-CPICH that is stronger than the current best P-CPICH.

During Stage 3, the handover algorithm is executed, and such an algorithm is implementation-specific.

A Question for You... Which network node or element implements the handover algorithm?[399]

The RNC processes the P-CPICH measurements received from the UE and decides whether to terminate the HSDPA operation in the previous cell and start the HSDPA operation in the new cell corresponding to the strongest pilot reported in the latest MEASUREMENT REPORT message. Note that the RNC handover algorithm for R99 channels maintains the Active Set because full soft and softer handover is available for R99 channels for an HSDPA-capable UE. As far as the HSDPA operation is concerned, the RNC selects only one of the cells from the Active Set for transmission and reception of HSDPA channels. Figure 11.18 illustrates two handover scenarios, intra–Node-B handover and inter–Node-B handover.

Assume that the UE's Active Set is {Cell 2, Cell 3, Cell 4}. The RNC has configured the UE for the HSDPA operation in Cell 3. Thus, the UE is monitoring the

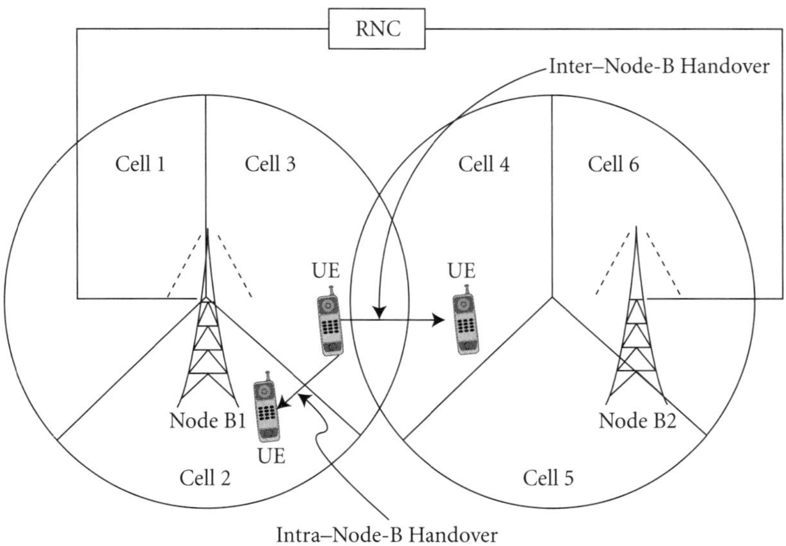

Figure 11.18: Handover Management in HSDPA

[399]The handover algorithm is implemented at the RNC.

HS-SCCH(s) and obtaining the data from the associated HS-DSCH in Cell 3 because Cell 3 is the strongest among the three cells.

Do You Recall... Which UE measurement can the RNC use to determine the strongest cell in the downlink?[400]

At the same time, Node B1 receives the HS-DPCCH from the UE in Cell 3. Furthermore, Node B1 receives the uplink Dedicated Physical Data channel from the UE in Cell 2 and Cell 3, and Node B2 receives it in Cell 4. Both Node B1 and Node B2 send the radio frame received on the uplink Dedicated Physical Data channel to the RNC so that the RNC can perform selection combining. Now we will consider two scenarios: in Scenario 1, the UE moves from Cell 3 into Cell 2, and in Scenario 2, the UE moves from Cell 3 into Cell 4.

In Scenario 1, the UE's event-1D–based reporting will indicate to the RNC the change of the best cell, and the RNC performs hard handover of the HSDPA channels. The RNC typically sends a TRANSPORT CHANNEL RECONFIGURATION or PHYSICAL CHANNEL RECONFIGURATION message to the UE, specifying a new cell in which the UE should resume the HSDPA operation. Since the cell that supports the HSDPA operation (i.e., Cell 2 in Scenario 1) for the UE has changed but the Node B remains the same, such a scenario is referred to as *intra*–Node-B hard handover. In Scenario 2, event 1D would indicate the change of the best cell, prompting the RNC to initiate *inter*–Node-B hard handover because the new cell (i.e., Cell 4 in our example) belongs to the new Node B. Note that the basic Active Set in both scenarios remains the same. Hence, the uplink Dedicated Physical Data channel would still be processed the same way as prior to the hard handover of the HSDPA channels.

Ever Wondered... Why is only hard handover used for HSDPA channels? HSDPA operation occurs between the UE and a single cell. If we were to support soft handover, we would have to rely upon the RNC for scheduling. The RNC would need to coordinate scheduling among multiple Node Bs. The transport delay between the Node B and the RNC would be too long (e.g., about 100 ms) to adapt to the changing channel conditions the fast CQI feedback from the UE would reflect. Channel-sensitive scheduling benefits would be diminished due to the long delay between the CQI feedback and the coordinated scheduling the RNC performs. Furthermore, backhaul bandwidth requirements would be increased significantly due to the need to transport data between the RNC and multiple Node Bs. In summary, due to factors such as fast channel-sensitive scheduling and backhaul bandwidth and due to the simplicity of scheduler implementation, HSDPA supports only hard handover.

[400]The P-CPICH E_c/N_0 and/or P-CPICH received signal code power can be used to determine the strongest cell.

In summary, the HSDPA channels are transmitted and received in only one cell for a given UE. The selection of the cell with the strongest P-CPICH ensures that the UE is communicating via the best possible downlink, thereby increasing the achievable downlink throughput.

11.8 NETWORK AND UE ARCHITECTURE

HSDPA relies upon the basic system architecture defined in UMTS 99. Section 11.8.1 discusses how HSDPA has enhanced the air-interface protocol stack between the UE and the UTRAN. Section 11.8.2 specifies main characteristics of the HSDPA UE categories. Section 11.8.3 describes the interaction between the Node B and the RNC to support the HSDPA operation. Finally, Section 11.8.4 summarizes how the Node B constructs a packet for transmission to a UE.

11.8.1 Radio Interface Protocol Stack Changes from Release 99 to Release 5

Figure 11.19 focuses on the radio network architecture in HSDPA and omits details of the core network. Similar to Release 99, the RNC is connected to the circuit-switched domain's MSC (mobile switching center) and the packet-switched domain's 3G-SGSN. *Let's discuss the protocol changes of Figure 11.19 from the perspectives of the RNC, the Node B, and the UE.*

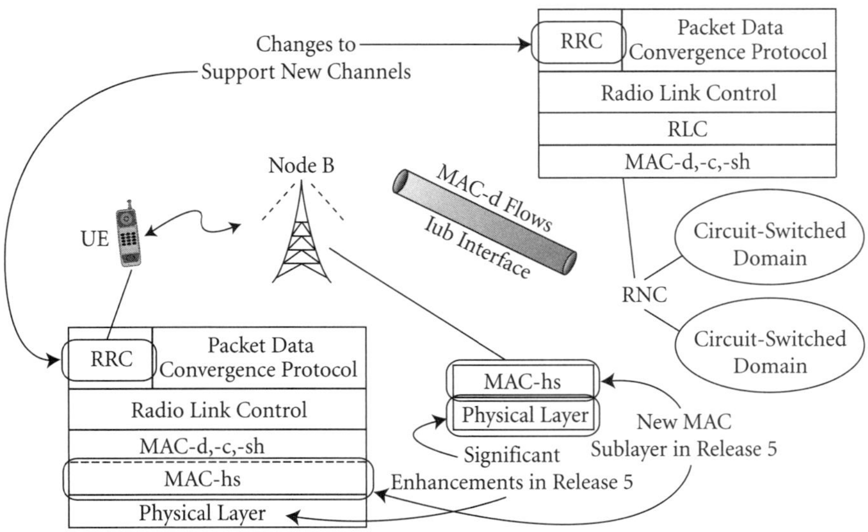

Figure 11.19: Radio Network Architecture in HSDPA

In an R99 RNC, the RNC typically implements RRC, Packet Data Convergence Protocol, and MAC sublayers, such as the MAC-c (common), MAC-d (dedicated), and MAC-sh (shared). During soft handover, the RNC also carries out some physical layer processing, such as selection combining, of the UE's uplink frames received from different the Node Bs in the UE's Active Set. Since the HSDPA introduces three new radio channels and since the RRC layer constructs the layer 3 RRC signaling messages, the RRC protocol undergoes some changes to support new channels. To support high-speed downlink transmission to the UE, the RNC sends MAC-d PDUs on MAC-d flows to the Node B over the Iub interface.

The Node B implements most of the physical layer functions. The physical layer is responsible for over-the-air transmission. Since HSDPA is primarily an air-interface feature, the physical layer of the Node B undergoes significant changes, such as support for 16-QAM and H-ARQ. The Node B also has a new MAC sublayer called Mac-hs. This sublayer performs several functions, such as scheduling and H-ARQ, which are briefly summarized here. The Node B scheduler chooses one or more users for data transmission during the upcoming 2 ms TTI. Such a scheduler is vendor-specific, and TDM or CDM scheduling can be implemented. For a user selected for new transmission during a given 2 ms TTI, the Node B determines the TBS and calculates a specific TFRI value based on the selected modulation scheme and the number of HS-PDSCH codes [291]. For example, $TFRI = 0$ corresponds to QPSK and one HS-PDSCH code, and $TFRI = 29$ corresponds to 16-QAM and fifteen HS-PDSCH codes. Note that the TBS for retransmission must be the same as that for the first transmission. The selected TFRI is a five-bit field transmitted on the HS-SCCH.

As far as the H-ARQ is concerned, the Node B maintains up to eight H-ARQ processes for a given UE. Multiple H-ARQ processes enable the Node B to continuously (or frequently) send data to the UE. One H-ARQ process transmits or retransmits one transport block during a TTI. One transport block is equivalent to one MAC-hs PDU that the MAC-hs sublayer submits to the physical layer for further processing and eventual over-the-air transmission. Section 11.8.4 describes the construction of a MAC-hs PDU. The identity of such an H-ARQ process is conveyed to the UE via the HS-SCCH. The HSDPA H-ARQ is considered fully adaptive as the Node B chooses a suitable redundancy version for a given UE during a TTI. Furthermore, the HSDPA H-ARQ is asynchronous because the time between the H-ARQ process's retransmissions is non-constant and the Node B can dynamically determine these times. Since the Node B is in charge of HS-DSCH scheduling, flow control of HS-DSCH data between the Node B and the RNC becomes important to prevent buffer overflow at the Node B.

The UE's physical layer requires significant enhancements to support the downlink transmission. The new MAC-hs layer at the UE performs the major functions of H-ARQ, packet re-ordering, and disassembly. The UE attempts to decode the packet received during a TTI and sends an ACK to the Node B if it succeeds and a NACK to the Node B if it fails. The Node B utilizes such an H-ARQ feedback to determine whether to perform retransmission or start a new packet transmission. Each MAC-hs PDU has a transmission sequence number, and the UE utilizes the transmission sequence number to re-order the MAC-hs PDUs for submission of consecutive PDUs to the disassembly function. The disassembly function removes the MAC-hs header and submits the MAC-d PDUs to the MAC-d sublayer.

11.8.2 High-Speed Downlink Packet Access UE Categories

Table 11.5 summarizes the physical layer characteristics of the frequency-division duplexing HS-DSCH categories in HSDPA. Let's summarize the main characteristics of the HSDPA UE category by discussing Category 8, which supports a maximum of ten simultaneous HS-PDSCHs during a given 2 ms subframe. Recall that each HS-PDSCH corresponds to one OVSF code at $SF = 16$. The minimum inter-TTI is 1, meaning that the Category 8 UE can receive HS-DSCH transmission in consecutive TTIs. Hence, if the Category 8 UE has received a transport block during subframe 5, it can receive another transport block in subframe $5 + \textit{inter-TTI} = 5+1 = 6$. Observe that the Category 4 UE has the minimum inter-TTI of 2. If the Category 4 UE has received a transport block during subframe 5, it can receive the next transport block in subframe $5 + \textit{inter-TTI} = 5 + 2 = 7$, implying that a gap of one TTI occurs between two successive transport blocks.

The Category 8 UE can receive a maximum of 14411 bits in a transport block. The total number of soft channel bits that can be stored in the UE's buffer is 134400. Recall that the H-ARQ enables the receiver to combine all transmissions (i.e., first transmission and all retransmissions) of a given transport block. The code symbols of a given transmission are referred to as *soft channel bits*, which must be shared among multiple H-ARQ processes. All UE categories except 10 and 11 can support both QPSK and 16-QAM modulation schemes. Categories 11 and 12 can support only QPSK.

Ever Wondered... If you are wondering why UE capabilities improve from 1 to 10 and suddenly degrade for categories 11 and 12, you are not alone! The reason for such an anomaly is a matter of history. In an earlier draft of the standard, only ten categories of UEs were defined, requiring all UEs to support 16-QAM. UE vendors had prior experience with QPSK but no experience with 16-QAM since QPSK was used in UMTS R99. To expedite the time-to-market for HSDPA, two additional categories, 10 and 11, were added later. Category 12 UEs were the first UEs to hit the market. So, the mystery of categories 11 and 12 has been solved!

11.8.3 Interaction Between the Node B and the RNC for High-Speed Downlink Packet Access

The radio resource management in HSDPA is distributed between the Node B and the RNC. In general, the RNC determines the configuration of HSDPA operation, and the Node B dynamically manages the radio resources for multiple users. Two main areas of the interaction between the Node B and the RNC are setup of the HSDPA operation

Table 11.5: Characteristics of Release 5 HSDPA UE Categories

UE (HS-DSCH) Category	Maximum Number of Simultaneous PDSCHs	Minimum Inter-TTI	Maximum TBS	Total Number of Soft Channel Bits	Supported Modulation Schemes
1	5	3	7298	19200	QPSK & 16-QAM
2	5	3	7298	28800	QPSK & 16-QAM
3	5	2	7298	28800	QPSK & 16-QAM
4	5	2	7298	38400	QPSK & 16-QAM
5	5	1	7298	57600	QPSK & 16-QAM
6	5	1	7298	67200	QPSK & 16-QAM
7	10	1	14411	115200	QPSK & 16-QAM
8	10	1	14411	134400	QPSK & 16-QAM
9	15	1	20251	172800	QPSK & 16-QAM
10	15	1	27952	172800	QPSK & 16-QAM
11	5	2	3630	14400	QPSK
12	5	1	3630	28800	QPSK

for a UE during the call setup (discussed in Section 11.8.3.1) and flow control of the data transfer from the RNC to the Node B (discussed in Section 11.8.3.2).

11.8.3.1 High-Speed Downlink Packet Access Operation Setup During the HSDPA call setup, the RNC sends a RADIO BEARER SETUP message to the UE to configure it to operate using HSDPA (as explained in Section 11.5). However, before sending this message to the UE, the RNC exchanges Node B Application Part messages with the Node B to reserve radio resources for the UE as illustrated in Figure 11.20.

The RNC provides numerous HSDPA configuration parameters to the Node B in the RADIO LINK SETUP REQUEST message so the Node B can reserve and configure resources for the UE. Examples of such parameters include *HS-DSCH Frequency-Division Duplexing Information*, *H-RNTI*, and *HS-PDSCH Id*. *HS-DSCH Frequency-Division Duplexing Information* includes information pertaining to MAC-d flows (e.g., the flow identity, priority queue identity, scheduling priority indicator, timer T1, discard timer, guaranteed rate, information related to MAC-d PDU size, and radio link control mode); to UE capabilities (e.g., the UE's HSDPA category); and to HS-DPCCH parameters (e.g., CQI feedback cycle k, CQI repetition factor, ACK-NACK repetition factor, CQI power offset, ACK power offset, NACK power offset, optional measurement power offset, and optional HS-SCCH power offset).

These parameters enable the Node B to perform suitable scheduling for the UE's packets and to properly demodulate, decode, and interpret the UL HS-DPCCH. The scheduling priority indicator ranges from 0 to 15, with 0 being the lowest priority and 15 being the highest priority. The Node B can utilize timer T1 to decide how many H-ARQ retransmissions to perform for the UE's MAC-hs PDU. The discard timer helps the Node B discard out-of-date MAC-hs SDUs from the queue.

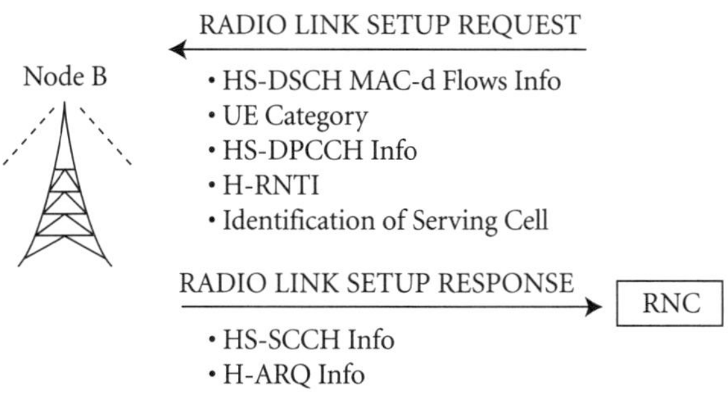

Figure 11.20: Dedicated Radio Connection Setup

Do You Recall... Do you remember the significance of H-RNTI?[401]

Note that the RNC is in charge of assigning the H-RNTI to the UE. *HS-PDSCH Id* specifies which of the Active Set cells is the HSDPA serving cell for the UE.

The Node B responds with a RADIO LINK SETUP RESPONSE message[402] and provides the RNC with several parameters it has determined for the UE in addition to the response to the RNC-dictated parameters in the RADIO LINK REQUEST SETUP message. Examples of response parameters are MAC-d flow specific response (e.g., flow identification and the initial amount of data that can be sent to the Node B without causing a buffer overflow) and HS-SCCH specific response (e.g., the specific HS-SCCH code(s) assigned to the UE). An example of the parameters the Node B autonomously determines is H-ARQ memory partitioning, which could be implicit or explicit. In an implicit configuration, the total number of soft channel bits is equally distributed among the H-ARQ processes. In an explicit configuration, the Node B specifies how much memory is reserved for each H-ARQ process.

Upon completion of the radio link setup, several HSDPA parameters have been configured for a given UE. However, the RNC may provide general HSDPA parameters to the Node B for all UEs via a PHYSICAL SHARED CHANNEL RECONFIGURATION REQUEST message as shown in Figure 11.21. The RNC specifies for a cell the following parameters: the total HS-PDSCH and HS-SCCH power; the downlink scrambling code for the HS-PDSCH and HS-SCCH (which is typically the primary scrambling code of the cell but may be a secondary scrambling code as well); HS-PDSCH OVSF codes; and HS-SCCH OVSF code(s) to the Node B. Now the Node B operates within the code and power constraints the RNC has given. For example, if the RNC has specified five codes at $SF = 16$ for the HS-PDSCHs, the Node B can now dynamically assign three codes to one user and the remaining two codes to another user and change these assignments possibly every 2 ms.

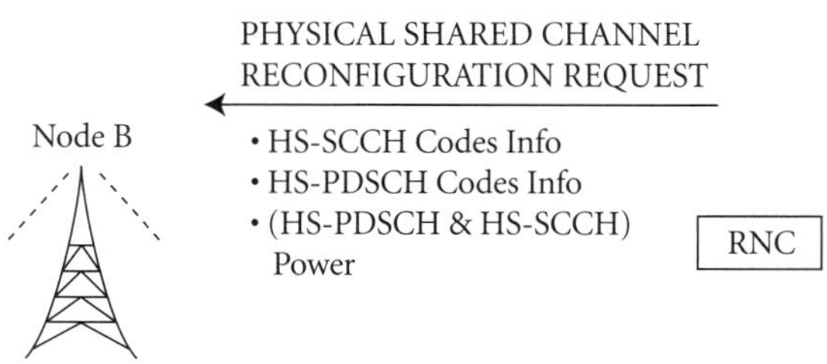

Figure 11.21: Configuration of the Node B for Generic HSDPA Operation

[401] H-RNTI is a sixteen-bit field that uniquely identifies an HSDPA UE in the downlink of an HSDPA serving cell.

[402] The standard [271] offers a discussion of RADIO LINK SETUP REQUEST, RADIO LINK SETUP RESPONSE, and PHYSICAL SHARED CHANNEL RECONFIGURATION REQUEST messages.

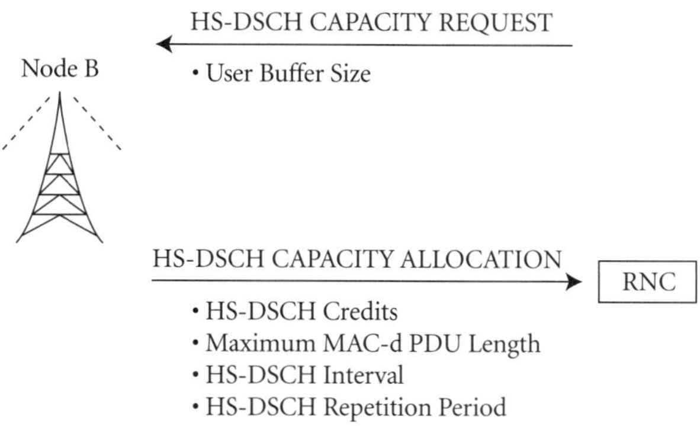

Figure 11.22: Flow-Control–Related Message Exchange for HSDPA

11.8.3.2 Flow Control Between the RNC and the Node B Control of the HSDPA data transfer from the RNC to the Node B prevents buffer overflow and the subsequent packet loss at the Node B. As shown in Figure 11.22, the RNC informs the Node B about the amount of data in the user's buffer via an HS-DSCH CAPACITY REQUEST message so that the Node B learns about the need to transmit such data over the air to minimize latency. The Node B now issues an HS-DSCH CAPACITY ALLOCATION message to the RNC to give the RNC permission to send a certain amount of data to the Node B to prevent overflowing the user's buffer at the Node B. This message includes parameters such as MAC-d PDU length, HS-DSCH credits, and HS-DSCH repetition period.

MAC-d PDU length is the maximum MAC-d PDU size allowed among the configured values. HS-DSCH credits is the number of MAC-d PDUs that the RNC can send during the HS-DSCH interval. An HS-DSCH credit value of 0 means that the RNC should stop transferring MAC-d PDUs to the Node B for the UE, and a value of 2047 means that no limit is placed. The HS-DSCH repetition period is the number of HS-DSCH intervals over which the value of HS-DSCH credits is valid; a value of 0 means that the time period is infinite (i.e., unlimited). The standard [294] offers additional details of the capacity request and allocation messages.

11.8.4 Construction of a Packet at the Node B

The Node B transmits one transport block or one MAC-hs PDU during a TTI of 2 ms. Recall that the Node B receives MAC-d PDUs from the RNC over the Iub interface and constructs a MAC-hs PDU. Figure 11.23 specifies the structure of a MAC-hs PDU.

One MAC-hs PDU can carry multiple logical channels that have the same priority. A MAC-hs PDU consists of a MAC-hs header and a MAC-hs payload. The MAC-hs header includes several fields: VF (i.e., a one-bit version flag), Queue ID, transmission sequence number, and one or more sets of the {SID, N, and F} fields where SID is

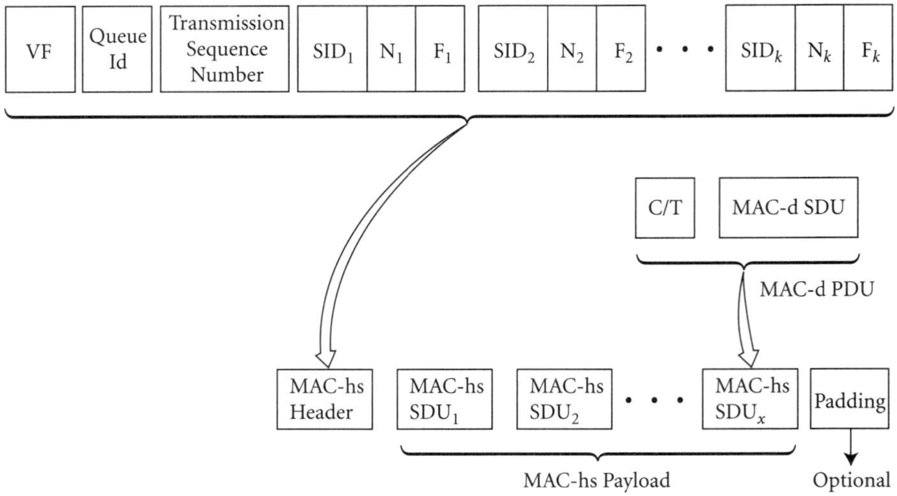

Figure 11.23: Construction of a MAC-hs PDU

size index descriptor, N is the number of MAC-d PDUs of the same size, and F stands for flag. The MC-hs payload incorporates MAC-hs SDUs, which consist of a header specifying the C/T (i.e., control/traffic) field and the MAC-d SDU containing radio link control PDUs from logical channels. VF is set to zero and included for future extensions. Queue ID is a three-bit field that specifies the identity of the re-ordering queue. Hence, when a packet is received, the receiver knows to which queue to forward this packet. Transmission sequence number is a six-bit field that indicates the transmission sequence number of the MAC-hs PDU and helps the receiver re-order the received packets.

One set of {SID, N, F} parameters is associated with one MAC-hs SDU. Since the MAC-hs payload may contain several MAC-hs SDUs, several sets of {SID, N, F} parameters may be required. SID specifies the size of a set of consecutive MAC-d PDUs. $F = 0$ means that the next set of bits represents the {SID, N, F} tuple, and $F = 1$ means that the next set of bits represents an actual MAC-hs SDU. C/T is a four-bit field that specifies the identity of the logical channel the MAC-hs SDU is carrying. C/T is required only if two or more logical channels are multiplexed on the same MAC-d flow. Padding is optional and is included if the sum of the MAC-hs header and MAC-hs payload is less than the TBS. Note that data from only one priority queue can be transmitted during a given TTI because only one Queue ID is present in a MAC-hs PDU. One MAC-hs PDU can carry multiple logical channels of the same priority by multiplexing them on the same MAC-d flow and by using the same priority Queue ID.

Let's understand the process of the MAC-hs PDU construction using two examples. For the first example, assume two MAC-d flows, MAC-d Flow 1 and MAC-d Flow 2.

MAC-d Flow 1 carries just one logical channel (e.g., DCCH). No C/T field is required because no multiplexing of logical channels occurs on MAC-d Flow 1. Assume that the messages to be sent on the DCCH are of high priority. So, MAC-d Flow 1 is assigned a higher-priority class and the Queue ID of 1. Now, MAC-d Flow 2 carries two logical channels, DTCH1 (e.g., for email) and DTCH2 (e.g., for web-browsing). The C/T field is required to differentiate between DTCH1 and DTCH2. Since both are given the same priority due to relaxed latency requirements, they can be assigned same priority class and hence the same Queue ID. The Queue ID of 2 can be assigned to these channels. Both DTCH1 and DTCH2 can be transmitted in the same TTI through multiplexing because they belong to the same Queue ID.

For the second example of the MAC-hs PDU construction, assume two priority queues within the same MAC-d flow. For example, we could have a higher priority for web-browsing and a lower priority for email. In such a case, two separate priority queues for two logical channels, DTCH1 (Queue ID 2) and DTCH2 (Queue ID 3), can be used. However, only one of the logical channels (either DTCH1 or DTCH2) can be transmitted from MAC-d Flow 2 in a given TTI because different priority queues exist for these two logical channels.

11.9 POINTS TO REMEMBER

- HSDPA is an air-interface enhancement specified in 3GPP Release 5. The focus of HSDPA is on the downlink with the peak user data rate reaching about 14 Mbps in the 5 MHz bandwidth for a Category 10 UE.
- Twelve UE categories are defined for the HS-DSCH frequency-division duplexing physical layer.
- HSDPA makes no alteration to peak uplink data rates and maintains Release 99 uplink data rates.
- The overall system architecture remains the same for Release 5 from the perspective of HSDPA.
- The majority of upgrades required for transitioning to HSDPA from R99 are in the RAN, especially the Node B.
- The interfaces, such as Iub, Iu-PS, and Iur, need to be properly provisioned to carry a significantly larger amount of data relative to R99. The UE also needs significant hardware and software upgrades.
- The overall call setup in HSDPA is similar to that in R99 because the basic messages are preserved. Additional fields are included in the original R99 messages to support the HSDPA operation.
- The main air-interface enhancements HSDPA exploits are fast channel feedback from the UE in the form of CQI, the Node B's fast and adaptive scheduling, higher-order modulation of 16-QAM, parallel downlink transmission on

multiple OVSF codes, physical layer H-ARQ, maximal downlink power utilization from the serving cell (instead of power-controlled transmission from the traditional Active Set cells), and flexibility of TDM or CDM scheduling.

CHAPTER 12

1x EVOLUTION–DATA-OPTIMIZED REVISION A

1xEV-DO (1x Evolution–Data-Optimized) Revision A (Rev. A) is an evolution of 1xEV-DO Revision 0 (Rev. 0). Enhancements incorporated in 1xEV-DO Rev. A relative to 1xEV-DO Rev. 0 are higher peak data rates and higher average throughput in the reverse link (RL) and forward link (FL), support for tighter control of latency, support for intra-user quality-of-service (QoS) in addition to the inter-user QoS, more efficient use of the air interface, lower-latency handoff and paging, smoother and faster interworking between 1xEV-DO and CDMA2000 1xRTT,[403] and inherent support for Voice-over-IP (VoIP, and IP is Internet Protocol), push-to-talk, and broadcast and/or multicast services.

In this chapter, a brief overview of 1xEV-DO Rev. A is first given. Different types of physical layer, FL, and RL radio channels are discussed. Rev. A makes session configuration more comprehensive and adds features, such as multiple access terminal (AT) personalities. Session configuration and associated protocols are presented. Overall FL data transmission is described along with enhancements, such as multi-user packets and improvements related to the Data Rate Control (DRC) index and

[403]CDMA represents code-division multiple access. CDMA2000 1xRTT is often referred to as 1xRTT or simply 1x.

Medium Access Control Index (MACIndex).[404] Rev. 0 uses an implicit interference control in the RL, but Rev. A more directly controls RL interference to achieve higher RL throughput. RL data transmission is illustrated by incorporating details of Traffic-to-Pilot power ratio (T2P) based data rate and interference control. Though Rev. 0 is suitable for best-effort services, Rev. A can be used to support real-time services. QoS control mechanisms that enable real-time services are summarized. Compared to Rev. 0, Rev. A reduces traffic interruption because, as the AT moves from the source base station (BS) to the target BS, the target BS is prepared for the AT's imminent arrival, and details of such handoff are given. Enhancements for an AT in the Idle State are described. 1xEV-DO Rev. A must work with legacy technologies, such as 1xEV-DO Rev. 0 and 1xRTT, and such interworking among 1xEV-DO Rev. A, 1xRTT, and 1xEV-DO Rev. 0 is explained.

12.1 INTRODUCTION

A 1xEV-DO Rev. A system is an evolution of 1xEV-DO Rev. 0, one of the stages of evolution of CDMA2000-based[405] systems. *In the rest of the chapter, 1xEV-DO Rev. A is simply referred to as Rev. A.* 1xEV-DO Rev. A is also referred to as TIA-856-A. *Chapter 10 discusses 1xEV-DO Rev. 0 and is a prerequisite for this chapter!* This chapter discusses the enhancements of Rev. A relative to 1xEV-DO Rev. 0 [277,284].

Rev. A is essentially an air-interface standard that focuses on the wireless link between the access terminal and the radio access network.[406] Recall that the radio access network consists of BSs and the radio network controller (RNC). Overcoming 1xEV-DO Rev. 0 deficiencies, increasing the reverse link rate, providing stricter and enforceable QoS control, and offering new services are major forces driving the design and specification of Rev. A. Both software and hardware upgrades are required to enable the Rev. 0 infrastructure to support Rev. A.

Rev. A supports peak user data rates of 3.072 Mbps in the forward link and 1.2288 Mbps or optionally 1.8432 Mbps in the reverse link. Rev. A. is fully backward-compatible with 1xEV-DO Rev. 0. In other words, a 1xEV-DO Rev. 0 AT works transparently in a Rev. A network, and a Rev. A AT utilizes Rev. 0 functions while communicating with a Rev. 0 BS.

Rev. A can efficiently support both asymmetric and symmetric services even when these services are bandwidth-intensive. Assume that you want to share with your friends a video file that has captured a funny moment of your vacation. When you send email with the video file attachment, you will be waiting for a long time after hitting the SEND button on the 1xEV-DO Rev. 0 device because you will get a maximum data rate of just 153.6 kbps. (You will be lucky to get 153.6 kbps. In a typical scenario, the average data rate would be in the tens of kilobits per second.)

[404] We do recognize a bit of inconsistency with the formatting of these two terms, but this is how they appear in the standard, so we want to follow that precedent.

[405] Code-Division Multiple Access 2000.

[406] Relevant changes would also be needed in the core network to support new Rev. A features, such as QoS support.

If you sent the same video file using a Rev. A device, the email would leave the device much sooner. The reverse link data rate of hundreds of kilobits per second (or perhaps even 1.8432 Mbps!) significantly reduces the wait time. Recall that Rev. 0 is designed to support asymmetric and delay-tolerant services with higher forward link bandwidth needs and lower reverse link bandwidths, such as web-browsing.

Rev. A can support delay-sensitive applications, such as voice and streaming, and intra-AT QoS due to various QoS control mechanisms. In a Rev. A system, the QoS for two services with different QoS requirements, such as web-browsing and video streaming, and running on the same AT can be individually controlled per service. In a 1xEV-DO Rev. 0 system, the QoS cannot be (inherently) customized for different services running in the same AT.

Just like Rev. 0, Rev. A is a packet data system that provides connectivity with the Internet and with the services network. Rev. A has no circuit-switched core network. Hence, a voice call made using the Rev. A air interface utilizes VoIP. Rev. A's latency-control features efficiently and feasibly support voice. 1xEV-DO Rev. 0's premise is that voice and data on the same carrier frequency cannot be optimized simultaneously; hence, 1xEV-DO utilizes a carrier frequency separate from the CDMA2000 1x carrier frequency. Now, we are placing voice onto the packet data system! From one perspective, voice in 1xEV-DO appears to be an afterthought. Looking at VoIP in Rev. A from another perspective, voice in 1xEV-DO follows the general trend in next-generation systems (see Chapter 1) where a unified packet-switched core network replaces the disjointed circuit-switched and packet-switched core networks.

Rev. A also enables an operator to offer advanced services, such as broadcast and multicast services in which users can watch regular TV programs and pay-per-view programs such as sports and movies. Rev. A improves packing efficiency by defining short packet sizes to minimize the amount of zero padding and by defining a multi-user packet that can carry packets from multiple ATs. Rev. A also adds numerous features to significantly enhance the reverse link. Examples of these features include Hybrid Automatic Repeat Request (H-ARQ), higher-order modulation, and explicit control of reverse link interference. An equivalent technology is a Third Generation Partnership Project (3GPP) Release 6 feature called high-speed uplink packet access.

A new radio channel means the inter-cell handoff experiences shorter traffic interruption in Rev. A than in Rev. 0. Such enhanced handoff results in reduced latency and higher throughput and facilitates support of delay-sensitive applications. Transition from the Idle State to the Connected State is accelerated in Rev. A by enhancements such as increased monitoring of the control channel. Idle State enhancements aim to reduce paging delay and to conserve the AT's battery power. Interworking between 1xEV-DO and CDMA2000 1x is also smoother in Rev. A due to features such as paging masks and 1x circuit services notification.[407]

Active users share the forward link traffic channel through time-division multiplexing (TDM), and the reverse link utilizes CDMA to enable multiple users to transmit

[407]The 1x circuit services notification was later desginated as a separate standard.

simultaneously. The access network (AN) determines the instantaneous forward link data rate based on the range of data rates the AT estimates to be supportable. The AT determines instantaneous reverse link data rate based on the RL resource the AN allocates.

12.2 OVERVIEW OF REV. A ENHANCEMENTS

To highlight Rev. A's enhancements from the operational perspective, Figure 12.1 depicts typical operations the AT carries out. Assume that our AT has been sleeping (i.e., powered-off) for a long time. Hence, the AN has no record of the AT, and the AT is consuming no radio or core network resources. Let's turn on the AT's power and wake it up. Typical operations the AT carries out are presented here.

The Rev. A AT utilizes the same Rev. 0 method of power-on initialization. The AT looks for the pilot channel in every 1.67 ms[408] slot at a given carrier frequency and waits for a Sync message on the control channel thereafter. The specific carrier frequency the AT seeks depends on the preferred roaming list stored in the AT. After the AT begins to operate in the Rev. A mode upon completing session negotiation, the Rev. A AN can reduce the AT's wait time by using a subsynchronous capsule of Rev. A instead of the typical synchronous capsule of Rev. 0. For example, the AT can observe overhead messages being transmitted on the control channel, such as the AccessParameters message. The subsynchronous capsules help reduce the overall time the AT requires to retrieve the system information that the overhead messages provide and to facilitate faster paging.

For the Rev. A session start-up, the AT and the AN negotiate various protocols and associated parameters for various layers of the seven-layer air-interface protocol stack. Upon power-up, the AT starts using the default Rev. 0 parameters to facilitate such session negotiation. In Rev. A, the concept of multiple *personalities* is intro-

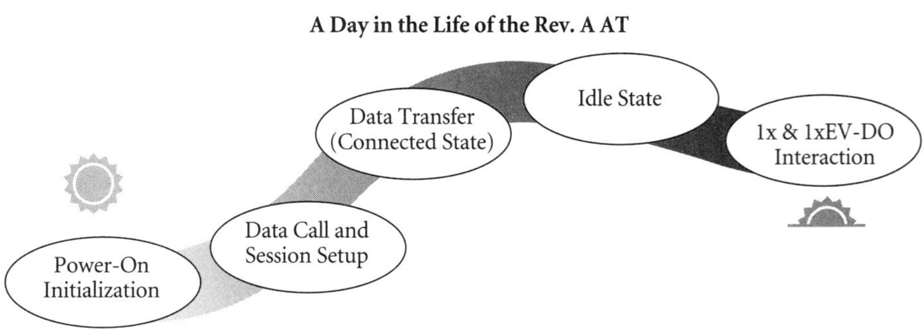

Figure 12.1: Overview of Rev. A AT Operations

[408] The exact amount is $1.\bar{6}$ ms, and the standard uses the term 1.66 . . . ms.

duced; different sets of protocols and/or parameters can be negotiated, and each set corresponds to a different *personality*. Such provisioning of multiple personalities facilitates smoother operation across Rev. 0 and Rev. A boundaries and faster activation of new services or applications. Since Rev. A offers numerous enhancements, multiple personalities can be negotiated to exploit these enhancements. Rev. A also allows changes in some session parameters without tearing down the dedicated radio connection. Section 12.6 discusses session-related enhancements.

The packing efficiency for FL and RL data transmission is improved by allowing transmission of shorter packets, thereby avoiding the need for a significant amount of zero padding. Rev. 0 allows FL data transmission to just one user in a slot, but Rev. A allows packets for up to eight users to be included in a given slot. Such multi-user packets reduce latency and improve packing efficiency. A larger packet size and a suitable amount of redundancy enable Rev. A to increase the peak user data rate to 3.072 Mbps from Rev. 0's 2.4576 Mbps. Rev. A also provides additional flexibility to the scheduling algorithm. For example, in a Rev. 0 system, when the AT conveys a DRC value on the DRC channel, the BS must use the data rate associated with this data rate control if it decides to send data to this AT. A Rev. 0 BS can neither upgrade nor downgrade the data rate for a given reported DRC. In a Rev. A system, one DRC value could correspond to multiple data rates, giving the BS an additional degree of freedom in deciding the data rate for a given AT.

Do You Recall... Do you recall from Rev. 0 how we identify a dedicated radio connection between the AT and the sector? (Hint: The name of the identity that the AN gives the AT starts with an M.)[409]

The number of connected users in a sector can be approximately doubled in Rev. A because Rev. A supports 128 MAC indices. (Rev. 0 supports only 64 MAC indices.) Section 12.7 discusses at length the FL data-transmission–related enhancements. Let's turn our attention to the RL. The RL of Rev. A undergoes a major overhaul compared to the Rev. 0 RL! The RL peak data rate is increased from 153.6 kbps to 1.8432 Mbps as a result of larger packet sizes and higher-order modulation. The average RL throughput of Rev. A can be expected to be much higher than that of Rev. 0 due to the presence of RL H-ARQ. The data rate can be changed in Rev. A as fast as every 6.67 ms.[410] Though the Rev. 0 data rate can be increased by just one step (e.g., from 19.2 kbps to the next step of 38.4 kbps), Rev. A allows sudden changes in the RL data rate.[411] Worried about the sudden increase in RL interference? Relax! Compared to Rev. 0, Rev. A has a more direct method of controlling the RL interference. An active AT is assigned a budgeted amount of RL interference, and,

[409]The MACIndex is the answer!

[410]Again, the exact amount is $6.\bar{6}$ ms, and the standard uses the term 6.66 . . . ms.

[411]According to the standard, the jumps in data rates can be set up as configurable parameters with the default parameter settings corresponding to gradual data rate changes.

as long as the AT stays within this budget, the interference stays within the limit the AN provisioned at that time.

Do You Recall... How fast can the data rate be changed in the Rev. 0 RL—once every frame or every 26.67 ms?[412]

See Section 12.8 for the RL data-transmission–related enhancements. Now, assume that our AT now travels toward another BS and that two BSs are in the AT's Active Set. The AT decides to discontinue getting data from the currently serving sector or BS and wants to resume FL data transfer from the new BS. In Rev. 0, the resumption of FL data transfer with a change in the FL serving BS requires a long time period (e.g., 200 ms or so) because the new BS lacks the AT's FL data. Rev. A introduces a new radio channel in the RL (called the Data Source Control channel) so that the new BS is ready to transmit data to the AT as soon as the AT selects its sector as the new FL serving sector. Handoff improvements are the focus of Section 12.13.

Switching gears to QoS control, Rev. 0 can easily provide inter-user QoS through suitable FL and RL scheduling. However, the Rev. 0 standard fails to adequately address the issue of intra-user QoS in which multiple applications (e.g., VoIP and web-browsing) running on a given AT need different QoS levels. Rev. A overcomes Rev. 0's shortcomings by defining new Radio Link Protocol (RLP) and MAC features. The Rev. A QoS control is discussed in Section 12.12.

Assume that our AT has worked pretty hard, uploading and downloading a lot of data. Now the AT is resting. In the absence of any data activity for a certain time period, the radio connection is closed and the AT enters the Idle State. In the Idle State, the AT periodically wakes up and looks for the Page message on the control channel. In Rev. 0, the AT wakes up every 5.12 seconds. When a data packet destined for the AT arrives at the AN, the AN must send a page to the AT. However, in Rev. 0, the paging delay could be as long as 5.12 seconds. Rev. A reduces such paging delay by using subsynchronous capsules and defining a three-tiered monitoring cycle. The AT wakes up more frequently after initially entering the Idle State and less frequently as time goes by. Sub-synchronous capsules are needed to implement the three-tiered monitoring cycle. Another Idle State enhancement is the ability to transfer a small amount of data without establishing a dedicated radio connection. See Section 12.14 for Idle-State–related enhancements.

Let's assume that the AT is leaving the Rev. A. coverage area and entering the Rev.-0–only coverage area. The multiple personality feature enables the AT to seamlessly move across Rev. 0 and Rev. A boundaries. Since a suitable Rev. 0 personality has already been negotiated, re-negotiation of the 1xEV-DO session is unnecessary even while moving across Rev. 0 and Rev. A boundaries. The interworking between Rev. A and CDMA2000 1xRTT is improved with the introduction of the paging mask concept. The AT can jump from the 1xEV-DO carrier frequency to the 1xRTT

[412]The data rate in Rev. 0 can be changed every 26.67 ms, or, more specifically, $26.\bar{6}$ ms (and the standard uses 26.66 . . . ms).

carrier frequency during the paging mask period.[413] The 1xEV-DO AN sends no page to the AT during the paging mask duration. Such provisioning of the paging mask reduces the page loss probability, translating into shorter paging delay. Section 12.15 illustrates technology-interworking enhancements of Rev. A.

12.3 CHARACTERISTICS OF PHYSICAL LAYER SUBTYPES

Rev. A enables higher peak and average data rates than Rev. 0 rates through numerous enhancements, some of which are designed at the physical layer. We will first provide an overview of the main physical layer subtypes and then dig into the details of the reverse link, in which Rev. A has made major physical layer changes compared to Rev. 0.

12.3.1 Overview of Physical Layer Subtypes

Though Rev. 0 defines just one type of physical layer, Rev. A defines *three subtypes of the physical layer*. Subtype 0 is nothing but the Rev. 0 physical layer! After power-on, the AT begins using this default physical layer subtype.

Do You Recall... Do you remember the peak data rates of the Access channel, Forward Traffic channel, and Reverse Traffic channel in Rev. 0?[414]

The lowest nominal data rates for the Forward Traffic channel and the Reverse Traffic channel (RTC) are 38.4 kbps and 9.6 kbps, respectively. The forward link transmission occurs in the units of 1.67 ms time-slots, and the reverse link transmission occurs in the units of 26.67 ms frames.

Subtype 1 enhances the access channel operation by supporting three data rates—9.6, 19.2, and 38.4 kbps—and by providing means of optimizing the transmit power of the access probe. The maximum Access channel data rate of a given AT depends on the maximum data rate, *SectorAccessMaxRate*, that the sector supports and the AT's capability specified by *TerminalAccessRateMax*. The SectorAccessMaxRate is assumed to be 9.6 kbps unless the AccessParameters message specifies a different value. *TerminalAccessRateMax* is a configurable parameter negotiated as part of session configuration and has a default value of 9.6 kbps. The peak data rates on the Forward Traffic channel and the Reverse Traffic channel for this subtype are the same as those for Subtype 0. As in Subtype 0, the forward link transmission occurs in 1.67 ms time-slots, and the reverse link transmission occurs in 26.67 ms frames.

The operation of the Access channel for Subtype 2 is the same as that for Subtype 1. For example, possible Access channel data rates are 9.6 kbps, 19.2 kbps,

[413] Though Rev. 0 allows the AT to suggest a paging offset to the AN so that the AN would not page the AT when the AT is on 1x, Rev. A makes this mechanism more sophisticated to optimize the tune-away intervals.

[414] 9.6 kbps, 2.4576 Mbps, and 153.6 kbps.

and 38.4 kbps. The AT transmits the Reverse Traffic channel at a variable data rate ranging from 4.8 kbps to 1.2288 Mbps (or 1.8432 Mbps). If the AT supports 8-PSK[415] modulation, the highest RL data rate is 1.8432 Mbps. Otherwise, the peak supportable RL rate is 1.2288 Mbps because the AT supports at least QPSK[416] modulation. The attribute, *Rate1M8Supported*, can be configured during the session negotiation to indicate the AT's support for 1.8432 Mbps. The default value of Rate1M8Supported is 0, indicating that the AT provides *no* support for 1.8432 Mbps on the Reverse Traffic channel. The nominal data rate on the Forward Traffic channel ranges from 4.8 kbps to 3.072 Mbps. In practice, Subtype 0 and Subtype 2 are widely used.

12.3.2 Reverse Link Frame Structure of Subtype 2 Physical Layer

The reverse link frame structure of this subtype is quite different from the Rev. 0 frame structure.

Do You Recall... Do you remember the transmission time unit in the reverse link of Rev. 0?[417]

Rev. A introduces the concept of a subframe as shown in Figure 12.2. One subframe consists of four slots with each slot occupying 1.67 ms. In other words, one subframe is 6.67 ms long. One reverse link physical layer frame has four subframes. A physical layer packet is represented by one, two, three, or four subpackets, and one subframe carries just one subpacket. The Subtype 2 Physical Layer enables the reverse link data rate to change every subframe (i.e., 6.67 ms) instead of every frame (i.e., 26.67 ms). When a physical layer packet is transmitted in the form of multiple subpackets, a three-subframe interlacing is used; two intervening subframes exist between the successive subframes carrying a subpacket of a given packet.

Assume that the AT has three physical layer packets, P, Q, and R. In Figure 12.2, the physical layer packet P is represented by four subpackets. The first subpacket of P is sent during subframe 1 of the RL Frame 1, and the second subpacket of P is sent in subframe 4 of the RL frame 1, assuming that the AT receives a H-ARQ negative acknowledgment (NAK) from the AN. The subframe interlacing provides the AN with adequate time to respond to the received RL subpacket via H-ARQ acknowledgments (ACKs) or NAKs. If the AT receives an ACK, no remaining subpackets of the physical layer packet P are transmitted because the AN has successfully decoded the packet P. The intervening two subframes carrying subsequent subpackets of a given physical layer packet P can be used to send subpackets of different physical layer packets Q and R.

A configurable attribute, MaxNumSubPackets *I* where *I* is the RL interlace offset, constrains the maximum number of subpackets. There are three RL interlaces. For

[415] 8-ary Phase Shift Keying
[416] Quadrature Phase Shift Keying
[417] The transmission time unit is 26.67 ms in the reverse link of Rev. 0.

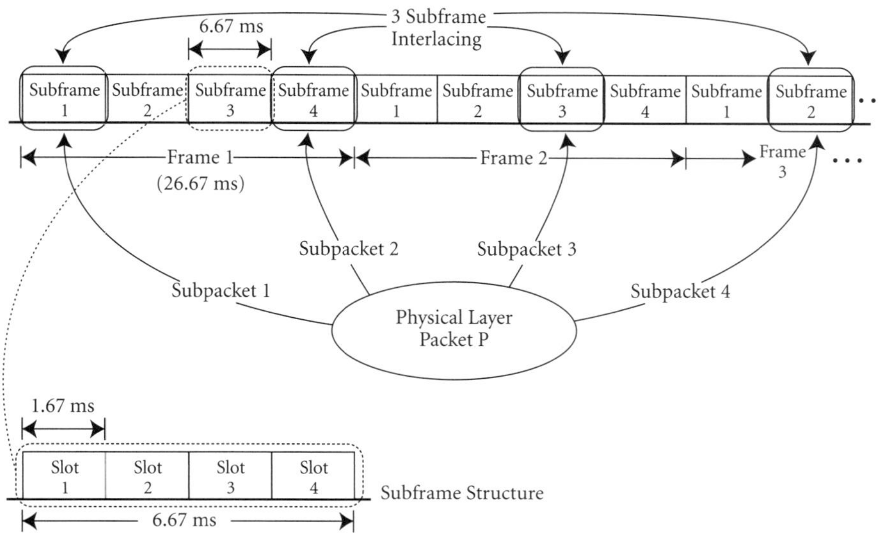

Figure 12.2: Reverse Link Frame Structure and Packet Transmission

example, in Figure 12.2, RL Interlace 0 consists of subframe 1 of RL Frame 1, subframe 4 of the RL Frame 1, subframe 3 of the RL Frame 2, and so on. Similarly, RL Interlace 1 consists of subframe 2 of RL Frame 1, subframe 1 of the RL Frame 2, subframe 4 of the RL Frame 2, and so on. The default value of MaxNumSubPackets$I$ is 4 subpackets for all the interlace offsets I. Furthermore, each active MAC flow is associated with a specific termination target for a set of payload sizes.

12.4 REV. A FORWARD LINK CHANNELS

We will discuss FL channels here. Section 12.4.1 provides an overview of all FL channels that are available in both Rev. 0 and Rev. A. Section 12.4.2 focuses on the new FL channel defined in Rev. A.

12.4.1 Overview of FL Channels

Figure 12.3 summarizes the four FL channels. As in Rev. 0, the Forward Pilot channel is used for coherent detection, synchronization, sector identification, FL channel quality determination, and handoff measurements. The Forward Control channel carries overhead messages, such as the SectorParameters message, and some AT-specific messages, such as the UATIAssignment message (where UATI represents the Unicast Access Terminal Identifier). The Forward Traffic channel represents a dedicated radio connection for one or more ATs in a given slot and carries user traffic

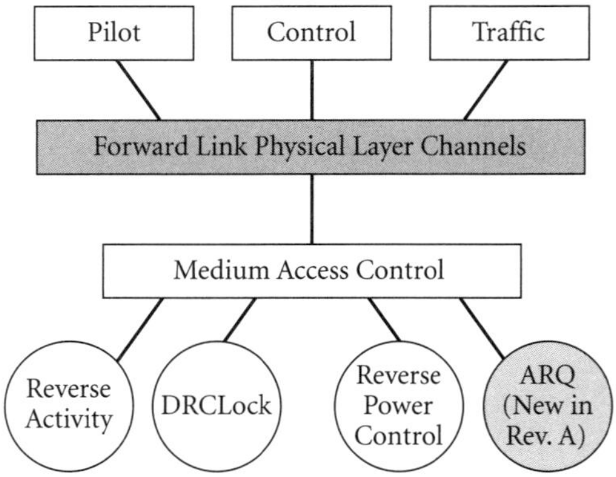

Figure 12.3: Rev. A Forward Link Channels

packets and user-specific signaling messages. The Forward MAC channel consists of four channels: the Reverse Activity (RA) channel, DRCLock channel, Reverse Power Control channel, and ARQ channel. The RA channel indicates the presence or absence of high RL interference. The DRCLock channel indicates the sector's current ability to read the AT's DRC channel. The Reverse Power Control channel carries the power-control commands for the connected ATs to influence their transmit power levels. The ARQ channel is newly defined in Rev. A to support the RL H-ARQ. In other words, the ARQ channel carries an ACK or NAK in response to the packet received from the AT.

Figure 12.4 shows a high-level view of how various FL channels are combined for FL transmission. See the standard [295] for a detailed view of the simplified block diagram shown in Figure 12.4, which includes the in-phase (I) channel and the quadrature-phase (Q) channel.

Recall from Chapter 10 that the FL in Rev. 0 utilizes TDM in a 1.67 ms slot to combine the Traffic or Control channel, the Pilot channel, the preamble (if required), and the MAC channel. The Rev. 0 FL also uses CDM to combine various component channels of the MAC channel, such as reverse power control (RPC) or DRCLock channels and the RA channel. Figure 12.4 shows how various channels are combined in a slot carrying the Forward Traffic channel.

12.4.1.1 Forward Traffic Channel Processing In a given 1.67 ms slot, the Forward Traffic channel (FTC) carries a subpacket that the H-ARQ process constructs for a given physical layer packet.

Do You Recall... Is the preamble sent in *every* forward link slot in Rev. 0?[418]

[418] In Rev. 0 and Rev. A, the preamble is sent in the slot only if the slot is carrying a new packet. If the slot carries the continuing retransmission of a previous physical layer packet, no preamble is sent.

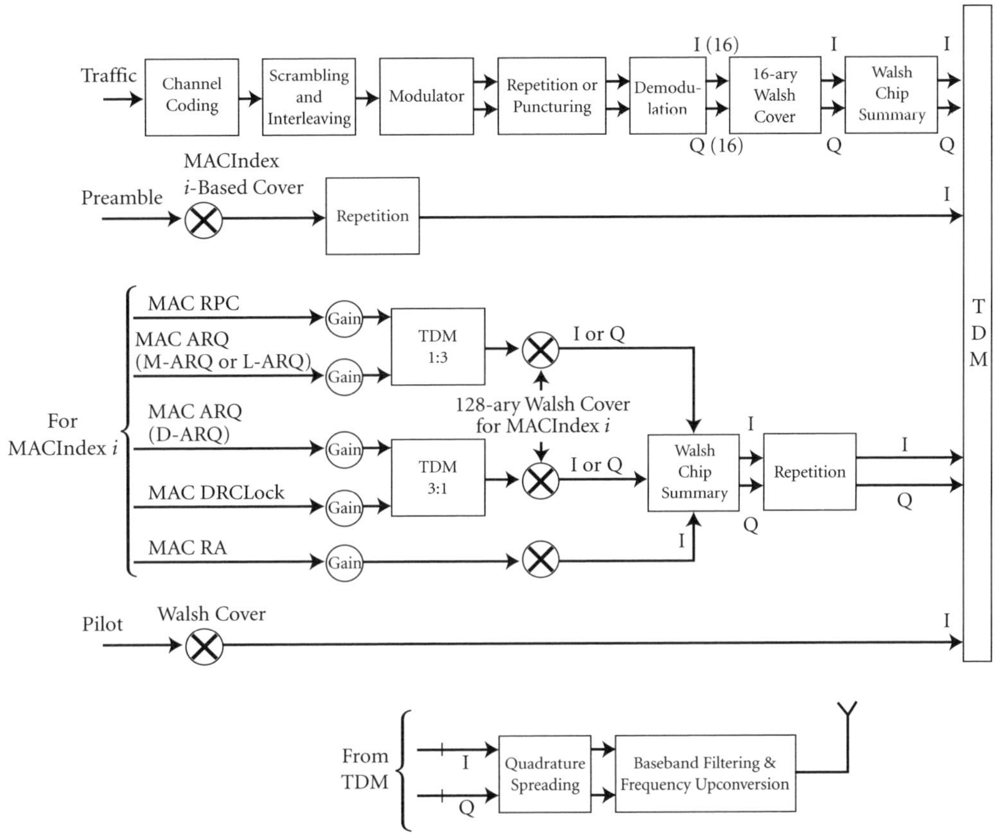

Figure 12.4: Combining of the Rev. A Forward Link Channels

The H-ARQ process constructs the FTC chips for transmission by performing the following basic steps according to the transmission format the AN scheduler selects. (In essence, the transmission format specifies the physical layer packet size, the coding rate, the modulation scheme, and the amount of repetition or puncturing. See Section 12.7 for details of the transmission format.)

The first step is channel coding. Turbo coding with a rate of 1/3 or 1/5 is used. The second step involves scrambling the turbo code symbols with a scrambling sequence to randomize data. The scrambling itself is simply an XOR operation between the scrambling sequence and the code symbols. The scrambling sequence is generated using the generator polynomial, $h(D) = D^{17} + D^{14} + 1$. The shift register implementing the generator polynomial is initialized with a value that is a function of the MACIndex and the transmission format associated with the packet. Such initialization occurs at the beginning of the transmission of a new physical layer packet.

In the next step, the scrambled code symbols are interleaved using channel interleaving. Two basic channel interleaving structures are defined, one for the rate 1/3 channel coding and another for rate 1/5 channel coding. Three modulation schemes are supported for the FTC in Rev. A, just like Rev. 0.

Do You Recall. . . Do you remember those modulation schemes?[419]

The next step is repetition or puncturing. In a 1xEV-DO forward link, the modulation symbol is equivalent to one chip because no real spreading of the modulation symbols occurs. The number of modulation symbols to be placed in the FTC portion of the slot could be more or less than the number of modulation symbols the digital modulator provided in the previous step. Hence, the H-ARQ process suitably repeats or punctures the modulation symbols.

Finally, the set of modulation symbols are now demultiplexed into sixteen in-phase and sixteen quadrature-phase parallel streams. Recall that a modulation symbol is simply a complex number, $(a + jb)$, a being the in-phase component and b being the quadrature-phase component. Each stream has the rate of 76.8 ksps and is covered with a distinct sixteen-chip Walsh function, resulting in the per-stream chip rate of 1.2288 Mcps. The Walsh chips from all sixteen in-phase streams are summed to create one in-phase stream, and the Walsh chips from all sixteen quadrature-phase streams are summed to create one quadrature-phase stream. During the chip period, all sixteen streams contribute one chip to the summing operation for a given phase combined stream. Hence, the per-stream chip rate and the combined stream chip rate are both 1.2288 Mcps.

[419]The supported FTC modulation schemes are QPSK, 8-PSK, and 16-Quadrature Amplitude Modulation.

12.4.1.2 Preamble Processing The preamble is sent during a 1.67 ms slot for a new physical layer packet being sent on the FTC or the Control channel. The preamble consists of all 0s with 0 mapped onto +1. The preamble is subsequently covered (i.e., multiplied) by a 64-chip bi-orthogonal Walsh cover. The resulting sequence is repeated as necessary depending upon the transmission format of the associated physical layer packet. The 64-chip Walsh cover is defined as

$$\begin{array}{ll} W_{i/2}^{64} & \text{for } i = 0, 2, \ldots, 126 \\ \overline{W_{(i-1)/2}^{64}} & \text{for } i = 1, 3, \ldots, 127 \end{array} \quad (12.1)$$

where i is the MACIndex value and $\overline{W_j^{64}}$ is the bit-by-bit complement of the Walsh function W_j^{64}.

12.4.1.3 MAC Channel Processing Recall that the MAC channel consists of the RPC channel, the DRCLock channel, the RA channel, and the ARQ channel.

Do You Recall... Do you remember which channel is absent from Rev. 0? What benefit does this channel offer Rev. A? [420]

The Forward MAC channel is constructed using the Walsh channels that are orthogonally covered by MACIndex-dependent Walsh functions. Table 12.1 summarizes the specifications with which the MAC channels are transmitted. These specifications include the Walsh function on which the channel is transmitted, the time-slot during which the channel is transmitted, and the branch for the channel. In Table 12.1, the slots T_1, T_2, T_3, and T_4 are four consecutive slots, each with the duration 1.67 ms. The slot T_1 is defined as slot T such that $(T - \mathit{FrameOffset}) \mod 4 = 3$ where T is the CDMA system time in slots and $\mathit{FrameOffset}$ is the time offset of the Reverse Traffic Data channel relative to the system time-aligned reverse link frame boundary.

Assume that a Rev. A AT is assigned MACIndex $i = 11$. The RPC channel for this AT is sent on the Q channel during the time-slot T_1 using the Walsh function

$$W_{((i-1)/2)+32}^{128} = W_7^{128}.$$

The DRCLock channel for this AT is sent on the I branch during the time-slot T_1 using the Walsh function

$$W_{((i-1)/2)+32}^{128} = W_7^{128}.$$

The RA channel is intended for all ATs and is transmitted in all slots on the I branch using the Walsh function

$$W_{i/2}^{128} = W_2^{128}.$$

[420]Rev. 0 lacks the ARQ channel, which allows Rev. A to support H-ARQ for the RL.

Table 12.1: MAC Channel Transmission Specifications

Channel	Associated Walsh Function for MACIndex i	Transmission Slot	Branch
RPC	$W^{128}_{((i-1)/2)+32}$ for $i = 5, 7, \ldots, 63$	T_1	Q
	$W^{128}_{i/2}$ for $i = 6, 8, \ldots, 62$	T_1	I
	$W^{128}_{((i-1)/2)+64}$ for $i = 73, 75, \ldots, 127$	T_1	Q
	$W^{128}_{(i/2)+32}$ for $i = 72, 74, \ldots, 126$	T_1	I
DRCLock	$W^{128}_{((i-1)/2)+32}$ for $i = 5, 7, \ldots, 63$	T_1	I
	$W^{128}_{i/2}$ for $i = 6, 8, \ldots, 62$	T_1	Q
	$W^{128}_{((i-1)/2)+64}$ for $i = 73, 75, \ldots, 127$	T_1	I
	$W^{128}_{(i/2)+32}$ for $i = 72, 74, \ldots, 126$	T_1	Q
RA	$W^{128}_{i/2}$ for $i = 4$	T_1, T_2, T_3, T_4	I
ARQ (H-ARQ or L-ARQ bits)	$W^{128}_{((i-1)/2)+32}$ for $i = 5, 7, \ldots, 63$	T_2, T_3, T_4	Q
	$W^{128}_{i/2}$ for $i = 6, 8, \ldots, 62$	T_2, T_3, T_4	I
	$W^{128}_{((i-1)/2)+64}$ for $i = 73, 75, \ldots, 127$	T_2, T_3, T_4	Q
	$W^{128}_{(i/2)+32}$ for $i = 72, 74, \ldots, 126$	T_2, T_3, T_4	I
ARQ (P-ARQ bits)	$W^{128}_{((i-1)/2)+32}$ for $i = 5, 7, \ldots, 63$	T_2, T_3, T_4	I
	$W^{128}_{i/2}$ for $i = 6, 8, \ldots, 62$	T_2, T_3, T_4	Q
	$W^{128}_{((i-1)/2)+64}$ for $i = 73, 75, \ldots, 127$	T_2, T_3, T_4	I
	$W^{128}_{(i/2)+32}$ for $i = 72, 74, \ldots, 126$	T_2, T_3, T_4	Q

When required, the H-ARQ and Last ARQ (L-ARQ) bits of the ARQ channel for this AT are sent on the Q branch during the slots T_2, T_3, and T_4 using the Walsh function

$$W^{128}_{((i-1)/2)+32} = W^{128}_{7}.$$

Furthermore, when required, the Packet ARQ (P-ARQ) bits of the ARQ channel for this AT are transmitted on the I-branch during the slots T_2, T_3, and T_4 using the

Walsh function

$$W^{128}_{((i-1)/2)+32} = W^{128}_{7}.$$

Though the transmit power of a given MAC Walsh channel can vary as a function of time, the cumulative transmit power of the channels forming the MAC channel must be constant. This total MAC channel power is the same as the pilot power (i.e., equal to the target maximum sector power).

12.4.1.4 RPC Channel The RPC channels for the active users that have been allocated the even-numbered MACIndex values are mapped onto the I-branch or I-signal. Conversely, the RPC channels for the active users that have been allocated the odd-numbered MACIndex values are mapped onto the Q-signal. The RPC channel is time-division-multiplexed with the ARQ channel such that the RPC channel is sent in one of the four slots, and the ARQ channel is sent in the remaining three of the four slots. Since one slot carries one new power-control command for an active AT, the power-control frequency is $1/1.67$ ms $= 600$ Hz. In other words, the data rate of the RPC channel is 600 bits per second.[421] One power-control command is represented by one RPC bit. The RPC bit of `0` is mapped onto $+1$ for spreading with a suitable Walsh function and corresponds to the *power-up* command for the AT. The RPC bit of `1` is mapped onto -1 for spreading with a suitable Walsh function and corresponds to the *power-down* command for the AT. The actual power step size is a session-configurable parameter. Since a Walsh function consisting of 128 chips represents one RPC bit and since the MAC RPC channel occupies 256 chips, the same RPC bit is sent twice in a given slot.

12.4.1.5 DRCLock Channel The DRCLock channel is time-division-multiplexed with ARQ channel such that one of four slots carries the DRCLock channel and three of four slots carry the ARQ channel. The DRCLock channel for the ATs assigned even-numbered MACIndex values are mapped to the Q-phase, and the DRCLock channel for the ATs assigned odd-numbered MACIndex values are mapped to the I-signal. During a slot associated with the DRCLock channel transmission (i.e., T_1 from Table 12.1), the DRCLock bit is sent twice because the 128-chip Walsh function spreads the DRCLock bit and the MAC channel is 256 chips long. The DRCLock channel's data rate is $600/\mathit{DRCLockLength}$ bits per second. DRCLockLength specifies the number of times the same DRCLock bit must be sent and assumes one of four values, eight, sixteen, thirty-two, or sixty-four slots, sixteen slots being the default value. Assume that $\mathit{DRCLockLength} = 16$ is used. If a new DRCLock bit value X is sent in slot N, the next set of $\mathit{DRCLockLength} - 1 = 16 - 1 = 15$ slots, $(N+4), (N+8), (N+12), \ldots, (N+60)$, all contain the same value X on the DRCLock channel.

[421] The exact data rate of the RPC channel is less than 600 Hz because the DRCLock and ARQ channels steal some RPC bits.

Do You Recall... How does the AT use the value of the DRCLock bit?[422]

Note that the SofterHandoff field denotes whether the sectors are transmitting the same DRCLock bit. The AT should perform maximal ratio combining of the DRCLock channels being sent from sectors of the same cell because those sectors are carrying the same DRCLock bit for a given AT. Formally, the DRCLock bit of 0 means "Out of Lock" and is mapped onto value $+1$ for spreading by a relevant Walsh channel function. The DRCLock bit of 1 means "In Lock" and is mapped onto value -1 for spreading by a relevant Walsh channel function.

As shown in Figure 12.4, each channel or subchannel of the MAC channel is Binary Phase Shift Keying (BPSK) or OOK (On-Off Keying) modulated on the in-phase or quadrature-phase of one of 128 128-ary Walsh covers. The resulting MAC symbol Walsh chips are transmitted twice in a slot in four bursts of sixty-four chips. Though the total transmit power of the MAC channel is the same as that of the Pilot channel, relative power levels of the subchannels of the MAC channel may be constant or variable based on vendor-specific implementations. Variability in relative power levels of the MAC subchannels may be achieved per channel and/or per user.

12.4.1.6 ARQ Channel The ARQ channel is the only new channel introduced in the FL of Rev. A. See Section 12.4.2 for details of this channel.

12.4.1.7 Pilot Channel Processing The Pilot channel consists of all 0s with 0 mapped to $+1$. It is covered by Walsh Cover 0 and is transmitted on the I channel only.

12.4.1.8 TDM, Quadrature Spreading, Filtering, and Frequency Upconversion The resulting in-phase and quadrature-phase chips of the preamble, the FTC, the Pilot channel, and the MAC channel are time-division-multiplexed as shown in Figure 12.4 for the quadrature spreading operation. Let's denote the combined in-phase chips by I and the quadrature-phase chips by Q, respectively. The I-channel and Q-channel generator polynomials for the quadrature spreading are defined to construct the pilot pseudo-noise (PN) sequences, PN_I and PN_Q. The PN sequences have a length of $2^{15} = 32768$ chips. In other words, they are repeated every 26.67 ms frame. The Pilot channel is defined with an offset relative to the zero offset PN sequence. The pilot PN offset is specified by an integer 0 through 511 with the units of sixty-four chips. For example, a pilot PN offset of 2 corresponds to a pilot PN sequence that lags the zero-offset pilot PN sequence by $2 * 64 = 128$ chips. The

[422]If the AT receives $DRCLock = 0$ from an Active Set cell, the AT should avoid pointing its DRC at any of that cell's Active Set sectors. The AN attempts to decode the DRC values the AT sends on the DRC channel. If the AN determines that it is unable to reliably detect the DRC values, it considers the RL to be of poor quality. The AN uses the DRCLock channel to convey to the AT the RL's status. $DRCLock = 1$ implies a reliable RL, and the AT can then start putting non-null DRC values toward the appropriate sector.

quadrature spreading is a complex multiply operation as specified in Eq. (12.2).

$$
\begin{aligned}
I' &= I\,PN_I - Q\,PN_Q \\
Q' &= I\,PN_Q - Q\,PN_I
\end{aligned} \tag{12.2}
$$

After the quadrature spreading, the resulting I′ and Q′ signals are processed by the I- and Q-baseband filters. Finally, they are upconverted to the RF.

12.4.2 ARQ Channel: A New Forward Link Channel in Rev. A

Rev. A introduces *only one new FL physical channel*, the ARQ channel. The AN transmits the ARQ channel to the AT for which the Subtype 2 Physical Layer has ben enabled. The basic function of the ARQ channel is to support RL H-ARQ. In other words, the AN sends a physical layer positive or negative acknowledgment on the ARQ channel in response to a packet received from the AT. However, instead of using just one type of ACK or NAK, the Rev. A RL H-ARQ uses three types of ARQ bits, the H-ARQ, L-ARQ, and P-ARQ, yielding higher reliability and more flexibility. Recall that the Subtype 2 Physical Layer includes four subframes, and one subframe carries one subpacket of a physical layer packet. Also recall that soft and softer handoff is supported in the RL in 1xEV-DO Rev. 0 and Rev. A. Each sector in the AT's Active Set transmits the ARQ channel. Two sectors belonging to two different cells may quite possibly carry different ARQ values. Assume that physical layer packet P's subpackets are identified as subpacket 0, subpacket 1, subpacket 2, and subpacket 3. The H-ARQ bit is sent in response to the received subpackets 0, 1, and 2. The L-ARQ bit is sent in response to the received subpacket 3. The P-ARQ bit is sent in response to the received physical layer packet P. Consider the scenario illustrated in Figure 12.5.[423]

The AT starts the transmission of a physical layer packet P during slot m of subframe n, which belongs to RL interlace 0. In Figure 12.5, $MaxNumSubPackets0 = 3$. MaxNumSubPackets0 is assumed to be 3, meaning that $MaxNumSubPackets0 + 1 = 4$ subpackets are present for the physical layer packet P. In other words, the AT will transmit a maximum of four subpackets. The AN receives subpacket 0 but is unable to correctly decode packet P. Hence, it sends a NAK as the H-ARQ bit on the ARQ channel. Observe the timing of the H-ARQ bit relative to the timing of the subpacket. If a subpacket transmission begins in the RL slot m, the H-ARQ bit transmission starts in the FL slot $m + 8$. Since the subpacket is transmitted in the entire subframe, the subpacket occupies the slots $\{m, m + 1, m + 2, m + 3\}$. The H-ARQ bit transmission occupies three FL slots, $\{m + 8, m + 9, m + 10\}$. Since the H-ARQ bit is a NAK, the AT continues transmission of packet P by sending subpacket 1 in subframe $n + 3$ in RL Interlace 0.

[423]Tx represents "transmit."

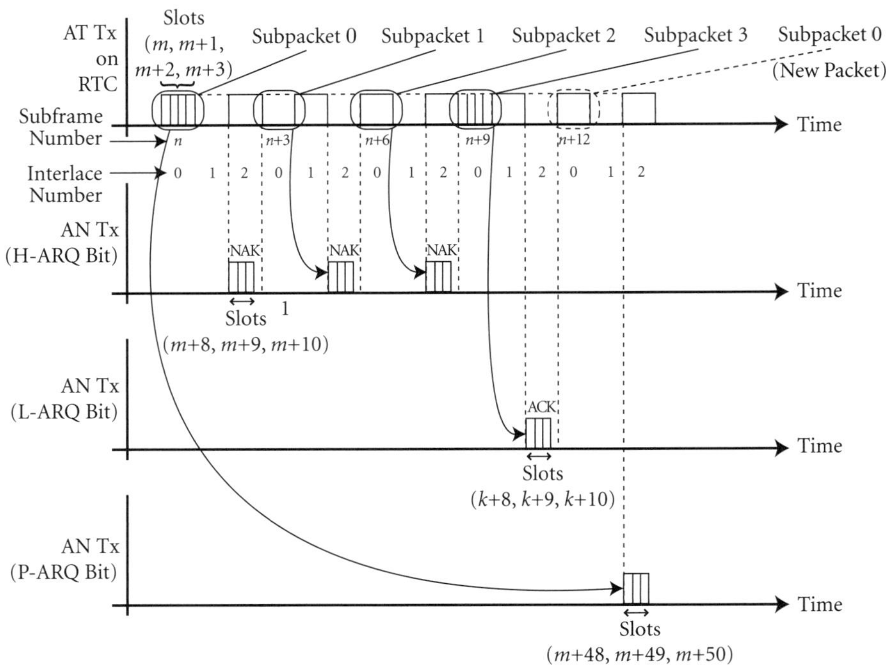

Figure 12.5: Transmission Timings of the RL Hybrid ARQ Bits

A Question for You... Why does the AT transmit packet P's subpacket 1 in the subframe $n + 3$ instead of the subframe $n + 1$?[424]

The AN now combines subpacket 0 with subpacket 1 in an attempt to recover packet P. Since it fails to decode packet P, it sends a NAK as the H-ARQ bit. The AT now sends subpacket 2. Again, the AN is unable to decode packet P and sends a NAK as the H-ARQ bit. Now, the AT sends the last subpacket, subpacket 3, to the AN. Observe that the AN sends no H-ARQ bit! Remember the rule: the H-ARQ bit is sent in response to all the subpackets *except* the last subpacket! Since the AT has indeed sent the LAST subpacket, the AN sends the L-ARQ bit. If the last subpacket is transmitted in the slots $\{k, k+1, k+2, k+3\}$, the L-ARQ bit is sent in the FL slots $\{k+8, k+9, k+10\}$. Since the AN has now successfully decoded packet P, an ACK is sent as the L-ARQ bit. Finally, the P-ARQ bits are sent in the FL slots $\{m+48, m+49, m+50\}$ in response to the packet transmission that started in the slot m.

[424] The delay provides adequate time for the feedback to support H-ARQ.

12.4.2.1 H-ARQ Bit of the ARQ Channel

An Active Set sector sends the H-ARQ bit to the AT in response to all the subpackets of the physical layer packet except the last one. Assuming that MaxNumSubPackets$I$ is 3, the sector sends the H-ARQ bit if the AT sends subpacket 0, 1, or 2 to the AN. If the AT sends no given subpacket number (because the AN successfully received the packet), then the AN sends no H-ARQ bit. If the Active Set *cell* correctly decodes the physical layer packet, the Active Set sectors of the cell send an ACK as the H-ARQ bit. If the Active Set cell is unable to decode the packet, the Active Set sectors of the cell send a NAK as the H-ARQ bit. No H-ARQ bit is transmitted in response to the last subpacket, subpacket 3.

The value of the H-ARQ bit a sector transmits depends on whether the sector is part of the cell that includes the serving sector and on the parameter ARQMode. A sector transmits the H-ARQ bit using ACK-oriented OOK if the sector is excluded from the serving cell. The ACK-oriented OOK is implemented by transmission of $+1$ to imply an ACK or transmission of 0 (i.e., no transmission at all) to imply a NAK.

An ARQMode of 0 means that the H-ARQ bit transmitted by a sector that is part of the serving cell uses Bi-Polar Keying, and an ARQMode of 1 means that the H-ARQ bit transmitted by a sector that is part of the serving cell uses ACK-oriented OOK. The Bi-Polar Keying involves transmission of $+1$ to convey an ACK and transmission of -1 to convey a NAK.

When the H-ARQ bit is transmitted for an even-numbered MACIndex, it uses the I-branch, and for an odd-numbered MACIndex, the Q-branch is used.

12.4.2.2 L-ARQ Bit of the ARQ Channel

The L-ARQ bit is transmitted only if the last subpacket is received. An ACK is sent if the physical layer packet is correctly received. A NAK is sent if the packet is incorrectly received. When the AT transmits the last (i.e., $MaxNumSubPacketsI$ + 1th) subpacket of a physical layer packet in slots $\{k, k+1, k+2,$ and $k+3\}$, an Active Set sector transmits the ACK or NAK in slots $\{k+8, k+9, k+10\}$. The sector transmits the L-ARQ bit using NAK-oriented OOK where 0 indicates an ACK and -1 represents a NAK.

The L-ARQ bit for the AT with an odd-numbered MACIndex is sent on the Q-branch, and the L-ARQ bit for the AT with an even-numbered MACIndex is sent on the I-branch.

12.4.2.3 P-ARQ Bit of the ARQ Channel

The P-ARQ bit is sent in response to the correct or incorrect reception of the physical layer packet. An Active Set sector transmits the P-ARQ bit in the slots $\{m+48, m+49, m+50\}$ in response to the first subpacket transmission in slots $\{m, m+1, m+2, m+3\}$; four subpackets are assumed to be present in the given interlace. In more generic terms, an Active Set sector transmits the P-ARQ bit in slots $\{m+((MaxNumSubPacketsI+1)\times 12), m+1+((MaxNumSubPacketsI+1)\times 12), m+2+((MaxNumSubPacketsI+1)\times 12), m+3+((MaxNumSubPacketsI+1)\times 12)\}$ where $MaxNumSubPacketsI$ is the maximum number of subpackets for interlace I. The P-ARQ bit is transmitted using NAK-oriented OOK, with `0` implying an ACK and `1` implying a NAK.

The P-ARQ bit for an even-numbered MACIndex is assigned to the quadrature branch, and the P-ARQ bit for an odd-numbered MACIndex is assigned to the in-phase branch. The AT should perform diversity combining of the ARQ bits for the Active Set sectors corresponding to the same cell because those sectors are transmitting the same ARQ bit (i.e., H-ARQ, L-ARQ, or P-ARQ).

A Question for You. . . How do you think the AT knows if two sectors belong to the same cell?[425]

12.5 OVERVIEW OF REV. A REVERSE LINK CHANNELS

This section discusses the RL channels for Subtype 2 Physical Layer. First, we summarize the main functions of the RL physical layer channels and illustrate how various RL channels are combined. Then, in Section 12.5.1, we discuss the enhancements Rev. A makes in Rev. 0 channels. The new RL channels that Rev. A introduces are discussed in detail in Section 12.5.2.

Figure 12.6 lists the RL physical layer channels. As in Rev. 0, the AT transmits on either the Access channel or the Reverse Traffic channel. When the AT has no dedicated radio connection with the AN, it uses the Access channel to contact the AN. The Access channel contains two channels, the Pilot channel and the Data channel. The Pilot channel provides a reference signal to the AN for the AT's transmission and serves as a preamble for the access probe, and the Data channel helps carry messages, such as the RouteUpdate message and ConnectionRequest message. As in Rev. 0, each Access channel transmission must have a RouteUpdate message. Though Rev. 0 allows only signalling messages on the Access channel, Rev. A enables the AT to send user traffic (up to 113 bytes) inside the DataOverSignaling message on the Access channel.

An AT with a dedicated radio connection with the AN transmits the RTC. The RTC consists of the Pilot channel (also referred to as Primary Pilot channel), Auxiliary Pilot channel, MAC channel, Data channel, and ACK channel. The MAC channel, in turn, is made up of the Reverse Rate Indicator (RRI) channel, DRC channel, and Data Source Control (DSC) channel. The new channels that Rev. A introduces are the Auxiliary Pilot channel and the DSC channel.

The Pilot channel provides the reference signal for the AT to the AN. The Data channel carries user traffic data, such as email and signaling messages. The ACK channel supports FL H-ARQ. The RRI channel essentially indicates to the AT the data rate the AN selected. Rev. A modifies the interpretation of the RRI channel, the details of which are discussed in Section 12.5.1. The DRC channel identifies the FL data rate that the AT can support under current channel conditions and the sector

[425]The SofterHandoff field within the TrafficChannelAssignment message indicates whether two sectors belong to the same cell or to different cells.

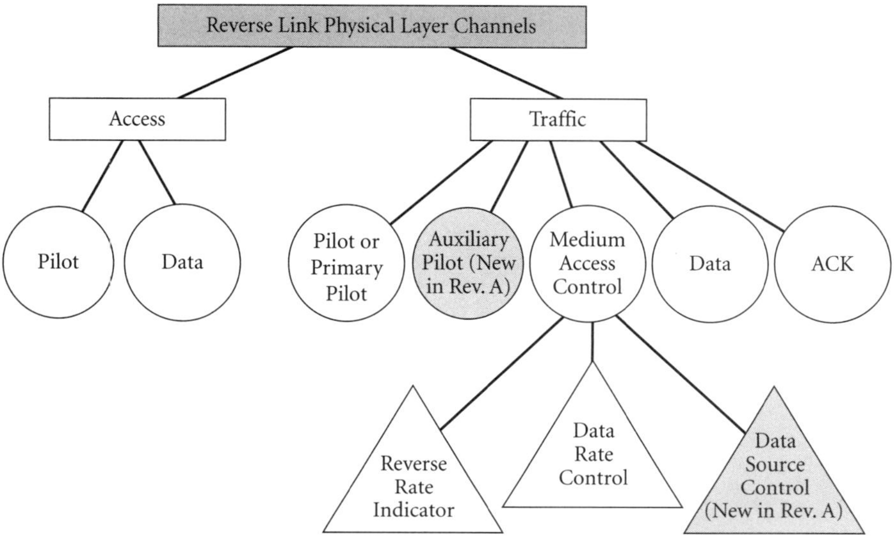

Figure 12.6: Rev. A Reverse Link Channels

the AT selected for FL data transmission. The Auxiliary Pilot channel provides an additional reference signal to the AN when the AT transmits at a very high data rate (e.g., 500 kbps). The RL packet size above which the AT sends the Auxiliary Pilot channel is a session-configurable parameter. The DSC channel identifies the forward link source (i.e., the cell) from which the AT prefers to receive data in the near future. In other words, the DSC channel helps the RNC prepare to send packets to a *new* cell when inter-cell handoff is expected to occur shortly. Additional details of the new Rev. A channels are available in Section 12.5.2.

Figure 12.7 depicts how various RL channels are combined to construct the RTC. The RL in 1xEV-DO Rev. A, just like 1xEV-DO Rev. 0, uses CDMA as the multiple access technique. Hence, orthogonal Walsh codes are used to combine various channels being transmitted from a given AT. Two ATs are differentiated by the sector via AT-specific user long codes. The Pilot channel, Auxiliary Pilot channel, RRI channel, ACK channel, DSC channel, and DRC channel are spread by the Walsh functions W_0^{16}, W_{28}^{32}, W_4^{16}, W_{12}^{32}, and W_8^{16}, respectively. The chips from the ACK channel and DSC channel are time-division-multiplexed such that the first half-slot carries ACK channel and the second half-slot carries the DSC channel. Once the Walsh spreading occurs, the resulting chip rate is 1.2288 Mcps.

The transmit power of the chips for a given channel effectively depends upon some power offset relative to the Pilot channel. More specifically, the chips of the RRI, DSC, DRC, ACK, and Data channels are transmitted with the gain of RRIChannelGain, DSCChannelGain, DRCChannelGain, and Transmit Traffic to Pilot ratio (TxT2P) relative to the transmit power of the Pilot channel, respectively. The chips of the

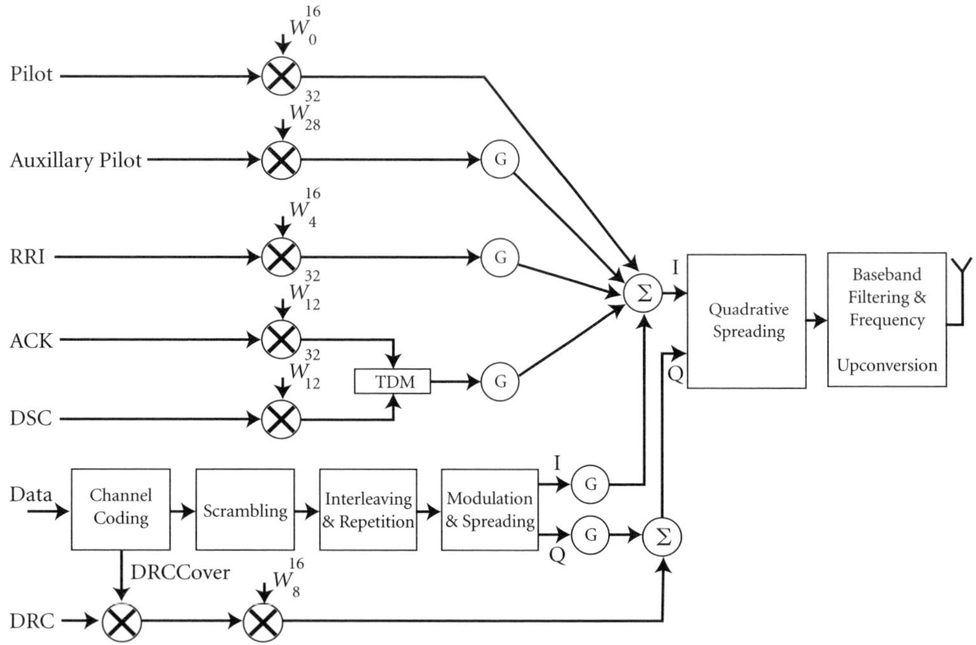

Figure 12.7: Combining of Rev. A RL Channels for the Reverse Traffic Channel

Auxiliary Pilot channel have a gain, AuxiliaryPilotChannelGain, relative to the Data channel. The relative gain value for the ACK channel transmitted in response to a single-user packet is specified by the parameter ACKChannelGain, and the relative gain value for the ACK channel transmitted in response to a multi-user packet is given by ACKChannelGain plus DeltaACKChannelGainMUP.

The quadrature spreading is a complex multiply operation:

$$\begin{aligned} I' &= I\,PN_I - Q\,PN_Q \\ Q' &= I\,PN_Q - Q\,PN_I \end{aligned} \tag{12.3}$$

In Eq. (12.3), PN_I depends on the I-channel short PN sequence and the I-channel user long code, and PN_Q depends on the Q-channel short PN sequence and the Q-channel user long code. The I- and Q-channel short PN sequences are short-code PN sequences with a period of 2^{15} chips and are common to all ATs. The I- and Q-channel long PN sequences also have a period of 2^{15} chips and are a function of user long codes. The user long codes are AT-specific and depend upon the assigned UATI. After the quadrature spreading, the resulting I′ and Q′ signals are processed by the I- and Q-baseband filters. Finally, they are upconverted to the RF frequency.

12.5.1 Access Channel and RRI Channel Enhancements in Rev. A

The Access channel carries the physical layer packet of size 256, 512, or 1024 bits over the duration of the 26.67 ms frame, yielding data rates of 9.6 kbps, 19.2 kbps, and 38.4 kbps. The MAC protocol controlling the Access channel is enhanced to provide more flexibility in the access operation compared to Rev. 0. The specific enhancements include pilot-strength–based adjustments in the access probe power level and flexibility of the Access channel rate selection.

The transmit power of an ith access probe is given by

$$X_i = X_0 + (i-1)PowerStep \tag{12.4}$$

where $PowerStep$, defined in the AccessParameters message, is the power difference between successive access probes in a probe sequence and X_0 is the open-loop mean output power of the Pilot channel. PowerStep is a four-bit value in units of 0.5 dB. For example, $PowerStep = 2$ means a $2 * 0.5 = 1.0$ dB step size. The AT calculates X_0 as

$$X_0 = -MeanRxPower + OpenLoopAdjust + ProbeInitialAdjust \tag{12.5}$$

where $MeanRxPower$ is the mean power the AT receives,[426] $OpenLoopAdjust$ is specified in the AccessParameters message, and $ProbeInitialAdjust$ is determined as

$$ProbeInitialAdjust = ProbeInitialAdjustEACMAC + \min\left\{\begin{array}{c} PilotStrengthCorrectionMax, \\ \max\big(PilotStrengthNominal - PilotStrength, \\ PilotStrengthCorrectionMin\big) \end{array}\right\}. \quad (12.6)$$

In Eq. (12.6), PilotStrength is the pilot strength of the sector to which the AT is sending the access probe and the parameters PilotStrengthCorrectionMax, PilotStrengthNominal, and PilotStrengthCorrectionMin are specified in the Access-Parameters message. These parameters are assumed to be 0 dB if they are absent in the AccessParameters message. ProbeInitialAdjustEACMAC is also specified in the AccessParameters message. The pilot-strength–based parameters enable the AN to provide additional adjustments to the AT's transmit power to reduce the access delay. For example, if the AT is far away from the cell site, the pilot strength would be weaker and Eq. (12.6) would yield a higher access probe transmit power. However, if the AT is close to the cell site, the pilot strength is stronger and Eq. (12.6) results in a smaller pilot-strength–based adjustment and a lower probe transmit power. The AT chooses the lowest Access channel data rate such that it can transmit the required payload (e.g., a message) *and* satisfy the transmit power constraints.

Do You Recall... Let's perturb the memory cells that have stored your Rev. 0 knowledge! What does the AT's RRI channel convey to the AN?[427]

Rev. A re-defines the RRI for the Subtype 2 Physical Layer. The RRI specifies the payload size of the RL Data channel in bits rather than the data rate of the RL Data channel in bits per second. The RRI also indicates the subpacket identity of the physical layer packet so that the AN knows how to combine the subpackets associated with the same RL physical layer packet. Furthermore, the RRI enables the AN to re-send an ACK the AT missed earlier.

The RRI symbol being conveyed on the RRI channel consists of six bits, with four bits representing payload size and two bits specifying the subpacket identity. The two-bit subpacket index j of 0, 1, 2, and 3 corresponds to the subpacket identity of 0, 1, 2, and 3, respectively. The relationship between the four-bit payload index i and the payload size is summarized in Table 12.2. A thirty-two–dimensional bi-orthogonal Walsh function encodes the six-bit RRI symbols as follows using the payload index i and the subpacket index j:

$$\begin{array}{ll} \overline{W}^{32}_{2i+\lfloor j/2 \rfloor} & \text{for odd } j \\ W^{32}_{2i+\lfloor j/2 \rfloor} & \text{for even } j. \end{array} \quad (12.7)$$

[426]Rx represents "receive."

[427]The RRI channel the AT transmits specifies the current data rate of the Data channel of the RTC the AT has selected.

Table 12.2: Interpretation of the Four-Bit Payload Size for the RRI Symbol

Payload Size (bits)	Four-Bit Payload Index i
0	0
128	1
256	2
512	3
768	4
1024	5
1536	6
2048	7
3072	8
4096	9
6144	10
8192	11
12288	12

The RRI channel is transmitted on the I channel. The null-rate RRI is specified by the payload index of 0 and subpacket index of 0.

12.5.2 Auxiliary Pilot Channel and DSC Channel: New Rev. A Channels

The Auxiliary Pilot channel is transmitted in addition to the usual Primary Pilot channel in Rev. A. This channel is available only for the Subtype 2 Physical Layer. Though the Primary Pilot channel facilitates coherent demodulation of the Data channel, the Auxiliary Pilot channel acts as an additional reference signal to further improve demodulation. The Auxiliary Pilot channel is transmitted during the half-slot n if the size of the transmitted payload during half slots $n-1$ or $n+1$ is greater than or equal to a configurable threshold AuxiliaryPilotChannelMinPayload. The default value of AuxiliaryPilotChannelMinPayload corresponds to the payload size of 3072 bits. The auxiliary pilot consists of all 0s with 0 mapped onto $+1$. As shown in Figure 12.7, the Walsh function W_{28}^{32} carries the Auxiliary Pilot channel on the I-branch. The transmit power of the Auxiliary Pilot channel is controlled by the parameter AuxiliaryPilotChannelGain, which specifies the gain of the Auxiliary Pilot channel relative to the Traffic channel gain that the AT uses. AuxiliaryPilotChannelGain ranges from 0 to -15 dB, with the default being -12 dB.

The DSC channel is like an early warning system in which the AT gives a heads-up to the AN about its intention to switch the FL data source (i.e., the cell) for FL packet reception. In such a case, the AN can prepare to send the FL data for the AT to the new source so that, when the DRCCover associated with the AT's DRCChannel points to a sector controlled by the new cell, the new cell is ready to send data to the AT. The

FL data can also be bi-cast to the currently serving cell and the new cell the AT has selected, which would be an implementation-dependent approach. In the absence of the DSC channel, the sectors in the Active Set need to wait for the new DRCCover to know what new sector the AT has selected, leading to longer transmission gaps in FL transmission and lower throughput.

A *cell cover* identifies the data source (i.e., the cell) from which the AT wants to receive the FL data. Each distinct cell (*not* the sector) in the AT's Active Set is assigned a unique cell-cover in the TrafficChannelAssignment message. The FL data source is assumed to maintain a common FL data queue for the AT for all the Active Set sectors with which it is associated. The DSC channel carries a three-bit DSC symbol index that ranges from 0 to 7. The DSC symbol index 0 is the null-cover, and other DSC symbol indices are regular cell-covers representing cells. In other words, the AT transmits either a null-cover or a cell-cover on the DSC channel.

The three-bit DSC symbol index maps onto a thirty-two–bit codeword and is then spread by the Walsh function W_{12}^{32}. The DSC channel is transmitted on the I-channel at the gain of DSCChannelGain relative to the Pilot channel. The value of DSCChannelGain depends on the parameters DSCChannelGainBase and DSC-ChannelGainBoost, which are two of the four configurable parameters (the other two are DSCLength and DSCBoostLength) that control the operation of the DSC channel.

The DSCLength parameter is the number of consecutive slots for which the same DSC symbol index is repeated. The DSC symbol index takes effect one slot after the end of its transmission and remains valid for DSCLength slots. The DSC symbol boundaries coincide with DRC symbol boundaries. The DSCLength is expressed in units of eight slots (the same as the maximum value of DRCLength), ranging from eight slots to 256 slots with the default value of sixty-four slots.

Assume that the AT starts transmitting a new cell-cover value on the DSC channel in slot T. The DSCChannelGain is set to $DSCChannelGainBase + DSCChannelGainBoost$ for the slots T through $T + DSCBoostLength - 1$. For subsequent slots, the DSCChannelGain is set to DSCChannelGainBase. DSCChannel-GainBoost can vary from 0 to 6 dB with 0 dB being the default value. DSCBoost-Length is expressed in units of eight slots with the numerical value ranging from 0 to 15 and from 17 to 64. Thus, its effective value in slots ranges from 0 to $15 \times 8 = 120$ slots and from $17 \times 8 = 136$ slots to $64 \times 8 = 512$ slots. The default value of DSC-BoostLength is 128 slots.

Section 12.13 describes how the DSC channel along with the DRC channel can be used to improve handoff in Rev. A systems. More specifically, the DSC channel minimizes data interruption time during handoff from one BS to another BS.

12.6 REV. A ENHANCEMENT FOR CALL SETUP AND SESSION SETUP

The basic call or session setup in Rev. A is similar to that in Rev. 0. See Section 10.8 of Chapter 10 for call setup details in Rev. 0. This section discusses 1xEV-DO session-related enhancements that Rev. A introduces.

Do You Recall. . . A 1xEV-DO session is a logical link (or *a shared state*) between which elements—the AT and the BS or the AT and the RNC?[428]

Any time the AT has a 1xEV-DO session with the RNC, the AT has a UATI.

12.6.1 Session Configuration and Multiple Personalities

One of the main session configuration enhancements is the idea of a *personality*. Recall that a 1xEV-DO air interface has a seven-layer protocol stack with the Physical layer at the bottom and the Application layer at the top. When a session is configured, the set of negotiated protocols and parameters is stored at the AT and the RNC. When one or more parameters or protocol subtypes need a change, the session has to be reconfigured. The session reconfiguration could take a while, perhaps a few hundreds of milliseconds, depending upon how much needs to be configuredor re-configured. Rev. A aims to reduce such processing delays by using the multiple personalities feature. *A personality is a set of negotiated protocols and parameters.* A 1xEV-DO session between the AT and the AN can support up to PersonalityCount personalities. The maximum value of PersonalityCount is sixteen because PersonalityCount is a four-bit number. PersonalityCount is an attribute of the Default Session Configuration Protocol, and its default value of 1 can be overwritten via session negotiation.

When the 1xEV-DO session between the AT and the AN is being established, several personalities can be negotiated. When a new set of parameters and protocols is desired, the personality can be changed quickly, avoiding processing delays while the AT is connected to the AN. One personality may correspond to non-real-time services with the RLP retransmission on. Another personality may correspond to a real-time and delay-sensitive application, such as streaming or VoIP, with the RLP retransmission off.[429]

Time for our AT's make-over! Let's see how to make a *simple* Rev. A AT assume multiple personalities. Figure 12.8 illustrates how two personalities are created. Recall that the AT and the AN (specifically, the RNC) exchange a series of Configuration-Request and ConfigurationResponse messages during the session configuration to negotiate a set of protocols and associated parameters. (We will discuss all basic messages, but Figure 12.8 shows only some of these messages.)

The session configuration in Rev. 0 and Rev. A consists of two stages, AT-led negotiation and AN-led negotiation. As in Rev. 0, the AN can kick-off the configuration by sending the ConfigurationStart message; typically, the AT starts sending a ConfigurationRequest message to the AN after obtaining the UATI from the AN. The AT-led negotiation involves a series of ConfigurationRequest and ConfigurationResponse messages. The ConfigurationRequest message going from the AT to the AN contains possible values for a given attribute (e.g., a parameter),

[428]The 1xEV-DO session is between the AT and the RNC.

[429]When physical layer H-ARQ retransmissions are inadequate for an error-free packet delivery, the RLP layer can retransmit the packet. See Chapter 10 for details.

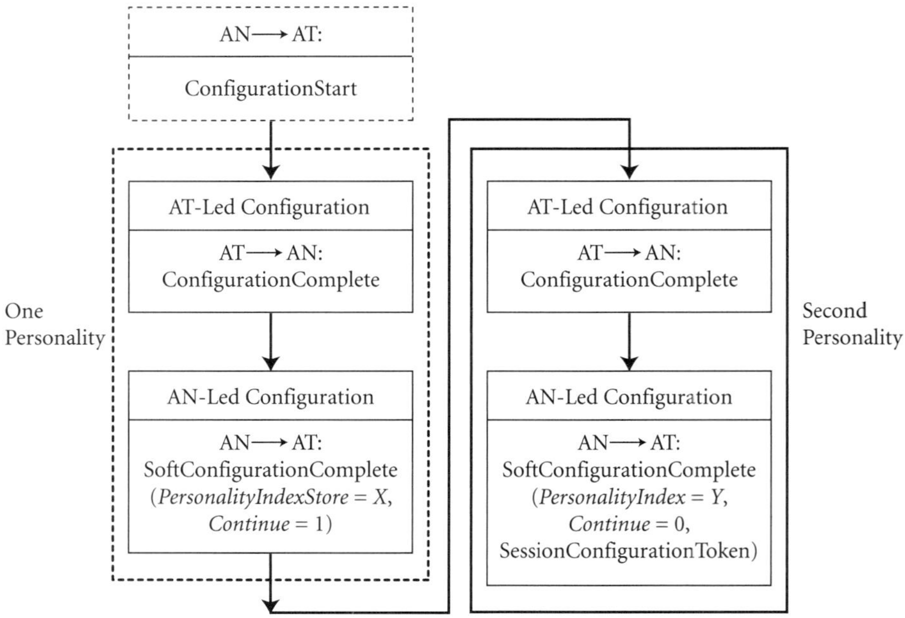

Figure 12.8: Creation of Multiple Personalities

and the ConfigurationResponse message going from the AN to the AT identifies the selected value of an attribute or indicates a reject decision. Once the AT has completed the negotiations, it sends the ConfigurationComplete message to the AN. Now, the AN-led configuration starts with ConfigurationRequest messages going from the AN to the AT and the ConfigurationResponse messages going from the AT to the AN.

So far, the session negotiation is the same for Rev. 0 and Rev. A. Here comes the Rev. A twist! When the AN is done with one set of negotiations, equivalent to one instance of the entire protocol stack, it sends a SoftConfigurationComplete message to the AT instead of the Rev. 0 message of ConfigurationComplete!

The value of the parameter PersonalityIndexStore, ranging from 0 to $PersonalityCount - 1$, indicates the personality number associated with the just negotiated set of AT-led and AN-led configurations. If the AN wants to negotiate multiple personalities, it makes the value of the parameter Continue equal to 1, indicating to the AT to start the negotiation for another personality. Again, AT-led and AN-led configurations are carried out. If the AN is finished constructing personalities, the SoftConfigurationComplete message contains the usual PersonalityIndex-Store parameter, $Continue = 0$, and a new parameter, SessionConfigurationToken. The four most significant bits of the SessionConfigurationToken parameter indicates which Personality number will be used after the session negotiation ends. Recall that the Personality index is a four-bit integer. The remaining twelve bits of Session-

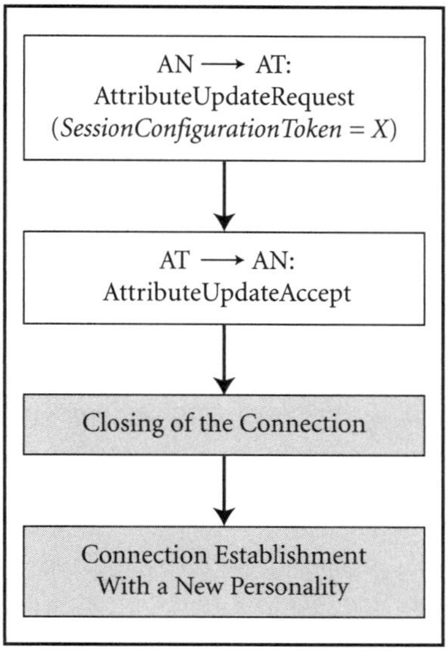

Figure 12.9: Changing the AT's Personality

ConfigurationToken are implementation-specific. For example, one set of twelve bits may correspond to a specific instance of the air-interface protocol stack where the delay-tolerant non-real-time traffic is supported with RLP retransmission on. Such use of the twelve bits can facilitate radio resource management in the AN.

Assume that the AT and the AN have negotiated several personalities, and the currently active personality must be changed. Perhaps, the existing personality supports just the non-real-time applications and another previously negotiated personality supports the real-time applications. Figure 12.9 shows one way of changing the AT's personality. The AN sends an AttributeUpdateRequest message indicating specifying some value for SessionConfigurationToken.

Do You Recall... Do you remember the relationship between the Personality and the SessionConfigurationToken?[430]

The AT responds with the AttributeUpdateAccept message. Some personality switch may require tearing down the current radio connection. For example, when we change the physical layer from a Rev. A physical layer to a Rev. 0 physical layer as a result of entering a Rev. 0 system from a Rev. A system, the connection with

[430]The four most significant bits of the SessionConfigurationToken represent the Personality index! Hence, the twelve least significant bits of the SessionConfigurationToken remain the same; just the Personality Index represented by the four most significant bits is changed when the AT's personality is to be changed.

the Rev. A system must be closed and reopened. Changes in certain attributes of a personality, however, do not require the closing of the connection. For example, if we are changing just the RLP retransmission approach from on to off and vice versa, the connection need not be closed. In case the personality change requires closing the connection, the AT and the AN exchange ConnectionClose messages, and a new connection that reflects the new personality is established.

12.6.2 Generic Attribute Update Protocol

The Generic Attribute Update Protocol is used to modify protocol attributes. Generic Attribute Update Protocol can be invoked at connection setup. In general, any protocol in the seven-layer protocol stack can use the Generic Attribute Update Protocol to update its attribute. The standard explicitly specifies which of the configurable protocol attributes can be updated using the Generic Attribute Update Protocol. In other words, not all attributes of a protocol can be updated using the Generic Attribute Update Protocol. The AT or the AN can make a request to update an attribute by sending an AttributeUpdateRequest message. One or many attributes may be present in an AttributeUpdateRequest message. The AN may accept or reject the updated value of the attribute, but the AT cannot reject an attribute update. The acceptance is indicated by an AttributeUpdateAccept message, and the rejection is indicated by an AttributeUpdateReject message.

12.6.3 Multimode Capability Discovery Protocol

This protocol residing at the session layer in the seven-layer protocol stack enables the AN to discover certain capabilities of the AT. Such knowledge of the AT's capabilities can be exploited in implementing better radio resource management techniques for a given AT. Table 12.3 summarizes the attributes of the Multimode Capability Discovery Protocol along with their default values. By default, the AT is assumed to support both CDMA2000 1x and 1xEV-DO, and the AT is assumed to utilize receive diversity. Of course, the default values can be overwritten during session configuration.

12.7 REV. A FORWARD LINK DATA TRANSMISSION: A DETAILED VIEW

The Subtype 2 Physical Layer is assumed throughout Section 12.7. Section 12.7.1 provides an overview of the FL data transmission, and, details of various steps are given in the sections 12.7.2 through 12.7.6.

12.7.1 Overview of Enhancements in Forward Link Transmission

Figure 12.10 illustrates the basic steps in the FL data transmission along with a list of related Rev. A enhancements. Let's take a look at each step.

Table 12.3: Attributes of the Multimode Capability Discovery Protocol

Attribute	Interpretation	Default
SimultaneousCommonChannelTransmit	0: AT cannot simultaneously transmit 1x and 1xEV-DO common channels 1: otherwise	0
SimultaneousDedicatedChannelTransmit	0: AT cannot simultaneously transmit 1x and 1xEV-DO dedicated channels 1: otherwise	0
SimultaneousCommonChannelReceive	0: AT cannot simultaneously receive 1x and 1xEV-DO common channels 1: otherwise	0
SimultaneousDedicatedChannelReceive	0: AT cannot simultaneously receive 1x and 1xEV-DO dedicated channels 1: otherwise	0
HybridMSAT	0: AT does not support hybrid operation with 1x and 1xEV-DO 1: Otherwise	1
ReceiverDiversity	0: AT does not support receiver diversity 1: otherwise	1

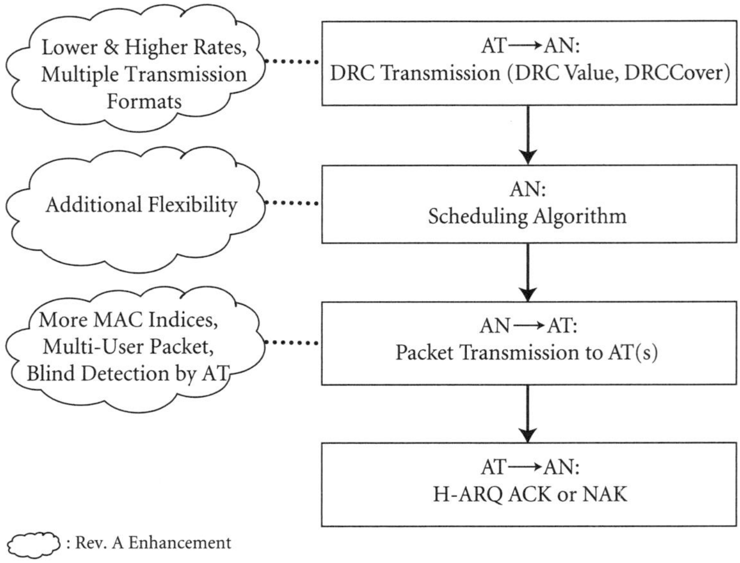

Figure 12.10: Overview of Forward Link Packet Transmission

In the DRC transmission step, the AT observes C/I ratios of the FL pilots in the Active Set and selects an Active Set sector as the FL serving sector. The AT conveys the FL serving sector selection to the AN via DRCCover. The DRC value specifies the nominal data rate the AT has requested. In a Rev. 0 system, a given DRC value corresponds to just one data rate. For example, the DRC value of 1 is associated with the nominal data rate of 38.4 kbps in Rev. 0. In a Rev. A system, a given DRC value could be associated with multiple nominal data rates represented by different transmission formats. Such an association gives the AN scheduler more flexibility during the packet transmission. The lowest FL data rate is 4.8 kbps instead of the Rev. 0 rate of 38.4 kbps, and the highest FL data rate is 3.072 Mbps rather than Rev. 0's 2.4576 Mbps. In addition to the DRC channel, the AN must decode the DSC channel before the FL data transmission can begin.

In the scheduling algorithm step, the Rev. 0 scheduling algorithm must choose just one user for transmission for an available FL slot. Since the DRC value of Rev. 0 maps onto a fixed data rate and predetermined transmission parameters, such as the coding rate and modulation scheme, the Rev. 0 scheduling algorithm is relatively simple. However, the Rev. A scheduling algorithm is much more comprehensive because numerous options are available to the scheduler. For example, for a given FL slot, the schedule can create a packet for just one user or it can create a multi-user packet. Furthermore, for a given user, the scheduler can typically choose from multiple transmission formats since a given DRC value may map onto multiple formats. Furthermore, intra-AT QoS control is facilitated by the multi-flow RLP protocol (see Section 12.12), and the scheduler can suitably create a packet for a user to reflect QoS requirements of various applications running on the AT.

In the packet transmission step, a Rev. 0 system can support a maximum of 59 unicast traffic connections in a sector due the availability of 59 MAC indices, but a Rev. A system can support 115 connections in a sector. Recall that a Rev. 0 system has a total of 64 MAC indices, and a Rev. A system supports 128 MAC indices in a sector. A multi-user packet can help transmit traffic bits to multiple users. An AT performs blind rate detection of its packet by comparing the contents of the received packet with a set of transmission formats compatible with the DRC value that the AT sent to the AN. Furthermore, broadcast and/or multicast services are also supported using a set of MAC indices.

Finally, in the H-ARQ ACK or NAK step, the AT that receives the FL packet sends an H-ARQ ACK or NAK to the AN just as in Rev. 0. The received packet may be a single-user packet or a multi-user packet in Rev. A. Regardless of the type of the received packet, the AT responds with an H-ARQ ACK or NAK just as if it were a Rev. 0 AT.

12.7.2 DRC Enhancements

Table 12.4 summarizes the transmission formats for different values of the DRC index for a single-user packet (i.e., a packet destined for just one user as opposed to a multi-user packet). The rate metric in Table 12.4 corresponds to the largest (canonical) packet format except for $DRC\ index = 0$.

A transmission format is specified a triple (a, b, c) where a specifies the physical layer packet size in bits, b denotes the span (the maximum number of slots for transmission of a packet), and c indicates the number of preamble chips. As shown in Table 12.4, a given value of DRC index could correspond to multiple compatible formats. For a given DRC index, the transmission format associated with the largest packet size is called the *canonical transmission format* for that DRC index. For example, in Table 12.4, the transmission format (1024, 16, 1024) is the canonical transmission format for $DRC\ index = 1$.

Do You Recall... What are the nominal data rate and the physical layer packet size associated with $DRC\ index = 1$ for Rev. 0?[431]

Let's calculate the nominal data rate for the canonical transmission format of $DRC\ index = 14$. This data rate is 5120 bits$/(1 * 1.67$ ms$) = 3.072$ Mbps, which is the rate metric shown in Table 12.4.

[431] The nominal data rate is 38.4 kbps and the packet size is 1024 bits!

Table 12.4: Interpretation of the DRC Index for a Single-User Packet

DRC Index	Compatible Transmission Formats	Rate Metric (kbps)
0	(128, 16, 1024), (256, 16, 1024), (512, 16, 1024), (1024, 16, 1024)	0
1	(128, 16, 1024), (256, 16, 1024), (512, 16, 1024), (1024, 16, 1024)	38.4
2	(128, 8, 512), (256, 8, 512), (512, 8, 512), (1024, 8, 512)	76.8
3	(128, 4, 256), (256, 4, 256), (512, 4, 256), (1024, 4, 256)	153.6
4	(128, 2, 128), (256, 2, 128), (512, 2, 128), (1024, 2, 128)	307.2
5	(512, 4, 128), (1024, 4, 128), (2048, 4, 128)	307.2
6	(128, 1, 64), (256, 1, 64), (512, 1, 64), (1024, 1, 64)	614.4
7	(512, 2, 64), (1024, 2, 64), (2048, 2, 64)	614.4
8	(1024, 2, 64), (3072, 2, 64)	921.6
9	(512, 1, 64), (1024, 1, 64), (2048, 1, 64)	1228.8
10	(4096, 2, 64)	1228.8
11	(1024, 1, 64), (3072, 1, 64)	1843.2
12	(4096, 1, 64)	2457.6
13	(5120, 2, 64)	1536
14	(5120, 1, 64)	3072

A Question for You. . . What is the lowest nominal data rate that a Rev. A system can support in the forward link?[432]

To appreciate the usefulness of the concept of compatible transmission formats, assume that a single AT is present in a sector and that the AT sends $DRC\ index = 1$ on the DRC channel. Further assume that the AN has just 100 bits in the AT's forward link buffer. In a Rev. 0 system, the AN would need to construct a packet of 1024 bits (because it is the minimum packet size) by appending about 900 zeros! That would be a significant waste of scarce radio interface resources. In a Rev. A system, the AN could choose any of the compatible transmission formats for $DRC\ index = 1$, and the allowed packet size ranges from 128 bits to 1024 bits. The selected packet size can now be matched to the available data for the AT in the buffer. Not only

[432]The lowest nominal data rate corresponds to the combination of the smallest packet size and the longest span. From Table 12.4, such a combination is 128 bits and sixteen slots, leading to the lowest rate of $128\ \text{bits}/(16 * 1.67\ \text{ms}) = 4.8\ \text{kbps}$.

are we using the air-interface resources efficiently, we are also incorporating a good degree of redundancy in the multi-slot packet transmission to increase the likelihood of successful and early decoding of the packet.

Let's re-visit the problem of an *optimistic* AT in a Rev. 0 system. Recall that the AT chooses a DRC index such that it can guarantee to the AN 1% packet error rate when the AN uses the transmission parameters associated with the DRC index (e.g., packet size, modulation, amount of coding, and the span). Assume that the AT's current channel conditions quantified by best sector C/I ratios are such that it can support the data rate associated with $DRC\ index = 2$. However, our AT is optimistic and perceives the channel conditions to be better than they actually are, so our AT sends $DRC\ index = 3$ to the AN. The AN utilizes the transmission parameters corresponding to $DRC\ index = 3$, but the packet is received in error because the channel conditions cannot support higher data rate. This is the error caused by an optimistic AT!

Something to Think About. . . Let's talk about optimism! Murphy's Law states, "Anything that can go wrong will go wrong." Lest you think Murphy was too pessimistic, O'Toole's Corollary states, "Murphy was an optimist."

Recall that the AN cannot upgrade or downgrade the data rate in Rev. 0. Hence, in general, our Rev. 0 system has to live with such optimistic ATs, leading to lower than optimal sector throughput. A Rev. A system forces our optimistic AT to face reality! Rev. A introduces the concept of DRCOffset. The actual DRC value the AT transmits on the DRC channel is $N - DRCOffsetN$ where N is the tentative DRC index that the AT thinks that it can support and $DRCOffsetN$ is the corresponding parameter of the DRCTranslationOffset attribute. For example, assume that the AT thinks that it can support $DRC\ index\ N = 3$. It then consults a table containing DRC offsets. Assume that the value of the DRC offset for DRC index 3 is 1. In other words, $DRCOfffset3 = 1$. Then, the actual value of the DRC index the AT sends on the DRC channel is $N - DRCOffsetN = 3 - 1 = 2$. So, the AT sends $DRC\ index = 2$ instead of $DRC\ index = 3$, and, with any luck, the packet will experience an error rate of less than 1%! Note that the AT expects the FL packet to correspond to $DRC\ index = 2$.

In summary, the Rev. A system can use the mechanism of DRCOffset to make adjustments to the AT's DRC determination algorithm and thus achieve the target packet error rate. The values of the DRCOffset$N$ can be set during the session configuration. These values, however, cannot be arbitrary and must follow these constraints: the rate metric of the DRC index corresponding to the transmitted DRC value shall be less than or equal to that of the DRC index corresponding to N, and the span of the DRC index corresponding to the transmitted DRC value shall be greater than or equal to that of the DRC index corresponding to N.

12.7.3 Multi-User Packets

In contrast to a single-user packet, a multi-user packet allows multiple users to receive their packets in the forward link. Information for up to eight users can be placed in a multi-user packet.

A Question for You... How many ATs can receive a packet from the AN during a 1.67 ms slot in a Rev. 0 system?[433]

If many ATs are running low-rate but delay-sensitive applications (e.g., VoIP or gaming), a significant delay could occur between two successive transmissions to a given AT. A Rev. A system allows the creation of a multi-user packet (i.e., a packet that carries packets for multiple users), and as many as eight users can now simultaneously receive a packet during a 1.67 ms slot. The multi-user capability is off by default and can be turned on by setting the value of the attribute MultiUserPacketsEnabled attribute to 1 during the session configuration.

A Rev. 0 system supports just a single-user packet, but a Rev. A system supports three types of MAC packets: single-user simplex packets, single-user multiplex packets,[434] and multi-user packets. A single-user simplex packet is addressed to one AT and contains just one security layer packet, which could be a Format A or Format B connection layer packet. In other words, the single-user simplex packet is like an AT's packet in a Rev. 0 system. A multi-user packet consists of zero or more (up to eight) security layer packets, addressed to zero or more ATs, with each security layer packet containing a Format A or Format B connection layer packet.

The last two bits of the MAC layer packet are termed the MAC Trailer. The AT infers the MAC packet type based on the preamble index and the MAC Trailer. A preamble MACIndex of 66 through 70 indicates a multi-user packet, and the MAC Trailer is set to `00`. The preamble MACIndex assigned to an AT corresponds to a single-user packet, and, in such case, the MAC trailer of `01` or `11` indicates a single-user simplex packet and `10` indicates a single-user multiplex packet.

Figure 12.11 shows the format of a multi-user packet. The size of the MAC layer packet excluding the MAC Trailer is 12, 24, 60, 124, 252, 380, 508, or 636 octets. The multi-user packet contains various fields; let's take a look at them.

The MAC Layer Payload field consists of n security layer packets, with each such packet containing a Format A or Format B connection layer packet. The value of n ranges from 0 to 8. The MAC Layer Header field consists of n two-octet fields, leading to the total length of $2n$ octets. Each two-octet field consists of a *PacketInfo* field and a *Length* field. The PacketInfo field contains a one-bit format field that

[433] Just one!

[434] Single-user multiplex packets were later removed from the standard [296]. Format B connection layer packets can provide similar function. A single-user multiplex packet was to carry one or more security layer packets, and each security layer packet contained one Format A connection layer packet. The single-user multiplex packet was intended for transmission of packets from different applications in a single slot.

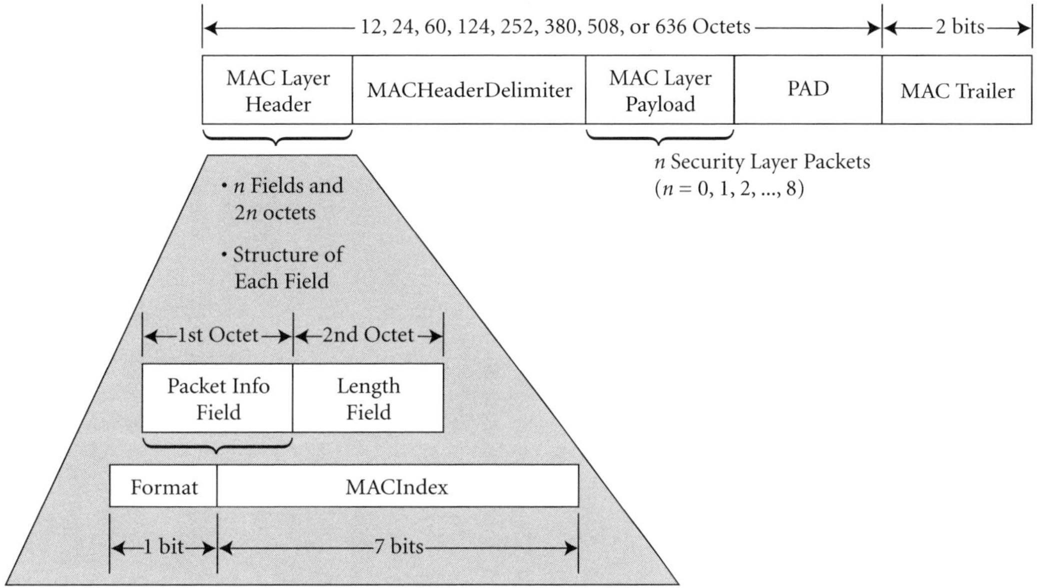

Figure 12.11: Structure of a Multi-User Packet

specifies whether the ith security layer packet is a Format A connection layer packet (0) or a Format B connection layer packet (1) and a seven-bit MACIndex field that specifies the AT for which the ith security layer packet is being sent. The Length field specifies the length of the ith security layer packet in octets.

If present, the MACHeaderDelimiter field consists of eight 0s. This field is present if the MAC layer packet size exceeds, by one or more octets, the sum of the length of the security layer packets plus the MAC Layer Header plus the MAC Trailer. The PAD field consists of all zeros, and its size is set such that the total MAC layer packet size is in the set of the allowed number of octets. Finally, for the multi-user packet, the MAC Trailer field is set to 00.

When the AN sends a multi-user packet to several ATs, it must follow the constraints of the DRC values these ATs have sent. Table 12.5 summarizes the compatible transmission formats for the entire set of DRC values.

Table 12.5: Interpretation of the DRC Index for a Multi-User Packet

DRC Index	Compatible Transmission Formats	Rate Metric (kbps)
0	(128, 4, 256), (256, 4, 256), (512, 4, 256), (1024, 4, 256)	0
1	(128, 4, 256), (256, 4, 256), (512, 4, 256), (1024, 4, 256)	38.4
2	(128, 4, 256), (256, 4, 256), (512, 4, 256), (1024, 4, 256)	76.8
3	(128, 4, 256), (256, 4, 256), (512, 4, 256), (1024, 4, 256)	153.6
4	(128, 4, 256), (256, 4, 256), (512, 4, 256), (1024, 4, 256)	307.2
5	(128, 4, 256), (256, 4, 256), (512, 4, 256), (1024, 4, 256), (2048, 4, 128)	307.2
6	(128, 4, 256), (256, 4, 256), (512, 4, 256), (1024, 4, 256)	614.4
7	(128, 4, 256), (256, 4, 256), (512, 4, 256), (1024, 4, 256), (2048, 4, 128)	614.4
8	(128, 4, 256), (256, 4, 256), (512, 4, 256), (1024, 4, 256), (2048, 4, 128), (3072, 2, 64)	921.6
9	(128, 4, 256), (256, 4, 256), (512, 4, 256), (1024, 4, 256), (2048, 4, 128)	1228.8
10	(128, 4, 256), (256, 4, 256), (512, 4, 256), (1024, 4, 256), (2048, 4, 128), (3072, 2, 64), (4096, 2, 64)	1228.8
11	(128, 4, 256), (256, 4, 256), (512, 4, 256), (1024, 4, 256), (2048, 4, 128), (3072, 2, 64)	1843.2
12	(128, 4, 256), (256, 4, 256), (512, 4, 256), (1024, 4, 256), (2048, 4, 128), (3072, 2, 64), (4096, 2, 64)	2457.6
13	(128, 4, 256), (256, 4, 256), (512, 4, 256), (1024, 4, 256), (2048, 4, 128), (3072, 2, 64), (4096, 2, 64), (5120, 2, 64)	1536
14	(128, 4, 256), (256, 4, 256), (512, 4, 256), (1024, 4, 256), (2048, 4, 128), (3072, 2, 64), (4096, 2, 64), (5120, 2, 64)	3072

12.7.4 MACIndex Enhancements

Rev. A includes several enhancements related to MAC indices. We discuss these enhancements next.

Do You Recall... Let's prod your Rev. 0 memory! What is the maximum number of MAC indices in a sector? What is the maximum number of traffic connections that a sector can support?[435]

A Rev. A system increases the total number of MAC indices from 64 to 128, and the maximum number of (unicast) Rev. A traffic connections is 115. Table 12.6 summarizes the use of the MACIndex in the preamble of a slot and on the forward link MAC channel.

As in Rev. 0, the preamble consists of all 0s the in-phase component only and is time multiplexed with the Forward Traffic channel stream. The preamble sequence is covered by a sixty-four–chip bi-orthogonal sequence, and such a sequence is repeated a suitable number of times based on the transmission format of the packet. The Rev. 0 preamble sequence and the Rev. A preamble sequence are different due to the availability of the total of 128 MAC indices rather than the 64 MAC indices in Rev. 0. The Rev. A bi-orthogonal sequence is defined as

$$W_{i/2}^{64} \qquad \text{for } i = 0, 2, \ldots, 126 \tag{12.8}$$

and

$$\overline{W_{(i-1)/2}^{64}} \qquad \text{for } i = 1, 3, \ldots, 127 \tag{12.9}$$

where i is the MACIndex and $\overline{W_j^{64}}$ is the bit-by-bit complement of the sixty-four–chip Walsh function of order j. As seen from Table 12.6, if the boadcast multicast services are supported in a sector, the total number of (unicast) traffic connections that can be supported is 114. If the boadcast multicast services are unsupported in a sector, a total of 115 unicast traffic connections can be supported in a sector.

In summary, the Rev. A offers following MACIndex–related enhancements.

- More unicast traffic connections are supported per sector.
- Multi-user packets at different data rates are supported by MACIndex 66 through MACIndex 70.
- Broadcast and/or multicast services are supported using MACIndex 5.
- In addition to the Rev. 0 Control channel rates of 38.4 and 76.8 kbps, the 19.2 kbps data rate is also supported for the control channel using MACIndex 71.

[435] Sixty-four is the maximum number of MAC indices in a sector, and a sector can support a maximum of 59 traffic connections.

Table 12.6: Use of MACIndex in the Preamble and on the MAC Channel

MACIndex	MAC Channel Use	Preamble Use
0 and 1	Unused	Unused
2	Unused	76.8 kbps Control channel with transmission format (1024, 8, 512)
3	Unused	38.4 kbps Control channel with transmission format (1024, 16, 1024)
4	RA channel	Unused
5	Available for RPC, DRCLock, and ARQ channels	Available for Broadcast and/or Multicast channel or Forward Traffic channel for single-user packets if broadcast is unnegotiated
64 and 65	Unused	Unused
66	Unused	Multi-user packet with one of these transmission formats: (128, 4, 256), (256, 4, 256), (512, 4, 256), or (1024, 4, 256)
67	Unused	Multi-user packet with transmission format (2048, 4, 128)
68	Unused	Multi-user packet with transmission format (3072, 2, 64)
69	Unused	Multi-user packet with transmission format (4096, 2, 64)
70	Unused	Multi-user packet with transmission format (5120, 2, 64)
71	Unused	19.2 kbps Control channel with (128, 4, 1024), 38.4 kbps Control channel with (256, 4, 1024), or 76.8 kbps Control channel with (512, 4, 1024) transmission formats
6 to 63 and 72 to 127	Available for RPC, DRCLock, and ARQ channels	Available for Forward Traffic channel for single-user packets

- Some MAC indices are used to support reverse link H-ARQ (e.g., 6 through 63) on the MAC channel. Section 12.8 describes the RL H-ARQ.

12.7.5 The AT's Detection of the FL Packet

In a Rev. 0 system, the AN, when sending a new packet on the Control channel or the Forward Traffic channel, includes a suitable MACIndex in the preamble part of the packet in the first slot of a (potentially) multi-slot transmission. For example, when the AN starts the transmission of a packet P at the nominal data rate of 38.4 kbps, it reserves sixteen slots using a four-slot interlacing method and includes the preamble corresponding to the MACIndex of the AT in the first slot; the remaining fifteen slots lack the preamble. Hence, an AT that has sent $DRC = 1$ (corresponding to 38.4 kbps) watches for the preamble corresponding to its own MACIndex and the MAC indices of the Control channel. If the AT finds a match with the preamble MAC index in a slot, it continues to receive additional chips of the multi-slot transmission of a packet until H-ARQ succeeds.

In a Rev. 0 system, the AT always knows the parameters of the FL transmission (e.g., packet size and modulation scheme) because the AN must follow the parameters associated with the DRC value the AT has requested. However, in a Rev. A system, a given DRC value could correspond to multiple compatible transmission formats, and the AN could choose any of these formats for a packet transmission to the AT. Hence, the AT must perform blind detection of the received packet, which is a new requirement in Rev. A. Though multiple transmission formats improve efficiency and provide flexibility to the AN scheduler, the AT's complexity is increased.

Another aspect of the AT's increased complexity stems from its need to match the received preamble chips with not only the Control channel MACIndex and its own (unicast) MACIndex, but also the MAC indices of the multi-user packets because the AN could choose to send data to an AT via a normal single-user packet or a multi-user packet if the multi-user packet has been enabled for that AT. For example, assume that the AT has sent out $DRC\ index = 5$ on the DRC channel. As specified in Table 12.5, this DRC index corresponds to the following multi-user packet transmission formats: (128, 4, 256), (256, 4, 256), (512, 4, 256), (1024, 4, 256), and (2048, 4, 128). Furthermore, the $DRC\ index = 5$ also corresponds to the following transmission formats for a single-user packet as shown in Table 12.4: (512, 4, 128), (1024, 4, 128), and (2048, 4, 128). The AN could thus choose any of these transmission formats. Thus, the AT tries to match the received preamble chips with its own unicast MACIndex and the multi-user packet MACIndex associated with one of the multi-user transmission formats. In our example, the AT will look for the multi-user MACIndex of 66 and 67 because Table 12.6 indicates that these MAC indices could accompany one or more of the multi-user transmission formats.

12.7.6 Forward Link H-ARQ

The basic H-ARQ process in support of the FL packet transmission remains the same in Rev. A as in Rev. 0. The main difference is that the Rev. 0 AN sends just the single-user packet, and the Rev. A AN may send a single-user packet or a multi-user packet to the AT. Regardless of the type of the received traffic packet, the Rev. A AT sends a positive or negative acknowledgment on the RL ARQ channel that is part of the Reverse Traffic channel. The AT that has successfully decoded a multi-user packet no longer processes the remaining (potential) FL H-ARQ retransmissions. Such an AT now looks for a suitable MACIndex in the (potential) preamble of the remaining multi-slot transmission of the multi-user packet. The AT that has been unsuccessful in decoding the packet sends an H-ARQ NAK and continues to receive the remaining retransmissions.

12.8 OVERVIEW OF REV. A REVERSE LINK DATA RATE CONTROL

Recall from Rev. 0 that the RTC consists of the Data channel carrying user packets and signaling messages and other channels, such as the Pilot channel and DRC channel. This section discusses at length how the rate of the RL Data channel is determined in Rev. A.

The Rev. 0 AT determines its instantaneous, specifically for a 26.67 ms frame, RL data rate probabilistically based on parameters such as transition probabilities and rate limit and on feedback from the AN, such as the reverse activity bit. Furthermore, the increase in the data rate from one frame to the next can be only gradual, such as from 19.2 kbps to the next higher level of 38.4 kbps. Rev. A offers potential enhancements, such as an increase in the peak data rate, a data rate change every 6.67 ms, and more direct control over the RL interference. *Four ways of operating the RTC are specified by defining four subtypes of the RTC MAC protocols*, and these four subtypes of the RTC MAC protocol utilize one or more of the three physical layer subtypes. The characteristics of these protocols are summarized in Figure 12.12.

The Subtype 0 RTC MAC Protocol is the default MAC subtype that operates with the Subtype 0 Physical Layer Protocol and Subtype 1 Physical Layer Protocol. This subtype is the same as the default Rev. 0 RTC MAC and provides backward-compatibility with Rev. 0. This subtype offers no new Rev. A functionality for the RTC. In other words, the basic Rev. 0 RL characteristics, such as 9.6 kbps minimum data rate, 153.6 kbps maximum peak data rate, and the probabilistic data rate control, remain intact. This protocol subtype can configure two attributes, PowerParameters and RateParameters. The PowerParameters attribute includes parameters for power offsets for various data rates (e.g., DataOffsetNom and DataOffset9k6 with default values of all these parameters being 0 dB) as well as the power step size, RPC-Step, with 0.5 dB and 1.0 dB as potential values and 1.0 dB as the default value. The RateParameters attribute contains all the transition probabilities (e.g., Transition009k6_019k2 and Transition153k6_076k8).

Subtype 1 RTC MAC Protocol works with the Subtypes 0 and 1 Physical Layer Protocols. Just like the Subtype 0 RTC MAC Protocol, this protocol can configure two attributes, PowerParameters and RateParameters. *However, the main benefit of the Subtype 1 RTC MAC Protocol compared to the Rev. 0 Subtype RTC MAC Protocol is that it allows the RateParameters attribute to be configurable via the Generic*

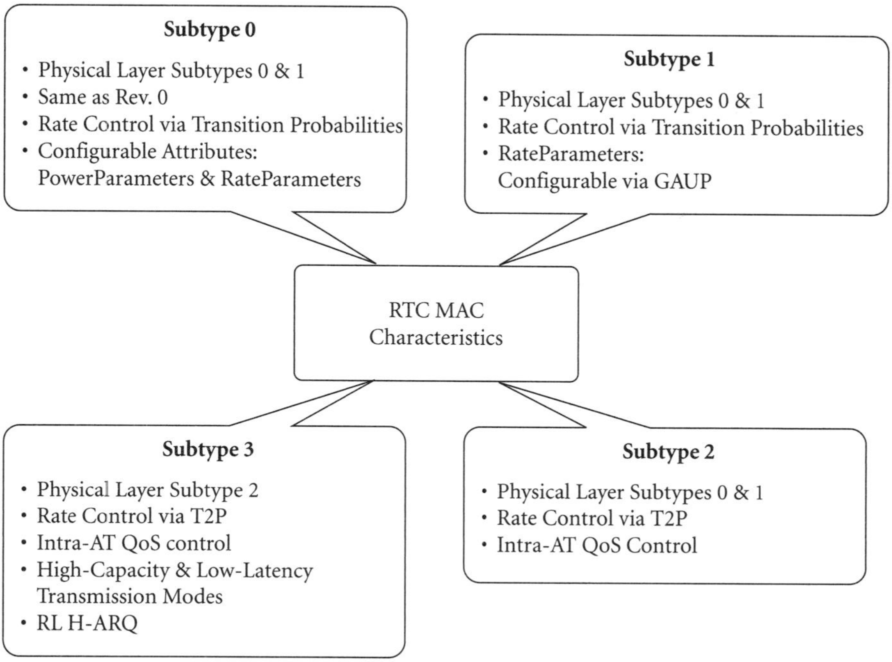

Figure 12.12: Characteristics of RTC MAC Protocols

Attribute Update Protocol. Other than such additional configuration flexibility, the basic functionality of this subtype is the same as that in Rev. 0.

Subtype 2 RTC MAC Protocol continues to rely upon the Rev. 0 physical layer.[436] However, it drastically changes the RL operation. Instead of using probabilistic data rate control via transition probabilities to control RL interference, this subtype introduces a more direct and more efficient way of controlling RL interference, which is T2P-based rate control. In addition, it facilitates implementation of intra-AT QoS control. In other words, Rev. A allows a standard mechanism to provide different QoS treatments for different applications (e.g., VoIP vs. email). Though this MAC subtype cannot provide higher peak rates than Rev. 0 due to its use of Subtype 0 Physical Layer Protocol, the T2P-based rate control helps achieve a higher average RL throughput because the RL interference is controlled more tightly and the rise-over-thermal (RoT) operating point can now be closer to the target RoT threshold.

[436]We are describing Subtype 2 RTC MAC Protocol for the sake of completion from the perspective of the original Rev. A standard. A newer version of the 1xEV-DO standard removed this MAC protocol.

A Question for You... If the AT has transmitted the packet at 19.2 kbps in a RL frame, can it transmit the packet at 153.6 kbps in the very next RL frame—yes or no? (Hmmm, that's easy!)[437]

The T2P-based rate control allows the AT to quickly change the rate (e.g., from 9.6 kbps in one frame to 153.6 kbps in the very next frame). The intra-AT QoS control is implemented by defining multiple MAC flows and (eventually) associating them with applications. The concept of T2P-based rate control is introduced in Section 12.9.2 and intra-AT QoS is discussed in Section 12.12. The RL rate control methodology in the Subtypes 2 and 3 RTC MAC Protocols are based on the philosophy of distributed rate selection at the AT and the AN's centralized (scheduled) resource allocation. The AN allocates a certain amount of RL resources (equal to T2P resources), and the AT is responsible for the rate selection by adhering to the rules of consuming the allocated T2P resources.

Subtype 3 RTC MAC Protocol provides the best RL performance and works with the Subtype 2 Physical Layer Protocol. Recall that the Subtype 2 Physical Layer Protocol supports peak RL data rates greater than 153.6 kbps, up to 1.2 Mbps or 1.8 Mbps. When QPSK is the highest-order modulation supported in RL, the maximum achievable RL rate is 1.2 Mbps. When 8-PSK is supported in RL, the highest RL peak rate becomes 1.8 Mbps. The data rate control is achieved using T2P as in the case of Subtype 2 RTC MAC Protocol. Furthermore, intra-AT QoS control is supported via MAC flow-based T2P allocation just as in Subtype 2 RTC MAC Protocol. Two transmission modes are supported by this MAC subtype, *high-capacity mode* and *low-latency mode*. High-capacity mode supports more users at a lower throughput, and low-latency mode allows lower latency for fewer users. The H-ARQ is used for the RL packet data transmission.

Subtype 2 and Subtype 3 are obviously quite different from Rev. 0 Subtype MAC. These two subtypes completely change the way 1xEV-DO RL operates and are discussed at length in Sections 12.10 and 12.11. Both subtypes use T2P-based rate control and QoS control. *Subtype 3 RTC MAC protocol is the most commonly used subtype for Rev. A ATs in commercial deployments.*

12.9 REVERSE LINK INTERFERENCE CONTROL VIA T2P

Rev. A aims to significantly enhance the RL throughput compared to Rev. 0 by using a T2P-based rate control instead of a probablistic rate control. Section 12.9.1 discovers the motivation for using T2P-based rate control. The T2P is formally defined in Section 12.9.2.

[437]No, the highest data rate cannot suddenly change from one frame to the next because the AT must meet the transition probability requirements.

12.9.1 Motivation for T2P-Based Rate Control

The RL rate control in Rev. 0 is slow and relies on transition probabilities. The transition probabilities and the value of RAB (reverse activity bit) govern how fast the data rate could change from one 26.67 ms frame to the next frame. If Rev. A continues to rely upon probabilistic and slow control, the full potential of the RL physical layer (i.e., Subtype 2 Physical Layer Protocol) cannot be exploited.

Consider probabilistic rate control for a low peak rate as shown in Figure 12.13. Assume that the network design has a certain target RoT threshold. The AN can explicitly quantify RoT noise based on the measured received signal strength indicator or estimate RoT noise based on factors such as the number of users and the data rates. Since the ATs in a sector probabilistically and slowly change the data rate (e.g., the fastest rate change could occur once every 26.67 ms based on the probabilities), some margin, such as X dB in Figure 12.13, is typically required to account for short-term variations in the RL interference. Thus, the target RL RoT operating threshold could be set so that the interference can be controlled around such a threshold. When the measured RoT noise exceeds such an operating threshold, an implementation-specific algorithm of the AN may set RAB to 1, indicating a higher RL interference. When RAB is set to 1, the ATs in a sector gradually decrease their RL rates based on the transition probabilities. When the average RL interference drops below the target RoT operating threshold, the AN algorithm may set RAB to 0. In response to $RAB = 0$, the ATs now begin to increase their rates gradually.

Assume that the similar approach is used for Rev. A where the peak rate is high. If the rate control is probabilistic and slow, the margin between the target RoT threshold and the RoT operating threshold must be larger, such as $(X + Y)$ dB, to account for larger variations caused by a wider range of data rates. Such a larger margin necessitates the RL RoT operating threshold that is lower than that for a lower peak rate, reducing the achievable RL throughput.

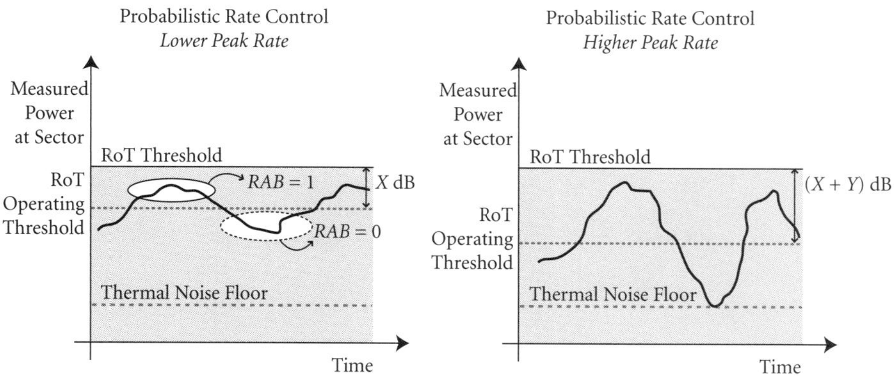

Figure 12.13: Probabilistic Rate Control for Lower and High Peak Rates

In summary, the probabilistic rate control is inadequate for higher throughput. T2P-based rate control is a more efficient alternative to the transition probabilities-based rate control.

12.9.2 What Is T2P, Anyway?

T2P is the ratio of the Traffic channel transmit power to the Pilot channel transit power. Figure 12.14 illustrates the concept of T2P. Assume that the Pilot channel transmit power is X dBm and that the T2P is 3 dB. Then, the transmit power of the Traffic channel becomes $X + 3$ dBm, i.e., the Traffic channel is transmitted at a power level that is 3 dB above the Pilot channel. Control of T2P helps control the Traffic channel power relative to the current Pilot channel power.

An active MAC flow of a given AT contributes to the sector loading or the RoT noise of the sector, and the average transmit T2P of such a flow is significantly correlated with its contribution to the sector's RoT noise. The power control in the RL ensures that the received power levels from different ATs are similar when the ATs use the same T2Ps for a given packet size. The absolute transmit power levels of the ATs' Pilot channels can be significantly different based on the path loss and the prevailing channel conditions, but the T2Ps can be strongly correlated with the interference these ATs caused. For example, when two ATs use $T2P = 3$ dB while transmitting the RTC, they will contribute a similar amount of interference to the sector's RoT noise. The higher the T2P value, the larger the interference the ATs cause.

Since T2P values are strongly correlated with the sector's RL loading, the AN can make reasonably accurate estimates of the interference the T2P values caused. The sector then allocates T2P resources to the ATs, such as $T2P = 4$ dB for one AT and $T2P = 6$ dB for another AT. Due to the reliability of T2P interference estimates, an aggressive (i.e., smaller) margin between the target RoT threshold and the operating RoT threshold can be used, effectively enabling a higher RL throughput. As long as the average T2P the AT uses stays below its allocated T2P resource, the AN can rest assured that the average RoT noise will be below the operating RoT threshold. The

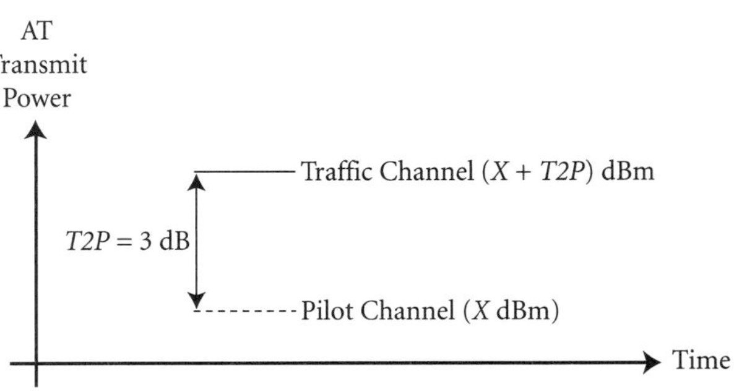

Figure 12.14: T2P Channel Transmit Power Ratio

use of T2P instead of transition probabilities enables direct and explicit control of RL interference and allows the AT to increase the data rates quickly instead of step-by-step. The AT, for example, can quickly increase the data rate as long as it stays within the allocated T2P resource and it adheres to the data rate change constraints conveyed during the session configuration. Since the AN has already accounted for the expected interference due to the T2P resource allocated to the AT, such quick data rate changes can be tolerated. In summary, an implementation-specific algorithm at the AN can allocate suitable T2P values for ATs in a sector in an attempt to control interference around the operating RoT threshold.

12.10 SUBTYPE 2 RTC MAC PROTOCOL DATA RATE CONTROL

The Subtype 2 RTC MAC Protocol allows per MAC flow T2P-based rate control. The AN allocates the T2P resource (e.g., $T2P = X$ dB) to an AT per active MAC flow. A total of $MaxNumMACFlows - 1$ MAC flows can be present with the maximum number of active MAC flows limited to MaxNumActiveMACFlows. A MAC flow numbered NN is considered to be active if the value of the parameter BucketLevelMaxNN is not null. The default value of MaxNumMACFlows is 4, with the highest supported value of 16. By default, two MAC flows numbered 00 and 01 are active because the default values of BucketLevelMax00 and BucketLevelMax01 are set to 20 dB and 27 dB, respectively. The AT can send signaling messages using only MAC flow 00; other MAC flows cannot be used for signaling. MaxNumMACFlows is the total number of activated and deactivated MAC flows. Both parameters, MaxNumMACFlows and MaxNumActiveMACFlows, are part of the MaxMACFlows attribute. Only the AT can initiate the negotiation of MaxMACFlows.

12.10.1 Terminology for Subtype 2 and 3 RTC MAC Protocol

Let's introduce some terms related to the T2P-based rate control before diving into the gory details of how the AT determines the data rate. Proceed with caution! The rate determination process defined in the standard is quite complex. The simplicity of Rev. 0 is giving way to the higher performance and increased complexity of Rev. A.

12.10.1.1 T2PInflow, T2POutflow, and BucketLevel Figure 12.15 provides a simplified view of the T2P resource that the AT uses to determine the instantaneous rate during a given time unit (e.g., a 26.67 ms frame in the case of Subtype 2 RTC MAC Protocol using Subtype 0 or 1 Physical Layer Protocol).

The AN allocates to the AT a T2P resource called *T2PInflow*,[438] which can be thought of as the average T2P that the AT can use over a time period. The higher the T2PInflow, the higher the data rate that the AT can use because the AT can transmit more power on the Data channel. In Figure 12.15, T2PInflow is 7 dB, so the AT can use an average of 7 dB of T2P for data transmission.

[438] In the absence of any T2PInflow received from the AN, the AT starts with T2PInflowmin, which is about 3.75 dB.

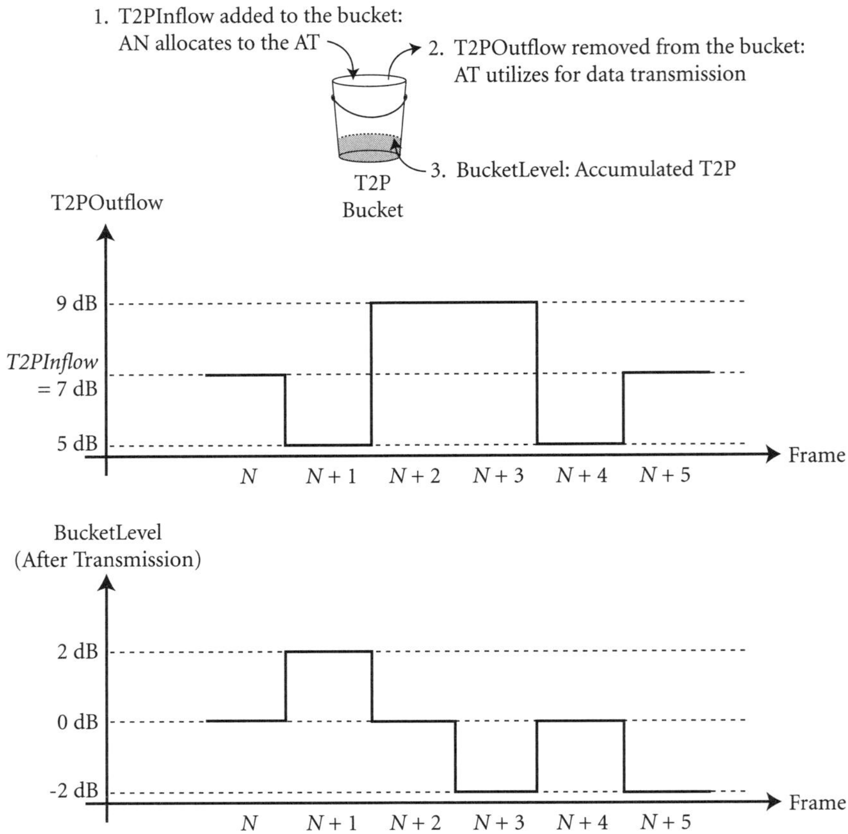

Figure 12.15: Processing of the T2P Resource

The goal here is to determine how much T2P the AT can use during frame N. At the beginning of the frame, T2PInflow is *added* to the T2P bucket. The T2P bucket may or may not be empty depending upon what has happened in the past. For simplicity, assume that the bucket is initially empty at the beginning of frame N. In other words, BucketLevel is 0 dB at the beginning of the frame N. The BucketLevel can be considered as an accumulated T2P resource. In Figure 12.15, the AT uses 7 dB T2P for data transmission during frame N. The consumed T2P resource is termed T2POutflow. T2POutflow is thus the resource the AT utilizes and is removed from the T2P bucket.

The remaining resource is 7 dB $-$ 5 dB $=$ 2 dB, which is the BucketLevel at the end of frame N or the beginning of frame $N+1$. In the next frame, frame $N+1$, again $T2PInflow = 7$ dB enters the bucket, leading to the approximate value of consumable T2P of $T2PInflow + BucketLevel = 7 + 0 = 7$ dB. The AT uses just

5 dB during frame $N + 1$. Since T2POutflow is 5 dB, the remaining T2P resource is 7 dB − 5 dB = 2 dB. Now, the BucketLevel at the end of frame $N + 1$ is 2 dB.

At the beginning of frame $N + 2$, $T2PInflow = 7$ dB enters the bucket that already has $BucketLevel = 2$ dB, yielding the consumable T2P of $7 + 2 = 9$ dB. Since $T2POutflow = 9$ dB for frame $N + 2$, BucketLevel at the end of frame $N + 2$ is $9 - 9 = 0$ dB.

T2PInflow of 7 dB enters the bucket at the beginning of frame $N + 3$. The AT now uses $T2POutflow = 9$ dB. Note that the AT uses more T2P (i.e., 9 dB here) than the available 7 dB! In other words, the AT can exceed the T2P because the AT is required to use an *average* T2PInflow; the instantaneous T2P the AT uses can be above or below T2PInflow. The BucketLevel now becomes 7 dB − 9 dB = −2 dB.

Finally, in Figure 12.15, $T2PInflow = 7$ dB enters the bucket with $BucketLevel = -2$ dB, yielding an approximate T2P resource of $7 - 2 = 5$ dB. The AT uses $T2POutflow = 5$ dB, giving the $BucketLevel = 0$ dB at the end of frame $N + 4$. In summary, the AN allocates T2PInflow, and the AT can use an average of T2PInflow for data transmission.

12.10.1.2 Making T2PInflow Dynamic Using Reverse Activity Bit, Soft Reverse Activity Bit, Quick Reverse Activity Bit, and Filtered Reverse Activity Bit Keeping T2PInflow constant for an AT's given MAC flow is simple, but dynamic T2PInflow can yield better performance by considering the prevailing RL interference environment. That's where RAB come into the picture.

Do You Recall. . . What role does the RAB play in the RL rate control in Rev. 0?[439]

If the RL interference in a sector is high, the ATs decrease their T2PInflows, leading to a reduction in data rates and, hence, a reduction in RL interference. However, if the RL interference in a sector is low, the ATs increase their T2PInflows, leading to an increase in data rates and, subsequently, better RL throughput.

Figure 12.16 summarizes how the quick reverse activity bit (QRAB) and filtered reverse activity bit (FRAB) are derived and subsequently utilized to make T2PInflow dynamic. QRAB indicates short-term, or instantaneous, RL loading, and FRAB specifies long-term RL loading.

12.10.1.3 From Reverse Activity Bit to Soft Reverse Activity Bit All sectors send out RABs on the RA channel that is part of the FL MAC channel. Every 1.67 ms FL slot contains the RAB, and the same RAB value, which could be 0 or 1, is repeated in consecutive RABLength slots. RABLength is specified to the AT in the TrafficChannelAssignment message. In Subtype 0 and 1 RTC MAC Protocol, RABLength can be set to eight, sixteen, thirty-two, or sixty-four slots. However, for Subtype 3 RTC MAC Protocol, RABLength is fixed at eight slots. The AT observes

[439]RAB indicates whether a given sector is heavily loaded with RL interference. Based on RAB and other parameters, the AT would decrease or increase its RL rate.

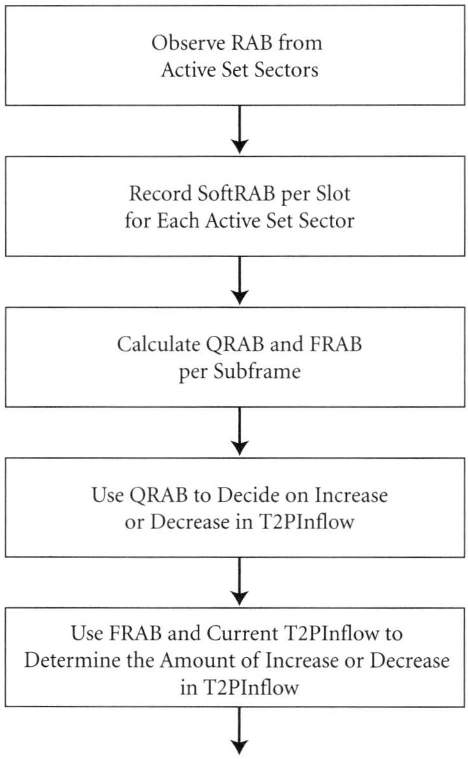

Figure 12.16: Making T2PInflow Dynamic at the AT

the RAB from each of its Active Set sectors, and the received real value of the RAB is termed SoftRAB. For example, when the decoder yields a real value of 0.8, a binary hard-limit decision-making entity interprets the RAB to be $+1$, while the real value of -0.7 would be interpreted to be -1 by such a binary hard-limit decision-making entity. The real (as opposed to binary) value of the received RAB, such as 0.8 or -0.7, is called SoftRAB and ranges from -2.5 to 2.5. The RAB is received every slot, and SoftRAB is calculated every time-slot. SoftRAB is then used to calculate the QRAB and FRAB.

12.10.1.4 From Soft Reverse Activity Bit to Quick Reverse Activity Bit

An IIR filter with the default time constant of four slots processes SoftRAB. QRAB for a given slot is determined by hard-limiting the output of the IIR filter to $+1$ or -1. The value of QRAB is sampled every subframe for each Active Set sector. If QRAB for any of the Active Set sectors is $+1$, the final QRAB for that subframe is considered to be $+1$, indicating a heavily loaded RL. Otherwise, the final QRAB for the given subframe is considered to be -1, corresponding to a lightly loaded RL. Since the QRAB is calculated using a short observation window, the QRAB reflects the instantaneous, or most recent, view of the prevailing RL interference environment.[440]

A Question for You... Let's do an easy yet important exercise. If QRAB is $+1$, should the AT try to increase or decrease T2PInflow?[441]

In summary, the AT uses the QRAB value to determine whether to increase or decrease T2PInflow. Now, *how much* should T2PInflow be increased or decreased? That's where the FRAB enters the scene!

12.10.1.5 From Soft Reverse Activity Bit to Filtered Reverse Activity Bit Like the QRAB, the FRAB per slot is obtained by using an IIR filter to process the SoftRAB. However, the time constant of the IIR filter for the FRAB is longer than that for the QRAB; its default value is 384 slots or 640 ms. The FRAB reflects the long-term loading in the recent past and is sampled every subframe for each Active Set sector. The final FRAB value is given by the largest FRAB among the FRABs for all Active Set sectors. For example, if the FRAB for a given subframe is -0.3, 0.1, and 0.8 for the AT's three Active Set sectors, the final FRAB for the subframe is determined to be 0.8.

12.10.1.6 Using Quick Reverse Activity Bit and Filtered Reverse Activity Bit for Updating T2PInflow As shown in Figure 12.16, the QRAB dictates whether to increase or decrease T2PInflow. The FRAB influences the amount by which the T2PInflow is changed. The Rev. A standard defines two transition functions, T2PUp and T2PDn (i.e., T2P Up and T2P Down), that specify the change in T2PInflow. In general, these transition functions depend upon the currently used value of T2PInflow and FRAB.

If $QRAB = +1$ (i.e., significant current, or instantaneous, interference), the current value of T2PInflow is large (meaning a high RL data rate), and the FRAB is a large positive number, such as 0.9 (meaning large long-term interference in the area), then both the long- and short-term interference levels are high and T2PInflow should be decreased by a large amount. However, if $QRAB = +1$, the current value of T2PInflow is small (corresponding to lower data rates), and FRAB is a large negative number (e.g., -0.9, associated with very little long-term loading), then T2PInflow requires only a small decrease (since the long-term interference is low) to alleviate the instantaneous high RL interference.

If $QRAB = -1$ (i.e., low current interference), the current value of T2PInflow is small (meaning a low RL data rate), and the FRAB is a large negative number, such as -0.9 (meaning low long-term interference in the system), then both the long-

[440] The determination of QRAB as discussed here is widely used in commercial deployments. Another way of calculating QRAB uses pilot strength measurements.

[441] Since QRAB is $+1$, interference is high and the AT should help the system by transmitting a packet with lower power, which means that the AT should *reduce* its data rate. The smaller the value of T2PInflow, the lower the data rate. Hence, the AT would *decrease* its current T2PInflow.

and short-term interference levels are low and T2PInflow can be increased by a large amount. However, if $QRAB = -1$, T2PInflow is large (corresponding to high data rates), and FRAB is a large positive number (e.g., 0.9, associated with large long-term loading), then T2PInflow should be increased by only a small amount to benefit from the instantaneous low RL interference.

The overall process of RL data transmission for Subtype 2 RTC MAC Protocol is summarized in Section 12.10.2, and, more details are available in Section 12.10.3

12.10.2 Overview of RL Data Transmission

Figure 12.17 provides an overview of the RL data transmission.[442] The AT sends on the Data channel the Request message conveying to the AN the amount of data in its buffer, the QoS requirements, and the maximum T2P that it can transmit.[443] The

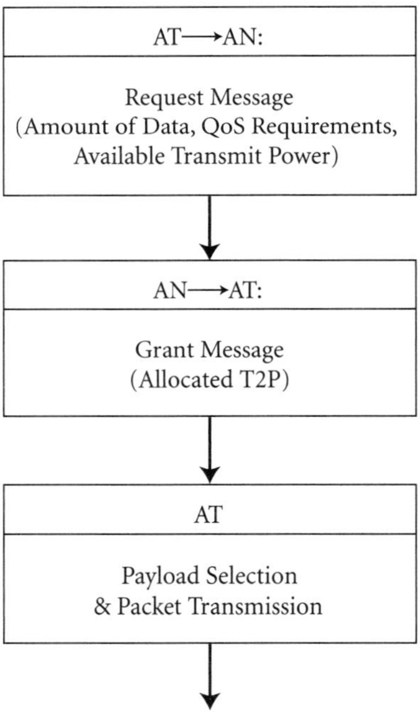

Figure 12.17: Reverse Link Data Transmission for Subtype 2 RTC MAC Protocol

[442] In another implementation approach that is common in commerical deployments, the AN does not explicitly allocate a T2P target to the AT. The AT starts with the default value of T2PInflowMin and autonomously decides the actual T2P target based on QRAB and FRAB calculations discussed earlier.

[443] Whether and how often the AT sends the Request message is a session-configurable parameter called MaxReqInterval under the control of the AN. The default value of such a parameter is "never," implying that the AT sends the Request message only if the AN configures the AT for this message during the session configuration.

AN responds to the AT by sending a Grant message on the Forward Traffic channel, which allocates the T2P target to the AT. The AT chooses a payload based on factors such as the allocated T2P, the prevailing RL interference, and the amount of data to be transmitted. The AT transmits on the Data channel the packet with the selected payload. As shown in Fig. 12.17, the Subtype 2 RTC MAC Protocol operation differs from the Rev. 0 operation in that it uses T2P instead of transition probabilities.

12.10.3 A Comprehensive View of RL Data Transmission

The AT obtains a T2P allocation by using default values defined in the standard and/or via a Grant message from the AN. The AN sends a Grant message autonomously or in response to the Request message received from the AT. For a MAC flow numbered NN, the T2PInflowRangeNN attribute specifies the minimum and maximum values of T2PInflow as $T2PInflowmin = 4$ dB and $T2PInflowmax = 10.75$ dB. Hence, in the absence of a Grant message from the AN, the AT can use these default values for the MAC flows activated during the session configuration. Recall that two MAC flows are defined as default by the standard and can be used without an explicit session configuration. Figure 12.18 summarizes the major contents of the Request and Grant messages.

The AT sends a Request message to the AN to indicate the maximum transmit T2P that the AT can support at this time and the queue length for each MAC flow for which the resources are being requested. The AT indicates the maximum T2P that the AT can currently transmit. Recall that the power control is carried out for the reverse link, and hence the AT may have varying amounts of *remaining* power usable for the RTC based on propagation path loss and prevailing channel conditions.

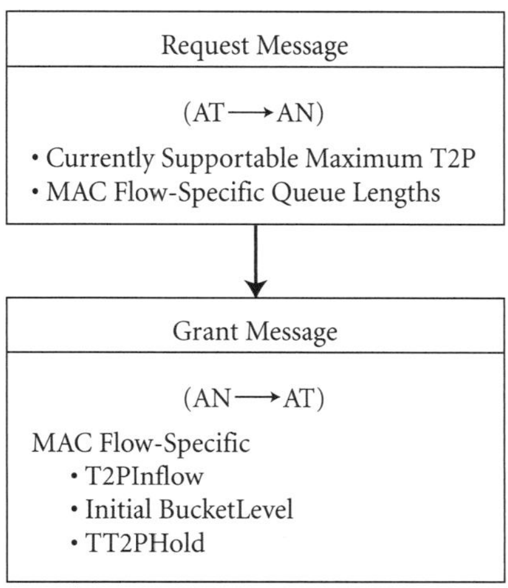

Figure 12.18: Allocation of the T2P Resource

The queue length is represented by a four-bit parameter, QueueLength, that maps onto a specific amount of data in AT's buffer. For example, the queue length of 0 indicates that $0 \leq QueueLength < 32$ bytes. A queue length of 15 corresponds to $QueueLength \geq 16384$ bytes. The AT can send the Request message any time. However, the AT must send the Request message if the AN has configured it for periodic transmission of the message or if it has sent a significant amount of bits since the last transmission of the message. The periodicity of the Request message transmission is controlled by the MaxReqInterval parameter, and the data-volume–based message transmission is controlled by the ReqRatio and RequestPktSize parameters. Both MaxReqInterval and ReqRatio are part of the RequestParameters attribute, and both of them have the default value of 0, implying that the AT need not send a Request message. The MaxReqInterval parameter specifies the *maximum* number of frames for which the AT *can wait* between the successive Request messages and ranges from 0 to 15 frames. The ReqRatio parameter ranges from 0 to 0.075, and the AT is required to send a Request message if the AT has transmitted $(1/ReqRatio) - 1$ RequestPktSize bits on the Data channel since transmitting the last Request message. The RequestPktSize parameter is the number of bits in the Request message.

The AN can send a Grant message anytime. The Grant message specifies the T2PInflow, BucketLevel, and TT2PHold parameters per active MAC flow for one or more flows. The T2PInflow parameter is the average T2P resource that the AT can use. The T2PInflow parameter is an eight-bit parameter that yields a range from 0 to 63.5 dB with the value 255 representing $-\infty$ dB or 0 in normal units. This is the single most critical parameter that gives the AN the ability to control the RL rate, curb RL interference, and provide intra- and inter-AT QoS.

The AT uses the BucketLevel parameter to initialize the bucket level. It is an eight-bit field representing a range of 0 to 63.5 dB, with the value 255 representing $-\infty$ dB. The AN can use this parameter to provide an initial T2P boost with no long-lasting impact on the average T2P resource the T2PInflow specified. For example, if the T2PInflow is 5 dB and the BucketLevel is specified to be 2 dB, the AT can use 7 dB while trying to determine the data rate. Choosing a non-zero BucketLevel thus helps the AT send data at an initially higher rate. Over time, the AT could use the T2P resource corresponding to the T2PInflow.

The TT2PHold parameter is the time duration for which the AT needs to maintain the T2PInflow value received in the current Grant message. In other words, if TT2PHold specifies four frames, the AT cannot update the value of T2PInflow for four frames. TT2PHold is a six-bit field and hence allows the longest time period of thirty-two frames or $32 * 26.67$ ms $= 1.7$ seconds.

A Question for You. . . How can the AN ensure that the AT would never update the value of T2PInflow?[444]

[444]The AN could send a new Grant message before the expiration of the TT2PHold timer corresponding to the previously sent Grant message. The AN can continue sending Grant messages one after another in this manner.

Since Subtype 2 RTC MAC Protocol relies upon Subtypes 0 and 1 Physical Layer Protocol, the data rate selection is completed once every 26.67 ms frame and the packet is constructed accordingly. The AT selects the data rate based on various factors, such as the allocated T2P resource, the available amount of data, the available transmit power, and the prevailing RL interference. A sector measures the RL interference and quantifies it as RAB as in Rev. 0. The AT observes its Active Set sectors transmitting the RAB and calculates two RAB-dependent variables, QRAB and FRAB. The basic idea is that the AT can dynamically update its initially allocated T2PInflow. However, some processing is done every slot and some processing is done every subframe (i.e., every 6.67 ms). Figure 12.19 summarizes the main steps that the AT follows to determine the data rate for a given frame and to construct a packet based on the selected rate.

The AT updates the variables PilotStrength, SlotQRAB, and SlotFRAB at the beginning of every slot for each Active Set sector. PilotStrength represents the IIR-filtered pilot E_c/I_0 in linear domain. PilotStrength can be used in the transition functions T2PUp and T2PDn to adjust the change in T2PInflow based on the FL

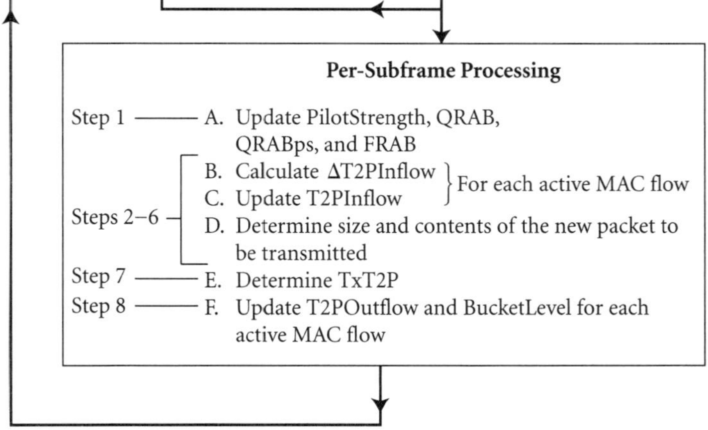

Figure 12.19: Subtype 2 RTC MAC Protocol Data Rate Determination at the AT

quality of the FL serving sector and to derive pilot-strength–based QRAB (discussed later). However, PilotStrength is essentially unused based on the default values in the standard.

SlotQRAB is derived by filtering the Soft RA bit received during this slot and hard-limiting the filter output such that positive input values are mapped to $+1$ and non-positive values are mapped to -1. SlotFRAB is obtained by filtering the Soft RA bit received during this slot.

Though per-slot processing is carried out to update relevant variables, this processing is done *at the start of a subframe* to determine the final values of the QRAB, FRAB, and T2PInflow for a given subframe and to construct a packet for transmission (if required). In the description that follows, n is the current subframe number and s is an Active Set sector.

Step 1 Sampling of Relevant Variables. The outputs of the IIR filters for PilotStrength, QRAB, and FRAB are sampled for each Active Set sector to determine $PilotStrength_{n,s}$, $QRAB_{n,s}$, and $FRAB_{n,s}$, respectively.

Step 2 Determination of $QRAB_n$. If any of the Active Set sectors is loaded, which is indicated by $QRAB_{n,s} = +1$, $QRAB_n$ is set to $+1$. If all Active Set sectors are unloaded (i.e., $QRAB_{n,s} = -1$), $QRAB_n$ is set to -1.

Step 3 Determination of $QRABps_n$. If the FL serving sector is loaded (i.e., $QRAB_{n,s} = +1$), $QRABps_n$ is set to $+1$. If an Active Set sector with a sufficiently strong $PilotStrength_{n,s}$ is loaded, $QRABps_n$ is set to $+1$. In all other cases, $QRABps_n$ is set to -1. The main idea here is to minimize weak sectors from influencing the AT's RL data rate since the AT is relatively far away from such weak sectors and cannot significantly alter the RL interference environment in such sectors.

Step 4 Determination of $FRAB_n$. It represents the worst-case RL interference and hence is set to the largest value of $FRAB_{n,s}$ among all Active Set sectors.

Step 5 Decision on T2PInflow Increase or Decrease (per MAC flow i). If the configurable parameter QRABSelect I is $+1$, $QRABps$ is used to decide whether to increase or decrease T2PInflow. If QRABSelect I is 0, $QRAB_n$ is used. By default, $QRAB_n$ is used. If $QRAB_n$ (or $QRABps$) is $+1$, T2PInflow is decreased to alleviate RL interference. If $QRAB_n$ (or $QRABps$) is -1, T2PInflow is increased to increase throughput since the interference is light.

Step 6 Update of T2PInflow. The amount by which T2PInflow is changed, ΔT2PInflow, is influenced primarily by the current T2PInflow and FRAB and is calculated using two-dimensional transition functions T2PUp and T2PDn. The basic idea is that T2PInflow is increased by a large amount if FRAB is low and T2PInflow is small because the potential for causing excessive interference is low. On the other extreme, T2PInflow is increased by a small amount if FRAB is high and T2PInflow is large because the potential for causing significant interference is high. See Section 12.11 for examples of numerical values that quantify ΔT2PInflow as a function of T2PInflow and FRAB.

Step 7 Determination of Packet Size and Packet Contents. The T2PInflow that the AT can use, PotentialT2POutflow, is a function of the T2PInflow, BucketLevel, and BurstFactor. For example, if T2PInflow is 2 dB, the AT may be able to use 2.5 dB where 2.5 dB – 2 dB $= 0.5$ dB is obtained by considering BucketLevel and BurstFactor. The basic idea is that the instantaneous T2PInflow that the AT uses can be smaller or greater than the target T2PInflow, and, on average, the T2PInflow the AT uses should be target T2PInflow. Though T2PInflow is updated every subframe, the packet is constructed only per 26.67 ms frame in Subtype 2 RTC MAC. Since one frame consists of four subframes, the AT gets an opportunity to construct a new packet using the latest updated T2PInflow every four subframes. The standard provides conditions, which the AT needs to satisfy for the packet size it selects. Let PS be the selected packet size and T2PPS be the T2P associated with PS. Now let's look at those conditions.

- The AT cannot have significant differences in packet sizes in consecutive frames. For example, if the AT has transmitted a packet of size 512 bits in frame $n - 1$, it can potentially transmit a packet of size 1024 bits but it cannot transmit a packet of size 4096 bits. This condition allows gradual (rather than sudden) changes in RL interference. However, such a change is not probabilistic, which is the case for Rev. 0.
- T2PPS cannot exceed the larger of the sum of T2POutflows for all MAC flows or the T2PInflowmin. This condition ensures the availability of at least some T2P for transmission and prevents the use of unnecessarily large T2P.
- T2PPS cannot exceed TxT2Pmax, which is a function of the Pilot-Strength of the serving sector. This condition places an upper limit on T2P.
- A packet size smaller than PS should be unable to carry the sum of the smaller of the available bits in the buffer and the T2P-transmittable bits for all MAC flows. The T2P-transmittable bits is the maximum number of bits that could be transmitted for a given PotentialT2POutflow.
- The available transmit power should be able to transmit the packet with T2PPS.

Step 8 Determination of TxT2P and Update of T2POutflow and BucketLevel. T2PPS is expressed in decibels, and T2PPS in normal units is referred to as TxT2P. TxT2P defines the power level at which the Data channel carrying the packet of size PS is transmitted relative to the Pilot channel. T2POutflow$_i$ is the T2P consumed by a MAC flow i and is proportional to the number of bits in the packet. For example, for a packet size of 400 bits and TxT2P of 4 (in normal units), if MAC flow 1 has 100 bits in the packet and MAC flow 2 has 300 bits in the packet, T2POutflow$_1$ is $4 * (100/400) = 1$ and T2POutflow$_2$ is $4 * (300/400) = 3$. BucketLevel$_i$ is increased by T2PInflow$_i$ and reduced by T2POutflow$_i$.

The AT transmits the packet of size PS on the data channel at TxT2P. Since the AT has selected the packet size, the AN has no idea about the packet size. Relying on blind packet size detection at the AN would mean a lot of work! To avoid that, the AT specifies to the AN the selected packet size.

Do You Recall... Let's see if you still remember Rev. 0, or if you need to re-visit Chapter 10! In Rev. 0, which channel does the AT use to convey to the AN the chosen data rate?[445]

The Subtype 2 RTC MAC Protocol utilizes the Subtype 0 Physical Layer Protocol and Subtype 1 Physical Layer Protocol, and its supported data rate ranges from 9.6 kbps (corresponding to the packet size of 256 bits) to 153.6 kbps (corresponding to the packet size of 4096 bits). The RRI value, therefore, indicates such data rates.

12.11 SUBTYPE 3 RTC MAC PROTOCOL DATA RATE CONTROL

Subtype 3 RTC MAC Protocol shares several similarities with Subtype 2 RTC MAC Protocol, such as the use of a T2P-based rate control instead of a probabilistic rate control. However, compared to Subtype 2 RTC MAC Protocol, Subtype 3 RTC MAC Protocol uses an enhanced physical layer to achieve higher peak and average throughput in the reverse link. A high-level view of RL data transmission is provided in Section 12.11.1, and a comprehensive view is offered in Section 12.11.2. As mentioned earlier, Subtype 3 RTC MAC Protocol is the most prevalent subtype in commercial deployments.

12.11.1 Overview of RL Data Transmission

Figure 12.20 provides a high-level view of RL data transmission for Subtype 3 RTC MAC Protocol. First, the AT sends a Request message to the AN specifying the amount of data in the buffer per active MAC flow, the QoS requirements, and the available transmit power. The AN allocates T2P per active MAC flow via a Grant message. The AT determines the payload size and constructs subpackets to implement H-ARQ. It sends a subpacket in a subframe. The AN tries to retrieve the packet by decoding the first subpacket. The AN sends an ACK or NAK to the AT to indicate successful or failed packet decoding. If the AN sends an ACK, then the AT sends subpackets of a new packet. Otherwise, it continues to transmit the subpackets of the current packet.

Let's compare Subtype 2 and Subtype 3 RTC MAC Protocol. Both use the T2P-based mechanism to determine the packet size and hence the data rate. Since Subtype 3 RTC MAC Protocol utilizes Subtype 2 Physical Layer Protocol, it controls the data rate ranging from 4.8 kbps to 1.8 Mbps (or 1.2 Mbps); Subtype 2 RTC MAC Protocol supports the data rates from 9.6 kbps to 153.6 kbps. Subtype 3 RTC MAC Protocol

[445]The RRI channel, which is time-multiplexed with the Pilot channel, carries the RRI. The RRI specifies the data rate ranging from 9.6 kbps to 153.6 kbps.

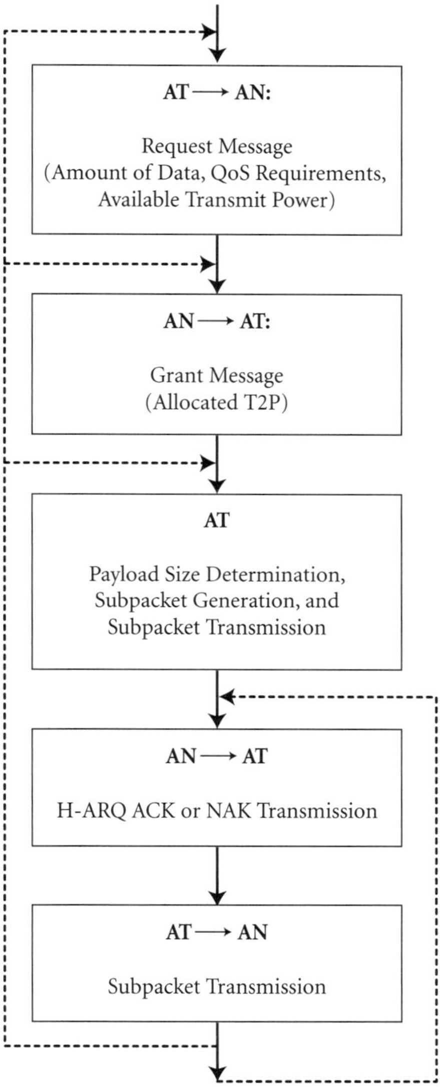

Figure 12.20: High-Level View of RL Data Transmission (Subtype 3 RTC MAC Protocol)

can change the data rate every subframe, and Subtype 2 RTC MAC Protocol can change the data rate every frame. RL H-ARQ is supported only in Subtype 3 RTC MAC Protocol, and, hence, the AT creates subpackets from a packet for transmission in interlaced subframes.

Table 12.7: Influence of FRAB on the Increase in T2PInflow

FRAB	T2PUp (dB)	Increase in T2PInflow (Normal Units)
−1.0	−1.25	0.75
−0.625	−6.0	0.25
−0.75	−7.0	0.13
0.875	−9.0	0.13

12.11.2 Detailed View of RL Data Transmission

Since both Subtype 2 and Subtype 3 RTC MAC Protocol utilize T2P-based rate determination, the process of allocating the T2P resource is the same for both. Section 12.10.3 has relevant details. In summary, the AT could send a Request message, and the AN allocates T2P via a Grant message.

The AT updates its T2PInflow by using the transition functions T2PUp and T2PDn per subframe. Such an update process is the same as that for Subtype 2 RTC MAC Protocol and is discussed in Section 12.10.3. If the RL interference is high (specified by $QRAB = +1$ and a large FRAB), T2PInflow is decreased, and T2PDn governs the amount of the T2PInflow decrease. Similarly, if the RL interference is low (specified by $QRAB = -1$ and a small FRAB), T2PInflow is increased, and T2PUp governs the amount of the T2PInflow increase. The standard defines default values for a two-dimensional T2PUp and T2pDn, with current T2PInflow and FRAB as two dimensions. Based on these values, Tables 12.7 and 12.8 are constructed for T2PUp to show numerical examples.

Assume that the current value of T2PInflow is 8.5 dB or 7.08 in normal units. As shown in Table 12.7, as long-term interference increases (shown by an increasing FRAB), the increase in T2P is smaller to prevent excessive interference.

Table 12.8 provides the trend of an increase in T2PInflow as a function of T2PInflow for the current value of FRAB $= -1.0$. As current T2PInflow increases, the amount of the increase in T2PInflow decreases to avoid excessive interference. If a subframe is available, the AT constructs a new packet to be transmitted in a predetermined number of interlaced subframes (e.g., four subframes).

Table 12.8: Influence of Current T2PInflow on the Increase in T2PInflow

Current T2PInflow (dB)	T2PUp (dB)	Increase in T2PInflow (Normal Units)
0.0	7.25	5.31
8.5	−1.25	0.75
17.75	−1.25	0.75

A Question for You... Actually, we have two questions for you! How long is the gap between successive subframes carrying a given packet, and why do we have such a gap?[446]

Subtype 3 RTC MAC Protocol offers two unique features, *transmission modes* and *power-boosted transmission*. Consider a sector in which forty ATs are sending packets with the payload size of 4000 bits. Assume that the average packet error rate of 1% is achieved when the ATs transmit all four subpackets using T2P of X dB. If we allocate $X + 2$ dB to the ATs and force the ATs to use the payload size of 4000 bits, what can we say about the average number of ATs that can be supported and the average packet latency? Will we be able to support more users or fewer users? Will the packet latency be shorter or longer? Let's think about this.

Since a given AT is transmitting a larger transmit power at the same payload size, the AN would now have somewhat higher interference per AT and hence would be unable to sustain forty ATs. Hence, the number of ATs supported in the sector will be less than forty to maintain the average RoT noise at the fixed target level. Now, since the ATs are transmitting the same size packet at a higher transmit power level, the AN can likely decode the packet before all four subpackets are transmitted. Thus, the packet latency is likely to be shorter. In summary, smaller T2P values can be used to support more simultaneous users, and such an operating mode is called high-capacity mode. When larger T2P values are used, fewer users are supported but latency decreases, and such an operating mode is called low-latency mode. The operating mode can be chosen per MAC flow for a given user.

Consider Figure 12.21 in which the AT uses $T2P = 5$ dB during each of the four subframes. Assume that the average packet error rate of 1% is obtained. Now, let's try to maintain the average T2P of 5 dB but use a little larger T2P for the first two subpackets and a little smaller T2P for the last two subpackets. In other words, we are implementing *power boosting* for the initial transmissions of the subpackets. If the AT transmits all four subpackets, the average T2P the AT consumes remains the same as before. However, since the AT is boosting its power during the initial transmissions of the subpackets, the AN is more likely to detect the packet early, essentially reducing the average packet latency. In summary, power boosting helps reduce latency while using a similar average T2P as non-power-boosted transmission.

The Rev. A standard combines the features of transmission modes and power boosting. For a given packet size PS (e.g., $PS = X$ bits), parameters, such as T2PHiCapPreTransitionPS, T2PHiCapPostTransitionPS, T2PLoLatPreTransitionPS, T2PLoLatPostTransitionPS, HiCapTerminationTargetPS, and LoLatTerminationTargetPS, are compatible.

[446]The gap is two subframes. If one subpacket is transmitted during the subframe n, the next subpacket is transmitted during the subframe $n + 3$. This gap enables H-ARQ because it gives the AT enough time to receive H-ARQ feedback from the AN and to decide on the need for transmission of the remaining subpackets.

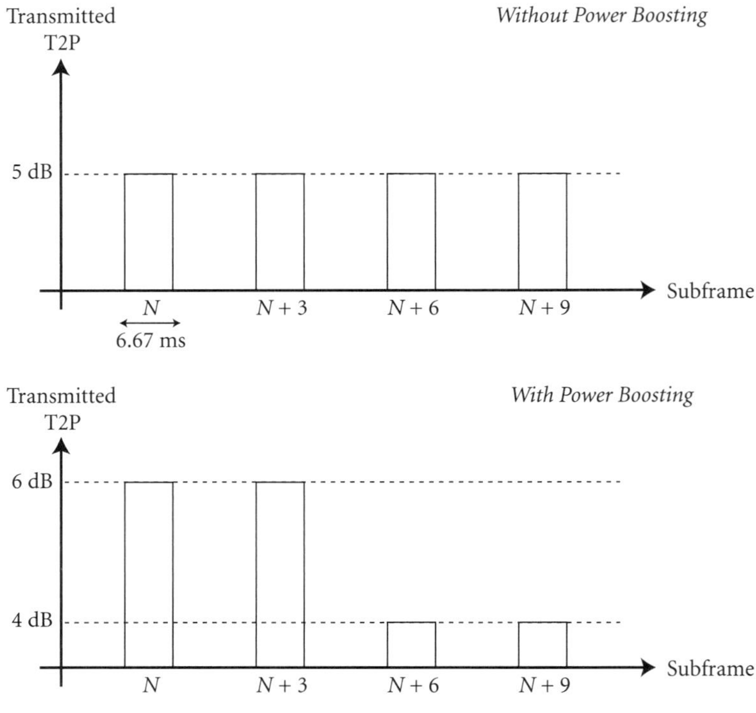

Figure 12.21: Power-Boosted Transmission

HiCapTerminationTargetPS is the termination target (e.g., four subpackets) for the high-capacity transmission mode, and LoLatTerminationTargetPS is the termination target (e.g., two subpackets) for the low-latency transmission mode. T2PHiCapPreTransitionPS is the T2P value used for the first $HiCapTerminationTargetPS + 1$ subpackets, and T2PHiCapPostTransitionPS is the T2P value used for the remaining subpackets for the high-capacity mode. T2PLoLatPreTransitionPS is the T2P value used for the first $LoLatTerminationTargetPS + 1$ subpackets, and T2PLoLatPostTransitionPS is the T2P value used for the remaining subpackets for the low-latency mode.

The AT forms a packet and starts transmitting the first of the set of the packet's subpackets. The RRI channel (see Section 12.5.1 for details) specifies the payload size of the packet and the identity of the subpacket (e.g., zeroth, first, second, or third) the AT formed so that the AN can avoid blind payload size detection and will know how to combine the received subpacket with the previously received packet's subpackets to recover the packet properly. The subpackets are transmitted once and in sequence. In other words, if required, the AT transmits subpacket 2 after transmitting subpacket 1. The AT also transmits the Auxiliary Pilot channel if the payload size is greater than or equal to a configurable threshold, AuxiliaryPilotChannelMinPayload. See Section 12.5.2 for details of the Auxiliary Pilot channel.

As discussed in Section 12.4.1, the RL H-ARQ used in Subtype 3 RTC MAC Protocol is quite comprehensive. The AN sends the H-ARQ bit in response to the all subpackets except the last one, the L-ARQ bit in response to the last subpacket, and the P-ARQ bit in response to the overall packet.

If the AT receives a positive acknowledgment for a packet P (e.g., $H\text{-}ARQ\ bit = ACK$), it discontinues the transmission of packet P's subpackets and starts sending subpackets of another packet, Q. If the AT receives no positive acknowledgment for packet P, it continues transmission of packet P's subpackets.

12.12 REV. A QoS CONTROL

Rev. A enables both intra- and inter-AT QoS control. Intra-AT QoS control means that two applications or services running in the AT can obtain different types of QoS. Inter-AT QoS control indicates that two different ATs can get a different QoS (e.g., a Platinum versus Gold subscription). Consider Figure 12.22, which shows an AT in a web-browsing session.

The end-to-end link is between the AT and the web server. A true end-to-end QoS would then involve control of various links between the AT and the web server, such as the radio interface between the AT and the BS, the (proprietary) link between the BS and the combined RNC and Packet Control Function systems, the link between the Packet Control Function and the combined Packet Data Serving Node (PDSN) and Foreign Agent systems, the link between the Foreign Agent and the Home Agent, and the link between the Home Agent and the web server. The logical link between the AT and the Foreign Agent or the Packet Data Serving Node is a Point-to-Point tunnel, and the logical link between the Foreign Agent and the Home Agent is a Mobile IP tunnel.

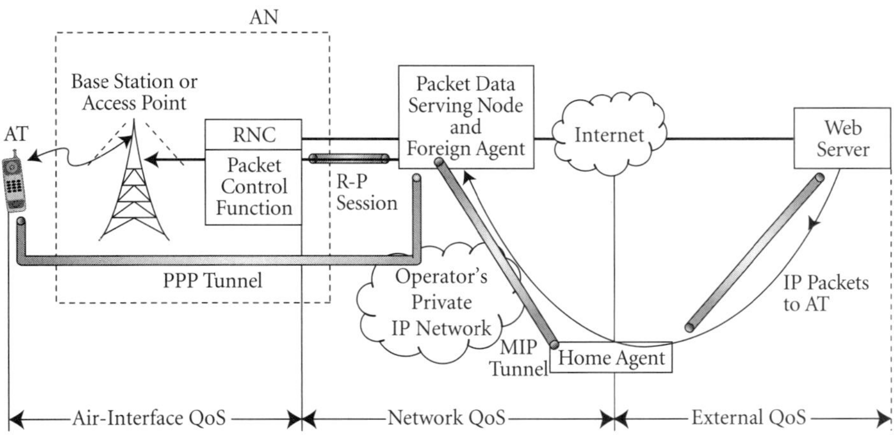

Figure 12.22: End-to-End QoS Control

The end-to-end QoS control can then be thought of having three main components—air-interface QoS, network QoS, and external QoS. The air-interface QoS is between the AT and the AN, the network QoS is between the AN and the Home Agent, and the external QoS is between the Home Agent and the web server. The QoS within an operator's control is the air-interface QoS and the network QoS. When the packet is in the public Internet, the QoS can be only influenced, not explicitly guaranteed.

The air-interface QoS control is achieved by managing the RLP flows via a multi-flow packet application, and the network QoS relies upon Radio-Packet (R-P) interface-specific QoS control and a generic IP-based QoS mechanism such as Diff-Serv (Differentiated Services). We'll discuss the air-interface QoS and R-P interface QoS control in just a moment; DiffServ details are given in Chapter 4. For right now, we'll just remind you that the basic idea of DiffServ is that IP routers can treat different services differently by mapping the services onto different DiffServ code points (DSCPs). One DSCP may have higher packet forwarding priority than another. Hence, an IP router would first handle delay-sensitive packets. If a delay-sensitive packet were found abandoned in the IP router's buffer for a long time, the router could discard such packets.

Assume that an AT is participating in web-browsing. The flow of IP packets from the web server to the AT is an IP flow, and the flow of IP packets from the AT to the web server is another IP flow. If the AT is also using a VoIP application, we have two unidirectional IP flows. In a given direction, the web-browsing IP flow has different requirements than the VoIP IP flow.

12.12.1 Air-Interface QoS

The PDSN obtains the subscribed QoS profile for a given user from the Authentication, Authorization, and Accounting Server and provides its copy to the RNC. When a service or application is activated for a given user (e.g., setting up a VoIP call), such a service could be associated with a QoS profile identity that defines a set of parameters, such as required data rate and latency. The RNC provides such QoS information to the BS so that the BS scheduler can consider QoS while allocating resources to the AT. Forward link QoS and reverse link QoS are somewhat different, although both rely upon manipulation of RLP flows.

The main mechanism for FL QoS control is management of multiple RLP flows and the BS scheduler's selective treatment of these flows. One IP flow in a given direction is established for one application and is mapped onto an RLP flow. Remember that one RLP flow gets one type of QoS treatment, and another RLP flow can get another type of QoS treatment. So, if the AT is participating in web-browsing and VoIP, it has two distinct QoS requirements, and, hence, two RLP flows would be required in the FL. The RLP flow for web-browsing can benefit from RLP retransmissions. However, the RLP flow for the VoIP application should omit RLP retransmissions.

Do You Recall... How many RLP retransmissions are used in Rev. 0 and Rev. A for a given packet—two or four?[447]

The IP flow for the web-browsing is assigned one value of ReservationLabel, and the IP flow for the VoIP application is assigned another. Furthermore, the FL scheduler will give higher priority to VoIP and lower priority to web-browsing. Additionally, short VoIP packets for multiple ATs can be grouped together to construct a multi-user packet to achieve higher resource utilization efficiency.

The air-interface QoS control in the RL relies on RLP flows and the RL scheduler. An IP flow in the RL is allocated a ReservationLabel and is mapped onto an RLP flow, which, in turn, is mapped onto a MAC flow. The RL scheduler allocates the T2P resource per MAC flow, enabling intra-AT QoS control. Furthermore, each MAC flow can be configured for the high-capacity or low-latency mode. Inter-AT QoS control is achieved by allocating different T2PInflows for different ATs.

12.12.2 Network QoS Control on R-P Interface

For the network QoS control on the R-P interface, an R-P (i.e., radio-packet) session occurs between the RNC and Packet Control Function and the PDSN for a given AT. One R-P session contains multiple R-P or A10 connections, with each connection providing one type of QoS. For example, one R-P connection could be present for web-browsing and another for VoIP. A certain bandwidth could be reserved for each R-P connection, and a Generic Routing Encapsulation key can be used to identify a given R-P connection. (See Chapter 4 for Generic Routing Encapsulation details.) Each R-P connection can be given a certain QoS level by establishing a suitable A10 tunnel between the PDSN and the AN.

When a packet for a given AT arrives at the PDSN from the outside world (e.g., Internet), the PDSN must know which R-P connection should carry such a packet. A Traffic Flow Template is defined to facilitate such packet forwarding. One example is the use of the combination of {Source IP Address, Destination IP Address, Port Numbers} to place an arriving packet onto an R-P connection. The R-P connection can then be mapped onto an IP flow via the ReservationLabel, which subsequently maps onto an RLP flow. The RLP flow now facilitates implementation of the air-interface QoS.

When a new application is activated, the AT and the PDSN exchange Resource Reservation Protocol messages, such as Resv and ResvConf, to establish the QoS for the application. The QoS itself is quantified in the form of a Flow Profile ID that describes parameters such as the data rate, packet delay budget, and packet error rate. For example, $Flow\ Profile\ SID = 0x100$ corresponds to conversational Rate Set 1 speech, 100 ms packet delay between the AT and the PDSN, and 1% packet error rate [49]. The PDSN and the RNC set up a suitable A10 tunnel for the required

[447] It's a trick! The answer is one! Only one NAK-based RLP retransmission occurs in Rev. 0 and Rev. A.

QoS, and the RNC works with the AT to set up an RLP flow that can meet the target QoS that the Flow Profile ID defines. Each R-P connection can be given a certain level of QoS by establishing a suitable A10 tunnel between the PDSN and the RNC. Furthermore, the IP flow or the application is associated with RerservationLabel that is valid between the AT and the RNC. This ReservationLabel is then mapped onto a specific RLP flow ID. The A10 tunnel for default best effort flow is referred to as default A10; additional A10 tunnels are called auxiliary A10s, which can have assured forwarding and expedited forwarding flows. Multiple reservations can be mapped onto a single RLP flow.

12.13 HANDOFF WITHIN 1xEV-DO

Handoff in Rev. A involves shorter traffic interruption and is more efficient than that in Rev. 0 because Rev. A utilizes *fast cell-switching*. Let's say that an AT has an Active Set {PN 100, PN 104, PN 300} where pilot PN offsets 100 and PN 104 belong to the sectors BS1 controls and pilot PN offset 300 is used in a sector that belongs to BS3. Such an AT is traveling, as shown in Figure 12.23.

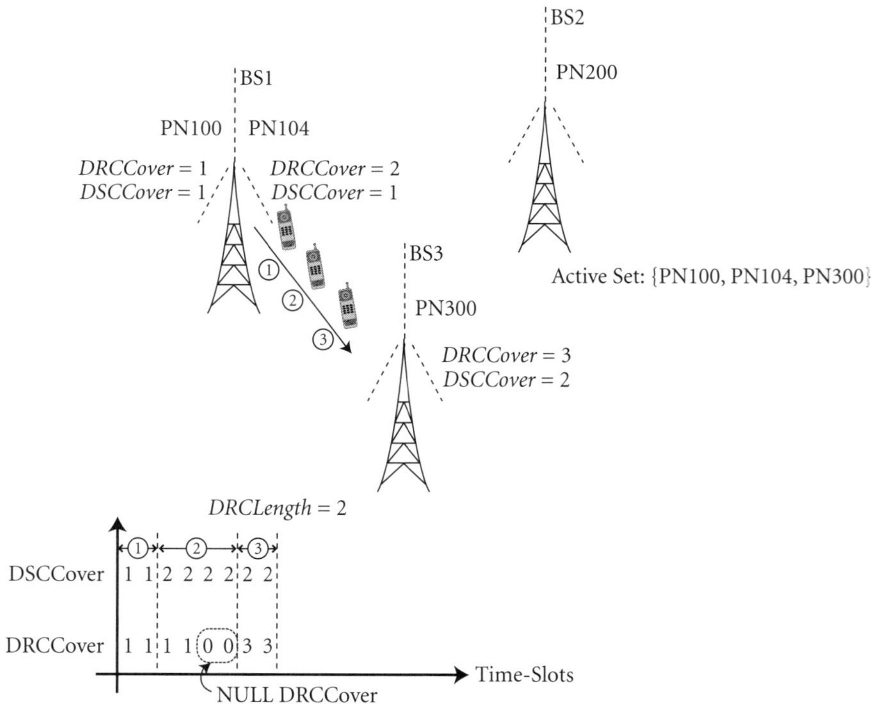

Figure 12.23: Fast Cell-Switching in Rev. A

Each sector in the Active Set is associated with DRCCover and DSCCover. The DRCCover identifies the sector within the Active Set, and the DSCCover identifies the Active Set source (i.e., the BS) that controls the sector. For example, in Figure 12.23, a sector that is transmitting a Pilot channel with PN offset of 100 is assigned the DRCCover of 1 and a sector that is transmitting a Pilot channel with the PN offset 104 is assigned $DRCCover = 2$. Since the same BS controls both of these sectors, both sectors have the same identity of the source, $DSCCover = 1$. A sector transmitting a Pilot channel with the PN Offset of 300 belongs to another BS, BS2, and, hence, the DSCCover for this sector is $DSCCover = 2$.

Do You Recall... Which message conveys the DRCCover to the AT? (Hint: The message name starts with T.)[448]

This message in Rev. A specifies both the DRCCover and DSCCover for every sector in the Active Set. The AT utilizes the DRCCover on the DRC channel and the DSCCover on the DSC channel. The *DRCCover* specifies the *sector* from which the AT wants to receive the FL packets *now*, and the *DSCCover* specifies the *BS* from which the AT would like to get data *in the near future*.

Let's see what the AT transmits on the RL as it travels from BS1 to BS3 as depicted in Figure 12.23. When the AT receives the best E_c/I_0 from PN 100, it chooses the sector with Pilot PN Offset 100 to receive data by specifying $DRCCover = 1$. The DSCCover during Stage 1 is 1, corresponding BS1, which controls the current FL serving sector.

During Stage 2, the AT predicts that it is moving toward a new BS, BS3, and, hence, it changes DSCCover to 2. At this time, the AN knows that the AT will soon be switching to a sector that belongs to BS3. The AN can prepare BS3 to be ready for transmission to the AT in the near future. The AN can, for example, start sending FL packets to BS3 in addition to sending the packets to the currently serving BS1. During Stage 2, the AT is still sending $DRCCover = 1$ on the DRC channel, and it can continue receiving FL packets from the current sector.

The AT enters Stage 3 where it gets the best pilot E_c/I_0 from the sector with Pilot PN Offset 300. To convey its intention to switch sectors, the AT sends NULL DRCCover on the DRC channel and then starts indicating $DRCCover = 3$, which is associated with the sector with the pilot PN offset 300. Since BS3 already has data for the AT, it can immediately start sending packets to the AT. In summary, fast cell-switching is implemented in Rev. A by using the DSCCover in addition to the DRCCover. The DSCCover essentially minimizes the interruption on the FL data transfer and facilitates QoS control for real-time services.

[448]The answer is the TrafficChannelAssignment message.

12.14 IDLE STATE ENHANCEMENTS AND SIGNALING ENHANCEMENTS

Rev. A introduces Idle State enhancements, such as three-tiered monitoring[449] and subsynchronous capsules, and supports efficient transmission of short data bursts via Data Over Signaling Protocol. Rev. 0 defines just two types of capsules that are used on the Control channel to send overhead messages, such as the QuickConfig, SectorParameters, and Page messages.

Do You Recall... Do you remember the names of these capsules? (Hint: One starts with an S and the other starts with an A).[450]

Synchronous capsules are sent every 256 slots. Subsynchronous capsules can be sent every sixty-four slots, reducing the delay by a factor of $256/64 = 4$. The AT can thus significantly reduce the time to receive a message, even a Page message, on the Control channel.

In the slotted mode operation in the Idle State in Rev. 0, the AT wakes up once every 5.12 seconds to look for a possible Page message. However, such a delay may be excessive for real-time services, such as VoIP and interactive gaming. To reduce the time to enter the Connected State from the Idle State, three-tiered monitoring is available. Instead of waking up every 5.12 seconds, the AT wakes up more frequently. Three tiers or levels of monitoring are available. After entering the Idle State, the AT wakes up quite frequently (e.g., once every A ms with A ms < 5.12 sec) for a certain time duration (e.g., P seconds). Once this time duration elapses, the AT enters the second level of monitoring in which it wakes up less frequently (e.g., once every B ms with 5.12 sec $> B$ ms $> A$ ms) during a certain time period (e.g., Q seconds). Finally, the AT enters the third level of monitoring in which it wakes up even less frequently (e.g., once every C ms with 5.12 sec $> C$ ms $> B$ ms $> A$ ms). In summary, more waking of the AT enables a faster response to a Page message and hence a faster transition to the Connected State.

Data Over Signaling Protocol is new to Rev. A and allows short messages to be sent in the FL or RL without establishing a dedicated traffic connection. Typically, this protocol allows an AT to send a short message on the Access channel and an AN to send a short packet on the Control channel. A short packet (e.g., a short-message-service) can be reliably sent using this protocol by requiring the receiver to send an acknowledgment. Duplicate detection is also possible by using a sequence number. A packet from any RLP flow can be carried using this protocol. In summary, Data Over Signaling Protocol facilitates the transfer of short packets with no need for establishing traffic channels.

[449] Unless this feature is activated via session configuration, Rev. A continues to use Rev. 0's monitoring method.

[450] If you said synchronous and asynchronous capsules, then you are correct!

12.15 1xEV-DO REV. A INTERWORKING: REV. 0, REV. A, AND CDMA2000 1x

Rev. A is fully backward-compatible with Rev. 0. A Rev. 0 sector can be treated like a Rev. A sector, and the Active Set can contain both Rev. 0 and Rev. A sectors. The RNC will then use Rev. 0 features for communications with the AT. A new session need not be negotiated while moving across the Rev. 0 and Rev. A boundary due to the multiple personalities feature. For example, one personality could be a Rev. A personality and another personality a Rev. 0 personality.

The interworking between Rev. A and CDMA2000 1x is similar to the interworking between Rev. 0 and CDMA2000 1x with some enhancements. As with a Rev. 0 AT, the Rev. A AT periodically jumps to the 1x carrier frequency to look for a possible Page message. In an enhancement Rev. A introduces, the AT in the Idle State jumps to 1x during the paging mask duration so that the Rev. A system sends no Page message and the AT loses no Rev. A Page message. If our Rev. A AT has no paging mask mechanism, the Rev. A system has no idea that the AT is on the 1x carrier frequency and may send a Page message to the AT. Such a Page message would be lost, effectively increasing the paging delay. The provisioning of the paging mask mechanism reduces the paging delay and results in better interworking between Rev. A and 1x. The AT implements a proprietary system selection algorithm to make a handoff to the 1x system if 1x is found to be a better system that Rev. A in a given geographical area. Once the AT is performing data transfer on 1x, it stays on 1x until the data transfer is over and until it enters the dormant mode in 1x when it can then search for a Rev. A system.

12.16 POINTS TO REMEMBER

- Rev. A is fully backward-compatible with Rev. 0.
- Rev. A makes numerous enhancements relative to Rev. 0 while keeping Rev. 0's network architecture.
- The FL performance is improved by the availability of the peak data rate of 3.07 Mbps for large packets, the low data rate of 4.8 kbps for small packets, and multi-user packets.
- The number of traffic connections is increased to 114 in Rev. A from 59 in Rev. 0.
- The RL performance is improved because of H-ARQ, a higher peak rate of 1.8 Mbps, and direct and faster interference control via T2P.
- Both inter- and intra-AT QoS control is possible using the multi-flow application protocol and a suitable scheduler.
- Handoff is improved by using fast cell-switching via the DSC channel.

- The use of subsynchronous capsules, three-tiered monitoring, the paging mask mechanism, and the Data Over Signaling Protocol improve the Idle State operations and inter-technology handoffs.
- The multiple personalities feature enables seamless handoff across Rev. 0 and Rev. A boundaries and reduces latency in handoff.

CHAPTER 13

HIGH-SPEED UPLINK PACKET ACCESS

HSUPA (High-Speed Uplink Packet Access) is a Third Generation Partnership Project (3GPP) Release 6 (R6) feature. The main goal of HSUPA is to significantly increase the uplink data rates from 384 kbps practical data rates of Release 99 (R99) and Release 5 (R5) to 5.7 Mbps. HSUPA is primarily an air-interface enhancement similar to HSDPA (High-Speed Downlink Packet Access). Unique characteristics and benefits of HSUPA are summarized first. A brief overview of the downlink and uplink data transmission is given. We'll see that the downlink data transmission follows the R99 and R5 philosophy, but the uplink transmission is quite different in R6 than in R99 and R5. Call setup for HSUPA, which is similar to R99 and R5, is discussed. New HSUPA radio channels are introduced, and the uplink data transmission is discussed in detail. HSUPA introduces no new elements in the existing R5 system architecture but defines new Medium Access Control (MAC) sublayers in the air-interface protocol stack. These architectural enhancements are briefly touched upon. Six categories of user equipments (UEs) for HSUPA are specified and their characteristics are outlined. Finally, HSUPA's key highlights are summarized.

13.1 INTRODUCTION TO HSUPA: AN EXECUTIVE SUMMARY

HSUPA is one of the features introduced in R6 of the 3GPP family of standards. Examples of other R6 features include Multimedia Broadcast-Multicast Services and enhancements to the HSDPA feature. The HSDPA feature was introduced in 3GPP R5. HSUPA promises higher data rates and shorter latency in the uplink. HSUPA is also referred to as the Enhanced Dedicated channel (E-DCH) feature or enhanced uplink. The E-DCH is a new uplink transport channel that has been added in R6 to transport high-speed data. See the standard [297] for an overall description of HSUPA.

Let's summarize the main characteristics of an HSUPA-based system. HSUPA increases the peak user data rate in the link from the UE to the Node B from a theoretical 2 Mbps in R5 to 5.7 Mbps in R6. The peak user data rate of 384 kbps is typically supported in the uplink in R99 and R5 deployments. The average throughput per user and per cell in the uplink is much higher in HSUPA than in R5, but HSUPA offers no increase in the peak user data rate in the downlink; hence, an HSUPA-based system offers R5's peak downlink data rate.

Do You Recall... Do you remember the peak downlink data rate for HSDPA?[451]

HSUPA can share the same spectrum with HSDPA. For example, when HSDPA is using a certain 5 MHz channel in the downlink and a certain 5 MHz channel in the uplink, HSUPA can use these same downlink and uplink channels. In other words, HSUPA requires no separate, dedicated spectrum. HSUPA is fully backward-compatible with previous standard releases, such as R99 and R5. For example, an R99 UE can transparently access and obtain services from an R6 system.

An HSUPA UE is required to support R99, R5, and R6. Major factors that lead to higher throughput and shorter delay in the uplink include direct and fast control of uplink interference, a UE's simultaneous data transmission on multiple radio channels, fast data rate changes, Hybrid Automatic Repeat Request (HARQ), and a shorter transmit time interval (TTI). As in R99 and R5, HSUPA continues to use CDMA's[452] fast power control and soft handover in the uplink.

One transport channel and five physical channels are added in R6 to support HSUPA. The new transport channel for the uplink is the E-DCH. Two new HSUPA physical channels in the uplink are the E-DCH Dedicated Physical Data channel (E-DPDCH) and the E-DCH Dedicated Physical Control channel (E-DPCCH). Three new HSUPA channels in the downlink are the E-DCH Absolute Grant channel (E-AGCH), the E-DCH Relative Grant channel (E-RGCH), and the E-DCH HARQ Indication channel (E-HICH). The E-DCH is mapped onto one or more E-DPDCHs to carry packets at a high speed. The E-DPCCH conveys the control information to

[451] The highest peak data rate in HSDPA is 14 Mbps for an HSDPA Category 10 UE.

[452] Remember that CDMA is Code-Division Multiple Access.

the Node B so that the Node B knows how to retrieve a packet from the E-DPDCH(s). The E-AGCH and E-RGCH perform fast and direct control of uplink interference. The E-HICH supports HARQ. [A new sixth physical channel, the Fractional Dedicated Physical channel (F-DPCH) is also introduced in R6 to enhance HSDPA performance.]

Pure R5-based UEs and networks need, in general, hardware and software upgrades.[453] Both the UE and the Node B need more processing power and buffering capabilities, and a new baseband processing block is required at the UE and the Node B to support new channels. Software upgrades are also required to manage new radio interface channels. The UTRAN (Universal Terrestrial Radio Access Network) configures an HSUPA terminal to use a scheduled or non-scheduled mode for a given logical channel. In the non-scheduled mode, the UE is allowed to transmit a maximum number of bits during a TTI. In the scheduled mode, the Node Bs use the E-AGCH and E-RGCH to dynamically control uplink data rates via fast and direct control of uplink interference. The non-scheduled mode is suitable for delay-sensitive applications, such as Signaling Radio Bearers and Guaranteed Bit Rate services. The scheduled mode helps achieve better performance than the non-scheduled mode at the expense of complexity and potentially longer latency. The scheduled mode is suitable for non-real-time, bandwidth-intensive applications, such as high-speed uplink transfer of video and other files. Using both scheduled and non-scheduled modes is possible during a given TTI. A UE sending a video email and Multimedia Message Service or uploading a file via FTP would benefit from HSUPA's high uplink data rates.

A Bit of Cellular History... You may wonder why, if HSUPA is providing good performance, HSDPA and HSUPA were designed in separate 3GPP releases. Well, the engineers creating the standard assumed that a typical user would use asymmetric services; a large amount of data would be transmitted in the downlink but little data in the uplink. Web-browsing is an example of this type of data service. Hence, to provide a good user experience, the standard focused on the downlink. Furthermore, time is a non-renewable resource (we have all wished for a day that is longer than 24 hours!), and the creators of the standard prioritized the downlink over the uplink. Thus, the uplink performance enhancement was designed in Release 6 rather than in Release 5.

13.2 BASICS OF DATA TRANSMISSION: A BRIEF OVERVIEW

Before getting into the minute details of how HSUPA achieves superior uplink performance, let's first have a brief overview of overall data transmission. The Universal Mobile Telecommunication System (UMTS) network establishes a packet data call (or session) with the UE so that data transfer can occur in the downlink and uplink.

[453] In practice, a typical R5 Node B was equipped with R6-ready hardware.

The basic call setup in an HSUPA system is similar to that in R99 and R5. The basic Radio Resource Control (RRC), i.e., layer 3, messages exchanged between the UE and the network during the call setup are the same in R99, R5, and R6. Additional fields have been incorporated in the R6 messages to support HSUPA and other R6 features. The downlink data transfer in an HSUPA system could be, in principle, based on an R99 approach or an HSDPA approach. In practice, the HSDPA approach is commonly used for downlink transmission due to its higher efficiency. The uplink data transfer in HSUPA is completely different from that in R99 and R5.

Section 13.2.1 offers a high-level view of a call setup in HSUPA. Section 13.2.2 will illustrate a scenario and thus introduce some terminology. Major stages of uplink data transmission are summarized in Section 13.5.1.

13.2.1 Simplified HSUPA Call Setup

The packet data call setup involves resource allocation for the UE throughout the network including radio resources and bandwidths on various interfaces. See Chapter 9 for a detailed packet data call setup in an R99 system. Figure 13.1 provides a high-level overview of the HSUPA call setup.

The UE first establishes an RRC signaling connection with the Radio Network Controller (RNC). During this process, the RNC learns about the HSUPA UE's specific capabilities, such as its maximum supported uplink data rate. A low-data-rate dedicated radio channel is typically established between the UE and the Node B to reliably exchange call setup messages. The efficiency of the call setup is better in R6 than in R5 through enhancements, such as the UE's early indications of HSDPA and HSUPA support. The UE is assigned an HSUPA identity called E-RNTI (E-DCH Radio Network Temporary Identifier).

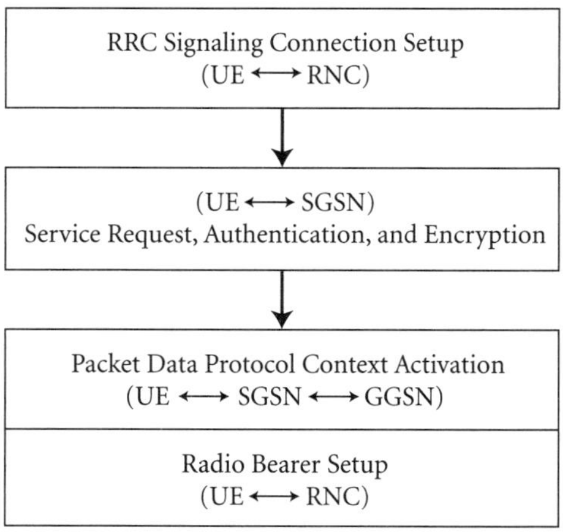

Figure 13.1: Simplified Call Setup

Upon completion of the RRC connection setup, the UE requests packet data service from the packet-switched core network. The Serving General Packet Radio Service Support Node (SGSN) authenticates the UE and could activate security mechanisms, such as integrity protection and ciphering. The UE then requests Packet Data Protocol context activation with two major purposes: obtaining an Internet Protocol address and requesting a specific quality-of-service (QoS). The Gateway General Packet Radio Service Support Node (GGSN), SGSN, and RNC work together to establish a specific QoS for the UE. The GGSN assigns the UE a dynamic Internet Protocol address. The RNC sets up a radio bearer with the UE with specific dedicated radio channels and configures the radio interface channels for high-speed downlink and uplink data transmission. For example, the RNC configures the UE to monitor the E-AGCH and E-RGCH so that the UE can determine an appropriate data rate for high-speed uplink transmission.

13.2.2 Some Terminology

Since the HSUPA uplink is based on CDMA,[454] traditional CDMA mechanisms, such as soft handover and power control, are used in the uplink as they are in R99 and R5. Figure 13.2 depicts a scenario useful for introducing some terminology here.

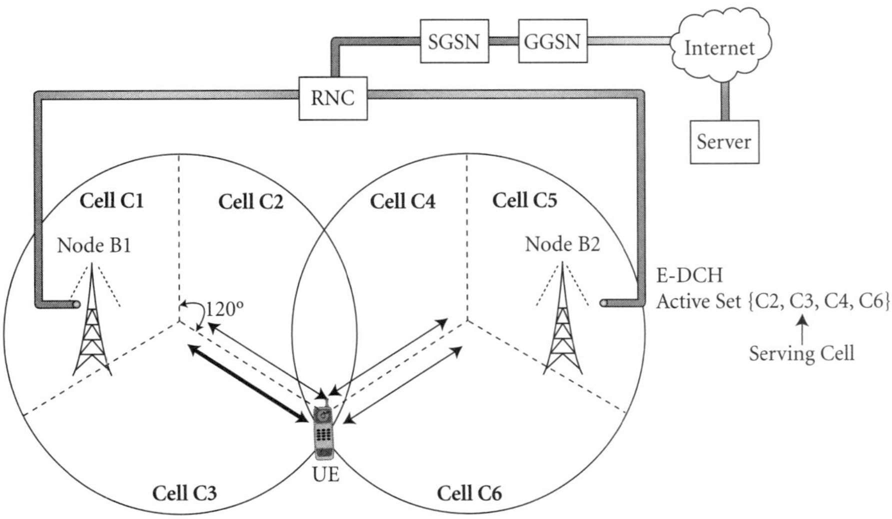

Figure 13.2: HSUPA Scenario

[454]The term CDMA here refers to the generic multiple access technique. UMTS and its evolutionary systems, such as HSDPA and HSUPA, use a wider channel than other CDMA-based systems, such as 1x. These UMTS-based systems are often said to be using wideband CDMA or WCDMA.

The UE is in the handover region with four cells in the Active Set, {C2, C3, C4, C6}. Node B1 controls cells C2 and C3, and Node B2 controls C4 and C6. When the UE transmits data on the high-speed channel, all the cells in the Active Set receive the packet. Node B1 and Node B2 perform maximal ratio combining, and the RNC performs selection combining of the packets it receives from Node B1 and Node B2. The RNC chooses one of the cells in the Active Set to be the serving or controlling cell. Cell C3 is the serving cell in Figure 13.2. The serving Radio Link Set (RLS) is the set of Active Set cells associated with the Node B that is in charge of the serving cell. In Figure 13.2, the serving RLS is {C3, C2}, Node B1 controls serving cell C3, and cells C3 and C2 are associated with Node B1.

Fast and dynamic interference control is achieved through Absolute Grants (AGs) and Relative Grants (RGs). The serving cell sends AGs to the UE, and all the Active Set cells send RGs to the UE. The AG directly controls the transmit power of the E-DPDCH relative to the Dedicated Physical Control channel (DPCCH),[455] and the RG adjusts the AG. The UE transmits data using a 2-ms or 10-ms TTI. During a 2-ms TTI, the UE sends scheduled data, non-scheduled data, or *both* scheduled and non-scheduled data. The scheduled data represents the data that the UE can send based on the received AG and RGs. The non-scheduled data is the data the UE can transmit based on prior configuration. The AG and RGs have no influence on non-scheduled data. Fast and dynamic interference control is possible via a scheduled mode of operation, but data rates can be guaranteed via a non-scheduled mode of operation.

13.2.3 Major Stages of Uplink Data Transmission: An Overview

The downlink data transmission in an HSUPA system relies on R99 or R5 philosophy. (See Chapter 9 on UMTS R99 and Chapter 11 on HSDPA for details.) However, the uplink data transmission in HSUPA is completely different from that in R99 and R5. HSUPA transmission has two basic modes: scheduled mode and non-scheduled mode. In the non-scheduled mode, the UE can transmit up to a certain number of bits during the TTI, which the RNC configures. Figure 13.3 outlines the basic steps involved in HSUPA data transmission in the uplink (UL). (We'll get into the *details* of these steps later on in Sections 13.4 through 13.6.)

When a packet data call is set up, the UE is configured with the HSUPA channels and various operational parameters. Such a configuration primarily takes place during RRC connection setup and radio bearer setup. At the end of this step, the UE knows which downlink channels to monitor and how to use the uplink channels. The UE makes a scheduling request to the serving Node B by sending *Scheduling Information* (SI) on the E-DPDCH and the *Happy Bit* on the E-DPCCH. (What?! Happy Bit?! Oh, yes, this is not a typo!)

[455] Please refer to Chapter 8 on UMTS Release 99 for DPCCH details.

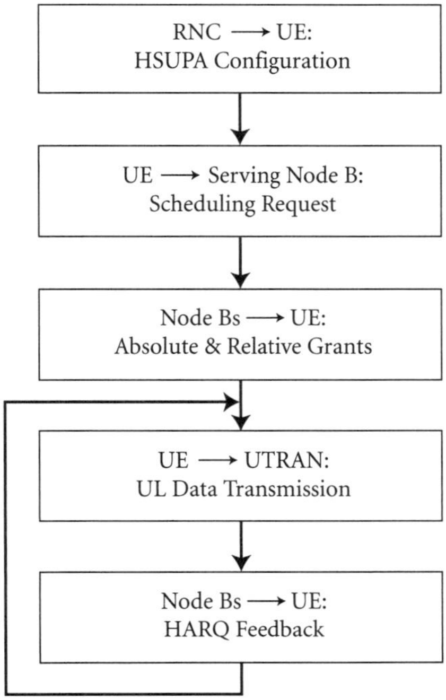

Figure 13.3: HSUPA Steps

A Funny Detail... How did the UMTS standard become so poetic? Here's our theory. You are likely aware of Happy Hour, a designated time in which alcohol, often beer, reigns supreme in restaurants and bars. We suspect the standard's authors continued their work after normal office hours and into Happy Hour at a restaurant. Once they had enough beer, they all became happy, relaxed, and poetic, and they called the bit the *Happy Bit*.

When Dr. Tripathi was teaching an HSUPA course and was reviewing the prior day's concepts, he asked the students about the name of the bit that the UE sends to the Node B. Multiple students replied at once, "The beer bit!" In keeping with the spirit (which spirit?! just choose your favorite!) of the standard, the author was *happy* to accept the answer!

The initial Serving Grant (SG) assigned to the UE during the RADIO BEARER SETUP message enables the UE to transmit the Scheduling Information. *The SG is the maximum allowed ratio of the E-DPDCH transmit power to the DPCCH transmit power.* Scheduling Information includes the amount of user traffic data in the UE buffer. The Happy Bit indicates if the UE is happy with the currently allocated uplink resource, i.e., the maximum resource the UE can use while transmitting the scheduled data on the E-DPDCH.

The serving Node B can send an AG to the UE from the serving cell on the E-AGCH. The AG is the maximum SG (i.e., ratio of E-DPDCH transmit power to DPCCH transmit power) allocated to the UE. All cells in the E-DCH Active Set can send RGs to the UE on the E-RGCHs. The RGs adjust the previously used ratio of the E-DPDCH power to DPCCH power. The UE calculates the SG to make a decision on the supportable uplink data rate on the E-DCH. The SG represents the influence of both the AG and RGs. Assume that the last received AG is 5 dB, and the actual E-DPDCH-power-to-DPCCH-power ratio the UE used in the last transmission is 4 dB. If the RG corresponds to a grant increase of 1.5 dB, the value of the SG is

$$\textit{actually used ratio} + \textit{adjustment} = 4 \text{ dB} + 1.5 \text{ dB} = 5.5 \text{ B}.$$

Now, the UE can use up to a 5.5 dB power ratio for the next transmission.

The UE calculates the supportable data rate in the uplink based on the calculated SG. It starts transmitting data on the E-DPDCH by multiplexing the non-scheduled data (if configured) with the scheduled data. Note that the number of bits in the scheduled data depends on the SG. The E-DPCCH is transmitted in parallel with the E-DPDCHs. The E-DPCCH indicates the data rate being transmitted on the E-DPDCHs, the HARQ configuration[456] used on the E-DPDCHs, and the Happy Bit.

The Node Bs that control the cells in the E-DCH Active Set attempt to retrieve the information bits from the packet received on the E-DPDCHs. If a Node B succeeds in decoding the packet, a positive acknowledgment (ACK) is transmitted on the E-HICHs from all of the Node B's E-DCH Active Set cells. Otherwise, a negative acknowledgment (NACK) is sent from all the cells in that Node B. As shown in Figure 13.2, assume that the E-DCH Active Set contains cells C2, C3, C4, and C6, with Node B1 controlling C2 and C3 and Node B2 controlling C4 and C6. Node B1 performs maximal ratio combining of the packet received at C2 and C3, and, if it successfully retrieves the original information bits of the packet the UE transmitted, it sends an ACK on both the E-HICH in C2 and the E-HICH in C3. Similarly, Node B2 performs maximal ratio combining of the packet received at C4 and C6. If Node B2 fails to recover the original information bits, it sends a NACK on both the E-HICH in C4 and the E-HICH in C6. Node B1 now forwards the correctly decoded packet to the RNC.

If the UE receives an ACK for the transmitted packet, it starts sending a new packet. However, if the UE receives NACKs from all the E-DCH Active Set cells, it retransmits the packet, possibly using a different HARQ configuration. For example, if the UE used redundancy version 0 in the first transmission, it could now use redundancy version 1.

Let's now dive into the HSUPA details! The HSUPA channels and their timing relationships are discussed in Section 13.3. Once we are familiar with the HSUPA channels, we will be ready to take an in-depth look at the major stages of uplink data transmission in Sections 13.4 through 13.6.

[456]The HARQ configuration specifies which turbo coding symbols the UE selected to form a packet. A parameter called Retransmission Sequence Number (RSN) dictates the HARQ configuration. See Section 13.3.2 for RSN details.

13.3 HSUPA CHANNELS

The uplink data transmission in HSUPA can be used in conjunction with downlink data transmission following either the R99 approach, which uses traditional soft handover, or the R5 approach of HSDPA. Figures 13.4 and 13.5 specify the basic channels required for uplink and downlink transmissions when R99 downlink transmission and R6 HSUPA uplink transmission are used. (Recall that DPDCH is the Dedicated Physical Data channel.) The RNC determines the R99 Active Set containing the cells that transmit the DPCCH (or possibly the F-DPCH[457]) to the UE and receive the DPCCH from the UE. The RNC also creates an E-DCH Active Set that could be the same as the R99 Active Set or a subset of the R99 Active Set.

As shown in Figure 13.4, the R99 UL DPCCH is essential because the transmit power levels of the HSUPA channels, such as the E-DPDCH and E-DPCCH, are defined relative to the DPCCH. Note that the cells in the R99 Active Set send power control commands to the UE so that the UE can adjust the DPCCH's transmit power. All other uplink channels maintain their power offsets with the DPCCH.[458] The UE transmits an E-DCH that is mapped onto one or more E-DPDCHs depending upon the amount of data to be transferred. The E-DCH is accompanied by the E-DPCCH.

In the downlink, R99 DPCCH and DPDCH are transmitted from every cell in the R99 Active Set. Only the E-DCH serving cell can transmit E-AGCH. The cells in the E-DCH Active Set transmit E-RGCH and E-HICH.

13.3.1 The E-DCH and E-DPDCHs (Uplink)

The E-DCH is the only new transport channel introduced in R6 to support HSUPA. Recall that it maps onto one or more E-DPDCHs to transport both the user traffic data as well as the Scheduling Information (discussed in Section 13.3.1.2). *Let's summarize the basic characteristics of the E-DCH or E-DPDCH.* One transport block is transmitted on the E-DCH during one TTI, which could be 2 ms or 10 ms. The UE determines the size of this block based on the received grants and the amount of data in the buffer. A given E-DPDCH is mapped onto either the in-phase channel (I-channel) or the quadrature channel (Q-channel). Multi-code Binary Phase Shift Keying (BPSK) is used for E-DCH transmission with one modulation symbol representing one turbo code symbol. Synchronous HARQ is used, and both incremental redundancy and Chase combining are allowed. The number of HARQ processes is four for 10-ms TTI and eight for 2-ms TTI. One HARQ process is responsible for the reliable transmission of one packet.

[457] See Chapter 11 on HSDPA.

[458] For example, E-DPDCH power offset would be the ratio of the E-DPDCH transmit power to the DPCCH transmit power, and E-DPCCH power offset would be the ratio of the E-DPCCH transmit power to the DPCCH transmit power.

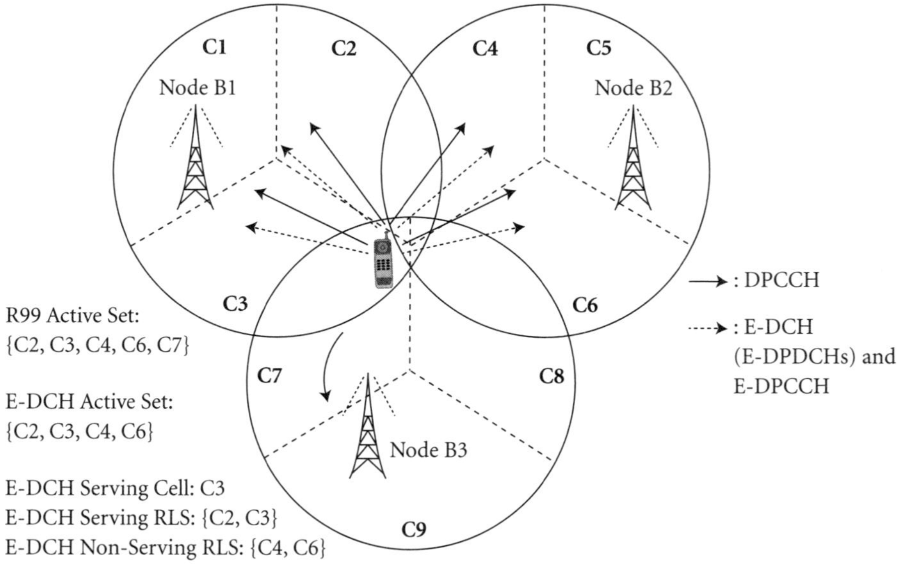

Figure 13.4: Uplink Channels for HSUPA

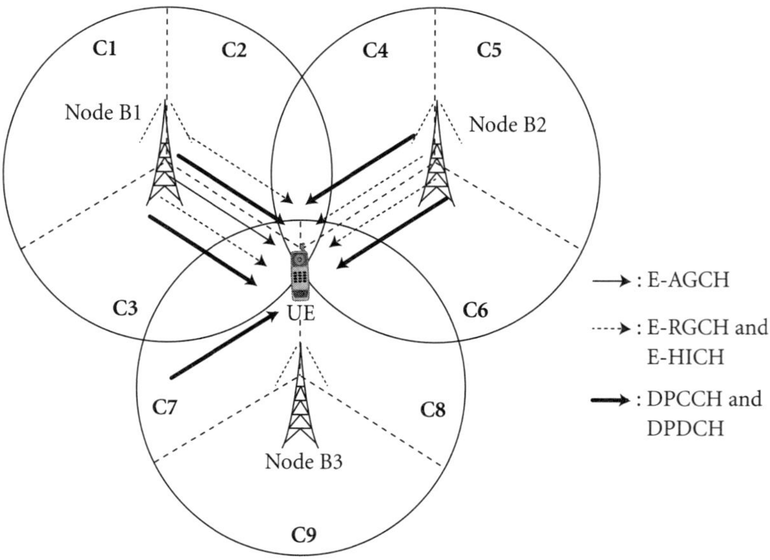

Figure 13.5: Downlink Channels for HSUPA

Do You Recall... Do you remember from HSDPA (Chapter 11) the benefit of multiple HARQ processes?[459]

A given TTI is *implicitly* associated with one HARQ process based on the timing. In other words, a given TTI has a deterministic relationship with the number (i.e., identity) of the HARQ process, which ranges from 0 to 3 for the 10-ms TTI and 0 to 7 for the 2-ms TTI. The E-DCH can be operated in a non-scheduled mode, a scheduled mode, or a combined non-scheduled and scheduled mode. In the non-scheduled mode, only non-scheduled data can be sent during a given TTI. The scheduled mode allows transmission of only scheduled data. The combined non-scheduled and scheduled mode, as the name implies, enables transmission of any type of data during a given TTI. The designation of a specific TTI for the scheduled or non-scheduled mode occurs implicitly through the association of a HARQ process with a scheduled or non-scheduled mode. For example, if HARQ process number 2 is configured to carry scheduled data and is also configured later to carry non-scheduled data, a TTI that maps onto the HARQ process number 2 now uses the E-DCH in the combined scheduled and non-scheduled mode.

Full soft and softer handover is supported. In other words, each Node B in the E-DCH Active Set performs maximal ratio combining, and the RNC performs selection combining on the packets received from multiple Node Bs in the E-DCH Active Set. R99-based closed-loop power control is used for the E-DCH. The SG puts a limit on the allowed ratio of the E-DPDCH power and the DPCCH power. A suitable ratio or power offset is used to transmit a transport block on the E-DCH. This power offset is maintained even when the DPCCH's transmit power is modified to reflect the power control commands received from the Active Set cells.

13.3.1.1 Physical Layer Processing of the E-DCH Figure 13.6 summarizes the overall physical layer processing carried out for the E-DCH [298–300]. Only one transport block is transmitted during a TTI. The transport block contains Scheduling Information and/or user traffic data. The standard defines four sets of transport block sizes (TBSs) [301]: TBS Table 0 for the 2-ms TTI and optimized for a 336-bit radio link control (RLC) protocol data unit (PDU); TBS Table 1 for a 2-ms TTI and exponentially distributed TBSs; TBS Table 0 for the 10-ms TTI and optimized for a 336-bit RLC PDU; and TBS Table 1 for the 10-ms TTI and exponentially distributed TBSs. The UE uses a specific table based on the UTRAN-controlled configuration. A maximum of 127 non-zero TBS values are supported, ranging from eighteen bits to 11484 bits for the 2-ms TTI. The minimum TBS of eighteen bits supports transmission of Scheduling Information (which is eighteen bits) on the E-DCH, and the maximum TBS of 11484 bits enables a peak data rate of $11484\ \text{bits}/2\ \text{ms} = 5.7$ Mbps. The supported TBS for the 10-ms TTI varies from eighteen bits to 20000 bits. A twenty-

[459]Multiple HARQ processes enable continuous data transmission, leading to shorter latency and higher throughput.

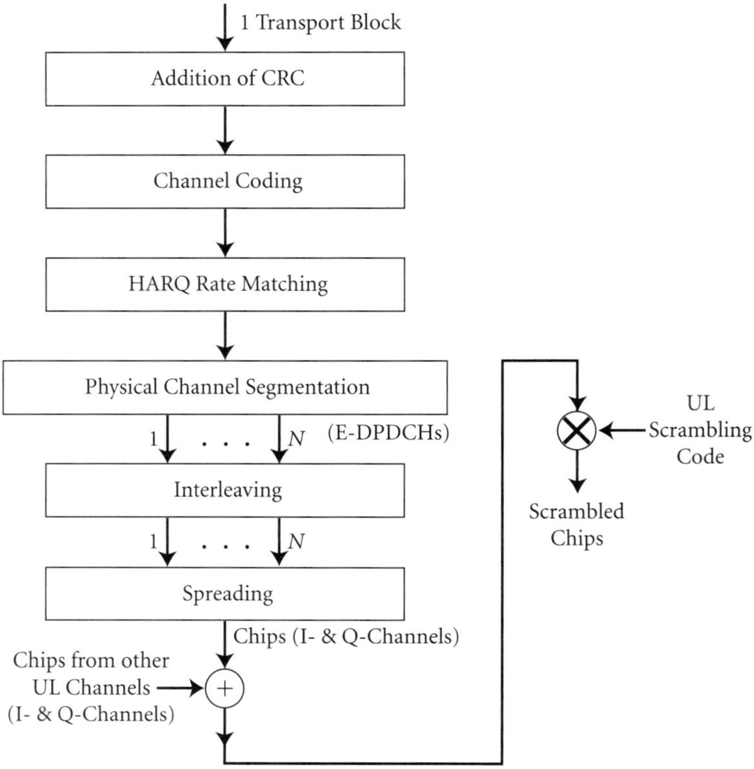

Figure 13.6: Physical Layer Processing of the E-DCH

four–bit cyclic redundancy check (CRC) is added to the transport block to facilitate error detection at the receiver.

Only turbo coding with the 1/3 rate is used to achieve better performance when a large amount of data must be transmitted. The UE selects a specific number of systematic and/or parity bits coming from the turbo coding block based on the Redundancy Version (RV) index. (See Sections 13.5.5 and 13.6 for details.) The selected turbo code symbols must be transmitted using a suitable spreading factor (SF) and a suitable number of physical channels. The standard specifies an algorithm that determines the SF and the number of physical channels [299]. One physical channel corresponds to one E-DPDCH at a given spreading factor on a given I-channel or Q-channel. Each E-DPDCH is separately interleaved. A given interleaved E-DPDCH undergoes BPSK modulation and is spread by a suitable SF to reach a certain chip rate.

A Question for You... What is the chip rate in UMTS—1.2288 Mcps or 5 Mcps?[460]

The turbo code symbol rate or the modulation symbol rate ranges from 15 ksps to 1920 ksps, and the SF ranges from 256 to 2 on a given E-DPDCH. The chips of the spread E-DPDCHs are added to the chips of other physical channels (e.g., DPCCH), and the combined chips are then scrambled using the UE-specific UL scrambling code.

13.3.1.2 Scheduling Information: Contents and Reporting The Scheduling Information conveys to the serving Node B how much grant the UE can use. Figure 13.7 illustrates the contents of the Scheduling Information. Assume that two logical channels have been configured for scheduled data transmission via RRC signaling; logical channel 2 has a priority of 0 and logical channel 3 has a priority of 1 where a lower number represents a higher priority. Further assume that the number of logical channel 2 bits is 700 and logical channel 3 bits is 300.

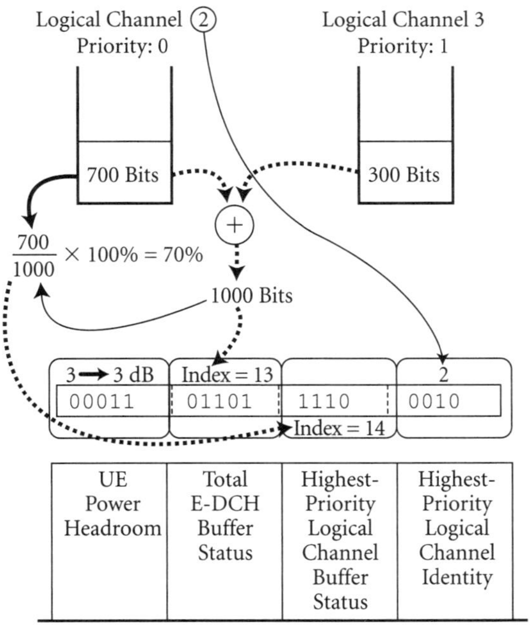

Figure 13.7: Scheduling Information

[460]The correct answer is neither 1.2288 Mcps nor 5 Mcps; it is 3.84 Mcps! We were just trying to ensure that you are fully awake and that you have unshakeable faith in your answer. The reserved channel bandwidth is 5 MHz including the guard bands.

Scheduling Information consists of four fields: UE Power Headroom, Total E-DCH Buffer Status, Highest-Priority Logical Channel Buffer Status, and Highest-Priority Logical Channel Identity. UE Power Headroom is a five-bit field representing the ratio of the maximum UE transmit power and the current DPCCH transmit power [301, 302]. If UE Power Headroom is just 3 dB, the Node B would give a small (e.g., $\leq$ 3 dB) AG to the UE because the UE would be unable to handle it due to the limited power headroom. The six-bit Total E-DCH Buffer Status field specifies an index corresponding to the total number of bits for the scheduled data. The total number of scheduled data bits is 700 bits + 300 bits = 1000 bits. According to the table specified in the standard [301], $index = 13$ corresponds to $848 < \mathit{Total\ E\text{-}DCH\ Buffer\ Status} \leq 1051$. Hence, $\mathit{Total\ E\text{-}DCH\ Buffer\ Status}$ will be $index = 13$.

Highest-Priority Logical Channel Buffer Status, a four-bit field, specifies the percentage of the total number of scheduled data bits that belongs to the highest-priority logical channel. In our example, 700 of 1000 bits belong to the logical channel 2, which has the highest priority. Hence, Highest-Priority Logical Channel Buffer Status should represent $(700/1000) * 100\% = 70\%$. In the Highest-Priority Logical Channel Buffer Status table in the standard [301], $index = 14$ is associated with $68 < \mathit{Highest\text{-}Priority\ Logical\ Channel\ Buffer\ Status} \leq 82$. Hence, the UE reports an Highest-Priority Logical Channel Buffer Status of 14. Since the logical channel *2* is the highest-priority channel, the four-bit Highest-Priority Logical Channel Identity field would be *2*.

SI is always sent when triggered [301], even when the SG is 0. *Periodic*, *event-based*, and *space-based* reporting are the three basic ways to trigger SI reporting. In periodic reporting, the UTRAN configures the UE to send SI *periodically*, such as one report every X time units when the SG is 0 and one report every Y time units when the SG is non-zero. The values of X and Y can be set independently and range from 2 ms to 1 second. In the case of event-based reporting, when some *event* condition is satisfied, SI reporting is triggered. For example, assume that the UE has no SG (i.e., $\mathit{Serving\ Grant} = 0$ dB) and some data arrives at the buffer. In this case, the UE sends SI so that the Node B can provide some SG to the UE for uplink transmission.

After sending SI corresponding to the prior data arrival, the UE sends SI again if new higher-priority data arrives. A list of all the conditions under which SI is triggered is available in the standard [301]. In space-based reporting, SI is included along with user traffic data anytime some *space* is available [301]. For example, if the UE's selected transport block size is 372 bits, corresponding to an E-DCH Transport Format Combination Indicator (E-TFCI) value of 4 and the amount of data available for transmission is 354 bits or less, an eighteen-bit Scheduling Information is appended at the end of the data bits so that $354 + 18 = 372$ bits can be transmitted during a TTI of 2 ms with an E-TFCI of 4.

13.3.1.3 *Achieving the Highest Data Rate of 5.7 Mbps* Here is a million dollar question... How can we achieve the highest data rate of 5.7 Mbps that HSUPA promises? Figure 13.8 provides a simplified view of the answer to this question.

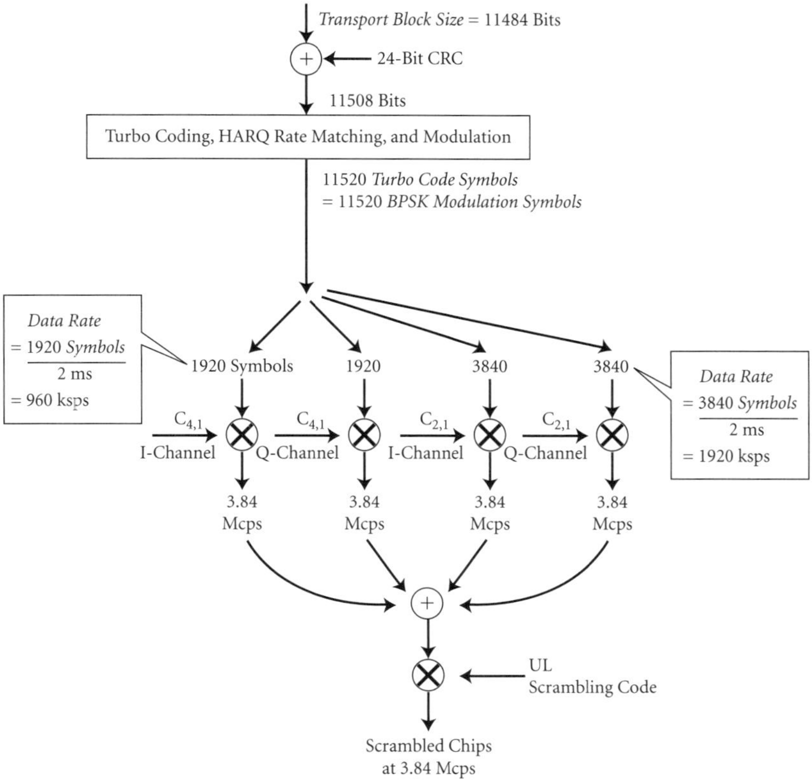

Figure 13.8: Achieving the Peak Data Rate of 5.7 Mbps in the Uplink

The maximum transport block size of 11484 bits is used for the 2-ms TTI. After the twenty-four–bit CRC is added, 11508 bits are obtained, which are then passed through turbo coding, HARQ rate matching, and BPSK modulation. A total of 11520 turbo code symbols or BPSK modulation symbols are obtained. Note that although the input to the turbo coding is 11508 bits, the effective number of code symbols to be transmitted over the air is just 11520. In other words, virtually no coding or protection is present in this case. *Our UE had better have a wireline (not a wireless) connection with the Node B to be able to decode this packet in a single shot!*

These code symbols are distributed over four E-DPDCHs. The Orthogonal Variable Spreading Factor (OVSF) code $C_{4,1}$ spreads a set of 1920 symbols; the channel at the rate of $1920/2$ ms $=$ 960 ksps is spread to 960 ksps $* \, 4 =$ 3.84 Mcps on the I-channel. The same procedure is repeated for the Q-channel for another set of 1920 symbols. Now, $C_{2,1}$ spreads a set of 3840 symbols on the I-channel to reach 3.84 Mcps from $3840/2$ ms $=$ 1920 ksps. Similarly, $C_{2,1}$ spreads another set of 3840 symbols on the Q-channel. The peak data rate can be approximately calculated as the ratio of the transport block size to the TTI, which is 11484 bits$/2$ ms $= 5.742$ Mbps.

At the physical layer, the raw data rate before spreading is $11520/2\text{ms} = 5.76$ Mbps. The user-perceived peak throughput would be slightly less than 5.742 Mbps since the transport block contains several upper-layer headers, such as MAC and RLC headers.

13.3.2 The E-DPCCH (Uplink)

The E-DPCCH conveys the control information from the UE to the Node B to facilitate the decoding of the E-DCH at the Node B. The E-DPCCH accompanies the E-DCH; hence, no E-DPCCH is transmitted during a TTI if no E-DCH is transmitted. As shown in Figure 13.9, the E-DPCCH contains a Happy Bit, Retransmission Sequence Number (RSN), and E-TFCI.

The Happy Bit is a one-bit field that conveys to the Node B whether the grant allocated to the UE is adequate, so 1 indicates that the UE is happy with the grant and 0 indicates that the UE is unhappy with the grant. The UE sets the bit to 0 if *all* of the following are true: the UE is transmitting as much data as possible with the current SG *and* the UE has sufficient transmit power headroom to transmit at a higher E-TFCI than the currently selected E-TFCI *and* the current SG would require more than X ms to empty the buffer where X is the *Happy Bit delay condition parameter*, which the RNC coveys to the UE. If *any* of the three criteria is unmet, the UE sets the bit to 1.

RSN is a two-bit field that dictates the RV index when the UE is configured to use a dynamically calculated E-DCH RV index. The MAC-e or MAC-es[461] layer at the UE sets the value of the RSN field. RSN is set to 0 for a new packet transmission. For subsequent transmissions (i.e., retransmissions) of the same packet, RSN is

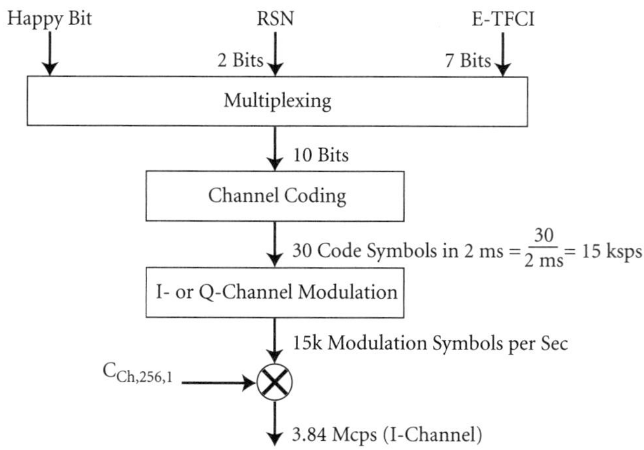

Figure 13.9: Physical Layer Processing of E-DPCCH

[461] The standard fails to explicitly define MAC-e or MAC-es, but MAC-e implies MAC for E-DCH and MAC-es stands for MAC-e selection.

incremented by 1. Since RSN is a two-bit field, its maximum value is 3. Beyond four transmissions of the same packet, the RSN value remains 3. The mapping between RSN and the E-DCH RV index remains unique and unambiguous (as we'll see later in Table 13.7 in Section 13.6).

E-TFCI is a seven-bit field that indicates the transport block size the E-DCH is carrying, and the value of E-TFCI ranges from 0 to 127. For example, in E-TFCI Table 0 for the 2-ms TTI, the value 0 represents the TBS of eighteen bits, and the value 127 corresponds to the TBS of 11484 bits. Four E-TFCI tables are defined in the standard [301], and the E-TFCI value points to the index in the table the RNC has configured.

As shown in Figure 13.9, one Happy Bit, two RSN bits, and seven E-TFCI bits are multiplexed together to create a set of ten bits. These ten bits pass through the channel coding, and thirty code symbols are generated. The channel coding scheme uses a subcode of the second-order Reed-Muller code. Thirty code symbols in a 2-ms subframe imply a symbol rate of $30/2$ ms $= 15$ ksps. An OVSF code spreads these symbols at $SF = 256$ to yield 3.84 Mcps. In the case of a 10-ms TTI, the contents of a 2-ms TTI are transmitted five times on the E-DPCCH.

13.3.3 The E-AGCH (Downlink)

Before discussing new downlink (DL) channels of HSUPA, let's refresh your memory. HSUPA defines new DL channels to convey grants to the UE and to support UL HARQ.

Do You Recall... Name the two types of grants that can be issued to the UE to control the UL data rate.[462]

The E-AGCH conveys the AG to the UE to directly adjust the maximum data rate at which the UE can transmit. Only the serving E-DCH cell can send the AG to the UE. Thus, a UE is assigned just one E-AGCH at $SF = 256$ in the serving E-DCH cell. The Node B sends the AG to the UE in the serving E-DCH cell when the Node B intends to directly and immediately change the uplink data rate. The E-AGCH typically operates in discontinuous mode. Recall that the AG is the maximum SG (i.e., ratio of E-DPDCH transmit power to DPCCH transmit power) that the UE can use. The UE considers both the AG and RGs to calculate the SG that controls the maximum uplink data rate.

Figure 13.10 outlines the basic physical layer processing carried out for the E-AGCH. The AG contents include the AG value and the AG scope. An AG scope of 1 indicates that the AG value is applicable to one specific process, depending upon the timing. An AG scope of 0 means that the AG value applies to all the HARQ processes. An AG value is represented by five bits, ranging from 0 to 31. An AG value of 0 corresponds to INACTIVE, which inactivates a HARQ process. An AG

[462] The two types of grants are the Absolute Grant and the Relative Grant.

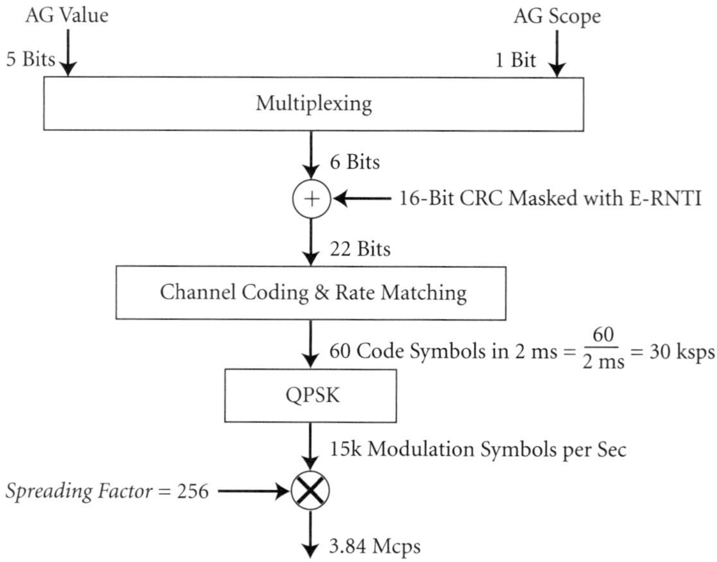

Figure 13.10: E-AGCH Physical Layer Processing

value of 1 is a grant of zero. An AG value of 2 represents $(7/15)^2$, which is the minimum non-zero grant. The maximum AG is $16 * (168/15)^2$, corresponding to the AG value of 31 [299, 301]. Six bits of AG contents are multiplexed, and a sixteen-bit CRC is calculated.

The CRC is then masked with the sixteen-bit E-RNTI to specify the UE(s) for which the AG is applicable. A total of twenty-two bits—six bits for the AG and sixteen bits for the masked CRC—are passed through rate 1/3 convolutional coding. After channel coding and puncturing, sixty bits or convolutional code symbols are generated. Sixty symbols in a 2-ms subframe represent the data rate of $60 \text{ symbols}/2 \text{ ms} = 30$ ksps. QPSK modulation converts sixty code symbols into thirty QPSK modulation symbols, leading to the rate of 15000 modulation symbols per second. Spreading at $SF = 256$ results in $15 \text{ ksps} * 256 = 3.84$ Mcps.

The Node B could use *primary E-RNTI* or *secondary E-RNTI* to mask the CRC; the Node B could assign a UE just one E-RNTI, called the primary E-RNTI, or two E-RNTIs, called the primary E-RNTI and the secondary E-RNTI. The Node B determines these identities, and the RNC conveys these identities to the UE via RRC signaling. The UE monitors the E-AGCH and looks for both the primary and secondary E-RNTI [303]. When the UE receives the AG via the primary E-RNTI, the primary AG influences the calculation of the SG. The UE uses this SG to determine the amount of scheduled data for the next transmission. When the UE receives the AG via the secondary E-RNTI, it stores such an AG as Stored_Secondary_Grant. When the Node B wants the UE to use Stored_Secondary_Grant, the Node B uses the primary E-RNTI on the E-AGCH and sends the AG value of INACTIVE. Now the UE switches from the primary AG to the secondary AG and starts using Stored_Secondary_Grant. The Node B instructs the UE to switch from the secondary AG to the primary AG by sending a suitable non-zero, non-INACTIVE AG using the primary E-RNTI on the A-AGCH.

13.3.4 The E-RGCH (Downlink)

The E-RGCH adjusts the SG for the UE's use for uplink data transmission via an RG. Any cell in the E-DCH Active Set can send an RG to the UE. Hence, the UE is assigned one E-RGCH at $SF = 128$ in every E-DCH Active Set cell. The cells in the E-DCH serving RLS send an identical RG to the UE so that the UE can soft-combine the RGs received from these types of cells. One non-serving E-DCH radio link sends an RG independently from another non-serving E-DCH radio link, and, hence, the UE does not soft-combine RGs received from non-serving E-DCH radio links even when the same Node B is sending an RG from two cells in the E-DCH Active Set. The serving RLS cells can send UP, DOWN, or HOLD (HOLD is equivalent to a "no recommendation" condition) commands as an RG. A non-serving radio link can send DOWN or HOLD commands as an RG. In general, when the UE receives at least one $RG = DOWN$, it attempts to *decrease* the SG to reduce the uplink interference. On the other hand, if the UE receives just $RG = UP$, it attempts to *increase* the SG. (See Section 13.5.3 for more details on how grants are processed.)

Figure 13.11 summarizes the physical layer processing of the E-RGCH. The E-RGCH helps the Node B convey different RGs to different UEs during the same TTI via the use of forty-bit orthogonal signature sequences that are transmitted during

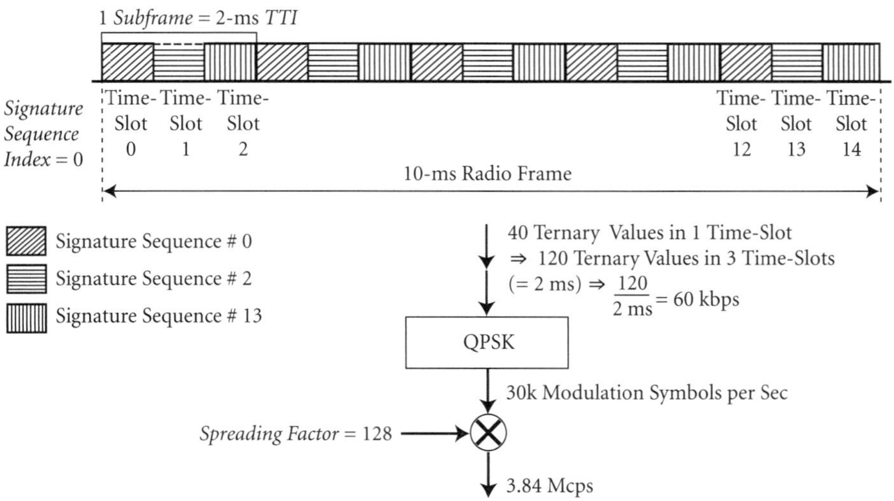

Figure 13.11: E-RGCH and E-HICH Physical Layer Processing

each of the time-slots within a given TTI. The UE is assigned a signature sequence index through RRC signaling, which ranges from 0 to 39. A signature sequence consists of a forty-bit sequence of $+1$ and -1. Figure 13.11 illustrates a scenario in which the UE has been assigned the signature sequence index of 0, which corresponds to the signature sequence of 0 in the zeroth time-slot, 2 in the first time-slot, and 13 in the second time-slot. The same signature sequences are repeatedly used throughout the radio frame in 10-ms TTI. During a time-slot, the specific information sent is $a * signature\ sequence$. Here, a is the RG, which is $+1$ for $RG = UP$, -1 for $RG = DOWN$, and 0 for $RG = HOLD$. When a is 0, the E-RGCH experiences discontinuous transmission and is not actually transmitted. The use of signature sequence hopping within the subframe ensures good cross-correlation properties among the sequences even under higher Doppler frequencies. Assume that the UEs are moving and that the UEs have been assigned *static* signature sequence indices. Predefined signature hopping would enable a UE to easily detect its own RG with no significant interference or confusion arising from the other sequences being sent to other UEs. In the absence of signature hopping, some UEs that have been assigned specific signature sequences would have difficulty detecting their own RGs due to poor cross-correlation among such sequences at higher velocities [304].

A set of forty ternary values (i.e., 0, $+1$, and -1 based on multiplication of a with a sequence of $+1$ and -1) in one time-slot means $40 * 3 = 120$ bits in a 2-ms subframe, leading to the rate of 120 bits/2 ms $= 60$ kbps in Figure 13.11. After these bits pass through QPSK modulation, the rate becomes 30000 modulation symbols per second. Spreading by a factor of 128 achieves the chip rate of 30 ksps $* 128 = 3.84$ Mcps.

13.3.5 The E-HICH (Downlink)

The Node B uses the E-HICH at $SF = 128$ to send to the UE a positive or negative acknowledgment in response to a correctly or incorrectly decoded packet, respectively. All the cells in the E-DCH Active Set must transmit the HARQ acknowledgment after receiving a packet. Since the Node B is combining the packet received from its own cells, the same HARQ acknowledgment is transmitted from all the cells belonging to the Node B. When different Node Bs exist in the E-DCH Active Set, they could potentially be sending different HARQ acknowledgments based on their success in decoding the packet received from the UE. Just like E-RGCH, the UE is assigned one E-HICH in every E-DCH Active Set cell. In fact, the OVSF code at $SF = 128$ is identical for the E-RGCH and the E-HICH for a given UE. The UE differentiates between the E-RGCH and the E-HICH based on the signature sequence index.

The basic physical layer structure is the same for the E-HICH and the E-RGCH. For example, during a subframe, three different signature sequences are transmitted in each time-slot. The set of these signature sequences is associated with a specific signature sequence index. The specific pattern of bits transmitted in each time-slot is $a * signature\ sequence$. The same value of a is transmitted during the entire TTI. Here, a is the HARQ acknowledgment, which is $+1$ for a positive acknowledgment, -1 for a negative acknowledgment from a cell that is part of the serving RLS, and 0 for a negative acknowledgment from a non-serving radio link. Recall that

$a = 0$ corresponds to discontinuous transmission and that the E-HICH is not actually transmitted. The UE associates the received HARQ acknowledgment with a specific HARQ process based on the timing. In other words, HARQ is synchronous. In the next section, we will look at more details of the timing relationships among all the HSUPA channels.

13.3.6 A Really Close Look at Timing Relationships Among HSUPA Channels

The standard defines strict timing relationships among the HSUPA channels. These timing relationships enable the UE and the Node B to interact with each other without ambiguity and with a clearly defined scope of physical layer indicators, such as AGs, RGs, and HARQ acknowledgments. Awaken your long-term memory cells and recall from Chapter 9 that the Node B transmits the Primary Common Control Physical Channel (P-CCPCH) in units of 10-ms radio frames, with each radio frame associated with a System Frame Number (SFN). The SFN represents the basic timing reference. The UE learns about the SFN by observing the SYSTEM INFORMATION message that carries System Information Blocks [305]. This message contains a parameter called SFNprime. SFN is determined as SFNprime for the first 10-ms frame of a 20-ms TTI and as $SFNprime + 1$ for the second, i.e., last, 10-ms frame of a 20-ms TTI.

Recall that each radio frame consists of fifteen time-slots, and each time-slot contains 2560 chips to achieve the chip rate of 3.84 Mcps. Relative timing between any two channels is specified with the P-CCPCH as the basic timing reference. Figure 13.12 illustrates an example of the timing relationships among various physical channels assuming certain parameter settings.[463]

Figure 13.12 specifies the relationship between the E-DCH and the E-HICH for a 10-ms TTI.[464] *The ACK or NACK that the UE receives in the E-HICH frame associated with SFN i corresponds to the data the UE transmits in the E-DPDCH frame associated with SFN $i - 3$.*

The downlink DPCH frame is transmitted τ_{DPCH} chips later than the P-CCPCH frame. $\tau_{DPCH} = 512$ chips is used here as an example. The UE receives the DL DPCH after some propagation delay. The UE transmits the E-DCH at $T_0 = 1024$ chips, a fixed value, after receiving the start of the DL DPCH frame. In Figure 13.12, the UE transmits a packet P in the frame associated with $SFN = X$ using the HARQ process 0. The timing offset between the P-CCPCH and E-HICH for a user depends on τ_{DPCH}; E-HICH is about seven slots earlier than the P-CCPCH frame in Figure 13.12 when $\tau_{DPCH} = 512$ chips. In the case of a 10-ms E-DCH TTI, the E-HICH frame offset relative to PCCPCH is given by

$$\tau_{\text{E-HICH},n} = 5210 + \left(7680 \times \left\lfloor \frac{(\tau_{DPCH,n}/256) - 70}{30} \right\rfloor \right) \quad (13.1)$$

[463] See the standards [298, 306] for additional details on the channel timings.

[464] Recall that Tx represents "transmit" and Rx represents "receive."

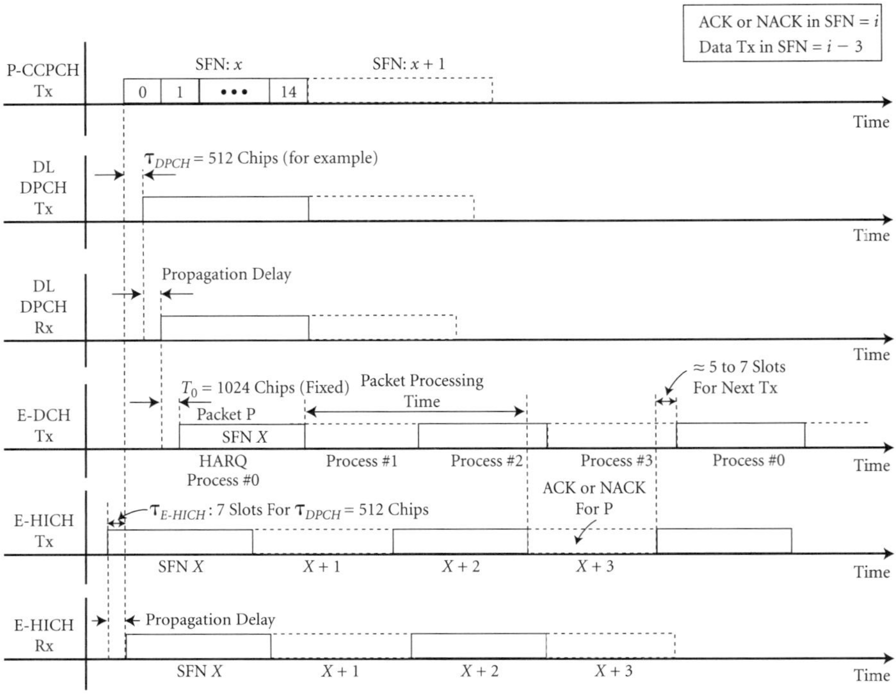

Figure 13.12: E-HICH and E-DCH Timing (10-ms TTI)

where the subscript n refers to the UE number n. The UE receives the ACK or NACK corresponding to packet P in the E-HICH frame associated with $SFN = X + 3$. The UE now has about five to seven slots to prepare for new transmission or retransmission in the TTI corresponding to HARQ process 0.

13.4 PRIOR TO DATA TRANSMISSION: CONFIGURATION DURING CALL SETUP AND E-DCH TRANSMISSION GUIDELINES

The RNC configures the UE for the HSUPA operation during the HSUPA call setup so that the UE can transmit data using the high-speed data channel [305]. (See Chapter 9 on UMTS for the comprehensive call setup. The focus here is on the HSUPA aspects of the call setup.) HSUPA reuses the basic call setup of R99 and R5 and makes some enhancements relative to R5. Figure 13.13 outlines the basic HSUPA configuration carried out during the call setup. Note that the RNC controls configuring the UE to operate in HSUPA. The Node B actually determines some configuration parameters and conveys them to the RNC so that the RNC can then configure the UE with these

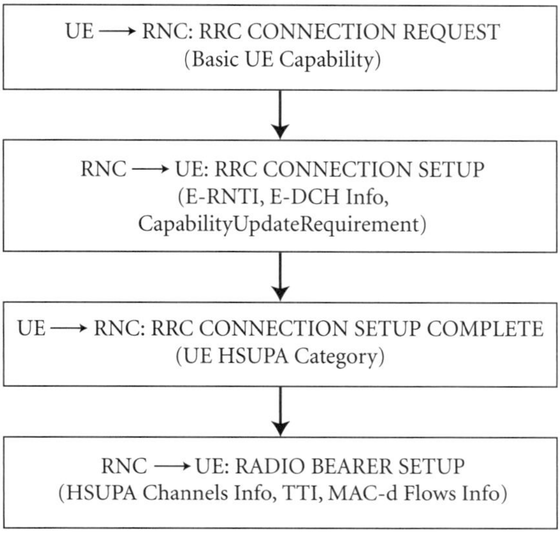

Figure 13.13: HSUPA Configuration During Call Setup

parameters. We will *mention* some configuration parameters in this section; their *meaning and use* will be discussed in Sections 13.5 and 13.6.

The UE sends an RRC CONNECTION REQUEST to the RNC and informs the RNC of whether it supports only HSDPA or both HSDPA and HSUPA. This message represents a signaling enhancement in R6 compared to R5. In R5, the UE provides no HSDPA support information to the RNC in the RRC CONNECTION REQUEST message. Note that the UE indicates only support for HSDPA and HSUPA but not the actual category. In other words, the RNC has no knowledge of the specific HSDPA or HSUPA category of the UE; it knows only that this UE supports HSDPA and HSUPA.

The RNC replies with an RRC CONNECTION SETUP message. Since the RNC knows that the UE supports HSUPA, it starts configuring selected parameters related to HSDPA and HSUPA. The RNC specifies a sixteen-bit E-RNTI to the UE. The RNC specifies one or two E-RNTIs to an HSUPA UE; i.e., primary E-RNTI and secondary E-RNTI. The main use of these E-RNTIs is to convey the AG to the UE. The RNC also assigns an H-RNTI to an HSDPA-capable UE. The RNC usually sets the CapabilityUpdateRequirement flag to TRUE to obtain more specifics regarding the UE's capabilities. The RNC includes an E-DCH Info field that contains the information elements of E-DPCCH info and E-DPDCH info, and E-DPCCH info provides the power offset of the E-DPDCH relative to the R99 DPCCH and the Happy Bit delay condition.

Do You Recall... What is the importance of the Happy Bit?[465]

E-DPDCH info includes the selected E-TFCI table index, reference E-TFCIs, periodicity and power offset of SI, identification of the HARQ processes where the scheduled mode of operation is used for a 2-ms TTI, (initial) SG, and two- and three-step thresholds.[466] The RNC can place an upper limit on the maximum data rate by specifying the maximum channelization codes that the UE can use as part of the E-DPDCH Info. For example, if the RNC sets the maximum channelization codes to sf4, the maximum channel rate is just 960 kbps (i.e., 960000 modulation symbols per second). The UE identities and E-DCH info can be provided to the UE in the RADIO BEARER SETUP message instead of the RRC CONNECTION SETUP message. The UE provides its HSUPA category to the RNC in the RRC CONNECTION SETUP COMPLETE message. Section 13.8.3 will offer details of the six UE categories.

As part of the radio bearer setup process, the RNC configures the rest of the parameters to complete the UE's configuration for HSUPA operation. As part of the DL radio resources IE, the RNC provides parameters such as the identity of the serving cell (or radio link), the E-AGCH's channelization code, the E-HICH's channelization code and signature index, and the E-RGCH's signature sequence and RG combination index.

Furthermore, as part of the added or reconfigured uplink transport channel info IE (which is a subset of the uplink transport channel info IE), the RNC sets parameters such as E-DCH TTI, HARQ info for E-DCH (e.g., the use of fixed redundancy version 0 or dynamic redundancy version, based on a table), and MAC-Dedicated (MAC-d) flow parameters (e.g., flow identity, power offset and maximum number of retransmissions to control flow-specific QoS, a multiplexing list containing identities of other flows, scheduled or non-scheduled mode, identities of the HARQ processes that can carry the flow for the non-scheduled mode in the case of a 2-ms TTI, and the maximum MAC-e PDU size that can be carried during one TTI for the non-scheduled mode).

The UE makes use of several guidelines to transmit data on E-DCH. First, only one transport block is transmitted during a given TTI. The UTRAN can change the TTI via RRC signaling. Secondly, a transport block could carry information from multiple MAC-d flows that contain relevant logical channels. The mapping between a MAC-d flow and a logical channel is established during the call setup. A given TTI is associated with a specific HARQ process. The HARQ Process Identity is given by $CFN \bmod 8$ for the 10-ms TTI and by $(5 * CFN + subframe\ number) \bmod 8$ for

[465]The Happy Bit conveys to the UTRAN whether the UE is *happy* with the current SG or the UE can use a larger grant.

[466]These thresholds are formally referred to as two-index-step threshold and three-index-step threshold.

the 2-ms TTI where CFN is the connection frame number the UE calculates based on the SFN. CFN is calculated as $(SFN - (DOFF \operatorname{div} 38400)) \bmod 256$ where $DOFF$ is the offset of the DPCH relative to the P-CCPCH.

Thirdly, one HARQ process attempts to deliver one transport block successfully until it reaches the worst-case maximum number of transmissions allowed for the transport block. Assume that a transport block contains two MAC-d flows, one with four as the maximum number of transmissions and another with seven as the maximum number of retransmissions. Then, the HARQ process will attempt up to seven times to deliver such a transport block. Each MAC-d flow is associated with a HARQ profile, which defines two attributes, the maximum number of retransmissions and a power offset. The power offset facilitates implementation of MAC-d flow-specific QoS. Assume that 0 dB power offset for MAC-d flow 1 yields a block error rate of 2%. A 2 dB power offset, for example, can be specified for MAC-d flow 2 to obtain a lower block error rate, such as 1%, for MAC-d flow 2.

Fourthly, the number of bits that can be transmitted in one transport block during a given TTI depends on the configuration of the associated HARQ process. E-TFCI represents the size of the transport block. For example, RRC signaling could have configured a process to carry scheduled data or non-scheduled data or both. The amount of non-scheduled data within one TTI is directly specified via RRC signaling during the call setup. The amount of scheduled data depends on the SG relevant to the specific TTI. An E-TFCI selection algorithm determines the E-TFCI and forms a packet.

Lastly, the 2-ms TTI provides the per-HARQ process configuration capability, and the 10-ms TTI configuration applies across all four HARQ processes. For example, in a 2-ms TTI operation, processes 0 through 6 can be configured for scheduled transmission for a scheduled MAC-d flow 1, and process 7 can be configured to carry a non-scheduled MAC-d flow 2. However, in a 10-ms TTI operation, if MAC-d flow 1 is configured for scheduled transmission, all four HARQ processes would correspond to the scheduled transmission. The per-HARQ process configuration for the 2-ms TTI is under the control of the serving Node B, and the RNC informs the UE about such a configuration via RRC signaling.

13.5 STEPS OF UPLINK DATA TRANSMISSION

Now that we know all about the HSUPA channels, we are ready to have an in-depth discussion of the steps involved in HSUPA uplink transmission. These steps include the UE's transmission of a scheduling request, the Node B's determination of grants, the UE's calculation of the SG, E-DCH transmission, Node B's packet decoding, and the UE's HARQ retransmission. These steps are described in Sections 13.5.1 through 13.5.5.

13.5.1 Scheduling Requests from the UE to the Node Bs

The UE sends a scheduling request to the Node B in the form of SI on the E-DPDCH(s) and of the Happy Bit on the E-DPCCH. SI is sent on the E-DCH either stand-alone or along with the user traffic data. When the Scheduling Information is sent stand-alone, the power offset to be used on the E-DPDCH is given by the Power Offset for Scheduling Info IE. When the SI is sent with the user traffic data, the E-TFCI–dependent power offset that the UE calculates gives the power offset to be used on the E-DPDCH. One way to facilitate SI transmission while starting the HSUPA transmission is via configuration of HSUPA channels during the call setup. According to this approach, the RNC provides the Power Offset for Scheduling Info IE and the UE can use the minimum TBS of eighteen bits. In another approach, the serving E-DCH cell provides an initial SG to the UE. The UE can use this grant to send the initial SI along with some user traffic data. The E-DCH transmission is always accompanied by the transmission on the E-DPCCH, which carries the Happy Bit and the E-TFCI corresponding to the E-DCH transmission as discussed earlier.

13.5.2 The Node B's Determination of Grants

The Node Bs allocate the AG and RG to the UE to directly and quickly control uplink interference. The AG and RG allocation algorithms must be implemented at the Node B and are non-standardized. The transmission of AG and RG is discontinuous. In other words, a grant is transmitted to the UE only when the Node B grant allocation algorithm decides so. The AG applies to *one or all* HARQ processes in the case of a 2-ms TTI, depending upon the RRC configuration, and the AG applies to *all* HARQ processes in the case of a 10-ms TTI. The RG applies to a specific TTI as a result of the strict timing relationship between the E-RGCH and the uplink E-DPDCH. *The main goals of these grant allocation algorithms are to maintain the average value of the total uplink rise-over-thermal (RoT)*[467] *noise below a suitable RoT threshold while maximizing the uplink throughput.* The closer the algorithms can operate to the RoT threshold, the higher the throughput can be.

The controlling RNC could, if the RNC vendor so chooses, send to the Node B some parameters, e.g., the maximum RTWP (Received Total Wideband Power) and reference RTWP, and thus play an important role in managing the uplink interference at the Node B. The RTWP is the total received power measured at the antenna in the suitable bandwidth, such as 5 MHz for the 3.84 Mcps Frequency-Division Duplexing air interface. The reference RTWP represents the thermal noise floor. The difference between the maximum RTWP and reference RTWP is the RoT threshold. For example, a system may target 4 dB as the target RoT threshold, which means that the maximum total received power that the Node B would allow in a given cell would

[467] RoT noise quantifies how much interference exists relative to the thermal noise power. For example, assume that the Node B measures X dBm in an unloaded system (i.e., a system with no users at all). If the total measured power is Y dBm in the presence of active users, RoT noise is said to be $Y - X$ dB. A typical CDMA system aims for an RoT threshold of 3 dB (e.g., for voice) to 6 dB (e.g., for higher data rates).

be 4 dB above the reference RTWP. The Node B may or may not use RTWP parameters coming from the controlling RNC depending upon the implementation [297].

Let's take a closer look at AG and RG. Consider a UE in the traditional handover region as shown in Figures 13.4 and 13.5. (Flip back to page 778.) The E-DCH Active Set for the UE is {Cell C2, Cell C3, Cell C4, Cell C6} in Figure 13.4.

A Question for You... Which types of cells can send AGs and RGs?[468]

The RNC selects Cell C2 as the E-DCH serving cell. Node B1 controls cells C2 and C3, and Node B2 controls cells C4 and C6. Let's call Node B1 the serving Node B since it controls the serving cell C2, and we'll call Node B2 the non-serving Node B since it offers no control of C2.

The serving Node B allocates the initial SG to the UE. The RNC conveys the initial SG to the UE as part of the HSUPA call setup and can be thought of as an initial AG value to facilitate immediate data transfer on the E-DCH. The initial SG obviates the need for immediate AG transmission on the E-AGCH. As shown in Figure 13.5, the serving Node B (i.e., Node B1) transmits the AG on the E-AGCH in only serving Cell C3, and it transmits the RG on the E-RGCH in both C2 and C3.

The non-serving Node B transmits RGs on E-RGCHs in C4 and C6. The same RG must be sent on C2 and C3 since they are members of the serving RLS. Due to the configuration of the RG Combination Index during the call setup, the UE soft-combines the RGs from C2 and C3 through maximal ratio combining because the UE is aware of the serving RLS members. For example, the RG Combination Index is set to the same value (e.g., 0) for both C2 and C3. The RG Combination Index is set to different values for different non-serving radio links (e.g., 1 for C4 and 2 for C6). When the RG Combination Index is the same for multiple cells or radio links, the UE soft-combines the RG commands.

The serving Node B executes an AG and RG allocation algorithm to determine AG and/or RG for the UEs. The basic goal of the AG is to directly allocate the uplink resource (i.e., the power ratio between the E-DPDCH and E-DPCCH). As shown in Figure 13.14, potential inputs to the AG and/or RG allocation algorithm at the serving Node B are the RoT noise measured at the receive antenna, scheduling requests from the UEs, QoS, and priorities of UEs or of services within the UE. Recall that this algorithm must be executed for each cell independently for AG allocation. For example, Node B1 considers only UEs 1, 2, and 3 and the RoT noise measured in Cell C1 upon allocating the AG to these UEs because it is the serving Node B for these UEs. Similarly, Node B1 considers only UEs 4 and 5 for AG allocation to these UEs in Cell C2. The situation becomes more complex when a UE's E-DCH serving RLS contains more than one cell. In such a scenario, the same RG command must be sent from all the cells in the E-DCH serving RLS. As a simple solution, the serving

[468] The serving E-DCH cell sends the AG, and all the cells in the E-DCH Active Set send RGs.

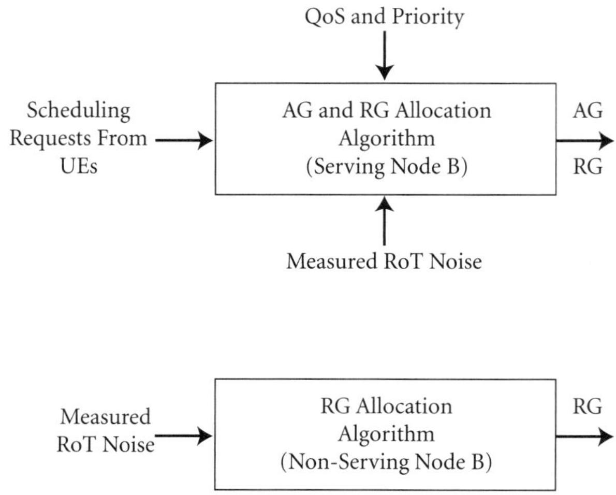

Figure 13.14: AG and RG Allocation Algorithms

Node B can consider the cell with the worst uplink interference as the RG-dictating cell within the E-DCH serving RLS for a given UE.

The grant allocation algorithm represents an area of product differentiation. When a UE indicates a large amount of data and large UE Power Headroom, the Node B is likely to allocate a large grant via a large AG value or an $RG = UP$ command. When the measured RoT noise is quite close to the RoT target threshold, smaller AG and/or $RG = DOWN$ commands are more likely. A platinum-level QoS subscriber is likely to receive a larger grant than a bronze-level QoS subscriber. (Capitalism!) A UE that has high-priority data waiting in the buffer would get a higher priority than a UE with low-priority data.

The non-serving Node B can send only $RG = HOLD$ or $RG = DOWN$ commands. As shown in Figure 13.14, the main input to the RG allocation algorithm executed at the non-serving Node B is the measured RoT noise in a cell. Though the standard specifies no exact algorithm, it does recommend that that the non-serving Node B refrain from sending an $RG = DOWN$ command unless the measured RTWP exceeds the maximum RTWP *and* the ratio of the received non-serving E-DCH power to the total received E-DCH power exceeds a target ratio the controlling RNC defines. Note that the Node B can use information such as the E-TFCI transmitted on the E-DPCCH and the UE-specific reference power offset to calculate the required ratio.

The serving Node B must execute the AG and RG allocation algorithm for each cell it controls. For example, assume that Cell C1 is the serving cell for UE1, UE2, and UE3, Cell C2 is the serving cell for UE4 and UE5, and Cell C3 is the serving cell for UE6 and UE7. Node B1 acts as the serving Node B for *all* these UEs. Though the standard allows enormous flexibility regarding the use of the AG and the RG, *the AG can be viewed as a long-term grant, and the RG can be viewed as a fast mechanism*

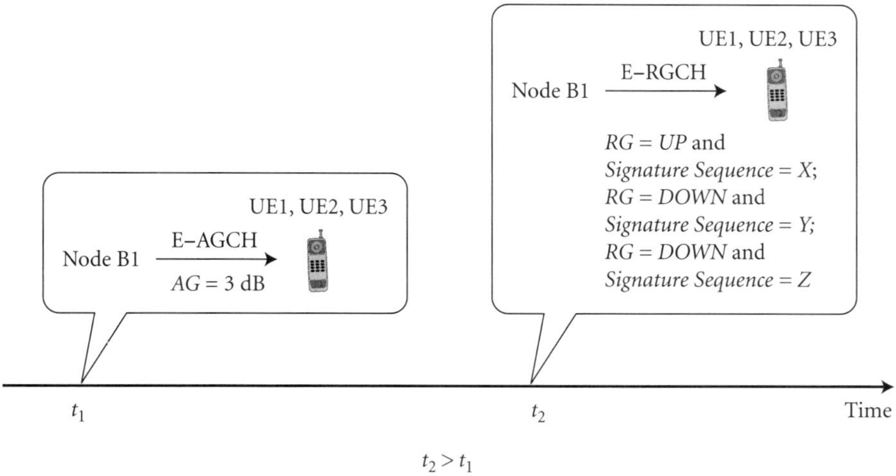

Figure 13.15: Using AG and RG to Influence the Uplink Data Rate

for adjusting the previously allocated grant, such as the AG. Furthermore, the RG helps extend the SG's effective range.

Figure 13.15 illustrates how the AG and RG can be used. Please note that this is only one possible approach; the many ways of using the AG and RG depend only upon your creativity! Assume that all the UEs in a given cell have been assigned the same primary E-RNTI and have been asked to monitor the E-AGCH with the same OVSF code. Further, assume that each UE has been assigned a different signature sequence index for the E-RGCH in a given cell.

Node B1 sends the same AG, such as 3 dB, to all the UEs by masking the CRC of the E-AGCH contents with the same primary E-RNTI in each of the three cells, C1, C2, and C3. All the UEs now have the same AG, which now becomes the SG. Node B1 can now adjust the previously allocated grant of 3 dB by sending different RGs to different UEs. For example, if Node B1 wants to allocate a larger grant to UE1 and smaller grants to UE2 and UE3, it can quickly do so by sending an $RG = UP$ command to UE1 and $RG = DOWN$ commands to UE2 and UE3 *at the same time* using the E-RGCH. UE1, UE2, and UE3 have been assigned the signature sequence indices X, Y, and Z, respectively. The UEs adjust the SG to reflect the received RG commands. If Node B1 wanted to use only E-AGCH to make such adjustments, it would have to assign different E-RNTIs to different UEs and send one AG to one UE at a given time because the UEs would monitor the E-AGCH with the same OVSF code. The initial SG could be considered a special AG case since it is sent to the UE just once as part of the HSUPA call setup.

13.5.3 Determination of the Serving Grant

The UE can transmit scheduled and non-scheduled data during a given TTI based on the RRC configuration. The UE determines the maximum amount of scheduled data for a given TTI based on the most recent value of the SG applicable to that TTI. The SG is a function of the AG and/or RGs received for a specific TTI. The relevance of the AG and RG to a specific TTI is based on strict timing relationships among the channels. Recall that the scope of the AG could be *one* or *all* HARQ processes for a 2-ms TTI. The scope of the AG is always *all* HARQ processes for a 10-ms TTI. *The UE updates the value of the SG every TTI regardless of whether a packet is to be transmitted.* When a new packet is to be transmitted during a TTI, the most recent SG applicable to that TTI is used. This section discusses in three stages how the UE updates the SG, using Figure 13.16 as an example. (We have used the three-stages concept only for the purpose of explanation; the standard contains no such term or description.)

The first stage involves the consideration of the AG. When the UE receives an AG value other than INACTIVE, the UE records it as the SG if the AG is received with primary E-RNTI or as the Stored_Secondary_Grant if the AG is received with secondary E-RNTI. When the UE receives INACTIVE as the AG value and the AG scope is for *all* HARQ processes, the UE switches to use a secondary grant by setting SG as Stored_Secondary_Grant. However, in this case, the AG value is INACTIVE but the AG scope is *per* HARQ processes, so just *one* HARQ process is deactivated. Once the AG is conveyed to the UE, the UE ignores the RG from the serving RLS for the duration of a timer, which is set to 40 ms for a 10-ms TTI and 16 ms for a 2-ms TTI.

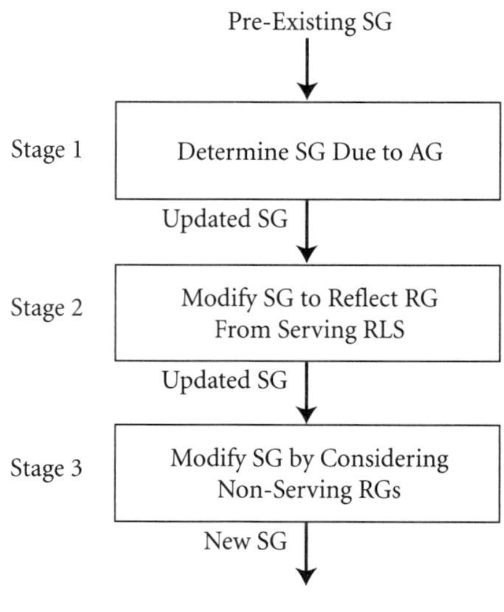

Figure 13.16: Stages of Serving Grant Update

During the second stage, the RG received from the serving RLS is evaluated to adjust the value of the SG coming from the first stage. If no new AG has been received for the TTI under consideration, the first stage is transparent and the SG is the same as the pre-existing SG. The RG value from the serving RLS can be UP, DOWN, or HOLD. If the RG value is HOLD, the second stage is transparent and the SG from the first stage remains unchanged. When the RG value is UP or DOWN, *the second stage then consists of three steps, A, B, and C.*

In Step A, the table index associated with the power ratio of the E-DPDCH to DPCCH is determined using a Scheduling Grant table that contains thirty-eight different values [301]. A variable called $reference\_ETPR$ (reference E-DCH traffic-to-pilot ratio) is calculated as the ratio of the E-DPDCH transmit power to the DPCCH transmit power in the last TTI corresponding to the HARQ process under consideration. Next, a reference index in the Scheduling Grant table is found for $reference\_ETPR$. For example, assume that the UE transmitted a packet on the E-DCH with $reference\_ETPR = 11.74$ dB, which translates into $(58/15)^2$ in normal units. In other words, $10 * \log_{10}((58/15)^2) = 11.74$ dB. In the table specified in the standard [301], $table\ index = 19$ corresponds to a Scheduling Grant of $(53/15)^2$, and $table\ index = 20$ corresponds to a Scheduling Grant of $(60/15)^2$. Then, the reference index is chosen to be $table\ index = 20$.

In Step B, a potential SG value is determined solely based on the RG with the help of the Scheduling Grant table. For example, $table\ index = 0$ in the table corresponds to a Scheduling Grant of $(5/15)^2$ and $table\ index = 37$ corresponds to a Scheduling Grant of $(168/15)^2 * 6$. Assume RRC signaling has set the *2-index-step threshold* parameter to 16 and *3-index-step threshold* parameter to 8. The RG value could be DOWN or UP. Upon completion of Step A, the minimum SG is constrained to be $(5/15)^2$ and the maximum SG is constrained to be $(168/15)^2 * 6$. When the RG value is DOWN, $SG = Scheduling\ Grant$ corresponds to the $table\ index = reference\ index - 1$. In our example, since the reference index is 20, $SG = Scheduling\ Grant$ corresponds to a table index of $20 - 1 = 19$, which is $(53/15)^2$. When the RG value is UP, the SG is increased based on an algorithm defined in the standard [301]. Figure 13.17 summarizes the operation of this algorithm.

If $reference\ index <$ *3-index-step threshold*, then $SG = Scheduling\ Grant$ corresponds to the $table\ index = (reference\ index\ + 3)$. If *3-index-step threshold* $\leq reference\ index <$ *2-index-step threshold*, then $SG = Scheduling\ Grant$ corresponds to $table\ index = (reference\ index + 2)$. If *2-index-step threshold* $\leq reference\ index$, then $SG = Scheduling\ Grant$ corresponds to $table\ index = (reference\ index + 1)$. In our example, since $reference\ index = 20 > 16 =$ *2-index-step threshold*, then $SG = Scheduling\ Grant$ corresponds to $table\ index = (reference\ index + 1) = 20 + 1 = 21$, which is $(67/15)^2$. Finally, in Step C, the SG is calculated as the minimum of the SG from the first stage or the SG from Step B in the second stage.

During the third stage, the RGs received from the non-serving radio links are considered to adjust the value of the SG arriving from the second stage. If no new RG has been received from the serving RLS for the TTI under consideration, the second stage is transparent and the SG entering the third stage is the same as the SG from

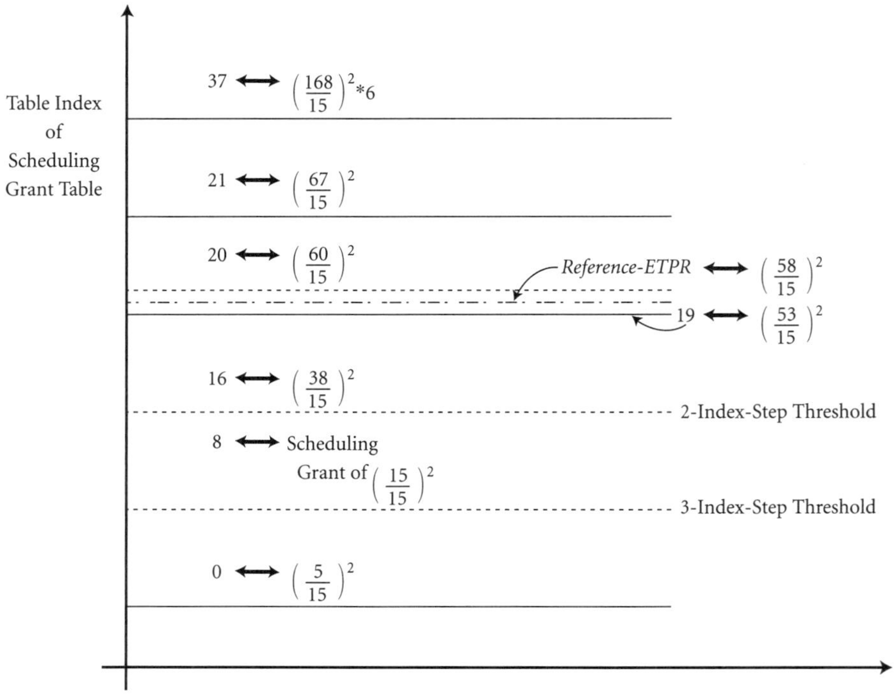

Figure 13.17: Influence of *Relative Grant* = *UP* on the Serving Grant

the first stage. The RG from the non-serving radio link can be DOWN or HOLD. If all the RG values are HOLD, the third stage is transparent and the SG from the second stage remains unchanged. When at least one RG value is DOWN, the SG is calculated as it is in the second stage. First, a potential value for the SG is determined solely based on the RG with the help of a Scheduling Grant table. The final SG is then calculated as the minimum of the SG from the second stage or the calculated SG in the third stage. This final SG cannot be exceeded during the next 40 ms for the 10-ms TTI and the next 16 ms for the 2-ms TTI.

In summary, once all the three stages are carried out, the final SG value is obtained. This SG value is used to determine how much scheduled data can be transmitted during a given TTI.

13.5.4 E-DCH Transmission

Figure 13.18 illustrates the basic steps for the determination of the E-TFCI for the uplink E-DCH. *This is where the UE determines the UL data rate based on SG and the allowed amount of non-scheduled data!* These six steps are very briefly summarized here and discussed in detail in Sections 13.5.4.1 through 13.5.4.6.

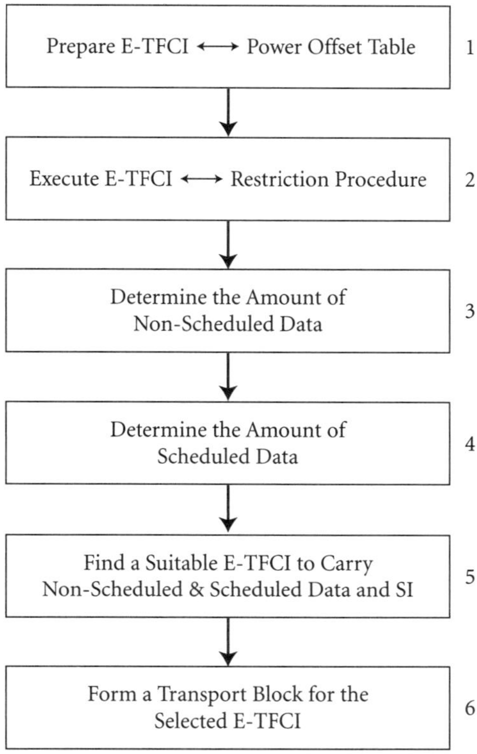

Figure 13.18: Overview of E-TFCI Determination at the UE

In Step 1, the UE is configured via RRC signaling to use one of four specific E-TFCI tables. The table specifies the allowed transport block sizes. RRC signaling also provides a reference E-TFCI set that contains the E-TFCI(s) and related power offset(s). The UE can now calculate the required power offset between the E-DPDCH and the DPCCH for each E-TFCI in the E-TFCI table. The infrastructure vendor must determine a reference E-TFCI set to ensure a certain block error rate for the E-DCH at various data rates. The UE prepares such a table for each configured MAC-d flow since the MAC-d flow-specific power offset is also configured via RRC signaling to provide a different QoS to different MAC-d flows. For example, an additional power offset of 2 dB may be used to reduce the block error rate for a MAC-d flow that needs a lower block error rate. In Step 2, the UE learns of the largest supportable E-TFCI based on the available power for the E-DCH.

In Step 3, the UE determines the amount of non-scheduled data for transmission during a given TTI based on the RRC configuration that specifies the amount of data for a given HARQ process. In Step 4, the UE considers the most recent SG applicable to the given TTI to calculate the amount of scheduled data that can be sent during a TTI. The UE now knows the supportable E-TFCIs from Step 1 and the amount of scheduled and non-scheduled data from Steps 3 and 4. The UE may need to transmit SI based on specific triggers mentioned earlier (see Section 13.3.1). In Step 5, the UE determines the most suitable E-TFCI that can carry the sum of scheduled data, non-scheduled data, and SI. In Step 6, the UE fills the transport block of the selected E-TFCI size with the data from the scheduled and/or non-scheduled MAC-d flows and SI.

13.5.4.1 Preparation of E-TFCI versus Power Offset Table The RNC conveys to the UE several parameters so that the UE can construct a table that maps every E-TFCI in the E-TFCI table to a power offset. Assume that one entry in such a table is (*E-TFCI* $=$ 37 corresponding to 517 bits, $Power\ Offset$ $=$ 2 dB) for a UE configured with a 10-ms TTI and E-DCH transport block size table 0. If the UE selects *E-TFCI* $=$ 37, it will transmit E-DPDCH at a transmit power level 2 dB above the DPCCH transmit power. These parameters, signaled via RRC, enable the UE to prepare the (E-RFCI, Power Offset) table: E-DCH TTI, E-TFCI table index, Reference E-TFCIs, and MAC-d-flow–specific power offset. These parameters are contained in the RRC messages, such as RADIO BEARER SETUP. The E-DCH TTI parameter indicates if the UE should use a 2 ms TTI or a 10 ms TTI. The E-TFCI table index is 0 or 1 since two tables are defined for each E-DCH TTI. The combination of the E-DCH TTI and E-TFCI table index indicates to the UE which of the four E-TFCI tables should be used. Reference E-TFCIs and MAC-d-flow–specific power offsets help the UE in forming the (E-TFCI, Power Offset) table for each MAC-d flow.

Example: Creation of an E-TFCI Table in the UE
Assume that reference E-TFCI includes just one entry, (120 bits, X dB) and that the MAC-d-flow–specific power offset is 0 dB for flow 1. The RNC has determined X dB such that a specific block error rate (e.g., 5%) on the E-DPDCH can be achieved for the transport block size of 120 bits. The UE has been configured to use a specific E-TFCI table. For flow 1, a suitable power offset is calculated for each E-TFCI in the E-TFCI table. The UE prepares a table such as the one shown in Table 13.1. If the same signal-to-interference ratio (to be more accurate, signal-to-interference-plus-noise ratio) is assumed across all the E-TFCIs, the required transmit power increases linearly with the amount of data. In other words, if the number of bits is doubled, the required transmit power doubles (i.e., increases by 3 dB). Based on such a simple assumption, the UE prepares a table similar to the one in Table 13.1. A more comprehensive analysis would yield a more accurate table. In Table 13.1, the UE has calculated the power offset of $(X + 9.5)$ dB corresponding to the TBS of 1015 bits as $10 * \log_{10}(1015/120) = 9.5$ dB.

Example: Creation of an E-TFCI Table at the UE including the Impact of a Power Offset
If a MAC-d flow 2 has a power offset of 1 dB, Table 13.1 is modified so that the required power offset for each E-TFCI is increased by 1 dB. Table 13.2 shows such

Table 13.1: (E-TFCI, Required Power Offset) Table for MAC-d Flow 1

E-TFCI	TBS (Bits)	Required Power Offset (dB)
0	18	$X - 8.2$
1	120	X
⋮	⋮	⋮
60	1015	$X + 9.5$
⋮		
79	2020	$X + 12.2$
⋮	⋮	⋮
127	11484	$X + 19.8$

a modified table. The power offset corresponding to the E-TFCI of 1015 bits is now $(X + 10.5)$ dB for flow 2 as opposed to $(X + 9.5)$ dB for flow 1. The RNC can determine the MAC-d-flow–specific power offset to achieve a better (or worse!) error rate or reduced latency. For example, compared to flow 1, which has a 0 dB power offset, flow 2 has a 1 dB power offset. Hence, when a transport block of a given size is transmitted for flow 2, it is transmitted at 1 dB higher power, leading to a lower error rate for flow 2. If flow 1 achieves a 5% error rate, flow 2 would achieve a lower error rate, such as 2%. A lower error rate translates into a lower average latency. The (E-TFCI, Power Offset) table can be reconfigured via RRC signaling.

Table 13.2: (E-TFCI, Required Power Offset) Table for MAC-d Flow 2

E-TFCI	TBS (Bits)	Required Power Offset (dB)
0	18	$X - 7.2$
1	120	$X + 1$
⋮	⋮	⋮
60	1015	$X + 10.5$
⋮	⋮	⋮
79	2020	$X + 13.2$
⋮	⋮	⋮
127	11484	$X + 20.8$

13.5.4.2 E-TFCI Restriction Procedure *The goal of this procedure is to identify which E-TFCIs are supportable based on the available transmit power at the UE during a given TTI.* Figure 13.19 summarizes the basic steps executed during the E-TFCI restriction procedure. These steps are carried out for each E-TFCI index i in the configured E-TFCI table for each configured MAC-d flow. In Step 1, the UE calculates the maximum allowed transmit power P_{max-i} as a minimum of the maximum UE transmit power based on its power class with E-TFCI–dependent power back-off and the maximum UE transmit power the UTRAN sets through RRC signaling (e.g., via the RRC CONNECTION SETUP message). In Step 2, the UE estimates the power available for the E-DCH, P_{rem}, as $P_{max-i} - (Power\ for\ DPCCH + Power\ for\ DPDCH +$ *Power for HS-DPCCH* $+$ *Power for E-DPCCH*). The UE in Step 3 considers an E-TFCI i as *supported* if P_{rem}/P_{DPCCH} is greater than or equal to the E-TFCI–dependent ratio of the E-DPDCH power to the DPCCH power (which is calculated as discussed in Section 13.5.4.1).

Figure 13.20 illustrates an example of the E-TFCI restriction procedure. The UE transmit power for the DPCCH is 2 mW or $10 * \log_{10}(2\ \text{mW}) = 3$ dBm. Other non–E-DCH channels, such as HS-DPCCH, occupy additional power. The total transmit power excluding the E-DCH channel is 97.64 mW or 19.89 dBm. The maximum UE transmit power is 125 mW or 21 dBm because the UE belongs to Power Class 4 [307]. The available power for the E-DCH is the difference between 125 mW and 97.64 mW, which is 28.25 mW or 14.5 dBm. The ratio of the maximum transmit power available to E-DCH to the DPCCH transmit power is $10 \log_{10}(28.25\ \text{mW}/2\ \text{mW}) = 11.5$ dB.

Now, let's identify which E-TFCIs can be supported. In Table 13.3, an E-TFCI index of 78 corresponds to 11.4 dB, and an E-TFCI index of 79 corresponds to

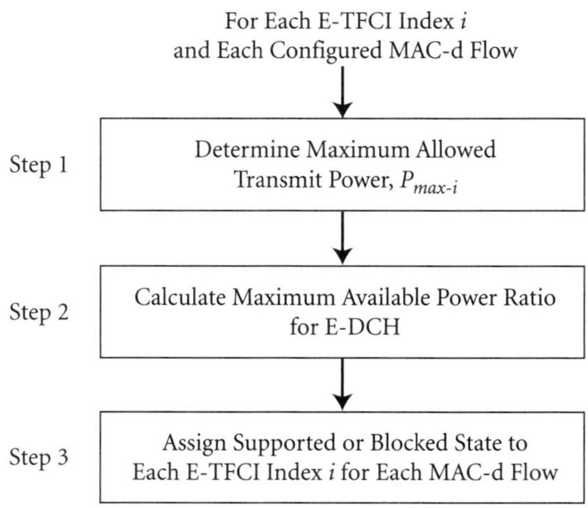

Figure 13.19: E-TFCI Restriction Procedure

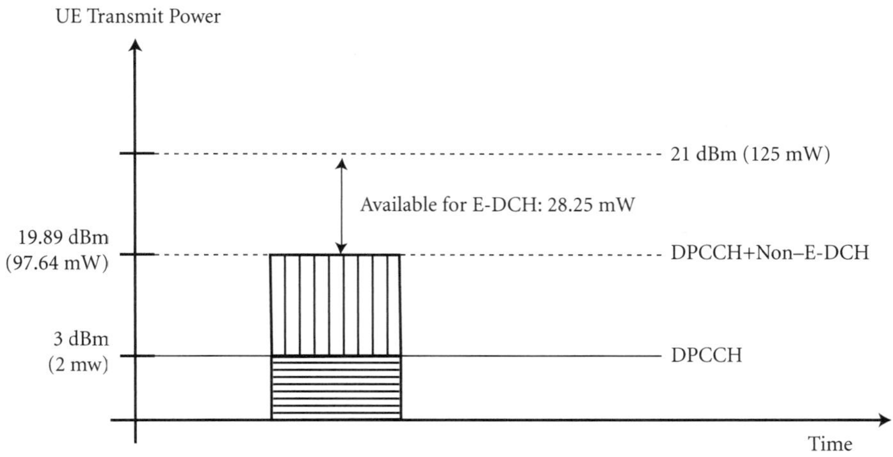

Figure 13.20: Transmit Power Levels and Margins at the UE

11.7 dB. (Please note that these are sample numbers.) Then, all E-TFCIs from 0 to 78 are assigned the supported state, and the E-TFCIs from 79 to 127 are assigned the blocked state.

13.5.4.3 Determination of Non-Scheduled Data

The UE calculates the maximum amount of non-scheduled data by considering the chosen MAC-d flow and

Table 13.3: Example E-TFCI Table after Execution of the E-TFCI Restriction Procedure

E-TFCI Index	TBS Size	Required Power Offset	State
0	18	−9.2	Supported
1	120	−1	Supported
⋮			⋮
60	1015	8.5	Supported
⋮	⋮	⋮	⋮
78	1948	11.4	Supported
79	2020	11.7	Blocked
80			Blocked
⋮	⋮	⋮	⋮
127	11484		Blocked

compatible[469] MAC-d flows. Assume that the chosen MAC-d flow is Flow 1 and the compatible MAC-d flows are Flow 2, Flow 3, and Flow 4. Further assume that all the MAC-d flows are configured for transmission in the HARQ process corresponding to the current TTI and that Flow 1 and Flow 2 belong to scheduled transmission and Flow 3 and Flow 4 belong to non-scheduled transmission. If the maximum numbers of bits configured via RRC signaling are 2000 bits for Flow 3 and 1000 bits for Flow 4, the maximum non-scheduled data is $2000 + 1000 = 3000$ bits for the TTI under consideration. The actual number of non-scheduled bits must be less than or equal to the number of bits calculated for a given TTI (e.g., 3000 bits in our example).

13.5.4.4 Determination of Scheduled Data Assume that the SG applicable to the TTI under consideration is 5 dB. The UE uses the E-TFCI table corresponding to the chosen MAC-d flow to find the maximum supportable scheduled payload. For example, Table 13.4 indicates that the required power offset is X dB corresponding to an E-TFCI index of N and Y dB for an E-TFCI index of $N + 1$. Assume that X dB $< SG = 5$ dB $< Y$ dB. The maximum scheduled payload is Z bits associated with the E-TFCI index N. If the total number of available scheduled bits, z' is less than Z, Z is set to z'. The actual scheduled payload for the current TTI must be Z bits or less.

13.5.4.5 E-TFCI Selection The UE selects a suitable value of E-TFCI for a given TTI. Figure 13.21 provides a simplified overview of the E-TFCI selection

Table 13.4: Determination of Maximum Scheduled Payload

E-TFCI Index	Required Power Offset	State	Number of Bits
0		Supported	
1			
⋮	⋮	⋮	⋮
$N-1$			
N	X dB	Supported	(Z)
$N+1$	Y dB	Supported	M
$N+2$	P dB	Blocked	R
⋮	⋮	⋮	⋮
127		Blocked	

[469]A compatible MAC-d flow can accompany a given MAC-d flow and has similar QoS requirements. A MAC-d flow multiplexing list containing such compatible flows is given to the UE via RRC signaling.

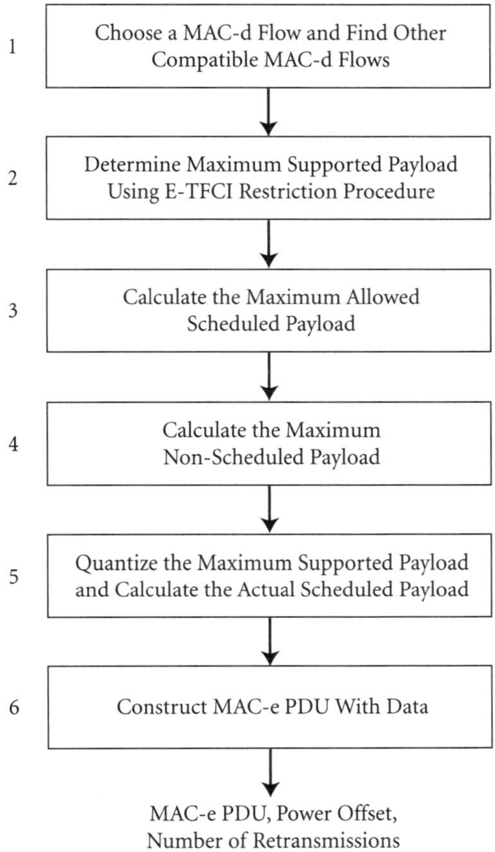

Figure 13.21: E-TFCI Selection: Overall Steps

process. The standard does provide a pseudo-code for this process. In Step 1, the UE chooses a specific MAC-d flow among the configured flows so that the highest-priority data is transmitted during a given TTI.

Do You Recall... How does the UE know which MAC-d flow to transmit during a given TTI?[470]

Each MAC-d flow is also associated with a logical channel that has a certain priority. In summary, the UE considers all the MAC-d flows that could potentially

[470]Every TTI is associated with a specific HARQ process. A given MAC-d flow, scheduled or non-scheduled, is configured to use one or more HARQ processes. Thus, a TTI is associated with one or more MAC-d flows.

be transmitted during a given TTI and selects the MAC-d flow associated with the highest-priority logical channel. The UE also knows about other MAC-d flows that can accompany the selected MAC-d flow based on the MAC-d multiplexing list specified via RRC signaling.

In Step 2, the UE uses the E-TFCI table corresponding to the selected MAC-d flow to determine the *upper limit on the payload* that can potentially be transmitted during the given TTI. Assume that there are two MAC-d flows, one mapped onto a high-priority logical channel 1 and the other mapped onto a low-priority logical channel 2. The UE will use a table, such as Table 13.4, corresponding to the MAC-d flow associated with logical channel 1. The largest E-TFCI index in the supported state is $N + 1$, and the E-TFCI index $N + 2$ is in the blocked state. Hence, the UE will not transmit more than M bits corresponding to the E-TFCI index of $N + 1$. In this example, M bits represent the upper limit on the payload size.

In Step 3, the UE calculates the *maximum allowed scheduled payload* based on the SG calculated for this TTI (as discussed in Section 13.5.4.4). The UE calculates the amount of *non-scheduled data* in Step 4 based on the configuration of MAC-d flows done via RRC signaling. (See Section 13.5.4.3 for an example.) The UE calculates the sum of non-scheduled data and the maximum scheduled data and adds SI bits if required and possible in Step 5. The total number of the data bits is quantized to the available transport block sizes. The scheduled data bits have the least priority. *The actual amount of scheduled bits is determined during this step. Furthermore, the specific value of the E-TFCI index is also determined at this stage.* We'll look at two examples in just a moment.

In Step 6, the UE constructs a MAC-e PDU using the number of bits associated with the selected E-TFCI index. Section 13.5.4.6 discusses the construction of a MAC-e PDU. The MAC-e PDU is a transport block that is given to the physical layer for transmission. The relevant pieces of information, such as the required power offset related to the selected E-TFCI index and the number of retransmissions for the highest-priority MAC-d flow, are also conveyed to the physical layer.

Let's discuss how the UE determines the value of a specific E-TFCI index to carry the sum of scheduled data, non-scheduled data, and possibly SI. If required, the number of scheduled bits is reduced to accommodate the SI. The E-TFCI index should be a minimum value that enables transmission of the total number of bits calculated in Step 5. Let's look at two examples for the E-TFCI table shown in Table 13.5.

Example: Selection of the E-TFCI Index (Example 1)

Assume that the total number of bits for transmission is M bits. From Table 13.5, $M_1 < M < M_2 < M_3$, and the E-TFCI index for both M_1 and M_2 is in the supported state. Then, the UE chooses the E-TFCI index of I_1 corresponding to the transport block size of M_1 bits.

Table 13.5: Determination of the E-TFCI Index

E-TFCI Index	Transport Block Size (Bits)	State	Required Power Offset
0		Supported	
1		Supported	
2			
⋮	⋮	⋮	⋮
$59 = I_1$	M_1		
60	M_2		
61	M_3		
⋮	⋮	⋮	⋮
$80 = I_2$	N_1	Supported	
81	N_2	Blocked	
82	N_3	Blocked	
⋮		⋮	
127		Blocked	

Example: Selection of the E-TFCI Index (Example 2)
Assume that the total number of bits for transmission is N bits with $N_2 < N < N_3$. However, N_2 and N_3 correspond to the E-TFCI indices in the blocked state. The highest E-TFCI index associated with the number of bits less than N_2 is I_2 corresponding to N_1 where $N_1 < N_2$. The UE chooses I_2 as the E-TFCI index.

13.5.4.6 Transport Block Formation The UE creates a MAC-e PDU, which is simply a transport block to be sent over the air on the E-DPDCH(s). The UE considers the logical channels in the order of their priorities while forming a MAC-e PDU. The MAC-e PDU includes a MAC-e header and a MAC-e Service Data Unit (SDU). Figure 13.22 illustrates an example of MAC-e PDU construction, and the standard [301] offers additional details.

Logical Channel 1 has a higher priority and is mapped onto MAC-d Flow 1 that is configured for non-scheduled transmission. Logical Channel 2 has a lower priority and is mapped onto MAC-d Flow 2 that has been configured for scheduled data transmission. To create a MAC-e PDU, the UE first considers Logical Channel 1 since it has the highest priority. The MAC-e SDU now contains data bits from

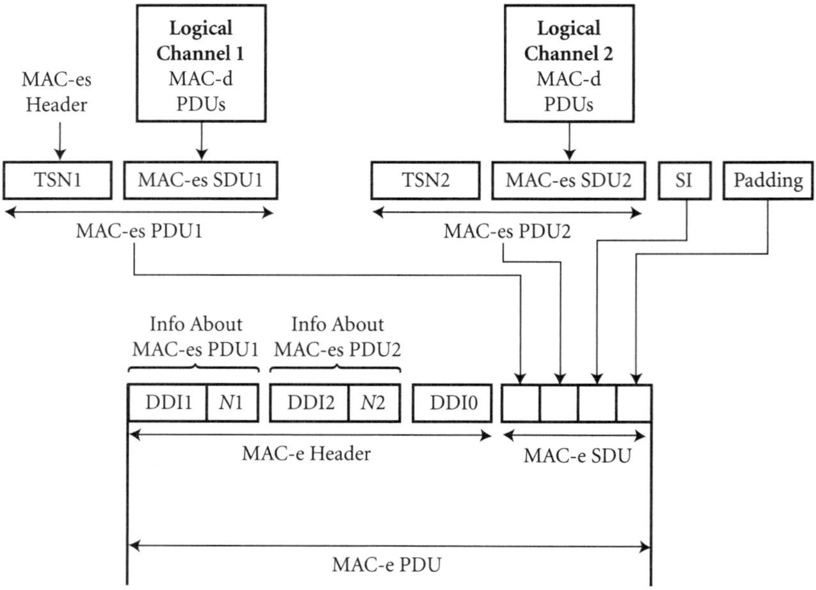

Figure 13.22: Construction of a MAC-e PDU

Logical Channel 1 or equivalently MAC-d Flow 1. Specifically, one MAC-es PDU that contains the transmission sequence number (TSN) and the actual data (called MAC-es SDU) is placed into the MAC-e SDU. The TSN helps the receiver re-order the MAC-es PDUs that might have arrived non-sequentially. The information about this MAC-es PDU must be inserted into the MAC-e header and includes the DDI (data descriptor indicator) and N. The DDI specifies the logical channel identity (1 in our example), MAC-d flow identity (1 in our example), and the size of the MAC-d PDUs concatenated in the MAC-es PDU. N specifies the number of MAC-d PDUs in the MAC-es PDU. Both DDI and N are six-bit fields. The mapping between a specific value of DDI and the *combination* of the logical channel, MAC-d flow, and MAC-d PDU size is configured via RRC signaling.

Now that the first logical channel is processed, the UE considers the next lower-priority Logical Channel 2. The MAC-es PDU from Logical Channel 2 or MAC-d Flow 2 is placed into the MAC-e SDU, and the relevant DDI and N for this MAC-es SDU are placed into the MAC-e header. If SI is to be included, it is appended at the end of the MAC-e SDU. If space allows, DDI0 (111111) is inserted into the MAC-e header; otherwise, just SI exists in the MAC-e SDU without DDI0.

The MAC-e PDU constructed at the MAC layer is passed to the physical layer along with the control information, such as the E-TFCI index, power offset of the MAC-d flow associated with the highest-priority logical channel, and the maximum number of retransmissions X, which is the maximum value among all the MAC-d flows included in the MAC-e PDU. Note that the power offset between the DPCCH and the E-DPDCH is maintained for the first transmission and all retransmissions of a given MAC-e PDU. The uplink closed-loop power control is still operational and directly adjusts the transmit power of the UL DPCCH, and the power offset between the UL DPCCH and the E-DPDCH remains constant at a value associated with the chosen E-TFCI index. For example, if the chosen E-TFCI index is N in Table 13.4, the E-DPDCH is transmitted at X dB higher relative to the DPCCH. The physical layer at the UE accepts the MAC-e PDU and related control information and forms a data packet to be transmitted on the E-DPDCH(s). A review question is now due...

Do You Recall... On which channel does the UE send control information, such as the selected E-TFCI index, RSN, and Happy Bit?[471]

13.5.5 Node B's Packet Decoding and HARQ Transmission

All the cells in the E-DCH Active Set listen to the UE's E-DPDCH(s) and E-DPCCH. E-DPCCH carries adequate information for the Node B in the Active Set to decode the packet on the E-DPDCH(s). The Node B performs maximal ratio combining of the signals received from the E-DCH Active Set cells. If the Node B successfully decodes the packet, it sends a positive acknowledgment on the E-HICH in all the E-DCH Active Set cells under its control. Otherwise, it sends a negative acknowledgment. After decoding a packet correctly or after receiving all the retransmissions of a given E-DCH packet, the Node B forwards the packet to the RNC along with the quality indicator for the received packet (e.g., whether or not the packet is decoded correctly). More specifically, the Node B disassembles the E-DCH packet and sends MAC-es PDUs on the I_{ub} interface for each MAC-d flow. The serving RNC performs selection combining of the packets received from the Node Bs and re-orders them separately for each logical channel based on the TSN included in each MAC-es PDU. The MAC-es sublayer in the serving RNC sends the packets to the MAC-d sublayer, which conveys the packets to the RLC sublayer. The RLC sublayer, in turn, forwards the packets to the Packet Data Convergence Protocol sublayer in the serving RNC. From the serving RNC, the packets reach the Third-Generation SGSN (3G-SGSN), which sends the packets to the GGSN. Finally, the GGSN enables the packets to enter the vast ocean we know as the Internet!

13.6 PHYSICAL LAYER RETRANSMISSIONS AND HARQ

HSUPA exploits synchronous HARQ for the uplink transmissions on the E-DCH to increase average throughput. The first transmission of a packet uses $RSN = 0$, which corresponds to $RV\ index = 0$. Table 13.6 summarizes the association between the

[471] The E-TFCI index, RSN, and Happy Bit are sent on the E-DPCCH.

Table 13.6: Mapping Between the RV Index and RV Parameters

RV Index	RV Parameters	
	s	r
0	1	0
1	0	0
2	1	1
3	0	1

RV index and the RV parameters. Chapter 11 offers interpretation of different values of s and r, but we'll help you refresh your memory.

Do You Recall... When $s = 1$ and $r = 0$, which type of bits at the output of the turbo coding block are getting a higher priority?[472]

HARQ rate matching[473] is carried out with the parameters specified in Table 13.6. The UE uses $RV\ index = 0$ for the first transmission and all retransmissions *or* the RSN-based RV index. Such configuration is controlled by setting the HARQ Info for E-DCH field in the Added or Reconfigured Uplink Transport Channel Info information element in an RRC message, such as the RADIO BEARER SETUP message. The RSN-based E-DCH RV index is derived using Table 13.7.

RSN starts at 0 for the first transmission, is incremented by one for each retransmission, and saturates at 3. *TTIN* represents the TTI number (TTIN) and is the same as the CFN for a 10-ms TTI and $5 * CFN + subframe\ number$ for a 2-ms TTI. The

Table 13.7: Mapping Between the E-DCH RV Index and the RSN

RSN	E-DCH RV Index for $((N_{sys})/(N_{e,data,j})) < (1/2)$	E-DCH RV Index for $((N_{sys})/(N_{e,data,j})) \geq (1/2)$
0	0	0
1	2	3
2	0	2
3	$(\lfloor(TTIN/N_{ARQ})\rfloor \bmod 2) \times 2$	$\lfloor(TTIN/N_{ARQ})\rfloor \bmod 4$

[472]The combination of ($s = 1, r = 0$) means that we are giving higher priority to systematic bits and lower priority to redundancy or parity bits.

[473]HARQ rate matching refers to the process whereby a set of code symbols from all the available code symbols at the output of the turbo coding block is chosen to form a packet for transmission during a TTI.

CFN is calculated as [305]

$$CFN = \big(SFN - (DOFF \operatorname{div} 38400)\big) \bmod 256 \tag{13.2}$$

where SFN is the system frame number specified in the SYSTEM INFORMATION message carrying System Information Blocks, $DOFF$ is the offset of the DL DPCH relative to the P-CCPCH, and div is the floor function. Note that the subframe number ranges from 0 to 4 in a 10-ms radio frame. N_{ARQ} is the number of HARQ processes, which is 4 for a 10-ms TTI and 8 for a 2-ms TTI.

The UE uses an E-DCH RV index value of 0 or 2 when the packet to be transmitted already includes numerous redundancy bits, which is the case when $(N_{sys}/N_{e,data,j}) < 1/2$. In such a case, two redundancy versions to choose redundancy bits are considered adequate. When $(N_{sys}/N_{e,data,j}) \geq 1/2$, fewer overall redundancy bits are available and multiple ways of including different sets of redundancy bits would enable the UE to cover these available redundancy bits.

Figure 13.23 illustrates how HARQ retransmissions occur in HSUPA for a dynamic E-DCH RV index with $(N_{sys}/N_{e,data,j}) < 1/2$. The UE intends to transmit packet P as a transport block. The UE chooses a HARQ process corresponding to the upcoming TTI. Recall that packet P could include multiple MAC-d flows corresponding to multiple logical channels.

In Step 1, the UE sets $RSN = 0$ since P is transmitted for the first time. An RV index of 0 is used, which means $(s = 1, r = 0)$. The rate matching process uses these RV parameters, and physical layer processing for the E-DCH is carried out. In Step 2, the Node B receives the E-DCH and attempts to retrieve packet P by decoding the received packet P1 that represents packet P. Since the Node B fails to correctly decode packet P, it sends in Step 3 a NACK on the E-HICH. In Step 4, the UE prepares for the physical layer HARQ retransmission by incrementing RSN by 1, which maps to an RV index value of 2 according to Table 13.7. The UE uses the same HARQ process as before to transmit packet P2. Note that packet P2 is formed by choosing $(s = 1, r = 1)$ corresponding to $RSN = 1$.

In Step 5, the Node B combines packets P1 and P2 in an attempt to retrieve packet P; again, the Node B is unsuccessful. The Node B sends in Step 6 a NACK on the E-HICH. The UE, in turn, increments RSN to 2 and uses an RV index value of 0 to construct packet P3 in Step 7. Packet P3 is transmitted on the E-DCH. In Step 8, the Node B combines previously stored packets P1 and P2 with the newly received packet P3. Good news! This time the Node B succeeds! In Step 9, the Node B sends a positive acknowledgment to the UE and the UE no longer needs to retransmit packet P. The UE can, in Step 10, start transmission of a new packet Q in the HARQ process previously used for packet P.

Figure 13.23 illustrates the operation of one HARQ process. Multiple such HARQ processes can be used to send different packets continuously on the E-DCH during different TTIs.

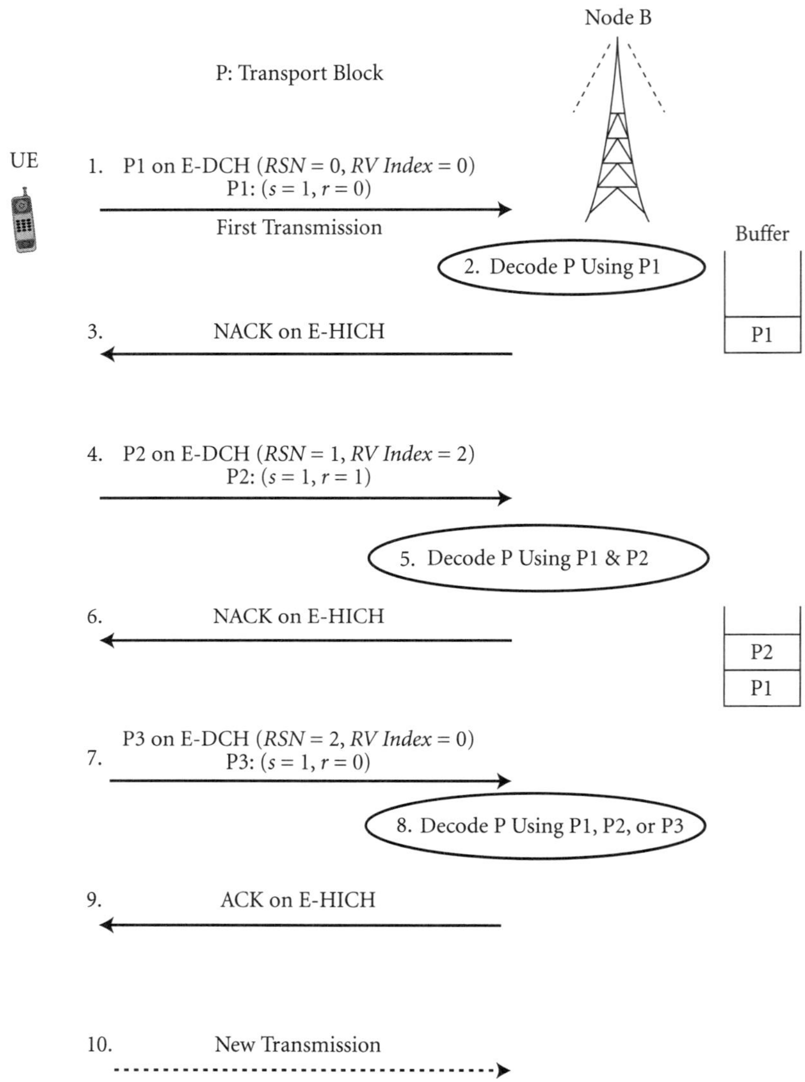

Figure 13.23: Physical Layer Retransmissions in an HSUPA HARQ Process

13.7 SUPPORT FOR USER MOBILITY

HSUPA supports traditional soft and softer handover in the uplink. *The basic HSUPA handover process is the same as that in R99*; however, one additional measurement reporting event, Event 1J, is introduced to support HSUPA. This new event is said to occur when a non-active E-DCH (but active Dedicated channel) Primary Common Pilot channel becomes better than an active E-DCH Primary Common Pilot channel.

This event helps the UTRAN update the E-DCH Active Set to add a strong cell to increase UL throughput.

The UE makes intra-frequency measurements of the pilot E_c/N_0 of the pilots in the Active Set and the Neighbor Set. It reports such measurements to the RNC via a measurement report message as a result of periodic or event-based triggering. The RNC executes the soft handover algorithm as discussed in Chapter 9 and creates an R99 Active Set. Then, it creates an E-DCH Active Set, which could be the same as or a subset of the R99 Active Set. The algorithm that the RNC uses to create the E-DCH Active Set is beyond the scope of the standard and is implementation-specific. Such an algorithm should attempt to maximize the uplink throughput while using a minimum number of cells in the E-DCH Active Set. This algorithm also needs to select one member of the E-DCH Active Set as the E-DCH serving Cell.

All the cells in the E-DCH Active Set process the packets received from the UE on the E-DPDCHs and send a HARQ ACK or NACK to the UE on the E-HICH. Furthermore, a cell in the E-DCH Active Set sends AGs and/or RGs. In the uplink, each Node B in the E-DCH Active Set performs maximal ratio combining, and the RNC performs selection combining of the uplink blocks received from multiple Node Bs. Furthermore, the cells in the R99 Active Set send uplink power control commands to the UE so that the UE can adjust the transmit power of the DPCCH. Transmit power levels of all the other uplink channels including the E-DPDCH are determined to maintain the power offsets of these channels relative to the UL DPCCH. User mobility across RNCs and across core networks in the Connected Mode and in the Idle Mode would be handled in the same way as in R99.

13.8 SYSTEM ARCHITECTURE

Our HSUPA discussion so far has primarily focused on the air interface because HSUPA indeed is an air-interface enhancement. Now, in Section 13.8.1, we'll discuss the impact of supporting HSUPA on the R99 and R5 network architecture. The major changes in the interaction between the Node B and the RNC required for HSUPA are summarized in Section 13.8.2. Finally, the characteristics of the UE categories are briefly touched upon in Section 13.8.3.

13.8.1 Enhancements to the Existing Architecture

The majority of the required architectural changes are primarily in the UE and the UTRAN, and the packet-switched core network requires only minor changes. As the HSUPA focuses on packet-switched services, the circuit-switched core network remains intact. The main air-interface–related enhancement in the system architecture is the introduction of a new MAC layer at the UE, the Node B, and the RNC, as illustrated in Figure 13.24.

The Node B introduces a new layer, MAC-e, above the physical layer. The RNC incorporates a new layer, MAC-es, below the MAC-d layer. A new layer, which is a combination of the MAC-e and MAC-es layers and is called, fittingly, the MAC-e/es,

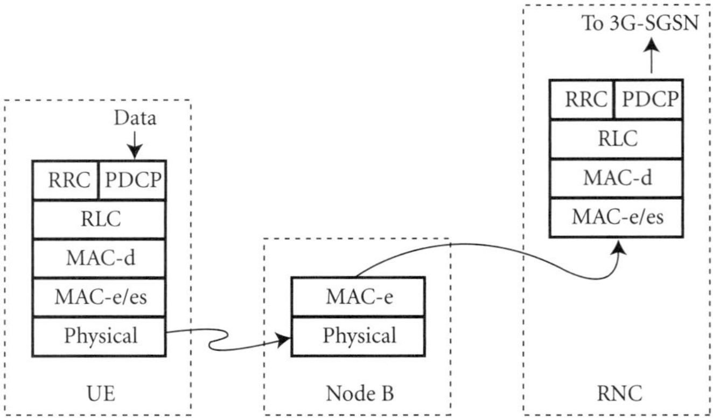

Figure 13.24: HSUPA-Specific Simplified System Architecture

is now placed between the MAC-d and physical layers at the UE. The split between MAC-e and MAC-es in the UE is unspecified in the standard.

The MAC-e and MAC-es layer at the UE multiplexes MAC-d PDUs into MAC-es PDUs and multiplexes MAC-es PDUs into one MAC-e PDU. One MAC-e PDU represents one transport block to be carried during a given TTI. Since one MAC-es PDU corresponds to one logical channel, a TSN is included in the MAC-es header so that the non-sequential packets can be re-ordered at the receiver. The MAC-e/es layer also selects the specific value of E-TFCI. Furthermore, the MAC-e/es layer processes the HARQ ACKs or NACKs received from the E-DCH Active Set cells to provide E-TFCI, RSN, and power offset to the physical layer for the next packet transmission or retransmission. The packet retransmission involves a (potentially) new RSN value, but the E-TFCI and power offset are maintained for all transmissions, including retransmissions, of a given MAC-e PDU.

The MAC-e layer at the Node B receives MAC-e PDUs from the physical layer and generates an ACK or NACK by using the CRC attached to the MAC-e PDU. This layer also processes the received scheduling requests and determines the AGs and RGs for the UEs. This layer performs functions such as determination of the value and scope of the AG and the selection of primary versus secondary E-RNTI. This layer also sets RGs. The Node B demultiplexes the received MAC-e PDUs into MAC-es PDUs and sends them on MAC-d flows to the RNC.

The MAC-es layer at the RNC receives multiple MAC-es PDUs from different Node Bs when the UE's E-DCH Active Set contains more than one Node B. After performing selection combining, this layer re-orders the MAC-es PDUs if required and extracts the MAC-d PDUs for submission to the upper MAC-d layer. The MAC-d layer forwards the packets to the RLC layer, which sends them to the PDCP layer. The PDCP layer at the RNC would then send the packets to the 3G-SGSN on the I_u-PS interface.

The physical layers at the UE and Node B undergo changes that are primarily related to support for new HSUPA channels. The physical layer at the UE must support the use of $SF = 2$ and of multiple E-DPDCHs to achieve higher peak data rates.

13.8.2 Interaction Between the Node B and the RNC for the Management of HSUPA Channels and Resources

The Node B and the RNC jointly manage the HSUPA channels. When a UE is configured for HSUPA operation during the call setup or when the E-DCH Active Set is updated, various parameters are exchanged between the Node B and the RNC. The RNC uses a Node B Application Part message, such as a RADIO LINK SETUP message, to direct the Node B to use the following parameters while communicating with a given UE:

- power offsets of the E-AGCH, E-RGCH, and E-HICH relative to the DL DPCCH or F-DPCH[474] for each radio link,
- serving E-DCH radio link,
- E-DPCH information, such as the E-TFCI table index, reference E-TFCIs and associated power offsets, E-DCH TTI (2 ms or 10 ms), E-DPCCH power offset, E-RGCH two-step and three-step thresholds, and
- E-DCH Frequency-Division Duplexing information, such as MAC-d flows information (e.g., MAC-d-flow–specific contents, such as the maximum number of retransmissions, power offset, number of non-scheduled bits per TTI, DDI, scheduling priority, guaranteed rate, and MAC-d flow multiplexing list), and HARQ process allocation for 2-ms TTI.

The Node B provides the following parameters to the RNC via messages, such as the RADIO LINK SETUP RESPONSE:

- the E-DCH Frequency-Division Duplexing information response, which includes MAC-d-flow–specific information, such as MAC-d flow identity, I_{ub} related identities and addresses, and HARQ process allocation for 2-ms TTI (same as above) for scheduled or non-scheduled transmission, and
- E-DCH Frequency-Division Duplexing DL control channel information that consists of primary or secondary E-RNTI, the E-RGCH and HICH channelization code, E-RGCH and E-HICH signature sequence indices, the initial SG, and the E-AGCH channelization code.

[474]The F-DPCH replaces the DPCCH for a data-only service so that the OVSF codes are used more efficiently. One F-DPCH can be shared among multiple UEs to convey UL power control commands. See Chapter 11 for more details.

Once the Node B and the RNC have exchanged information, the RNC conveys some of the parameters to the UE via an RRC message, such as the RADIO BEARER SETUP message, during the HSUPA call setup.

13.8.3 HSUPA UE Categories

"All men are created equal," but all UEs are not! Six categories of UEs are defined for HSUPA [308]. An HSUPA category is characterized by the maximum number of simultaneous codes, the minimum supported SF, the maximum transport block size, and the supported TTI as summarized in Table 13.8.

The Maximum Number of Simultaneous Codes column specifies the maximum number of codes or physical channels onto which one E-DCH maps. For example, the E-DCH maps onto just one E-DPDCH associated with one OVSF code for Category 1. In case of a Category 6 UE, one E-DCH can be mapped onto four E-DPDCHs, with two E-DPDCHs at $SF = 2$ and two E-DPDCHs at $SF = 4$. The Minimum Spreading Factor column indicates the maximum possible peak data rate for a given E-DPDCH. When the minimum SF is 4, the modulation symbol rate is 3.84 Mcps$/4 = 960$ ksps. However, the minimum SF of 2 enables a higher rate of 1920 ksps.

A UE supports a 10-ms TTI *or* both a 10-ms TTI and a 2-ms TTI. For example, a Category 1 UE supports just a 10-ms TTI and a maximum of 7110 bits can be transmitted during one 10-ms TTI. A Category 6 UE supports transmission of up to 20000 bits in a 10-ms TTI and 11484 bits in a 2-ms TTI. The highest peak rate of 5.7 Mbps is supported only by a Category 6 UE. Transmission of 11484 bits as a transport block in a 2-ms TTI yields the data rate of 11484 bits$/2$ ms $= 5.7$ Mbps. As of mid-2011, E-DCH Category 5 devices have been widely available and Category 6 devices have begun to appear as well.

13.9 POINTS TO REMEMBER

- HSUPA is a Release 6 feature that provides significant throughput and latency enhancements in the uplink relative to previous releases such as R99 and R5.
- The highest peak user data rate is 5.7 Mbps.
- Major factors contributing to higher uplink throughput include the use of multiple OVSF codes in parallel, support for a shorter 2 ms TTI, implementation of HARQ, optimal use of UE transmit power, fast uplink interference control by the Node B, instantaneous and fast data rate selection by the UE based on received AGs and RGs, and support for soft handover in the uplink.
- A UE can be configured to use non-scheduled transmission for delay-sensitive MAC-d flows and scheduled transmission for delay-tolerant MAC-d flows.
- Scheduled transmission enables optimal use of the available uplink interference, and non-scheduled transmission supports delay-sensitive applications.

Table 13.8: HSUPA UE Categories

UE Category	Maximum Number of Simultaneous Codes	Minimum Spreading Factor	Maximum Transport Block Size (10 ms, 2 ms)	Peak Data Rate (10 ms, 2 ms)
1	1	4	(7110, –)	(711 kbps, –)
2	2	4	(14484, 2798)	(1.4 Mbps, 1.4 Mbps)
3	2	4	(14484, –)	(1.4 Mbps, –)
4	2	2	(20000, 5772)	(2 Mbps, 2.9 Mbps)
5	2	2	(20000, –)	(2 Mbps, –)
6	4	2	(20000, 11484)	(2 Mbps, 5.7 Mbps)

- The overall system architecture to support HSUPA is the same as previous releases with the majority of HSUPA-related enhancements in the UE, Node B, and RNC.
- The main architectural change is the addition of a MAC-e/es layer at the UE, a MAC-e layer at Node B, and a MAC-es layer at the RNC.
- Six categories of UEs are defined with Category 1 as a low-end UE and Category 6 as a high-end UE. A Category 6 UE can achieve the peak rate of 5.7 Mbps in favorable channel conditions and a lightly loaded system.
- Though the peak data rate of 5.7 Mbps cannot be achieved all the time and in all locations in commercial deployments, average user-perceived throughput in the uplink can be expected to be much better than that in R99 and R5.

PART III

IP MULTIMEDIA SUBSYSTEM AND 4G CELLULAR TECHNOLOGIES

CHAPTER 14

IP MULTIMEDIA SUBSYSTEM

Internet Protocol Multimedia Subsystem (IMS, also called IP Multimedia Core Network Subsystem) is a service delivery platform that could potentially yield numerous benefits to consumers and service operators. The consumer can access a variety of IP-based services at a lower cost than a non-IMS solution. The operator can cost-effectively introduce new services and obtain additional revenue from those new services. This chapter provides an overview of IMS. Motivation for IMS is given, followed by the history of IMS. The basic components of an IMS network are described. The IP-based protocols IMS exploits are briefly discussed. Various call scenarios are illustrated to explain how IMS acts as a call control protocol. Examples of services IMS offers are given. Implementation of quality-of-service (QoS) using IMS is discussed.

14.1 INTRODUCTION TO IMS: WHAT AND WHY?

IMS enables a service operator to offer numerous IP-based services to its subscribers. *We focus on the wireless aspects in this chapter, but IMS also benefits wireline networks, such as cable networks.* Figure 14.1 shows where IMS exists in a service operator's network.

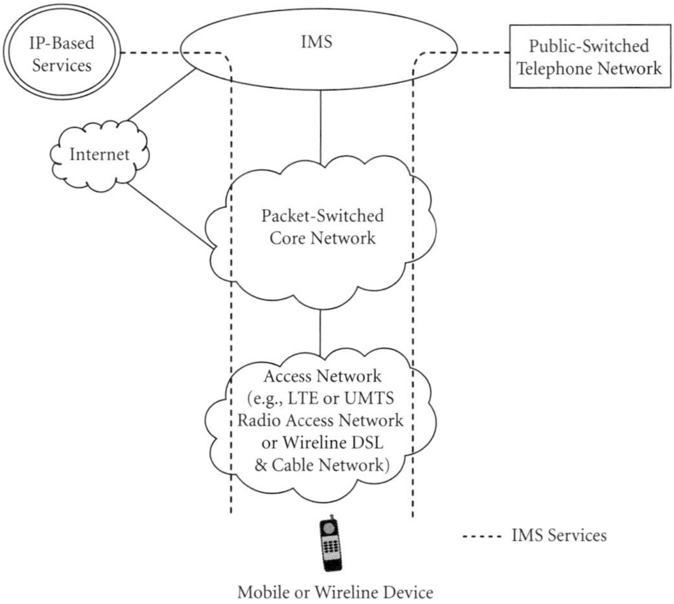

Figure 14.1: IMS for Services in the Next-Generation Network

As the communications industry prepares to offer a variety of IP-based services, IMS plays a vital role as the service delivery platform in the next-generation network. As shown in Figure 14.1, *IMS controls access to operator-provided IP services* [e.g., Voice over IP (VoIP), instant messaging, presence, and push-to-talk over cellular]. IMS is independent of the underlying access technologies and core networks. For example, a user may access IMS services via a wireline access, such as a digital subscriber line (DSL) or cable modem, or a wireless access, such as cellular technologies [e.g., Long-Term Evolution (LTE) and Universal Mobile Telecommunication System (UMTS)]. Furthermore, the IMS utilizes only an IP-based packet-switched core network and needs no circuit-switched core network. IMS can work with any suitable packet core network (e.g., a UMTS-based core network and a 1x core network). Since IMS focuses on only IP-based services, even IMS voice services are VoIP rather than traditional circuit-switched services.

Let's discuss the benefits that IMS provides and thus we'll discover the motivation for using it [309, 310]. Figure 14.2 summarizes the potential benefits of IMS; only time will tell what benefits IMS will eventually realize in practice!

IMS facilitates *Fixed Mobile Convergence* (FMC) and *Quad Play*. Regarding FMC, IMS acts as a common unified services platform regardless of how the user accesses the services—using a *fixed* system (e.g., a cable modem or fixed wireless access system) or a *mobile* system (e.g., a cellular network). Quad Play refers to the bundling of voice, video or TV, data or Internet access, and mobility. FMC

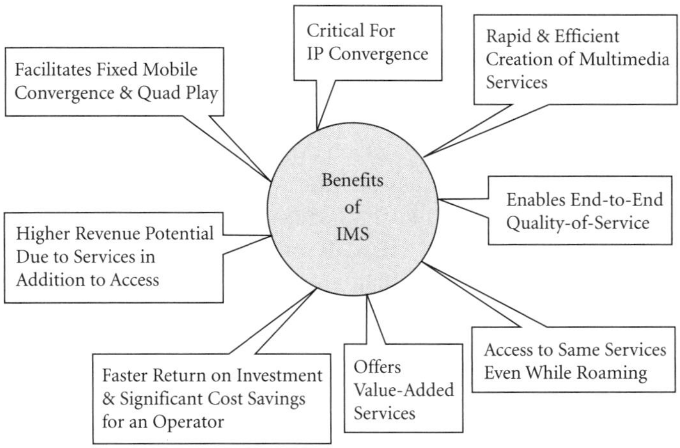

Figure 14.2: Motivation for IMS: (Potential) Benefits Galore!

and Quad Play allow a service operator to integrate all user communication and entertainment needs into a personalized and seamless user experience, increasing customer satisfaction and loyalty.

Both wireline and wireless service operators can offer the same IMS services to both wireless and wireline users, thereby expanding their customer bases. See Section 14.2 for additional information on the role of IMS in FMC.

IMS also facilitates *IP Convergence*, which allows the operator to use an IP-converged network rather than maintaining diverse or divergent radio and core networks for different services. Such a converged network can then rely upon IMS to provide services. The user can then enjoy a ubiquitous and personalized communication experience at home, at work, or on the move! Furthermore, IMS requires no circuit-switched core network, but it supports seamless Voice Call Continuity by defining interworking between the circuit and packet core networks. See Section 14.2 for additional information on the role of IMS in IP Convergence.

IMS enables the *rapid and efficient creation of multimedia-rich services*, e.g., VoIP, push-to-talk, video sharing, and location-based services. Because voice revenue is declining and data revenue growth is registering double-digit gains, service operators must offer an array of services to attract and retain subscribers. A user can access multiple IMS services during the same session. For example, a user can initiate a VoIP call and then add video or text to the ongoing VoIP session without adversely affecting the VoIP service (e.g., the addition of a video to an ongoing Yahoo or Skype voice call)! IMS offers users access to services that different vendors may have created, encouraging competition and innovation in the services arena. IMS also allows a smooth evolution of existing services with no adverse impact to the existing circuit-switched services. New Session-Initiation-Protocol–based user devices can benefit from IMS services, and existing devices can continue to use existing circuit- and packet-switched core networks for services.

In addition, IMS helps implement end-to-end QoS even when the user experiences a change in the access technology [e.g., from a fourth-generation (4G) network to a third-generation (3G) network]. IMS also allows the home network to provide services even when the user is roaming. The home network chooses the best route for the bearer traffic (e.g., an email or web page), which could be completely different from the route that the control traffic (e.g., session setup signaling) takes. IMS-based services are *value-added services* that add a significant value compared to the services that are delivered via the open Internet. The IMS services add value with interoperability between service providers, seamless roaming, secure connectivity, payment security, consolidated billing, compliance with regulatory requirements, and QoS.

IMS promises to accelerate service integration, allowing the service provider to respond promptly to the market and to customer requirements, yielding a time-to-market advantage. This faster response also results in a *faster return on investment* and a service differentiator. IMS becomes a single point of service control for various applications, yielding *cost savings* to the service provider.

IMS enables a service operator to become more than a bit pipe provider, in which the user is using the service provider's access and core network to transport bits but is providing more revenue to a different entity. For example, if the wireless user is watching a high-definition game or movie from a video service provider, such as ESPN or HBO, the user is paying relatively more to ESPN or HBO for the video service than to the service operator for the wireless access. Perhaps the user is paying a flat-rate fee to the wireless service provider but paying the video service provider per program. Since the user would be willing to pay more for services than for access, the wireless service provider shoulld be motivated to *offer services* (including the video service that we mentioned) *in addition to access*. The wireless service provider can, of course, partner with content providers, such as ESPN and HBO.

We discussed IMS' numerous benefits. Although "there is no free lunch" is a cliché, it is also usually accurate! The major drawback of IMS is the initial cost of deployment. An operator must invest in a brand-new architecture with numerous new components (even though limited numbers of off-the-shelf IP components can be used). Since operators have already invested in the existing mechanisms to offer some services, they may be reluctant to invest in IMS until they see some *killer applications* (i.e., highly popular applications), which, of course, are infrequent visitors!

Now, let's address an important question head on: Is IMS an absolute necessity or could an operator do something else to offer a variety of services to its subscribers? Well, it is possible to use other solutions to offer the services that IMS provides, but it is quite difficult to reuse the components of such alternative solutions (e.g., databases and gateways) as the service environment expands. Furthermore, implementation and maintenance costs of such individual solutions for different services can be expected to be higher than those for IMS. Additionally, interworking among service operators becomes more complex and difficult with the individual solutions.

14.2 IMS, IP CONVERGENCE, AND FIXED MOBILE CONVERGENCE

Let's discuss the role IMS plays in two scenarios, FMC and IP Convergence. Service providers are evolving their networks to reap benefits that IP Convergence and FMC offer.

Do You Recall... Do you remember the main benefits of IP Convergence and FMC? The footnote[475] summarizes the answer.

FMC enables an operator to reuse the core network for fixed as well as mobile access. IP Convergence allows the user access to a variety of services anywhere and anytime in a fixed or mobile environment via multiple access technologies and different types of devices using the service operator's all-IP infrastructure.

FMC is often viewed as the first step toward network and service convergence [311]. When fixed and mobile networks converge, the user experiences seamless services, and the operator can provide bundled and value-added services. For example, Unlicensed Mobile Access enables the consumer to move between WLAN and cellular networks seamlessly, and the consumer gets a single bill in spite of using two different types of networks. FMC facilitates a trend toward converged devices since mobile phones are becoming more powerful and "smart." Such *smart phones* are increasingly performing the duties of traditional computing devices (e.g., laptops and personal digital assistants). The consumer can thus access services from a single converged device. For example, a user may have started watching a high-definition news video on a hand-held device while riding the bus and can continue to watch the news on an LCD or plasma TV screen once the user arrives home! Implementation of end-to-end QoS provides a seamless user experience across fixed and mobile networks.

Also facilitating FMC are emerging broadcast and multicast technologies, such as Digital Video Broadcasting–Hand-held; emerging 4G cellular technologies, such as Worldwide Interoperability for Microwave Access and LTE; and emerging deployment strategies, such as femtocells. These emerging technologies can cost-effectively offer a living room entertainment experience in the mobile environment. As far as a service operator is concerned, the FMC requires migration from the existing disjointed circuit-switched and packet-switched core networks to a unified core network that supports existing fixed and mobile access technologies [312]. Such network convergence would require some CapEx but would reduce OpEx in due course due to demand for service bundles and for operational efficiency. The use of IMS should reduce the cost of service development and deployment.

Let's now briefly discuss *IP Divergence* from the consumer's and service provider's perspectives so that you can appreciate IP Convergence! Imagine a typical day for a typical consumer, Mr. Gonzalez, who has subscribed to different services from

[475]The main benefits of FMC and IP Convergence are the service operator's reduced cost of offering services and the consumer's seamless and ubiquitous service experience.

different service providers. Mr. Gonzalez gets several bills every month from his communication service providers: one from his local phone service provider, one from his cell-phone operator, one from an Internet service provider, and one from the cable operator. Depending upon which service has problems, Mr. Gonzalez needs to call a specific customer service number. Since Mr. Gonzalez's contacts have their own preferred method of communication, Mr. Gonzalez needs to access all kinds of storage—voice mails, emails, and short messages—at different locations. Mr. Gonzalez needs to access several different websites to learn about the prevailing weather and traffic conditions for the commute between his home and work and about the impact of the latest stock quotes on his retirement nest egg!

Now, consider a service operator that offers local telephone service, long-distance telephone service, cellular service, and Internet access service. In the absence of convergence, the operator would have separate access networks (e.g., different physical links for different services between the subscriber's address and the operator's networks) and separate Operations Support Systems, which configure and maintain the network, and Business Support Systems, which support functions such as customer service, inventory management, and billing. The network management and introduction of new services (e.g., audio conferencing) become cumbersome and inefficient when a service operator offers multiple services using divergent networks. Replication of Operations Support Systems and Business Support Systems increases the cost, adversely affecting the business bottom line. More real estate and more personnel are required to maintain dissimilar networks.

IP convergence addresses the user's needs as well as the operator's needs. Figure 14.3 shows the basic concept of IP convergence. A unified packet-switched and IP-based core network is used, which allows a variety of access technologies to procure the same services. IMS serves as a services delivery platform. Examples of access technologies include the wireline networks serving DSL modem and cable modem networks and wireless networks, such as wireless local area networks (WLANs) based on IEEE 802.11, and cellular technologies, such as UMTS and 1xEV-DO. As with FMC, both wireline and wireless access technologies can be used. The consumer has access to a variety of services at a low cost and gets a single bill at the end of the month. Even when using different access technologies as a result of mobility, the subscriber has a seamless service experience. The service operator can now have a common Operations Support System and Business Support System. Furthermore, the operator must maintain only one core network rather than multiple core networks. Additionally, since IMS is used to control access to the services, the service operator can easily introduce (and remove) services.

14.3 HISTORY OF IMS

Various organizations started their efforts in parallel to realize the vision of IMS. Furthermore, some organizations facilitate the standardization of IMS. Examples of the organizations that have facilitated the creation of an IMS eco-system include the Third Generation Partnership Project (3GPP), the Internet Engineering Task Force (IETF),

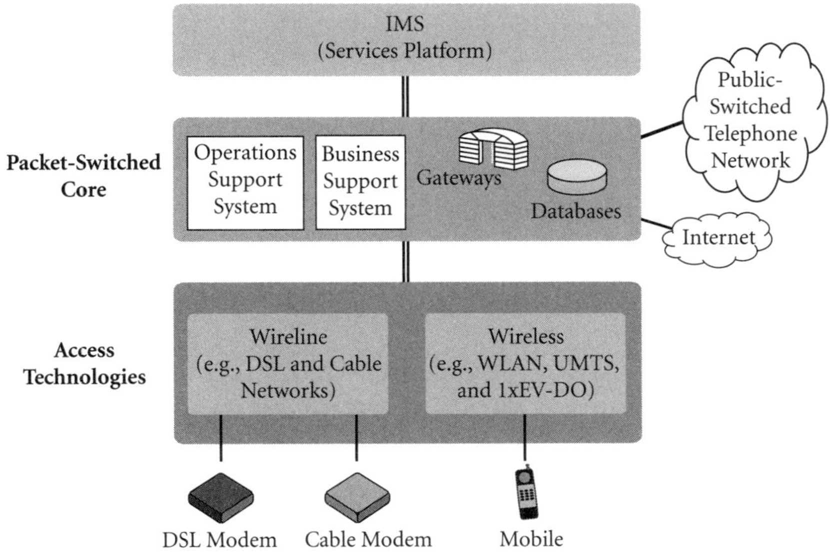

Figure 14.3: Network Infrastructure for IP Convergence

the International Telecommunications Union (ITU), the Open Mobile Alliance, the Third Generation Partnership Project 2 (3GPP2), CableLabs or PacketCable, the Alliance for Telecommunications Industry Solutions Next Generation Network Focus Group, Telecoms and Internet Converged Services and Protocols for Advanced Networks, the European Telecommunications Standards Institute, and the MultiService Forum [309, 310, 313].

The 3GPP became a leader in defining the IMS architecture. *IMS uses Session Initiation Protocol (SIP) as a call control protocol*[476] as well as many other IETF–defined protocols to avoid re-inventing the wheel and to help create an open services network that encourages competition and innovation. The 3GPP and the IETF work together to ensure harmonization of standards [153, 314]. The ITU defines some of the protocols IMS uses (e.g., H.248[477] for media control and Q.1912.SIP for SIP–ISUP[478] inter-working) [315]. The Open Mobile Alliance defines services for the IMS architecture (e.g., instant messaging and push-to-talk).

Various organizations embraced the basic 3GPP IMS architecture with some modifications and extensions to suit their individual needs. For example, the 3GPP2 calls the combination of the modified version of IMS and its packet-switched core network

[476] Recall that SIP is pronounced as the word *sip*, rather than as the individual letters. Refer to Chapter 4 for details on SIP.

[477] Megaco is an older term, and H.248 is the newer term for a protocol that enables a Media Gateway Controller to configure a Media Gateway.

[478] ISUP stands for ISDN (Integrated Services Digital Network) User Part.

Multimedia Domain. PacketCable supports 3GPP IMS beginning with version 2.0. The Alliance for Telecommunications Industry Solutions addresses end-to-end solutions over wireline and wireless access and has embraced 3GPP IMS. As yet another example of the acceptance of 3GPP IMS, Telecoms and Internet Converged Services and Protocols for Advanced Networks Release 1 uses an architecture similar to 3GPP IMS but specifically includes support for non-SIP–based applications, such as gaming, peer-to-peer applications, IP television, and video on demand. The European Telecommunications Standards Institute has also adopted the IMS core architecture.

Figure 14.4 tracks the evolutionary path that 3GPP IMS has taken [316]. Although the original 3GPP IMS assumed wireless access and mandated the use of IP version 6 alone, such constraints no longer exist; 3GPP IMS is now access-agnostic and both IP version 4 and IP version 6 are supported. 3GPP first defined UMTS in Release 99, and the IMS architecture and its main logical components were first introduced in Release 5. Examples of other Release 5 IMS features include the definition of various QoS-enabled session and service control call flows; charging; and operations, administration, maintenance, and provisioning. Release 5 focused on call control and roaming for a UMTS network.

Release 6 enhanced IMS's scope by incorporating the interworking with IP networks (e.g., wireless local area networks and the Internet) and circuit-switched networks as well as including complete real-time capabilities to support streaming and conversational services. Example IMS services introduced in Release 6 include group

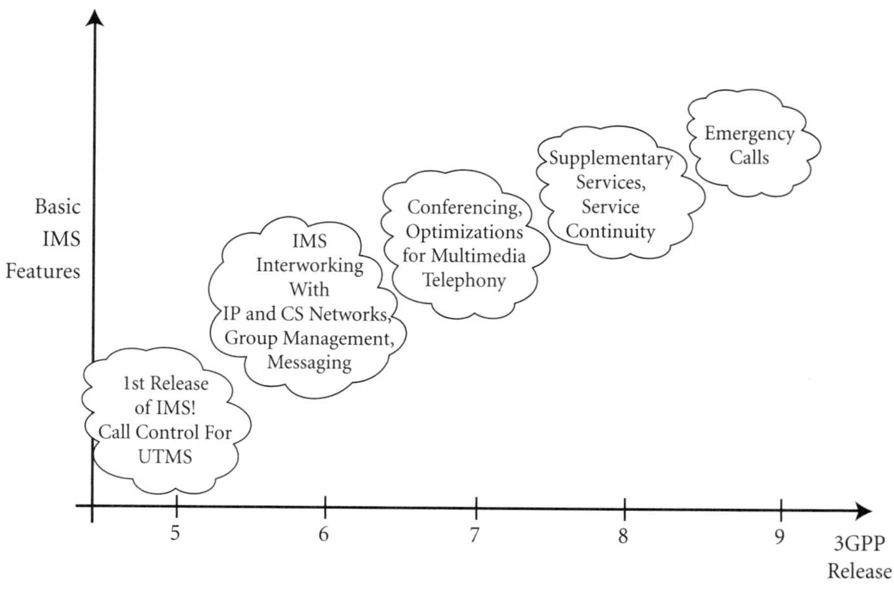

Figure 14.4: In the Footsteps of 3GPP IMS

management and messaging (e.g., instant messaging similar to Yahoo's). Release 7 supports conferencing and dynamic and interactive multimedia scenes and discusses optimizations for multimedia telephony over IMS.

Release 8 focuses on features such as multimedia telephony and supplementary services, communication wait, enhancements for corporate network access, service continuity, and security enhancements. Release 9 addresses protection against unsolicited communication, services centralization and continuity, and support for IMS emergency calls over General Packet Radio Service (GPRS) as well as Evolved Packet System, which is a 4G system based on LTE.

14.4 IMS ARCHITECTURE

IMS defines a comprehensive architecture with several logical components [317]. This architecture is access-agnostic and interfaces with various networks, such as a packet-switched core network of a traditional 3G cellular system, a public-switched telephone network, and application servers (ASs). Section 14.4.1 introduces major components of an IMS network using an example of a VoIP call. Section 14.4.2 depicts a comprehensive IMS network architecture and provides the details of various components.

14.4.1 High-Level View of the IMS Architecture

Until some *killer app* is found, VoIP can be considered a force-to-be-reckoned-with in the domain of IMS applications. However, before a VoIP call can be initiated from or received at an IMS device, the device needs to register with the IMS network to obtain IMS services. Let's first understand the roles of various IMS entities that we encounter as we attempt to perform service registration. We will later discuss additional IMS entities through a high-level discussion on how to set up a simple VoIP call. The goal here is to get a basic understanding of the functions of various IMS components; the architecture details are awaiting you in Section 14.4.2! *We will discuss detailed call flows for various types of call or session setups in Section 14.7.*

Figure 14.5 illustrates a *simplified view* of the basic communication steps that occur during the service registration [317]. The mobile device needs to establish an IP traffic bearer with an IP Connectivity Access Network (IP CAN) in Step 1. Examples of IP CANs are 1xRTT, 1xEV-DO, and UMTS. Think of Step 1 as a prerequisite for the IMS registration. As part of Step 1, the mobile device learns about the IP address of the Proxy–Call Session Control Function (P–CSCF). For example, with a UMTS system, the mobile device obtains the P–CSCF's IP address as part of (primary) packet data protocol Context Activation. As another example, with a 1xRTT or 1xEV-DO system, the mobile device obtains the P–CSCF's IP address during Mobile IP Registration. The P–CSCF is the first IMS element with which the IMS device comes into contact. The P–CSCF is always present regardless of the device's roaming status. If the device is roaming, the P–CSCF typically resides in the visited network; otherwise, it resides in the home network.

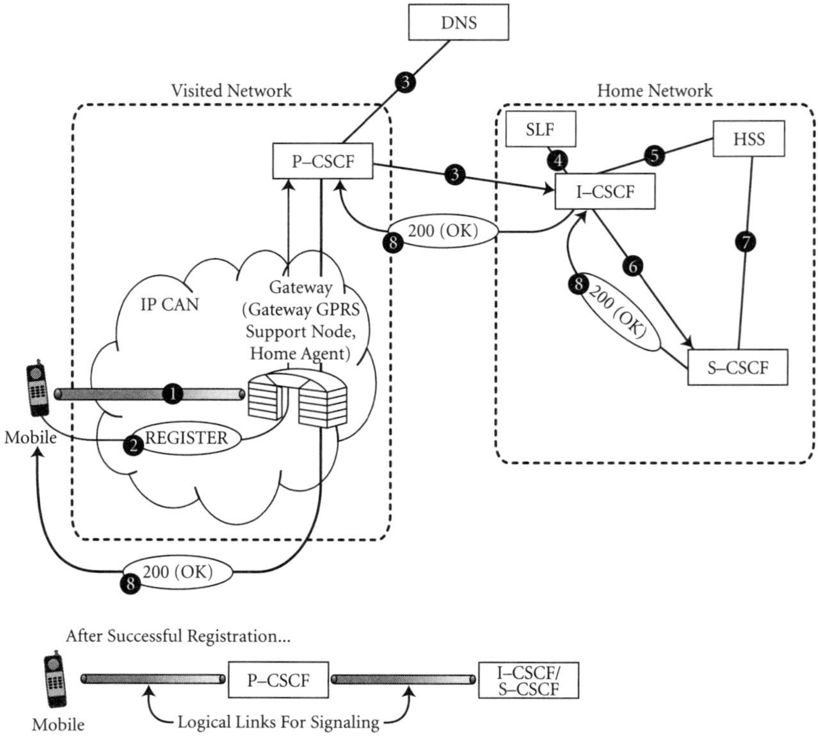

Figure 14.5: Using the IMS Architecture for IMS Service Registration (without Authentication)

Do You Recall. . . Do you remember the name of the call control protocol IMS uses? Hints: The name of the protocol starts with an S and its acronym could be used as a verb that helps you quench your thirst! See the footnote[479] to confirm your answer!

The mobile sends a SIP REGISTER message to the P–CSCF using the IP bearer in Step 2. Note that the user plane is used to send a SIP signaling message because SIP signaling is considered just like some application data, e.g., FTP data or HTTP data. The P–CSCF acts on the IMS device's behalf (hence, the name proxy!) to communicate with the home network. The P–CSCF also works with the IP CAN that the IMS device is accessing to implement QoS after a multimedia session is established. In Step 3, the P–CSCF queries the Domain Name System (DNS) using the home

[479]IMS uses SIP as the basic call control protocol, and the IETF defines SIP.

network domain name (e.g., ims_HomeOperator.com) received in the REGISTER message and finds the IP address of the I–CSCF (Interrogating–Call Session Control Function).

The I–CSCF always resides in the home network, and one of its functions is *topology hiding* (THIG), which means that the outside network cannot observe SIP messages to infer the home network's network topology. Furthermore, THIG enables an operator to reconfigure its internal network behind-the-scenes (or should we say behind-the-I–CSCF!) without announcing the reconfiguration to the whole world. In Step 4, the I–CSCF uses the subscriber identity (e.g., sip:JohnDoe@ims_HomeOperator.com) and the home network domain name to query the Subscription Locator Function (SLF) to retrieve the IP address of the Home Subscriber Server (HSS). The HSS is like a Home Location Register or Authentication, Authorization, and Accounting Server and is essentially a database containing subscriber profiles. The HSS in Step 5 uses the visited network name received from the I–CSCF to determine if the IMS device is allowed to use the visited network based on the subscription profile. If the device is allowed to access the (visited) IP CAN, the HSS provides the name of the Serving–Call Session Control Function (S–CSCF) to the I–CSCF.

The S–CSCF is in charge of the IMS device and controls the IMS device's access to applications or services. When the IMS device wants to initiate a multimedia session with some target device or server, it registers with the S–CSCF as an active user, and the S–CSCF handles the IMS device's communication with the target device or server. In Step 6, the I–CSCF contacts the S–CSCF using the S–CSCF's name (and potentially a DNS query to associate a name with an IP address) received from the HSS. The S–CSCF then retrieves the subscriber profile from the HSS in Step 7 so that it can control access to services when the IMS device sets up a session in the future.

If the S–CSCF accepts the service registration,[480] it replies with a SIP 200 (OK) response in Step 8. This SIP 200 (OK) response goes from the S–CSCF to the I–CSCF, from the I–CSCF to the P–CSCF, and from the P–CSCF to the IMS device. When the P–CSCF receives the SIP 200 (OK) response, it knows which IMS component in the home network to contact in the future. If the I–CSCF has decided to use THIG, the P–CSCF would contact I–CSCF for a multimedia session setup; otherwise, it would directly communicate with the S–CSCF. In summary, at the end of IMS service registration, the IMS device has been registered in the home network and a logical signaling link exists between the P–CSCF in the visited or home network and the I–CSCF/S–CSCF in the home network.

[480]If the REGISTER message from the IMS device contains no authentication information, the S–CSCF indicates an authentication failure as the SIP response. The IMS device then sends a new REGISTER message containing the authentication parameters. The S–CSCF then works with the HSS to authenticate the IMS device. See Chapter 4 for additional details on the authentication process.

Something to Think About... We encountered three incarnations of CSCF: P–CSCF, I–CSCF, and S–CSCF. Our colleague, Dr. Bob Best, has found a link between Hollywood and IMS. No, not downloading Hollywood movies using IMS! He has discovered a link between the CSCF and Tom Hanks' roles in three Hollywood movies: "Forrest Gump," "Philadelphia," and "Apollo 13." Just as Forrest Gump seems to be present in every major event of that era, the P–CSCF is present in every visited or home network. Tom Hanks plays the role of an interrogator (i.e., a lawyer) in "Philadelphia," just like the I–CSCF interrogates the HSS about the IMS device. In "Apollo 13," Tom Hanks' character is in charge of the space ship, just like S–CSCF is in charge of the IMS device for access to services. Only one Tom Hanks plays all three different roles, and, similarly, one CSCF also plays three different roles—as a proxy, as an interrogator, and as a controller!

The IMS device essentially deals with four logical IMS components during the service registration phase, i.e., the P–CSCF, I–CSCF, S–CSCF, and HSS. Let's understand the roles of a few more IMS components before facing a comprehensive IMS network architecture. A little disclaimer is warranted. We'll focus on the IMS components here; detailed call flows are illustrated in Section 14.5.

Assume that we have an IMS-capable mobile device, and we want to call someone who is using a landline phone. Furthermore, assume that we have already taken care of IMS service registration. The mobile device constructs a SIP INVITE message, which finds its way to the P–CSCF, the I–CSCF, and finally the S–CSCF. The S–CSCF learns from the INVITE message that the destination is a public-switched telephone network (PSTN) phone. The Breakout Gateway Control Function (BGCF) helps the S–CSCF locate a Media Gateway Control Function (MGCF) suitable for the destination phone (e.g., close to the destination).

The MGCF and the IMS device negotiate the type of media (e.g., 12.2 kbps adaptive multirate vocoder speech or 8k enhanced variable-rate codec vocoder speech). Although the MGCF takes care of the signaling aspects of the call setup, we need an entity to translate between the speech codecs used in the two different systems, i.e., the IP CAN and the PSTN. That's where a Media Gateway (MGW)[481] comes into the picture. The MGCF determines a suitable MGW (e.g., geographically closest to the destination) for the voice call to and from the PSTN. The MGCF configures such an MGW using H.248 signaling for speech conversion, such as the conversion between the adaptive multirate or enhanced variable-rate codec speech that the IP CAN uses and the 64 kbps Pulse Coded Modulation that the PSTN uses. We need to ensure that the destination phone is ready for communication. The MGCF works with the T–SGW (Trunk–Signaling Gateway) to set up the trunk between the MGW and the destination phone in the PSTN.

[481] The Media Gateway is also sometimes referred to as the IMS Media Gateway (IMS-MGW). We will use simply MGW.

14.4.2 IMS Architecture: A Closer Look

Now that we have a basic understanding of the roles of several major IMS elements, let's dig deeper into the IMS architecture. Figure 14.6 provides a comprehensive view of the IMS architecture defined in 3GPP Release 7.[482]

An IMS network consists of a CSCF, HSS, BGCF, MGCF, MGW, SLF, AS, Media Resource Function Controller (MRFC), and Media Resource Function Processor (MRFP). The IMS network can interface with other networks, as shown in Figure 14.6, such as an IP CAN, IP multimedia networks, a circuit-switched core network, and legacy mobile signaling networks. Let's first introduce new IMS components that we have yet to encounter. Then, we will discuss details of the architecture. The SLF helps the I–CSCF locate a suitable HSS (i.e., an HSS with the information about the subscriber of interest) when multiple HSSs are present in an operator's network. The AS provides one or more services, such as a presence service. The MRFC and MRFP together constitute the Media Resource Function.

An AS, e.g., a SIP-based AS, provides to the subscriber one or more valued-added and operator-controlled services. Examples of AS applications include presence and conference control. Any innovative applications or services can be developed and then deployed using a suitable AS. The subscribers of a service provider can now access

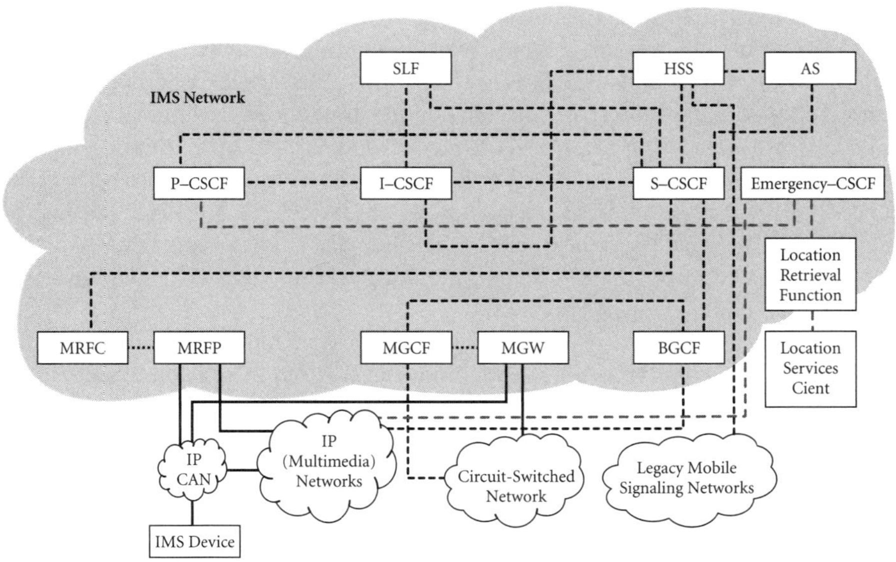

Figure 14.6: A Comprehensive View of the IMS Architecture

482 See the standards [318] for additional details on user and network identities as well as the IMS architecture.

such services. ASs enable the IMS to rapidly deploy new services. (See Section 14.8 for a discussion of various IMS services.) The AS can reside in the home network, a visited network, or a trusted external network. The AS can communicate with the HSS via Sh and Si interfaces to retrieve the subscriber information. The S–CSCF uses the interface with the AS to provide the user with AS services. The S–CSCF provides no authentication and security functionality for secure, direct, third-party access to the IMS. The I–CSCF uses the interface with the AS to forward SIP requests destined for a Public Service Identity that the AS hosts.

The CSCF is considered the IMS's brain. We have looked at three types of CSCFs so far. Release 7 defines one more type of CSCF, the Emergency–CSCF, which handles certain aspects of emergency sessions (e.g., forwarding of an emergency request to a suitable Public Safety Access Point or emergency center) [319]. Let's discuss the various roles of the CSCF.

The S–CSCF controls the session for a subscriber. During the IMS device's service registration, the S–CSCF communicates with the HSS to obtain the subscriber's service profile. This service profile includes initial filter criteria that the S–CSCF uses to involve one or more ASs to provide services [317]. The information that can be used to construct the filter criteria includes the type of the SIP METHOD (e.g., INVITE), the presence or absence of a message header, the content of a message header (e.g., Request URI[483]), and the Session Description Protocol (SDP) portion of a SIP message [320]. The S–CSCF performs different duties depending on whether it is serving the originating subscriber or the destination subscriber [320]. The S–CSCF ensures that the originating endpoint has subscribed to the IMS communication service being requested. When the S–CSCF is serving an originating subscriber, the S–CSCF uses the terminating subscriber's destination identity (e.g., E.164 phone number or SIP URI as discussed in Chapter 4) and contacts the DNS to find the destination subscriber's I–CSCF's IP address. The S–CSCF then sends the SIP messages to the I–CSCF. However, if the DNS query indicates to the S–CSCF that the destination identity is a PSTN address, the S–CSCF sends the SIP message to the BGCF within the operator's network. The S–CSCF communicates with a BGCF for call routing to the PSTN or circuit-switched domain.

When the S–CSCF is serving a terminating or destination subscriber (i.e., the one who is receiving a request for a multimedia session), it sends the SIP request to a P–CSCF using the registered location for the subscriber. A given network may have multiple S–CSCFs with perhaps different capabilities that the service operator defines. Since the S–CSCF for a given subscriber resides in the home network, it gives the subscriber access to the home network's services even when the subscriber is roaming.

During the service registration, the S–CSCF acts as a registrar and the HSS acts as a location server as defined in the IETF SIP architecture [153]. In other words, the S–CSCF accepts registration requests and makes the registration information available through the HSS. Hence, when an I–CSCF or an AS contacts the HSS looking for

[483] Uniform Resource Identifier

a particular IMS subscriber, the HSS knows which S–CSCF is responsible for that subscriber. During registration, the S–CSCF also allocates some identities when the IMS device requests that it do so. An AS may also originate a session request. In such a case, the S–CSCF determines the served user and processes the request using suitable logic even if the served user is unregistered. Examples of such applications are periodic stock updates[484] and weather updates. The S–CSCF also generates Call Detail Records to facilitate charging and to monitor resource utilization.

The I–CSCF is the *initial* point of contact for an IMS subscriber within the home network (e.g., during service registration and not for every multimedia session setup). As mentioned earlier, during the service registration, the I–CSCF, with help from the HSS, informs the visited network's P–CSCF about the subscriber's S–CSCF location. Of course, the I–CSCF can implement THIG to hide the configuration of the home network. In such a case, the I–CSCF acts as a single point of contact for the home network and encrypts and decrypts the SIP headers that reveal the network topology, e.g., a Route header (see Chapter 4). One or more I–CSCFs may exist in an operator's network. An I–CSCF can also generate Call Detail Records.

As discussed earlier, the P–CSCF is the first IMS component that the IMS device encounters during registration or a session setup. The P–CSCF may be in a visited or home network based on the IMS device's location. During registration, the P–CSCF finds the home network's I–CSCF. During a session setup, the P–CSCF communicates with the I–CSCF or the S–CSCF.

Do You Recall... Do you recall the name of the feature used if the P–CSCF talks to the I–CSCF (instead of the S–CSCF) for a session setup? Hint: The feature name starts with a T and rhymes with "pig!" See the footnote[485] to check your answer.

The P–CSCF can also generate Call Detail Records. Note that SIP is a text-based protocol, and every character in the message text needs bytes! Hence, SIP messages can consume quite a few bytes. The P–CSCF can control such byte consumption through SIP message compression (especially for over-the-air transmission) and decompression. To authorize bearer resources based on the agreed-upon SDP and to control QoS, the P–CSCF works with the IP CAN that the IMS device is using. The P–CSCF has an interface with the Policy and Charging Rules Function (PCRF) to facilitate QoS implementation.[486] The P–CSCF detects a request for an emergency session establishment and forwards it to the Emergency–CSCF, which retrieves a PSAP URI based on the location of the user equipment (UE) [322]. The P–CSCF can also provide security by establishing an IPsec tunnel with the IMS device, such as a mobile or another operator.

[484] On second thought, when the economy is in recession and stocks are being consistently hammered, you may want to avoid periodic stock quotes!

[485] The feature is THIG, topology hiding. See Section 14.4.1 if you missed it.

[486] See Section 14.9 as well as the standard [321] for additional details on QoS.

The HSS is the master database for the subscribers of an operator just like the Home Location Register of a circuit-switched core network. It stores the subscription-related information to facilitate operations of the entities, such as the I–CSCF and the S–CSCF. An operator's network may contain one or more HSSs based primarily on the number of subscribers and the capacity of the equipment. A few examples of the type of information stored in the HSS are user identities, security-related information, and user profiles [323]. The HSS generates user security information to implement mutual authentication, communication integrity check (referred to as *integrity protection* in the UMTS family of standards) and ciphering.

A Question for You... What is the difference between integrity check and ciphering? The footnote[487] provides the answer.

The BGCF plays an important role in the interworking between the IMS or IP domain and the PSTN or circuit-switched domain. When the S–CSCF determines (using DNS) that the session cannot be routed to an IP network, it contacts the BGCF within the network. Consider Figure 14.7; two scenarios describe the BGCF's function. In Scenario 1, both the originating endpoint, IMS Device A, and the terminating endpoint, Landline Phone B, belong to Network X. BGCF1 in Network X has jurisdiction or control over its MGCFs and chooses a suitable MGCF, MGCF2 in our example, which connects to the PSTN where the terminating endpoint resides.

In Scenario 2, the originating endpoint, IMS Device C, belongs to Network X, but the terminating endpoint, Landline Phone D, belongs to another network, Network Y. Since the BGCF in Network X knows about the MGCFs within Network X (and knows nothing of those in Network Y), it contacts BGCF2 in the destination network, Network Y. Now, BGCF2 in Network Y selects a suitable MGCF, MGCF3 in our example, which is closest to the terminating endpoint Landline Phone D. Network X and Network Y may be in different countries (e.g., the U.S. and India). The operators who own Network X in the U.S. and Network Y in India must have some type of interoperability agreement. We will see later that the transport of voice traffic between two endpoints becomes quite efficient due to the combined efforts of the BGCF, the MGCF, and the MGW.

The BGCF determines the next hop for routing the INVITE message based on information stored in some operator-specific database. For a PSTN– or circuit-switched–terminated call, the BGCF determines the network (e.g., Network X or Network Y in Figure 14.7) in which the breakout (i.e., leaving some network) is to occur. If the BGCF determines, based on the database, that the breakout is to occur in the same network in which the BGCF resides, the BGCF selects a suitable MGCF to serve as a bridge between the IMS domain and the circuit-switched domain for the signaling messages. This is what happens in Scenario 1. However, if the BGCF determines, based on the database, that the breakout is to occur in a different network, the

[487] Integrity check ensures that the message has come from a valid sender and that the message has not been altered during its journey from the sender to the receiver.

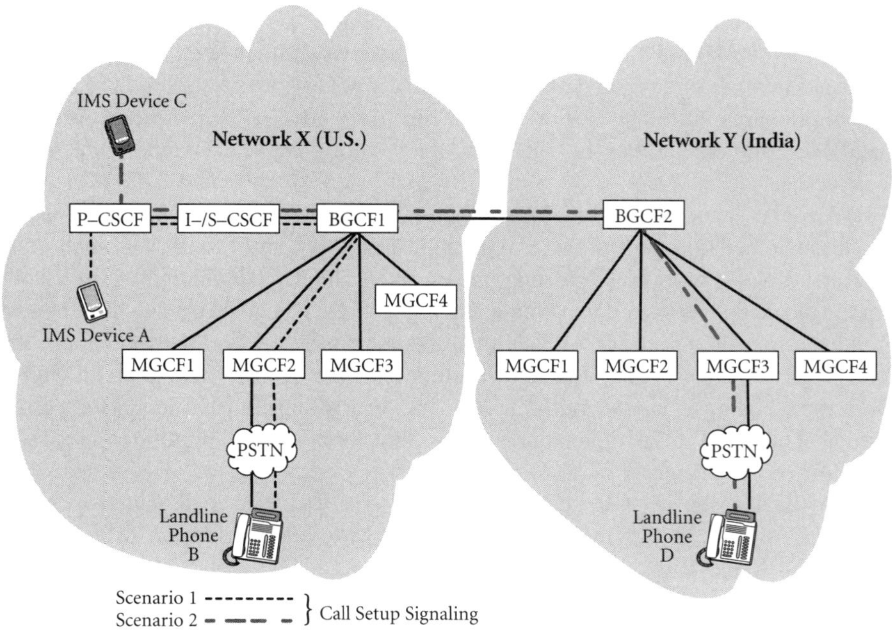

Figure 14.7: Role of the Breakout Gateway Control Function

BGCF selects a BGCF that serves that different network. Scenario 2 illustrates this situation. Several BGCFs may be present within an operator's network. The BGCF can also generate Call Detail Records and can provide security between two networks through authorization. Example criteria that the BGCF can use to select the PSTN access point include the current location of the originating IMS device, the location of the destination PSTN, the operator's local policies, and the business agreements between the source and destination networks' operators.

The MGCF participates in the signaling part of the interworking between the IMS domain and the PSTN or circuit-switched domain. It communicates with the BGCF, the CSCF, and the PSTN. The T–SGW converts an ISUP message that runs over IP into an ISUP message that runs over Message Transport Protocol. The MGCF converts a SIP message into an ISUP message [324], and the T–SGW converts the lower layers from IP to Message Transport Protocol, which is understood by Signaling System 7. (Appendix A offers an overview of Signaling System 7.) As an example, the MGCF converts the SIP INVITE message into an ISUP Initial Address message. The MGCF sends the ISUP Initial Address message to the T–SGW over IP. The T–SGW replaces the IP and lower layers with Message Transport Protocol but makes no modification to the ISUP Initial Address message. The SS7 network can then forward the ISUP message to the destination. This protocol conversion is reversed when the PSTN sends an ISUP message to the T–SGW. An MGCF controls one or more MGWs, and one MGW can be shared among multiple MGCFs. Such flexibility allows the operator to buy MGCFs and MGWs based on their capabilities (e.g., ability to handle a certain number of sessions) and the expected signaling and traffic load in a given geographic area, and then to provision them independently. When the MGCF and MGW are implemented in the same box, the resulting entity is called an *integrated gateway*.

An MGW handles the media part of the interworking between the PSTN or circuit-switched domain and the IP domain. The MGW's most important function is to convert one type of media format into another. The MGCF configures the MGW for the media conversion. For example, assuming a UMTS network using the adaptive multirate speech codec and a PSTN that understands Pulse Coded Modulation, the MGW may convert a speech packet traveling from the UMTS network to the PSTN from the 12.2 kbps adaptive multirate format to the 64 kbps Pulse Coded Modulation format. The MGW may implement an echo canceller and may support features such as a conference bridge.

Specifically, the MGCF instructs the MGW to create a context between two terminations (i.e., bearers) with the Pulse Coded Modulation format for the PSTN bearer and some IP-friendly format (e.g., 12.2 kbps adaptive multirate codec or 8k enhanced variable-rate codec) for the IP bearer. The MGW may participate in the QoS implementation by marking the IP packets destined for the IMS device with a suitable DiffServ Code Point.

The Media Resource Function is divided into the MRFC and the MRFP.[488] The MRFC controls the media stream resources in the MRFP based on its communication with an AS and S–CSCF. Specifically, the S–CSCF sends an INVITE message to the MRFC, and the MRFC configures the MRFP using H.248. The MRFC may be co-located with an AS for services, such as conferencing. Furthermore, the MRFC can also be co-located with the MRFP. The MRFC also generates Call Detail records. The MRFP processes the actual media streams, such as audio transcoding. To support an audio conference among multiple parties, the MRFP mixes incoming media streams and outputs the multiplexed streams to multiple parties. Another example of the MRFP function is the playing of multimedia announcements, e.g., "The conference will start when the chairperson arrives." The MRFP also executes floor control during the conference call by granting access rights to the shared resources. An AS facilitates the Multimedia Resource Function by managing the reservation configuration or booking of the conference call (e.g., start time, duration, and the list of participants). When an operator supports ad-hoc multiparty conferences, the IMS device can establish a session directly with the MRFC and then invite others to join the conference.

Let's look at additional IMS entities that support various interworking between IMS and another system. Two entities, i.e., an Interconnection Border Control Function and a Transition Gateway, are included to support interworking between IP ver-

[488] An entity called the Media Resource Broker enables multiple IMS applications to share a pool of heterogeneous MRF resources. The MRB allocates and releases suitable MRF resources according to the requests the IMS applications make [325].

sion 4 and IP version 6. The Border Control Function provides application-specific functions at the SIP and SDP layer to enable interconnection between two operator domains and between IP version 6 and IP version 4 SIP applications. Examples of other Border Control Function tasks include hiding the network topology and generating Call Detail Records. The Border Control Function controls the Transition Gateway. The Transition Gateway works with the media and carries out tasks such as network address and port translation and IP version 4 and IP version 6 protocol translation. An entity called the Location Retrieval Function facilitates the identification of the user's location, retrieves location information for the IMS device, and may interact with a gateway mobile location center or other type of location server function to obtain the location [319].

14.5 IMS COMMUNICATION PROTOCOLS FOR SIGNALING AND MEDIA TRANSPORT

As we mentioned earlier, IMS reuses protocols, especially IETF–defined protocols, instead of defining its own, which creates an open system and allows faster specification development. SIP is one of the prime examples of these protocols. Figure 14.8 specifies the protocols used on various interfaces. Most of the protocols are discussed in detail in Chapter 4; the remaining protocols are discussed here.

IMS uses SIP, H.248, DNS, and Diameter for signaling messages and Real-Time Transfer Protocol for traffic. As noted earlier, SIP is the primary protocol IMS uses for signaling among IMS components. For example, SIP is used for the signaling message exchange between the IMS device and the P–CSCF; the P–CSCF and the I–CSCF/S–CSCF; the I–CSCF and the S–CSCF; the S–CSCF and the BGCF; the BGCF and the MGCF; the S–CSCF and the MGCF; the S–CSCF and the AS; and the S–CSCF and the MRFC. SIP is considered an application layer protocol and usually runs over User Datagram Protocol and IP but could also run over Transmission Control Protocol and IP. The MGCF and MRFC use H.248 to configure the MGW and the MRFP, respectively. H.248 is also considered an application layer protocol that runs over User Datagram Protocol and IP or Transmission Control Protocol and IP [326]. DNS is used between the P–CSCF and the DNS server and between the I–CSCF and DNS server so that the IP address of the target node can be found. DNS runs over Transmission Control Protocol and IP or User Datagram Protocol and IP. Diameter is used for communication between the I–CSCF and the SLF; the HSS and the I–CSCF; and the HSS and the S–CSCF.

As far as transporting media (e.g., a VoIP packet) is concerned, Real-Time Transfer Protocol is used over User Datagram Protocol and IP in the IP domain. In Figure 14.8, speech is carried in the IP domain between the IMS device and the MGW (through the MRFP if required). The MGW[489] converts the speech between the IP format (e.g.,

[489]The UMTS standard uses the term IMS-MGW to distinguish it from the Media Gateway introduced in the Release 4 bearer-independent circuit-switched core network. The Release 4 Media Gateway is used by the MSC (mobile switching center) Server, and this Media Gateway does the traffic processing. *Remember that we use MGW to refer to the IMS Media Gateway* (not the gateway used by the Mobile Switching Center server).

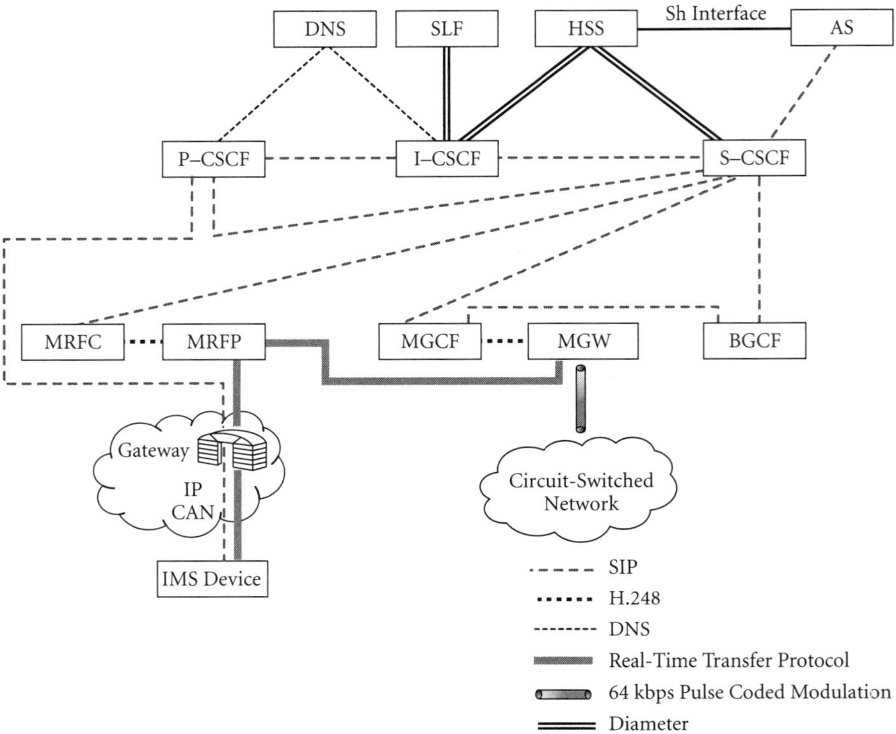

Figure 14.8: Signaling and Media Transport Protocols for IMS

12.2 kbps adaptive multirate or 8 kbps enhanced variable-rate codec speech) on the IP CAN side and 64 kbps PCM format on the PSTN side.

Though the basics of SIP are discussed in Chapter 4, the way IMS uses SIP is discussed in Section 14.5.1. H.248 is briefly touched upon in Section 14.5.2.

14.5.1 Using SIP in IMS

Figure 14.9 summarizes several characteristics of how IMS uses SIP. Since the IMS requires the basic SIP entities, IMS maps the SIP entities into the IMS components. For example, the SIP proxy server is mapped to the S–CSCF, the SIP registrar is the SLF, and the HSS implements the location and redirect server functionalities. SIP is quite flexible and allows many messages to be optional, but IMS develops a more definitive message sequence for a session setup. Several messages that are optional

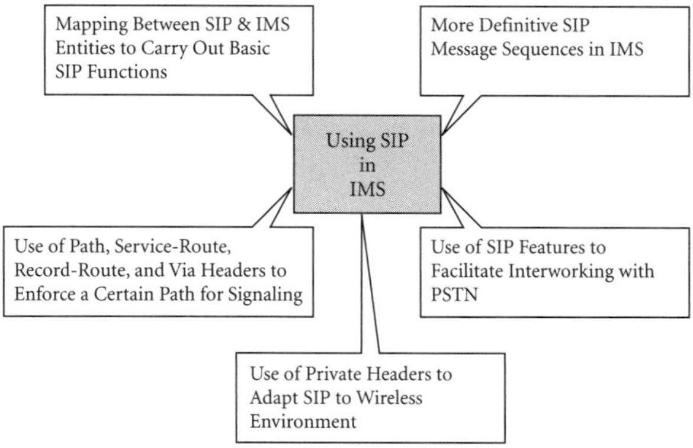

Figure 14.9: Use of SIP in IMS

in the basic SIP session setup are mandatory in IMS to facilitate implementation of features, such as QoS. For example, PRACK and UPDATE messages are optional in a basic SIP session setup but required in IMS. IMS uses certain SIP features to facilitate interworking between the IP domain and the circuit-switched domain (e.g., PSTN).

We can make a voice call from an IP device to a PSTN phone and address the destination landline phone with identities such as sip:+12345678901@gateway.com; user=phone or tel:+1-234-567-8901. As part of such a voice call setup, the IMS network configures the MGW to translate between the speech in an IP packet and the speech in a 64 kbps Pulse Coded Modulation format. IMS introduces new private headers to adapt SIP to the wireless environment [327]. For example, the P–CSCF inserts a private header, P-Visited-Network-ID, into a SIP message so that the I–CSCF and HSS in the home network can determine whether to allow roaming in the specific visited network.

In a pure SIP environment, each IP packet carrying a SIP message is treated independently of other IP packets carrying a SIP message or of SIP responses. The SIP requests and responses must travel along different paths. Furthermore, one set of requests and responses and another set of requests and responses may also traverse completely different paths. Note that SIP and IMS intend that the signaling messages and the user traffic can (and do!) follow different paths. Such independence of traffic from signaling makes the system more efficient; operators can independently provision these two paths and buy both the network nodes that are best for signaling and those that are best for transporting traffic. Such a philosophy is in direct contrast to the concept of traditional circuit-switched core network's monolithic Mobile Switching Center, which may have extremely under-utilized signaling or traffic capabilities. IMS uses the SIP headers, such as Via, Path, Service-Route, and Record-Route, to enforce a certain signaling path.

A Question for You... Can you think of any benefits of forcing the same path for the SIP signaling messages? See the footnote[490] for our take on that!

The reverse path is guaranteed to be the same as the forward path through the use of the Via SIP header. Let's take a quick look at how to use the Path and the Service-Route headers. During the IMS registration, the IMS device sends a REGISTER message to P–CSCF that adds its address in the Path header of the message. The REGISTER message finds its way to the S–CSCF as explained in Section 14.4.1. *The S–CSCF now learns about the P–CSCF's address by observing the Path header.* In the future, the S–CSCF can contact the P–CSCF in the case of an incoming session for the IMS device. The S–CSCF responds to the REGISTER message with a SIP 200 (OK) response that contains the Service-Route header, which includes the S–CSCF's address. The P–CSCF also includes its address in the Service-Route header after receiving this SIP 200 (OK) response. The SIP 200 (OK) response eventually reaches the UE. *The UE observes the Service-Route header and learns about the P–CSCF and S–CSCF addresses.* In the future, the UE includes the identities of the P–CSCF and the S–CSCF in the Route header when constructing an INVITE message to initiate a session so that the INVITE message can be properly routed. The information contained in the Service-Route header is considered fresh as long as the registration remains valid.

Now, let's see how to use the Record-Route header to enforce a specific route for the signaling messages. Assume that an IMS Device A in Network X initiates a multimedia session with IMS Device B in Network Y. The IMS device sends an INVITE message to the P–CSCF. Since the P–CSCF wants to be in the signaling path, it adds its address, `proxy@here.com`, in the Record-Route header. The P–CSCF sends the INVITE to S–CSCF. Since S–CSCF also wants to be part of the signaling path, it adds another Record-Route header with its own address, `serving@here.com`. Network X's S–CSCF queries a DNS server to find Network Y's I–CSCF and sends the INVITE message to that I–CSCF. Network Y's I–CSCF queries Network Y's HSS to find IMS Device B's S–CSCF. The I–CSCF then forwards the INVITE message to the S-SCCF. It does not implement topology hiding so it does not include its own address in the Record-Route header. The S–CSCF adds its address, `serving@there.com`, to the Record-Route header in the received INVITE message and forwards the message to IMS Device B's P–CSCF. The P–CSCF then adds its address in Record-Route header and sends the resulting message to IMS Device B. Now, the INVITE message has finally reached its destination!

IMS Device B must acknowledge the reception of this message and send the SIP 200 (OK) response all the way back to IMS Device A. More specifically, the overall path is traced back to the originating mobile using the Via SIP header and

[490]Forcing SIP signaling messages to travel the same path facilitates the implementation of various features, such as the generation of charging data records (e.g., to determine session start time and session end time) and topology hiding (encrypting and decrypting relevant headers). Furthermore, it also allows the service operator to provision the required capacity for various logical links.

includes both the nodes that *did not* and the nodes that *did* insert their own Record-Route headers. The information in the Via header is valid within the transaction, which terminates when the final SIP response to a SIP request is 200 or higher, e.g., a SIP 200 (OK) response. The originating mobile device processes the Record-Route headers and creates a Route header that defines the required nodes in subsequent messages.

14.5.2 H.248 in IMS

H.248 enables the MGCF to control the operation of the MGW. The MGCF is also referred to as the Media Gateway Controller and the Mobile Switching Center Server in some technologies. The mapping between the MGCF and the MGW is many-to-many, i.e., one MGCF can control multiple MGWs, and one MGW can be shared among multiple MGCFs. We will first introduce several fundamental concepts of H.248,[491] followed by an example of how H.248 is used in a VoIP call setup and teardown. Cooperation between the ITU and the IETF has resulted in these two bodies jointly creating H.248.1. The IETF no longer maintains relevant updates and corrections; the ITU will handle all the future work [329].

The MGCF configures the MGW by defining *terminations* and *contexts*. Figure 14.10 illustrates the terminations and contexts concepts using a voice call between an IMS device, such as an IMS-capable mobile, and a PSTN phone. Termination T1 and Termination T2 are associated with Context C. The termination originates or terminates one or more media streams. A termination may represent a physical entity or an ephemeral flow. A DS0 trunk on a T1 link, which is a time-division multiplexing channel, is an example of a physical termination. A Real-Time Transfer Protocol flow is an example of an ephemeral termination, and such an ephemeral termination is specified by parameters such as the IP address and the port number.

The Media descriptor must describe the ephemeral terminations and includes a Stream descriptor that is really just the SDP. The SDP for the ephemeral termination includes type of codec, data rate, and type of bearer (e.g., IP for ephemeral terminations). Physical terminations require no Media Descriptor. The MGW allocates identities to the terminations during their creation. In general, the MGW can choose any arbitrary schema (e.g., an integer or some text name such as A4444 [326]) as a *TerminationID*. The TerminationIDs of physical terminations are provisioned in the MGW. For DS0 physical terminations, the TerminationID may be chosen as a combination of a trunk group number and a trunk number within the trunk group. The MGCF defines a termination for the MGW in terms of descriptors, which are groups of *properties* that characterize a termination. MGWs may have terminations that have quite different characteristics. Such characteristics are grouped into *packages*. A termination implements one or more packages.

The MGW may process multiplexed media streams. More specifically, a bearer channel that carries a stream has a physical or ephemeral *bearer termination*, and

[491] Groves et al. [328] offer details on H.248.

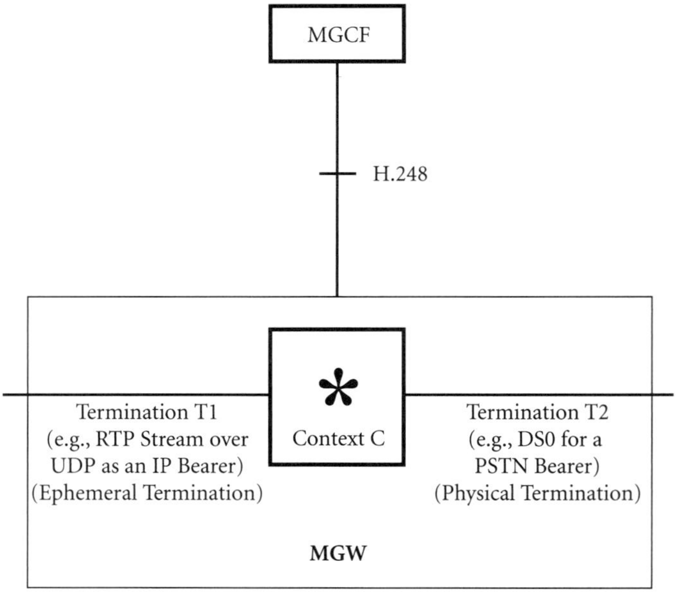

Figure 14.10: H.248 and an Inside Look at MGW

multiple bearer terminations are then connected to an ephemeral termination called the *multiplexing termination*. Within a set of terminations, a *context* is an association between the terminations. In our example, Context C has two terminations. A null context includes the terminations unassociated with any other termination. In Figure 14.10, the asterisk represents the association between the terminations the context specifies. Attributes such as ContextID, topology, and priority specify the context. The topology specifies how media flow occurs between the terminations of a context. The priority attribute enables the MGCF and the MGW to prioritize a given context relative to other contexts.

Table 14.1 summarizes H.248 commands that carry out specific tasks [328]. The MGCF sends an *Add* command to the MGW to add a termination to a context. However, if the MGCF specifies no existing context in the *Add* command, the MGW creates a new context. Not surprisingly, a *Subtract* command is used to remove a termination from a context! Another auxiliary function of the Subtract command is to provide statistics on the termination's role in the context. The Add and Subtract commands create and remove ephemeral terminations, respectively. An Add command adds a physical termination to a *null context* (i.e., a context that contains no terminations), and a Subtract command subtracts a physical termination. A *Modify* command is used to change the parameter settings of an existing context. From the perspectives of call setup and teardown, the Add command is used to establish a context while a call is being set up. The Modify command is used to activate the

Table 14.1: Summary of Selected H.248 Commands

H.248 Command	Purpose
Add	123... Create a context; Add a termination to an existing context
Subtract	Remove a termination from a context
Modify	Change the parameters of a context
AuditValue	Obtain currently used parameters and statistics of a termination
AuditCapabilities	Get all possible parameter settings on a termination
Move	Move a termination from one context to another
ServiceChange	Inform about a termination being taken out of or returned to service

context when the called party answers. Finally, the Subtract command is used to tear down the context when the call terminates.

The MGCF can also obtain statistics on a termination via the *AuditValue* command. The AuditValue command provides the *currently used values* of the parameters (e.g., properties and statistics) associated with terminations, and the *AuditCapabilities* command provides *all possible values* (not just the currently used values) of the parameters associated with terminations. The MGCF can configure the MGW's terminations to detect events (e.g., a phone off-the-hook) and to send Notify messages to the MGCF. The *Move* command is used to move a termination from the current context to another context. When one or more terminations is to be taken out of or returned to service, the MGW uses the *ServiceChange* command to inform the MGCF.

Let's understand how the processing for media streams is specified. The MGCF specifies certain characteristics of the *Media* descriptors for the terminations. The Media descriptor specifies the parameters for the media streams using *TerminationState* and *Stream* descriptors. The *TerminationState* descriptor yields information on the state of a termination, which could be "test," "out of service," or "in service." The state "in service" means that a termination either is being used or can be used for normal traffic. One or more media streams may be present, with a *Stream* descriptor associated with one bidirectional stream. The Stream descriptor includes the StreamID and three descriptors: *LocalControl*, *Local*, and *Remote*.

The MGCF allocates a StreamID to each stream. The media streams with the same StreamID are connected within a context. The LocalControl descriptor specifies properties, such as the Mode, the Reserve properties, and the stream-specific properties of a termination. The Mode may be send-only, receive-only, send and receive, inactive, or loop-back. The terms *send* and *receive* are used relative to the exterior of the context. As an example, a stream with `mode=sendOnly` passes no received media into the associated context. The Reserve properties of a termination, *ReserveValue* and *ReserveGroup*, dictate the MGW's action upon receiving a Local and/or Remote descriptor. If *ReserveValue* is true, the MGW reserves the resources for all the alternatives specified in the Local and/or Remote descriptors for which it currently has resources available. If ReserveValue is false, the MGW chooses an alternative from the Local descriptor and an alternative from the Remote descriptor. Similarly, if *ReserveGroup* is true, the MGW reserves the resources required to support the *property group alternatives* the MGCF requests. For example, assume that the Local descriptor specifies two groups of properties, one corresponding to packetized G.711 A-law audio and the other corresponding to G.723.1 audio. The MGW reserves resources so that it can decode an audio stream that has been encoded in G.711 A-law or G.723.1 format [328].

The MGW replies to the MGCF, specifying the alternatives it has chosen for the Local and Remote descriptors. If the MGW runs out of resources, it replies to the MGCF with error number 510, which corresponds to "insufficient resources." The MGCF uses Local and Remote descriptors to reserve (and commit) MGW resources for processing the media (e.g., decoding and encoding for the relevant media streams and terminations). "Local" refers to the media the MGW receives and "Remote" refers to the media the MGW sends.

The Local and Remote descriptors can take the form of session descriptions that the SDP defines and uses [155]. Certain exceptions to the basic SDP are allowed. Examples of such exceptions include the `s=`, `t=`, and `o=` lines being optional[492] and the use of alternatives instead of a single parameter value. Since a Stream descriptor specifies a single bidirectional media stream, a given session description includes just one media description (i.e., `m=` line). The Stream descriptor can specify multiple session descriptions (and hence multiple media descriptions for a given media stream) as alternatives, using `v=` lines as delimiters. However, note that different media streams for a given termination must be specified in different Stream descriptors. The Local and Remote descriptors consist of two parts, one specifying the descriptor itself and the other dictating what the MGW is required to do after receiving the descriptors. An example of the MGW action is to reserve resources. (Groves et al. [328] provide additional details on the contents of the Local and Remote descriptors.) Now let's move on to call setup to understand how the MGCF configures the MGW for a mobile-originated and PSTN-terminated VoIP call.

14.6 IDENTIFYING THE IMS SUBSCRIBER AND THE IMS ELEMENTS

Let's briefly discuss how to identify or address an IMS subscriber and the IMS elements. An IMS subscriber has one or more private user identities and one or more public user identities. The IMS Subscriber Identity Module *securely* stores a private user identity and at least one public user identity.

The home network operator permanently (rather than dynamically) assigns a globally unique private user identity to the subscriber. Hence, the private user identity remains valid for the duration of the user's subscription with the home network. Such

[492] See Section 4.5.1 of Chapter 4 for further details of these parameters.

an identity is used for purposes such as registration, authorization, administration, and accounting. To be more accurate, the private user identity identifies the subscription, not the individual user, because the identity is associated with the services to which the user has subscribed. An example of a private user identity is the Network Access Identifier [330]. The Network Access Identifier may include a representation of the International Mobile Subscriber Identity (IMSI). Examples of Network Access Identifiers include `JohnDoe_Private@HomeOperatorName.com` or `IMSI-of-JohnDoe@HomeOperatorName.com`.

The private user identity is like a social security number; it uniquely identifies a person, but very few people know it. (At least, let's hope so!) The public user identity is how people are known and may differ based on association. For example, a person may be "Jennifer" as a formal contact name, "Jen" as a less formal contact name, and "Jenny" for those inside the "inner circle." The public user identity may change with each session and determines how the network will treat the session. Within the IMS network, the HSS stores the private user identity and the S–CSCF obtains it from the subscriber during registration. This identity is authenticated during registration. The public user identity is used to request communications. The public user identity or identities may be a SIP URI (e.g., sip: tom@xyz.com) [153, 331] or a tel:-URI (e.g., tel: +1-234-567-8901) [332]. A public user identity is registered either explicitly or implicitly before an IMS session can be originated. A group of alias public user identities may also be defined, and such information is stored in the HSS, which can provide such information to the AS and the S–CSCF. All public user identities may be registered via a single request from the subscriber.

Note that two different IMS devices may correspond to the same public user identity; the public user identity need not be unique. That's when a Globally Routable User Agent URI (GRUU) enters the scene. The GRUU specifies a unique combination of public user identity and the IMS device or instance. A device may have multiple instances or applications, such as a VoIP application and a text application, each with a different GRUU. When two or more devices or applications are sharing the same public user identity, the GRUU ensures that the SIP request is sent to the correct mobile device. There are two types of GRUUs, Public GRUUs and Temporary GRUUs. The Public GRUUs reveal the user's public user identity and are intended to be long-lasting. On the contrary, Temporary GRUUs conceal the user's public user identity and are valid until the contact is explicitly de-registered or the current registration expires. The IMS instances can obtain both Temporary GRUUs and Public GRUUs during registration and can exchange GRUUs using SIP requests and responses [318, 333].

An IMS subscriber has one or more IMS service profiles. The IMS service profile refers to the collection of service- and user-related data [318]. The service profile includes an initial filter criteria, which essentially determines when the associated service is invoked. This service logic is implementable via processing of the SIP messages. The ASs can utilize more complex and dynamic service logic that uses the information unavailable in typical SIP messages (e.g., location, time, and day). The IMS service profile is defined and stored in the HSS. The S–CSCF retrieves the service profile from the HSS during the registration. As shown in Figure 14.11, multiple service profiles may be defined for an IMS subscription. Each public user identity is associated with just one service profile, but a given service profile could be associated with one or more public user identities. Furthermore, a given public user identity may be shared across multiple private user identities for the same IMS subscription.

A given IMS subscription may be shared among multiple IMS users or UEs. We will use the UMTS standard's term UE to refer to a mobile terminal, but the actual designation of the terminal would depend on the technology (e.g., Access Terminal for 1xEV-DO). UEs with different private user identities and different contact addresses may share a particular public user identity. However, the sharing of a public user identity across numerous private user identities requires that these UEs be associated with the same IMS subscription and hence the same S–CSCF. Figure 14.11 illustrates the relationships among the IMS subscription, private user identities, public user identities, and service profiles. Observe that `sip:` precedes a public user identity but not the private user identity.

Assume that John Doe has enrolled in IMS services from a cellular service operator, Operator.com. John's IMS Subscriber Identity Module has stored two private user identities, `JohnDoe_Personal@Operator.com` and `JohnDoe_Professional@Operator.com`. The outside world can communicate with John Doe using one of three public user identities, `sip:JohnAtHome@Operator.com`, `sip:JohnDaddyOfTwins_SOS_FromHomeToWork@Operator.com`, and `sip:JohnAtWork@Operator.com`. Two service profiles are associated with the three public user identities. Service Profile 1 says that all the terminating (i.e., destined for John Doe) INVITE requests act as a service trigger and that two public user identities are associated with the service profile. Furthermore, the AS `sip:CallScreeningAndPhotoID@operator.com` is designed to provide a service. This server associates a digital photograph (stored as part of the service data) with the source SIP URI in the INVITE message so that John knows who is trying to reach him. Furthermore, John Doe has specified some contacts in the database to automatically get a voice mail when they call. John can call them back at a more convenient time.

Calls that require immediate attention, e.g., calls from home or certain work-related calls, are treated as part of Service Profile 2. The AS `sip:CallScreeningAndFastAck@Operator.com` serves these calls. For example, this AS associates the public ID `sip:JohnDaddyOfTwins_SOS_FromHomeToWork@Operator.com` with the image of the family and sends a quick (it *must* be quick!) response that John will call them back as soon as possible. Furthermore, this AS sends a text message to all locations John has registered, indicating that the situation at home needs urgent attention. When a call associated with the public ID `sip:JohnAtWork@Operator.com` comes, the AS responds with John's professional photograph, and, if John does not answer the call, the server allows the caller to leave a message on a specific voice mailbox.

Each public user identity may be related to one or more GRUUs, and such a relationship is established during registration. A set consisting of a Permanent GRUU and a Temporary GRUU is associated with one public user identity and one IMS instance during registration. However, during subsequent re-registrations, the UE is allocated

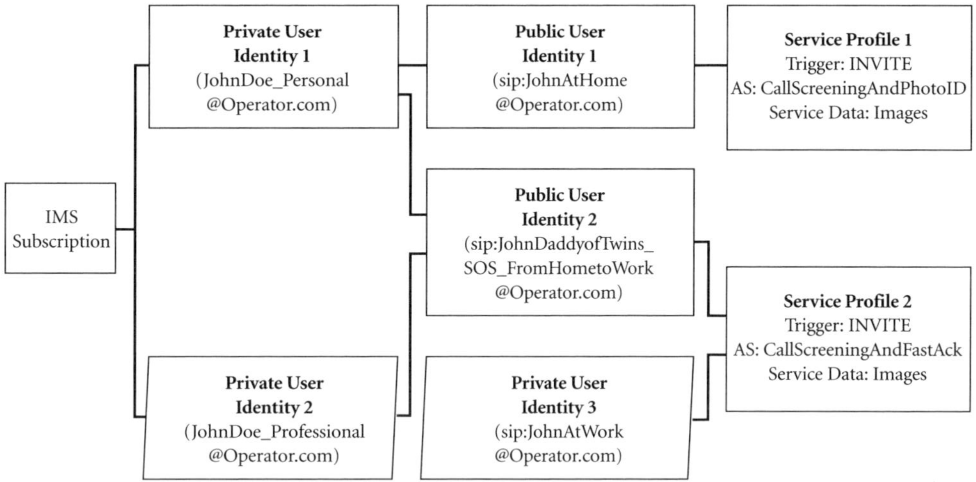

Figure 14.11: IMS Subscription, User Identities, and Service Profiles

the same Permanent GRUU but a new and different Temporary GRUU, with all the previous Temporary GRUUs still valid during the period of this registration. The UE may replace some or all of the previous Temporary GRUUs with the new Temporary GRUU. The current set of the Permanent GRUU and all the valid Temporary GRUUs during the registration period is called a GRUU set. When a UE registers, a separate GRUU set is associated with each public user identity. Of course, if different UEs register with the same public user identity, each UE is allocated a different GRUU set.

Now that we know how an IMS device is addressed, let's summarize how the IMS network nodes are addressed. Since the CSCF, BGCF, and MGCF all use SIP, they are identifiable via valid SIP URIs containing the host domain name or network address.

14.7 IMS SESSION SETUP SCENARIOS

We will take a closer look at certain call setup scenarios. Section 14.7.1 describes how an IMS device can call a PSTN phone. The opposite scenario of a PSTN phone calling an IMS device is illustrated in Section 14.7.2.

14.7.1 IMS Device to PSTN Call Setup: A Closer Look

In this scenario, we are using an IMS-capable UE to make a VoIP call to a landline phone. Since the call is connecting an IMS device and a PSTN landline phone, we would need the MGCF's and the MGW's help. Before the MGCF and the MGW are ready to participate in this specific VoIP call, they have exchanged some messages. When the MGCF and the MGW come into service, then such an interaction or message exchange occurs. The MGCF and the MGW interactions include transaction requests from the originator and reply acknowledgments from the receiver. For example, the MGW registers with the MGCF by sending a ServiceChange command with some transaction ID and root as the termination to represent the entire MGW. The MGCF responds with a ServiceChange reply message. The MGCF learns about the capabilities of the MGW (e.g., supported speech codecs) by sending an AuditCapabilities command to the MGW and receiving a command response message from the MGW.

Figure 14.12 illustrates the end-to-end message flow for an IMS device-initiated VoIP call setup when the destination is a PSTN phone [318, 334, 335]. Prior to the initiation of the VoIP session, IMS service registration has already occurred. Hence, the UE knows about the P–CSCF, and the P–CSCF knows about the S–CSCF. In Step 1, the UE sends an INVITE message to the P–CSCF, which replies with a 100 (Trying) response (Step 2) and forwards the INVITE message to S–CSCF (Step 3).

This INVITE message includes[493] the identity of the destination device and the initial SDP offer containing media descriptions (e.g., a list of one or more voice

[493] Many other parameters are included and are discussed in Chapter 4.

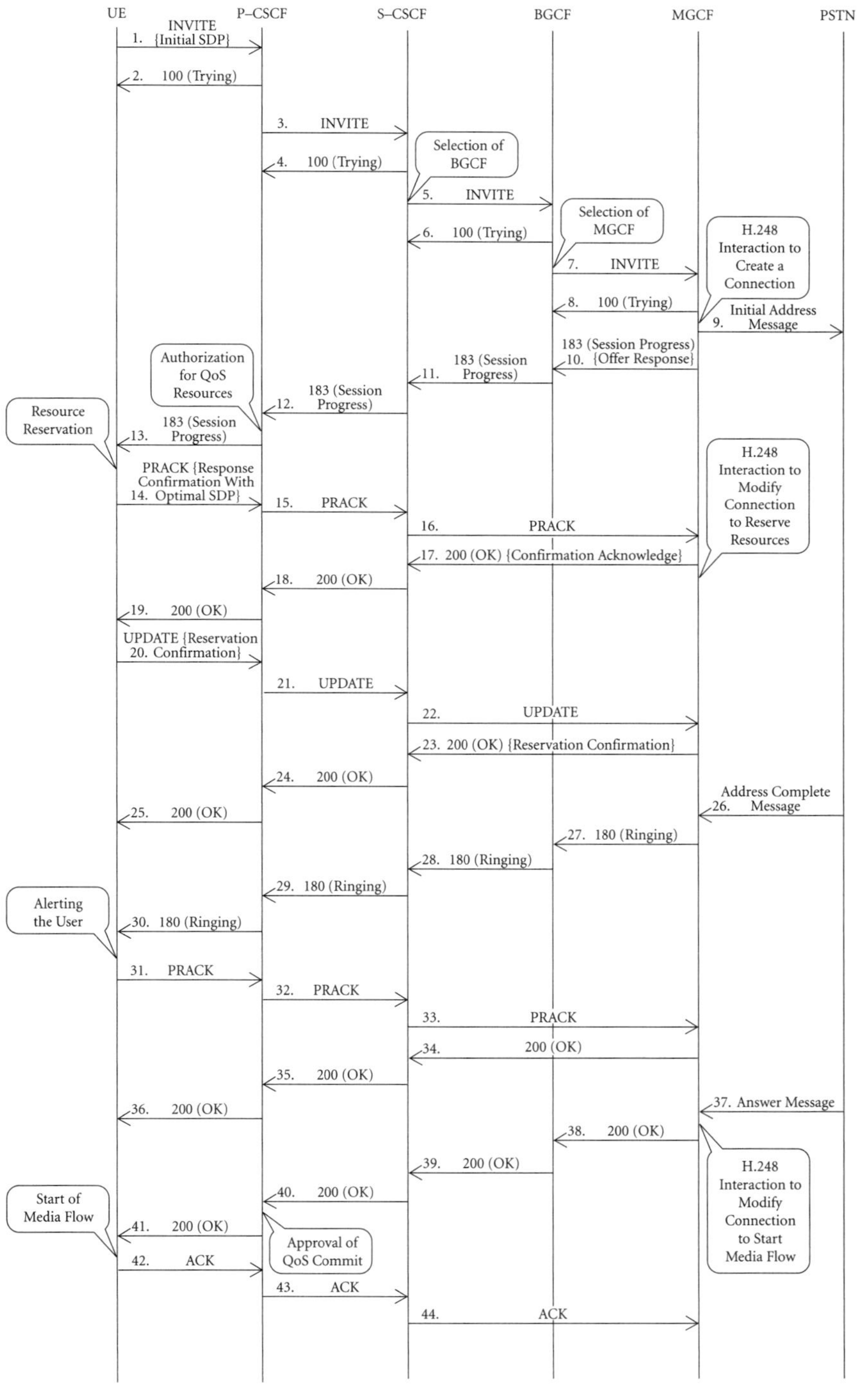

Figure 14.12: IMS Device-to-PSTN VoIP Call Setup

codecs). Since our destination phone is a PSTN phone, the identity may be expressed as sip:+12345678901@gateway.com;user=phone. The S–CSCF validates the service profile and may invoke any origination service logic. It also authorizes the requested SDP based on the user's subscription profile.

The S–CSCF in Step 4 sends a SIP 100 (Trying) response to the P–CSCF. The S–CSCF determines from the destination address that the destination is a PSTN address and hence the call needs to break out to the PSTN. Hence, the S–CSCF sends the INVITE message to a suitable BGCF in Step 5. The S–CSCF may make such a determination based on the public ID, the destination address, or some other parameters that indicate that special treatment is required prior to proceeding. The BGCF in Step 6 sends a 100 (Trying) response to the S–CSCF. The BGCF chooses a suitable MGCF (e.g., closest to the destination phone) and forwards the INVITE message in Step 7. In Step 8, the MGCF sends a 100 (Trying) response to the BGCF.

H.248 interaction occurs between the MGCF and the MGW the MGCF has chosen, and the MGCF sends an Add transaction command to the MGW to create a new context for the VoIP call. It specifies certain characteristics of the Media descriptors for the terminations. An outgoing channel (e.g., a 64 kbps TDM trunk designated by a circuit identification code) is selected for the PSTN-side of the MGW.

The MGCF sends an ISUP Initial Address message to the PSTN via T–SGW (Step 9[494]) as explained in Section 14.4. The MGCF responds to the INVITE message by sending a 183 (Session Progress) message in Step 10. The headers in this SIP message contain the IP addresses for SIP signaling, and the SDP message body present in the SIP message contains the IP addresses for the media. This message includes the Offer Response from the MGCF in the form of the destination's media stream capabilities (which is MGW on the behalf of the landline phone). More specifically, this Offer Response specifies the media characteristics the MGCF prefers based on the MGCF's prior interaction with the MGW. In our example, the MGCF may specify the same codec as that the UE suggested (e.g., 12.2 kbps adaptive multirate codec or 8k enhanced variable-rate codec). The Session Progress message passes through the BGCF (Step 11), the S–CSCF (Step 12), and the P–CSCF to finally reach the UE in Step 13.

The P–CSCF authorizes the resources required for the session. The UE responds to the MGCF by sending a SIP PRACK message that serves as the Response Confirmation (Step 14). The Response Confirmation may contain the SDP, which may be the same SDP as in the Offer Response or its subset. Please note that the originating UE can offer new media during this step or later using the SIP UPDATE method. Of course, the P–CSCF would need to authorize the required QoS when the SDP contains new media descriptions. The UE initiates resource reservation procedures for the offered media after sending the PRACK message. For example, the UE works with the IP CAN to create a secondary PDP context for this QoS when the IP CAN

[494]While we have shown the MGCF sending the Initial Address message in Step 9, the MGCF may do so after Step 22. Some standard documents show the Initial Address message in Step 9, but others show the Initial Address message between Step 22 and Step 23.

is a 3G UMTS-based network. The IP-CAN can also initiate the reservation of the required resources.

The PRACK message passes through the P–CSCF and the S–CSCF (Step 15) and through the BGCF so it reaches the MGCF (Step 16). The MGCF sends a SIP 200 (OK) response as a Confirmation Acknowledge in Step 17. If an optional SDP is included in the PRACK Response Confirmation, the Confirmation Acknowledge in the SIP 200 (OK) response would include an SDP response in Step 17. As mentioned earlier, the P–CSCF would need to authorize the media if the SDP had changed. This acknowledgment message traverses the S–CSCF and the P–CSCF to reach the UE (Steps 18 and 19).

After sending the SIP 200 (OK) response in Step 17, the MGCF reserves the resources to meet the QoS. The MGCF uses a H.248 Modify command to the MGW modify the context to reserve the MGW resources for media processing (e.g., reservation of an adaptive multirate speech codec for speech toward the IP CAN). Note that the resource reservation itself is beyond the scope of the SIP environment. In other words, the UE and IP CAN work together for resource reservation, and the MGCF and the MGW work together for resource reservation. The SIP UPDATE message in Step 20 indicates that this external resource reservation function has been completed and that session setup can continue. This message from the UE serves as a successful resource Reservation Confirmation message. The UPDATE message reaches the MGCF after passing through the P–CSCF (Step 21) and the S–CSCF (Step 22). The MGCF responds in kind with a SIP 200 (OK) message (Steps 23–25).

In the meantime, the PSTN rings the destination landline phone and sends an ISUP Address Complete message to the MGCF via the T–SGW (Step 26). The T–SGW translates the Address Complete message over Message Transport Protcol into an Address Complete message over IP. The MGCF now creates a SIP 180 (Ringing) message, which passes through the BGCF, S–CSCF, and P–CSCF (Steps 27–29) and finds its way to the UE (Step 30). The reception of the 180 (Ringing) message at the UE in Step 30 results in alerting the user. The UE sends a PRACK message to acknowledge the reception of the 180 (Ringing) message. The PRACK and 200 (OK) responses are exchanged in Steps 31–36.

Once the destination party answers the call, the PSTN sends the Answer message to the MGCF via the T–SGW in Step 37. The MGCF now sends an H.248 Modify command to the MGW to start the media flow. The MGCF sends a final SIP 200 (OK) response in Step 38 through the BGCF, S–CSCF, and P–CSCF and to the UE (Steps 39–41), which is the final response to the very first INVITE message the UE sent in Step 1! When the P–CSCF receives this final SIP 200 (OK) response from the S–CSCF, it commits to the approved QoS and indicates to the Policy and Charging Enforcement Function (PCEF)[495] that the media flows authorized for the session should now be enabled. The UE acknowledges the reception of the SIP 200 (OK) response by sending a SIP ACK message in Step 42. The VoIP call setup is complete, and speech flows in both directions!

[495]We will discuss the PCEF in Section 14.9.

Let's say that the VoIP call is over, and the UE ends the call. The UE sends a SIP BYE message to the MGCF, and the MGCF replies with a SIP 200 (OK) response. The MGCF sends an ISUP REL (Release) message to the PSTN via T–SGW, and the PSTN responds with an ISUP RLC (Release Complete) message. The MGCF sends the H.248 Subtract command to the MGW to release the IP CAN side termination and the TDM termination. The MGW releases both terminations and deletes the context.

14.7.2 PSTN-Originated and IMS Device-Terminated VoIP Call Setup: An Overview

Figure 14.13 summarizes the main steps of the VoIP call setup when the originating party is a PSTN phone and the destination party is an IMS device. The PSTN in Step 1 sends an ISUP Initial Address message to a suitable MGCF via the T–SGW based on the destination phone number. The MGCF creates a context in a suitable MGW using H.248 and constructs an INVITE message containing an initial SDP offer based on the MGW capabilities. In Step 2, the MGCF sends the INVITE message to the I–CSCF, and the I–CSCF responds with a 100 (Trying) response (Step 3). The I–CSCF learns from the HSS the identity of the S–CSCF responsible for the IMS device and provides this identity to the MGCF. The I–CSCF is now able to forward the INVITE message to the S–CSCF in Step 4. The S–CSCF sends a 100 (Trying) response to the P–CSCF in Step 5, evaluates the filter criteria, and decides to forward the INVITE message to the P–CSCF accordingly (Step 6). The S–CSCF may communicate with an AS based on the identifiers in the INVITE message and the filter criteria before forwarding the INVITE message to the P–CSCF. For example, an AS may help forward the message to the correct final destination depending upon a specific public user identity and the time of the day and may support recurring conference calls.

The P–CSCF in Step 7 sends the INVITE message to the UE in Step 8. The UE replies with a 100 (Trying) response to the P–CSCF in Step 9 and responds to the MGCF by sending a 183 (Session Progress) message containing the SDP offer response (Step 10). This SDP indicates the UE's preferences for the media stream characteristics (e.g., 12.2 kbps adaptive multirate speech codec). This 183 (Session Progress) message finds its way back through the P–CSCF. When the P–CSCF receives this 183 (Session Progress) message from the UE, it authorizes the resources required to support the QoS for the session. The 183 (Session Progress) message then passes back through the S–CSCF and the I–CSCF and to the MGCF in Steps 10–13.

The MGCF sends a SIP PRACK message as a Response Confirmation message in Step 14 and may include an optional SDP as discussed in Section 14.7.1. The SIP PRACK passes through the S–CSCF and P–CSCF to reach the UE in Steps 14–16. The UE answers with a SIP 200 (OK) response in Step 17 as a confirmation acknowledge message, which reaches the MGCF via the P–CSCF and the S–CSCF in Steps 17–19. Furthermore, the UE also works with its IP CAN to reserve resources to support the VoIP session.

When the SIP 200 (OK) response reaches the MGCF in Step 19, it interacts with the MGW to modify the connection using H.248 to reserve the resources. After

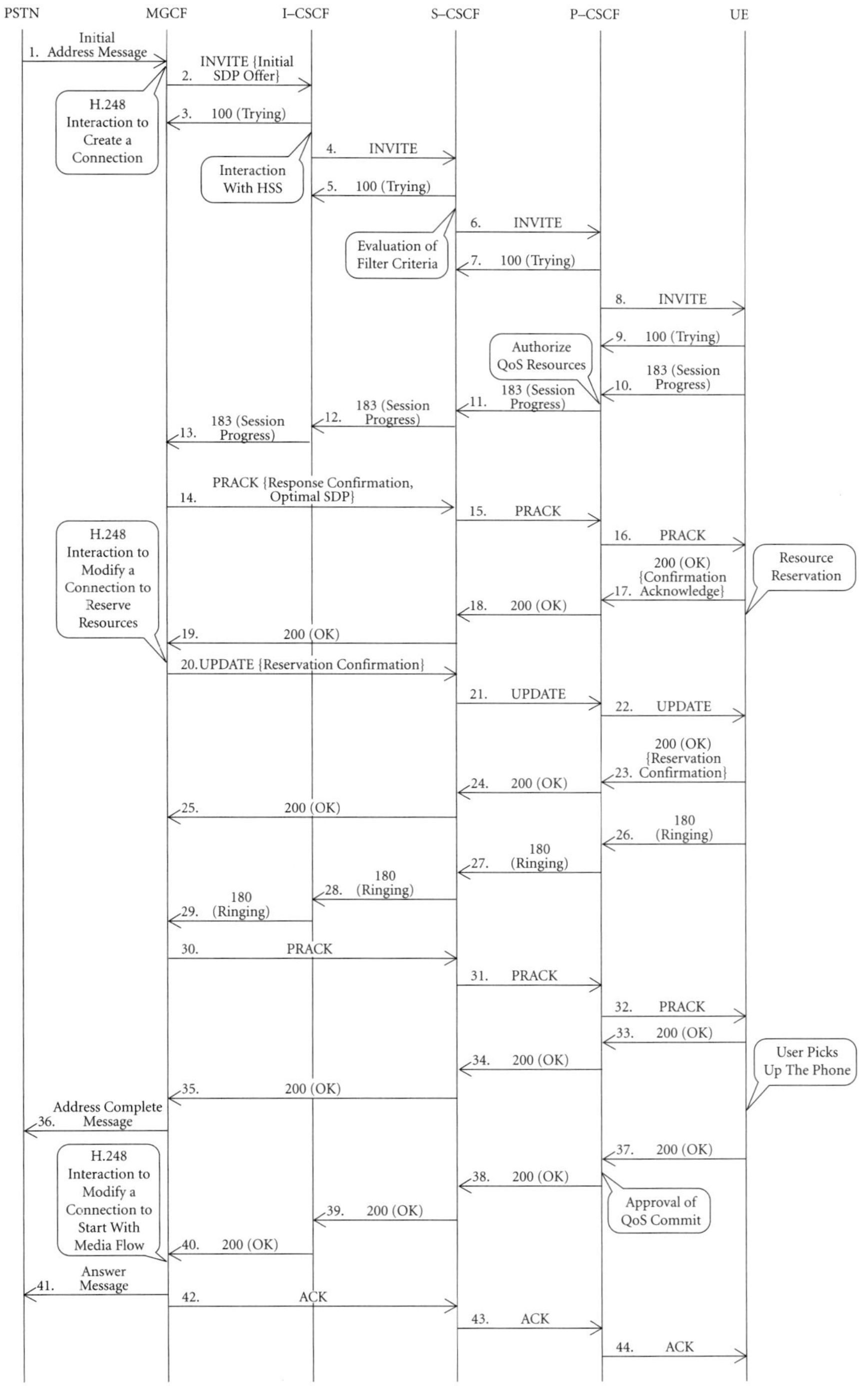

Figure 14.13: PSTN-to-IMS Device VoIP Call Setup

reserving the resources, the MGCF sends a SIP UPDATE message in Step 20 as a Reservation Confirmation message, which travels through the S–CSCF and P–CSCF to reach the UE in Steps 20–22. The UE replies with a SIP 200 (OK) response as a Reservation Confirmation message in Step 23, which reaches the MGCF via the P–CSCF and the S-CSCF in Steps 23–25.

The UE also alerts the user about the incoming call and sends a SIP 180 (Ringing) message in Step 26, which passes through the P–CSCF, S–CSCF, and I–CSCF and reaches the MGCF in Steps 26–29. The MGCF translates the SIP 180 (Ringing) message into an ISUP (Ringing) message and forwards the ISUP (Ringing) message to the PSTN via the T–SGW. The MGCF and the UE exchange SIP PRACK and SIP 200 (OK) responses via the S–CSCF and the P–CSCF to acknowledge the SIP 180 (Ringing) message and the PRACK message, respectively (Steps 30–35). In Step 36, the MGCF sends the ISUP Address Complete message to the PSTN after receiving the SIP 200 (OK) response in Step 35 in response to its PRACK message of Step 30.

Once the user answers the phone call, the UE sends a SIP 200 (OK) response in Step 37, which is the final response to the INVITE message. This 200 (OK) response finds its way from the UE to the MGCF via the P–CSCF, S–CSCF, and I–CSCF in Steps 37–40. The P–CSCF commits the QoS resources for the session and indicates that the media flows authorized for the session should be enabled. The MGCF receives the final SIP 200 (OK) response in Step 40 and uses H.248 to modify the connection so that the media can start flowing. Specifically, the MGCF modifies the context to allow bidirectional transfer of speech; the MGCF sets the Mode to "send receive" in the LocalControl descriptor, which resides within the Stream descriptor, for the relevant media between the UE and the PSTN phone. Furthermore, the MGCF constructs an ISUP Answer message in Step 41 for the PSTN to let the calling party know that all is clear for voice communications. The MGCF sends the UE a SIP ACK message in Step 42 in response to the received SIP 200 (OK) response to culminate the call setup process.

14.7.3 Multimedia Session Setup for IMS Endpoints: A Brief Overview

We have discussed the details of session setup for IMS-to-PSTN and PSTN-to-IMS voice calls. Now let's take a high-level look at the session setup between two IMS devices, as shown in Figure 14.14. The details of the messaging would be similar to those discussed in Sections 14.7.1 and 14.7.2. Note that the session could be a VoIP session or any other type of data session (e.g., a video call). In our example, IMS Device A (referred to as A hereafter) initiates a session with IMS Device B (not surprisingly, referred to as B hereafter) by sending an INVITE message in Step 1. The INVITE message passes through A's P–CSCF and reaches A's S–CSCF.

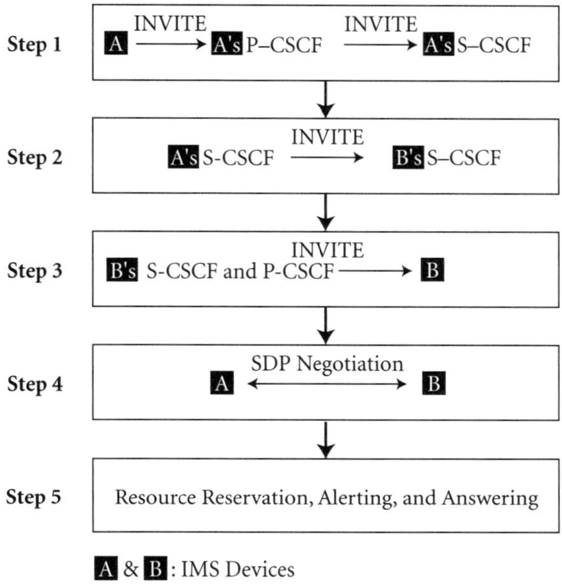

Figure 14.14: Session Setup between Two IMS Devices

A Question for You... How does the P–CSCF know which S–CSCF is responsible for A? See the footnote[496] to verify your answer!

In Step 2, the S–CSCF determines based on the target SIP URI that the destination is a SIP-capable device and asks the DNS server about B's I–CSCF. A's S–CSCF then contacts B's I–CSCF, which, in turn, communicates with B's HSS to learn which S–CSCF is assigned to B. A's S–CSCF can now forward the INVITE message to B's S–CSCF (through the I–CSCF if THIG is active). In Step 3, B's S–CSCF forwards the INVITE message to B via the P–CSCF. A and B now negotiate the SDP for the session in Step 4, using SIP messages such as Session Progress, PRACK, and UPDATE as explained in Sections 14.7.1 and 14.7.2. Since the SIP messages pass through A's S–CSCF (among several other entities), the S–CSCF knows if media conversion or translation is needed.

If A and B have no common ground for media processing (e.g., if each supports a different codec), we need to construct a bridge between the two. In such a case, A's IMS network determines that the signaling with B's IMS domain needs to occur via the Interconnection Border Control Function [318], which adds its list of codecs to those A specified in the INVITE message. When B responds with an SDP Offer

[496]During the IMS device's service registration process, the P–CSCF determines the S–CSCF's identity and stores it in the database.

Response message, the Interconnection Border Control Function determines if B has selected a codec from A's list or from the Interconnection Border Control Function's list. If B has chosen a codec from the Interconnection Border Control Function list, the Interconnection Border Control Function invokes a suitable Transition Gateway to enable media conversion between A and B. The media between A and B would then pass through the Transition Gateway. Note that a Transition Gateway is unnecessary if A and B have a common codec.

Each IMS device works with its own IP CAN to reserve resources in Step 5. When A successfully reserves resources for the session, it sends a SIP UPDATE message to B. After B has also reserved resources, it replies with a SIP 200 (OK) response. Once the destination party is alerted, B sends a SIP 180 (Ringing) message to A, A replies with a SIP PRACK message, and B completes the transaction by sending a SIP 200 (OK) response. When the destination party answers the call (i.e., accepts the session), B sends a SIP 200 (OK) response to the original INVITE request. *The mission (of call setup) is accomplished when A sends a SIP ACK to B in response to the received SIP 200 (OK) response!*

14.8 IMS SERVICES AND APPLICATIONS

As mentioned earlier, IMS enables a service operator to offer a variety of services. *IMS enablers* are reusable components of the services network that multiple applications or services can leverage to acquire specific information (e.g., the location of the IMS device) or to carry out a specific task (e.g., to insert an image) [309]. Such enablers reduce services' capital and operating expenditures, provide scalability, and facilitate implementation of complex and rich services via combining multiple enablers. The limit on services will be set only by the limits of imagination [309].

The *Presence enabler* obtains user-presence–related updates from suitable entities (e.g., the S–CSCF) and provides such information to Presence-based applications. An application that intends to use Presence information must register with the Presence enabler to be notified of Presence-related changes (e.g., a change in the user's presence and availability state). Examples of states include "Present and Available for Communications," "Present but Momentarily Stepped Out," and "Not Present." The user of a service can also use the Presence enabler; for example, the presence of video or audio conference call participants can be confirmed before initiating the conference call.

An XML Document Manager Server enabler makes user-specific data (e.g., a list of URIs) available to other services and service enablers. A service that can exploit such an XML Document Manager Server enabler is instant messaging. For example, assume that you want to arrange a potluck with some of your close friends. You can then create a "Potluck List" using the contacts in your mobile phone and submit the list to the XML Document Manager Server enabler. The "Instant Messaging, Chat, and Call" AS (in the spirit of potluck, we just cooked up the name for this new service!) fetches the resource list from the XML Document Manager Server enabler and sends an INVITE message to all of your friends. Your friends also have access to the Potluck List and can use the list to decide, for example, what food to bring. The list can also include ways of contacting the list members (e.g., email addresses and phone numbers). Another service that can make use of the XML Document Manager Server enabler is video or audio conferencing in which the list includes the conference participants.

A *Location enabler* provides a user's physical location to relevant entities while conforming to certain security and privacy rules. Note that a position server can indeed provide the user's location; however, the Location enabler offers a much richer set of location-related capabilities. The Location enabler periodically obtains the users' physical locations and notifies a given application when predefined events occur. For example, imagine a TraceMyFriend application; the Location enabler alerts the user's TraceMyFriend application when a friend arrives in the user's city, and the user can then call the friend to inquire about getting together for lunch.

A *Messaging enabler* provides enhanced or enriched instant messaging and group chat in a fixed as well as a mobile environment regardless of the end devices the clients use. For example, one user may have Yahoo instant messaging and another user may have MSN instant messaging, but the Messaging enabler will act as a bridge so these two users can interact via instant messaging or chat. The Messaging enabler can be exploited by services such as gaming, two-way paging, and content sharing.[497]

Now that we have discussed several application enablers, let's look at a few examples of services and applications. VoIP is indeed considered a major service that IMS can offer. Since we have already discussed VoIP quite a bit so far, we will focus on other applications.

Video Share is a service that allows a user to show a unidirectional real-time video to a user at the other end of the communications session. For example, AT&T in the U.S. has been offering the Video Share service. Potential enhancements to the basic Video Share service may be bidirectional Video Share and a full-fledged multi-party video conferencing.

Push-to-Talk Over Cellular (also called push-to-talk) is a walkie-talkie service but on a grand scale. Push-to-Talk Over Cellular allows a one-to-one and one-to-many VoIP service over cellular networks. A user can initiate a call with one user or a group of users. Push-to-Talk Over Cellular can exploit the Presence enabler to determine the availability of one or more group members before initiating a call. The main difference between a regular VoIP call and a Push-to-Talk Over Cellular call is Push-to-Talk Over Cellular's very short call setup delay; only one person can offer the media at any given time. To gain a right to speak, the user presses a designated button or key on the device (i.e., the push-to-talk button). A Push-to-Talk Over Cellular user may be a member of several groups or communities. The user can hear when someone is talking in any of the groups and can even reply immediately due to Push-to-Talk Over Cellular's promise of an always-on connection. Push-to-Talk Over Cellular is expected to be a very popular SIP-based service [309, 337].

[497] See the standard [336] for details on the implementation of SMS (Short Message Service) in 3GPP networks.

Gaming in IMS is no longer limited to simple or pre-downloaded games. A user can download complex games as well as compete with other users in real-time.

14.9 IMPLEMENTATION OF QUALITY-OF-SERVICE USING IMS

IMS can use the Policy and Charging Control architecture, illustrated in Figure 14.15,[498] to facilitate the implementation of QoS [338]. The Policy and Charging Control architecture includes the PCRF, PCEF, Application Function, and Subscription Profile Repository. The PCRF determines the policy rules that will allow or disallow certain IMS or SIP flows (e.g., VoIP may be allowed but VoIP plus video may be prohibited) as well as charging rules that will enable flow-based charging (e.g., the price per minute for a VoIP session). The PCRF, upon constructing the rules governing service flows and charging, forwards the rules to the PCEF using the Gx interface.

The PCEF is typically a gateway in the IP CAN in which the IMS device is currently located (e.g., Packet Data Serving Node or Home Agent in 1x and 1xEV-DO core networks and GGSN[499] or its equivalent in UMTS-based core network). The

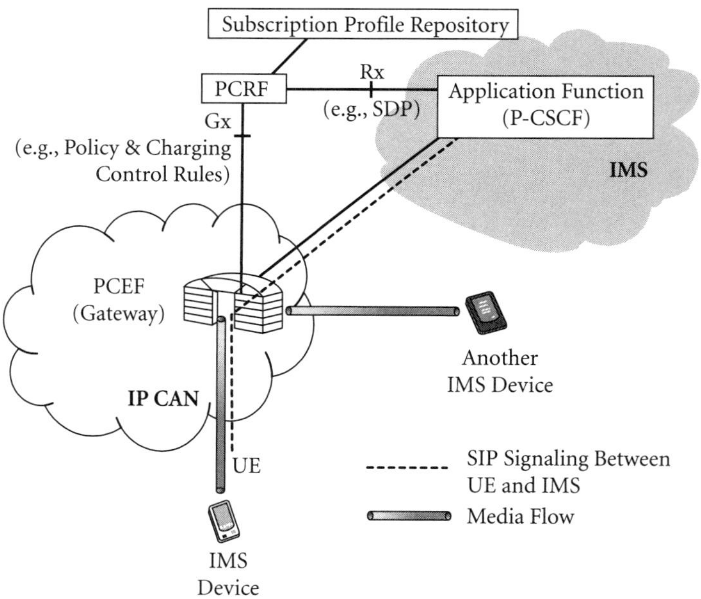

Figure 14.15: Policy Charging and Control Network Architecture for Quality-of-Service

[498] Rx represents "receive."

[499] Gateway GPRS (General Packet Radio Service) Support Node.

PCEF is able to enforce the rules because the user traffic flows through it. When the UE is roaming, the home PCRF controls the visited PCEF, and the service operator who owns the PCEF may place additional constraints. The Application Function helps establish the sessions for the IMS device.

A Question for You... The Application Function formally resides in the IMS and is the first IMS element that the IMS device encounters. Who, do you think, is such an IMS element? See the footnote[500] for the answer.

The Subscription Profile Repository (SPR) contains QoS-related parameters (e.g., average packet latency less than y ms and aggregate maximum bit rate of X Mbps) for the subscribers. The PCRF can retrieve subscriber-specific QoS parameters from the SPR when needed. Since all the elements we have been discussing are logical nodes, the Subscription Profile Repository could be part of the HSS or could be a separate database.

Now that we know the Policy and Charging Control network elements' basic functions, let's discuss how QoS can be implemented for a SIP-based session using Figure 14.16. See the standard [338] for details of the QoS-related signaling flows. Let's assume that a UE initiates a SIP-based multimedia session in Step 1. After some initial SIP message exchange (i.e., INVITE and ACK) between the UE and the destination party, the SIP PRACK message from the UE reaches the Application Function, which is the P–CSCF with IMS. The Application Function extracts the final SDP negotiated between the UE and the destination party from the PRACK message, which contains policy-related information in the form of a media description (e.g., the required bandwidth for the service flow). The Application Function provides SDP and other pieces of information, such as the source and destination IP addresses and Transmission Control Protocol or User Datagram Protocol ports,[501] to the PCRF in Step 2.

The PCRF develops the policy and charging rules for the service flow the UE is requesting and specifies these rules to the PCEF in Step 3. Recall that the PCRF can contact the SPR to retrieve the QoS subscription information for a subscriber. The policy rules dictate whether the flow is authorized and the granted QoS for an authorized service flow. In Step 4, the IP CAN executes admission control to evaluate the availability of resources (e.g., radio resources in the radio access network) to support the required QoS. Assuming that all goes well, the UE and the destination party exchange remaining SIP signaling messages to complete the session setup (Step 5).

For QoS implementation, the IP CAN reserves the required resources, such as tunnels with suitable bandwidths. Between the PCEF and the destination party, the

[500] The P–CSCF is the first point of contact for the IMS device and acts as the Application Function.

[501] This information eventually finds its way to the PCEF that constructs a Traffic Flow Template. The Traffic Flow Template facilitates detection (and thereby policy and charging control) of a service flow at the PCEF for an incoming IP packet.

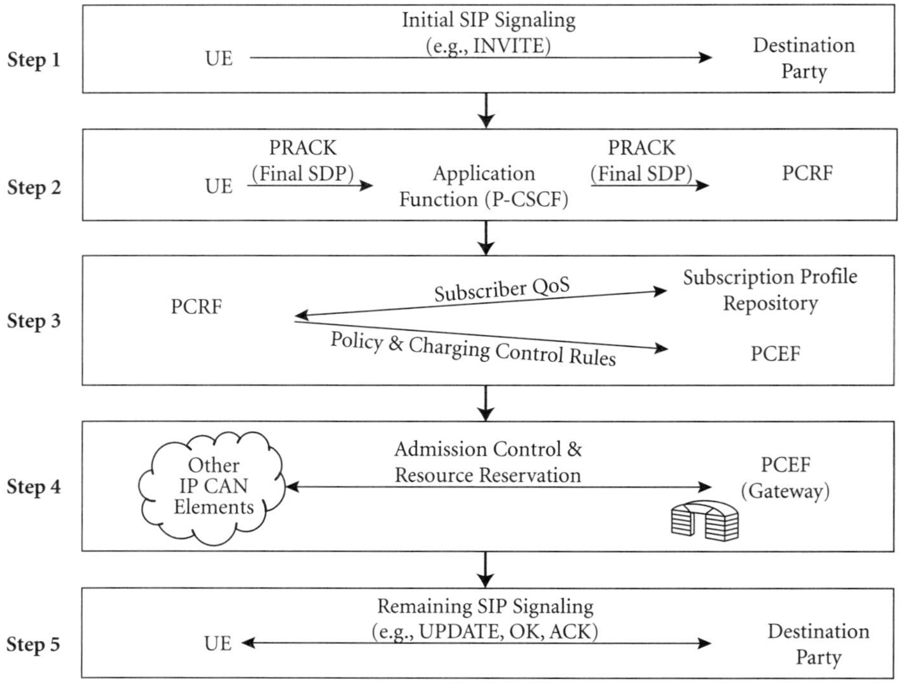

Figure 14.16: QoS for a SIP Session: An Overview

specific approach to implement QoS depends on the type of connectivity. For example, if the destination party is an IP phone connected to the regular Internet, DiffServ can be used. In such a case, the PCEF would map the QoS parameters into a DiffServ Code Point.

Here we have discussed *dynamic Policy and Charging Control*, in which the PCRF dynamically determines the Policy and Charging Control rules in real-time. Another approach is *static Policy and Charging Control*, in which the PCRF is provisioned with certain predefined rules [321].

14.10 POINTS TO REMEMBER

- IMS is beneficial to both consumers and service operators.
- Various organizations started working toward the vision of IMS, with 3GPP's work gaining wide acceptance.
- IMS was formally defined in 3GPP Release 5. Subsequent 3GPP releases have focused on enhancing various aspects of IMS, such as support for emergency calls or sessions.

- The basic components of an IMS network include the P–CSCF, I–CSCF, S–CSCF, HSS, BGCF, MGCF, T–SGW, and MGW.
- Examples of the IP-based protocols that IMS exploits are SIP, DNS, Diameter, and H.248.
- IMS provides session control and hence can be used to set up different types of sessions between two endpoints, such as a VoIP call between an IP phone and a landline phone.
- IMS can offer a variety of services, such as video share and interactive multiplayer gaming.
- IMS plays a vital role in the realization of Fixed Mobile Convergence and IP Convergence.

CHAPTER 15

FOURTH-GENERATION TECHNOLOGIES

As voice revenues decline and data revenues grow, cellular systems must proactively plan to efficiently and cost-effectively meet increasing data demands. Fourth-generation (4G) technologies aim to reduce the cost per bit by using an efficient air interface that exploits OFDM (Orthogonal Frequency-Division Multiplexing) and advanced antenna techniques, a flat all-IP (Internet Protocol) network architecture, a converged packet-switched core network, and a scalable service delivery platform. This chapter first provides the motivation behind 4G technologies. After discussing the essential elements of 4G technologies, the chapter dives into the specifics of two 4G system contenders, WiMAX (Worldwide Interoperability for Microwave Access) and LTE (Long-Term Evolution). Finally, the challenges 4G technologies face are summarized.

15.1 WHY 4G TECHNOLOGIES?

So far, we have discussed major second-generation (2G), second-and-a-half–generation (2.5G), and third-generation (3G) cellular systems. Though various 3G technologies are being deployed in co-existence with 2G and 2.5G technologies, efforts are underway to design and deploy next-generation cellular technologies. We will refer to such next-generation systems as fourth-generation or 4G systems. We will provide an overview of two 4G technologies, WiMAX and LTE. Before we dive into it, let's understand the driving forces behind 4G technologies (see Figure 15.1).

4G systems are expected to be at least three times more efficient (i.e., spectral efficiency measured in bits per second per Hertz) than 3G systems, thus reducing the cost per bit. 4G technologies' higher efficiency will benefit consumers and operators alike; it would (or should!) effectively reduce the bill for the consumer, and service operators will be able to reduce the cost of supporting a target data rate for a subscriber. Lower monthly bills may entice more subscribers, leading to a potentially huge economies-of-scale benefit. Please note that there will be initial capital expenditures to deploy a new 4G network; however, efficiency and economies-of-scale would significantly reduce operating expenditures. Since 4G technologies' peak and average data rates are quite high [e.g., tens of Megabits per second (Mbps) and often exceeding 100 Mbps] and since special attention is given to reduce latency or packet delay, the user will experience services at a new level. The user will have an enhanced experience of existing services. such as email and web-browsing, due to higher download and upload speeds, and new services, such as video sharing and interactive multiplayer gaming, can be effectively and efficiently offered to subscribers.

The network infrastructure and user devices are also evolving and are pushing a move toward 4G systems. To enable the operator to execute IP convergence and fixed mobile convergence, a new network architecture is required. The user device is no longer a plain old telephone; it is a powerful multimedia device that is becoming a miniature version of a laptop.

The wireless business is changing; the earlier business focus was voice services, but now data services are becoming more important. Though voice is still king in terms of absolute earnings, revenue growth is coming from data services. For example, many service operators are experiencing double-digit growth (e.g., 25% to 45%) in data

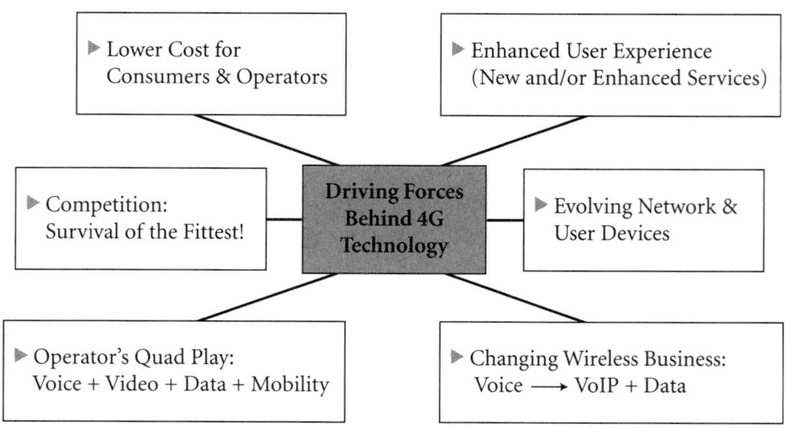

Figure 15.1: Motivation for 4G Technologies

average revenue per user [339]. 4G technologies' high-speed data rates are needed to keep up with the increasing demand for data services. Since voice will still be quite important, it will be supported in a more efficient form—VoIP (Voice-over-IP). Operators are increasingly offering bundled services and are aiming for a *quad play*, in which they can offer voice (e.g., VoIP), video (e.g., video streaming and TV), data (e.g., Internet access), and mobility (e.g., the ability to communicate while on the move). 4G technology promises large bandwidth to facilitate a quad play.

Finally, the communications industry is a cutthroat competition. Fixed operators (e.g., DSL and cable operators) that have saturated the home and office sector are trying to offer mobility to broaden their product portfolio, and cellular operators want to enter the home and office sector by offering the broadband experience over wireless. Operators, fixed or mobile, that have embraced 4G technologies are likely to be more profitable than those who have not.

15.2 ESSENTIAL ELEMENTS OF 4G TECHNOLOGIES

A 4G system includes certain essential elements to fulfill its promises of superior performance (e.g., latency of less than 5 to 10 ms and peak data rates exceeding 100 Mbps) [340]. Figure 15.2 summarizes the *basic* elements of a 4G system. (The intent here is to provide only *general* traits of 4G technologies. We will discuss specific features of WiMAX and LTE later.)

The air interface transitions from Code-Division Multiple Access (CDMA) in 3G systems to scalable OFDMA (Orthogonal Frequency-Division Multiple Access) in 4G technologies.[502] OFDMA involves the creation of numerous narrowband radio channels and the allocation of multiple radio channels to users. (We will take a closer look at OFDMA in Section 15.3.) Scalable OFDMA means that the number of radio channels varies as a function of spectrum bandwidth. The narrower the spectrum bandwidth, the fewer radio channels. OFDMA has performance and complexity advantages over CDMA in cases involving very high data rates (e.g., those exceeding 100 Mbps).

4G systems would typically use a larger bandwidth than 3G systems, even though they support deployment using narrow spectrum bandwidths, such as 1.4 MHz. 4G systems would need a total of 10 MHz or 20 MHz bandwidth to achieve stellar performance. Though 10 MHz deployment may occur when spectrum availability is an issue, 20 MHz deployment is expected to be more common.[503]

16-QAM (Quadrature Amplitude Modulation) is the highest-order modulation scheme supported in a typical 3G system (e.g., HSPA and 1xEV-DO Rev. A[504]), but 4G technologies support 64-QAM in both the uplink (UL) and downlink (DL). Typical cellular systems use space or spatial diversity in the UL (and sometimes in the

[502] A 4G system called HSPA+ (High-Speed Packet Access +) uses CDMA (not OFDMA).

[503] In frequency-division duplexing (FDD) deployment, thie 10 MHz bandwidth includes 5 MHz downlink bandwidth and 5 MHz uplink bandwidth.

[504] 1xEvolution—Data Optimized Revision A.

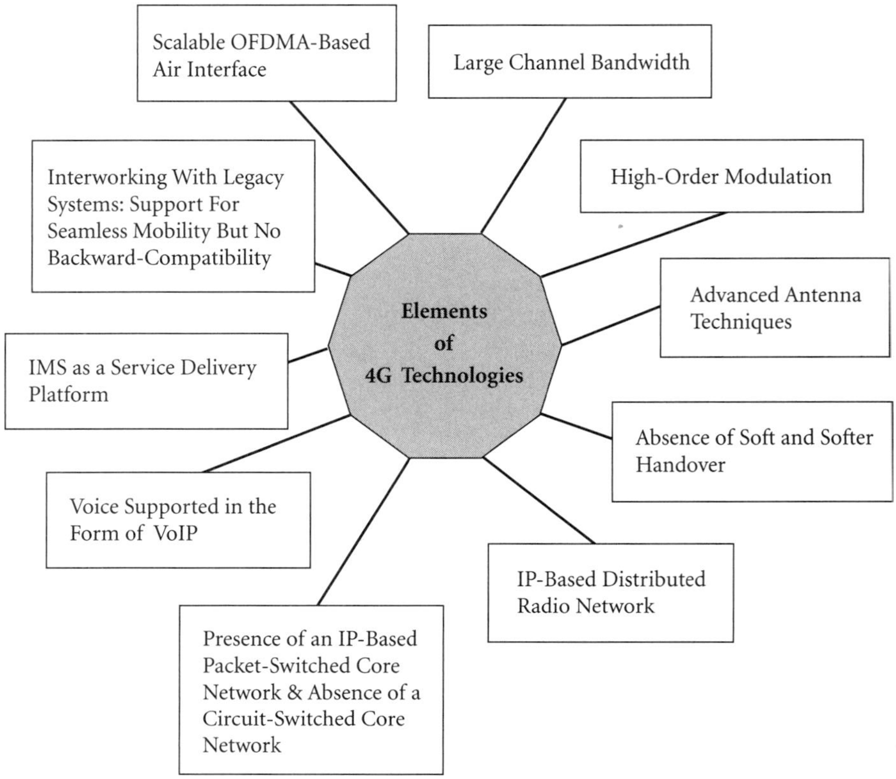

Figure 15.2: Elements of a 4G System

DL, too). 4G systems exploit advanced antenna techniques, such as beamforming and MIMO (Multiple-Input Multiple-Output), to achieve higher reliability and spectral efficiency.

To facilitate fast and dynamic scheduling of radio resources in a cell or sector, soft and softer handover are unsupported. Cell-edge reliability is addressed by the nature of OFDMA and by some interference management methods.

The architecture of the radio access network is distributed as opposed to hierarchical. There is no radio network controller in the radio access network; the radio access network consists of base stations that can communicate with one another. The radio access network may or may not have a gateway.

A Question for You... In typical 3G systems, can two base stations directly communicate?[505]

IP is used to transfer signaling messages and data traffic between two base stations. 4G technologies use no circuit-switched core network. Recall that 3G systems, such as 1x Radio Transmission Technology and Universal Mobile Telecommunication System (UMTS), have both circuit-switched and packet-switched core networks, with the circuit-switched core handling voice services and the packet-switched core providing basic Internet access. The IP-based packet-switched core network serves as a unified common framework that enables seamless mobility across different radio access technologies (e.g., handover from LTE to 1xEV-DO). 4G technologies are not backward-compatible with legacy technologies and require new radio and core networks.

As new services increasingly rely upon IP, IP Multimedia Subsystem (IMS)[506] is a suitable service delivery platform in 4G technologies. In the absence of a circuit-switched core network, voice service is supported in another form, VoIP. The media gateway of the IMS is used to translate voice between the IP and Public-Switched Telephone Network formats. IMS facilitates fast and cost-effective development and deployment of IP-based services. Though IMS is optional in 4G technologies, the industry trend is to increasingly rely upon the IMS for services. Several operators have already deployed IMS in conjunction with LTE. A service operator may have its own specific way of offering IP-based services.

15.3 FUNDAMENTALS OF OFDM AND OFDMA

As mentioned earlier, OFDMA is the multiple access technique exploited in 4G technologies. OFDMA builds upon OFDM. We will explain the motivation for OFDM and OFDMA in Section 15.3.1. An overview of OFDM is given in Sections 15.3.2 and 15.3.3. OFDM encounters challenges such as inter-symbol interference (ISI) and intercarrier interference (ICI); these challenges and their solutions are discussed in Section 15.3.4. Finally, Section 15.3.5 provides a simplified view of an OFDMA transceiver.

15.3.1 OFDM and OFDMA: Why and Why Now?

Although OFDM and OFDMA are often used interchangeably, they are actually slightly different. OFDM involves the use of multiple narrowband radio channels for data transfer between the source (e.g., the base station) and the destination (e.g.,

[505] In 3G systems, two base stations cannot directly communicate. The radio network controller controls the base stations. When coordination between the base stations is required (e.g., during handover), the radio network controller takes control by communicating with the base stations individually.

[506] See Chapter 14 for an overview of IMS.

the mobile device). For example, when a base station sends data to *one user* using multiple radio channels in parallel, it is using OFDM. Furthermore, the base station is said to be using OFDMA when it sends information to *multiple users simultaneously*, with OFDM between the base station and each user.

Though the basic OFDM technique has been around for several years, recent advances in digital signal processing have made OFDM more attractive for widespread deployment in *mobile* communication systems. OFDMA is considered a *very efficient multiple access technique for high-speed data transfer* and is thus well-suited for bandwidth-intensive applications, such as broadcast-multicast multimedia services. Hence, OFDMA is viewed as an essential component in any next-generation wireless system, e.g., IEEE 802.16-based WiMAX, 3GPP[507] LTE, and IEEE 802.11n-based WLAN (wireless local area network). Other examples of existing OFDM-based commercially deployed systems include Digital Audio Broadcast, Digital Video Broadcast–Terrestrial, HiperLAN, and IEEE 802.11-based WLAN systems [341].

15.3.2 OFDM: A High-Level View

The next time you take a shower, think OFDM![508] As shown in Figure 15.3 [342], a large amount of water flows through a big pipe, and the showerhead distributes water among numerous tiny streams. What is the end result? A nice user experience, assuming the water is warm! (If the water is cold, we assume no responsibility!) Similarly, OFDM distributes a large amount of data onto numerous low-data-rate radio channels, and parallel transmission of data on these narrowband channels yields a very high throughput. For example, when a base station sends data to a user on one hundred radio channels in parallel with each channel transmitting just 20 kbps, the effective user-perceived data rate is $100 * 20$ kbps $=$ 20 Mbps, which leads to a nice user experience, just as in the case of a shower! *In essence, OFDM*

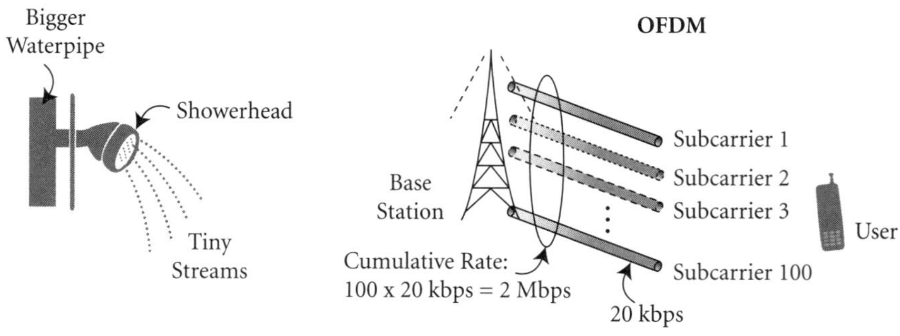

Figure 15.3: OFDM and Showerhead Analogy

[507] Third Generation Partnership Project
[508] The showerhead example is discussed in Langton [342].

enables parallel data transmission on multiple narrow-bandwidth radio channels, increasing the effective user data rates. Each narrow bandwidth radio channel is called a *subcarrier.*

In an OFDM system, the radio channels or subcarriers are orthogonal; such orthogonality makes OFDM very efficient. OFDM can be used with TDD (Time-Division Duplexing) or FDD (Frequency-Division Duplexing).

OFDM is a special case of Frequency-Division Multiplexing. In a traditional Frequency-Division Multiple Access (FDMA) system, two users are distinguished via two different radio channels, with one radio channel per user. Furthermore, in an FDMA system, interference between two adjacent channels (i.e., ICI) is minimized by separating them by guard bands, as illustrated in Figure 15.4.[509]

In a traditional FDMA system, ICI is minimized through guard bands. Similar to FDMA, OFDMA can use different channels to distinguish different users. However, contrary to FDMA, OFDMA can assign multiple subcarriers to the same user. *Furthermore (and this is quite important!), OFDMA does not require a guard band between the adjacent subcarriers due to the orthogonality of the subcarriers.* In Figure 15.4, User 1 is assigned the subcarriers f_1, f_3, f_5, f_7, User 2 is assigned the subcarrier f_2, and User 3 is given the subcarriers f_4, f_6, f_8.[510] The absence of guard

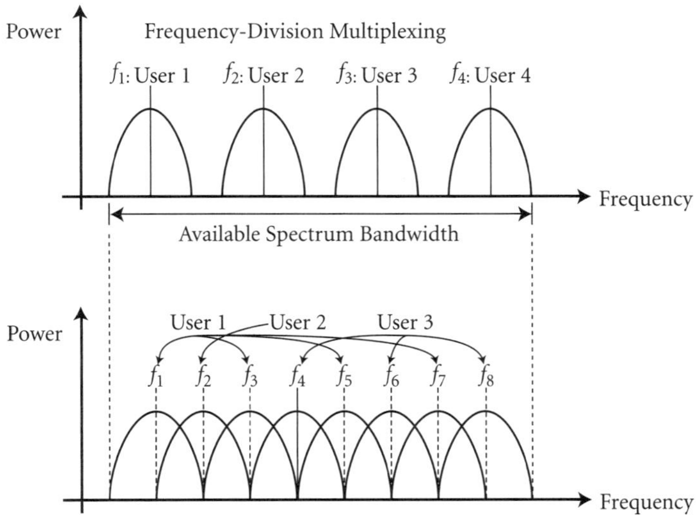

Figure 15.4: OFDMA versus Frequency-Division Multiple Access

[509]In the interest of simplicity, Figure 15.4 shows a given subcarrier crossing the zero-power points where the two adjacent subcarriers reach their highest power levels. A more comprehensive diagram would represent a given subcarrier in the form of a sync function such that this subcarrier crosses the zero-power points where *all* the other subcarriers reach their highest power levels.

[510]A user may be allocated adjacent or non-adjacent subcarriers in a generic OFDMA system. The exact allocation of subcarriers depends on technology.

bands between the adjacent subcarriers enables more subcarriers to exist within the available spectrum bandwidth, increasing the system capacity and throughput.

In a traditional Time-Division Multiple Access (TDMA) system, different users use the same frequency radio channel, such as f_1, at different times, achieving orthogonality in the time domain. For example, a given 200 kHz radio channel can be shared among eight users in Global System for Mobile Communications when each user uses the channel only during the assigned time-slot. Similarly, an OFDMA system can assign the same subcarrier to different users at different times, achieving orthogonality in the time domain. Contrary to traditional TDMA, the OFDMA system could assign many time-slots (as opposed to one or two in typical 2G and 2.5G deployments) to the same user during a radio frame. A radio frame is a time unit that encompasses several time-slots. Both TDMA and OFDMA systems need special mechanisms to counteract ISI caused by multipaths.

Both OFDMA and CDMA use a carrier frequency with a wide-spectrum bandwidth. In fact, when fast frequency-hopping is used in conjunction with OFDMA, it derives benefits of a frequency-hopping spread-spectrum system [343]. Code-Division Multiple Access 2000 (CDMA2000) and UMTS Release 99 systems use a direct-sequence spread-spectrum technique to use a wide-spectrum bandwidth. In a typical CDMA system, such as 1x and UMTS, the UL is non-orthogonal and about two-thirds of the total UL interference comes from the users within the same sector or cell. In a typical OFDMA system, the users are allocated orthogonal subcarriers for transmission within a sector, leading to minimal (ideally, zero) in-sector interference.

An OFDMA system allows multiple users to simultaneously access the system. For example, User 1 can use subcarrier f_1 and User 2 can use the subcarrier f_2 to receive data from the base station at the same time instant t_1. A pure OFDM system allows DL transmission to just one user at a time, using all the available subcarriers for the selected user. For example, the 802.16-based fixed WiMAX system that uses the configuration of OFDM-256 typically sends data to one user at a given time instant using many of the 256 subcarriers.

When the spacing between adjacent subcarriers is properly determined, each subcarrier experiences a flat-fading propagation channel. This facilitates the implementation of various multiple antenna techniques. The receiver design also becomes simpler in the case of flat fading as opposed to frequency-selective fading.

15.3.3 Construction of Orthogonal Subcarriers

Orthogonality of subcarriers in an OFDM system means (ideally) zero interference among these subcarriers. Assume that User 1 is assigned the subcarrier f_1 and User 2 is assigned the subcarrier f_2. Orthogonality ensures that User 1 experiences no interference from f_2 while attempting to retrieve information from f_1. Figure 15.5[511] shows how subcarrier frequencies and the useful symbol period are selected to achieve orthogonality among subcarriers. The available spectrum bandwidth is divided into N

[511] All the subcarriers shown here are at the baseband.

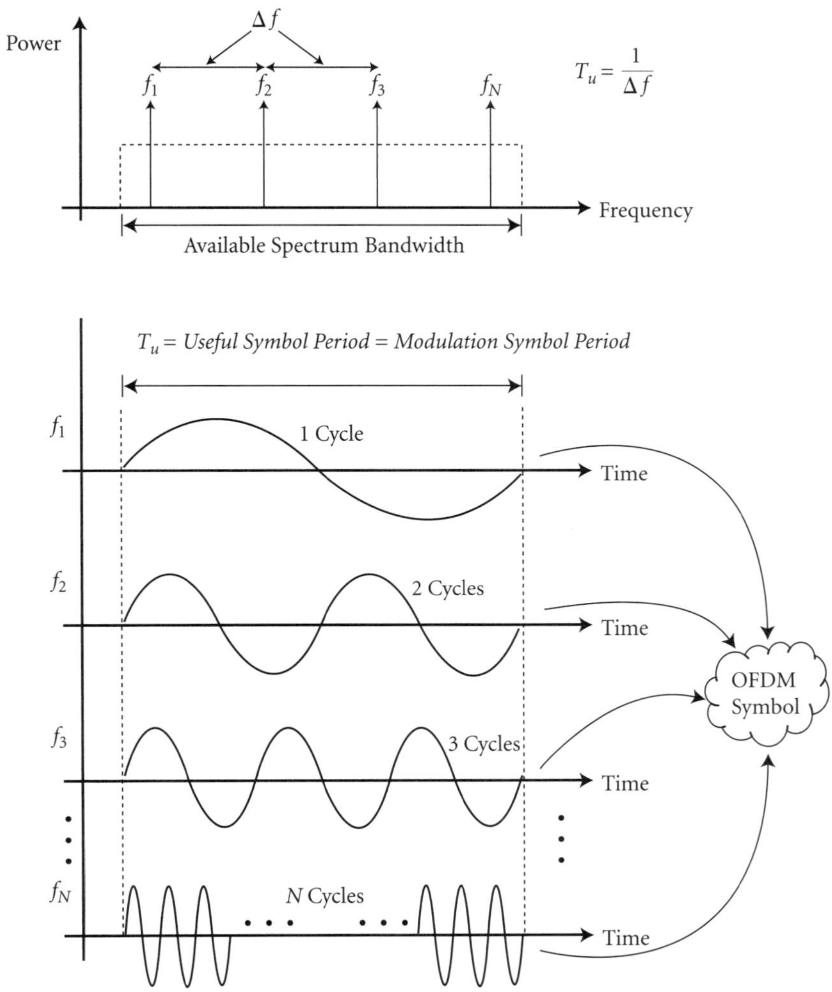

Figure 15.5: Creation of Orthogonal Subcarriers

equally spaced subcarriers with Δf as the spacing between the adjacent subcarriers. The inverse of Δf is defined to be T_u, which is called the *useful symbol period. One subcarrier carries one modulation symbol during the time period T_u.* In other words, T_u is equivalent to the modulation symbol period. A total of N modulation symbols are transmitted during T_u with one modulation symbol per subcarrier as shown in Figure 15.5. The frequency of the first subcarrier is $f_1 = \Delta f$. The frequency of the second subcarrier is $f_2 = 2\Delta f$ or $f_2 = 2f_1$. The frequency of the third subcarrier is $f_3 = 3\Delta f$ or $f_3 = 3f_1$, and so on.

In other words, an integer multiple of cycles exists for each subcarrier during the time interval T_u. *The modulation symbols on all the subcarriers during the time interval T_u constitute one OFDM symbol.* Section 15.3.5 illustrates where modulation symbols and OFDM symbols appear in an OFDMA transmitter and receiver. Figure 15.5 illustrates all subcarriers as pure sine waves with $0°$ phase shift and the same amplitude to represent binary phase shift keying (BPSK) modulation symbols (e.g., $A * \sin(2 * \pi * f_i * t)$ where $A = +1$ represents the bit 0 and $i = subcarrier\ number$). However, in practice, any modulation scheme, such as quadrature phase shift keying (QPSK), 16-QAM, and 64-QAM, can be used, which would result in different phases and/or amplitudes for different subcarriers. Orthogonality is still maintained in such cases. Let's discuss orthogonality from two perspectives.

Orthogonality: A Conceptual Example. Once the users are assigned specific subcarriers, they attempt to match the received signal with the reference waveforms at their assigned subcarriers. The users will find a match only with their individual subcarriers, and orthogonality will prevent interference during the matching process. For example, assume that the base station has assigned f_1 to User 1 and f_2 to User 2 as shown in Figure 15.6. The base station sends bit 0 to User 1 on f_1 and bit 0 to User 2 on f_2. In other words, the base station sends two different waveforms simultaneously during the time interval T_u. The first waveform is a pure sine wave at

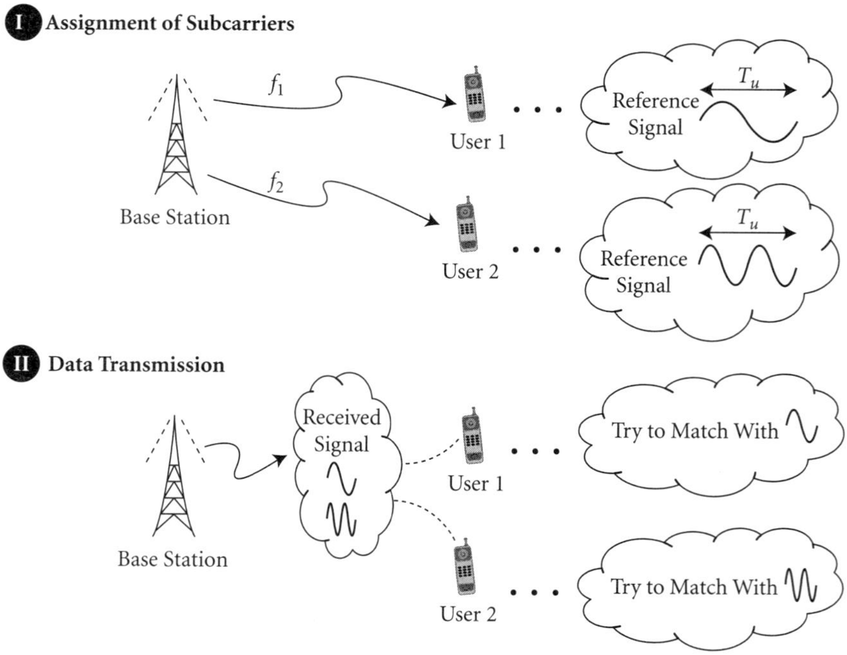

Figure 15.6: Orthogonal: A Conceptual Example

the frequency f_1: $A * \sin(2 * \pi * f_1 * t) = \sin(2 * \pi * f_1 * t)$. The second waveform is also a pure sine wave but at frequency f_2: $\sin(2 * \pi * f_2 * t)$. Since User 1 has been assigned f_1, User 1's reference signal consists of just one cycle of a sine wave during the time interval T_u. This reference signal will match with the received signal $\sin(2 * \pi * f_1 * t)$. However, User 1's reference signal does not match the received signal $\sin(2 * \pi * f_2 * t)$ because this received signal has two cycles during time interval T_u.

Orthogonality: A Mathematical Example. When the received signal is multiplied by the reference signal and integrated over the time period T_u, the signal at the correct subcarrier frequency is retrieved successfully, and all other subcarriers disappear. Assume two subcarrier frequencies, f_n and f_m, where $m \geq n$, f_n is the base frequency such that $f_n = 1/T_u$, and f_m is a multiple of f_n. Further assume that User 1 has been assigned f_n and that the base station sends two waveforms, $A_1 * \sin(2 * \pi * f_n * t)$ and $A_2 * \sin(2 * \pi * f_m * t)$ during the time period T_u. The goal for User 1 is to retrieve bit A_1 that is being carried on subcarrier f_1. The reference signal for User 1 becomes $\sin(2 * \pi * f_n * t)$. Now, User 1 performs the mathematical operation shown in the following example.

Example: Mathematical Processing and Orthogonality

Received Signal: $s(t) = A_1 \sin(2\pi f_n t) + A_2 \sin(2\pi f_m t)$
Reference Signal: $r(t) = \sin(2\pi f_n t)$ for User 1
Goal: Retrieve A_1
Processing at User 1: Retrieved bit or modulation symbol

$$\begin{aligned} Result &= \int_0^{T_u} (A_1 \sin(2\pi f_n t) + A_2 \sin(2\pi f_m t)) * (\sin(2\pi f_n t))\, dt \\ &= A_1 \int_0^{T_u} \sin 2\pi f_n t * \sin 2\pi f_n t dt + A_2 \int_0^{T_u} \sin 2\pi f_m t * \sin 2\pi f_n t dt. \end{aligned}$$

Now,

$$-2\sin(a_1)\sin(a_2) = \cos(\alpha_1 + \alpha_2) - \cos(\alpha_1 - \alpha_2)$$

where $\alpha_1 > \alpha_2$. In our case, $f_m > f_n$.

$$f_n = \frac{1}{T_u}, \qquad f_m = i * \frac{1}{T_u}, \qquad i = integer$$

$$\begin{aligned} Result = &-\frac{A_1}{2}\left[\int_0^{T_u} \cos(2\pi f_n + 2\pi f_n) t dt - \int_0^{T_u} \cos(2\pi f_n - 2\pi f_n) t dt\right] \\ &-\frac{A_2}{2}\left[\int_0^{T_u} \cos(2\pi f_m + 2\pi f_n) t dt - \int_0^{T_u} \cos(2\pi f_m - 2\pi f_n) t dt\right] \end{aligned}$$

Since integration of a sine or cosine wave over an integer multiple of cycles is always zero,

$$Result = -\tfrac{A_1}{2}[0 - T_u] - \tfrac{A_2}{2}[0 - 0]$$
$$Result = \tfrac{A_1}{2} * T_u.$$

Since T_u is known, multiplying the result by $2/T_u$ will yield A_1.

Orthogonality enables users to retrieve their own information bits or modulation symbols from the assigned subcarriers with no interference from other subcarriers. Each subcarrier is processed separately. In other words, when a user is assigned two frequencies, f_i and f_j, one reference signal at f_i would help retrieve one modulation symbol and another reference signal at f_j would help recover another modulation symbol.

Let's summarize the construction of orthogonal subcarriers at an OFDM transmitter. Once the subcarrier spacing is determined, the useful symbol time T_u is fixed. During the time period T_u, subcarriers with an integer multiple of cycles are constructed. Each subcarrier carries one modulation symbol during time period T_u.

15.3.4 OFDM Challenges and Solutions

Multipaths in a wireless system cause problems. Signal fading can occur due to multipath signals with different amplitudes and phases combining together. ISI is another risk because the previous symbol overlaps with the current symbol. Different systems rely on various methods for overcoming the impact of multipaths.

In a CDMA system, we typically welcome multipaths! CDMA uses a RAKE[512] receiver in which different fingers lock onto different multipaths so that their energies can be combined together to reconstruct a more reliable signal at the receiver. As long as we have an adequate number of RAKE fingers,[513] we are in a good shape!

A traditional TDMA system uses an equalizer to estimate the impact of multipaths via channel estimation, and the system subsequently corrects for potential errors. An OFDM system uses a cyclic prefix or cyclic extension to solve the problem of orthogonality loss multipaths cause.

Figure 15.7 illustrates how multipaths destroy orthogonality. The mobile receives two copies of the signal, one via a direct path and the other via a reflected path. The time difference in the arrival of multipaths depends on the environment (e.g., about ten microseconds). The modulation symbol period can be from seventy to one hundred microseconds in commercial 4G systems. Since OFDM offers no RAKE receiver and for the sake of illustration and discussion, we have shown two distinct paths in Figure 15.7. In reality, the multipaths automatically combine at the receive antenna itself. When the mobile tries to recover the current modulation symbol, the

[512] Recall that RAKE is not an acronym but a description of the receiver structure's appearance.
[513] Six to eight fingers in a RAKE receiver are common these days.

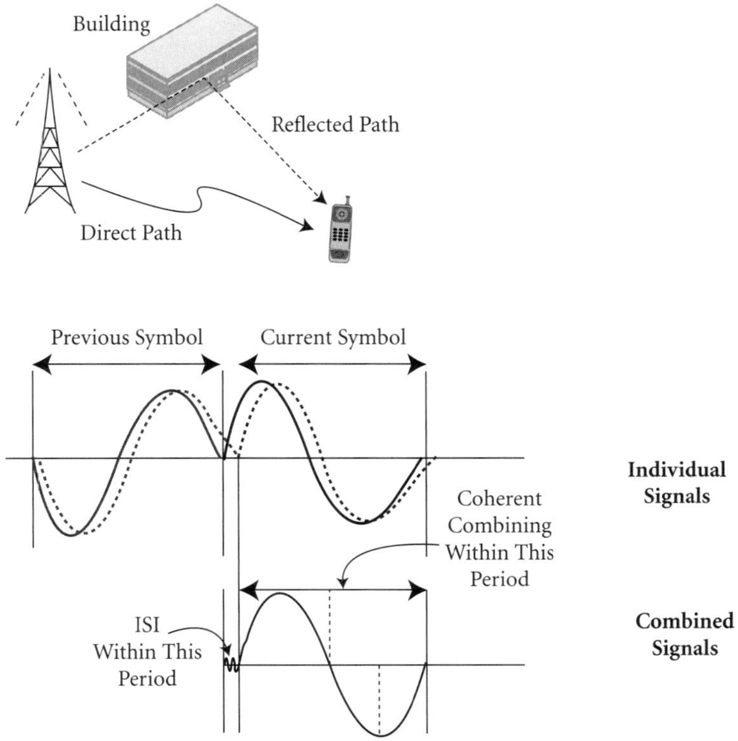

Figure 15.7: ISI Problem Faced by OFDM

previous modulation symbol riding on the reflected path causes ISI at the symbol boundary. Once the overlap between the two successive symbols multipaths cause is over, the multipaths combine coherently within the symbol period.[514] ISI exists at the symbol boundary, and we lose orthogonality due to the inability to recover full cycles of subcarriers. But no need to worry! We have a simple and elegant solution to the ISI problem: *Cyclic Prefix (CP)* or *Cyclic Extension*!

Figure 15.8 depicts how the Cyclic Prefix or Extension helps solve the ISI problem. The basic idea is to transmit the symbol for a slightly longer time period. The tail-end of the signal is copied and pasted at the front of the signal, giving the appearance of an extended cycle (hence the name cyclic *extension*!). The transmitter is now sending the extended symbol in the time domain. At the receiver, we will again see ISI at the symbol boundary. However, in a simple receiver implementation, we can ignore the initial portion of the received symbol (which the ISI has corrupted) and recover

[514]Within the symbol period, multipaths carry the *same* modulation symbol and hence energy from different multipaths adds up at the receive antenna. In contrast, at the symbol boundary, two different modulation symbols overlap when multipaths are present.

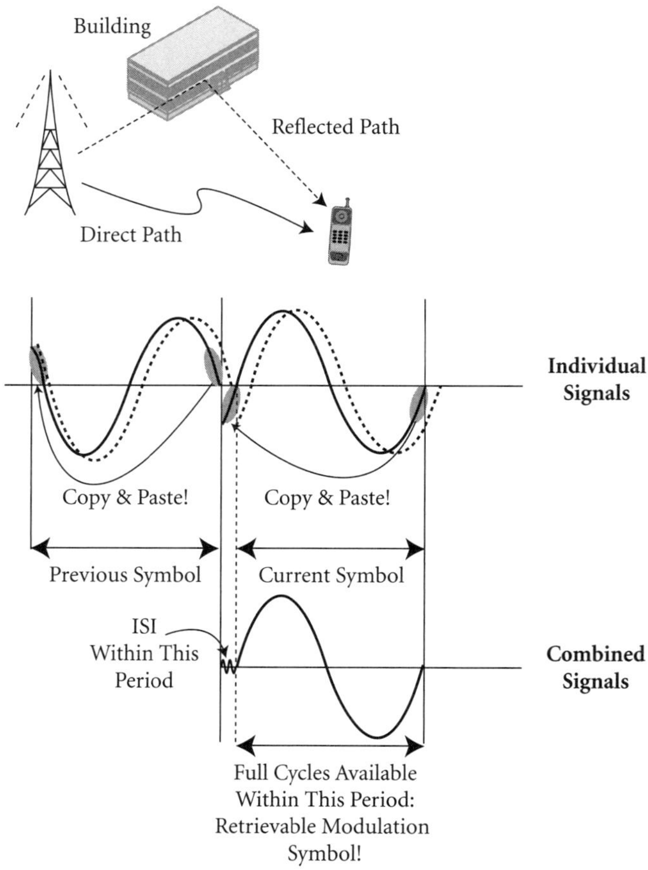

Figure 15.8: Solving the ISI Problem Using Cyclic Prefix

the modulation symbol using the remainder of the symbol period. An attempt should be made to choose the length of the CP such that it exceeds the delay spread[515] in the geographical area in which the system is being deployed. The sum of the useful symbol time (i.e., $modulation\ symbol\ period = 1/subcarrier\ spacing$) and the CP is referred to as the *total OFDM symbol time*. The subcarrier spacing remains the same regardless of the CP value. If the time period in the case of delay spread exceeds the CP, some ISI will remain within the symbol period. An alternative solution to the ISI problem would be simply a gap or guard time between two successive symbols. However, the benefits of using the CP instead of such a guard time are that energy from

[515]Delay spread quantifies the time difference in arrival instants of the first arriving path and the last arriving strong multipaths. See Chapter 3 on radio frequency (RF) propagation to find the exact definition of delay spread.

the initial and ending portions of the symbol can be added for more reliable detection, the power amplifier's preference for continuous transmission, and conversion of the linear channel convolution to the circular channel convolution for enhanced receiver operation [344].

Now let's take care of another challenge OFDM faces: ICI. Consider Figure 15.9 in which the base station (BS) is observing signals coming from three different users. We will discuss two cases, one with narrow subcarrier spacing, Δf_n, and one with wide subcarrier spacing, Δf_w. For now, *assume that all users are stationary.*

Due to the orthogonality among subcarriers, all other subcarriers have no power at the frequency of a given subcarrier, leading to zero ICI regardless of the subcarrier

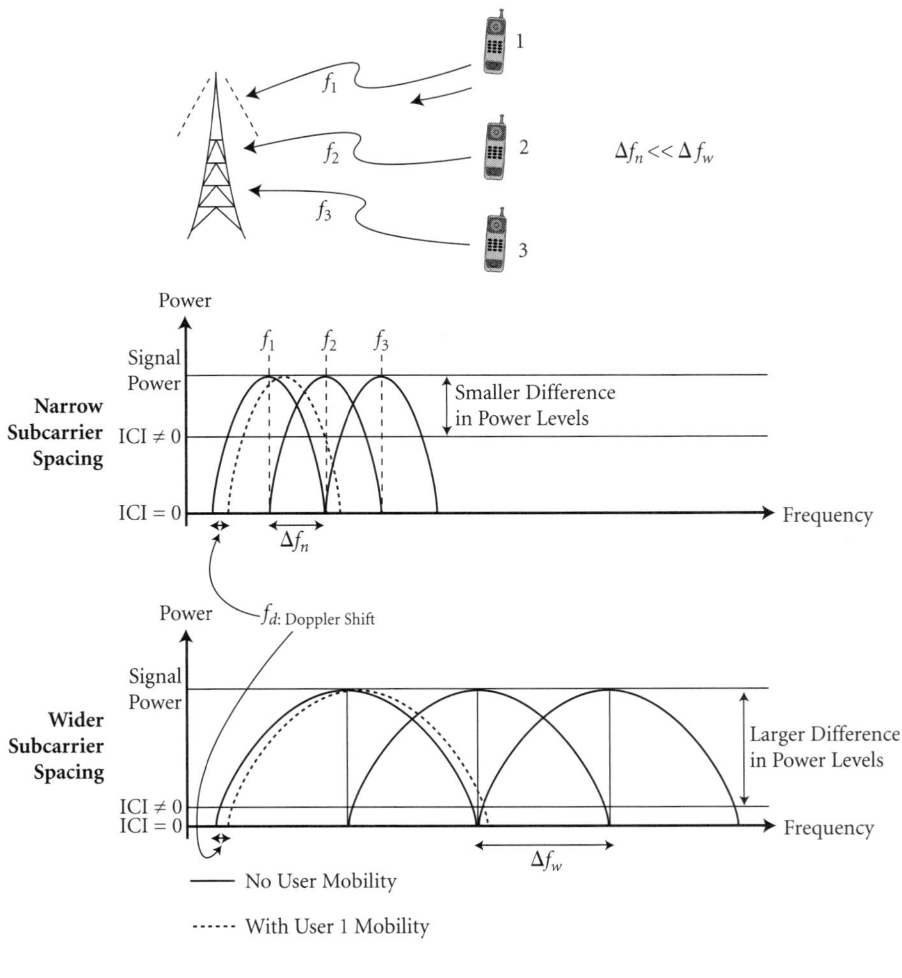

Figure 15.9: Solving the ICI Problem Using Adequate Subcarrier Spacing

spacing. In our example, each user is allocated one subcarrier.[516] Assume that the base station is trying to detect User 2's subcarrier f_2.

The BS sees that f_2 has pretty good signal power. Due to the orthogonality among subcarriers, subcarriers f_1 and f_3 contribute no interference at subcarrier f_2's frequency. In other words, ICI is zero when the users are stationary. Now let's discuss the impact of user mobility on ICI. Assume that User 1 starts moving toward the BS. If the subcarrier spacing is narrow, as the upper portion of Figure 15.9 depicts, the interfering subcarrier's power could be close to the desired subcarrier's power, making it difficult for the receiver to retrieve the information the desired subcarrier carries. In contrast, if the subcarrier subspacing is wide, the interfering subcarrier's power would be much smaller than the desired subcarrier's power, enabling the receiver to easily retrieve the information the desired subcarrier carries.

A Question for You... As the mobile moves at a faster velocity, would the magnitude of the Doppler shift become smaller or larger?[517]

As an example, at the carrier frequency of 3.5 GHz, the Doppler shift is about 408 Hz for the speed of 125 km/hr and the ICI power is about 27 dB weaker than the signal power on a subcarrier for the 10-kHz subcarrier spacing [345, 346]. Since signal power is much stronger than ICI power, the receiver is able to work reliably.

OFDM systems also face the challenge of peak-to-average-power ratio (PAPR). In an OFDM or OFDMA system, different subcarriers carry different modulation symbols in parallel. Since the Inverse Fast Fourier[518] Transform (IFFT) combines all independent modulation symbols[519] and since modulation symbols can be considered random numbers, it is as if we were adding random numbers.

Figure 15.10 illustrates an example of the required transmit power as a function of time samples within one OFDM symbol period when four sinusoidal signals are added. As the figure shows, the required peak power can be quite large compared to the average transmit power. When PAPR is high, the average transmit power, rather than the maximum transmit power, must be used to estimate the supportable data rate so that we can meet the peak power needs when required without entering the non-linear region of the power amplifier operation. Note that operating in the non-linear region would distort the transmitted signal. In other words, we need some power back-off relative to the maximum transmit power to account for high PAPR.

[516]This is just an example! In practice, a user is usually allocated more than just one subcarrier.

[517]The higher the (relative) velocity is, the larger the magnitude of the Doppler shift. The actual received frequency would be the transmit frequency plus the Doppler shift. If the transmitter is moving toward the receiver, the Doppler shift is positive and the actual received frequency is greater than the transmitted frequency. However, if the transmitter is moving away from the receiver, the Doppler shift is negative and the actual received frequency is less than the transmitted frequency.

[518]Pronounced as FOOR-ee-ay (not like "furrier") since this is the surname of Jean Baptiste Joseph Fourier, a French mathematician and physicist.

[519]Section 15.3.5 discusses IFFT details.

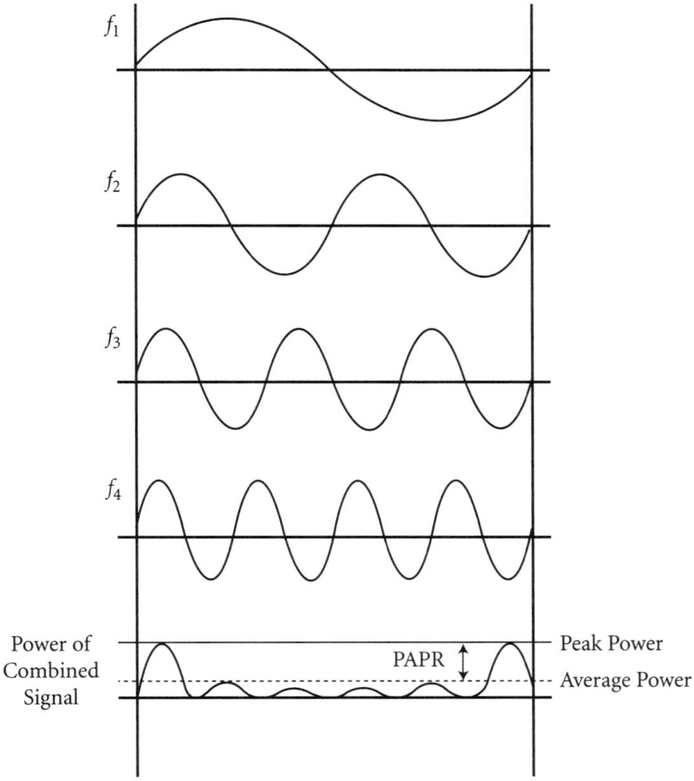

Figure 15.10: Peak-to-Average-Power Ratio Problem in OFDM

As an example, when the maximum transmit power is 23 dBm, we can use a 3 dB power back-off, essentially leading to $23-3 = 20$ dBm as the average power available for use. Observe that when peak power must be transmitted, power back-off will give us the required cushion or buffer. Since the provisioning of the power back-off results in less-than-maximum available power, it reduces the peak supportable cell-edge data rate. Hence, several PAPR reduction techniques have been proposed [347]. The simplest technique is peak-clipping, in which the instantaneous transmit power is clipped at the maximum transmit power, avoiding the use of the non-linear region. However, peak clipping may reduce the overall throughput due to packet errors and retransmissions.

Another PAPR reduction technique is channel coding, in which suitable codes or bit sequences are used to represent information. Such codes reduce the PAPR; however, coding adds redundancy, leading to lower peak data rates. Some techniques such as Selective Mapping, Partial Transmit Sequences, and Random Phasor modify amplitudes and/or phases to reduce the PAPR [348–353]. The main drawback of these techniques is increased complexity and increased signaling overhead to convey phase manipulation to the receiver.

A technique called tone reservation or peak cancellation transmits anti-peak signals in unused or reserved subcarriers [354] to reduce the PAPR of the transmitted signal. Single-Carrier—Frequency-Division Multiple Access (SC-FDMA), a variation of OFDMA, relies upon Discrete Fourier Transform signal processing to modify the nature of the transmitted signal, leading to a reduction in the PAPR. SC-FDMA's main drawback is increased complexity of the transmitter and receiver due to extra processing related to Discrete Fourier Transform. WiMAX allows the use of the peak cancellation technique and LTE uses SC-FDMA in the UL.

15.3.5 Simplified OFDMA Transceiver

Let's take a look at how OFDMA transmitters and receivers work. Consider Figure 15.11 where basic elements of an OFDMA transmitter at the base station are shown. Assume that the BS wants to send data to two users during some time period, such as a 1-ms or 5-ms interval.

For each user, the transmitter scrambles the information bits to provide some level of privacy and to randomize the stream of bits to facilitate the decoding operation at the receiver. The scrambled bits pass through FEC coding, such as turbo coding,

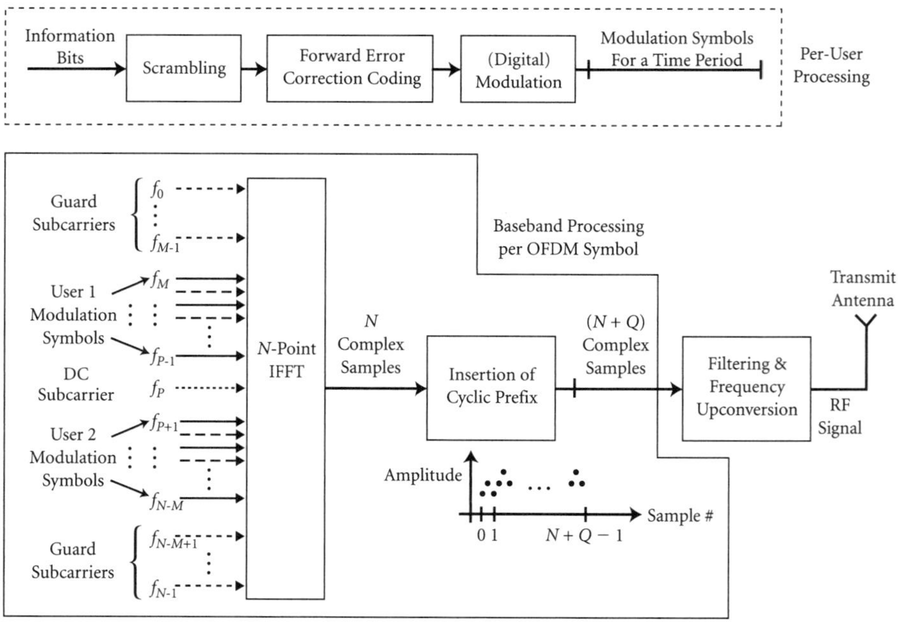

Figure 15.11: A High-Level Look at an OFDMA Transmitter

and the code symbols may be interleaved to improve decoding and rate-matched (e.g., punctured) to support Hybrid Automatic Repeat Request (HARQ).[520] The code symbols are then converted into modulation symbols by the digital modulation technique.

A Question for You... What is the highest-order modulation scheme supported in 4G technologies, and how many bits (or code symbols) does one modulation symbol represent for such technique?[521]

Once the modulation symbols for transmission during a specific transmission period (e.g., 1 ms or 5 ms) for all users are determined, the OFDMA signal is generated for each OFDMA symbol. Such processing includes IFFT, addition of the Cyclic Prefix, Filtering, digital-to-analog conversion, and frequency upconversion.

IFFT combines the modulation symbols that orthogonal subcarriers carry into a time-domain signal that can be transmitted over the air. Using IFFT instead of actually sending numerous subcarriers in parallel makes the implementation efficient, cost-effective, and fairly simple. IFFT is an algorithm that can carry out the mathematical calculations required to implement Inverse Discrete Fourier Transform.

Now let's take a closer look at what goes into the IFFT block. Certain numbers of subcarriers at the left side and the right side of the spectrum, f_0 to f_{M-1} and f_{N-M+1} to f_{N-1} in Figure 15.11, are null subcarriers, meaning that they are not transmitted, and they are called *guard subcarriers*. Recall that no guard band is required between two adjacent subcarriers due to the subcarriers' orthogonality. However, our guard subcarriers are between our OFDM system and the neighboring systems to the left and to the right of our OFDM spectrum.

A Question for You... What may be the purpose of the guard subcarriers?[522]

Another type of null subcarrier is the DC subcarrier, which is located in the center of the DL spectrum bandwidth. Historically, the DC subcarrier was required in the analog system implementation, but its existence in modern digital implementations facilitates detection of the center of the spectrum that corresponds to the system's carrier frequency. The center carrier frequency is around the spectrum bandwidth's center, wherever the mobile finds the frequency with minimum power or energy.

The remaining subcarriers contain data subcarriers and pilot (or reference signal) subcarriers. The data subcarriers carry modulation symbols to convey a signaling message or a traffic packet (e.g., email bits). The pilot subcarriers help implement

[520] See Section 11.3.2 and Figure 11.5 in Chapter 11 for an example of HARQ operation.

[521] It is 64-QAM, and each 64-QAM symbol represents six code symbols. Observe that $2^6 = 64$.

[522] Since the guard subcarriers are not transmitted, the total power the system transmits gradually decays to acceptable levels outside the spectrum bandwidth. This will minimize interference caused to (and by!) the adjacent system and will facilitate meeting regulatory emission requirements (e.g., FCC requirements in the U.S.).

coherent demodulation at the receiver by acting as a reference signal, facilitating fine time and frequency synchronization, and supporting handoff or handover.

In Figure 15.11, the subcarriers f_M through f_{P-1} are allocated to User 1 and $f_P + 1$ through f_{N-M} are assigned to User 2. Some of the allocated subcarriers carry pilot symbols for the entire cell or sector, and the rest of the symbols carry data modulation symbols. The unallocated or unused subcarriers are not transmitted. N-point IFFT generates N complex-valued samples; for example, 1024-point IFFT in a 10-MHz deployment generates 1024 complex-valued samples during the useful symbol period of $1/(subcarrier\ spacing)$.

Once the IFFT provides the time samples for transmission, the CP is added. The last few samples are appended to the N samples generated by the IFFT to yield a total of $N + Q$ samples where Q samples are for the CP.

Do You Recall... Do you remember what radio propagation phenomenon the CP addresses? Hint: The name of this phenomenon starts with an M.[523]

The next step is to ensure that the signal remains within the intended system bandwidth (e.g., 10 MHz). Baseband filtering will do that job. So far, all operations have been performed at the baseband, and most of the energy of our signal stays from 0 to 10 MHz (as an example). However, we lack permission to send the baseband signal over the air. We need to convert the signal from the baseband to the RF to remain emission-law–abiding citizens! Examples of frequency bands in which 4G systems are deployed are 700 MHz, 1.7 GHz, and 2.5 GHz. Frequency upconversion will translate the baseband signal into an RF signal. After the high-power amplifier offers suitable amplification, the antenna will radiate the signal into the air.

We discussed the OFDMA transmitter for the DL in which the base station must multiplex information for different users via the frequency domain. In the UL, each user sends modulation symbols, along with predefined pilot or reference signals, on subcarriers allocated to the specific user.

Let's turn our attention to the receiver. Consider the mobile receiver shown in Figure 15.12. The main function of the receiver is reverse-processing relative to the transmitter. The antenna receives the RF signal, which is downconverted to the baseband and filtered so that the received signal can be processed further in the intended bandwidth. Once the analog signal is sampled and analog-to-digital conversion is carried out, $N + Q$ samples are taken within the total OFDMA symbol duration with the time between two successive sampling instants being $1/(N * \Delta f)$ where N is the size of the IFFT at the transmitter and Δf is the subcarrier spacing. The receiver next removes the samples that are suspected of having been affected by ISI. Of course, no more than Q samples should be removed because the next processing block of Fast Fourier Transform (FFT) needs N samples as its input.

FFT processing is opposite to that of IFFT and provides modulation symbols on orthogonal subcarriers as the output. Pretend that you are a user who has been allo-

[523]Multipaths! The CP alleviates the ISI that multipaths cause.

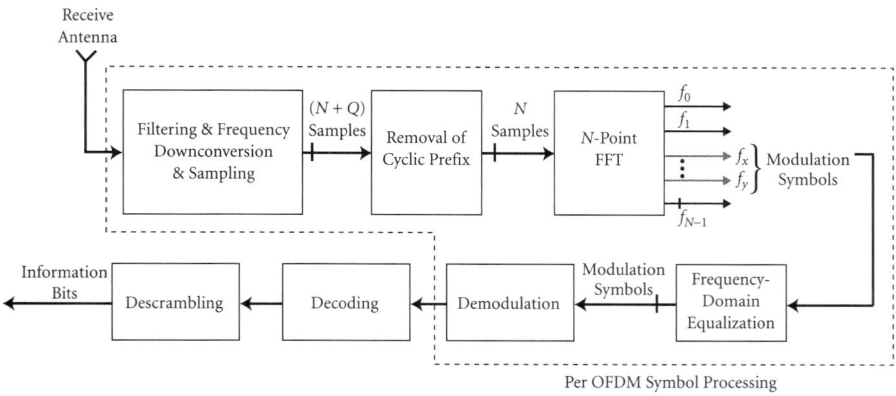

Figure 15.12: A High-Level Look at an OFDMA Receiver

cated the subcarriers f_x through f_y. You will process the modulation symbols on only those subcarriers to receive your signaling message or a traffic packet. Specifically, you need not (and should not![524]) demodulate the data subcarriers allocated to other users. Such processing involves frequency-domain equalization and demodulation per OFDMA symbol and decoding and descrambling for the overall transmission period (e.g., 1 ms or 5 ms). In frequency-domain equalization, the receiver looks at the pilot subcarriers in the frequency domain to estimate how the radio environment has affected the transmitted signal and then uses such knowledge to undo the damage caused to the data subcarriers. For example, if the pilot subcarrier f_{10} undergoes a 90° phase shift, the adjacent data subcarrier f_{11} is likely to have undergone a similar phase shift. Frequency-domain equalization facilitates the retrieval of the transmitted modulation symbols. Demodulation estimates the modulation symbols and provides code symbols to the decoding block. The decoder processes the code symbols and estimates the scrambled bits. The next step of descrambling will yield the original information bits.

We briefly discussed the DL receiver implemented at the mobile. Now, let's take a quick look at the UL receiver implemented at the base station. The base station receiver processes the received signal in pretty much the same way as the mobile receiver. The basic steps are identical to the ones shown in Figure 15.12. The first few steps of frequency downconversion, filtering, removal of the CP, and FFT are done just once for the combined signal containing signals from all users. Once the FFT is performed, the modulation symbols for each user will be treated separately. The BS knows which subcarriers are for which user and hence will perform frequency-domain

[524]Because each user's data is scrambled, we need not worry about security or privacy. Furthermore, higher-layer encryption can also be implemented for security.

equalization for each user individually. Such user separation in the UL is important for frequency-domain equalization because different users experience different channel conditions.

15.4 MULTIPLE ANTENNA TECHNIQUES

OFDMA alone is inadequate for delivering the spectacular performance that 4G technologies have promised. Advanced antenna techniques are needed to help achieve performance targets. Let's take a look at the basic concepts of these antenna techniques, such as receive diversity, transmit diversity, beamforming, spatial multiplexing or MIMO, and SDMA (Space- or Spatial-Division Multiple Access).

15.4.1 Diversity Techniques: Some Simple and Some Sophisticated!

Let's first discuss the three basic diversity techniques—space diversity, time diversity, and frequency diversity—as illustrated in Figure 15.13. Then, we will look at two sophisticated diversity techniques, which involve some type of redundancy.

Space diversity can be applied at the transmitter to implement *transmit diversity* or at the receiver to implement *receive diversity*. In the case of transmit diversity, the same information is sent from two (or more) different antennas (i.e., spaces). In Figure 15.13, two transmit antennas, Tx1 and Tx2,[525] send the same modulation symbol S_1 at the same frequency f_1. Now we have two distinct paths between the transmitter and the receiver. Since two paths usually experience independent fading and since the probability of two paths experiencing fading at the same time is quite low, the receiver is likely to get at least one of the modulation symbols successfully. Recall from Chapter 3 that short-term fading is typically modeled as a Rayleigh or Ricean fading and long-term fading is typically modeled as a log-normal distribution.

A Question for You... What kind of fading would be different between the transmitter-to-receiver paths—short-term or long-term?[526]

In the case of receive diversity in Figure 15.13, the modulation symbol S_1 is sent from only one antenna, Tx, but is received by two receive antennas, Rx1 and Rx2. Again, two paths between the transmitter and the receiver experience different short-term fading.

In a simple example of *time diversity*, symbol S_1 is sent during two different time periods, t_1 and t_2, and symbol S_2 is sent during two different time periods, t_3 and t_4. Hence, if we lose a symbol during one time period, perhaps we can still recover

[525] Recall that Tx represents "transmit" and Rx represents "receive."

[526] Short-term fading allows very little separation between the two transmit antennas. Long-term fading would be different only if two paths experience widely different environments (e.g., one path encounters numerous buildings and another path encounters no buildings).

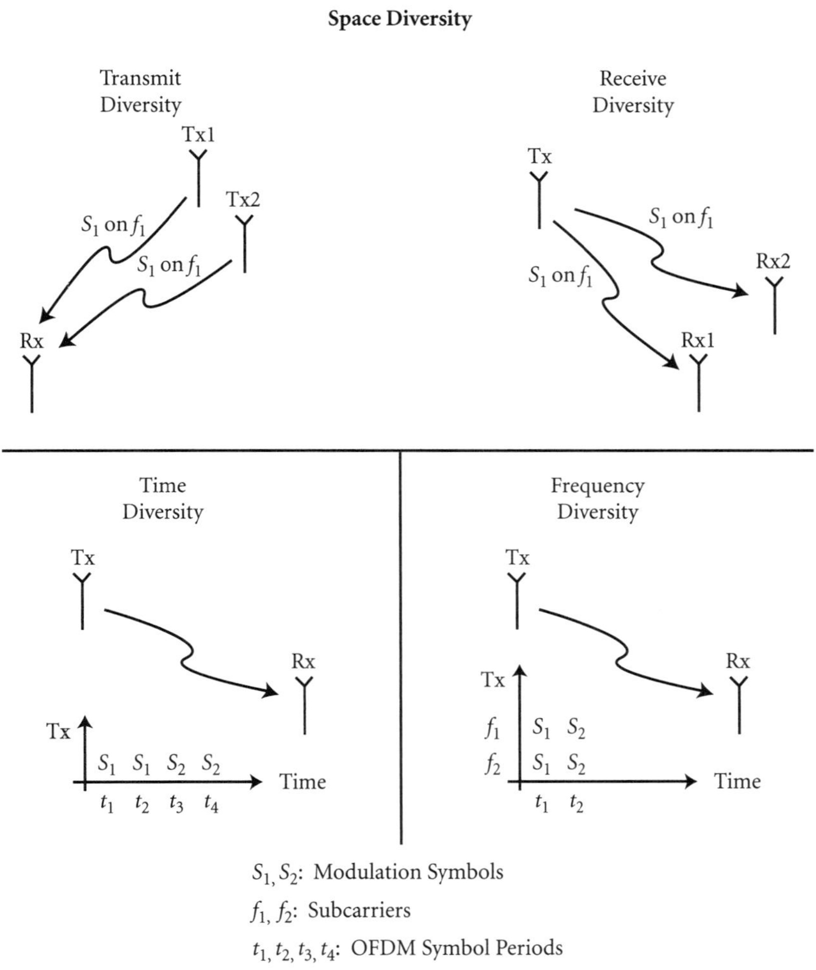

Figure 15.13: Basic Diversity Techniques

the information by processing the same symbol during another time period. Another example of time diversity is interleaving in which the same information is separated by a sufficient time interval so that the information bit (usually the code symbol) experiences different fading.

Frequency diversity involves transmission of the same information (e.g., symbol S_1) on two different frequencies (e.g., subcarriers f_1 and f_2). Since different parts of the spectrum experience different fading, f_1 and f_2 are unlikely to experience fading at the same time, allowing us to retrieve the information. *In summary, diversity techniques exploit signal variations in space, time, and frequency to incorporate redundancy and thereby improve reliability*. Receive diversity in the UL has been widely deployed in commercial 2G and 3G systems, and receive diversity in the DL is also widely used in 3G 1xEV-DO systems.

Now that we understand basic diversity techniques, let's discuss two sophisticated transmit diversity schemes, Space Time Coding (STC, also called *Alamouti coding*) and Space Frequency Block Coding (SFBC). STC is used in WiMAX, and SFBC is used in LTE. As the name implies, STC exploits both space diversity and time diversity.

Figure 15.14 gives the basic idea of how STC works. The antenna Tx1 transmits the modulation symbol S_1 during the OFDMA symbol period t_1, and the antenna Tx2 transmits the same symbol (although in a different form, S_1^*) during t_2. The same information is sent using two different antennas, thus space diversity is used. Furthermore, since the same information is sent during two different time periods, time diversity is also used.

In Figure 15.14, S_1^* is the complex conjugate of S_1. The same subcarrier is used to send the symbols and their conjugates. Since we are sending two modulation symbols on the same subcarrier during two symbol periods, we are providing redundancy but no increase in the data rate. The data rate would have increased if we were to send different modulation symbols (e.g., S_3 and S_4) during t_2, but STC aims to improve reliability by providing structured redundancy that exploits both space diversity and time diversity, not by increasing the data rate.

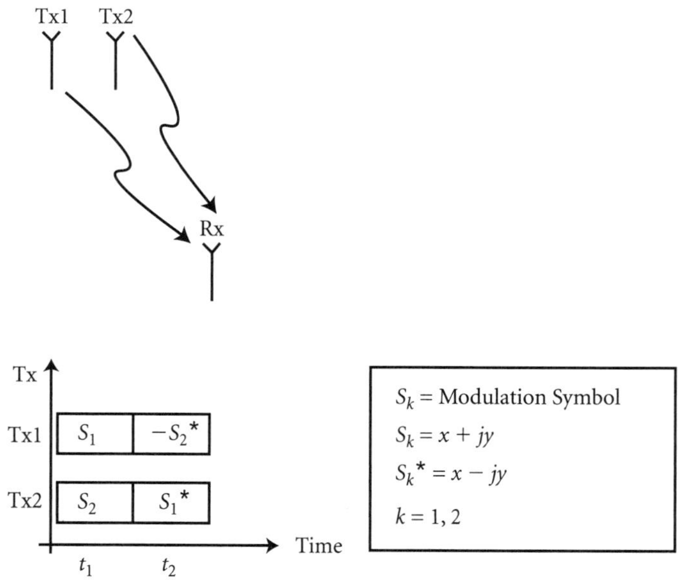

Figure 15.14: Space Time Coding

Ever Wondered... If you have wondered why we have used complex conjugates for diversity rather than pure pristine modulation symbols, many of us have asked the same question! Using complex conjugates and negative complex conjugates makes the receiver's job easier. The receiver performs addition and subtraction for the signals received during t_1 and t_2 to extract S_1 and S_2. Take a leap of faith on this one. If you insist on seeing for yourself, try adding and subtracting r_1 and r_2 with and without complex conjugates for r_1 and r_2, and you will then quiet your inner skeptic. Here, $r_1 = S_1 + S_2$ and $r_2 = S_1^* - S_2^*$. Alamouti provides the formal way of retrieving S_1 and S_2 [355].

Figure 15.15 summarizes SFBC, which aims to combine the benefits of space diversity with frequency diversity. Antenna Tx1 transmits S_1 on subcarrier f_1, and antenna Tx2 transmits S_1 (to be precise, S_1^*) from subcarrier f_2. Since the same information is sent using two different antennas, space diversity is being used. Furthermore, since two different subcarriers carry the same information, frequency diversity is also being used. As in the case of STC, the use of complex conjugates facilitates decoding at the receiver. SFBC enhances reliability at the expense of throughput, just like STC.

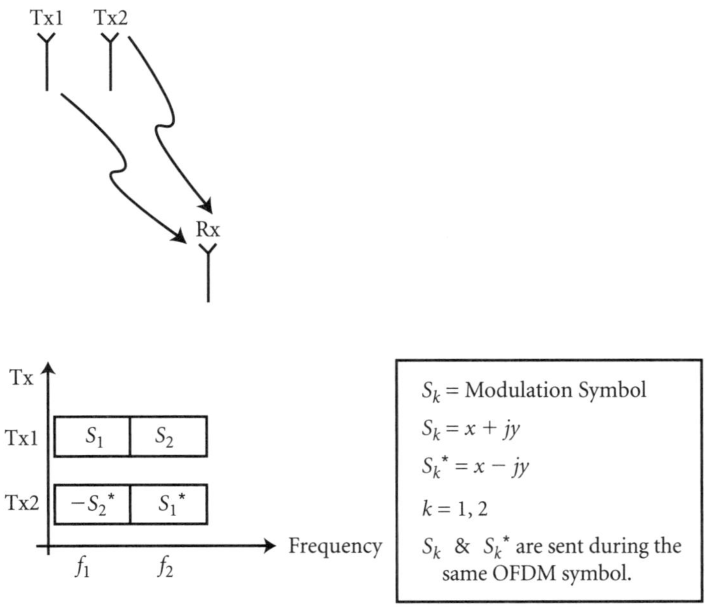

Figure 15.15: Space Frequency Block Coding

15.4.2 Spatial Multiplexing

Though diversity techniques require *either* multiple transmit antennas *or* multiple receive antennas (go back to the beginning of Section 15.4 if this point went unnoticed!), spatial multiplexing requires multiple antennas at both the transmitter and the receiver. MIMO is often used to imply spatial multiplexing.[527] Multiple-input refers to the use of multiple transmit antennas, and multiple-output refers to the use of multiple receive antennas. 2×4 MIMO for the DL means that we have two transmit antennas at the base station and four receive antennas at the mobile.

Figure 15.16 depicts how a 2×2 MIMO works. The transmitter sends two modulation symbols, A from Tx1 and B from Tx2, using the same subcarrier during a given OFDMA symbol period. Two receive antennas, Rx1 and Rx2, are used to obtain the received signal. The radio environment consists of four distinct propagation paths. In our example, the path from Tx1 to Rx1 and from Tx1 to Rx2 makes no change to the phase of the signal transmitted from Tx1. The signal transmitted from Tx2 is different. The Tx2 to Rx1 path makes no alteration to the phase of the transmitted signal, but the Tx2 to Rx2 path causes a 180° phase shift.

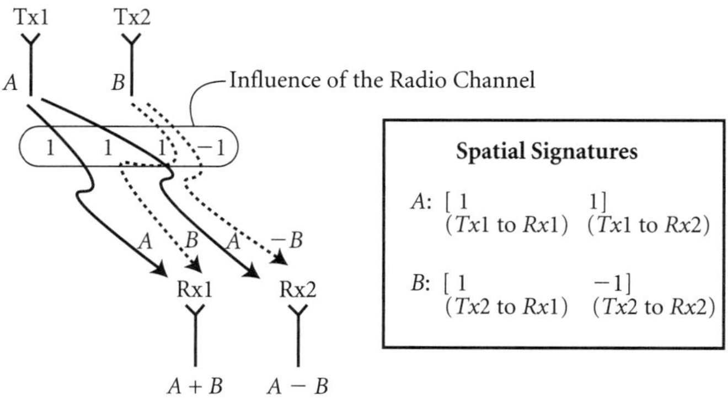

Information Retrieval

For A: $(Rx\ Signal) * [1 \quad 1]^T = [A + B \quad A - B]\ [1 \quad 1]^T = (A + B) + (A - B) = 2A$

For B: $(Rx\ Signal) * [1 \quad -1]^T = [A + B \quad A - B]\ [1 \quad -1]^T = (A + B) + (-A + B) = 2B$

Figure 15.16: MIMO Spatial Multiplexing

[527]The term *MIMO* may be used differently by different technologies. WiMAX uses MIMO as a generic term that may actually mean transmit diversity or spatial multiplexing. LTE uses the term *Single-User–MIMO* to refer to spatial multiplexing.

Since both A and B are being sent on the same subcarrier, Rx1 receives $A+B$, and Rx2 receives $A-B$. The influence of the radio environment on the transmitted signals is quantified in the form of the *spatial signature* [356], $[1\ 1]$ for Tx1 to Rx1 and Tx2 to Rx2 paths and $[1\ -1]$ for Tx2 to Rx1 and Tx2 to Rx2 paths, respectively. Assume that the receiver has perfect knowledge of the propagation channel (of course, in practice, the receiver would need to estimate the spatial signatures using the help of the pilot subcarriers!) and thus knows the spatial signatures. To retrieve A, the receiver would multiply the received signal by its spatial signature $[1\ 1]$. In other words, the receiver will calculate $\{[A+B\ A-B][1\ 1]^T = (A+B)+(A-B) = 2A\}$ where T is the transpose operation. Similarly, to retrieve B, the receiver will calculate $\{[A+B\ A-B][1\ -1]^T = (A+B)-(A-B) = 2B\}$.

As long as the receiver can adequately estimate the spatial signatures and as long as these signatures are distinguishable, spatial multiplexing will provide a significant increase in throughput. In theory, a 2×2 MIMO will double the throughput and a 4×4 MIMO will quadruple the throughput. In practice, about 60% to 70% of the theoretical throughput is considered feasible. In our case, spatial signatures are orthogonal. The greater the correlation between two signatures, the less the throughput gain due to interference between the paths.

15.4.3 Beamforming

Beamforming is yet another advanced antenna technique. Nearly everyone has heard of *smart antennas*, which focus energy in a given direction by forming a beam as opposed to spreading energy all over the place.

Something to Think About... The term *smart antennas* is a misnomer; the antennas are never "smart!" The antennas are passive elements that only transmit and receive energy. However, the *software* that manipulates the antenna signals enables the system to focus energy in a given direction, imparting intelligence to the antennas, but the antennas get all the credit!

In Figure 15.17, the transmitter has two antennas and the receiver has one antenna. User 1 experiences an X° phase shift difference between the two transmitted signals due to the structure of the transmitted signals. When these signals reach the receive antenna, they have identical phase, and the automatic addition of in-phase signals at the antenna increases the signal power. If two signals are received with each signal having an amplitude of N, the cumulative signal will have the amplitude $N+N = 2N$. Since power is proportional to the square of the amplitude, the receiver experiences a fourfold increase in power compared to the single-antenna case. For User 2, the signals transmitted from Tx1 and Tx2 have no phase difference, and, again, the automatic addition of the signals at the receive antenna increases the overall signal power.

In summary, suitable phase shifts between the transmitted signals form a beam with focused energy in a given direction. The only trick is to determine these *suitable*

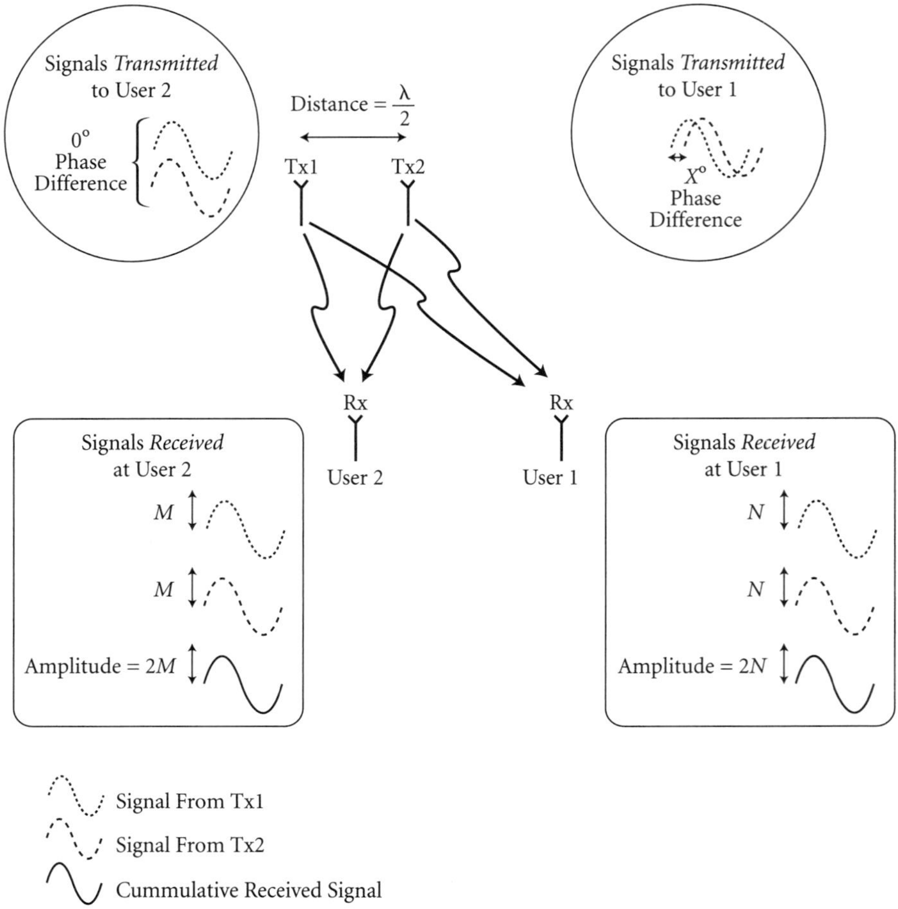

Figure 15.17: Beamforming

phase shifts. If incorrect phase shifts are chosen, the signals may arrive at the receive antenna out-of-phase, completely wiping out the received signal! In practice, the feedback from the receiver to the transmitter on *correct* amounts of phase shifts facilitates the implementation of beamforming. Transmit beamforming involves forming a beam using suitable phase shifts at different transmit antennas. Receive beamforming can also be implemented where just one transmit antenna exists, but the receiver manipulates the phases of the signals received on different receive antennas.

15.4.4 Space-Division Multiple Access

SDMA is the combination of spatial multiplexing with the concept of beamforming. Figure 15.18 contrasts a traditional non-SDMA system with an SDMA system.

In a traditional non-SDMA system, two users in a sector or cell are allocated different subcarriers such as f_1 through f_M for User 1 and f_{M+1} through f_N for User 2. Since the subcarriers are orthogonal to one another, there is (ideally) no interference within the sector between the two users. In an SDMA system, two users are allocated the same set of subcarriers such as f_1 through f_N. Since each user is getting more subcarriers compared to a non-SDMA system, user throughput and sector throughput increase.

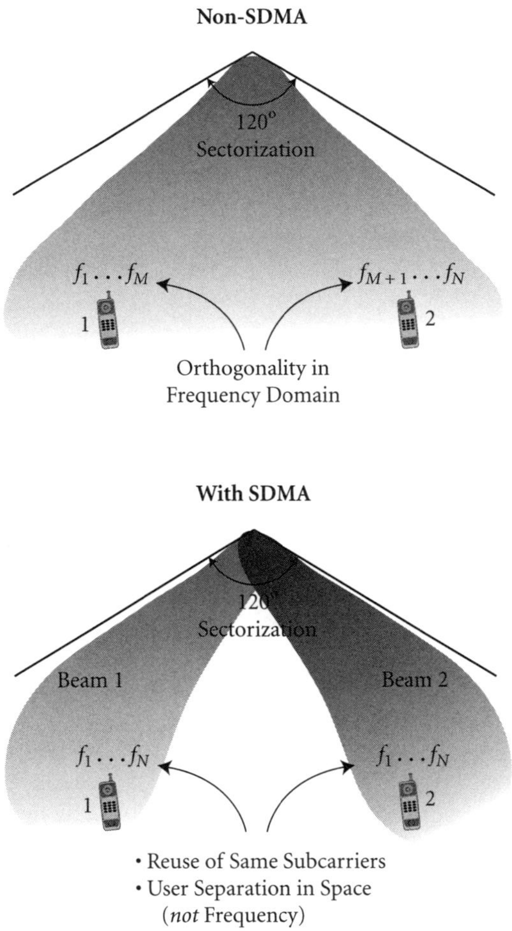

Figure 15.18: Space-Division Multiple Access

The only issue that we need to resolve is how to separate these two users. We'll separate the users spatially by forming beams, one beam for each user. We use *space* as a dimension to allow multiple users to access the same resources, and thus this technique is called *space*-division multiple access. As long as the beams are constructed properly, this technique works well. If the beams are constructed improperly, significant interference can occur due to the reuse of subcarriers within the sector. LTE also calls SDMA *multi-user–MIMO* (MU-MIMO), and SDMA can be used in the DL and UL. WiMAX refers to its UL SDMA implementation as *collaborative spatial multiplexing*.

The various antenna techniques yield different benefits and are suitable under different channel conditions. Table 15.1 summarizes the main benefits and usage scenarios for various antenna techniques.

Diversity techniques are used for reliability, and MIMO is used for throughput. Beamforming in a non-SDMA case is used for larger coverage or higher cell-edge data rates for the same coverage as a non-beamforming case. Beamforming in an SDMA case increases the capacity. Beamforming relies on phase adjustments and is thus unsuitable for an area with too many multipaths. Transmit and receive diversity techniques in the DL will improve reliability throughout the cell coverage area. Receive diversity in the UL helps the link budget as well, increasing the cell radius.

Table 15.1: Decisions, Decisions... Where to Use Which Antenna Technique?

Antenna Technique	Primary Benefit	Usage Scenario
Transmit Diversity (STC, SFBC)	Increased Reliability	Any radio environment throughout the cell coverage area
Receive Diversity	Increased Reliability, Better link budget for diversity in the UL	More beneficial for cell-edge users
Spatial Multiplexing	Significant increase in peak data rates	DL implementation, good channel conditions (e.g., near cell site), multipath-rich environment (e.g., urban area)
Beamforming	Significant increase in coverage or cell-edge data rates	Rural environment with few multipaths
SDMA	Increase in throughput & coverage	Primary means of increasing UL throughput, areas with few multipaths

Since MIMO requires good estimation of spatial signatures, high signal-to-noise-plus-interference ratio is required. Furthermore, spatial signatures must be distinguishable, and, hence, a multipath-rich scattering environment is needed.

15.5 WiMAX OVERVIEW

WiMAX is an IEEE 802.16-based broadband wireless access system. There are two main flavors of WiMAX, fixed WiMAX and mobile WiMAX. Fixed WiMAX is based on the IEEE 802.16-2004 standard, and mobile WiMAX is based on the IEEE 802.16e-2005 standard. As the name suggests, fixed WiMAX offers no support for mobility. We will focus on mobile WiMAX. From this point on, we will use the term WiMAX to refer to mobile WiMAX. Section 15.5.1 provides a succinct history of WiMAX and mentions its salient features. The network architecture of WiMAX is discussed in Section 15.5.2, and the air-interface protocol stack defined for WiMAX is described in Section 15.5.3. Finally, Section 15.5.4 summarizes various operations, such as data session setup.

A Bit of Cellular History. . . WiMAX has not evolved from any pre-existing cellular system, so how did it come about? Well, rumor has it that Intel wanted an efficient wireless Internet access technology to be built into their laptops and initially explored the use of CDMA technology. The CDMA path led Intel to Qualcomm. Intel and Qualcomm disagreed on royalty terms, so Intel started looking for an alternative to CDMA; meanwhile OFDMA-based IEEE 802.16 was waiting for a savior. Intel's heavy research and development investments into IEEE 802.16 energized the standardization efforts, and WiMAX became a formidable 4G technology!

15.5.1 Salient Features of WiMAX

Two entities, IEEE and the WiMAX Forum, play a vital role in the design and implementation of a WiMAX system. IEEE focuses on the air interface between the mobile station and the base station and provides 802.16 specifications. The IEEE 802.16 group's stated goal was to make broadband wireless easily accessible. For example, 802.16e-2005 defines the air-interface protocol stack that consists of the physical layer and the MAC (Medium Access Control) layer. IEEE 802.16e-2005 also describes the process for handover from one BS to another BS.

The WiMAX Forum defines an end-to-end architecture that includes a radio network called an ASN (Access Service Network) and a core network called a CSN (Connectivity Service Network). The WiMAX Forum also specifies profiles that include a list of features for the mobile station (MS) and the BS to ensure interworking between the two.

The basic IEEE 802.16e-2005 is quite flexible and gives numerous choices, and, hence, the WiMAX Forum short-lists features to a manageable and limited list (in the form of a system profile) to ensure common ground between the BS and the

MS [357]. The WiMAX Forum defines certification profiles and also designates labs that test the MS and the BS and provide the WiMAX certification.[528]

The main advantage of WiMAX over the competing LTE technology was time-to-market. Initial LTE deployments occurred in 2010 in Europe, and the first WiMAX deployment occurred in 2006 in South Korea.[529] Both WiMAX and LTE offer a comprehensive ecosystem, leading to lower MS and BS cost. WiMAX's core network, however, can reuse many off-the-shelf IP-based components (rather than WiMAX-specific components) whereas LTE requires many LTE-specific components within the core network.

Figure 15.19 summarizes the main features of WiMAX. The air interface supports both FDD and TDD. Early commercial deployments have used TDD. Numerous channel bandwidths are allowed, e.g., 3.5 MHz, 5 MHz, 7 MHz, 8.75 MHz, 10 MHz,

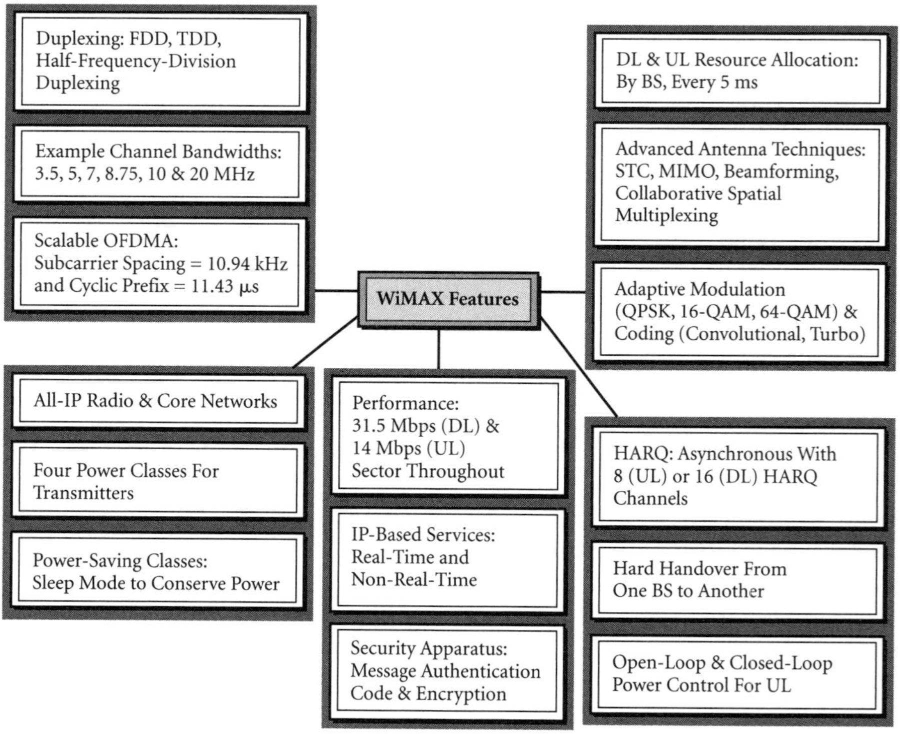

Figure 15.19: Salient Features of WiMAX

[528]See `http://www.wimaxforum.org` for the latest in the world of WiMAX.

[529]WiMAX service is offered under the brand name WiBro (Wireless Broadband) in South Korea.

and 20 MHz. Both the DL and the UL use scalable OFDMA with the typical subcarrier spacing of 10.94 kHz and the CP of 11.43 μs.

The basic timing unit for data transmission is a 5-ms frame, allowing resource allocation, and, hence, the data rate to change as fast as every 5 ms. The BS allocates DL and UL resources to the MS. Various advanced antenna techniques, such as STC, MIMO, beamforming, and collaborative spatial multiplexing, are supported. Adaptive modulation and coding is included. The modulation schemes are QPSK, 16-QAM, and 64-QAM for UL and DL, with 64-QAM optional in the UL. Convolutional coding and turbo coding are available.

Chase combining[530] is the required HARQ technique, and the number of HARQ processes or channels is negotiated between the BS and the MS. Based on to which of the four HARQ categories the MS belongs, the DL can have up to sixteen HARQ channels and the UL can have up to eight HARQ channels [357].[531] Asynchronous HARQ is used in both the DL and UL. Hard handover from one BS to another BS is supported. The MS makes measurements of neighboring BSs based on the configuration the serving BS sets up and provides reports to the serving BS. These measurements may be of the frame preambles and/or pilot subcarriers in a certain time zone of the frame. The serving BS instructs the MS to carry out the handover. Soft or softer handover (also called *macro diversity handover*) and fast BS switching are unsupported. Both open-loop and closed-loop power-control schemes are supported for the UL with the fastest power-control corresponding to one power-control command every 5-ms frame.

In the case of a 10 MHz channel bandwidth shared equally between the UL and the DL in a TDD fashion, sector throughput of 31.5 Mbps and 14 Mbps is feasible in the DL and the UL, respectively [358]. A variety of IP-based services, including VoIP, can be offered using five data delivery services: Unsolicited Grant Service, Real-Time Variable Rate, Non-Real-Time Variable Rate, Best Effort, and Extended Real-Time Variable Rate. Examples of security mechanisms include message authentication code and encryption using 128-bit Advanced Encryption Standard.

All-IP radio and core networks are defined. The radio network may have just the BSs or the BSs and a gateway with possible inter-BS communication. The core network primarily relies on traditional IP nodes, such as the home agent, authentication, authorization, and accounting (AAA) server,[532] and domain name system server. Four power classes are defined to specify the maximum transmit power of the transmitter, $P_{t,max}$. For example, power class 1 satisfies 20 dBm $\leq P_{t,max} <$ 23 dBm for QPSK modulation, and power class 2 satisfies 23 dBm $\leq P_{t,max} <$ 27 dBm. When the MS has a dedicated radio connection with the BS, it may be actively transferring data only some of the time. Power-saving classes are defined to conserve power during the data inactivity period.

[530] See Section 11.4.4 in Chapter 11 for details of Chase combining.

[531] See Chapter 6 on GPRS and EDGE for HARQ basics, Chapter 10 for synchronous HARQ specifics in 1xEV-DO, and Chapter 11 for asynchronous HARQ specifics in HSDPA.

[532] AAA is pronounced "triple A."

15.5.2 WiMAX Network Architecture

A question is needed to make sure you are alive and well...

Do You Recall... Who do you think defines the end-to-end network architecture for WiMAX—IEEE or the WiMAX Forum?[533]

Figure 15.20 illustrates the overall end-to-end architecture of a WiMAX network [359]. The overall network consists of an ASN and a CSN. A CSN can be connected to networks such as the Internet, an Application Service Provider network that provides specific services, such as video streaming and VoIP, or a full-fledged IMS.

Let's first look at an ASN and its various interfaces, and then we'll discuss the CSN. An ASN includes two major logical nodes, the BS and ASN gateway (ASN-GW). One logical BS takes care of one carrier frequency assignment in one sector, and physical implementation of a BS may include multiple (e.g., three for three sectors) logical BSs. The interface between two logical entities is called a *reference*

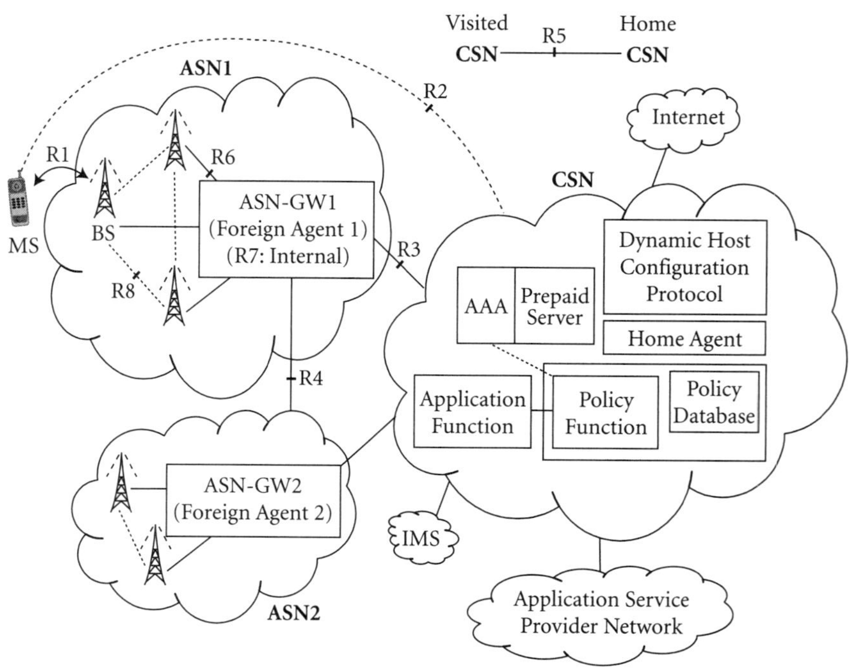

Figure 15.20: WiMAX Network Architecture

[533]The WiMAX Forum specifies the end-to-end network architecture.

point. Reference Point R1 is between the MS and the BS and is the air interface. IEEE 802.16e-2005 fully specifies the operations on R1. The logical link between the MS and the CSN is R2, which helps functions such as the network's authentication of the MS. R3 occurs between the ASN and the CSN and is the bridge between the radio network and the core network for signaling and data traffic.

Two ASNs are interconnected via R4 to support mobility across ASNs. If the MS is roaming, the visited CSN and home CSN communicate via R5 to verify the support for roaming and to do proper accounting of charges. R6 enables the BS and the ASN-GW to exchange signaling and traffic. For a given MS, the BS is connected to one ASN-GW. However, one BS can be connected to multiple ASN-GWs for functions such as load balancing and redundancy.

R7 is an internal reference point of the ASN-GW. *If* the ASN-GW is decomposed into two entities, a Decision Point and an Enforcement Point, R7 exists between these two entities. The Decision Point implements control plane functions, such as a AAA client for communication with a AAA server to facilitate authentication. The Enforcement Point handles the bearer plane functions, such as routing of IP packets between the BS and the CSN.

Two BSs can talk to each other via the R8 interface, which is new in 4G systems. R8 enables the exchange of signaling messages for faster and seamless handover and can be used (optionally) to carry data traffic.

The basic function of the BS is to communicate with the MS over the air interface (no surprise here!), and the example tasks of the ASN-GW are to act as a Foreign Agent to support mobile IP and to connect to the CSN so that the MS can access IP-based services. The three ASN profiles define the division of labor between the BS and the ASN-GW and specify which interfaces are exposed to conform to equipment interoperability requirements. Two different ASN profiles can work together using suitable exposed reference points. ASN Profile A resembles the centralized radio network architecture of 3G systems and allows the ASN-GW control. For example, the ASN-GW implements Radio Resource Control (RRC) to manage the radio resources among multiple BSs. The ASN-GW also controls the handover. The main features of Profile B are unexposed R6 and the possible implementation of BS and ASN-GW functions in a single (physical) entity. ASN Profile C lets the BS control handover and manage its own radio resources. Furthermore, it exposes the R6 interface to enable the BS and ASN-GW to exist as distinct identities.

Let's turn our focus to the CSN. As we mentioned earlier, CSN attempts to reuse many of the typical IP components. The AAA server authenticates the user with the help of the ASN-GW. The AAA server can obtain user data records from the Decision Point function of the ASN-GW to take care of accounting. The prepaid server is co-located with the AAA server to support prepaid services. The Home Agent helps implement mobile IP and allocates an IP address to the MS for client-based mobile IP. A Dynamic Host Configuration Protocol server can be used to allocate an IP address to the MS in the case of fixed or proxy-mobile IP access. The MS itself understands mobile IP (MIP) and can exchange MIP messages with the Foreign Agent and Home Agent in the case of client-based mobile IP. In proxy-mobile IP, the MS itself has no role to play in IP mobility; the network has a proxy server that acts on behalf of the MS to work with the Foreign Agent and the Home Agent to obtain an IP address for the MS.

The application function, policy function, and policy database enable implementation of the quality-of-service (QoS) for a special IP-based service. In such a case, the MS exchanges application layer messages with the application function, which queries the policy function to determine if the service is approved. The policy function consults the policy database and determines whether the user is allowed such service with the requested QoS parameters (e.g., a peak data rate of 1 Mbps). The policy function works with the ASN-GW to initiate the addition of a WiMAX service flow to support the requested service. The AAA server may also have some pre-provisioned service flows that can be downloaded to the ASN-GW during the authentication process.

15.5.3 WiMAX Air Interface

The MS and the BS communicate over the air interface using the protocol stack shown in Figure 15.21.[534] The protocol stack consists of just two layers, the physical layer and the MAC layer. The physical layer implements OFDMA transmitter and receiver functions, such as channel coding and decoding, modulation and demodulation, IFFT and FFT, Discrete Fourier Transform and Inverse Discrete Fourier Transform, addition and removal of cyclic prefix, filtering, and frequency upconversion and downconversion.

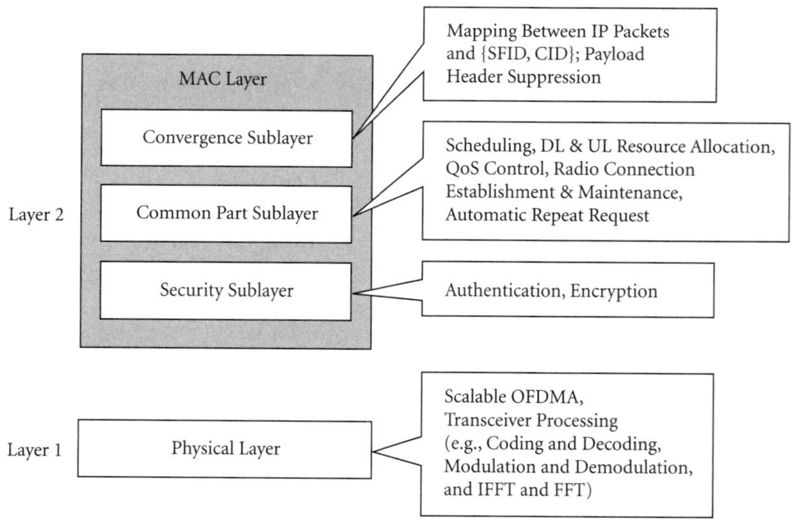

Figure 15.21: WiMAX Air-Interface Protocol Stack

[534]See the standard [360] for additional details.

Do You Recall. . . Do you remember which of these functions involves the combining of modulation symbols being carried on orthogonal subcarriers into a series of samples for (eventual) transmission?[535]

The MAC layer consists of three sublayers, the convergence sublayer, the common part sublayer, and the security sublayer. The MAC convergence sublayer maps an upper-layer data unit, such as an IP version 4 packet, onto a specific service flow identifier (SFID) and transport connection identifier (CID). When a new application or service is added for a given MS, the ASN-GW assigns an SFID and the BS needs to allocate a transport CID. The SFID facilitates implementation of QoS because the SFID is associated with a set of QoS parameters, and the CID uniquely identifies the MS (and a specific IP flow) over the air interface. The CID is valid only within one sector or BS. One-to-one mapping occurs between the SFID and the CID. The MAC convergence sublayer can also perform payload header suppression, such as robust header compression, for an IP header.

The MAC common part sublayer is where intelligence resides! The MAC common part sublayer carries out user scheduling and allocates resources to the MSs for both the UL and DL. It is also responsible for providing QoS over the air interface. The MAC common part sublayer constructs and interprets signaling messages, such as those associated with the establishment and teardown of radio connections. When the physical layer HARQ fails, the MAC common part sublayer can attempt to rectify the packet error by implementing Automatic Repeat Request.

The MAC security sublayer helps secure the air interface by including cipher-based message authentication code in signaling messages and by encrypting information using an algorithm such as 128-bit Advanced Encryption Standard. The security sublayer also participates in the authentication process.

15.5.4 WiMAX Operations

Let's take a look at what happens after the MS's power is turned on. The MS carries out initialization and network entry as summarized in Figure 15.22. In Step 1, the MS scans the WiMAX signal in the DL at a set of carrier frequencies based on its configured list. At a given carrier frequency, it looks for a match with one of the 114 preamble sequences defined in the standard. The BS sends such a sequence on every third subcarrier in the first OFDMA symbol of each 5-ms frame. Different BSs send out different preamble sequences within a local geographical area.

Once the MS finds a good match with a preamble sequence, it has already achieved frame (timing) synchronization and (coarse) frequency synchronization. The MS next processes the Frame Control Header that occupies predefined bandwidth in the next two (i.e., second and third) OFDMA symbols in the frame. The Frame Control Header informs the MS about the subchannels being used in the DL of the BS and the length of

[535] IFFT at the transmitter combines the modulation symbols on subcarriers into a time-domain waveform for transmission during the OFDMA symbol duration.

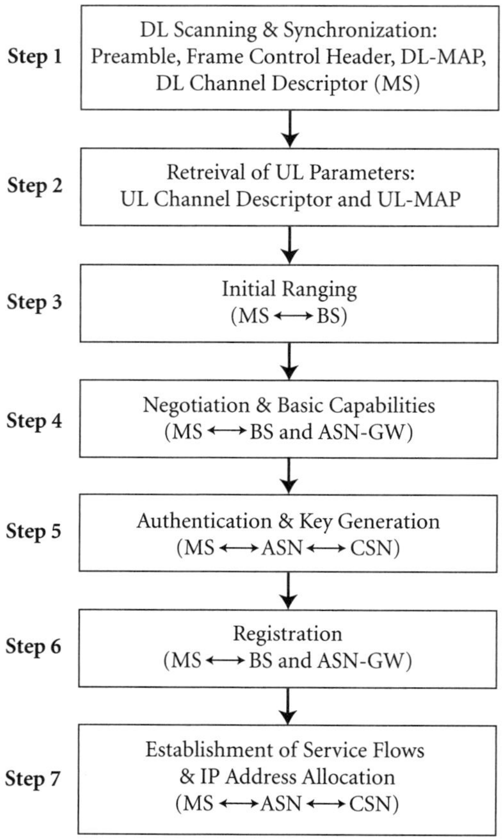

Figure 15.22: Network Entry upon MS Power-Up

a message called DL-MAP. A subchannel consists of a certain number of subcarriers (e.g., twenty-four subcarriers). The BS may or may not use all the subchannels of the channel bandwidth in the sector based on network planning.

The DL-MAP message is a key overhead message that allocates DL resources to the MSs and that specifies locations of other overhead messages, such as UL-MAP, DL Channel Descriptor, and UL Channel Descriptor. The Frame Control Header may, for example, indicate that the DL-MAP occupies all the subchannels that the Frame Control Header doesn't consume in the second and third OFDMA symbols in a given frame. The MS can now process DL-MAP and know the location (i.e., subchannels and symbol numbers) of the DL Channel Descriptor. The DL Channel Descriptor specifies the physical layer characteristics of the DL, such as the supported burst profiles (e.g., one profile being rate 3/4 turbo coding and 16-QAM modulation and another profile being rate 1/2 convolutional coding and QPSK modulation).

Once the MS learns about the DL parameters, it proceeds to learn the UL parameters in Step 2 by processing UL Channel Descriptor and UL-MAP messages. Just as the DL Channel Descriptor specifies DL characteristics, the UL Channel Descriptor specifies the UL characteristics. For example, the UL Channel Descriptor provides information on the subchannels used in the UL, *ranging codes* (i.e., the code that the MS uses to contact the network when it has no dedicated UL resources) available in the cell, and the supported UL burst profiles. The UL-MAP indicates UL resource allocation for user traffic and subchannels that constitute a ranging channel. Note that the MS uses the subchannels corresponding to the ranging channel to send a ranging code.

In Step 3, since the MS knows about the UL transmission parameters, it is able to perform *initial ranging*, a process whereby the MS uses a ranging code on the ranging channel to contact the BS and get adjustments in time, frequency, and transmit power. The MS then becomes in sync with other MSs in the BS. The MS also receives two CIDs, basic and primary management CIDs, from the BS. These CIDs help the MS exchange signaling messages with the BS using dedicated resources.

In Step 4, the MS and the BS exchange information regarding their capabilities and find common ground. For example, the MS may be supporting QPSK and 16-QAM, and the BS may be supporting QPSK, 16-QAM, and 64-QAM in the UL. The BS works with the ASN-GW to create a new context for the MS. The authorization policy (e.g., the Rivest-Shamir-Adelman–based or the Extensible-Authentication-Protocol–based approach) the ASN-GW and the MS support must be reconciled. The Extensible Authentication Protocol is mandatory in Release 1 of WiMAX.

In Step 5, mutual authentication occurs in which the MS and the AAA server authenticate each other. At the end of Step 5, several keys are generated, which are later used to provide security (e.g., the Traffic Encryption Key to encrypt traffic packets). Step 6 is registration, during which the MS and the BS agree on additional parameters, such as the use of hard handover. The BS informs the ASN-GW about MS registration so that ASN-GW can update its database regarding the MS. Next, the ASN-GW acts on the MS's QoS profile obtained from the AAA server and works with the BS in Step 7 to create the pre-provisioned service flows. The ASN-GW allocates SFIDs and the BS allocates CIDs for the service flows. If the MS supports client mobile IP, it exchanges MIP registration messages with the ASN-GW and the Home Agent to get an IP address. The MS may use a secondary management CID (absent from Release 1 of WiMAX) or a transport CID corresponding to some initial or default service flow to exchange mobile IP messages over the air. The MS may also use Dynamic Host Configuration Protocol to obtain an IP address. The only thing that the MS needs now is bandwidth to send and receive packets.

Figure 15.23 summarizes the main steps of the DL data transfer with the assumption that required service flows have been established and are associated with SFIDs and CIDs.[536] Step 1 involves quantifying the DL channel conditions. The MS estimates the carrier-to-interference-plus-noise ratio using the preamble or the pilot

[536]See the standard [360] for additional details.

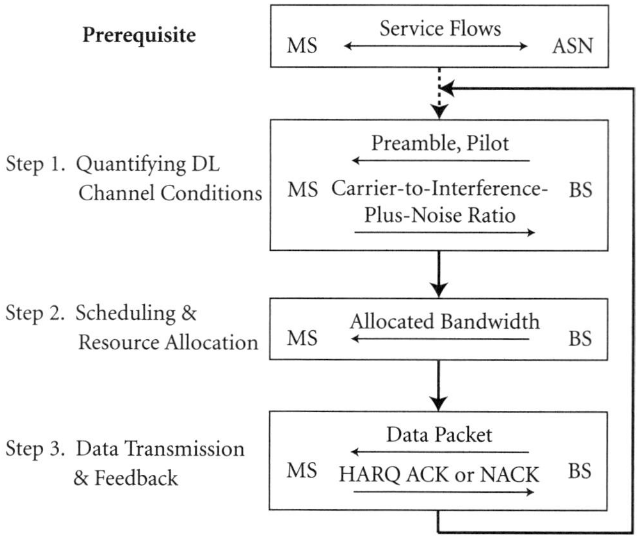

Figure 15.23: DL Data Transfer: An Overview

subcarriers during a certain part of the frame based on the configuration of the BS. The MS provides the measured carrier-to-interference-plus-noise ratio to the BS.

Step 2 involves the scheduling algorithm at the BS. Example inputs to the implementation-specific scheduling algorithm include the required QoS for the SFID, amount of service flow data in the BS buffer, availability of resources, number of MSs, history of data transmission to the MSs, past HARQ feedback from the MSs, and reported carrier-to-interference-plus-noise ratios. The scheduling algorithm determines the number of subchannels and the number of OFDMA symbol periods to be allocated to each MS chosen for DL transmission. The scheduler also decides the combination of the type and amount of coding and the modulation scheme for each chosen MS from the DL burst profiles advertised in the DL Channel Descriptor message. The DL-MAP specifies the resource allocation for the MS per CID. *The resource allocation in WiMAX is non-persistent or non-sticky*, meaning that it is valid for a specific 5-ms frame only. Once the frame is over, the resources are no longer available for that MS; the BS must re-allocate the resources for another transmission. Furthermore, since the DL-MAP can be sent as fast as every 5-ms frame, the data rate for an MS can change as quickly as every 5 ms.

In Step 3, the BS sends a packet to the MS on the allocated resources, and the MS responds with an HARQ acknowledgment (ACK) or negative acknowledgment (NACK). The BS retransmits the packet if it receives a NACK. If the BS receives an HARQ ACK, it can send a new packet. The BS decides the maximum number of retransmissions; the default number is four.

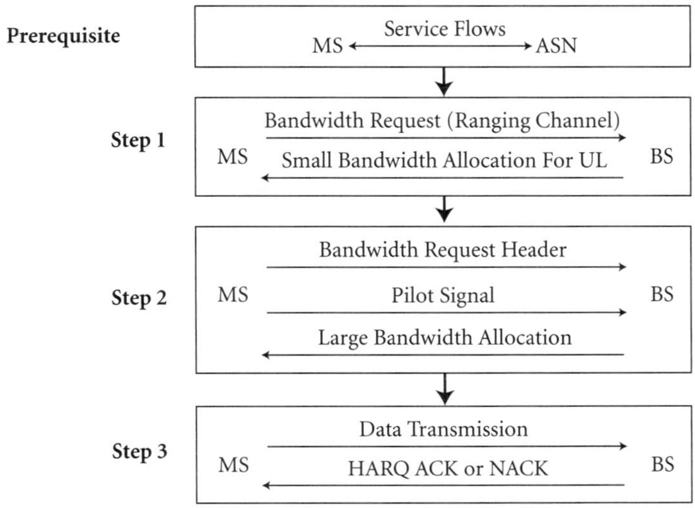

Figure 15.24: UL Data Transfer: An Overview

Now let's turn our attention to UL data transfer, which is shown in Figure 15.24, and, again, the service flows are assumed to be already established. Depending on the data delivery service (e.g., unsolicited grant versus non-real-time variable rate), slight variations occur in the overall mechanism.

The first step involves letting the BS know that new application data has arrived in the buffer at the MS. The MS sends a bandwidth request ranging code on the ranging channel. The BS then allocates adequate bandwidth in Step 1 to enable the MS to send a Bandwidth Request Header in Step 2, which informs the BS of the amount of data the MS has in the buffer. Since the pilot subcarriers are transmitted along with the data subcarriers carrying the header, the BS can estimate, to a certain degree, the prevailing UL channel conditions. The BS estimates the carrier-to-interference-plus-noise ratio using the UL pilot subcarriers in Step 2. The BS scheduler considers various inputs, such as the estimated carrier-to-interference-plus-noise ratio, amount of data in the MS's buffer, required QoS for the SFID, availability of UL resources, and number of MSs. The scheduling algorithm then determines the number of subchannels and the number of OFDMA symbol periods to be allocated to each MS chosen for UL transmission. The scheduler also specifies to the MS in Step 2 one of the UL burst profiles advertised in the UL Channel Descriptor message. The UL-MAP specifies the resource allocation for the MS in Step 2, which can be shared among multiple service flows and hence transport CIDs. As in case of the DL, *the resource allocation in the UL is also non-persistent.* The MS sends a packet to the BS in Step 3 using the allocated resources in Step 3, and the BS responds with an HARQ ACK or NACK in Step 3. The MS retransmits the packet if it receives a NACK. If the MS receives an HARQ ACK, it can send a new packet if the resources have been allocated. Just as in the DL scenario, the BS decides the actual number of HARQ retransmissions, and, again, the default is four.

15.6 OVERVIEW OF LTE

LTE is a 3GPP Release 8 feature [361]. Recall that the 3GPP is the organization that developed the UMTS specifications. Typical 3GPP-based commercial systems, such as high-speed downlink packet access and high-speed uplink packet access systems, and some 3GPP2[537] systems, such as 1xRTT and 1xEV-DO, are expected to evolve to LTE. For example, in the U.S., AT&T has deployed high-speed downlink packet access and high-speed uplink packet access and began deploying LTE in 2011, and Verizon Wireless has deployed 1xEV-DO and began deploying LTE in 2010. U.S. public safety entities have chosen LTE technology, and LTE's importance can be inferred from this choice. The Federal Communication Commission (FCC) has designated 758 MHz to 775 MHz and 788 MHz to 805 MHz for public safety entities.

Strictly speaking, the term *LTE* refers to evolution (or should we say *revolution* due to a significant jump in the target performance?) of the air interface and the radio network. The core network evolution is the responsibility of the System Architecture Evolution group of the 3GPP, which has defined the Evolved Packet Core (EPC). However, the term *LTE* is often used to encompass both the radio network and the core network, and we will use LTE that way as well. Section 15.6.1 lists the main features of LTE. The network architecture of LTE is illustrated in Section 15.6.2. The LTE air-interface protocol stack is described in Section 15.6.3. Operations such as initial attach, DL data transfer, and UL data transfer are briefly discussed in Section 15.6.4.[538]

A Funny Detail... What does LTE stand for? Well, the answer depends on your perspective! The formal definition (and one that cannot land you in any controversy) is Long-Term Evolution. Critics of LTE have often called it "Looooooooong-Term Evolution" and "Late-Term Evolution" due to the time the standardization process has taken. One of our colleagues (obviously a fan of WiMAX) even calls it "Long-Time Editing!"

When Dr. Tripathi was teaching LTE, the author mentioned that LTE meant Long-Term Evolution. One student protested and said that it was something else. Since LTE has been a work-in-progress for quite some time, Dr. Tripathi wondered if the 3GPP had redefined LTE. Dr. Tripathi grew concerned; as an LTE (and WiMAX) instructor, he should know the name of the technology! He was quite relieved when the protesting student revealed that LTE stands for "Long-Term Employment!" Who

[537] The 3GPP2 is the Third Generation Partnership Project 2.
[538] See LTE University [362] for numerous references on LTE.

would not appreciate this definition of LTE, especially during the time of a global economic slowdown?!

15.6.1 Salient Features of LTE

Figure 15.25 illustrates the basic features of LTE. You will observe quite a few similarities between LTE and WiMAX; despite their raging battle for dominance, they share more similarities than differences. The LTE air interface supports both FDD and TDD, and initial commercial deployments use FDD.

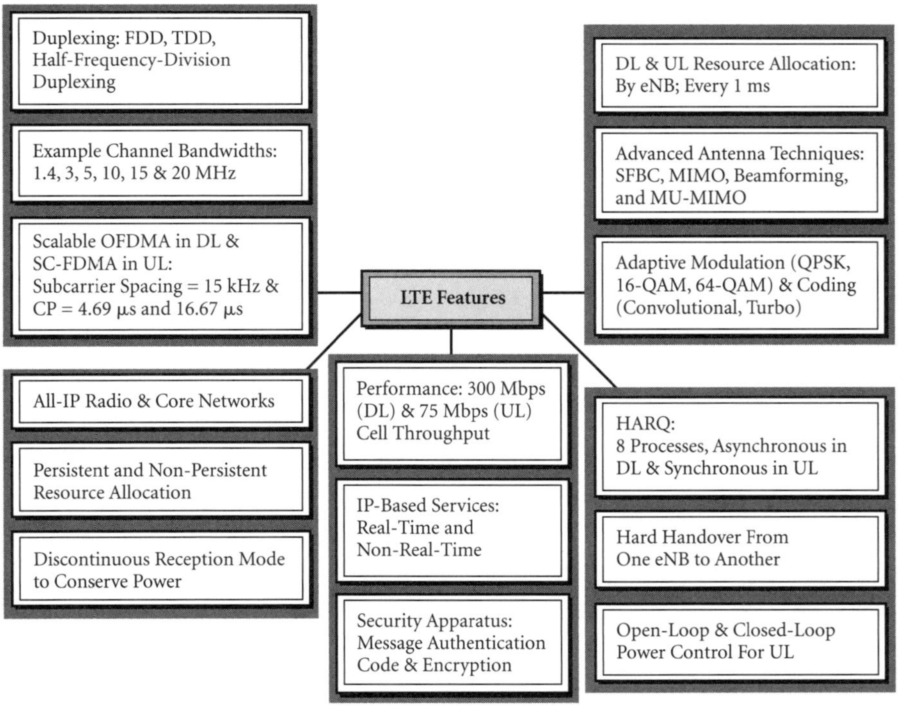

Figure 15.25: Salient Features of LTE

Do You Recall... Do initial WiMAX deployments focus on TDD or FDD?[539]

The channel bandwidths supported are 1.4 MHz, 3 MHz, 5 MHz, 10 MHz, 15 MHz, and 20 MHz.[540] For multiple access techniques, the DL uses scalable OFDMA and the UL uses an OFDMA with a twist, SC-FDMA. The typical subcarrier spacing is 15 kHz, and two options for the CP are 4.69 μs and 16.67 μs.

Do You Recall... Do you recall the main benefit and drawback of SC-FDMA?[541]

Something to Think About... Consider the output of the IFFT in an OFDMA transmitter. Recall that a subcarrier carries one modulation symbol during the useful symbol period. The IFFT combines multiple subcarriers, thereby combining multiple modulation symbols. During a given OFDMA symbol, the IFFT generates a certain number of samples (e.g., 1024). Now focus on a given sample or time instant: all the modulation symbols are present at the output of the IFFT. Such combining of modulation symbols results in a high PAPR for OFDMA. Let's contrast SC-FDMA with OFDMA. SC-FDMA first passes the modulation symbols through Discrete Fourier Transform (DFT), which yields DFT symbols. One subcarrier carries one DFT symbol, and the IFFT combines all the subcarriers carrying DFT symbols. Now let's again focus on a sample coming from the IFFT: the sample is influenced by a single modulation symbol! Since DFT and IFFT are inverse operations, we are effectively getting the modulation symbols sequentially at the IFFT output. Since a single modulation symbol affects the IFFT output sample in SC-FDMA, SC-FDMA reduces the PAPR. SC-FDMA can, thus, be viewed as "OFDMA with a twist," where the modulation symbols appear to be transmitted in sequence rather than in parallel within the symbol period. LTE uses SC-FDMA in the uplink for enhanced performance at the expense of increased complexity for the SC-FDMSA transmitter at the user equipment (UE) and the SC-FDMA receiver at the eNodeB. See Rumney's work [364] and Myung's work [365] for more SC-FDMA details.

[539]Initial WiMAX deployments focus on TDD primarily due to the availability of the type of spectrum in South Korea and the U.S. In the U.S., Sprint-Nextel made a decision to deploy WiMAX-TDD under the brand name XOHM due to the nature of their spectrum. Later, Clearwire assumed the responsibility of WiMAX deployments from Sprint-Nextel under the brand name Clear.

[540]LTE supports numerous frequency bands, including 700 MHz and Advanced Wireless Service (AWS) bands [363]. The advantage here is the flexibility of deployment, but the disadvantage is the challenge in designing the UEs (user equipments) that operate at multiple frequency bands to achieve economies-of-scale.

[541]SC-FDMA reduces PAPR compared to OFDMA, and this reduction can result in benefits such as larger cell radius or increased UL throughput near the cell edge. The main drawback of SC-FDMA is increased complexity of the transmitter and the receiver.

The basic timing unit for data transmission is a 1-ms subframe, enabling the data rate to potentially change every 1 ms. The BS is formally called the eNB (evolved Node B[542]). The eNB allocates both DL and UL resources to the UE. Examples of LTE's advanced antenna techniques are SFBC, Single-User–MIMO, and MU-MIMO. Adaptive modulation and coding is supported for the high-speed traffic channels in the UL and the DL with support for QPSK, 16-QAM, 64-QAM, and turbo coding. Five categories of UE are defined with different characteristics and capabilities [366]. For example, only Category 5 supports 64-QAM in the UL and 4×4 DL MIMO.

Chase combining and Incremental Redundancy are supported to execute HARQ. Eight HARQ processes are available. Asynchronous HARQ is used in the DL, and synchronous HARQ is used in the UL. Hard handover from one cell (i.e., sector) to another cell of the same or of a different eNB is supported. The UE makes measurements of neighboring eNBs and provides reports to the serving eNB. These measurements may be the received power of the reference signal or a quantity that is proportional to the ratio of the reference signal power to the total power. The serving eNB makes the handover decision after communicating with the target eNB using an inter–eNB interface. Soft or softer handover and fast sector or cell switching are unsupported. Open-loop and closed-loop power-control schemes are defined to control the UE's transmit power with the fastest power control corresponding to one power-control command every 1-ms subframe.

In the case of a 20 MHz DL, 20 MHz UL, and 4×4 DL MIMO, cell or user throughput of 300 Mbps in the DL and 75 Mbps in the UL is feasible for a Category 5 UE [366]. A variety of IP-based services including VoIP can be offered using one of nine QoS class indicators (QCIs). Examples of security mechanisms include message authentication code (called *integrity protection*) and ciphering using 128-bit Advanced Encryption Standard.

All-IP radio and core networks are defined. The radio network consists of only eNBs. A comprehensive core network provides an efficient packet-switched core network for the LTE and facilitates interworking with non-LTE technologies, such as HSPA and 1xEV-DO. Resource allocation can be non-persistent (like WiMAX) or persistent. Persistent allocation remains valid for a long time period as opposed to a single subframe. The services with fixed- and low-data-rate requirements (e.g., VoIP) can be supported efficiently using persistent allocation since the resource allocation overhead per subframe is reduced. Discontinuous Reception Mode is used so the UE can conserve power during the data inactivity period even when a dedicated radio connection exists.

15.6.2 LTE Network Architecture

LTE's overall network architecture is depicted in Figure 15.26. The network elements and the interfaces shown in Figure 15.26 are logical; in terms of physical implementation, one physical node may implement two logical nodes, and a logical interface

[542]Or the Evolved Universal Terrestrial Radio Access Network (E-UTRAN) Node B, i.e., an acronym within an acronym!

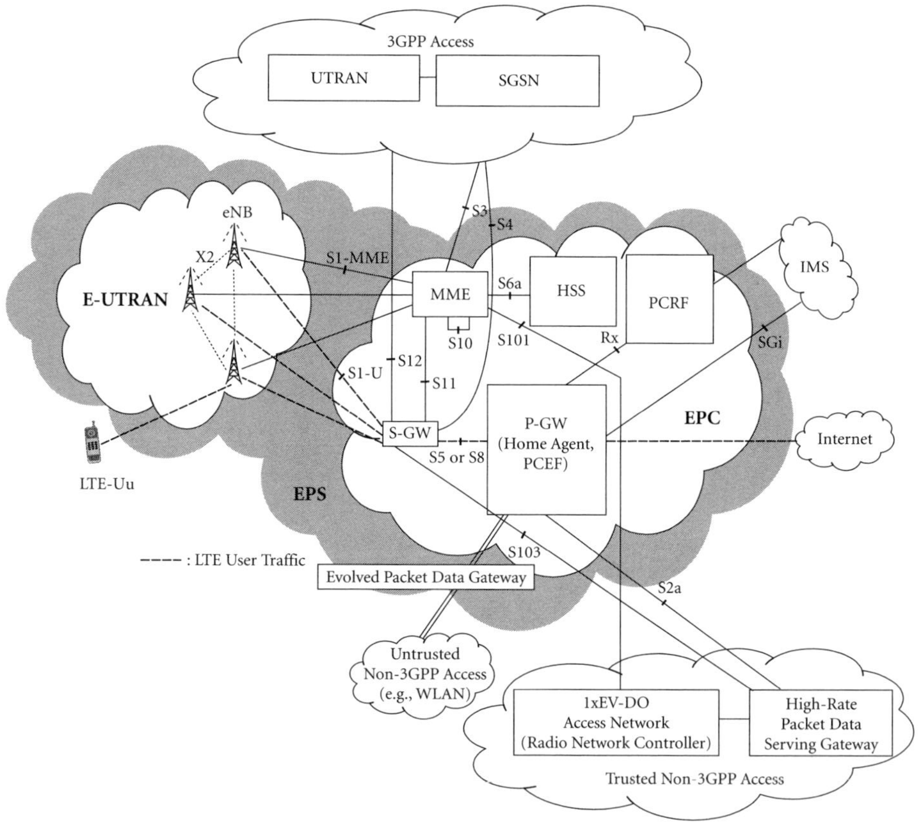

Figure 15.26: LTE Network Architecture

may pass through multiple physical elements (e.g., IP routers). Various interfaces are specified for the sake of curious minds only and are most certainly not intended to intimidate you! The radio access network is called the Evolved Universal Terrestrial Radio Access Network (E-UTRAN), and the packet-switched core network is the EPC. The E-UTRAN and EPC together constitute the EPS (Evolved Packet System), which is an end-to-end infrastructure network.

The E-UTRAN consists of eNBs. One eNB controls one or more (e.g., three) cells (i.e., sectors). eNBs in a given local area communicate with one another via an X2 interface. The connectivity between two network elements would typically be achieved using an internal or private IP network. The eNB and the UE communicate with each other using LTE-Uu, which is simply the air interface.

The eNB connects to the EPC using two distinct interfaces, one for the control plane (i.e., signaling) and one for the user plane. Specifically, the eNB and MME (Mobility Management Entity) connect via the S1-MME interface, which is used to exchange signaling messages, such as those related to authentication and paging. The MME knows the geographical area, called *tracking area*, in which the UE is currently located. Hence, for a UE-terminated session, the MME sends a page message to the eNBs in the tracking area.

Furthermore, during the data session setup, the MME and the UE authenticate each other. To authenticate the UE, the MME talks to the Home Subscriber Server, which stores the secret key (i.e., K) for authentication and other information about the subscriber, such as subscribed QoS. S10 is the interface between two MMEs, allowing transfer of UE information when the UE travels from one MME area to another.

The interface between the eNB and serving gateway (S-GW) is S1-U (S1-user plane). The S-GW is essentially a router that forwards the packets between the appropriate eNB and the P-GW (packet data network—gateway). The P-GW is a gateway to the outside world; IP packets destined for the UE arrive from elsewhere to the P-GW. In the reverse direction, IP packets from the UE leave the EPC from the P-GW for their outside destination.

When both the S-GW and P-GW are in the home network, S5 is the interface between the two; however, when S-GW is in a visited network and P-GW is in the home network, S8 is the interface between the two. The P-GW can also allocate an IP address to the UE. The P-GW acts as a home agent to support mobile IP. The P-GW is equivalent to the Gateway General Packet Radio Service Support Node used in pre-LTE 3GPP releases. The P-GW interfaces an IMS network via SGi to enable the UE to get IMS-based IP services.

The Policy and Charging Rules Function (PCRF) is a component of the Policy and Charging Control architecture used to facilitate implementation of QoS and flow-based charging. The PCRF determines the policy rules (e.g., which IP services are allowed) and provides the rules to the P-GW. The P-GW works as a Policy and Charging Enforcement Function (PCEF) to enforce the rules and is able to do so because the UE's IP packets must pass through the P-GW in both directions.

As far as user traffic is concerned, the P-GW primarily takes care of the interworking between an LTE and a non-LTE system. For example, when IP packets are sent to the hybrid UE that supports LTE and 1xEV-DO using the 1xEV-DO radio network, the packets pass to the P-GW, High-Rate Packet Data S-GW, radio network controller, and the 1xEV-DO base station. Similarly, when the IP packets are to be sent to a UE that is using a WLAN, packets traverse the P-GW, evolved packet data gateway, and WLAN Access Points. The evolved packet data gateway is a gateway used to interface an untrusted network with the P-GW and offers protection for the EPC against hackers who may use an unsecured WLAN air link to access the EPC!

To support interworking between LTE and non-LTE technology, a control plane is also needed. MME exchanges signaling messages with entities such as the 1xEV-DO radio network controller and the Serving General Packet Radio Service Support Node for LTE interworking with 1xEV-DO and pre-LTE UMTS, respectively. The interface between the S-GW and the High-Rate Packet Data Serving Gateway of 1xEV-DO and between the S-GW and the Serving General Packet Radio Service Support Node is used to forward the leftover packets when the UE makes a handover from LTE to pre-LTE UMTS and 1xEV-DO, respectively.

15.6.3 LTE Air Interface

Let's turn our attention to the air-interface protocol stack shown in Figure 15.27. Those familiar with the basic UMTS air-interface protocol stack will find that the LTE stack has a striking resemblance to the UMTS Release 99 protocol stack. The air-interface protocol stack consists of the physical, MAC, Radio Link Control (RLC), Packet Data Convergence Protocol (PDCP), and RRC layers.

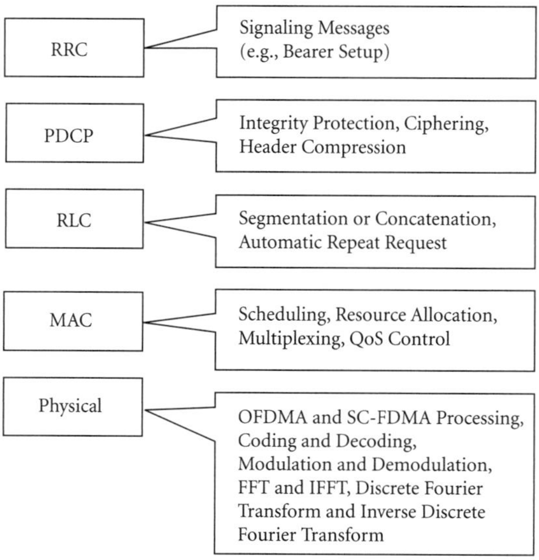

Figure 15.27: LTE Air-Interface Protocol Stack

The physical layer carries out basic OFDMA processing for the DL and SC-FDMA processing for the UL. Examples of transceiver functions implemented at the physical layer include scrambling and descrambling, coding and decoding, modulation and demodulation, FFT and IFFT, Discrete Fourier Transform and Inverse Discrete Fourier Transform, addition and removal of the CP, filtering, and frequency upconversion and downconversion. The MAC layer implements scheduling and allocates resources to meet QoS requirements. It can also multiplex data for different services (e.g., email and web-browsing) into a single packet for the physical layer's further processing.

The RLC layer increases reliability through retransmission as part of the Automatic Repeat Request mechanism. The RLC layer also breaks down a large IP packet into smaller RLC pieces for a low-data-rate transmission and puts together multiple IP packets into a large RLC segment for a high-data-rate transmission. The PDCP layer implements integrity protection for signaling messages and ciphering for signaling messages and user traffic. The PDCP layer can also carry out header compression using robust header compression. Finally, the RRC layer helps construct and interpret signaling messages, so it comes in handy during data session setup.

15.6.4 LTE Operations

Let's see what happens after the UE's power is turned on. Figure 15.28 summarizes the main steps for the UE. The UE looks for the existence of an LTE system in its neighborhood in Step 1. It searches for an LTE signal at certain carrier frequencies based on its configuration [e.g., one stored in the Subscriber Identity Module (SIM)]. At a given carrier frequency, it first tries to find an LTE signal called a *primary synchronization signal*. After detecting the primary synchronization signal, it tries to detect a *secondary synchronization signal*. Once the IE has successfully detected these two signals, it has achieved *time and frequency synchronization* with the DL.

The existence of two synchronization signals facilitates the UE's quest to find an LTE cell. These synchronization signals are sent on sixty-two subcarriers in the center of the channel bandwidth during specific symbols of specific subframes. The by-product of the synchronization operation is implicit determination of the physical layer cell identity of the cell the UE has been observing. Many of the signals the eNB transmits (e.g., reference signal and the broadcast channel) are a function of the cell identity. Hence, if the UE has no knowledge of the cell identifier, it cannot process such signals. In Step 2, the UE learns the most critical *system information* (e.g., the actual channel bandwidth being used in the cell) by observing the broadcast channel that is being transmitted on center 72 subcarriers during specific symbols of specific subframes. By observing a specific part of the subframe the eNB dictates, the UE learns about various types of system information, such as a list of 1xEV-DO or UMTS neighbor cells and information on how to use the UL to contact the E-UTRAN.

The UE is now ready to inform the network about its existence in Step 3 so it can receive page messages in the future. Remember that we recently powered-on the UE; when the UE was powered-off, the system had no idea of the UE's whereabouts! Since the UE has no dedicated radio connection yet, it must use a common channel,

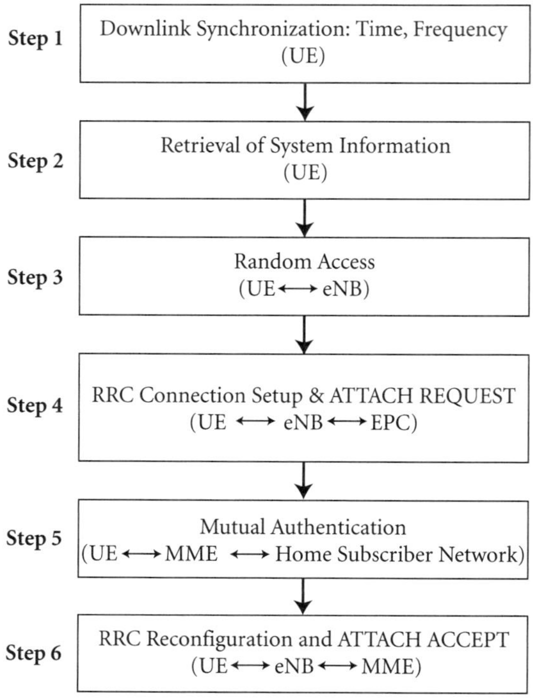

Figure 15.28: Initial Attach Upon UE Power-Up

which is the random access channel, to communicate with the eNB. The UE and the eNB establish an RRC signaling connection in Step 4 so that future signaling can be done reliably via dedicated UL and DL radio channels. The UE next sends an ATTACH REQUEST message to attach itself to the EPC (more specifically, to the MME). The MME works with the Home Subscriber Server to authenticate the UE in Step 5, and the UE also authenticates the MME. In other words, mutual authentication occurs.

The next step is to establish a default EPS bearer, which is a logical link between the UE and the P-GW and includes the LTE-Uu radio bearer, S1-U bearer, and an S5 or S8 bearer, and to allocate an IP address to the UE. The default bearer provides an always-on experience to the UE without guaranteeing any specific data rate or packet delays, and the P-GW allocates an IP address to the UE. The MME includes the IP address in the ATTACH ACCEPT message in Step 6, which is the response to the original ATTACH REQUEST the UE sends. The air interface encapsulates the ATTACH ACCEPT message in the RRC Reconfiguration message. When a service with new QoS requirements is invoked (e.g., a VoIP call or a video-streaming session), a dedicated EPS bearer is established.

Now that we have established one or more EPS bearers, the UE should transfer data! Let's first summarize the DL data transfer, and then we'll take a look at the UL data transfer. Again, you will indeed find similarities between WiMAX and LTE.

Figure 15.29 provides a example of the DL data transfer. In Step 1, the UE observes the reference signal the eNB sends during specific symbols and specific subcarriers and quantifies the DL channel conditions in the form of a channel-quality indicator (CQI), and the CQI corresponds to the combination of the amount of coding redundancy and the modulation scheme that the UE can support at a 10% instantaneous error rate. The UE sends the CQI feedback to the eNB. The eNB configures the UE with the CQI reporting parameters, e.g., how frequently the UE should send the CQI, perhaps every 1-ms subframe!

Now the ball is in the eNB's court. The eNB has an implementation-specific scheduler that considers the reported CQI and many other factors, such as the QCIs, amount of data in the eNB buffer, availability of resources, number of UEs, history of data transmission to the UEs, and past HARQ feedback received from the UEs, to determine OFDMA resources for the UE. The minimum allocation is two *resource blocks* in a subframe. One *resource block* spans twelve subcarriers and one slot. Each slot occupies 0.5 ms, and, of course, two slots occur in a 1-ms subframe. The scheduling algorithm determines the number of Resource Blocks for each selected UE in Step 2.

The UE retrieves its packet from the subcarriers in the allocated resource blocks in Step 3. The UE sends an HARQ feedback to the eNB. If the eNB receives an HARQ ACK, it can send a new packet in the future. Otherwise, it retransmits the packet using Chase combining or the Incremental Redundancy approach.

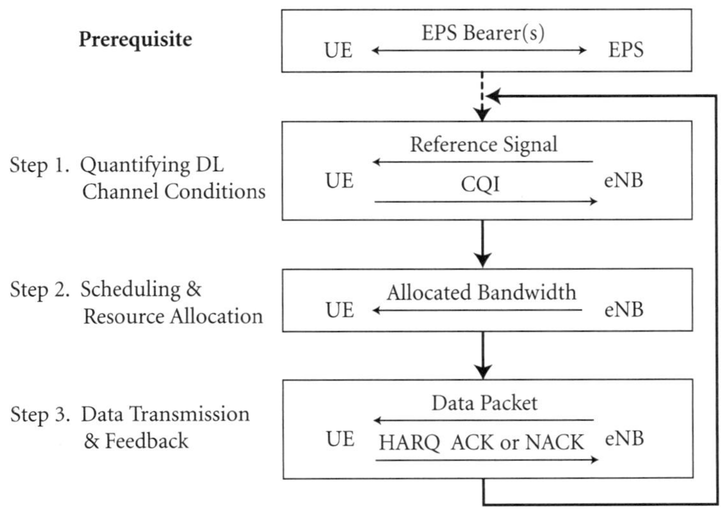

Figure 15.29: DL Data Transmission

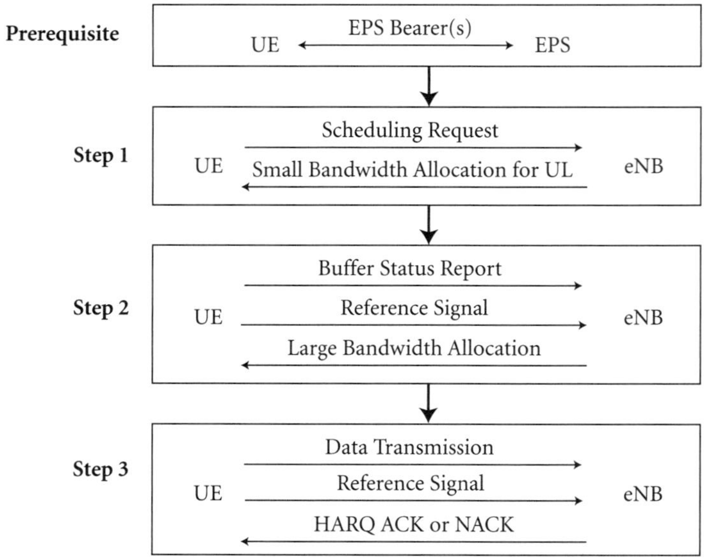

Figure 15.30: UL Data Transmission

Now, let's steer our attention to the UL data transfer, which is summarized in Figure 15.30. Again, it is assumed that the EPS bearers have already been established. Depending upon the data delivery service (e.g., unsolicited grant versus non-real-time variable rate), the overall mechanism will vary slightly.

The UE sends a Scheduling Request in Step 1 on a UL control channel to inform the eNB that new data has arrived in the UE's buffer. The eNB responds with resource allocation sufficient for the UE to send a buffer status report, which indicates the amount of data in the UE's buffer in Step 2.

The eNB can get an idea of the overall UL channel conditions by observing the reference signal that accompanies the data subcarriers carrying the buffer status report. The eNB scheduler considers factors such as the UL channel conditions, amount of data in the UEs buffer, relevant QCIs, availability of UL resource blocks, and number of UEs. The scheduling algorithm then determines the number of resource blocks to allocate to each selected UE. The scheduler also dictates to the UE the amount of coding and the modulation scheme as part of large bandwidth allocation in Step 2. The allocated UL resources can be shared among multiple EPS bearers. Resource allocation may be persistent or non-persistent. Using the allocated resource blocks, the UE sends a packet to the eNB in Step 3. The eNB provides HARQ feedback to the UE in Step 3, and, depending on this feedback, the UE may retransmit a packet or send a new packet.

Ever Wondered... Have you wondered how Release 8 LTE can achieve 300 Mbps in the downlink and 75 Mbps in the uplink? Let's take a simplified view of how to reach LTE's highest promised data rates. Here are the key parameters that will help us reach such data rates: 20 MHz channel bandwidth, 64-QAM modulation, and 4×4 MIMO in the downlink. Recall that one 64-QAM modulation symbol represents six bits. The 20 MHz channel bandwidth contains 1200 subcarriers, and one subcarrier carries one modulation symbol during the symbol time. The total symbol time including the normal cyclic prefix is about 71 microseconds. Hence, the data rate is

$$\begin{aligned}&\big(1200\ subcarriers * ((1\ modulation\ symbol)/(subcarrier))\\&\quad * ((6\ bits)/(modulation\ symbol))\big)/\big(72\ \text{microseconds}\big) = 100\ \text{Mbps.}\end{aligned}$$

Since we are using 4×4 MIMO in the downlink, each transmit antenna contributes 100 Mbps, and the total data rate is 100 Mbps $per\ antenna * 4\ antennas =$ 400 Mbps. However, we cannot ignore the overhead, such as channel coding, control channels, and reference signals. Assuming 25% overhead, we get 400 Mbps $*\ (100 - 25) =$ 300 Mbps! Once we are comfortable with 300 Mbps in the downlink, 75 Mbps in the uplink is easy. The uplink in Release 8 LTE supports a single-antenna transmission only, so the achievable data rate in the uplink is 300 Mbps$/4 =$ 75 Mbps.

15.6.5 A Brief Overview of LTE-Advanced[544]

3GPP Release 8 introduces LTE, and Release 10 introduces LTE-Advanced, which is fully backward-compatible with LTE. Both LTE and LTE-Advanced can share the same spectrum. A legacy LTE can work with both LTE and LTE-Advanced infrastructures. Additionally, an LTE-Advanced UE can work with both LTE and LTE-Advanced infrastructures. Table 15.2 summarizes LTE-Advanced's features, some of which are available in Release 11 and later releases.

The main features of LTE-Advanced are carrier aggregation, enhanced antenna techniques, relays, and Coordinated Multipoint (CoMP) transmission and reception. Carrier aggregation allows multiple carrier frequencies to be combined for data transmission between the eNodeB and a given UE, and each carrier is backward-compatible to a Release 8 carrier frequency. Up to five carrier frequencies, each with a maximum bandwidth of 20 MHz, can be combined. In the frequency domain, the carrier frequencies could be contiguous or non-contiguous (i.e., from the same or from different frequency bands). Carrier aggregation enables the operator to exploit

[544]The entirety of this section originally appeared in a prior work. Only minor, non-technical edits have been made to the material as it appears here.
SOURCE: N. Tripathi, "LTE Deployments in the LMS Band for ITS Radio Communications and Location" [367]. © 2012, Skybridge Spectrum Foundation and Telesaurus Holdings GB LLC (sometimes together known as "SkyTel"). Used with permission.

Table 15.2: Features of LTE-Advanced

Feature	Feature Specifics	Comments
Carrier Aggregation	Use of multiple LTE carrier frequencies for data transmission between the eNodeB and a given UE	Flexible combining of different frequency bands for higher throughput
Enhancements to Multiple Antenna Techniques	8×8 SU-MIMO in the downlink; 4×4 SU-MIMO in the uplink	High throughput
Relays	Layer 3 self-backhauling relay (advanced repeater)	Increased coverage at a lower cost
Coordinated Multipoint (CoMP) Transmission and Reception	Different flavors of soft handover with possible transmission from multiple cells and reception at multiple cells	Increased reliability and increased throughput near the cell edge
Heterogeneous Network (HetNet)	Deployment of large cells and small cells in a given geographic area	Capacity and throughput enhancements due to significant frequency reuse in small cells and coverage of wide areas due to large cells

fragmented spectrum (e.g., some spectrum at 700 MHz and some spectrum at the AWS spectrum). Carrier aggregation increases user throughput and cell throughput.

Though Release 8 supports up to 4×4 SU-MIMO in the downlink, Release 10's LTE-Advanced supports up to 8×8 SU-MIMO in the downlink. Release 8 allows the UE to transmit from a single antenna, and Release 10 supports 4×4 SU-MIMO in the uplink. Beamforming is also enhanced in Release 10 to achieve better performance.

LTE-Advanced also enhances the network architecture by introducing relays. A relay can be considered as an enhanced form of a repeater. The repeater is a low-cost solution to increase coverage without using regular full-fledged base stations. When the goal is to increase coverage at a lower cost without adding capacity, relays are useful. LTE has defined a *layer 3 self-backhauling relay* that increases coverage and uses LTE-based wireless backhaul between the donor eNodeB and the relay node. No direct connectivity exists between the relay node and the EPC; the relay node communicates only with the UEs and the donor eNodeB. The communication between the UEs and the relay node and between the donor eNodeB and the relay node can be achieved using the same or a different spectrum.

CoMP transmission comes in different flavors. In one flavor, multiple cells can transmit the same packet to the UE and the UE combines the signals from these cells for reliable reception. In another flavor, multiple cells are ready to transmit to the UE but only one cell transmits the packet. The specific cell that actually transmits the packet to the UE can be dynamically and quickly changed. The CoMP reception in the uplink involves multiple cells listening to the UE signal, and one of the eNodeBs processes the packets received in different cells. The main benefits of CoMP transmission and reception are increased reliability and increased throughput near the cell edge.

A heterogenous network (HetNet) uses the combination of large cells and small cells in a given geographic area to increase the achievable spectral efficiency per unit area. In particular, the use of small cells allows the use of the same radio resources numerous times within a given area. Furthermore, LTE-Advanced defines mechanisms to coordinate the use of radio resources between the large cells and the small cells to minimize interference.

15.7 4G TECHNOLOGY CHALLENGES

4G technology promises an excellent user-perceived experience for a variety of data services. Certain challenges need to be overcome, however, for 4G technology to reach its full potential. First, adequate spectrum bandwidth is required to reach high data rates. An operator's 5 MHz × 5 MHz FDD spectrum may be insufficient. At least 10 MHz in a given direction would be needed per sector or cell to outperform existing 3G systems.

3G UMTS initially lagged due to the unavailability of UEs. 4G systems must ensure the availability of high-performance and user-appealing smart phones to exploit the 4G system's efficiency to yield an awesome user experience.

Since this is the first introduction of OFDMA and complex advanced antenna technologies into a mobile cellular system, technical roadblocks may arise. Sophisticated design and troubleshooting tools will vastly help alleviate potential problems. Network planning and design is another challenge as the decision is made regarding frequency reuse of 1 or 3.

Finally, the backhaul is an issue that cannot be ignored since it implements the connectivity of the 4G system's base station to the core network and the inter-BS connectivity in the network architecture of WiMAX and LTE. Though air-interface data rates could reach hundreds of Megabytes per second, huge amounts of data must get from the core to the BS and vice versa. Combined DL and UL peak throughout on the BS-core backhaul could easily exceed 1 Gbps when one BS controls three sectors. *No business case can be made for supporting T1-based backhaul in 4G systems.* The lease fee for one T1 is a few hundred U.S. dollars per month, and one T1 can handle just 1.54 Mbps. An operator should not even consider deploying 4G technology if it intends to rely upon T1s! The backhaul must migrate to something else. The industry is indeed moving toward Multi-Protocol Label-Switching[545] backhaul [368,369], which

[545] See Chapter 4 to learn how IP and Multi-Protocol Label-Switching can be exploited in the operator's infrastructure network.

can exploit optical or wireless transport mechanisms to yield a scalable, efficient, and cost-effective backhaul; such backhaul *would* have a business case!

15.8 POINTS TO REMEMBER

- Emerging cellular technologies beyond 3G systems promise superior data rates and low latency efficiently, enabling service operators to cost-effectively offer VoIP and multimedia services. LTE and WiMAX are such competing 4G technologies.
- A 4G system is characterized by an all-IP network infrastructure, distributed radio network architecture, inter-BS communications, OFDMA-based air interface, use of advanced antenna techniques, large radio channel bandwidths, high-order modulation schemes, stringent QoS control, presence of a packet-switched core network, absence of a circuit-switched core network, and potential use of IMS as a service delivery platform.
- LTE and WiMAX share a lot of similarities, especially on the air interface.
- WiMAX reuses many traditional IP components, and LTE defines the EPC as a comprehensive core network. The EPC provides a common framework for interworking across various radio access technologies and for offering IMS-based IP services.

Just wait and see. Your life will be transformed by these exciting 4G technologies!

Appendix A

A Brief Overview of Signaling System 7

A Signaling System 7 (SS7) network is a message-based packet network in which messages are sent using dedicated channels that are different from the voice channels. The Public-Switched Telephone Network uses SS7 to carry out functions such as call setup and teardown, call forwarding, toll-free (e.g., a company's 800 or 888 customer service phone number) and toll (e.g., a weather service's 900 phone number) services, wireless roaming, and wireless subscriber authentication. Since SS7 uses a channel separate from the channel that carries voice, SS7 is said to use *out-of-band* (as opposed to *in-band*) signaling. The use of out-of-band signaling offers faster call setup and facilitates implementation of intelligent network features, such as 1-800 number services. An SS7 signaling link between the two network nodes typically utilizes bidirectional 64 or 56 kbps DS0 or DS0-A channels.[546]

Figure A.1 illustrates the architecture of an SS7 network, which contains three types of nodes: Service Switching Points (SSPs), Signal Transfer Points (STPs), and Service Control Points (SCPs). A telephone switch with SS7 capability is an SSP. The

[546]DS represents *digital signal.*

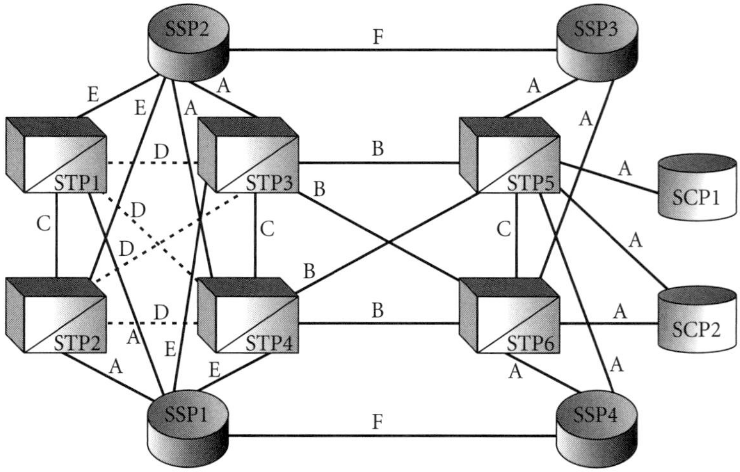

Figure A.1: Architecture of an SS7 Network

STPs route the arriving signaling messages toward their destination (e.g., one SSP to another SSP). The SCPs are databases that store information to support special call processing, such as 800 and 900 number services, credit card calls, local number portability, and cellular roaming.

Each SS7 node is identified by a unique address called a Point Code, which consists of twenty-four bits for an ANSI-based[547] SS7 network and fourteen bits for an ITU-based[548] SS7 network. The Point Code consists of three parts: the network number, cluster number, and member number. The network number is assigned on a nationwide basis. For example, Regional Bell Operating Companies and Interexchange Carriers in the U.S. have been assigned specific network numbers. One network number is associated with several clusters, and an SS7 node belongs to one of the clusters through the member number. Such hierarchy of networks and clusters provides scalability.

Now that we know a little about the nodes, let's look at the types of signaling links between the nodes. There are six types of links: A, B, C, D, E, and F. The A (for Access) link connects an STP to an SSP or an SCP. In Figure A.1, STP2 connects to SSP1 via the A link and STP5 connects to SCP1 via the A link. The B (for Bridge) link interconnects *mated STPs*; since STPs and SCPs are critical, they are deployed in *mated pairs*. For example, STP3 and STP4 constitute one mated pair, and STP5 and STP6 form another mated pair. The B links interconnect the STP3-STP4 *pair* and the STP5-STP *pair*. The C (for Cross) link connects two STPs of the mated pair. For example, STP3 and STP4 are connected via a C link, as are STP5 and STP6.

The D (for Diagonal) links are the quad (i.e., a set of four) links that interconnect two STP pairs at different hierarchical levels. In Figure A.1, STP1 and STP2 are at

[547] American National Standards Institute
[548] International Telecommunications Union

one hierarchy level and STP3 and STP4 are at another; hence, we have linked these STP pairs using D links. In practice, no clear hierarchy exists and the links connecting two STP pairs may be considered B, D, or B-D links. An SSP is connected to its *home* STP pair using A links. For example, in Figure A.1, STP1-STP2 is a home STP pair for SSP1 and STP3-STP4 is a home STP pair for SSP2.

To improve reliability, the SSP can be connected to a non-home STP pair via an E (for Extended) link. For example, SSP1 is connected to a non-home STP pair STP3-STP4 as a backup connection in case the home STPs are unreachable. The E links may be deployed at a service operator's discretion. Finally, the F (for Fully Associated) link directly connects two nodes (e.g., SSP1 and SSP4 in Figure A.1). Since the F links bypass the STP's security features, they are generally absent in practice.

Let's discuss how the SS7 architecture can be used to set up a voice call between two landline phones. Figure A.2 summarizes the messages associated in call setup and teardown [370]. Subscriber X connected to SSPA places a call to Subscriber Y connected to SSPB. SSPA observes the digits dialed by Subscriber X and constructs an Initial Address Message (IAM) in Step 1. The IAM includes information such as the initiating switch (i.e., SSPA), the destination switch (i.e., SSPB), the idle voice trunk SSPA selected, the calling party number, and the called party number. SSPA sends the IAM to its home STP, STP1, using an A link in Step 2. STP1 routes the IAM to SSPB based on the routing label using an A link. SSPB confirms that it is in charge of the called party number and that this number is idle.

In Step 3, SSPB creates the Address Complete Message (ACM), containing the sending switch (i.e., SSPB), the recipient switch (i.e., SSPA), and the selected trunk

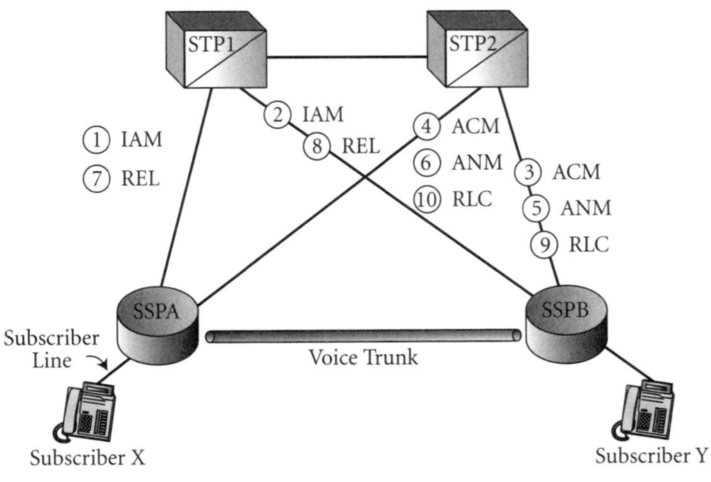

Figure A.2: Using SS7 for Call Setup and Teardown
SOURCE: "Signaling System 7 (SS7)," [370]. © International Engineering Consortium. Used with permission.

and sends the ACM to its home STP2. The ACM indicates that the IAM has found its way to the called party number. SSPB completes the voice call path in the backward direction (i.e., toward SSPA) by choosing the voice trunk toward SSPA, sends the ringing tone to SSPA using the voice trunk, and rings the subscriber line of the called party (i.e., Subscriber Y).

In Step 4, STP2 forwards the ACM to SSPA based on the routing label. SSPA now connects the subscriber line to the selected voice trunk so that the subscriber can hear the ringing tone SSPB has sent. In Step 5, once Subscriber Y picks up the phone, SSPB constructs an Answer Message (ANM) containing the sending switch, the recipient switch, and the selected trunk. SSPB also connects the line of Subscriber Y to the selected voice trunk to enable conversation. In Step 6, STP2 forwards the ANM to SSPA, and SSPA confirms that the conversation can indeed take place since the voice trunk is connected between the SSPs.

Let's assume that the conversation is over and that Subscriber X hangs up the phone first. In Step 7, SSPA generates a Release (REL) message, containing the identities of the voice trunk and SSPB, and sends it to STP1. In Step 8, STP1 forwards the REL to SSPB. SSPB disconnects the trunk from the subscriber line and makes the state of the trunk idle. In Step 9, SSPB creates a Release Complete (RLC) message, containing the identities of SSPA and the voice trunk, and sends it to STP2. In Step 10, STP2 forwards the RLC to SSPA, which makes the voice trunk idle.

Now that we understand SS7-based call setup signaling between two subscribers, let's discuss signaling units of an SS7 network. Three types of signal units are defined for an SS7 network: Message Signal Units, Link Status Signal Units, and Fill-In Signal Units. The Message Signal Unit carries information related to call control, database query and response, and network management and maintenance. The routing label within the Message Signal Unit helps route a message from the originating SSP to the destination SSP. The Link Status Signal Unit ensures the availability of a given link for the transfer of a signaling unit. The Fill-In Signal Unit fills up the signaling link in the absence of other signaling units.

Let's take a quick look at the SS7 protocol stack shown in Figure A.3. The bottom-most layer is the Message Transfer Part Level 1, which is analogous to the physical layer of the seven-layer Open System Interconnect model. The Message Transfer Part Level 1 typically uses a 56 kbps DS-0A or a 64 kbps DS0 channel, with 56 kbps being more common. Other examples of the Message Transfer Part Level 1 implementation are E1 (E for European) at the data rate of 2.048 Mbps and DS1 at the rate of 1.544 Mbps. The Message Transfer Part Level 2 is like the link layer of the Open System Interconnect model and implements functions such as flow control, packet sequencing, error detection, and error correction through retransmission. The Message Transfer Part Level 2 ensures that the two nodes associated with a signaling link can reliably exchange signaling messages. The Message Transfer Part Level 3 is analogous to the network layer of the Open System Interconnect model, routes the messages between the SSPs, and performs congestion control. The Message Transfer Part Level 3 enables the delivery of signaling messages regardless of the direct or indirect connectivity between the originating node and the destination node.

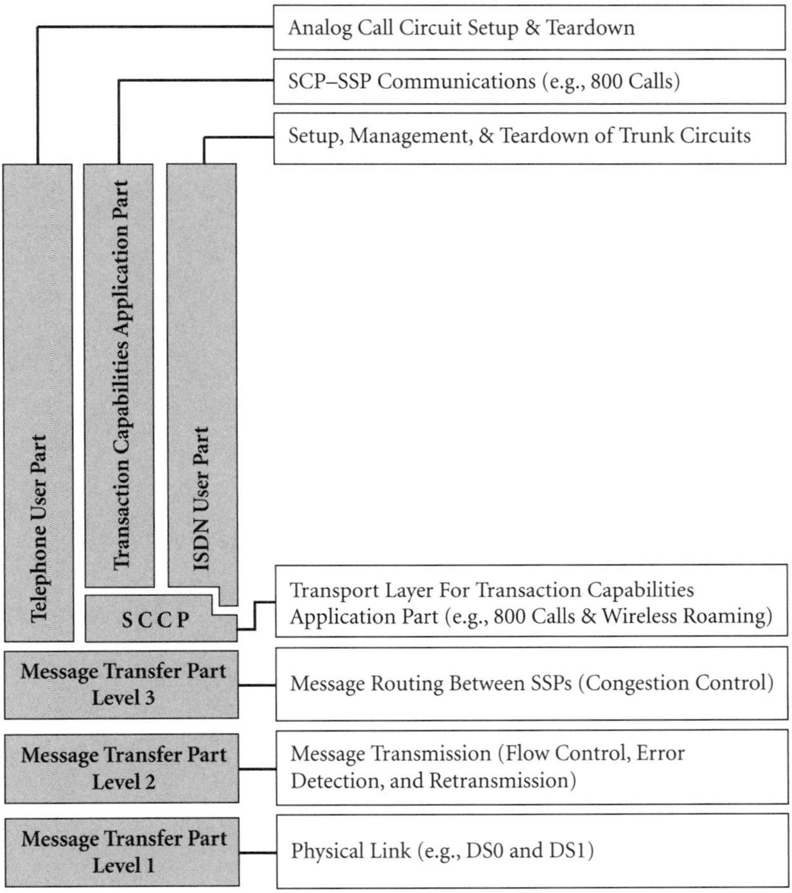

Figure A.3: SS7 Protocol Stack

The upper layers include the Telephone User Part, Transaction Capabilities Application Part, Integrated Services Digital Network User Part (ISDN in the figure), and Signaling Connection Control Part (SCCP in the figure). The Telephone User Part facilitates setup and teardown of the analog call circuit. The Transaction Capabilities Application Part helps route 800, 888, and 900 calls by allowing communications between the SCP and SSP and encapsulates Mobile Application Part messages, which contain the subscriber profile, to support roaming cellular subscribers. The Integrated Services Digital Network User Part defines messages (e.g., IAM, ACM, ANM, REL, and RLC) to set up, manage, and release the trunk circuits. Finally, the Signaling Connection Control Part acts as a transport layer for services based on the Transaction Capabilities Application Part (e.g., 800, 888, and 900 numbers and wireless roaming). The Signaling Connection Control Part enables SS7 to address specific software applications (e.g., 800 call processing and calling-card processing) at the destination SS7 nodes. Furthermore, the Signaling Connection Control Part provides the ability to perform incremental routing using Global Title Translation, which relieves the originating SS7 node from the burden of knowing every potential destination for message routing. An SSP node can query an STP with a request for Global Title Translation. Such an STP, in turn, can find another STP along the path from origination to destination. One STP would perform the final Global Title Translation and route the message to the destination.

Appendix B

Erlang-B Table

An Erlang-B table quantifies the relationship among the call blocking probability or loss probability (P_b), the number of trunked channels or capacity (N), and the Erlang traffic intensity (Erl). We introduced Erlang-B formulation in Chapter 1, and, in Table B.1, we summarize for quick reference a selected set of numerical values for P_b, N, and Erl. Other works [371–373] offer other values of these parameters.

Cellular Communications, First Edition. Nishith D. Tripathi and Jeffrey H. Reed

Table B.1: Erlang-B Table: Selected Parameters

Number of Trunked Channels or Capacity (N)	Erl for $P_b = 1\%$	Erl for $P_b = 2\%$
1	0.01	0.02
5	1.36	1.66
10	4.46	5.08
15	8.11	9.01
20	12.03	13.18
25	16.13	17.51
30	20.34	21.93
35	24.64	26.44
40	29.01	31.00
45	33.43	35.61
50	37.90	40.26
55	42.40	44.90
60	46.90	49.60
65	51.50	54.35
70	56.10	59.10
75	60.70	63.90
80	65.35	68.65
85	70.00	73.45
90	74.65	78.30
95	79.35	83.10
100	84.05	87.95

Appendix C

A High-Level Comparison of Third-Generation Technologies

Here, we provide via several tables a simplified and brief comparison of the UMTS (Universal Mobile Telecommunication System) family of standards [e.g., Release 99 (the first UMTS release), HSDPA (High-Speed Downlink Packet Access), and HSUPA (High-Speed Uplink Packet Access)] and the CDMA2000[549] family of standards [e.g., 1xRTT (1x Radio Transmission Technology or simply 1x) and 1xEV-DO (1x Evolution—Data-Optimized, referred to as DO in this appendix)]. If you are familiar with one 3G family and are trying to learn the other family, this comparison will accelerate your learning curve. Traditionally, UMTS and CDMA2000 have been competing technologies, but they are merging in fourth-generation systems since UMTS and CDMA2000 operators have chosen LTE (Long-Term Evolution) as their next-generation system.

[549]CDMA represents Code-Division Multiple Access.

Table C.1 lists the basic terms used in the standards, and Table C.2 compares UMTS Release 99 and 1x. Table C.3 highlights the similarities and differences between HSDPA and 1xEV-DO Rev. 0, and Table C.4 does the same for HSUPA and 1xEV-DO Rev. A.

Table C.1: Basic Terms

UMTS Family	CDMA2000 Family	
	1x	DO
UE[a]	Mobile Station	Access Terminal
Cell	Sector	Sector
Node B	Base Station or Base Transceiver Station	Base Station, Base Transceiver Station, or Access Point
Radio Network Controller	Base Station Controller	Radio Network Controller
Handover	Handoff	Handoff
Cell Reselection	Idle Handoff	Idle Handoff

[a] User Equipment.

Table C.2: UMTS and CDMA2000: Similarities and Differences

Topic or Area	UMTS Family	CDMA2000 Family
Equivalent or Similar Technologies	R99[a]	1x
	R5[b] HSDPA	1xEV-DO Rev. 0
	R6[c] HSUPA	1xEV-DO Rev. A
Radio Network Architecture	Node Bs control cells and carrier frequencies, and each Radio Network Controller controls hundreds of Node Bs.	Base stations control sectors and carrier frequencies, each Radio Network Controller controls hundreds of base stations, and Packet Control Functions provide an interface between the Radio Network Controller and the Packet-Switched Core Network. (A Packet Control Function typically resides inside the Radio Network Controller cabinet.)
Core Network Architecture	Major Circuit-Switched Domain Components: MSC[d]	Major Circuit-Switched Core Network Components of 1x: MSC, VLR,[e] HLR[f] (the MSC and VLR typically reside in the same physical cabinet)
	Major Packet-Switched Domain Components: SGSN[g] and GGSN[h]	Major Packet-Switched Core Network Components of 1x and DO: PDSN,[i] AAA[j] Server, Foreign Agent, and Home Agent
	Common Circuit-Switched and Packet-Switched Domain Components: HLR and VLR (the MSC and VLR typically reside in the same physical cabinet)	No elements common to circuit-switched and packet-switched core networks

Continued on next page

Table C.2: UMTS and CDMA2000: Similarities and Differences (continued from previous page)

Topic or Area	UMTS Family	CDMA2000 Family
Multiple Access Technique	Wideband CDMA	CDMA
Spectrum Requirements	3.84 Mcps chip rate with channel bandwidth of 5 MHz in a given direction (uplink or downlink); a total of 10 MHz bandwidth is required for FDD[k]	1.2288 Mcps chip rate with channel bandwidth of 1.25 MHz in a given direction; a total of 2.5 MHz bandwidth is required for FDD
Radio Frame Structure	10 ms frame with 15 slots, with power-control command per slot	20 ms frame with 16 Power-Control Groups, with power-control command per Power-Control Group
Closed-Loop Power Control	1500 Hz	800 Hz
Maximum Active Set Size	8 cells allowed by the standard; limited to about 3 to 4 in commercial networks primarily due to implementation choice	6 sectors

Continued on next page

Table C.2: UMTS and CDMA2000: Similarities and Differences (continued from previous page)

Topic or Area	UMTS Family	CDMA2000 Family
Layers of Air-Interface Protocol Stack	Physical, MAC,[l] RLC,[m] RRC,[n] and PDCP[o]	Physical, MAC, LAC,[p] and Upper-Layer (L3) Signaling
Cell or Sector Identification	Primary Scrambling Code (512 options, with all options usable in an area)	Pilot PN[q] Offset (just one code with 512 offsets; usable options are typically less than 512 to avoid confusion between two adjacent offsets)
Mobile Identification in a Cell or Sector in the Downlink	Typically, one OVSF[r] code (more than one allowed by the standard but uncommon)	One or two Walsh codes with one adequate for voice and two for data (one for a low-rate channel and one for a higher-rate channel)
Mobile Identification in a Cell or Sector in the uplink	(Uplink) Scrambling Code	User-specific long code
Concurrent Voice and Data Services	Supported	Usually unsupported because Release 0 of 1x lacks support for it and Release A of 1x, which supports this feature, is uncommonly deployed
Integrity Protection of Signaling Messages and Ciphering (i.e., Encryption) of Signaling and Traffic	Quite common	Uncommonly used
Circuit-Switched Voice Capacity per MHz	Higher than GSM[s] (speech codec: full-rate 12.2 kbps AMR[t])	Higher than UMTS primarily due to lower effective data rate (speech codec: 8 kbps EVRC[u] to carry voice)

Continued on next page

Table C.2: UMTS and CDMA2000: Similarities and Differences (continued from previous page)

Topic or Area	UMTS Family	CDMA2000 Family
Call Setup Delay	Longer due to extensive signaling	Shorter due to fewer signaling messages
Global Positioning System	Allowed but typically unused in commercial systems; asynchronous Node B transmissions	Used in commercial systems; synchronous transmissions from all the base stations on the planet!

[a] Release 99

[b] Release 5

[c] Release 6

[d] Mobile Switching Center

[e] Visitor Location Register

[f] Home Location Register

[g] Serving GPRS Support Node where GPRS is General Packet Radio Service

[h] Gateway GPRS Support Node

[i] Packet Data Serving Node

[j] Authentication, Authorization, and Accounting

[k] Frequency-Division Duplexing

[l] Medium Access Control

[m] Radio Link Control

[n] Radio Resource Control

[o] Packet Data Convergence Protocol

[p] Link Access Control

[q] Pseudo-Noise

[r] Orthogonal Variable Spreading Factor

[s] Global System for Mobile Communication

[t] Adaptive Multirate

[u] Enhanced Variable-Rate Codec

Table C.3: HSDPA and 1xEV-DO Rev. 0: Similarities and Differences

Topic or Area	HSDPA	1xEV-DO Rev. 0
Spectrum Sharing	Possible and typical; the same spectrum bandwidth can be shared among UMTS R99 and HSDPA UEs	Unavailable; need a spectrum bandwidth separate from 1x; two parallel networks, 1x and DO, exist in case of an overlay
Required Total Spectrum	10 MHz for FDD (5 MHz in UL[a] and 5 MHz in DL[b])	2.5 MHz for FDD (1.25 MHz in RL[c] and 1.25 MHz in FL[d]) (Overlay: Another 2.5 MHz total for 1x)
Focus of the Technology	Air interface for the high-speed data transmission in the DL	Air interface for the high-speed data transmission in the FL
Peak User Data Rate (DL or FL)	14 Mbps for a Category 10 UE; 7.2 Mbps for a Category 8 UE	2.4 Mbps
Modulation Schemes	QPSK[e] and 16-QAM[f]	QPSK, 8-PSK[g], and 16-QAM
Channel Coding	Turbo with rate of 1/3	Turbo with rates of 1/3 and 1/5
HARQ[h] Terminology	HARQ Process	Interlaces

Continued on next page

Table C.3: HSDPA and 1xEV-DO Rev. 0: Similarities and Differences (continued from previous page)

Topic or Area	HSDPA	1xEV-DO Rev. 0
HARQ Characteristics	- DL only - Asynchronous retransmissions with configurable number of retransmissions - Type of HARQ: Chase combining and Incremental Redundancy (configurable) - Configurable number of HARQ processes	- FL only - Synchronous retransmissions with data-rate–dependent predefined number of retransmissions - Type of HARQ (predefined per data rate): Incremental Redundancy (and repetition of code symbols beyond a certain number of retransmissions) - Four HARQ interlaces
Fastest Data Rate Change in DL or FL	Every 2 ms TTI[i]	Every 1.67 ms slot
Maximum Number of Users who can Receive DL or FL Data Simultaneously	Around 13 per TTI	One per slot
Radio Resources Available for High-Speed Data	Typically leftover resources after UMTS R99 UEs have been considered (e.g., transmit power allocated to a UE is less than the full power of the HPA[j])	All 1xEV-DO resources since the 1xEV-DO carrier frequency is different from a 1x carrier frequency

Continued on next page

Table C.3: HSDPA and 1xEV-DO Rev. 0: Similarities and Differences (continued from previous page)

Topic or Area	HSDPA	1xEV-DO Rev. 0
Data Transmission Philosophy	Wideband CDMA with constant power transmission during a 2 ms TTI for the UEs allocated to a high-speed data channel	TDM[k] with full-power transmission to one Access Terminal during the 1.6 ms slot

[a] Uplink
[b] Downlink
[c] Reverse Link
[d] Forward Link
[e] Quadrature Phase Shift Keying
[f] 16-Quadrature Amplitude Modulation
[g] 8-ary Phase Shift Keying
[h] Hybrid Automatic Repeat Request
[i] Transmit Time Interval
[j] High-Power Amplifier
[k] Time-Division Multiplexing

Table C.4: HSUPA and 1xEV-DO Rev. A: Similarities and Differences

Topic or Area	HSDPA	1xEV-DO Rev. A
Spectrum Sharing	Possible and typical, the same spectrum bandwidth can be shared among UMTS R99, HSDPA, and HSUPA UEs.	Rev. A, like Rev. 0, cannot share the same spectrum bandwidth with 1x; two parallel networks, 1x and DO, exist in case of an overlay. 1xEV-DO Rev. A is fully backward-compatible with 1xEV-DO Rev. 0, and the same spectrum bandwidth can be shared between the two revisions.
Required Total Spectrum	10 MHz for FDD (5 MHz in UL and 5 MHz in DL)	Like 1xEV-DO Rev. 0, 2.5 MHz for FDD (1.25 MHz in RL and 1.25 MHz in FL)
Primary Focus of the Technology	Air interface for the high-speed data transmission in the UL	Air interface for the high-speed data transmission in the RL and Support for QoS[a]
Peak User Data Rates	Same as HSDPA in DL; 5.76 Mbps in UL for a Category 6 UE and 2 Mbps for a Category 5 UE	3.1 Mbps in FL and 1.8 Mbps in RL
Modulation Schemes (UL or RL)	Multi-code BPSK[b](with the overall constellation diagram resembling QPSK)	BPSK, QPSK, and 8-PSK
Channel Coding	Turbo with rate of 1/3	Turbo with rates of 1/3 and 1/5
HARQ Terminology	HARQ Process	Interlace

Continued on next page

Table C.4: HSUPA and 1xEV-DO Rev. A: Similarities and Differences (continued from previous page)

Topic or Area	HSDPA	1xEV-DO Rev. A
HARQ Characteristics	- DL and UL - DL HARQ the same as HSDPA - UL HARQ: synchronous retransmissions with configurable number of retransmissions - Type of HARQ: Chase combining and Incremental Redundancy (configurable) - 8 HARQ processes for 2 ms TTI and 4 for 10 ms TTI	- DL and UL - DL HARQ the same as Rev. 0 - UL HARQ: synchronous retransmissions with configurable number of retransmissions (default value: 4) - Type of HARQ (predefined per new or retransmission): Incremental Redundancy (and Repetition of code symbols beyond a certain number of retransmissions) - Three HARQ interlaces
Fastest Data Rate Change in UL or RL	Every 2 ms or 10 ms TTI (configurable)	Every 6.67 ms slot
Radio Resources Available for High-Speed Data in UL	Function of the actual rise-over-thermal noise and the target rise-over-thermal margin	Function of the actual rise-over-thermal noise and the target rise-over-thermal margin
Data Transmission Philosophy (UL)	Wideband CDMA with power control	CDMA with power control
Mechanism to Influence UL Data Rate Selected by UE	Absolute and Relative Grants with Absolute Grant defining the ratio of data channel power and pilot channel power and Relative Grant adjusting the overall grant	Traffic-to-pilot ratio, which defines ratio of traffic channel power and pilot channel power to traffic channel including both data channel and pilot channel

[a] Quality-of-Service
[b] Binary Phase Shift Keying

Appendix D

HSPA+ Overview

HSPA (High-Speed Packet Access) implies support for HSDPA (High-Speed Downlink Packet Access) in the downlink and HSUPA (High-Speed Uplink Packet Access) in the uplink. HSPA+ or enhanced HSPA introduces several features to improve the basic HSPA operations. HSPA+ is introduced in the Third Generation Partnership Project (3GPP) Release 7. This appendix provides a brief overview of HSPA+. Section D.1 summarizes the main features of HSPA+ in Release 7, and Section D.2 discusses details of how data transmission occurs in the downlink (DL) and the uplink (UL) in HSPA+. Section D.3 provides a glimpse of HSPA+ enhancements targeted in 3GPP Releases 8 and 9.

D.1 SUMMARY OF HSPA+ FEATURES

Fourth-generation LTE (Long-Term Evolution) is a 3GPP Release 8 feature. HSPA+ is expected to provide similar performance to LTE when both use 5 MHz spectrum bandwidth and implement similar features [374, 375]. Most HSPA+ features are defined as part of 3GPP Release 7, and some features are deferred to Release 8 and 9. Just like HSPA, HSPA+ focuses on the air interface to enhance the radio network performance. The main benefits of HSPA+ relative to HSPA are higher user peak data rates, higher cell throughput, reduced latency, higher VoIP (Voice-over-IP[550]) capacity, higher spectral efficiency, and longer UE (User Equipment) battery life. Table D.1 summarizes the main HSPA+ features and their primary benefits. A service operator may choose to deploy a subset of these features based on the cost-to-benefit analysis, the feature availability at the UE and the UTRAN (Universal Terrestrial Radio Access Network), and the business strategy (e.g., accelerated or delayed LTE deployment).

The use of high-order modulation (HOM) schemes in both the downlink and uplink significantly increases peak user rates and throughput. The downlink can use 64-QAM (64-ary Quadrature Amplitude Modulation) in addition to the two modulation schemes, QPSK (Quadrature Phase Shift Keying) and 16-QAM, available in Release 5. The uplink can use 4-PAM (4 Pulse Amplitude Modulation) in addition to BPSK (Binary Phase Shift Keying) on a given Enhanced Dedicated channel (E-DCH) (or E-DPDCH[551]) OVSF (Orthogonal Variable Spreading Factor) code. The higher-order Release 7 (R7) E-DCH modulation scheme is often referred to as 16-QAM because two OVSF codes simultaneously using 4-PAM give the appearance of 16-QAM.

A Question for You. . . By what factors can the peak user data rates in the downlink and the uplink increase in R7 compared to Release 5 (R5) and Release 6 (R6) as a result of R7's higher-order modulation? See the footnote[552] for the answer.

Discontinuous transmission (DTX) of the uplink DPCCH (Dedicated Physical Control channel) saves precious battery power. Furthermore, DTX reduces the uplink interference. In Code-Division Multiple Access systems, capacity is inversely proportional to interference, and, hence, reduction in interference increases uplink capacity. In general, the uplink is usually the limiting link for VoIP capacity because the DL cell throughput is greater than the uplink cell throughput. An increase in the uplink capacity enables the system to support more VoIP users. For systems supporting data services, lower uplink interference translates into higher uplink throughput. A short DTX cycle is activated if a short period of data inactivity occurs, and a long DTX cycle is used if the inactivity period persists longer [376].

In R5 and R6 systems, the UE must continuously monitor one or more HS-SCCHs (High-Speed Shared Control channels) to determine if the PDSCH (Physical

[550] IP is, of course, Internet Protocol.

[551] E-DCH Dedicated Physical Data Channel

[552] The downlink peak rate increases by a factor of $6/4 = 1.5$ as a 64-QAM modulation symbol represents six bits (or code symbols) and a 16-QAM modulation symbol represents four bits. The uplink peak rate increases by a factor of $2/1 = 2$ because a 4-PAM modulation symbol represents two bits (as opposed to a BPSK modulation symbol, which represents one bit) and significantly increases the processing requirements.

Table D.1: HSPA+ Features and Associated Benefits

Feature	Primary Benefits
HOM	Higher peak data rates, larger throughput, and better spectral efficiency
DTX in the UL[a]	Longer UE battery life, potential for higher capacity and throughput in UL due to lower interference
DRX in the DL[b]	Longer UE battery life due to reduced processing requirements
Enhanced CELL_FACH[c] RRC[d] State	Reduced latency due to the use of HS-DSCH[e] in the CELL_FACH state
Enhanced CELL_PCH	Reduced paging delay
HS-SCCH–less[f] Operation	Potential for higher VoIP capacity and throughput due to efficient use of OVSF codes; part of Continuous Packet Connectivity
Transmit Diversity[g]	Improved reliability and potential for higher throughput
MIMO[h]	Higher peak data rates, larger throughput, and better spectral efficiency
Enhanced Receivers (Types 2i and 3i)[i]	Improved reliability and better throughput
Enhanced Layer 2 in the DL	Higher efficiency and throughput
Direct Tunnel Architecture	Reduced latency and more efficient interface provisioning

[a] This feature is considered part of an overall feature called Continuous Packet Connectivity.

[b] This feature is also considered part of Continuous Packet Connectivity.

[c] Forward Access channel.

[d] Radio Resource Control.

[e] High-Speed Downlink Shared channel.

[f] High-Speed Shared Control channel.

[g] Basic transmit diversity is defined in Universal Mobile Telecommunication System (UMTS) Release 99 but not implemented in commercial systems.

[h] Multiple-Input Multiple-Output.

[i] The letter i in Type 2i and Type 3i indicates that the receiver is interference-aware and can cancel out some interference to increase the effective signal-to-interference ratio at the receiver.

Downlink Shared Channel) has allocated code to it for downlink transmission in a given TTI (Transmit Time Interval). Such continuous information reception and subsequent processing drains the UE's battery, and the user may find, upon wanting to watch a high-definition video, that the battery has run out of "juice"! HSPA+ introduces DRX (Discontinuous Reception) in which the UE in the Connected Mode wakes up only periodically and processes the received HS-SCCH. Such HS-SCCH DRX increases the UE's battery life and lightens the UE's processing load. The network controls the activation and deactivation of DTX and DRX cycles through fast layer 1 signaling called *HS-SCCH orders* [377].

Release 99 and subsequent 3GPP releases have four RRC states: CELL_DCH, CELL_PCH, CELL_FACH, and URA_PCH. Transitions from one state to another, such as from CELL_PCH to CELL_DCH, may take a long time. HSPA+ allows the use of the HS-DSCH for paging and for data transfer. An HSPA+ UE monitors the PICH (Page Indicator channel) for an indication of a potential page message and retrieves the actual page message from the PDSCH associated with an HS-DSCH.

Do You Recall... Which physical channel does the UE use to collect the actual page message prior to R7? The footnote[553] contains the answer.

Shorter TTI, higher data rate, and fast repetitions make the paging operation more efficient in R7 than in previous 3GPP releases. Furthermore, the UE can receive data faster in the CELL_FACH mode before entering the CELL_DCH mode.

When UEs using different releases, such as Release 99 (R99), R5, R6, and R7 are sharing the same 5 MHz spectrum bandwidth, resources such as OVSF codes must be used efficiently to maximize voice capacity and data throughput. When HSDPA channels are used for VoIP users, the number of OVSF codes may quickly become a capacity and throughput bottleneck because the UE must be configured to monitor one or more HS-SCCH codes. HS-SCCH–less operation enables the UE to receive data on the HS-DSCH without the help of an HS-SCCH. The UE performs blind detection of the HS-DSCH packet using a preconfigured set of transport formats.

HSPA+ supports Space Time Transmit Diversity (STTD) in the downlink to improve reliability, especially for cell-edge users who experience challenging channel conditions. Two transmit antennas transmit the same information; this redundancy coupled with the independent fading on the two propagation paths between the transmitter and the receiver provide improved performance. STTD's open-loop flavor involves transmission of predefined STTD-encoded symbols [263], and its closed-loop flavor involves active feedback from the UE to the Node B regarding recommended phase adjustments for the two transmit antenna signals [376].

MIMO is available for use in the DL with two transmit antennas at the Node B and two receive antennas at the UE. Unlike transmit diversity, MIMO involves the transmission of different information from two different antennas, doubling the achievable peak data rate.

HSPA+ also introduces advanced receiver designs, such as Type 2i and Type 3i receivers, for UEs [378, 379]. The use of such receivers instead of traditional RAKE receivers improves the DL performance. A Type 1 receiver implements a two-branch receive diversity, and the RAKE receiver now has access to more paths to choose the best signals for its fingers. A Type 2 receiver performs chip-level linear minimum

[553] Prior to R7, the UE receives the actual page message from the S-CCPCH (Secondary Common Control Physical channel).

mean-squared error equalization to undo the effect of the radio channel. A Type 3 receiver exploits the concepts of receive diversity and chip-level equalization.

Layer 2 of the air-interface protocol stack is enhanced for the DL. Specifically, the radio link control (RLC) now supports both flexible and fixed-size PDUs (protocol data units). Additionally, the maximum RLC PDU size can be as large as 1503 bytes so that a 1500-byte Ethernet frame or packet no longer needs to be segmented into several RLC chunks [380]. A PDCP (Packet Data Convergence Protocol) header is accounted for in the 1503-byte RLC PDU size. A larger RLC PDU size can thus increase the packing efficiency and reduce the relative overhead the RLC layer adds. The MAC (Medium Access Control) layer has a new MAC sublayer called MAC-ehs (MAC–enhanced high speed) in HSPA+. MAC-ehs implements scheduling to support features such as multiple antenna techniques, HS-SCCH–less operation, and segmentation. The MAC-ehs can also multiplex RLC PDUs of different sizes and RLC PDUs from different priority queues [381]. Recall that HSDPA allows RLC PDUs from just one priority queue within a given transport block.

Direct tunnel architecture enables the UTRAN, specifically the RNC (radio network controller), to directly connect to the GGSN (gateway GPRS[554] support node), instead of SGSN (serving GPRS support node), for user traffic [382]. The user traffic no longer needs to travel via the SGSN. The signaling is still exchanged between the RNC and the SGSN. A service operator with an existing network may skip upgrading to the new direct tunnel architecture if the expected reduction in latency fails to justify the cost. However, the direct tunnel architecture may be attractive to a new service operator deploying a brand-new UMTS network. The direct tunnel architecture also allows more efficient provisioning of the required link capacities to accommodate signaling traffic and user traffic since SGSNs carry only signaling traffic.

Release 7 introduces six new HSDPA UE categories (bringing the total to eighteen) and one new HSUPA category (bringing the total to seven) [383]. The new HSDPA categories support features such as 64-QAM or MIMO, and the new HSUPA UE category supports 16-QAM in the uplink.

D.2 DATA TRANSMISSION IN HSPA+: A CLOSER LOOK

Let's discuss the DL data transmission first, and then we'll consider the UL data transmission. Figure D.1 illustrates the basic steps of the data transmission. The UE is assumed to have established primary PDP (Packet Data Protocol) context and 0 or more secondary PDP context and to have obtained an IP address as part of the PDP context activation.

In Step 1, the UE is configured to operate in either an HSDPA-like single antenna transmission mode, a transmit diversity mode, or a MIMO mode. We will focus on the latter two modes. The UE may also be configured to use the HS-SCCH–less operation and the DRX cycle. Furthermore, the UE may be configured to use 64-QAM when channel conditions are conducive. The RNC can perform such configuration using

[554]General Packet Radio Service

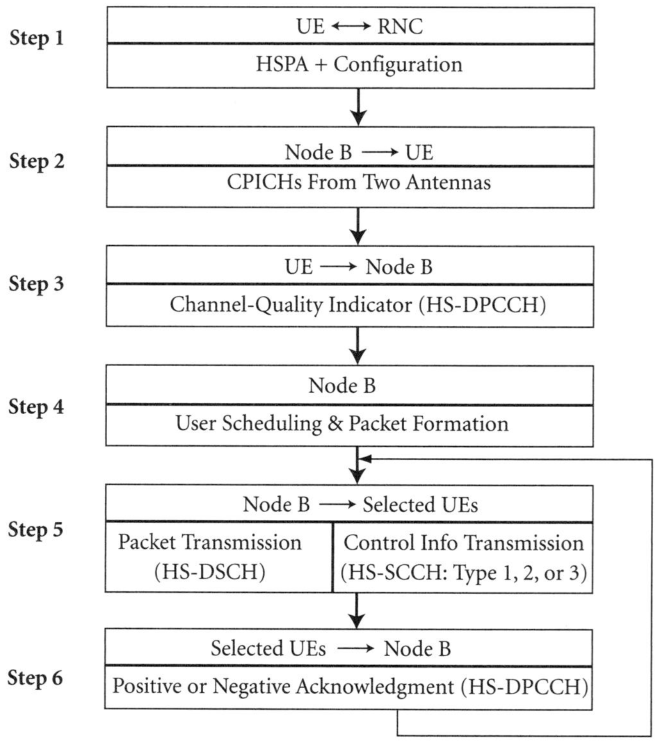

Figure D.1: Downlink Data Transmission using HSPA+

RRC messages, such as RRC CONNECTION SETUP, RADIO BEARER SETUP, RADIO BEARER RECONFIGURATION, and ACTIVE SET UPDATE [384].

In Step 2, the UE observes the CPICHs (Common Pilot channels) that a Node B transmits from two antennas in the cell. The CPICHs are orthogonal to help the UE estimate the channel conditions for the propagation paths between each transmit antenna and the receiver. The Node B can ensure orthogonality of the CPICHs using one of two approaches: transmission of the P-CPICH (Primary Common Pilot channel) with the same OVSF code and primary scrambling code but different (orthogonal) modulation patterns from different antennas or transmission of the P-CPICH on one OVSF code from one antenna and the S-CPICH on another OVSF code from the second transmit antenna, with the same modulation pattern (all 0s) and the same primary scrambling code on both antennas [263].

In Step 3, the UE provides feedback to the Node B using the HS-DPCCH (High-Speed Dedicated Physical Control channel) to facilitate the scheduler's decision-making process. If the UE is configured for closed-loop transmit diversity, the feedback includes a CQI (Channel-Quality Indicator) on the HS-DPCCH and feedback information on the DPCCH. The CQI conveys the supportable data rate at 10% or less instantaneous block error rate. The UE uses a specific CQI table based on its category, 64-QAM configuration, and the type of CQI report [376]. Two types of CQI reports, Type A and Type B, are defined in support of MIMO, with the UTRAN configuring the UE with a relative proportion of these reports (e.g., 50% of Type A CQI reports and 50% of Type B CQI reports) [376].

A Type A CQI report is for MIMO transmission and indicates the supportable data rate in case of a single transport block or two transport blocks; the UE's preference for one or two transport blocks is implicit in the CQI value. The Type A CQI report with CQI greater than 30 indicates the UE preference for a dual transport block transmission from the Node B; CQI occupies eight bits for this type of report. The Type B CQI report is for non-MIMO transmission and indicates the supportable data rate for a single transport block; CQI occupies five bits for this report. The feedback information conveys the UE-recommended phase difference, 0° or 180°, between the two transmit antenna signals that would maximize the received signal strength at the UE receiver. When the UE is configured with MIMO, the HS-DPCCH includes the CQI and PCI (Precoding Control Indication). The PCI helps the Node B construct primary and secondary precoding vectors in which the primary transport block is transmitted using the primary precoding vector and the secondary transport block is transmitted using the secondary precoding vector. These vectors specify the amplitude and phase adjustments[555] to the basic scrambled chips so that the receiver's task of separating the antenna signals is simplified when spatial multiplexing is exploited.

In Step 4, an implementation-specific scheduling algorithm at the Node B selects the UEs for data transmission in the next subframe and forms packets for them based on feedback, such as the CQI, PCI, and feedback information. The scheduler needs to consider the UE capabilities, amount of data for the UE, quality-of-service (QoS) promises made to the UE, and the transmission mode configured for the UE as well.

Of course, the transmission mode can be changed through RRC signaling based on reported CQIs over a longer time period. The Node B in Step 5 sends the control information on the HS-SCCH and the actual data on the HS-DSCH (which is mapped onto PDSCHs). HSPA+ contains three types of HS-SCCHs, called, funnily enough, Type 1, Type 2, and Type 3 [377]. The Node B may choose an HS-SCCH–less operation for a UE, and, in this case, the Node B directly sends an HS-DSCH. The UE uses preconfigured parameters, such as a set of transport block sizes, one or two OVSF codes, and a modulation scheme, to blindly decode the HS-DSCH for such operation. If the UE is configured for operation other than HS-SCCH–less, the Node B transmits HS-SCCH Type 1 or HS-SCCH Type 3 in Step 5 to specify the characteristics of the associated HS-DSCH. HS-SCCH Type 1 provides backward-compatibility with pre-R7 operation so that pre-R7 UEs can correctly interpret its contents. An R7 UE interprets HS-SCCH Type 1 differently if the UE is configured for 64-QAM. In such a case, the UE interprets the Modulation Scheme and Channelization and Code Set

[555]Let A be the scrambled chip from the primary transport block and X be the scrambled chip generated from the secondary transport block. Antenna 1 transmits signal $(W_1 * A + W_3 * X)$, and Antenna 2 transmits signal $(W_2 * A + W_4 * X)$. $W_1 = W_3 = 1/\sqrt{2}$, $W_4 = -W_2$, and W_2 takes one of four values: $(1+j)/2$, $(1-j)/2$, $(-1+j)/2$, and $(-1-j)/2$. The PCI specifies W_2 [376].

fields differently to determine the modulation scheme (QPSK, 16-QAM, and 64-QAM) and the OVSF codes associated with the HS-DSCH. HS-SCCH Type 2 is irrelevant to the first transmission or a new transmission of a packet. When a packet sent as part of the HS-SCCH–less operation is retransmitted, HS-SCCH Type 2 becomes relevant and specifies parameters such as the retransmission number and the transport block size. HS-SCCH Type 3 supports MIMO transmission and specifies parameters such as the number (which can be 1 or 2) and size of transport blocks, the precoding weight, and the modulation scheme.

If the UE successfully decodes the packet, it sends a positive acknowledgment (ACK) to the Node B in Step 6; otherwise, it sends a negative acknowledgment (NACK). Such an ACK or NACK is sent over the HS-DPCCH. The new HS-DPCCH structure helps send individual ACKs or NACKs in response to each transport block. Note that, in MIMO mode, the Node B may have sent (one or) two transport blocks and that the UE is able to respond to each transport block.

If the Node B receives a NACK from the UE, the Node B retransmits the transport block in the form of selected turbo code symbols. As in basic HSDPA, the HARQ (hybrid automatic repeat request) retransmissions are asynchronous, and the Node B performs a certain number of retransmissions before giving up at the physical layer. HS-SCCH Type 1 has a field called *New Data Indicator*, which helps the UE decide if the information received on the HS-DSCH is a new transmission or a retransmission. HS-SCCH Type 2 is actually transmitted for the retransmissions of a transport block; a total of three transmissions, i.e., the initial new transmission and two retransmissions, are allowed for the HS-SCCH–less operation. (It no longer stays HS-SCCH–less, does it?[556]) HS-SCCH Type 2 indicates the retransmission number. HS-SCCH Type 3 has no New Data Indicator field; the UE implicitly finds out about new transmission or retransmission by observing the Redundancy Version (RV) field of the HS-SCCH. $RV = 0$ implies a new transmission, and $RV \neq 0$ implies a retransmission. Simply put, Chase combining is no longer possible with the MIMO mode; Incremental Redundancy is required.

Now that we know how DL data transmission occurs in HSPA+, let's turn our attention to the UL data transmission. Most of the processing in HSPA+ is similar to that for HSUPA as shown in Figure D.2.

In Step 1, the UE is configured with a specific E-TFCI (E-DCH Transport Format Combination Indicator) table that corresponds to the larger transport block sizes available in HSPA+. Such configuration, in general, can be completed via traditional RRC messages, such as the RADIO BEARER SETUP message. The UE in Step 2 sends a scheduling request to the serving Node B by sending *Scheduling Information* on the E-DCH, indicating the amount of data in the buffer along with other information associated with the data to be transmitted, such as the power headroom and the priority of the logical channel. The Node B scheduler issues an Absolute and/or Relative Grant to the UE using the E-AGCH (E-DCH Absolute Grant channel) and/or E-RGCH (E-DCH Relative Grant channel) in Step 3. No HSPA+ enhancement is necessary for the

[556]For the first transmission of a packet, no HS-SCCH is transmitted; HS-SCCH Type 2 is transmitted only for packet retransmissions.

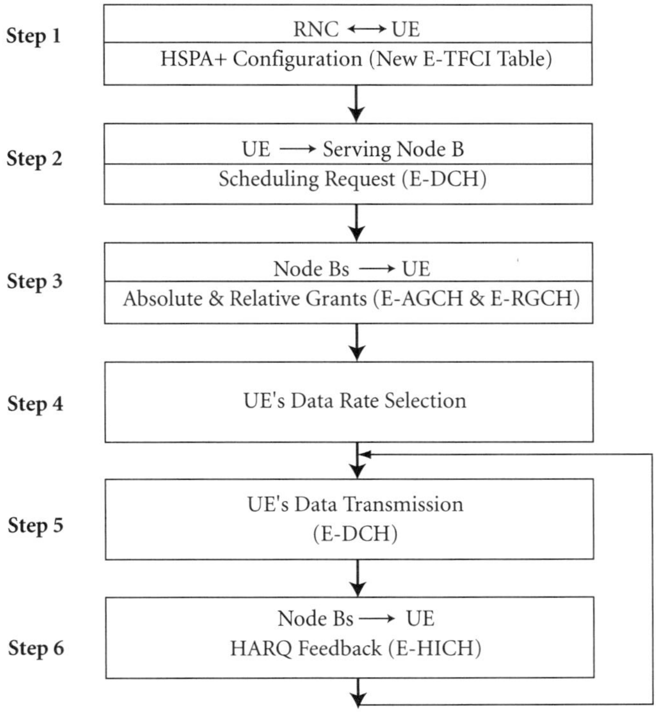

Figure D.2: Uplink Data Transmission using HSPA+

E-AGCH and E-RGCH. However, a larger grant would help the UE achieve a higher UL data rate. In HSUPA, even a very large grant would result in a 5.76 Mbps peak rate, and HSPA+ increases the peak rate to 11.5 Mbps.

In Step 4, the UE executes an algorithm to determine the overall Serving Grant it can use based on the Absolute and/or Relative Grants received from the cells in the E-DCH Active Set. The value of the usable Serving Grant along with the available amount of data and the power headroom dictates the data rate the UE has chosen. When the use of QPSK modulation is inadequate to achieve the selected data rate, the UE starts using the 16-QAM modulation scheme. The UE specifies the selected transport block size on the E-DPCCH (E-DCH Dedicated Physical Control channel) in the form of E-TFCI and transmits the actual data on the E-DCH mapped onto one or more E-DPDCHs in Step 5.

In Step 6, the Node B sends an HARQ ACK or NACK on the E-HICH (E-DCH HARQ Indication channel) from the Active Set cells. The UE performs synchronous retransmission if necessary, using the Serving Grant determined for the first transmission of a transport block. In summary, the overall UL data transmission in HSPA+ is quite similar to that in HSUPA, with the major HSPA+ enhancements being the use of a new E-TFCI table and of 16-QAM when QPSK is inadequate.

D.3 HSPA+: BEYOND RELEASE 7

Table D.2 summarizes main features available in Release 8 and Release 9 [385]. Circuit-switched (CS) speech over HSPA enables the UE to make a regular CS call with the CS core network but with the help of HSPA over the air interface rather than a CS radio bearer. Such enhancement avoids reservation of an OVSF code dedicated for the UE's voice call since HSPA channels shared among multiple UEs would be used to carry voice. A higher overall voice capacity can be expected as a result of this feature. The AMR (adaptive multirate) speech would still be traveling on the I_u interface between the RNC and the CS core network, just as in a regular CS call. The RNC would translate between the CS speech and HSPA speech. Specifically, the PDCP layer introduces a counter to help maintain the order of the speech frames.

Though Release 7 does support MIMO and 64-QAM in the DL, both cannot be used simultaneously. Release 8 removes such restriction, and, 2×2 MIMO in the DL can be used along with 64-QAM. Such a combination will allow the peak rate to reach 42 Mbps.

Dual-cell HSDPA allows two adjacent carrier frequencies to be used for HS-DSCH transmission to a given UE. 64-QAM can be used for data transmission, but MIMO is unavailable. The peak data rate can then be 21 Mbps $* 2 = 42$ Mbps where 21 Mbps is the peak data rate achievable by 64-QAM in the absence of MIMO. Enhanced

Table D.2: HSPA+ Features for Release 8 and Release 9

Feature	Primary Benefits	Release
Circuit-Switched Speech over HSPA	Higher voice capacity	8
MIMO with 64-QAM	Higher peak rate and throughput in the DL	8
Dual-Cell HSDPA	Higher peak rate and throughput in the DL	8
Enhanced HSPA Cell Change	Higher throughput and lower latency in the DL during inter-cell mobility	8
Enhanced CELL_FACH	Higher UL throughput	8
Improved Layer 2 for Uplink	Higher UL throughput	8
Dual-Cell HSDPA with MIMO	Higher peak rate and throughput in the DL	9
Dual-Band Dual-Cell HSDPA	Higher peak rate and throughput in the DL	9
Dual-Cell HSUPA	Higher peak rate and throughput in the UL	9

HSPA cell change facilitates faster handover transition from one cell to another, in which the UE monitors an HS-SCCH from the target cell to determine if HSDPA operation should be discontinued in the current cell and resumed in the target cell. Compared to a pure RRC signaling-based handover, such faster cell change results in a shorter gap between the transmission from the old serving cell and the transmission from the new serving cell.

The enhanced CELL_FACH state enables the use of the E-DCH instead of the RACH (Random Access channel), improving the packet delay and throughput in the uplink. Release 8 also improves layer 2 for the uplink by allowing larger RLC PDU size, flexible size RLC PDU, and MAC segmentation. Release 9 now allows the combination of the dual-cell HSDPA feature and MIMO, increasing the peak rate to 42 Mbps $* 2 = 84$ Mbps where 42 Mbps is the peak data rate achievable with the combined use of MIMO and 64-QAM on a single carrier frequency. Another feature being discussed for Release 9 is dual-band dual-cell HSDPA, in which the two carrier frequencies doing the HSDPA transmission no longer need to be adjacent. Dual-cell HSUPA allows the UE to use two adjacent carrier frequencies for E-DCH transmission, increasing the peak data rate to 11.5 Mbps$*2 = 23$ Mbps in the uplink. A total of twenty-four HSDPA UE categories exist in Release 8, with Categories 19 to 24 supporting features such as combined 64-QAM and MIMO as well as dual-cell HSDPA.

You are encouraged to refer to the 25.x series of documents for Releases 7, 8, and 9 to dive deeper into HSPA+ [386].

References

1. www.cnn.com/2005/TECH/01/03/cnn25.top25.innovations/.
2. Global Mobile Statistics 2012. mobithinking.com/mobile-marketing-tools/latest-mobile-stats#subscribers.
3. www.ctia.org/research_statistics/statistics/index.cfm/AID/10202.
4. www.medi-news.com/article2753.html.
5. www.cbsnews.com/stories/2004/07/22/tech/main631231.shtml.
6. www.ieee.org/web/aboutus/history_center/conferences/comsoc/timelines.html.
7. www.ctia.org/content/index.cfm/AID/101.
8. www.sri.com/policy/csted/reports/sandt/techin2/chp4.html.
9. Claude Shannon. Communication in the Presence of Noise. *Proceedings of the IRE*, 37(1):10–21, January 1949.
10. www.fcc.gov/Bureaus/OPP/working_papers/oppwp38chart.pdf.
11. www.ntia.doc.gov/osmhome/allochrt.pdf.
12. www.comsearch.com/bandrange/.
13. Spectrum Dashboard. reboot.fcc.gov/reform/systems/spectrum-dashboard.
14. W. C. Y. Lee. *Mobile Cellular Telecommunications: Analog and Digital Systems*. McGraw-Hill, 2nd edition, 1995.

Cellular Communications, First Edition. Nishith D. Tripathi and Jeffrey H. Reed

15. High-Capacity Mobile Telephone System. Technical report, Bell Laboratories, December 1971.
16. V. H. MacDonald. The Cellular Concept. *Bell System Technical Journal*, 58:15–42, January 1979.
17. en.wikipedia.org/wiki/Mobile_phone.
18. en.wikipedia.org/wiki/Radio_propagation.
19. www.nortel.com/solutions/wireless/features/2005/wireless_advanced_technologies.html.
20. Lal C. Godara. Applications of Antenna Arrays to Mobile Communications, Part I: Performance Improvement, Feasibility, and System Considerations. *Proceedings of the IEEE*, 85(7), July 1997.
21. www.cdg.org/technology/cdma_technology/WLL/wll_overview_v1.pdf.
22. W. C. Y. Lee. Elements of Cellular Mobile Radio Systems. *IEEE Transactions on Vehicular Technology*, VT-35(2):48–56, May 1986.
23. T. S. Rappaport and L. B. Milstein. Effects of Radio Propagation Path Loss on DS-CDMA Cellular Frequency Re-Use Efficiency for the Reverse Channel. *IEEE Transactions on Vehicular Technology*, 41(3):231–242, August 1992.
24. J. Jacobsmeter. Improving Throughput and Availability of Cellular Digital Packet Data (CDPD). In *Proc. Virginia Tech 4th Symposium on Wireless Personal Communications*, pages 18.1–18.12, June 1994.
25. wireless.fcc.gov/services/index.htm?job=operations&id=broadband_pcs.
26. hraunfoss.fcc.gov/edocs_public/attachmatch/FCC-02-247A1.pdf.
27. Rand McNally 1992 Commercial Atlas & Marketing Guide.
28. wireless.fcc.gov/auctions/default.htm?job=auction_factsheet&id=66.
29. www.ero.dk/fc2e8966-1db9-445b-a8d5-e5c7cf825cc2?mid=97605DCA-E7D9-4E5A-84B7-1E84586C7205&frames=no.
30. www.gsmworld.com/using/spectrum/frequencies.shtml.
31. Consultation Paper on Allocation and Pricing of Spectrum for 3G Services and Broadband Wireless Access, June 2006.
32. TRAI Issues Recommendations on Allocation and Pricing of Spectrum for 3G Services and Broadband Wireless Access, September 2006.
33. en.wikipedia.org/wiki/Bell_System_divestiture.
34. en.wikipedia.org/wiki/Cingular.
35. news.com.com/Protective+parents+Gold+for+cellular+services/2100-1039_3-6058756.html.
36. files.shareholder.com/downloads/QCOM/93727467x0x69425/d81d0439-9484-4e8a-ae5c-a33db348f60e/Q107%20ER%20FINAL.pdf.
37. files.shareholder.com/downloads/QCOM/93727467x0x69424/7c22ccec-1213-4fab-b64c-6ae55e6dcdf5/q107_web_final.pdf.
38. D. O. Reudink. Properties of Mobile Radio Propagation Above 400 MHz. *IEEE Transactions on Vehicular Technology*, 23(4):143–159, November 1974.

39. Error Function. en.wikipedia.org/wiki/Error_function.

40. Theodore S. Rappaport. *Wireless Communications: Principles and Practice*. Prentice Hall, 2nd edition, 2002.

41. Annual Report and Analysis of Competitive Market Conditions with Respect to Mobile Wireless, Including Commercial Wireless Services. Technical Report FCC 11-103, FCC, June 2011. Fifteenth Report.

42. www.phonemag.com/index.php/weblog/read_more/20061128nokia_games_study_reveals_changing_mobile_gaming_habits_worldwide/.

43. www.sprint.com/index.html.

44. news.com.com/Verizon+Wireless+takes+the+road+less+traveled/2008-1035_3-6116181.html.

45. www.qualcomm.com/mediaflo/index.shtml.

46. www.lge.com/about/press_release/detail/PRO|NEWS$∧PRE|MENU$∧PRER|MENU_20134_PRE|MENU.jhtml.

47. en.wikipedia.org/wiki/Audio_Frequency.

48. Jeremy Bradbury. Linear Predictive Coding, December 2000. ecaaser3.ecaa.ntu.edu.tw/jjchou/jjchouweb/Authorization/Progress Reports and Schedules/PhD Student/DavidHu/PDF/Algorithm/lpc_paper.pdf.

49. Administration of Parameter Value Assignments for CDMA2000 Spread Spectrum Standards, Release H. Technical Report 3GPP2 C.R1001-H, Third Generation Partnership Project 2, July 2011. Version 1.0.

50. Enhanced Variable Rate Codec, Speech Service Options 3, 68, and 70 for Wideband Spread Spectrum Digital Systems. Technical Report 3GPP2 C.S0014-C, Third Generation Partnership Project 2, January 2007. Version 1.0.

51. AMR Speech Codec; General Description. Technical Report 3G TS 26.071, Third Generation Partnership Project, August 1999. Version 3.0.1.

52. AMR Speech Codec; Transcoding Functions. Technical Report 3G TS 26.090, Third Generation Partnership Project, December 1999. Version 3.1.0.

53. Performance Characterization of the Adaptive Multi-Rate (AMR) Speech Codec. Technical Report TR 26.975, Third Generation Partnership Project, December 2004. Version 6.0.0.

54. AMR Speech Codec Frame Structure. Technical Report 3GPP TS 26.101, Third Generation Partnership Project, March 2002. Version 3.3.0.

55. R. Gallager. Low-Density Parity-Check Codes. *IRE Transactions on Information Theory*, 8(1):21–28, January 1962.

56. R. Gallager. Low-Density Parity-Check Codes, 1963. ms. published online at www.rle.mit.edu/rgallager/documents/ldpc.pdf.

57. Multiplexing and Channel Coding (FDD) (Release 1999). Technical Report 3GPP TS 25.212 (2002-09), Third Generation Partnership Project, Technical Specification Group Radio Access Network, September 2002. Version 3.11.0.

58. Chip Fleming. A Tutorial on Convolutional Coding with Viterbi Decoding. November 2006.

59. Joachim Hagenauer and Peter Hoeher. A Viterbi Algorithm with Soft-Decision Outputs and its Applications. In *IEEE Global Telecommunications Conference and Exhibition*, pages 1680–1686, 1989.

60. L. Bahl, J. Cocke, F. Jelinek, and J. Raviv. Optimal Decoding of Linear Codes for Minimizing Symbol Error Rate. *IEEE Transactions on Information Theory*, 20(2):284–287, March 1974.

61. M.C. Valenti and J. Sun. *Handbook of RF and Wireless Technologies*, chapter Turbo Codes, pages 375–399. Newnes Press, 2004. `csee.wvu.edu/~mvalenti/documents/DOWLA-CH12.pdf`.

62. Iterative Solutions. *CML Source Code, cml.1.10.zip*. `www.iterativesolutions.com/download.htm`.

63. P. Robertson, E. Villeburn, and P. Hoeher. A Comparison of Optimal and Suboptimal MAP Decoding Algorithms Operating in the Log Domain. In *Proceedings of the International Conference on Communications*, pages 1009–1013, 1995.

64. A. Viterbi. An Intuitive Justification and a Simplified Implementation of the MAP Decoder for Convolutional Codes. *IEEE Journal on Selected Areas in Communications*, pages 260–264, February 1998.

65. John G. Proakis. *Digital Communications*. McGraw-Hill, 3rd edition, 1995.

66. Siegmund M. Redl, Matthias K. Weber, and Malcolm W. Oliphant. *An Introduction to GSM*. Artech House Publishers, 1995.

67. CDMA2000 High Rate Packet Data Air Interface Specification. Technical Report 3GPP2 C.S0024-A, Third Generation Partnership Project 2, July 2005. Version 2.0.

68. `en.wikipedia.org/wiki/Friis_formulas_for_noise`.

69. `en.wikipedia.org/wiki/Tower_Mounted_Amplifier`.

70. `en.wikipedia.org/wiki/Radio_frequency`.

71. `people.deas.harvard.edu/~jones/es151/prop_models/propagation.html`.

72. M. J. Feuerstein, K. L. Blackard, T. S. Rappaport, S. Y. Seidel, and H. H. Xia. Path Loss, Delay Spread, and Outage Models as Functions of Antenna Height for Microcellular System Design. *IEEE Transactions on Vehicular Technology*, 43(3):487–498, August 1994. Parts 1–2.

73. W. C. Y. Lee. *Mobile Communications Engineering*. McGraw-Hill, 1985.

74. K. Bullington. Radio Propagation at Frequencies Above 30 Megacycles. In *Proceedings of the IEEE*, volume 35, pages 1122–1136, 1947.

75. S. Y. Seidel, T. S. Rappaport, S. Jain, M. Lord, and R. Singh. Path Loss, Scattering, and Multipath Delay Statistics in Four European Cities for Digital Cellular and Microcellular Radio Telephone. *IEEE Transactions on Vehicular Technology*, 40(4):721–730, November 1991.

76. Propagation Prediction Models. Technical report, COST-231.

77. Forsk. *Atoll—RF Planning and Optimisation Software: Technical Reference Guide*, 2009. Version 2.8.0.

78. Morten Tolstrup. *Indoor Radio Planning: A Practical Guide for GSM, DCS, UMTS, HSPA, and LTE*. Wiley, 2nd edition, 2011.

79. WiMAX's Technology for LOS and NLOS Environments. Technical report, WiMAX Forum, August 2004. `www.wimaxforum.org/technology/downloads/WiMAXNLOSgeneral-versionaug04.pdf`.

80. Masaharu Hata. Empirical Formula for Propagation Loss in Land Mobile Radio Services. *IEEE Transactions on Vehicular Technology*, VT-29(3):317–325, August 1980.

81. Y. Okumura, et al. Field Strength and Its Variability in UHF and VHF Land-Mobile Radio Service. *Rev. Elect. Commun. Lab.*, 16, 1968.

82. H. H. Xia, H. L. Bertoni, L. R. Maciel, A. Lindsay-Stewart, and R. Rowe. Microcellular Propagation Characteristics for Personal Communications in Urban and Suburban Environments. *IEEE Transactions on Vehicular Technology*, 43(3):743–752, August 1994.

83. H. Xia and H. L. Bertoni. Diffraction of Cylindrical and Plane Waves by an Array of Absorbing Half Screens. *IEEE Transactions on Antennas and Propagation*, 40(2):170–177, February 1992.

84. L. R. Maciel, H. L. Bertoni, and H. H. Xia. Unified Approach to Prediction of Propagation Over Buildings for All Ranges of Base Station Antenna Height. *IEEE Transactions on Vehicular Technology*, 43(1):35–41, 1993.

85. V. Graziano. Propagation Correlations at 900 MHz. *IEEE Transactions on Vehicular Technology*, 27(4):182–189, November 1978.

86. D. C. Cox, R. R. Murray, and A. W. Norris. 800 MHz Attenuation Measured in and Around Suburban Houses. Technical Report 6, AT&T Bell Laboratory Technical Journal, 1984.

87. R. C. Bernhardt. Macroscopic Diversity in Frequency Reuse Systems. *IEEE Journal on Selected Areas in Communications*, SAC-5:862–878, 1987.

88. M. Gudmundson. Correlation Model for Shadow Fading in Mobile Radio Systems. *Electronics Letters*, 27(23):2145–2146, November 1991.

89. D. Giancristofaro. Correlation Model for Shadow Fading in Mobile Radio Channels. *Electronics Letters*, 32(11):958–959, May 1996.

90. F. Graziosi and F. Santucci. A General Correlation Model for Shadow Fading in Mobile Radio Systems. *IEEE Communications Letters*, 6(3):102–104, March 2002.

91. CDMA2000 Evaluation Methodology. Technical Report 3GPP2 C.R1002-0 v1.0, Third Generation Partnership Project 2, December 2004. Version 1.0.

92. Selection Procedures for the Choice of Radio Transmission Technologies of the UMTS. Technical Report TR 101 112 V3.2.0 (1998-04), Universal Mobile Telecommunications System (UMTS). UMTS 30.03 Version 3.2.0.

93. Guidelines for Evaluation of Radio Transmission Technologies for IMT-2000. Technical Report ITU-R M.1225, International Telecommunication Union.

94. T. S. Rappaport, S. Y. Seidel, and R. Singh. 900 MHz Multipath Propagation Measurements for U.S. Digital Cellular Radio Telephone. *IEEE Transaction on Vehicular Technology*, pages 132–139, May 1990.

95. W. C. Y. Lee. *Mobile Cellular Telecommunications Systems*. McGraw-Hill, 1989.

96. R. Steele, editor. *Mobile Radio Communications*. IEEE Press, 1994.

97. R. H. Clark. A Statistical Theory of Mobile Reception. *Bell Systems Technical Journal*, 47:957–1000, 1968.

98. W. C. Jakes, editor. *Microwave Mobile Communications*. John Wiley & Sons, 1974.

99. T. S. Rappaport. Characterization of UHF Multipath Radio Channels in Factory Building. *IEEE Transactions on Antennas and Propagation*, 37(8):1058–1069, August 1989.

100. D. C. Cox. Delay Doppler Characteristics of Multipath Delay Spread and Average Excess Delay for 910 MHz Urban Mobile Radio Paths. *IEEE Transactions on Antennas and Propagation*, AP-20(5):625–635, September 1972.

101. K. Pahlavan and A. H. Levesque. *Wireless Information Networks*. John Wiley & Sons, 1995.

102. Michel C. Jeruchim, Philp Balaban, and K. Sam Shanmugan. *Simulation of Communication Systems*. Plenum Press, 1992.

103. C. Loo and N. Secord. Computer Models for Fading Channels with Applications to Digital Transmission. *IEEE Transactions on Vehicular Technology*, 40(4):700–707, November 1991.

104. M. J. Gans. A Power-Spectral Theory of Propagation in the Mobile-Radio Environment. *IEEE Transactions on Vehicular Technology*, 21(1):27–38, February 1972.

105. J. I. Smith. A Computer Generated Multipath Fading Simulation for Mobile Radio. *IEEE Transactions on Vehicular Technology*, 24(3):39–40, August 1975.

106. `en.wikipedia.org/wiki/Rayleigh_fading`.

107. S. Von Eckardstein and K. Isaksson. Kanalmodeller För Radiotransmission. Master's thesis, Royal Institute of Technology, 1991.

108. P. Dent, G. E. Bottomly, and T. Croft. Jakes Fading Model Revisited. *Electronics Letters*, 29(13):1162–1163, June 1993.

109. Brad Grimes. Vinton Cerf: The Search Continues. *Government Computer News*, January 2006.

110. www.isoc.org/internet/history/.

111. Vint Cerf. A Brief History of the Internet and Related Networks. `www.isoc.org/internet/history/cerf.shtml`.

112. Barry M. Leiner, Vinton G. Cerf, David D. Clark, Robert E. Kahn, Leonard Kleinrock, Daniel C. Lynch, Jon Postel, Larry G. Roberts, and Stephen Wolff. A Brief History of the Internet. `www.isoc.org/internet/history/brief.shtml`.

113. William F. Slater, III. Internet History and Growth, September 2002. `www.isoc.org/internet/history/`.

114. Robert Kahn and Vinton Cerf. Al Gore and the Internet. September 2000. `amsterdam.nettime.org/Lists-Archives/nettime-l-0009/msg00311.html`.

115. Al Gore. `en.wikipedia.org/wiki/Al_gore`.

116. Internet Protocol Suite. `en.wikipedia.org/wiki/Internet_protocol_suite`.

117. Lydia Parziale, David T. Britt, Chuck Davis, Jason Forrester, Wei Liu, Carolyn Matthews, and Nicolas Rosselot. *TCP/IP Tutorial and Technical Overview*. IBM RedBooks, December 2006.

118. James F. Kurose and Keith W. Ross. *Computer Networking—A Top-Down Approach Featuring the Internet.* Addison-Wesley, 2nd edition, 2003.

119. Information Sciences Institute (University of Southern California). Transmission Control Protocol. Technical Report RFC793, Internet Engineering Task Force, September 1981. www.ietf.org/rfc.html.

120. M. Allman, V. Paxson, and W. Stevens. TCP Congestion Control. Technical Report RFC2581, Internet Engineering Task Force, April 1999. www.ietf.org/rfc.html.

121. M. Allman, V. Paxson, and W. Stevens. Increasing TCP. Technical Report RFC3390, Internet Engineering Task Force, October 2002. www.ietf.org/rfc.html.

122. R. Stewart, Q. Xie, K. Morneault, C. Sharp, H. Schwarzbauer, T. Taylor, I. Rytina, M. Kalla, L. Zhang, and V. Paxson. Stream Control Transmission Protocol. Technical Report RFC2960, Internet Engineering Task Force, October 2000. www.ietf.org/rfc.html.

123. L. Ong and J. Yoakum. Introduction to the Stream Control Transmission Protocol (SCTP). Technical Report RFC3286, Internet Engineering Task Force, May 2002. www.ietf.org/rfc.html.

124. J. Postel. User Datagram Protocol. Technical Report RFC768, Internet Engineering Task Force, August 1980. www.ietf.org/rfc.html.

125. P. Srisuresh and K. Egevang. Traditional IP Network Address Translator (Traditional NAT). Technical Report RFC3022, Internet Engineering Task Force, January 2001. www.ietf.org/rfc.html.

126. P. Srisuresh and M. Holdrege. IP Network Address Translator (NAT) Terminology and Considerations. Technical Report RFC2663, Internet Engineering Task Force, August 1999. www.ietf.org/rfc.html.

127. Internet Protocol. Technical Report RFC791, Internet Engineering Task Force, September 1981. www.ietf.org/rfc.html.

128. P. Almquist. Type of Service in the Internet Protocol Suite. Technical Report RFC1349, Internet Engineering Task Force, July 1992. www.ietf.org/rfc.html.

129. J. Mogul and J. Postel (ISI). Internet Standard Subnetting Procedure. Technical Report RFC950, Internet Engineering Task Force, August 1985. www.ietf.org/rfc.html.

130. Jeffrey Mogul. Broadcasting Internet Datagrams. Technical Report RFC919, Internet Engineering Task Force, October 1984. www.ietf.org/rfc.html.

131. Jeffrey Mogul. Broadcasting Internet Datagrams in the Presence of Subnets. Technical Report RFC922, Internet Engineering Task Force, October 1984. www.ietf.org/rfc.html.

132. D. Grossman. New Terminology and Clarifications for Diffserv. Technical Report RFC3260, Internet Engineering Task Force, April 2002. www.ietf.org/rfc.html.

133. K. Ramakrishnan, S. Floyd, and D. Black. The Addition of Explicit Congestion Notification (ECN) to IP. Technical Report RFC3168, Internet Engineering Task Force, September 2001. www.ietf.org/rfc.html.

134. J. Postel. Internet Control Message Protocol. Technical Report RFC792, Internet Engineering Task Force, September 1981. www.ietf.org/rfc.html.

135. en.wikipedia.org/wiki/Ping_of_death. Wikipedia.

136. *Ethernet Technologies*. www.cisco.com/en/US/docs/internetworking/technology/handbook/Ethernet.html.

137. Ethernet. en.wikipedia.org/wiki/Ethernet.

138. Metro Ethernet. en.wikipedia.org/wiki/Metro_Ethernet.

139. Ralph Santitoro. Metro Ethernet Services—A Technical Overview. Technical report, Metro Ethernet Forum, 2003. metroethernetforum.org/PDF_Documents/metro-ethernet-services.pdf.

140. metroethernetforum.org.

141. Asynchronous Transfer Mode. en.wikipedia.org/wiki/Asynchronous_transfer_mode.

142. Mark Jeffrey. ATM—The Ultimate Broadband Solution? *IEE Colloquium on "What's New in Telecommunications?"*, pages 3/1–314, January 1994.

143. ATM. www.protocols.com/pbook/atm.htm#AAL2.

144. Martyn J. Riley and Iain E.G.Richardson. Quality of Service and the ATM Adaptation Layers. *IEE Colloquium on ATM in Professional and Consumer Applications*, pages 2/1–2/5, May 1997.

145. C. Hedrick. Routing Information Protocol. Technical Report RFC1058, Internet Engineering Task Force, June 1998. www.ietf.org/rfc.html.

146. J. Moy. OSPF Version 2. Technical Report RFC2328, Internet Engineering Task Force, April 1998. www.ietf.org/rfc.html.

147. Y. Rekhtera, T. Li, and S. Hares. A Border Gateway Protocol 4 (BGP-4). Technical Report RFC4271, Internet Engineering Task Force, January 2006. www.ietf.org/rfc.html.

148. Routing Protocol. en.wikipedia.org/wiki/Routing_protocol.

149. G. Malkin. RIP Version 2. Technical Report RFC2453, Internet Engineering Task Force, November 1998. www.ietf.org/rfc.html.

150. David C. Plummer. An Ethernet Address Resolution Protocol or Converting Network Protocol Addresses to 48.bit Ethernet Address for Transmission on Ethernet Hardware. Technical Report RFC826, Internet Engineering Task Force, November 1982. www.ietf.org/rfc.html.

151. Ross Finlayson, Timothy Mann, Jeffrey Mogul, and Marvin Theimer. A Reverse Address Resolution Protocol. Technical Report RFC903, Internet Engineering Task Force, June 1984. www.ietf.org/rfc.html.

152. E. Rosen, A. Viswanathan, and R. Callon. Multiprotocol Label Switching Architecture. Technical Report RFC3031, Internet Engineering Task Force, January 2001. www.ietf.org/rfc.html.

153. J. Rosenberg, H. Schulzrinne, G. Camarillo, A. Johnston, J. Peterson, R. Sparks, M. Handley, and E. Schooler. SIP: Session Initiation Protocol. Technical Report RFC3261, Internet Engineering Task Force, June 2002. www.ietf.org/rfc/rfc3261.txt.

154. J. Franks, P. Hallam-Baker, J. Hostetler, S. Lawrence, P. Leach, A. Luotonen, and L. Stewart. HTTP Authentication: Basic and Digest Access Authentication. Technical Report RFC2617, Internet Engineering Task Force, June 1999. www.ietf.org/rfc.html.

155. M. Handley and V. Jacobson. SDP: Session Description Protocol. Technical Report RFC2327, Internet Engineering Task Force, April 1998. www.ietf.org/rfc.html.

156. H. Schulzrinne, S. Casner, R. Frederick, and V. Jacobson. RTP: A Transport Protocol for Real-Time Applications. Technical Report RFC3550, Internet Engineering Task Force, July 2003. www.ietf.org/rfc.html.

157. H. Schulzrinne and S. Casner. RTP Profile for Audio and Video Conferences with Minimal Control. Technical Report RFC3551, Internet Engineering Task Force, July 2003. www.ietf.org/rfc.html.

158. R. Droms. Dynamic Host Configuration Protocol. Technical Report RFC2131, Internet Engineering Task Force, March 1997. www.ietf.org/rfc.html.

159. C. Perkins. IP Mobility Support for IPv4. Technical Report IETF RFC3344, Internet Engineering Task Force, August 2002. www.ietf.org/rfc.html.

160. S. Gundavelli, K. Leung, V. Devarapalli, K. Chowdhury, and B. Patil. Proxy Mobile IPv6. (RFC5213), August 2008. www.ietf.org/rfc.html.

161. P. Mockapetris. Domain Names—Concepts and Facilities. (RFC1034), November 1987. www.ietf.org/rfc.html.

162. P. Mockapetris. Domain Names—Implementation and Specification. (RFC1035), November 1987. www.ietf.org/rfc.html.

163. C. Rigney, S. Willens, A. Rubens, and W. Simpson. Remote Authentication Dial In User Service (RADIUS). (RFC2865), June 2000. www.ietf.org/rfc.html.

164. C. Rigney. RADIUS Accounting. Technical Report RFC2866, Internet Engineering Task Force, June 2000. www.ietf.org/rfc.html.

165. Diameter (protocol). en.wikipedia.org/wiki/Diameter_(protocol).

166. P. Calhoun, J. Loughney, E. Guttman, G. Zorn, and J. Arkko. Diameter Base Protocol. Technical Report RFC3588, Internet Engineering Task Force, September 2003. www.ietf.org/rfc.html.

167. R. Braden, D. Clark, and S. Shenker. Integrated Services in the Internet Architecture: An Overview. (RFC1633), June 1994. www.ietf.org/rfc.html.

168. R. Braden (Ed.), L. Zhang, S. Berson, S. Herzog, and S. Jamin. Resource ReSerVation Protocol (RSVP) - Version 1 Functional Specification. Technical Report RFC2205, Internet Engineering Task Force, September 1997. www.ietf.org/rfc.html.

169. S. Blake and D. Black and M. Carlson and E. Davies and Z. Wang and W. Weiss. An Architecture for Differentiated Services. (RFC2475), December 1998. www.ietf.org/rfc.html.

170. K. Nichols, S. Blake, F. Baker, and D. Black. Definition of the Differentiated Services Field (DS Field) in the IPv4 and IPv6 Headers. Technical Report RFC2474, Internet Engineering Task Force, December 1998. www.ietf.org/rfc.html.

171. Implementing Quality of Service Policies with DSCP. Technical Report 10103, Cisco Systems, Inc. www.cisco.com/application/pdf/paws/10103/dscpvalues.pdf.

172. J. Babiarz, K. Chan, and F. Baker. Configuration Guidelines for DiffServ Service Classes. Technical Report RFC4594, Internet Engineering Task Force, August 2006. www.ietf.org/rfc.html.

173. J. Heinanen, Telia Finland, F. Baker, W. Weiss, and J. Wroclawski. Assured Forwarding PHB Group. Technical Report RFC2597, Internet Engineering Task Force, June 1999. www.ietf.org/rfc.html.

174. V. Jacobson, K. Nichols, and K. Poduri. An Expedited Forwarding PHB. (RFC2598), June 1999. www.ietf.org/rfc.html.

175. G. Dommety. Key and Sequence Number Extentions to GRE. Technical Report RFC 2890, September 2000.

176. General Packet Radio System (GPRS) Tunnelling Protocol—User Plane (GTPv1-U). Technical Report TS 29.281, Third General Partnership Project, September 2009. Release 8, Version 8.3.0.

177. S. Kent and K. Seo. Security Architecture for the Internet Protocol. Technical Report RFC4301, December 2005.

178. Robust Header Compression. en.wikipedia.org/wiki/Robust_Header_Compression. Wikipedia.

179. C. Bormann, C. Burmeister, M. Degermark, H. Fukushima, H. Hannu, L-E. Jonsson, R. Hakenberg, T. Koren, K. Le, Z. Liu, A. Martensson, A. Miyazaki, K. Svanbro, T. Wiebke, T. Yoshimura, and H. Zheng. Robust Header Compression (ROHC): Framework and Four Profiles: RTP, UDP, ESP, and Uncompressed. Technical Report RFC309, Internet Engineering Task Force, July 2001. www.ietf.org/rfc.html.

180. S. Deering and R. Hinden. Internet Protocol, Version 6 (IPv6). (RFC2460), December 1998. www.ietf.org/rfc.html.

181. R. Hinden, S. Deering, and E. Nordmark. IPv6 Global Unicast Address Format. (RFC3587), August 2003. www.ietf.org/rfc.html.

182. R. Hinden and S. Deering. IP Version 6 Addressing Architecture. (RFC4291), February 2006. www.ietf.org/rfc.html.

183. S. Thomson and T. Narten. IPv6 Stateless Address Autoconfiguration. (RFC2462), December 1998. www.ietf.org/rfc.html.

184. R. Droms. Stateless Dynamic Host Configuration Protocol (DHCP) Service for IPv6. (RFC3736), April 2004. www.ietf.org/rfc.html.

185. L. Delgrossi and L. Berger. Internet Stream Protocol Version 2 (ST2) Protocol Specification—Version ST2+. Technical report, August 1995.

186. Mobile Subscriptions Market Share Worldwide. www.gsacom.com/downloads/charts/GSM_market_share_global.php4.

187. GSM Celebrates 20 Years, July 2011. www.nokiasiemensnetworks.com/news-events/press-room-press-releases/gsm-celebrates-20-years.

188. www.gsmworld.com/about/history.shtml.

189. Michel Mouly and Marie-Bernadette Pautet. The GSM System for Mobile Communication. *Telecom Publishing*, June 1992.

190. Network Architecture. Technical Report GSM 03.02, December 1996. Version 4.2.1.

191. Base Station Controller—Base Transceiver Station (BSC–BTS) Interface—Interface Principles. Technical Report GSM 08.52, August 1995. Version 4.2.0.

192. Base Station System—Mobile-Services Switching Centre (BSS–MSC) Interface—Interface Principles. Technical Report GSM 08.02, July 1995. Version 4.2.0.

193. Mobile-Services Switching Centre—Base Station System (MSC–BSS) Interface; Layer 3 Specification. Technical Report GSM 08.08, October 1998. Version 4.12.1.

194. Customised Applications for Mobile Network Enhanced Logic (CAMEL); Service definition—Stage 1. Technical Report GSM 02.78, March 1999. Version 5.6.0.

195. Physical Layer on the Radio Path: General Description. Technical Report 3GPP TS 05.01, December 1995. Version 4.6.0.

196. Channel Coding. Technical Report 3GPP TS 05.03, December 1997. Version 4.5.1.

197. Modulation. Technical Report 3GPP TS 05.04, September 1994. Version 4.0.3.

198. Radio Transmission and Reception. Technical Report 3GPP TS 05.05, December 1999. Version 4.23.1.

199. P. Laurent. Exact Approximation Construction of Digital Phase Modulations by Superposition of Amplitude Modulated Pukes (AMP). *IEEE Transactions on Communications*, 34(2), February 1986.

200. Radio Subsystem Link Control. Technical Report GSM 05.08, March 1998. Version 4.22.1.

201. Data Link (DL) Layer—General Aspects. Technical Report GSM 04.05, September 1994. Version 4.03.

202. Mobile Station—Base Station System (MS–BSS) Interface: Data Link (DL) Layer Specification. Technical Report GSM 04.06, September 1994.

203. `en.wikipedia.org/wiki/Pseudonym`.

204. Radio Subsystem Synchronization. Technical Report GSM 05.10, May 1996.

205. Security Related Network Functions. Technical Report GSM 03.20, August 1997. Version 4.4.1.

206. Multiplexing and Multiple Access on the Radio Path (Phase 2). Technical Report 3GPP TS 05.02, August 2001. Version 4.11.0.

207. Radio Network Planning Aspects. Technical Report GSM 03.30, October 1993.

208. Functions Related to Mobile Station (MS) in Idle Mode. Technical Report GSM 03.22, March 1996.

209. Mobility Management, Connection Management. Technical Report GSM 04.08.

210. Service Accessibility. Technical Report GSM 02.11, September 1996.

211. Discontinuous Reception (DRX) in the GSM System. Technical Report GSM 03.13, September 1994.

212. Layer 1: General Requirements. Technical Report GSM_04.04, September 1994.

213. Radio Subsystem Link Control. Technical Report 3GPP TS 45.008, Third Generation Partnership Project, November 2005. Version 4.17.0.

214. General Packet Radio Service (GPRS) Service Description: Stage 2. Technical Report 3GPP TS 23.060, Third Generation Partnership Project, December 2006. Version 3.17.0, Release 1999.

215. General Packet Radio Service (GPRS); Mobile Station (MS)–Serving GPRS Support Node (SGSN); Subnetwork Dependent Convergence Protocol (SNDCP). Technical Report 3GPP TS 04.65, Third Generation Partnership Project, September 2001-09. Version 8.2.0.

216. Charging Management; Call and Event Data for the Packet Switched (PS) Domain. Technical Report 3GPP TS 32.015, Third Generation Partnership Project, December 2003. Version V3.12.0.

217. www.gsmworld.com/technology/gprs/class.shtml.

218. General Packet Radio Service (GPRS); Base Station System (BSS)–Serving GPRS Support Node (SGSN) Interface; Network Service. Technical Report 3GPP TS 08.16, Third Generation Partnership Project, May 2002. Version 8.0.1.

219. General Packet Radio Service (GPRS); Mobile Station–Serving GPRS Support Node (MS-SGSN) Logical Link Control (LLC) Layer Specification. Technical Report 3GPP TS 04.64, Third Generation Partnership Project, December 2001. Version 8.7.0.

220. General Packet Radio Service (GPRS); Base Station System (BSS)–Serving GPRS Support Node (SGSN); BSS GPRS Protocol (BSSGP). Technical Report 3GPP TS 08.18, Third Generation Partnership Project, May 2004. Release 1999, Version V8.12.0.

221. General Packet Radio Service (GPRS): Mobile Station (MS)–Base Station System (BSS) Interface; Radio Link Control/ Medium Access Control (RLC/MAC) Protocol. Technical Report 3GPP TS 04.60, Third Generation Partnership Project, July 2002. Version 6.14.0.

222. Physical Layer on the Radio Path: General Description. Technical Report 3GPP TS 45.001, Third Generation Partnership Project, April 2005. Release 4, Version V4.5.0.

223. GSM/EDGE Radio Access Network; Channel Coding. Technical Report 3GPP TS 45.003, Third Generation Partnership Project, January 2005. Version 4.4.0.

224. General Packet Radio Service (GPRS); Overall Description of the GPRS Radio Interface; Stage 2. Technical Report 3GPP TS 43.064, Third Generation Partnership Project, April 2004. Version 4.5.0.

225. General Packet Radio Service (GPRS); Mobile Station (MS)–Base Station System (BSS) Interface; Radio Link Control/Medium Access Control (RLC/MAC) Protocol. Technical Report 3GPP TS 44.060, Third Generation Partnership Project, November 2005. Version 4.23.0.

226. GSM/EDGE Radio Access Network; Digital Cellular Telecommunications System (Phase 2+); Modulation. Technical Report 3GPP TS 45.004, Third Generation Partnership Project, November 2001. Version 4.2.0.

227. GSM/EDGE Radio Access Network; Multiplexing and Multiple Access on the Radio Path. Technical Report 3GPP TS 45.002, Third Generation Partnership Project, June 2003. Version 4.8.0.

228. Digital Cellular Telecommunications System (Phase 2); Functions Related to Mobile Station (MS) in Idle Mode. Technical Report GSM 03.22, March 1996.

229. Third Generation Partnership Project. Core Network Protocols. Technical Report 3GPP, Stage 3, TS 24.008, Release 4, V4.17.0, September 2007.

230. RFC 792: Internet Control Message Protocol, September 1981. www.ietf.org/rfc/rfc0792.txt?number=792.

231. The Case for Evolved EDGE. Technical report, 3G Americas, August 2008. www.4gamericas.org/userfiles/File/2008_3G_Americas_The_Case_for_Evolved_Edge_Aug08.pdf.

232. D. Mužić and D.Opatić. Capabilities and Impacts of EDGE Evolution Toward Seamless Wireless Networks. In *International Convention on Information and Communication Technology, Electronics, and Microelectronics*. Ericsson, Inc., May 2009. www.ericsson.com/hr/etk/dogadjanja/mipro_2009/03_1077_F.pdf.

233. www.cdg.org/worldwide/index.asp?h_area=0&h_technology=99.

234. www.cdg.org/.

235. Eric W. Weisstein. Walsh Function. *MathWorld–A Wolfram Web Resource*. mathworld.wolfram.com/WalshFunction.html.

236. J. L. Walsh. A Closed Set of Normal Orthogonal Functions. *Amer. J. Math.*, 45:5–24, 1923.

237. Mobile Station–Base Station Compatibility Standard for Wideband Spread Spectrum Cellular Systems. Technical Report TIA/EIA-95-B, Telecommunications Industry Association and Electronic Industries Association, March 1999.

238. www.cdg.org/.

239. Inter-Operability Specification (IOS) for CDMA2000 Access Network Interfaces. Technical Report TIA/EIA/IS-2001-A, Telecommunications Industry Association and Electronic Industries Alliance, June 2001.

240. Physical Layer Standard for CDMA2000 Spread Spectrum Systems, Version 1.13. Technical Report 3GPP2 C.S0002-0-2, Third Generation Partnership Project 2, April 2001.

241. Medium Access Control (MAC) for CDMA2000 Spread Spectrum Systems, Version 1.0. Technical Report C.S0003-0-2, Third Generation Partnership Project 2, April 2001.

242. Data Service Options for Spread Spectrum Systems: Radio Link Protocol Type 3. Technical Report PN-4692.10 (Ballot Version), Third Generation Partnership Project 2, January 2000. Also published as TIA/EIA/IS-707-A-2.10, March 2001, Telecommunications Industry Association and Electronic Industries Alliance.

243. Common Cryptographic Algorithms, Revision C. Technical report, CTIA—The Wireless Association, 1997. An EAR-controlled document subject to restricted distribution. Contact the Telecommunications Industry Association, Arlington, VA, U.S.

244. Interface Specification for Common Cryptographic Algorithms, Revision C. Technical report, CTIA—The Wireless Association, 1997. An EAR-controlled document subject to restricted distribution. Contact the Telecommunications Industry Association, Arlington, VA, U.S.

245. Signaling Link Access Control (LAC) for CDMA2000 Spread Spectrum Systems, 3GPP2 C.S0004-0-2, Version 1.0. Technical Report C.S0004-0-2, Third Generation Partnership Project 2, April 2001.

246. Upper Layer (Layer 3) Signaling for CDMA2000 Spread Spectrum Systems, Version 1.0. Technical Report C.S0005-0-2, Third Generation Partnership Project 2, April 2001.

247. Sandeep Sarkar, Tao Chen, Gilbert Leung, Luca Blessent, and Edward Tiedemann. CDMA2000 Reverse Link: Design and System Performance. In *IEEE Vehicular Technology Conference*, pages 2713–2719, 2000.

248. The CDMA Development Group. Delivering Voice and Data: Comparing CDMA2000 and GSM/GPRS/EDGE/UMTS. Technical report, December 2005.

249. Third Generation Partnership Project, Technical Specification Group Services and Systems Aspects. Network Architecture (Release 1999). Technical Report 3G TS 23.002 V3.6.0 (2002-09).

250. Third Generation Partnership Project, Technical Specification Group Radio Access Network. UTRAN Overall Description (Release 1999). Technical Report 3GPP TS 25.401 V3.10.0 (2002-06).

251. en.wikipedia.org/wiki/Mobile_Network_Code.

252. Guy Webster and Dwayne Brown. Mars Rovers Near Five Years of Science and Discovery. www.nasa.gov/mission_pages/mer/news/mer-20081229.html.

253. Service Description. Technical Report TS 23.060, Stage 2, V3.16.0, Release 1999, Third Generation Partnership Project, December 2003.

254. Third Generation Partnership Project. GPRS Tunnelling Protocol (GTP) Across the Gn and Gp Interface. Technical Report 3GPP, TS 29.060, V3.19.0, Release 1999, March 2004.

255. Third Generation Partnership Project. UTRAN Iu Interface RANAP Signaling. Technical Report 3GPP, TS 25.413, V3.14.0, Release 1999, September 2003.

256. Third Generation Partnership Project. Architectural Requirements for Release 1999. Technical Report 3GPP, TS 23.121, V3.6.0, June 2002.

257. Third Generation Partnership Project, Technical Specification Group Radio Access Network. Radio Interface Protocol Architecture (Release 1999). Technical Report 3GPP TS 25.301 V3.11.0 (2002-09).

258. Third Generation Partnership Project. Mobile Radio Interface Signalling Layer 3–General Aspects. Technical Report 3GPP, TS 24.007, Release 1999, V3.10.0, June 2004.

259. Third Generation Partnership Project, Technical Specification Group Terminals. Common Test Environments for User Equipment (UE); Conformance Testing (Release 5). Technical Report 3GPP TS 34.108 V5.2.0 (2004-09).

260. Third Generation Partnership Project, Technical Specification Group Radio Access Network. Medium Access Control (MAC) Protocol Specification (Release 1999). Technical Report 3GPP TS 25.321 V3.17.0 (2004-06).

261. Third Generation Partnership Project, Technical Specification Group Radio Access Network. Radio Link Control (RLC) Protocol Specification (Release 1999). Technical Report 3GPP TS 25.322 V3.18.0 (2004-06).

262. M. Degermark, B. Nordgren, and S. Pink. IP Header Compression. Technical Report Internet Engineering Task Force, RFC 2507, February 1999.

263. Physical Channels and Mapping of Transport Channels onto Physical Channels (FDD), V7.7.0. Technical Report TS 25.211, Third Generation Partnership Project, March 2009.

264. Third Generation Partnership Project, Technical Specification Group Radio Access Network. Spreading and Modulation (FDD) (Release 1999). Technical Report 3GPP TS 25.213 V3.9.0 (2003-12).

265. Third Generation Partnership Project, Technical Specification Group Radio Access Network. Radio Resource Control (RRC) Protocol Specification (Release 1999). Technical Report 3GPP TS 25.331 V3.21.0 (2004-12).

266. Third Generation Partnership Project, Technical Specification Group Radio Access Network. User Equipment (UE) Procedures in Idle Mode and Procedures for Cell Reselection in Connected Mode (Release 1999). Technical Report 3GPP TS 25.304 V3.14.0 (2004-03).

267. Third Generation Partnership Project, Technical Specification Group Radio Access Network. Physical Layer Procedures (FDD) (Release 1999). Technical Report 3GPP TS 25.214 V3.12.0 (2003-03).

268. thinkexist.com/quotation/persistence_is_the_twin_sister_of_excellence-one/11454.html.

269. User Equipment (UE) Radio Transmission and Reception (FDD) (Release 1999). Technical Report 3GPP TS 25.101 V3.17.0 (2004-03), Third Generation Partnership Project, Technical Specification Group Radio Access Networks.

270. R. Kreher and T. Rudebusch. *UMTS Signaling*. John Wiley & Sons Ltd., 2005.

271. UTRAN Iub Interface NBAP Signaling (Release 5). Technical Report TS 25.433 V5.11.0 (2004-12), Third Generation Partnership Project; Technical Specification Group Radio Access Network.

272. Third Generation Partnership Project. Physical Layer–Measurements (FDD). Technical Report 3GPP, TS 25.215,Release 1999, V3.13.0, March 2005.

273. Third Generation Partnership Project, Technical Specification Group Services and System Aspects. Quality of Service (QoS) Concept and Architecture (Release 1999). Technical Report 3GPP TS 23.107 V3.9.0 (2002-09).

274. China Mobile to Add 25 Min TD-SCDMA Subscribers in 2011, November 2010. www.cn-c114.net/576/a589011.html.

275. Siemens. TD-SCDMA: The Solution for TDD Bands. Technical report, March 2004.

276. P. Bender, P. Black, M. Grob, R. Padovani, N. Sindhushyana, and S. Viterbi. CDMA/HDR: A Bandwidth Efficient High Speed Wireless Data Service for Nomadic Users. *IEEE Communications Magazine*, 38(7), 2000.

277. CDMA2000 High-Rate Packet Data Air-Interface Specification, Version 4.0. Technical Report 3GPP2 C.S0024, Third Generation Partnership Project 2, October 2002.

278. R. Rezahfar, P. Agashe, and P. Bender. Macro-Mobility Management in EVDO. *IEEE Communications Magazine*, 44(2):65–72, February 2006.

279. Arnab Das, Farooq Khan, Ashwin Sampath, and Hsuan-Jung Su. Adaptive, Asynchronous, Incremental Redundancy (A2IR) with Fixed Transmission Time Intervals (TTI) for HSDPA. In *IEEE International Symposium on Personal, Indoor, and Mobile Communications*, volume 3, pages 1083–1087, 2002.

280. P. Frenger, S. Parkvall, and E. Dahlman. Performance Comparison of HARQ with Chase Combining and Incremental Redundancy for HSDPA. In *IEEE Vehicular Technology Conference Proceedings*, volume 3, pages 1829–1833, 2001.

281. Third Generation Partnership Project 2. Interoperability Specification (IOS) for High Rate Packet Data (HRPD) Radio Access Network Interfaces with Session Control in the Access Network, Version 2.0. Technical Report A.S0008-C v2.0, January 2009.

282. Third Generation Partnership Project 2. Interoperability Specification (IOS) for CDMA2000 Access Network Interfaces —Part 7 (A10 and A11 Interfaces), Version 2.0. Technical Report A.S0017-C v2.0, December 2005.

283. A. Jalali, R. Padovani, and R. Pankaj. Data Throughput of CDMA-HDR a High Efficiency-High Data Rate Personal Communication Wireless System. In *Vehicular Technology Conference Proceedings*, volume 3, pages 1854–1858, 2002.

284. Peter J. Black and Qiang Wu. Link Budget of CDMA2000 1xEV-DO Wireless Internet Access System. In *The 13th IEEE International Symposium on Personal, Indoor and Mobile Radio Communications*, volume 4, pages 1847–1852, September 2002.

285. 3G Americas. The Evolution of UMTS, 3GPP Release 5 and Beyond. Technical report, June 2004.

286. 3G Americas. The Global Evolution of UMTS/HSDPA—3GPP Release 6 and Beyond. Technical report, December 2005. `www.3gamericas.org/pdfs/UMTS_Rel6_Beyond-Dec2005.pdf`.

287. 3G Americas. `www.3g.co.uk/PR/April2007/4532.htm`.

288. High-Speed Downlink Packet Access (HSDPA); Overall Description; Stage 2. Technical Report 3GPP TS 25.308, Third Generation Partnership Project, December 2004.

289. Third Generation Partnership Project; Technical Specification Group Radio Access Network. Physical Channels and Mapping of Transport Channels onto Physical Channels (FDD), (Release 5). Technical Report 3GPP TS 25.211, V5.5.0 (2003-09).

290. Third Generation Partnership Project; Technical Specification Group Radio Access Network. Multiplexing and Channel Coding (FDD) (Release 5). Technical Report 3GPP TS 25.212 V5.9.0 (2004-06).

291. Third Generation Partnership Project; Technical Specification Group Radio Access Network. Medium Access Control (MAC) Protocol Specification (Release 5). Technical Report 3GPP TS 25.321 V5.9.0 (2004-06).

292. Third Generation Partnership Project; Technical Specification Group Radio Access Network. Physical Layer Procedures (FDD) (Release 5). Technical Report 3GPP TS 25.214 V5.9.0 (2004-06).

293. Third Generation Partnership Project; Technical Specification Group Radio Access Network. Spreading and Modulation (FDD) (Release 5). Technical Report 3GPP TS 25.213 V5.5.0 (2003-12).

294. UTRAN Iub Interface User Plane Protocols for Common Transport Channel Data Streams (Release 5). Technical Report 3GPP TS 25.435 V5.7.0 (2004-03), Third Generation Partnership Project; Technical Specification Group Radio Access Network.

295. CDMA2000 High-Rate Packet Data Air-Interface Specification, Version 3.0. Technical Report 3GPP2 C.S0024-A, Third Generation Partnership Project 2, September 2006.

296. CDMA2000 High Rate Packet Data Air Interface Specification. Technical Report 3GPP2 C.S0024-B, Third Generation Partnership Project 2, February 2012. Version 2.0.

297. FDD Enhanced Uplink—Overall Description, Stage 2, Release 6, V6.6.0. Technical Report 3GPP, TS 25.309, Third Generation Partnership Project, March 2006.

298. Physical Channels and Mapping of Transport Channels onto Physical Channels (FDD), V6.7.0. Technical Report 3GPP, TS 25.211, Third Generation Partnership Project, December 2005.

299. Multiplexing and Channel Coding (FDD), V6.7.0. Technical Report 3GPP, TS 25.212, Third Generation Partnership Project, December 2005.

300. Spreading and Modulation (FDD), V6.4.0. Technical Report 3GPP, TS 25.213, Third Generation Partnership Project, September 2005.

301. Medium Access Control (MAC) Protocol Specification, V6.7.0. Technical Report 3GPP, TS 25.321, Third Generation Partnership Project, December 2005.

302. Physical Layer Measurements (FDD), V6.4.0. Technical Report 3GPP, TS 25.215, Third Generation Partnership Project, September 2005.

303. Single Mode Scheduler—Way Forward. Technical Report R2-050929, TSG-RAN WG2 #46bis, Ericsson, 4–8 April 2005. www.3gpp.org/ftp/tsg_ran/WG2_RL2/TSGR2_46bis/Documents/.

304. Text Proposal on E-HICH and E-RGCH Signature Sequence Hopping. Technical Report R1-041424, TSG-RAN WG1 #39, Ericsson, November 15–19 2004. www.3gpp.org/ftp/tsg_ran/wg1_rl1/TSGR1_39/Docs/.

305. Radio Resource Control (RRC) Protocol Specification, V6.8.0. Technical Report 3GPP, TS 25.331, Third Generation Partnership Project, December 2005.

306. Physical Layer Procedures (FDD), V6.7.1. Technical Report 3GPP, TS 25.214, Third Generation Partnership Project, December 2005.

307. User Equipment (UE) Radio Transmission and Reception (FDD), V6.10.0, Release 6. Technical Report 3GPP, TS 25.101, Third Generation Partnership Project, December 2005.

308. UE Radio Access Capabilities, V6.7.0. Technical Report 3GPP, TS 25.306, Third Generation Partnership Project, December 2005.

309. IMS: Application Enabler and UMTS/HSPA Growth Catalyst. Technical report, 3G Americas, July 2006. www.3gamericas.org/documents/wp_IMS_UMTS-HSPA_Growth_Catalyst.pdf.

310. C.V. Chakravarthy. IP Multimedia Subsystems: A Tutorial. October 2006. www.imsexpo.com.

311. FMC Market View. Technical report, Qualcomm, December 2007. white paper.

312. Fixed-Mobile Convergence: Understanding the Marriage of Wireless and Wireline Technologies. Technical report, 3G Americas, July 2007. white paper.

313. www.msforum.org.

314. K. Rosenbrock, R. Sanmugam, S. Bradner, and J. Klensin. 3GPP-IETF Standardization Collaboration. Technical Report RFC 3113, June 2001. www.ietf.org/rfc/rfc3113.txt.

315. Interworking between Session Initiation Protocol (SIP) and Bearer Independent Call Control Protocol or ISDN User Part. Technical Report ITU-T Q.1912.5, International Telecommunication Union, March 2004.

316. 3GPP Work Plan. Technical report, October 2008. www.3gpp.org/.

317. IP Multimedia (IM) Session Handling—IM Call Model. Technical Report TS 23.218, December 2008. Stage 2, Release 8, Version 8.4.0.

318. IP Multimedia Subsystem (IMS). Technical Report TS 23.228, March 2009. Stage 2, Release 8, Version 8.8.0.

319. IP Multimedia Subsystem (IMS) Emergency Sessions. Technical Report TS23.167, December 2008. Version 8.2.0, Release 8.

320. Thomas T. Towle. IMS in Next Generation Networks. In *Next Generation Networks Technical Workshop*. ITU, March 2005.

321. Policy and Charging Control Architecture. Technical Report TS23.203, December 2008. Version 8.4.0.

322. IP Multimedia Call Control Protocol based on Session Initiation Protocol (SIP) and Session Description Protocol (SDP). Technical Report TS 24.229, March 2009. Stage 3, Release 8, Version 8.7.0.

323. Network Architecture. Technical Report TS 23.002, 2008-09. Release 7, Version 7.5.0.

324. G. Camarillo, A. B. Roach, J. Peterson, and L. Ong. Integrated Services Digital Network (ISDN) User Part (ISUP) to Session Integration Protocol (SIP) Mapping. Technical Report RFC3398, December 2002.

325. IP Multimedia (IM) Session Handling—IM Call Model. Technical Report TS 23.218, March 2011. Stage 2, Release 10, Version 10.0.0.

326. H.248.1 Version 1—Implementors' Guide. Technical report, ITU, November 2004.

327. M. Garcia-Martin, E. Henrikson, and D. Mills. Private Header (P-Header) Extensions to the Session Initiation Protocol (SIP) for the 3rd-Generation Partnership Project (3GPP). Technical Report RFC 3455, January 2003.

328. C. Groves, M. Pantaleo, T. Anderson, and T. Taylor. Gateway Control Protocol Version 1. Technical report, IETF, June 2003. `www.ietf.org/rfc/rfc3525.txt`.

329. T. Taylor. Reclassification of RFC 3525 to Historic. Technical Report RFC5125, February 2008.

330. B. Aboba and M. Beadles. The Network Access Identifier. Technical Report RFC 2486, January 1999.

331. T. Berners-Lee, R. Fielding, and L. Masinter. Uniform Resource Identifiers (URI): Generic Syntax. Technical Report RFC 2396, August.

332. H. Schulzrinne. The Tel URI for Telephone Numbers. Technical Report RFC 3966, December 2004.

333. Obtaining and Using Globally Routable User Agent (UA) URIs (GRUU) in the Session Initiation Protocol (SIP). Technical report, Internet Engineering Task Force, October 2007. draft-ietf-sip-gruu-15.

334. Signalling Flows for the IP Multimedia Call Control Based on Session Initiation Protocol (SIP) and Session Description Protocol (SDP). Technical Report TS 24.228, September 2006. Stage 3, Version 5.15.0.

335. Interworking Between the IP Multimedia (IM) Core Network (CN) Subsystem and Circuit Switched (CS) Networks. Technical Report TS 29.163, Third Generation Partnership Project, March 2009. Version 8.6.0.

336. Support for Short Message Service (SMS) Over Generic 3GPP Internet Protocol (IP) Access. Technical Report TS 23.204, September 2008. Stage 2, Release 8, Version 8.3.0.

337. Push To Services Take the Lead. Technical report, IMS Research, March 2006.

338. Policy and Charging Control Signaling Flows and QoS Parameter Mapping. Technical Report TS 29.213, March 2009. Release 8, Version 8.3.0.

339. Mobile Broadband: The Global Evolution of UMTS/HSPA, 3GPP Release 7 and Beyond. Technical report, 3G Americas, July 2006.

340. Requirements Related to Technical Performance for IMT-Advanced Radio Interface(s). Technical Report ITU-R M.2134, International Telecommunication Union, 2008.

341. Feasibility Study for Orthogonal Frequency Division Multiplexing (OFDM) for UTRAN Enhancement, Release 6. Technical Report 3GPP TR 25.892, v6.0.0, Third Generation Partnership Project, June 2004.

342. Charan Langton. Intuitive Guide to Principles of Communications—OFDM. `www.complextoreal.com`.

343. OFDM for Mobile Data Communications, March 2003. Whitepaper, Flarion Technologies, Inc.

344. Yasir Ahmed [John, pseud.]. Can We Do Without a Cyclic Prefix? October 2011. `www.raymaps.com/index.php/can-we-do-without-a-cyclic-prefix/`.

345. Hassan Yaghoobi. Scalable OFDMA Physical Layer in IEEE 802.16 WirelessMAN. *Intel Technology Journal*, 08(03), August 20, 2004. ISSN 1535-864X.

346. Y. Li and L.J. Cimini. Bounds on the Interchannel Interference of OFDM in Time Varying Impairments. *IEEE Transactions On Communications*, 49(3):401–404, March 2001.

347. Cristina Ciochina, Fabien Buda, and Hikmet Sari. An Analysis of OFDM Peak Power Reduction Techniques for WiMAX Systems. Sequans Communications, France.

348. C. Tellambura. Phase Optimization Criterion for Reducing Peak-to-Average Power Ratio in OFDM. *Electronics Letters*, 34:169–170, January 1998.

349. M. Breiling, S. H. Muller, and J. B. Huber. SLM Peak-Power Reduction Without Explicit Side Information. *IEEE Communications Letters*, 5:239–241, June 2001.

350. S. H. Muller and J. B. Huber. OFDM with Reduced Peak-to-Average Power Ratio by Optimum Combination of Partial Transmit Sequences. *Electronics Letters*, 33:368–369, February 1997.

351. B.S. Krongold and D.L. Jones. PAR Reduction in OFDM via Active Constellation Extension. *IEEE Transactions on Broadcasting*, 3:258–268, September 2003.

352. H.K. Kwok and D.L. Jones. PAR Reduction via Constellation Shaping. In *Proceedings of IEEE International Symposium on Information Theory*, June 2000.

353. S. Sezginer and H. Sari. Peak Power Reduction in OFDM Systems Using Dynamic Constellation Shaping. In *Proceedings of European Signal Processing Conference*, 2005.

354. J. Tellado. *Multicarrier Modulation with Low Peak to Average Power Applications to xDSL and Broadband Wireless*. Kluwer Academic Publishers, Boston/Dordrecht/London, 2000.

355. Siavash M. Alamouti. A Simple Transmit Diversity Technique for Wireless Communications. *IEEE Journal on Select Areas in Communications*, 16(8):1451–1458, October 1998.

356. IntelliCell: A Fully Adaptive Approach to Smart Antennas. Technical Report WP-ISA-031502-2.0, Arraycomm.

357. WiMax Forum Mobile System Profile, Release 1.0 Approved Specification, Revision 1.7.0. Technical Report 2008-09-18, WiMAX Forum.

358. Mobile WiMAX—Part I: A Technical Overview and Performance Evaluation. Technical report, WiMAX Forum, August 2006.

359. WiMAX Forum Network Architecture (Stage 2: Architecture Tenets, Reference Model and Reference Points), Release 1, Version 1.3.0. Technical report, WiMAX Forum, November 2, 2008.

360. Draft Standard for Local and Metropolitan Area Networks, Part 16: Air Interface for Broadband Wireless Access Systems. Technical Report P802.16Rev2/D5, IEEE Standards Association, June 2008. Revision of IEEE Std 802.16-2004 and consolidates material from IEEE Std 802.16e-2005, IEEE Std 802.16-2004/Cor1-2005, IEEE Std 802.16f-2005 and IEEE Std802.16g-2007.

361. LTE. www.3gpp.org/LTE.

362. www.lteuniversity.com/. LTE University.

363. User Equipment (UE) Radio Transmission and Reception, Release 8. Technical Report 3GPP TS 36.101 V8.7.0, Third Generation Partnership Project, September 2009.

364. Moray Rumney. 3GPP LTE: Introducing Single-Carrier FDMA. *Agilent Measurement Journal*, (5989-7898EN), January 2008. Issue 4, cp.literature.agilent.com/litweb/pdf/5989-7898EN.pdf.

365. Hyung G. Myung, Junsung Lim, and David J. Goodman. Single Carrier FDMA for Uplink Wireless Transmission. *IEEE Vehicular Technology Magazine*, pages 30–38, September 2006.

366. User Equipment (UE) Radio Access Capabilities, Release 8. Technical Report TS 36.306, Third Generation Partnership Project.

367. N. Tripathi. LTE Deployments in the LMS Band for ITS Radio Communications and Location. Technical report, Skytel, 2012. https://wireless2.fcc.gov/UlsEntry/attachments/attachmentViewRD.jsp?applType=search&fileKey=270750084&attachmentKey=18951929&attachmentInd=applAttach.

368. Esmael Dinan. Deploying MPLS in WiMAX and LTE. 10th Annual International Conference on MPLS and Related Technologies, 2007. www.slideshare.net/n888n/323-esmael-dinan-presentation.

369. MPLS in Mobile Backhaul Networks Framework and Requirements Technical Specification. Technical Report IP/MPLS Forum 20.0.0, IP/MPLS Forum (now Broadband Forum), October 2008. www.broadband-forum.org/technical/ipmplstechspec.php.

370. Signaling System 7 (SS7). Web ProForum Tutorials.

371. Erlang B-Table for 1 to 50 Channels, 0.7%–40%. www.umtsworld.com/resources/erlang.htm.

372. Erlang B Calculator. www.erlang.com/calculator/erlb/.

373. Erlang B Calculator. www.ansapoint.com/calculator/erlb/.

374. HSPA+ For Enhanced Mobile Broadband. www.qualcomm.com/documents/files/hspa-enhanced-mobile-broadband-whitepaper.pdf, February 2009.

375. VoIP Overview. www.qualcomm.co.in/common/documents/white_papers/VoIP_Overview.pdf, August 2008.

376. Physical Layer Procedures (FDD), V7.11.0. Technical Report TS 25.214, Third Generation Partnership Project, March 2009.

377. Multiplexing and Channel Coding (FDD), V7.9.0. Technical Report TS 25.212, Third Generation Partnership Project, September 2008.

378. C. Gessner. T&M Requirements for Mobile Radios with Receive Diversity and Advanced Receivers, 2006. Rohde & Schwarz GmbH & Co. KG, Application Note 1MA88.

379. Feasibility Study on Interference Cancellation for UTRA FDD: User Equipment (UE), V7.0.0. Technical Report 3GPP TR 25.963, Third Generation Partnership Project, April 2007.

380. Radio Link Control (RLC) Protocol Specification, V7.9.0. Technical Report TS 25.322, Third Generation Partnership Project, March 2009.

381. Medium Access Control (MAC) Protocol Specification, V7.12.0. Technical Report TS 25.321, Third Generation Partnership Project, March 2009.

382. GPRS Tunnelling Protocol (GTP) Across the Gn and Gp Interface, V7.12.0. Technical Report TS 29.060, Third Generation Partnership Project, December 2008.

383. UE Radio Access Capabilities, V8.5.0. Technical Report TS 25.306, Third Generation Partnership Project, December 2008.

384. Radio Resource Control (RRC) Protocol Specification, V7.12.1. Technical Report TS 25.331, Third Generation Partnership Project, April 2009.

385. Courau, Francois. HSPA+ Summit: HSPA+ Standardization Review. `www.ezcom.cn/meeting/hspa/ppt/3_4%E4%BD%9B%E6%9C%97%E5%93%A5%E7%A7%91%E5%8A%B3.pdf`, October 2007. Alcatel-Lucent, Beijing.

386. Index of /ftp/Specs/latest. `www.3gpp.org/ftp/Specs/latest/`.

Glossary

1G first-generation

1x CDMA2000 1xRTT

1xEV-DO 1xEvolution—Data-Optimized, or 1xEvolution—Data-Only

1xEV-DV 1xEvolution—Data-and-Voice

1xRTT CDMA2000 1xRTT

2G second-generation

2.5G second-and-a-half-generation

3G third-generation

3GPP Third Generation Partnership Project

3GPP2 Third Generation Partnership Project 2

3G-SGSN Third-Generation Serving GPRS Support Node

3x or 3xRTT CDMA2000 3xRTT

4G fourth-generation

8-PSK 8-ary phase shift keying

AAA [pronounced as "triple A"] authentication, authorization, and accounting [server]

AAL ATM adaptation layer

ACELP algebraic code excited linear prediction
ACAck Access Channel Acknowledgment
ACH Access channel
ACK acknowledgment, or positive acknowledgment
ADC analog-to-digital converter
AG Absolute Grant
AGCH Access Grant channel
AI acquisition indicator
AICH Acquisition Indicator channel
AMR adaptive multirate
AMPS Analog Mobile Phone Service
AN access network
AN-AAA access-network–authentication-authorization-and-accounting
ANSI American National Standards Institute
APN access point name
ARFCN absolute radio frequency channel number
ARP Address Resolution Protocol
ARQ automatic repeat request
AS autonomous system, or application server
ASIC application-specific integrated circuit
ASN access service network
ASN-GW access service network—gateway
AT access terminal
ATM Asynchronous Transfer Mode
AuC authentication center
AWGN additive white Gaussian noise
AWS Advanced Wireless Services
BCCH broadcast control channel
BCS block check sequence
BG border gateway
BGCF breakout gateway control function
BGP Border Gateway Protocol
BLER block error rate
BLOB block of bits
BMC broadcast/multicast control
BPSK binary phase shift keying
BS base station
BSC base station controller

BSIC base station identity code
BSN block sequence number
BSS base station system
BSSGP BSS GPRS Protocol
BTA basic trading area
BTS base transceiver station
CCCH Common Control channel
CCI co-channel interference
CDM code-division multiplexing
CDMA code-division multiple access
CDMA2000 code-division multiple access 2000
cdmaOne IS-95
CELP code excited linear prediction
CFN connection frame number
CGF charging gateway function
ch channelization code
C/I carrier-to-interference [ratio]
CID connection identifier
CIDR classless inter-domain routing
CM connection management
CoA care of address
CP cyclic prefix
CPFSK continuous phase frequency shift keying
CPICH Common Pilot channel
CQI channel-quality indicator
CRC cyclic redundancy check
CS coding scheme, or circuit-switched
CSN connectivity service network
C/T control/traffic (field name)
cwnd congestion window
DAC digital-to-analog converter
DAS distributed antenna system
DCS digital cellular system
DCCH Dedicated Control channel
DCH Dedicated channel
DDI data descriptor indicator
DHCP Dynamic Host Configuration Protocol
DiffServ differentiated services

DL downlink

DL-MAP downlink-MAP

DNS domain name system

DO [shortened form of] 1xEV-DO

DPCCH Dedicated Physical Control channel

DPCH Dedicated Physical channel

DPDCH Dedicated Physical Data channel

DRC data rate control

DRCLock Data Rate Control Lock [channel]

DRX discontinuous reception

DTX discontinuous transmission

DS digital signal [In other texts, DS can also represent differentiated services.]

DSC data source control

dsch dedicated signaling channel

DSCP differentiated services code point

DSL digital subscriber line

DTCH, or dtch dedicated traffic channel

DTM Dual Transfer Mode

DTX discontinuous transmission

E extension [bit]

E1 European

E-AGCH E-DCH Absolute Grant channel

E-DCH Enhanced Dedicated channel

EDGE Enhanced Data Rates for GSM Evolution, or Enhanced Data Rates for Global Evolution

E-DPCCH E-DCH Dedicated Physical Control channel

E-DPDCH E-DCH Dedicated Physical Data channel

EGPRS Enhanced GPRS

E-GSM Extended GSM 900 band

E-HICH E-DCH HARQ Indication channel

EIR equipment identity register

EIRP effective isotropic radiated power

eNB evolved Node B or E-UTRAN Node B

EPC evolved packet core

EPS evolved packet system

E-RGCH E-DCH Relative Grant channel

E-RNTI E-DCH radio network temporary identifier

ESCAM Extended Supplemental Channel Assignment message

ESN electronic serial number

ETACS European Total Access Communication System
E-TFCI E-DCH transport format combination indicator
ETPR E-DCH traffic-to-pilot ratio
E-UTRAN Evolved Universal Terrestrial Radio Access Network
EVRC enhanced variable-rate codec
F flag (parameter name)
FA foreign agent
FACCH Fast Associated Control channel
FACCH/F Fast Associated Control channel/full-rate
FACCH/H FACCH/H (Fast Associated Control channel/half-rate)
FACH Forward Access channel
FBI feedback information
FCC Federal Communications Commission
FCCH Frequency Correction channel
FCH fundamental channel
FCS frame check sequence
f-csch forward common signaling channel
FDD frequency-division duplexing
FDMA frequency-division multiple access
F-DPCH Fractional Dedicated Physical channel
f-dsch forward dedicated signaling channel
f-dtch forward dedicated traffic channel
FEC forward error correction
FER frame error rate
F-FCH forward fundamental channel
FFT fast Fourier transform
FIN final control bit
FL forward link
FMC fixed mobile convergence
FN frame number
FPC forward power control
FQI frame-quality indicator
FRAB filtered reverse activity bit
FSK frequency shift keying
FRF frequency reuse factor
F-SCH forward supplemental channel
FTC Forward Traffic channel
FTP File Transfer Protocol

GAUP Generic Attribute Update Protocol

GGSN gateway GPRS support node

GMSC gateway MSC

GMSK Gaussian minimum shift keying

GPRS General Packet Radio Service

GRE generic routing encapsulation

GRUU globally routable user agent URI

GRX GPRS roaming exchange

GSM Global System for Mobile Communications or Global System Mobile, formerly Groupe Speciale Mobile

GTP GPRS Tunneling Protocol

GTP SND GPRS Tunneling Protocol Sequence Number Downlink

GTP SNU GPRS Tunneling Protocol Sequence Number Uplink

GTP-U GPRS Tunneling Protocol–User Plane

HA home agent

HARQ or H-ARQ [pronounced "hark"] hybrid automatic repeat request

HCS header check sequence

HE (bits) header extension

H-FDD half-FDD

HLR home location register

HoA home address

HOM high-order modulation

HPA high-power amplifier

H-RNTI HS-DSCH radio network temporary identifier

HSDPA high-speed downlink packet access

HS-DPCCH High-Speed Dedicated Physical Control channel

HS-DSCH High-Speed Downlink Shared channel

HSPA high-speed packet access

HSPA+ high-speed packet access plus or high-speed packet access evolved

HS-PDSCH High-Speed Physical Downlink Shared channel

HSS home subscriber server

HS-SCCH High-Speed Shared Control channel

HTML hypertext markup language

HTTP Hypertext Transfer Protocol

HSUPA high-speed uplink packet access

I in-phase [channel or phase]

ICI intercarrier interference

ICMP Internet Control Message Protocol

I–CSCF interrogating–call session control function

IEEE Institute for Electronics and Electrical Engineers
IETF Internet Engineering Task Force
IFFT inverse fast Fourier transform
IIR infinite impulse response
IIS Internet information server
IMS IP multimedia subsystem
IMSI [pronounced as "eem-zee"] international mobile subscriber identity
IMTS Improved Mobile Telephone Service
IntServ integrated services
IP Internet Protocol
IP CAN IP Connectivity Access Network
IPsec IP security
IPv4 IP version 4
IPv6 IP version 6
IS integrated services
ISDN integrated services digital network
ISI inter-symbol interference
ISP internet service provider
ISUP integrated services digital network user part
ITU International Telecommunications Union
Iu-CS [interface name]
Iu-PS [interface name]
IWF interworking function
IWMSC interworking MSC
K secret key
L1 layer 1 (physical layer)
L2 layer 2
LAC link access control
LAN local area network
LAPDm Link Access Protocol for mobile D
L-ARQ last ARQ
LCR level crossing rate
LDPC low-density parity-check
LLC logical link control
LMA local mobility anchor
LNA low-noise amplifier
LOS line-of-sight
LPC linear predictive coding

LSP line spectral pairs
LTE Long-Term Evolution
MAC medium access control
MAC-c MAC-common
MAC-d MAC-dedicated
MAC-e MAC for E-DCH
MAC-ehs MAC–enhanced high speed
MAC-es MAC-e selection
MAC-hs MAC–high speed
MACIndex medium access control index
MAC-sh MAC-shared
MAG mobile access gateway
MAHO mobile-assisted (hard) handover
MAP maximum a posteriori
ME mobile equipment
MGCF media gateway control function
MGW media gateway
MIMO multiple-input multiple-output
MIP mobile IP
MM mobility management
MME mobility management entity
MOS mean opinion score
MPLS multi-protocol label switching
MRFC media resource function controller
MRFP media resource function processor
MS mobile station
MSC mobile switching center
MSID MS identity [field]
MSK minimum shift keying
MTS mobile telephone system
Mux multiplexing
MuxPDU multiplexing protocol data unit
N number (parameter name)
NACK negative acknowledgment
NAK negative acknowledgment
NAPT network address and port translation
NAS Non-Access Stratum
NAT network address translation

NID network identity
NSS network switching subsystem
OFDM orthogonal frequency division multiplexing
OFDMA orthogonal frequency division multiple access
OOK on-off keying
OVSF orthogonal variable spreading factor
P polling [field or bit]
PACCH Packet Associated Control channel
PAGCH Paging and Access Grant channel [Chapter 5] or Packet Access Grant channel [Chapter 6]
PAM pulse amplitude modulation
PAPR peak-to-average power ratio
P-ARQ packet ARQ
PBCCH Packet Broadcast Control channel
PCCCH Packet Common Control channel
P-CCPCH Primary Common Control Physical channel
PCF packet control function
PCH Paging channel
PCI precoding control indication
PCM pulse coded modulation
P-CPICH Primary Common Pilot channel
PCS Personal Communication System
P–CSCF proxy–call session control function
PDC Personal Digital Cellular
PDCH packet data channel
PDCP Packet Data Convergence Protocol
PDN packet data network
PDP Packet Data Protocol
PDSCH Physical Downlink Shared channel
PDSN packet data serving node
PDTCH Packet Data Traffic channel
PDU protocol data unit
P-GSM primary GSM 900 band
P-GW packet data network—gateway
PHB per-hop behavior
PHY physical (layer)
PICH Paging Indicator channel
PLMN public land mobile network
PMM packet mobility management

PN pseudo-noise
PNCH Packet Notification channel
PPCH Packet Paging channel
PPP Point-to-Point Protocol
PRACH Packet Random Access channel, or Physical Random Access channel
PS packet-switched
P-SCH Primary Synchronization channel
PSI packet system information
PSK phase shift keying
PSTN public-switched telephone network
PTCCH Packet Timing Advance Control channel
PTCCH/D Packet Timing Advance Control channel for the downlink
PTCCH/U Packet Timing Advance Control channel for the uplink
PTM-M Packet Transfer Mode–mobile-terminated
P-TMSI packet temporary mobile subscriber identity
PVC permanent virtual circuit
Q quadrature [phase or channel]
QAM quadrature amplitude modulation
QCI QoS class indicators
QoS quality-of-service
QPSK quadrature phase shift keying
QRAB quick reverse activity bit
R5 Release 5
R6 Release 6
R7 Release 7
R99 Release 99
RA routing area, or reverse activity
RAB radio access bearer, or reverse activity bit
R-ACH Reverse Access channel
RACH Random Access channel
RADIUS Remote Authentication Dial-In User Service
RAI routing area identity
RAN radio access network
RAND random number
RAT radio access technology
RC radio configuration
RCELP relaxed code excited linear prediction
r-csch reverse common signaling channel

RDA rate determination algorithm
R-DCCH Reverse Dedicated Control channel
r-dsch reverse dedicated signaling channel
r-dtch reverse dedicated traffic channel
REL Release [message]
Rev. revision
RF radio frequency
R-FCH Reverse Fundamental channel
RG Relative Grant
RIP Routing Information Protocol
RIP-1 RIP version 1
RIP-2 RIP version 2
RL reverse link
RLC radio link control, or Release Complete [message]
RLC/MAC radio link control/medium access control
RLP Radio Link Protocol
RLS radio link set
rms root mean square
RNC radio network controller
RoT rise-over-thermal [noise]
R-P radio-packet
RPC Reverse Power Control [channel]
RR radio resource
RRC radio resource control
RRH remote radio head
RRI Reverse Rate Indicator [channel]
R-SCH Reverse Supplemental channel
RSCP received signal code power
RSN retransmission sequence number
RTC Reverse Traffic channel
RTCAck Reverse Traffic Channel Acknowledgment
RTCP Real-Time Transport Control Protocol
RTP Real-Time Transport Protocol
RTT Radio Transmission Technology
RTWP received total wideband power
RV Redundancy and Constellation Version [field]
Rx receive
S1-U S1-user plane (an interface)

SACCH Slow Associated Control channel
SACCH/C4 Slow Associated Control channel/4 kbps
SACCH/C8 Slow Associated Control channel/Control channel 8 kbps
SACCH/TH Slow Associated Control channel/Traffic channel half-rate
SACCH/TF Slow Associated Control channel/Traffic channel full-rate
SACK selective acknowledgment
SB synchronization burst
SCCP signaling connection control part
S-CCPCH Secondary Common Control Physical channel
SC-FDMA single-carrier—frequency-division multiple access
SCH Synchronization channel [Chapter 5] or Supplemental channel [Chapter 8]
SCM station class mark
SCP service control points
S–CSCF serving–call session control function
SCTP Stream Control Transmission Protocol
SDCCH Stand-Alone Dedicated Control channel
SDCCH/4 Stand-Alone Dedicated Control channel/4 kbps
SDCCH/8 Stand-Alone Dedicated Control channel/8 kbps
SDMA space- or spatial-division multiple access
SDP Session Description Protocol
SDU service data unit
SF spreading factor
SFD start-of-frame delimiter
SFID service flow identifier
SFN system frame number
SG Serving Grant
SGi [an interface]
SGSN serving GPRS support node
S-GW serving gateway
SI scheduling information
SIB system information blocks
SID Silence Descriptor [Chapters 2 and 5], system identification [Chapter 7], or size index descriptor [field; Chapter 11]
SIM subscriber identity module
SIP [pronounced as "sip"] Session Initiation Protocol
SIR signal-to-interference ratio
SLF subscription locator function
SMS short message service
SM-SC short message service center

SMTP Simple Mail Transfer Protocol
SNDCP Subnetwork Dependent Convergence Protocol
SRB signaling radio bearer
SRBP Signaling Radio Burst Protocol
SS7 Signaling System 7
SSC secondary synchronization code
S-SCH Secondary Synchronization channel
SSP service switching points
ssthresh slow-start threshold
STC space time coding
STP signal transfer points
STTD space time transmit diversity
SYN synchronize (control bit)
Sync synchronization
T2P Traffic-to-Pilot [power ratio]
TBF temporary block flow
TBS transport block size
TCH traffic channel
TCH/F4.8 Traffic channel/full-rate 4.8 kbps
TCH/F9.6 Traffic channel/full-rate 9.6 kbps
TCH/FS Traffic channel/full-rate Speech
TCP Transmission Control Protocol
TDD time-division duplexing
TDM time-division multiplexing
TDMA time-division multiple access
TEID tunnel endpoint identifier
TFCI transport format combination indicator
TFI temporary flow identity
TFRI transport format resource indicator
TFT traffic flow template
THIG topology hiding
TMA tower mounted amplifier
TMSI [pronounced as "teem-zee"] temporary mobile subscriber identity
TN time-slot number
TPC transmit power control
TRAU transcoder rate adaptation unit
TRX transceivers
T–SGW trunk–signaling gateway

TSN transmission sequence number
TTI transmit time interval
TTIN TTI number
TTL time-to-live
Tx transmit
UAC user agent client
UAS user agent server
UATI unicast access terminal identifier
UDP User Datagram Protocol
UE user equipment
UI unnumbered information
UL uplink
UL-MAP uplink-MAP
UMTS Universal Mobile Telecommunication System
URA UTRAN registration area
URI uniform resource identifier
USF uplink state flag
UTRAN Universal Terrestrial Radio Access Network
VA Viterbi algorithm
VAD voice activity detection
VCI virtual channel identifier
VF version flag [field]
VLR visitor location register
VoIP Voice-over-IP
VPI virtual path identifier
WCDMA wideband CDMA
Wi-Fi wireless fidelity
WiMAX Worldwide Interoperability for Microwave Access
WLAN wireless local area network
WS-SUS wide-sense stationary uncorrelated scattering

INDEX

16-ary quadrature amplitude modulation, 90, 96
1xEV-DO Rev. 0
- Access Attempt, 598
- Call Setup, 623–630
- Channels, 593
- Connection Setup, 624–628
- Forward Link Data Transmission, 630–633
- Forward Link Slot Structure, 595–596
- Handoff, 640–641
- Link Budget, 646–649
- Network Architecture, 590–591
- Paging, 600–603
- Protocol Stack, 591–593
- Reverse Link Data Transmission, 633–640
- Reverse Link Timing Structure, 596

1xEV-DO Rev. A
- Forward Link Channels, 707–718
- Forward Link Data Transmission, 728–740
- H-ARQ, 739–740
- Interworking, 767
- Physical Layer Subtypes, 705–706
- QoS Control, 761–764
- Reverse Link Channels, 718–724
- Reverse Link Data Rate Selection, 637
- Reverse Link Frame Structure, 706
- Subtype 2 RTC MAC Protocol Data Rate Control, 745–756
- Subtype 3 RTC MAC Protocol Data Rate Control, 756–761

1xRTT and 1xEV-DO Interworking, 645–646
64-ary quadrature amplitude modulation, 90, 96
8-ary phase shift keying, 90, 96

AAA server, 428
Absolute Grant, 774
absorption, 120–121
Access Channel, 472–477
access probe, 382, 614
Access Service Network, *see* ASN
AccessParameters, 600–603
AccessSignature, 602–603
Active Set, 376, 559–569
active slot, 595–596
adaptive multirate codec, *see* AMR
adjacent channel interference, 26–27
Alamouti coding, 890
AMR, 66–69

antenna
  antenna azimuth, 104
  antenna beamwidth, 104–106
  antenna configuration, 103
  antenna down-tilt, 107–108
  antenna gain characteristics, 106–107
  antenna height, 104
  half-wave dipole, 114–115
  isotropic, 113–115
  smart antennas, 893–894
ARFCN, 260, 262
ARQ Channel, 711–713, 715–718
ASN, 897
ASN-GW, 900–902
Asynchronous Transfer Mode, *see* ATM
ATM, 188–193
  AAL, 191–193
  ATM Adaptation layer 2, 493–495
  ATM Adaptation layer 5, 495
  cell, 188–193
ATM Adaptation layer, *see* AAL
Auxiliary Pilot Channel, 723–724
average fade duration, 152–153

backward error correction, 338
Base Station Identity Code, 285
BCCH, 264
BCCH carrier, 244
BCJR algorithm, 85–86
beacon frequency, 271
beamforming, 893–894
Bessel function, 147–152
BGCF, 834
bi-orthogonal code, 608–609
binary phase shift keying, *see* BPSK
BLER, 47, 381
block error rate, *see* BLER
BMC, 497
Border Control Function, 841
BPSK, 90–91
Breakout Gateway Control Function, *see* BGCF
BSIC, 271
building penetration loss, 121
burst, 267

camping on the cell, 282
candidate set, 414
carrier aggregation, 919–920
CCO, 360–361
CDM scheduling, 678
CDMA, 25
CDMA2000 1x
  Access Attempt, 474
  Authentication, 469–472
  Data Call Setup, 459–460
  Forward Link Packet Data Transmission, 462–466
  Forward Link Radio Channels, 433–436
  Link Layer, 445–454
  Network Architecture, 426–431
  Physical Layer, 441–445
  Power Control, 481–484
  Protocol Stack, 431–433
  Radio Link Protocol, 447–450
  Registration, 469–472
  Reverse Link Packet Data Transmission, 467–468
  Reverse Link Radio Channels, 436–438
  Voice Call Setup, 455
cdmaOne, 365
cell allocation, 262
Cell Change Notification, 360
Cell Change Order, *see* CCO
cell reselection, 283, 359–361
cell search, 269–272
cell selection, 256
cell update procedure, 359–361
cell-area reliability, 48–49
cell-edge reliability, 48–49
cellular standards
  evolution, 28–34
channel coding, *see* FEC
channel sounding, 142–146
Charging Gateway Function, 318–319
Chase combining, 610, 680–681
CID, 903, 905
CIDR, 179, 181–183
ciphering, 277–278
classless inter-domain routing, *see* CIDR
cluster size, 14
clutter classes, 121
co-channel interference, 27–28
code-division multiple access, *see* CDMA
Coded Composite Transport channel, 501
coherence bandwidth, 134
coherence time, 138–140
coherent demodulation, 97–99
ColorCode, 602–603
combining
  equal-gain, 374
  maximal ratio, 374
  selection, 374
CoMP, 919
Compressed Mode, 569–573
Connectivity Service Network, *see* CSN
Control Channel, 600–603
convolutional coding, 72–74
convolutional decoding, 74–83

Coordinated Multipoint, *see* CoMP
correlated shadow fading, 129–131
COST–Hata model, 125
CQI, 667, 917
CSN, 897
cyclic extension, 879–881
cyclic prefix, 879–881

data subcarrier, 885
data transfer
  DL, 905, 917
  UL, 907, 918
dBd, 114–115
dBi, 115
DC subcarrier, 885
de-interleaving, 87–88
decibel dipole, *see* dBd
decibels isotropic, *see* dBi
delay jitter, 50
demodulation, 89–100
Diameter, 222–224
Differentiated Services, 224–228
Differentiated Services Code Point, 226–228
diffraction, 117–119
diffraction loss, 119
digital communication system
  block diagram, 58
digital modulation, 89–100
discontinuous reception, *see* DRX
discontinuous transmission, *see* DTX
distance-based path loss, 122–128
DL channel descriptor, 904
DL-MAP, 904
DNS, 220–222
Domain Name System, *see* DNS
Doppler shift, 131–132, 137–140
Doppler spectrum, 138, 141–146
Doppler spread, 138–140
downlink capacity, 389
DRCChannelGain, 617–619
DRCCover, 765
DRCGating, 617–619
DRCLength, 617–619
DRCLock Channel, 597–600, 711–714
DRCOffset, 733
drift RNC, 495
DRX, 281
DSC Channel, 723–724
DSCCover, 765
DTX, 502
duplexing, 22–23

E-DCH transmission, 790–793, 800–811
E-RNTI, 791
E-TFCI determination, 800–811
E-TFCI selection, 806–809
E-UTRAN, 913
EDGE, 305–362
  Acknowledged Mode, 340–341
  Cell Search, 348–349
  Frame Structure, 342–348
  Logical Channels, 344
  Modulation and Coding Schemes, 330–333
  Packet Transfer, 353–357
  Paging, 356–357
  Physical Channels, 346–348
  physical layer, 325–327
  Radio Block, 327–328
  RLC functions, 336–342
  RLC/MAC layer, 334–335
  Synchronization, 348–349
  Unacknowledged Mode, 341–342
effective radiated power, *see* ERP
electromagnetic wave, 111–113
eNB, 911
Enhanced Data Rates for GSM Evolution, *see* EDGE
enhanced variable-rate codec, 62–66
EPC, 913
EPS, 913
Equipment Identity Register, 248, 318–319
Erlang-B, 45–46, 929–930
ERP, 113–115
Ethernet, 185–188
  Carrier Ethernet, 187–188
  Metro Ethernet, 187–188
Event 1J, 814–815
evolved Node B, *see* eNB
evolved packet core, *see* EPC
evolved packet system, *see* EPS
excess delay, 133–137
excess delay spread, 133

Fast Associated Control Channel, 254
fast connect, 626
FDD, 22–23
FDMA, 24–25
FEC, 70–87
FER, 47, 381
Fixed Mobile Convergence, 824
flat fading, 134
foreign agent, 428
forward error correction coding, *see* FEC
Forward Fundamental channel, 435
Forward Supplemental channel, 436
Forward Traffic Channel, 604–610
ForwardTrafficValid, 602–603

four-slot interlacing, 610, 620
frame control header, 903–904
frame error rate, *see* FER
free-space path loss, 115
Frequency Correction Channel, 264
frequency reuse, 13–19
frequency reuse factor, 14
frequency shift keying, *see* FSK
frequency-division duplexing, *see* FDD
frequency-division multiple access, *see* FDMA
frequency-domain equalization, 887
frequency-selective fading, 134, 136
Fresnel zone, 117–119
FSK, 91–92

Gateway GPRS Support Node, *see* GGSN
Gaussian minimum shift keying, *see* GMSK
General Packet Radio Service, *see* GPRS
Generic Routing Encapsulation, *see* GRE
GGSN, 310–320
GMSK, 94, 255
GPRS, 305–362
    Acknowledged Mode, 338–340
    Attach, 314–318, 349–351
    Cell Search, 348–349
    Coding Schemes, 328–330
    Frame Structure, 342–348
    Logical Channels, 344
    multislot classes, 320
    Packet Transfer, 353–357
    Packet-Domain Core Network, 310
    Paging, 356–357
    Physical Channels, 346–348
    physical layer, 325–327
    Protocol Stacks, 323
    Radio Block, 327
    RLC functions, 336–342
    RLC/MAC layer, 334–335
    Roaming Exchange, 322
    Synchronization, 348–349
    Unacknowledged Mode, 341–342
GRE, 228–231
GSM
    Air-Interface Protocol Stack, 251
    Authentication, 276
    Base Station System, 244–247
    Bursts, 267
    Cell Search, 269
    cell selection, 282
    Core Network, 247–249
    Deployments, 300
    DTX, 297–298
    Frequency Reuse, 300–301
    Handover Process, 283
    IMSI Attach, 273
    Logical Channels, 264
    Logical-to-Physical Channel Mapping, 266–267
    MS-Originated Call Setup, 279–281
    MS-Terminated Call Setup, 281–282
    Paging, 281–282
    power control, 294–298
    Synchronization, 269
    System Architecture, 242–249
    Timing Structures, 260
    Voice Call Setup, 279–282
guard subcarriers, 885

H-RNTI, 660
H.248, 834, 845–848
half-FDD, 23
handoff, 375–381
    Channel Assignment into Soft Handoff, 477
    enhanced 1x handoff, 478
    hard, 375
    idle, 411
    make-before-break, 375
    soft, 365, 375, 377, 411–414
    softer, 377
    softer-soft, 377
handover, 283–294, 375
    algorithm, 287–290
    inter-cell handover, 290
    intra-cell handover, 290
    mobile-assisted hard handover, 284
    soft handover, 559–562
Happy Bit, 774–776
Hata's propagation model, 124–125
high-capacity mode, 742, 759
high-power amplifier, *see* HPA
High-Speed Downlink Packet Access, *see* HS-DPA
High-Speed Uplink Packet Access, *see* HSUPA
history of wireless communications, 4–11
HLR, 247, 318
home agent, 428
home location register, *see* HLR
Hopping Sequence Number, 301
HPA, 100
HS-SCCH–less operation, 950
HSDPA, 651
    Call Setup, 681–683
    Channels, 655–670
    Downlink Data Transmission, 654
    H-ARQ, 658
    Handover, 687
    handover management, 685–689

OVSF Code Tree, 683
OVSF Code Tree Utilization, 683–685
Radio Network Architecture, 689
Scheduling Algorithm, 675
Tx, 653–655
UE categories, 691
HSPA+, 943–953
HSUPA, 769, 774
Call Setup, 772–773
Downlink Channels, 777
Handover, 814–815
HARQ, 813
System Architecture, 815–818
Timing Relationships, 789–790
UE Categories, 818
Uplink Channels, 777
Uplink Data Transmission, 774–776, 793–811
hyperframe, 260
hysteresis, 566–567

I–CSCF, 833
ICI, 873, 881–882
ICMP, 183–184
Idle Mode mobility, 558–559
idle slot, 595–596
IEEE 802.16, *see* WiMAX
IEEE 802.16-2004, *see* WiMAX
IEEE 802.16e-2005, *see* WiMAX
IEEE 802.3, *see* Ethernet
IFFT, 882
IMS
Architecture, 831–841
Communication Protocols, 841–848
QoS, 862–864
Session Setup, 852–860
IMSI attach, 272
Incremental Redundancy, 340, 610, 680–681
initial attach, 915
initial filter criteria, 849
initial ranging, 905
Integrated Services, 224–228
inter-symbol interference, 136
interface, 591
A, 428
A10, 430
A11, 430
A12, 591
A13, 591
Abis, 246
AN-AAA, 591
Gb, 321
Gn, 322
Gp, 322
Gr, 322
Gs, 322
Iu-CS, 492
Iu-PS, 492
Iub, 491
Iur, 491
R-P, 429
Uu, 491
interlacing, 706
interleaving, 87–88
Internet Control Message Protocol, *see* ICMP
Interrogating–Call Session Control Function, *see* I–CSCF
IP, 173–184
forwarding, 193–200
header, 179–181
IPsec, 231
Authentication Header, 231
Encapsulating Security Payload, 231
IPv6, 232–235
Mobile IP, 428, 628–630
Robust Header Compression, 231–232
routing, 193–200
security, *see* IPsec
IP address, 174–179
IP Convergence, 825
IS-856 Rev. 0, 586–587
IS-95
Call Setup, 405–406
Downlink Channels, 396
IS-95B, 419
Multiplex Sublayer, 403–404
Network Acquisition, 407–408
Physical Layer, 399–403
Power Control, 414–418
Protocol Stack, 392–393
Registration, 410–411
System Architecture, 390–391
Uplink Channels, 398–399
ITU-R M.1225, 125–127
IWF, 248

Jakes' model, 151–152

key performance indicators, *see* KPI
knife-edge diffraction model, 118–119
KPI, 45–50

LAPDm, 256–257
large-scale fading loss, 122–123, 128–131
latency, 50
layers
LLC layer, 334
MAC, 450–452
sublayers

- LAC, 433, 452–454
- MAC, 432
- Mux, 432, 451–452
- QoS, 432
- SRBP, 432

LCR, 152–153
level crossing rate, *see* LCR
linear predictive speech coding, 61–62
location updating, 272
log-normal fading, 129
low-latency mode, 742, 759
LTE air-interface protocol stack, 914–915
LTE-Advanced, 919–921

MAC, 497–499
- functions, 335–336

MAC Channel, 597–600, 711–713
MAC-e, 815–817
MAC-e PDU, 809–811
MAC-e/es, 815–816
MAC-es, 815–817
MAC-hs PDU, 695–697
MACIndex, 587, 604–605, 737–739
MAHO, *see* handover
mean excess delay, 133
Media Gateway, *see* MGW
Media Gateway Control Function, *see* MGCF
Media Resource Function Controller, *see* MRFC
Media Resource Function Processor, *see* MRFP
MGCF, 834
MGW, 490, 834
milestones of wireless or cellular communications, 6
MIMO, 892–893
- MU-MIMO, 896

minimum shift keying, *see* MSK
MME, 913
mobile allocation, 262
mobile allocation index offset, 262, 301
mobility management entity, *see* MME
mobility management procedure, 315–318
mode
- dormant, 428
- non-slotted, 409–410
- slotted, 408–410

modulation symbol period, 875
MPLS, 188, 202–204
MRC, 381
MRFC, 835
MRFP, 835
MSC, 248
MSC server, 490
MSK, 92–94
Multi-Protocol Label Switching, *see* MPLS
Multi-User Packets, 734–736
multi-user–MIMO, *see* MIMO
multiframe, 260, 266
Multimode Capability Discovery Protocol, 728
multipath power delay profile, 133–137
multiple access, 24–25
multiplex sublayer, 403–404
multislot packet configuration, 344

NACC, 360–361
NAPT, 177–179
NAT, 177–179
near-far effect, 388
neighbor set, 414
Network Address and Port Translation, *see* NAPT
network architecture, 11–12
Network Assisted Cell Change, *see* NACC
network entry, 903
Node B Application Part, 493
Node B, 491–493
non-coherent demodulation, 97–99
non-scheduled mode, 774–776

OFDM, 871–888
OFDM symbol, 876
OFDMA, 25, 871–888
- receiver, 886–888
- transceiver, 884–888
- transmitter, 884–885

orthogonal code, 608–609
orthogonal frequency-division multiple access, *see* OFDMA
orthogonal subcarriers, 874–878
orthogonality, 367
overall access operation, 529
OVSF code, 366, 498
OVSF code tree, 519–523

P–CSCF, 831
P-GW, 913
packet control function, *see* PCF
packet data network—gateway, *see* P-GW
Packet Data Traffic Channel, *see* PDTCH
paging slot cycle, 408
PAPR, 882
Path, 843
PCEF, 862, 913
PCF, 429
PCRF, 862, 913
PDCP, 497
PDP Context Activation, 351–353
PDSN, 428
PDTCH, 326
persistence test, 475

personality, 725
Pilot Channel, 597
pilot subcarrier, 885
PN offsets, 396
PN sequence, 396
pole capacity, 389
Policy and Charging Control, 862
Policy and Charging Enforcement Function, *see* PCEF
Policy and Charging Rules Function, *see* PCRF
port, 160–173
power control, 381–388
 closed-loop, 381, 383–388, 573–580
 downlink, 416–418
 forward link, 482–484
 inner-loop, 381, 573–580
 open-loop, 381–383, 573
 outer-loop, 381, 573–580
 reverse link, 481–482
 uplink, 415–416
Power Control Group, 444
PPP Session, 591
prediction of received signal strength, 121–153
PreferredControlChannelCycle, 627
primary scrambling code, 518
processing gain, 388
propagation path loss models, 121–153
protocols
 Address Resolution Protocol, 173, 187, 200
 Border Gateway Protocol, 193–200
 Client Mobile IP, 216–220
 Dynamic Host Configuration Protocol, 213–214
 Generic Attribute Update Protocol, 728
 GPRS Tunneling Protocol, 228–231
 LLC, 323
 Mobile IP, 214–220
 Open Shortest Path First , 193–200
 Proxy Mobile IP, 216–220
 Real-Time Transport Control Protocol, 211–213
 Real-Time Transport Protocol, 211–213
 Routing Information Protocol, 193–200
 Session Description Protocol, 209–211
 Session Initiation Protocol, 206–209
 SNDCP, 323, 334
 Stream Control Transmission Protocol, 169–172
 TCP/IP stack, 159–193
 Transmission Control Protocol, 162–169
 User Datagram Protocol, 172–173
Proxy–Call Session Control Function, *see* P–CSCF

QoS
 classes, 580–581
 QoS class indicators, 911
QPSK, 94–96
Quad Play, 824
quadrature phase shift keying, *see* QPSK
QuickConfig, 600–603

RA Channel, 711–713
Radio Access Network Application Part, 493–495
radio bearer, 550–551
radio blocks, 327–330
radio configurations, 433–441
radio frame, 515–517
radio interface protocol stack, 495–515
radio network subsystem, 491–493
radio waves, 113
RADIUS, 222–224
RAKE receiver, 365, 374
ranging codes, 905
rate matching, 501
Rayleigh distribution, 146–152
Rayleigh fading, 136
Rayleigh fading simulator, 149
receive diversity, 888–891
received signal quality, 286
received signal strength, 286
Received Total Wideband Power, 794–795
Record-Route, 843
Redundancy and Constellation Version, 664
reference point, 901
reference signal, 886
reflection, 116–117
registration
 power-up, 542–544
Relative Grant, 774
relays, 919
remaining set, 414
remote electrical tilt, 108
remote radio head, 108–109
reporting
 event-based, 563–567
 inter-frequency measurement reporting events, 568–569
 inter-RAT reporting events, 569
 intra-frequency measurement reporting events, 563–567
 periodic, 563–567
resource block, 917
Reverse Activity Channel, 597–600

Reverse Fundamental channel, 437
Reverse Pilot channel, 441
Reverse Supplemental channel, 438
RF filter, 100
RF processor, 100–109
Ricean distribution, 147–152
rise-over-thermal noise, 389, 794–796
rms delay spread, 133
RNC, 491–493
root mean square delay spread, *see* rms delay spread
Routing Area Update, 358–359
RPC Channel, 597–600, 711–713
RPCCount, 602–603
RRC, 497
RRC states, 514–515
RTC MAC Protocols, 741

S–CSCF, 833
S-GW, 913
SC-FDMA, 884, 910
scattering, 119–120
scheduled mode, 774–776
Scheduling Information, 774–775, 781–782
scheduling request, 794
SDCCH, 264
SDMA, 895–897
sectorization, 20–22, 370–371
SectorParameters, 600–603
sectors per user, *see* SPU
SectorSignature, 602–603
selective ARQ, 338
selective Type I ARQ, 340
service option, 456
Service-Route, 843
serving gateway, *see* S-GW
Serving GPRS Support Node, *see* SGSN
Serving Grant, 775–776
serving RNC, 495, 546
Serving–Call Session Control Function, *see* S–CSCF
Session configuration, 626–628
SFBC, 890–891
SFID, 903
SGSN, 310–320
shadow fading, *see* large-scale fading loss
short data burst, 454
Signaling ATM Adaptation layer, 493–495
Signaling System 7, *see* SS7
signature sequence, 787–788
Silence Interval, 639–640
SLF, 833
slot, 515
Slow Associated Control Channel, 266
small-scale fading loss, 122–123, 131–153
Smith's Rayleigh fading simulator, 149
socket, 160–173
space diversity, 888–891
Space Frequency Block Coding, *see* SFBC
space time coding, *see* STC
Space Time Transmit Diversity, *see* STTD
Space-Division Multiple Access, *see* SDMA
spatial multiplexing, 892–893
spectrum allocation, 5–8
    Non-U.S., 34
    U.S., 32–34
speech vocoder, 61–62
SPR, 863
spreading, 368
SPU, 380
SRBP, 447
SS7, 923–928
    protocol stack, 926–928
STANDBY state, 348–349
STC, 890–891
STTD, 946
subcarrier subspacing, 882
subchannel, 904
Subscription Locator Function, *see* SLF
Subscription Profile Repository, *see* SPR
Sync, 600–603
Synchronization Channel, 264
SYSTEM INFORMATION message, 538
system profile, 897

T–SGW, 834
T2P-based rate control, 742–745
TAIL bits, 75
TDD, 22–23
TDM, 701
TDM scheduling, 677
TDMA, 24–25
TDMA frame, 260
TEID, 317
Temporary Block Flow, 335
TFCI, 504
THIG, 833
throughput, 49–50
time diversity, 888–891
time-division duplexing, *see* TDD
time-division multiple access, *see* TDMA
time-division multiplexing, *see* TDM
time-slot, 260
time-to-trigger, 566–567
timing advance, 260
TMA, 100–102
topology hiding, *see* THIG
tower mounted amplifier, *see* TMA

Tower Top Low Noise Amplifier, 100–102
tracking area, 913
Traditional Network Address Translation, *see* NAT
traffic classes, *see* QoS classes
Transition Gateway, 841
transition probabilities, 627
transmit diversity, 888–891
transmit time interval, *see* TTI
transport format combination indicator, *see* TFCI
Trunk–Signaling Gateway, *see* T–SGW
TRX, 244
TTI, 500–501
tunneling
  reverse, 431
turbo coding, 83–85
turbo decoding, 85–86
two-ray ground reflection model, 117
Type II H-ARQ, 340

UATI, 587
UMTS
  Cell Reselection, 558–559
  Cell Search, 539
  Data Call Setup, 553–557
  Frame Structure, 515
  Logical Channel, 499
  MAC Sublayer, 503–504
  Network Architecture, 489–495
  PDCP Sublayer, 510–511
  Physical Channel, 499
  Physical Layer, 499–503
  Power Control, 573–580
  Protocol Stacks, 493
  QoS, 580–581
  Radio Channels, 517
  Radio Interface Channels, 497
  Radio Interface Protocol Stack, 495–497
  Radio Link Control Sublayer, 504–510
  Radio Resource Control Layer, 512–515
  Registration, 542–543
  RLC, 497
  RLC Acknowledged Mode, 508–510
  RLC Transparent Mode, 506
  RLC Unacknowledged Mode, 506–508
  Soft Handover, 559–560
  Synchronization, 539
  Transmission Timing, 537–538
  Transport Channel, 499
  Voice Call Setup, 544–552
Unicast Access Terminal Identifier, *see* UATI
Universal Terrestrial Radio Access Network, *see* UTRAN
uplink capacity, 388–389
Uplink State Flag, *see* USF
useful symbol period, 875
user identity
  private, 848–852
  public, 848–852
USF, 335–336
UTRAN, 490
UTRAN Radio Network Temporary Identity, 504

Via, 843
Visitor Location Register, 247
Viterbi Algorithm, 74–83
Viterbi decoder, 78–83
VoIP, 204–213

Walsh codes, 366, 394–395
Walsh-Hadamard matrix, 394–395
wavelength of the signal, 112–113
wideband transmission, 369
WiMAX
  air interface, 902–903
  air-interface protocol stack, 902–903
  fixed, 897–908
  IEEE 802.16, 897
  IEEE 802.16-2004, 897
  IEEE 802.16e-2005, 897
  mobile, 897–908
  network architecture, 900–902
  operations, 903–908
  UL channel descriptor, 905
  UL-MAP, 905

IEEE PRESS SERIES ON DIGITAL AND MOBILE COMMUNICATION

John B. Anderson, *Series Editor*
University of Lund

1. *Wireless Video Communications: Second to Third Generation and Beyond*
   Lajos Hanzo, Peter Cherriman, and Jurgen Streit
2. *Wireless Communications in the 2lst Century*
   Mansoor Sharif, Shigeaki Ogose, and Takeshi Hattori
3. *Introduction to WLLs: Application and Deployment for Fixed and Broadband Services*
   Raj Pandya
4. *Trellis and Turbo Coding*
   Christian Schlegel and Lance Perez
5. *Theory of Code Division Multiple Access Communication*
   Kamil Sh. Zigangirov
6. *Digital Transmission Engineering,* Second Edition
   John B. Anderson
7. *Wireless Broadband: Conflict and Convergence*
   Vern Fotheringham and Shamla Chetan
8. *Wireless LAN Radios: System Definition to Transistor Design*
   Arya Behzad
9. *Millimeter Wave Communication Systems*
   Kao-Cheng Huang and Zhaocheng Wang
10. *Channel Equalization for Wireless Communications: From Concepts to Detailed Mathematics*
    Gregory E. Bottomley
11. *Handbook of Position Location: Theory, Practice, and Advances*
    Edited by Seyed (Reza) Zekavat and R. Michael Buehrer
12. *Digital Filters: Principle and Applications with MATLAB*
    Fred J. Taylor
13. *Resource Allocation in Uplink OFDMA Wireless Systems: Optimal Solutions and Practical Implementations*
    Elias E. Yaacoub and Zaher Dawy
14. *Non-Gaussian Statistical Communication Theory*
    David Middleton
15. *Frequency Stabilization: Introduction and Applications*
    Venceslav F. Kroupa
16. *Mobile Ad Hoc Networking: Cutting Edge Directions,* Second Edition
    Stefano Basagni, Marco Conti, Silvia Giordano, and Ivan Stojmenovic
17. *Techniques for Surviving the Mobile Data Explosion*
    Dinesh C. Verma and Paridhi Verma
18. *Cellular Communications: A Comprehensive and Practical Guide*
    Nishith D. Tripathi and Jeffrey H. Reed